S0-BKT-992

3rd Edition

TEXTBOOK
of
ORGANIC CHEMISTRY

CARL R. NOLLER
Professor of Chemistry, Stanford University

W. B. SAUNDERS COMPANY
Philadelphia and London

1966

QD
253
N66
1966

W. B. Saunders Company: West Washington Square,
Philadelphia, Pa. 19105

12 Dyott Street,
London, W.C.1.

Textbook of Organic Chemistry

© Copyright 1966 by W. B. Saunders Company. Copyright 1951 and 1958 by W. B. Saunders Company. Copyright under the International Copyright Union. All rights reserved. This book is protected by copyright. No part of it may be duplicated or reproduced in any manner without written permission from the publisher. Made in the United States of America. Press of W. B. Saunders Company. Library of Congress catalog card number: 65-10290.

Preface

In addition to the tremendous expansion of scientific knowledge in recent years, there has been a shift in emphasis in the various fields. Biology now is concerned more with the molecular aspects of the subject, chemistry with the physical aspects, and physics with the mathematical aspects.

Organic chemistry is no exception in this regard. Accordingly the material presented in this third edition of *Textbook of Organic Chemistry* not only has been revised in the light of new facts but has been largely rewritten. Before each class of compounds is considered in detail, an attempt is made to summarize the general properties to be expected from a knowledge of reactive sites, the polarity of bonds, and the relative electron densities at different points in the molecule. More attention is paid to thermodynamic properties and thermochemistry. The transition state is given more prominence, and the courses of reactions frequently are illustrated by energy profiles. Greater use is made of the current practice of indicating by curved arrows the probable flow of electrons in planar transition states that leads to the final products. Ionization potentials frequently are cited to support statements regarding relative stabilities, electronegativities, and electron-donating power. Use is made of the better understanding of the part played by solvents in organic reactions. Now that most educational institutions are equipped with relatively inexpensive infrared spectrometers, the discussion of characteristic infrared absorption of functional groups is introduced at an early stage. The text does not attempt, however, to make the student a specialist in the interpretation of spectra. A working knowledge can come only with actual practice in the laboratory.

Discussions of mass spectrometry, nuclear magnetic resonance, electron paramagnetic resonance, and electronic absorption spectra now are included, but their presentation is postponed until after the main types of compounds have been considered. In this way the various aspects of each topic can be considered as a whole. This material and its applications may be introduced earlier if the instructor so prefers.

Regardless of the shift in emphasis in organic chemistry, there is a basic core of physical and chemical properties that the student must assimilate at some time. The sooner this material is presented, the sooner the student can start accumulating a store of factual knowledge. For this reason this text retains the functional group approach and continues to

79724

stress the important specific reactions of organic compounds. At the same time, it tries to develop the subject in such a way that the student acquires a feeling for the current conception of what organic molecules are, of how they behave, and of why they behave the way they do.

Chapters 2 and 3 are new. Chapter 2 gives a general discussion of ionic and covalent bonding, bond strengths, polarity of covalent bonds, and intermolecular forces. Chapter 3 gives an elementary survey of chemical reactivity, including thermodynamics, chemical equilibrium, kinetics, and the main types of mechanisms of chemical reactions. The material in these chapters is illustrated almost exclusively by the properties of simple inorganic compounds, with the idea that these chapters will provide a bridge between the usual course in general chemistry and the course in organic chemistry. When the same principles are applied later to organic reactions, it can be assumed that the student is familiar with this material. Some well-prepared students should be able to skip these chapters altogether. Others who have been exposed to these topics in high school or freshman courses may find the discussions a useful review. Those for whom the material is new will need to study it thoroughly. Under most circumstances, the actual course work can begin with Chapter 4 or even with Chapter 5, and reference can be made to the earlier chapters as the need arises.

Numerous new problems have been added, and answers are provided to meet what appears to be a general demand for them. The additional space required has been obtained by omitting the review questions and the variations in the same type of problem, which were features of the second edition.

In order to keep within a limited number of pages, it is necessary to make a judicious selection from the vast array of material available. The attempt here has been to continue to present a well-rounded text by giving equal weight to the theoretical, experimental, and practical aspects of the subject, both classical and current. Many interesting topics, especially those of a more esoteric or sophisticated nature, have been omitted or touched on only briefly. Where a more complete text that can be used also in a second course is desired, or where the instructor thinks that a text should serve a student as ready reference for a number of years, *Chemistry of Organic Compounds,* of which this *Textbook* is an abridgment, is suggested.

CARL R. NOLLER

Stanford University

Contents

APPENDIX

INDEX

CHAPTER ONE

ORGANIC CHEMISTRY

What is organic chemistry? What does an organic chemist do? How has his work affected the well-being of the human race? What currently is exciting to scientists who are working in this field? Organic chemistry is defined as the *chemistry of the compounds of carbon.* Why has a separate branch of chemistry been devoted to the compounds of carbon? The answer to this last question is that carbon atoms are unique in being able to form up to four strong covalent bonds with other atoms, including like carbon atoms. Covalent bonding is a type of attractive force that, unlike ionic bonding, is localized between two atoms and that, for multivalent elements, has direction. Carbon, with four valences directed in space to the corners of a tetrahedron, is able to form complex three-dimensional structures that may contain any number of atoms. This ability explains the existence of the countless simple to complex compounds that make up all living matter.

Man has been curious about the composition of his environment since he first began to think, and those who sought to determine what constitutes matter derived from living organisms, as opposed to nonliving matter, came to be called organic chemists. The investigation of living matter from this viewpoint involves first the separation of pure individual compounds from the complex mixtures that occur in nature. Next it is necessary to determine by analysis what elements are present in the pure compound and in what ratio. Then the structure of the compound must be determined, that is, the exact order in which the various atoms are bonded to each other. Finally the configuration of the molecule or the arrangement of the atoms in space, and the conformation of the molecule or the shape that the molecule assumes as the result of internal and external forces, must be learned before all of the properties of the compound can be explained. Over the years a large body of information has been accumulated as to how particular groups of atoms behave, and this knowledge makes possible the identification of special arrangements of atoms in small molecules or of groups of atoms attached to large molecules. Often the more complex molecules must be broken into smaller parts, and these smaller parts identified. As with a cutout puzzle or a detective story, the various parts or bits of evidence are fitted together, and a structure for the original compound is proposed.

Next, as a proof of structure, the compound must be synthesized. The organic chemist not only has devised means of arriving at the structure of molecules; he also has learned how to join atoms in just the right way to build up desired structures. There is no more convincing evidence of the soundness of theories concerning how atoms are attached to each other in organic molecules than the ability of the chemist to arrive at the structure and configuration of a complex compound such as cholesterol, $C_{27}H_{46}O$, penicillin G, $C_{16}H_{18}N_2S$, or chlorophyll, $C_{55}H_{72}N_4O_5Mg$, from its chemical and physical properties and those of its degradation products, and then to rebuild the molecules step by step from small molecules and in the end obtain a compound identical with the original natural product.

From the knowledge that has been gained of the chemical properties of organic compounds, the organic chemist now is able to synthesize molecules of almost any structure. This ability has led him to try to improve on nature. Whereas at one time most medicinals were obtained from natural sources, for example, morphine, quinine, or digitalin, most now are synthetic compounds that do not exist as natural products. It is the organic chemist who has been responsible for the "wonder drugs" of modern times. He has conceived new structures and synthesized them for the pharmacologist, who determines their physiological properties. Those with promise of therapeutic value are turned over to the physician for clinical evaluation. While less than a generation ago linseed oil was the main vehicle for protective coatings, numerous synthetic resins now are available. Modern dyes, insecticides, and selective herbicides are chiefly synthetic organic compounds. Synthetic plastics are replacing metal and wood for machine parts and construction purposes; synthetic elastomers for natural rubber; synthetic fibers for cotton, wool, and silk.

Although the practical applications are spectacular, they all are based on knowledge acquired by fundamental investigations of the chemical behavior of organic compounds. Many scientists consider practical applications merely as by-products of their work. Many other scientists carry on fundamental investigations in the hope that they will lead to practical applications. The main objective of all is the acquisition of new knowledge.

Historical Development

Organic chemistry was late in becoming a separate branch of science, dating as such only from the early part of the nineteenth century. Although many organic compounds existed on earth countless years before life began, only in very recent times has much progress been made in their study. The investigation of organic compounds lagged considerably behind that of inorganic compounds because the naturally occurring organic compounds were more complex, their reactions were more difficult to understand, and they usually occurred as mixtures that were more difficult to separate than the mixtures of inorganic compounds.

Specific properties of a few organic compounds were known to the ancients. Noah was familiar with the effect of fermented grape juice on the human system. Acetic acid in the form of sour wine was well known; one of the proverbs of Solomon refers to the action of vinegar on chalk. Acetic acid was, in fact, the only acid known to the ancients. Indigo and alizarin have been identified as dyestuffs used on Egyptian mummy cloth, and royal purple was extracted from a Mediterranean mollusk by the Phoenicians.

The assignment of specific properties to individual compounds required the development of methods for isolating the compounds in a pure state. The process of distillation first was described in detail by the Alexandrians in the beginning of the fifth century A.D., and was developed and used by their successors, the Arabians, during the eighth century to concentrate acetic acid and alcohol. It remained for Lowitz,[1] however, to prepare acetic acid sufficiently pure to obtain it in crystalline form in 1789, and to obtain alcohol free of water in 1796. Cane sugar had been obtained in crystalline form in northeastern India about 300 A.D., but it did not become known until about 640 to the Arabians, who introduced the growing of cane into Egypt and southern Europe. Tartar from wine was known at an early date, and the older alchemists experimented with it. Benzoic acid and succinic acid were prepared in the sixteenth century, and wood alcohol, grape sugar, and milk sugar were isolated during the seventeenth century. Persistent **attempts to isolate pure organic compounds** came only with the rise of chemistry during the latter half of the

[1] Johann Tobias Lowitz (1757–1804), German-born Court Apothecary at St. Petersburg, who made several outstanding contributions to basic laboratory technique. He discovered the absorptive property of activated charcoal and used it in the purification of organic compounds. He also discovered the phenomenon of supersaturation, used ice and various salts to make freezing mixtures, and was the first to use calcium chloride to remove water from organic liquids.

eighteenth century. Scheele[2] was the first to isolate acetaldehyde, glycerol, and uric, oxalic, tartaric, lactic, citric, and malic acids. As late as the beginning of the nineteenth century, however, the view was held that acetic acid was the only vegetable acid, and that all others consisted of some combined form of acetic acid.

Along with the work on isolation, some **chemical transformations** were discovered. It was known as early as the thirteenth century that sulfuric acid reacts with ethyl alcohol, and later a mixture of ether and alcohol prepared by this reaction was used in medicine. It was not until 1730 that the properties of a relatively pure ether, incompletely miscible with water, were described, and not until 1800 that the absence of sulfur was proved, thus showing that the sulfuric acid was not incorporated into the ether molecule. During the seventeenth century, acetone was prepared by the thermal decomposition of lead acetate, and ethyl chloride was prepared from ethyl alcohol and hydrochloric acid. Several organic acids were converted to the ethyl esters during the latter half of the eighteenth century. Scheele made oxalic acid by the oxidation of cane sugar, and mucic acid by the oxidation of milk sugar.

Much less was known about **compounds present in animal organisms.** At the end of the eighteenth century, the only accomplishments of importance were the isolation of urea from urine in 1773, and of uric acid from urinary calculi in 1776. It generally was assumed that there was a fundamental difference between animal matter and vegetable matter since decomposition of the former by heat always yielded ammonia.

Of greatest importance for the advance of organic chemistry was the **development of methods for qualitative and quantitative analysis.** Lavoisier,[3] who was one of the earliest investigators to analyze organic compounds, showed that they contain carbon and hydrogen and usually oxygen. He showed also that products derived from animal sources frequently contained nitrogen. Berzelius,[4] Liebig,[5] and Dumas[6] made further improvements in organic analysis, particularly from the quantitative standpoint.

Originally no distinction was made between the chemistry of mineral compounds and the chemistry of those of biological origin. After a representative group of compounds had been isolated from natural sources, however, the differences became so apparent that a division soon was made. Bergman[7] in 1780 was the first to differentiate between organic compounds and nonorganic compounds. Berzelius first used the term *organic chemistry* in

---

[2] Carl Wilhelm Scheele (1742–1786), Swedish apothecary. Besides being the first to isolate and prepare many organic compounds, he was the first to prepare molybdic, tungstic, and arsenic acids, to show that graphite is a form of carbon, to characterize manganese and barium, and to note the insolubility of barium sulfate. He also discovered oxygen (independently and prior to Priestley) and chlorine.

[3] Antoine Lavoisier (1743–1794), French scientist and public servant, who was condemned to death by a judge of the Revolutionary Tribunal with the remark, "The Republic has no need for savants."

[4] Joens Jacob Berzelius (1779–1848), Swedish chemist, known chiefly for his accurate work on the combining weights of the elements and for his support of the dualistic theory, which held that elements were either electrically positive or negative in character and that only elements oppositely charged were capable of combining with each other.

[5] Justus Liebig (1803–1873), German professor who first introduced laboratory work into general instruction in chemistry. He perfected the combustion method of organic analysis, established the theory of radicals, laid the foundations of agricultural chemistry, and was the forceful editor who built up the prestige of the *Annalen der Chemie und Pharmacie*. The present title, *Justus Liebig's Annalen der Chemie*, was given to this publication after his death.

[6] Jean Baptiste André Dumas (1800–1884), French chemist, teacher, and public servant. He was noted for the accuracy of his experimental work and for his clear thinking. His name is associated chiefly with his analytical methods and with the phenomenon of substitution in organic compounds.

[7] Torbern Olof Bergman (1735–1784), professor of mathematics and later of chemistry at the University of Upsala, Sweden.

1808 and published the first independent treatment of organic chemistry in his textbook of 1827.

The slowness in the advance of organic chemistry before the nineteenth century has been ascribed to the belief that compounds produced by living organisms could not be synthesized without the intervention of a *vital force*. The rapid advance during the nineteenth century has been attributed to the renunciation of this theory after Woehler[8] reported in 1828 the synthesis of urea from ammonium cyanate. Actually this event had little immediate effect in combating the vitalistic theory, which persisted in one form or another for at least another twenty years.

Although arguments for and against the vitalistic theory continued, they did not deter chemists from trying to synthesize organic compounds. Their work, however, was leading them into chaos because of the confusion that existed concerning **atomic and molecular weights.** Avogadro's hypothesis that the same volume of any gas at the same temperature and pressure contains the same number of molecules had provided the basis for atomic and molecular weight determinations in 1811, but order did not begin to appear until after Cannizzaro[9] clearly showed in 1858 how the application of the hypothesis would resolve many of the difficulties. At about the same time, the **theories of valence and structure** expounded by Couper[10] and by Kekulé[11] and others paved the way for the extremely rapid advances of the next fifty years.

During the first half of the nineteenth century, it gradually was realized that *the essential difference between inorganic and organic compounds is that the latter always contain carbon.* Gmelin[12] was the first to state this fact in 1848. At the present time, the number of carbon compounds synthesized in the laboratory far exceeds the number isolated from organic products, and the phrase *chemistry of the compounds of carbon* is more accurate than *organic chemistry*. The latter term, however, is less cumbersome and is used almost universally. The study of the chemical reactions that take place in living organisms, usually under the influence of enzymes, now is termed *biochemistry*.

With the standardization of a system of atomic and molecular weights and the acceptance of concepts of valence and structure as applied to organic chemistry, this branch of the science made very rapid progress after 1860. Investigations of the properties of compounds present in coal tar led to the commercial production, from 1856 to the 1890's and after, of a vast variety of dyes. During this period fundamental knowledge of the stereo-

---

[8] Friedrich Woehler (1800–1882), professor at the University of Goettingen and friend of Berzelius and Liebig. Besides his work in organic chemistry, he discovered aluminum when he was only 27 years old and did important work on boron, aluminum, and titanium.

[9] Stanislao Cannizzaro (1826–1910), Sicilian revolutionist and politician, professor of chemistry at the Universities of Genoa, Palermo, and Rome. His *Summary of a Course of Chemical Philosophy,* which he wrote for his students, was published in 1858. An international congress of the most eminent chemists was called by Wurtz and Kekulé and presided over by Weltzien, Kopp, and Dumas at Karlsruhe in 1860 to try to bring order out of the confusion regarding atomic weights and chemical notation. At the close of the congress, which had failed to reach agreement, Cannizzaro's pamphlet was distributed to those present by his friend, Pavesi. Lothar Meyer was convinced by the exposition, and, largely through his influence, Cannizzaro's views soon were accepted.

[10] Archibald Scott Couper (1831–1892), brilliant young Scottish chemist whose career was cut short by illness in 1859. He was the first to publish formulas for organic compounds comparable to present-day structural formulas.

[11] Friedrich August Kekulé (1829–1896), professor at the Universities of Ghent and Bonn. He extended the type theory and laid the foundations of structural theory in organic chemistry. It seems significant that before becoming interested in chemistry, he was a student of architecture.

[12] Leopold Gmelin (1788–1853), professor of medicine and chemistry at the University of Heidelberg, and originator of *Gmelin's Handbuch der Chemie,* now in its eighth edition. His family is noted for the large number and long line of prominent chemists that it contained, beginning with Johann Georg Gmelin, born in 1674.

chemistry or spatial arrangement of atoms in organic compounds and of the chemistry of the more complex natural products such as carbohydrates and proteins also was being acquired. The years 1890 to 1910 were particularly fruitful. Synthetic analgesics, soporifics, and local anesthetics such as aspirin, Veronal, and Novocaine were introduced, and the anthraquinone vat dyes were discovered. In 1900 Gomberg prepared the first stable free radical, a compound with an unpaired electron, and Grignard, the versatile magnesium-containing reagent that bears his name. Since 1925 the use of petroleum as a raw material for the synthesis of organic compounds has been exploited and has led to the vast current production of solvents, plastics, rubbers, and textile fibers.

Organic chemistry still is a dynamic subject. New physical procedures have been developed for separating complex mixtures and handling minute amounts of material such as the sex attractants of insects or the substance from glands of the queen bee that sterilizes worker bees and inhibits the production of queen bee cells. Various types of electronic equipment developed or improved since 1950, such as ultraviolet, infrared, nuclear magnetic, and mass spectrometers, immediately have been put to use by the organic chemist to speed up his work. One hundred and fifty-seven years elapsed between the first isolation of cholesterol and the completion of the elucidation of its structure in 1932. It is safe to say that the structures of compounds of equal or greater complexity are being determined today in six months or less. Ultraviolet, infrared, nuclear magnetic resonance, electron paramagnetic resonance, and mass spectrometric methods not only are aiding new investigations. They have confirmed in every respect what the organic chemist, by painstaking investigation of chemical properties, has come to believe concerning the structure of organic molecules and the making and breaking of bonds between atoms.

Many unusual compounds whose possible existence has been predicted by theory are being synthesized in the laboratory. The investigation of the mechanism of organic reactions is being accelerated. More importance is being attached to the shapes that molecules assume under various conditions and their effect on the reactivity of the compound. More attention is being given to the interaction of solvent molecules with reacting molecules and the role that solvents play in the formation of the transition state. Many of the ideas developed by organic chemists concerning the importance of structure, the mechanisms of reactions, and the transfer of electrons have been applied to inorganic chemistry and have rejuvenated this discipline. Clearly and inevitably organic chemistry will continue to be an important branch of the sciences, and it will play an ever increasing role in the biological sciences as well.

It may be noted here that not only is the study of organic chemistry necessary as a background for certain scientific professions. It so affects man's life that it may be considered to be desirable also as part of the cultural training of an educated person.

CHAPTER TWO

CHEMICAL BONDS AND INTERMOLECULAR FORCES

It is assumed that students of organic chemistry will have studied general chemistry. It seems desirable, however, to review and possibly to elaborate somewhat on those principles that are used most in organic chemistry. This chapter considers the attractive and repulsive forces acting between atoms and between molecules. Chapter 3 summarizes some general concepts concerning chemical reactivity.

CHEMICAL BONDS

Electronic Structure of Atoms

Since organic chemistry is concerned chiefly with the elements in the first three periods of the periodic table, this discussion is limited to these elements. According to the electronic theory, atoms consist of positive nuclei surrounded by a number of negative electrons equal to the positive charge on the nucleus (the *atomic number*). These electrons group themselves in a first shell with a maximum of two electrons (*K* shell), a second shell with a maximum of eight electrons (*L* shell), and a third shell with a maximum of eighteen electrons (*M* shell).

A somewhat more detailed picture represents the electrons as distributed in a definite way in orbitals and groups of orbitals. Strictly speaking, an **orbital** is the mathematical expression describing the behavior of an electron moving in the vicinity of a positively charged nucleus. For a qualitative pictorial discussion of bonding, it is sufficient to think of orbitals as *preexisting regions in space about atomic nuclei in which electrons, if present, are most likely to be found;* that is, the system electrons-nucleus will be most stable when the electrons are in these regions (cf. p. 10). The groups of orbitals are known as **shells,** and within a shell there are **subshells.** The *K* shell consists of a single orbital, the 1*s* orbital. The *L* shell consists of four orbitals: a subshell known as the 2*s* orbital, and a subshell of three orbitals known as the $2p_x$, $2p_y$, and $2p_z$ orbitals. The numerals 1, 2, 3, 4, 5, 6, and 7, called the *principal quantum numbers,* designate the shell or principal energy levels of the electrons just as do the letters *K*, *L*, *M*, *N*, *O*, *P*, and *Q*. They indicate more, however, in that the number of subshells in a shell is equal to the principal quantum number, and the total number of orbitals in a shell is equal to the square of the principal quantum number. Thus the *M* shell with the principal quantum number 3 contains three subshells and nine orbitals—one 3*s*, three 3*p*, and five 3*d* orbitals. The *N* shell contains four subshells and sixteen orbitals—one 4*s*, three 4*p*, five 4*d*, and seven 4*f* orbitals.

Because of the attraction of the positive charge on the nucleus, electrons occupy an orbital as close to the nucleus as possible. Hence there is a greater tendency to occupy a 1*s* orbital than a 2*s* orbital, and a 2*s* orbital than a 2*p* orbital. No more than two electrons can occupy a given orbital, however, and even this condition is possible only if the electrons have opposite spin, that is, if they are affected in opposite ways by a magnetic field. If more than one electron is available for a given set of *p* orbitals that are equidistant from the nucleus, the electrons usually do not pair but occupy separate orbitals until each of the three *p* orbitals has at least one electron. Hence the pairing of electrons in itself does not contribute to the stability of the atom but merely is permissive if the electrons have opposite spins. This behavior of the electrons may be summarized by two important rules for the distribution of electrons in orbitals: (*1*) no orbital can contain more than two electrons and these electrons must have opposite spin (***Pauli exclusion principle***); (*2*) two electrons usually do not occupy a given orbital in a subshell until all of the orbitals of the subshell have at least one electron (***Hund rule***). Electrons with opposed spins usually are represented by arrows pointing in opposite directions, ↑↓, whereas electrons with parallel spins are represented by arrows pointing in the same direction, ↑↑.

The distribution of the electrons for the elements in the first three periods is given in Table 2–*1*. In accordance with the Hund rule, the carbon atom has two unpaired electrons in

TABLE 2–*1*. DISTRIBUTION OF ELECTRONS, FIRST IONIZATION POTENTIAL, AND ELECTRON AFFINITY FOR THE GROUND STATE OF ELEMENTS OF THE FIRST THREE PERIODS OF THE PERIODIC TABLE

PERIOD	ATOM	ORBITALS										IONIZATION POTENTIAL	ELECTRON AFFINITY
		K SHELL	*L* SHELL				*M* SHELL						
		1*s*	2*s*	$2p_x$	$2p_y$	$2p_z$	3*s*	$3p_x$	$3p_y$	$3p_z$	3*d**	e.v.†	e.v.‡
1	H	(↑)	○	○	○	○	○	○	○	○	○	13.60	0.75
	He	(↑↓)	○	○	○	○	○	○	○	○	○	24.58	—
2	Li	(↑↓)	(↑)	○	○	○	○	○	○	○	○	5.39	0.82
	Be	(↑↓)	(↑↓)	○	○	○	○	○	○	○	○	9.32	−0.19
	B	(↑↓)	(↑↓)	(↑)	○	○	○	○	○	○	○	8.30	0.33
	C	(↑↓)	(↑↓)	(↑)	(↑)	○	○	○	○	○	○	11.26	1.12
	N	(↑↓)	(↑↓)	(↑)	(↑)	(↑)	○	○	○	○	○	14.54	0.05
	O	(↑↓)	(↑↓)	(↑↓)	(↑)	(↑)	○	○	○	○	○	13.61	1.47
	F	(↑↓)	(↑↓)	(↑↓)	(↑↓)	(↑)	○	○	○	○	○	17.42	3.48
	Ne	(↑↓)	(↑↓)	(↑↓)	(↑↓)	(↑↓)	○	○	○	○	○	21.56	—
3	Na	(↑↓)	(↑↓)	(↑↓)	(↑↓)	(↑↓)	(↑)	○	○	○	○	5.14	0.47
	Mg	(↑↓)	(↑↓)	(↑↓)	(↑↓)	(↑↓)	(↑↓)	○	○	○	○	7.64	−0.32
	Al	(↑↓)	(↑↓)	(↑↓)	(↑↓)	(↑↓)	(↑↓)	(↑)	○	○	○	5.98	0.52
	Si	(↑↓)	(↑↓)	(↑↓)	(↑↓)	(↑↓)	(↑↓)	(↑)	(↑)	○	○	8.15	1.46
	P	(↑↓)	(↑↓)	(↑↓)	(↑↓)	(↑↓)	(↑↓)	(↑)	(↑)	(↑)	○	10.98	0.77
	S	(↑↓)	(↑↓)	(↑↓)	(↑↓)	(↑↓)	(↑↓)	(↑↓)	(↑)	(↑)	○	10.36	2.07
	Cl	(↑↓)	(↑↓)	(↑↓)	(↑↓)	(↑↓)	(↑↓)	(↑↓)	(↑↓)	(↑)	○	12.97	3.69
	A	(↑↓)	(↑↓)	(↑↓)	(↑↓)	(↑↓)	(↑↓)	(↑↓)	(↑↓)	(↑↓)	○	15.75	—

* Five orbitals of equal energy, all empty.
† e.v. = electron volts = 23.063 kcal./mole.
‡ Approximate values.

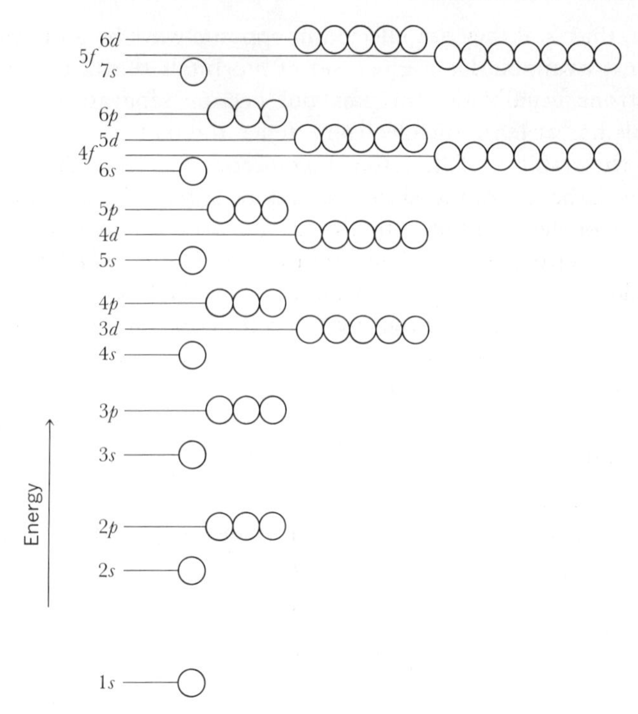

Figure 2–*1*. Relative energy levels of atomic orbitals.

two *p* orbitals, and nitrogen has three unpaired electrons in the three *p* orbitals. Pairing of electrons in the *p* orbitals begins with oxygen, which has one pair in one *p* orbital and an unpaired electron in each of the two remaining *p* orbitals. The number of unpaired electrons decreases to one for fluorine and none for neon. The elements of the third period have no electrons in the 3*d* orbitals, although these orbitals are available for bonding purposes and may be used in complex ion formation, or for electrons promoted into them in excited states. There are no 1*p* orbitals corresponding to the 2*p* and 3*p* orbitals, and no 1*d* or 2*d* orbitals corresponding to the 3*d* orbitals.

Because energy would be required to promote an electron from a 2*s* orbital to a 2*p* orbital or from a 2*p* orbital to a 3*s* orbital, orbitals frequently are referred to as being of lower or higher energy. Thus the 2*s* orbital is said to be of lower energy than the 2*p* orbital and the 3*s* orbital is of higher energy than the 2*p* orbital. Figure 2–*1* illustrates the approximate relative energy levels of atomic orbitals. For the first three periods the order follows the order of the principal quantum numbers and the *s,p,d* order of the subshells. For the higher periods, overlapping of energy levels of orbitals with different quantum numbers takes place, and there is much less difference in energy between adjacent levels. For example, the 4*s* orbital is of lower energy than the 3*d* orbital, and there is less difference in energy between the 4*s*, 3*d*, and 4*p* orbitals than between the 3*s*, 3*p*, and 3*d* orbitals.

The differences in the energies of the various orbitals and the effects of electron repulsion are reflected in the first ionization potential of the atoms, the **ionization potential** being the energy necessary to remove an electron from an atom or molecule. These ionization potentials are given in Table 2–*1*. The ionization potential of helium is greater than that of hydrogen because the positive charge on the nucleus has doubled. It is not twice that of hydrogen because of the coulombic repulsion of the two electrons in the same 1*s* orbital. Lithium has an ionization potential less than that of helium because, although the nuclear charge has increased, the 2*s* orbital is farther from the nucleus. The electrons in the 1*s* orbital exert a screening effect on the nucleus, and the effective nuclear charge on the 2*s* electrons is decreased. The potential increases again for beryllium because the nuclear charge increases without forcing the electron to occupy an orbital of higher energy, but decreases for boron because the outer electron now must be in the 2*p* orbital. As the *p* orbitals become filled, the ionization potential increases again for carbon and nitrogen but decreases for oxygen because of pairing of electrons. It increases again for fluorine and neon but decreases for sodium because the electron most easily removed now is in the 3*s* orbital. Similar considerations explain the order of **electron affinities** given in Table 2–*1*, the electron affinity being the energy *released* when an isolated atom acquires an additional electron.

Ionic Bonds

It is known that *atoms with filled electron shells,* such as helium and neon, *are extremely inert.* If the outside shell is incompletely filled, it is conceivable that an atom could acquire one or more electrons from another atom or atoms, provided that the transfer would lead to an energetically more stable arrangement. The transfer of an electron from one atom to another would leave a positive charge on the donor and a negative charge on the acceptor.

$$A\cdot + B\cdot \longrightarrow A^{+} + :B^{-}$$

The atoms involved in this process originally were electrically neutral; that is, the number of negative electrons equaled the number of positive charges on the nucleus. In the salt that is formed, the metallic atom has lost an electron and become positively charged, whence metals are called *electropositive* elements. The halogen has gained an electron and become negatively charged and is referred to as an *electronegative* element. It should be noted that these terms refer to the condition of the particle *after* the transfer of electrons has taken place. The charged particles are called ions, and the attractive force between ions of opposite charge is called an **ionic bond** or an **electrovalence.**

Actually there is no bond between ions because the charges are distributed equally about the ions, and the oppositely charged particles attract each other equally in all directions. The result is a crystal lattice in which the arrangement is governed chiefly by the size of the ions and by the number of charges on them. In lithium fluoride and sodium chloride, each metal ion is surrounded by six halogen ions and each halogen ion by six metal ions to give a crystalline solid belonging to the cubic system.

Covalent Bonds

In general the tendency to form ionic bonds is greatest between elements on opposite sides of the periodic table. For elements having positions close to each other in the periodic table, the transfer of electrons becomes highly endothermic. Yet the gas densities of hydrogen and fluorine show that they are diatomic molecules, and compounds that contain carbon bonded to nitrogen are known. Hence there must be a second way in which attractive forces can operate. In fact a considerable amount of energy is liberated when an unpaired electron from each of two atoms is attracted by two nuclei instead of one. This process, which is called a *sharing of electrons,* gives rise to what is known as a **covalence** or as a **covalent** or **electron-pair bond.** The covalent bond is described best in terms of molecular orbitals, which requires first a more detailed description of atomic orbitals.

Shapes of Atomic Orbitals. The atomic orbitals have a definite distribution in space. The *s* orbitals are spherically symmetrical about the nucleus as illustrated schematically in perspective in Fig. 2–*2a*, and in cross section through the nucleus in Fig. 2–*2b*. In these fig-

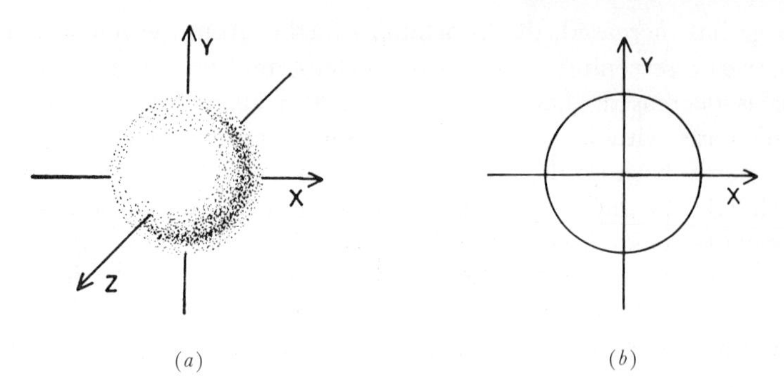

Figure 2–2. Representation of the atomic 1*s* orbital (*a*) in perspective and (*b*) in cross section through the nucleus.

ures no attempt is made to indicate the radial distribution of the electron within the orbital but only the relatively small region in space that the electron occupies most of the time. Several methods may be used to illustrate the distribution of the electron about the nucleus. Visually one may think of the electron as moving randomly in all directions but being more frequently in some places than in others. Suppose one could locate the electron at a particular moment and place a dot at that point in space and could do this repeatedly after equal intervals of time. Then after many such operations, the relative density of the dots would represent the probability of finding the electron in space. This relative density is called the *charge distribution* or *charge cloud* of the electron. Figure 2–*3a* represents the cross section of such a cloud formed by an electron in the 1*s* orbital of the hydrogen atom. The charge density is distributed spherically about the nucleus but decreases with increasing distance from the nucleus. This picture does not indicate *how much* of the charge is at a particular distance from the nucleus. Thus although the charge density is greatest close to the nucleus, the volume occupied by a thin shell is very small, and only a small fraction of the total charge is present. Similarly at great distances from the nucleus, a shell of equal thickness has a large volume, but the density is so small that again a negligible fraction of the charge is present. Figure 2–*3b* illustrates the variation in the amount of charge, *e*, on a uniformly charged shell with increasing distance, *r*, from the nucleus. The maximum of the curve is at 0.529 A, the radius of the old Bohr orbit. For most purposes it is sufficient to represent the orbital by a boundary surface or by a cross section of the surface. This surface is chosen to enclose the space in which the charge density is greatest and in which most of the charge is contained.

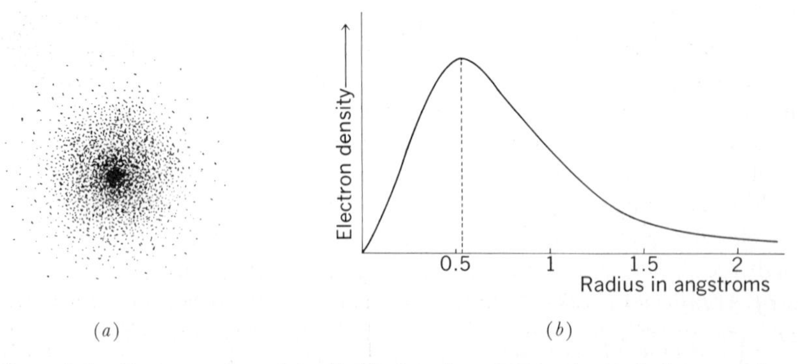

Figure 2–3. Representations of the distribution of an electron about a hydrogen nucleus.

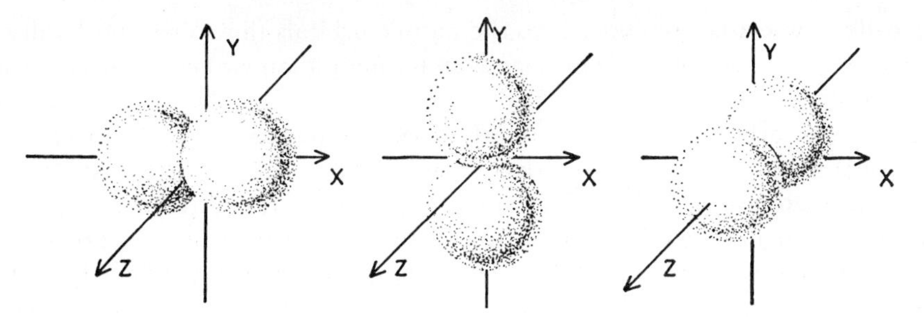

Figure 2–4. Perspective representation of the p_x, p_y, and p_z atomic orbitals.

For elements in the second period the 1*s* orbital shrinks to a size proportional to the size of the 1*s* orbital of hydrogen divided by the atomic number. Thus the 1*s* orbital of carbon is about one sixth the size of that of hydrogen. The 1*s* orbital is surrounded by the 2*s* orbital, which also is spherically symmetrical about the nucleus.

The electrons of the 2*p* orbitals are most likely to be in regions resembling a dumb-bell as illustrated in Fig. 2–4. These orbitals possess a *nodal plane,* that is, a plane in which the probability of finding the electron is zero. The three dumb-bells representing the three 2*p* orbitals are oriented perpendicularly to each other in the *x*, *y*, and *z* directions.

Molecular Orbitals and Covalent Bonds. The covalent or electron-pair bond, when first considered, seems to be anomalous. Electrons can only repel each other, and the pairing of electrons cannot account for the liberation of energy when a covalent bond is formed. Yet the hydrogen molecule is more stable than two hydrogen atoms by 103 kcal. per mole.[1] As two hydrogen atoms approach each other, their 1*s* orbitals begin to overlap (Fig. 2–5*a*). Since the orbitals are regions in space where the electron is most likely to be, the overlapping of the atomic orbitals increases the electron density and hence the negative charge between the two positive nuclei (Fig. 2–5*b*). Attraction between the positive and negative charges gives rise to the chemical bond. Each electron now is attracted to two positive nuclei or an effective charge approaching +2 and is held more strongly than when it is attracted by the +1 charge of a single hydrogen nucleus. Hence energy is evolved when two hydrogen atoms combine to form the hydrogen molecule.

In the hydrogen molecule, the electrons no longer occupy the separate atomic orbitals about the hydrogen nuclei but instead both electrons occupy an orbital that encompasses both nuclei. Because orbitals encompassing more than one nucleus give rise to molecules,

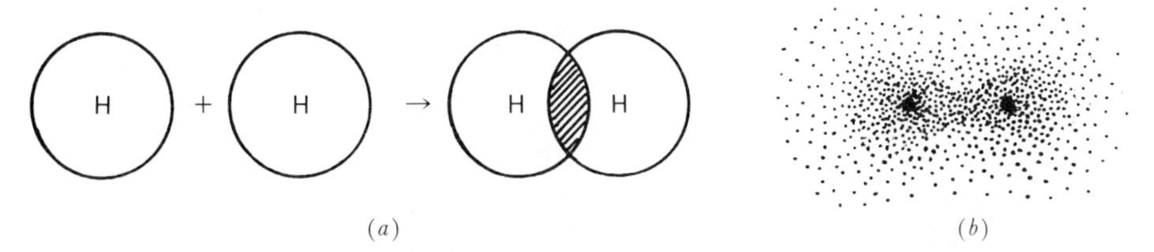

Figure 2–5. Schematic representation of (*a*) the bonding of two hydrogen atoms and (*b*) the electron density in a hydrogen molecule.

[1] Chemists usually express energies gained, lost, or required in a process as the energy per gram mole of average particles. A mole is defined as the amount of a substance that contains the same number of units (*moles, ions, atoms,* or *electrons*) equal to the number of atoms in 12 mass units of carbon-12. Hereafter when energies are given in this text, energy per gram mole or per indicated gram molar quantity is to be understood.

they are called **molecular orbitals.** Like the atomic orbitals they obey the Pauli exclusion principle in that no molecular orbital may contain more than two electrons and these must have opposite spin.

A molecular orbital of the type used in the formation of the hydrogen molecule is symmetrical about the line joining the two nuclei. It is called a *sigma* (σ)-*type orbital,* in analogy to the *s* atomic orbital, which is spherically symmetrical about the nucleus. Since the bond results from the overlapping of two *s* orbitals, it is called an *s*—*s* bond.

Hybridization of Atomic Orbitals. The ground state of the carbon atom has two unpaired electrons in two *p* orbitals (Table 2–*1*, p. 7). Hence it might be expected to form either two ionic bonds or two covalent bonds. Since the electrons are in *p* orbitals, the covalent bonds should make an angle of 90°. It is known, however, that carbon usually forms four covalent bonds and that in molecules such as methane, CH_4, or carbon tetrachloride, CCl_4, the bonds are directed to the corners of a regular tetrahedron and make angles of 109°28′ to each other. The explanation is that the 2*s* orbital can combine with the three 2*p* orbitals to give four sp^3 hybrid orbitals of equal energy. Each orbital holds one electron and all four electrons have parallel spins.[2] Promoting an electron from the 2*s* orbital to a 2*p* orbital requires 96 kcal. of energy, and the change from one *s* and two *p* orbitals to four sp^3 hybrids would require somewhat more. The formation of four *s*—sp^3 bonds in methane, however, liberates 397 kcal., whereas the formation of two *s*—*p* bonds would yield a maximum of 198 kcal. The difference is more than enough to bring about the hybridization of the atomic orbitals.

Actually the bond strength of an *s*—*p* bond is considerably less than that of an *s*—sp^3 bond. One lobe of an sp^3 hybrid is much larger and the other much smaller than the lobes of a *p* orbital (Fig. 2–*6a*). Hence an sp^3 orbital can overlap the orbital of another atom much better than can a *p* orbital and lead to a stronger bond. The lobes of the sp^3 hybrids are tetrahedrally distributed, thus accounting for bond angles of 109°28′ (Figs. 2–*6b* and 2–*7*).

The distribution of orbitals of like energy that have a common nucleus can be

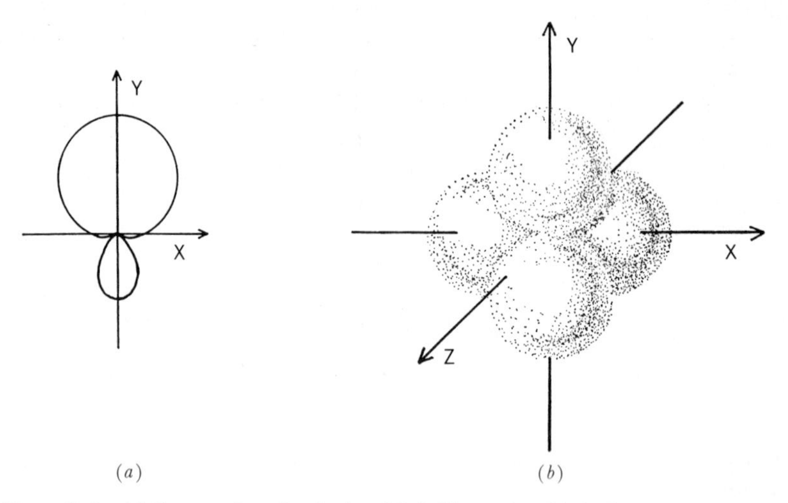

Figure 2–*6*. (*a*) Cross section of a single sp^3 hybrid atomic orbital; (*b*) perspective of four sp^3 hybrid orbitals.

---

[2] The symbol sp^3 merely means that the orbital is one of four made up of one *s* orbital and three *p* orbitals. It is pronounced *ess pee three* and not *ess pee cubed.* The symbol first was used by spectroscopists. A chemist would have written it sp_3.

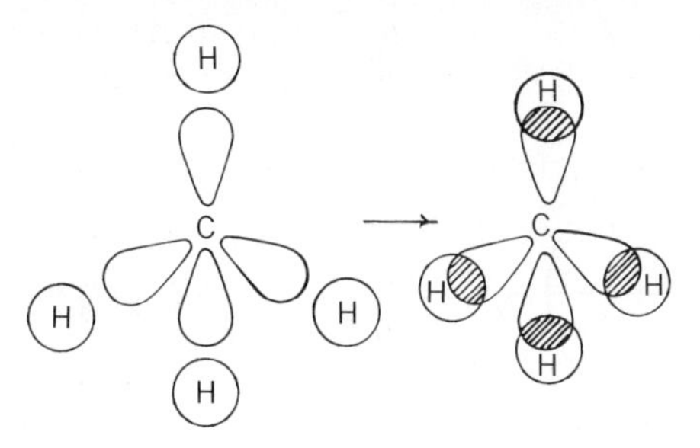

Figure 2–7. Tetrahedral bonding of hydrogen to carbon in methane.[3]

explained very simply on the basis of electrostatic repulsion. The lobes of the p orbitals, for example, are essentially negative charges equivalent to $\frac{1}{2}e$, and the repulsive forces cause them to be separated from each other as far as possible. Hence the axes of the p orbitals are mutually perpendicular. The situation is analogous to six like charges on the surface of a sphere. Mutual repulsion would cause four charges to be distributed at equal intervals about the equator and two at the poles (Fig. 2–8). Mutual repulsion of the lobes of four sp^3 hybrids causes them to be tetrahedrally distributed.

Hybrid orbitals other than sp^3 also are possible. The ground state of boron contains two electrons in the $2s$ orbital and one in a $2p$ orbital. Hence boron might be expected to form a mono- or trivalent ion or a monocovalent bond. Combination of the $2s$ orbital with two $2p$ orbitals, however, yields three sp^2 hybrid orbitals, each of which contains one electron. Electrostatic repulsion causes the three orbitals to lie in a plane and make angles of 120° with each other (Fig. 2–9*a*,*b*). Boron forms tricovalent compounds such as boron trichloride, and these molecules have been shown to be planar with 120° bond angles. Somewhat stronger bonds should result from sp^3 hybridization, but one sp^3 orbital would remain empty. The greater use of the s orbital when three sp^2 orbitals are formed, leaving an empty p orbital, makes the stability of sp^2 hybrids greater than that of sp^3 hybrids. Apparently the greater bond strength of the sp^3 hybrids is not sufficient to overcome the greater stability of the sp^2 hybrids. Hybridization of three atomic orbitals is called *trigonal hybridization.* The axis of the remaining p orbital is perpendicular to the plane of the sp^2 orbitals.

Beryllium with two electrons in the $2s$ orbital loses both electrons and forms divalent ions as do magnesium and the elements of the IIa group of the periodic table. Elements of the IIb group, however, form molecules that are largely dicovalent. Here hybridization of an s orbital with one p orbital provides two sp hybrid orbitals that can contain the two electrons, now with parallel spins. Electron repulsion causes these orbitals to be opposed

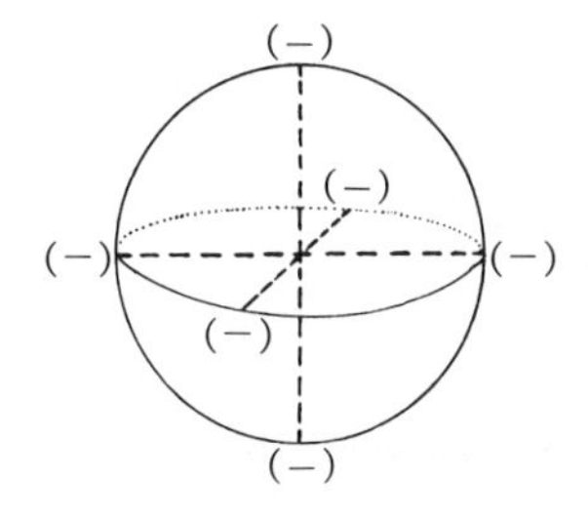

Figure 2–8. Distribution of six like charges on the surface of a sphere.

[3] In the illustrations used in Fig. 2–7 and later, the lobes of the hybrid orbitals have been elongated to make the space relationships clear and do not represent the actual shape of the orbitals.

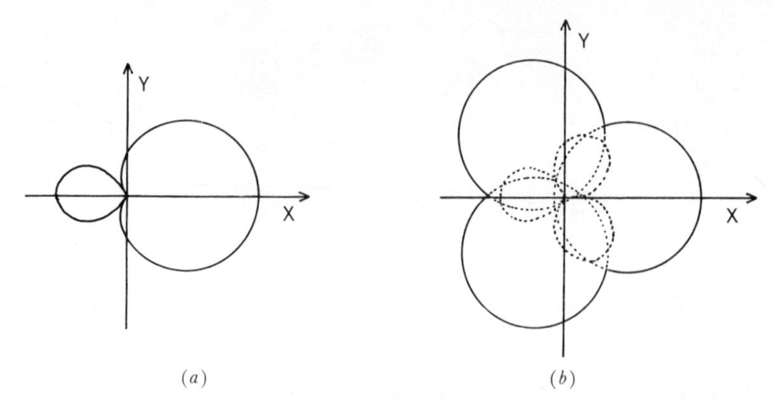

Figure 2–9. Cross section of (*a*) a single sp^2 hybrid atomic orbital and (*b*) three sp^2 hybrid orbitals.

to each other and leads to 180° bond angles (Fig. 2–*10*). The halides of zinc, cadmium and mercury have been shown to be linear in the vapor state, and mercurous chloride, Hg_2Cl_2, is linear in the crystal (Cl—Hg—Hg—Cl). Hybridization of an *s* orbital and a *p* orbital is called *digonal hybridization*. The axes of the remaining *p* orbitals and of the *sp* hybrid are mutually perpendicular.

For elements of the third and higher periods, *d* orbitals are available for bond formation and lead to other types of hybrid orbitals. Hybridization of one 3*s* orbital, three 3*p* orbitals, and one 3*d* orbital gives five sp^3d orbitals that are directed to the corners of a triangular bipyramid (Fig. 2–*11a*) and accounts for the molecule PCl_5. The four dsp^2 orbitals that result from the hybridization of one 3*d*, one 4*s*, and two 4*p* orbitals are directed to the corners of a square (Fig. 2–*11b*), and nickel complexes are planar with 90° bond angles. Hybridization of two 3*d*, one 4*s*, and three 4*d* orbitals gives six d^2sp^3 orbitals directed to the corners of an octahedron (Fig. 2–*11c*), the characteristic arrangement for cobalt complexes.

Hybridization of orbitals takes place only when their energy levels are close together (Fig. 2–*1*, p. 8) and would not be expected between orbitals that have principal quantum numbers of 2 and 3. Hence the octet rule of the older Lewis[4] theory that no atom may have more than four pairs of electrons in its valence shell must be obeyed for elements in the second period.

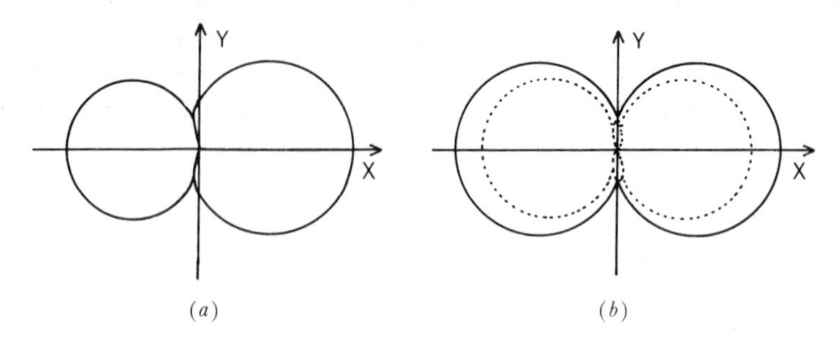

Figure 2–*10*. Cross section of (*a*) a single *sp* hybrid atomic orbital and (*b*) two *sp* hybrid orbitals.

[4] Gilbert Newton Lewis (1875–1946), professor of physical chemistry at the University of California. He is noted for his development of the electronic theory of valency, for his theory of acids and bases (p. 33), for his work in chemical thermodynamics, and for his investigations of deuterium and the absorption spectra of organic compounds.

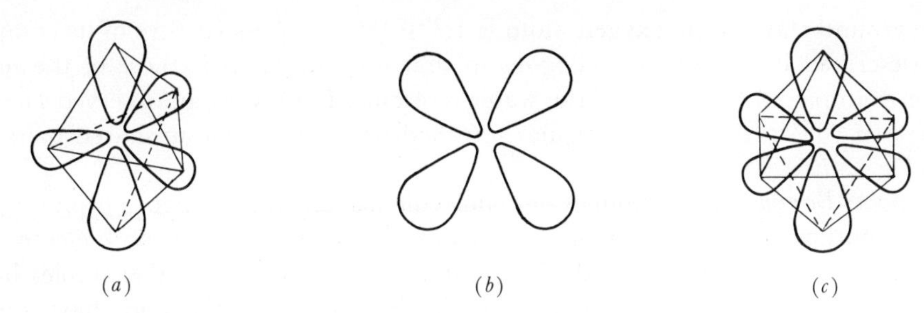

Figure 2–*11*. Spatial arrangement of (*a*) five sp^3d hybrid orbitals, (*b*) four dsp^2 hybrid orbitals, and (*c*) six d^2sp^3 hybrid orbitals.

Partial Hybridization of Atomic Orbitals. Hybridization need not be pure *sp*, sp^2, or sp^3 and have *s*:*p* ratios of only 1:1, 1:2, and 1:3. Instead *s* and *p* orbitals may mix in various proportions. Moreover, some mixing may take place with higher orbitals also, such as the *d* and *f* orbitals. The amount of *s* orbital in the hybrid is referred to as its *s character,* or the amount of *p* orbital as its *p character.*

In the ground state the nitrogen atom has three unpaired electrons in three 2*p* orbitals, and one reasonably might expect that combination with three hydrogen atoms would yield an ammonia molecule with bond angles of 90°. The observed H—N—H angle, however, is 107°. One possible explanation is that the larger angle is due to the electrostatic repulsion of the bonding electron clouds. Calculations indicate, however, that this repulsion would not widen the angle more than about 5°. Moreover, microwave spectroscopy and calculations of dipole moment indicate that the unshared pair of electrons on nitrogen is not symmetrically distributed about the nitrogen in the 2*s* orbital but extends from the nitrogen in a direction opposite to that of the resultant of the nitrogen-hydrogen bonds. In other words there is some sp^3 hybridization with widening of the bond angle. The orbital containing the unshared pair thus is not a pure *s* orbital but has some *p* character, and the hydrogen overlap is not with pure *p* orbitals, but with orbitals that have some *s* character. The resulting structure, though tetrahedral with the unshared pair occupying one corner of the tetrahedron, is not regular as for methane since the base of the tetrahedron has a greater area than the sides (Fig. 2–*12a*). When the ammonia molecule bonds with a proton to form the ammonium ion, complete sp^3 hybridization takes place with the formation of four equivalent sp^3—*s* bonds to give a regular tetrahedral structure.

The bond angles of phosphine (PH_3, 93°) and arsine (AsH_3, 92°) approach 90°. The greater distance of the valence electrons in the 3*s* and 3*p* orbitals from the nucleus and the still greater distance of the 4*s* and 4*p* electrons decrease the repulsive forces among the bonding electrons and permit the hydrogen nuclei to move closer together. The bonding orbitals become more like pure 3*p* and 4*p* orbitals with bond angles closer to 90°.

Figure 2–*12*. Spatial arrangement of atoms and unshared pairs of electrons in (*a*) the ammonia molecule and (*b*) the water molecule.

The ground state of the oxygen atom is $1s^2 2s^2 2p^4$ and the structure of its compounds may be described in a manner analogous to that used to describe those of the nitrogen atom. For example, the structure of the water molecule, H_2O, with an observed H—O—H bond angle of 105° is that of an irregular tetrahedron with the two unshared pairs at two corners (Fig. 2–*12b*).

Semipolar Bonds. If an atom of one molecule has an empty valence orbital and that of another contains an unshared pair of electrons, the latter can share both electrons with the former to give a strong σ bond. Trivalent boron compounds, for example, have an empty p orbital, and trivalent nitrogen and divalent oxygen compounds have unshared pairs of electrons. Hybridization of atomic orbitals takes place and the boron bonds become tetrahedral as indicated in Fig. 2–*13a* for the boron hydride-ammonia complex and in Fig. 2–*13b* for boron fluoride hydrate. The bond is covalent, but because the nitrogen or oxygen has supplied both electrons, they have in effect supplied boron with a charge of approximately one electron and have lost a corresponding amount. This fact is indicated by placing a plus sign on the *donor atom* and a minus sign on the *acceptor atom* and gives rise to the term **semipolar bond** or **dative bond.** Compounds involving several donor molecules usually are called *coordination complexes,* and the semipolar double bond frequently is called a *coordinate covalence.*

It should be noted that the formation of the ammonium ion or the fluoroborate ion does not involve the formation of a semipolar bond. Although the nitrogen or the fluorine has supplied both electrons, a charged ion has been transferred in the process.

$$H_3N\!: + H\!:\!Cl \longrightarrow [H_3\overset{+}{N}\!:\!H][\!:\!Cl^-]$$

$$H\!:\!\underset{\cdot\cdot}{\ddot{F}}\!: + BF_3 \longrightarrow [H^+][\!:\!\underset{\cdot\cdot}{\ddot{F}}\!:\!\bar{B}F_3]$$

All of the N—H and B—F bonds are ordinary covalent bonds, and the bonds between the ions are ionic bonds.

Electron-dot Formulas. For convenience in writing the formulas of compounds with ionic bonds the charges may be indicated as in Li^+F^- or $Mg^{2+}(Cl^-)_2$, although more frequently they are omitted. The electronic structures of molecules that have covalent bonds may be indicated by the Lewis convention in which dots or other marks represent the valence electrons (Fig. 2–*14*). More often an electron-pair bond is indicated by a long dash joining the atoms. Usually unshared pairs of valence electrons are indicated by two dots, although a bar sometimes is used for this purpose.

Double and triple bonds may be shown by placing two or three pairs of electrons or long dashes between the atoms involved. Semipolar bonds are best indicated by assigning the proper charges to the atoms. Pre-Lewis formulas showing double bonds between oxygen and sulfur or between oxygen and phosphorus as in the last formulas given for sulfuric acid and phosphorus oxychloride have again come into use because it has been shown that the sulfur-oxygen and phosphorus-oxygen interatomic distances are somewhat shorter than those expected for semipolar double bonds. These bonds, however, are not the same as ordinary π double bonds, and it is better to distinguish them from ordinary double bonds by the use of the semipolar representation (cf. pp. 257, 271).

(*a*) (*b*)

Figure 2–*13.* (*a*) Boron hydride-ammonia complex; (*b*) boron fluoride hydrate.

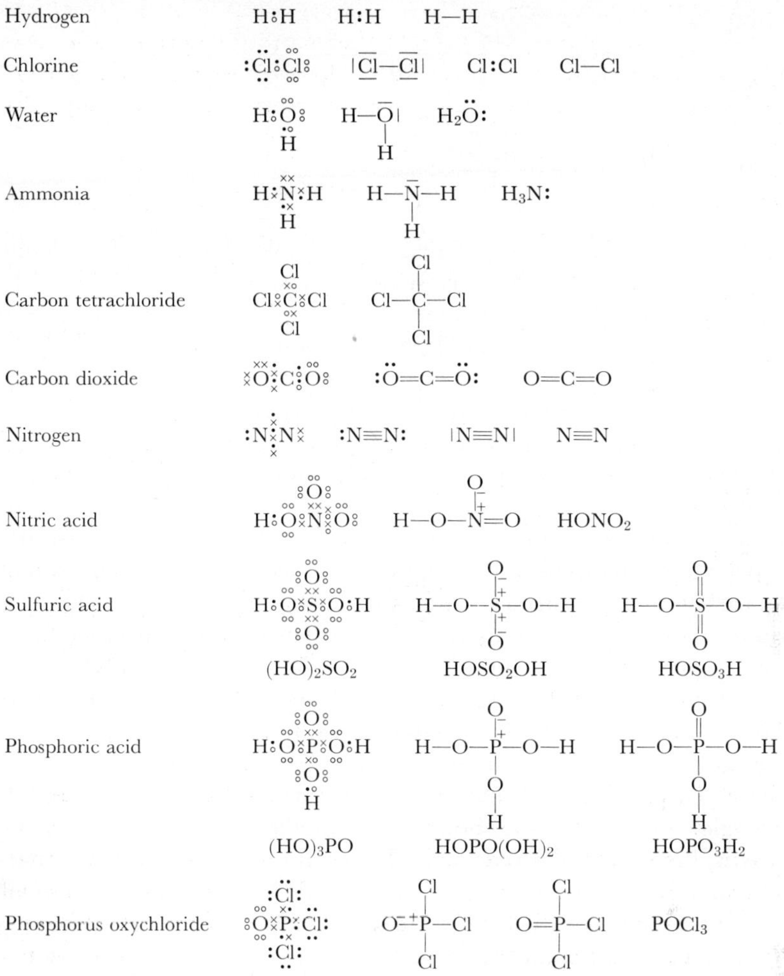

Molecule	Representations			
Hydrogen	H:H	H:H	H—H	
Chlorine	:Cl:Cl:	|Cl—Cl|	Cl:Cl	Cl—Cl
Water	H:O: (H)	H—O| (H)	$H_2\ddot{O}$:	
Ammonia	H:N:H (H)	H—N—H (H)	H_3N:	
Carbon tetrachloride	Cl:C:Cl (Cl, Cl)	Cl—C—Cl (Cl, Cl)		
Carbon dioxide	:O::C::O:	:O=C=O:	O=C=O	
Nitrogen	:N:::N:	:N≡N:	|N≡N|	N≡N
Nitric acid	H:O:N:O: (:O:)	H—O—N=O (O, with − and +)	$HONO_2$	
Sulfuric acid	H:O:S:O:H (:O:, :O:)	H—O—S—O—H (O, O, with − and +)	H—O—S—O—H (=O, =O)	
	$(HO)_2SO_2$	$HOSO_2OH$	$HOSO_3H$	
Phosphoric acid	H:O:P:O:H (:O:, :O:H)	H—O—P—O—H (O with − and +; O—H)	H—O—P—O—H (=O; O—H)	
	$(HO)_3PO$	$HOPO(OH)_2$	$HOPO_3H_2$	
Phosphorus oxychloride	:O:P:Cl: (:Cl:, :Cl:)	O$\overset{-+}{—}$P—Cl (Cl, Cl)	O=P—Cl (Cl, Cl)	$POCl_3$

Figure 2–*14*. Various representations of covalent molecules.

In some of the formulas in Fig. 2–*14* and occasionally throughout the text, different marks are used to differentiate between the sources of the bonding electrons. For example, in the ammonia molecule the crosses represent the five valence electrons that belonged originally to the nitrogen atom and the dots the electrons that belonged to the hydrogen atoms. Since electrons are all alike and constantly are interchanging (*electronic interaction*), there is no distinction between them once the bond is formed. The above representation is justified only in that it simplifies the procedure in keeping account of electrons when writing formulas representing electronic structures.

Bond Strengths. If sufficient energy can be supplied to either an ionic or a covalent compound, for example by heat or electromagnetic radiation, the attractive forces between the atoms can be overcome. Conceivably a bond can break in two ways. Either one electron can go to each atom to give neutral particles (*homolysis*), or both electrons can go with one atom (*heterolysis*). From a consideration of ionization potentials and electron affinities (Table 2–*1*, p. 7), it is clear that homolysis is energetically the more favorable, assuming that no other forces are present to stabilize the ions.

$$\begin{array}{lll} \mathrm{Na\cdot} \longrightarrow \mathrm{Na^+} + e & & -5.14 \text{ e.v.} \\ \mathrm{Cl\cdot} + e \longrightarrow \mathrm{Cl{:}^-} & & +3.69 \text{ e.v.} \\ \hline \mathrm{Na\cdot} + \mathrm{Cl\cdot} \longrightarrow \mathrm{Na^+} + \mathrm{Cl{:}^-} & & -1.45 \text{ e.v.} = -33 \text{ kcal.} \end{array}$$

$$\begin{array}{lll} \mathrm{H\cdot} \longrightarrow \mathrm{H^+} + e & & -13.60 \text{ e.v.} \\ \mathrm{H\cdot} + e \longrightarrow \mathrm{H{:}^-} & & +\ 0.75 \text{ e.v.} \\ \hline \mathrm{H\cdot} + \mathrm{H\cdot} \longrightarrow \mathrm{H^+} + \mathrm{H{:}^-} & & -12.85 \text{ e.v.} = -296 \text{ kcal.} \end{array}$$

Thus it requires 33 kcal. more energy to dissociate a mole of sodium chloride into ions than into atoms, and 296 kcal. for the corresponding process with hydrogen. Hence when hydrogen is passed through an electric arc, hydrogen atoms are formed. Since electron affinities ordinarily are small compared to ionization potentials (Table 2–*1*, p. 7), this behavior is the usual one.

The energy required to break bonds homolytically is referred to as *bond strength* or *bond energy*. Two methods of expressing bond energies are in general use. One method assumes that the bond energy between two given kinds of atoms is the same regardless of the structure of the rest of the molecule. It is calculated from the energies required to dissociate diatomic molecules into atoms and from thermochemical data that lead to the total bond energies for polyatomic molecules. By subtractive procedures that go from molecules that contain only carbon and hydrogen to molecules containing other elements, bond energies for the various types of bonds are arrived at that, when added together for a particular molecule, give approximately the bond energy for the molecule as a whole. They usually are referred to simply as **bond energies,** although they should be called **empirical bond energies.** Their chief use has been to estimate the heat liberated or absorbed during a reaction. The heat of reaction in the gas phase is the difference between the sum of the bond energies of the products and that of the reactants (cf. Problems 2–*4* and 2–*5*). Empirical bond energies are of no value, however, in predicting the relative rates of different reactions, that is, the course of a reaction. The second method recognizes that the energy required to break a bond between two given atoms depends on the structure of the rest of the molecule. This energy is called the **bond dissociation energy** and refers to the energy required to break a specific bond. For example, the empirical bond energy for the N—H bond is taken as 93 kcal., which is one third of the energy necessary to convert the ammonia molecule into one atom of nitrogen and three atoms of hydrogen. The energy required to break the first nitrogen bond in ammonia, however, is 107 kcal., and is the bond dissociation energy for the reaction $H_2N{-}H \longrightarrow H_2N\cdot + \cdot H$. Breaking the next bond requires only 88 kcal. and the third only 85 kcal. Similarly the average bond energy for the oxygen-hydrogen bond in water is 111 kcal., but breaking the first bond requires 119 kcal. and the second 103 kcal. Empirical bond energy is the same as average bond energy for molecules such as water and ammonia, but not the same for molecules that contain more than one kind of bond.

Some bond dissociation energies and empirical bond energies are listed in Table 2–*2*. Next to hydrogen, carbon atoms form stronger covalent bonds with each other and with other atoms than do most other elements.

Polarity of Covalent Bonds, Electric Dipole Moments, and Polar Molecules. In a diatomic molecule that has like atoms such as hydrogen or chlorine, each atom has the same ionization potential and the same electron affinity, and the bonding pair of electrons spends the same amount of time about both nuclei. For bonds between unlike atoms, however, it would be expected that the bonding electrons would be held more strongly by one atom than by the other. For example, the transfer of an electron from a fluorine to a chlorine atom would require 13.7 e.v., whereas the transfer from chlorine to fluorine would

TABLE 2–2. SOME BOND DISSOCIATION ENERGIES (D) AND EMPIRICAL BOND ENERGIES (E) IN KILOCALORIES PER MOLE*

BOND	D	E	BOND	D	E	BOND	D	E
F—F	36	37	C—O		84	H_3C—I	53	
Cl—Cl	57	58	H_3C—OH	89		C_6H_5—I	57	
Br—Br	45	46				Benzyl—I	39	
I—I	35	36	H_2C=O		164	Allyl—I	36	
			RCH=O		171			
H—H	103	104	R_2C=O		174	$O_2 \longrightarrow 2\,O$	118	119
H—F	134	135				O—O		33
H—Cl	102	103	N—H		93	HO—OH	51	
H—Br	87	88	H_2N—H	107		*t*-BuO—OH	39	
H—I	70	71	CH_3NH—H	105				
						$N_2 \longrightarrow 2\,N$	225	226
C—H		99	C—N		70	N—N		38
n-C_4H_9—H	101		H_3C—NH_2	80		H_2N—NH_2	66	
s-C_4H_9—H	95		C=N		147	MeNH—NH_2	70	
t-C_4H_9—H	90							
OHC—H	76		C—F		105	S—H		81
			H_3C—F	107		HS—H	95	
C—C		83	C—Cl		79	CH_3S—H	90	
Et—Et	80		H_3C—Cl	80		C_2H_5S—H	96	
			C_6H_5—Cl	86				
C=C		147	C—Br		66	C—S		62
C≡C		194	H_3C—Br	67		H_3C—SH	70	
			C_6H_5—Br	71		H_3C—SCH_3	73	
O—H		111	Benzyl—Br	51				
HO—H	119		Allyl—Br	46		S—S		51
CH_3O—H	100		C—I		57	CH_3S—SCH_3	73	

* Bond dissociation energies, D, in kilocalories per mole, are selected from Cottrell, "The Strengths of Chemical Bonds" (1958) and from Mortimer, "Reaction Heats and Bond Strengths" (1962). Empirical bond energies, E, are taken from Pauling, "The Nature of the Chemical Bond" (1960). The values for D are at 0°K., whereas those for E are at 25°C. Roughly 1 kcal. must be added to D at 0°K. to convert it to D at 25°C.

require only 9.5 e.v. Hence the shared electron pair in chlorine fluoride, Cl**:**F, would be expected to be closer to the fluorine nucleus on the average than to the chlorine, thus making the fluorine end of the molecule negative with respect to the chlorine end. The bond is said to be polarized, which frequently is indicated by the symbol $\overset{\longleftrightarrow}{\text{Cl—F}}$ or $\overset{\delta+}{\text{Cl}}\text{—}\overset{\delta-}{\text{F}}$. Such a molecule has a positive and negative end and is called a *dipole*. Because there is a difference in charge separated by a distance, the molecule has an electric moment, called the *electric dipole moment*.

The electric moment of a molecule, which is the product of the size of the charge by the distance between the centers of the positive and negative charges, can be measured. If a gas whose molecules have an electric dipole is placed between the plates of a condenser and a potential difference is applied to the plates, the molecules tend to line up with the electric field and change the capacity of the condenser. The change in capacity as compared with a vacuum is the *dielectric constant* of the substance. The alignment of the molecules is opposed by thermal agitation, and from the variation of dielectric constant with the temperature it is possible to calculate the dipole moment of the molecule, which is designated by the symbol μ. An electron has a charge of 4.80×10^{-10} electrostatic units, and hence the charge at either end of the bond is less than this but of the order of 10^{-10} esu. Since the distances between the centers of charge are of the order of 10^{-8} cm., the usual magnitude of the moment is of the order of 10^{-18} esu. $\times$ cm. This quantity is known as a *debye unit* or simply a *debye,* after P. Debye, who first measured electric moments, and is given the symbol D. Chlorine fluoride has a dipole moment of 0.88 D.

It should be noted here that the polarized bond is not the only contributor to the dipole moment. The energy required to form a positive chloride ion and a negative hydride ion is 12.2 e.v., and the energy required to form a positive hydrogen ion and a negative chloride ion is 9.9 e.v. The difference of 2.3 e.v. is less than the difference of 4.2 e.v. calculated from similar considerations for chlorine and fluorine. Hence hydrogen chloride might be expected to have a lower dipole moment than chlorine fluoride. The observed dipole moment of hydrogen chloride, however, is 1.08 D. The reason for the higher moment is the unshared electrons in the valence shell of chlorine, which produce a chlorine nucleus–*p* cloud dipole that is added vectorially to the σ bond dipole and may be represented as $\overset{+\!\!\longrightarrow\!+\!\!\rightarrow}{\text{H—Cl}}\colon$. A more striking demonstration of the effect of unshared electrons is shown by the dipole moments of the ammonia molecule and nitrogen fluoride. Bond moments are vector quantities and the dipole moment of ammonia and of nitrogen fluoride could be considered as arising from the vector sum of the individual bond moments. From the ionization potentials and electron affinities of the atoms, the dipole moment for ammonia should be small and that of nitrogen fluoride rather large. The observed moments are 1.47 D and 0.23 D respectively. The explanation of this discrepancy is that the moment of ammonia is due almost entirely to the unshared pair of electrons. The low moment for nitrogen fluoride results from the fact that the effect of the unshared electrons is opposed by the strong nitrogen-fluorine bond moments, which are in the opposite direction, namely $\overset{+\!\!\longrightarrow}{\text{N—F}}$. The dipole moments of a few compounds are summarized in Table 2–3.

Hydrogen and chlorine have zero moment because the bonds themselves are nonpolar. The bonds in mercuric chloride, boron fluoride, methane, and carbon tetrachloride are polarized, but the digonal, trigonal, and tetrahedral arrangement of the bonds leads to symmetrical molecules and vectorial addition of the individual bond moments leads to zero moment for the molecule.

Molecules that have a permanent dipole are known as *polar* molecules. It is only in this sense that the term polar should be used. Substances such as sodium chloride that have ionic bonds should be called ionic or electrovalent compounds. Liquids such as water should be called polar only if one is referring to their properties arising from the dipole in the molecule, and not when referring to their abnormal properties such as association caused by hydrogen bonding (p. 23), or to their power to dissolve salts with the formation of electrically conducting solutions, which is dependent on ion-dipole interaction and a high dielectric constant (p. 25). Although a high dielectric constant is related to a high dipole moment, other factors also are involved. Thus water, which is the best ionizing solvent, has a dipole moment of 1.8 and a dielectric constant of 81, whereas ethyl bromide,

TABLE 2–3. ELECTRIC DIPOLE MOMENTS OF SOME SIMPLE MOLECULES

FORMULA	μ IN D	FORMULA	μ IN D	FORMULA	μ IN D
H_2	0	HBr	0.80	$CHCl_3$	1.02
Cl_2	0	HI	0.42	CCl_4	0
BrF	1.29	$HgCl_2$	0	NH_3	1.47
ClF	0.88	BF_3	0	NF_3	0.23
BrCl	0.57	CH_4	0	H_2O	1.85
HF	1.91	CH_3Cl	1.87	H_2S	0.92
HCl	1.08	CH_2Cl_2	1.55	HCN	2.95

which does not dissolve salts, has the same dipole moment, but a dielectric constant of only 9. Although liquid hydrogen cyanide has a higher dipole moment (2.9) and dielectric constant (115) than water, it is a poorer ionizing solvent. The loose use of the term "polar" to include all of these factors, as in the common expression, "Water is a polar solvent," is so frequent, however, that it seems desirable to drop the term "polar" and to use the term "dipolar" when referring to a molecule that has a dipole moment.

Electronegativity. The power of a chemically bonded atom or group to attract electrons is referred to as its *electronegativity*.[5] Electronegativity of atoms in the ground state should be proportional to the sum of the energy necessary to remove an electron (its ionization potential) and the energy evolved if the atom acquires an electron (its electron affinity). The values given in Table 2–*1*, p. 7, indicate that the electronegativity increases from left to right in a given row of the periodic table, and decreases from top to bottom in a given column. Values for the ionization potentials and electron affinities for atoms in the ground state, however, are not expected to be the same for atoms in compounds, since in molecules the atoms are not in the ground state but in what is called the *valence state*. The valence state of an atom is the state it would have if all the other atoms with which it is joined were removed with their electrons without any rearrangement of the electrons belonging to the atom.

Numerous attempts have been made to arrive at quantitative values for relative electronegativities in the valence state. Such assignments are made on the assumption that electronegativity is a constant quantity regardless of the nature of the other groups to which the atoms are attached. This assumption certainly is not true. In this respect the assignment of valence-state electronegativities suffers from the same defect as attempts to assign values to atom and group refractivities, empirical bond energies, and individual bond dipole moments. Nevertheless all of these concepts are useful at times for qualitative comparisons. Insofar as valence-state electronegativities are concerned, the rules stated at the beginning of this section generally hold; namely, that electronegativity increases from left to right in a row and decreases from top to bottom in a column of the periodic table. An additional useful rule is that the greater the *s* character of a hybrid orbital, the more electronegative the atom because *s* orbitals are closer to the nucleus than are *p* orbitals. Thus the electronegativity of carbon in the valence state increases as the hybridization of bond orbitals changes from tetrahedral to trigonal to digonal; that is, $sp^3 < sp^2 < sp$.

INTERMOLECULAR FORCES

It has been indicated that the forces holding ions together in crystals of ionic compounds are due to electrostatic attraction between the positive and negative charges on the ions (p. 9). The high melting points of ionic compounds are associated with these strong attractive forces. Molecules that have only covalent bonds are neutral and have only very weak attraction for each other. Hence they are gases, liquids, or low-melting solids. Yet weak attractive forces between the molecules do exist as evidenced by the fact that most such compounds are liquids or solids. Even the inert gas helium can be liquefied when cooled under pressure, indicating that attractive forces exist between the helium atoms when they are close enough together and not moving too rapidly.

---

[5] The term *electronegativity* is confusing because the property that it describes is the extent to which an atom or group is positive with respect to other groups. Originally the term was used to describe the electrical state of the group attached to hydrogen in acids; the stronger the acid, the more negative the group. With the realization that the strength of acids varied with the ability of a group to take an electron away from a hydrogen atom, electronegativity came to be used to indicate the relative ability of groups to attract electrons (cf. also p. 9).

Attractive forces between uncharged molecules are known as **van der Waals forces**[6] because they are responsible for the a term in the van der Waals equation, $\left(P + \frac{a}{V^2}\right)(V - b) = RT$ for one mole of gas. The law for a perfect gas, $PV = RT$, holds only if the molecules of a gas have no attraction for each other and occupy no volume. By introducing a constant a to take care of the attraction between molecules and a constant b to offset the volume occupied by the molecules, the equation fits more closely the experimental data. Constants a and b are characteristic for each compound. Van der Waals forces may be divided into three groups: (*1*) those that are due to a transient polarization, which usually are called *dispersion* or *London forces*;[7] (*2*) those that are due to a permanent polarization, known as *dipole-dipole attraction;* and (*3*) those that are due to so-called *hydrogen bonding.*

London Forces. All attractive forces are electrical; that is, they are due to attraction between positive and negative charges. If there is a difference in charge between two parts of a molecule, the molecules behave much like an aggregation of bar magnets, with the positive and negative ends of one molecule attracting the negative and positive ends of other molecules.

Although the negative electrons in a neutral molecule are balanced by the positive charges on the nucleus, London forces arise because the electrons are in motion, and the center of density of the electrons does not coincide continuously with the center of density of the positively-charged nuclei. Hence the molecules sometimes have a positive end and a negative end; that is, they acquire an *electric dipole*. In this condition they are *polarized* and can exert an attraction for other molecules having a dipole. In helium, for example, the particles are nonpolar only when the electrons are on opposite sides of the nucleus and in line with it (Fig. 2–*15a*). In all other positions the atoms are dipolar (Fig. 2–*15b*). If two properly oriented polarized molecules approach each other closely enough, a slight attraction between the molecules results (Fig. 2–*15c*). A polarized molecule even can induce a dipole in a nonpolarized molecule and lead to some attraction. Because such molecules continually revert to the nonpolarized state, the dipole may be said to be *transient*.[8]

The ease of polarization or *polarizability* of a molecule increases as the number of electrons in the molecule increases and as their distance from the nucleus increases. For example, in a given column of the periodic table, the atomic polarizability should increase from top to bottom. Since the refraction of light is caused by the interaction of the electric

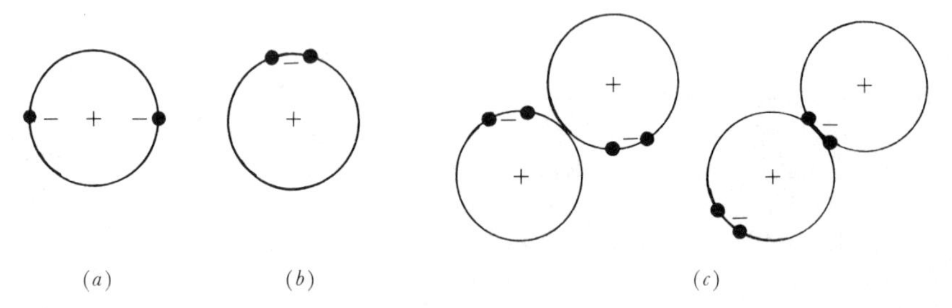

Figure 2–*15*. Representation of source of attractive forces (*London forces*) between helium atoms: (*a*) nonpolar atom; (*b*) polarized atom; (*c*) attraction between polarized atoms.

---

[6] Johannes Diderik van der Waals (1837–1923), professor of physics at the University of Amsterdam, introduced his equation in 1873. He was awarded the Nobel Prize in Physics in 1910.

[7] Fritz London (1900–1954), professor of chemical physics at Duke University. He is noted for his application of theoretical physics to superconductivity and other low-temperature phenomena.

[8] Frequently it is only these forces that are called van der Waals forces. The a term in the van der Waals equation, however, includes all attractive forces between molecules.

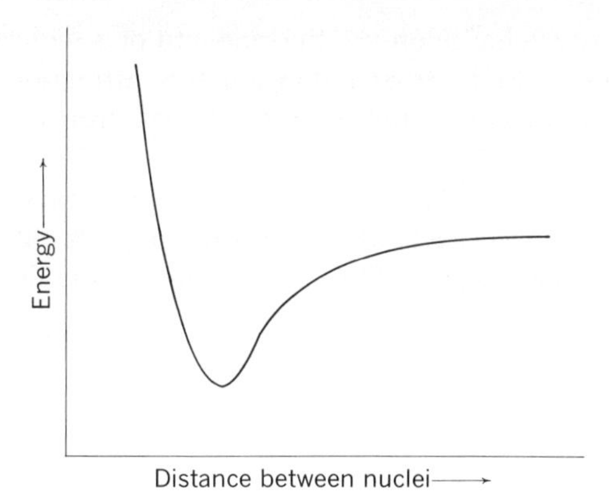

Figure 2–*16*. Energy diagram for van der Waals attraction between two molecules.

vectors of the light ray with the electrons, the atomic refraction is a direct measure of polarizability. The atomic refractions for the halogen atoms are F, 1.00; Cl, 5.97; Br, 8.87; and I, 13.90.

The greater the polarizability of a molecule, the stronger should be the London forces and the higher the boiling points. Hydrogen, fluorine, chlorine, bromine, and iodine boil at −253°, −187°, −35°, 59°, and 184° respectively. Similarly helium, neon, argon, krypton, xenon, and radon boil at −270°, −246°, −186°, −152°, −109°, and −62°.

The attraction caused by London forces varies inversely with the seventh power of the distance. Hence they are operative only over a very short range. The behavior of two molecules approaching each other may be represented by the usual diagram shown in Fig. 2–*16* in which the potential energy of the system is plotted against the distance between the two molecules. When the molecules get close enough, the attractive forces resulting from the polarization of the molecules begin to be effective and the potential energy decreases. This attraction is opposed by the repulsion between the electron shells and is rapidly overcome on closer approach. The minimum in the potential energy curve is at the internuclear distance of closest approach and leads to the calculation of what are known as *van der Waals radii* for individual atoms. These radii are larger than atomic radii, which are calculated from interatomic distances in molecules, because the attractive forces between bonded atoms are greater than the attractive forces between molecules.

Dipole-dipole Attraction. Both neon and hydrogen fluoride have ten electrons and are about the same size, argon and hydrogen chloride have eighteen electrons each, krypton and hydrogen bromide have thirty-six electrons, and xenon and hydrogen iodide have fifty-four electrons. The members of each pair might be expected to have about the same polarizability and hence the same boiling points. The hydrogen halides, however, boil 265°, 101°, 85°, and 73° higher respectively than their partners. The difference is that the hydrogen halides are polar molecules. They have permanent dipole moments (Table 2–*3*, p. 20) that lead to **dipole-dipole attraction.**

Hydrogen Bonding. The effectiveness of the dipole moment in increasing the attraction between molecules is dependent on how well it can operate. For example, hydrogen fluoride with a dipole moment of 1.9 D boils at +19° but ethyl fluoride with an almost identical dipole moment boils at −32°. The reason for this difference is that the positive end of the dipole in hydrogen fluoride is relatively exposed on the hydrogen nucleus, whereas that in ethyl fluoride is buried in the ethyl group and screened by the surrounding electrons. The result is that the positive end of a hydrogen fluoride molecule can approach the concentration of negative charge on the fluorine end of a molecule closely and give rise to a strong attractive force. In hydrogen fluoride this attraction is so strong (37 kcal.) that the HF_2^- ion exists in aqueous solution and polymers from $(HF)_2$ to $(HF)_6$ have been detected in the vapor state.

Similar effects are observed when hydrogen is bonded to oxygen or nitrogen and when a concentrated negative charge, such as the unshared pairs of electrons on oxygen or nitrogen, is present on another atom. Thus water and ammonia are associated liquids.[9]

```
                    H
                     \
                      O------H—O
                     /         \
                    H           H
      H            /                     H          H  H
       \          /                       \          \ |
        O------H—O                       H—N------H—N
       /          \                       /            \
      H            H                     H          H  H  H
                                                     \ | /
                                                       N
```

Because this type of attraction always involves hydrogen and is stronger than dipole association not involving hydrogen, it is referred to as **hydrogen bonding.** It should not be confused with the valence bonds between hydrogen and other elements. Hydrogen bonding plays an important role in the physical properties of many organic molecules.

The attractive forces between hydrogen-bonded molecules, other than hydrogen fluoride, have been estimated to be from 5 to 8 kcal. The length of the hydrogen bond usually is between one and one half and two times the length of the covalent hydrogen-oxygen bond or the covalent hydrogen-nitrogen bond.

The question may be asked why hydrogen united to carbon does not form hydrogen bonds, and why the unshared pairs of electrons on elements such as sulfur in hydrogen sulfide or chlorine in hydrogen chloride do not act as electron donors for hydrogen bonding. Apparently the usual carbon-hydrogen bond is not sufficiently polar to provide a large enough positive charge for hydrogen bonding. If the carbon atom has several electronegative groups attached to it as in chloroform, $HCCl_3$, the $\overset{+\!\longrightarrow}{\text{H—C}}$ dipole is sufficiently increased to permit weak hydrogen bonding. For elements other than nitrogen, oxygen, or fluorine, the negative charge of the unshared electrons is too widely dispersed, because of the larger size of the valence shell, to permit strong hydrogen bonding. Hydrogen bonding is the result of a balance of forces. If the valence bond to hydrogen is not sufficiently polar or if the negative charge on the other molecule is not sufficiently concentrated, little attraction results. If the valence bond to hydrogen is too polar or if an unshared pair of electrons is too readily available, the proton is transferred completely from one molecule to another and an acid-base reaction results.

Solubility. Solubility is the intermingling of the particles that make up different kinds of matter, whether matter is in the gaseous, liquid, or solid state. The extent to which the different kinds of matter are soluble in each other depends on how nearly the attractive forces between like particles are the same as the attractive forces between unlike particles. Carbon tetrachloride and chloroform are miscible in all proportions because the attractive forces between carbon tetrachloride molecules and chloroform molecules are little different from those of carbon tetrachloride molecules for each other or those of chloroform molecules for each other. On the other hand, neither carbon tetrachloride nor chloroform

[9] *Normal liquids* are those that obey certain empirical rules, such as the Trouton rule that the molal heat of vaporization divided by the boiling point in °K. equals 21. At the molecular level a liquid may be defined as normal if the internal degrees of freedom of each molecule are similar to those for gaseous molecules; that is, they are not seriously disturbed by the close proximity of other molecules. In *associated liquids* the degrees of freedom of the molecules are seriously modified.

is appreciably soluble in water. Water molecules are held to each other much more strongly by hydrogen bonding than water molecules are attracted to carbon tetrachloride or chloroform molecules by London forces, or by dipole-dipole attraction.

Although water molecules have a strong attraction for each other, and ammonia molecules have a strong attraction for each other, ammonia is very soluble in water. Not only do water molecules hydrogen bond with water molecules and ammonia molecules with ammonia molecules; water and ammonia molecules hydrogen bond with each other thus minimizing the difference in the attractive forces between the various entities and permitting intermingling.

```
H—O···H—O        H—O···H—N—H        H—N···H—O        H—N···HNH
  |     |          |     |           /|     |          /|   |
  H     H          H     H          H H     H         H H   H
```

Ionic compounds are said to have a high **crystal lattice energy;** that is, a large amount of energy is required to separate the ions from each other. Some energy, though a much smaller amount, is required to separate water molecules from each other. Nevertheless many salts are soluble in water. Ions on the surface of a crystal not only are held less strongly than those in the interior, but their charge is partially exposed on the surface. When a crystal, for example sodium chloride, is placed in water, the polar water molecules are attracted to the surface of the crystal. The negative oxygen ends are attracted to the sodium ions and the positive hydrogen ends to the chlorine ions (Fig. 2–*17a*). Partial neutralization of the charges on the surface ions by the water molecules decreases their attraction for other ions in the crystal.

The high dielectric constant of water ($\epsilon = 78.5$) means that it is a good insulator of electric charge, and as the water molecules come between the sodium ions and the chlorine ions, the attraction between them is decreased further, and they become more heavily hydrated. Thus the energy liberated on hydration is more than enough to overcome the lattice energy, and the hydrated ions (Fig. 2–*17b* and *c*) mingle with the water complexes. Although only single molecules are represented in Fig. 2–*17*, each water molecule is associated with other water molecules by hydrogen bonding. The solvation of ions by dipolar molecules is said to be due to **ion-dipole interaction.** The lattice energies of ionic compounds that are not soluble in water, such as silver chloride, are too high to permit appreciable solution.

Water molecules do not form hydrogen bonds with covalently bonded halogen. The

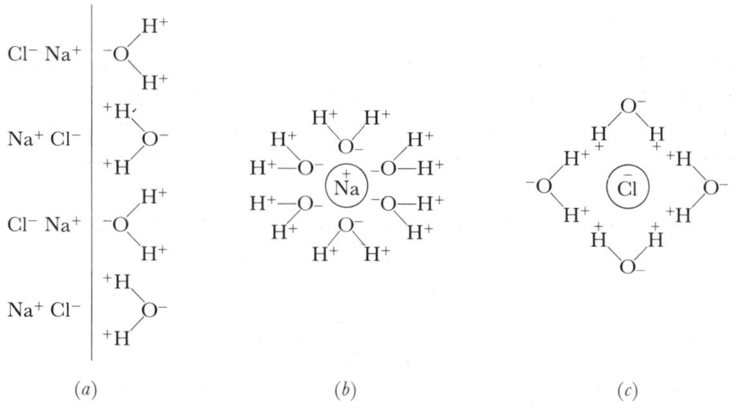

(*a*) (*b*) (*c*)

Figure 2–*17*. Dissolution of sodium chloride in water.

79724

high solubility of halogen acids in water results from transfer of a proton to the water molecule (p. 31) followed by hydration and solution of the ions.

$$\mathrm{H}:\underset{\underset{\mathrm{H}}{\cdot\cdot}}{\ddot{\mathrm{O}}}: + \mathrm{H}:\mathrm{X} \longrightarrow \mathrm{H}:\underset{\underset{\mathrm{H}}{\cdot\cdot}}{\overset{+}{\ddot{\mathrm{O}}}}:\mathrm{H} + {}^{-}:\mathrm{X}$$

It may be noted here that although an indefinite number of water molecules are associated with each other in liquid water, there is evidence that each of the hydrogens of the H_3O^+ ion is strongly bound to a water molecule to give the ion $H_9O_4{}^+$. This ion is bound to an indefinite number of other molecules by hydrogen bonds.

*PROBLEMS**

2–*1*. Calculate the energy released or absorbed (*a*) when an electron is transferred from a carbon atom to a sodium atom, and from a sodium atom to a carbon atom; (*b*) from a carbon atom to a fluorine atom, and from a fluorine atom to a carbon atom.

2–*2*. Write electron-dot formulas for hydrogen sulfide, carbon monoxide, carbonic acid, nitrous acid, nitrous oxide, nitric oxide, nitrogen dioxide, dinitrogen trioxide, dinitrogen tetroxide, dinitrogen pentoxide, phosphorous acid, phosphorus trichloride, sulfur dioxide, sulfur trioxide, thionyl chloride, and sulfuryl chloride.

2–*3*. Which of the compounds in Fig. 2–*14* and in Problem *2* have (*a*) one or more semipolar bonds, (*b*) double bonds, and (*c*) an unpaired electron?

2–*4*. From empirical bond energies, estimate the heat evolved or absorbed per mole of product in the reaction of hydrogen with (*a*) iodine, (*b*) bromine, (*c*) chlorine, and (*d*) fluorine to give the corresponding halogen acid; with oxygen to give (*e*) hydrogen peroxide and (*f*) water; with nitrogen to give (*g*) ammonia and (*h*) hydrazine.

2–*5*. Calculate the energy evolved per gram of total reactants in the combustion with oxygen of (*a*) hydrogen, (*b*) ammonia, and (*c*) hydrazine.

2–*6*. From the ionization potentials and electron affinities given in Table 2–*1*, p. 7, predict the positive end of the dipole in (*a*) N—H, (*b*) N—F, (*c*) B—H, (*d*) B—Cl, (*e*) C—Si, (*f*) C—B, and (*g*) C—N.

2–*7*. (*a*) Compare the magnitude of the dipole moments for the C—H bond for the various valence states of carbon *i.e.*, *sp*, sp^2, and sp^3. (*b*) In which valence state for carbon would hydrogen bonded to carbon be removed most readily as a proton?

---

* The purpose of problems is to provide practice in applying the principles and in making use of the information supplied by the text. Students should work as many problems as possible. When an instructor wishes to limit the amount of material for which the student is held responsible, applicable problems may be assigned. The student should make every effort to solve the problem first without referring to the text, and then, if necessary, with the aid of the text. Although answers are given beginning on p. 671, they should be consulted only to confirm the student's answer or to supply the answer if the student has been unable to do so. The answers given frequently are not detailed. Only enough information may be given that should enable the student to work out the correct solution.

CHAPTER THREE

CHEMICAL REACTIVITY

Chemistry is the science of chemical change; that is, it is the science of the breaking and making of bonds between atoms that results in new atomic arrangements. This chapter reviews those principles of chemical change that are most important to organic chemistry.

Chemical Equilibrium

The extent to which a reaction takes place depends on the equilibrium constant, K, for the reaction. For a bimolecular reaction such as

$$A + B \rightleftharpoons C + D$$

$K = [C][D]/[A][B]$, where the symbols in brackets indicate the concentrations of the reactants and of products in moles per liter after equilibrium is reached. Strictly speaking, activities should be used in this equation instead of concentrations. Since activities cannot be measured directly, the results for various concentrations may be extrapolated to infinite dilution where activities and concentrations become equal. The larger the value of K, the greater the amount of products formed.

Most organic reactions are carried out in such a way that the products at the end of the reaction are at the same temperature and pressure as the reactants were at the beginning of the reaction. Under these conditions

$$-\ln K = -2.303 \log K = \frac{\Delta G}{RT}$$

where ΔG[1] is the change in free energy for the reaction at constant temperature and pressure, R is the gas constant, and T is the absolute temperature. From this relationship, if $\Delta G = 0$, $K = 1$. In the special case of the above reaction, this means that starting with equal concentrations of A and B, their concentrations will decrease and those of C and D will increase until all concentrations are equal. If ΔG is negative, K is greater than one and at equilibrium the concentrations of products will be greater than the concentrations of re-

[1] The symbol G is used for the Gibbs[2] free energy, which is concerned with processes that take place at constant pressure and temperature. Helmholtz[3] free energy is concerned with processes that take place at constant volume and temperature and is denoted by the symbol A. Formerly F frequently was used for the Gibbs free energy, but this symbol has been used also for the Helmholtz free energy. Hence it seems desirable to drop the use of F.

[2] Josiah Willard Gibbs (1839–1903), American professor at Yale College, was the most distinguished mathematical physicist of his period and a contemporary of Helmholtz. Gibbs is considered to be the father of chemical thermodynamics. In 1901 he was awarded the Copley medal of the Royal Society as being "the first to apply the second law of thermodynamics to the exhaustive discussion of the relation between chemical, electrical, and thermal energy, and capacity for external work."

[3] Hermann Ludwig Ferdinand von Helmholtz (1821–1894), German philosopher and scientist who occupied the chair of physiology at Heidelberg and of physics at Berlin. His contributions ranged from the discovery of nerve cells in ganglia through the invention of the ophthalmoscope, the explanation of color vision and color blindness, the mechanism of hearing and the perception of tone, theories of electricity, and the abstract principles of dynamics. He is regarded as one of the founders of the first law of thermodynamics, the law of the conservation of energy.

actants, whereas if ΔG is positive, K is less than one and the concentrations of products will be less than those of the reactants.

ΔG is a thermodynamic function defined by the equation

$$\Delta G = \Delta H - T\Delta S$$

ΔH is the change in enthalpy and is equal to the heat evolved or absorbed during a reaction at constant temperature and pressure, and hence frequently is called the change in heat content. ΔS is the change in entropy, and T is the absolute temperature. Thus the maximum amount of product that can be formed under a given set of conditions is governed only by energy relationships between the initial and final states of the reacting system, that is, by *chemical thermodynamics*.

Enthalpy, H, is a measure of the extent to which the energy of a system can be transferred to its surroundings under constant temperature and pressure. A chemical reaction has a tendency to take place if the enthalpy of the reacting system can decrease by the evolution of heat. The enthalpy may be considered to reside in the bond energies (p. 18). If the sum of the bond energies of the products is less than the sum of the bond energies of the reactants, heat is evolved and there has been a decrease in the enthalpy of the system. Since ΔH for a process such as a chemical reaction is defined as the enthalpy of the final state minus the enthalpy of the initial state, ΔH is negative when heat is evolved. The products are said to be more stable than the reactants because energy would be required to put them back in their original condition.

A decrease in enthalpy, however, is not the only driving force for a reaction. Many natural processes exhibit a property that everyone recognizes but that actually is inexplicable. Water flows spontaneously downhill; heat flows from a hotter body to a colder body; gas flows from a higher pressure to a lower pressure; a clock runs down. None of these processes takes place spontaneously in the opposite direction. These phenomena are generalized by saying that all systems tend to a state of higher probability. To deal with these phenomena quantitatively, the concept of **entropy,** S, has been introduced, which may be defined as a measure of the probability for the existence of a given state. If state A of a sys-system is more probable than state B and all other things are equal, state A has a greater entropy than state B. Hence the ability of a process to take place with an increase in entropy is a driving force that causes the process to take place spontaneously. At the molecular level there is an increase in entropy as a gas passes from a higher pressure to a lower pressure because the molecules are less confined. Liquid water has a lower entropy than water vapor and ice a lower entropy than liquid water because the molecules are losing degrees of freedom and are becoming more ordered and less random and hence in a less probable state. Similarly if a system changes with an increase in the total number of molecules, an increase in entropy results because there has been an increase in the degrees of freedom for the system.

To summarize, the direction and extent of a chemical reaction are determined by two driving forces: (*1*) the tendency to reach a state of lowest enthalpy and (*2*) the tendency to reach a state of highest entropy. Keeping in mind that for a decrease in enthalpy ΔH is negative and for an increase in entropy ΔS is positive, both tendencies can be expressed simultaneously by the concept of a new function of state, called the **free energy,** G, where at constant temperature and pressure

$$-\Delta G = -\Delta H + T\Delta S$$
$$\text{or } \Delta G = \Delta H - T\Delta S$$

Thus reactions can take place spontaneously when there can be a decrease in free energy of the system (ΔG *negative*).

It should be noted that a positive change in free energy does not mean that a reaction cannot take place in the forward direction, nor does a negative change in free energy mean that a reaction will go to completion in the forward direction. The extent to which the reaction takes place, that is, how large the equilibrium constant K is, depends on the *extent* to which ΔG is negative. Moreover since ΔS is temperature dependent, temperature will have an important effect on the size of ΔG if ΔS is large. Also, since K is dependent on the concentration of both reactants and products, changes in concentration influence the extent to which a reaction takes place.

In order to measure and calculate changes in enthalpy, entropy, and free energy, elements in their most stable physical form at 760 mm. of mercury pressure and 298°K. have been assigned arbitrarily a value of zero enthalpy, and perfect crystals of elements or compounds at the absolute zero are assumed to have zero entropy. From these reference points standard enthalpies of formation, ΔH_f°, in kcal. per mole at 298° K., standard entropies, ΔS°, in entropy units (e.u. = calories per mole per degree) at 298° K., and standard free energies of formation, ΔG_f°, in kcal. per mole at 298°K. have been determined or calculated for many compounds. The most complete compilation of these values is Circular 500 of the U.S. National Bureau of Standards (1952) entitled "Selected Values of Chemical Thermodynamic Properties." Additional data have been published by the American Petroleum Institute and the Manufacturing Chemists' Association.

At this point it may be well to review the different ways in which the term *reversible reaction* is used. In thermodynamic parlance, a reversible reaction is one for which the change in free energy is zero; that is, the reaction has reached equilibrium and is proceeding in both directions at the same rate without the intervention of any outside force. On the other hand, chemists frequently say that a reaction is reversible if ΔG is small enough that the reaction can be forced in either direction by a moderate change in conditions. A reaction is said to be irreversible if ΔG for the forward reaction is so negative that for all practical purposes it is not possible to make the products react to give the original reactants.

Rates of Reactions

Since time does not enter into the equations involving free energy, there is no apparent relation between how fast a reaction takes place and the change in free energy. For example, the conversion of nitrogen dioxide to dinitrogen tetroxide is extremely fast at room temperature although the change in free energy is only −14.6 kcal., whereas the reaction of hydrogen and oxygen at room temperature to give one mole of water is immeasurably slow, although the change in free energy of −56.6 kcal. is much larger. Clearly the velocity or rate of a reaction, that is, the speed with which it takes place, is as important to the chemist as the equilibrium constant for the reaction. First, a reaction must be rapid enough to be practical and yet not so fast as to be uncontrollable. Second, a given organic compound or mixture of compounds may undergo many different thermodynamically possible reactions. The products formed in largest amounts will result from the fastest reaction, which will not necessarily be that with the largest equilibrium constant. By modification of conditions and use of specific catalysts, it frequently is possible to bring about a desirable reaction or suppress an undesirable one. Finally, a study of the effect of variation in the concentrations of reactants or of modifications in the structure of a reactant on the rate may permit the determination of the steps by which a reaction takes place, that is, the mechanism of the reaction. The study of rates of reactions belongs to the realm of **chemical kinetics.**

Since the nature of the reactants, the concentrations of reactants, the temperature of the reaction, and the presence of catalysts or inhibitors all influence the rate of a reaction, chemical kinetics obviously is a complicated subject. In regard to the effect of concentration, the concept of **order** is important. An attempt is made to keep all variables constant except the concentration of a single reactant, A, and to measure the rate of reaction for different concentrations of this reactant. It may be found that the rate doubles if the concentration of A is doubled and triples as A is tripled; in other words, the rate of reaction is directly proportional to the concentration of A. The reaction then is said to be *first order in A*. Should the rate be proportional to the square of the concentration of A, the reaction is said to be *second order in A*. Next the effect of varying the concentration of reactant B is studied, keeping the other variables constant, and the order, n_B, of the reaction with respect to B is determined. If the rate is proportional to $[A]^{n_A}$ and to $[B]^{n_B}$, it is proportional to their product and may be expressed by the equation

$$\text{Rate} = k[A]^{n_A}[B]^{n_B}\cdots$$

where k is a proportionality constant known as the *rate constant* or *specific reaction rate* for the reaction, and the dots represent terms for any other additional reactants such as C, D, or E.

For a rate that can be expressed in the form of a product of powers of concentrations, the reaction itself is said to be of the order n, where $n = n_A + n_B + \cdots$. Thus a reaction that is first order in A and zero order in B is an *over-all first order reaction;* one that is first order in both A and B is an *over-all second order reaction.* The exponents n_A, n_B, . . . usually are small positive integers but may be fractions or even negative.

Since the rate is the change in concentration per unit time, the dimensions of k are

$$\frac{\text{concentration}}{\text{time} \times \text{concentration}^n} = \text{concentration}^{1-n}\ \text{time}^{-1}$$

Concentrations ordinarily are expressed in moles per liter and time in seconds. Hence k for a first order reaction has the units, $\left(\frac{\text{moles}}{\text{liters}}\right)^0 \text{sec.}^{-1} = \text{sec.}^{-1}$. For a second order reaction the units are $\left(\frac{\text{moles}}{\text{liters}}\right)^{-1} \text{sec.}^{-1} = \text{liters moles}^{-1}\ \text{sec.}^{-1}$.

The order of a reaction is an empirical fact and is not connected with the stoichiometry of the reaction. Thus the equation

$$A + 2B \longrightarrow C + D$$

may be first order in A or B, first order in A and B, second order in A or B, or of a more complex order. The number of molecules necessarily consumed in the reaction as represented by the equation is the **molecularity** of the reaction. Thus the above reaction is termolecular but may be of any order. Although a bimolecular reaction may be second order, a second order reaction is not necessarily bimolecular. For reactions that take place in more than one step, molecularity refers to the *rate-determining step* rather than to the reaction as a whole (p. 36).

Reactions Between Ions. The reactions of hydrated ions in solution are relatively simple. Since the attraction between the ions and the water molecules is very weak, the activation energies are low and reactions take place on almost every collision. The position of equilibrium depends on the change in free energy. Thus a solution containing silver ions when mixed with one containing chloride ions gives an immediate precipitate of silver chloride. Proton transfer reactions are extremely rapid as in the reaction of hydroxide ion with hydronium ion.

$$HO{:}^- + H_3O^+ \longrightarrow H_2O + H_2O$$

The second order rate constant for this reaction is 10^{11} liter mole^{-1} sec.$^{-1}$ and appears to be governed only by the rates of diffusion. Because of the large decrease in free energy, the reaction goes practically to completion.

Acids and Bases. Not only can charged ions react to give neutral molecules; neutral molecules can react with each other to give ions. Hydrogen chloride and water are typical covalent nonionic compounds, yet a solution of hydrogen chloride in water is a strong electrolyte. A reaction has taken place to give hydronium ions and chloride ions, both of which are heavily hydrated.

$$H_2O\!: + HCl \longrightarrow H_3O^+ + Cl^-$$

This behavior is considered to be the reaction of an acid, hydrogen chloride, with a base, water.

Several types of acid-base reactions exist, and different definitions of acids and bases are given to satisfy each type. From the standpoint of organic chemistry, the most useful definitions are that *a base is an entity having an unshared pair of electrons that can be shared with a proton,* and *an acid is an entity that can transfer a proton to a base.* Either acid or base or both may be neutral molecules, positive ions, or negative ions. These concepts essentially are those proposed almost simultaneously by Lowry[4] and by Brønsted.[5] Although in their first publications Brønsted was the more specific in his definitions, Lowry's views are more in accord with current usage. Brønsted defined acids and bases by the equilibrium

$$\text{Acid} \rightleftharpoons \text{Base} + H^+$$

It was Lowry who stated specifically that protons under ordinary conditions are incapable of independent existence and that the reaction of a base with an acid is a *proton transfer reaction* and yields a new acid and a new base.

$$\begin{aligned}
\text{Base } Y + \text{Acid } Z &\rightleftharpoons \text{Acid } Y + \text{Base } Z \\
H_2\ddot{O}\!: + H\!:\!OSO_3H &\rightleftharpoons H_3O\!:^+ + {}^-\!:\!OSO_3H \\
H_3N\!: + H_3O\!:^+ &\rightleftharpoons NH_4^+ + H_2\ddot{O}\!: \\
H_2\ddot{N}\!:^- + H\!:\!OH &\rightleftharpoons H_3N\!: + {}^-\!:\!OH \\
HO\!:^- + H\!:\!CN &\rightleftharpoons H_2O + {}^-\!:\!CN \\
H_3N\!: + H\!:\!OSO_3^- &\rightleftharpoons NH_4^+ + SO_4^{2-}
\end{aligned}$$

Each acid and its corresponding base are said to be *conjugate* to each other. Ammonia is the *conjugate base* of ammonium ion and ammonium ion is the *conjugate acid* of ammonia. Similarly, ammonia and amide ion, water and hydroxide ion, and sulfuric acid and bisulfate ion are conjugate pairs.

The position of equilibrium of a proton transfer reaction depends on the ease with which the acid can give up a proton and on the tendency of the base to share its pair of electrons. These properties are referred to as the *strength* of the acid or base. Within a row of the periodic table, they depend on the relative electronegativity of the atoms to which the proton or the pair of electrons is attached. Just as the electronegativity of atoms increases in going from lower to higher atomic number within a given period (p. 21), so the acidity of the hydrides increases and the basicity of their conjugate bases decreases. For example, the order of increasing acidity is $CH_4 < NH_3 < H_2O < HF$ and of decreasing basicity is $H_3C\!:^- > H_2N\!:^- > HO\!:^- > F\!:^-$.

Relative electronegativity is the dominating influence only if the unshared electrons of the atoms being compared are at approximately the same distance from the nucleus. With increasing size of the halogen atoms in going from the top to the bottom of the periodic table, there is increasingly poorer overlap of a $3p$, $4p$, or $5p$ orbital with the $1s$ orbital of

[4] Thomas Martin Lowry (1878–1936), professor of physical chemistry at Cambridge University. He is noted for his investigations in the fields of optical rotatory power, of catalysis by acids and bases, and of proton transfer reactions.

[5] J. N. Brønsted (1879–1947), director of the Fysisk-Kemiske Institute of Copenhagen. His chief contributions were in the fields of electrolytes and reaction kinetics.

hydrogen, which leads to a progressive decrease in the ability of the proton to share the pair of electrons with the conjugate base. This decreased tendency more than compensates for the decrease in electronegativity, with the result that the order of acidity for the halogen acids is HI > HBr > HCl > HF. Similarly the order of acidity for water and its analogs is $H_2Te > H_2Se > H_2S > H_2O$, and the order of basicity for ammonia and its analogs is $H_3N{:} > H_3P{:} > H_3As{:}$.

For the oxy acids, the proton is bound to oxygen, and electronegativity again controls the relative acidity. Thus sulfurous acid, $(HO)_2SO$, is a stronger acid than selenious acid, $(HO)_2SeO$. Addition of electronegative elements increases the electronegativity of the elements to which they are attached. Sulfuric acid is stronger than sulfurous acid, and the order for the oxygenated halogen acids is $HOX < HOXO < HOXO_2 < HOXO_3$.

Quantitative comparisons of acidity usually have been made by determining the position of equilibrium for the reaction

$$\text{Acid} + H_2O \rightleftharpoons \text{Base} + H_3O^+$$

which is called the degree of ionization of the acid. The equilibrium constant for this reversible reaction is

$$K = \frac{[\text{Base}][H_3O^+]}{[\text{Acid}][H_2O]}$$

where the symbols in brackets indicate the concentration in moles per liter (p. 27). Because it is so large compared with the concentration of the ions, the concentration of water is essentially constant and may be incorporated in the equilibrium constant to give the expression

$$K_a = \frac{[\text{Base}][H_3O^+]}{[\text{Acid}]}$$

where K_a is called the *acidity constant* or the *ionization constant* of the acid. Similarly for the equilibrium

$$\text{Base} + H_2O \longrightarrow \text{Acid} + HO^-$$

$$K_b = \frac{[\text{Acid}][HO^-]}{[\text{Base}]}$$

where K_b is the *basicity constant* of the base.

Since the extent to which proton transfer takes place is called the *strength* of the acid or base, it is common practice to consider the relative strengths of acids and bases as proportional to their acidity or basicity constants. These constants range from very small to very large numbers. In order to simplify working with them, it has become customary to convert them to logarithmic functions defined by the equations

$$pK_a = -\log K_a \quad \text{and} \quad pK_b = -\log K_b$$

In water solution a useful relationship exists between the pK_a of an acid and the pK_b of its conjugate base. Multiplication of both the numerator and the denominator of the expression for K_a by the concentration of hydroxide ions gives

$$K_a = \frac{[\text{Base}][H_3O^+][HO^-]}{[\text{Acid}][HO^-]}$$

But $[H_3O^+][HO^-]$ is the ion product of water, which is approximately equal to 10^{-14} $(\text{moles/liter})^2$ at 25°. Since

$$\frac{[\text{Base}]}{[\text{Acid}][HO^-]} = \frac{1}{K_b}, \quad K_a = \frac{10^{-14}}{K_b} \quad \text{and} \quad K_b = \frac{10^{-14}}{K_a}, \quad \text{or} \quad pK_a = 14 - pK_b \quad \text{and} \quad pK_b = 14 - pK_a.$$

TABLE 3–*1*. ACIDITY CONSTANTS IN WATER AT 18–25°

ACID	pK_a	CONJ. BASE	ACID	pK_a	CONJ. BASE
(H^+)*	$\ll -10$		H_2CO_3	6.46†	HCO_3^-
$HClO_4$	< -10	ClO_4^-	H_2S	7.04	HS^-
HI	~ -10	I^-	HCN	9.14	CN^-
HBr	~ -9	Br^-	NH_4^+	9.21	NH_3
HCl	~ -7	Cl^-	H_3BO_3	9.24	$H_2BO_3^-$
H_2SO_4	~ -3	HSO_4^-	H_2O_2	11.62	HO_2^-
HNO_3	-2.3	NO_3^-	H_2O	15.74	HO^-
H_3O^+	-1.74	H_2O	NH_3	~ 35	H_2N^-
HSO_4^-	1.70	SO_4^{2-}	H_2 }	> 37	{ H^-
H_2SO_3	1.77	HSO_3^-	CH_4 }		{ CH_3^-
H_3PO_4	2.12	$H_2PO_4^-$	(HO^-)* }	$\gg 37$	{ (O^{2-})*
H_2SeO_3	2.52	$HSeO_3^-$	(H_2N^-)* }		{ (HN^{2-})*
HF	3.14	F^-	(HN^{2-})*	$\ggg 37$	(N^{3-})*
HNO_2	3.35	NO_2^-			

* Hypothetical.

† This value results from including dissolved CO_2 molecules as nonionized H_2CO_3. The value for H_2CO_3 itself is approximately 3.7.

Because of this interdependence of pK_a and pK_b, acidity and basicity now commonly are expressed as pK_a, this value referring to the conjugate acid of any base. The strengths of acids *decrease* and of their conjugate bases *increase* with increasing pK_a. Moreover each unit difference in pK corresponds to a tenfold difference in the values of K. Thus the conjugate acid of a base with a pK_a of 3 has a K_a one tenth that of one with a pK_a of 2, and the K_b of the first base is ten times that of the second base. A difference of 3 in pK corresponds to a thousandfold difference in K.

Table 3–*1* lists the acidity constants of the more common inorganic acids and of the conjugate acids of entities usually thought of as bases. No values can be given for the first seven acids of the table because water is a sufficiently strong base to ionize completely the weakest of these acids. Similarly, water is too weak as a base to determine the acidity of acids weaker than water. This limitation is referred to as the *leveling effect* of the solvent. To determine the relative acidity of the strong acids, it is necessary to use a solvent that is a much weaker base than water. Likewise, to determine the relative acidity of the very weak acids, it is necessary to use a solvent that is much more basic than water. Such measurements as have been made agree with the order expected from the relative electronegativities of the atoms involved.

In 1923, the same year in which Lowry and Brønsted discussed acids and bases, Lewis (p. 14) stated in his book entitled *Valence* that "the definition of an acid or a base as a substance which gives up or takes up hydrogen ion would be more general than the one used before, although it will not be universal." "The one used before" refers to the Arrhenius concept that an acid yields hydrogen ions in water and a base yields hydroxide ions. Thus Lewis expressed the conjugate pair concept as early as Lowry and Brønsted, but he discarded it in favor of what he considered to be a more general one. He looked on the proton with its empty valence shell as the acid and then classed with it any ion or molecule that lacks a pair of electrons in its valence shell. Just as a proton combines with an ammonia molecule to give an ammonium ion, so boron fluoride forms a stable complex with ammonia.

$$H_3N\!: + H^+ \longrightarrow H_3\overset{+}{N}\!:\!H$$

$$H_3N\!: + BF_3 \longrightarrow H_3\overset{+}{N}\!:\!\overset{-}{B}F_3$$

Accordingly Lewis classed boron fluoride as an acid and defined an acid as a substance that can fill the valence shell of one of its atoms with an unshared pair of electrons from another molecule. In favor of this concept is the fact that substances such as boron fluoride, aluminum chloride, zinc chloride, or stannic chloride can catalyze the same types of reaction, for example polymerization of olefins (p. 87) or the formation of ethers (p. 139), as can a proton.

According to the Lewis definition, however, hydrogen chloride, sulfuric acid, acetic acid, and in fact any of the countless number of substances that from the beginning of chemistry have been called acids, are not acids. The acid as defined by Lewis is the bare unsolvated proton, which is practically incapable of existence. Moreover substances such as cupric ion, which seldom are considered as acids, are acids in the Lewis sense, because they can fill their valence shell with unshared pairs from other molecules, as when cupric ion reacts with ammonia molecules to give the cupric-ammonia complex ion.

$$Cu^{2+} + 4\,NH_3 \longrightarrow \left[\begin{array}{c} NH_3 \\ \ddot{} \\ H_3N{:}Cu{:}NH_3 \\ \ddot{} \\ NH_3 \end{array}\right]^{2+}$$

To obviate this difficulty, Sidgwick,[6] in his book on valency published in 1927, called substances that have a vacant orbital *electron-acceptors,* and those that have an unshared pair of electrons *electron-donors.* In this text the terms *acid* and *base* refer only to the Lowry-Brønsted concept. In discussions of electron acceptor properties analogous to those of boron fluoride, the popular idiom *Lewis acid* is used.

Other Reactions Between Covalent Molecules. Proton transfer reactions between acids and bases usually are rapid because the base has an exposed pair of electrons and the hydrogen nucleus of the acid is practically bare of electrons on the exposed side. The nucleus is exposed because it is encompassed by only two electrons, which spend most of the time between the proton and the bonding atom. Moreover the nucleus is further deshielded from these electrons by the electronegativity of the group to which it is bonded, that is, the weakness of its conjugate base. Most other reactions between covalent molecules are slow. Molecules must collide before reaction takes place and many collisions are ineffective because the repulsions between the electron clouds about the molecules cause them to bounce away from each other without reacting, or because they are surrounded by solvent molecules that interfere with collision. Often only certain orientations of the molecules lead to reaction (p.100), an effect that is known as the *steric factor.* Bonds must be stretched and molecules brought into a reactive state before new bonds can form; that is, energy must be supplied to molecules before reaction will take place. This energy is called the **activation energy,** $E^{\ddagger}$, for the reaction, and the reactive state is called the **transition state** (TS) or the **activated complex.** Once molecules have reached the transition state, the atoms may rearrange to the new products or revert to the original compounds with the evolution of energy.

The relation between rate and activation energy is given by the equation

$$\ln k = -E^{\ddagger}/RT + C$$

where C is a temperature independent constant. The activation energy for a reaction is determined experimentally by measuring the rate at different temperatures. A plot of ln k against $1/T$ gives a straight line, the slope of which is equal to $-E^{\ddagger}/R$.

Processes requiring energy may be thought of as going uphill, leading to systems of higher energy, whereas those evolving energy go downhill and lead to more stable systems

[6] Nevil Vincent Sidgwick (1873–1952), professor of chemistry at Oxford University. His writings did much to disseminate ideas arising from the application of physical methods to organic chemical problems.

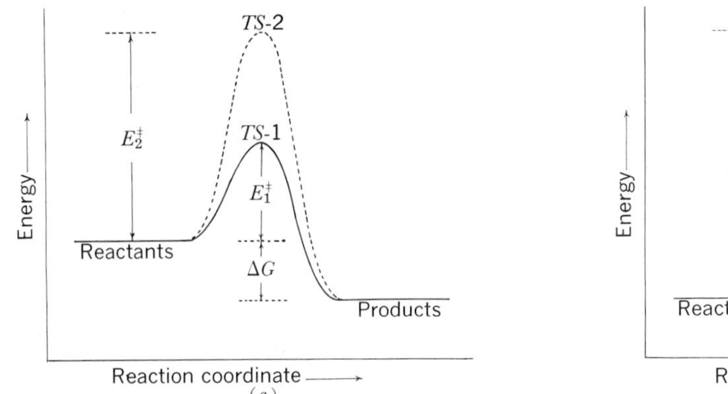

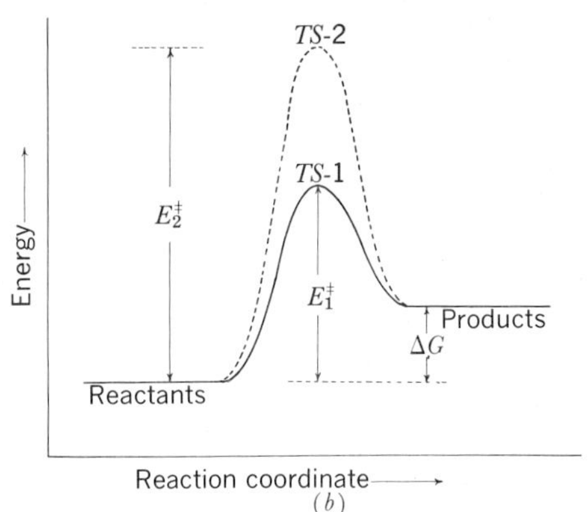

Figure 3–*1*. Diagrams illustrating energy changes during the course of a reaction: (*a*) ΔG negative; (*b*) ΔG positive.

of lower energy. These considerations frequently are represented schematically by energy diagrams such as those illustrated in Fig. 3–*1*. Here the ordinate represents the free energy of the system, and the abscissa represents the progressive change in spatial relationships as reacting molecules approach each other, form the transition state, and separate into product molecules. It usually is called the *reaction coordinate*. If a process requires an activation energy, it must pass over a hump or energy barrier before it can reach the lower level of greater stability. The dotted lines in Fig. 3–*1* represent a higher activation energy than the solid lines. The height of the energy barrier determines the rate of the reaction because the lower this activation energy, the greater the number of molecules that will have sufficient kinetic energy to form the activated complex on collision and the faster the reaction will take place.

The difference, ΔG, between the free energy of the reactants and the free energy of the products determines the position of equilibrium for the reaction (p. 27). For a reaction having an energy profile like that of Fig. 3–*1a*, the products would predominate at equilibrium, whereas for one having an energy profile like Fig. 3–*1b*, the starting materials would predominate.

Frequently the same reactants may undergo two or more separate reactions to yield different products.

$$A + B \rightleftharpoons C + D$$
$$A + B \rightleftharpoons E + F$$

Each reaction has its own activation energy and its own change in free energy. The reaction with the lowest activation energy is the fastest because at any given temperature there are more molecules with sufficient kinetic energy to produce the activated complex. The position of equilibrium at the end of the reaction depends, however, on the free energy change, the products with the lowest free energy predominating. If the energy diagram for the reaction giving C and D is represented by Fig. 3–*2a* and that giving E and F by Fig. 3–*2b*, and if the reaction is stopped before the reverse reactions become important, C and D will exceed E and F in the reaction mixture. If, however, the reaction is allowed to proceed to equilibrium, E and F will predominate. Under these circumstances C and D are called the *kinetically controlled products* and E and F the *thermodynamically controlled products*. The reaction that leads to unwanted products or products of lesser values usually is referred to as a *side reaction*.

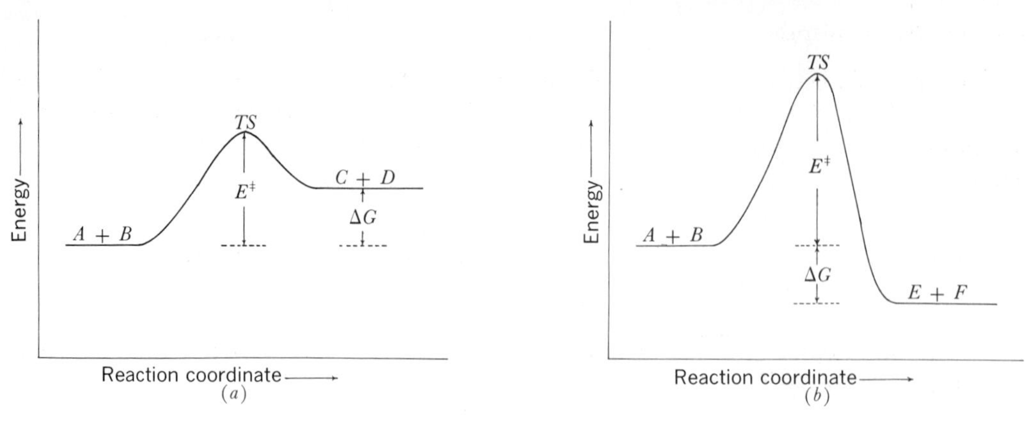

Figure 3–2. Energy profiles for (*a*) reaction $A + B \longrightarrow C + D$ and (*b*) reaction $A + B \longrightarrow E + F$.

Another type of reaction that often is encountered is one that takes place by a series of steps. Each step is really a separate reaction and has its own $E^{\ddagger}$ and ΔG. The intermediates may be so unstable, however, that they cannot be isolated, and the reaction goes to the final products. The position of equilibrium is determined by the over-all change in free energy. Figure 3–*3a* illustrates a two-step reaction in which the first step is much slower than the second ($E_1^{\ddagger} > E_2^{\ddagger}$). The first step determines the over-all rate of reaction and is called the *rate-controlling step*. Figure 3–*3b* illustrates a two-step reaction in which the second step is rate-controlling ($E_2^{\ddagger} > E_1^{\ddagger}$).

It has been found that if an atom involved in a bond-breaking step of a reaction is replaced by one of its isotopes, the effect on the rate of the reaction is much greater than the effect on the position of equilibrium, a phenomenon known as the *kinetic isotope effect*. This fact can be used in assigning the rate-determining step of a reaction. If replacement of an atom by one of its isotopes has an appreciable effect on the over-all rate of the reaction, the step involving a bond to this atom probably is the rate-controlling step. The greatest effects are observed when hydrogen is replaced by deuterium or tritium, which may reduce the rate at room temperature to as low as one sixteenth of that for hydrogen, if hydrogen is being transferred in the reaction.

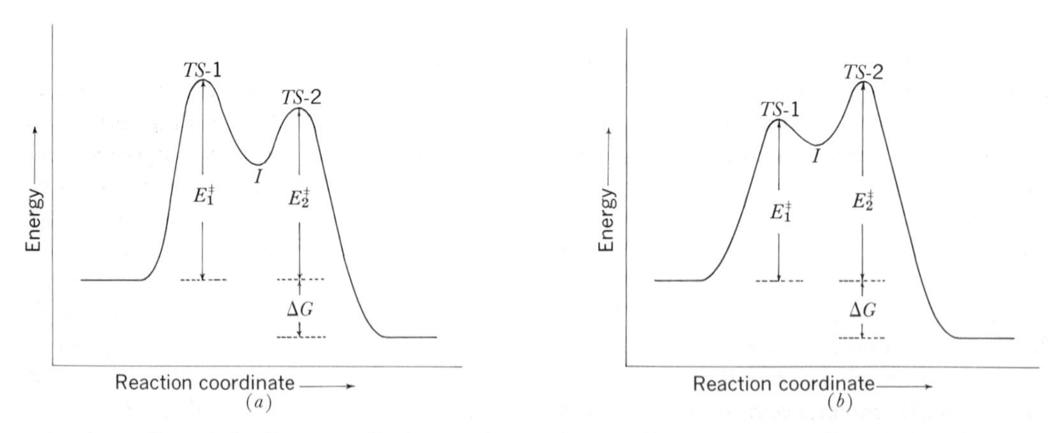

Figure 3–3. Energy profiles for two-step reactions: (*a*) first step rate-controlling; (*b*) second step rate-controlling.

Reaction Mechanisms

One possible way for covalent molecules to react would be for them to dissociate first into atoms. Recombination of the atoms could give either the starting compounds or the new products. For example, in the decomposition of hydrogen iodide to give hydrogen and iodine,

$$2\ HI \rightleftharpoons H_2 + I_2$$

the hydrogen iodide might dissociate into hydrogen atoms and iodine atoms.

$$HI \rightleftharpoons H\cdot + I\cdot$$

The atoms then could either recombine to hydrogen iodide, or like atoms could combine to hydrogen and iodine.

$$2\ H\cdot \longrightarrow H_2$$
$$2\ I\cdot \longrightarrow I_2$$

The activation energy for this process would be expected to be equal to the dissociation energy of hydrogen iodide or 71 kcal. ($E_2^\ddagger$ of Fig. 3–*1a*). The measured activation energy, however, is only 39 kcal. ($E_1^\ddagger$ of Fig. 3–*1a*). Hence some process that does not require scission into free atoms is operating. This particular reaction is thought to be of the *four-center type* in which, as the two molecules approach each other, bonds are being made and broken simultaneously at all four nuclei. As bonding begins between the two hydrogen nuclei and the two iodine nuclei, the hydrogen-iodine bonds stretch and weaken. In the transition state or activated complex, bonding is represented by dotted lines. Dissociation of the activated complex can take place either to give two hydrogen iodide molecules or one molecule each of hydrogen and iodine.

$$\begin{matrix} H{-}I \\ + \\ H{-}I \end{matrix} \rightleftharpoons \begin{matrix} H\cdots I \\ \vdots \quad \vdots \\ H\cdots I \end{matrix} \rightleftharpoons \begin{matrix} H \\ | \\ H \end{matrix} + \begin{matrix} I \\ | \\ I \end{matrix}$$

Activated complex

At no point in the reaction have free hydrogen atoms or free iodine atoms been formed. Hence the activation energy necessary to form the complex is less than that required to form free atoms. Experimentally the velocity of this reaction is proportional to the square of the concentration of hydrogen iodide in conformity with the postulated simple reaction.

$$\text{Velocity of reaction} = k[HI]^2$$

It should be noted that the transition state for the decomposition of hydrogen iodide can form only when two colliding molecules are oriented in a particular way. Actually another reaction undoubtedly takes place more frequently in which the colliding molecules are oriented in opposite directions, an orientation that should lead to dipole-dipole attraction. This transition state, however, can lead only to regeneration of the same products.

$$\begin{matrix} H{-}I \\ + \\ I{-}H \end{matrix} \rightleftharpoons \begin{matrix} H\cdots I \\ \vdots \quad \vdots \\ I\cdots H \end{matrix} \rightleftharpoons \begin{matrix} H \\ | \\ I \end{matrix} + \begin{matrix} I \\ | \\ H \end{matrix}$$

For the reaction

$$H_2 + Br_2 \longrightarrow 2\ HBr$$

the velocity of the reaction might be expected to be proportional to the product of the concentration of hydrogen and the concentration of bromine. Instead, the relation found by experiment is

$$\text{Velocity of reaction} = \frac{k[H_2][Br_2]^{3/2}}{[Br_2] + k'[HBr]}$$

Since this complicated rate expression cannot be explained by the simple reaction indicated by the over-all equation, the reaction must be complex. Further evidence that balanced equations seldom tell the whole story is the fact that catalysts and solvents markedly affect the rates of reactions but do not even appear in the balanced equations. The *path that molecules are assumed to follow in order to account for the effect of various factors on the course and rate of a reaction* is called the **mechanism of the reaction.** For any reversible reaction, the mechanism of the reverse reaction is exactly the reverse in all details of the mechanism for the forward reaction. This fact is known as the *principle of microscopic reversibility.*

The reactions of covalent compounds usually are divided into two main types according to the way the bonds are broken. Either one electron of the bonding pair goes with each atom as in (*a*), or both electrons stay with one or the other of the two atoms as in (*b*) or (*c*).

$$A \cdot\!/\!\cdot B \qquad A\,|\!:B \qquad A:\!|\,B$$
$$(a) \qquad\quad (b) \qquad\quad (c)$$

Process (*a*) is referred to as *homolysis,* and process (*b*) or (*c*) is called *heterolysis* (Gr. *homos* same, *heteros* other, and *lysis* a loosing). The products of process (*a*) have unpaired electrons and are either atoms or groups called *free radicals.* Reactions involving process (*a*) are said to take place by a *free radical mechanism.* Process (*b*) or (*c*) would lead to ionic intermediates, and reactions involving heterolysis are said to take place by *polar* or *ionic mechanisms.* Since ions need never actually be present in measurable amounts in these processes, the term *polar mechanism* seems preferable.

Free Radical Reactions. It has been noted (p. 17) that less energy is required to break a bond homolytically than heterolytically. Reactions in the gas phase commonly take place in this way. If the bond dissociation energy is large, considerable activation energy must be supplied to initiate the reaction. This may be done by raising the temperature to increase the kinetic energy of the molecules or by promoting electrons to orbitals of higher energy by the absorption of light. The particular wavelength absorbed depends on the difference in quantized energy levels of the molecules. One mole of photons corresponds to $6.024 \times 10^{23}\, h\nu$, where h is the Planck constant and ν is the frequency of the radiation. The range of wavelength, λ, from the near infrared through the ultraviolet is from about 1 micron (0.0001 cm.) to 0.2 micron (200 mμ), or in frequency from 3×10^{14} to 1.5×10^{13} per second, corresponding to 28.6 to 143 kcal. per mole of photons.[7] Since activation energies are in the range of 10 to 100 kcal., the absorption of visible or near ultraviolet light can initiate chemical reactions, either by causing dissociation of a molecule into free radicals, or by producing an excited molecule that may retain the energy long enough to collide with another molecule and form the activated complex. Reactions activated by light are called *photochemical reactions.*

The reaction of hydrogen and chlorine, initiated by light of wavelength less than 400 mμ or by heat, takes place by a free radical mechanism. The first step is the dissociation of chlorine into free chlorine atoms.

---

[7] Spectroscopists prefer to describe radiation in terms of frequency rather than wavelength because frequency is directly proportional to the energy whereas wavelength is inversely proportional. The frequency of a vibration usually is characterized by its **wavenumber,** which is the number of cycles per centimeter of length along the path of the wave. Hence the wavenumber in reciprocal centimeters is 10,000 divided by the wavelength in microns. Thus a wavelength of 1μ (0.0001 cm.) corresponds to a wavenumber of 10,000 cm.$^{-1}$. The energy of 10,000 cm.$^{-1}$ is 28.6 kcal. or 1.24 e.v. Light with a wavelength of 200 mμ = 50,000 cm.$^{-1}$ = 28.6 $\times$ 5 = 143 kcal.

$$\text{Cl:Cl} \xrightarrow[h\nu]{\text{Heat or}} 2\ \text{Cl}\cdot \quad -57\ \text{kcal.} \qquad (\Delta H = +57\ \text{kcal.})^{8} \qquad \textit{Initiation}$$

A **chain reaction** then occurs.

$$\begin{array}{lll} \text{Cl}\cdot + \text{H:H} \longrightarrow \text{H:Cl} + \text{H}\cdot & -1\ \text{kcal.} & (\Delta H = +\ 1\ \text{kcal.}) \\ \text{H}\cdot + \text{Cl:Cl} \longrightarrow \text{H:Cl} + \text{Cl}\cdot & +45\ \text{kcal.} & (\Delta H = -45\ \text{kcal.}) \\ \hline \text{H}_2 + \text{Cl}_2 \longrightarrow 2\ \text{HCl} & +44\ \text{kcal.} & (\Delta H = -44\ \text{kcal.}) \end{array} \qquad \textit{Propagation}$$

Thus the main reaction is exothermic by 44 kcal. for each mole of hydrogen and chlorine that reacts to give two moles of hydrogen chloride.

The chain may be terminated by combination of any two free radicals. The amount of energy liberated in the formation of a bond is equal to that required to dissociate it. Hence the formation of diatomic molecules from atoms can take place only in the presence of a third body such as the wall of the vessel or another molecule, M, that can absorb the energy liberated.[9]

$$\begin{array}{l} \text{Cl}\cdot + \text{Cl}\cdot + \text{wall} \longrightarrow \text{Cl}_2 + (\text{wall} + 57\ \text{kcal.}) \\ \text{or Cl}\cdot + \text{Cl}\cdot + M \longrightarrow \text{Cl}_2 + M^* \ (M^* = M + 57\ \text{kcal.}) \end{array} \qquad \textit{Termination}$$

Combination of two hydrogen atoms or a hydrogen atom and a chlorine atom by similar processes can terminate the chain also. The three steps *initiation, propagation,* and *termination* are characteristic of chain reactions.

The energy profile for the activation step is illustrated in Fig. 3–*4a*. As a chlorine atom thus formed approaches a hydrogen molecule, the orbitals begin to overlap with simultaneous stretching of the H—H bond until the activated complex is formed as illustrated schematically in Fig. 3–*4b*. Figures 3–*5a* and *b* represent the energy profiles for the propagation steps. Since the activation energies for reactions involving free radicals are probably less than 5 kcal., these steps are extremely rapid. The first propagation step probably is endothermic by about 1 kcal., but the second step is exothermic by about 45 kcal. In an adiabatic closed system, the heat evolved increases the kinetic energy of the remaining reactants with corresponding increase in the rate of reaction until eventually an explosion results.

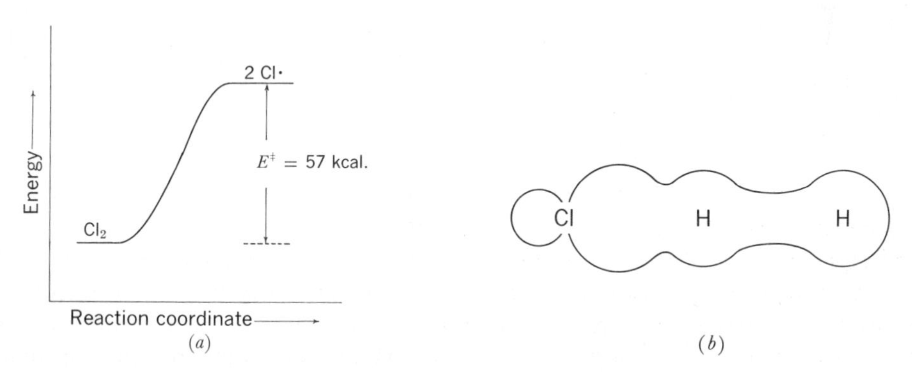

Figure 3–*4*. (*a*) Activation of chlorine molecules; (*b*) schematic representation of probable transition state (*activated complex*) in reaction of a chlorine atom with a hydrogen molecule.

[8] Heat changes, when written as part of an equation, are the changes in the heat content of the surroundings and are considered to be negative if heat is absorbed (*endothermic reactions*) and positive if heat is evolved (*exothermic reactions*). When expressed as ΔH, however, they are the changes in the heat content of the reactants (*changes in enthalpy*) and have the opposite sign.

[9] If polyatomic molecules are involved, a third body is not necessary because the energy may be absorbed by vibrations in other parts of the molecule. Similarly in the initiation of photochemical reactions energy absorbed by one part of a molecule may be transmitted to and activate another part of the molecule.

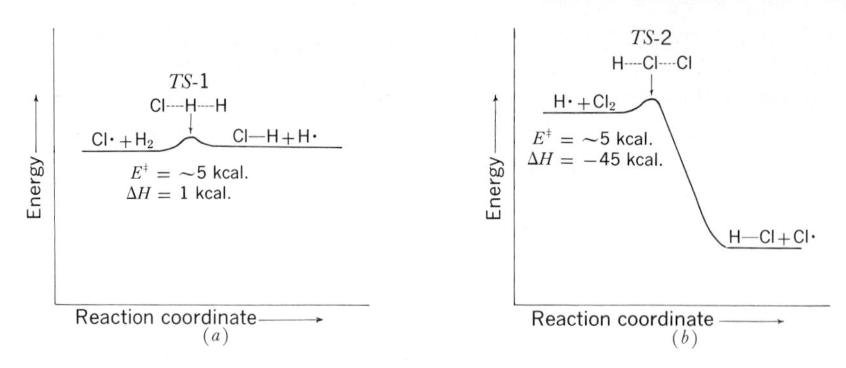

Figure 3–5. Energy profiles for the propagation steps in the reaction of hydrogen with chlorine.

Free radical reactions may take place also in the liquid phase. An example is the reaction of ferrous ion with hydrogen peroxide to give ferric ion, hydroxide ion, and free hydroxyl radicals.

$$\mathrm{Fe}\cdot^{2+} + \mathrm{HO{:}OH} \rightleftharpoons \underbrace{\mathrm{Fe^{3+} + HO{:}^-}}_{(\mathrm{FeOH})^{2+}} + \cdot\mathrm{OH}$$

Curved arrows, first used by Lowry[4], indicate the movement of the electrons in the over-all process. In this text dotted curved arrows will indicate one-electron transfers, whereas solid curved arrows will be used for the transfer of a pair of electrons. Many organic reactions such as autoxidation (p. 205) and polymerization (p. 88) take place by free radical mechanisms in the liquid phase.

Polar Mechanisms. Except for proton transfer reactions, polar reactions do not take place in the gas phase because it is easier to break a bond homolytically than heterolytically (p. 17). In the liquid phase, however, polar reactions between polar molecules in polar solvents are more frequent than free radical reactions. The bonds are already polarized, and the solvent can assist the heterolysis of the bond by solvation of the transition state and of the products of the reaction.

A typical polar reaction is the formation of bromine when chlorine is passed into an aqueous solution of sodium bromide.

$$\mathrm{Br{:}^-} + \mathrm{Cl}{:}\mathrm{Cl} \longrightarrow \mathrm{Br{:}Cl} + {}^-{:}\mathrm{Cl}$$

$$\mathrm{Br{:}^-} + \mathrm{Br}{:}\mathrm{Cl} \longrightarrow \mathrm{Br{:}Br} + {}^-{:}\mathrm{Cl}$$

The Cl—Cl bond is stronger than the Br—Br bond, as regards homolysis, by 12 kcal. Yet the reaction takes place because chlorine is more electronegative than bromine and can better accept a negative charge. Moreover the heterolysis of the chlorine bond is aided by hydration of the leaving chlorine in the transition state. Hence, although not shown in the reaction, the solvent actually takes part, and a more complete representation would be

$$(\mathrm{H_2O})_x\mathrm{Br}^- + \mathrm{Cl{-}Cl} + (\mathrm{H_2O})_y \longrightarrow [(\mathrm{H_2O})_x\text{----}\mathrm{Br}\text{----}\mathrm{Cl}\text{----}\mathrm{Cl}\text{----}(\mathrm{H_2O})_y]^- \longrightarrow (\mathrm{H_2O})_x + \mathrm{Br{-}Cl} + {}^-\mathrm{Cl}(\mathrm{H_2O})_y$$

The dielectric constant of the solvent (p. 102), its ability to form hydrogen bonds, and even its shape play an important role whenever the solvent is involved in the transition state.

In reactions of the general type

$$A{:} + B{:}C \longrightarrow A{:}B + {:}C$$

the reagent A**:** that provides the pair of electrons for the new bond is called a **nucleophile** (Gr. *philos* loving; literally *nucleus-loving* or *nucleus-seeking*) because it becomes attached to an electron-poor group. Because another group leaves, or is displaced, the reaction is called a nucleophilic displacement reaction or an S_N-type reaction.[10] The group displaced has come to be called the *leaving group,* a rather awkward expression.

All nucleophilic reagents are bases, but the order of reactivity does not necessarily follow the order of basicity. For example, iodide ion is a much weaker base than hydroxide ion. Yet iodide ion frequently reacts much faster than hydroxide ion with the same substrate. Relative basicities are based on equilibrium and measured in terms of equilibrium constants. Relative rates are governed by relative activation energies, which may be influenced by several factors. The term **nucleophilicity** has been introduced to indicate the relative rates with which nucleophiles react with a given substrate. Thus in nucleophilic reactions in which iodide ion reacts faster than hydroxide ion, iodide ion is said to have a higher nucleophilicity.

If a reagent is thought of as being attracted to a point of high electron density, it is said to be **electrophilic.** Whether a reaction is called electrophilic or nucleophilic depends on which reacting substance is considered to be the attacking agent. Thus the reaction between hydrogen chloride and water may be considered as a nucleophilic attack on hydrogen by a water molecule with displacement of chloride ion, or as an electrophilic attack of hydrogen chloride on oxygen with the transfer of a proton.

$$H_2O\!:\; + \;H\,|\,\!:\!Cl \longrightarrow H_2\overset{+}{O}\!:\!H + {}^{-}\!:\!Cl$$

$$Cl\!:\,|\,H \;+\; :\!OH_2 \longrightarrow Cl\!:^{-} + H\!:\!\overset{+}{O}H_2$$

*PROBLEMS**

3–*1*. For H_3PO_4(aq.), $\Delta H_f^\circ = -308$ kcal. mole^{-1} and $\Delta S^\circ = 42.1$ e.u. For $H_2PO_4^-$(aq.), $\Delta H_f^\circ = -311$ kcal. mole^{-1} and $\Delta S^\circ = 21.3$ e.u. Calculate the change in free energy, ΔG, for the ionization of the first hydrogen of phosphoric acid, and the first ionization constant at 25°C.

3–*2*. Convert (*a*) $K_a = 5.9 \times 10^{-2}$ to pK_a; (*b*) $K_b = 1.5 \times 10^{-14}$ to pK_b; (*c*) $K_b = 3.8 \times 10^{-10}$ to pK_a; (*d*) $pK_a = -7.1$ to pK_b; (*e*) $pK_b = 14.5$ to pK_a.

3–*3*. Which is the stronger acid: (*a*) $K_a = 4.5 \times 10^{-5}$ or $K_a = 8.5 \times 10^{-6}$; (*b*) $pK_a = -3.2$ or $pK_a = 2.5$?

3–*4*. Which is the stronger base: (*a*) $K_b = 1.4 \times 10^{-4}$ or $K_b = 8.3 \times 10^{-4}$; (*b*) $pK_b = 15.5$ or $pK_b = 8.6$; (*c*) $pK_a = -1.5$ or $pK_a = 5.4$?

---

[10] Many terms introduced with the development of theories of mechanism are unnecessary jargon and frequently are not as descriptive as they should be. Thus a nucleophilic group is not seeking a nucleus but merely becomes attached at a point of low electron-density. The symbol S_N stands for *substitution-nucleophilic,* but S_N reactions properly are called displacement reactions. The term *substitution reaction* always has referred only to the substitution of another group for hydrogen attached to carbon. Once terms and symbols come into common usage, however, it is not possible to replace them.

* See footnote, p. 26.

CHAPTER FOUR

ISOLATION AND ANALYSIS OF ORGANIC COMPOUNDS

Before any study can be made of an unknown compound, it first must be isolated in a pure state. During this process a number of its physical properties ordinarily will be determined. Next a qualitative and quantitative analysis for the elements present must be made and the molecular weight of the compound determined. These operations precede the investigation of the structure of a natural product and follow the production of a new compound by synthesis.

Since all organic compounds have covalent bonds and most of them lack ionic bonds, their physical properties are markedly different from those of most inorganic compounds, which usually are ionic. These differences are summarized in Table 4–*1* and have a distinct bearing on the experimental procedures used when organic compounds are investigated.

Isolation and Purification of Organic Compounds

Distillation. Because of the small attractive forces between most individual organic molecules, many organic compounds are converted into the gaseous state, without decomposition, when heated to a temperature at which the vapor pressure becomes equal to the external pressure. This temperature is known as the **boiling point** of the liquid and is a characteristic property of a pure compound. The vapors may be forced into a cooling device where they condense to a liquid again. This process is called **distillation.**

Because the attraction between the molecules of one compound differs from that between the molecules of another, different organic compounds usually distill at different temperatures. Hence distillation can be used not only to separate volatile organic compounds from nonvolatile materials but also to separate mixtures of organic compounds that have different boiling points. To separate substances whose boiling points are not far apart, it may be necessary to use fractional distillation or an efficient fractionating column.

In **fractional distillation,** the mixture is distilled into a number of fractions. These fractions usually differ in composition from the original material, the lower-boiling fractions

TABLE 4–*1*. USUAL PHYSICAL PROPERTIES OF IONIC AND COVALENT COMPOUNDS

IONIC COMPOUNDS	COVALENT COMPOUNDS
High melting point (above 700°)	Low melting point (under 300°)
Nonvolatile	Distill readily
Insoluble in nonaqueous liquids	Soluble in nonaqueous liquids
Soluble in water	Insoluble in water
Conduct electric current in molten state and in solution	Solutions and melts are nonconducting

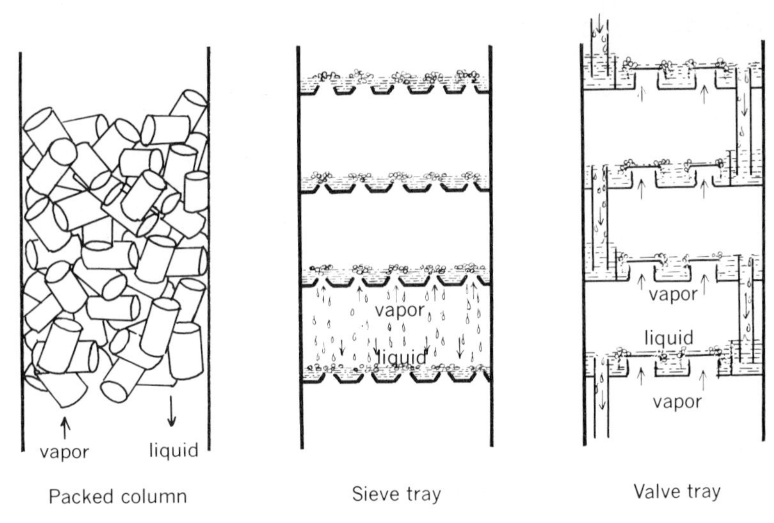

Figure 4–*1*. Sections through three types of fractionating columns.

containing a higher percentage of the lower-boiling substance. By a systematic redistillation of the fractions a further separation takes place, and the process is repeated until the desired degree of separation has been achieved. A **fractionating column** is a device for bringing about a more efficient separation than can be obtained in an ordinary distilling flask. It consists of a tall column through which the vapors rise and a certain amount of liquid, known as *reflux,* descends. As the hot vapors from the still come into contact with the cooler descending liquid, a heat interchange takes place. The vapors are cooled slightly and some of the higher-boiling components condense. The heat of condensation given up to the reflux vaporizes a small amount of its lower-boiling components. Hence there is a gradual enrichment of the lower-boiling components in the vapors as they rise in the column, and an enrichment of the higher-boiling components in the reflux as it descends.

The efficient operation of a fractionating column depends on intimate contact of the distilling vapors and the reflux and on the establishment of equilibrium between them, and many types of columns have been devised to achieve this. There are two main classes of columns, the **packed columns** and the **plate** or **tray columns,** sections of which are illustrated in Fig. 4–*1*. Stacked grids may be used instead of bulk packing in packed columns, and sieve trays may have devices (*down-comers*) for the return of liquid from tray to tray as with the valve trays. The valve trays may have circular or rectangular openings. The older bubble cap trays rarely are used now in new columns. Theoretically it is not possible to separate two volatile compounds completely by distillation. Even when the difference in their boiling point is large, the higher-boiling component always exerts its vapor pressure at the boiling point of the other component and some molecules of the higher-boiling member distill. For example, when a mixture of equal amounts of ethyl alcohol, a colorless liquid boiling at 78°, and of azobenzene, an orange solid boiling at 297°, is distilled from an ordinary distilling flask, even the first distillate is colored yellow by the azobenzene molecules that have come over with the alcohol, despite the difference in boiling points of over 200°. For all practical purposes, however, it is possible to separate by means of efficient fractionation, liquids boiling as little as 3° apart.

Distillation under reduced pressure, which frequently is called **vacuum distillation,** is used for high-boiling compounds to avoid the decomposition that might take place at the high temperatures necessary to produce distillation at atmospheric pressure. Since substances distill as soon as their vapor pressures exceed the pressure at the surface of the liquid, reduction of the pressure enables the substance to distill at a lower temperature. **Steam distillation,** in which steam is passed through the substance or the substance is distilled in the presence of water, accomplishes the same purpose. Distillation takes place when the sum of the vapor pressures of the compound and of water exceeds the pressure in the distillation apparatus. This condition occurs at a lower temperature than for either component alone.

Crystallization. This procedure is useful for the purification of organic solids, just as it is an important method for salts, but the technique is different. The solubilities of organic compounds usually increase greatly with increasing temperature. Hence in the usual method of crystallization, a saturated solution is made at the boiling point of the solvent and then cooled to room temperature or below. The pure compound crystallizes, and the more soluble or less abundant impurities remain in solution. A large variety of solvents is available, and the one most suitable for the material being purified can be chosen. Because of the highly selective forces involved in fitting molecules into a crystal lattice, crystallization, where applicable, is one of the most effective purification processes available.

Extraction. Extraction with solvents also is an important method for the isolation and purification of organic compounds. Solid mixtures are extracted with volatile solvents that remove certain components and leave other components behind. Organic compounds can be separated from aqueous solutions by extraction with a water-immiscible solvent that dissolves more of the desired material than does water. The organic compound passes into the organic solvent, from which the compound can be recovered by evaporation or distillation of the solvent. Organic or inorganic salts can be separated from nonionic compounds by means of a mixture of water and an organic solvent, the ionic compound dissolving in the water and the nonionic compound in the organic solvent.

In general when a compound is shaken with a mixture of two immiscible liquids, the ratio of the concentration in one layer to that in the other layer is a constant. The value of the constant is called the *distribution coefficient* for the two liquids and is characteristic of the compound. Furthermore if two or more species are present, each usually behaves independently of the other. If the different species have different distribution coefficients, the composition of the mixtures in each layer will be different from that in the original mixture and a partial separation will have occurred. The extent of separation is greater, the greater the difference in the distribution coefficients of the components. If the layers are separated and fresh upper-layer solvent is added to the lower layer and fresh lower-layer solvent is added to the upper layer and the partitions are repeated, a further separation is effected. When the process is carried out countercurrently, it is known as *countercurrent distribution.* To separate two substances that have nearly the same distribution coefficients, a complex apparatus has been devised to perform a large number of these operations automatically. For most separations the same result can be obtained more conveniently by partition chromatography.

Adsorption and Chromatography. Selective adsorption at the surfaces of solids is important for the removal of impurities and for the separation of mixtures of compounds having similar physical properties. Activated charcoal (*decolorizing carbon*) is used widely to remove colored impurities of high molecular weight from liquids or solutions. In another procedure a solution of the mixture of compounds is passed through a column of activated aluminum oxide, silica gel, or one of numerous other solid adsorbents, and the column is eluted with suitable solvents. The components of the mixture usually pass down the column at different rates, dependent on their relative adsorbability, and are separated. This process first was applied to a mixture of leaf pigments (p. 662), which separated as colored bands on the white alumina. Hence the procedure was called *chromatography,* a term now used for all types of separations using a moving phase and a stationary phase. In *thin layer* chromatography, the immobile phase is a thin layer of silica or alumina on a glass or other inert surface, and the developing solvent moves by capillary attraction. Separation of mixtures of ions in aqueous solution can be made by use of a column of an *ion-exchange* resin (pp. 356, 484) as the adsorbent.

In *partition chromatography,* a liquid is held stationary at the surface of an adsorbent packed in a column. A second liquid, insoluble in the first, is passed through the column. Separation of a mixture of solutes that have different distribution coefficients in the two liquids is effected by continuous partition between the two phases. *Paper chromatography* is a type of partition chromatography in which cellulose in the form of filter paper is used as the support for the immobile phase and in which elution occurs by capillary attraction or by capillarity and gravity flow. The immobile phase is either adsorbed water or a liquid such as *N,N*-dimethylformamide with which the paper has been impregnated.

Since 1952 a process called *gas-liquid partition chromatography* (GLC) or *vapor phase chromatography* (VPC) has been used to separate complex mixtures of compounds that have an appreciable vapor pressure at any temperature below that at which decomposition begins. The vaporized mixture in a carrier gas such as nitrogen or helium is passed through a tube packed with fire brick granules or a capillary tube lined with a porous material, the support being impregnated with a nonvolatile liquid. The components are carried through the column at rates dependent on their distribution coefficients for the carrier gas and the stationary liquid. Amazingly sharp separations can be obtained using very small samples. A four-foot column can be as effective as a 1200-plate fractionating column and capillary columns rated at 400,000 theoretical plates are in use.

The various types of detectors used respond in proportion to the concentrations of the separated compounds in the effluent gas. The signals are amplified electronically, and the results are recorded on a moving chart. Since the areas under the curves are proportional to the quantity of each component present, this method yields quantitative as well as qualitative results and is the preferred method for the analysis of mixtures of volatile compounds. Coupled with combustion procedures (p. 47), it can be used in the ultramicro quantitative determination of carbon, hydrogen, and nitrogen in organic compounds.

Electrophoresis and Ionophoresis. For a given potential gradient, the rate at which charged particles migrate in an electric field depends on the net charge on the particle and on its resistance to movement in the surrounding medium, which in turn depends on the size and shape of the particle, and on the attractive forces between the particle and the medium. Both colloidal and ionic particles are charged, and mixtures of substances that have different rates of migration can be separated by application of the proper electrical field to an aqueous solution. Convection is inhibited by conversion of the solution to a gel or by impregnation of a horizontal sheet of filter paper with the solution. The separation of colloidal particles in this way is called *electrophoresis,* and of ionic particles, *ionophoresis.*

Zone Melting. If a crystalline substance is heated in such a way that a narrow molten zone is created, and if this zone is then moved through the bulk of the material at a rate such that the material through which the zone has passed immediately recrystallizes, the crystallization causes the lower-melting mixtures with impurities to move along with the molten zone to the end of the mass. The process has had only limited application to organic compounds because of their tendency to form supercooled melts.

Criteria of Purity. Pure substances distill at a constant temperature; hence *constancy of boiling point* is a good criterion of purity. Pure solid compounds usually *melt over a very narrow range of temperature* that is characteristic of the compound. Since these melting points are relatively low, they may be determined easily and form a second criterion of purity. A substance that moves as a single spot on a paper chromatogram usually is pure. The ratio of the rate of movement of the spot to the rate of movement of the solvent front is known as the R_F value and is characteristic of the substance. It is so dependent on conditions, however, that direct comparison with a known compound must be made on the same strip of paper used for identification purposes. Other useful criteria are *density, refractive index, solubilities in various solvents,* and *absorption spectra* (pp. 133 and 541).

Analysis of Organic Compounds for the Elements

Qualitative Tests. After a compound has been isolated in a pure state, any investigation of it must begin with the identification of the elements present. Qualitative tests for the elements depend almost exclusively on reactions of ions. Since the elements in organic compounds usually are not present in ions, other types of tests must be devised, or the organic compounds must be converted into ionizable salts. Both methods of attack are used. **Carbon** and **hydrogen** may be detected readily if the unknown is heated in a test tube with dry copper oxide powder and the evolved gas passed into a solution of barium hydroxide. The hydrogen is burned to water, which collects in droplets on the cooler portion of the test tube and may be observed visually. The carbon is burned to carbon dioxide, which reacts with the barium hydroxide solution to give a precipitate of barium carbonate.

Several methods are available for the conversion of other elements into detectable ions, but the procedure most commonly used is to heat the compound with molten sodium. This process is known as *sodium fusion.* **Sulfur** is converted to sodium sulfide, **halogen** to sodium halide, and **nitrogen,** in the presence of carbon, to sodium cyanide. The symbol X is used to represent chlorine, bromine, or iodine.

$$\begin{array}{c}\text{Organic compound containing}\\ \text{C, H, N, S, X}\end{array} \xrightarrow{\text{Na fusion}} \begin{array}{l}NaCN\\ Na_2S\\ NaX\end{array}$$

An aqueous solution of the ions is obtained by decomposition of the melt with water. If a portion of the aqueous solution is heated with ferrous and ferric sulfate in alkaline solution and *cyanide ion* is present, insoluble ferric ferrocyanide (Prussian blue) is formed. Addition of hydrochloric acid dissolves the iron hydroxide and leaves the blue precipitate.

$$18\ NaCN + 3\ FeSO_4 + 2\ Fe_2(SO_4)_3 \longrightarrow Fe_4[Fe(CN)_6]_3 + 9\ Na_2SO_4$$

Sulfide ion is detected readily if the solution is acidified with acetic acid, heated to drive off hydrogen sulfide, and the vapors allowed to come in contact with a solution of lead acetate on filter paper. A dark, lustrous spot of lead sulfide forms if sulfide ion is present.

$$Na_2S + 2\ HC_2H_3O_2 \longrightarrow H_2S + 2\ NaC_2H_3O_2$$
$$Pb(C_2H_3O_2)_2 + H_2S \longrightarrow PbS + 2\ HC_2H_3O_2$$

To test for *halide ion,* the solution is boiled with nitric acid to remove cyanide ion and sulfide ion, and then aqueous silver nitrate is added. If halogen is present, a white, pale yellow, or deep yellow precipitate forms, depending on whether chlorine, bromine, or iodine is present.

$$NaX + AgNO_3 \longrightarrow AgX + NaNO_3$$

Quantitative Analysis. The determination of the *kind* of elements present in a new compound under investigation is followed by the determination of the *relative amounts* of the elements. **Carbon** and **hydrogen** are estimated by a modification of a procedure developed by Liebig, which is based on the same principle used for qualitative detection, namely combustion to carbon dioxide and water. The apparatus and the functions of the various parts are given in Fig. 4–2.

Figure 4–2. Apparatus for the determination of carbon and hydrogen: *a*, source of oxygen; *b*, tube packed with granular sodium hydroxide on asbestos for removing carbon dioxide, and a second solid absorbent such as anhydrous magnesium perchlorate for removing water; *c*, fused silica combustion tube; *d*, platinum boat containing weighed sample and heated by a sectional combustion furnace (dotted lines); *e*, packing of quartz granules heated to 950–1000°; *f*, silver wool to remove sulfur and halogen; *g*, weighed tube containing water absorbent; *h*, tube containing manganese dioxide to remove oxides of nitrogen; *i*, weighed tube containing carbon dioxide absorbent backed by water absorbent.

In the course of an analysis, the tube is swept with oxygen until the weights of the absorption tubes *g* and *i* become constant. The sample then is heated in a slow stream of oxygen until it is burned completely, and the tube is swept with oxygen again until all the combustion products are passed through the absorption train. Finally the absorption tubes *g* and *i* are weighed again. The increases in weight give the weights of water and carbon dioxide respectively. From these data and the weight of the original substance, the per cent of hydrogen and of carbon can be calculated (p. 49).

The *Dumas method* is the most important for the determination of **nitrogen** because it can be used for any type of nitrogen compound. The apparatus is illustrated in Fig. 4–3. In this procedure the weighed sample is burned to carbon dioxide and nitrogen, the carbon dioxide removed, and the volume of the remaining nitrogen measured. The combustion is carried out in a quartz tube packed with copper oxide, cobalt oxide, and clean copper. The weighed sample, covered with copper oxide powder, is placed in the tube and the tube is swept with pure carbon dioxide. When the section containing the sample and copper oxide is heated to 1000°, the sample is burned to carbon dioxide, water, and nitrogen. The gases are passed over copper oxide and cobalt oxide at 800–850° to ensure complete combustion.

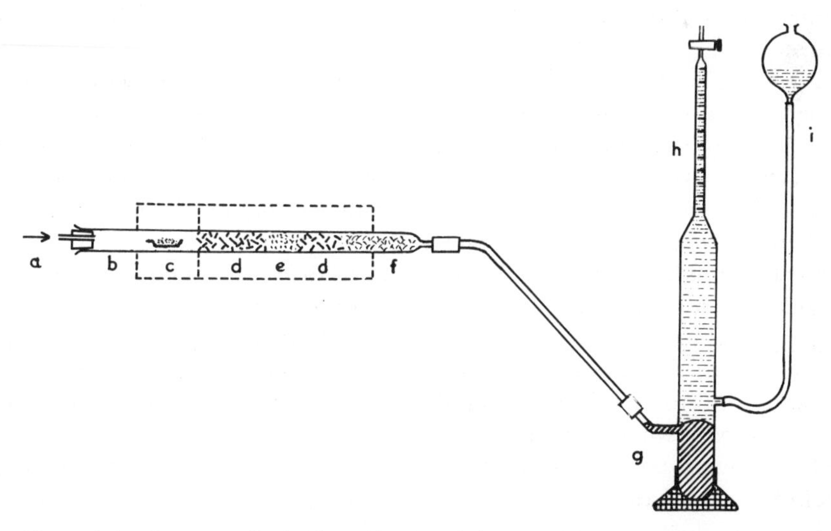

Figure 4–3. Apparatus for the determination of nitrogen (Dumas): *a*, source of pure carbon dioxide; *b*, fused silica tube; *c*, weighed sample covered with copper oxide powder and heated by sectional combustion furnace (dotted lines); *d*, copper oxide wire and *e*, cobalt oxide heated to 800–850°; *f*, clean copper to reduce oxides of nitrogen to nitrogen; *g*, mercury valve; *h*, nitrometer containing 50 per cent aqueous potassium hydroxide solution; *i*, leveling bulb and potassium hydroxide reservoir.

Any oxides of nitrogen that are formed are reduced to nitrogen by means of a packing of clean copper in the form of short thin rods, which is partly at 500° and partly at lower temperatures where it extends beyond the furnace. At the end of the combustion, the tube is swept again with carbon dioxide. The gases are passed into a nitrometer tube filled with 50 per cent potassium hydroxide solution,[1] which absorbs the carbon dioxide and leaves the nitrogen. The volume of nitrogen is measured and the temperature and barometric pressure noted. The vapor pressure of the potassium hydroxide solution is obtained from a table and subtracted from the barometric pressure. From these data the weight of nitrogen and the per cent in the sample can be calculated (p. 49).

In the *Kjeldahl method* a weighed sample is heated with concentrated sulfuric acid and an oxidation catalyst such as mercuric oxide, or copper sulfate and selenium. The nitrogen in the sample is converted to ammonium sulfate. At the end of the digestion, sodium hydroxide is added to liberate ammonia, which is distilled into a measured volume of standard acid. Back titration with standard alkali and with methyl red as an indicator gives the amount of ammonia formed. The procedure lends itself to routine analyses of large numbers of samples and is very useful for those types of compounds for which it gives accurate results. Some compounds, however, evolve nitrogen or do not decompose completely during the digestion, and the results are low.

The only special feature in the quantitative determination of **other elements** is the necessity to convert them into ionized compounds. In the determination of **sulfur, halogen,** or **phosphorus,** for example, the compound may be fused in a Parr bomb with sodium peroxide, which burns the carbon and hydrogen to carbon dioxide and water, and converts halogen, sulfur, and phosphorus into the alkali halide, sulfate, and phosphate. Alternatively the compound may be oxidized with nitric acid in a sealed tube at elevated temperature (*Carius method*). When the conversion to the inorganic salts has been effected, the usual quantitative methods, gravimetric or volumetric, may be employed. In recent years procedures have been developed for the direct determination of **oxygen,** but these methods are satisfactory only if the apparatus is used almost continuously. As a result this element usually is determined as the difference between 100 per cent and the sum of the per cents of the other elements.

The *quantity of material* needed for an analysis depends on the experience of the operator, on the apparatus, and on the sensitivity of the balance available. Macroanalyses use from 0.2 to 0.1 g. of substance; semimicroanalyses, 0.02 to 0.01 g.; microanalyses, 5 to 3 mg.; and submicroanalyses, also called ultramicroanalyses, 1 to 0.05 mg. For each type a balance must be used that is accurate to four significant figures. Ordinarily analyses now are carried out by specialists in microanalysis.

Empirical Formulas. From the analyses the relative proportion by weight of the various elements present may be calculated, but for convenience the results are expressed as an *empirical formula,* which shows the relative number of different atoms in the molecule. To calculate the empirical formula, the per cent of each element is divided by its atomic weight. The result usually is in the form of fractions, and it is customary to convert these fractions into ratios of whole numbers.

The following example illustrates the general method for the calculation of an empirical formula from analyses.

---

[1] Potassium hydroxide is used instead of sodium hydroxide because potassium carbonate is soluble in concentrated potassium hydroxide solution, whereas sodium carbonate is insoluble in concentrated sodium hydroxide solution and would precipitate.

A sample weighing 0.1824 g. gave on combustion 0.2681 g. of carbon dioxide and 0.1090 g. of water.

$$\text{Weight of carbon in sample} = 0.2681 \times \frac{12}{44} = 0.07312 \text{ g.}$$

$$\text{Weight of hydrogen in sample} = 0.1090 \times \frac{2}{18} = 0.01211 \text{ g.}$$

$$\text{Per cent carbon in sample} = \frac{0.07312}{0.1824} \times 100 = 40.09$$

$$\text{Per cent hydrogen in sample} = \frac{0.01211}{0.1824} \times 100 = 6.64$$

$$\text{Per cent oxygen in sample} = 100 - (40.09 + 6.64) = 53.27$$

Element	Per cent by weight		Atomic weight		Atomic ratio				Atomic ratio in integers
C	40.09	÷	12	=	3.36	÷	3.33	=	1
H	6.64	÷	1	=	6.64	÷	3.33	=	2
O	53.27	÷	16	=	3.33	÷	3.33	=	1

The empirical formula therefore is CH_2O.

Molecular Formulas. Usually a determination of the empirical formula identifies an inorganic compound. For most organic compounds, however, the empirical formula is less definitive. One reason for this difference is that several organic compounds may have the same empirical formula but differ in molecular weight. Thus formaldehyde, acetic acid, lactic acid, and glucose all have the empirical formula CH_2O, but their molecular formulas, that is, formulas indicating not only the ratio but also the number of atoms in the molecule, are respectively CH_2O, $C_2H_4O_2$, $C_3H_6O_3$, and $C_6H_{12}O_6$. Molecular formulas represent the *composition* of the compound. The importance of determining molecular weights was brought to the attention of chemists by Cannizzaro in 1860, although the basis for their determination had been provided by Avogadro in 1811 (p. 4).

The two methods for the determination of molecular weights that most generally have been used are known as the *cryoscopic* or *freezing point method,* and the *ebullioscopic* or *boiling point method.* Nonvolatile solutes lower the melting point and raise the boiling point of a pure solvent. Since the change, Δt, in melting point or boiling point of the solvent depends on the number of molecules of solute dissolved, one gram molecular weight of any nondissociating substance dissolved in 1000 grams of a solvent lowers the melting point $K_{f.p.}$ degrees, called the *molal lowering of the freezing point* of the solvent, or raises the boiling point $K_{b.p.}$ degrees, the *molal elevation of the boiling point.* If W grams of a substance of molecular weight[2] M is dissolved in G grams of solvent,

$$\Delta t = K\frac{W}{M} \times \frac{1000}{G}$$

$$\text{or } M = \frac{KW}{\Delta t} \times \frac{1000}{G}$$

[2] Throughout this text, the term *molecular weight* means *gram molecular weight,* a convention commonly adopted by chemists. It should be noted also that strictly speaking the terms *atomic mass* and *molecular mass* should be used because their values are determined by means of a chemical balance and are not dependent on the force of gravity. Since the reference units used with a chemical balance are called *weights,* however, the terms atomic weight and molecular weight are in common use, weight in this sense being synonymous with mass.

TABLE 4–2. CRYOSCOPIC AND EBULLIOSCOPIC CONSTANTS

SOLVENT	M.P.	$K_{f.p.}$	B.P.	$K_{b.p.}$ (760 MM.)
Water	0	1.86	100	0.51
Acetic acid	16	3.86	118	3.07
Benzene	5	5.12	81	2.53
Exaltone	66	21.3		
Camphene	49	31.0		
Camphor	178	39.7		

The experimentally determined values of K for a few of the common solvents are given in Table 4–2.

In the estimation of the molecular weights of unknown compounds, the boiling point or freezing point is determined in a special apparatus, first on the pure solvent and then on a solution of known concentration. With K, W, G, and Δt known, the molecular weight M can be calculated. Because the changes in the boiling point or melting point usually are small, a special differential thermometer that measures small differences in temperature, such as the *Beckmann thermometer,* or the *Menzies-Wright thermometer,* is used, and relatively large amounts of material are needed. The molecular lowering of the freezing point of camphor, however, is large, and since it is a solid, the melting point lowering may be determined in a capillary tube by means of an ordinary thermometer. This procedure is known as the *Rast camphor method* for determining molecular weights.

More recently, commercial *vapor pressure osmometers,* priced around $2500, have become available for the determination of molecular weights with an accuracy of one per cent. In an atmosphere saturated with solvent vapor, a drop of solvent and a drop of sample solution as dilute as 0.01 molar are placed on two thermistors capable of detecting a difference in temperature of 0.0001°. Since the solution has a lower vapor pressure than the solvent, vapor condenses on the solvent drop, which causes a difference in temperature between the two drops that is proportional to the molar concentration of the solution. The instrument is calibrated with pure compounds of known molecular weight.

The relative error of molecular weight determinations by most methods rarely is less than ±5 per cent, and the result of a combustion analysis usually is considered acceptable if the absolute error is less than ±0.5 per cent.[3] Hence it is difficult to determine by analyses alone the exact molecular formulas of saturated hydrocarbons (compounds containing carbon and the maximum amount of hydrogen) having molecular weights greater than about 150. If other elements are present, the differences in the per cent composition of compounds differing by one carbon atom are considerably greater. Thus if a compound contains carbon and hydrogen and one oxygen atom per molecule, empirical formulas can be determined for compounds having molecular weights up to about 400. Analyses for the elements usually are more accurate than ordinary molecular weight determinations, and the latter are used merely to determine whether the molecular formula is a multiple of the empirical formula. Since about 1955, however, *mass spectrographs* capable of handling organic compounds other than hydrocarbons of low molecular weight have become increasingly

[3] Accuracy is expressed in terms of absolute error or relative error. The error of a measurement is the deviation from what is thought to be the true value. The *absolute error* is the difference between the numerical value observed and the true value and is expressed in the units used to express the numerical value. The *relative error* usually is expressed as the per cent deviation and is the absolute error divided by the true value and multiplied by 100. For example, if a compound has a molecular weight of 200 g. and the observed value is 210 g., the absolute error is 10 g. and the relative error is 5 per cent. If a compound contains 5.5 per cent hydrogen and the value found is 5.3 per cent, the absolute error is 0.2 per cent and the relative error is 3.6 per cent.

available in the larger laboratories and provide a means for the determination of molecular weights of less than 500 to within one mass unit (p. 543).

Classification of Organic Compounds

The number of organic compounds known by July of 1961 has been estimated to be around 1,750,000 as compared to 500,000 inorganic compounds. Moreover the number of organic compounds was increasing at the rate of 90,000 per year and of inorganic, 20,000. The reasons for the large number of organic compounds have been given, namely that a carbon atom is able to combine with four other atoms or groups of atoms, and that carbon atoms are able to combine with each other indefinitely to produce stable compounds (p. 62). The possible number of compounds of carbon is infinite. Because of this large number, it might seem that the study of organic chemistry would be hopeless, but fortunately it is possible to divide these compounds into a comparatively small number of groups or *families of compounds*. Most members of each family may be prepared by similar methods, and most members exhibit similar chemical behavior. Moreover the individual members of each family differ from each other more or less regularly in their physical properties. The families, on the other hand, usually differ from each other markedly in their chemical properties and are prepared by different types of chemical reactions. Accordingly it is possible to concentrate on the methods of preparation and reactions of the families of compounds, and it is not necessary to discuss each member of a family. Special methods of preparation and special reactions of individual compounds then can be considered as additions or exceptions to the generalizations.

The numerous families may be grouped according to similarities in structure. The three main classifications are (*1*) *acyclic* compounds, which contain no ring structures of the atoms; (*2*) *carbocyclic* compounds, which contain rings made up solely of carbon atoms; and (*3*) *heterocyclic* compounds, which contain rings made up of more than one kind of atom. The acyclic compounds more commonly are called *aliphatic* compounds from the Greek word *aleiphatos* meaning *fat,* because the fats have this type of structure. The carbocyclic compounds are subdivided into benzenoid carbocyclic or aromatic compounds and non-benzenoid carbocyclic compounds. The organization of this text largely follows this classification.

*PROBLEMS**

4–*1*. Calculate the per cent composition of the compounds having the following molecular formulas: (*a*) C_3H_8, (*b*) C_2H_6O, (*c*) C_2H_3ClO, (*d*) $C_2H_8N_2$, (*e*) $C_2H_5NO_2$, (*f*) $C_3H_9O_4P$, (*g*) $C_4H_8Cl_2S$, (*h*) CH_6BrN, (*i*) C_2H_6Hg, (*j*) C_3H_9ClSi.

4–*2*. Calculate the differences in the per cents of carbon and of hydrogen for the following pairs of compounds: (*a*) $C_{17}H_{36}$ and $C_{18}H_{38}$, (*b*) $C_{17}H_{36}O$ and $C_{18}H_{38}O$, (*c*) $C_{30}H_{48}O$ and $C_{31}H_{50}O$.

4–*3*. Calculate the volume of nitrogen that would be obtained at S.T.P.† when 0.044 g. of C_2H_7N is burned in a Dumas apparatus.

4–*4*. Calculate the empirical formula for a compound that contains 38.65 per cent carbon, 9.68 per cent hydrogen, 51.62 per cent sulfur, and possibly oxygen.

4–*5*. A compound contains 54.55 per cent carbon, 9.02 per cent hydrogen, and possi-

---

* See footnote, p. 26.

† Standard temperature and pressure; that is, 0° and 760 mm. of mercury. When designating pressure, chemists ordinarily omit "of mercury," this qualification being understood. Recently the unit, *torr* = 1 mm. of mercury, has come into use.

bly oxygen, and has the approximate molecular weight 84. Calculate the molecular formula of the compound and the exact molecular weight.

4–6. Calculate the approximate molecular weight of a compound that gave a freezing point depression of 1.582° when 0.310 g. was dissolved in 15.0 g. of benzene.

4–7. What is the minimum molecular weight of a compound that contains (*a*) 59.2 per cent chlorine; (*b*) 4.25 per cent sulfur; (*c*) 11.6 per cent nitrogen; (*d*) 4.24 per cent copper; (*e*) 0.33 per cent iron?

4–8. A compound contains carbon, hydrogen, bromine, and possibly oxygen. Combustion of 0.2001 g. gave 0.1902 g. of carbon dioxide and 0.0907 g. of water. Fusion of 0.1523 g. with sodium peroxide, acidification with nitric acid, and precipitation with silver nitrate gave 0.2058 g. of silver bromide. Calculate the empirical formula of the compound.

4–9. Combustion of 0.1908 g. of a substance gave 0.2895 g. of carbon dioxide and 0.1192 g. of water. Combustion of 0.1825 g. in a Dumas apparatus gave 40.2 cc. of nitrogen measured over 50 per cent aqueous potassium hydroxide solution at 25° (vapor pressure of solution = 9 mm. of mercury) and at 735 mm. atmospheric pressure. When 1.082 g. was dissolved in 25 g. of benzene, the freezing point of the benzene was lowered 1.72 degrees. Calculate the molecular formula of the compound.

CHAPTER FIVE

ALKANES

Several families of organic compounds exist that contain only hydrogen and carbon in the molecule. They are known as *hydrocarbons*. The simplest family of this group is known as the *alkanes*. For reasons that will become apparent, they are known also as *saturated hydrocarbons* or *paraffin hydrocarbons*, or the *methane series* of hydrocarbons. All other families of the acyclic or aliphatic series can be considered as derived from the alkanes, and hence a knowledge of their properties and constitution is the best preparation for further study.

The simplest member of this family of hydrocarbons contains only one carbon atom and is called *methane*. It is a gas that liquefies at $-161°$ and has the composition and molecular weight corresponding to the molecular formula CH_4. The next member boils at $-89°$, has the molecular formula C_2H_6, and is called *ethane*. The third member, *propane*, C_3H_8, boils at $-42°$. It is apparent already that this family is a series of compounds having the general formula C_nH_{2n+2}, in which one member differs from each adjacent member by one carbon atom and two hydrogen atoms, that is, by CH_2. Such a series is known as a **homologous series,** and the members of the series are known as **homologs** (Gr. *homos* same, *logos* speech; that is, related or similar).

Structural Formulas

The electronic structure of the carbon atom explains the composition of these compounds and the existence of a homologous series. The carbon atom has four electrons that it can pair with four electrons of other atoms (p. 12). Since only compounds of carbon and hydrogen are being considered, the simplest compound is that in which a single carbon atom has paired its four electrons with the electrons of four hydrogen atoms to give the molecule $\begin{array}{c} H \\ \times\bullet \\ H\overset{\bullet}{\times}C\overset{\times}{\bullet}H \\ \bullet\times \\ H \end{array}$, or CH_4. If two carbon atoms are in the molecule, they must be united to each other by a mutual sharing of electrons, $\cdot\dot{\underset{\cdot}{C}}\overset{\circ}{\underset{\bullet}{}}\overset{\circ}{\underset{\circ}{C}}\circ$, because the hydrogen atom with only one valence electron cannot combine with more than one carbon atom. Six electrons remain to be paired by combination with six hydrogen atoms to give the molecule $\begin{array}{c} H\;\;H \\ \times\bullet\;\;\times\circ \\ H\overset{\bullet}{\times}C\overset{\bullet}{\underset{\circ}{}}C\overset{\times}{\circ}H \\ \bullet\times\;\;\circ\times \\ H\;\;H \end{array}$, or C_2H_6. This process amounts to the insertion of a CH_2 group between one of the hydrogen atoms and the carbon atom in the methane molecule, or to the replacement of a hydrogen atom by a CH_3 group. If three atoms of carbon are in the molecule, the formula is

$$\begin{array}{c} H\;H\;H \\ \cdot\cdot\;\;\cdot\cdot\;\;\cdot\cdot \\ H:C:C:C:H \\ \cdot\cdot\;\;\cdot\cdot\;\;\cdot\cdot \\ H\;H\;H \end{array}, \quad \text{or} \quad C_3H_8.$$

In the last electronic formula no distinction has been made between the electrons supplied by different atoms. This representation is less artificial than the previous ones, since all electrons are identical. The important points to observe when electronic formulas are written are that the total number of valence electrons is correct, that all are paired, and that the valence shells of the atoms are filled (two electrons for hydrogen and usually eight electrons for all elements in the next two periods).

The operation of building hydrocarbons by application of the rules for valence can be carried on indefinitely. Hence compounds should exist with the molecular formulas C_4H_{10}, C_6H_{14}, or in fact any compound with the general formula C_nH_{2n+2}; that is, n times CH_2 plus two hydrogen atoms to satisfy the remaining two unshared electrons. For alkanes having four or more carbon atoms, however, the situation becomes somewhat more complicated because two compounds are known that have the same molecular formula, C_4H_{10}. One boils at $-0.5°$, and the other boils at $-12°$. Two or more compounds having the same molecular formula but differing in at least one chemical or physical property are known as **isomers** (Gr. *isos* equal, *meros* part; that is, having equal or like parts), and the phenomenon is known as **isomerism.** It can be explained readily on the basis of electronic formulas. As soon as more than three carbon atoms are present, they can be arranged in more than one way. Four carbon atoms, for example, can be joined consecutively to each other, or one carbon atom can be united to the central atom of a chain of three atoms.

```
 . . . .            . . .
·C:C:C:C·    or    ·C:C:C·
 . . . .            . .. .
                     ·C·
                      .
```

The remaining unpaired electrons can pair with those of hydrogen atoms to give the formulas

```
  H H H H             H    H    H
  .. .. .. ..         ..   ..   ..
H:C:C:C:C:H    and  H:C  : C  : C:H
  .. .. .. ..         ..   ..   ..
  H H H H             H  H:C:H  H
                           ..
                           H
```

Ordinarily the electron pair bond is represented by a dash, and the above formulas may be written

```
   H  H  H  H               H     H     H
   |  |  |  |               |     |     |
H—C—C—C—C—H      and     H—C—————C—————C—H
   |  |  |  |               |     |     |
   H  H  H  H               H     |     H
                               H—C—H
                                  |
                                  H
```

Since the atoms are arranged differently in the two molecules, they should have different chemical and physical properties. The above two compounds are known as *butanes.* The compound with the carbon atoms linked consecutively in a chain is known as *normal butane,* and its isomer is called *isobutane.* Hydrocarbons having the carbon atoms linked in a continuous chain are called *normal* or *straight-chain* hydrocarbons. If *side chains* or *branches* are present, as in isobutane, they are known as *branched-chain* hydrocarbons. Formulas that show not only the number of atoms in the molecule but also the way in which they are united to each other are known as *structural* or *graphic formulas.* The arrangement of the atoms in the molecule is referred to as the *constitution* of the compound as distinguished from its *composition* as represented by its molecular formula (p. 49).

If the formulas are written in a plane with the carbon bonds directed to the corners of a square, more than one propane should be possible also; that is,

```
       H
       |
H3C—C—H
       |
      CH3
```

might be

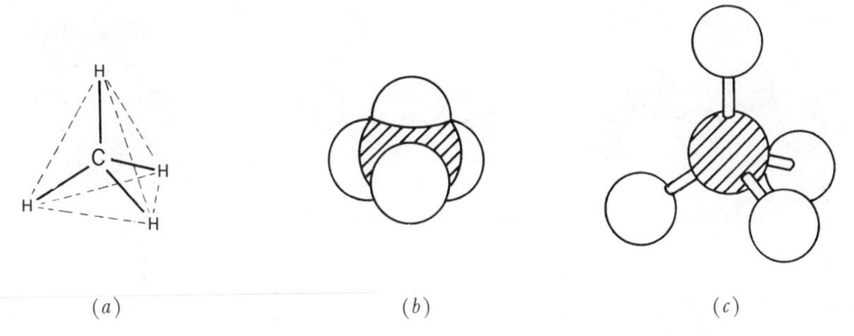

Figure 5–*1*. Space representations of methane: (*a*) tetrahedral; (*b*) Stuart model; (*c*) ball and stick model.

expected as well as $H_3C-\overset{\displaystyle H}{\underset{\displaystyle H}{\overset{|}{\underset{|}{C}}}}-CH_3$. Similarly a considerably larger number of butanes should exist. Additional isomers, however, are not known, and the existence of optical isomers, which can be explained only by different arrangements in three dimensions (p. 302), led to the postulation that the bonds of a carbon atom are directed to the corners of a regular tetrahedron, as shown by the formula for methane (Fig. 5–*1a*) (cf. p. 12). Stuart, Hirschfelder, or Briegleb models (Fig. 5–*1b*) give the correct interatomic distances and the distances of closest approach of different molecules as determined by X-ray diffraction studies. It is difficult, however, to see the mode of linkage of atoms in complicated molecules with this type of model, and the ball and stick type (Fig. 5–*1c*) more commonly is used. On paper or on the lecture board the plane formulas, which show only the order in which the atoms are linked to each other, are most convenient. It is important, however, to acquire the habit of visualizing molecules according to the spatial arrangement of the atoms, rather than as symbols in which letters are connected by short lines.

With a tetrahedral arrangement of the carbon valences, all of the hydrogen atoms in the ethane molecule are alike; that is, each bears exactly the same space relationship to all the other atoms in the molecule (Fig. 5–*2a*). Hence replacement of any hydrogen atom by another atom or group of atoms gives only one new compound, and the two plane formulas for propane become identical in the space formula (Fig. 5–*2b*).

For compounds of more than three carbon atoms, the tetrahedral distribution of the valences might be thought capable of giving isomeric compounds even for normal hydrocarbons. A five-carbon chain might be expected to exist in numerous arrangements. In the

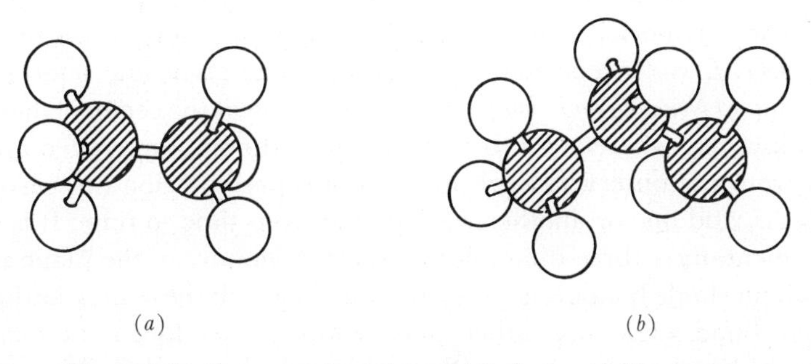

Figure 5–*2*. Space representations of (*a*) ethane and (*b*) propane.

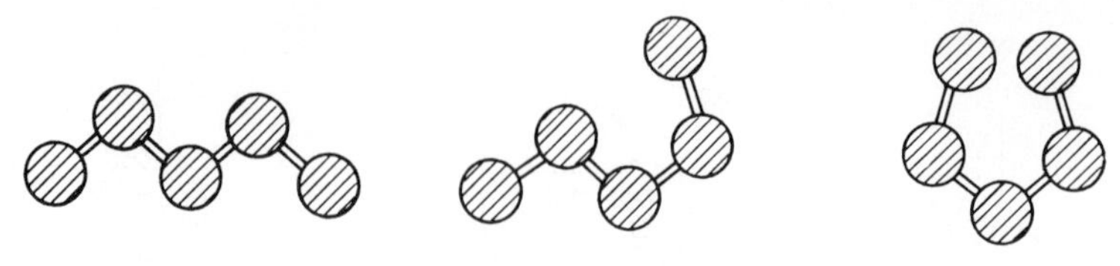

Figure 5–*3*. Some possible conformations of a five-carbon chain.

examples illustrated in Fig. 5–*3*, only the carbon skeleton is indicated. Nonplanar conformations are even more probable. Actually no isomers corresponding to these different arrangements have been detected. It is believed that all of the above arrangements and an infinite number of others exist in gases and liquids, but that, because of essentially free rotation about a single bond, the various forms interconvert at room temperature. The actual relative positions of the atoms at any instant is referred to as the *conformation* of the molecule. Thus the same molecule may have an infinite number of conformations, and a gas or liquid of the same molecular species consists of a mixture of molecules of different and continually changing conformations.

The molecules do have preferred conformations, however, in which they exist most of the time. Gaseous ethane, for example, is believed to prefer the staggered position (Fig. 5–*2a*) because the electron concentrations in the carbon-hydrogen bonds repel each other. It has been calculated that the barrier to free rotation about the carbon-carbon bond in the ethane molecule is about 3 kcal. per mole. This so-called *energy barrier* is the energy required to rotate the methyl groups with respect to each other sufficiently to bring the hydrogen atoms from the staggered position to the opposed position. Once the hydrogen atoms reach this position, further rotation of the methyl groups in the same or in the reverse direction returns the hydrogen atoms to a more stable position with the liberation of the same amount of energy. An energy barrier of 3 kcal. per mole is sufficient to ensure that most of the molecules at any instant have the more stable conformation. An energy barrier of at least 20 kcal. is required to prevent thermal interconversions, that is, to permit the independent existence of isomers, at room temperature.

Microwave spectroscopy has shown that practically all propane molecules have both methyl groups staggered with respect to the methylene (CH_2) group (Fig. 5–*2b*). In *n*-butane at a temperature of 14°, 60 per cent of the molecules have methyl groups as far apart as possible with the four carbon atoms in a plane (Fig. 5–*4 I*), and 40 per cent have a *skew* or *gauche* conformation (Fig. 5–*4 II* and *III*). In other words on the average the molecules have the conformation represented by *I* three fifths of the time, and by each of the conformations *II* and *III* one fifth of the time. Figures *a*, *d*, and *g* are views from the side. The common convention is that the solid lines represent bonds in the plane of the paper; the dotted lines, bonds directed behind the plane of the paper; and the dark wedges, bonds in front. The "sawhorse" representation in figures *b*, *e*, and *h* views the molecule from an angle. Figures *c*, *f*, and *i* represent end views looking along the bond joining the two central carbon atoms (*Newman projection*). The lines meeting at the center of the circle are bonds nearer to the observer, whereas the bonds touching the circumference are bonds to the groups farther from the observer. Each method of representation has certain advantages and disadvantages, and one or another will be used from time to time. It is necessary to be able to picture mentally a three-dimensional model from any of the plane representations.

Rotation about single bonds is not the only way in which the conformation of a molecule may be changed. Bond angles are rather "soft" and repulsive forces can change the degree of hybridization of bond angles (p. 15). Thus although the bond angles in methane are all

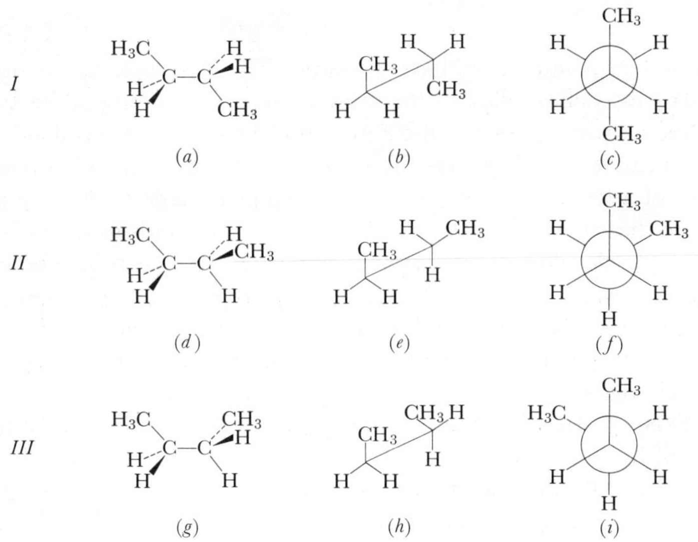

I (a) (b) (c)

II (d) (e) (f)

III (g) (h) (i)

Figure 5–4. Conformations of molecules of *n*-butane.

109°28′, the H—C—H angles in the methyl groups of propane are 108° and in the methylene group 106°, indicating a greater amount of *p* character. The C—C—C angle is 112°, indicating a greater amount of *s* character. In isobutane, the C—C—C angle is 111°. The C—C—C angle of *n*-butane also is 112° as determined by electron diffraction.

In the solid state the molecules are held in more or less fixed positions. Thus X-ray diffraction has shown that in crystals of long-chain paraffin hydrocarbons the molecules have the extended zigzag conformation (Fig. 5–7, p. 64).

With these facts available, it is possible to predict the number of isomeric alkanes having five carbon atoms. Three arrangements of the carbon atoms, called *carbon skeletons,* are possible.

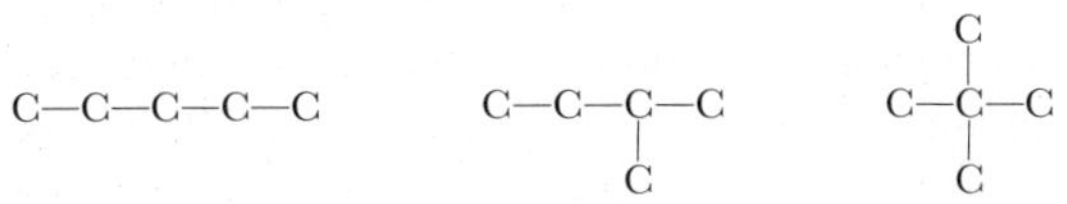

```
                                               C
                                               |
C—C—C—C—C          C—C—C—C                   C—C—C
                         |                     |
                         C                     C
```

They give rise to three alkanes.

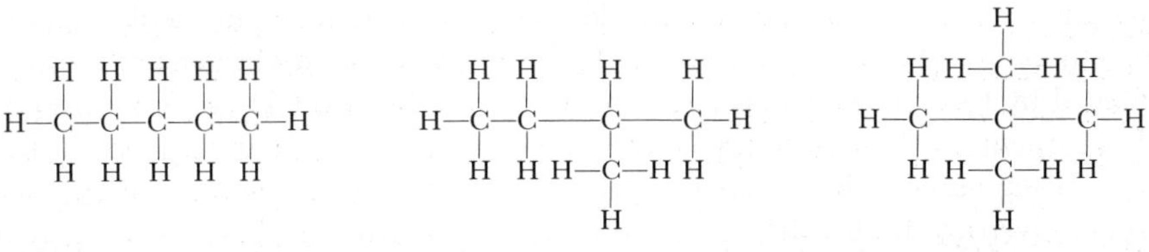

```
                                                                    H
                                                                    |
  H  H  H  H  H          H  H     H     H              H  H—C—H  H
  |  |  |  |  |          |  |     |     |              |     |     |
H—C—C—C—C—C—H          H—C—C—————C—————C—H          H—C—————C—————C—H
  |  |  |  |  |          |  |     |     |              |     |     |
  H  H  H  H  H          H  H  H—C—H  H              H  H—C—H  H
                                  |                          |
                                  H                          H
```

Actually three and only three compounds that have the molecular formula C_5H_{12} are known. The compound to which the first structure has been assigned boils at 36° and is called *normal pentane,* that having the second structure boils at 28° and is called *isopentane,* and that having the third structure boils at 9.5° and is called *neopentane* (Gr. *neos* new).

The above formulas are known as *extended structural formulas*. In order to conserve space and time in writing, they usually are replaced by *condensed structural formulas,* for example, $CH_3(CH_2)_3CH_3$, $CH_3CH_2CH(CH_3)_2$ or $C_2H_5CH(CH_3)_2$, and $C(CH_3)_4$.

Nomenclature

Each organic compound must have a name. The development of nomenclature in organic chemistry always has followed the same pattern. When only a few compounds of a particular type are known and before their structures have been determined, they are given names usually indicative of their source. These names are called *trivial* or *common* names. As more compounds of a series become known, an attempt is made to develop a more systematic nomenclature that shows the relation of the new compounds to the older ones. Finally an attempt is made to develop a truly rational system. In the meantime, however, the older names have become established, and it is difficult to replace them by new ones, with the result that the simpler compounds usually have two or three names. All of these names are in more or less general use, and it is necessary to become familiar with them and to be able to use them interchangeably.

Common Names. In all systems names of the members of the alkane series end in *ane*. The names of hydrocarbons having four carbon atoms or less are derived from the common names of the alcohols having the same number of carbon atoms (p. 116). Above C_4 the names become a little more systematic in that the Greek (or Latin)[1] prefix is used to indicate the number of carbon atoms in the molecule.

C_5	pentanes	C_{16}–C_{19}	etc.
C_6	hexanes	C_{20}	eicosanes
C_7	heptanes	C_{21}	heneicosanes
C_8	octanes	C_{22}	docosanes
C_9	nonanes	C_{23}	tricosanes
C_{10}	decanes	C_{24}–C_{29}	etc.
C_{11}	undecanes	C_{30}	triacontanes
C_{12}	dodecanes	C_{31}	hentriacontanes
C_{13}	tridecanes	C_{32}–C_{39}	etc.
C_{14}	tetradecanes	C_{40}	tetracontanes
C_{15}	pentadecanes		etc.

These names do not indicate the structures of the many isomers existing in each case. When a second butane was found it was called *isobutane,* and the third pentane became *neopentane,* but new prefixes, if coined indefinitely, would become difficult to remember. The term *iso* still is retained for all compounds having a single carbon atom branch at the end of a straight chain. Thus isohexane is $(CH_3)_2CH(CH_2)_2CH_3$ and isononane is $(CH_3)_2CH(CH_2)_5CH_3$. The term *neo* is used in this series only for neopentane and neohexane.

As Derivatives of Methane. The necessity for a system that could be used for a larger number of isomers led to the naming of compounds as derivatives of a simple compound. Thus any alkane can be considered as derived from methane by replacement of the hydrogen atoms by other groups of atoms. In isobutane three hydrogens of methane have been replaced by three CH_3 groups, and in isopentane by two CH_3 groups and one C_2H_5 group. In neopentane four hydrogens have been replaced by four CH_3 groups. If these groups are given names, the compounds can be named as derivatives of methane. The groups themselves are hydrocarbons less one hydrogen atom and are called *alkyl* groups. To name them, the *ane* of the alkane having the same number of carbon atoms is dropped and *yl* added. Thus CH_3 is *methyl* and C_2H_5 is *ethyl.* Isobutane then becomes trimethyl-

---

[1] Most of the names are of Greek derivation, but the Latin combining form *nona* is used almost universally instead of the Greek *ennea.* Similarly *undecane* is used more frequently than *hendecane.* Purists object to these inconsistencies as they do to the use of terms such as *mono-, di-, tri-,* and *tetravalent,* where prefixes of Greek origin are used with a word of Latin origin. Once a word becomes a part of a language, however, it is impractical, even if it were desirable, to try to bring about changes in terminology merely to be consistent.

TABLE 5–*1*. ALKYL GROUPS CONTAINING UP TO FOUR CARBON ATOMS

STRUCTURE OF GROUP	COMMON NAME	INTERNATIONAL NAME *
CH_3—	methyl	methyl
CH_3CH_2— or C_2H_5—	ethyl	ethyl
$CH_3CH_2CH_2$—	*n*-propyl	propyl
$CH_3\underset{\mid}{C}HCH_3$, $(H_3C)(H_3C)$CH—, or $(CH_3)_2CH$—	*i*-propyl	(methylethyl)
$CH_3CH_2CH_2CH_2$—	*n*-butyl	butyl
$CH_3CH_2\underset{\mid}{C}HCH_3$, $CH_3CH_2\underset{CH_3}{CH}$—, or $C_2H_5\underset{CH_3}{CH}$—	*s*-butyl	(1-methylpropyl) *not* (2-butyl)
$CH_3\underset{CH_3}{CH}CH_2$— or $(CH_3)_2CHCH_2$—	*i*-butyl	(2-methylpropyl)
$CH_3\overset{CH_3}{\underset{CH_3}{C}}$— or $(CH_3)_3C$—	*t*-butyl	(dimethylethyl)

* Rules 54.2 and 54.3 now state that isoalkyl names through isohexyl and also *s*-butyl, *t*-butyl, and neopentyl "are preferred to the systematic names."

methane, isopentane becomes dimethylethylmethane, and neopentane is tetramethylmethane. The prefix always is attached directly to the name to give one word.

It is customary to make use also of groups containing three and four carbon atoms in order to name more complicated compounds. Since there are *two kinds of hydrogen*[2] in propane, there are two propyl groups. The structures of these groups are given in Table 5–*1*. In order to indicate more clearly the carbon atom that lacks a hydrogen atom, it is customary to attach a long dash. In a molecule, the group must be bonded to some other atom or group of atoms at this point. To distinguish between the two propyl groups

[2] This expression arises from the fact that the properties of an atom in a molecule depend not only on the atom itself but also on the atom or atoms to which it is bonded. Thus it is possible to distinguish a hydrogen atom bonded to carbon from one bonded to oxygen or nitrogen. Likewise a hydrogen atom bonded to a carbon atom that in turn is joined to one carbon atom and two other hydrogen atoms is different from a hydrogen atom bonded to a carbon atom joined to two carbon atoms and one other hydrogen atom.

by name, they are called *normal propyl* or *n-propyl,* and *isopropyl* or *i-propyl,* respectively. Two groups each can be derived from *n*-butane and isobutane. *n*-Butane gives rise to the *n-butyl* group and a group that is called *secondary butyl* or *s-butyl* because the carbon atom that lacks a hydrogen atom is united directly to two other carbon atoms. Such a carbon atom is known as a *secondary carbon atom.* Isobutane gives rise to the *isobutyl* or *i-butyl* group and to a group that is known as *tertiary butyl* or *t-butyl,* because the carbon atom lacking a hydrogen atom is a *tertiary carbon atom;* that is, it is united to three other carbon atoms. These eight groups form the basis for the systematic naming of a large number of compounds. Their names and their structures must be memorized. Once this has been done, the nomenclature of most organic compounds is remarkably simple.

It is customary to name only branched alkanes as derivatives of methane. In general the most highly branched carbon atom is considered to be the methane carbon unless a less highly branched atom permits naming a smaller group. Thus compound (*a*) would be

```
           CH3                      CH3
           |                        |
CH3—CH—C—CH3          CH3—CH—C—CH3
     |     |                  |     |
    CH3  CH3          CH3—CH  CH3
                             |
                            CH3
        (a)                    (b)
```

called trimethyl-*i*-propylmethane rather than dimethyl-*t*-butylmethane. On the other hand compound (*b*) would require a name for a five-carbon group if the more highly branched carbon atom were chosen to represent methane but could be named methyl-*i*-propyl-*t*-butylmethane if the less highly branched carbon atom were considered to be the methane carbon atom.

International System. The system that names compounds as derivatives of methane has the disadvantage that the eight groups given in Table 5–*1* still are not sufficient for naming the more complicated compounds. This disadvantage is overcome to a considerable extent in the system adopted by the International Congress held at Geneva, Switzerland, in 1892. This method became known as the *Geneva system.* It was extended at the meeting of the International Union of Chemistry at Liège in 1930 and by the International Union of Pure and Applied Chemistry at Amsterdam in 1949 and subsequently. It attempts to cover all the more important phases of the nomenclature of organic chemistry.[3] A summary of the rules for alkanes follows:[4]

(*1*) The ending for alkanes is *ane.*

(*2*) The common names for the normal (*straight-chain*) hydrocarbons are used.

(*3*) Branched-chain hydrocarbons are regarded as derivatives of normal hydrocarbons, the longest normal chain in the molecule being considered as the parent hydrocarbon. If there are two or more chains of equal length, the chain having the most branches is selected, that is, the one that is the most highly substituted.

(*4*) The carbon atoms of the parent hydrocarbon are numbered from the end that gives the branched atoms the smaller numbers.

(*5*) The names of the branches or side chains are attached as prefixes directly to the

[3] The original rules were called the *Geneva system.* After the revisions, the rules were referred to by some as the *I.U.C. system* and then as the *I.U.P.A.C. system.* This text refers to the latest published rules as the *international system.*

[4] The rules laid down by the International Union are not without ambiguity, and a certain amount of flexibility has been allowed. What may be the best system for general use may not be the most suitable for indexing. The rules given are considered to be those most widely followed by chemists in the United States.

TABLE 5–*2*. NOMENCLATURE OF ALKANES

STRUCTURE	COMMON NAME	AS A DERIVATIVE OF METHANE	INTERNATIONAL SYSTEM
$CH_3(CH_2)_4CH_3$	normal hexane	not used	hexane
$CH_3CH(CH_2)_2CH_3$ | CH_3	isohexane	dimethyl-*n*-propylmethane	isohexane
$CH_3CH_2CHCH_2CH_3$ | CH_3	none	methyldiethylmethane	3-methylpentane
$CH_3CH—CHCH_3$ | | CH_3 CH_3	(diisopropyl*)	dimethyl-*i*-propylmethane	2,3-dimethylbutane
CH_3 | $CH_3—C—CH_2CH_3$ | CH_3	neohexane	trimethylethylmethane	2,2-dimethylbutane
9 5 $CH_3(CH_2)_3CH(CH_2)_3CH_3$ |4 $CH_3—CH$ |3 2 1 $CH_2—CHCH_3$ | CH_3	none	not possible using the first eight groups	2,4-dimethyl-5-butylnonane

* Occasionally a hydrocarbon is given a common name indicating that it may be divided into two like groups.

parent name. Their position is indicated by the number of the atom to which they are attached. If two branches are on the same carbon atom, the number is repeated. The numbers precede the groups and are separated from the groups by hyphens. Consecutive numbers are separated from each other by commas.

(*6*) Alkyl groups containing more than four carbon atoms are named in the same way as the alkanes except that the ending is *yl*, and the point of attachment is numbered 1. The full name of the group is enclosed in parentheses (Table 5–*1*, p. 59).

These rules and the other methods of nomenclature are illustrated in Table 5–*2* by the names for the five isomeric hexanes and for a still more complicated compound. The advantages of naming compounds as derivatives of parent hydrocarbons are that it is easy to assign a name to every compound and to write the formula from the name. Thus the structure for 2,2,5-trimethyl-3-ethylhexane is written by joining six carbon atoms in a row, numbering them from 1 to 6, attaching three methyl groups and an ethyl group at the proper positions, and satisfying the remaining valences of the carbon atoms with hydrogen atoms.

A useful check on the correctness of a one-word name is that the sum of the carbon atoms in the various groups and in the parent hydrocarbon must equal the total number of carbon atoms in the molecule. In the last example of Table 5–*2* each methyl group has one carbon atom, the butyl group has four, and the parent name nine, making a total of fifteen carbon atoms. This number checks with the number of carbon atoms in the formula. It should be noted also in this example that there are two different nine-carbon chains, but that when the most highly substituted chain is chosen, it is not necessary to name a complicated six-carbon group. If it had been necessary to name this group, it would have been called the (1,3-dimethylbutyl) group. It should *not* be called the (3-methyl-2-pentyl) group.

The parentheses are necessary to prevent confusion concerning the portion of the molecule to which the numbers refer. The compound

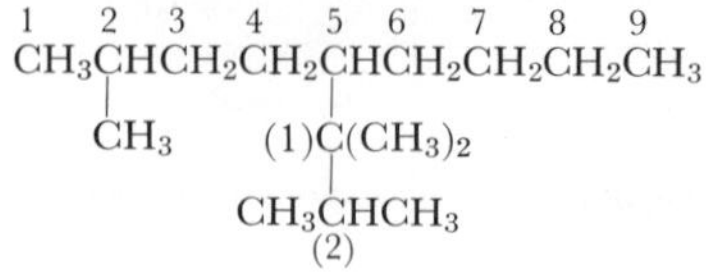

```
  1   2  3   4   5  6   7   8   9
CH3CHCH2CH2CHCH2CH2CH2CH3
     |          |
    CH3   (1)C(CH3)2
                |
          CH3CHCH3
               (2)
```

for example, would be called 2-methyl-5-(1,1,2-trimethylpropyl)nonane. The first five compounds in Table 5–*2* have six carbon atoms, and all are hexanes regardless of the fact that they are named as derivatives of methane, butane, and pentane. The last compound of the list is a pentadecane even though it is named as a derivative of nonane.

In the naming of compounds as derivatives of another compound, it is conventional to consider the name as one word rather than to write the names of groups and parent compound as separate words or to use an unnecessary number of hyphens. It is preferable to limit the use of hyphens to the attachment of position numbers and symbols.

As the number of carbon atoms increases, the number of structural isomers possible soon reaches astronomical proportions as shown by the following figures which were arrived at by rather complicated mathematical formulas.

C_7—9
C_8—18
C_9—35
C_{10}—75
C_{15}—4347
C_{20}—366,319
C_{30}—4,111,846,763
C_{40}—6.25×10^{13}

By 1947 all of the predicted alkanes through the nonanes, and over half of the decanes were known. Only a few isomers of each of the higher groups of compounds are known, chiefly because there has not been a sufficiently good reason for organic chemists to attempt to synthesize them or because those who are interested in them have not yet had time to do so. The largest alkane of known structure synthesized up to 1952 is *n*-hectane, $C_{100}H_{202}$, with a molecular weight of 1405.

Physical Properties

The physical properties of organic compounds depend in general on the number and kind of atoms in the molecule and on the way in which the atoms are linked together. At 25° and 760 mm.,* the normal hydrocarbons are gases from C_1 to C_4, liquids from C_5 to C_{17}, and solids for C_{18} and above.

Boiling Points. The boiling points of the normal hydrocarbons increase with increasing molecular weight. When plotted against the number of carbon atoms, they fall on a smooth curve as shown in Fig. 5–*5*. The rise in boiling point is due to the increased attraction between molecules. As the number of atoms and hence the number of electrons in the molecule increases, the polarizability increases with resulting increase in the attractive London forces (p. 22). For normal alkanes the energy required to convert the liquid to the vapor phase amounts to about 1.0 kcal. per carbon atom per mole. It is not possible to distill without decomposition hydrocarbons having more than around 80 carbon atoms, no matter how perfect the vacuum, because the energy of about 80 kcal. per mole required to separate the molecules is approximately the same as that necessary to break a carbon-carbon bond.

Branching of the chain always results in a lowering of the boiling point. Thus *n*-pentane boils at 36°, isopentane at 28°, and neopentane at 9.5°. Branching tends to decrease the magnitude of the transient dipole and prevents the optimum proximity of the molecules to each other.

---

* See footnote, page 51.

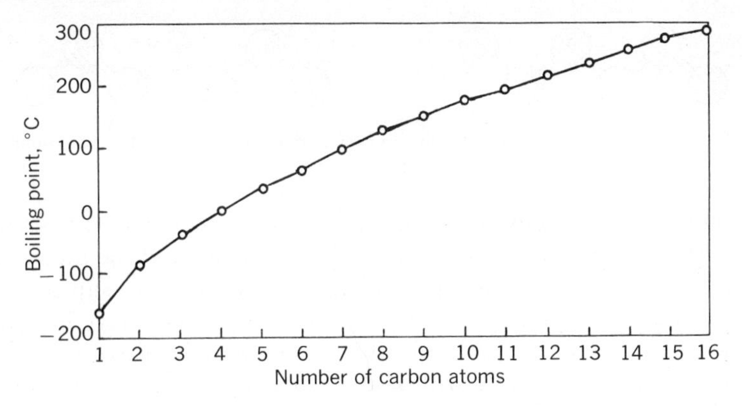

Figure 5–5. Boiling points of normal alkanes.

Melting Points. The melting points of normal alkanes do not fall on a smooth curve but show alternation (Fig. 5–6). With the exception of methane they fall on two curves, an upper one for the hydrocarbons having an even number of carbon atoms and a lower one for those with an odd number of carbon atoms.

Analyses of the X-ray diagrams of solid normal alkanes show that the chains are extended, the carbon atoms having a zigzag arrangement (Fig. 5–7). Compounds with an even number of carbon atoms have the end carbon atoms on opposite sides of the chain, whereas those with an odd number of carbon atoms have the end carbon atoms on the same side of the chain. The chains with an even number of carbon atoms are packed more closely, which apparently makes the van der Waals forces more effective and leads to a higher melting point.

Unlike the variation in boiling points, there is no regularity in the change in melting point with branching because the effectiveness of the attractive forces depends on how well the molecule fits into the crystal lattice. Thus *n*-pentane melts at −129.7°, isopentane at −160°, and neopentane at −20°. In general, however, the more symmetrical and compact the molecule, the higher its melting point. Hexamethylethane, $(CH_3)_3CC(CH_3)_3$, is interesting in that it melts at 100.7° and boils at 106.3°.

Other Properties. The *density* of the normal alkanes gradually increases from 0.626 g. per cc. at 20° for pentane to 0.769 for pentadecane. Branching may cause a decrease or an increase in density. All hydrocarbons float on water. They are the lightest of all classes of

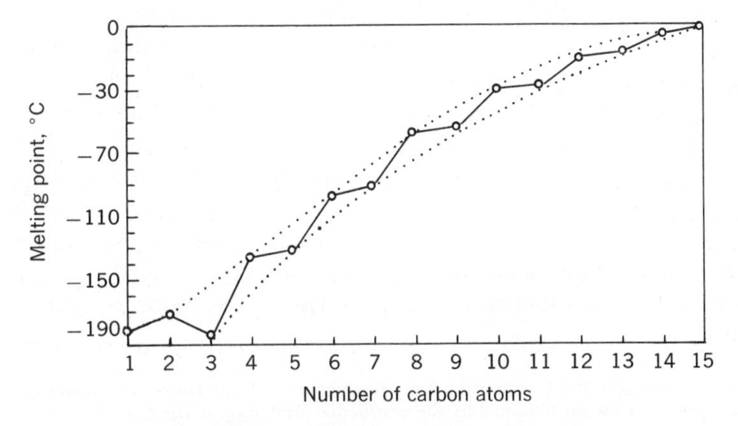

Figure 5–6. Melting points of normal alkanes.

Figure 5–7. Extended chains having an even and an odd number of carbon atoms.

organic compounds. The *viscosity* of the normal alkanes increases with increasing chain length, because the greater attraction between the molecules and the increased possibility for entanglement decreases the ease with which the molecules can slip past one another. The alkanes are almost completely *insoluble in water* because they have little attraction for water molecules, whereas water molecules have considerable attraction for each other (p. 24). Liquid alkanes are miscible with many other organic liquids because the attractions between like molecules and unlike molecules are of the same order of magnitude.

Sources of the Alkanes

The commercial sources of the saturated hydrocarbons are *natural gas* and *petroleum.* Paraffin hydrocarbons are present also in the *products of the destructive distillation of coal.* Natural gas usually is composed chiefly of **methane** and smaller amounts of **ethane, propane** and the **butanes** (p. 93). Methane also is a product of the action of anaerobic organisms on cellulose and other organic matter. For example, it is formed during the decomposition of vegetable matter under water in marshes, whence the common name *marsh gas.* Large amounts are formed during the treatment of sewage by the activated sludge process. It also is present in coal mines and, because it is one of the causes of explosions, it is known as *fire damp.* **n-Heptane** is present in the volatile oil of the fruit of *Pittosporum resiniferum,* and in the turpentine of the digger pine (*Pinus sabiniana*) and the Jeffrey pine (*Pinus jeffreyi*) of the mountain forests of the Pacific Coast. It readily can be obtained pure in quantity from the last source. Alkanes having odd and even numbers of carbon atoms have been isolated from many plant and insect waxes. Tobacco leaf wax, for example, contains all of the odd and even carbon *n*-alkanes from C_{25} through C_{33} and *i*-alkanes from C_{27} through C_{33}. For both the normal and the iso series, however, the odd-carbon alkanes predominate and account for 83 per cent of the hydrocarbon fraction.

Individual alkanes of known structure usually are obtained by synthesis, that is, by making them from certain other compounds. These methods of synthesis are given in Chapter 10.

Chemical Properties

1. ***Inertness to Most Reagents.*** High chemical reactivity is associated with ions (p. 30), with free radicals (p. 38), or with compounds that contain an unshared or weakly held pair of electrons (p. 31), an incomplete valence shell (p. 16), or a partially polarized covalent bond (p. 19). None of these factors is present in the alkanes. They are not ionic, all electrons are paired, all pairs are shared, and all valence shells are filled. The carbon-carbon bonds of alkanes are nonpolar and the carbon-hydrogen bonds are practically so. Hence it is not surprising that the alkanes are relatively unreactive. At room temperature they are not affected by concentrated acids or alkalies, strong oxidizing or reducing agents, or most other reagents, and they frequently resist reaction under more drastic conditions. It is this inertness that gave rise to the name *paraffin* (L. *parum* little, *affinitas* affinity).[5]

[5] The term *paraffin* first was used by Karl Reichenbach (1788–1869), an Austrian, for a wax that he isolated from the tar obtained by the destructive distillation of beechwood.

2. ***Decomposition at High Temperature.*** At a sufficiently high temperature, homolytic bond breaking takes place with subsequent recombination of the free radicals (p. 39). This process is known as **cracking** or **pyrolysis** (Gr. *pyr* fire, *lysis* loosing[6]). Methane yields carbon and hydrogen as final products, a process that is endothermic by 190 kcal.[7]

$$CH_4 \longrightarrow C(s) + 4\,H\cdot(g) \qquad -396 \text{ kcal.}$$
$$4\,H\cdot(g) \longrightarrow 2\,H_2(g) \qquad +206 \text{ kcal.}$$

For the reaction to proceed rapidly with methane, temperatures above 1200° are required.[8] Other hydrocarbons decompose rapidly at considerably lower temperatures (500–600°).

For the decomposition of ethane, two primary reactions are possible, dissociation of the carbon-carbon bond to give two methyl radicals, or dissociation of the carbon-hydrogen bond to give an ethyl radical and a hydrogen atom.

$$CH_3CH_3 \longrightarrow 2\,CH_3\cdot \qquad -83 \text{ kcal.} \qquad (1)$$
$$CH_3CH_3 \longrightarrow C_2H_5\cdot + H\cdot \qquad -96 \text{ kcal.} \qquad (2)$$

Since in these decompositions the activation energy is approximately equal to the bond dissociation energy (p. 18), the first reaction is faster. The carbon-carbon bond dissociation energy of 83 kcal., compared with 101 kcal. for the carbon-hydrogen bond in methane, accounts for the fact that ethane decomposes at a lower temperature than methane.

Once free methyl radicals are formed, several other reactions can take place readily because relatively small energy changes are involved.

$$H_3C\cdot + H{:}CH_2CH_3 \rightleftharpoons H_3C{:}H + \cdot CH_2CH_3 \qquad +5 \text{ kcal.} \qquad (3)$$
$$\dot{C}H_2{:}CH_2{:}H \rightleftharpoons CH_2{=}CH_2 + \cdot H \qquad -40 \text{ kcal.} \qquad (4)$$
$$H\cdot + H{:}CH_2CH_3 \rightleftharpoons H{:}H + \cdot CH_2CH_3 \qquad +7 \text{ kcal.} \qquad (5)$$

The product $CH_2{=}CH_2$ is a new hydrocarbon, called *ethylene,* that contains a carbon-carbon double bond. It belongs to another homologous series, C_nH_{2n}, known as the *alkenes, olefins,* or *unsaturated hydrocarbons* (Chapter 6).

In the decomposition of ethane, reaction *1* initiates the process. Reaction *3*, which has a very low activation energy, takes place very rapidly to give ethyl radicals. It makes possible the self-perpetuating reactions *4* and *5* that yield the main products. The over-all reaction of ethane giving ethylene and hydrogen (addition of reactions *4* and *5*) is endothermic by 33 kcal. In the transition states for the various steps, only partial bonding exists between the reacting species (p. 39). For example, the transition state for reaction *3* may be represented by Fig. 5–*8* in which the hybrid orbitals of the carbon atoms overlap almost equally well the *s* orbital of the hydrogen. The energy profiles for the several steps are

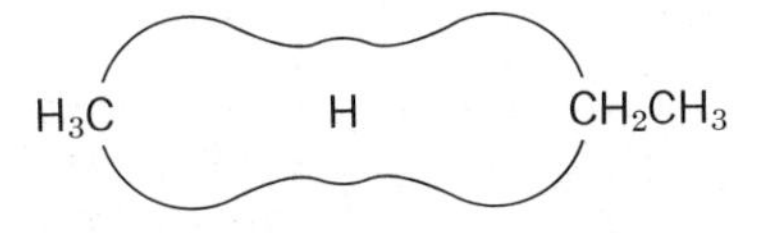

Figure 5–*8*. Schematic representation of transition state for the reaction of a methyl radical with an ethane molecule or of an ethyl radical with a methane molecule.

[6] The combining form *lysis* in chemical terms refers to a scission of the molecule by the combining form that precedes. *Hydrolysis* means scission by water and *pyrolysis* mean scission by heat.

[7] See footnote 1, page 11.

[8] When methane is subjected to higher temperatures for very short periods, other products such as ethylene, acetylene, and benzene also are formed (p. 151).

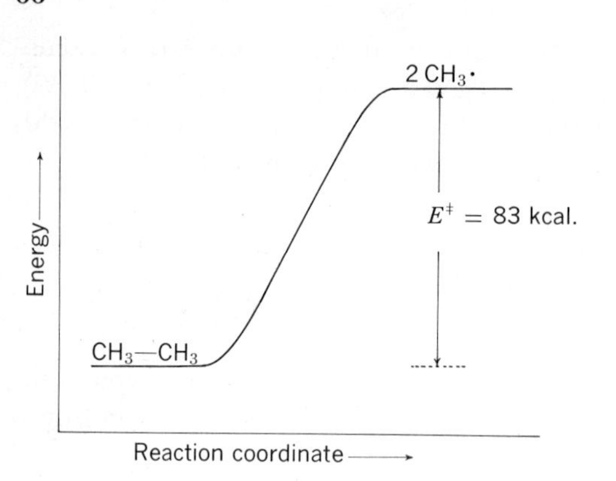

Figure 5–*9a*. Dissociation of ethane into methyl radicals (*reaction 1*).

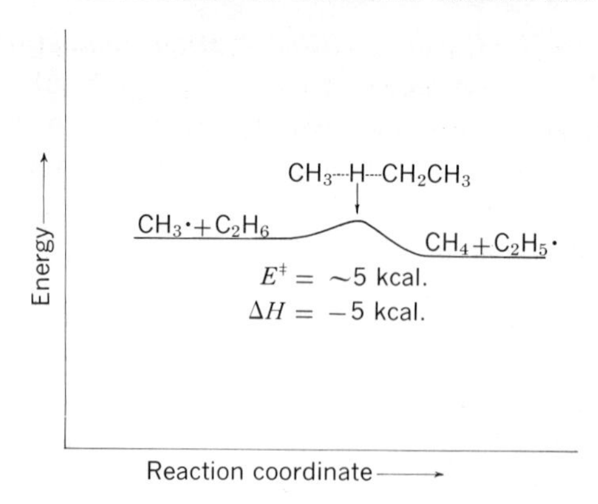

Figure 5–*9b*. Transfer of hydrogen atoms (*reaction 3*).

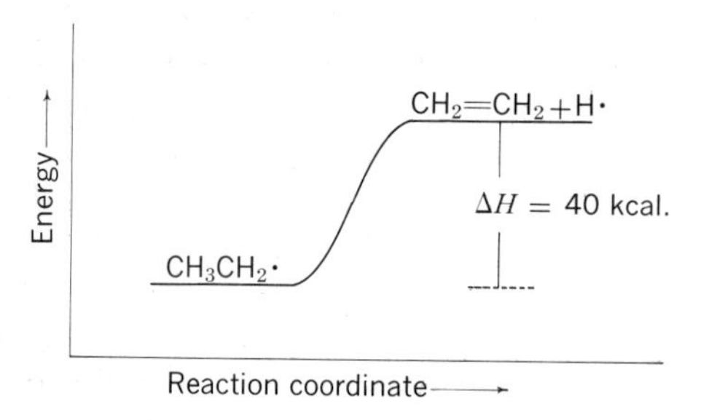

Figure 5–*9c*. Decomposition of ethyl radicals into hydrogen atoms and ethylene (*reaction 4*).

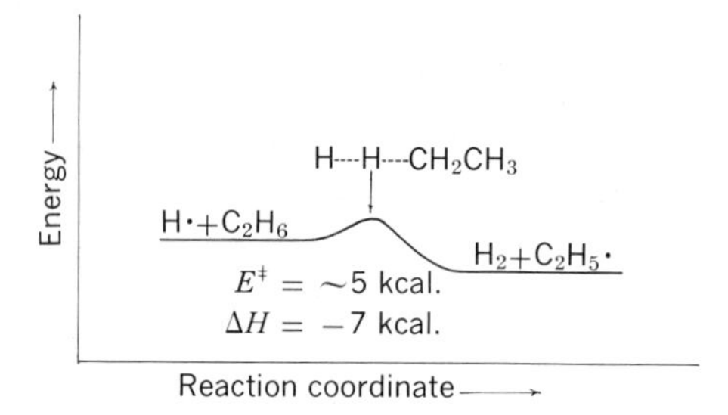

Figure 5–*9d*. Reaction of hydrogen atoms with ethane (propagation step, *reaction 5*).

given in Fig. 5–9.[9] The kinetics of the reaction can be accounted for if the chief chain-breaking reaction is the combination of an ethyl radical and a hydrogen atom to regenerate a molecule of ethane.

$$C_2H_5\cdot + H\cdot \longrightarrow C_2H_6 \qquad +96 \text{ kcal.}$$

Pyrolysis of higher alkanes gives hydrogen and mixtures of alkenes and alkanes, both having varying numbers of carbon atoms. This cracking process is of great importance to the petroleum industry (p. 94) because it provides a means for making smaller, more volatile hydrocarbon molecules from the larger, less volatile molecules that predominate in most natural petroleum.

3. ***Oxidation at High Temperature.*** Hydrocarbons in the presence of strong oxidizing agents such as chromium trioxide in concentrated sulfuric acid at 150° or oxygen at 600° burn to carbon dioxide and water.

$$CH_4 + 2\,O_2 \longrightarrow CO_2 + 2\,H_2O$$

$$C_2H_6 + 3\tfrac{1}{2}\,O_2 \longrightarrow 2\,CO_2 + 3\,H_2O$$

$$C_3H_8 + 5\,O_2 \longrightarrow 3\,CO_2 + 4\,H_2O$$

$$C_nH_{2n+2} + \frac{3n+1}{2}O_2 \longrightarrow n\,CO_2 + (n+1)\,H_2O$$

[9] When values are given for energy changes in these and subsequent energy profiles, they are enthalpies rather than free energies. To be correct the ordinate should represent free energy, but it is not possible to measure the entropy of the transition state, and even the entropies of the reactants and products usually are not known (p. 70). The use of enthalpies, however, is qualitatively satisfactory because chemists ordinarily are concerned with the relative rates of similar reactions under the same conditions, for which entropy changes are approximately the same.

The changes in volume that take place on combustion of gaseous hydrocarbons formerly were used as the basis for the analysis of mixtures, but gas analysis now is done by gas-liquid chromatography (p. 45).

The oxidation or combustion of alkanes to carbon dioxide and water takes place with the evolution of heat amounting to 11 to 13 kcal. per gram. The amount of energy liberated per mole of compound is known as the *heat of combustion* of the compound. The heat of combustion increases with increasing number of carbon atoms. Most of the heat arises from the difference between the sum of the bond energies (cf. p. 18) of the reactants and that of the products. Therefore the greater the number of bonds broken and reformed to more stable structures, the greater the heat evolved. For any homologous series the increment per CH_2 group is constant at approximately 156 kcal. Isomers have slightly different heats of combustion. The heat of combustion of a liquid compound is less than that of the gaseous form by the heat of vaporization of the compound.

Although the energy relationships (*thermodynamics*) for the oxidation of alkanes are highly favorable, the activation energy is high (cf. p. 34). Thermal energy equal to the activation energy must be supplied to start the reaction. Once the reaction starts, sufficient energy is liberated to permit the reaction to continue. Catalysts, such as finely divided platinum, may provide a pathway with a much lower activation energy and hence lower the temperature at which ignition takes place.

The fact that the combustion of hydrocarbons is exothermic is the basis for their use as fuels.Combustion of gasoline yields about 11.5 kcal. per gram. The flammability depends on the volatility of the hydrocarbon. Mixtures of air and hydrocarbon vapors in the proper proportions explode on ignition, which gives rise to their use in the internal combustion engine. In a deficient oxygen supply, carbon monoxide, or elementary carbon in the form of carbon black, may be produced.

4. ***Halogenation.*** Alkanes, like hydrogen (p. 38), can be made to react with fluorine, chlorine, or bromine. Hydrogen gives two moles of hydrogen halide, whereas alkanes give one mole of hydrogen halide and one mole of halogenated alkane. The bond dissociation energy of fluorine is so low (36 kcal.)[10] that at ordinary temperatures the kinetic energy is sufficiently high to supply the necessary activation energy, and violent reaction takes place on mixing.

$$R \cdot\!\!\cdot H + F \cdot\!\!\cdot F \longrightarrow R\cdot + HF + \cdot F \qquad \textit{Initiation}$$

$$\left.\begin{array}{rcl} R\cdot + F_2 & \longrightarrow & RF + F\cdot \\ F\cdot + HR & \longrightarrow & FH + R\cdot \end{array}\right\} \qquad \textit{Propagation}$$

In these reactions and wherever else it is used, R stands for any alkyl group, whence RH is the general formula for any alkane, RF is an alkyl fluoride (*fluoroalkane*), and RX is any alkyl halide. The propagation steps are exothermic by approximately 104 kcal. (Table 2–2, p. 19). Since this energy is sufficient to break carbon-hydrogen and carbon-carbon bonds, the reaction is difficult to control. Control may be brought about in two ways. In one the reactants are diluted with nitrogen and passed over a large heat-conducting surface such as copper gauze or turnings. In the other a less reactive fluorinating agent is used, such as cobalt fluoride. This reaction is carried out as a cycle. Fluorine is passed over cobaltous fluoride at 250° to give cobaltic fluoride, the excess fluorine is swept out with nitrogen, and the hydrocarbon vapor passed over the cobaltic fluoride.

$$2\,CoF_2 + F_2 \longrightarrow 2\,CoF_3$$

$$2\,CoF_3 + RH \longrightarrow CoF_2 + RF + HF$$

[10] The anomalous bond dissociation energy of fluorine is ascribed to a large electron repulsion between the kernels because of the small size of the atoms.

Each step evolves about half of the over-all heat of reaction. Even under these conditions all of the hydrogen is replaced by fluorine and only about 60 per cent of the product has the original carbon skeleton.

The bond dissociation energy of chlorine is sufficiently high (57 kcal.) that activation energy must be supplied by heat or light (p. 38).

$$Cl_2 \xrightarrow[h\nu]{\text{Heat or}} 2\,Cl\cdot \qquad \textit{Initiation}$$

$$\left.\begin{array}{l} Cl\cdot + HR \longrightarrow Cl{-}H + \cdot R \\ R\cdot + Cl_2 \longrightarrow R{-}Cl + Cl\cdot \end{array}\right\} \qquad \textit{Propagation}$$

Since the heat of reaction is only around 25 kcal., the reaction is fairly easily controlled. Atomic chlorine is so reactive, however, that all types of hydrogen are attacked with almost equal ease, and the replacement of one hydrogen does not appreciably affect the ease of replacement of a second hydrogen atom. The result is that a mixture of all the possible monosubstitution products is obtained and some polysubstitution products as well. The reaction finds use only in a few commercial processes where all the products can be used.

Methane gives a mixture of four products.

$$CH_4 + Cl_2 \longrightarrow CH_3Cl + HCl$$
Methyl chloride[11] (*chloromethane*)

$$CH_3Cl + Cl_2 \longrightarrow CH_2Cl_2 + HCl$$
Methylene chloride (*dichloromethane*)

$$CH_2Cl_2 + Cl_2 \longrightarrow CHCl_3 + HCl$$
Chloroform (*trichloromethane*)

$$CHCl_3 + Cl_2 \longrightarrow CCl_4 + HCl$$
Carbon tetrachloride (*tetrachloromethane*)

The reaction is carried out industrially either thermally or photochemically. The mixture of methane and chlorine is passed through hot tubes or into vessels fitted with mercury arc lamps. Each of the products has important technical uses, and they are separated by fractional distillation.

In the commercial chlorination of pentanes the reaction is carried out in the gas phase at 250–300°. A volume ratio of pentane to chlorine of 15 to 1 is used in order to form chiefly monosubstitution products. Even at this ratio about 5 per cent of disubstitution products are formed. The rate of flow of the gases is about 60 miles per hour. This rate is faster than the rate of propagation of the chlorine-hydrocarbon flame, and the unreacted material behind the flame cannot explode. The relative amounts of the monosubstitution products formed from *n*-pentane at 300° are 24 per cent 1-chloropentane, 49 per cent 2-chloropentane, and 27 per cent 3-chloropentane. From *i*-pentane there is obtained 33 per cent 1-chloro-2-methylbutane, 22 per cent 2-chloro-2-methylbutane, 17 per cent 1-chloro-3-methylbutane, and 28 per cent 2-chloro-3-methylbutane. It is possible to separate the mixture by distillation, but for most uses separation is unnecessary.

Bromination of alkanes also can be carried out thermally or photochemically but has no practical use. Although iodine dissociates readily into iodine atoms, the over-all reaction with methane to form methyl iodide is endothermic by about 14 kcal.

The direct replacement of hydrogen by a halogen atom or other monovalent group is known as a **substitution reaction** and is of considerable importance for other compounds as well as for the saturated hydrocarbons.[12] One of the *fundamental concepts of substitution and of*

---

[11] For the nomenclature of halogen compounds see p. 98.

[12] Substitution of hydrogen by halogen was investigated first by Dumas after an episode at the Tuileries. On the occasion of a ball during the reign of Charles X of France, the guests were driven from the ballroom by choking fumes given off by the burning candles. Brongiart, the chemical advisor to the king, called in his son-in-law Dumas, who found that the candles had been bleached by a new process using chlorine. The bleaching had given rise to chlorinated fat acids, which on burning gave off hydrogen chloride. Dumas then made an extensive investigation of the substitution reaction.

all replacement reactions is that the new atom or group takes the position formerly occupied by the replaced atom or group of atoms. *Substitution, chlorination, halogenation,* and other terms for analogous reactions should be used only for the direct replacement of hydrogen.

5. ***Isomerization.*** The conversion of a compound into an isomeric compound is called *isomerization*. The resistance of most organic compounds to a change in structure, a phenomenon that van't Hoff (p. 301) referred to as the "inertness of the carbon bond," is strikingly exhibited by the fact that although branched-chain alkanes are more stable thermodynamically than normal alkanes, some petroleums that are millions of years old contain mainly straight-chain hydrocarbons. The reason is that the conditions under which the less stable isomers were formed and have since existed have not been such as to provide sufficient activation energy for the isomerization. What is needed is a catalyst that provides a pathway or mechanism for which the activation energy of the rate-controlling step is sufficiently low to permit isomeric change to take place until thermodynamic equilibrium is established. A complex of anhydrous aluminum chloride with an alkyl chloride, RCl, is able to provide such a pathway. The chlorine of an alkyl chloride has unshared pairs of electrons, and the aluminum of aluminum chloride lacks a pair of electrons in its valence shell. These conditions permit the formation of what is essentially a semipolar bond (p. 16).

$$R:\ddot{\underset{\cdot\cdot}{Cl}}: + AlCl_3 \longrightarrow R:\overset{+}{\underset{\cdot\cdot}{\ddot{Cl}}}:\overset{-}{A}lCl_3$$

The withdrawal of electrons from chlorine also causes withdrawal from the R group, which thus acquires a positive charge. Usually the complex is written as if it were a salt, $R^{+-}AlCl_4$, of the hypothetical chloroaluminic acid. R^+ is called a *carbonium ion* because if it were free, the carbon atom to which the chlorine originally was attached would lack a pair of electrons and bear a positive charge.

In the presence of *n*-butane, this complex is viewed as being able to abstract a hydride ion, $H:^-$, from the more highly substituted secondary carbon atom to give another alkane and a secondary butyl cation.

$$CH_3CH_2\underset{\ddot{H}}{C}HCH_3 + R^{+-}AlCl_4 \rightleftharpoons CH_3CH_2\overset{+}{C}HCH_3 + H:R + {}^-AlCl_4$$

Migration of a methide ion to the electron-deficient carbon atom gives a new carbonium ion, which removes a hydride ion from another *n*-butane molecule to give *i*-butane and a new *s*-butyl cation.

$$\underset{\ddot{C}H_3}{CH_2}\overset{+}{C}HCH_3 \rightleftharpoons \overset{+}{C}H_2\underset{\ddot{C}H_3}{C}HCH_3 \xrightleftharpoons{CH_3CH_2CH_2CH_3} CH_3\underset{CH_3}{\underset{|}{C}}HCH_3 + CH_3CH_2\overset{+}{C}HCH_3$$

Thus the isomerization is self-propagating. None of these processes requires a high activation energy because all of the carbonium ions are highly reactive intermediates, and the energy required to break the bonds heterolytically is regained with the formation of bonds of essentially the same energy. Since the steps all are reversible, equilibrium is established between the isomers. The isomerization of alkanes is important in the production of high octane gasoline (p. 95).

Reactions that take place with the migration of a group from one position in a molecule to another position with a change in the carbon skeleton are called *molecular rearrangements*. Frequently the migration is from a carbon atom to an adjacent atom and is referred to as a *1,2 shift*. 1,2 Shifts in which a group migrates with its bonding pair of electrons to an electron-deficient atom are among the more common types of molecular rearrangements.

Kinds of Stability

At this point it is desirable to discuss the various meanings of the words "stable" and "stability" as used by chemists. Both *n*-butane and *i*-butane usually are thought of as being stable compounds. By this it is meant that each can be isolated in a pure state and can be kept in a pure state indefinitely under atmospheric conditions of temperature and pressure. In this sense *n*-butane is for all practical purposes just as stable as *i*-butane. Some compounds are said to be "relatively stable" or "relatively unstable" to indicate that they are subject to slow or rapid spontaneous change. Still other compounds are said to be "unstable" meaning that, though structurally they should be capable of existence or may exist as transient intermediates, they cannot be isolated in a pure state but rearrange or decompose under ordinary conditions. As has been noted (p. 65), compounds such as the butanes that are stable indefinitely at 25° decompose at around 500°, and some compounds that are unstable at room temperature may be stable at −80° or −195° or at some still lower temperature. Sometimes a substance is said to be stable to acids or bases or to oxidizing or reducing agents. When used in any of the above ways "stability" usually means *lack of reactivity* under particular conditions.

When one says as on page 69 that *i*-butane is more stable than *n*-butane at 25°, one refers to an entirely different type of stability, namely *thermodynamic stability*. The reason that there is more *i*-butane than *n*-butane at equilibrium is that there is a decrease in free energy when *n*-butane is converted to *i*-butane. For the reaction *n*-butane → *i*-butane at 25°, $\Delta G = -0.90$ kcal., which leads (p. 27) to an equilibrium constant for the reaction of 4.6, and hence 4.6 times as much *i*-butane as *n*-butane.

It should be remembered, however, that the change in free energy is related to the change in two other thermodynamic properties, the enthalpy and the entropy, by the equation, $\Delta G = \Delta H - T\Delta S$ (p. 28). For the reaction *n*-butane → *i*-butane, $\Delta H^\circ = -2000$ cal. and $\Delta S^\circ = -3.69$ cal. per degree. At 25° $T\Delta S^\circ = (25 + 273)(-3.69) = -1100$ cal., whence $\Delta G = -2000 + 1100 = -900$ cal. Accordingly at 25° the negative value of ΔH is sufficient to offset the entropy term $-T\Delta S$. At a temperature of 269° or 542°*K*, however, the value of ΔG becomes zero and the concentrations of *n*-butane and *i*-butane at equilibrium are equal. Above 269°, ΔG becomes positive and *n*-butane becomes thermodynamically more stable than *i*-butane.

The fact that the enthalpy and entropy of *n*-butane are not the same as the enthalpy and entropy of *i*-butane illustrates two points made previously. (*1*) The enthalpy resides in the bond energies, and the bond energies depend not only on the kinds of atoms bonded to each other but also on the structure of the molecule (p. 19). Since ΔH° is negative for the conversion of *n*-butane to *i*-butane, the sum of the bond energies for *i*-butane is slightly greater than that for *n*-butane despite the fact that each has three carbon-carbon bonds and ten carbon-hydrogen bonds. (*2*) A decrease in the degrees of freedom leads to a decrease in entropy (p. 28). Here the entropy of *n*-butane is greater than that of *i*-butane because the *n*-butane molecule has greater flexibility than the *i*-butane molecule. The four-carbon chain can be extended or can coil back and assume various conformations, whereas the conformation of the carbon skeleton of *i*-butane is fixed.

The determination of the free energy for a reaction involves the determination of the change in enthalpy and the change in entropy. The change in enthalpy is merely the heat evolved or absorbed in a reaction at constant temperature and pressure and is readily determined, but the determination of the change in entropy is more difficult experimentally, and relatively few data concerning changes in entropy are available. It can be seen, however, that the difference in entropy between *n*-butane and *i*-butane is not large enough to outweigh the change in enthalpy at 25°, and the change in free energy follows the change in enthalpy. In general the difference in entropy for isomers is sufficiently small for this to be true at room temperature, and the isomer that has the lower enthalpy is the more stable.

Usually some guess can be made as to whether a large or small change in entropy for a reaction is probable, and heats of reaction frequently are used to predict relative stabilities. To distinguish between true thermodynamic stability and stability based on heats of reaction, the latter has been called *thermochemical stability*.

Production of Alkanes for Chemical Conversion

In addition to their use as fuel, large quantities of the lower alkanes are used to make other chemicals. Production of ethane in 1964[13] for chemical conversion was 1.2 billion pounds, of propane, 3.7 billion pounds, of *n*-butane, 1.7 billion pounds, and of *i*-butane, 0.6 billion pounds. All sold for around 1 cent per pound. They were used chiefly to make the chemically reactive alkenes by cracking, but considerable amounts were used directly to make other chemicals (pp. 107, 168). In addition large amounts of methane were converted into products as diverse as carbon blacks, hydrogen, chlorinated methanes, methanol, and acetylene (pp. 93, 68, 120, 151).

*PROBLEMS**

5–*1*. Write a skeleton structure for each of the compounds that meet the following descriptions, name each isomer by the international system, and name branched-chain compounds also as derivatives of methane: (*a*) all of the heptanes; (*b*) all of the octanes having a chain of five carbon atoms.

5–*2*. Write condensed structural formulas for the following compounds: (*a*) isodecane, (*b*) methyldiethyl-*i*-butylmethane, (*c*) 6-(1,2-dimethylpropyl)dodecane, (*d*) hexadecane, (*e*) 2,3,4-trimethylpentane, (*f*) 5-neopentylnonane, (*g*) 4-*i*-propyl-5-*t*-butyloctane, (*h*) 5-(2,2-dimethylpropyl)-7-(1-methyl-2-ethylbutyl)-dodecane.

5–*3*. Give the international name and one other name for the following compounds:

(*a*) $(CH_3)_2CH(CH_2)_7CH_3$, (*b*) $(CH_3)_2CHCH_2CH(CH_3)_2$,
(*c*) $(CH_3)_3CCH(CH_3)C_2H_5$, (*d*) $(CH_3)_2CHCH_2CH(C_2H_5)CH_2CH_2CH_3$,
(*e*) $CH_3(CH_2)_3C(CH_3)(C_2H_5)CH(CH_3)C_2H_5$

5–*4*. Give the international name for each of the following alkyl groups:

(*a*) $C_2H_5CH(CH_3)CH(CH_3)\underset{|}{C}HCH_3$, (*b*) $(CH_3)_3C\underset{|}{C}(CH_3)_2$,

(*c*) $(CH_3)_3CC(CH_3)_2CH_2—$

5–*5*. Give a suitable name for each of the following compounds:

(*a*) $CH_3(CH_2)_4\underset{\underset{C(CH_3)_2C_2H_5}{|}}{C}H(CH_2)_3CH_3$

(*b*) $(CH_3)_3CC(CH_3)_2CH_2\underset{\underset{(CH_2)_4CH_3}{|}}{C}HCH_2CH(CH_3)CH(CH_3)_2$

5–*6*. Tell why the following names are objectionable and give a suitable name to each: (*a*) 3-propylhexane, (*b*) 4-methylpentane, (*c*) methyl-3-pentane, (*d*) methyl-*i*-propylmethane, (*e*) 3-*i*-propylhexane, (*f*) 4-(1,2-dimethylethyl)heptane.

5–*7*. Write structural formulas and give the international names for all of the alkyl groups that can be derived from the following compounds and indicate whether the alkyl group is primary, secondary, or tertiary: (*a*) 2-methylpentane, (*b*) methyldiethyl-*i*-propylmethane.

---

[13] Figures for commercial production as given in this text are for the United States as reported by the U.S. Tariff Commission or in the *Chemical Economics Handbook* published by the Stanford Research Institute.

* See footnote, page 26.

5–*8*. How many different monobromo derivatives of each of the following alkanes are possible: (*a*) isohexane, (*b*) tetramethylbutane, (*c*) 2,2,4-trimethylpentane, (*d*) 3-ethylpentane, (*e*) 3-methylhexane?

5–*9*. Give the number of primary, secondary, tertiary, and quaternary carbon atoms in (*a*) 2,2,4-trimethylpentane, (*b*) tetramethylbutane, (*c*) 3-ethylpentane, and (*d*) 2,3-dimethyl-3-*i*-propyl hexane.

5–*10*. Place the following compounds in the order of increasing boiling point: (*a*) *n*-octane, (*b*) isooctane, (*c*) *n*-hexane, (*d*) 2,4-dimethylhexane, and (*e*) 2,4-dimethylpentane.

5–*11*. (*a*) How many moles of oxygen are required for the complete combustion of one mole of butane? (*b*) What is the total number of moles of products formed?

5–*12*. (*a*) Would the wavelength of the light necessary for initiation of photochemical bromination be longer or shorter than that required for chlorination? (*b*) In the bromination of methane to methyl bromide, what is the approximate activation energy? What is the heat of reaction for each of the propagation steps and for the over-all reaction? (*c*) Why would the change in entropy for the reaction be expected to be small for monohalogenation? (*d*) Assuming the change in entropy to be small at room temperature, calculate the approximate equilibrium constants for chlorination, bromination, and iodination. (*e*) What values of ΔH would correspond approximately to 90 per cent and to 99 per cent conversion starting with one mole each of methane and halogen?

5–*13*. From the data on p. 70, calculate the equilibrium composition of a mixture of *n*-butane and *i*-butane at 500°.

5–*14*. For *n*-pentane, $\Delta G_f^\circ = -2.00$ kcal. and for *i*-pentane, -3.50 kcal. Calculate the equilibrium composition of the mixture at 25°.

CHAPTER SIX

ALKENES. CYCLIC HYDROCARBONS. NATURAL GAS AND PETROLEUM

ALKENES

Like the alkanes, members of the second family or homologous series of organic compounds also are hydrocarbons. They are known as the *alkenes.* They also are called *ethylenes* from the name of the first member of the series, or *olefins* from *olefiant gas,* an old name for ethylene (p. 80). The members of this second series have the general formula C_nH_{2n}; that is, each member has two less hydrogen atoms than the corresponding alkane. Because they do not contain the maximum number of hydrogen atoms, they frequently are called *unsaturated* hydrocarbons in contrast to the alkanes, which are called *saturated* hydrocarbons.

Structure

The simplest member of this homologous series, ethylene, has the molecular formula C_2H_4. No stable compound having a single carbon atom is known for this series. Three electronic formulas suggest themselves. Either one carbon atom has two unshared electrons (*I*), each has one unshared electron (*II*), or the carbon atoms are joined by two pairs of electrons (*III*).

```
  H H          H H          H   H
  •× ×o        •× ×o        •+  o+
H×C:C∘       H×C:C×H         C::C
  ×• o          •  o        •+  o+
  H                         H   H
  I            II           III
```

If *I* or *II* were capable of existence,

```
 H         H
 •×        •×
H×C•  or  H×C•
 •         ×•
           H
```

should be stable also, but these groups have been detected only as highly reactive, short-lived species (p. 248). Formula *I* is ruled out also on chemical grounds because its reactions would involve only one carbon atom, whereas all known reactions of the olefins involve both carbon atoms. Formula *III* satisfies all requirements. Atoms sharing two pairs of electrons are said to be joined by a **double bond.**

The carbon-carbon double bond and its properties are explained best by means of molecular orbitals. When a carbon atom is united to only three other atoms, it makes use of sp^2 hybrid orbitals, the three bonds being planar and making angles of approximately 120 degrees with each other (p. 13).[1] The remaining unpaired electron is in a p orbital (p. 11)

[1] In molecules the experimentally measured angles between single bonds are somewhat smaller than 120° and those between single and double bonds somewhat larger. Thus for ethylene the measured H—C—H angle is 117.6° ± 0.5° and the H—C—C angle 121.2°.

perpendicular to the plane of the three bonds. In ethylene, overlapping of one sp^2 orbital from each carbon atom forms a single sp^2—sp^2 bond between the carbon atoms. The remaining four sp^2 orbitals overlap with the s orbitals of four hydrogen atoms to form four s—sp^2 bonds. All of these bonds result from the formation of σ-type molecular orbitals (p. 12). These molecular orbitals accommodate three of the valence electrons of each carbon atom, leaving one electron in a p orbital on each carbon atom. This orbital is perpendicular to the plane of the carbon-hydrogen bonds. Figure 6–*1a* represents a cross section through this plane and Fig. 6–*1b* represents a cross section through the carbon atoms and perpendicular to this plane. The molecule represented by Fig. 6–*1* would acquire added stability if the two electrons that are occupying monocentric p orbitals (electrons encompassing a single nucleus) could occupy a dicentric molecular orbital (encompass two nuclei). The formation of a molecular orbital could result if the p orbitals overlapped sufficiently. Since the p orbitals are perpendicular to the planes of the CH_2 groups, these orbitals can overlap best when the planes of the two CH_2 groups are coincident. In this position the two p orbitals coalesce to form a molecular orbital that resembles two fat sausages above and below the plane of the molecule and can accommodate two electrons. Because it has a nodal plane, it is called a *pi* (π) *orbital,* by analogy to the p atomic orbital, which also has a nodal plane. Figure 6–*2a* is the same cross section as Fig. 6–*1b* but after the formation of the molecular orbital. Figure 6–*2b* is a perspective representation, and Fig. 6–*2c* is a convenient schematic representation in which the figure eights indicate the p orbitals and the light lines joining them indicate the overlapping to form the molecular π orbital.

Because the overlapping of the two p orbitals is relatively poor, the π-type molecular orbital is not as stable as a σ-type molecular orbital. The energy associated with a carbon-carbon σ bond is around 85 kcal., whereas that associated with a carbon-carbon π bond is about 58 kcal. The lower stability of the π bond (higher energy content of the electrons) accounts for the greater reactivity of the π bond, that is, the tendency to form the more stable σ bonds with other atoms. Hence the π electrons also are called *unsaturation electrons.* The energy of 58 kcal. that is evolved when the π bond is formed by overlapping of the p orbitals is quite sufficient, however, to provide a strong barrier to rotation about the double bond (p. 76) and ensures that all the atoms in the ethylene molecule lie in the same plane.

Another observable effect of the π bond is the decrease in the distance between the carbon atoms. The interatomic distance of a carbon-carbon single bond in hydrocarbons is 1.53 A, but that of a carbon-carbon double bond is 1.33 A; that is, the nuclei are bound together more strongly when encompassed by an additional pair of electrons.

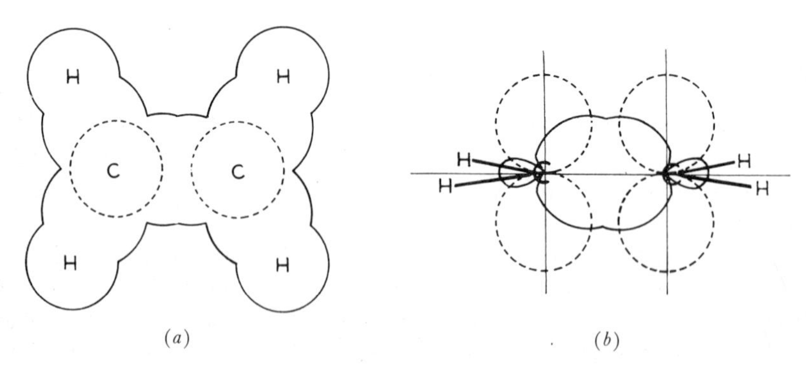

Figure 6–*1*. Cross sections of the ethylene molecule before π bond formation: (*a*) through the carbon and hydrogen atoms; (*b*) through the carbon atoms and perpendicular to the plane of the hydrogen atoms.

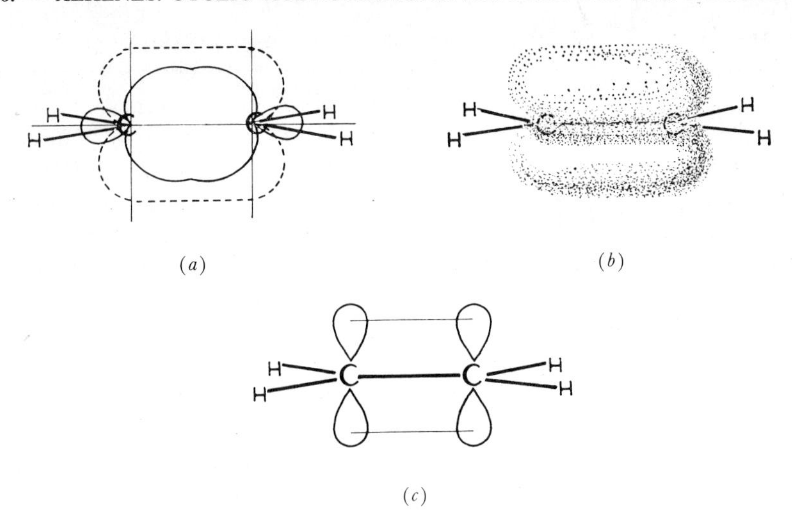

Figure 6–2. The ethylene molecule after π bond formation: (*a*) cross section through the carbon atoms and perpendicular to the plane of the hydrogen atoms; (*b*) perspective representation of the π bond; (*c*) schematic representation of the π bond.

This theory amounts to a reinterpretation of the much older partial valence theory proposed by Thiele.[2] He assumed that the valence forces, the nature of which was unknown, were satisfied only partially at the carbon atoms of a double bond. The second bond was represented by a dotted line to indicate that it was more reactive than the single bond. Thiele's theory lost adherents because it could not be explained by the Lewis electron-pair bond theory.

Formulas for the homologs of ethylene can be constructed in the same way as formulas for the homologs of methane, that is, by the insertion of CH_2 groups (*methylene groups*) between hydrogen and carbon atoms or by the replacement of hydrogen atoms by alkyl groups. For example, the next member of the series, propylene, C_3H_6, would have the formula $CH_3CH{=}CH_2$. Since the propylene molecule has three kinds of hydrogen atoms, there should be three structurally isomeric butylenes, all of which are known.

$CH_3CH_2CH{=}CH_2$ $\quad$ $CH_3CH{=}CHCH_3$ $\quad$ $CH_3\underset{\displaystyle |\atop\displaystyle CH_3}{C}{=}CH_2$

The same formulas would have been obtained if the double bond had been put in all possible positions in normal butane and in isobutane.

Nomenclature

Common Names. The common names of the alkenes are derived by replacement of the *ane* of the common name for the saturated hydrocarbon by *ylene,* but *n-* is omitted for straight-chain compounds. Greek letters sometimes are used to distinguish between isomers (Table 6–*1*, p. 76).

As Derivatives of Ethylene. In this system the alkyl groups replacing the hydrogen atoms of ethylene are named and the word *ethylene* added. Since ethylene has two hydrogens on each of two carbon atoms, the replacement of two hydrogens by two groups can give rise to two isomers depending on whether the new groups are on different carbon

[2] Johannes Thiele (1865–1918), professor at the University of Strassburg. He is noted chiefly for his work on the addition reactions of unsaturated compounds (p. 582).

TABLE 6–*1*. NOMENCLATURE OF ALKENES

FORMULA	COMMON NAME	AS DERIVATIVE OF ETHYLENE	INTERNATIONAL SYSTEM
$CH_2{=}CH_2$	ethylene	none	ethene
$CH_3CH{=}CH_2$	propylene	methylethylene	propene
$CH_3CH_2CH{=}CH_2$	α-butylene	ethylethylene	1-butene
$CH_3CH{=}CHCH_3$	β-butylene	*sym*-dimethylethylene	2-butene
$CH_3C{=}CH_2$ CH_3	isobutylene*	*unsym*-dimethylethylene	methylpropene
$C_2H_5CH{=}CHCH_3$	β-amylene†	*sym*-methylethylethylene	2-pentene
6 5 4 $CH_3CH_2CHCH_3$ $CH_3C{=}CHCH_3$ 3 2 1	none	*sym*-dimethyl-*s*-butylethylene	3,4-dimethyl-2-hexene

* The name isobutene, which frequently is used, is undesirable because it mixes two systems.
† Olefins having five carbon atoms are known as *amylenes* instead of *pentylenes* (see footnote, p. 121).

atoms or both on the same carbon atom. In order to distinguish between these two isomers, the one with the two groups on different carbon atoms is called *symmetrical* (prefix *sym-*); that is, the groups are placed symmetrically with respect to the double bond. The isomer with both groups on the same carbon atom is called *unsymmetrical* (prefix *unsym-*). These designations apply only to the distribution of the two groups that give rise to the isomerism and have nothing to do with the symmetry of the molecule as a whole. In the last two examples given in Table 6–*1*, the molecules as a whole are unsymmetrical, but the methyl and ethyl groups in the first and the two methyl groups in the second are symmetrically placed with respect to the double bond. Hence they are called *sym*-methylethylethylene and *sym*-dimethyl-*s*-butylethylene.

International System. The international rules for naming alkenes resemble the rules for naming alkanes (p. 60): (*1*) the ending *ane* of the corresponding saturated hydrocarbon is replaced by the ending *ene;* (*2*) the parent compound is considered to be the longest chain *containing the double bond;* (*3*) the chains are numbered from the end nearest the double bond, and the position of the double bond is indicated by the number of the lower-numbered carbon atom to which it is attached; (*4*) side chains are named and their position indicated by a number.

Geometric Isomerism

Although only one 1-butene and one methylpropene are known, two 2-butenes have been isolated. One boils at 0.96° and the other at 3.73°. Similarly 1,1-dichloroethylene exists in only a single form, but two forms of 1,2-dichloroethylene are known, one boiling at 48.3° and the other at 60.5°. The existence of these isomers results from a lack of free rotation about a double bond (p. 74). If each of the doubly bound carbon atoms is attached to two different atoms or groups, two isomers are possible that are structurally identical; that is, the order of attachment of the atoms is the same in both. In one, like groups are on the same side of the double bond and in the other, on opposite sides (Fig. 6–*3*). Because the isomerism is due to differences in the shapes of the molecules, it is called **geometric isomerism.** The isomer with like groups on the same side is known as the *cis* form, and that with like groups on opposite sides is the *trans* form (L. *cis* on this side, *trans* across).

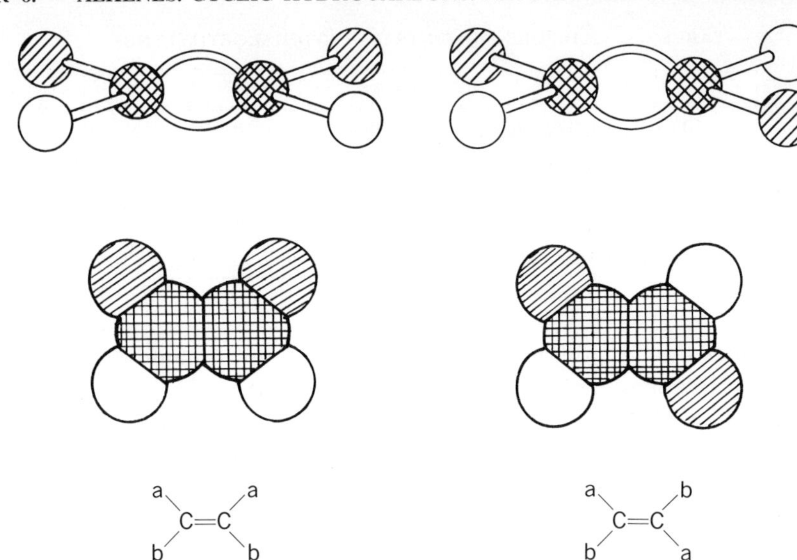

Figure 6–3. Representation of *cis* and *trans* forms of geometric isomers by molecular models and by projection formulas.

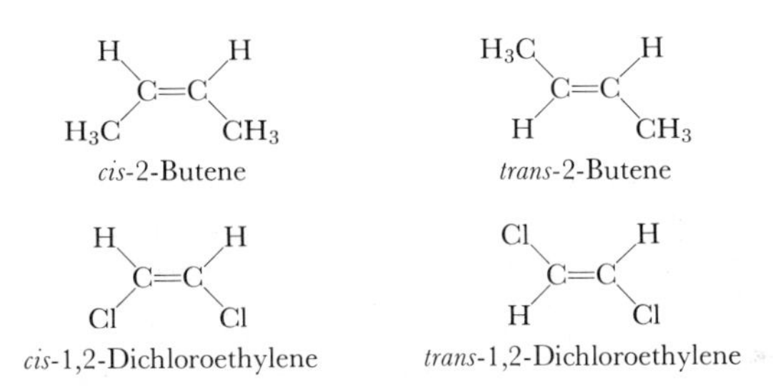

The theory that bond formation results from the overlap of atomic orbitals explains the lack of free rotation, since the *p* orbitals can overlap only when they are parallel. Rotation about the double bond cannot take place without breaking the π bond. Thus the barrier to rotation has been raised from 3 kcal. for a single bond (p. 56) to about 58 kcal. for a double bond.

Isomers are compounds that have the same kinds and numbers of atoms (the same *composition*) but different properties. The isomerism of the alkanes is ascribed to a difference in structure (a difference in *constitution*). The relative positions of the atoms and groups that are permitted by free rotation about single bonds and by the flexibility of bond angles is called the *conformation* of the molecule (p. 56). The difference between conformers and isomers is that conformers are too readily convertible each into the other to be isolable as separate entities. Geometric isomers are structurally stable compounds that have the same constitution but differ in the spatial arrangement of the groups. They make up one division of **stereoisomers** (Gr. *stereos* solid), and are said to differ in *configuration.* Each of the four terms, composition, constitution, conformation, and configuration, has a distinct meaning, and they should not be used interchangeably.

It is of interest to determine which of a pair of geometric isomers has the *cis* configuration and which the *trans* configuration. The assignment is relatively easy for the 1,2-dihaloethylenes. The carbon-hydrogen bond is practically nonpolar, but the carbon-halogen bond is strongly polar. An examination of the proposed configurations of the two isomers indicates

TABLE 6–2. DIPOLE MOMENTS OF 1,2-DIHALOETHYLENES

	μ	M.P.	B.P.
1,2-Dichloroethylene (*trans*)	0	−50	47
(*cis*)	1.85	−80	60
1,2-Dibromoethylene (*trans*)	0	− 6	108
(*cis*)	1.35	−53	112
1,2-Diiodoethylene (*trans*)	0	+72	190
(*cis*)	0.75	−14	188

that the *trans* isomer should have zero dipole moment (p. 20) because the carbon-halogen bond moments are in opposite directions and cancel each other. The *cis* compound, on the other hand, should have a resultant moment for the molecule as a whole.

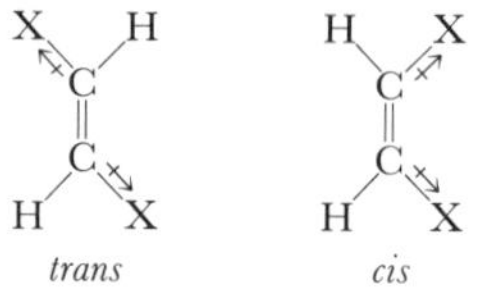

trans *cis*

The dipole moments and physical properties of the 1,2-dihaloethylenes are given in Table 6–2. By a rather complicated procedure, it has been shown that *cis*-2-butene is the isomer boiling at 3.73° and that *trans*-2-butene boils at 0.96°. The configuration of some geometric isomers can be determined from differences in their chemical reactions (p. 634).

Sources and Uses of the Alkenes

Alkenes are produced commercially by the cracking of natural gas and petroleum hydrocarbons (p. 65). **Ethylene** is by far the most important olefin as a raw material for the synthesis of other chemicals. Production in 1964 was over 8.6 billion pounds, and the contract price less than 5 cents a pound. About 35 per cent was used for the synthesis of the plastic, polyethylene (p. 88), 25 per cent for ethylene oxide (p. 605), 20 per cent for ethyl alcohol (p. 121), and 10 per cent for ethylbenzene (p. 391).

Production of **propylene** in 1964 was 3.6 billion pounds, and the selling price was 2 cents per pound. About 40 per cent was used for the synthesis of *i*-propyl alcohol (p. 123), 30 per cent for propylene trimer and tetramer (p. 87), and 12 per cent for propylene oxide (p. 606). Combined production of the **butylenes** was 1.4 billion pounds. They are used chiefly for the production of butadiene (p. 587) and synthetic rubbers (p. 588). Methods for the synthesis of individual olefins, for example from alcohols or from compounds containing halogen, are given in Chapter 10.

Physical Properties

The general physical properties of the alkenes are much the same as those of the corresponding saturated hydrocarbons. The solubility of the lower alkenes in water, though slight, is considerably greater than that of the alkanes, because the greater availability of electrons in the double bond leads to a greater attraction for the positive end of the water dipole (p. 25).

General Chemical Properties

In contrast to the alkanes, the alkenes are very reactive. The π bond is weaker than most σ bonds between carbon and other elements. The empirical bond energy assigned to a carbon-carbon double bond is 146 kcal. and that to a single bond is 83 kcal. The difference of 63 kcal. should be the energy required to unpair the electrons of the π bond. The bond dissociation energy of the carbon-carbon single bond in ethane is 84 kcal., but that of the double bond in ethylene has been estimated to be only 125 kcal., indicating that the energy required to break the π bond in ethylene is around 41 kcal. The generally accepted value for a dialkyl-substituted alkene is about 58 kcal. Not only is the π bond weaker than a σ bond, but the electrons of the π bond are more exposed than those in a σ bond and are subject to attack by electrophilic reagents. In this respect the π bond resembles an unshared pair of electrons.

Reaction of a π bond may take place in several ways. In a reaction involving free radicals, one electron goes with each carbon.

$$A\cdot + RCH{:}CHR \longrightarrow RCH(A)\text{—}\dot{C}HR$$

$$RCH(A)\text{—}\dot{C}HR + A{:}A \longrightarrow RCH(A)CH(A)R + A\cdot$$

In a polar reaction the pair of electrons goes with one or the other carbon atom.

$$A{:}\,|\,A + RCH{:}CHR \longrightarrow A{:}^- + RCH(A)\text{—}\overset{+}{C}HR$$

$$RCH(A)\text{—}\overset{+}{C}HR + A{:}\,|\,A \longrightarrow RCH(A)\text{—}CH(A)R + A^+$$

The polar reaction may be thought of as an initial electrophilic attack by the reagent $A{:}A$ on the double bond followed by a nucleophilic attack on the positive ion intermediate, which is called a *carbonium ion.* Alternatively the reaction may be considered as a nucleophilic attack by the olefin on the reagent followed by an electrophilic attack by the carbonium ion. It is probable that in these reactions, ions never are free in the sense that sodium and chlorine ions are free in aqueous solution. The reaction probably takes place by a *concerted mechanism* with the transition state involving at least three different molecules. Although ordinarily written as a one-step process, the formation of the transition state undoubtedly takes place in two steps.

$$\begin{array}{ccccc} A{:}A + CHR & \longrightarrow & A\text{---}A\text{---}CHR & \longrightarrow & A^+ + A\text{—}CHR \\ \| & & \| & & | \\ RCH + A{:}A & & RCH\text{---}A\text{---}A & & RCHA + {:}A^- \\ \text{Reagents} & & \text{Transition state} & & \text{Products} \end{array}$$

Another possibility is a concerted reaction that takes place by way of a cyclic transition state similar to that for the decomposition of hydrogen iodide (p. 37).

$$\begin{array}{ccccc} RCH{=}CHR & & RCH{=}CHR & & RCH\text{—}CHR \\ + & \longrightarrow & \vdots \quad \vdots & \longrightarrow & | \quad | \\ A{:}A & & A\text{------}A & & A \quad A \end{array}$$

This type is called a *multicenter reaction.* Usually four atoms are involved in the breaking of old bonds and the making of new bonds during the reaction.

In either free radical or polar mechanisms the *driving force* of the reaction is the energy evolved because the sum of the energies of the two new σ bonds in the product is greater than the sum of the energy of the π bond and the σ bond in the reactants. Energy lost in the formation of A^+ and A^- in the polar reactions is regained either by reaction with olefin molecules by steps similar to those above, or by reaction with each other to regenerate $A{:}A$. The former would be more frequent at the beginning of the reaction, whereas the latter would predominate towards the end of the reaction.

Because one portion of a reagent appears to have added to one carbon atom and the other portion to the other carbon atom, the reactions of alkenes are known as *addition reactions* and the reagent is said to *add* to the double bond. Numerous reagents add to carbon-carbon double bonds. It is convenient to classify them as *identical addenda* and *non-identical addenda.* Two other types of reaction of alkenes are *oxidation* and *isomerization.*

ADDITION OF IDENTICAL ADDENDA

Halogen. The most characteristic reaction of the double bond is the rapid polar addition, in the liquid phase or in solution, of chlorine or bromine to those alkenes in which each of the doubly bound carbon atoms is united to at least one hydrogen atom.

$$CH_2{=}CH_2 + Cl_2 \longrightarrow ClCH_2CH_2Cl + {\sim}36 \text{ kcal.}$$

Ethylene chloride[3]
(*1,2-dichloroethane*)

$$CH_3CH{=}CH_2 + Br_2 \longrightarrow CH_3CHBrCH_2Br + {\sim}22 \text{ kcal.}$$

Propylene bromide
(*1,2-dibromopropane*)

The reaction in general may be expressed by the equation

$$RCH{=}CHR + X_2 \longrightarrow RCHXCHXR$$

where R is any alkyl group or a hydrogen atom, and X is any halogen atom. In practice the reaction is limited to chlorine and bromine, because fluorine reacts too violently to be controllable ($\Delta H = -110$ kcal.), and iodine does not give stable 1,2-diiodo derivatives except with a few simple olefins. If both hydrogen atoms on the same carbon atom of ethylene are replaced by alkyl groups as in isobutylene, addition of chlorine does not take place as readily as another reaction, *substitution,* in which a hydrogen atom is replaced by a chlorine atom (pp. 68, 593). Bromine, however, adds regularly, forming isobutylene bromide.

An energy profile for the addition of bromine is shown in Fig. 6–4. The first transition state, *a*, results from the electrophilic attack by a bromine molecule on the unsaturation electrons. The transition state then goes over to the positive intermediate *b*.

$$R_2C{\stackrel{..}{=}}CR_2 + Br{:}Br \longrightarrow \text{Transition state } a \longrightarrow R_2C(\overset{+}{Br})\text{—}CR_2 + {:}Br^-$$

Intermediate *b*

It has been postulated that this intermediate is stabilized somewhat by the formation of a three-membered ring to account for the fact that nucleophilic attack of the second mole-

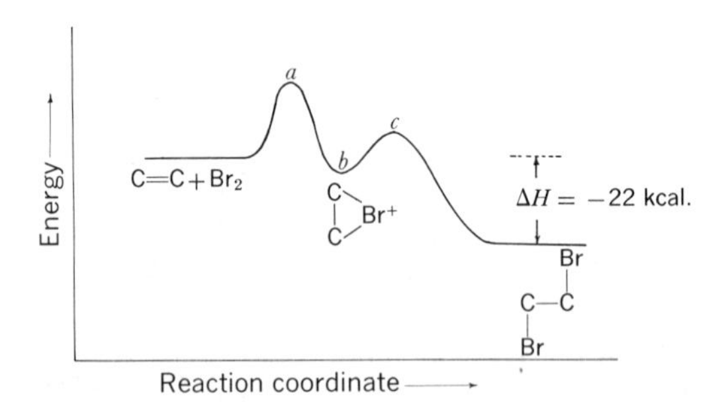

Figure 6–4. Energy profile for the addition of bromine to a double bond.

[3] It is this reaction that gave rise to the old name for ethylene, namely *olefiant gas* (meaning *oil-making gas*) because the reaction of gaseous ethylene with gaseous chlorine gives liquid ethylene chloride. "Olefiant gas" then was contracted to "olefin" as a general term for unsaturated hydrocarbons. For the nomenclature of halogen compounds see page 98 and footnote 1, page 138.

cule, to give the second transition state, *c*, and the final product, always is from the side opposite to that of the electrophilic attack.[4]

$$R_2C(-Br^+)-CR_2 \cdots :Br:\,Br: \longrightarrow \text{Transition state } c \longrightarrow R_2C(Br)-C(Br)R_2 + :Br:^+$$

These halogen derivatives are colorless liquids. Hence decolorization of a solution of bromine in water, or better in a mutual solvent such as carbon tetrachloride or acetic acid, may be used as a *test for unsaturation,* provided that no other group is present that reacts with bromine. To determine the amount of a known olefin in a mixture with other substances, the mixture can be titrated with a standardized solution of bromine. A similar titration of a pure unknown olefin would give the equivalent weight of the compound, that is, the weight associated with one double bond. The number of double bonds would be equal to the molecular weight divided by the equivalent weight.

The addition of halogen to a double bond should not be referred to as halogenation. This term should be used only for substitution reactions in which hydrogen is displaced by halogen (p. 69).

Hydrogen. From empirical bond dissociation energies, the addition of a molecule of hydrogen to a double bond should be exothermic by about 32 kcal.

$$\underset{64}{RCH{=}CHR} + \underset{104}{H_2} \longrightarrow \underset{2\times 99}{RCH_2{-}CH_2R} \quad +30$$

Nevertheless no detectable reaction takes place between hydrogen and an alkene at 25°. In the presence of certain finely divided metals, however, addition takes place rapidly with the evolution of heat. Metal surfaces have unpaired electrons, and if the lattice spacing is right, these electrons can pair with those of the hydrogen molecule. The result is a weak bonding of hydrogen atoms on the metal surface as in (*a*). This type of adsorption is known as *chemisorption* in contrast to *physical adsorption,* which is due to van der Waals attraction.

$$M\cdot\,M\cdot + H{:}H \longrightarrow \underset{(a)}{M\cdot\cdot H,\ M\cdot\cdot H} \xrightarrow{CR_2{:}CR_2} \underset{(b)}{M\cdot\text{---}H\text{---}CR_2 \,\|\, M\cdot\text{---}H\text{---}CR_2} \longrightarrow M\cdot\,M\cdot + HCR_2{-}HCR_2$$

Collision of an unsaturated molecule with adsorbed hydrogen (*a*) probably gives a multicenter transition state (*b*) that goes over to the saturated alkane in which hydrogen has added *cis* to the double bond. Because the various steps in this process involve unpaired electrons and weak bonds, none has a high activation energy.

Substances that increase the rate of a desired reaction are called **catalysts.** They operate by providing a path having a lower activation energy than that for the uncatalyzed reaction. Increases in rate brought about by agents that supply energy, such as heat or light, should not be classed as catalytic. When catalysts are present as a distinct solid phase as in catalytic hydrogenation, the process is called *heterogeneous catalysis.* If the process takes place in a single phase, it is referred to as *homogeneous catalysis.*

---

[4] It is customary to indicate by means of curved arrows the plausible movement of electrons during bond breaking and bond formation (p. 40). The convention regarding the use of solid lines, dotted lines, and dark wedges for bonds is given on page 57.

Finely divided platinum,[5] palladium, rhodium, ruthenium, or nickel are active heterogeneous catalysts for hydrogenation at 25°. Their activity depends on the conditions under which they are prepared. Platinum and palladium catalysts made by reduction of the finely divided oxides with hydrogen are known as *Adams catalysts.* Nickel catalysts made by removing aluminum from a powdered nickel-aluminum alloy by reaction with hot aqueous sodium hydroxide is called *Raney nickel.* Active catalysts containing 5 to 10 per cent of one of the platinum metals on a powdered charcoal carrier are widely used also.

The over-all process may be represented by the reaction

$$R_2C{=}CR_2 + H_2 \xrightarrow{\text{Pt, Pd, or Ni}} R_2CHCHR_2$$

If the olefin is a gas, a mixture with hydrogen may be passed over the catalyst. Liquid olefins, or solid olefins dissolved in an inert solvent, may be shaken with hydrogen in the presence of a suspension of the finely divided catalyst. The reaction can be used for analytical as well as for preparative purposes. From the volume of hydrogen absorbed, the amount of an unsaturated compound of known structure in a mixture with saturated compounds, or the number of double bonds in a pure unknown compound whose molecular weight has been determined, can be calculated.

ADDITION OF NONIDENTICAL ADDENDA

Alkyl groups are more polarizable than hydrogens because they have a larger number of electrons. Hence when called upon to do so, they can relieve a deficiency of electrons better than can a hydrogen. An alkene is a nucleophile, and the first step of a polar reaction is the combination of the π electrons with the attacking electrophilic reagent. The greater polarizability of an alkyl group compared to hydrogen has two effects on the reaction. First, the ease of reaction is increased as hydrogens attached to the doubly bound carbon atoms are replaced by alkyl groups because the positive charge resulting from the combination of the electrophile with the π electrons is relieved by polarization of the alkyl groups. This view is supported by the ionization potentials of the alkenes given in Table 6–3. The energy necessary to remove an electron within any group of isomers decreases as the number of alkyl groups attached to the doubly bound carbon atoms increases, thus reflecting the electron-releasing power of the alkyl groups.

The second effect of the greater polarizability of alkyl groups compared to hydrogen, is that it determines the point of attack of a polarized molecule on an unsymmetrically substituted alkene. Thus the intermediate carbonium ion in the following reaction will have either structure (*a*) or (*b*).

$$RCH\overset{..}{=}CH_2 + \overset{\delta+}{A}{:}\overset{\delta-}{B} \longrightarrow \underset{(a)}{R{-}\overset{+}{C}H{-}\underset{A}{\underset{|}{C}}H_2} + B^- \quad \text{or} \quad \underset{(b)}{R{-}\underset{A}{\underset{|}{C}}H{-}\overset{+}{C}H_2} + B^-$$

Because the R group is more polarizable than hydrogen, the positive charge on (*a*) can be

[5] The first reference to the use of catalytic hydrogenation appears to be that by Debus in 1863. He found that hydrogen added to hydrogen cyanide in the presence of platinum to give methylamine (p. 221). Scattered references to similar reactions were made during the next thirty-four years. In 1897 Paul Sabatier (1854–1941) and Jean Baptiste Senderens (1856–1937) at the University of Toulouse found that ethylene and hydrogen when passed over cobalt, iron, copper, or platinum at 300° gave ethane, and that benzene and hydrogen over nickel gave cyclohexane (p. 374). Sabatier pursued the subject until his retirement in 1930. He was awarded the Nobel Prize in Chemistry, jointly with Victor Grignard (p. 108), in 1912. Senderens entered industry and introduced contact catalysis into technical organic chemistry. He was a very religious man and became an abbé and later a canon in the Roman Catholic Church.

TABLE 6–3. IONIZATION POTENTIALS OF ALKANES, ALKENES, AND ALKYL FREE RADICALS IN ELECTRON VOLTS*

ALKANE	E.V.	ALKENE	NUMBER OF SUBSTITUENTS					ALKYL FREE RADICALS	E.V.
			0	1	2	3	4		
Methane	13.12							Methyl	9.95
Ethane	11.65	Ethylene	10.56					Ethyl	8.87
Propane	11.21	Propylene		9.80				*n*-Propyl	8.69
								i-Propyl	7.90
n-Butane	10.80	1-Butene		9.72				*n*-Butyl	8.64
		cis-2-Butene			9.34			*s*-Butyl	7.93
		trans-2-Butene			9.27			*t*-Butyl	7.42
i-Butane	10.79	Methylpropane			9.26			*i*-Butyl	8.35
n-Pentane	10.55	1-Pentene		9.67					
		cis-2-Pentene			9.11				
		trans-2-Pentene			9.06				
i-Pentane	10.60	2-Methyl-1-butene			9.20				
		3-Methyl-1-butene		9.60					
		2-Methyl-2-butene				8.89			
n-Hexane	10.43	1-Hexene		9.59					
		trans-2-Hexene			9.16				
		trans-3-Hexene			9.12				
i-Hexane	10.34								
3-Methylpentane	10.30								
2,3-Dimethylbutane	10.24	2,3-Dimethyl-2-butene					8.53		

*1 e.v. = 23 kcal./mole.

distributed better than that on (*b*), and (*a*) should be the intermediate most readily formed. In other words a secondary alkyl cation is more stable than a primary alkyl cation. Similarly a tertiary alkyl cation is more stable than a secondary alkyl cation. Ionization potentials again may be cited as confirmation of this view. Removal of an electron from a free radical leads to a carbonium ion. The more readily this electron is removed, the greater the stability of the carbonium ion. The ionization potentials for any group of isomeric free radicals decrease with increasing number of alkyl substituents (Table 6–3).

To summarize, *the reaction of a polar molecule with a carbon-carbon double bond results in the addition of the positive end of the polar molecule to the less alkyl-substituted end of the polarized double bond.* Markovnikov[6] formulated essentially this rule in 1870 from experimental observations. He stated that in the addition of halogen acid to a double bond, "the halogen adds itself to the less hydrogenated carbon atom." In 1875 he modified the rule to "the more negative element or group adds to the less hydrogenated carbon atom." Regardless of how it is stated, the mode of addition in polar reactions is referred to as the **Markovnikov rule.**

Sulfuric acid. Alkenes add sulfuric acid according to the Markovnikov rule to give *alkyl hydrogen sulfates.*

$$R_2C{=}CHR + H{:}OSO_3H \longrightarrow R_2\underset{\displaystyle OSO_3H}{\underset{|}{C}}{-}CH_2R$$

[6] Vladimir Vasil'evich Markovnikov (1838–1904), director of the Chemical Institute of the University of Moscow. Although his ideas concerning the effect of structure on the course of chemical reactions were published in Russian in 1869, they went unnoticed in Europe until 1899, because he refused to publish them in a foreign language. After 1881 he did important work on the chemistry of petroleum hydrocarbons.

$$CH_3CH_2CH{=}CH_2 \text{ (1-Butene) or } CH_3CH{=}CHCH_3 \text{ (2-Butene)} + H_2SO_4 \longrightarrow CH_3CH_2\underset{\displaystyle OSO_3H}{\underset{|}{C}}HCH_3$$

s-Butyl hydrogen sulfate

$$(CH_3)_2C{=}CH_2 + H_2SO_4 \longrightarrow (CH_3)_2\underset{\displaystyle OSO_3H}{\underset{|}{C}}CH_3$$

Methyl propene (*isobutylene*) — *t*-Butyl hydrogen sulfate

The ease of addition of sulfuric acid depends on the number of alkyl substituents at the double bond (p. 82). For example, ethylene reacts slowly at room temperature with concentrated sulfuric acid, requiring a catalyst such as silver sulfate for rapid addition; propylene reacts with 85 per cent sulfuric acid in the absence of a catalyst; isobutylene reacts with 65 per cent sulfuric acid at room temperature.

It is difficult to separate the free alkyl hydrogen sulfates from the excess sulfuric acid because neither compound is volatile and both have about the same solubility characteristics. The mixture can be neutralized, however, for example with potassium hydroxide, and the potassium alkyl sulfate separated from potassium sulfate by crystallization. The calcium and barium alkyl sulfates can be isolated even more readily because they are soluble in water, whereas calcium or barium sulfate is insoluble in water.

In the presence of an excess of olefin, a second molecule can add to the alkyl hydrogen sulfate to give an alkyl sulfate.

$$CH_3CH_2OSO_2OH + CH_2{=}CH_2 \longrightarrow CH_3CH_2OSO_2OCH_2CH_3$$

Ethyl hydrogen sulfate — Ethyl sulfate

Halogen Acids. Olefins add halogen acids to give *alkyl halides.*

$$R_2C{=}CHR + HX \longrightarrow R_2\underset{\displaystyle X}{\underset{|}{C}}{-}CH_2R$$

$$CH_3CH{=}CH_2 + HBr \longrightarrow CH_3\underset{\displaystyle Br}{\underset{|}{C}}HCH_3$$

i-Propyl bromide (*2-bromopropane*)

Halogen acids also add more readily the greater the amount of alkyl substitution at the double bond. For example, the order of ease of addition of hydrogen chloride is isobutylene > propylene > ethylene. The ease of addition varies also with the acidity (electrophilicity, pp. 41, 101) of the halogen acid, the order from concentrated aqueous solutions being hydrogen iodide > hydrogen bromide > hydrogen chloride > hydrogen fluoride.

In contrast to the behavior of bromine, which has been shown to add *trans* to a double bond, apparently by way of a bromonium ion intermediate (p. 80), addition of hydrogen bromide is primarily *cis*. It is postulated that protonation of the π electrons yields not a free carbonium ion but an ion pair with the bromide ion retained on the same side as the hydrogen. Collapse of the ion pair yields the *cis* addition product.

$$\text{HRC}{=}\text{CHR} + \text{H:Br} \longrightarrow \text{H(R)(H)C}{-}\overset{+}{\text{C}}\text{(H)(R)} \cdots \text{Br}^- \longrightarrow \text{H(R)(H)C}{-}\text{C(Br)(H)(R)}$$

Trans addition takes place only if an ion pair survives long enough for migration of bromide ion to the other side or for attack by a second molecule of hydrogen bromide. Presumably the other halogen acids also add chiefly *cis* (cf. pp. 106, 648).

The behavior of hydrogen bromide is anomalous if oxygen or peroxides are present. Under these conditions the mode of addition to unsymmetrically substituted ethylenes is contrary to the Markovnikov rule. For example, when propylene reacts with hydrogen bromide in the presence of peroxides, *n*-propyl bromide is formed instead of *i*-propyl bromide.

$$CH_3CH{=}CH_2 + HBr \xrightarrow{\text{(Peroxides)}} CH_3CH_2CH_2Br$$

The mode of addition of hydrogen chloride or hydrogen iodide is not affected by the presence of peroxides.

Whenever the same reactants give different products under different conditions, different mechanisms are involved. The fact that oxygen or peroxides are necessary indicates that in the abnormal addition, a free radical mechanism is operating. The reaction is initiated by the attack of a free radical $Z\cdot$ ($\cdot O{-}O\cdot$, $HO\cdot$, or $RO\cdot$) on a molecule of hydrogen bromide to give bromine atoms. Initiation is followed by the propagation steps that lead to the product.

$$Z\cdot + H{:}Br \longrightarrow Z{:}H + Br\cdot \qquad \textit{Initiation}$$

$$\left.\begin{array}{l} Br\cdot + CH_2{=}CHR \longrightarrow BrCH_2{-}\dot{C}HR \\ BrCH_2\dot{C}HR + H{:}Br \longrightarrow BrCH_2CH_2R + Br\cdot \end{array}\right\} \quad \textit{Propagation}$$

The intermediate bromo free radical is electron deficient. Of the two possible structures for the intermediate, that having the greater number of alkyl groups on the carbon with the unpaired electron is the more stable and the more readily formed. These facts lead to an extension of the rule for addition to carbon-carbon double bonds, namely that *free radical attack, like electrophilic attack, is at the less alkyl-substituted carbon atom.* It commonly is called *anti-Markovnikov addition.*

Of the halogen acids, only hydrogen bromide shows the peroxide effect. The bond dissociation energies of hydrogen fluoride (134 kcal.) and hydrogen chloride (102 kcal.) are considerably higher than that of hydrogen bromide (87 kcal.). Although iodine atoms are formed readily, the attack on alkene by iodine atoms is endothermic and 10 to 14 kcal. less favorable than the attack by bromine atoms. Accordingly the free radical route for the reaction of hydrogen fluoride, chloride, and iodide is much slower than addition by the polar mechanism. Various other reagents, however, such as aldehydes, hydrogen sulfide, bisulfite (p. 268), and organic halogen compounds, undergo free radical anti-Markovnikov additions to olefins in the presence of peroxides or light.

Water or Hydrogen Peroxide. The initial attack on a carbon-carbon double bond by an electrophilic agent must be accompanied by attack of the intermediate by a nucleophilic agent. Water is a fair nucleophile but is not a sufficiently good electrophile to bring about the first step. Hydronium ion is, however, and the presence of a strong acid can catalyze the addition of water to the double bond.

$$H_2O + H_2SO_4 \rightleftharpoons H_3O^+ + HSO_4^- \overset{H_2O}{\rightleftharpoons} H_3O^+ + SO_4^{2-}$$

$$RCH{=}CH_2 + H_3O^+ \rightleftharpoons R\overset{+}{C}H{-}CH_3 + H_2O$$

$$R\overset{+}{C}H{-}CH_3 + {:}OH_2 \rightleftharpoons RCH(\overset{+}{O}H_2){-}CH_3$$

$$R{-}CH(\overset{+}{O}H_2){-}CH_3 + H_2O \rightleftharpoons \underset{\text{An alcohol}}{RCH(OH){-}CH_3} + H_3O^+$$

Because of the high concentration of water molecules, nucleophilic attack by them predominates over attack by a bisulfate or sulfate ion. Moreover the nucleophilicity of these ions is greatly reduced because they are heavily hydrated (p. 25). The over-all reaction for the hydration of propylene may be represented by the simple equilibrium

$$CH_3CH{=}CH_2 + H_2O \xrightleftharpoons{H_2SO_4} CH_3CHOHCH_3$$
Isopropyl alcohol

Alcohols react with alkenes by an entirely analogous process to give ethers (p. 139). Thus the formation of isopropyl alcohol always is accompanied by the formation of isopropyl ether.

$$(CH_3)_2CHOH + CH_2{=}CHCH_3 \xrightleftharpoons{H_2SO_4} (CH_3)_2CH{-}O{-}CH(CH_3)_2$$
Isopropyl ether

Similarly, the acid-catalyzed addition of hydrogen peroxide to isobutylene yields *t*-butyl hydroperoxide and *t*-butyl peroxide.

$$(CH_3)_2C{=}CH_2 + H_2O_2 \xrightleftharpoons{H_2SO_4} (CH_3)_3C{-}O{-}OH$$
t-Butyl hydroperoxide

$$(CH_3)_2C{=}CH_2 + (CH_3)_3C{-}O{-}OH \xrightleftharpoons{H_2SO_4} (CH_3)_3C{-}O{-}O{-}C(CH_3)_3$$
t-Butyl peroxide

The properties of alcohols are discussed in Chapter 8 and those of ethers in Chapter 9. Like hydrogen peroxide, the alkyl hydroperoxides are stronger acids (pK_a = 11.5 to 12.7) than water (pK_a = 15.7) and form salts with strong bases in aqueous solution. Methyl and ethyl hydroperoxide decompose violently when heated above 60° or when subjected to shock. The higher hydroperoxides decompose smoothly to give complex mixtures of products. Because of the low bond dissociation energy of around 35 kcal. for the O—O bond, alkyl hydroperoxides and alkyl peroxides decompose first to free radicals.

$$RO{-}OH \longrightarrow RO\cdot + \cdot OH$$
$$RO{-}OR \longrightarrow 2\,RO\cdot$$

They are used as initiators for free radical chain reactions (p. 88). Production of *t*-butyl peroxide amounted to over 1 million pounds in 1964.

Borane. Diborane, B_2H_6, is a toxic gas, b.p. −92°, that in ether solutions probably is in equilibrium with solvated borane molecules.

$$B_2H_6 + 2\,R_2O{:} \rightleftharpoons 2\,R_2O{:}BH_3$$

The latter can be generated in the presence of the compound with which it is to react by the reduction of boron fluoride with lithium aluminum hydride in ether solution.

$$3\,LiAlH_4 + 4\,BF_3 \longrightarrow 4\,BH_3 + 3\,LiAlF_4$$

Borane also can be purchased as a 1 *M* solution in the cyclic ether, tetrahydrofuran (p. 512).

Borane readily adds to olefins to give **alkylboranes.**

$$3\,RCH{=}CH_2 + BH_3 \longrightarrow (RCH_2{-}CH_2{-})_3B$$

The process is called *hydroboration.* The direction of addition is that expected, since the boron lacks a pair of electrons in its valence shell. The alkylboranes are useful for the synthesis of other compounds (p. 118).

Self Addition (Polymerization). An alkene molecule in the presence of an acid catalyst can add another molecule to give a new alkene. The product has twice the molecular weight of the original alkene and is called a *dimer* (Gr. *meros* part, hence *two parts*). The original alkene is called a *monomer.*

The initial step is the same as that for the addition of an acid or of water to the double bond. The catalyst adds a proton to the monomer according to the Markovnikov rule to yield a carbonium ion. This intermediate carbonium ion is a Lewis acid and adds to the second molecule to give a new carbonium ion.

$$CH_3CH{=}CH_2 \underset{B^-}{\overset{HB}{\rightleftharpoons}} (CH_3)_2\overset{+}{C}H \xrightleftharpoons{CH_2=CHCH_3} (CH_3)_2CH{-}CH_2{-}\overset{+}{C}H{-}CH_3$$

The second carbonium ion can be stabilized by loss of a proton to a base with the formation of a double bond, but since the proton can be lost from either carbon adjacent to the carbonium carbon, a mixture of two olefins results.

$$(CH_3)_2CHCH_2\overset{+}{C}H{-}\underset{H}{C}H_2 \underset{HB}{\overset{B^-}{\rightleftharpoons}} (CH_3)_2CHCH_2CH{=}CH_2$$

$$(CH_3)_2CH\underset{H}{C}H{-}\overset{+}{C}HCH_3 \underset{HB}{\overset{B^-}{\rightleftharpoons}} (CH_3)_2CHCH{=}CHCH_3$$

In this representation of steps and in subsequent equilibrium reactions illustrating stepwise mechanisms, a **convention** is adopted that eliminates the necessity of rewriting formulas. The catalyst or reagent is placed above the double arrow and the minor product that is eliminated is placed below the double arrow. The interpretation of the first reaction is that the acid HB transfers a proton to the methylene (CH_2) group of 1-butene with loss of the base B^- to give the intermediate carbonium ion. In the second and third equations, a base B^- removes a proton from a carbon atom adjacent to the carbonium carbon with regeneration of the acid HB and the formation of the hexenes. In the reverse reactions, the reagent is below the arrow and the minor product is above the arrow. It will be observed that reactants that appear both above and below the arrows cancel each other and do not appear in the the final products.

Reaction of the dimeric carbonium ion with a third molecule of alkene gives a trimeric carbonium ion, which can give a mixture of *trimers*. It can react also with a fourth molecule of alkene to give a mixture of *tetramers*. It should be noted that the product always contains one double bond and that all of the steps are reversible. This process can repeat itself until products of high molecular weight are obtained. The reaction is strongly exothermic because at each step a π bond is replaced by a carbon-carbon bond with the liberation of about 25 kcal. of energy. The products may be considered to be made up of many small parts and are called *polymers* (Gr. *polys* many, *meros* part). The process by which they are formed is called *polymerization.* If, as with the alkenes, polymerization takes place by addition, it is known as *addition polymerization,* and the product is an *addition polymer.* Although the acid catalyst was symbolized by the general formula HB, and although phosphoric or sulfuric acid is the usual catalyst, Lewis acids such as boron fluoride or aluminum chloride also initiate polymerization.

$$F_3B + CH_2{=}CHCH_3 \rightleftharpoons F_3B{:}CH_2{-}\overset{+}{C}HCH_3 \xrightleftharpoons{x\,CH_2=CHCH_3} \text{Polymer}$$

The acid-catalyzed polymerization of alkenes is important commercially for the production of alkenes of medium molecular weight and of polymers of high molecular weight. **Propylene trimer** and **propylene tetramer** have been the chief intermediates in the manufacture of synthetic detergents (pp. 406, 444). The polymers result when propylene is passed over a catalyst consisting of a film of phosphoric acid on quartz granules. If isobutylene is passed into cold 60 per cent sulfuric acid and the solution is heated to 100°, a mixture of dimers and trimers (about 4:1), together with smaller amounts of higher polymers, is formed. The mixture of dimers is known as **diisobutylene** and consists of four parts of 2,4,4-trimethyl-1-pentene and one part of 2,4,4,-trimethyl-2-pentene.

$$2\,(CH_3)_2C{=}CH_2 \xrightleftharpoons{H_2SO_4} \underset{\text{20 per cent}}{(CH_3)_3CCH{=}C(CH_3)_2} \quad \text{and} \quad \underset{\text{80 per cent}}{(CH_3)_3CCH_2C(CH_3){=}CH_2}$$

Catalytic hydrogenation of the mixed diisobutylene yields **2,2,4-trimethylpentane,** the standard 100-octane motor fuel (p. 94).

The trimers and higher polymers are formed by reaction of the dimers with more isobutylene. If boron fluoride or anhydrous aluminum chloride is used as a catalyst at −100°, **polyisobutylenes** having from 400 to 8000 C_4H_8 units and varying from sticky viscous resins to elastic rubber-like solids are obtained.

$$x\,(CH_3)_2C{=}CH_2 \xrightleftharpoons[-100^\circ]{AlCl_3} [-C(CH_3)_2CH_2-]_x$$

The acid-catalyzed polymerization of olefins gives as products only viscous semisolids of high molecular weight. In 1938 a patent was issued in Great Britain for the production of a solid polymer of ethylene. The polymerization is carried out above 100° and at pressures above 15,000 p.s.i. in the presence of 0.01 per cent of oxygen. Molecular oxygen has two unpaired electrons of like spin and acts as a free radical initiator. Free radicals generated by the thermal decomposition of organic peroxides (p. 86) also may be used.

The free radical, RO·, starts a chain reaction (p. 39), which is terminated by the union with any other free radical, $Z\cdot$, with which the growing chain happens to collide in the proper way. Usually $Z\cdot$ is another growing chain.

$$RO\cdot + CH_2{=}CH_2 \longrightarrow RO{-}CH_2{-}\dot{C}H_2 \qquad \textit{Initiation}$$

$$ROCH_2CH_2\cdot + x\,CH_2{=}CH_2 \longrightarrow RO(CH_2CH_2)_xCH_2CH_2\cdot \qquad \textit{Propagation}$$

$$RO(CH_2CH_2)_xCH_2CH_2\cdot + Z\cdot \longrightarrow RO(CH_2CH_2)_xCH_2CH_2{-}Z \qquad \textit{Termination}$$

The product, called **polyethylene,** is a linear polymer but contains about one methyl branch for every 8 to 10 methylene (CH_2) groups. The branching results from the occasional 1,2 shift of a hydrogen atom during the polymerization.

$$\underset{\displaystyle H}{RCHCH_2\cdot} \longrightarrow \underset{\displaystyle CH_3}{RCH\cdot}$$

Various catalysts, such as the so-called Ziegler catalysts, for example titanium tetrachloride and an alkylaluminum (p. 113), have been developed that not only permit polymerization to take place at essentially atmospheric pressure but also reduce the amount of branching and yield a denser and more rigid polymer called *high density polyethylene.*

The polyethylenes are *thermoplastic;* that is, they soften and flow when heated because the molecules can slip past each other. In this condition they can be extruded into sheets and various shapes, which solidify when cooled. Because polyethylenes are saturated hydrocarbons, they are very inert, and their high molecular weight makes them insoluble in most solvents. During 1964, United States production reached 2.6 billion pounds. Polyethylenes are used chiefly in the manufacture of thin film for wrapping and bags, of insulation for wire and cable, of plastic pipe, and of household utensils and toys.

In the polymerization of substituted olefins such as propylene, the usual catalysts give a random spatial arrangement of the polymeric units as in (*a*) with the result that the attractive forces between the chains are not strong, and the products are oils and sticky semisolids. The catalysts that produce high density polyethylene (p. 88), however, give an ordered arrangement of the polymeric units as in (*b*) with the result that a useful product is obtained. Polymers with a random arrangement of units are said to be *atactic* (Gr. *an* not,

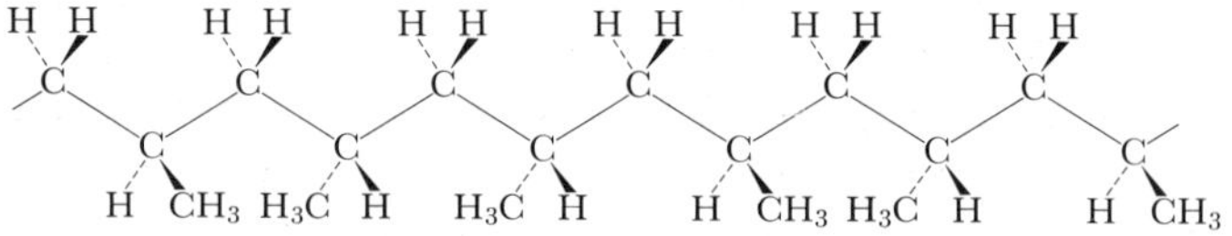

H H　H H　H H　H H　H H　H H
C　C　C　C　C　C
C　C　C　C　C　C
H CH3　H3C H　H3C H　H CH3　H3C H　H CH3

(*a*) Section of atactic polypropylene chain.

H H　H H　H H　H H　H H　H H
C　C　C　C　C　C
C　C　C　C　C　C
H CH3　H CH3　H CH3　H CH3　H CH3　H CH3

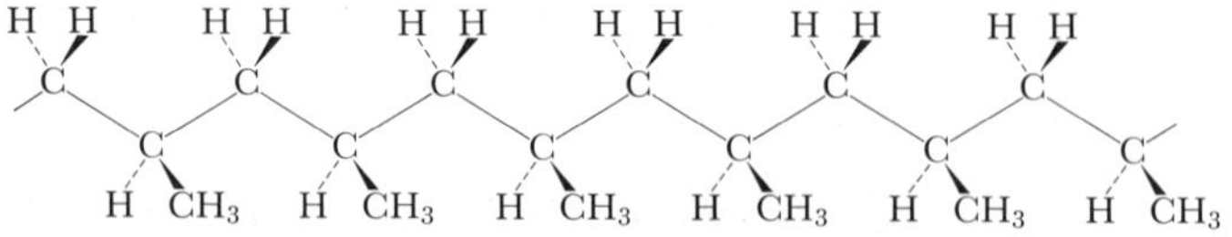

(*b*) Section of isotactic polypropylene chain.

taktika arrangement), whereas those with an ordered arrangement are said to be *isotactic.*

Catalysts capable of producing isotactic polymers are said to be *stereoselective* (p. 113) (commonly, but less properly called *stereospecific*).

Polypropylene produced by means of stereoselective catalysts rapidly is becoming of importance not only for film and molded articles but also for synthetic fiber. Production in the United States amounted to 270 million pounds in 1964. Certain synthetic rubbers such as butyl rubber (IIR, p. 589) and ethylene-propylene rubber (*EPT*, p. 589) are predominantly alkene polymers.

Isomerization

At 25° in the absence of catalysts, the four butenes do not undergo detectable interconversion. Yet the available thermodynamic data indicate that the standard free energies of formation of 2-methylpropene, *trans*-2-butene, *cis*-2-butene, and 1-butene are 13.88, 15.05, 15.74, and 17.09 kcal. From these values the calculated percentages at equilibrium at 25° (p. 27) should be 84.3, 11.7, 3.6, and 0.4. As with *n*-butane and *i*-butane (p. 69), various catalysts have been found that bring about interconversion of the isomers. They do not necessarily bring about thermodynamic equilibrium, however. Thus acids readily catalyze the interconversion of 1-butene and the *cis* and *trans* 2-butenes at 25° to give approximately the expected equilibrium ratio of 2.6:22.9:74.5, but no 2-methylpropene, the most stable isomer, is formed. It is postulated that in the isomerization a carbonium ion is an intermediate.[7]

$$CH_3CH_2CH{=}CH_2 \underset{B^-}{\overset{HB}{\rightleftharpoons}} CH_3CH_2\overset{+}{C}HCH_3 \underset{HB}{\overset{B^-}{\rightleftharpoons}} CH_3CH{=}CHCH_3$$

The base removes a proton only from carbon atoms adjacent to the carbonium carbon because it is only these hydrogens that are made sufficiently acidic by the electron-attracting effect of the positive charge. Since the base can remove a proton from either side of the carbonium carbon, equilibrium between 1-butene and the 2-butenes is established. Moreover rotation about the bond between the methylene carbon and the planar (sp^2 hybridized) carbonium ion is not restricted, and the equilibrium ratio of *cis* and *trans* forms (23.5:76.5) should result.

Oxidation

Ozonolysis. If a stream of ozonized air or ozonized oxygen is passed through a liquid olefin or a nonaqueous solution of an unsaturated compound, the ozone is removed rapidly and quantitatively with the formation of a product that usually is called an *ozonide*. The course of the reaction and the structure of the ozonide vary with the structure of the alkene and the nature of the solvent. The ozonide may be one of several types of cyclic or polymeric peroxides, in all of which both the π bond and the σ bond have been broken with formation of bonds between carbon and oxygen. They are represented here by the noncommittal formula, $R_2C(O_3)CR_2$. The ozonides of low molecular weight are very unstable. It is reported that a sample of ethylene ozonide exploded when poured from one vessel to another. Many polymeric peroxides are viscous, explosive oils. On the other hand, some ozonides are stable crystalline solids or distillable liquids.[8]

The importance of the reaction is that reduction of the products, for example with zinc

---

[7] The convention in regard to reactants that appear above and below the arrows is discussed on page 87.

[8] Ozone was discovered by Schoenbein (1799–1868) in 1840, and he reported its reaction with ethylene in 1855. Little further study was made of the reaction of ozone with organic compounds until 1903 when Carl Dietrich Harries (1866–1923), professor at the University of Kiel, began his extensive work on the ozonation of unsaturated compounds. Not until 1953 was any real progress made in understanding the course of the reaction.

and acetic acid or with hydrogen and platinum, gives two molecules of aldehydes or ketones (p. 193). In these molecules, oxygen is doubly bound to the carbon atoms that originally were joined by the carbon-carbon double bond.

$$R_2C{=}CR_2 + O_3 \longrightarrow R_2C(O_3)CR_2 \xrightarrow[\text{or } H_2/Pt]{Zn,\ H^+} R_2C{=}O + O{=}CR_2 + H_2O$$

The over-all process starting with the unsaturated compound is known as *ozonolysis* because the molecule is split by means of ozone.

Ozonolysis of unsaturated compounds is a useful method for making aldehydes and ketones. Moreover, since the aldehydes and ketones can be isolated and analyzed, the reaction can be used to locate the position of the double bond in organic molecules. For example, the structures of the three butenes can be determined by ozonolysis. 1-Butene yields a one-carbon aldehyde and a three-carbon aldehyde; 2-butene yields two moles of a two-carbon aldehyde; and methylpropene (*isobutylene*) yields a one-carbon aldehyde and a three-carbon ketone.

$$CH_3CH_2CH{=}CH_2 \xrightarrow{O_3} [CH_3CH_2CH(O_3)CH_2] \xrightarrow{H_2/Pt} CH_3CH_2CH{=}O + O{=}CH_2 + H_2O$$

$$CH_3CH{=}CH{-}CH_3 \xrightarrow{O_3} [CH_3CH(O_3)CHCH_3] \xrightarrow{H_2/Pt} CH_3CH{=}O + O{=}CHCH_3 + H_2O$$

$$(CH_3)_2C{=}CH_2 \xrightarrow{O_3} [(CH_3)_2C(O_3)CH_2] \xrightarrow{H_2/Pt} (CH_3)_2C{=}O + O{=}CH_2 + H_2O$$

Oxidation by Aqueous Potassium Permanganate. Strong oxidizing agents such as permanganate ion and chromic acid also attack the double bond, but they are not as selective as ozone in that they may oxidize other groups in the molecule as well as the double bond (pp. 132, 205). Aqueous potassium permanganate frequently is used to test for the presence of a double bond in the absence of other easily oxidized groups because the reaction takes place rapidly at room temperature and can be detected visually. When a dilute solution of potassium permanganate acidified with sulfuric acid is shaken with an unsaturated compound, the solution is decolorized, the inorganic product being manganous sulfate, $MnSO_4$. If a neutral solution of permanganate is used, the purple color disappears and a brown precipitate of hydrated manganese dioxide is formed. If the permanganate solution is alkaline, reaction with the double bond converts the purple permanganate ion, MnO_4^-, to the green manganate ion, MnO_4^{2-}. The reaction with permanganate usually is referred to as *Baeyer's test for a double bond.* It should be pointed out, however, that like the reaction with bromine and with hydrogen, *this test is not specific for a double bond.* Reaction with permanganate indicates a double bond only in the absence of other easily oxidized groups.

The organic products vary with the conditions of the reaction. The initial step is believed to be the *cis* addition of permanganate ion, Mn(VII), to the double bond to give a cyclic ester (p. 126) of a manganese(V) acid, H_3MnO_4. When dilute permanganate and strongly alkaline conditions (pH 12–13) are used, the ester reacts with hydroxide ion to give a practically quantitative yield of a dihydroxy compound known as a 1,2 diol or glycol (p. 602).

$$\text{RHC{=}CHR} + Mn^{VII}O_4^- \longrightarrow \text{cyclic ester } (Mn^{V}) \xrightarrow{2\ ^-OH} \underset{\text{OH}}{\text{RCH}}{-}\underset{\text{OH}}{\text{CHR}} + MnO_4^{3-}$$

As fast as the manganese(V) ion is formed, it is oxidized by permanganate in the alkaline solution to manganate ion.

$$MnO_4^{3-} + MnO_4^- \longrightarrow 2\,MnO_4^{2-}$$

All of the reactions of the alkenes that have been discussed are dependent on the presence of the double bond. The rest of the molecule is inert like the paraffin hydrocarbons. *A reactive group such as the double bond is called a* **function** or a **functional group.** Most of the homologous series have their characteristic functional groups. The methods of introducing this group are the methods of preparation for members of the series, and the effects of other agents on this group are the general reactions of the series. The number of functional groups is small in comparison to the number of organic compounds. By emphasizing the functional groups of each homologous series, the facts of organic chemistry are systematized and made easier to remember.

CYCLIC HYDROCARBONS

In addition to the alkenes, a second series of hydrocarbons that also has the general formula C_nH_{2n} is known. The lowest member of the series has three carbon atoms. The only way in which three carbon atoms can be joined other than in a linear fashion is in the form of a closed triangle as in (*a*). Each of the six unshared electrons on the carbon atoms

(*a*) (*b*) (*c*)

can bond with a hydrogen atom to give structure (*b*), or as usually written (*c*). This molecule has the molecular formula C_3H_6 and is isomeric with propylene. No double bond is present, however, which is consistent with the fact that the compound does not decolorize alkaline permanganate solution.

If four, five, six, or *n* carbon atoms are joined in similar fashion, the homologous series C_4H_8, C_5H_{10}, C_6H_{12}, C_nH_{2n} results. Carbon atoms joined in this way are thought of as forming a closed ring and compounds having this feature are called *cyclic compounds.* Because the members of this series are saturated and hence resemble the alkanes, they are called *cyclanes.* Another name is *alicyclic hydrocarbons,* that is, cyclic compounds having aliphatic properties. Petroleum chemists frequently call them *naphthenes* because they occur in the naphtha fraction of petroleum and are isomeric with the alkenes. Names for the individual compounds are formed by adding the prefix *cyclo* to the name of the saturated hydrocarbon having the same number of carbon atoms. Thus C_3H_6 is cyclopropane; C_4H_8, cyclobutane; C_5H_{10}, cyclopentane; C_6H_{12}, cyclohexane; and C_7H_{14}, cycloheptane.

Each of the cyclanes can give rise to a separate homologous series if hydrogen atoms are replaced by alkyl groups. Thus methylcyclopropane, ethylcyclopropane, several dimethylcyclopropanes, and higher homologs are known. The cyclic atoms are numbered consecutively around the ring to distinguish among the various possible isomers.

Methylcyclopropane 1,1-Dimethylcyclopropane 1,2-Dimethylcyclopropane

Of the various types of cyclanes, only those having five and six carbon atoms in the rings are abundant. Consideration of those having fewer or more carbon atoms and of the derivatives of cyclanes is deferred until Chapter 38.

The cyclopentane ring is less stable at room temperature than the cyclohexane ring (p. 639), and enlargement of carbon-substituted cyclopentane rings to cyclohexane rings can take place. Thus methylcyclopentane isomerizes in the presence of anhydrous aluminum chloride to an equilibrium mixture containing 80 per cent of cyclohexane. 1,2-Dimethylcyclopentane isomerizes to a mixture containing 97 per cent of methylcyclohexane.

H_2C—$CHCH_3$, H_2C, CH_2, C, H_2 $\underset{}{\overset{AlCl_3}{\rightleftharpoons}}$ H_2, C, H_2C, CH_2, H_2C, CH_2, C, H_2

Methylcyclopentane — Cyclohexane

H_2C_5—$_1CHCH_3$, H_2C_4, $_2CHCH_3$, C_3, H_2 $\overset{AlCl_3}{\rightleftharpoons}$ H_2, C_6, H_2C_5, $_1CHCH_3$, H_2C_4, $_2CH_2$, C_3, H_2

1,2-Dimethylcyclopentane — Methylcyclohexane

Just as alkenes of the type RCH=CHR can have like groups on the same side of the plane of the bond or on opposite sides and give rise to *cis* and *trans* geometric isomers (p. 76), so also can substituted cyclanes have like groups on the same side or on opposite sides of the plane of the ring to give *cis-trans* isomers.

H_3C, CH_3, H, H — *cis*-1,4-Dimethylcyclohexane

H_3C, H, H, CH_3 — *trans*-1,4-Dimethylcyclohexane

Alicyclic hydrocarbons having double bonds in the ring also are known. The compound with one double bond in a six-membered ring is called cyclohexene. Those with two double bonds are called 1,3-cyclohexadiene and 1,4-cyclohexadiene.

H_2, C, H_2C, CH, H_2C, CH, C, H_2 — Cyclohexene

H, C_1, H_2C_6, $_2CH$, H_2C_5, $_3CH$, C_4, H — 1,3-Cyclohexadiene

H_2, C_6, HC_5, $_1CH$, HC_4, $_2CH$, C_3, H_2 — 1,4-Cyclohexadiene

These compounds undergo all of the addition reactions of the alkenes (p. 80).

When a third double bond is present in a six-membered ring, the properties differ radically from those of either the cyclanes or the cyclenes. The members of this series are called *aromatic hydrocarbons.* The parent compound is benzene, C_6H_6. The lower homologs are toluene, $C_6H_5CH_3$, ethylbenzene, $C_6H_5C_2H_5$, and three dimethylbenzenes, known as the xylenes.

```
      H                 H                  H
      C                 C                  C
  HC     CH         HC     CCH3        HC     CC2H5
  HC     CH         HC     CH          HC     CH
      C                 C                  C
      H                 H                  H
   Benzene           Toluene          Ethylbenzene
```

```
      CH3                    CH3                    CH3
      C1                     C                      C
  HC6    2CCH3           HC     CH              HC     CH
  HC5    3CH             HC     CCH3            HC     CH
      C4                     C                      C
      H                      H                      CH3
1,2-Dimethylbenzene    1,3-Dimethylbenzene    1,4-Dimethylbenzene
  (ortho xylene)         (meta xylene)          (para xylene)
```

The special properties of the aromatic hydrocarbons are discussed in Chapter 20. It is sufficient here to state that when three double bonds alternate with three single bonds in a six-membered ring, the unsaturation electrons are almost as unreactive as the electrons in single bonds. For example, benzene does not decolorize a solution of bromine and does not react with aqueous permanganate solution.

Aromatic hydrocarbons occur in coal tar (p. 390) and petroleum (p. 94). They can be prepared by dehydrogenating cyclohexanes over a platinum catalyst or by cyclization of alkanes to cyclanes, followed by dehydrogenation.

```
       H2                          H
       C                           C
  H2C     CH2      Pt         HC      CH
  H2C     CH2   <====>        HC      CH     + 3 H2
       C           500°            C
       H2                          H
   Cyclohexane                  Benzene
```

```
     CH3                       CH3                     CH3
     CH2                       CH                      C
 H2C    CH3               H2C     CH2             HC      CH
 H2C    CH2    <===>  H2 + H2C     CH2    <===>    HC      CH    + 3 H2
     CH2                       C                       C
                               H2                      H
   Heptane             Methylcyclohexane            Toluene
```

NATURAL GAS AND PETROLEUM

Natural gas and petroleum are the chief sources of hydrocarbons. **Natural gas** varies greatly in composition. Unprocessed gases contain 60 to 80 per cent methane, 5 to 9 per cent ethane, 3 to 18 per cent propane, and 2 to 14 per cent higher hydrocarbons. Most of the natural gas is used for fuel, although an increasing amount is being used as raw material for the synthesis of organic compounds (p. 71). Partial combustion of natural gas yields **carbon blacks,** about 95 per cent of which are used as reinforcing agents for rubber and synthetic rubbers (p. 585). U.S. production in 1964 was over 2 billion pounds. Reaction of natural gas with steam and with air over suitable catalysts is the chief method for the production of synthesis gas, the hydrogen-nitrogen mixture used for the synthesis of ammonia.

Petroleum is a complex liquid mixture of organic compounds. The chief components are hydrocarbons, which may be aliphatic, alicyclic (p. 91), or aromatic (p. 92) in varying proportions. In addition to carbon and hydrogen, petroleum contains 1 to 6 per cent of sulfur, nitrogen, and oxygen present as organic compounds.

The separation of the individual components is very difficult, but between 1927 and 1960, at the U.S. Bureau of Standards and at the Carnegie Institute of Technology, 145 pure hydrocarbons were isolated from the gas, gasoline, and kerosene fractions of a midcontinent petroleum. Of the 145 hydrocarbons, 49 were paraffinic, 48 were alicyclic, and 48 were aromatic. It is estimated that the portion of petroleum boiling up to 200° contains at least 500 compounds.

Petroleum is *refined,* that is, separated into useful products, by distillation into fractions of different boiling ranges, conversion of the less desirable components into more valuable products, and treatment of fractions in various ways to remove undesirable components. The **liquefied petroleum gases** (*LPG*), methane, propane, and butanes, are used chiefly as fuel. The **petroleum ethers, ligroins, naphthas,** and **mineral paint spirits** boil from 30° to 140° and are used as solvents. **Gasoline** is the fraction boiling between 0° and 200°. Qualities other than volatility also are important, especially the *octane number.* Everyone is familiar with *knocking* in an automobile gasoline engine, the *ping* that develops when pulling up a long grade or attempting to accelerate a car too rapidly. Knocking varies with the fuel and depends on the structures of the hydrocarbon molecules.

As a means of measuring the knocking properties of a fuel, two pure hydrocarbons were selected as standards. One is *n*-heptane, which is worse in its tendency to cause knocking than any ordinary gasoline, and the other is 2,2,4-trimethylpentane, which was better than any gasoline known at the time. Blends of these two pure hydrocarbons can be made to match the knocking characteristics of ordinary fuels. The knocking property of the fuel then is described by its **octane number,** the per cent of 2,2,4-trimethylpentane in the synthetic blend that matches the gasoline in knocking properties. Investigation of a large number of pure hydrocarbons has shown that in general the octane number increases with an increase in branching, and that olefins and aromatic hydrocarbons are better than saturated hydrocarbons.

Besides the fraction obtained from the first distillation of petroleum, referred to as *straight-run* gasoline, large amounts of gasoline are obtained by **cracking** the higher-boiling fractions. At first, cracking was purely thermal, but as new cracking units were built, purely thermal cracking was replaced by *catalytic cracking* in which the oil vapors are heated in the presence of aluminum silicate. Not only is the yield of gasoline increased by cracking; the octane number of straight-run gasoline may be 50–55, that of thermally cracked gasoline 70–72, and of catalytically cracked gasoline 77–81. More recently the trend is to crack in the presence of hydrogen and special catalysts to yield a saturated product, free of oxygen, sulfur, and nitrogen.

$$R-R + H_2 \longrightarrow 2\,RH$$
$$ROH + H_2 \longrightarrow RH + H_2O$$
$$R_2S + 2\,H_2 \longrightarrow 2\,RH + H_2S$$
$$R_2NH + 2\,H_2 \longrightarrow 2\,RH + NH_3$$

Moreover, because no coke is produced and because the density of hydrocarbons decreases with decreasing molecular weight, up to 115 barrels of products result from 100 barrels of feed stock. Although the octane number of gasoline produced in this way is low, it can be raised to as high as 100 by **reforming.** In this process gasolines are heated in the presence of platinum catalysts. The acyclic and alicyclic hydrocarbons are converted by cyclization and dehydrogenation to aromatic hydrocarbons (p. 93). Hydrogen from the reforming operation is used in the cracking process.

Octane number is improved also by the addition of *tetraethyllead,* $Pb(C_2H_5)_4$ (p. 113). One of the disadvantages of tetraethyllead when used alone is that the product of combustion is lead oxide, which is reduced to lead and causes pitting of the cylinder walls. If ethylene halides are added with the tetraethyllead, lead halides are formed, which are more resistant to reduction and are volatile at the temperature of combustion. Commercial *Ethyl Fluid,* which is added to gasoline, consists of approximately 59 per cent tetraethyllead, 13 per cent ethylene bromide, 24 per cent ethylene chloride, and 4 per cent kerosene and dye. *Tetramethyllead* is more effective than tetraethyllead in gasolines having a high aromatic content.

In the cracking operations considerable amounts of gaseous hydrocarbons are produced. The olefins may be converted by *polymerization* (p. 86) to higher-boiling olefins that can be used as gasoline, or they may be used to synthesize other organic chemicals. The saturated compounds may be *dehydrogenated* to olefins or dienes (p. 587).

Two other operations are important in the efficient utilization of the C_4 fraction from the cracking operations. Despite the usual lack of reactivity of alkanes, isobutane has been added to butenes to give a mixture of highly branched octanes, a process known in the industry as **alkylation.** The reaction is thermodynamically feasible only at relatively low temperatures and is carried out below 20° in the presence of 85 to 100 per cent sulfuric acid or of anhydrous hydrogen fluoride. The initial steps are the same as those in the acid-catalyzed polymerization (p. 86), namely the formation of a carbonium ion and addition to a second molecule of olefin to give the dimeric ion.

$$(CH_3)_2C{=}CH_2 \xrightarrow{H_2SO_4} HSO_4^- + (CH_3)_3C^+ \xrightarrow{H_2C{=}C(CH_3)_2} (CH_3)_3C{-}CH_2{-}\overset{+}{C}(CH_3)_2$$

In the presence of isobutane, however, the dimeric ion does not lose a proton to give the unsaturated dimers nor does it undergo further polymerization. Instead the alkane rapidly transfers a hydride ion, H:⁻, from a tertiary carbon atom to the dimeric ion to give the saturated addition product with regeneration of a *t*-butyl carbonium ion.

$$(CH_3)_3C{-}CH_2{-}\overset{+}{C}(CH_3)_2 + (H{:})\,C(CH_3)_3 \longrightarrow \underset{\text{2,2,4-Trimethylpentane}}{(CH_3)_3CCH_2CH(CH_3)_2} + {}^{+}C(CH_3)_3$$

Although 1-butene or the 2-butenes also can react, *n*-butane cannot replace isobutane. The reason is that only the *t*-butyl carbonium ion is sufficiently stabilized by the electron-donating properties of enough methyl groups (p. 83) to permit transfer of a hydrogen with a pair of electrons from the alkane to the dimeric ion. In practice the mixed butenes are used, and the product is a mixture of isomeric octanes that has an octane number of 92 to 94. Alkylation has the advantage over polymerization in that saturated as well as unsaturated gases are used, and the product is saturated.

The second operation of importance is the **isomerization** of alkanes. *n*-Butane can be isomerized to the more valuable *i*-butane (p. 69) for use in the alkylation reaction. A liquid catalyst consisting of anhydrous aluminum chloride, hydrogen chloride, and some unsaturated hydrocarbon, or a solid catalyst of aluminum chloride on alumina is used. Isomerization also converts straight-chain hydrocarbons in gasolines to branched-chain hydrocarbons and thereby improves the octane number of the gasoline.

Other commercial fractions of petroleum are **kerosene, turbine fuel,** and **jet fuel** (b.p. 175–275°), **gas oil, fuel oil,** and **diesel oil** (b.p. 250–400°), and **lubricating oil.** A desirable characteristic of diesel fuel is a low spontaneous ignition temperature. Straight-chain paraffin hydrocarbons are superior to branched-chain hydrocarbons and aromatics. Cetane (*n-hexadecane*), $CH_3(CH_2)_{14}CH_3$, which ignites rapidly, is rated as 100 and α-methylnaphthalene (p. 491), which ignites slowly, is rated as zero. The **cetane number** of a fuel is the per cent of cetane in a cetane-α-methylnaphthalene mixture that has the same ignition qualities as the fuel. Lubricating oil is purified by distillation at reduced pressure and by extraction with solvents to remove undesirable components. It then is mixed with various *additives* to improve the lubricating properties of the oil. Polyisobutylene (p. 87) may be added to improve the viscosity index, that is, to decrease the change in viscosity with change in temperature. **Paraffin** is the mixture of solid saturated hydrocarbons that crystallizes from various high-boiling fractions. **Greases** are made by dispersing metallic soaps (p. 191) in hot lubricating oils. **Pitch** and **asphalt** are residual products used as protective coatings and as binding agents for fibers and crushed rock. If these residues are distilled to dryness, **petroleum coke** is obtained, which can be calcined to a practically pure carbon that is valuable for the manufacture of carbon electrodes.

The conversion of coal, lignite, or other carbonaceous material into liquid fuels has been of interest to countries that do not have petroleum resources. Germany was the pioneer in this development, but other countries have built experimental plants as a safeguard against the day when petroleum supplies have become exhausted. The *Bergius process* is carried out in two stages. In the liquid phase treatment, powdered coal is mixed with heavy oil residues and a small amount of catalyst, presumably iron oxide, and converted to heavy and middle oils at 450° and hydrogen pressures of 10,000 p.s.i. In the second stage, the middle oils are vaporized and passed with hydrogen over a fixed catalyst. The products are separated by distillation.

Processes based on the *Fischer-Tropsch synthesis* of liquid fuels have been investigated extensively. Coal or other carbonaceous material is converted into carbon monoxide and hydrogen by the water-gas reaction.

$$C + H_2O \longrightarrow CO + H_2$$

The mixture of carbon monoxide and hydrogen is enriched with hydrogen and passed over a cobalt-thoria catalyst on a suitable support at 200–250° and atmospheric pressure.

$$n\,CO + (2n+1)H_2 \longrightarrow C_nH_{2n+2} + n\,H_2O$$

The gasoline fraction must be reformed to improve the octane number.

PROBLEMS

6–*1*. Write condensed structural formulas and give a suitable name to the members of the following groups of compounds: (*a*) all of the isomeric pentenes; (*b*) all of the octenes with a central double bond.

6–*2*. Write structural formulas for the following compounds: (*a*) 2,2-dimethyl-3-hexene, (*b*) *s*-butylethylene, (*c*) *sym*-ethyl-*t*-butylethylene, (*d*) 3-(1-methylpropyl)-2-heptene, (*e*) (*sym*-dimethyl)ethylethylene.

6–3. Name the following compounds by the international system and as derivatives of ethylene:

(*a*) $(CH_3)_3CCH{=}CHCH_3$ (*b*) $CH_3CH_2CH(CH_3)CH{=}CHC_2H_5$

(*c*) $(CH_3)_2CHC{=}C(CH_3)_2$ with C_2H_5 on the first doubly bonded carbon:

$$\begin{array}{l} (CH_3)_2CH\underset{\displaystyle C_2H_5}{\underset{|}{C}}{=}C(CH_3)_2 \end{array}$$

(*d*) $CH_3CH_2C{=}C(CH_2)_4CH_3$ with CH_3 on each doubly bonded carbon:

$$CH_3CH_2\underset{\displaystyle CH_3}{\underset{|}{C}}{=}\underset{\displaystyle CH_3}{\underset{|}{C}}(CH_2)_4CH_3$$

6–4. Indicate the objections to the following names and give suitable ones: (*a*) 2-ethyl-3-hexene, (*b*) 3-methyl-4-heptene, (*c*) methyl-*t*-butylethylene, (*d*) 2,3-dimethylpentene-2, (*e*) (1-methylethyl)ethylene.

6–5. Write equations for the following reactions and give a suitable name to the organic product: (*a*) chlorine and 2-pentene; (*b*) sulfuric acid and 1-hexene; (*c*) hydrogen chloride and 2-methyl-1-pentene; (*d*) hydrogen, platinum, and 3-ethyl-1-pentene; (*e*) hydrogen bromide, peroxides, and isobutylethylene; (*f*) borane and *s*-butylethylene.

6–6. Give reactions illustrating the preparation of the following compounds from the proper alkene: (*a*) 2,3-dimethylpentane, (*b*) 2-chloro-2-methylbutane, (*c*) 1,1,2-trimethylpropyl hydrogen sulfate, (*d*) 2,3-dibromo-2-methylhexane, (*e*) 2-methylpentyl bromide.

6–7. Deduce the structure of the olefin that gives the following products on ozonolysis: (*a*) a one-carbon aldehyde and a three-carbon ketone, (*b*) a two-carbon aldehyde and a four-carbon ketone, (*c*) two moles of a four-carbon ketone, (*d*) a two-carbon aldehyde and a four-carbon aldehyde having an isopropyl group, (*e*) a three-carbon ketone and a four-carbon ketone.

6–8. What are the names of the olefins that give on ozonolysis: (*a*) $(CH_3)_2CHCHO$ and $CH_3COC_2H_5$, (*b*) $(CH_3)_3CCHO$ and CH_2O, (*c*) only CH_3CH_2CHO, and (*d*) $(CH_3)_2CO$ and $(CH_3)_2CHCOCH_3$.

6–9. Write the possible structural formulas for an olefin, C_8H_{16}, that gives two different ketones on ozonolysis.

6–10. Pair the following terms with the related phrases: (*a*) homologs, (*b*) free radical, (*c*) carbonium ion, (*d*) isomers, (*e*) carbanion, (*f*) polymers, (*g*) ionization potential; (*1*) repeating units, (*2*) loss of an electron, (*3*) unpaired electron, (*4*) differ by CH_2, (*5*) negative charge on carbon, (*6*) same molecular formula, (*7*) positive charge on carbon.

6–11. Without referring to Table 6–3, place the following compounds in the order of decreasing ionization potential: (*a*) 1-butene, (*b*) *n*-butane, (*c*) propane, (*d*) 2-methyl-2-butene, (*e*) 2-methyl-1-butene.

6–12. Place the following compounds in the order of increasing ease of addition of sulfuric acid to the double bond: (*a*) propylene, (*b*) 2-butene, (*c*) 2,2-dimethyl-2-butene, (*d*) ethylene, (*e*) 2-methyl-2-butene.

6–13. Arrange the following compounds in the order of increasing octane number: (*a*) 2,2-dimethylpentane, (*b*) *n*-heptane, (*c*) 2,2,3-trimethylbutane, and (*d*) isoheptane.

6–14. How many grams of bromine will react with 35 g. of 2-pentene?

6–15. If 1.0 g. of an olefin decolorized 14 cc. of a standard solution of bromine in carbon tetrachloride containing 20 g. of bromine per 100 cc. of solution, calculate the equivalent weight of the compound.

6–16. If 8.9 g. of a pure olefin having one double bond decolorized 14.5 g. of bromine, calculate the molecular formula of the olefin.

6–17. When 0.235 g. of a pure hydrocarbon having one or more double bonds was shaken with hydrogen in the presence of a platinum catalyst, 168 cc. of hydrogen (S.T.P.) was absorbed. If 90 ± 8 was the molecular weight, calculate the number of double bonds in the compound and its molecular formula.

6–18. When 4.2 g. of a mixture of pentenes and pentanes was titrated with a standard bromine solution, 5.76 g. of bromine was absorbed. Calculate the per cent of pentenes in the mixture.

6–19. If 10 g. of a mixture of hexenes and octanes, when shaken with hydrogen and a platinum catalyst, absorbs 1280 cc. of hydrogen (S.T.P.), what is the per cent of hexenes in the mixture?

6–*20*. What volume of hydrogen (S.T.P.) would be required to prepare one liter of 2,2,4-trimethylpentane (sp. gr. 0.69) from mixed diisobutylenes?

6–*21*. Write the separate steps of the mechanism for the acid-catalyzed conversion of (*a*) propylene to *i*-propyl ether, and (*b*) isobutylene to *t*-butyl hydroperoxide.

6–*22*. Write structural formulas and give names for the chief compounds present in propylene tetramer.

6–*23*. Show how 2,2,3-, 2,2,4-, and 2,3,4-trimethylpentane can arise from *t*-butyl cation, 2-butene, and *i*-butane.

CHAPTER SEVEN

ALKYL HALIDES. ORGANOMETALLIC COMPOUNDS

ALKYL HALIDES

Alkyl halides are important intermediates for the synthesis of other compounds. Not only do they undergo a variety of reactions, but they are prepared readily in good yields.

Structure and Nomenclature

From the analysis for the elements and from molecular weight determinations, alkyl halides are known to have the general formula $C_nH_{2n+1}X$. Hence the only possible structure that obeys the rules of valence is one in which a halogen atom replaces a hydrogen atom of an alkane.

Alkyl halides usually are named as if they were salts of alkyl groups. Thus CH_3Cl is called methyl chloride and $(CH_3)_3CI$ is *t*-butyl iodide. They may be considered also as halogen derivatives of saturated hydrocarbons and often are so named. Ethyl chloride may be called chloroethane and isobutyl bromide may be called 1-bromo-2-methylpropane. The alkyl halides are classed as primary, secondary, and tertiary. In *primary* halides the halogen is attached to a primary carbon atom, that is, a carbon atom united directly to only *one* other carbon atom. *Secondary* halides have the halogen bonded to a secondary carbon atom, that is, one joined to *two* other carbon atoms. In *tertiary* halides the halogen is attached to a tertiary carbon atom, that is, one united directly to *three* other carbon atoms. Of the above named compounds ethyl chloride and isobutyl bromide are primary halides and *t*-butyl iodide is a tertiary halide. Isopropyl bromide (*2-bromopropane*) is a secondary halide. Methyl halides ordinarily are grouped with the primary halides although they clearly belong to a separate class.

Preparation

Methyl and ethyl chloride are made commercially by the chlorination of methane and ethane (p. 68). Ethyl chloride and secondary and tertiary alkyl halides can be obtained by the addition of the halogen acids to alkenes (p. 84). Iodides can be obtained by the reaction of alkyl sulfates or alkyl sodium sulfates with concentrated aqueous sodium iodide.

$$R_2SO_4 + 2\,NaI \longrightarrow 2\,RI + Na_2SO_4$$

Primary alkyl chlorides or bromides react with sodium iodide in acetone solution to give primary iodides (*Finkelstein reaction*).

$$RCH_2Cl + NaI \longrightarrow RCH_2I + NaCl$$

This reaction goes to completion because sodium chloride and sodium bromide, in contrast to sodium iodide, are insoluble in acetone, and the equilibrium is shifted to the right. The

TABLE 7–*1*. COMPARISON OF THE BOILING POINTS OF *n*-ALKANES WITH THOSE OF *n*-ALKYL HALIDES HAVING APPROXIMATELY THE SAME NUMBER OF ELECTRONS

ALKANES			FLUORIDES			CHLORIDES			BROMIDES			IODIDES		
	ΣZ*	B.P.		ΣZ	B.P.		ΣZ	B.P.		ΣZ	B.P.		ΣZ	B.P.
C_2	18	−89	C_1	18	−78									
C_3	26	−45	C_2	26	−32	C_1	26	−24						
C_4	34	+0.6	C_3	34	− 3	C_2	34	+13						
C_5	42	36	C_4	42	34	C_3	42	46	C_1	44	5			
C_6	50	69	C_5	50	62	C_4	50	78	C_2	52	38			
C_7	58	98	C_6	58	93	C_5	58	108	C_3	60	71	C_1	62	42
C_8	66	126	C_7	66	120	C_6	66	133	C_4	68	102	C_2	70	75
C_9	74	150	C_8	74	142	C_7	74	157	C_5	76	130	C_3	78	103
C_{10}	82	174				C_8	82	183	C_6	84	156	C_4	86	130
C_{11}	90	194				C_9	90	202	C_7	92	178	C_5	94	157
C_{12}	98	214				C_{10}	98	223	C_8	100	202	C_6	102	180

* ΣZ = sum of the atomic numbers of the atoms = total number of electrons outside of the nuclei.

reaction of secondary and tertiary chlorides and bromides is too slow to be practical. The procedure most generally used to prepare alkyl halides, however, is by the reaction of alcohols with halogen acids or with inorganic acid halides (p. 127).

Physical Properties

In physical properties the alkyl halides resemble the alkanes. They are insoluble in water and are good solvents for many organic compounds. Because of the sizable dipole moment of the alkyl halides ($\sim$2 D), they might be expected to boil higher than hydrocarbons that have the same number of electrons.[1] Actually the fluorides boil at approximately the same temperature as the alkanes, and the chlorides only around 10° higher (Table 7–*1*). Bromides boil lower than chlorides, and iodides boil lower than bromides for compounds expected to have the same polarizability. It is possible that the marked increase in size of the bromine and iodine have much the same effect as branching of the chain in decreasing the van der Waals forces between molecules.

Alkyl halides are more dense than hydrocarbons, the order being hydrocarbons $<$ alkyl chlorides $<$ bromides $<$ iodides, if compounds with the same number of carbon atoms are compared. The density of hydrocarbons increases with increasing length of chain because of the increasing van der Waals forces, but because carbon atoms are less dense than elements of higher atomic number, increasing the size of the alkyl group in alkyl halides causes a decrease in density. Halogenated alkanes that have more than around 65 per cent halogen by weight are heavier than water.

Physiological Action

Volatile halogenated hydrocarbons when breathed at high concentration cause general anesthesia (footnote 4, p. 142). Halogenated hydrocarbons in general are toxic compounds. Prolonged exposure to even low concentrations leads to fatty degeneration of the liver. The acute toxicity of carbon tetrachloride is said to be as high as that of hydrogen sulfide or hydrogen cyanide.

[1] Comparison is made of compounds having approximately the same number of electrons because they might be expected to have approximately the same polarizability (p. 22), and hence, in the absence of other factors, the same attractive forces between molecules.

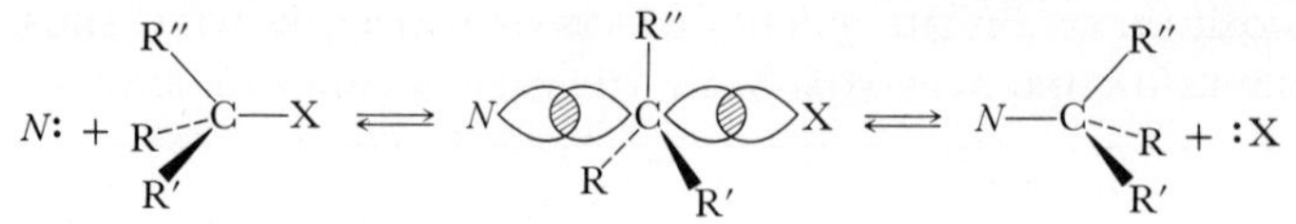

Figure 7–*1*. Progress of an S_N2 reaction.

Reactions

Alkyl halides are dipolar molecules, the carbon-halogen bond being the bond most polarized. The negative end of the dipole is on halogen and the positive end on carbon (p. 18). This condition has two effects on reactivity: (*1*) the electronegativity makes it relatively easy for the halogen to leave as a halide ion, and (*2*) the positive charge on the α carbon tends to pull electrons away from adjacent β carbon atoms,[2] which in turn increases the ease of removal of a proton from a β carbon atom. The first effect permits displacement of halide ion by other nucleophiles (p. 41). Two kinetically distinguishable mechanisms operate. One gives rise to second order kinetics; that is, the rate is proportional to the concentration of both the nucleophile and the substrate (p. 30). This type is designated as S_N2 (*substitution, nucleophilic, bimolecular*). The other shows first order kinetics, the rate being dependent only on the concentration of the substrate. It is referred to as an S_N1 reaction (*substitution, nucleophilic, unimolecular*). The second effect leads to the elimination of halogen acids by means of bases. Here again both first order and second order kinetics are observed, the mechanisms for which are designated as *E*1 and *E*2 (p. 104).

S_N2 Displacement Reaction. In the so-called S_N2 mechanism, the nucleophile *N*: approaches the molecule from the side opposite the leaving group, in this case the halide ion. The reason that attack from the back is more effective than from any other direction is that it yields the best transition state for the reaction. The sp^3-hybridized orbitals of the carbon atom are suitable for the formation of bonds with one other atom but not with two. If the orbitals are sp^2-hybridized, however, a *p* orbital remains that has lobes of equal size on opposite sides of the plane of the three sp^2 orbitals (p. 11). One lobe then can overlap with the orbital of the nucleophile and the other with that of the leaving group to give a transition state in which both the nucleophile and the leaving group are bonded partially to the carbon (Fig. 7–*1*). The transition state can either revert to the reactants or lose the halide ion to give the products. In the latter event the process resembles an umbrella being turned inside out and is known as a *Walden inversion.*[3]

Because the nucleophile may be any ion or neutral molecule that has an unshared pair of electrons, the displacement of halogen can give rise to numerous types of organic compounds as the following examples indicate.

$$HO:^- + R:X \longrightarrow HO:R + :X^-$$
An alcohol

$$H_2N:^- + R:X \longrightarrow H_2N:R + :X^-$$
An amine

$$H_3N: + R:X \longrightarrow [H_3\overset{+}{N}:R]:X^-$$
An alkylammonium salt

---

[2] The position of carbon atoms with respect to a particular group, in this instance halogen, is designated by lower-case letters of the Greek alphabet, e.g., $\underset{\delta}{C}—\underset{\gamma}{C}—\underset{\beta}{C}—\underset{\alpha}{C}—X$.

[3] Paul Walden (1863–1957), Russian-born and educated German physical chemist, whose principal work was in the field of electrolytes. He was awarded life pensions by each of three governments but at the end of World War II, at the age of 82, found it necessary to support himself.

$$:N{\equiv}C:^- + R:X \longrightarrow N{\equiv}C:R + :X^-$$

An alkyl cyanide

$$HS:^- + R:X \longrightarrow HS:R + :X^-$$

A thiol
(thioalcohol, mercaptan)

$$I:^- + R:X \longrightarrow I:R + :X^-$$

An alkyl
iodide

It should be noted that neutral molecules as well as negative ions can react as nucleophiles. Neutral molecules yield positive ions, whereas negative ions yield neutral molecules. Other displacement reactions that yield ethers (p. 139), esters (p. 166), sulfides (p. 261), and thiocyanates (p. 293), as well as other types of compounds, will be discussed as the occasion arises.

Reactivity in S_N2 Displacement Reactions. The *extent* to which a reaction takes place depends only on the change in free energy for the reaction (p. 27), which in general parallels the heat of the reaction. The latter can be estimated from the sum of the bond dissociation energies of the bonds that are formed and the sum of the bond dissociation energies of the bonds being broken. If bond dissociation energies are not known, a less accurate estimate may be made from the difference between the sum of the empirical bond energies of the products and that of the reactants (Table 2–2, p. 19). The *rates* of reactions, however, depend on the size of the activation energy, that is, on the ease of formation of the transition state (p. 34), which in turn depends on a number of factors. For the reaction

$$N: + S:L \longrightarrow N:S + :L$$

it depends on *(1)* the ability of the nucleophile, $N:$, to supply a pair of electrons for the new bond, which is called its **nucleophilicity** (p. 41); *(2)* the ability of the atom attacked by the nucleophile, the substrate atom *S*, to accept a pair of electrons, which may be termed its **electrophilicity** (p. 41) towards *N*; *(3)* the attraction of the substrate atom for the pair of electrons of the leaving group, *L*, which may be called its electrophilicity towards *L*; and *(4)* the nucleophilicity of the leaving group. The greater the first two factors, the easier it is to form the transition state and the lower the activation energy for the reaction, whereas the greater the last two factors, the more difficult it is to form the transition state and the higher the activation energy. These four properties are those of the ground state and influence each other in the transition state. Thus as *N* approaches *S*, the electrophilicity of *S* towards *N* increases and the nucleophilicity of *N* towards *S* decreases. Simultaneously the electrophilicity of *S* towards *L* decreases and the nucleophilicity of *L* towards *S* increases.

Another factor that influences the activation energy is the **steric condition in the transition state.** Any structural features that lead to greater crowding in the transition state, that is, greater repulsion among the electrons of the bonding orbitals, increases the activation energy, whereas reduction of repulsive forces decreases the activation energy. Cyclic planar structures that permit the ready transfer of electrons in the transition state also may decrease the activation energy.

Finally, the nucleophiles either are dipolar liquids, or they must be dissolved in a dipolar solvent that is also a solvent for the substrate. These dipolar solvents are capable of solvating the nucleophile, the substrate, the transition state, and the leaving group. Since **solvation** always involves the liberation of energy, solvation always increases stability. Thus the nucleophilicity of *N* and the electrophilicity of *S* decrease with increasing solvation, which leads to an increase in activation energy. Solvation of the transition state, however, increases the reactivity because the solvated transition state is at a lower energy level, and the activation energy necessary to reach it is less than that to reach the unsolvated state. Solvation of the leaving group L decreases its nucleophilicity and makes the reaction energetically more favorable.

Solvents may be classified into two main types: those that have hydrogen on oxygen or nitrogen and hence are capable of hydrogen bonding (p. 23) are called *protic solvents;* those that lack hydrogen attached to oxygen or nitrogen are called *aprotic solvents.* Each class may be subdivided further according to the dielectric constant, ϵ, of the solvent. Dielectric constants of liquids at room temperature (20–25°) range from about 2 to 115 (Table 7–2). Solvents that have dielectric constants below 10 arbitrarily will be called *low dielectric* or *LD solvents* and those with dielectric constants above 20 will be called *high dielectric* or *HD solvents.* Thus water, the lower alcohols, liquid ammonia, and concentrated sulfuric acid are protic high dielectric solvents, and acetic acid is a protic low dielectric solvent.

$$\underset{\text{Acetic acid}}{CH_3—\overset{}{\underset{\underset{O}{\|}}{C}}—O—H} \qquad \underset{\textit{N,N}\text{-Dimethylformamide}}{H—\underset{\underset{O}{\|}}{C}—\underset{\underset{CH_3}{|}}{N}—CH_3} \qquad \underset{\text{Methyl sulfoxide}}{CH_3—\underset{\underset{O^-}{|^+}}{S}—CH_3}$$

N,N-Dimethylformamide (*DMF*) and methyl sulfoxide (more frequently called dimethyl sulfoxide, *DMSO*) are used widely as aprotic high dielectric solvents. Other useful aprotic HD solvents are acetone, acetonitrile, nitrobenzene, and sulfolane. Hexane, benzene, and carbon tetrachloride are aprotic low dielectric solvents. For reactions involving polar molecules and ions, solvents of high dielectric constant are most useful because they are good insulators of electric charge (p. 25). The following discussion refers only to such solvents.

Solvation of nucleophiles by protic solvents results from the attraction between the positive hydrogen of solvent molecules and the unshared electrons on the nucleophile (Fig. 2–*17c*, p. 25). In aprotic solvents the attraction between a dipolar solvent molecule and the nucleophile is due merely to the association of the dipole with the dipole of the nucleophile and is much weaker than that between nucleophile and a protic molecule. Since the nucleophile must lose some of its solvent sphere before the transition state can be formed, and since the aprotic solvent molecules are held less strongly than protic solvent molecules, the activation energy is less in aprotic solvents and the rate is much faster than in protic solvents. For example, the rate of displacement of iodide ion by chloride ion is over a million times faster in dimethylformamide than in methanol.

For a given aprotic solvent, the attraction between solvent molecules and nucleophile increases with the polarizability of the nucleophile and hence with its size. Thus the relative reactivity of the nucleophile, which is referred to as its nucleophilicity, decreases from top to bottom in a given column of the periodic table. For example, in acetone the order

TABLE 7–2. DIELECTRIC CONSTANTS OF SOME COMMON SOLVENTS

COMPOUND	ϵ at (t°)	COMPOUND	ϵ at (t°)
Acetic acid (p. 168)	6 (20)	*n*-Hexane	2 (20)
Acetone (p. 216)	21 (25)	Hydrogen cyanide	115 (20)
Acetonitrile (p. 240)	38 (20)	Methyl sulfoxide (p. 270)	45 (25)
Ammonia	22 (−33)	Nitrobenzene (p. 416)	35 (25)
Benzene (p. 368)	2 (20)	Sulfolane (p. 507)	44 (25)
Carbon tetrachloride (p. 591)	2 (20)	Sulfur dioxide	14 (20)
Dimethylformamide (p. 235)	38 (25)	Sulfuric acid	100 (25)
Ethyl alcohol (p. 120)	24 (25)	Water	78 (25)

of nucleophilicity is $Cl^- > Br^- > I^-$. Since formation of the transition state requires partial bonding with a pair of electrons, the reactivity should increase also with increasing localization of negative charge, which leads to the same order. The latter effect is less important, however, because the solvent shell must be broken before bonding with the substrate can take place.

Solvation of the leaving group is determined by the same factors. Here, however, a decrease in concentration of negative charge and an increase in polarizability assist rather than retard formation of the transition state. Hence the order of reactivity of alkyl halides is iodides > bromides > chlorides. It should be noted, however, that for the leaving group, localization of charge is more important than polarizability, since the bond to the substrate must be broken before solvation becomes important.

The third participant in the equation for an S_N2 reaction is the substrate, *S*. For an alkyl halide, it is a saturated carbon atom joined to three other groups that are tetrahedrally distributed with respect to each other and to the leaving group, *L*. The transition state requires the accommodation of five groups (Fig. 7–*1*, p. 100). The smaller the size of the three groups R, R′, and R″ that must remain within bonding distance to carbon, the easier the transition state is formed and the more reactive the substrate. For this reason the order of reactivity for S_N2 reactions of alkyl halides is methyl > primary > secondary > tertiary. This order holds if all of the groups replacing the hydrogens of methyl are methyl groups. If any are further substituted, especially at the β position, it becomes more difficult to accommodate them in the transition state, and the reactivity is decreased. Thus ethyl bromide may be twice as reactive as *n*-propyl bromide, 100 times more reactive than *i*-butyl bromide, and 100,000 times more reactive than neopentyl bromide, although all are primary halides. The presence on the substrate carbon atom of groups that are electron-attracting or electron-repelling makes the carbon atom either more or less electrophilic towards both the nucleophile and the leaving group and affects the rate of reaction.

Most of the earlier work on the determination of rates of reaction and equilibrium constants for S_N2-type reactions has been carried out in aqueous or alcoholic solution, or in water-organic solvent mixtures that have a sufficient concentration of water to make them essentially protic solvents. Under these conditions, the *order of nucleophilicity* of some common nucleophiles for a variety of substrates *is* HS^-, $CN^- > I^- > SCN^- > NH_3 > HO^- > N_3^- > Br^- > Cl^- > F^-$, $NO_3^- > H_2O$. Although any explanation of the relative order of reactivity in protic solvents of all the nucleophiles in this series would be open to question, certain trends may be noted. Since protic solvents are held most strongly by the smallest ions and since solvation outweighs concentration of negative charge, the order of nucleophilicity for halide ions is $I^- > Br^- > Cl^- > F^-$, the reverse of that for aprotic solvents. Similarly HS^- is more nucleophilic than HO^-. For elements in a given period, the size of the molecules is approximately the same, and solvation has about the same effect. Here the relative reactivity depends on the availability of a pair of electrons for bonding, that is, on the basicity. Thus $NH_3 > H_2O > HF$ and $H_2N^- > HO^- > F^-$.

As would be expected, steric effects in the transition states are the same in protic as in aprotic solvents. Similarly the dependence of reactivity on the leaving group is the same in both solvents. Although solvation is in the opposite order to that in aprotic solvents, its effect is outweighed by the effect of concentration of negative charge, since the group must leave with the pair of electrons before it can be heavily solvated. Protic solvents frequently still are used for organic reactions because the aprotic solvents that are available may not dissolve one or more of the reactants.

S_N1 Displacement Reactions. In the second type of nucleophilic displacement, the rate constant is first order. It is independent of the concentration of the nucleophile and is dependent only on the concentration of the substrate. Moreover in certain reactions the initial rate is independent of the nature of the nucleophile. To explain these and other facts it is postulated that the rate-controlling step is the formation of a carbonium ion intermediate. The carbonium ion then reacts rapidly with the nucleophile.[4]

$$R_3C{:}X \underset{}{\overset{\text{Slow step}}{\rightleftharpoons}} R_3C^+ + {:}X^-$$

$$R_3C^+ + {:}N^- \underset{}{\overset{\text{Fast step}}{\rightleftharpoons}} R_3C{:}N$$

In the S_N1 reactions of alkyl halides, both the intermediate carbonium ion and the leaving group are charged, and solvation plays an even more important role than in S_N2 reactions. It should be noted that the carbon of the carbonium ion is sp^2-hybridized and that the carbonium ion is planar. Attack by the nucleophile may be from either side; that is, with or without Walden inversion.

Reactivity by S_N1 mechanisms is dependent primarily on the substrate. In the transition state the C—X bond is stretched and the R groups have become more nearly planar with the central carbon atom, which has acquired a greater positive charge than in the original polar molecule. Since in forming the transition state the molecule is on the way towards forming the carbonium ion, it is assumed that whatever stabilizes the carbonium ion also stabilizes the transition state and increases the rate of reaction. Replacement of the three hydrogens of a methyl cation successively by three alkyl groups leads to increasing stability of the carbonium ion. Two factors are responsible for this effect. One is that the planar carbonium ion with 120° bond angles can accommodate the larger groups better than can the tetrahedral carbon with 109.5° bond angles; that is, there is less steric interference of the groups attached to the central carbon in the planar carbonium ion, which is more important to large alkyl groups than to small hydrogen atoms. The second factor is electronic in that the greater polarizability of the alkyl groups can neutralize the positive charge better (distribute it more widely) than can the less polarizable hydrogen atoms. The result is that the reactivity for an S_N1 mechanism is tertiary halides > secondary > primary > methyl. The order of reactivity of leaving groups is the same as that for S_N2 reactions, namely iodides > bromides > chlorides > fluorides.

Since for the S_N2 mechanism the order of reactivity is methyl > primary > secondary > tertiary, the S_N2 mechanism predominates for methyl and primary halides, and the S_N1 mechanism is favored for tertiary halides. Many intermediate cases exist, however, where it is reasonable to assume that the mechanism is neither purely S_N2 nor S_N1 but something in between. The energy profile for the S_N2 mechanism (Fig. 7–2) may be represented by a curve with a single peak at the half-way point, which is the transition state for the reaction. The curve for the S_N1 mechanism has a dip at the half-way point where the carbonium ion exists, and two peaks at the two transition states. The mechanism for intermediate cases may be thought of as having energy profiles in which the dip for the dotted portion of the curve is something between zero for the S_N2 mechanism and the maximum for the S_N1 mechanism.

---

[4] Considerable controversy has arisen over whether ionization actually precedes reaction, but much experimental evidence favors ionization as a first step. That simple alkyl cations are stable entities under certain conditions is indicated by studies of the electrical conductivity of aluminum chloride in ethyl chloride (cf. p. 69). More recently a variety of properties show that alkyl fluorides in antimony pentafluoride are converted completely to stable tertiary alkyl hexafluoroantimonates.

$$RF + SbF_5 \longrightarrow R^{+-}SbF_6$$

In particular, the tertiary alkyl cations are characterized by strong down-field shifts in the proton magnetic resonance spectra (p. 549) owing to deshielding by the positively charged sp^2-hybridized carbonium carbon.

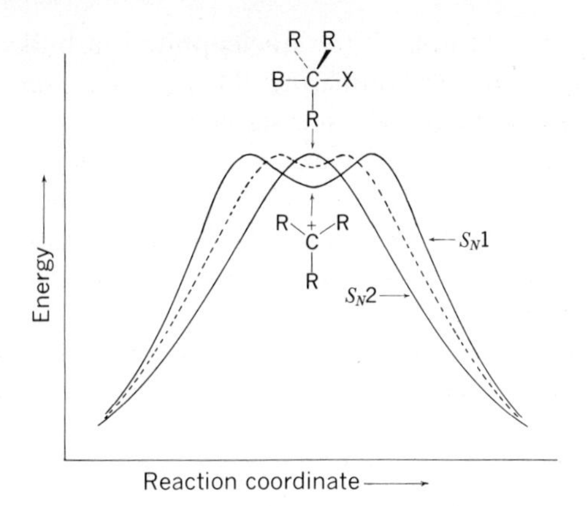

Figure 7–2. Section of the energy profiles for S_N1 and S_N2 reactions in the region of the transition states.

Elimination Reactions. Thus far only displacement reactions have been discussed. The second type of reaction, namely the removal of a proton from the β carbon atom (p. 100), leads to the elimination of halogen acid. Reactions in which two groups are eliminated from adjacent carbons are called *beta eliminations*. As with displacement reactions, two principal types are recognized. $E\,2$ reactions show second order kinetics, whereas $E\,1$ reactions show first order kinetics. As with displacement reactions, the $E\,2$ reactions are assumed to proceed by a concerted mechanism with a single transition state. Elimination takes place readily only when the molecule is in the conformation such that the groups being eliminated are *trans* to each other and planar with the two carbon atoms, a condition in which the electronic repulsions in the transition state are at a minimum.

$$\mathrm{B:^-} + \mathrm{R_2C(H)\!-\!CR_2(X)} \rightleftharpoons \mathrm{R_2C(H\cdots B)\!=\!=\!CR_2(\cdots X)} \rightleftharpoons \mathrm{B\!:\!H} + \mathrm{R_2C\!=\!CR_2} + \mathrm{:X^-}$$

The $E\,1$ reactions are assumed to be two-stage reactions with carbonium ions as intermediates.

$$\underset{\;\;\mathrm{R}\qquad\mathrm{R}}{\mathrm{H\!:\!CH\!-\!CH\!:\!X}} \xrightleftharpoons{\text{Slow step}} \underset{\;\;\mathrm{R}\qquad\mathrm{R}}{\mathrm{H\!:\!CH\!-\!\overset{+}{C}H}} + \mathrm{:X^-}$$

$$\mathrm{B\!:^-} + \underset{\;\;\mathrm{R}\qquad\mathrm{R}}{\mathrm{H\!:\!CH\!-\!\overset{+}{C}H}} \xrightleftharpoons{\text{Fast step}} \mathrm{B\!:\!H} + \underset{\mathrm{R}\qquad\mathrm{R}}{\mathrm{CH\!=\!CH}}$$

Because elimination reactions take place under neutral or alkaline conditions, no rearrangement takes place, in contrast to the acid-catalyzed dehydration of 1-alkanols (p. 130). If the halogen is terminal, the 1-alkene is formed. If the halogen is not terminal, two products may be possible. Since in the transition state the double bond is partly formed, it would be expected that in the absence of other factors the more stable olefin would be the one that would be obtained in larger amount. The more highly substituted olefin is usually the more stable (p. 89), and the order of removal of β hydrogens usually is tertiary in preference to secondary and secondary in preference to primary **(Zaĭtsev rule[5]).** Other factors, however,

---

[5] Alexander Zaĭtsev (1841–1910), student of Butlerov and successor to Markovnikov at the University of Kazan.

may change this order. For example, if the nucleophile is a bulky group, then one of the less hindered, more acidic, and more numerous β hydrogens (*primary* > *secondary* > *tertiary*) may be removed more rapidly to give the less stable olefin.

The factors that govern the reactivity of the *leaving group* in $E2$ reactions operate in the same way as for S_N2 reactions, and those for $E1$ reactions in the same way as for S_N1 reactions. For both $E2$ and $E1$ the reactivity of the *nucleophile* increases with increasing basicity because its function is to remove a proton. Nucleophiles are more basic and hence more reactive in aprotic than in protic solvents because solvation is less in the former. As would be expected, the rate of an $E2$ reaction is more dependent on the basicity of the nucleophile than is that of the S_N2 reaction. Hence the greater the basicity, the higher the ratio of elimination to displacement.

The rates of $E2$ reactions decrease with increasing branching of the alkyl group but not to the same extent as for S_N2 reactions. Thus the rate for the $E2$ reaction of *i*-propyl bromide with ethoxide ion is one twentieth of that for *n*-propyl bromide whereas the S_N2 reaction for *i*-propyl bromide is only one thousandth of that for *n*-propyl bromide.

Since the same reagent can bring about both nucleophilic displacement and elimination, the two reactions are competitive, and the relative amounts of displacement product and olefin depend on the relative rates of the two reactions. The S_N2 reaction for *n*-propyl bromide is ten times faster than the $E2$ reaction, whereas the S_N2 reaction for *i*-propyl bromide is only one fourth that for the $E2$ reaction. The result is that *n*-propyl bromide gives about 91 per cent displacement and 9 per cent propene, whereas *i*-propyl bromide gives 20 per cent displacement and 80 per cent propene. For tertiary halides the elimination predominates to such an extent that even mildly basic nucleophiles such as cyanide ion give only olefin and no displacement product.

In the above discussion the assumption has been made that in the S_N1 and $E1$ mechanisms the carbonium ion intermediate is entirely free from the influence of the anion with which it was associated originally. If such were the case the relative amounts of displacement and elimination should be independent of the leaving group. This is approximately so in aqueous solvents, but in ethyl alcohol or in acetic acid, the ratio of elimination to substitution decreases in the order $Cl > Br > I > \overset{+}{S}(CH_3)_2$. Since there are strong arguments for ionization preceding both S_N1 and $E1$ reactions, the variation indicates that the intermediate for the $E1$ reaction is an ion pair. By the principle of microscopic reversibility (p. 38), these results support the view that an ion pair is an intermediate in the *cis* addition of hydrogen halides to olefins (p. 84).

Solvolysis. Because even mildly basic nucleophiles bring about $E1$ elimination, S_N1 reactions of tertiary halides take place only with neutral or acidic nucleophiles, and even then only in protic solvents. Thus fluoride ion is a sufficiently strong base in aprotic solvents to bring about even $E2$ eliminations. The protic solvents themselves, however, are sufficiently good nucleophiles to complete S_N1 reactions of tertiary halides.

$$RX \rightleftharpoons R^+ + X^-$$

$$R^+ + H_2O \rightleftharpoons R\overset{+}{O}H_2 \overset{H_2O}{\rightleftharpoons} ROH + H_3O^+$$

Reaction with the solvent is called *solvolysis,* and the term solvolysis or solvolytic reaction is used almost synonymously with S_N1 reaction.

Alkyl Halides of Commercial Importance

It is evident from the variety of compounds obtainable from alkyl halides that they are extremely important to the organic chemist. Only a few of the simple halides are of industrial importance, however. In the laboratory the chemist usually uses bromides or iodides because of their greater reactivity, but industrially, the chlorides are used because of their lower cost. Many of the polyhalogen compounds are commercially important, and they are discussed in Chapter 34. **Methyl chloride,** b.p. −24°, is made by heating methyl alcohol with hydrogen chloride and a catalyst (p. 128). It is formed also as one product of the reaction of chlorine with methane (p. 68). Around 134 million pounds was consumed in the United States in 1964, of which about 70 per cent was used for the synthesis of silicones (p. 278), 15 per cent as a solvent for the production of butyl rubber (p. 589), 10 per cent for the synthesis of tetramethyllead (p. 113), and 5 per cent each for the manufacture of methylcellulose (p. 344), and of miscellaneous chemicals. The price was around 8 cents per pound. **Methyl bromide,** b.p. 4.5°, is made from methyl alcohol and hydrogen bromide. Production was 17 million pounds and the price 42 cents per pound. It is highly toxic and is used as a poison gas for rodent control. It is toxic to other forms of life also and is used as a soil and grain fumigant. Its use in fire extinguishers probably depends on dissociation into free radicals, which act as chain breakers (p. 39) and inhibit the propagation of flame. Because of its high toxicity to human beings, it should be used for this purpose only with proper precautions to prevent inhalation of the vapors. The same precautions should be taken when using methyl bromide in chemical reactions. **Ethyl chloride,** b.p. 13°, is made from ethyl alcohol and hydrochloric acid in the presence of zinc chloride, from ethylene and hydrogen chloride by passing an equimolar mixture into a solution of anhydrous aluminum chloride in liquid ethyl chloride (p. 84), or from ethane and chlorine (p. 68). Over 666 million pounds was produced in 1964, and sold for 7 cents per pound. It is used chiefly for the manufacture of tetraethyllead (p. 113) and of ethylcellulose (p. 344). Physicians carry a small cylinder of ethyl chloride in their medical kits for use as a quickly acting general anesthetic (footnote 4, p. 142) for minor operations. Evaporation of the liquid sprayed on the skin also produces local anesthesia by cooling. The **amyl chlorides** are made by the chlorination of *n*-pentane and of *i*-pentane (p. 68). The mixed monochlorides are hydrolyzed to the mixed alcohols (*Pentasol*). Reaction with ammonia gives amylamines, and with sodium hydrosulfide, amyl mercaptans and amyl sulfides.

ORGANOMETALLIC COMPOUNDS

Organometallic compounds are those compounds in which carbon is linked directly to a metallic element. Many organometallic compounds are highly reactive and useful in the synthesis of other organic compounds. Others have uses as antiseptics, medicinals, fungicides, and antiknock compounds. In general they are made either directly or indirectly from organic halides.

Of the alkali metals, the lithium compounds are most useful, with sodium compounds being next in importance. By far the most important organometallic compounds are the magnesium derivatives known as *Grignard reagents.* Zinc and cadmium compounds are of some interest, and certain mercury compounds are important. Of the third group of the periodic table, boron and aluminum compounds are most useful. Both tin and lead compounds are of technical importance, while arsenic, antimony, and bismuth compounds have been used as war gases and medicinals. Only alkyl derivatives are discussed in this chapter. Organometallic compounds containing functional groups and many of the reactions of organometallic compounds are considered from time to time throughout the text. Other types of organometallic compounds are discussed in Chapter 38, p. 644.

The simplest method for the preparation of organometallic compounds is the direct reaction of the metal with an alkyl halide, usually in the presence of a solvent. Since the metals have free electrons, adsorption of the alkyl halide on the metal surface can result in the simultaneous formation of the organometallic compounds and the metallic halide.

$$\begin{matrix} M\cdot \\ M\cdot \end{matrix}\left|\begin{matrix} R \\ \colon \\ X \end{matrix}\right. \longrightarrow \begin{matrix} M\colon R \text{ (or } M^+\colon R^-) \\ + \\ M^+\colon X^- \end{matrix}$$

The carbon-metal bond may be largely ionic as in alkylsodium compounds or largely covalent as in alkylmercury compounds, with all gradations between the extremes.

Grignard Reagents

Zinc or magnesium reacts with alkyl halides to give alkylzinc or alkylmagnesium compounds. The organozinc compounds found use in early synthetic work but have the disadvantage that only methyl and ethyl iodides give good yields, and that even methylzinc and ethylzinc[6] are difficult to use because they are spontaneously flammable in air. In 1899 Barbier[7] announced that in many cases the alkylzincs could be replaced by a mixture of alkyl halide and magnesium in ether (p. 140) in the presence of the compound with which the reaction was to take place. This reaction was improved by Grignard,[8] a student of Barbier's, who reported in 1900 the separate preparation of alkylmagnesium halide solutions and their reaction with a variety of compounds. Subsequent work by Grignard and a large number of other investigators has shown that the use of alkylmagnesium halide solutions, now commonly known as **Grignard reagents,** is of more practical importance for the laboratory synthesis of organic compounds than is any other single synthetic method. By the time of Grignard's death, the literature contained about 6000 articles dealing with this reagent. Only recently, however, has the reaction been used on a large scale in commercial syntheses.

Preparation. Grignard reagents are made by the direct action of alkyl halides on magnesium turnings.

$$RX + Mg \longrightarrow RMgX$$

Reaction takes place in the absence of a solvent but soon stops because the alkylmagnesium halide is a solid that is insoluble in the alkyl halide and coats the surface of the magnesium. In the presence of a solvent for the alkylmagnesium halide, the reaction goes to completion. Ethyl ether (p. 142) is the solvent most commonly used, although other ethers such as *n*-butyl ether, the dimethyl ether of diethylene glycol (*diglyme,* p. 142), or tetrahydrofuran (p. 512) may be advantageous.

Sometimes the reaction is slow to start. The drier the reagents and apparatus and the less oxide film on the surface of the magnesium, the easier the reaction starts. Usually a small crystal of iodine or a small amount of ethylene bromide is added to initiate the reaction. Alkyl iodides react more readily than bromides, and bromides than chlorides, but the yields are in the reverse order, namely chlorides > bromides > iodides. The higher the dilution with ether and the greater the purity and state of subdivision of the magnesium, the higher are the yields.

Structure and Reactions. Attempts to establish the structure of Grignard reagents have been made ever since their discovery. Despite numerous investigations since 1958 that have used the most modern techniques, the question still has not been settled completely. It appears, however, that when ethylmagnesium bromide is made in ethyl ether from the purest reagents, the only species present is solvated $C_2H_5—Mg—Br$. Certain impurities, or the products of reaction with oxygen, or other complexing agents such as dioxane, cause an equilibrium to be established that has been represented by one or the other of two series of reactions.

---

[6] Regarding nomenclature, see footnote 1, p. 138.

[7] François Philippe Antoine Barbier (1848–1922), professor at the University of Lyon, France, who is known not only for his discovery of the usefulness of magnesium in organic synthesis but also for his work on the constitution of the terpenes. In his later years he was interested in mineralogy and the analysis of minerals.

[8] Victor Grignard (1871–1935), professor at the University of Nancy and later successor to Barbier at Lyon. In 1912 he was awarded the Nobel Prize in Chemistry jointly with Paul Sabatier (p. 82).

$$2\,RMgX \rightleftharpoons R{:}Mg\begin{matrix}X\\ \\X\end{matrix}MgR \rightleftharpoons R_2Mg + MgX_2$$

or

$$2\,RMgX \rightleftharpoons \begin{matrix}R\\ \\R\end{matrix}Mg\begin{matrix}X\\ \\X\end{matrix}Mg \rightleftharpoons R_2Mg + MgX_2$$

Since the reactions of R_2Mg in the presence of MgX_2 are the same as those of RMgX, it seems desirable to continue the long-established use of the formula RMgX.

The fact that methylmagnesium is a nonvolatile solid insoluble in hydrocarbon solvents indicates that the carbon-magnesium bond is largely ionic. Accordingly it is expected that the alkyl group of Grignard reagents acts as a nucleophile. As such it attacks polarized molecules at points of low electron density. Thus the Grignard reagents evolve hydrocarbon with all compounds having **reactive hydrogen.** By reactive hydrogen the organic chemist means hydrogen that is more acidic (i.e., more polarized) than that in alkanes.

$$H\overset{\delta-}{O}{:}\overset{\delta+}{H} + \overset{\delta-}{R}{:}\overset{\delta+}{Mg}X \longrightarrow H{:}R + HO^{-+}Mg^{+-}X$$

$$H_2NH + RMgX \longrightarrow HR + H_2NMgX$$

$$XH + RMgX \longrightarrow HR + MgX_2$$

These reactions may be viewed as acid-base reactions, the stronger acid liberating the weaker acid, the alkane, from its salt. Other types of compounds that have reactive hydrogen, such as alcohols (p. 115), phenols (p. 435), thiols (p. 258), amines (p. 221), and acetylenes (p. 150), undergo analogous reactions.

This type of reaction is valuable for the estimation of groups containing replaceable hydrogen, for example hydroxyl groups. The procedure was proposed first by Chugaev[9] and further developed by his student Zerevitinov and by others. It commonly is referred to as the *Zerevitinov determination.* A simple form of the apparatus is shown in Fig. 7–*3*. A solution of methylmagnesium iodide in a

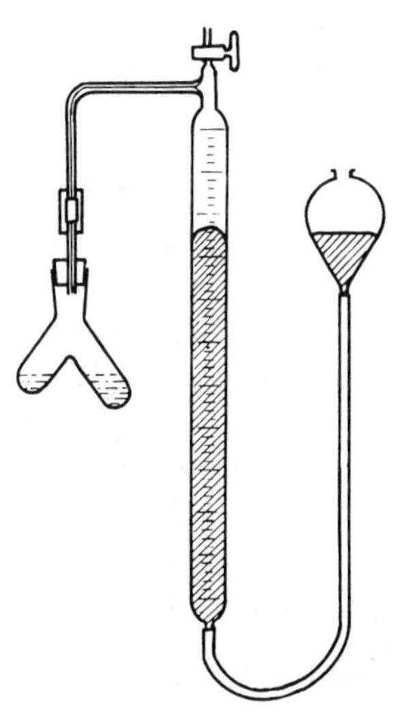

Figure 7–*3*. Zerevitinov apparatus for the determination of reactive hydrogen.

[9] Leo A. Chugaev (1872–1922), professor at the University of St. Petersburg, Russia.

high-boiling ether, such as *i*-amyl ether (p. 142), is placed in one leg of the bifurcated tube, and a weighed amount of the compound dissolved in a suitable solvent is placed in the other leg. The apparatus is brought to temperature equilibrium with the stopcock open. The cock then is closed, and the height of the mercury column is read. The two solutions are mixed, and after again reaching temperature equilibrium, the volume of methane formed in the reaction is determined. From these data and the molecular weight of the compound, the number of reactive hydrogen atoms in the molecule can be calculated. Various refinements in the above procedure have been introduced to increase the accuracy of the determination.

The concentration of Grignard solutions can be determined by adding water to a known volume of the reagent and titrating the magnesium hydroxide formed. The method is inaccurate to the extent that the reagent has reacted with air, since magnesium alkoxides also react with water to give magnesium hydroxide (p. 125).

Grignard reagents also react with **halogen.**

$$X{:}X + R{:}MgX \longrightarrow X{:}R + MgX_2$$

This reaction can be used to determine the concentration of Grignard solutions by titration with standard iodine solution.

Oxygen with its two unpaired electrons reacts rapidly with Grignard reagents at $-80°$ to give the halomagnesium salts of alkyl hydroperoxides.

$$:\ddot{O}{-}\ddot{O}: + R{:}MgX \longrightarrow XMg^{+-}{:}\ddot{O}{-}\ddot{O}{:}R$$

Further reaction with Grignard reagent takes place at room temperature to give the halomagnesium salt of an alcohol (p. 125), from which the alcohol may be liberated by acidification.

$$R{-}O{-}O^{-+}Mg^{+-}X + RMgX \longrightarrow 2\,R{-}O^{-+}Mg^{+-}X \xrightarrow{2\,HX} 2\,ROH + 2\,MgX_2$$

Just as magnesium reduces the halides of all metals below it in the electromotive series to the metal, alkylmagnesium halides reduce the **inorganic halides** of all elements below magnesium to the alkyl derivative of the element.

$$M^{2+} + {:}Mg \longrightarrow M{:} + Mg^{2+}$$

$$M^{2+} + 2\,R{:}Mg^{+} \longrightarrow MR_2 + 2\,Mg^{2+}$$

The following reactions are typical and usually are the best methods for preparing the indicated compounds.

$$ZnX_2,\ CdX_2,\text{ or }HgX_2 + 2\,RMgX \longrightarrow ZnR_2,\ CdR_2,\text{ or }HgR_2 + 2\,MgX_2$$

$$BX_3\text{ or }AlX_3 + 3\,RMgX \longrightarrow BR_3\text{ or }AlX_3 + 3\,MgX_2$$

$$SiX_4,\ SnX_4,\text{ or }PbX_4 + 4\,RMgX \longrightarrow SiR_4,\ SnR_4,\text{ or }PbR_4 + 4\,MgX_2$$

$$PX_3 + 3\,RMgX \longrightarrow PR_3 + 3\,MgX_2$$

If less reagent is used than is required to react with all of the halogen, intermediate reaction products are obtained. For example, silicon tetrachloride and methylmagnesium chloride can give rise to methyltrichlorosilane, CH_3SiCl_3, dimethyldichlorosilane, $(CH_3)_2SiCl_2$, trimethylchlorosilane, $(CH_3)_3SiCl$, and tetramethylsilane, $(CH_3)_4Si$. Mercurous chloride and the chlorides of iron, copper, silver, and gold cause chiefly coupling of the alkyl groups, presumably by way of the organometallic compound.

$$2\,RMgX + CuX_2 \longrightarrow R_2Cu + 2\,MgX_2$$

$$R_2Cu \longrightarrow R{-}R + Cu{:}$$

The carbon-oxygen double bond is strongly polarized with the negative end on oxygen. Nucleophilic attack by Grignard reagents on carbon leads to addition to the double bond.

Thus they react readily with **carbon dioxide** to give the halomagnesium salt of a carboxylic acid (p. 159).

$$\overset{\delta-}{O}{=}\overset{\delta+}{C}{=}\overset{\delta-}{O} + \overset{\delta-}{R}{-}\overset{\delta+}{Mg^{+}}{}^{-}X \longrightarrow O{=}\underset{\displaystyle R}{\underset{|}{C}}{-}O^{-+}Mg^{+-}X$$

Addition of a mineral acid liberates the free carboxylic acid (p. 161).

$$RCOOMgX + HX \longrightarrow RCOOH + MgX_2$$

Simple **olefins** ordinarily do not add Grignard reagents because the double bond is not sufficiently polarized. In the presence of titanium tetrachloride, however, olefins having a terminal double bond in effect abstract HMgX from Grignard reagents to give a new Grignard reagent and olefin.

$$RCH{=}CH_2 + CH_3CH_2CH_2MgCl \xrightarrow{TiCl_4} RCH_2CH_2MgCl + CH_3CH{=}CH_2$$

The reaction probably takes place by way of an unstable organotitanium compound and amounts to an olefin exchange reaction (cf. *i*-butylaluminum, p. 112). The many important reactions of Grignard reagents with **other organic compounds** are considered as new functional groups are discussed.

Lithium and Sodium Compounds

A few of the alkyl derivatives of lithium and sodium can be made directly from alkyl halides and the metal using ether, benzene, or cyclohexane as solvent. Lithium containing 0.02 to 2.0 per cent of sodium reacts readily with *n*-butyl chloride or bromide to give good yields of ***n*-butyllithium.**

$$n\text{-}C_4H_9Br + 2\,Li \longrightarrow n\text{-}C_4H_9Li + LiBr$$

n-Butyllithium is used as a catalyst for the polymerization of 1,3-butadiene and of isoprene to synthetic rubbers (p. 588). Organosodium compounds can be prepared from alkyl halides if the sodium is highly dispersed (10–30μ) in a liquid hydrocarbon.

$$RX + 2\,Na \longrightarrow RNa + NaX$$

Alkyl derivatives of the alkali metals can be made also by heating the alkylmercury with the alkali metal (p. 112).

$$R_2Hg + 2\,Li \longrightarrow 2\,RLi + Hg$$

The alkyl derivatives of the strongly electron-donating alkali metals undergo reactions analogous to those of the Grignard reagents but are somewhat more reactive. Thus alkyllithiums add to both double bonds of carbon dioxide to give lithium salts of ketone hydrates, which hydrolyze to ketones.

$$O{=}C{=}O + 2\,LiR \longrightarrow R{-}\underset{\displaystyle R}{\underset{|}{\overset{\displaystyle OLi}{\overset{|}{C}}}}{-}OLi \xrightarrow{H_2O} R{-}\underset{\displaystyle O}{\underset{\|}{C}}{-}R + 2\,LiOH$$

A ketone

They also undergo other addition reactions more readily than Grignard reagents (p. 197). Although alkyllithiums are soluble in liquid hydrocarbons, they are more reactive in solvating liquids such as the ethers (p. 140).

Zinc, Cadmium, and Mercury Compounds

Methylzinc and **ethylzinc** can be made by the direct reaction of methyl or ethyl iodide with a zinc-copper couple.[10] They are volatile and can be distilled from the reaction mixture.

$$2\,CH_3I + 2\,Zn(+Cu) \longrightarrow (CH_3)_2Zn + ZnI_2(+Cu)$$

Air must be excluded during preparation and manipulation because they are spontaneously flammable. The higher alkylzinc compounds are prepared best from anhydrous zinc halides and Grignard reagents. The alkylcadmium and alkylmercury compounds also are prepared best from the metal halides and Grignard reagents, although the mercury compounds can be made also from the alkyl halide and sodium amalgam.

$$2\,RX + Na_2Hg \longrightarrow R_2Hg + 2\,NaX$$

Ethylmercuric chloride, which is an important fungicide used to prevent sap stain in lumber and to disinfect seeds, is made from mercuric chloride and tetraethyllead (p. 113).

$$4\,HgCl_2 + Pb(C_2H_5)_4 \longrightarrow 4\,C_2H_5HgCl + PbCl_4$$

The ionic character of the carbon-metal bond decreases in the order $RMgX > R_2Zn > R_2Cd > R_2Hg$ and is paralleled by decreasing ease of hydrolysis and addition to carbon-oxygen double bonds, the mercury compounds being practically unreactive. A general method for the preparation of alkyl derivatives of any metal above mercury in the electromotive series consists of heating the alkylmercury, prepared from the Grignard reagent (p. 110), with the metal in a sealed tube.

$$2\,M\cdot + HgR_2 \longrightarrow 2\,MR + Hg:$$

Organomercury compounds are cleaved by halogen acids or halogen.

$$R_2Hg \xrightarrow{HX} RH + RHgX \xrightarrow{HX} RH + HgX_2$$

$$R_2Hg \xrightarrow{X_2} RX + RHgX \xrightarrow{X_2} RX + HgX_2$$

$$R_2Hg \xrightarrow{(SCN)_2} RSCN + RHgSCN \xrightarrow{(SCN)_2} RSCN + Hg(SCN)_2$$

The last two reactions are useful for the preparation of halogen compounds and thiocyanates when the mercury compound can be prepared by direct mercuration (pp. 441, 506, 513).

Boron and Aluminum Compounds

Alkylboron and alkylaluminum compounds can be made from the trihalide and Grignard reagents (p. 110). Alkylboranes can be made readily by the addition of borane to alkenes (p. 86). Primary alkylaluminums can be made from alkenes by the addition of aluminum hydride, which is formed *in situ*. Thus isobutylene reacts with hydrogen and clean finely divided aluminum at 150° and 3000 p.s.i. to give **isobutylaluminum.**

$$6\,(CH_3)_2C{=}CH_2 + 3\,H_2 + 2\,Al \longrightarrow 2\,[(CH_3)_2CHCH_2]_3Al$$

Isobutylaluminum undergoes an exchange reaction at 100° with other olefins having a terminal double bond to give other alkylaluminums.

$$Al(C_4H_9)_3 + 3\,C_2H_4 \longrightarrow Al(C_2H_5)_3 + 3\,C_4H_8$$

$$+ 3\,RCH{=}CH_2 \longrightarrow Al(CH_2CH_2R)_3 + 3\,C_4H_8$$

---

[10] A pair of metals in contact, one of which is above hydrogen and the other below hydrogen in the electromotive series, is known as a *couple*. Usually couples are more reactive than the pure metal.

Ethylaluminum cannot be made directly from ethylene, hydrogen and aluminum because at the temperature required, ethylaluminum adds to ethylene to give a mixture of the higher alkylaluminums.

$$Al(C_2H_5)_3 + 3x\ C_2H_4 \longrightarrow Al[CH_2(CH_2)_{2x}CH_3]_3$$

A cyclic process can be operated, however, in which ethylaluminum is converted to **diethylaluminum hydride** at 120°, and the latter allowed to react with ethylene at 70°.

$$4\ Al(C_2H_5)_3 + 2\ Al + 3\ H_2 \xrightarrow{120°} 6\ AlH(C_2H_5)_2$$

$$3\ AlH(C_2H_5)_2 + 3\ C_2H_4 \xrightarrow{70°} 3\ Al(C_2H_5)_3$$

Alkylaluminums are not sufficiently reactive to be as generally useful as the alkyllithiums (p. 111) or Grignard reagents. All three alkyl groups can be oxidized to alkoxyl groups, and hydrolysis of the alkoxides gives alcohols.

$$AlR_3 \xrightarrow{O_2} Al(OR)_3 \xrightarrow{H_2O} Al(OH)_3 + 3\ HOR$$

Bromine gives the primary alkyl bromide.

$$AlR_3 \xrightarrow{Br_2} 3\ RBr + AlBr_3$$

Thus a process is available for converting any olefin with a terminal double bond into a primary alcohol or halide, in contrast to the formation of secondary or tertiary alcohols or halides by the addition of water or halogen acids (pp. 100, 99).

Technically, isobutylaluminum and ethylaluminum have become of considerable importance. In the presence of titanium tetrachloride they act as stereoselective catalysts (*Ziegler catalysts*) for the polymerization of ethylene to polyethylene and of propylene to polypropylene (p. 88). Ethylaluminum is used in the synthesis of the higher alcohols (p. 122). A plant for this purpose began operation in 1962 with a capacity of 20 million pounds of ethylaluminum per year. The spontaneously flammable methylaluminum, ethylaluminum, and methylboron are used as jet fuel igniters, and when added to fuel help to prevent flame-out of jet engines.

Tin and Lead Compounds

Dibutyltin dichloride and **tributyltin chloride** are made from tin tetrachloride and butylmagnesium chloride.

$$SnCl_4 + 2\ C_4H_9MgCl \longrightarrow (C_4H_9)_2SnCl_2 + 2\ MgCl_2$$

$$(C_4H_9)_2SnCl_2 + C_4H_9MgCl \longrightarrow (C_4H_9)_3SnCl + MgCl_2$$

They are intermediates in the production of other derivatives used chiefly as stabilizers for poly(vinyl chloride) plastics (p. 593) and as fungicides and biocides in general. Hydrolysis, for example, yields **dibutyltin oxide** and **bis(tributyltin) oxide,** which are used to prevent slime in paper mills.

$$(C_4H_9)_2SnCl_2 + H_2O \longrightarrow (C_4H_9)_2SnO + 2\ HCl$$

$$2\ (C_4H_9)_3SnCl + H_2O \longrightarrow (C_4H_9)_3Sn{-}O{-}Sn(C_4H_9)_3 + 2\ HCl$$

The organometallic compounds produced in by far the largest amounts are **tetramethyllead** (TML) and **tetraethyllead** (TEL) used as knock inhibitors in motor fuels (p. 94). The principal method for the manufacture of tetraethyllead has been the reaction of sodium-lead alloy with ethyl chloride.

$$4\ C_2H_5Cl + Na_4Pb \longrightarrow Pb(C_2H_5)_4 + 4\ NaCl$$

Other processes, including electrolysis of Grignard or alkylaluminum complexes using a lead anode, also are being used.

$$4\ CH_3MgCl + 4\ e \longrightarrow 2\ Mg + 2\ MgCl_2 + 4\ CH_3{:}^- \quad \textit{Cathode process}$$

$$4\ CH_3{:}^- + Pb \longrightarrow 4\ e + (CH_3)_4Pb \quad \textit{Anode process}$$

United States production of tetraethyllead in 1964 amounted to 587 million pounds. The tetraalkylleads accounted for 15 per cent of the total consumption of lead in the United States, being second only to storage batteries, which consumed 38 per cent.

PROBLEMS

7–1. Give a skeleton structure and a suitable name for each member of the monobromopentanes.

7–2. How many primary, secondary, and tertiary alkyl bromides have the same carbon skeleton as 3-ethylhexane?

7–3. Write equations for the following reactions: (*a*) *s*-butyl bromide with silver hydroxide, (*b*) *i*-octyl chloride with sodium ethoxide, (*c*) 1-iodobutane with sodium sulfide, (*d*) *t*-pentyl bromide with magnesium in ether, (*e*) *i*-propyl iodide with sodium cyanide, (*f*) *n*-hexyl chloride with sodium iodide in acetone, (*g*) *t*-butyl chloride with water, (*h*) *n*-butyl bromide with sodium, (*i*) 2-bromopentane with sodium hydrosulfide, (*j*) ethylmercury with sodium.

7–4. List the following compounds in the order of increasing rate of reaction with aqueous silver nitrate: (*a*) $CH_3CH_2CHBrCH_3$, (*b*) $CH_3CH_2CH_2CH_2Br$, (*c*) $(CH_3)_3CBr$, (*d*) $(CH_3)_2CClCH_2CH_3$, (*e*) $CH_3CH_2CH_2CH_2Cl$.

7–5. Predict the order of increasing reactivity of the following compounds to S_N2 displacement: (*a*) $CH_3CH_2CCl(CH_3)_2$, (*b*) $CH_3(CH_2)_4Cl$, (*c*) $CH_3(CH_2)_4I$, (*d*) $(CH_3)_3CCH_2Cl$, (*e*) $CH_3CH_2CH_2CHClCH_3$.

7–6. During the preparation of a tertiary alkyl bromide from the pure alcohol, the product was washed with cold conc. sulfuric acid to remove unreacted alcohol, separated, and then washed with water and dilute carbonate solution to remove the remaining sulfuric acid, and dried over anhydrous magnesium sulfate. The product boiled over a sufficiently wide range to indicate that it was grossly impure. Explain.

7–7. Predict the direction of the dipole moment of iodine monochloride and its mode of addition to 1-butene.

7–8. What reagent can be used to convert a Grignard reagent into (*a*) an alkane, (*b*) an alcohol, (*c*) an alkyl iodide, (*d*) a carboxylic acid, (*e*) an alkylmercury, and (*f*) a trialkylphosphine?

7–9. Specify the type of mechanism for each of the following reactions: (*a*) chlorination of methane, (*b*) *n*-butyl chloride with sodium iodide, (*c*) *t*-butyl chloride with water, (*d*) *t*-butyl chloride with ammonia, (*e*) *n*-butyl bromide with sodium hydroxide to give 1-butene.

7–10. If 10 cc. of a solution of ethylmagnesium bromide reacts with 25.4 g. of iodine, what is the concentration of Grignard reagent in moles per liter?

7–11. Calculate the volume (S.T.P.) of methane that would be evolved when an excess of methylmagnesium iodide reacts with 0.1 g. of 2,3-dihydroxybutane.

7–12. If 0.1 g. of an unknown compound having a molecular weight of 62 ± 5 g. reacts with an excess of methylmagnesium iodide in the Zerevitinov apparatus to give 72 cc. of methane (S.T.P.), calculate the number of hydroxyl groups that are in the compound.

CHAPTER EIGHT

ALCOHOLS. ESTERS OF INORGANIC ACIDS. INFRARED SPECTRA

ALCOHOLS

The simplest alcohols contain oxygen and have the general empirical formula $C_nH_{2n+2}O$. The first member, methyl alcohol, has the molecular formula CH_4O. Since oxygen has six valence electrons, it requires two more to complete its valence shell. It does this by pairing two of its electrons with two electrons from one or two other atoms or groups, thus forming two covalent bonds. In the alcohol molecule, CH_4O, the oxygen atom must be joined to both carbon and hydrogen because if it shared both of the electrons with carbon, carbon would have only two valence electrons left to share with hydrogen and the formula would be CH_2O. If both electrons of oxygen were shared with hydrogen, a water molecule would result. Hence only one structural formula is possible, namely,

$$\begin{array}{c} \text{H} \\ | \\ \text{H—C—O—H} \\ | \\ \text{H} \end{array} \text{ or } CH_3OH.$$

This formula is in accord with the general methods of preparation and the reactions of methyl alcohol.

The number of possible structures for the other members of this homologous series can be predicted in much the same way as was done for the olefins, that is, from a consideration of the possible positions that the functional group, in this case the *hydroxyl group* (OH), can occupy in the carbon skeletons of the alkanes. This procedure leads to one C_2 alcohol, two C_3 alcohols, four C_4 alcohols, and eight C_5 alcohols.

Nomenclature

The word *alcohol* has been applied to the active principle of intoxicating beverages since the eighteenth century. The usually accepted derivation is from *al-kuhl,* the Arabic word for finely powdered antimony sulfide, which was used by Arabian women as a cosmetic to darken eyelids. Later the term was applied to any finely divided substance and then during the sixteenth century in the sense of "essence". Thus *alcool vini* was the essence or spirit of wine. Gradually *vini* was dropped, but it was not until the early part of the nineteenth century that the term *alcohol* came to be used generally for wine spirits. It now is used also as a family name for the homologous series of hydroxyalkanes.

Three systems of nomenclature are in general use. In the first the alkyl group attached to the hydroxyl group is named and the separate word *alcohol* is added. In the second system the higher alcohols are considered as derivatives of *carbinol,* another name for methyl alcohol.[1] The third method is the international system in which (*1*) the longest carbon chain containing the hydroxyl group determines the surname, (*2*) the ending *e* of the corresponding saturated hydrocarbon is replaced by *ol*, (*3*) the carbon chain is numbered from the end that gives the hydroxyl group the smaller number, and (*4*) the side chains are named and their positions indicated by the proper number. The following examples illustrate these various systems.

Formula	Names
CH_3OH	methyl alcohol carbinol[2] methanol[2]
C_2H_5OH	ethyl alcohol[3] methylcarbinol[3] ethanol
$CH_3CH_2CH_2OH$	normal propyl alcohol (*n-propyl alcohol*) ethylcarbinol 1-propanol
CH_3CHCH_3 $\quad\,\vert$ $\quad OH$	isopropyl alcohol (*i-propyl alcohol*) dimethylcarbinol 2-propanol
$CH_3CH_2CH_2CH_2OH$	normal butyl alcohol (*n-butyl alcohol*) *n*-propylcarbinol 1-butanol
$CH_3CH_2CHCH_3$ $\qquad\quad\vert$ $\qquad\quad OH$	secondary butyl alcohol (*s-butyl alcohol*) methylethylcarbinol 2-butanol
CH_3CHCH_2OH $\quad\,\vert$ $\quad CH_3$	isobutyl alcohol (*i-butyl alcohol*) *i*-propylcarbinol 2-methyl-1-propanol
$\qquad CH_3$ $\qquad\,\vert$ $CH_3—C—OH$ $\qquad\,\vert$ $\qquad CH_3$	tertiary butyl alcohol (*t-butyl alcohol*) trimethylcarbinol 2-methyl-2-propanol
$\qquad CH_3 \quad \overset{1}{C}H_3$ $\qquad\quad ^{2}CH$ $\qquad\qquad\vert$ $CH_3—CH_2—{}^{3}C—OH$ $\qquad\qquad\vert$ $\qquad\quad ^{4}CH$ $\overset{6}{C}H_3—\overset{5}{C}H_2 \quad CH_3$	ethyl-*i*-propyl-*s*-butylcarbinol 2,4-dimethyl-3-ethyl-3-hexanol

[1] It has been recommended that the term *carbinol* be replaced by the international name *methanol*, but the use of the former term is so well established that it will be difficult to replace it. Moreover a name such as trimethylmethanol is as undesirable as isobutene or isopropanol (see footnote to Table 6–*1*, p. 76).

[2] With the exception of the word *alcohol*, the ending *ol* is pronounced with a long *o*. Thus *carbinol* and *methanol* are pronounced as if they were spelled *carbinole* and *methanole*.

[3] Whether a chemical name for an organic compound is written as one word or more than one depends on whether the compound is being named as a derivative of a chemical entity. Thus *carbinol* is a definite compound, CH_3OH, and names for compounds considered to be derivatives of it are written as one word, for example *methylcarbinol*. On the other hand *alcohol* is not the name of any compound but is the name of a class of compounds. Hence *ethyl alcohol* is written as two words.

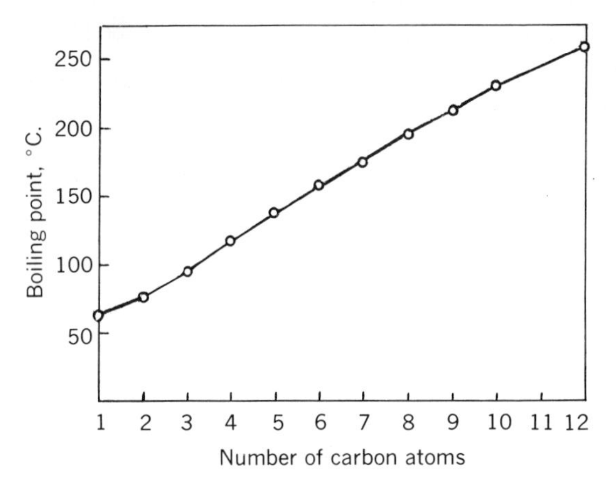

Figure 8–*1*. Boiling points of normal alcohols.

Alcohols, like the alkyl halides (p. 98), are classified as primary, secondary, or tertiary. The three classes can be represented by the general formulas RCH_2OH, R_2CHOH, and R_3COH. In the above list ethyl, *n*-propyl, *n*-butyl, and *i*-butyl alcohols are primary alcohols, *i*-propyl and *s*-butyl alcohols are secondary alcohols, and *t*-butyl alcohol and ethyl-*i*-propyl-*s*-butylcarbinol are tertiary alcohols. Ordinarily methyl alcohol is grouped with the primary alcohols, although really it is unique in that it is the only alcohol having only hydrogen atoms attached to the carbon atom that bears the hydroxyl group.

Physical Properties

Methyl alcohol, the first member of the series, is a liquid boiling at 65°. Ethane, which has the same number of electrons and presumably about the same polarizability (p. 22), boils at −88°. The alcohols in general have abnormally high boiling points when compared with the alkanes. The reason for the high boiling points is the same as that for the high boiling points of water and ammonia, namely, association by hydrogen bonding (p. 24). The attractive forces between alcohol molecules, however, are not so great as those between water molecules, because the hydrogen atoms of the alkyl groups do not form hydrogen bonds, and the bulk of the alkyl groups decreases the chance that collision of two alcohol molecules will form a hydrogen bond between two hydroxyl groups.

The rise in boiling point of the straight-chain primary alcohols with increasing molecular weight is about 18° for each additional CH_2 group, as shown by Fig. 8–*1*. For a given molecular weight, branching of the carbon chain lowers the boiling point just as it does in hydrocarbons. The boiling points of *n*-, *i*-, and *t*-butyl alcohols are respectively 117°, 107°, and 83°. The boiling point of *s*-butyl alcohol is 100° indicating that the hydroxyl group acts like a branch as might be expected.

Dodecyl alcohol is the first straight-chain alcohol that is solid at room temperature, although the more nearly spherical branched alcohols with fewer carbon atoms, for example *t*-butyl alcohol, also may be solids at room temperature. There is no evidence for alternation in melting point (Fig. 8–*2*) as is exhibited by saturated hydrocarbons (Fig. 5–*6*, p. 63).

The alcohols containing three carbon atoms or less and *t*-butyl alcohol are miscible with water at 20°, but *n*-butyl alcohol is soluble to the extent of only about 8 per cent, and primary alcohols with more than five carbon atoms are less than 1 per cent soluble in water. Solubility in general has been discussed previously (p. 24). From the differences in boiling points, it may be concluded that the attraction between water molecules is considerably greater than between methyl alcohol molecules, and it might be predicted that

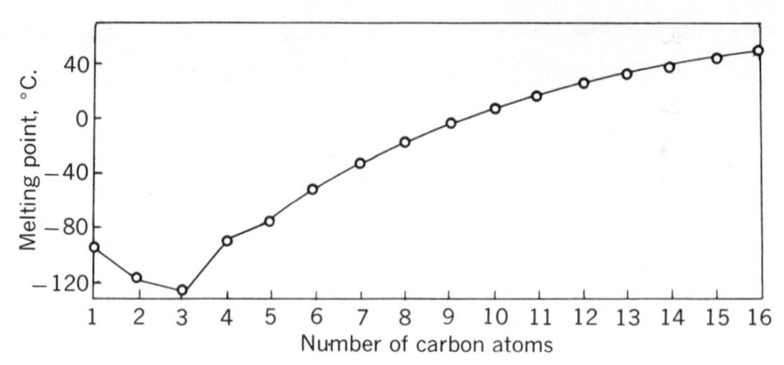

Figure 8–2. Melting points of normal alcohols.

water and methyl alcohol should not mix. Alcohol molecules, however, can hydrogen-bond to water molecules and water molecules to alcohol molecules with the result that the differences in attractive forces are decreased and intermingling takes place.

$$\begin{array}{c} R \\ | \\ H{-}O \end{array}\cdots\cdots \begin{array}{c} H \\ | \\ H{-}O \end{array} \qquad \begin{array}{c} H \\ | \\ H{-}O \end{array}\cdots\cdots \begin{array}{c} R \\ | \\ H{-}O \end{array}$$

As the length of the hydrocarbon chain increases, the attractive forces resulting from the polarizability of the molecule (p. 22) increase until a point is reached at which the association with water molecules no longer is sufficient to prevent alcohol molecules from being attracted more strongly to each other than to water complexes, and two phases result. Thus 100 g. of *n*-butyl alcohol dissolves only 37 g. of water, and 100 g. of water dissolves 7.3 g. of *n*-butyl alcohol at 15°. Another factor that appears to be important is the decrease in entropy for the system as the hydrocarbon chain is lengthened. There is some attraction of the hydrocarbon chain for water molecules, which restricts the motions of the water molecules. As the length of the hydrocarbon chain increases, the loss of entropy (p. 28) outweighs the decrease in enthalpy of solution, and the solubility of the alcohol becomes negligible. The miscibility of *t*-butyl alcohol with water agrees with the fact that it boils lower than *n*-butyl alcohol, since both the higher solubility and lower boiling point are dependent on the lower attractive forces among *t*-butyl alcohol molecules.

General Methods of Preparation

Secondary and **tertiary alcohols** are made by the catalytic addition of water to alkenes (p. 85), or by the acid hydrolysis of alkyl hydrogen sulfates or alkyl sulfates formed by the addition of sulfuric acid to alkenes (p. 83). The last reaction is a nucleophilic attack on sulfur with displacement of an alcohol molecule from the protonated sulfate by a water molecule.

$$H_2O\!: + R\overset{\overset{H}{\cdot\cdot}}{\underset{+}{O}}SO_3H \rightleftharpoons ROH + H_2\overset{+}{O}\!:\!SO_3H$$

The equilibrium is shifted to the right by a high concentration of water or by removing the alcohol by distillation.

Because addition of water or sulfuric acid follows the Markovnikov rule, ethyl alcohol is the only primary alcohol that can be obtained from an alkene by these reactions. Alkenes that yield secondary or tertiary alcohols by these reactions, however, can be converted to **primary alcohols** through the primary alkylboranes (p. 86), which yield primary alcohols by oxidation with alkaline hydrogen peroxide.

$$R_3B + 3\,H_2O_2 + 3\,NaOH \longrightarrow 3\,ROH + Na_3BO_3 + 3\,H_2O$$

In this way the Markovnikov rule is circumvented.

The attacking agent in this reaction undoubtedly is hydroperoxide anion, which provides a pair of electrons for the unfilled orbital of the boron atom.

$$\text{HO:}^- + \text{HO—OH} \longrightarrow \text{H}_2\text{O} + {}^-\text{:O—OH}$$
$$\text{R}_3\text{B} + {}^-\text{:O—OH} \longrightarrow \text{R}_3\overset{-}{\text{B}}\text{:O—OH}$$

A 1,2 shift of an alkyl group with its pair of electrons from boron to oxygen with displacement of a hydroxide ion gives an ester of borinic acid, which hydrolyzes to the alcohol (p. 127).

$$\text{R}_2\text{B}(\text{R})\text{—O:OH} \longrightarrow \text{R}_2\text{BOR} + {}^-\text{:OH}$$

$$\text{R}_2\text{BOR} + {}^-\text{:OH} \longrightarrow \text{R}_2\text{BOH} + {}^-\text{:OR} \xrightarrow{\text{H}_2\text{O}} \text{HOR} + {}^-\text{:OH}$$

A repetition of these steps completes the reaction. Primary alkylaluminums can be oxidized by air to aluminum alkoxides, which yield primary alcohols on hydrolysis (pp. 113, 122).

Alcohols can be obtained by the hydrolysis of alkyl halides (pp. 100, 106), but the reaction is of little practical use. Other important methods for the synthesis of primary alcohols are by the reduction of aldehydes (p. 206) or of esters of carboxylic acids (p. 175), or by the reaction of Grignard reagents with formaldehyde (p. 197) or with ethylene oxide (p. 601). Secondary alcohols can be made readily by the reduction of ketones (p. 206) or by the reaction of Grignard reagents with aldehydes other than formaldehyde (p. 197). Tertiary alcohols result from the reaction of Grignard reagents with ketones (p. 197) or with esters (pp. 176, 455).

Alcohols of Commercial Importance

Methyl Alcohol. The word *methyl,* first used by Dumas and Peligot in 1834, was derived from the Greek word *methy* meaning wine and *yle* meaning wood or material, and refers to the fact that it is the chief alcohol formed by the destructive distillation of wood. Previous to 1919 it commonly was called *wood alcohol* or *Columbian spirits* in the United States. During the operation of the Volstead Act (1917–1933), which outlawed alcoholic beverages in the United States, anything called alcohol was used by many for intoxicating drinks. Since methyl alcohol is highly poisonous, the result was an alarming number of deaths and cases of blindness. As one combative measure the use of the international name *methanol* was urged, and it now largely has displaced the older term.

Some methyl alcohol still is produced by wood distillation, although most is made by synthesis. In the **wood distillation** process dried hardwood such as beech, birch, hickory, maple, or oak is decomposed in an oven or a vertical retort at a temperature increasing from about 160° to 450°. If the wood is dry when placed in the retort, the reaction is exothermic, and heat is supplied only to start the reaction. Barrel valves at the top and bottom of the retort for the introduction of wood and the removal of charcoal permit continuous operation. The products are gases, which are burned as fuel, a liquid condensate, and a residue of charcoal. The liquid condensate separates into an aqueous layer called *pyroligneous acid,* and a tarry layer. The pyroligneous acid, of which 200 to 250 gallons per cord of wood is obtained, is mostly water but contains 1 to 6 per cent of methyl alcohol, 4 to 10 per cent acetic acid, 0.1 to 0.5 per cent of acetone, and smaller amounts of methyl acetate and numerous other organic compounds.

During the year 1924 the total importation of methyl alcohol by the United States was 48 gallons. This material was of high purity, the import duty of 12 cents per gallon being sufficient to prevent importation of the commercial grade of foreign wood alcohol. During the first five months of 1925, the Badische Company of Germany shipped into the United States almost a quarter of a million gallons of methyl alcohol. Wood alcohol was selling for 88 cents per gallon, but the imported material was being made by a synthetic process perfected in 1923 that could produce methanol at a cost of around 20 cents per gallon. Even raising the import duty to 18 cents per gallon, the maximum permitted by law, was of little help to the wood distillers. Not enough methanol was shipped into the country to drive the price below 57 cents per gallon in 1925, and for some reason the price rose again to 82 cents per gallon in February 1927. With the advent of synthetic production in the United States, however, the price became stabilized in 1930 at about 40 cents per gallon. Many of the wood distillation plants survived because of more efficient operation and revision of their procedures, particularly the development of new methods for separating acetic acid by extraction or by special distillation processes (p. 168). It is of some interest that the United States exported around 65 million gallons of synthetic methanol in 1964, about half of which went to West Germany.

The **synthetic process** starts with carbon monoxide and hydrogen, which combine over a zinc oxide catalyst containing other promoter oxides, for example 10 per cent chromium oxide, at a temperature of 300–400° and at pressures of 200 to 300 atmospheres.

$$CO + 2\,H_2 \xrightarrow{ZnO-Cr_2O_3} CH_3OH$$

The synthesis frequently is carried out in conjunction with some other operation such as the synthesis of ammonia or acetylene (p. 151), but most of the carbon monoxide and hydrogen is produced by the partial oxidation of methane with oxygen prepared from liquid air.

$$2\,CH_4 + O_2 \longrightarrow 2\,CO + 4\,H_2$$

The total production of synthetic methanol in the United States in 1964 was over 2.6 billion pounds, or 400 million gallons, exceeding the production of any other synthetic organic chemical. About 47 per cent was used for the manufacture of formaldehyde (p. 212), 28 per cent for the synthesis of other chemicals, 8 per cent for jet fuels, 4 per cent as radiator antifreeze, and 7 per cent as solvent[4] and as a denaturant for ethyl alcohol. Production of wood alcohol was around 1.5 million gallons. Practically all of it was used to denature ethyl alcohol. Whereas pure synthetic methanol sold for 30 cents per gallon, impure wood alcohol sold for 85 cents per gallon, because by law it is a required ingredient of certain formulas for denatured ethyl alcohol (p. 122).

Ethyl Alcohol. Although the suffix *yl* was used in 1834 by Dumas and Peligot in the word *methyl* to indicate its derivation from wood, Liebig and Woehler had used the same suffix in 1832 in the term *benzoyl* in the sense of stuff or material. It is in the latter sense that it ordinarily is used. *Ethyl* means the material that gives rise to ether (p. 139).

One important source of ethyl alcohol is the *fermentation of sugars.* **Fermentation** is the decomposition of organic compounds into simpler compounds through the agency of enzymes. **Enzymes** are protein catalysts (p. 352) produced by living cells. The name *enzyme* (Gr. *en* in, *zyme* leaven) was given because the earliest known enzymes were those secreted by yeast cells. Pasteur,[5] who discovered the nature of fermentation, thought that the living cell was necessary, but this view was disproved by Buchner in 1897. He showed that the juice expressed from yeast cells that had been completely destroyed still was capable of bringing about fermentation.

The chief sources of sugars for fermentation are the various starches and the molasses residue from sugar refining. Corn (*maize*) is the chief source of starch in the United States, and ethyl alcohol made from corn commonly is known as *grain alcohol.* Potatoes are the chief source of starch in Europe, and rice in Asia. In preparing alcohol from corn, the grain, with or without the germ, is ground and cooked to give the mash. The enzyme *diastase* is added in the form of malt (sprouted barley that has been dried in air at 40° and ground) or of a mold such as *Aspergillus oryzae,* and the mixture is kept at 40° until all of the starch has been converted into the sugar maltose. This solution is known as the *wort.*

$$\underset{\text{Starch}}{2\,(C_6H_{10}O_5)_n} + n\,H_2O \xrightarrow[\text{in malt}]{\text{Diastase}} \underset{\text{Maltose}}{n\,C_{12}H_{22}O_{11}}$$

The wort is cooled to 20°, diluted to 10 per cent maltose, and a pure yeast culture added, usually a strain of *Saccharomyces cerevisiae* (or *ellipsoidus*). The yeast cells secrete two enzymes, *maltase,* which converts the maltose into glucose, and *zymase,* which converts the glucose into carbon dioxide and alcohol.

$$\underset{\text{Maltose}}{C_{12}H_{22}O_{11}} + H_2O \xrightarrow{\text{Maltase}} \underset{\text{Glucose}}{2\,C_6H_{12}O_6}$$

$$C_6H_{12}O_6 \xrightarrow{\text{Zymase}} 2\,CO_2 + 2\,C_2H_5OH + 26\text{ kcal.}$$

Heat is liberated, and the temperature must be kept below 32° by cooling, to prevent destruction of the enzymes. Oxygen in large amounts is necessary initially for the optimum reproduction of

[4] The term *solvent* refers to the use of organic liquids for the purpose of dissolving other organic substances. Solvents may be used as reaction media, or for extraction of organic compounds from solids or from aqueous solutions. Following the operation the solvent is removed by distillation or evaporation. Hence they usually are low-boiling liquids. Solvents may be used also to dissolve solid organic materials in order to form the latter into sheets or fibers or to coat other materials with them. They comprise an important segment of industrial organic chemicals.

[5] Louis Pasteur (1822–1895), French chemist and microbiologist whose studies of fermentation led to the germ theory of disease and to immunization by inoculation with attenuated organisms and viruses. The "Life of Pasteur" written by his son-in-law, Vallery-Radot, is a most interesting account of the life and work of an outstanding scientist.

yeast cells, but the production of alcohol is anaerobic. During fermentation the evolution of carbon dioxide soon establishes anaerobic conditions. If oxygen were freely available, only carbon dioxide and water would be produced. After 40 to 60 hours, fermentation is complete, and the product is run through beer stills (*sieve tray type,* Fig. 4–*1*, p. 43) to remove the alcohol from solid matter. The distillate is fractionated by means of an efficient column of the valve tray type. A small amount of acetaldehyde, b.p. 21°, distills first and is followed by 95 per cent alcohol. **Fusel oil** is withdrawn from the center of the column. The fusel oil consists of a mixture of higher alcohols, chiefly *n*-propyl, *i*-butyl, *i*-amyl (*3-methyl-1-butanol*), and active amyl[6] (*2-methyl-1-butanol*). The exact composition of fusel oil varies considerably, being dependent particularly on the type of raw material that is fermented. These higher alcohols are not formed by fermentation of glucose but arise from certain amino acids (p. 353) derived from the proteins present in the raw material and in the yeast.

Industrial alcohol is ethyl alcohol used for nonbeverage purposes. Previous to the development of efficient synthetic processes, the chief source of industrial alcohol was the fermentation of blackstrap molasses, the noncrystallizable residue from the refining of sucrose (*cane sugar*). The waste liquors from the production of wood pulp by the sulfite process (p. 342) contain a low concentration of fermentable sugars and can yield about 20 gallons of alcohol per ton of wood pulp produced. Numerous plants utilizing waste sulfite liquors are operated in the Scandinavian countries and in Germany, and since 1954 a few plants have been operating in Canada and in the United States. Glucose formed by the hydrolysis of wood cellulose (pp. 317, 342) also is a possible source of carbohydrate for fermentation and has been utilized in Germany. Experimental plants have been built in the United States, but they have not been able to compete with other processes under normal economic conditions. The flowers of *Bassia latifolia* and other species have a high content of glucose and fructose and are used extensively for the production of alcohol in India.

About 15 per cent of U.S. production in 1964 was by fermentation, chiefly from grain but also from the sugars in fruit juices, wood pulp liquors, and molasses. The distribution among these sources varies greatly from year to year depending on the availability and price of the raw material.

Several reactions can lead to the **synthesis of ethyl alcohol.** In 1826 Hennel[7] had reported the isolation of potassium ethyl sulfate (p. 84) from a sample of sulfuric acid that had absorbed 80 volumes of ethylene, and that had been given to him by Faraday[8] for investigation. In 1828, the same year in which Woehler described his synthesis of urea (p. 282), Hennel reported the hydrolysis of potassium ethyl sulfate to ethyl alcohol. Hennel's discovery was overlooked, however, and the synthesis was rediscovered in 1855 by Berthelot,[9] who absorbed the ethylene from coal gas in concentrated sulfuric acid, and diluted and distilled the solution.

$$CH_2{=}CH_2 + HOSO_3H \longrightarrow CH_3CH_2OSO_3H$$

$$CH_3CH_2OSO_3H + HOH \longrightarrow CH_3CH_2OH + H_2SO_4$$

Although the possibility of industrial synthesis by this process was discussed in the following year and a claim made in France in 1862 that the cost of synthetic alcohol was about one third that of fermentation alcohol, the first continuously successful process began operation in the United States in 1930. The ethylene used as the raw material is produced by the cracking of hydrocarbons. It is absorbed in concentrated sulfuric acid at 100° to give a mixture of ethyl hydrogen sulfate and ethyl sulfate (p. 84). Dilution with water brings about hydrolysis to ethyl alcohol, which is removed by distillation. The dilute sulfuric acid is concentrated for reuse. Since 1949 ethyl alcohol has been manufactured also by direct hydration of ethylene in the vapor phase. The conditions reported are a large excess of water, high temperature (300°) and pressure (1000–4000 p.s.i.), and a solid catalyst such as phosphoric acid on a carrier, hydrogen fluoride-treated clays, or promoted tungsten oxide. By 1940, the synthetic process accounted for 25 per cent of the total U.S. production of industrial alcohol and in 1964 for 85 per cent or 2.3 billion pounds. Ethyl alcohol also is one of the products of the butyl alcohol fermentation of starch (p. 122) and of the Fischer-Tropsch synthesis of

---

[6] The five-carbon alcohols commonly are referred to as *amyl alcohols,* because they first were obtained from the products of fermentation (L. *amylum* starch). The use of *pentyl* instead of *amyl* is being advocated. The term *active* refers to the effect of the compound on plane-polarized light (p. 298). The higher alcohols are toxic even in small amounts. The term *fusel oil* comes from the German word *fusel* meaning *bad liquor.*

[7] Henry Hennel, English apothecary and contemporary of Michael Faraday. He was killed in 1842 by the explosion of a large quantity of mercury fulminate that he had prepared for the East India Company for military purposes.

[8] Michael Faraday (1791–1867). Although best known for his work on electricity and magnetism, he made outstanding contributions to chemistry as well. He was the first to liquefy a number of gases and was the discoverer of benzene (p. 368). He established the laws of electrolysis and discovered the phenomenon of magnetic optical rotation.

[9] Marcellin Pierre Eugene Berthelot (1827–1907), French chemist and statesman. He is noted especially for his high temperature syntheses of organic compounds, for his study of esterification, and for his work in thermochemistry.

liquid hydrocarbon fuels (p. 93). A rough estimate indicates that approximately the same amount of alcohol was consumed in beverages as was produced for industrial use.

Various grades of ethyl alcohol are produced. **Ordinary alcohol** is 92–95 per cent ethyl alcohol by weight, the remainder being chiefly water. Anhydrous alcohol cannot be obtained by simple distillation because a constant-boiling mixture (also called an *azeotrope*) containing 95.6 per cent alcohol by weight boils lower (78.15°) than pure alcohol (78.3°).

Absolute alcohol (anhydrous, 99.9+ per cent) usually is prepared in the laboratory by removing the water by chemical means, for example by heating with calcium oxide, which reacts with the water, and distilling the dried alcohol from the calcium hydroxide. The 5 per cent water in ordinary alcohol has a marked effect on its solvent properties, and there is a considerable large-scale demand for the anhydrous product. One commercial method of dehydration makes use of the fact that a ternary mixture consisting of 18.5 per cent alcohol, 74.1 per cent benzene (C_6H_6, p. 93), and 7.4 per cent water by weight boils at 64.85°. Since the ratio of water to alcohol in the ternary mixture is 1:2.5, enough benzene can be added to remove all of the water in the low-boiling distillate, and anhydrous alcohol can be withdrawn from the still pot.

Because all countries derive a sizeable portion of their revenue by taxing alcohol for beverage purposes, ethyl alcohol for industrial use first must be converted into **denatured alcohol** if the payment of tax is to be avoided. Denaturing is the addition of substances that render the alcohol unfit to drink. Many formulas are available to the manufacturer in order that he may choose one suitable for his purpose. Only denatured alcohols containing obnoxious mixtures that are difficult to remove may be sold to the general public without payment of tax. On January 1, 1940, the price of tax-free alcohol was 31 cents per gallon, compared with $4.55 per gallon for taxed grain alcohol; on January 1, 1964, the prices were 68 cents and $20.63 respectively. The difference in each case is approximately the tax per gallon of 190 proof alcohol. The United States tax is based on "100 U.S. proof" alcohol, which contains half of its volume of absolute alcohol. Since there is a contraction in volume on mixing alcohol and water, one volume of proof spirits contains 0.537 volumes of water. Proof spirits is 42.5 per cent alcohol by weight. Ordinary so-called 95 per cent alcohol is 190 proof. One volume contains 0.95 volume of alcohol and 0.062 volume of water and is 92.4 per cent alcohol by weight. The term **proof spirit** has its origin in an old method of testing whiskey by pouring it on some gunpowder and lighting it. Ignition of the gunpowder after the alcohol burned away was considered proof that the whiskey did not contain too much water.

Higher Alcohols. Previous to World War I the chief source of higher alcohols was fusel oil from which *n*-propyl, *i*-butyl, *i*-amyl, and active amyl alcohols were separated by distillation (p. 121). Since then many other alcohols have become available. ***n*-Butyl alcohol** is made by a special bacterial fermentation process. Corn mash or black-strap molasses is inoculated with a pure culture of one of several strains of *Clostridium acetobutylicum* in closed tanks under anaerobic conditions. The products of fermentation are *n*-butyl alcohol, acetone, and ethyl alcohol in proportions varying from 60:30:10 to 74:24:2. The evolved gas contains hydrogen and carbon dioxide in the ratio of 1 volume:2 volumes. The process was developed during World War I (1914–1918) as a source of acetone when the supply from the decomposition of calcium acetate from pyroligneous acid was insufficient to meet the requirements of the British for the production of cordite (pp. 343, 608). At that time there was no large-scale use for *n*-butyl alcohol, and huge stocks accumulated. Since then uses for *n*-butyl alcohol have been developed, and now it is the most valuable product, and acetone has become the by-product. *n*-Butyl alcohol is manufactured also by synthetic processes starting with acetaldehyde or *n*-butyraldehyde (p. 215). In 1964 the total U.S. production of *n*-butyl alcohol was over 388 million pounds of which about one fourth was produced by fermentation and three fourths by synthesis. The chief uses of *n*-butyl alcohol are as a solvent, for the manufacture of esters (p. 177), and for the synthesis of *n*-butyraldehyde (p. 215) and *n*-butyric acid (p. 168).

1-Alkanols having **10 to 18 carbon atoms** have been made by the reduction of the esters of acids obtained from fats (p. 187). Synthesis from ethylene (*Alfol process*) provides a source with greater price stability. Ethylaluminum is prepared by a cyclic two-stage process from aluminum powder, ethylene, and hydrogen (p. 113). In this reaction two moles of ethylaluminum yield three moles. Reaction of ethylaluminum with ethylene at 125° gives a mixture of higher alkylaluminums, which are oxidized by air to the aluminum alkoxides. Up to this point all of the reactions are conducted in a paraffin hydrocarbon solvent to minimize the fire hazard and permit temperature control of the reaction. Hydrolysis with water or aqueous sulfuric acid yields the alcohols and high purity aluminum hydroxide or aluminum sulfate (p. 113). The mixture of alcohols is purified and separated by distillation. The C_8–C_{10} alcohols are used for the synthesis of plasticizers (p. 474), and the C_{12}–C_{16} alcohols for the synthesis of detergents (p. 191). A plant with an alcohol capacity of 100 million pounds per year began operation in 1962. Higher branched-chain primary alcohols are produced from alkenes by the "oxo" process (pp. 215, 216).

Secondary and tertiary alcohols are manufactured on a large scale by hydration of lower olefins obtained by the cracking of saturated hydrocarbons (p. 78).

$$\underset{\text{Propylene}}{CH_3CH{=}CH_2} \xrightarrow{H_2SO_4} \underset{\displaystyle OSO_3H}{CH_3\underset{|}{C}HCH_3} \xrightarrow{H_2O} \underset{i\text{-Propyl alcohol}}{\underset{\displaystyle OH}{CH_3\underset{|}{C}HCH_3}}$$

$$\left.\begin{matrix}\underset{\text{2-Butene}}{CH_3CH{=}CHCH_3}\\ \text{or}\\ \underset{\text{1-Butene}}{CH_3CH_2CH{=}CH_2}\end{matrix}\right\} \xrightarrow{H_2SO_4} \underset{\displaystyle OSO_3H}{CH_3CH_2\underset{|}{C}HCH_3} \xrightarrow{H_2O} \underset{s\text{-Butyl alcohol}}{\underset{\displaystyle OH}{CH_3CH_2\underset{|}{C}HCH_3}}$$

$$\underset{\text{Isobutylene}}{(CH_3)_2C{=}CH_2} \xrightarrow{H_2SO_4} \underset{\displaystyle OSO_3H}{(CH_3)_2\underset{|}{C}CH_3} \xrightarrow{H_2O} \underset{t\text{-Butyl alcohol}}{(CH_3)_3COH}$$

Similarly amylenes (*pentenes*) are converted into secondary and tertiary amyl alcohols.

Production of *i*-propyl alcohol in 1964 was over 1.5 billion pounds, and the price was around 40 cents per gallon. It is used chiefly as a solvent and for the synthesis of acetone and hydrogen peroxide (p. 216).

Chemical Properties

In alcohol molecules, both the carbon-oxygen bond and the hydrogen-oxygen bond are polarized with the negative ends of the bond dipoles at oxygen. Moreover the oxygen has two unshared pairs of electrons. It is these features that determine the chemical reactions of the saturated alcohols. Insofar as the features are common to water and alcohols, the latter resemble water in their chemical behavior. The alkyl group lacks strongly polarized bonds, unshared electrons, and vacant valence orbitals and, with the exception of the somewhat increased reactivity of β hydrogen atoms, which is due to the electron-attracting effect of oxygen (p. 130), is as unreactive as the saturated hydrocarbons.

Despite the polarity of the carbon-oxygen bond, no displacement reactions analogous to those of the alkyl halides take place in neutral or alkaline solution because the high basicity of the hydroxide ion makes it a poor leaving group (p. 41). Reactions involving scission of the C—O bond are catalyzed by acids (p. 130). The protonated hydroxyl group is a much better leaving group, the water molecule being much less basic than the hydroxide ion.

1. ***Formation of Complexes with Inorganic Salts.*** Both alcohol molecules and water molecules have unshared pairs of electrons on oxygen, and both can coordinate with metal ions. Thus magnesium chloride forms a complex with six molecules of methanol or of ethanol. Although calcium chloride forms a hexahydrate, the reported solid complexes for methanol and ethanol are $CaCl_2 \cdot 4\ CH_3OH$ and $CaCl_2 \cdot 3\ C_2H_5OH$. Some salts that form hydrates, such as calcium sulfate and copper sulfate, do not form complexes with alcohols. These differences in the behavior of alcohols from water and from each other are not surprising, since the size and shape of the molecule would be expected to influence the formation of the complex, especially in the solid state.

In general, salts are not so soluble in alcohols as in water because alcohols have a much lower dielectric constant (methanol = 34, water = 78) (p. 25). If a salt, such as calcium chloride, forms a definite coordination complex with the alcohol, the complex may be highly soluble.

2. ***Oxonium Salt Formation.*** When a strong acid is dissolved in water, there is practically complete transfer of the proton from the acid radical to the water molecule. The product from hydrogen chloride is called *hydronium chloride,* and the H_3O^+ ion is the *hydronium ion.* The proton is transferred to a water molecule because the water molecule is a stronger base than the chloride ion (p. 31).

Alcohols also are stronger bases than the anions of strong acids. Hence alcohols dissolve in strong acids with the formation of alkonium salts.

$$R{:}\ddot{\underset{..}{O}}{:}\overset{H}{} + H{:}\ddot{\underset{..}{Cl}}{:} \longrightarrow R{:}\overset{H}{\underset{+}{\ddot{\underset{..}{O}}}}{:}H + {:}\ddot{\underset{..}{Cl}}{:}^-$$

The general term for this type of compound is *oxonium salt.* Concentrated sulfuric acid dissolves practically all organic compounds containing oxygen, frequently without chemical change other than salt formation. This fact is used in qualitative organic analysis *to distinguish oxygen-containing compounds from saturated hydrocarbons.*

Alcohols and all other oxygen-containing compounds are more soluble in aqueous solutions of strong acids than in water because of the competition of alcohol and water molecules for the proton and the greater degree of hydration of ions as compared to neutral molecules.

$$R{:}\overset{H}{\underset{..}{\ddot{O}}}{:} + H{:}\overset{H}{\underset{+}{\ddot{\underset{..}{O}}}}{:}H \rightleftharpoons R{:}\overset{H}{\underset{+}{\ddot{\underset{..}{O}}}}{:}H + H_2O$$

n-Butyl alcohol, for example, is miscible with concentrated aqueous hydrochloric acid although it is soluble to the extent of only 8 per cent in water. It should be noted that oxonium salt formation differs from hydrogen bonding. In the latter the proton bonds two molecules together by electrostatic attraction, whereas in the former a proton is transferred completely from one molecule to another. In the oxonium salt the proton is engulfed in the electron shell of the new molecule and carries a positive charge with it. Water molecules and alcohol molecules are hydrogen bonded in the liquid state, but the liquids have a very low electrical conductivity. A solution of hydrogen chloride in water or methanol, on the other hand, is a strong electrolyte.

Proton transfer reactions usually will be indicated merely by the symbol H^+ without designating the base with which the proton is associated. Thus the formation of an oxonium salt may be written

$$ROH + H^+ \rightleftharpoons R\overset{+}{O}H_2$$

If anhydrous hydrogen chloride is used, H^+ stands for HCl; if an aqueous solution of hydrogen chloride is used, H^+ stands for H_3O^+. If a stronger base than an alcohol is being neutralized with an alcoholic solution of hydrogen chloride, H^+ stands for $ROH_2{}^+$. An objection to the use of the symbol H^+ is that it represents a proton, and free protons under ordinary conditions are nonexistent. In organic reactions, however, where a number of bases may be involved, there usually is no point in trying to indicate the base with which the proton is associated. Moreover in aqueous solutions the H_3O^+ ion is associated strongly with three more molecules of water to give $H_9O_4{}^+$, which is less strongly associated with an indefinite number of water molecules (p. 26).

3. ***Deuterium Exchange.*** The polarity of the hydrogen-oxygen bonds in water molecules and in alcohol molecules permits rapid exchange of hydrogen, which can be detected by use of its isotope, deuterium. Merely mixing a water-soluble alcohol or shaking a water-insoluble alcohol with deuterium oxide establishes equilibrium.

$$R{:}\underset{\underset{H}{|}}{\ddot{O}}{:} + D{:}\underset{\underset{D}{|}}{\ddot{O}}{:} \rightleftharpoons R{:}\underset{\underset{H}{|}}{\overset{+}{\ddot{O}}}{:}D + {}^-{:}\underset{\underset{D}{|}}{\ddot{O}}{:} \rightleftharpoons R{:}\underset{..}{\ddot{O}}{:}D + H{:}\underset{..}{\ddot{O}}{:}D$$

$$ROH + DOH \rightleftharpoons ROD + H_2O$$

As would be expected, the exchange is catalyzed by either acids or bases.

4. ***Reaction with Metals.*** The polarity of the oxygen-hydrogen bonds in water makes water sufficiently acidic to react with the active metals. Thus water reacts with sodium to give hydrogen and sodium hydroxide. Alcohols react in the same way to give hydrogen and sodium alkoxide.

$$HOH + Na \longrightarrow HO^{-+}Na + \tfrac{1}{2} H_2$$
Sodium hydroxide

$$CH_3OH + Na \longrightarrow CH_3O^{-+}Na + \tfrac{1}{2} H_2$$
Sodium methoxide

$$ROH + Na \longrightarrow RO^{-+}Na + \tfrac{1}{2} H_2$$
Sodium alkoxide

The reaction with alcohols is slower than with water and decreases with increasing molecular weight of the alcohol, probably because of the decreasing solubility of the alkoxide in the alcohol.

In the above equations the ionic bond between the sodium atom and the hydroxyl, methoxyl, and alkoxyl groups is indicated by (+) and (−) signs. Usually ionic bonds will not be indicated for inorganic salts, because it is assumed that the reader understands that inorganic salts are completely ionized and that although the formula is written NaOH, for example, two types of bonds are involved, the ionic bond between sodium ion and hydroxide ion, and the covalent bond between the hydrogen and the oxygen. Also it is assumed that the reader understands that the molecule represented by the formula CH_3OH involves only covalent bonds. When an organic molecule contains both ionic and covalent bonds, however, the presence of the ionic bond frequently will be emphasized by indicating the ionic bond with (+) and (−) charges.

The reaction product of methanol with sodium is known as *sodium methoxide*. The term *sodium methylate* also is used but is less satisfactory because the term *alcoholate* frequently is used for complexes of alcohols with inorganic salts analogous to the hydrates (p. 123).

Other active metals also react with alcohols. Anhydrous methanol reacts with magnesium to give hydrogen and magnesium methoxide, $Mg(OCH_3)_2$, and amalgamated aluminum reacts with ethanol, *i*-propyl alcohol, or *t*-butyl alcohol, to give respectively aluminum ethoxide, $Al(OC_2H_5)_3$, *i*-propoxide, $Al(OC_3H_7\text{-}i)_3$, or *t*-butoxide, $Al(OC_4H_9\text{-}t)_3$. Since no more than one hydrogen atom per mole of alcohol ever is liberated, these reactions indicate that one hydrogen is different from all the rest and is linked to oxygen.

Because the acidity of water and of alcohols is about the same, metallic alkoxides undergo a reversible reaction with water.

$$RO^- + H_2O \rightleftharpoons ROH + {}^-OH$$

The position of equilibrium is dependent on the relative concentrations of the reactants (cf. pp. 27, 146). Sodium ethoxide is made commercially from ethanol and sodium hydroxide by azeotropic removal of the water with benzene (p. 93). Magnesium or aluminum alkoxides may be used to remove small amounts of water from alcohols because the extreme insolubility of magnesium or aluminum hydroxide in alcohols shifts the equilibrium to the right.

5. ***Conversion to Esters of Inorganic Acids.*** The hydroxide ion is a poor leaving group because it is not highly polarizable and is a good nucleophile (p. 103). If it is protonated, however, the positive charge enables it to be displaced as a water molecule even by weak nucleophiles. Hence alcohols can displace protonated hydroxyl groups from inorganic oxygen acids such as **sulfuric acid.**

$$HO{:}\overset{O^-}{\underset{O^-}{\overset{|+}{\underset{|+}{S}}}}{-}OH + HOSO_3H \rightleftharpoons H_2\overset{+}{O}{:}\overset{O^-}{\underset{O^-}{\overset{|+}{\underset{|+}{S}}}}{-}OH + {}^-OSO_3H$$

$$\underset{\underset{H}{|}}{RO}{:} + H_2\overset{+}{O}{:}\overset{O^-}{\underset{O^-}{\overset{|+}{\underset{|+}{S}}}}{-}OH \rightleftharpoons H_2O + \underset{\underset{H}{|}}{R\overset{+}{O}}{:}SO_2OH \rightleftharpoons H_3O^+ + ROSO_3H$$

An alkyl hydrogen sulfate

In agreement with this mechanism, there is evidence that the R—O bond is not broken during the reaction. The second stage yielding the alkyl sulfate, however, appears to take place by displacement of a water molecule from a protonated alcohol molecule with scission of the R—O bond.

$$ROSO_3H + ROH \longrightarrow ROSO_2O{:}^- + R{:}\overset{+}{O}H_2 \longrightarrow ROSO_2OR + H_2O$$

The products, in which an alkyl group is present instead of the acidic hydrogen of an oxygen acid, are called *esters,* and the reaction is called *esterification*. Esters are named as if they were alkyl salts of the acid even though they are strictly covalent compounds.

The mechanism postulated for the esterification of sulfuric acid is of the S_N2 type described for alkyl halides (p. 100). **Nitrous acid** definitely is known to react with alcohols in this way.

$$HONO + HONO \underset{NO_2^-}{\rightleftharpoons} H_2\overset{+}{O}NO \underset{H_2O}{\overset{ROH}{\rightleftharpoons}} \underset{\underset{H}{|}}{R\overset{+}{O}NO} \underset{HNO_2}{\overset{NO_2^-}{\rightleftharpoons}} RONO$$

Nitric acid, on the other hand, reacts by way of an S_N1 type of mechanism, even in aqueous solutions.

$$HONO_2 + HONO_2 \underset{NO_3^-}{\rightleftharpoons} H_2\overset{+}{O}NO_2 \underset{H_2O}{\rightleftharpoons} \overset{+}{N}O_2 \overset{ROH}{\rightleftharpoons} R{-}\underset{\underset{H}{|}}{\overset{+}{O}}{-}NO_2 \underset{HNO_3}{\overset{NO_3^-}{\rightleftharpoons}} RONO_2$$

Sulfuric acid esterifies less readily than nitrous acid or nitric acid, probably because attack on the sulfur atom is more sterically hindered than attack on the nitrogen atom. The failure of **phosphoric acid** to esterify directly may be due to the fact that its phosphorus atom is less positive than is sulfur in sulfuric acid (p. 21). **Boric acid,** though a weak acid, esterifies very easily, undoubtedly because the boron atom lacks a pair of electrons in its valence shell and is very susceptible to nucleophilic attack.

$$R{-}\underset{\underset{H}{|}}{O}{:} + B(OH)_3 \rightleftharpoons \underset{H}{R\overset{+}{O}}{:}\underset{OH}{\overset{-}{B}}(OH)_2 \rightleftharpoons RO\underset{\underset{^+OH_2}{|}}{\overset{-}{B}}(OH)_2 \rightleftharpoons H_2O + ROB(OH)_2$$

Repetition of these steps yields the trialkyl ester, $(RO)_3B$.

Esters of inorganic acids can be made also by nucleophilic displacement of halogen from **inorganic acid halides.**

$$\underset{\underset{H}{|}}{RO}{:} + \underset{\underset{Cl}{|}}{\overset{..}{P}}Cl_2 \rightleftharpoons \underset{\underset{H}{|}}{R\overset{+}{O}}{-}PCl_2 + {}^-{:}Cl \rightleftharpoons RO{-}PCl_2 + HCl$$

Further reaction yields the alkyl phosphite.

$$ROPCl_2 + 2\,ROH \rightleftharpoons (RO)_3P + 2\,HCl$$

Phosphorus oxychloride gives alkyl phosphates.

$$3\,ROH + POCl_3 \longrightarrow (RO)_3PO + 3\,HCl$$

Usually reactions with acid halides are carried out in the presence of an organic tertiary amine such as pyridine (p. 514). Like ammonia, pyridine is a stronger base than chloride ion and can remove a proton better than can a chloride ion, giving pyridinium chloride instead of hydrogen chloride.

$$Cl:^- + R\overset{+}{\underset{H}{O}}—PCl_2 + :NC_5H_5 \longrightarrow RO—PCl_2 + C_5H_5NH^{+-}Cl$$

Pyridine; Pyridinium chloride

Combination with the proton decreases the concentration of protonated ester and in this way reduces the amount of alkyl halide formed by attack of the halide ion on the R group.

In esters of oxygen acids the hydrocarbon group always is linked to the acid group through oxygen. Although the esters are named as if they were salts, they are covalent compounds. Esters of low molecular weight are volatile liquids, and the higher esters can be distilled under reduced pressure. Most esters are insoluble in water and soluble in organic solvents, although ethyl phosphate is soluble in water. Anhydrous ethyl nitrate and ethyl perchlorate can decompose explosively.

The esters are hydrolyzed by water with varying degrees of ease. Ethyl borate and ethyl silicate, for example, are hydrolyzed readily at room temperature.

$$(C_2H_5O)_3B + 3\,H_2O \longrightarrow 3\,C_2H_5OH + B(OH)_3$$
$$(C_2H_5O)_4Si + 4\,H_2O \longrightarrow 4\,C_2H_5OH + Si(OH)_4$$

One alkyl group of sulfates and phosphates is hydrolyzed much more readily than a second group. Ethyl nitrate and ethyl perchlorate are hydrolyzed only very slowly by water even at elevated temperatures.

Nitrites relax the smooth muscles of the body and produce a rapid lowering of the blood pressure. ***i*-Amyl nitrite** is used for the relief of pain in acute angina pectoris although the normal brevity of the attacks in the absence of medication makes it difficult to determine whether amyl nitrite actually assists in the relief of pain. **Ethyl, butyl,** and ***i*-amyl nitrites** react with hydrogen chloride to give the alkyl chloride and nitrous acid.

$$RONO + HCl \longrightarrow RCl + HONO$$

They are used extensively as a source of nitrous acid in organic reactions when it is necessary or desirable to carry the reaction out in an anhydrous solvent (p. 430). The **sulfites, sulfates,** and **phosphates** find important use as alkylating agents, that is, reagents for the introduction of alkyl groups (pp. 139, 329). **Methyl sulfate,** which is used extensively as a methylating agent to convert hydroxyl to methoxyl groups (p. 437), is highly toxic and care must be exercised to avoid exposure to its vapors. Other volatile esters of inorganic acids such as ethyl pyrophosphate (p. 272) and ethyl silicate also are noxious chemicals. The formation and hydrolysis of phosphates and pyrophosphates are involved in vital biochemical processes (p. 533). Related compounds may be highly toxic to the organism, and numerous esters of the phosphoric acids and their sulfur analogs are important commercial insecticides (p. 445). Phosphites and phosphates are important also as intermediates for the synthesis of organophosphorus compounds (p. 274).

6. ***Conversion to Alkyl Halides.*** The reactions of **inorganic acid halides** to yield esters may be followed by nucleophilic attack of the halide ion on carbon of the alkyl group, rather than on the proton of the protonated ester, to yield alkyl halides instead of the ester.

$$X:^- + R:\overset{+}{\underset{H}{O}}—PX_2 \longrightarrow X:R + HOPX_2$$

Like the reactions of alkyl halides, these displacements may be of the S_N2 type with inversion of the alkyl group (p. 100) or of the S_N1 type, which may or may not involve inversion.

The ease of formation of alkyl halides is tertiary > secondary > primary. In fact the reaction of acid halides with tertiary alcohols gives exclusively alkyl halides. Since the nucleophilicity of halide ions is iodide > bromide > chloride, phosphorus tribromide and phosphorus triiodide give alkyl halides even with primary alcohols, whereas primary alcohols give chiefly esters with phosphorus trichloride. These reactions constitute one of the best preparative methods for alkyl bromides and iodides.[10]

The reactions of thionyl chloride with alcohols in the presence of pyridine require special mention. When one mole each of alcohol, thionyl chloride, and pyridine are allowed to react, excellent yields of alkyl chloride are obtained, even from primary alcohols. Because of the electron-attracting properties of the two oxygen atoms and the chlorine atom, the OSOCl group leaves readily and the pyridine removes the proton from the protonated chlorosulfite, leaving the necessary chloride ion for the S_N2 reaction.

$$\underset{\mathrm{H}}{\mathrm{R\ddot{O}}}\!: + \mathrm{SOCl(Cl)} \rightleftharpoons \underset{\mathrm{H}}{\mathrm{R\overset{+}{O}}}\mathrm{-SOCl} + {}^{-}\!:\!\mathrm{Cl} \xrightarrow{C_5H_5N} C_5H_5NH^+ + \mathrm{ROSOCl}$$

$$\mathrm{Cl^-} + \mathrm{ROSOCl} \longrightarrow \mathrm{ClR} + [{}^{-}\mathrm{OSOCl}] \longrightarrow SO_2 + \mathrm{Cl^-}$$

If two moles of alcohol, one of thionyl chloride, and two of pyridine are allowed to react, an excellent yield of alkyl sulfite is produced. Apparently the replacement of the chlorine by the less electronegative OR group makes the OSOOR group a sufficiently poorer leaving group than the OSOCl group to make attack by the weakly nucleophilic chloride ion ineffective.

$$\mathrm{ROH} + \mathrm{ROSOCl} \longrightarrow \mathrm{ROSO(\overset{+}{H}OR)} + \mathrm{Cl^-} \xrightarrow{C_5H_5N} \mathrm{ROSO(OR)} + C_5H_5NH^+Cl^-$$

Halogen acids also convert alcohols to alkyl halides. The reaction involves nucleophilic displacement on carbon by halide ion. Since hydroxide ion is a poor leaving group, the reaction is assisted by protonation of the alcohol or by complex formation with a Lewis acid (p. 34). Where an additional catalyst is needed zinc chloride usually is used as the Lewis acid because it is highly soluble in concentrated hydrochloric acid and does not react with the water present. As with displacement reactions of the alkyl halides, either an S_N2 or an S_N1 mechanism may operate (p. 100).

$$\underset{\mathrm{H}}{\mathrm{R\ddot{O}}}\!: + \mathrm{H\!:\!X} \longrightarrow \underset{\mathrm{H}}{\mathrm{R\overset{+}{O}}}\mathrm{-H} + {}^{-}\!:\!\mathrm{X}$$

$$\mathrm{X\!:^-} + \mathrm{R\overset{+}{O}H_2} \longrightarrow \mathrm{X\!:\!R} + H_2O \qquad S_N2 \text{ reaction}$$

$$\mathrm{R\overset{+}{O}H_2} \longrightarrow H_2O + R^+ \xrightarrow{{}^{-}:X} \mathrm{R\!:\!X} \qquad S_N1 \text{ reaction}$$

The conditions necessary for reaction depend both on the particular halogen acid and on the type of alcohol. Since the order of nucleophilicity in protic solvents is fluoride < chloride < bromide < iodide, and the order of increasing acidity of the acids is in the same direction (p. 32), the reactivity of the halogen acids is hydrogen fluoride < hydrogen chloride < hydrogen bromide < hydrogen iodide. The factors influencing the reactivity of

[10] Reactions for the conversion of an alcohol to a halide should not be referred to as the *halogenation* of an alcohol or the *substitution* of hydroxyl by halogen (cf. p. 69) but as the *replacement* of hydroxyl by halogen. To avoid this rather cumbersome phrase, it has been suggested that such reactions be designated by the name of the entering group followed by *de*, the name of the departing group, and the suffix *ation*. Thus the replacement of hydroxyl by bromine would be called *bromodehydroxylation*.

alcohols are the same as those that influence the reactivity of alkyl halides (pp. 101–104). The order of reactivity of the alcohols by the S_N2 mechanism is methyl > primary > secondary > tertiary. Steric factors greatly influence the reactivity, β methyl groups being particularly effective in decreasing the rate of reaction. Isobutyl alcohol reacts about one tenth as fast as the straight-chain primary alcohols and neopentyl alcohol is practically nonreactive. For the S_N1 mechanism, the order is tertiary > secondary > primary. The intermediate secondary and primary carbonium ions are subject to molecular rearrangement. Thus neopentyl alcohol reacts with hydrogen bromide at a reasonable rate under conditions favorable to the S_N1 mechanism, but the product is *t*-amyl bromide (cf. footnote 4, p. 104).

$$(CH_3)_2C(CH_3)CH_2\overset{+}{O}H_2 \longrightarrow H_2O + (CH_3)_2C(CH_3)-\overset{+}{C}H_2 \longrightarrow (CH_3)_2\overset{+}{C}(CH_3)-CH_2 \xrightarrow{Br^-} (CH_3)_2C(Br)CH_2CH_3$$

Neopentyl iodide can be prepared in 74 per cent yield[11] by refluxing the alcohol with a mixture of phenyl phosphite (p. 272) and methyl iodide. The hydroxyl is converted to a good leaving group without the aid of strong acids. In the following reactions R = neopentyl.

$$(C_6H_5O)_3P\!: + CH_3I \longrightarrow (C_6H_5O)_3\overset{+}{P}CH_3 + I^- \xrightarrow{ROH}$$

$$C_6H_5OH + (C_6H_5O)_2\overset{+}{P}(CH_3)OR + I^- \longrightarrow (C_6H_5O)_2P(CH_3)\overset{\pm}{=}O + RI$$

From a practical preparative viewpoint the following summary may be made. In general the over-all order of reactivity of the alcohols is tertiary > secondary > primary. Primary and secondary alkyl chlorides are prepared best from concentrated hydrochloric acid along with zinc chloride. They may be made also by reaction of the alcohol with thionyl chloride and pyridine using the mole ratio of 1:1:1. Tertiary alkyl chlorides are formed readily from concentrated hydrochloric acid alone. Primary and secondary bromides are obtained from alcohols and phosphorus tribromide at 0°, anhydrous hydrogen bromide, or concentrated aqueous hydrobromic acid and sulfuric acid. Aqueous hydrobromic acid suffices for tertiary bromides. All iodides are made best from phosphorus triiodide, the actual reagents used being iodine and phosphorus, which react to give phosphorus triiodide.

$$2\,P + 3\,I_2 \longrightarrow 2\,PI_3$$

Although hydrogen iodide reacts readily with alcohols, it has the disadvantage that it is a strong reducing agent and may reduce the alkyl iodide to the hydrocarbon (cf. p. 37).

$$R{-}I + H{-}I \longrightarrow \begin{matrix} R\cdots H \\ \vdots \quad \vdots \\ I\cdots I \end{matrix} \longrightarrow R{-}H + I_2$$

The *Lucas test* for **distinguishing among primary, secondary, and tertiary alcohols** is based on their relative rates of reaction with hydrogen chloride. The reagent is a solution of zinc chloride in concentrated hydrochloric acid. The lower alcohols all dissolve in this reagent because of oxonium salt formation (p. 123), but the alkyl chlorides are insoluble. Tertiary alcohols react so rapidly that it is difficult to detect solution, the chloride separating immediately. Secondary alcohols give a clear solution at first, which becomes cloudy within five minutes with eventual separation into two layers. Primary alcohols dissolve, and the solution remains clear for several hours.

[11] Because organic reactions rarely go to completion and usually take several courses simultaneously, the amount of desired product actually obtained is expressed as *per cent yield,* that is, as the per cent of the amount of product that would have been obtained if no side reactions had taken place, and if the reaction had gone to completion. If unreacted starting material is recovered, the yield is based on the amount consumed in the reaction.

7. ***Dehydration to Olefins.*** Bases do not cause elimination of water from a simple alcohol analogous to the elimination of hydrogen halide from an alkyl halide (p. 105). The hydroxide ion is a poor leaving group because of its nucleophilicity, and the removal of a proton from the hydroxyl by a base gives the alkoxide ion $RO{:}^-$ from which the oxygen has still less tendency to leave as the doubly charged oxide ion. Moreover the hydroxyl group is not as electronegative as halogen and does not increase the acidity of the β hydrogen as much as does halogen. On the other hand, acids catalyze dehydration to olefins. Protonation of the hydroxyl group permits it to leave as a water molecule. The positive charge on the resulting carbonium ion greatly increases the acidity of a β hydrogen making it readily removable by another alcohol molecule or a negative ion. A nonvolatile acid such as sulfuric, phosphoric, or *p*-toluenesulfonic acid (p. 407) is used as a catalyst, and the olefin either is distilled from the reaction mixture or separates as an insoluble layer, thus driving the reaction to completion. If the alcohol is volatile, the process may be made continuous by passing the vapor through a tube packed with pumice impregnated with sulfuric or phosphoric acid and heated to the required temperature.

$$RCH_2CH(OH)R + H_2SO_4 \rightleftharpoons {}^-SO_4H + RCH_2CH(\overset{+}{O}H_2)R \underset{}{\overset{150^\circ}{\rightleftharpoons}} H_2O + RCH_2\overset{+}{C}HR \overset{ROH}{\rightleftharpoons} RCH{=}CHR + R\overset{+}{O}H_2$$

The order of reactivity for the alcohols is tertiary > secondary > primary, reflecting the order of stability of carbonium ions (p. 83).

In the removal of halogen acid from alkyl halides by bases, only the nonterminal alkyl halides give a mixture of alkenes (p. 105) because carbanion intermediates ordinarily do not undergo molecular rearrangements. In the acid-catalyzed dehydration of alcohols, however, the intermediate is a carbonium ion, and rearrangements may take place (p. 69). In the acid-catalyzed dehydration of either 1-butanol or 2-butanol, one of the carbonium ion intermediates is the same as that formed in the acid-catalyzed isomerization of 1-butene (p. 89).

$$\underset{\text{1-Butanol}}{CH_3CH_2CH_2CH_2OH} \underset{B^-}{\overset{HB}{\rightleftharpoons}} CH_3CH_2CH_2CH_2\overset{+}{O}H_2 \underset{H_2O}{\rightleftharpoons} CH_3CH_2CH_2\overset{+}{C}H_2$$

$$CH_3CH_2CH_2\overset{+}{C}H_2 \underset{B^-}{\overset{HB}{\rightleftharpoons}} \underset{\text{1-Butene}}{CH_3CH_2CH{=}CH_2}$$

$$\underset{\text{2-Butenes}}{CH_3CH{=}CHCH_3} \underset{HB}{\overset{B^-}{\rightleftharpoons}} CH_3CH_2\overset{+}{C}HCH_3 \underset{B^-}{\overset{HB}{\rightleftharpoons}} \underset{\text{1-Butene}}{CH_3CH_2CH{=}CH_2}$$

$$CH_3CH_2\overset{+}{C}HCH_3 \overset{H_2O}{\rightleftharpoons} CH_3CH_2CH(\overset{+}{O}H_2)CH_3$$

$$\underset{\text{2-Butanol}}{CH_3CH_2CH(OH)CH_3} \underset{B^-}{\overset{HB}{\rightleftharpoons}} CH_3CH_2CH(\overset{+}{O}H_2)CH_3$$

As expected, either 1-butanol or 2-butanol yields a mixture of olefins of approximately the same composition as that obtained from 1-butene, that is, chiefly the 2-butenes.

Because of the electron-donating properties of alkyl groups, methide ions would be expected to migrate more readily from tertiary or quaternary carbon atoms than from secondary carbon atoms. Thus although no 2-methylpropene has been found in the products from 2-butanol, the acid-catalyzed dehydration of 3,3-dimethyl-2-butanol yields very little of the expected *t*-butylethylene, the chief products being the result of a methide ion shift.

$$(CH_3)_2\underset{\displaystyle CH_3}{C}-\underset{\displaystyle \overset{+}{O}H_2}{C}HCH_3 \longrightarrow H_2O + (CH_3)_2-\underset{\displaystyle CH_3}{C}-\overset{+}{C}HCH_3 \rightleftharpoons \underset{\text{3 per cent}}{(CH_3)_3CCH{=}CH_2} + H^+$$

$$\Big\updownarrow \text{Methide ion 1,2 shift}$$

$$(CH_3)_2\overset{+}{C}-\underset{\displaystyle CH_3}{C}HCH_3 \rightleftharpoons CH_2{=}\underset{\displaystyle CH_3}{C}-\underset{\displaystyle CH_3}{C}HCH_3 + H^+ \quad \text{31 per cent}$$

$$\Big\updownarrow$$

$$\underset{\text{61 per cent}}{(CH_3)_2C{=}C(CH_3)_2} + H^+$$

The percentages given are those reported from an analysis of the products and indicate ratios of 3:33:64. From the recorded free energies of formation of the hexenes, the ratios should be 0.5:17.8:81.7. Although the structure of neopentyl alcohol does not appear to permit an elimination reaction, it can be dehydrated with molecular rearrangement to trimethylethylene (cf. p. 69).

$$CH_3\overset{\displaystyle CH_3}{\underset{\displaystyle CH_3}{C}}CH_2\overset{+}{O}H_2 \longrightarrow H_2O + CH_3\overset{\displaystyle CH_3}{\underset{\displaystyle CH_3}{C}}-\overset{+}{C}H_2 \longrightarrow CH_3\overset{\displaystyle CH_3}{\underset{+}{C}}-CH_2CH_3 \longrightarrow (CH_3)_2C{=}CHCH_3 + H^+$$

It often is stated that rearrangements of carbonium ions take place because the order of stability is tertiary > secondary > primary. Although it is true that the ease of formation of carbonium ions is in this order (p. 104), the final composition depends on the relative thermodynamic stability of the products. The isomerization of *n*-butane to *i*-butane, for example involves a methide ion shift that converts a secondary carbonium ion to a primary carbonium ion (p. 69).

Molecular rearrangements of alkenes, alkyl halides, or alcohols that involve a carbonium ion and a change in the carbon skeleton usually are called *Wagner-Meerwein* rearrangements. Because of the possibility of molecular rearrangements during chemical reactions, the structure of new compounds must be arrived at by synthetic as well as degradative procedures. The structures of the above olefins can be determined by ozonolysis and identification of the products (p. 89).

Alcohols may be converted to olefins also by passing the vapors over activated[12] alumina at 350°. Pure alumina is a Lewis acid and yields a mixture of olefins. If alumina that has been treated with alkali is used as the catalyst, rearrangement largely is avoided and yields of 1-alkenes as high as 97 per cent can be obtained from 1-alkanols. Even 2-alkanols yield chiefly 1-alkenes if thoria is used as the catalyst.

8. ***Dehydrogenation (Oxidation) to Aldehydes and Ketones.*** Many metals absorb hydrogen directly at temperatures ranging from 25° to 700°. The alkali metals react with hydrogen to give hydrides such as sodium hydride, $Na^{+-}H$, and lithium hydride, $Li^{+-}H$, which definitely are salts. Other metals, such as copper, iron, nickel, and palladium, absorb hydrogen with little change in the metal lattice. Apparently the hydrogen occupies holes in the crystal lattice. These solutions of hydrogen have strong reducing properties, indicating that the hydrogen is present in atomic form; that is, the bond of the hydrogen molecule has been broken, a process that presumably has been aided by interaction of the

[12] Some forms of solid catalysts are more effective than others. When prepared in such a way that they have the maximum activity, they are called *activated* catalysts. Activated alumina has an optimum porosity or surface structure, and an optimum amount of adsorbed water or water of constitution.

free electrons of the metal with the electrons of the hydrogen molecule (p. 81). The adsorbed hydrogen can be driven off as molecular hydrogen by heating to a higher temperature.

In the same way these metals are able to break carbon-hydrogen and oxygen-hydrogen bonds, adsorb the hydrogen atoms and then give them up as molecular hydrogen. The removal of a hydrogen atom leaves an unpaired electron on the atom to which it was attached, and if two such atoms are adjacent, their unpaired electrons can pair to form a double bond. In this way the metal can catalyze the removal of hydrogen from the hydroxyl group and from an adjacent CH group of a primary or secondary alcohol and produce a double bond between carbon and oxygen (cf. the hydrogenation of alkenes and dehydrogenation of alkanes, pp. 81, 587).

$$\begin{array}{c} R_2C{:}H \\ | \\ O{:}H \end{array} + \left|\begin{array}{l} \cdot Cu \\ \\ \cdot Cu \end{array}\right. \longrightarrow \begin{array}{c} R_2C\cdot \\ | \\ O\cdot \end{array} + \begin{array}{c} H\cdot \\ \\ H\cdot \end{array}\left|\begin{array}{l} \cdot Cu \\ \\ \cdot Cu \end{array}\right.$$

$$\downarrow \qquad \downarrow$$

$$\begin{array}{c} R_2C \\ \| \\ O \end{array} + \begin{array}{c} H \\ | \\ H \end{array} + \left|\begin{array}{l} \cdot Cu \\ \\ \cdot Cu \end{array}\right.$$

In practice the alcohol vapor is passed over a copper-zinc alloy known as *brass spelter* at 300–325°. Primary alcohols yield aldehydes and secondary alcohols yield ketones (p. 193).

$$RCH_2OH \underset{325^\circ}{\overset{Cu}{\rightleftarrows}} \underset{\text{An aldehyde}}{\overset{H}{\overset{|}{R C}}{=}O} + H_2$$

$$R_2CHOH \underset{325^\circ}{\overset{Cu}{\rightleftarrows}} \underset{\text{A ketone}}{R_2C{=}O} + H_2$$

The reactions are reversible and do not go to completion. The aldehyde or ketone is removed from the unreacted alcohol by fractional distillation and the alcohol is recycled. If air is used along with the alcohol and at a higher temperature, the hydrogen is burned to water, and the reaction goes to completion.

$$RCH_2OH + \tfrac{1}{2}\,O_2 \xrightarrow[600^\circ]{\text{Cu or Ag}} RCHO + H_2O$$

$$R_2CHOH + \tfrac{1}{2}\,O_2 \xrightarrow[600^\circ]{\text{Cu or Ag}} R_2CO + H_2O$$

Tertiary alcohols, which do not have a hydrogen atom on the carbon atom united to the hydroxyl group, do not undergo these reactions but may dehydrate if the temperature is sufficiently high.

The same type of dehydrogenation can be brought about by chemical oxidizing agents. The reagent most commonly employed is chromic acid in aqueous sulfuric acid (*sodium dichromate and sulfuric acid*), or chromic anhydride (*chromium trioxide*) dissolved in glacial acetic acid.

$$3\,RCH_2OH + Na_2Cr_2O_7 + 4\,H_2SO_4 \longrightarrow 3\,RCHO + Na_2SO_4 + Cr_2(SO_4)_3 + 7\,H_2O$$

$$3\,R_2CHOH + 2\,CrO_3 + 6\,HC_2H_3O_2 \longrightarrow 3\,R_2CO + 2\,Cr(C_2H_3O_2)_3 + 6\,H_2O$$

The formation of aldehydes by this method is successful only if the aldehyde is removed rapidly from the oxidizing mixture by distillation; otherwise it is oxidized further to an organic acid (p. 205).

The mechanism of the oxidation of primary and secondary alcohols to aldehydes and ketones in acidic solution appears to follow the same course for most oxidizing agents, probably through the formation of an unstable ester that decomposes by a two-electron transfer mechanism. Protonation of the chromate portion of the ester makes it a better leaving group and accounts for acid catalysis of the reaction.

$$R_2CHOH + H_2CrO_4 + H^+ \longrightarrow H_2O + R_2\underset{H}{C}{-}O{-}\overset{:\ddot{O}:}{\underset{:\ddot{O}:H}{\overset{+}{Cr}}}{:}OH \longrightarrow H_3O^+ + R_2C{=}O + OCr(OH)_2$$

Disproportionation of Cr(IV) to Cr(V) and Cr(III) takes place rapidly, and $HCrO_3$ can oxidize a second mole of alcohol by the same mechanism as that for the oxidation by H_2CrO_4.

$$2\,H_2CrO_3 \longrightarrow HCrO_3 + HCrO_2 + H_2O$$

$$R_2CHOH + HCrO_3 \longrightarrow R_2C{=}O + HCrO_2$$

The formation of chromic sulfate in the presence of sulfuric acid accounts for the over-all stoichiometry of the reaction.

$$2\,HCrO_2 + 3\,H_2SO_4 \longrightarrow Cr_2(SO_4)_3 + 4\,H_2O$$

Tertiary alkyl esters of oxidizing acids cannot decompose by this mechanism because they lack the necessary hydrogen on carbon. Hence they are fairly stable to oxidation. When oxidation does occur, the carbon chain is broken and the products have fewer carbon atoms than the original alcohol. Presumably the tertiary alcohol is converted to olefins by dehydration, and the final products are formed by oxidation of the olefins with breaking of the carbon chain. The stability of tertiary alcohols to oxidation permits the use of a solution of chromium trioxide in *t*-butyl alcohol (*t-butyl chromate*) to oxidize higher primary alcohols to aldehydes in good yields. Milder procedures for oxidizing alcohols to aldehydes and ketones have been developed.

VIBRATIONAL OR INFRARED SPECTRA

The positions of atoms in molecules are mean equilibrium positions, and the bonds between atoms may be looked upon as springs subject to stretching, bending, or twisting. Each atom and group of atoms in a molecule oscillates about a point at which the attraction of nuclei for electrons balances the repulsion of nuclei for nuclei and electrons for electrons. These oscillations have natural periods, which depend on the mass of the atoms and the strength of the bonds involved. The amplitude of the oscillations can be increased by supplying energy through the action of an external force. Because nuclei and electrons bear electric charges, the force necessary can be supplied by the oscillating electric vector of an electromagnetic wave (p. 542) of frequency and phase matching those of a particular molecular vibration. For energy to be transferred from the electromagnetic wave to the molecular vibration it is necessary also that a change in amplitude of that vibration results in a change of molecular dipole moment. Under these conditions some of the radiant energy is absorbed, and the intensity of the radiation at this particular wavelength is decreased in passing through the substance.

The motions of the atoms in molecules may be divided into stretching and bending vibrations. The **stretching frequency** as derived from Hooke's law is given by the equation

$$\tilde{\nu} = \frac{1}{2\pi c}\left(\frac{f}{M}\right)^{1/2}$$

where $\tilde{\nu}$ = vibrational frequency (cm.$^{-1}$),[13] c = velocity of light (cm. sec.$^{-1}$), and f = force constant (dynes cm.$^{-1}$). M is the reduced mass as defined by the equation $M = \frac{M_1 M_2}{M_1 + M_2}$ where M_1 and M_2 are the masses of atoms 1 and 2. The force constant is a measure of the resistance of the bond to stretching and is roughly proportional to the strength of the bond.

The frequencies of stretching vibrations of bonds between carbon, hydrogen, oxygen, nitrogen, sulfur, and halogen atoms lie in the infrared region in the range 4000 to 500 cm.$^{-1}$. The energy absorption pattern of a molecule in this region commonly is referred to by the organic chemist as the *infrared spectrum* of a compound. Although the absorption of energy is quantized, each vibrational energy change is accompanied by a number of smaller quantized rotational energy changes. Hence with the resolution ordinarily available, absorption appears as a band rather than as a line.

The infrared spectrum can be determined by means of commercially available instruments known as *infrared spectrometers*. The mechanical operation consists of placing a sample of the substance, in the form of a film or solution, or supported in a potassium bromide disc, in the path of the source of infrared radiation, usually a rod of carborundum heated to 1200°. As little as a few micrograms of sample may suffice. The instrument automatically scans the transmitted radiation over the desired range and measures and records on a chart the per cent transmission as a function of the frequency. In such spectra the absorption maxima or peaks appear as minima or valleys.

From the equation relating frequency to force constant and masses, it is clear that bonds between hydrogen and carbon, oxygen, or nitrogen will absorb at higher frequency than bonds between the three heavier elements. Not only are the bond energies and thus the force constants larger, but the reduced masses are smaller. Since the reduced masses of OH, NH, and CH are nearly the same, the order of decreasing frequency of the bonds will be the same as the order of decreasing bond strength, namely O—H > N—H > C—H. The observed frequency ranges for nonhydrogen-bonded molecules containing these groups are respectively 3650–3590, 3500–3300, and 3300–2840 cm.$^{-1}$. For single bonds between carbon and the heavier elements, the stretching absorption lies between 1500 and 500 cm.$^{-1}$. The effect of an increase in the reduced mass is illustrated further by the shift in frequency from 3600 cm.$^{-1}$ for the stretching vibration of the O—H bond to 2630 cm.$^{-1}$ for the O—D bond, both of which have the same bond energy.

The intensity of absorption depends on the magnitude of the change in dipole moment of the bond during the transition and also is directly proportional to the number of bonds in the molecule responsible for the absorption band. Thus hydrogen or carbon bonded to oxygen or nitrogen results in strong infrared absorption because of the polarity of the bonds. Although the carbon-hydrogen bond is only slightly polar, the absorption band usually is strong because most organic compounds have a large number of carbon-hydrogen bonds. The carbon-carbon double bond has a much larger force constant than the carbon-carbon single bond and absorbs at higher frequency (1680–1620 cm.$^{-1}$). The absorption usually is weak, however, because of the low polarity of the bond, and it often is difficult to detect. No absorption at all results from stretching vibrations if the double bond is symmetrically substituted.

---

[13] **Frequency** is the number of cycles passing a given point in space per unit time as the wave travels through space. Spectroscopists usually characterize the frequency of a vibration by its **wave-number,** which is the number of cycles per centimeter of length along the path of the wave and is expressed as reciprocal centimeters or cm.$^{-1}$ Hence frequency expressed as reciprocal centimeters is 10,000 divided by the wavelength in microns.

The total number of independent vibrational motions possible for a nonlinear molecule of n atoms is $3n - 6$, and each has a characteristic fundamental band frequency. Motions that are not stretching vibrations are classed as **bending vibrations** or **deformations.** The various types are described as scissoring, rocking, twisting, or wagging vibrations. The energy required for bending vibrations falls in the 1600–500 cm.$^{-1}$ range. Because of the large number of bands and their tendency to overlap, it often is difficult to use this spectral region to identify structural features. The region is, however, extremely characteristic of the molecule as a whole. It is called the *profile* or *fingerprint region* and is used to establish the identity or nonidentity of two compounds, for example, a synthetic and a natural product.

In addition to the bands resulting from fundamental modes of vibration, **overtones** may appear at wavelengths corresponding to frequencies that are approximately an integral multiple of the fundamental frequency. These bands usually are very weak, but the intensity may amount to as much as 10 per cent of that of the fundamental frequency. If the latter is a strong band, the overtone may be as strong as some of the weaker bands caused by fundamental vibrations (p. 195).

Figure 8–*3* is the infrared spectrum for liquid *n*-propyl alcohol. The band at 3330 cm.$^{-1}$ is broad and at lower frequency than that characteristic for unassociated alcoholic hydroxyl because of strong hydrogen bonding. In the vapor phase or in sufficiently dilute solutions in a nonpolar solvent, a sharp band is observed at around 3600 cm.$^{-1}$. The weak absorption at 3660 cm.$^{-1}$ is due to a small amount of hydrogen-bonded water present in the alcohol. The partially resolved band around 2900 cm.$^{-1}$ results both from methyl and from methylene C—H stretching vibrations. The band at 1230 cm.$^{-1}$ may be due to OH deformation and that at 1050 cm.$^{-1}$ to C—O stretching absorption, but assignments in this region are uncertain.

Infrared absorption spectrometry is of value not only for the identification of structural features in an unknown compound and for the establishment of identity. It is used routinely in the laboratory before working up a reaction mixture to determine whether a desired reaction has taken place. It is used also to detect impurities in a compound. Since the intensity of absorption is proportional to the concentration, it can be used for the quantitative estimation of the components of a mixture.

In discussions of the physical properties of other functional groups, mention will be made of the regions in the infrared in which absorption takes place. The interpretation of infrared spectra requires much practice and no attempt is made in this text to go into the subject in detail.

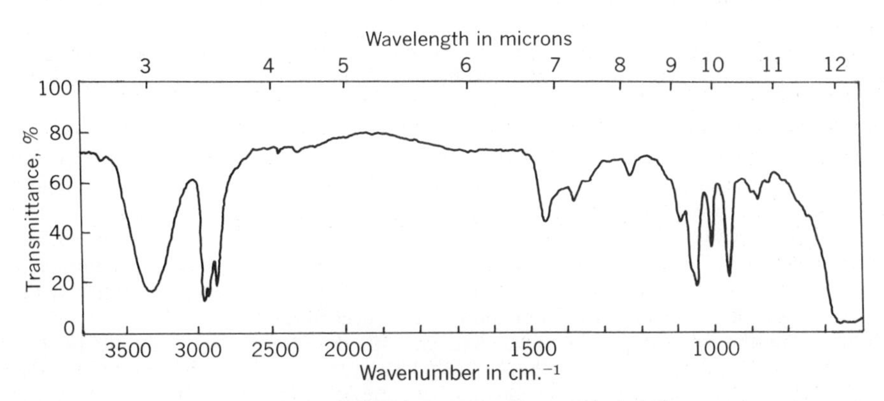

Figure 8–*3*. Infrared spectrum of *n*-propyl alcohol.

PROBLEMS

8–*1*. Write skeleton structures for all of the secondary hexyl alcohols and name them by the international system.

8–*2*. How many isomeric alcohols have the same carbon skeleton as (*a*) 2,3,3-trimethylpentane, (*b*) isooctane, (*c*) neohexane, (*d*) *n*-hexane, and (*e*) 2,5-dimethylhexane?

8–*3*. How many primary, secondary, and tertiary alcohols are in each of the above sets?

8–*4*. Write condensed structural formulas for the following alcohols: (*a*) *n*-amyl alcohol, (*b*) dimethylethylcarbinol, (*c*) 2,2-dimethyl-3-pentanol, (*d*) *i*-hexyl alcohol, (*e*) diethyl-*s*-butylcarbinol, (*f*) 3-ethyl-2-hexanol, (*g*) neopentyl alcohol, (*h*) di-*i*-propylcarbinol, (*i*) 4-methyl-2-*i*-propyl-1-pentanol, (*j*) active amyl alcohol.

8–*5*. Which of the following compounds do not belong to the group of isomeric octyl alcohols: (*a*) *n*-propyl-*i*-butylcarbinol, (*b*) 2,6-dimethyl-4-heptanol, (*c*) 4-ethyl-3-hexanol; (*d*) 2,3,4-trimethyl-2-pentanol, (*e*) diethyl-*i*-propylcarbinol, (*f*) 2-methyl-1-octanol?

8–*6*. Tell why each of the following names is unsatisfactory as written and give an acceptable name: (*a*) isopropanol, (*b*) 3-*i*-propyl-1-pentanol, (*c*) methyl ethyl carbinol, (*d*) dimethyl-(2-methylbutyl)carbinol, (*e*) 2-methyl-hexanol-1, (*f*) *s*-amyl alcohol, (*g*) *i*-amylalcohol, (*h*) *s*-butanol, (*i*) 4,4-dimethyl-3-pentanol, (*j*) 3-ethyl-2-butanol.

8–*7*. Place the following compounds in the order of increasing boiling point: (*a*) water, (*b*) ethanol, (*c*) *n*-butane, (*d*) *i*-pentane, (*e*) *n*-pentane.

8–*8*. Place the following compounds in the order of decreasing solubility in water: (*a*) 2-pentanol, (*b*) *n*-butyl alcohol, (*c*) *t*-butyl alcohol, (*d*) 2-pentene, (*e*) 1-octanol, (*f*) *i*-pentane.

8–*9*. List the following compounds in the order of increasing density: (*a*) water, (*b*) *i*-hexane, (*c*) methylene iodide, (*d*) ethyl bromide, (*e*) ethylene bromide, (*f*) bromoform, (*g*) 2-hexanol.

8–*10*. Give reactions illustrating the preparation of the following alcohols from suitable olefins: (*a*) 2-pentanol, (*b*) dimethyl-*n*-propylcarbinol, (*c*) *t*-amyl alcohol, (*d*) triethylcarbinol.

8–*11*. Give a reaction for the preparation of (*a*) an alkoxide from *n*-nonyl alcohol, (*b*) an alkene from methyl-*i*-butylcarbinol, and (*c*) an inorganic ester from 2,3-dimethyl-1-pentanol. Name the organic product.

8–*12*. Give a series of reactions for the following conversions: (*a*) *n*-butyl alcohol to *s*-butyl alcohol, (*b*) 2-methyl-1-butene to 2-methyl-2-butene, (*c*) ethyl-*i*-propylcarbinol to 2-methyl-2-chloropentane, (*d*) 2-propanol to 1-propanol.

8–*13*. List the following alcohols in the order of (*a*) increasing ease of oxidation by chromium trioxide and (*b*) increasing ease of acid-catalyzed dehydration: (*1*) 1-butanol, (*2*) 2-butanol, (*3*) 2-methyl-2-propanol, (*4*) 3-methyl-2-butanol, (*5*) 2-methyl-1-propanol.

8–*14*. Give equations for the following reactions and name the organic product: (*a*) sodium *n*-butoxide and phosphorus oxychloride, (*b*) ethyl nitrite and dry hydrogen chloride, (*c*) methyl borate and water.

8–*15*. Give and explain a chemical test that could be used to distinguish between the members of each of the following pairs of compounds: (*a*) *n*-propyl alcohol and *i*-propyl alcohol, (*b*) ethyl nitrite and ethyl nitrate, (*c*) sodium *n*-butoxide and aluminum *n*-butoxide, (*d*) methanol and hexane, (*e*) *n*-propyl sulfate and *n*-undecane, (*f*) *n*-butyl iodide and *n*-octane, (*g*) 1-tridecene and ethyl phosphate.

8–*16*. When 2.22 g. of a compound containing only hydroxyl groups reacted completely with metallic sodium, 336 cc. of hydrogen (S.T.P.) was evolved. What is the equivalent weight of the compound with respect to hydroxyl?

8–*17*. Pair the following types of mechanism with the appropriate reaction: (*a*) free radical, (*b*) S_N1, (*c*) S_N2, (*d*) *E* 1, and (*e*) *E* 2; (*1*) ammonia with *n*-butyl chloride, (*2*) dehydration of *n*-butyl alcohol, (*3*) autoxidation, (*4*) hydrolysis of *t*-butyl chloride, and (*5*) action of sodium cyanide on *t*-butyl chloride.

8–*18*. If a 20 per cent excess of sodium bromide and sulfuric acid is used and if the yield is 90 per cent, calculate the amount of alcohol and sodium bromide that must be used to prepare 100 g. of *i*-propyl bromide.

8–*19*. Balance the reactions proposed for the various steps of the oxidation of a secondary alcohol to a ketone by sodium dichromate and sulfuric acid, and show that they account for the over-all stoichiometry of the reaction.

8–*20*. Indicate by reactions the probable mechanistic steps for the hydrolysis of boric acid esters to alcohols.

CHAPTER NINE

ETHERS

The simplest ethers have the general molecular formula $C_nH_{2n+2}O$, and hence are isomeric with the alcohols. Unlike the alcohols, no ether is known containing only one carbon atom. The first member of the series has the molecular formula C_2H_6O. If the formula CH_3CH_2OH is assigned to ethyl alcohol, the only other structural formula possible for an isomer is that in which the oxygen atom joins two carbon atoms. Hence the first ether should contain the structure C—O—C. By placing the hydrogen atoms on the two carbon atoms to give the compound CH_3OCH_3, all of the rules of valence are satisfied. The general formula for the ethers is ROR. The methods of preparation and chemical properties of the ethers confirm this structure.

Nomenclature

The word *ether* (Gr. *aither* applied to the material filling heavenly space) was given to ethyl ether because of its volatility. The ethers commonly are divided into two groups, the simple ethers, ROR, in which both R groups are alike, and the mixed ethers, ROR′, in which the R groups are different. They generally are designated by naming the alkyl groups and adding the word *ether*. For the simple ethers no indication need be given that two alkyl groups are present in the molecule; $(CH_3)_2O$ is methyl ether, $CH_3OC_2H_5$ is methyl ethyl ether, and $(C_2H_5)_2O$ is ethyl ether.[1]

The group RO is known as an *alkoxyl group*, although as a substituent it is called *alkoxy*. In the international system the ethers are named as alkoxy derivatives of the longest hydrocarbon chain. Methyl ether is methoxymethane, and methyl ethyl ether is methoxyethane. This system usually is used only when other functional groups are present. For example, $CH_3OCH_2CH_2CH_2OH$ may be called 3-methoxy-1-propanol.

Physical Properties

The ethers do not contain hydrogen united to oxygen. Hence there is no tendency for molecules to associate with each other by hydrogen bonding, and the boiling points are nearly normal. Methyl ether boils at $-24°$ and propane at $-42°$, methyl ethyl ether boils at 6° and *n*-butane at $-0.5°$, and ethyl ether boils at 35° and *n*-pentane at 36°.

Ethers have, however, unshared pairs of electrons on oxygen and can form hydrogen bonds with water molecules. As would be expected, the solubilities in water are roughly the same as those of alcohols having the same number of carbon atoms. A saturated solution of ethyl ether in water at 25° contains 5.7 per cent by weight of the ether, and a saturated solution of *n*-butyl alcohol at 25° contains 7.3 per cent by weight of the alcohol. Water is

[1] Since there is no ambiguity in the term methyl ether, it is redundant to call it dimethyl ether. The situation is comparable to naming Na_2SO_4 sodium sulfate rather than disodium sulfate. Similarly the addition product of chlorine to ethylene was called ethylene chloride rather than ethylene dichloride (p. 80), just as $MgCl_2$ is called magnesium chloride rather than magnesium dichloride. This same principle is followed in naming derivatives of other polyvalent compounds such as the ketones (p. 194), sulfides (p. 262), sulfoxides and sulfones (p. 265), esters of polycarboxylic acids (p. 628), and peroxides (p. 86).

soluble in both but only to the extent of 1.3 per cent in ether at 25° compared with 20.3 per cent in *n*-butyl alcohol. Because the C—O—C angle is not linear, ethers are dipolar molecules with dipole moments of 1.2–1.3 D.

The infrared absorption for the C—O stretching vibration of alkyl ethers is in the range 1150–1060 cm.$^{-1}$. It is the only band in the profile region that ordinarily can be assigned with certainty in an unknown compound.

Preparation

Ethers as a rule are made by either of two general methods. In one an alcohol is converted into a good nucleophile that can displace a good leaving group from some other substrate. Thus reaction with metallic sodium gives the alkoxide ion, which can displace halogen from an alkyl halide or alkyl sulfate ion from an alkyl sulfate by an S_N2 reaction. For primary alkyl halides or sulfates, a sufficiently high concentration of alkoxide ion results from the reaction of the alcohol with sodium hydroxide (p. 146).

$$ROH + Na \longrightarrow RO^{-+}Na + \tfrac{1}{2}H_2$$

$$\text{or} \quad ROH + NaOH \rightleftharpoons RO^{-+}Na + H_2O$$

$$RO{:}^- + RX \longrightarrow ROR + {}^-{:}X$$

$$RO{:}^- + ROSO_2OR \longrightarrow ROR + {}^-{:}OSO_2OR$$

This procedure is known as the **Williamson synthesis.**[2] The alkyl groups in the alkoxide ion and in the alkyl halide may be alike or different to give either simple or mixed ethers. Ethers with one tertiary alkyl group may be made by this process using a tertiary alkoxide, but ditertiary ethers cannot be prepared because the basic alkoxide ion removes halogen acid from tertiary alkyl halides to give olefin (p. 106). Tertiary butyl ether has been obtained by the reaction of *t*-butyl chloride with dry silver carbonate.

$$2\,(CH_3)_3CCl + Ag_2CO_3 \longrightarrow (CH_3)_3C{-}O{-}C(CH_3)_3 + 2\,AgCl + CO_2$$

In the second general method the hydroxyl of the alcohol is converted into a better leaving group by formation of the oxonium salt or the acid sulfate.[3]

$$ROH \underset{B^-}{\overset{HB}{\rightleftharpoons}} R\overset{+}{O}H_2 \underset{H_2O}{\overset{ROH}{\rightleftharpoons}} R\underset{\substack{|\\H}}{\overset{+}{O}}R \underset{HB}{\overset{B^-}{\rightleftharpoons}} ROR$$

$$ROH \underset{^-OSO_3H}{\overset{HOSO_3H}{\rightleftharpoons}} R\overset{+}{O}H_2 \underset{H_2O}{\overset{^-OSO_3H}{\rightleftharpoons}} ROSO_3H \underset{^-OSO_3H}{\overset{ROH}{\rightleftharpoons}} R\underset{\substack{|\\H}}{\overset{+}{O}}R \underset{H_2SO_4}{\overset{^-OSO_3H}{\rightleftharpoons}} ROR$$

All strong acids, including Lewis acids such as zinc chloride and boron fluoride, catalyze the reaction, but sulfuric acid is most often used. Because dehydration is a competing reaction, the yields are best with the lower primary alcohols, which give ethers that can be distilled from the reaction mixture. Tertiary alcohols yield only olefins under these conditions, but olefins can add primary or secondary alcohols to give mixed ethers (cf. p. 86).

The reaction of sulfuric acid with ethanol is a good example of the way in which the course of an organic reaction may be modified by conditions. At room temperature the chief reaction is the formation of the oxonium salt.

$$C_2H_5OH + H_2SO_4 \rightleftharpoons \left[C_2H_5\overset{+}{O}H_2\right]\left[^-OSO_3H\right]$$

---

[2] Alexander William Williamson (1824–1904), professor at the University College, London. His synthesis of ethyl ether in 1850 cleared up the confusion existing then concerning the constitution of the alcohols and ethers. He also was the first to synthesize ortho esters (p. 177), and to assign the correct structure to acetone (p. 193).

[3] For the convention used in these consecutive reactions, see page 87.

If ethanol is warmed with an excess of sulfuric acid, ethyl hydrogen sulfate is formed.

$$C_2H_5OH + H_2SO_4 \rightleftharpoons C_2H_5OSO_3H + H_2O$$

If ethyl hydrogen sulfate is heated to the point where it decomposes (above 150°), ethylene is formed.

$$C_2H_5OSO_3H \rightleftharpoons CH_2{=}CH_2 + H_2SO_4$$

If sulfuric acid is allowed to react with an excess of alcohol and the mixture heated under reduced pressure, ethyl sulfate distills.

$$C_2H_5OSO_3H + C_2H_5OH \rightleftharpoons (C_2H_5)_2SO_4 + H_2O$$

If ethyl hydrogen sulfate is heated to 140–150° and alcohol added below the surface, ethyl ether distills.

$$C_2H_5OSO_3H + C_2H_5OH \rightleftharpoons (C_2H_5)_2O + H_2SO_4$$

Since all of these reactions are reversible, all of the reactants and products are in equilibrium with each other. To prepare any particular compound, the optimum conditions must be chosen for obtaining it rather than the unwanted compounds.

Reactions

Lacking hydrogen attached to oxygen, the ethers do not react with active metals or strong bases and cannot be dehydrogenated. Otherwise their reactions are qualitatively the same as those of alcohols. Like hydroxide ion, alkoxide ion is a poor leaving group (p. 130) and is displaced readily by nucleophiles only after it has been protonated or complexed with a Lewis acid. Because the oxygen atom contains unshared pairs of electrons, it would be expected that ethers should form oxonium salts. The basicity of the aliphatic acyclic ethers ($pK_a \sim -4$) is only one hundredth that of water, however, and solutions of hydrogen chloride in ethers do not conduct the electric current. Hydrogen chloride dissolves in ethers because of strong hydrogen bonding, and the greater solubility of ethers in concentrated aqueous solutions of strong acids than in water probably results from strong hydrogen bonding with hydronium ions. Nevertheless, low concentrations of oxonium ions may be present and usually are postulated to account for acid-catalyzed reactions of ethers (p. 141).

Molecules that have an empty orbital in their valence shells form definite coordination complexes with ethers.

$$\begin{array}{c}\mathrm{R{:}\ddot{\underset{..}{O}}{:}} \\ \mathrm{R}\end{array} + BF_3 \rightleftharpoons \begin{array}{c}\mathrm{R{:}\ddot{\underset{..}{O}}{:}BF_3} \\ \mathrm{R}\end{array}$$

$$2\,R_2O + MgX_2 \rightleftharpoons \begin{array}{c}\mathrm{R} \\ \mathrm{R{:}\ddot{O}{:}} \\ \mathrm{X{:}\ddot{\underset{..}{Mg}}{:}\ddot{\underset{..}{O}}{:}R} \\ \mathrm{X\ \ R}\end{array} \qquad 2\,R_2O + RMgX \rightleftharpoons \begin{array}{c}\mathrm{R} \\ \mathrm{R{:}\ddot{O}{:}} \\ \mathrm{R{:}\ddot{\underset{..}{Mg}}{:}\ddot{\underset{..}{O}}{:}R} \\ \mathrm{X\ \ R}\end{array}$$

It is because of the last reaction that some ethers dissolve alkylmagnesium halides and are used as solvents in preparing Grignard reagents (p. 108). Not all ethers are suitable solvents because the complexes sometimes are insoluble in an excess of the solvent, for example those with *i*-propyl ether or with dioxane (p. 527).

Trialkyloxonium salts have been prepared from ethers. Thus ethyl ether reacts with ethyl bromide in the presence of silver fluoroborate to give **triethyloxonium fluoroborate** and silver bromide.

$$(C_2H_5)_2O + C_2H_5Br + AgBF_4 \longrightarrow (C_2H_5)_3O^{+-}BF_4 + AgBr$$

Other reactions of the ethers involve scission of a carbon-oxygen bond. The usual reagents are strong acids or Lewis acids, and it seems likely that under the conditions of high concentration of acid and elevated temperatures, oxonium salt formation precedes or accompanies nucleophilic attack. An alcohol is one of the initial products, but the conditions necessary to cause the ether to react with the reagent also cause the reagent to react with the alcohol (p. 128).

$$\text{ROR} \underset{I^-}{\overset{HI}{\rightleftharpoons}} \text{R}\overset{+}{\underset{\substack{|\\ \text{H}}}{\text{O}}}\text{R} \overset{I^-}{\rightleftharpoons} \text{RI} + \text{ROH}$$

$$\text{ROH} \underset{I^-}{\overset{HI}{\rightleftharpoons}} \text{R}\overset{+}{\text{O}}\text{H}_2 \overset{I^-}{\rightleftharpoons} \text{RI} + \text{H}_2\text{O}$$

Hydrogen iodide is the most reactive of the halogen acids because it is the strongest acid and iodide ion is the best nucleophile. Hot sulfuric acid also splits ethers.

$$\text{ROR} \underset{^-OSO_3H}{\overset{H_2SO_4}{\rightleftharpoons}} \text{R}\overset{+}{\underset{\substack{|\\ \text{H}}}{\text{O}}}\text{R} \overset{^-OSO_3H}{\rightleftharpoons} \text{ROSO}_3\text{H} + \text{ROH}$$

$$\text{ROH} \underset{^-OSO_3H}{\overset{H_2SO_4}{\rightleftharpoons}} \text{R}\overset{+}{\text{O}}\text{H}_2 \overset{^-OSO_3H}{\rightleftharpoons} \text{ROSO}_3\text{H} + \text{H}_2\text{O}$$

In these reactions the final products are likely to be olefins (p. 130) and their polymers (p. 86). Tertiary alkyl ethers are cleaved easily by an S_N1 mechanism.

$$(\text{CH}_3)_3\text{COC}_2\text{H}_5 \overset{H^+}{\rightleftharpoons} (\text{CH}_3)_3\text{C}\overset{+}{\underset{\substack{|\\ \text{H}}}{\text{O}}}\text{C}_2\text{H}_5 \rightleftharpoons \text{C}_2\text{H}_5\text{OH} + (\text{CH}_3)_3\text{C}^+ \overset{H_2O}{\rightleftharpoons}$$

$$(\text{CH}_3)_3\text{C}\overset{+}{\text{O}}\text{H}_2 \underset{H_3O^+}{\overset{H_2O}{\rightleftharpoons}} (\text{CH}_3)_3\text{COH}$$

Ethers can be cleaved also by heating with anhydrous aluminum chloride.

$$\text{ROR} \overset{AlCl_3}{\rightleftharpoons} \text{R}\underset{\substack{\cdot\cdot\\ \text{AlCl}_3}}{\text{O}}\text{R} \overset{Cl^-}{\rightleftharpoons} \text{RCl} + {}^-\underset{\substack{\cdot\cdot\\ \text{AlCl}_3}}{\text{O}}\text{R} \longrightarrow \text{ROAlCl}_2 + \text{Cl}^-$$

Ethers can be cleaved without subjecting them to strong acids by reaction with triphenylphosphine dibromide.

$$\text{R}_2\text{O} + (\text{C}_6\text{H}_5)_3\text{PBr}_2 \xrightarrow{125^\circ} 2\ \text{RBr} + (\text{C}_6\text{H}_5)_3\text{PO}$$

The reaction of ethers and of other compounds containing alkoxyl groups with hydrogen iodide is important, chiefly because it is the basis of the *Zeisel procedure for the estimation of methoxyl and ethoxyl.* Methyl or ethyl groups linked to oxygen are converted to the volatile methyl or ethyl iodide, which is distilled. The amount of alkyl iodide in the distillate is determined by allowing it to react with alcoholic silver nitrate solution and collecting and weighing the precipitate of silver iodide.

A reaction of ethers that has not found a use, but that always should be kept in mind because of its potential danger, is the attack by atmospheric oxygen (*autoxidation*) to form peroxides.

$$\text{R}_2\text{CHOR}' + \text{O}_2 \longrightarrow \text{R}_2\underset{\substack{|\\ \text{O—OH}}}{\text{C}}\text{OR}'$$

These peroxides or further reaction products are unstable and decompose violently when heated. An instance has been reported in which a can containing *i*-propyl ether that had been exposed to air for some time exploded on being moved. Hence ethers should not be exposed to air unnecessarily and always should be tested for peroxides with acidified potassium iodide or titanium sulfate before distillation or use. In another convenient test for peroxides in liquids a portion is shaken with a globule of clean mercury. In the presence of peroxides a black film of oxidized mercury is formed. Peroxides may be removed by passing the ether through a column of activated alumina or of the hydroxide form of a strong anion-exchange resin (p. 484). Autoxidation can be inhibited by the addition of antioxidants (p. 445).

Uses of Ethers

Methyl ether boils at −24° and can be used as a solvent and extracting agent at low temperatures and as a propellant for aerosol sprays. The preparation of **ethyl ether,** b.p. 35°, first was recorded in the sixteenth century. Because it was made by the sulfuric acid process, it usually contained sulfur compounds as impurities, and it was not until 1800 that it was proved that sulfur was not part of the ether molecule. Since 1846 ethyl ether has been used as a general anesthetic[4] and still is the most widely used substance for this purpose. It is used also as a solvent for fats. It frequently is used in the laboratory as a solvent for extracting organic compounds from aqueous solutions, although it is not always the best solvent for this purpose. The objections are that ether is highly flammable and that it is very soluble in strongly acidic solutions. Moreover it emulsifies readily making separation of the ether and water layers difficult. Finally it usually contains peroxides, which should be removed both for safety and because the peroxides may oxidize the product being extracted. One of the lower-boiling chlorinated hydrocarbons, such as methylene chloride (p. 68) or ethylene chloride (p. 80), may be preferable. One advantage of ether is that the equilibrium distribution of a compound between the water and the ether is attained rapidly because ether and water are appreciably soluble in each other.

Ordinary ether usually contains some water and alcohol, which are detrimental for some purposes such as the preparation of Grignard reagents. Both impurities can be removed if the ether is allowed to stand over metallic sodium. The sodium hydroxide and sodium ethoxide formed are insoluble in ether and are nonvolatile. The pure product is known as **absolute ether.**

***i*-Propyl ether** is a coproduct of the manufacture of *i*-propyl alcohol from propylene (p. 86). ***n*-Butyl ether** (b.p. 142°) and ***i*-amyl ether** b.p. 173°) are made from the alcohols and are used as higher-boiling solvents for Grignard reagents. **Tetrahydrofuran** (p. 512) and **dioxane** (p. 527) are cyclic ethers used as solvents. **Diglyme,** another important solvent, is the methyl ether of β-hydroxyethyl ether (*diethylene glycol,* p. 605). The structural formula $CH_3OCH_2CH_2OCH_2CH_2$-OCH_3 shows that it has three ether-type oxygen atoms.[5]

PROBLEMS

9–*1*. How many compounds can have the molecular formula $C_4H_{10}O$?

9–2. Give the ways in which ethyl *n*-propyl ether can be made from alcohols and alkyl chlorides, bromides, and iodides by the Williamson reaction.

9–*3*. From a handbook, obtain the boiling points of the alcohols and halides used in Problem *2* and tell (*a*) which pair of reagents would permit the easiest separation of the product by distillation and (*b*) which pair would permit easy separation and be easiest to handle.

9–*4*. What reagents yield the same product when they react with either ethyl alcohol or ethyl ether?

9–*5*. Which of the following compounds reacts rapidly with methylmagnesium bromide and slowly with Lucas reagent:

(*a*) $CH_3CH_2CH_2CH_2OH$, (*b*) $(CH_3)_2C{=}CH_2$, (*c*) $CH_3CH_2CHOHCH_3$, (*d*) $(CH_3)_3COH$, (*e*) $(CH_3)_3COCH_3$?

---

[4] A general anesthetic is one that acts on the brain and produces unconsciousness as well as insensitivity to pain. In local anesthesia and spinal anesthesia, only portions of the body are rendered insensitive to pain, and the patient retains consciousness.

[5] As in other fields, frequently used complex words or phrases often have been replaced colloquially by acronyms or by key letters of various parts of the word or phrase. The inconvenience that results for persons not actually working in a particular specialized field is compounded by the fact that these shorthand notations may have been coined from names that already are colloquial rather than systematic. Thus *diglyme* is the acronym for the common name di(ethylene glycol) methyl ether, and *DDT* stands for dichlorodiphenyltrichloroethane. A correct structure for the latter compound can be written only from the systematic name 1,1-*bis*(*p*-chlorophenyl)-2,2,2-trichloroethane.

9–6. What simple chemical tests could be used to distinguish between the members of the following groups of compounds: (*a*) ethanol, *i*-propyl alcohol, and *i*-propyl ether; (*b*) 2-hexene, ethyl ether, and *n*-hexane; (*c*) *n*-butyl chloride, *t*-butyl chloride, and *n*-propyl ether; (*d*) *i*-amyl ether, *n*-amyl bromide, and *n*-octane?

9–7. If 0.1178 g. of a compound gave 0.3137 g. of silver iodide in a Zeisel determination, calculate (*a*) the per cent methoxyl and (*b*) the equivalent weight for the compound.

9–8. Two compounds have the molecular formula $C_5H_{12}O$. One reacts with metallic sodium with evolution of hydrogen, and the other does not. Both react readily with concentrated hydrochloric acid to give a water-insoluble product. What is the structure of each compound?

9–9. Two compounds have the molecular formula $C_4H_{10}O$. The boiling point of one is above 80° and of the other under 50°. Each gives a single product when heated with hydrogen iodide. When passed over neutral alumina at 350°, one gives a gas that is not condensed at the temperature of an ice-salt freezing mixture whereas the other gives a liquid at this temperature. Careful ozonolysis of the liquid gives a one-carbon aldehyde and a three-carbon aldehyde. What are the two compounds?

CHAPTER TEN

SYNTHESIS OF ALKANES AND ALKENES. ALKYNES (ACETYLENES)

ALKANE AND ALKENE SYNTHESIS

As indicated in Chapters 7 and 8, some of the reactions of alcohols and alkyl halides lead to the production of paraffin and olefin hydrocarbons. It now is possible to summarize some of the general methods for the synthesis of paraffin hydrocarbons and olefins having a desired structure.

Syntheses of Alkanes

1. ***Reduction of an Alkyl Halide.*** The direct reduction of alkyl halides may be brought about by a variety of reagents such as zinc and hydrochloric acid in alcoholic solution, sodium and alcohol, sodium amalgam and water, hydrogen iodide, or hydrogen and a catalyst such as platinum. The reaction may be represented by the equation

$$RX + 2\,[H] \longrightarrow RH + HX$$

where [H] represents any of the above reducing agents. If hydrogen iodide is used (p. 129), the reaction is

$$RI + HI \longrightarrow RH + I_2$$

In this reaction red phosphorus frequently is used along with aqueous hydrogen iodide. The phosphorus reacts with the iodine to form phosphorus triiodide, which is hydrolyzed by the water, regenerating hydrogen iodide. In this way the concentration of hydrogen iodide does not decrease, and only red phosphorus is consumed in the reduction. One of the best methods for preparing pure alkanes is the indirect reduction of halides through the Grignard reagents (p. 109).

$$RX + Mg \longrightarrow RMgX$$
$$RMgX + HOH \longrightarrow RH + Mg(OH)X$$

2. ***Wurtz Reaction.*** When a large surface of metallic sodium is in contact with a dilute solution of alkyl halide, the alkylsodium is formed (p. 111). If, however, bulk sodium is allowed to react with pure alkyl halide, the alkyl sodium reacts with excess alkyl halide by an S_N2 displacement of halogen by alkide ion to give a saturated hydrocarbon in which the two alkyl groups are joined to each other.

$$Na^{+}\,{}^{-}{:}R + R{:}X \longrightarrow Na^{+} + R{-}R + X^{-}$$

The over-all reaction, $2\,RX + 2\,Na \longrightarrow R{-}R + 2\,NaX$, is known as the *Wurtz reaction* or *Wurtz*[1] *synthesis*. The best yields are obtained with primary halides. Tertiary halides yield almost exclusively olefin by elimination of halogen acid.

3. ***Reduction of Olefins.*** The double bond of an olefin adds hydrogen quantitatively in the presence of finely divided platinum, nickel, palladium, or other hydrogenation catalysts, with the formation of alkanes (p. 81).

$$RCH{=}CHR + H_2 \xrightarrow[\text{or Pd}]{\text{Pt, Ni,}} RCH_2CH_2R$$

If the unsaturated hydrocarbon is available, for example by the dehydration of an alcohol, this method is an excellent one for alkane synthesis. To convert an alcohol to an alkane, it is easier in general, provided there is no skeletal rearrangement, to use the series of reactions alcohol → alkene → alkane, rather than the series alcohol → alkyl halide → alkane.

Syntheses of Alkenes

1. ***Pyrolysis of Saturated Hydrocarbons.*** Refinery gases from the industrial cracking of petroleum (p. 94) and the gases formed by the cracking of propane or the dehydrogenation of butane (p. 587) are the chief source of the lower alkenes. Cracking of alkanes is not a useful laboratory procedure, however, because the reaction cannot be controlled to give a single product and because the separation of pure products from the complex mixture that is formed is too unrewarding in small-scale operations.

2. ***Dehydration of Alcohols.*** Generally the best procedure for converting alcohols to olefins consists of passing them in the vapor state over hot activated alumina (p. 130).

$$RCH_2CHOHR \xrightarrow[350\text{–}450^\circ]{Al_2O_3} RCH{=}CHR + H_2O$$

This method is less likely to give rearranged products than acid catalysts. For example, if alumina that has been treated with alkali is used, *n*-butyl alcohol gives largely 1-butene with only small amounts of 2-butene.

When acid catalysts are used, molecular rearrangement frequently takes place (p. 89). When rearrangement does not take place, or when the constitution of the rearranged product is known, heating the alcohol to a sufficiently high temperature in the presence of a strong acid such as sulfuric, phosphoric, or *p*-toluenesulfonic acid (p. 407) may be a satisfactory procedure. The last two acids have the advantage that they are relatively nonoxidizing. If the alcohol boils above the temperature required for decomposition, a trace of the acid may be sufficient; otherwise enough acid must be present to keep the alcohol in the form of the nonvolatile oxonium salt. Alternatively, the alcohol vapor may be passed over hot pumice impregnated with phosphoric acid. Some alcohols such as diacetone alcohol (p. 202) dehydrate so easily that a trace of iodine is sufficient to catalyze the decomposition. Alcohols may be dehydrated indirectly by the pyrolysis of their acetates (p. 176) or xanthates (p. 292).

3. ***Removal of Halogen Acid from Alkyl Halides.*** The order of ease of removal of halogen acid from alkyl halides is tertiary > secondary > primary for both the halogen

[1] Charles Adolphe Wurtz (1817–1894), successor to Dumas on the latter's resignation from the faculty of the École de Médecine in 1853, and first occupant of the chair of organic chemistry established at the Sorbonne in 1875. He was the first to synthesize amines in 1849. His synthesis of alkanes was published in 1855. Among his many other contributions are the synthesis of ethylene glycol and ethylene oxide in 1859, the reduction of aldehydes to alcohols in 1866, and the synthesis of aldol in 1872. In a work on the history of chemical principles, he antagonized many chemists of other nations with his opening statement, "Chemistry is a French science. It was founded by Lavoisier of immortal memory."

atom and the hydrogen atom. The usual reagent for secondary halides is an alcoholic solution of potassium hydroxide.

$$RCH_2CHXR + KOH\,(\text{alcoholic}) \longrightarrow RCH{=}CHR + KX + H_2O$$

Alcohol is used as a mutual solvent for the base and the alkyl halide, thus permitting the reaction to take place in a single phase. Potassium hydroxide is preferred to sodium hydroxide because the former dissolves in alcohol much more easily than does sodium hydroxide. Tertiary alkyl halides lose halogen acid so readily that a higher-boiling organic base such as pyridine (p. 514) or quinoline (p. 519) may be used.

Alcoholic sodium or potassium hydroxide is not a suitable reagent for removing halogen acid from primary alkyl halides. In a tenth molar solution of sodium hydroxide in 99 per cent ethyl alcohol, 96 per cent of the base is present as ethoxide ions (p. 125).

$$C_2H_5OH + {}^{-}{:}OH \rightleftharpoons C_2H_5O{:}^{-} + H_2O$$

Hence the primary alkyl halide undergoes nucleophilic attack by ethoxide ions to give chiefly the ether.

$$C_2H_5O{:}^{-} + RX \longrightarrow C_2H_5OR + {}^{-}{:}X$$

Better yields of olefin are obtained by passing the alkyl halide over hot soda lime, a solid mixture of calcium oxide and sodium hydroxide.

4. ***Removal of Two Halogen Atoms from Adjacent Carbon Atoms.*** If a compound contains two halogen atoms on adjacent carbon atoms, the halogen may be removed readily by heating with zinc dust in alcohol, leaving a double bond in the molecule.

$$Zn \; \begin{array}{l} R_2 \\ C{-}X \\ | \\ C{-}X \\ R_2 \end{array} \longrightarrow Zn^{2+} + \begin{array}{c} R_2 \\ C \\ \| \\ C \\ R_2 \end{array} + 2\,X^{-}$$

This reaction is not of much use as a preparative method because the best way to obtain 1,2 dihalides is by adding halogen to an olefin. It occasionally is valuable, however, for the purification of unsaturated compounds if the dihalide can be purified more readily than the olefin itself. The olefin usually is converted to the dibromide, the dibromide purified, and the olefin regenerated by means of zinc dust. Zinc also removes halogen and methoxyl from β-bromoalkyl methyl ethers.

$$\underset{\displaystyle CH_3O \;\; Br}{RCHCHR} + Zn \longrightarrow RCH{=}CHR + Zn(OCH_3)Br$$

5. ***Other Methods of Preparation.*** Olefins can be prepared by the pyrolysis of esters (pp. 176, 292), of amine oxides (p. 231), and of quaternary ammonium hydroxides (p. 231), or by the partial catalytic reduction of acetylenes (p. 148). The double bond can be introduced also by way of the Wittig reaction (p. 428).

ALKYNES (ACETYLENES)

In addition to the olefins and cyclenes, another homologous series of unsaturated hydrocarbons exists, called **acetylenes** from the first member of the series, or **alkynes.** Acetylene has the molecular formula C_2H_2. The acetylenes add two moles of halogen whereby two halogen atoms are united to each of two adjacent carbon atoms. Using the same arguments advanced for the structure of the olefins, the only logical structure for the acetylenes is one in which two carbon atoms are joined by a triple bond. Acetylene has the structure $HC{\equiv}CH$, and acetylenes in general the structure $RC{\equiv}CR$.

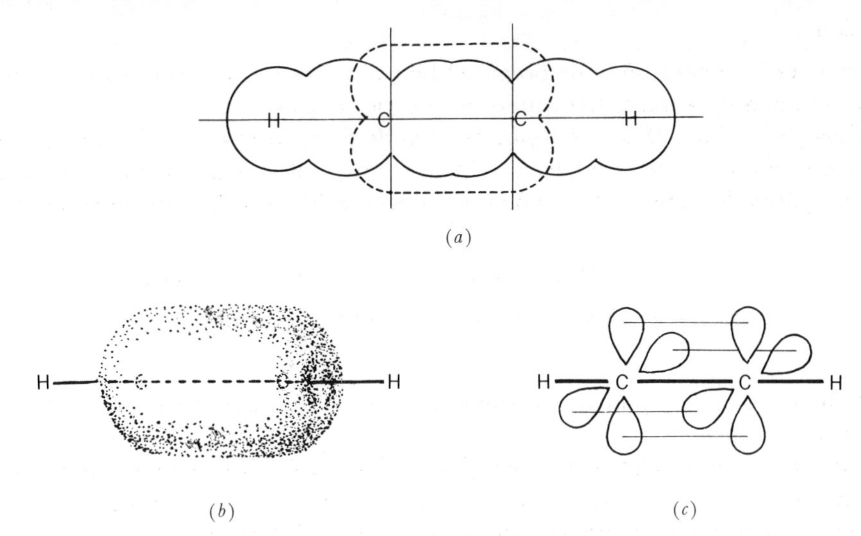

Figure 10–*1*. The acetylene molecule: (*a*) cross section through the molecular orbitals; (*b*) perspective representation of the π bonds; (*c*) schematic representation of the π bonds.

The description of the triple bond follows the same reasoning as that for the double bond (p. 74). In acetylene each carbon atom is joined to only two other atoms. Hence *sp* hybridization takes place on bonding, and the bond directions are in a straight line (p. 13). The primary carbon-carbon bond results from *sp—sp* overlapping of orbitals, and the carbon-hydrogen bonds result from *sp—s* overlapping, giving rise to a linear molecule. Both carbon atoms still have two *p* orbitals perpendicular to each other, each containing one electron. Overlapping of the four *p* orbitals gives two bonding π molecular orbitals, both having cylindrical symmetry. Figure 10–*1a* illustrates a cross section through the σ orbitals and the π orbitals. Figure 10–*1b* gives a perspective view of the π orbitals, and Fig. 10–*1c* is a schematic representation of the overlapping of *p* orbitals. The formation of the second π bond increases the carbon-carbon bond energy by around 55 kcal. and decreases the internuclear distance to 1.20 A from 1.33 A for ethylene.

Physical Properties and Nomenclature

The physical properties of the alkynes closely resemble those of the alkanes and the alkenes, although the boiling points are somewhat higher and the solubility in water is greater. Monosubstituted acetylenes show strong C≡C stretching absorption in the infrared at 2140–2100 cm.$^{-1}$. The frequency is higher than that for the alkenes (1680–1620 cm.$^{-1}$) because the triple bond is stronger than the double bond. Like the alkenes, symmetrically substituted acetylenes have zero transition dipole moment and show no absorption. The absorption of other disubstituted acetylenes may be very weak.

Alkynes may be named (*1*) as derivates of acetylene, and (*2*) by the international system. In the latter the ending is *yne*. These methods are illustrated by the following examples.

$(CH_3)_2CHC{\equiv}CH$	$CH_3CH_2\underset{\displaystyle CH_3}{\underset{\vert}{C}}HC{\equiv}CC_2H_5$
i-Propylacetylene (*methylbutyne*)	Ethyl-*s*-butylacetylene (*5-methyl-3-heptyne*)

Preparation

Acetylenes in general may be prepared by two procedures analogous to the preparation of olefins, and by a third that cannot be used for olefins.

1. ***Removal of Two Moles of Hydrogen Halide from Dihalides.*** If a dihalide has the halogen atoms on the same or adjacent carbon atoms, boiling with an alcoholic solution of potassium hydroxide removes two moles of halogen acid with the introduction of a triple bond.

$$RCH_2CX_2R + 2\,KOH\ (\text{alcoholic}) \longrightarrow RC{\equiv}CR + 2\,KX + 2\,H_2O$$

$$R\underset{X}{\underset{|}{C}}H\underset{X}{\underset{|}{C}}HR + 2\,KOH\ (\text{alcoholic}) \longrightarrow RC{\equiv}CR + 2\,KX + 2\,H_2O$$

Since 1,2 dihalides can be prepared easily from olefins, this method can be used to convert olefins into acetylenes.

2. ***Removal of Four Halogen Atoms from Adjacent Carbon Atoms.*** If a tetrahalide has two halogen atoms on each of two adjacent carbon atoms, boiling with zinc dust in alcoholic solution removes the halogen, and an alkyne is formed.

$$RCX_2CX_2R + 2\,Zn\ (\text{in alcohol}) \longrightarrow RC{\equiv}CR + 2\,ZnX_2$$

This reaction has the same limitations as the similar one used for the preparation of olefins, namely that the halides required are best made by adding halogen to the unsaturated compound.

3. ***From Other Acetylenes.*** These procedures are discussed under the reactions of acetylenes (p. 150) and of other compounds (p. 200).

Reactions

Because of the triple bond, acetylenes undergo addition reactions analogous to those of the olefins but are capable of adding two moles of reagent instead of one. The large amount of *s* character to the *sp* σ orbitals of the carbon atoms, however, brings any bonding pairs of electrons closer to the carbon nuclei than with alkenes (sp^2 hybrids) or alkanes (sp^3 hybrids). Hence hydrogen is removed as a proton much more readily from an acetylenic carbon atom than from an ethylenic or saturated carbon atom. This greater attraction of the carbon nuclei for electrons even makes rates of electrophilic addition to the triple bond slower than addition to the double bond. An alternative explanation of the greater reactivity of π electrons in alkenes is that torsional vibrations about the double bond, which are absent in the acetylenes, assist the breaking of the π bond. The greater tightness with which the π electrons are held in acetylene is reflected in its higher ionization potential (11.25 e.v.) compared to that of ethylene (10.52 e.v.).

1. ***Addition to the Triple Bond.*** Hydrogen can be added to the triple bond in the presence of a suitable catalyst.

$$RC{\equiv}CR + 2\,H_2 \xrightarrow{Pt} RCH_2CH_2R$$

If finely divided palladium partially poisoned with lead and quinoline (p. 519, *Lindlar catalyst*) is used, it is possible to stop the reaction when one mole of hydrogen has been added.

$$RC{\equiv}CR + H_2 \xrightarrow{Pd} RCH{=}CHR$$

Reduction may be brought about also by means of sodium in liquid ammonia.

$$RC{\equiv}CR + 2\,Na + 2\,NH_3 \longrightarrow RCH{=}CHR + 2\,Na^{+-}NH_2$$

Because of adsorption on the surface of the metal, catalytic reduction yields *cis* olefins. Reduction by sodium in liquid ammonia is a stepwise process from opposite sides of the molecule, and *trans* olefins result.

One or two moles of halogen or halogen acid can be added to the triple bond.

$$RC{\equiv}CH \begin{cases} + \ X_2 \longrightarrow RCX{=}CHX \\ + \ 2X_2 \longrightarrow RCX_2CHX_2 \\ + \ HX \longrightarrow RCX{=}CH_2 \\ + \ 2HX \longrightarrow RCX_2CH_3 \end{cases}$$

The polar addition of nonidentical addenda follows the Markovnikov rule (p. 83). As with the alkenes, free radical additions take place contrary to the Markovnikov rule (p. 85).

$$RC{\equiv}CH + HBr \xrightarrow{\text{Peroxides}} RCH{=}CHBr$$

Acetylene adds water in the presence of sulfuric acid and mercurous sulfate (mercury-mercuric sulfate mixture). Instead of a hydroxy compound being formed, however, as results from the reaction of olefins with water and sulfuric acid (p. 85), acetylene yields acetaldehyde. The initial complex of acetylene with the mercury salt probably yields a hydroxy olefin in which a hydroxyl group is on a carbon atom joined to another carbon atom by a double bond.

$$HC{\equiv}CH + H_2O \xrightarrow[\text{Hg—HgSO}_4]{H_2SO_4,} [H_2C{=}CHOH]$$

Compounds having such structures are called *enols*. The simple enols are unstable[2] with respect to the isomer having an additional carbon-hydrogen bond and a carbon-oxygen double bond, and rearrange to the more stable isomer. A simple calculation using empirical bond energies (Table 2–2, p. 19) indicates that the conversion of the enol to the aldehyde is exothermic by about 11 kcal. The rearrangement involves electrophilic attack by a proton on the unsaturated methylene (CH_2) group with loss of a proton from the hydroxyl group.

$$H^+ + CH_2{=}\underset{\displaystyle H}{\underset{|}{C}}{-}O{-}H \longrightarrow CH_3\underset{\displaystyle H}{\underset{|}{C}}{=}O + H^+$$

Acetaldehyde

Alkynes other than acetylene yield ketones because addition takes place according to the Markovnikov rule.

$$RC{\equiv}CH + HOH \xrightarrow[\text{Hg—HgSO}_4]{H_2SO_4,} \left[R\underset{\displaystyle OH}{\underset{|}{C}}{=}CH_2 \right] \longrightarrow R\underset{\displaystyle O}{\underset{\|}{C}}CH_3$$

A ketone

2. ***Reactions of Acetylenic Hydrogen.*** The increased acidity of hydrogen attached to a carbon atom bearing a triple bond (p. 148) is shown by the ease with which metallic salts are formed. Acetylene reacts with molten sodium to form a monosodium derivative **(sodium acetylide)** at 110°.

$$HC{\equiv}CH + Na \text{ (molten at 110°)} \longrightarrow HC{\equiv}C^{-+}Na + \tfrac{1}{2} H_2$$

Sodium acetylides can be formed readily also by the reaction of acetylenes with highly dispersed sodium (particle size 10–25μ) in xylene or by reaction with sodium amide in liquid ammonia.

$$RC{\equiv}CH + Na^{+-}NH_2 \longrightarrow RC{\equiv}C^{-+}Na + NH_3$$

---

[2] To indicate that a structure is unstable and that a compound having this structure cannot be isolated, the formula is enclosed in brackets.

The complex of lithium acetylide with ethylenediamine (p. 611), in the form of a dry powder, is convenient to handle.

The acetylide ion is a strong nucleophile and is useful for the introduction of the triple bond into other molecules (p. 200). Thus it can displace halide ion from a primary alkyl halide to give higher acetylene homologs.

$$RC{\equiv}C{:}^- + R'X \longrightarrow RC{\equiv}CR' + {}^-{:}X$$

$$\underset{\text{Sodium methylacetylide}}{CH_3C{\equiv}C^{-+}Na} + C_2H_5Br \longrightarrow \underset{\text{Methylethylacetylene } (\textit{2-pentyne})}{CH_3C{\equiv}CC_2C_5} + Na^{+-}Br$$

A mixture of dimethylformamide (p. 235) and xylene (p. 93) is a good medium for the reaction.

Acetylene reacts readily with aqueous ammoniacal silver nitrate or aqueous ammoniacal cuprous chloride solutions to give water-insoluble carbides.

$$HC{\equiv}CH + 2\,Ag(NH_3)_2NO_3 \longrightarrow Ag^{+-}C{\equiv}C^{-+}Ag + 2\,NH_4NO_3 + 2\,NH_3$$

$$HC{\equiv}CH + 2\,Cu(NH_3)_2Cl \longrightarrow Cu^{+-}C{\equiv}C^{-+}Cu + 2\,NH_4Cl + 2\,NH_3$$

Compounds of the type $RC{\equiv}CH$ yield acetylides, $RC{\equiv}C^{-+}Ag$ and $RC{\equiv}C^{-+}Cu$, whereas compounds of the type $RC{\equiv}CR$ do not react. These reactions are useful to distinguish acetylene and monosubstituted acetylenes from olefins and disubstituted acetylenes.

The heavy metal carbides are thermodynamically unstable and when dry may be exploded by heat or shock with the formation of the metal and carbon. Silver carbide, for example, explodes at 140–150°. Sodium carbide, on the other hand, is stable up to 400°, and calcium carbide, which is made in the electric furnace (p. 150), melts without decomposition at 2300°.

Since the acetylides and carbides are salts of the very weak acid, acetylene, they are hydrolyzed by water.

$$HC{\equiv}C^{-+}Na + H_2O \longrightarrow NaOH + C_2H_2$$

$$Ca^{2+}({}^-C{\equiv}C^-) + 2\,H_2O \longrightarrow Ca(OH)_2 + C_2H_2$$

A mineral acid is required to decompose the heavy metal salts.

$$Ag_2C_2 + 2\,HNO_3 \longrightarrow 2\,AgNO_3 + C_2H_2$$

Acetylenic hydrogen (the hydrogen in $HC{\equiv}CH$ or $RC{\equiv}CH$) is sufficiently more acidic (p. 148) than alkane hydrogen (the hydrogen in RCH_3) to liberate alkanes from alkyl Grignard reagents with the formation of acetylenic Grignard reagents.

$$RC{\equiv}CH + R'MgX \longrightarrow RC{\equiv}CMgX + R'H$$

Since Grignard reagents undergo a great variety of reactions, the acetylenic Grignard reagents are valuable for the preparation of other compounds containing the triple bond.

Industrial Preparation and Use of Acetylene

Acetylene is the only compound of the series produced commercially in large amounts. Although it can be made by any of the general methods, it is made much more cheaply from coke by way of calcium carbide, or from natural gas.

$$\underset{\text{Coke}}{3\,C} + \underset{\text{Lime}}{CaO} \xrightarrow[(2500°)]{\text{Electric furnace}} \underset{\substack{\text{Calcium}\\\text{carbide}}}{CaC_2} + CO$$

$$CaC_2 + 2\,H_2O \xrightarrow[\text{temp.}]{\text{Room}} HC{\equiv}CH + Ca(OH)_2$$

When methane is heated to a very high temperature, acetylene is one of the products.

$$2\,CH_4 \rightleftharpoons C_2H_2 + 3\,H_2 - 95.5\text{ kcal.}$$

The initial step in the decomposition of methane is the formation of methyl radicals and hydrogen atoms (p. 65). The methyl radicals can give rise to ethane, ethylene, and acetylene. Combination of hydrogen atoms gives molecular hydrogen. Whereas the order of thermodynamic stability at 25° is ethane > ethylene > acetylene, at 1000° they are about equally stable and at higher temperatures the order is reversed. None of these products really is stable at this temperature. All decompose rapidly to carbon and hydrogen. If, however, the time of exposure to the high temperature (1400–1600°) is short enough (0.1 to 0.01 second) and the products are cooled quickly enough to prevent further decomposition to carbon and hydrogen, a reasonably high yield of acetylene can be obtained. Several processes for attaining these conditions are used such as passing the methane or higher hydrocarbons through an electric arc (*Huels process*) or over a refractory heated to a high temperature (*Wulff process*). The process operated most generally in the United States has been one developed in Germany (*Sachsse* or *BASF process*). The necessary heat is supplied by partial combustion of methane with oxygen. Natural gas and oxygen of 95 per cent purity are preheated separately to 600°, mixed, ignited in a special burner, and the products immediately cooled by a water spray. The cooled gases are compressed and the acetylene is absorbed in the water, from which it can be recovered in 99.8 per cent purity. The off-gases consist of carbon monoxide and hydrogen in the ratio of 7 to 15, a satisfactory composition for the synthesis of methanol (p. 120). More recently, practical processes have been developed for the production of acetylene and ethylene by the cracking of petroleum naphtha fractions (p. 390). Of the estimated 1.4 billion pounds of acetylene produced in the United States in 1964, over 30 per cent was made from natural gas. The price was about 10 cents per pound.

Acetylene is a colorless gas that boils at −84°. The garlic odor of the crude product from calcium carbide and water is due to the presence of phosphine derived from calcium phosphide. Acetylene cannot be liquefied safely because it is thermodynamically unstable and explodes from shock with the formation of the elements.

$$C_2H_2 \longrightarrow 2\,C + H_2 + 56 \text{ kcal.}$$

Procedures have been developed for handling acetylene safely at pressures up to 30 atmospheres (450 p.s.i.). Contact with all copper and copper alloys is avoided, and the free space kept to a minimum. No pipes for carrying the gas under pressure are over 35 mm. in diameter, all larger pipes being filled with small pipes. Higher pressures can be used without explosion by adding an inert gas such as water vapor or nitrogen.

At atmospheric pressure acetylene is soluble to the extent of one volume of the gas in one volume of water, four volumes in one of benzene (p. 93), six volumes in one of ethyl alcohol, twenty-five volumes in one of acetone (p. 216), and 33 volumes in one of methyl sulfoxide (p. 270). At 180 p.s.i., 300 volumes of acetylene dissolve in one volume of acetone. Since this solution is stable, acetylene is transported under pressure in tanks filled with a porous material saturated with acetone.

About half of the acetylene consumed in the United States is used for the welding, cutting, and cleaning of iron and steel by means of the oxyacetylene flame, which has a temperature in the neighborhood of 2800°. The other half is used for the preparation of other organic chemicals. Acetylene tetrachloride is made by the addition of chlorine to acetylene and is used as a solvent and for the preparation of other compounds (p. 592). Catalytic addition of water to acetylene gives acetaldehyde from which acetic acid and a large number of other organic compounds can be made (p. 215). Catalytic addition of acetic acid yields vinyl acetate (p. 599), and addition of alcohols yields vinyl ethers (p. 600), which are used in the preparation of plastics and other useful products. Addition of hydrogen cyanide gives acrylonitrile used for the production of synthetic fibers (p. 619) and synthetic rubber (p. 589); pyrrolidone gives vinylpyrrolidone (p. 618). Dimerization of acetylene gives vinylacetylene, the intermediate for the manufacture of neoprene (p. 588). Reaction of acetylene with carbon monoxide and water in the presence of nickel carbonyl, $Ni(CO)_4$, gives acrylic acid (p. 619); carbon monoxide and methyl alcohol give methyl acrylate (p. 619). Addition of acetylene to aldehydes and ketones gives acetylenic alcohols (p. 200).

PROBLEMS

10–*1*. Give reactions for the following preparations: (*a*) 2,3-dimethylbutane from a three-carbon alkyl halide, (*b*) 3-methyl-1-pentene from a dihalide, (*c*) *n*-heptane from an alkyl halide, (*d*) *i*-butylene from an alcohol.

10–*2*. Write skeleton structures for and give a name to each alkyne having the molecular formula C_6H_{10}.

10–*3*. Predict for $HC{\equiv}C{-}CH_2{-}CH{=}CHCH_2OH$ the approximate sizes of the bond angles.

10–*4*. Arrange the following compounds in the order of increasing stability to heat: (*a*) methane, (*b*) calcium carbide, (*c*) ethyl peroxide, (*d*) isobutane.

10–*5*. Give reactions for the following syntheses: (*a*) 1-pentyne from *n*-propyl bromide, (*b*) 2,2-dibromobutane from ethylacetylene, (*c*) 3-hexene from 3-hexyne, (*d*) 2-butyne from acetylene, (*e*) *n*-butylacetylene from 1-hexene.

10–*6*. Give a series of reactions for carrying out the following conversions: (*a*) 1-pentene to 2-pentene, (*b*) ethyl-*s*-butylcarbinol to 3-methylhexane, (*c*) *n*-hexyl bromide to 1-hexyne, (*d*) *i*-butyl alcohol to 2,5-dimethylhexane, (*e*) 2,3-dibromobutane to 2,2-dibromobutane, (*f*) 2,2-dibromobutane to 2,3-dibromobutane.

10–*7*. If all of the following compounds were available, which should be chosen to make 2-methyl-1-butene: (*a*) 2-bromo-2-methylbutane, (*b*) 2-bromo-3-methylbutane, (*c*) 1,2-dibromo-2-methylbutane, (*d*) 2,3-dibromo-2-methylbutane, (*e*) 1-bromo-2-methylbutane?

10–*8*. How many steps are involved in the conversion of acetylene to tetrachloroethylene?

10–*9*. Without resorting to quantitative determinations, how may one distinguish by chemical tests among the members of the following groups of gases: (*a*) ethyl chloride, butane, and 2-butyne; (*b*) propane, propene, and propyne; (*c*) ethylacetylene, *i*-butane, and dimethylacetylene?

10–*10*. Making use of chemical reactions, describe a procedure for obtaining in a relatively pure form the components of the following mixtures: (*a*) ethane, ethylene, and acetylene; (*b*) methyl chloride, 2-butyne, and 1-butyne; (*c*) 1-pentyne, 2-pentyne, and pentane; (*d*) ethylacetylene, diethylacetylene, and 3-hexene.

10–*11*. From empirical bond energies, calculate the heat of reaction for the conversion of hydroxyethylene to acetaldehyde.

CHAPTER ELEVEN

CARBOXYLIC ACIDS AND THEIR DERIVATIVES. ORTHO ESTERS.

CARBOXYLIC ACIDS

Numerous types of organic compounds transfer protons to bases more readily than does the water molecule. These compounds are referred to as organic acids or acidic compounds. The carboxylic acids constitute one of the most important groups of compounds having this property.

Structure

The simplest carboxylic acids have the general formula $C_nH_{2n}O_2$, and since a carboxylic acid having only one carbon atom is known, namely formic acid, CH_2O_2, the number of possible structures is limited. If unlikely structures containing unpaired electrons are omitted, only two formulas, *I* and *II*, are reasonable from the theory of electronic structure.

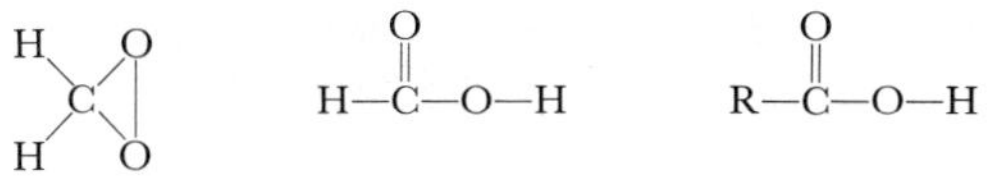

$$\underset{I}{H_2C\langle{}^{O}_{O}|} \qquad \underset{II}{H-\overset{\overset{O}{\|}}{C}-O-H} \qquad \underset{III}{R-\overset{\overset{O}{\|}}{C}-O-H}$$

Formula *I* has two oxygen atoms joined to each other; that is, it contains a peroxide structure. Carboxylic acids, however, do not show any of the properties of peroxides, such as the liberation of iodine from hydrogen iodide solutions. Moreover all monocarboxylic acids contain only one ionizable hydrogen atom. If the hydrogen atoms in formula *I* ionized, both would be expected to do so. Formula *II* in which one hydrogen is joined to oxygen and one to carbon accounts for the difference in behavior of the hydrogen atoms. Formula *III* is the general formula for the homologous series. The methods of synthesis and reactions of carboxylic acids confirm the view that both the doubly bound oxygen atom and the hydroxyl group are combined with a single carbon atom to give the group —COOH, which is known as the *carboxyl group*. The C=O group is called a *carbonyl group*, and the $R-\underset{|}{C}=O$ group is called an *acyl group*.

General Methods of Preparation

1. ***From the Grignard Reagent and Carbon Dioxide*** (p. 111).

$$RMgX + O=C=O \longrightarrow R-\overset{\overset{O}{\|}}{C}-OMgX \xrightarrow{HX} R-\overset{\overset{O}{\|}}{C}-OH + MgX_2$$

This reaction is not only one of the best general methods for synthesizing carboxylic acids, but it proves that both oxygen atoms are united to the same carbon atom.

2. ***By the Oxidation of Primary Alcohols.*** When primary alcohols are oxidized by an excess of a strong oxidizing agent such as sodium dichromate and sulfuric acid, chromium trioxide in glacial acetic acid, potassium permanganate, or nitric acid, carboxylic acids are produced without loss of carbon.

$$RCH_2OH + 2\,[O] \longrightarrow RCOOH + H_2O$$

An aldehyde is an intermediate in this reaction (p. 132). The symbol [O] merely means any suitable oxidizing agent. It is assumed that the student has learned how to balance oxidation and reduction reactions and is able to write balanced equations when necessary (see below). For example, when sodium dichromate and sulfuric acid is used as the oxidizing agent, the balanced equation is

$$3\,RCH_2OH + 2\,Na_2Cr_2O_7 + 8\,H_2SO_4 \longrightarrow 3\,RCOOH + 2\,Cr_2(SO_4)_3 + 2\,Na_2SO_4 + 11\,H_2O$$

The mechanism of the oxidation of alcohols to aldehydes and ketones is discussed on page 133. The oxidation of aldehydes to acids is discussed on page 205. Mixtures of acids of lower molecular weight are formed on vigorous oxidation of alkenes (p. 90), tertiary alcohols (p. 154), or secondary alcohols or ketones (p. 205).

3. ***Other Methods.*** Monocarboxylic acids result also from the hydrolysis of alkyl cyanides (p. 238) or nitroalkanes (p. 245), from the oxidation of aldehydes (p. 205), and by the decarboxylation of malonic acids (p. 628). Acid derivatives such as anhydrides (p. 171), acyl halides (p. 169), esters (p. 173), and amides (p. 234) yield acids on hydrolysis.

Balancing Oxidation-Reduction Equations

By Inspection. When sodium dichromate is reduced to chromic sulfate, three oxygen atoms are available for oxidation. When a primary alcohol is oxidized to an acid, two atoms of oxygen are required.

$$Na_2Cr_2O_7 + 4\,H_2SO_4 \longrightarrow Na_2SO_4 + Cr_2(SO_4)_3 + 4\,H_2O + 3\,[O]$$

$$RCH_2OH + 2\,[O] \longrightarrow RCOOH + H_2O$$

Therefore if the first equation is multiplied by two and the second by three and these equations are added, a balanced equation results.

Potassium permanganate in neutral or alkaline solution is reduced to manganese dioxide. For each two moles of permanganate, three atoms of oxygen are available for oxidation.

$$2\,KMnO_4 + H_2O \longrightarrow 2\,KOH + 2\,MnO_2 + 3\,[O]$$

In acid solution the reduction product is a manganous salt, and five atoms of oxygen are available from two moles of permanganate.

$$2\,KMnO_4 + 3\,H_2SO_4 \longrightarrow K_2SO_4 + 2\,MnSO_4 + 3\,H_2O + 5\,[O]$$

When an excess of nitric acid is used as an oxidizing agent, the reduction product ordinarily is considered to be nitrogen dioxide and each two moles of nitric acid provide one atom of oxygen.

$$2\,HNO_3 \longrightarrow 2\,NO_2 + H_2O + [O]$$

If the reduction product is dinitrogen trioxide, each mole of nitric acid provides one atom of oxygen.

$$2\,HNO_3 \longrightarrow N_2O_3 + H_2O + 2\,[O]$$

If the reduction product is nitric oxide, two moles of nitric acid provide three atoms of oxygen.

$$2\,HNO_3 \longrightarrow 2\,NO + H_2O + 3\,[O]$$

By Change in Polar Number. **(Oxidation Number).** Since oxidation is the loss of electrons and reduction the gain of electrons, balancing the loss and gain of electrons leads to the balancing of oxidation and reduction equations. Here certain atoms of the molecules involved are considered to be oxidized or reduced in the reaction. The degree of oxidation or reduction of an atom in a molecule or ion is considered to be proportional to the electron concentration about the atom compared with that of the free atom. Each bond to the atom is assigned a unit polarity, the direction of polarity depending on which atom has the greater attraction for electrons. Hydrogen always is considered positive with respect to other atoms. For other elements the attraction for electrons increases from left to right in a given period and decreases from top to bottom in a given column of the periodic table. Thus in a primary alcohol the R—C bond is between two carbon atoms and is assigned zero polarity. Each C—H bond and the O—H bond are assigned polarities with the negative end at carbon or oxygen and the positive end at hydrogen. The C—O bond is assigned a polarity with the positive end at carbon and the negative end at oxygen. The same is true for each bond of the double bond in the carboxyl group of the acid.

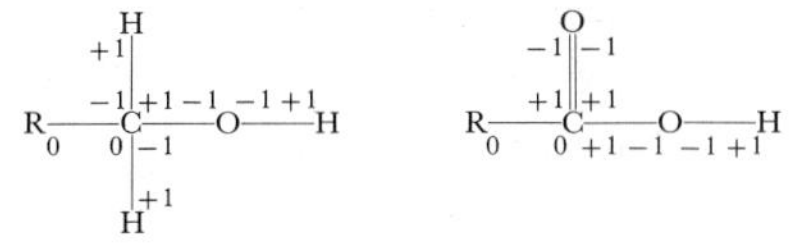

```
        H                            O
      +1|                         -1||-1
        |                           ||
      -1|+1 -1  -1 +1             +1||+1
  R-----C-----O-----H        R------C------O------H
  0    0|-1                  0      0 +1 -1  -1 +1
        |
        |+1
        H
```

The algebraic sum of the charges on any atom is its **polar number** or **oxidation number.** Thus the polar number of carbon in a primary alcohol group is $0 - 1 + 1 - 1 = -1$, and that in a carboxyl group is $0 + 1 + 1 + 1 = +3$. If the polar number becomes more positive in a reaction, indicating a decrease in electron density, the atom is oxidized, and if it becomes less positive, indicating an increase in electron density, it is reduced. The change in polar number of an atom in an oxidizing agent must balance the change in the polar number of an atom in the molecule being oxidized.

In the unbalanced reaction

$$RCH_2OH + Na_2Cr_2O_7 + H_2SO_4 \longrightarrow RCOOH + Cr_2(SO_4)_3 + Na_2SO_4 + H_2O$$

the atoms undergoing a change in polar number are the carbon atom in the primary alcohol, which changes from -1 to $+3$ or a change of $+4$, and the chromium atoms in the dichromate which change from $+6$ to $+3$ or a change of -3 for each chromium atom, a total change for the two atoms of -6. To balance $+4$ against -6, $+4$ may be multiplied by 6 and -6 by 4, or more simply $+4$ by 3 and -6 by 2. Hence three molecules of alcohol can be oxidized by two molecules of dichromate. The moles of sulfuric acid required and the moles of water formed follow from the number of sulfate ions required and the amount of hydrogen available.

It is unfortunate that the expression *change in valence* sometimes is used in balancing oxidation-reduction reactions when *change in polar number* is meant. The term *valence* should be reserved for the number of covalent bonds by which an atom is united to other atoms or for the number of charges that an ion bears. It is obvious in this example that the valence of carbon has not changed on being oxidized from a CH_2 group to a C=O group but remains four throughout, whereas the polar number has changed from -1 to $+3$.

By Means of Half-Reactions. Oxidation and reduction equations may be balanced by the method of half-reactions. In the present example one half of the reaction is the oxidation of alcohol molecules to acid molecules, and the other half is the reduction of dichromate ion to chromic ion. The source of oxygen for the oxidation half may be considered to be water molecules.

$$RCH_2OH + H_2O \longrightarrow RCOOH + 4\,H^+ + 4\,e^-$$

$$Cr_2O_7^{2-} + 14\,H^+ + 6\,e^- \longrightarrow 2\,Cr^{3+} + 7\,H_2O$$

Since the first half-reaction must supply as many electrons as are used by the second half-reaction, the first half-reaction must be multiplied by 3 and the second by 2. Addition then gives

$$3\,RCH_2OH + 2\,Cr_2O_7^{2-} + 16\,H^+ \longrightarrow 3\,RCOOH + 4\,Cr^{3+} + 11\,H_2O$$

All of the above schemes for balancing oxidation and reduction equations are empirical and arbitrary and none is more scientific than the others. In the first method oxygen is assumed to be available from the oxidizing agent; in the second method the atoms are assumed to differ in polarity by unit charges; in the third method the reaction is assumed to take place by the elimination of electrons and the consumption of electrons. These assumptions merely are devices for arriving at the desired result, and the method used should be that which seems preferable to the individual. The oxidation of organic compounds usually is limited to the addition of oxygen or the removal of hydrogen, and reduction to the addition of hydrogen or the removal of oxygen. Since the oxygen or hydrogen requirement in organic oxidations and reductions can be determined at a glance, the first method given for balancing oxidation-reduction equations is the simplest. Whatever method is used, it should be practiced sufficiently to be carried out correctly and with facility.

Nomenclature

Common Names. The straight-chain or normal carboxylic acids were isolated first from natural sources, particularly by the hydrolysis of natural fats and waxes (Chapter 12). Hence they frequently are called *fatty acids.* Since nothing was known about their structure, they were given common names indicating their source. These names and the derivations of the names, together with some physical properties, are given in Table 11–*1*.

International names are given for the odd-carbon acids above C_{10} rather than common names. The reason is that only acids with an even number of carbon atoms had been found in fats,[1] the odd-carbon acids having been prepared by synthesis. The name *margaric* appears to be an exception. It has been shown, however, that the material isolated from fats and thought to be a C_{17} acid was actually a mixture of palmitic and stearic acids. When the true C_{17} normal acid was synthesized, the common name was retained.

As with other homologous series, those compounds having an isopropyl group at the end of a normal hydrocarbon chain may be named by adding the prefix *iso* to the common name, for example $(CH_3)_2CHCOOH$, isobutyric acid, or $(CH_3)_2CH(CH_2)_6COOH$, isocapric acid. If the methyl branch occurs at any other portion of the chain, the designation *iso* may not be used. Acids having a secondary butyl group at the end of a straight chain, e.g. $CH_3CH_2CH(CH_3)(CH_2)_xCOOH$, have been called *anteiso* acids.

As Derivatives of Normal Acids. Acetic acid frequently is chosen as the parent compound, and derivatives are named as compounds in which the hydrogen atoms of the methyl group are replaced by other groups.

$$\overset{5}{\underset{\delta}{C}}H_3\overset{4}{\underset{\gamma}{C}}H_2\overset{3}{\underset{\beta}{C}}H(CH_3)-\overset{2}{\underset{\alpha}{C}}H(CH_3)\overset{1}{C}OOH$$

This compound may be called methyl-*s*-butylacetic acid. It also may be considered to be derived from valeric acid and called α,β-dimethylvaleric acid or 2,3-dimethylvaleric acid. When Greek letters are used to designate the positions of substituents, the α carbon atom of the chain is the carbon atom adjacent to the functional group. When numerals are used, the carbon atom of the functional group is numbered 1.

International System. The final *e* is dropped from the name of the hydrocarbon having the same number of carbon atoms as the longest chain containing the carboxyl group, and *oic acid* is added. The carbon atom of the carboxyl group always is numbered 1 when numbering the atoms of the longest chain. For example, $CH_3CH_2CH(CH_3)CH(CH_3)COOH$ is called 2,3-dimethylpentanoic acid.

Physical Properties

The **boiling points** of carboxylic acids (Table 11–*1*) rise more or less uniformly with increase in molecular weight. The increase for those listed averages about 18° per additional methylene group, the same as for the alcohols. The magnitude of the boiling points, however, is even more abnormal than for the alcohols. Ethyl alcohol boils at 78°, but formic acid, which has about the same number of electrons, boils at 101°; *n*-propyl alcohol boils at 98°, but acetic acid boils at 118°. The explanation of the abnormal boiling points

[1] The presence of small amounts of normal odd-carbon acids in natural fats has been reported. Thus tri-, penta-, and heptadecanoic acids have been obtained from butter fat. Heptadecanoic acid has been isolated also from mutton fat and from shark liver oil. Moreover branched-chain acids with a methyl group on the second or third carbon from the hydrocarbon end of the chain (*iso* and *anteiso* acids) also have been isolated. The total amount of odd-carbon and branched-chain acids may amount to as much as 2 per cent of the total fat acids.

TABLE 11–*1*. COMMON NAMES OF NORMAL CARBOXYLIC ACIDS

NO. OF CARBON ATOMS	NAME OF ACID	DERIVATION OF NAME	BOILING POINT	MELTING POINT	DENSITY 20°/4°
1	Formic	L. *formica* ant	100.7	8.4	1.220
2	Acetic	L. *acetum* vinegar	118.2	16.6	1.049
3	Propionic	Gr. *proto* first, *pion* fat	141.4	−20.8	0.993
4	Butyric	L. *butyrum* butter	164.1	−5.5	0.958
5	Valeric	valerian root (L. *valere* to be strong)	186.4	−34.5	0.939
6	Caproic	L. *caper* goat	205.4	−3.9	0.936
7	Enanthic	Gr. *oenanthe* vine blossom	223.0	−7.5	0.918
8	Caprylic	L. *caper* goat	239.3	16.3	0.909
9	Pelargonic	pelargonium	253.0	12.0	
10	Capric	L. *caper* goat	268.7	31.3	
11	Undecanoic		280	28.5	
12	Lauric	laurel		43.2	
13	Tridecanoic			41.6	
14	Myristic	*Myristica fragrans* (nutmeg)		54.4	
15	Pentadecanoic			52.3	
16	Palmitic	palm oil		62.8	
17	Margaric	Gr. *margaron* pearl		61.2	
18	Stearic	Gr. *stear* tallow		69.6	
19	Nonadecanoic			68.7	
20	Arachidic	*Arachis hypogaea* (peanut)		75.4	
21	Heneicosanoic			74.3	
22	Behenic	behen oil		79.9	
23	Tricosanoic			79.1	
24	Tetracosanoic			84.2	
25	Pentacosanoic			83.5	
26	Cerotic	L. *cera* wax		87.7	

of the acids is the same as that for the alcohols, namely association by hydrogen bonds, but the acids are able to form double molecules that are more stable than the association complexes formed by the alcohols.

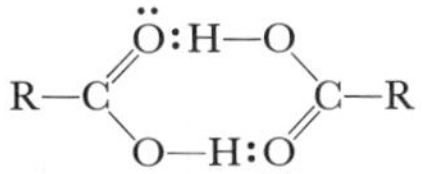

```
        ..
        O:H—O
       //     \
   R—C         C—R
       \     //
        O—H:O
            ..
```

It has been shown by vapor density measurements that the double molecules of acetic acid persist even in the vapor state. Hence it is not surprising that the boiling point of acetic acid (118°, $Z = 32 \times 2 = 64$) is of the same order of magnitude as *n*-octane (126°, $Z = 66$).

An interesting characteristic of the normal carboxylic acids is the alternation in **melting points.** Acids with an even number of carbon atoms always melt at a higher temperature than the next higher member of the series (Fig. 11–*1*). X-ray diffraction has shown that in the solid state the carbon atoms of the hydrocarbon chain assume an extended zigzag arrangement in which the carboxyl groups of the odd carbon acids are on the same side of the chain as the terminal methyl groups, whereas those of the even carbon acids are on the opposite side of the chain (see p. 64). Although all of the acids are double molecules, the arrangement of those with an even number of carbon atoms gives a more symmetrical molecule and a more stable crystal lattice.

Because of partial ionization in water, carboxylic acids are somewhat more heavily hydrated than alcohols and hence show somewhat greater **solubility** in water. In general they have about the same solubility in water as the alcohol with one less carbon atom. For example, *n*-butyric acid is miscible with water as is *n*-propyl alcohol, whereas *n*-butyl

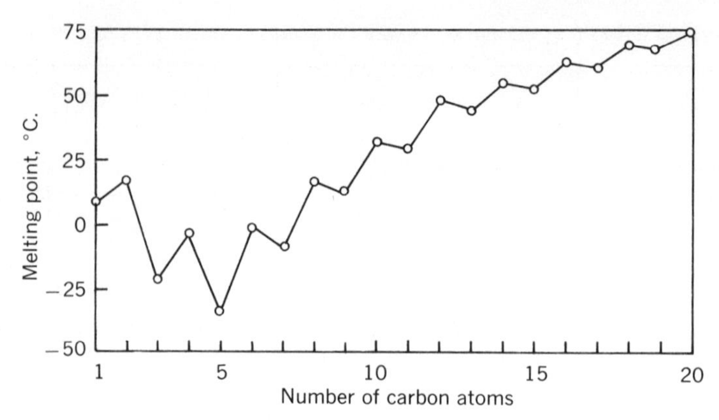

Figure 11–*1*. Melting points of normal carboxylic acids.

alcohol dissolves only to the extent of about 1 volume in 11 volumes of water. Monocarboxylic acids usually are soluble in other organic solvents.

The characteristic infrared absorption of saturated aliphatic carboxylic acids is that of the stretching vibration of the strong carbon-oxygen double bond at 1725–1700 cm.$^{-1}$. The C—O stretching absorption at 1320–1211 cm.$^{-1}$ also is prominent. It occurs at somewhat higher frequency than that for ethers (1150–1060 cm.$^{-1}$) because the bond is strengthened by the electronegativity of the doubly bound oxygen. The O—H stretching absorption is a very broad band extending over the region 3500–2500 cm.$^{-1}$ with various submaxima. It includes the absorptions of monomeric molecules at 3560–3500 cm.$^{-1}$ and those of dimeric molecules at 3000–2500 cm.$^{-1}$ as well as the C—H stretching absorptions at 2960–2850 cm.$^{-1}$. These features are illustrated in the infrared spectrum for propionic acid given in Fig. 11–*2*. The broad O—H stretching absorption reaches a maximum around 3100 cm.$^{-1}$ and the C—H maxima show at 2980 and 2940 cm.$^{-1}$. The strong carbonyl absorption maximum is at 1710 cm.$^{-1}$, and the C—O stretching maximum at 1230 cm.$^{-1}$.

Odor and Taste

The carboxylic acids that are sufficiently soluble in water to give an appreciable hydronium ion concentration have a sour taste. The lower members have a sharp acrid odor, and the acids from butyric through caprylic have a disagreeable odor. The odor of rancid butter and strong cheese is due partially to volatile acids (pp. 186, 195), and caproic,

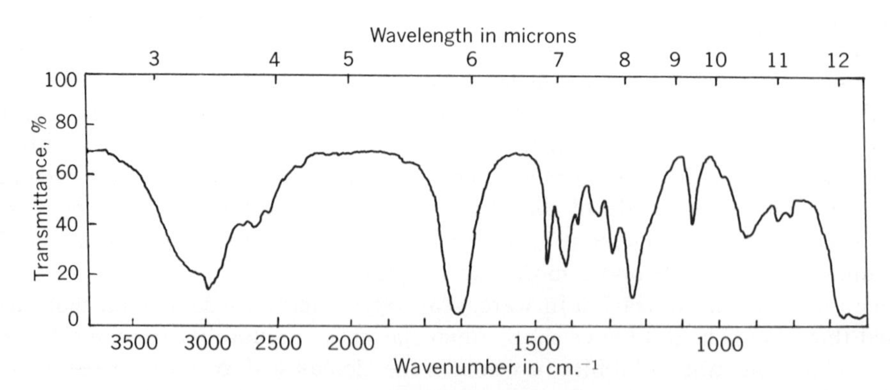

Figure 11–*2*. Infrared spectrum of propionic acid.

capric, and caprylic acids were so named because they are present in the skin secretions of goats. The higher acids are practically odorless because of their low volatility.

General Reactions of Free Acids

1. ***Salt Formation.*** The acidity of simple alcohols is of the same order of magnitude as that of water ($pK_a = 15.74$), whereas carboxylic acids ($pK_a = 4$ to 5) are about as strong as carbonic acid ($pK_a = 3.7$; apparent $pK_a = 6.76$).

The greater acidity of carboxylic acids compared to alcohols may be ascribed to two factors. First the additional oxygen atom, because of its electronegativity, increases the electron-withdrawing effect on the pair of electrons binding the hydrogen to the carboxyl group and permits the hydrogen to leave more readily as a proton. The action of groups in bringing about changes in electron density because of differences in electronegativity or in polarizability is called an **electrostatic** or **inductive effect.**

The second factor increasing acidity is **resonance** in the carboxylate ion. The primary bonds between the carbon atom and the three groups to which it is attached are formed by the presence of two electrons in each of three σ-type orbitals (p. 12). Because of the sp^2 hybridization of the atomic orbitals of the carbon atom, the molecular orbitals make angles of approximately 120 degrees with each other and lie in a plane (p. 14). These bonds, which are indicated in Fig. 11–*3* by the usual solid lines, utilize three of the four valence electrons of the carbon atom. Each oxygen atom contributes one of its six valence electrons to these bonds. Two more of the valence electrons of each oxygen atom are in a $2s$ orbital and two are in a p orbital, leaving one unassigned electron for each oxygen atom. Since one electron is acquired in the ionization of the proton, there is a total of four electrons that have not been assigned. Three p orbitals remain, one at each of the oxygen atoms and one at the carbon atom. Their axes can become parallel to each other and perpendicular to the plane of the σ orbitals. If the p orbital of carbon overlaps that of one of the oxygen atoms to form a molecular π orbital, the latter can hold two electrons. The remaining pair can occupy the empty p orbital of the second oxygen atom. This type of overlapping is indicated by Fig. 11–*3a* or by the conventional valence bond structure (*b*). Alternatively the p orbital of carbon can overlap with that of the other oxygen atom as indicated by (*c*) or (*d*). Either structure (*a*)

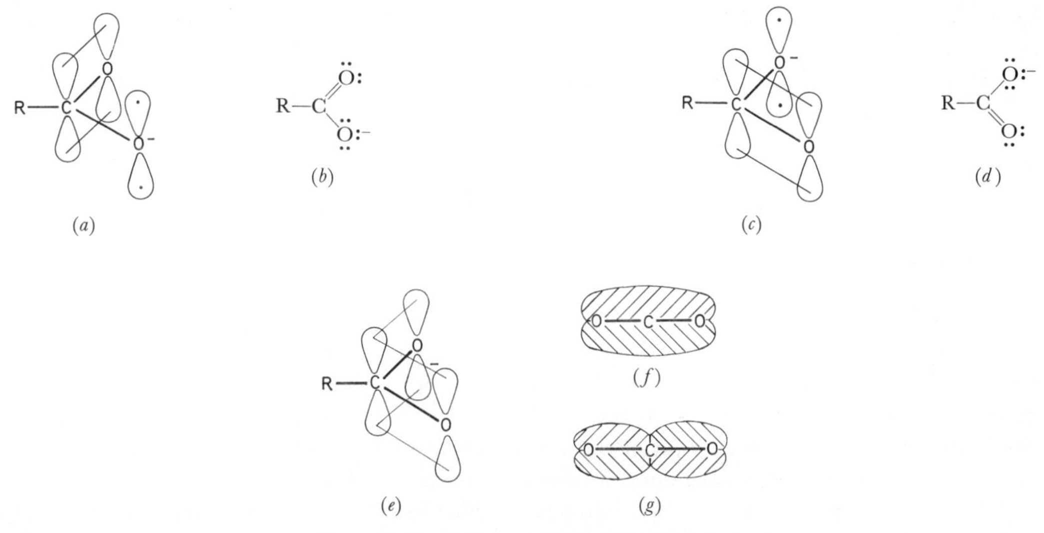

Figure 11–*3*. Resonance in the carboxylate ion.

or structure (*c*) is more stable than that with an unpaired electron occupying each of two *p* orbitals, because in either case the electrons can encompass two positive nuclei (p. 11). Since the *p* orbital of carbon can overlap the *p* orbital of either oxygen atom equally well, however, a molecular orbital can be formed encompassing all three nuclei as indicated in (*e*) in which the electrons have a still lower energy than in either (*a*) or (*c*) because now the electrons encompass three nuclei. Two such tricentric molecular orbitals are possible, one having a single nodal plane (*f*) and the other having two nodal planes perpendicular to each other (*g*). Hence two pairs of electrons can be accommodated. Although the orbital (*g*) with two nodal planes has a higher energy than that with a single nodal plane (*f*), the total energy of two electrons in (*g*) and two in (*f*) is less than the total energy of two electrons in a dicentric orbital and two electrons in a *p* orbital as in (*a*) or (*c*). This ability to utilize molecular orbitals involving more than two positive nuclei is the phenomenon that is called *resonance.* The difference in energy between the tricentric and the dicentric states is called the *resonance energy* of the molecule. This type of interaction may be indicated by placing a double-headed arrow between the conventional valence structures and enclosing the whole in braces.

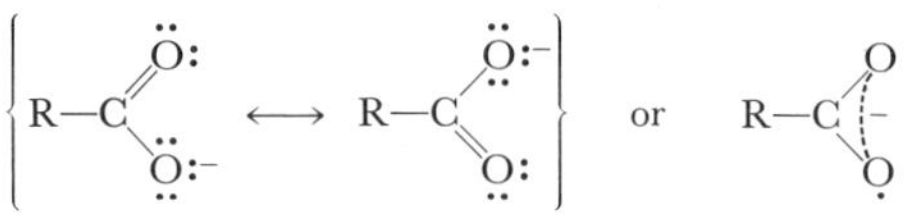

$$\left\{ R-C\begin{matrix} \nearrow \ddot{O}: \\ \searrow \ddot{O}:^- \end{matrix} \longleftrightarrow R-C\begin{matrix} \nearrow \ddot{O}:^- \\ \searrow \ddot{O}: \end{matrix} \right\} \quad \text{or} \quad R-C\begin{matrix} O \\ O \end{matrix}^-$$

It should be clear from the above discussion that these separate valence structures do not exist in the ground state of the molecule. It usually is said that the actual molecule is a *resonance hybrid* of the two structures. The hybrid concept is a convenient alternate symbolism for resonance.

The **source of resonance energy** is the same as that of any other type of bonding energy. Thus the energy of a single bond or of a double bond arises from the ability of an electron to encompass two positive nuclei in a molecule instead of one in an atom (p. 11). Resonance energy is merely the additional stabilization resulting from the ability of an electron to encompass three or more nuclei instead of two. Electrons encompassing more than two nuclei frequently are called *mobile electrons,* and are said to be *delocalized* in contrast to the relatively localized electrons in the σ bonds. Hence resonance energy is referred to also as *delocalization energy.*

The delocalization of electrons increases the stability of the carboxylate ion over that of an ethoxide or hydroxide ion by the amount of the resonance energy of the carboxylate ion. When a proton combines with one of the oxygen atoms of a carboxylate ion, the total energy for the reaction is the energy evolved in forming the oxygen-hydrogen bond less the energy necessary to destroy the resonance of the carboxylate ion. Hence combination of a proton with a carboxylate ion evolves considerably less energy than combination with an alkoxide or hydroxide ion, neither of which is stabilized by resonance. Conversely a proton can leave the carboxyl group of a carboxylic acid more readily than it can leave the hydroxyl group of water or the alkoxyl group of an alcohol. Hence both the electrostatic (*inductive*) effect and the resonance effect operate to increase the ease of ionization of the proton from a carboxyl group. It is not known which effect is the more important.

Although the opportunity for overlapping of *p* orbitals might appear to be identical in the carboxyl group and in the carboxylate ion (Fig. 11–4), the resonance energy of the carboxyl group is relatively small. In the undissociated carboxyl group (*a*) the *p* orbital of the hydroxyl oxygen overlaps that of the carbon atom much less than does the *p* orbital of the other oxygen atom because bond formation with the additional positive nucleus causes a deficiency of electrons about the hydroxyl oxygen. In other words because the hydroxyl oxygen has one less unshared pair of electrons, it is less able than the other oxygen atom to

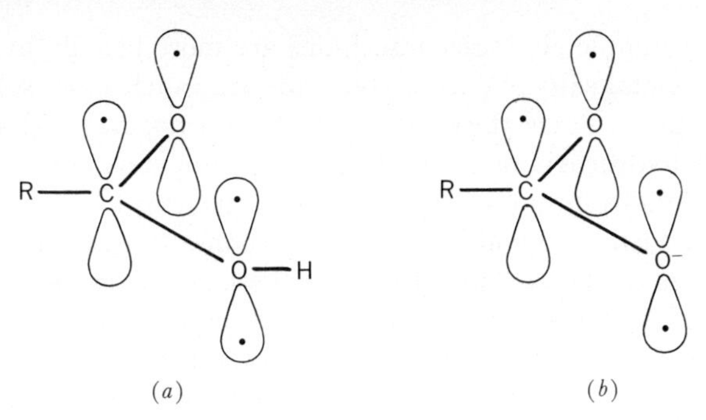

Figure 11–4. p Orbitals available for π bond formation (*a*) in the undissociated carboxyl group and (*b*) in the carboxylate ion.

contribute electrons to a π orbital. Therefore there is a greater tendency to form a dicentric orbital, that is, an ordinary double bond, than to form a tricentric orbital. If the molecule is considered as a resonance hybrid of the ordinary electronic valence structures, the

$$\left\{ \mathrm{R{-}C}\begin{matrix}\nearrow \ddot{\mathrm{O}}: \\ \searrow \ddot{\underset{..}{\mathrm{O}}}\mathrm{{-}H}\end{matrix} \longleftrightarrow \mathrm{R{-}C}\begin{matrix}\nearrow \ddot{\underset{..}{\mathrm{O}}}:^- \\ \searrow \overset{+}{\underset{..}{\mathrm{O}}}\mathrm{{-}H}\end{matrix} \right\}$$

second structure involving a separation of charge within the molecule is much less stable than the first. Since one of the requirements for resonance is that the conventional structures have the same or nearly the same stability, the resonance energy for the undissociated carboxyl group is much lower than that of the carboxylate ion.

These views receive experimental support from the measured interatomic distances. Thus the distance between a carbon and oxygen atom linked by a single bond is 1.43 A, whereas that between doubly linked carbon and oxygen is 1.24 A. Two carbon-oxygen distances are observed in monomeric formic or acetic acid, one at 1.43 A and one at 1.24 A. Only one carbon-oxygen distance of 1.27 A is observed, however, in the formate ion, the distances between carbon and the two oxygen atoms being equal. Moreover the observed distance is less than the average distance for a single and a double bond because the resonance energy causes the atoms to be bound more strongly.

Because of their acidity, carboxylic acids react with hydroxides to form salts that are hydrolyzed only slightly by water. They react also with carbonates and bicarbonates because the evolution of carbon dioxide, owing to its limited solubility in aqueous acid, shifts the equilibrium.

$$\mathrm{RCOOH + NaHCO_3 \longrightarrow RCOO^{-+}Na + CO_2 + H_2O}$$
$$\mathrm{2\,RCOOH + Na_2CO_3 \longrightarrow 2\,RCOO^{-+}Na + CO_2 + H_2O}$$
$$\mathrm{RCOOH + NaOH \longrightarrow RCOO^{-+}Na + H_2O}$$

Reaction of the salts with mineral acids ($pK_a = <1$) regenerates the carboxylic acid.

$$\mathrm{RCOO^{-+}Na + HCl \longrightarrow RCOOH + NaCl}$$

The neutralization of an acid by a standard solution of a base is the customary procedure for estimating acids. When a weak acid is neutralized by a strong base, the equivalence point is on the alkaline side because of hydrolysis of the salt. Therefore an indicator changing color at the proper acidity is necessary. For carboxylic acids, phenolphthalein usually is satisfactory. The equivalent weight of an acid as determined by neutralization with a standard base is known as the *neutralization equivalent* of the acid.

Because salts are completely ionized and ions are more heavily hydrated than neutral molecules, the alkali metal salts of carboxylic acids are much more soluble than the acids themselves. For example, whereas the solubility of the free normal acids in water approaches that of the saturated hydrocarbons above C_5, the sodium salts are very soluble up to C_{10} and form colloidal solutions from C_{10} to C_{18}. This fact is used to separate acids from water-insoluble compounds such as alcohols or hydrocarbons. Extraction of the mixture with dilute alkali causes the acid to go into the aqueous layer as the salt. The aqueous layer then can be separated, and the free acid liberated from its salt by the addition of a mineral acid. It is necessary to add at least the calculated amount of mineral acid, or if the amount of salt present is unknown, to add mineral acid until a universal indicator, such as Hydrion paper, shows that a hydrogen ion concentration of 10^{-2} to 10^{-1} (pH = 2 to 1) has been reached. Merely making the solution acid to litmus does not free the carboxylic acid from its salt completely because a mixture of the salt and organic acid is acid to litmus.

Another result of the ionic nature of salts is that they are nonvolatile. This property is utilized in the recovery of volatile water-soluble acids from aqueous solutions. The solution is neutralized with base and evaporated to dryness. The free acid may be liberated from its salt by means of a nonvolatile strong acid such as concentrated sulfuric acid.

Just as the inductive effect of oxygen influences the ease with which a proton may leave a molecule, so the electron-releasing or electron-attracting properties of other groups attached to a carboxyl group can affect its acidity. Thus acetic acid, $pK_a = 4.76$, is weaker than formic acid, $pK_a = 3.75$, because the methyl group has greater electron-releasing ability than hydrogen. The replacement of hydrogen in the methyl group of acetic acid by electron-attracting halogen atoms increases the acidity. The pK_a's for chloroacetic, dichloroacetic, and trichloroacetic acids are respectively 2.86, 1.26, and 0.64. Determinations of the change in enthalpy and in entropy for the ionization of acids indicates that the entropy change is the more important factor governing the extent of ionization. Apparently the inductive effect of the electronegative groups decreases the effectiveness of the negative charge in holding water molecules, and the lesser restriction of the movement of water molecules results in less decrease in entropy for the ionization (cf. solubility of alcohols, p. 118).

2. ***Replacement of Hydroxyl by Halogen.*** Since alcohols react with halogen acids to give alkyl halides (p. 128), it might be expected that a similar reaction would take place with carboxylic acids. For the latter, however, the position of equilibrium is so far on the side of carboxylic acid and halogen acid that the formation of acyl halide cannot be detected. A calculation from empirical bond energies indicates that the reaction of hydrogen chloride with a carboxylic acid is endothermic by 8 to 11 kcal., whereas the reaction with a primary alcohol is endothermic only by around 1 kcal. Carboxylic acids do bring about nucleophilic displacement of halide ion from inorganic acid chlorides, which is followed by nucleophilic attack by halide ion to give the *acyl halide.*

$$\mathrm{RC(=O){-}O(H) + PCl_3 \rightleftharpoons RC(=O){-}\overset{+}{O}(H){-}PCl_2 + {}^{-}Cl}$$

$$\mathrm{Cl^- + R{-}C(=O){-}\overset{+}{O}(H){-}PCl_2 \longrightarrow Cl{-}C(=O){-}R + HOPCl_2}$$

An acyl halide

Repetition of these steps completes the reaction.

$$\mathrm{2\ RCOOH + HOPCl_2 \longrightarrow 2\ RCOCl + (HO)_3P}$$

The positive charge on the oxygen of the intermediate makes the carbonyl carbon strongly electrophilic, and the stability of dichlorophosphinous, chlorophosphonous, and phosphorous acids makes them good leaving groups.

Thionyl chloride as a reagent has the advantage that the initially formed hydroxysulfinyl chloride, HOSOCl, decomposes to sulfur dioxide and hydrogen chloride, which are readily removed from the product.

$$RCOOH + SOCl_2 \longrightarrow RCOCl + SO_2 + HCl$$

For acids that do not react readily with thionyl chloride alone, the reaction can be facilitated by the addition of dimethylformamide.

3. ***Conversion to Esters.*** *(a)* BY DIRECT ESTERIFICATION. In the presence of strong acids, carboxylic acids react with alcohols to give esters and water. In these reactions it is possible for bonds to be broken in two ways, which usually are referred to as *acyl-oxygen scission* and *alkyl-oxygen scission.*

$$RCO \mid OH + H \mid OR' \rightleftharpoons RCOOR' + H_2O \qquad \textit{Acyl-oxygen scission}$$

$$RCOO \mid H + HO \mid R' \rightleftharpoons RCOOR' + H_2O \qquad \textit{Alkyl-oxygen scission}$$

It has been shown by several procedures that the usual esterification of *primary* alcohols takes place by acyl-oxygen scission. For example, if the oxygen in the alcohol is labeled by incorporating the oxygen isotope O^{18}, the isotope appears in the ester.

$$RCO \mid OH + H \mid O^{18}R' \longrightarrow RCOO^{18}R' + H_2O$$

The catalysis by acids therefore cannot result from protonation of the alcohol. Instead protonation of the carboxyl group increases the electrophilicity of the carbonyl carbon atom, because of the positive charge on oxygen, and makes possible an effective nucleophilic attack by the weakly nucleophilic alcohol molecule. An unstable intermediate addition product (*IV*) is formed that can lose water and establish equilibrium between the reactants and products.

$$\underset{I}{R{-}C(OH){=}O} \underset{B^-}{\overset{HB}{\rightleftharpoons}} \underset{II}{R{-}C(OH){=}\overset{+}{\ddot{O}}{:}H} \overset{R'OH}{\rightleftharpoons} \underset{III}{R{-}C(OH)(\overset{+}{:}\ddot{O}(H){-}R'){-}OH} \underset{HB}{\overset{B^-}{\rightleftharpoons}}$$

$$\underset{IV}{R{-}C(O{-}R')(OH){-}OH} \underset{B^-}{\overset{HB}{\rightleftharpoons}} \underset{V}{R{-}C(OR')(\overset{+}{:}\ddot{O}{-}H(H)){-}OH} \underset{HOH}{\rightleftharpoons} \underset{VI}{R{-}C(OR'){=}\overset{+}{O}H} \underset{HB}{\overset{B^-}{\rightleftharpoons}} \underset{VII}{R{-}C(OR'){=}O}$$

The energy profile for this mechanism is illustrated in Fig. 11–5. Each of the steps has its own transition state. They occur at the maxima in the diagram. The minima are labeled with numerals to indicate the intermediates of the reaction. The proton transfer reactions are expected to have low activation energies, whereas the formation of the carbon-oxygen bond should require a much higher activation energy.

As would be expected, the over-all energy changes in esterification are small. Starting with one mole of a straight-chain acid and one mole of a primary alcohol, equilibrium is established when the reaction has gone approximately two thirds of the way to ester and

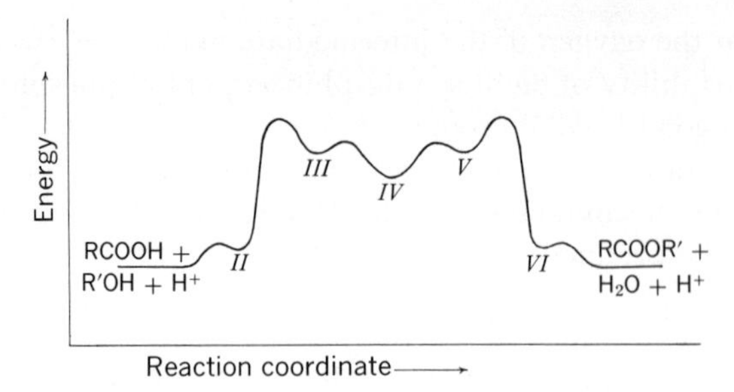

Figure 11-5. Energy profile for the acid-catalyzed esterification of a carboxylic acid by a primary alcohol.

water ($K_E = \sim 4$). In practice the equilibrium is shifted in favor of the ester by using a large excess of one of the reactants or by removing the water and ester by distillation, especially if a low-boiling azeotrope can be formed (p. 122). The acids generally used to catalyze the reaction are sulfuric acid, hydrogen chloride, boron fluoride, or the acid form of a strong acid cation-exchange resin (p. 484).

Since the transition states involve the carbonyl carbon atom, it is expected that branching of the alkyl group near its point of attachment will make the formation of the transition state more difficult and will increase the activation energy of the reaction. Thus the rate of esterification of aliphatic carboxylic acids with methyl alcohol decreases with increasing substitution by alkyl groups on the α, β, and even the γ carbon atoms of the acid. Sterically hindered acids can be converted to esters by pyrolysis of their tetraalkylammonium salts (p. 231).

(b) BY REACTION WITH CARBONIUM IONS FROM TERTIARY ALCOHOLS OR OLEFINS. If a tertiary alcohol or an olefin that can yield a stable carbonium ion is dissolved in concentrated sulfuric acid and a carboxylic acid is added, an ester of a tertiary alcohol is produced. Electrophilic attack by the carbonium ion on the carbonyl oxygen atom yields an intermediate that can lose a proton and give the ester.

$$\left.\begin{array}{l}(CH_3)_3COH \underset{H_2O,\,^-OSO_3H}{\overset{H_2SO_4}{\rightleftharpoons}} \\ (CH_3)_2C{=}CH_2 \underset{^-OSO_3H}{\overset{H_2SO_4}{\rightleftharpoons}}\end{array}\right\} (CH_3)_3C^+ \overset{CH_3COOH}{\rightleftharpoons} CH_3C(OH){=}\overset{+}{O}{:}C(CH_3)_3 \underset{H_2SO_4}{\overset{^-OSO_3H}{\rightleftharpoons}} CH_3C({=}O){-}O{-}C(CH_3)_3$$

t-Butyl acetate

When a tertiary alcohol is used, the reaction involves alkyl-oxygen scission in contrast to the more common acyl-oxygen scission of esterification reactions.

(c) BY REACTION WITH DIAZOMETHANE. Diazomethane (p. 249) is represented best by the resonance structure $\{CH_2{=}\overset{+}{N}{=}N{:}^- \longleftrightarrow {}^-{:}CH_2{-}\overset{+}{N}{\equiv}N{:}\}$. Carboxylic acids are sufficiently acidic to transfer a proton to the methylene group to give a methyldiazonium salt. Nucleophilic attack by the carboxylate ion to give the methyl ester then is very easy because the great stability of molecular nitrogen makes it an excellent leaving group.

$$RCOOH + {}^-{:}CH_2{-}\overset{+}{N}{\equiv}N{:} \longrightarrow RCOO{:}^- + CH_3{-}\overset{+}{N}{\equiv}N \longrightarrow RCOOCH_3 + N_2$$

The reaction takes place spontaneously when an ether solution of diazomethane is added to a solution of the carboxylic acid and provides one of the most convenient methods for the conversion of acids into methyl esters. Complete reaction of the acid is observed easily by the cessation of the evolution of nitrogen and the persistence of the yellow color of diazomethane.

4. ***Reaction with Hydrogen Peroxide.*** When carboxylic acids are mixed with 90 per cent hydrogen peroxide in the presence of sulfuric acid or methanesulfonic acid (p. 263), **peroxy acids** (*per acids*) are formed. The process undoubtedly is analogous to esterification.

$$RCOOH + HO{-}OH \xrightarrow{H_2SO_4} RC(=O){-}O{-}OH + H_2O$$

To avoid the presence of a strong mineral acid in the product, the reaction can be brought about by passing the mixture of organic acid and hydrogen peroxide over the acid form of a strong acid cation-exchange resin (p. 484). **Peroxyacetic acid** is made commercially both from acetic acid and hydrogen peroxide, and from acetaldehyde (p. 205).

Peroxy acids readily add oxygen to unsaturated compounds to give the three-membered *oxirane* ring. Such compounds commonly are called **epoxides** (p. 601) and the process is known as **epoxidation.** The reaction is first order in both olefin and peroxy acid. The rate increases with increasing acidity of the peroxy acid and with increasing electron-releasing ability of substituents at the double bond. Since the reaction goes faster in nonpolar than in polar solvents, it is assumed that proton transfer takes place intramolecularly.

$$RCH{=}CHR + R'C(=O){-}O{-}O{-}H \longrightarrow RCH\overset{O}{\frown}CHR + R'COOH$$

Epoxidized oils (p. 186) have found important commercial uses.

5. ***Decomposition to Ketones.*** Ketones are formed when carboxylic acids are heated in the presence of thoria or manganous oxide to the point where decomposition takes place, or when salts of polyvalent metals such as calcium, lead, or thorium are pyrolyzed.

$$2\,R{-}C(=O){-}OH \xrightarrow[400\text{–}450^\circ]{ThO_2 \text{ or } MnO} \underset{\text{A ketone}}{R{-}C(=O){-}R} + CO_2 + H_2O$$

$$(R{-}C(=O){-}O^-)_2\,Ca^{2+} \xrightarrow[\text{decomposition}]{\text{Heat to}} R{-}C(=O){-}R + CaCO_3$$

If the acid is volatile the vapor-phase decomposition over thoria is the preferred method.

6. ***Reduction to Primary Alcohols.*** Carboxylic acids are reduced less readily either catalytically or by most chemical reducing agents than are other compounds containing double bonds. They are reduced more readily than such compounds, however, by borane (p. 86). The relative rates of reduction by this reagent are carboxylic acids > olefins > ketones > nitriles > epoxides > esters > acyl chlorides. Hence borane can be used to reduce carboxyl groups selectively in the presence of other reducible functional groups.

$$RCOOH + BH_3 \longrightarrow H_2 + [RCOOBH_2] \longrightarrow RCH_2OBO \xrightarrow{2\,H_2O} RCH_2OH + B(OH)_3$$

General Reactions of Salts of Carboxylic Acids

1. ***Electrolysis* (Kolbe[2] Hydrocarbon Synthesis).** If a methanol solution of a salt of a carboxylic acid is electrolyzed, the carboxylate ions migrate to the anode and are discharged to give the free carboxylate radical. This unstable radical decomposes to give carbon dioxide and free alkyl radicals, which combine to give the saturated hydrocarbon.

$$RCOO^- \longrightarrow e + RCOO\cdot \longrightarrow CO_2 + R\cdot$$
$$2\,R\cdot \longrightarrow R{-}R$$

Hydroxide ion and hydrogen are formed at the cathode.

$$H_2O + e \longrightarrow {}^-OH + H\cdot$$
$$2\,H\cdot \longrightarrow H_2$$

2. ***Reaction with Alkyl Halides.*** Reaction of a metallic salt with an alkyl halide produces an ester and metallic halide.

$$RCOO^{-+}M + R'X \longrightarrow RCOOR' + M^{+-}X$$

This reaction follows the same course as other displacement reactions of alkyl halides (p. 100).

Although sodium salts frequently give satisfactory yields, the use of the salt of a metal such as silver or mercury usually improves the yield. Coordination of the halogen with the metal ion in the transition state makes the halogen more easily displaced by the carboxylate ion; that is, the metal ion acts as an electrophilic catalyst for the reaction.

$$RCOO^- + R'{:}X{:} + Ag^+ \longrightarrow RCOO{\cdots}R{\cdots}X{\cdots}Ag \longrightarrow RCOOR' + XAg$$

Salts of carboxylic acids with tertiary amines such as triethylamine (p. 226) also have been used, since they are more soluble in organic solvents than are the alkali metal salts.

3. ***Pyrolysis of Ammonium Salts.*** Like other ammonium salts, ammonium carboxylates, when heated, dissociate into the acid and ammonia. At the elevated temperature, addition of the ammonia to the carbonyl group followed by loss of water yields an *amide*.

$$RCOO^{-+}NH_4 \rightleftharpoons RCOOH + NH_3$$

$$R{-}\overset{\overset{\displaystyle O}{\|}}{C}{-}OH \underset{}{\overset{NH_3}{\rightleftharpoons}} R{-}\underset{\underset{\displaystyle {}^+NH_3}{|}}{\overset{\overset{\displaystyle O^-}{|}}{C}}{-}OH \underset{\text{transfers}}{\overset{\text{Proton}}{\rightleftharpoons}} R{-}\underset{\underset{\displaystyle NH_2}{|}}{\overset{\overset{\displaystyle O{-}H}{|}}{C}}{-}OH \rightleftharpoons R{-}\underset{\underset{\displaystyle NH_2}{|}}{\overset{\overset{\displaystyle O}{\|}}{C}} + H_2O$$

Yields are increased by heating in the presence of a high concentration of either the free acid or ammonia and by removal of the water by distillation. The mechanism of the reac-

[2] Hermann Kolbe (1818–1884), student of Woehler, successor to Bunsen at Marburg, and after 1865, professor at the University of Leipzig. He reported in 1845 the first synthesis from the elements of a natural product, namely acetic acid. Kolbe recognized the significance of this work and was the first to use the term *synthesis* in connection with organic chemistry. He also synthesized methanesulfonic acid, taurine, and salicylic acid for the first time, and made nitromethane in the same year that Victor Meyer made nitroethane (p. 242). He showed that fatty acids could be synthesized from alkyl cyanides and was the first to synthesize malonic acid. He demonstrated also the functional relationship of ethyl alcohol, acetaldehyde, and acetone to acetic acid, and of lactic acid to alanine.

Kolbe early recognized the importance of functional groups and had a clear understanding of their transformations. He firmly believed that his formulas represented molecules, and they enabled him to predict the possible existence of then unknown compounds such as secondary and tertiary alcohols, which soon were synthesized by Friedel and Butlerov. Unfortunately Kolbe cast his formulas in terms of the old Berzelius idea of coupled compounds and used the old equivalent weights. Although he is called the "father of structural theory," his refusal to accept the ideas of Kekulé and van't Hoff became an obsession with him and led him to make vitriolic attacks that undoubtedly did much to prevent his receiving the recognition that he deserved.

tion is strictly analogous to that of direct esterification. Ammonium chloride can be used as an acid catalyst.

4. ***Reaction of Silver Salts with Bromine.*** Dry silver carboxylates undergo various reactions with halogen depending on the ratio of the reactants and on whether alkenes, alkynes, or aromatic compounds also are present. If, however, the silver salt is allowed to react with one mole of bromine in boiling carbon tetrachloride, the alkyl halide is obtained in from poor to good yield (*Hunsdiecker reaction*). The initial product is the acyl hypobromite, which apparently undergoes decomposition and yields the alkyl halide by a series of free radical reactions.

$$\text{RCOO}^- + \text{Br—Br} + {}^+\text{Ag} \longrightarrow \text{RCOO—Br} + \text{AgBr}$$

$$\text{RCOO—Br} \longrightarrow \text{Br}\cdot + \text{RCOO}\cdot \longrightarrow \text{R}\cdot + \text{CO}_2$$

$$\text{R}\cdot + \text{Br—OOCR} \longrightarrow \text{RBr} + \text{RCOO}\cdot$$

The over-all reaction provides a method for decreasing the length of a carbon chain by one carbon atom.

Industrially Important Acids

Compounds of commercial importance frequently are made by special reactions that are not applicable to other members of the homologous series. The reason for this situation is that industry tries to use low-cost raw materials and to find new ways of reducing the cost of indispensable substances. Hence a compound may be of commercial importance because a special method has been discovered by which it can be made cheaply. On the other hand the intrinsic value of a compound may be such that a great deal of effort and money have been spent to discover new ways of making it more cheaply than by the general procedures.

Sodium formate is made by a special process from carbon monoxide and caustic soda.

$$\text{CO} + \text{NaOH} \xrightarrow[\text{6–10 atm.}]{200^\circ} \text{HCOO}^{-+}\text{Na}$$

This reaction was one of the earliest used commercially for the synthesis of an organic compound from carbon and salt as the raw materials. It results from attack on carbon by hydroxide ion.

$$\text{H}\overset{-}{\text{O}}\text{:} + \left\{\underset{\cdot\cdot}{\text{C}}\text{=}\underset{\cdot\cdot}{\text{O}}\text{:} \longleftrightarrow \text{:}\overset{-}{\text{C}}\equiv\overset{+}{\text{O}}\text{:}\right\} \longrightarrow \text{H:O:}\overset{-}{\underset{\cdot\cdot}{\text{C}}}\text{=O} \longrightarrow {}^{-}\text{:O—}\underset{\text{H}}{\underset{|}{\text{C}}}\text{=O}$$

Formic acid can be liberated from the sodium salt by the addition of mineral acid. With the increased uses for pentaerythritol, more than enough formic acid is available as a coproduct than is needed to meet the demand (p. 609).

Because the carboxyl group is united to a hydrogen atom, rather than to a carbon atom as in all subsequent members of the series, formic acid undergoes a number of *special reactions.* When mixed with concentrated sulfuric acid, it decomposes into carbon monoxide and water.

$$\text{HCOOH} + \text{H}_2\text{SO}_4 \longrightarrow \text{CO} + \text{H}_3\text{O}^+ + {}^-\text{OSO}_3\text{H}$$

When sodium formate is pyrolyzed, hydrogen is evolved and sodium oxalate remains.

$$\begin{matrix}\text{HCOO}^{-+}\text{Na}\\ \text{HCOO}^{-+}\text{Na}\end{matrix} \xrightarrow{400^\circ} \text{H}_2 + \begin{matrix}\text{COO}^{-+}\text{Na}\\ | \\ \text{COO}^{-+}\text{Na}\end{matrix}$$

Like other compounds containing the $\text{H—}\underset{|}{\text{C}}\text{=O}$ group (p. 206), formic acid is a mild reducing agent.

$$\text{HCOOH} + [\text{O}] \longrightarrow [\text{HOCOOH}] \longrightarrow \text{CO}_2 + \text{H}_2\text{O}$$

Because the acidity of formic acid is approximately ten times that of its homologs (p. 162), it is used when an acid stronger than acetic acid, but not so strong as a mineral acid, is desired. It is used also for the manufacture of its esters and salts. Sodium formate is used to make formic acid and oxalic acid, and as a reducing agent.

Acetic acid is by far the most important organic acid from the standpoint of quantity used. Production was 185 million pounds in 1940 and 1.1 billion pounds in 1964, exclusive of the amount produced in the form of vinegar or salts. The unit value was 7 cents per pound. It appears in the market largely as *glacial acetic acid* of about 99.5 per cent purity, so-called because on cold days it freezes to an ice-like solid. The melting point of pure acetic acid is 16.7°.

Several methods for the preparation of acetic acid are in use. **(*a*)** ENZYME-CATALYZED OXIDATION OF ETHYL ALCOHOL. Acetic acid is the chief component of **vinegar.** The alcohol in fermented fruit juices or fermented malt (*beer*) in the presence of various species of *Acetobacter* and air is oxidized to acetic acid.

$$CH_3CH_2OH + O_2 \text{ (air)} \xrightarrow{Acetobacter} CH_3COOH + H_2O$$

Vinegar produced by fermentation is used almost exclusively as a preservative and condiment, its value for this purpose being enhanced by flavors present in the cider, wine, or malt.

(*b*) FROM PYROLIGNEOUS ACID. Pyroligneous acid from the destructive distillation of wood (p. 119) contains 4–10 per cent of acetic acid, which may be recovered by extraction from the vapor state with tar oil (*Suida process*), or by azeotropic distillation (p. 122) using ethylene chloride, propyl acetate, or butyl acetate to form a constant-boiling mixture with the water (*Clarke-Othmer process*). It may be recovered as ethyl acetate by esterification with ethyl alcohol.

(*c*) FROM ACETALDEHYDE. In the United States, prior to 1952, most of the synthetic acetic acid was made by the oxidation of acetaldehyde, which may be produced either by the hydration of acetylene (p. 149), by the catalytic dehydrogenation or air oxidation of ethyl alcohol (p. 132), by the partial oxidation with air of propane and butane (p. 213), or by the indirect air-oxidation of ethylene (p. 214). In the presence of manganese and cobalt acetates, acetaldehyde absorbs oxygen rapidly to give acetic acid.

$$2\,CH_3CHO + O_2 \longrightarrow 2\,CH_3COOH$$

When 99–99.8 per cent acetaldehyde is used, 96 per cent acetic acid is obtained, which may be rectified to 99.5 per cent.

(*d*) FROM BUTANE. Since 1952 increasing amounts of acetic acid have been made by the air oxidation of butane in the liquid phase. A mixture of air and a large excess of butane is passed into acetic acid containing dissolved manganese and cobalt acetates maintained at 165° under 300 p.s.i. The rate of flow of gases is such that the nitrogen and unreacted butane carry off the products as fast as they are formed. The oxygenated products, chiefly acetic acid and methyl ethyl ketone (p. 216), are extracted with water and separated by fractional distillation. Methyl ethyl ketone in excess of demand is recycled.

$$CH_3CH_2CH_2CH_3 + O_2 \longrightarrow \underset{\text{Methyl ethyl ketone}}{CH_3CH_2COCH_3} + H_2O$$

$$2\,CH_3CH_2COCH_3 + 3\,O_2 \longrightarrow 4\,CH_3COOH$$

The largest single use for acetic acid is for the preparation of acetic anhydride (p. 171). It is used also wherever a cheap organic acid is required; for the preparation of metallic salts and esters (p. 117); in the manufacture of cellulose acetates (p. 343) and white lead; as a precipitating agent for casein from milk, and for rubber or synthetic rubber from their aqueous emulsions (p. 583); and for numerous other purposes. **Sodium acetate** is used to reduce the acidity of mineral acids. **Lead acetate,** known as **sugar of lead,** and **basic lead acetate,** $Pb(OH)(OCOCH_3)$, are used to prepare other lead salts. **Aluminum acetate** is used to impregnate cotton cloth or fibers with aluminum hydroxide prior to dyeing, a process known as *mordanting* (p. 561).

Propionic acid and **butyric acids** may be made by the oxidation of the corresponding alcohols or aldehydes, or by special fermentation processes from starch. They are used in the manufacture of cellulose acetate-propionates and acetate-butyrates (p. 344). **Calcium propionate** is used in bread to prevent molding and ropiness. The higher normal acids having an even number of carbon atoms are obtained by the hydrolysis of fats (p. 184).

ACYL HALIDES

Preparation and Structure

The preparation of acyl halides by reaction of carboxylic acids with inorganic acid halides (p. 162) involves the replacement of a hydroxyl group by halogen. Accordingly the acyl halides are assigned the general structure $R{-}\overset{\overset{\displaystyle O}{\|}}{C}{-}X$. Although acyl fluorides, bromides, and iodides can be prepared, only the acyl chlorides commonly are used. They are highly reactive compounds and are useful only as intermediates for the preparation of other acyl derivatives.

Nomenclature

Acyl halides are named by dropping the ending *ic acid* from the name of the corresponding acid and adding *yl halide;* for example, CH_3COCl is acetyl chloride or ethanoyl chloride, and $(CH_3)_2CHCOBr$ is *i*-butyryl bromide or 2-methylpropanoyl bromide. If common names are used for acyl halides having more than five carbon atoms in the chain, the ending is *oyl*. Thus lauric acid gives rise to lauroyl chloride. Where confusion may arise from common names, systematic names should be used.

Physical Properties

Since the acyl halides do not contain hydrogen united to oxygen, no proton bonding can occur; hence they have normal boiling points. For example, propionic acid with 42 electrons boils at 141°, but acetyl chloride with 40 electrons boils at 51°. This value is higher than that of pentane ($Z = 42$, b.p. 36°) because acetyl chloride has a high dipole moment ($\mu = 2.7$ D). Acyl halides are insoluble in water, the covalently bound halogen atom having the effect of about two or three methylene groups in reducing the water solubility caused by the carbonyl group. The volatile acyl halides have a sharp odor and an irritating action on the mucous membranes. The carbonyl and C—Cl stretching absorptions lie in the 1815–1770 cm.$^{-1}$ and 750–700 cm.$^{-1}$ ranges respectively, but the acid halide function rarely is encountered in unknown compounds.

Reactions

1. ***Nucleophilic Displacement of Halogen.*** Because of the electronegative character of the doubly bound oxygen atom, the electrophilic character of the carbon atom to which halogen is attached is much greater than that in alkyl halides. Accordingly nucleophilic displacement of the halogen of acyl halides takes place much more readily than that of alkyl halides. For example, acyl halides undergo **hydrolysis, alcoholysis,** and **ammonolysis** to give acids, esters, or amides at room temperature. Usually the S_N2 mechanism (p. 100) is involved.

$$H_2O\!: + RCO{-}Cl \longrightarrow Cl^- + RCO\overset{+}{O}H_2 \longrightarrow RCOOH + HCl$$

$$R'\underset{H}{\underset{|}{O}}\!: + RCOCl \longrightarrow RCOOR' + HCl$$

$$H_3N\!: + RCOCl \longrightarrow RCONH_2 + HCl \xrightarrow{NH_3} NH_4^+Cl^-$$

Reactions of this type often are called *acylations,* and the nucleophile is said to be *acylated.* If the R group in the acyl halide is sufficiently electron-donating, displacement of halogen may take place by the S_N1 mechanism.

2. ***Halogenation.*** Acyl halides chlorinate and brominate more readily than hydrocarbons or carboxylic acids. Moreover only α hydrogen atoms are replaced readily. The combined electron-attracting effect of the carbonyl group and the halogen is sufficient to permit loss of a proton from the α carbon atom and enolization. Reaction of the enol with halogen gives the halogenated acyl halide.

$$\underset{H}{\underset{|}{RCH}}{-}\overset{O}{\overset{\|}{C}}{-}X \underset{H^+}{\rightleftharpoons} RCH{=}\overset{O^-}{\overset{|}{C}}{-}X \overset{H^+}{\rightleftharpoons} RCH{=}\overset{O-H}{\overset{|}{C}}{-}X \;\; X{-}X \rightleftharpoons \underset{X}{\underset{|}{RCH}}{-}\overset{O}{\overset{\|}{C}}{-}X + HX$$

In actual practice free acids are used, the halogenation being carried out in the presence of a small amount of phosphorus trihalide. The reactions then are

$$3\,RCH_2COOH + PX_3 \longrightarrow 3\,RCH_2COX + P(OH)_3$$

$$RCH_2COX + X_2 \longrightarrow RCHXCOX + HX$$

$$RCHXCOX + RCH_2COOH \rightleftharpoons RCHXCOOH + RCH_2COX$$

Because of the last reaction a small amount of acid halide is sufficient to permit the direct halogenation of a large amount of acid. This procedure for making halogen acids is known as the *Hell-Volhard-Zelinsky reaction.*

3. ***Reduction.*** ***(a)*** TO ALCOHOLS. Hydride ion is a strong nucleophile and not only displaces halide ion from acyl chlorides but also a pair of electrons from the double bond of the intermediate aldehyde.

$$RCOCl \underset{X:^-}{\overset{H:^-}{\rightleftharpoons}} R\underset{H}{\underset{|}{C}}{=}O \overset{H:^-}{\rightleftharpoons} R{-}\overset{H}{\overset{|}{\underset{H}{\underset{|}{C}}}}{-}O^-$$

Lithium aluminum hydride, which is soluble in ether, is a very satisfactory source of hydride ion. The product is the mixture of lithium and aluminum alkoxides from which the alcohol is liberated by addition of hydrochloric acid.

$$2\ RCOCl + LiAlH_4 \longrightarrow (RCH_2O)_2LiAlCl_2$$
$$(RCH_2O)_2LiAlCl_2 + 2\ HCl \longrightarrow 2\ RCH_2OH + LiAlCl_4$$

(b) TO ALDEHYDES. The reducing power of lithium aluminum chloride can be decreased if the hydrogen is replaced partially by alkoxyl groups. Acyl chlorides are reduced only to the aldehyde stage by lithium aluminum triethoxide hydride, made by adding three moles of ethanol to one of lithium hydride in tetrahydrofuran (p. 512) or in the dimethyl ether of diethylene glycol (*diglyme,* p. 142).

$$LiAlH_4 + 3\ C_2H_5OH \longrightarrow LiAl(OC_2H_5)_3H + 3\ H_2$$
$$RCOCl + LiAl(OC_2H_5)_3H \longrightarrow RCHO + LiAl(OC_2H_5)_3Cl$$

Acyl chlorides can be reduced catalytically to aldehydes by means of a palladium catalyst (*Rosenmund reduction*). The catalyst usually is poisoned partially by the addition of sulfur compounds, which renders it inactive for the catalysis of the further reduction of aldehydes to alcohols (p. 206).

$$RCOCl + H_2 \xrightarrow{Pd} RCHO + HCl$$

ACID ANHYDRIDES

Preparation and Structure

The preparation of carboxylic acid anhydrides by the nucleophilic displacement of chloride ion from acyl chlorides by carboxylate ion clearly defines the structure of the anhydrides as $R\overset{O}{\overset{\|}{C}}{-}O{-}\overset{O}{\overset{\|}{C}}R$. A second method of preparation depends on an exchange reaction that results from nucleophilic attack by carboxylic acids on acid anhydrides, which permits equilibrium to be established with acetic anhydride.

$$R\underset{O-H}{C}{=}\ddot{O}: + CH_3\underset{\|}{\underset{O}{C}}{-}O{-}\underset{\|}{\underset{O}{C}}CH_3 \rightleftharpoons HOCOCH_3 + R\underset{\|}{\underset{O}{C}}{-}O{-}\underset{\|}{\underset{O}{C}}CH_3 \xrightleftharpoons{RCOOH} R\underset{\|}{\underset{O}{C}}{-}O{-}\underset{\|}{\underset{O}{C}}R + HOCOCH_3$$

Since acetic acid boils at a lower temperature than any other component in the system, it can be removed by careful distillation and the reaction forced to completion. The first reaction may be used for preparing anhydrides in which the R groups are either alike or different. Although the second method is suitable only for the preparation of anhydrides in which both R groups are the same, it is preferred for this purpose because of the availability and low cost of acetic anhydride and the excellent yields.

Nomenclature

A member of this class of compounds is named by adding the word *anhydride* to the name of the acid or acids from which it is derived. For example, $(CH_3CO)_2O$ is acetic anhydride or ethanoic anhydride, and $(CH_3CO)O(COCH_2CH_2CH_3)$ is acetic butyric anhydride or ethanoic butanoic anhydride. Those anhydrides having both R groups alike are called *simple anhydrides* whereas those with different R groups are known as *mixed anhydrides.* The former are the more important.

Physical Properties

Anhydrides boil higher than hydrocarbons but lower than alcohols of comparable polarizability. Despite the relatively large amount of oxygen present, acetic anhydride and its higher homologs are insoluble in water. Apparently the high resonance energy (30 kcal.) arising from the interaction of the electrons in the *p* orbitals of oxygen and carbon prevents the localization of negative charge necessary for hydrogen bonding.

Anhydrides show two carbonyl absorption bands about 60 cm.$^{-1}$ apart in the 1850–1800 cm.$^{-1}$ and 1790–1740 cm.$^{-1}$ regions. The reason for two bands is not known but the doubling is useful for identification. Strong C—O stretching absorption occurs in the 1175–1040 cm.$^{-1}$ range.

Reactions

In acylation reactions, the reactivity of the acylating agent depends on the electrophilicity of the carbonyl carbon atom, which in turn depends on the electronegativity of the group attached to the acyl group. Thus the order of reactivity as acylating agents is $RCOR \cong RCHO < RCO{-}NR_2 < RCO{-}OR < RCO{-}OCOR < RCO{-}X < RCO{-}OSO_3H < RCO{-}OClO_3 < RCO^{+-}BF_4$. Aldehydes and ketones usually are not considered to be acylating agents, and the strongest acylating agents are acyl perchlorates and acyl fluoroborates. Anhydrides are good acylating agents and readily undergo hydrolysis, alcoholysis, and ammonolysis, but the reactions are not as violent as those of acyl halides.

$$H_2O\!: + \underset{\underset{O}{\|}}{\overset{\overset{R}{|}}{C}}{-}OCOR \longrightarrow H_2\overset{+}{O}{-}\underset{\underset{O}{\|}}{\overset{\overset{R}{|}}{C}} + {}^{-}OCOR \longrightarrow RCOOH + HOCOR$$

$$R'OH + (RCO)_2O \longrightarrow R'OCOR + HOCOR$$

$$2\,H_3N + (RCO)_2O \longrightarrow H_2NCOR + NH_4^{+-}OCOR$$

Like the acyl halides, the anhydrides are used only as acylating agents. Ammonia and primary and secondary amines are much better nucleophiles than water (p. 103). Hence even though anhydrides react slowly with water, they react so much more rapidly with ammonia or amines that aqueous solutions of the latter can be used for the preparation of amides.

Acetic anhydride is the only important member of the series. Several processes have been used for its commercial production. Currently the two most important methods are by the air oxidation of acetaldehyde (p. 168)

$$2\,CH_3CHO + O_2 \longrightarrow (CH_3CO)_2O + H_2O$$

and by the addition of acetic acid to ketene.

$$CH_3COOH + CH_2{=}C{=}O \longrightarrow (CH_3CO)_2O$$

The former process uses a cobalt acetate–copper acetate catalyst at 50° and an entrainer such as ethyl acetate to remove the water. In this way it is possible to convert two thirds of the acetaldehyde to anhydride and one third to acetic acid (p. 168). In the second process ketene is prepared by decomposing acetic acid vapor carrying a small amount of ethyl phosphate at 700–720°.

$$CH_3COOH \longrightarrow \underset{\text{Ketene}}{CH_2{=}C{=}O} + H_2O$$

The mixture of ketene, water, and acetic acid vapor is cooled quickly to condense the water and unreacted acetic acid. The ketene then is absorbed in acetic acid in a scrubbing tower, giving acetic anhydride. In an alternative procedure for the manufacture of ketene, acetone is decomposed non-catalytically at 700°.

$$CH_3COCH_3 \longrightarrow CH_2{=}C{=}O + CH_4$$

Acetic anhydride is used chiefly to make esters that cannot be made by direct esterification of alcohols with acetic acid. The most important of these is cellulose acetate. Production of acetic anhydride in the United States rose from 250 million pounds in 1940 to 1.4 billion pounds in 1964, when it was valued at 15 cents per pound.

ESTERS

Preparation

Most of the procedures for making esters have been described, namely direct esterification between an acid and an alcohol (p. 163), the addition of an acid to an olefin (p. 164), the reaction of a metallic salt with an alkyl halide (p. 166), the reaction of an acyl halide with an alcohol (p. 169), and the reaction of an acid anhydride with an alcohol (p. 171). They can be prepared also by the pyrolysis of quaternary ammonium salts (p. 231). A few generalizations should be made about these reactions. In the first place all except direct esterification proceed practically to completion. Side reactions may take place, however, particularly with tertiary alcohols and tertiary alkyl halides, which make the reactions useless unless special conditions can be found that will produce satisfactory yields. In general the ease of loss of water from alcohols or of halogen acid from alkyl halides is tertiary > secondary > primary (pp. 130, 106). Therefore when a tertiary alcohol reacts with an acid anhydride, which is a dehydrating agent, or when a tertiary alkyl halide reacts with a sodium salt, which acts as a base, the rate of olefin formation is much greater than the rate of ester formation and only olefin results.

Nomenclature

Esters are named as if they were alkyl salts of the organic acids because the early investigators assumed that esterification is analogous to neutralization. Thus $CH_3COOC_2H_5$ is ethyl acetate or ethyl ethanoate. It is necessary to be careful to recognize the portion of the molecule derived from the acid and that from the alcohol, particularly in condensed structural formulas. For example, both $(CH_3)_2CHOCOCH_2CH_3$ and $CH_3CH_2COOCH(CH_3)_2$ are *i*-propyl propionate and not ethyl *i*-butyrate. No difficulty should be encountered if it is remembered that the oxygen of a carbonyl group usually follows immediately the carbon atom to which it is attached and that the alkyl group from the alcohol portion of the ester is joined to an oxygen atom. If it is necessary to name esters as substitution products, the ester group is called an *alkoxycarbonyl* group. For example, $COOC_2H_5$ is the ethoxycarbonyl group.

Physical Properties

The esters have normal boiling points, but their solubility in water is less than would be expected from the amount of oxygen present. Ethyl acetate with four carbon atoms and two oxygen atoms dissolves to about the same extent as *n*-butyl alcohol, which has four carbon atoms and one oxygen atom (cf. acid anhydrides, p. 171). The infrared absorption bands for the C=O and C—O stretching vibrations lie in the 1750–1735 $cm.^{-1}$ and the 1300–1180 cm.$^{-1}$ respectively. The volatile esters have pleasant odors that usually are described as fruity.

Reactions

1. ***Displacement of the Alkoxyl Group.*** Alkoxyl, like hydroxyl, is a poor leaving group. Moreover it is not as electronegative as acetate or chlorine and hence does not contribute as much to the electrophilicity of the carbonyl carbon atom. Hence displacement of alkoxyl by weak nucleophiles occurs only when it is catalyzed by acids or bases.

Acid-catalyzed hydrolysis of primary alkyl esters is exactly the reverse of acid-catalyzed esterification (p. 163) because of the principle of microscopic reversibility (p. 38). **Alcoholysis** takes place by a strictly analogous mechanism.

$$H_2O + RCOOR' \overset{H^+}{\rightleftharpoons} RCOOH + HOR'$$

$$R''OH + RCOOR' \overset{H^+}{\rightleftharpoons} RCOOR'' + HOR'$$

Hydrolysis and alcoholysis of tertiary alkyl esters and of esters of strong acids take place by a carbonium ion mechanism with alkyl-oxygen scission rather than acyl-oxygen scission (p. 163). Hence reaction with alcohol leads to the formation of an ether rather than another ester.

$$CH_3\underset{\underset{O}{\|}}{C}{-}O{-}C(CH_3)_3 \underset{B^-}{\overset{HB}{\rightleftharpoons}} CH_3{-}\underset{\underset{\overset{+}{O}H}{\|}}{C}{-}O{-}C(CH_3)_3 \rightleftharpoons$$

$$CH_3\underset{\underset{OH}{|}}{C}{=}O + (CH_3)_3C^+ \overset{ROH}{\rightleftharpoons} (CH_3)_3C{-}\underset{\underset{H}{|}}{\overset{+}{O}}{-}R \underset{HB}{\overset{B^-}{\rightleftharpoons}} (CH_3)_3C{-}O{-}R$$

Since hydroxide ion and alkoxide ion are stronger nucleophiles than water or alcohol molecules, they bring about **base-catalyzed hydrolysis** and **alcoholysis.**

$$RCOOR' \overset{^-OH}{\rightleftharpoons} R{-}\underset{\underset{OH}{|}}{\overset{\overset{O^-}{|}}{C}}{-}OR' \underset{H_2O}{\overset{^-OH}{\rightleftharpoons}} R{-}\underset{\underset{O^-}{|}}{\overset{\overset{O^-}{|}}{C}}{-}OR' \longrightarrow RCOO^- + {}^-OR' \underset{^-OH}{\overset{H_2O}{\rightleftharpoons}} HOR'$$

$$H_2O \Big\| {}^-OH \qquad\qquad H_2O \Big\| {}^-OH$$

$$R{-}\underset{\underset{OH}{|}}{\overset{\overset{OH}{|}}{C}}{-}OR' \underset{H_2O}{\overset{^-OH}{\rightleftharpoons}} R{-}\underset{\underset{O^-}{|}}{\overset{\overset{OH}{|}}{C}}{-}OR'$$

These mechanisms for catalysis by either acid or base agree with the observation that if the hydrolyses are carried out in water rich in the O^{18} isotope, O^{18} enters the acid molecule formed on hydrolysis but not the alcohol. This result involves the rapid proton transfer reactions between hydroxide ions and water molecules.

$$H{:}\ddot{\underset{..}{O}}{:}^- + H{:}\overset{..18}{\underset{..}{O}}{:}H \rightleftharpoons H{:}\ddot{\underset{..}{O}}{:}H + {}^-{:}\overset{..18}{\underset{..}{O}}{:}H$$

Alkaline hydrolysis frequently is referred to as **saponification,** because it is the type of reaction used in the preparation of soaps (p. 184). The hydrolysis goes to completion and requires one equivalent of alkali for each equivalent of ester because the acid formed in the base-catalyzed equilibrium reacts irreversibly with the catalyst to form a salt and water. The over-all reaction becomes

$$RCOOR' + NaOH \longrightarrow RCOO^-Na^+ + R'OH$$

Because this reaction goes to completion, alkaline saponification is used as a quantitative procedure for the estimation of esters. A weighed sample of the unknown is refluxed with an excess of a standardized aqueous or alcoholic solution of alkali, and the excess base at the end of the reaction is titrated with standard acid and a suitable indicator, usually phenolphthalein (p. 579). The equivalent weight as determined by saponification is called the **saponification equivalent** of the ester.

Since in the acid-catalyzed hydrolysis the same position of equilibrium is attained as in esterification, the effect of the constitution of the alcohol and of the acid on the rate of hydrolysis is the same as their effect on the rate of esterification (p. 164). In alkaline hydrolysis the substitution of hydrogen atoms of both acetic acid and methyl alcohol by alkyl groups greatly decreases the rate of reaction. Thus the rate of saponification for ethyl trimethylacetate is about one hundredth of that for ethyl acetate, and the rate for *t*-butyl acetate one hundredth of that for methyl acetate.

Because hydroxide ion brings about the irreversible saponification of an ester, it cannot be used to catalyze the alcoholysis of esters. Under anhydrous conditions alkoxide ions, however, are very effective catalysts for alcoholysis.

$$\text{RCOOR}' \underset{}{\overset{\text{R}''\text{O:}^-}{\rightleftharpoons}} \text{R—}\underset{\text{OR}''}{\overset{\text{O}^-}{\text{C}}}\text{—OR}' \underset{\text{R}''\text{O:}^-}{\overset{\text{R}''\text{OH}}{\rightleftharpoons}} \text{R—}\underset{\text{OR}''}{\overset{\text{O—H}}{\text{C}}}\text{—OR}' \rightleftharpoons \text{HOR}' + \text{RCOOR}''$$

As with acid-catalyzed alcoholysis, all of the steps are easily reversible. Hence high conversion of one ester into another can be attained only by using a large excess of the alcohol that supplies the new alkyl group. This alcohol is also the solvent for the reaction. Preformed alkoxide ion can be added, but usually it is supplied by allowing a small amount of sodium to react with the alcohol prior to addition of the ester.

Alkoxide ion can bring about also the exchange of alkyl groups between different esters, a process called **transesterification.**

$$\text{R}''\text{O}^- + A\text{—}\overset{\text{O}}{\overset{\|}{\text{C}}}\text{OR}' \rightleftharpoons A\text{—}\underset{\text{OR}''}{\overset{\text{O}^-}{\text{C}}}\text{—OR}' \rightleftharpoons A\text{—}\overset{\text{O}}{\overset{\|}{\text{C}}}\text{—OR}'' + {}^-\text{OR}'$$

$$\text{R}'\text{O}^- + B\text{—}\overset{\text{O}}{\overset{\|}{\text{C}}}\text{OR}'' \rightleftharpoons B\text{—}\underset{\text{OR}'}{\overset{\text{O}^-}{\text{C}}}\text{—OR}'' \rightleftharpoons B\text{—}\overset{\text{O}}{\overset{\|}{\text{C}}}\text{—OR}' + {}^-\text{OR}''$$

Both alcoholysis and transesterification have practical uses (p. 188).

The **ammonolysis** of simple esters also requires catalysis. In alcohol solution the reaction is accelerated by alkoxide ion. The base apparently improves the nucleophilicity of the ammonia molecule by bonding with the proton in the transition state.

$$\text{RO:}^- + \text{H}_3\text{N:} + \underset{\text{R}''}{\overset{\text{O}}{\overset{\|}{\text{C}}}}\text{OR}' \rightleftharpoons \text{RO---H---NH}_2\text{---}\underset{\text{R}''}{\overset{\text{O}}{\overset{\|}{\text{C}}}}\text{---OR}' \rightleftharpoons \text{ROH} + \text{H}_2\text{N—}\underset{\text{R}''}{\overset{\text{O}}{\overset{\|}{\text{C}}}} + {}^-\text{:OR}'$$

Ammonolysis in liquid ammonia solution is accelerated by ammonium salts, the ammonium ion acting as a proton donor. The mechanism then is analogous to acid-catalyzed hydrolysis.

2. ***Displacement of Carboxylate Ion.*** Dry hydrogen bromide, concentrated aqueous hydrogen iodide, or concentrated sulfuric acid can split esters just as they do ethers (p. 141). Protonation of the alkoxyl group permits the carboxylate ion to be displaced as the carboxylic acid.

$$\mathrm{RC(=O){-}O{-}R'} \underset{\mathrm{Br^-}}{\overset{\mathrm{HBr}}{\rightleftharpoons}} \mathrm{RC(=O){-}\overset{+}{O}(H){-}R'} \overset{\mathrm{Br^-}}{\rightleftharpoons} \mathrm{RCOOH + R'Br}$$

The reaction is important chiefly for the determination of methoxyl or ethoxyl by the Zeisel procedure (p. 141).

3. ***Reduction.*** Esters, and thus indirectly carboxylic acids, can be converted to primary alcohols by reduction with agents such as lithium aluminum hydride or sodium and an alcohol, or by catalytic reduction. The chemical reducing agents react through the hydride ion, either preformed as with lithium aluminum hydride or as an intermediate on the surface of the metal as with sodium and alcohol (cf. p. 132).

$$2\,\mathrm{Na\cdot} + \mathrm{HOR} \longrightarrow 2\,\mathrm{Na^+} + \mathrm{H{:}^-} + \mathrm{^-{:}OR}$$

$$\mathrm{RC(=O){-}OR} + \mathrm{{}^-{:}H} \longrightarrow \mathrm{R{-}C(O{:}^-)(H){-}OR} \underset{\mathrm{:OR^-}}{\rightleftharpoons} \mathrm{R{-}C(=O){-}H} \xrightarrow{\mathrm{H{:}^-}} \mathrm{RC(O^-)(H){-}H} \underset{\mathrm{^-{:}OH}}{\overset{\mathrm{H_2O}}{\rightleftharpoons}} \mathrm{RCH_2OH}$$

The over-all reactions are

$$2\,\mathrm{RCOOR'} + \mathrm{LiAlH_4} \longrightarrow \mathrm{(RCH_2O^-)_2Li^+Al^{3+}(^-OR)_2}$$

$$\mathrm{RCOOR'} + 4\,\mathrm{Na} + 3\,\mathrm{R'OH} \longrightarrow \mathrm{RCH_2O^{-+}Na} + 3\,\mathrm{Na^{+-}OR'}$$

Addition of water at the end of the reduction hydrolyzes the alkoxides to the alcohols. Reduction stops at the aldehyde stage when di-*i*-butylaluminum hydride (p. 112) is used as a reducing agent at $-70°$.

Lithium aluminum hydride reductions are carried out in ether solution. In the sodium-alcohol reductions the methyl, ethyl, or *n*-butyl esters ordinarily are used along with an excess of the corresponding alcohol as solvent and source of hydrogen. Improved yields are obtained if an equivalent amount of a higher-boiling secondary alcohol, such as the secondary hexyl alcohols or methylcyclohexanols (p. 647), is used as the source of hydrogen in an inert solvent such as toluene or xylene (p. 93). One half mole or 19 g. of lithium aluminum hydride is equivalent in reducing power to four atoms or 92 g. of sodium. The former is, however, a considerably more expensive reagent, since it is made from lithium hydride.

$$4\,\mathrm{LiH} + \mathrm{AlCl_3} \longrightarrow \mathrm{LiAlH_4} + 3\,\mathrm{LiCl}$$

It is reported that sodium aluminum hydride can be made directly from sodium, aluminum, and hydrogen and that it is as good as lithium aluminum hydride as a reducing agent.

In the *catalytic reduction* of esters, a copper oxide-chromium oxide catalyst is used with hydrogen in the absence of a solvent at 200° and 200 atmospheres.

$$\mathrm{RCOOR'} + 2\,\mathrm{H_2} \xrightarrow[200°,\ 200\ \mathrm{at.}]{\mathrm{CuO{-}Cr_2O_3}} \mathrm{RCH_2OH} + \mathrm{HOR'}$$

Catalytic reduction of esters is preferred to sodium-alcohol reduction for large-scale laboratory or industrial preparations.

The reduction of esters to alcohols is important for preparing higher alcohols from natural fat acids. Thus octyl, decyl, dodecyl (*lauryl*), tetradecyl (*myristyl*), hexadecyl (*cetyl*), and octadecyl (*stearyl*) alcohols can be made readily from the corresponding fat acids. The mixed alcohols formed by the reduction of coconut oil are used in the manufacture of synthetic detergents (p. 191). The reaction is important also because it is the last step in a series of reactions that permits the increase in length of a hydrocarbon chain by one carbon atom at a time.

$$ROH \longrightarrow RX \longrightarrow RMgX \longrightarrow RCOOH \longrightarrow RCOOCH_3 \longrightarrow RCH_2OH$$

A repetition of the series produces the next higher homolog.

4. ***Reaction with Grignard Reagents.*** Alkylmagnesium halides add to a carbonyl group (p. 153). When they react with an ester, the initial addition product loses alkoxide ion and halomagnesium ion to give a ketone.

$$R{-}\overset{O}{\overset{\|}{C}}{-}OR' + R''MgX \longrightarrow R{-}\underset{R''}{\underset{|}{\overset{O^{-+}MgX}{\overset{|}{C}}}}{-}OR' \longrightarrow R{-}\underset{R''}{\underset{|}{\overset{O}{\overset{\|}{C}}}} + RO^{-+}Mg^{+-}X$$

Loss of the alkoxyl group is facilitated by coordination with magnesium halide (p. 140), which always is present. The resulting ketone reacts at once with more Grignard reagent to give the halomagnesium salt of a tertiary alcohol from which the alcohol can be obtained by hydrolysis.

$$R{-}\underset{R''}{\underset{|}{\overset{O}{\overset{\|}{C}}}} + R''MgX \longrightarrow R{-}\underset{R''}{\underset{|}{\overset{O^{-+}MgX}{\overset{|}{C}}}}{-}R'' \xrightarrow{H_2O} R{-}\underset{R''}{\underset{|}{\overset{OH}{\overset{|}{C}}}}{-}R'' + HOMgX$$

5. ***Acyloin Formation.*** In solvents such as ether that cannot supply protons, esters react with sodium to give bimolecular reduction products. The reaction undoubtedly occurs on the surface of the metal. The product is the disodium salt of an enolized α-hydroxy ketone. Addition of water hydrolyzes it to the enol, which spontaneously ketonizes.

$$4\,Na\cdot + 2\,RC(=O)OR \longrightarrow 2\,Na^+ + 2\,{}^-{:}OR + \begin{matrix}Na\cdot\ O{=}C{-}R\\ Na\cdot\ O{=}C{-}R\end{matrix} \longrightarrow 2\,Na^+ + \begin{matrix}{}^-O{-}C{-}R\\ \|\\ {}^-O{-}C{-}R\end{matrix} \underset{2\,{}^-{:}OH}{\overset{2\,H_2O}{\rightleftharpoons}} \begin{matrix}HO{-}C{-}R\\ \|\\ HO{-}C{-}R\end{matrix} \longrightarrow \begin{matrix}O{=}C{-}R\\ |\\ H{-}C{-}R\\ |\\ OH\end{matrix}$$

These α-hydroxy ketones commonly are called **acyloins** or **α-ketols.** *Methylacetylcarbinol,* $CH_3COCHOHCH_3$, is made in this way from ethyl acetate.

6. ***Pyrolysis to Olefins.*** When esters that have hydrogen on the β carbon atom of the alkoxy group are heated to 500°, a molecule of acid is lost with the formation of an olefin. Esters of primary alcohols give 1-alkenes.

$$RCOOCH_2CH_2R' \xrightarrow{500^\circ} RCOOH + CH_2{=}CHR'$$

When esters of secondary or tertiary alcohols are decomposed, a mixture of olefins is obtained. The reaction is believed to go by way of a cyclic transition state.

$$\begin{matrix}H\ \ \ \ \ \ \ \ \ \\ R_2C\ \ \ \ O\\ |\ \ \ \ \ \ \|\\ R_2C\ \ \ C{-}R'\\ \ \ \ O\ \ \ \end{matrix} \longrightarrow \begin{matrix}H\ \ \ \ \ \ \ \ \ \\ R_2C\ \ \ \ O\\ |\ \ \ \ \ \ \|\\ R_2C\ \ \ C{-}R'\\ \ \ \ O\ \ \ \end{matrix} \longrightarrow \begin{matrix}R_2C\\ \|\\ R_2C\end{matrix} + \begin{matrix}H\ \ \\ O\\ |\\ O{=}C{-}R'\end{matrix}$$

Tertiary butyl esters decompose very easily to isobutylene and the acid. In this way carboxyl groups can be regenerated from the ester without resorting to alkaline or acid hydrolysis (p. 173).

Uses

By far the most important general use for esters of aliphatic acids is as solvents, especially for cellulose nitrate in the formulation of lacquers (p. 343). For this purpose ethyl acetate and butyl acetates are used to the greatest extent. Production of these esters in the United States in 1964 was 100 and 116 million pounds valued at 12 and 10 cents per pound respectively. Ethyl formate is used as a fumigant and larvicide for grains and food products. Higher-boiling esters are used as softening agents (*plasticizers*) for resins and plastics (p. 474), and numerous resins and plastics are themselves esters, such as poly(methyl methacrylate) (p. 620), poly(vinyl acetate) (p. 599), cellulose acetate (p. 343), Dacron (p. 476), alkyd and polyester resins (pp. 474, 635), and polyurethans (p. 605).

Some of the volatile esters have specific fruit odors. For example, the odors of *i*-amyl acetate, *i*-amyl valerate, butyl butyrate, and *i*-butyl propionate resemble the odors of banana, apple, pineapple, and rum respectively. Hence they are used to a limited extent in synthetic flavors or perfumes. Natural odors and flavors are the result of complex mixtures of organic compounds. Very careful blending of synthetic compounds is necessary to imitate the natural product. Table 11–2 summarizes the results of an analysis of the substances responsible for the odor and flavor of the pineapple. Both the amount of volatile oil and its components vary with the time of the year at which the fruit is harvested.

A more exhaustive investigation of strawberry extract and of raspberry extract has shown that both are complex mixtures. Over twenty different compounds have been isolated and identified from each, together with other unidentified compounds and unseparated mixtures. Moreover there is considerable variation in composition depending on factors such as the strain of plant, ripeness, soil nutrients, and climatic conditions. The bouquet of fine wines has been ascribed to esters produced by the slow esterification of organic acids during the ageing process.

Because of its easy decomposition to isobutylene and acetic acid, *t*-butyl acetate (p. 164) is used as an additive for gasoline. It has been known for some time that acetic acid prevents the decrease in effectiveness of tetraethyllead as an antiknock agent (p. 94) as its concentration is increased, but the addition of acetic acid would lead to corrosion problems. Addition of *t*-butyl acetate has the same effect and allows an increase in tetraethyllead concentration from 3 cc. to 4 cc. per gallon with a proportionate increase in antiknock rating.

ORTHO ESTERS

Ortho acids are acids in the highest state of hydration; that is, they contain the maximum number of hydroxyl groups. For example, orthoformic acid should have the formula $HC(OH)_3$. Compounds having more than one hydroxyl group on the same carbon, however, usually are unstable and lose water.

$$\left[\text{H—C(OH)}_2\text{—O—H}\right] \longrightarrow \text{H—}\overset{\text{O}}{\overset{\|}{\text{C}}}\text{—OH} + H_2O$$

If orthoformic acid exists at all, it exists only in aqueous solutions. When alkyl groups are united to oxygen instead of hydrogen, this type of reaction cannot take place. The 1,1,1-trialkoxyalkanes are stable compounds known as *ortho esters*.

TABLE 11–2. COMPOSITION OF THE VOLATILE OIL OF THE PINEAPPLE

WINTER FRUIT		SUMMER FRUIT	
CONSTITUENT	MG. PER KG.	CONSTITUENT	MG. PER KG.
Total volatile oil	15.6	Total volatile oil	190.0
Ethyl acetate	2.91	Ethyl acetate	119.6
Ethyl alcohol	0.0	Ethyl alcohol	60.5
Acetaldehyde	0.61	Acetaldehyde	1.35
Methyl *n*-valerate	0.49	Ethyl acrylate	0.77
Methyl *i*-valerate	0.60	Ethyl *i*-valerate	0.39
Methyl *i*-caproate	1.40	Ethyl *n*-caproate	0.77
Methyl caprylate	0.75		

Two methods commonly are used for the synthesis of ortho esters. One method, to be discussed later (p. 236), involves the reaction of a nitrile with an alcohol. The other method, discovered by Williamson (p. 139), is the reaction of a trihalide with sodium alkoxide. For example, **ethyl orthoformate** usually is prepared by the reaction of chloroform with sodium ethoxide.

$$HCCl_3 + 3\,NaOC_2H_5 \longrightarrow HC(OC_2H_5)_3 + 3\,NaCl$$

Although formally this reaction is analogous to the reaction of alkyl halides with sodium ethoxide to give ethers, chloroform is about 1000 times more reactive than either carbon tetrachloride or methylene chloride. It has been proposed that reactions involving chloroform and a strongly alkaline reagent take place through the intermediate formation of dichloromethylene (*dichlorocarbene*). The first step is the removal of a proton from chloroform by the alkoxide ion, a reaction made possible by the presence of three electronegative chlorines on the same carbon.

$$CHCl_3 + C_2H_5O^- \longrightarrow {}^-CCl_3 + C_2H_5OH$$
$${}^-CCl_3 \longrightarrow :CCl_2 + Cl^-$$
$$:CCl_2 + 3\,C_2H_5O^- \longrightarrow (C_2H_5O)_3C:^- + 2\,Cl^-$$
$$(C_2H_5O)_3C:^- + C_2H_5OH \longrightarrow (C_2H_5O)_3CH + C_2H_5O^-$$

Much evidence has accumulated that methylene and derivatives of methylene are intermediates in many reactions (p. 250). One question that has arisen is whether these intermediates have paired electrons (*singlet state*) and an empty orbital or two unpaired electrons (*triplet state*). In the singlet state the intermediate could act either as an electron-pair donor or as an electron-pair acceptor, whereas in the triplet state it would be a biradical. It appears that these intermediates react both by free radical and by polar mechanisms. In most of their reactions, however, they act initially as electron-pair acceptors (p. 250).

Originally there was little or no justification for coining the new term *carbene* to replace the well-established *methylene.* Since the term *carbene* has become popularized, however, it would be well to limit its use to intermediates that appear to react in the singlet state and to reserve the term *methylene* for the biradical.

The lack of a carbonyl group makes the ortho esters resemble acetals (p. 200) rather than carboxylic esters in some of their chemical properties. Thus they are very stable to aqueous alkali but are hydrolyzed readily and irreversibly by dilute acid solutions.

$$RC(OC_2H_5)_3 + H_2O \xrightarrow{H^+} RCOOC_2H_5 + 2\,C_2H_5OH$$

The carboxylic ester goes to reversible equilibrium with the alcohol and the free acid. Grignard reagents replace one of the alkoxy groups by an alkyl group.

$$H\!-\!\underset{\displaystyle OC_2H_5}{\overset{\displaystyle OC_2H_5}{\underset{|}{\overset{|}{C}}}}\!-\!OC_2H_5 + RMgX \longrightarrow H\!-\!\underset{\displaystyle OC_2H_5}{\overset{\displaystyle R}{\underset{|}{\overset{|}{C}}}}\!-\!OC_2H_5 + C_2H_5OMgX$$

Displacement of the ethoxide ion by alkide ion undoubtedly is assisted by coordination of the oxygen of the ethoxy group with magnesium halide. Use is made of this reaction for the synthesis of aldehydes (p. 193).

PROBLEMS

11–*1*. Write skeleton structures for members of the saturated carboxylic acids that have six carbon atoms. Name each by the international system and where convenient by a common name or as a derivative of acetic acid.

11–*2*. Using accepted structural theory, write formulas for all of the possible compounds having the molecular formula $C_2H_4O_2$. From empirical bond energies (let O=C = 180 kcal.), estimate the heats of formation for each and arrange in the order of decreasing stability.

11–*3*. List the following groups in the order of increasing ability to confer water solubility when they replace hydrogen of an alkane: (*a*) chlorine, (*b*) hydroxyl, (*c*) methoxyl, (*d*) carboxyl.

11–*4*. Give balanced equations for the oxidation of the following compounds to carboxylic acids: (*a*) 2-pentene ozonide with sodium dichromate and sulfuric acid, (*b*) *i*-butyl alcohol with alkaline potassium permanganate, (*c*) 1-octanol with excess nitric acid.

11–*5*. (*a*) Which of the following compounds would be the most convenient starting point for the preparation of *n*-propyl cyanide: (*1*) propylene, (*2*) propane, (*3*) *n*-propyl chloride, (*4*) *n*-propyl bromide, (*5*) 1-propanol, (*6*) *n*-propyl ether? (*b*) Which of the above compounds could be converted readily to the desired starting material and by what reagent?

11–*6*. Place the following compounds in the order of increasing strength as acids: (*a*) propionic acid, (*b*) hydrochloric acid, (*c*) carbonic acid, (*d*) acetylene, (*e*) 1-propanol, (*f*) pentane, (*g*) dimethylacetic acid, (*h*) 2-chloropropionic acid, (*i*) 3-chloropropionic acid.

11–*7*. Write reactions for the preparation of the following compounds from any other organic compound: (*a*) heptanoyl bromide, (*b*) *n*-butyric anhydride, (*c*) *i*-caproic acid, (*d*) *n*-valeramide, (*e*) *s*-butyl propionate, (*f*) calcium acetate.

11–*8*. What is the minimum number of steps in the usual laboratory procedure for the conversion of a carboxylic acid to the next higher homolog?

11–*9*. In the acid-catalyzed esterification of acetic acid with ethyl alcohol, which of the following procedures will lead to the maximum conversion of acetic acid to ethyl acetate: (*a*) starting with equal moles of acid and alcohol, (*b*) using a two-fold excess of ethyl alcohol, (*c*) conducting the reaction under pressure, (*d*) removing the water as it is formed, (*e*) using a high temperature?

11–*10*. In the esterification of a carboxylic acid by a primary alcohol, which of the following processes is involved: attack of (*a*) alcohol on the carboxylate anion, (*b*) alcohol on the protonated acid, (*c*) catalyst on the alcohol, (*d*) carboxylic acid on the protonated alcohol, or (*e*) alcohol on the free acid?

11–*11*. Give equations for the reaction of water, of ethyl alcohol, and of ammonia with each of the following compounds, assuming, where necessary, that the reaction with water or alcohol is acid-catalyzed: (*a*) 2-methylpentanoic anhydride, (*b*) *i*-valeryl chloride, (*c*) *s*-butyl propionate.

11–*12*. List the following types in the order of decreasing rate of reaction with ammonia: (*a*) primary alkyl chlorides, (*b*) esters, (*c*) anhydrides, (*d*) ethers, (*e*) carboxylic acids.

11–*13*. Devise a simple chemical procedure for distinguishing among the members of the following groups of compounds: (*a*) *n*-butyl ether, ethyl butyrate, and acetic anhydride; (*b*) cetyl chloride, palmitoyl chloride, and palmitic anhydride; (*c*) water, ethyl ether, and acetic acid; (*d*) aluminum hydroxide, aluminum acetate, and aluminum ethoxide; (*e*) *i*-butyl alcohol, *i*-propyl ether, and *n*-valeric acid; (*f*) *s*-butyl ether, *n*-propyl butyrate, *n*-octane, and 2-octene.

11–*14*. Which of the indicated materials (*a*) aluminum oxide, (*b*) thoria, (*c*) copper-zinc alloy, (*d*) Raney nickel, or (*e*) zinc oxide is used to catalyze (*1*) methanol synthesis, (*2*) reduction of alkenes, (*3*) dehydrogenation of alcohols, (*4*) dehydration of alcohols, (*5*) conversion of acids to ketones.

11–*15*. Give a series of reactions for each of the following conversions: (*a*) *n*-hexyl bromide to heptanoic acid, (*b*) *i*-amyl alcohol to ethyl *i*-valerate, (*c*) methyl caprate to *n*-decyl bromide, (*d*) 1-dodecanol to *n*-docosane, (*e*) stearic acid to 1-octadecanol, (*f*) ethyl myristate to 1-tetradecene, (*g*) *n*-hexadecyl alcohol to *n*-pentadecyl bromide, (*h*) lauroyl chloride to *n*-dodecyl iodide, (*i*) propionic acid to triethylcarbinol, (*j*) 2-butene to 2-methylbutanoic acid, (*k*) *n*-butyl alcohol to *n*-propyl ketone, (*l*) 1-octene to 2-bromoheptanoic acid, (*m*) lauric acid to lauroyl peroxide, (*n*) isobutylene to *t*-butyl peroxyacetate.

11–*16*. (*a*) Calculate the change in free energy for the esterification reaction, assuming that $K_E = 4$. (*b*) Starting with 10 moles of ethyl alcohol and 1 mole of acetic acid, calculate the per cent conversion of acetic acid to ethyl acetate at equilibrium. (Hint: let X = moles of acetic acid at equilibrium, set up quadratic equation, and solve.)

11–*17*. If 0.2410 g. of a carboxylic acid required 23.61 cc. of 0.1 N sodium hydroxide when titrated to the phenolphthalein end-point, calculate the equivalent weight of the acid.

11–*18*. Calculate the equivalent weight of the alcohol with which an acid was esterified if the saponification equivalent of the ester is 116 g. and the neutralization equivalent of the recovered acid is 74 g.

11–*19*. If in Problem *18* the acid is monobasic and the alcohol monohydric, what are the possible structures for the ester?

11–*20*. How many grams of sodium theoretically is required to reduce 22.8 g. of ethyl laurate in ethyl alcohol solution to lauryl alcohol?

11–*21*. Tell why the following procedure did not give the expected result. During the purification of an ether-soluble, water-insoluble acid, it was dissolved in 5 per cent sodium hydroxide, and the water-insoluble impurities were removed by extraction with ether. To recover the acid, the aqueous layer was acidified to litmus, extracted with ether, and the ether evaporated. Only a small amount of acid was recovered.

11–*22*. A compound was insoluble in water but became soluble on boiling with alkali. When the solution was acidified, an insoluble acid separated. Quantitative saponification of the original compound indicated an equivalent weight that was nine units less than that of the purified acid as determined by titration with standard base. What can be said about the nature of the original compound?

11–*23*. Compound *A* ($C_7H_{14}O_2$), only very slightly soluble in water and dilute aqueous acids and bases, dissolves in concentrated sulfuric acid; it dissolves also when refluxed with aqueous sodium hydroxide. When *A* is refluxed with ethyl alcohol and sodium, then treated with water in the cold, it gives a new compound *B* only slightly soluble in water and in dilute aqueous base. With acetic anhydride *B* gives *C* ($C_7H_{14}O_2$) which is *different from A*. Give the possible structures for *A* and show the reactions that it undergoes.

11–*24*. An unknown compound is insoluble in water but dissolves in dilute sodium carbonate solution. It decolorizes bromine readily without the evolution of hydrogen bromide. No other characteristic groups can be detected. Oxidation with strong permanganate gives two acids, a liquid having an equivalent weight of 158 as shown by titration, and a solid acid having an equivalent weight of 94. Give the likely structure of the original compound and the reactions it undergoes.

11–*25*. (*a*) When 0.1568 g. of a compound containing no elements other than carbon, hydrogen, and possibly oxygen was burned in a combustion train, there was obtained 0.3630 g. of carbon dioxide and 0.1222 g. of water. Calculate the empirical formula of the compound. (*b*) A solution of 0.0116 g. of the above compound in 1.000 g. of camphor melted 4° lower than pure camphor. What is the molecular formula of the compound and its exact molecular weight? (*c*) The original compound was only slightly soluble in water but dissolved readily in dilute alkali. Strong oxidation gave among other products another acidic substance that was volatile with steam and that had an equivalent weight of 89 ± 1 by titration. Of the two possible isomers satisfying these data, the new compound corresponded to that with the lower boiling point. Write the structural formula for the new acid and for the original compound.

11–*26*. Give a probable mechanism for (*a*) the cleavage of a tertiary alkyl ester by hydrogen bromide; (*b*) the ammonium ion-catalyzed ammonolysis of esters; (*c*) the catalysis of the reaction of Grignard reagents with *ortho* esters by magnesium halide.

CHAPTER TWELVE

WAXES, FATS, AND OILS

Waxes, fats, and oils are naturally occurring esters of higher straight-chain carboxylic acids. They usually are classified on a mixed basis including source, physical properties, and chemical properties.

- Waxes
  - Vegetable
  - Animal
- Fats
  - Vegetable
  - Animal
- Oils
  - Vegetable
    - Nondrying
    - Semidrying
    - Drying
  - Animal
    - Terrestrial
    - Marine

Work on the composition of waxes, fats, and oils has been concerned largely with the separation and identification of the alcohols (*unsaponifiable fraction*) and of the acids (*saponifiable fraction*) obtained by saponification (p. 173). Modern practice for the separation of these complex mixtures into pure compounds involves chiefly physical methods including amplified fractional distillation, low temperature crystallization, column and gas-liquid chromatography, counter-current distribution, and the formation of urea adducts (p. 283). At various stages alcohols may be converted into acetates and acids into methyl esters to prevent hydrogen bonding and the formation of double molecules, and so increase the ease of separation.

WAXES

A practical definition of a wax might be that it is anything with a waxy feel and a melting point above body temperature and below the boiling point of water. Thus the term *paraffin wax* is used for a mixture of solid hydrocarbons, *beeswax* for a mixture of esters, and *Carbowax* for a synthetic polyether. Chemically, however, waxes have been defined as *esters of long-chain* (C_{16} and above) *monohydric* (one hydroxyl group) *alcohols with long-chain* (C_{16} and above) *fatty acids* (Table 11–*1*, p. 157). Hence they have the general formula of a simple ester, RCOOR′. Actually the natural waxes are mixtures of esters and frequently contain hydrocarbons as well. Esters of dibasic acids, of hydroxy acids, and of diols also may be present. Moreover sugarcane wax does not contain esters, but consists chiefly of long-chain alcohols and polymeric aldehydes.

Spermaceti crystallizes and separates when the oil from the head of the sperm whale (*Cetacea*) is chilled. It is mainly cetyl palmitate, $C_{15}H_{31}COOC_{16}H_{33}$, and melts at 42–47°. **Beeswax,** m.p. 60–82°, is the material from which the bee builds the cells of the honeycomb. Saponification yields chiefly the C_{26} and C_{28} acids and C_{30} and C_{32} alcohols. About 25 per cent of the total fatty acids are hydroxy acids from which 14-hydroxypalmitic acid has been isolated. Beeswax contains also 10–14 per cent of hydrocarbons with C_{31} predominating. **Carnauba wax** is the most valuable of the natural waxes. It occurs as a coating on the leaves of a Brazilian palm, *Corypha cerifera,* from which it is removed by shredding and beating the leaves. Because of its high melting point of 80–87°, its hardness, and its impervious-

TABLE 12–1. COMPOSITION OF SAPONIFIED WOOL WAX

SAPONIFIABLE PORTION		UNSAPONIFIABLE PORTION	
PRODUCT	PER CENT	PRODUCT	PER CENT
Alkanoic acids	60	Lanosterol (p. 662)	44
α-Hydroxy acids	30	Cholesterol (p. 663)	31
ω-Hydroxy acids	5	Alkanols	9.5
Resinous acids	1	α,β-Diols	6.5
		Hydrocarbons	1
		Resinous material	3

ness to water, carnauba wax is a valuable ingredient of automobile and floor polishes and of carbon paper coatings. A careful investigation has shown that it is a mixture of the esters of chiefly C_{24} and C_{28} normal fatty acids with C_{32} and C_{34} 1-alkanols, together with a considerable quantity of ω-hydroxy acids, $HO(CH_2)_xCOOH$, where x is 17–29, and smaller amounts of esterified α,ω-diols, $HO(CH_2)_xOH$, where x is 22 to above 28. It is believed that the ω-hydroxy acids are responsible for the unique physical properties of carnauba wax, probably because they are present as polymeric esters of high molecular weight. **Ouricuri wax** from the leaves of the Brazilian palm, *Scheelia martiana,* has about the same composition and properties as carnauba wax. **Degras,**[1] **wool grease,** or **wool wax** is recovered from the scouring of wool. It has the unusual property of forming a stable semisolid emulsion containing up to 80 per cent water. A purified product known as **lanolin** or **lanum** is used as a base for salves and ointments in which it is desired to incorporate both water-soluble and fat-soluble substances. The commercial product is a complex mixture of esters, alcohols, free fatty acids, and hydrocarbons, together with their various autoxidation products. The results of detailed investigations of the composition of the saponification products of freshly secreted wool wax are given in Table 12–1.

Over 140 compounds have been isolated and identified. All classes of open-chain compounds listed in the table contain members of the normal, iso, and anteiso series. Compounds with an even number and compounds with an odd number of carbon atoms are present in the normal series, the members of the iso series all have an even number of carbon atoms, and those of the anteiso series all have an odd number.

FATS AND OILS

The constitution of fats and oils was investigated systematically first by Chevreul.[2] They are *esters of higher fat acids and the trihydric alcohol, glycerol,* $HOCH_2CHOHCH_2OH$. Esters of glycerol frequently are called *glycerides.* They have the general formula $RCOOCH_2CH(OCOR')CH_2OCOR''$. The difference between fats and oils is merely that

[1] The material known as *degras, moellon degras,* or *moellon* in Europe is a by-product of the tanning of chamois skins with fish oils. When tannage is complete, the excess autoxidized oil is removed by pressing.

[2] Michael Eugène Chevreul (1786–1889), professor at the Collège de France and director of the dye plant at the Gobelin Tapestry Works. He was among the first to study the chemistry of complex natural products and isolated the glycoside quercitrin (p. 577) in 1810. He is noted chiefly for his work on fats begun in 1811. He was the first to show the nature of saponification, and from the hydrolysis products of fats he isolated caproic, capric, palmitic, stearic, and oleic acids. He reported in 1816 that cholesterol from gall stones (p. 663) was not saponifiable and hence not a fat. In 1818 he discovered cetyl alcohol in the unsaponifiable fraction of spermaceti and in 1834 isolated creatine (p. 289) from urine. In 1864 he published an extensive criticism of spiritualism and of the pseudoscientific psychic research then in vogue. He was an active investigator until his death at the age of 102 years and 7 months.

fats are solid or semisolid at room temperature, whereas oils are liquids. Vegetable fats and oils usually occur in the fruits and seeds of plants and are extracted (*1*) by cold pressing in hydraulic presses or continuous expellers, (*2*) by hot pressing, and (*3*) by solvent extraction. Cold pressing gives the blandest product and is used for producing the highest grade food oils such as olive, cottonseed, and peanut oils. Hot pressing gives a higher yield, but larger quantities of undesirable components are expressed, and the oil has a stronger odor and flavor. Solvent extraction gives the highest recovery, and in recent years the process has been so improved that even food oils may be prepared that are free from undesirable odors and flavors. Animal fats are recovered by heating fatty tissue to a high temperature (*dry-rendering*) or by treating with steam or hot water and separating the liberated fat.

Fat Acids

Since all fats and oils are esters of glycerol, their differences must be due to the acids with which the glycerol is esterified. These acids are both saturated and unsaturated. Of the saturated acids the most important are **lauric acid,** $CH_3(CH_2)_{10}COOH$, **palmitic acid,** $CH_3(CH_2)_{14}COOH$, and **stearic acid,** $CH_3(CH_2)_{16}COOH$. The most important unsaturated acids have eighteen carbon atoms, and one double bond usually is at the middle of the chain. If other double bonds are present they lie further removed from the carboxyl group. **Oleic acid,** $CH_3(CH_2)_7CH{=}CH(CH_2)_7COOH$, has only one double bond; **linoleic acid (*linolic acid*),** $CH_3(CH_2)_4CH{=}CHCH_2CH{=}CH(CH_2)_7COOH$, has two double bonds separated by one methylene group; **linolenic acid,** $CH_3CH_2CH{=}CHCH_2CH{=}CH$-$CH_2CH{=}CH(CH_2)_7COOH$, has three double bonds all separated by methylene groups; **eleostearic acid,** $CH_3(CH_2)_3CH{=}CH{-}CH{=}CH{-}CH{=}CH(CH_2)_7COOH$, also has three double bonds, but they are conjugated;[3] **licanic acid** is 4-oxoeleostearic acid, $CH_3(CH_2)_3CH{=}CH{-}CH{=}CH{-}CH{=}CH(CH_2)_4CO(CH_2)_2COOH$.

Several other unsaturated acids frequently are encountered. **Ricinoleic acid,** CH_3-$(CH_2)_5CHOHCH_2CH{=}CH(CH_2)_7COOH$, is 12-hydroxyoleic acid. In **palmitoleic acid,** $CH_3(CH_2)_5CH{=}CH(CH_2)_7COOH$, **petroselenic acid,** $CH_3(CH_2)_{10}CH{=}CH(CH_2)_4$-$COOH$, **vaccenic acid,** $CH_3(CH_2)_5CH{=}CH(CH_2)_9COOH$, and **erucic acid,** $CH_3(CH_2)_7$-$CH{=}CH(CH_2)_{11}COOH$, the double bond is not at the middle of the chain. It is of interest that for all of the common naturally occurring unsaturated acids, with the exception of eleostearic acid, all of the double bonds have the thermodynamically less stable *cis* configuration (p. 76). In eleostearic acid the double bond at position 9 is *cis,* but the other two double bonds are *trans.* Of all of the fat acids palmitic acid is the most abundant, and oleic acid is the most widely distributed.

The most abundant fat acids from the higher plants and animals have an even number of carbon atoms. The reason is that in their biosynthesis they are built up two carbon atoms at a time from acetate ion, which is derived chiefly from carbohydrate. Coenzyme-A (p. 534) and reductive enzymes are involved in the synthesis. The small amounts of the odd-carbon acids and the branched-chain acids isolated from fats are believed to be formed by the accidental incorporation of methyl groups. It has been shown that enzymic dehydrogenation of stearic acid gives rise stereospecifically to oleic acid. With the exception of a zooflagellate, *Leishmania enrietti,* only plants are known to synthesize oleic, linoleic, and linolenic acids. Animals, however, after ingesting fats containing these acids can increase the length of the chain and introduce further unsaturation to give the acids characteristic of fish oils (p. 185) and organ phosphatides (p. 187).

Natural products other than waxes, fats, and oils, such as the phosphatides (p. 610)

---

[3] If double bonds and single bonds alternate successively in a molecule, the double bonds are said to be *conjugated.*

and cholesterol esters (p. 187), also yield fatty acids on hydrolysis. The biochemist groups fatty acids and substances that yield fatty acids, as well as numerous other fat-soluble compounds, under the general term *lipide* or *lipid.*

Fats and oils are not a mixture of simple glycerides having all three hydroxyl groups esterified with the same fat acid. They are a mixture chiefly of mixed glycerides, each molecule of glycerol being esterified with more than one fat acid. In the vegetable fats the different fat acids are not wholly randomly distributed. Partial enzymatic hydrolyses have shown that the secondary hydroxyl groups of the glycerol molecules are esterified with C_{18} unsaturated acids to the extent that they are available. The primary hydroxyl groups are esterified with the remaining fat acids, the distribution being statistical. Simple glycerides occur in quantity only if more than two thirds of the acyl groups are of one kind. In the fats from mammals, the fat acids appear to be randomly distributed except that the amount of the glycerides that are solid at body temperature does not exceed the amount that would render the mixture of fats nonfluid.

Unsaturation lowers the melting point. Hence saturated acyl groups predominate in fats and unsaturated acyl groups predominate in oils. Another factor that affects the melting point is molecular weight. The acids obtained from low-melting fats such as coconut oil, palm oil, and butter contain relatively small amounts of unsaturated acids but considerable amounts of lower fat acids. Although classified as fats because they are solid in temperate zones, coconut oil and palm oil are called oils because they are liquids in the tropics where they are produced. The approximate amounts of the different acids obtained by hydrolysis of fats are given in Table 12–*2*.

Reactions of Fats and Oils

The characteristic chemical features of the fats are the ester linkages and the unsaturation. As esters they may be hydrolyzed in the presence of acids, enzymes, or alkali to free fat acids, or their salts, and glycerol.

$$\begin{array}{c}\mathrm{RCOOCH_2}\\ |\\ \mathrm{R'COOCH}\\ |\\ \mathrm{R''COOCH_2}\end{array}\left\{\begin{array}{l}+\,3\,\mathrm{H_2O}\xrightarrow[\text{enzymes}]{\mathrm{H^+}\text{ or}}\begin{array}{c}\mathrm{RCOOH}\\ \mathrm{R'COOH}\\ \mathrm{R''COOH}\end{array}+\begin{array}{c}\mathrm{CH_2OH}\\ |\\ \mathrm{CHOH}\\ |\\ \mathrm{CH_2OH}\\ \text{Glycerol}\end{array}\\[2ex] +\,3\,\mathrm{M^{+-}OH}\longrightarrow\begin{array}{c}\mathrm{RCOO^{-+}M}\\ \mathrm{R'COO^{-+}M}\\ \mathrm{R''COO^{-+}M}\end{array}+\begin{array}{c}\mathrm{CH_2OH}\\ |\\ \mathrm{CHOH}\\ |\\ \mathrm{CH_2OH}\end{array}\end{array}\right.$$

Because the fat acids differ in molecular weight, and because substances that do not react with alkali, such as alcohols and hydrocarbons, may be present, different fats require different amounts of alkali for saponification. Hence the amount of alkali required to saponify a given weight of fat may be used as a characteristic of the particular fat. An arbitrary unit known technically as the **saponification value** is used, which is the *number of milligrams of potassium hydroxide required to saponify one gram of fat.* Table 12–*2* shows that the fats containing chiefly C_{18} acids have almost identical saponification values and the determination is useful only to identify or detect the presence of coconut oil and butter fat, or to determine whether these fats have been adulterated with others having a lower saponification value, or with mineral oils or greases.

The extent of unsaturation likewise is characteristic of a fat and may be determined by the amount of halogen that the fat can add. Iodine does not ordinarily form stable addition products with the double bond (p. 80), and chlorine or bromine replace hydrogen as

TABLE 12–2. SAPONIFICATION AND IODINE VALUES OF FATS AND OILS AND THE COMPOSITION OF THE FAT ACIDS OBTAINED BY HYDROLYSIS

FAT OR OIL			SAPONIFICATION VALUE	IODINE VALUE	COMPOSITION OF FAT ACIDS (PER CENT)						
					MYRISTIC	PALMITIC	STEARIC	PALMITOLEIC	OLEIC	LINOLEIC	OTHER COMPONENTS
VEGETABLE FATS		Coconut	250–60	8–10	17–20	4–10	1–5		2–10	0–2	*a*
		Babassu	245–55	10–18	15–20	6–9	3–6		12–18	1–3	*b*
		Palm	196–210	48–58	1–3	34–43	3–6		38–40	5–11	
ANIMAL FATS		Butter	216–35	26–45	7–9	23–26	10–13	5	30–40	4–5	*c*
		Lard	193–200	46–66	1–2	28–30	12–18	1–3	41–48	6–7	*d*
		Tallow	190–200	31–47	2–3	24–32	14–32	1–3	35–48	2–4	
VEGETABLE OILS	NONDRYING	Castor	176–87	81–90		0–1			0–9	3–7	*e*
		Olive	185–200	74–94	0–1	5–15	1–4	0–1	69–84	4–12	
		Peanut	185–95	83–98		6–9	2–6	0–1	50–70	13–26	*f*
		Rape	172–5	94–106	0–2	0–1	0–2		20–38	10–15	*g*
	SEMI-DRYING	Corn	188–93	116–30	0–2	7–11	3–4	0–2	43–49	34–42	
		Sesame	187–93	104–16		8	4	1	45	41	*h*
		Cottonseed	191–6	103–15	0–2	19–24	1–2	0–2	23–33	40–48	
	DRYING	Soybean	189–94	124–36	0–1	6–10	2–4		21–29	50–59	*i*
		Sunflower	190–2	122–36		10–13			21–39	51–68	
		Safflower	186–94	130–50		5–10			14–21	73–78	
		Hemp	190–3	149–67		4–10			13	53	*j*
		Linseed	189–96	170–204		4–7	2–5		9–38	3–43	*k*
		Tung	189–95	160–80		2–6			4–16	1–10	*l*
		Oiticica	186–94	139–55		11			6		*m*
ANIMAL OILS	TER-RES-TRIAL	Lard oil	190–95	46–70		22–26	15–17		45–55	8–10	
		Neat's-foot	192–7	67–73		17–18	2–3		74–77		
	MA-RINE	Whale	188–94	110–50	4–6	11–18	2–4	13–18	33–38		*n*
		Fish (sardine)	185–95	120–90	6–8	10–16	1–2	6–15			*o*

(*a*) 5–10 caprylic, 5–11 capric, and 45–51 lauric acids.
(*b*) 4–7 caprylic, 3–8 capric, 44–46 lauric acids.
(*c*) 3–4 butyric, 1–2 caproic, 1 caprylic, 2–3 capric, and 2–3 lauric acids.
(*d*) 2 of C_{20} and C_{22} unsaturated fat acids.
(*e*) 80–92 ricinoleic acid.
(*f*) 2–5 arachidic and 1–5 tetracosanoic acids.
(*g*) 1–2 tetracosanoic, 1–4 linolenic, and 43–57 erucic acids.
(*h*) 1 arachidic acid.
(*i*) 4–8 linolenic acid.
(*j*) 24 linolenic acid.
(*k*) 25–58 linolenic acid.
(*l*) 74–91 eleostearic acid.
(*m*) 70–78 licanic acid.
(*n*) 11–20 C_{20} and 6–11 C_{22} unsaturated acids.
(*o*) 70 unsaturated acids, C_{16}–C_{22}, having one to six double bonds.

well as add to the double bond. In practice standardized solutions of iodine monochloride (*Wijs solution*) or of iodine monobromide (*Hanus solution*) in glacial acetic acid are used. The Wijs or Hanus solution is standardized, however, by adding potassium iodide and titrating the liberated iodine with standard thiosulfate solution. The amount of reagent remaining after reaction with a fat is determined in the same way. *The difference expressed in terms of grams of iodine* (as if iodine had added) *per 100 grams of fat* is known as the **iodine value** of the fat. The increase in iodine value with increasing amounts of unsaturated acids is apparent in Table 12–2.

Thiocyanogen, $(SCN)_2$, shows many of the properties of halogens. It liberates iodine from potassium iodide, and also adds to double bonds. It has been found, however, that whereas one mole adds to monoethylenic acids such as oleic acid, only slightly more than one mole adds to the diethylenic linoleic acid, and somewhat less than two moles to the triply unsaturated linolenic acid. Solutions of thiocyanogen in glacial acetic acid are prepared by adding bromine to lead thiocyanate and are standardized and used like Wijs and Hanus solutions. *The amount of thiocyanogen added expressed in terms of grams of iodine per gram of fat* is known as the **thiocyanogen value.** Since Wijs or Hanus solution saturates all double bonds, it is possible to estimate the composition of mixtures of oleic, linoleic, and saturated acids from the iodine and thiocyanogen values or the amounts of oleic, linoleic, and linolenic acids if the amount of saturated acids is known.

The **hydrogenation of oils** is carried out technically on a large scale by bubbling hydrogen through the hot oil containing a suspension of finely divided nickel. In this process the double bonds of unsaturated glycerides are hydrogenated and the oils converted into the hard waxy tristearin. By controlling the amount of hydrogen added, any consistency desired may be obtained. People of temperate zones prefer fat to oil for cooking purposes, and fats are more useful than oils as soap stocks. Hence hydrogenation greatly increases the value of an oil.

Rancidity of oils is caused chiefly by autoxidation of the unsaturation, which gives peroxides that degrade the molecule to a complex mixture of volatile aldehydes, ketones, and acids. In some cases rancidity may be caused by microorganisms. Fats that have been freed of odor and undesirable tastes are stabilized by the addition of antioxidants such as butylated hydroxytoluene (BHT, p. 445) or butylated hydroxyanisole (BHA, p. 447). Vegetable oils are more resistant to autoxidation than animal oils because of the presence of the naturally occurring antioxidant, α-tocopherol (p. 522).

Waste or rags containing unsaturated oils are subject to **spontaneous combustion** if air is not excluded, or if not enough ventilation is possible to prevent a rise in temperature as the oil oxidizes. Any rise in temperature increases the rate of oxidation, and the process is accelerated until the material bursts into flame.

Epoxidation of highly unsaturated oils with peroxy acids (p. 165) is carried out industrially. The products are used as plasticizers for vinyl resins (p. 593) and in the formulation of epoxy resins (p. 608).

Castor oil is characterized by the high percentage of ricinoleic acid, which contains a hydroxyl group. The hydroxyl group may be acetylated and the *number of milligrams of potassium hydroxide that is needed to neutralize the acetic acid liberated from one gram of acetylated product* is known as the **acetyl value.** Castor oil has an acetyl value of 142 to 150, but other common fats and oils range from 2 to 20. Substances other than ricinoleic acid, for example high molecular weight alcohols and partially hydrolyzed glycerides, also acetylate and may account for all or part of the acetyl value.

A second double bond may be introduced into the ricinoleic portion of castor oil by **dehydration,** giving rise to a mixture of linoleic and the 9,11-diunsaturated acyl groups, thus changing it from a nondrying to a drying oil. Moreover the high percentage of conjugation that results makes the dehydrated oil resemble tung oil in its properties.

Transesterification is important technically for modifying the properties of a fat. It is carried out under anhydrous conditions using sodium methoxide as a catalyst. In the presence of methoxide ion the various acyl groups rapidly exchange places within and between molecules (p. 174).

$$\begin{array}{l} RCO\!:\!OCH_2 \\ \quad | \\ CH_3\!:\!\ddot{O}\!:^- \quad CHOCOR' \\ \quad | \\ \quad CH_2OCOR'' \end{array} \longrightarrow \begin{array}{c} R \\ | \\ CO \\ | \\ OCH_3 \end{array} + \begin{array}{l} CH_2{-}O\!:^- \quad R' \\ | \qquad\qquad C \\ CH{-}O \qquad\quad O \\ | \\ CH_2{-}OCOR'' \end{array} \longrightarrow \begin{array}{l} CH_2{-}O{-}COR' \\ | \qquad\qquad R'' \\ CH{-}O\!:^- \quad C \\ | \\ CH_2{-}O \qquad O \end{array} \longrightarrow$$

$$\begin{array}{l} CH_2{-}O{-}COR' \\ | \\ CH{-}O{-}OCR'' \\ | \\ CH_2{-}O\!:^- \end{array} + \begin{array}{l} \quad R''' \\ \quad | \\ \rightarrow CO\!:\!O{-}CH_2 \end{array} \begin{array}{l} CH_2OCOR \\ | \\ CHOCOR \\ | \\ \end{array} \longrightarrow \begin{array}{l} CH_2OCOR' \\ | \\ CHOCOR'' \\ | \\ CH_2OCOR''' \end{array} + \begin{array}{l} CH_2OCOR \\ | \\ CHOCOR \\ | \\ CH_2{-}O\!:^- \end{array}$$

This redistribution changes the properties of the fat. Lard, for example, is unsuitable for baking cake because it has a grainy texture resulting from the crystallization of solid glycerides. It is greatly improved by redistribution of the acyl groups among the glyceride molecules.

The equilibrium is so mobile that transesterification can be carried out at a temperature such that the higher-melting saturated triglycerides crystallize from the mixture. As they crystallize, the equilibrium shifts until most of the saturated acyl groups are removed from the liquid portion. After separation of the crystallized material, the oil remaining is much more highly unsaturated. In this way the drying properties of the oils are improved.

Fats in the presence of glycerol and sodium methoxide undergo alcoholysis (p. 173). The triglycerides are thus converted to mono- and diglycerides, useful as emulsifying agents (p. 191).

Reduction of fats **to primary alcohols** and glycerol can be carried out using sodium and a secondary alcohol (p. 175). Alternatively the fats may be converted by methanolysis to glycerol and methyl esters, and the latter reduced catalytically (p. 175).

Oxides of nitrogen from the reaction of mercury with nitric acid (*Poutet reagent*) convert liquid unsaturated acids having the *cis* configuration to an equilibrium mixture with the *trans* isomers (p. 76). Thus oleic acid, m.p. 16°, is converted into an equilibrium mixture containing 67 per cent of the *trans* form, m.p. 51°, which is called *elaidic acid.* Oils may be solidified partially in the same way, and the process is known as **elaidinization.**

Uses of Fats and Oils

Food. From 25 to 50 per cent of the caloric intake of man consists of fats. When oxidized to carbon dioxide and water by the organism, they produce about 9.5 kcal. of energy per gram compared to 4 kcal. per gram for carbohydrates or proteins. Fats in their ordinary form cannot be absorbed through the walls of the intestine because of their insolubility in water and their large particle size. In the small intestine they are partially hydrolyzed in the presence of the enzyme catalyst, **steapsin,** to di- and monoglycerides, which, together with the bile salts (p. 664), can bring about emulsification. Sufficient reduction in particle size takes place to permit passage through the intestinal walls into the lymph ducts. Before leaving the walls of the intestine, mono- and diglycerides are recombined with fat acids to form triglycerides again. The lymph discharges the fats into the blood stream as a highly dispersed emulsion. The blood stream transports them to the various tissues where they are burned for energy, converted into phosphatides (p. 610) or into esters of cholesterol (p. 663), or stored as fat deposits for future use.

From the standpoint of digestion, one fat is as useful as another unless its melting point is so high that when mixed with other fats it does not melt or emulsify with the bile at body temperature, in which case it passes into the feces unchanged. Either linoleic or linolenic acid, however, must be supplied by the ingested fats to insure a healthy condition of the skin. Moreover, the ingestion of oils containing large amounts of linoleic or linolenic groups markedly lowers the cholesterol content of the blood. Saturated fats raise the cholesterol content. Current interest in the replacement of saturated fats by corn oil and safflower oil in diets is due to the possible correlation of hardening of the arteries and heart disease with high blood cholesterol. Linoleic acid is more effective in lowering blood cholesterol than linolenic acid and is the probable precursor of **arachidonic acid,** $CH_3(CH_2)_4(CH{=}CHCH_2)_4(CH_2)_2COOH$, which is believed to be formed by lengthening of the chain of linoleic acid and dehydrogenation. Arachidonic acid has several times the biological activity of linoleic acid and is the chief fat acid obtained when the phosphatides (p. 610) of the liver and other organs are hydrolyzed. The C_{18}–C_{22} polyenoic acids obtained from organ phosphatides resemble those from fish oils (p. 185) except that the latter appear to be derived from linolenic acid.

Although some of the fat consumed is mixed with the other foodstuffs, the proteins and carbohydrates, a considerable portion first is isolated in a relatively pure state and consumed as such in bread spreads, or used for frying or as salad oils and salad dressings such as mayonnaise. The relatively high cost of butter has led to the development of a substitute generally referred to as **oleomargarine.** Selected vegetable or animal fats and oils that have been highly refined and properly hydrogenated to give the desired melting point and consistency, are emulsified with about 17 per cent by weight of milk that has been cultured with certain microorganisms to give it flavor. An emulsifying agent such as a monoglyceride (p. 191) or a vegetable lecithin (p. 610) usually is added as well. Butter consists of droplets of water suspended in oil (water-in-oil type emulsion). Oleomargarines may be either water-in-oil or oil-in-water types of emulsions depending on the method of manufacture. Diacetyl, $CH_3COCOCH_3$, and methylacetylcarbinol, $CH_3CHOHCOCH_3$ (p. 176), which account for the characteristic taste of butter, also may be added along with vitamins A and D. For many years oleomargarine manufacturers in the United States were not permitted to add color to their product unless an additional Federal tax of ten cents per pound was paid. By changing from coconut oil to soybean oil as a raw material, the makers of oleomargarine were able to enlist the aid of soybean growers to combat the dairy lobby, and this law was repealed in 1950. Since 1953 the production of oleomargarine has exceeded that of butter. In 1962, soybean oil accounted for 75 per cent of the fats used in the manufacture of margarine.

Protective Coatings. Glycerides of fat acids containing two or more double bonds absorb oxygen on exposure to air and become solid or semisolid oxidized polymers, a process referred to as "drying." If exposed in thin layers, tough elastic waterproof films are formed. **Paint** is a mixture of drying oil, pigment, thinner, and drier. The pigment is an opaque material having a refractive index different from that of the oil film. It provides color and covering power, and protects the oil film from the destructive action of light. The thinner is a volatile solvent, either turpentine (p. 661) or a petroleum fraction called mineral paint spirits (p. 94). It permits spreading and on evaporation leaves a thin even film of oil and pigment that does not run. The drier is a solution of cobalt, manganese, or lead salts of organic acids, usually naphthenic acids from petroleum (p. 645), which catalyzes the oxidation and polymerization of the oil. The drying oil is known as the *vehicle* because after polymerization it holds or carries the pigment. Linseed oil is the most widely used drying oil, although a certain amount of tung oil in a paint gives it superior properties, which seem to be due to the conjugation of the unsaturation. Tung oil has been imported chiefly from China, but the tree can grow in a rather limited belt extending across southern United States, and plantations are in production. Dehydrated castor oil (p. 186) has properties resembling tung oil and is used to replace tung oil to a certain extent, especially when supplies from China are not available. The amount of conjugation of the double bonds can be determined from the ultraviolet absorption spectrum of the oil (p. 553). White paints made from safflower oil have less tendency to become yellow with age than those made from other drying oils.

Although soybean oil has in recent years been classed as a drying oil, it formerly was classed as a semidrying oil, and as indicated by Table 12–*2* its composition is such that its properties would be expected to be intermediate between those of cottonseed oil and linseed oil. Its drying properties can be enhanced, a procedure known as *up-grading,* by extraction with solvents that separate the oil into fractions having varying degrees of unsaturation, or by removing saturated acid residues by transesterification (p. 174). Another method of up-grading an oil consists of liberating the mixed acids by hydrolysis and re-esterifying with alcohols having a larger number of hydroxyl groups such as pentaerythritol, $C(CH_2OH)_4$ (p. 609), or sorbitol, $CH_2OH(CHOH)_4CH_2OH$ (p. 349). In this way molecules having higher molecular weights than the glycerides are formed, and less polymerization is necessary to give a solid film. Such "synthetic" vehicles "dry" much faster and may give superior films. In still another process the saturated acids and oleic acid are removed as inclusion compounds with urea (p. 283), and the remaining highly unsaturated acids are re-esterified with a polyhydric alcohol. Many of the synthetic resins used for baking enamels contain unsaturated acid residues in the molecule and depend on polymerization for the solidification of the film (p. 474).

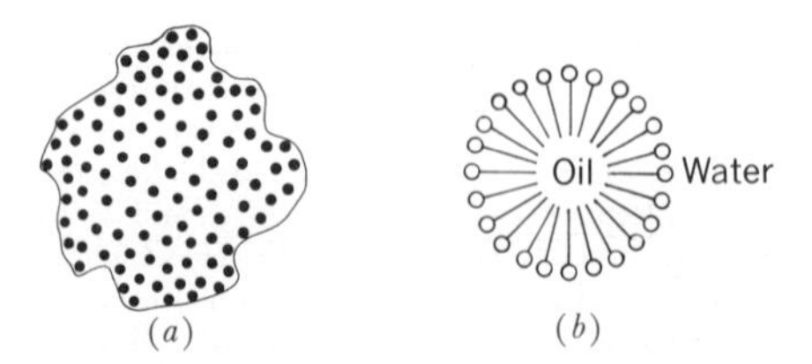

Figure 12–*1*. (*a*) Oil in water emulsion; (*b*) emulsified oil droplet.

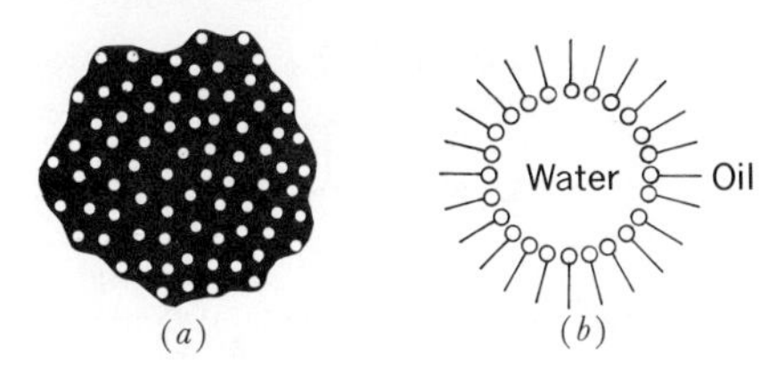

Figure 12–2. (a) Water in oil emulsion; (b) emulsified water droplet.

Varnish is a mixture of drying oil, rosin (p. 662), and thinner. The rosin imparts hardness and high gloss to the dried film and may be replaced by other natural or synthetic resins. **Enamels** are pigmented varnishes. **Oil cloth** is made by coating cotton cloth with a mixture of partially oxidized oil and a pigment, and drying in warm chambers. If the oil is more highly oxidized to a thick viscous mass, mixed with rosin and ground cork or other filler, and rolled into a continuous sheet, **linoleum** is produced. The sheet travels slowly through warm chambers to complete the polymerization.

Wetting Agents, Emulsifying Agents, and Detergents. If the attractive force between water molecules and some solid surface is not sufficient to overcome the surface tension of water, water does not wet the surface. If some third substance is added that is adsorbed on the surface more strongly than water and provides a new surface that strongly adsorbs water, the surface then can be wet by water. Such substances are known as **wetting agents.**

Oil and water do not mix because there is a greater attraction of water molecules for each other and of oil molecules for each other than of water molecules for oil molecules. If the molecules of a third substance have a portion that has a strong attraction for oil molecules (*oleophilic group*) and a portion that has a strong attraction for water molecules (*hydrophilic group*), the substance can disperse in water to give a colloidal solution, if the effect of the water-attracting portion of its molecules is sufficiently great. The dispersed substance, however, has an attraction for oil molecules also, and if an oil is shaken with the colloidal solution, the oil may be dispersed into tiny droplets and an oil-in-water type of emulsion result (Fig. 12–*1*). Each droplet has a surface film of the dispersing agent with the oleophilic group in the oil phase and the hydrophilic group in the water phase. In Fig. 12–*1* the dispersing agent is indicated by the symbol o— in which the small circle represents the hydrophilic group and the line the oleophilic group. On the other hand, if the effect of the oil-soluble portion of the molecule having mixed properties is sufficiently great, the substance can dissolve in oil. The water-solubilizing group, however, has an attraction for water, and if the solution is shaken with water, the latter may be dispersed in the oil to give a water-in-oil type of emulsion (Fig. 12–*2*). Substances having the property of facilitating the production of emulsions are called **emulsifying agents.**

Dirt adheres to fabric and other surfaces chiefly by means of films of oil, and **detergent action** is largely the result of wetting and emulsification. The adsorption of the detergent at the solid surface permits wetting of the surface by the water and allows the oil film to roll up into small droplets (Fig. 12–*3*). Agitation of the oil droplets with the detergent solution causes emulsification of the oil, which permits the oil and adhering dirt to be washed away from the surface. If the detergent foams, the oil may be emulsified more readily when the system is agitated because it is spread out into thin lines at the junctures of the bubbles (Fig. 12–*4*). It should be noted, however, that detergent action is not dependent on the formation of foam, and that numerous nonfoaming detergents

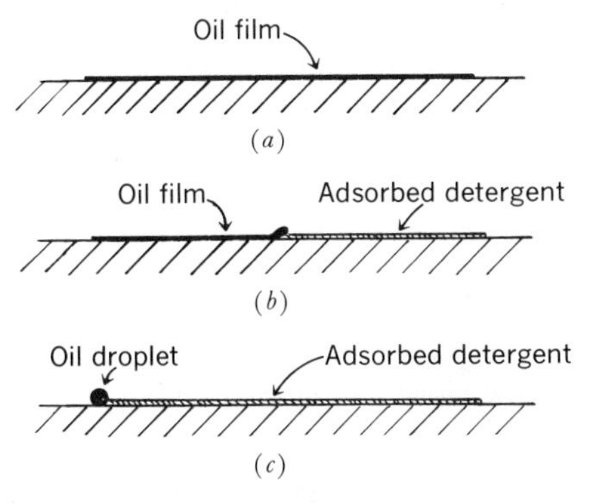

Figure 12–*3*. Adsorbed oil film (a) before, (b) during, and (c) after action of detergent.

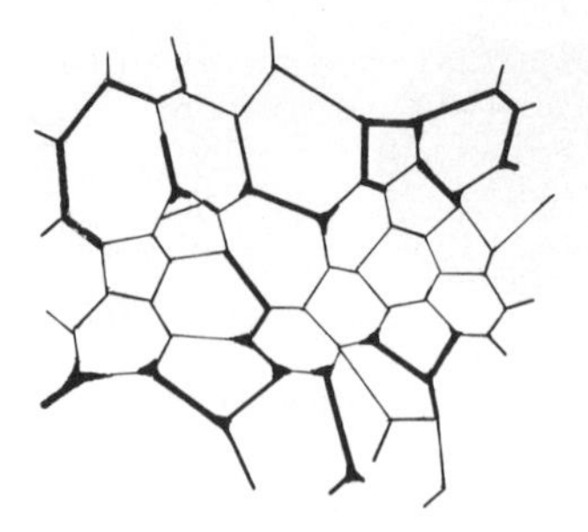

Figure 12–4. Oil spread out in the interfaces of foam bubbles.

are available and have distinct advantages over foaming detergents in equipment such as automatic dishwashers.

The alkali metal salts of fat acids having from ten to eighteen carbon atoms are known as **soaps.** They have a long oil-soluble hydrocarbon chain attached to a water-soluble carboxylate ion, and hence act as wetting agents, emulsifying agents, and detergents. Strictly speaking, only such salts should be referred to as soaps. If the hydrocarbon chain of the alkali metal salts is less than ten carbon atoms long, it does not cause emulsification of oil. If more than eighteen carbon atoms are present, the salt is too insoluble in water to form a sufficiently concentrated colloidal solution. The alkaline earth and heavy metal salts are insoluble in water and hence useless as detergents. Hard water, containing calcium, magnesium, and iron ions, precipitates insoluble salts when soap solutions are added, and detergent action does not take place until these ions are completely removed. Moreover the scum of insoluble salts produced from the soap and inorganic salts may be difficult to remove from the article being washed. The insoluble salts have other uses, however (p. 191), and frequently are called soaps.

Ordinary soaps usually are sodium salts. Stocks containing mostly saturated fats give hard soaps, whereas highly unsaturated fats give soft soaps. Soaps of low molecular weight, for example from coconut oil, are most soluble in water and give a loose lather consisting of large unstable bubbles, whereas soaps of higher molecular weight, such as those from tallow, give a close lather consisting of fine stable bubbles. Potassium soaps are more soluble than sodium soaps but more costly. The fats and hydrogenated oils used in soap manufacture are blended carefully for the particular type of soap desired.

About three fourths of the hard soap is made by the *full boiled* or *settled* process. The melted fat and the solution of sodium hydroxide are run simultaneously into a large tank with a conical bottom and boiled by steam in both open and closed coils in the bottom of the tank. When saponification is complete, the soap is *grained,* that is, precipitated by the addition of salt. After washing by re-solution and reprecipitation, it is processed into the various types available on the market. The brine solutions are evaporated to recover the glycerol and salt. In the *cold process* the fats and lye are agitated in a mechanical mixer at 30–40° until a homogeneous emulsion is formed, which then is run into molds where completion of the saponification takes place.

In addition to direct saponification of fats, soaps are made also by hydrolyzing the fat continuously and countercurrently in coils with water at 260°, separating the aqueous glycerol from the fatty acids, and neutralizing the fatty acids with alkali. Extraction of glycerol is more efficient and recovery is easier than when direct saponification is used. Moreover the acids can be purified by distillation, permitting the use of lower grade fats.

Synthetic Detergents and Emulsifying Agents. In recent years numerous synthetic compounds have become available that meet the general structural requirements for a detergent or emulsifying agent, namely a water-soluble portion and an oil-soluble portion. Their chief advantage is that their alkaline earth and heavy metal salts are sufficiently soluble to avoid precipitation in hard water. Over 2000 commercial products are on the market. Since 1954 the sale of synthetic detergents has exceeded that of soaps and in 1963 accounted for 77 per cent of total sales of detergents, which amounted to 5.3 billion pounds.

The earliest of the semisynthetic wetting agents were the *sulfated fats and oils* (formerly called sulfonated), which have been in commercial use for over one hundred years. When an unsaturated fat is treated with sulfuric acid, the double bonds add sulfuric acid to give the hydrogen sulfate of the hydroxy acid. For example, the oleic acid portion reacts according to the equation

$$\begin{array}{l} CH_2OCO(CH_2)_7CH{=}CH(CH_2)_7CH_3 \\ CHOCOR \\ CH_2OCOR' \end{array} + H_2SO_4 \longrightarrow \begin{array}{l} CH_2OCO(CH_2)_7CH_2CH(CH_2)_7CH_3 \\ CHOCOR \qquad\qquad\quad OSO_3H \\ CH_2OCOR' \end{array}$$

Neutralization gives the sodium salt, which provides the water-solubilizing portion of the molecule. In the sulfation of *castor oil,* the hydroxyl group of the ricinoleic acid portion of the molecule reacts more readily than the double bond. If sulfation is carried out at a sufficiently low temperature, the double bond is largely unaffected.

$$C_3H_5\left[OCO(CH_2)_7CH{=}CHCH_2\underset{OH}{CH}(CH_2)_5CH_3\right]_3 + 3\ H_2SO_4 \longrightarrow C_3H_5\left[OCO(CH_2)_7CH{=}CHCH_2\underset{OSO_3H}{CH}(CH_2)_5CH_3\right]_3$$

Sulfated castor oil is known as *Turkey Red oil* because it was used in the application of the dye alizarin to an aluminum mordanted cloth to give the color known as Turkey Red (p. 572). Turkey Red oil should not be confused with *red oil,* which is the technical name given to commercial oleic acid. During 1964 over 31 million pounds of sulfated oils were produced in the United States.

Since 1930 the catalytic reduction of fat acids or their methyl esters, or the sodium-alcohol reduction of fats (p. 187), especially coconut oil, has been used to produce mixtures of higher alcohols, which can be sulfated with sulfuric acid or chlorosulfonic acid to produce detergents.

$$RCH_2OH + H_2SO_4 \longrightarrow H_2O + RCH_2OSO_3H \xrightarrow{NaOH} RCH_2OSO_3Na + H_2O$$

$$RCH_2OH + ClSO_3H \longrightarrow HCl + RCH_2OSO_3H \xrightarrow{NaOH} RCH_2OSO_3Na + H_2O$$

1-Alkanols now are being made for this purpose by the oxidation of long-chain alkylaluminums made by the polymerization of ethylene with ethylaluminum (p. 122).

The synthetic detergents used in largest amount previous to 1966 were the sodium salts of aromatic sulfonic acids that have highly branched alkyl side chains (p. 406). This widespread use, however, has caused serious difficulty because the branched chains are not destroyed by microorganisms as are the straight chains of soaps, and they have appeared in domestic water supplies. Threatened by Government action, detergent manufacturers agreed to convert to "biodegradable" straight-chain alkyl sulfonates or sulfates by 1966.

Surface active soaps, sulfates, and sulfonates belong to the class of **anionic surfactants** because their activity resides in the anion. **Cationic surfactants** (p. 232), **nonionic surfactants** (p. 606), and **amphoteric surfactants** (p. 618) also are produced commercially.

In contrast to the triglycerides, monoglycerides have free hydroxyl groups, which act as water-solubilizing groups, and hence they become emulsifying agents. They have found extensive use particularly in the manufacture of oleomargarine and in other food industries, and in the preparation of cosmetic creams. They are mixtures prepared either by the partial esterification of glycerol with free fat acids or by the glycerolysis of fats (p. 187). Over 67 million pounds of glycerol esters were produced in 1964. Acetylation of the monostearate gives a product that is used as a flexible, non-greasy, edible coating for foods.

Other Uses for Fat Acids. Free fat acids are used as softening agents for rubber. Commercial stearic acid is a mixture of stearic and palmitic acids used in the manufacture of candles, cosmetics, and shaving soaps. The aluminum, calcium, lead, and other metallic soaps, when heated with petroleum oils, form a gel and are used to thicken oils in the manufacture of lubricating greases. Magnesium and zinc stearates are used in face powders and dusting powders, and as lubricants to prevent sticking in the molding of plastics. Long alkyl chains are incorporated into the molecules of antiseptics, drugs, dyes, resins, and plastics to modify their solubility or setting characteristics. Fatty acids are converted on an industrial scale into esters, amides, nitriles, and amines, the last being particularly valuable as a starting point for the synthesis of cationic detergents and antiseptics (p. 232), and for amphoteric surfactants (p. 618).

PROBLEMS

12–*1*. Which of the following reactions may be used to distinguish readily vegetable oils from unsaturated petroleum oils: (*a*) reaction with halogen, (*b*) polymerization, (*c*) saponification, (*d*) hydrogenation, (*e*) oxidation?

12–*2*. Of the following fats or oils, (*1*) linseed oil, (*2*) olive oil, (*3*) tallow, (*4*) castor oil, (*5*) tung oil, and (*6*) coconut oil, which has (*a*) the highest saponification value, (*b*) the lowest saponification value, (*c*) the highest iodine value, (*d*) the lowest iodine value, (*e*) conjugated double bonds, and (*f*) the highest acetyl value?

12–*3*. Of the following compounds, (*1*) zinc stearate, (*2*) ammonium palmitate, (*3*) potassium caproate, (*4*) glycerol monostearate, and (*5*) sodium docosanoate, which (*a*) would be a useful detergent and (*b*) is used as an emulsifying agent in foods?

12–*4*. Calculate the volume (S.T.P.) of hydrogen that would be required to convert 500 g. of an oil having an iodine value of 90 into the completely saturated fat.

12–*5*. An acid isolated by the hydrolysis of a fat had a neutralization equivalent of 254. Vigorous oxidation with permanganate gave two new acids with neutralization equivalents of 94 and 130. (*a*) What acid probably was isolated? (*b*) How could identity be established?

12–*6*. How many pounds of sodium hydroxide would be required to saponify 10 pounds of a fat having a saponification value of 195?

12–*7*. What would be the maximum yield of (*a*) glycerol and (*b*) anhydrous soap obtainable in the saponification indicated in Problem *6*?

12–*8*. (*a*) Calculate the weight of metallic sodium required to reduce to alcohols 100 g. of a fat that has a saponification value of 250. (*b*) What would be the maximum amount of anhydrous sodium alkyl sulfate that could be obtained from the reduction product?

CHAPTER THIRTEEN

ALDEHYDES AND KETONES

Both alcohols and ethers have the general formula $C_nH_{2n+2}O$. A less saturated group of compounds is known having the formula $C_nH_{2n}O$. Because the first member of the series, CH_2O, has only one carbon atom, the unsaturation cannot be due to a carbon-carbon double bond. Therefore the only logical structure satisfying the rules of valence is one in which the carbon atom is united to the oxygen atom by a double bond. This functional group, C=O, which is known as a *carbonyl group,* is present in the carboxyl group of carboxylic acids, but its properties are masked by the presence on the same carbon atom of a hydroxyl group. If only hydrogen or carbon atoms are united to the carbonyl group, the group is characteristic of the compounds known as *aldehydes* and *ketones.* The **aldehydes** have at least one hydrogen atom united to the carbonyl group and are represented by the general formula RCHO. The **ketones** have two carbon atoms united to the carbonyl group and are represented by the formula RCOR. The methods of preparation and reactions are in agreement with these structures.

Preparation

1. ***Aldehydes or Ketones.*** Several reactions that yield aldehydes or ketones from other compounds have been described, namely, the oxidation or dehydrogenation of primary or secondary alcohols (p. 131), the pyrolysis of carboxylic acids (p. 165), the reduction of acyl chlorides (p. 170) or esters (p. 175), the addition of water to alkynes (p. 149), and the ozonolysis of alkenes (p. 90.) One of the best laboratory procedures for the synthesis of aldehydes involves the nucleophilic displacement of ethoxide ion from ethyl orthoformate (p. 178) by an alkide ion by means of a Grignard reagent derived from an alkyl halide. An acetal is formed, which can be hydrolyzed easily in acid solution to the aldehyde (p. 200) (*Chichibabin-Bodroux reaction*).

$$\mathrm{HC(OC_2H_5)_3 + RMgX \longrightarrow \underset{\text{An acetal}}{RCH(OC_2H_5)_2} + C_2H_5OMgX}$$

$$\mathrm{RCH(OC_2H_5)_2 + H_2O \xrightarrow{H^+} C_2H_5OH + \left[\begin{array}{c}\mathrm{RCHOH}\\ |\\ \mathrm{OC_2H_5}\end{array}\right] \longrightarrow C_2H_5OH + RCHO}$$

An additional advantage of the procedure is that acetals can be stored as such until the aldehyde is needed, whereas aldehydes do not keep well because they polymerize (p. 201) and autoxidize easily (p. 205).

Grignard reagents displace chloride ion from acyl chlorides at low temperature in the presence of ferric chloride to give good yields of ketones.

$$\mathrm{RCOCl + R'MgCl \xrightarrow[-65^\circ]{FeCl_3} RCOR' + MgCl_2}$$

The ferric chloride probably catalyzes the reaction by coordinating with the chlorine of the acyl chloride.

$$\mathrm{RCOCl + FeCl_3 \longrightarrow R\underset{\underset{O}{\|}}{C}-\overset{+}{Cl}:\overset{-}{Fe}Cl_3}$$

If the reaction is carried out at room temperature, the ketone reacts further to give a tertiary alcohol (p. 197). Methyl ketones have been obtained in good yields from Grignard reagents and acetic anhydride at low temperatures.

$$(CH_3CO)_2O + RMgX \xrightarrow{-70^\circ} CH_3COR + CH_3COOMgX$$

Alkylcadmiums (p. 112) in the presence of magnesium halide also replace the halogen of acyl halides by alkyl, but react less readily with the carbonyl group. The alkylcadmium solutions are prepared from the Grignard reagents and cadmium chloride.

$$2\,RMgCl + CdCl_2 \longrightarrow CdR_2 + 2\,MgCl_2$$
$$2\,R'COCl + CdR_2 \longrightarrow 2\,R'COR + CdCl_2$$

The magnesium halide formed in the preparation of the reagent is necessary for rapid reaction and probably functions in this reaction in the same way at room temperature as does the more electrophilic ferric chloride at low temperature in the Grignard reaction. Acyl chlorides react also with alkylaluminums (p. 112) to give ketones.

Nitriles and *N,N*-dimethyl amides can be reduced to aldehydes or ketones (pp. 240, 235) or converted to higher aldehydes or ketones (p. 235), and ketones can be prepared by decarboxylation of β-keto acids (p. 622). Either aldehydes or ketones can be made by way of the Wittig reaction (p. 428).

Nomenclature

Aldehydes. The **common names** of aldehydes are derived from the acids that would be formed on oxidation, that is, the acids having the same number of carbon atoms. In general the *ic acid* is dropped and *aldehyde* added. For example, formaldehyde, acetaldehyde, and *i*-butyraldehyde correspond to the fact that they yield formic, acetic, and *i*-butyric acids on oxidation. Just as more complex acids may be named as derivatives of acetic acid, branched-chain aldehydes may be named **as derivatives of acetaldehyde;** thus $C_2H_5CH(CH_3)CHO$ may be called methylethylacetaldehyde. In the **international system** the aldehydes are named by dropping the *e* of the hydrocarbon corresponding to the longest chain containing the aldehyde group and adding *al.* The compound having the above formula would be called 2-methylbutanal. The position of the aldehyde group is not indicated because it always is at the end of the chain, and its carbon atom is numbered one.

Ketones. The **common names** of ketones are derived from the acid that on pyrolysis would yield the ketone. For example, acetone, CH_3COCH_3, may be derived from two molecules of acetic acid, and *i*-butyrone, $(CH_3)_2CHCOCH(CH_3)_2$, from *i*-butyric acid. A second method, especially useful for naming **mixed ketones,** simply names the alkyl groups and adds the word *ketone.* For example, $CH_3COC_2H_5$ is methyl ethyl ketone. The name is written as three separate words since the compound is not a substitution product of a substance *ketone.* In the **international system** the ending is *one,* and the position of the carbonyl group must be indicated by a number, unless the name is unambiguous without the number. Thus methyl ethyl ketone may be called simply butanone, but ethyl ketone, $CH_3CH_2COCH_2CH_3$, must be called 3-pentanone to distinguish it from methyl *n*-propyl ketone, $CH_3COCH_2CH_2CH_3$, which is 2-pentanone. Side chains are named and numbered as usual.

If it is necessary to designate a carbonyl function when another functional group is present, the doubly bound oxygen is called *oxo.* Thus $CH_3COCH_2CH_2COOH$ is 4-oxopentanoic acid. Generically, however, it is called a keto acid rather than an oxo acid.

Physical Properties

The aldehydes and ketones resemble the ethers in their solubility characteristics, but aldehydes and ketones boil somewhat higher than ethers having the same number of electrons. For example, methyl ether boils at $-24°$ and acetaldehyde at $+20°$; methyl ethyl ether boils at 6°, whereas propionaldehyde boils at 49° and acetone at 56°. Aldehydes and ketones have a higher dipole moment (~2.7 D) than ethers (~1.2 D) because of the greater polarizability of the double bond, and hence the attractive force between molecules owing to dipole-dipole association (p. 23) is greater.

The C=O stretching frequency for saturated aliphatic aldehydes is in the region 1740–1720 $cm.^{-1}$ and for saturated open-chain ketones, 1725–1705 $cm.^{-1}$. Much is known about the effect of substituents and structure in the immediate vicinity of the carbonyl group on the frequency of absorption, and this information can be used for the identification of specific structures. Figure 13–*1* gives the infrared spectrum for liquid acetone. When compared with that for propionic acid (Fig. 11–*2*, p. 158), it is seen that the OH absorption is lacking and that the strong carbonyl absorption at 1710 $cm.^{-1}$ is present in both. The spectrum for acetone shows a sharp weak peak at 3400 $cm.^{-1}$, which is the overtone of the carbonyl absorption (p. 135). It is not detectable in the spectrum for propionic acid, probably because it is wiped out by the broad OH absorption. Aldehydes and ketones show characteristic absorption also in the ultraviolet at around 280 mμ resulting from an electronic transition (p. 554).

Odor and Taste

Aldehydes and ketones have characteristic odors and flavors and are largely responsible, along with the volatile fatty acids, for the flavor of rancid and stale foods. From the volatile components of blue cheese 29 neutral substances and 18 acids have been isolated and identified. The neutral components are chiefly C_5 to C_{11} methyl *n*-alkyl ketones, with methyl *n*-amyl ketone making up 50 per cent of the neutral fraction. Capric acid is the most prominent acid. The C_8 to C_{14} normal aldehydes are used in the formulation of perfumes.

Reactions

In carboxylic acids and their derivatives, attack by nucleophiles on the electron-deficient carbon atom usually leads to displacement reactions, rather than addition to the carbon-oxygen double bond, because of the presence of good or potentially good leaving

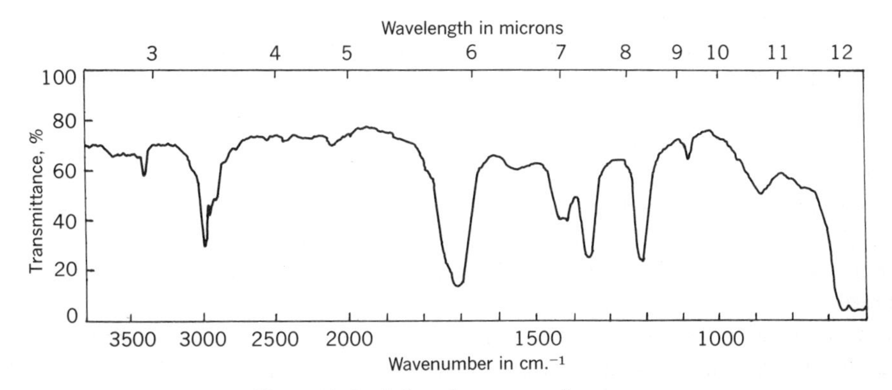

Figure 13–*1*. Infrared spectrum of acetone.

groups. Since hydrogen and alkyl groups attached to carbon are poor leaving groups, aldehydes and ketones undergo chiefly addition reactions. The carbon-oxygen double bond is much more highly polarized than even a 1,1-dialkyl-substituted carbon-carbon double bond. Hence the carbonyl group reacts more readily and with a greater variety of reagents than does a carbon-carbon double bond.

The polarity of the carbonyl group and the unshared electron pairs on oxygen determine the direction of addition of polar reagents. Nucleophilic attack occurs at the carbon atom of the carbonyl group with displacement of a pair of electrons from the double bond. Usually the reaction is completed by the transfer of a proton to the negatively charged oxygen atom.

$$N{:}^- + \overset{\overset{\displaystyle O}{\|}}{C}R_2 \rightleftharpoons N{-}\overset{\overset{\displaystyle O^-}{|}}{C}R_2 \underset{B^-}{\overset{HB}{\rightleftharpoons}} N{-}\overset{\overset{\displaystyle OH}{|}}{C}R_2$$

Electrophilic attack occurs on an unshared pair of electrons on the oxygen atom. Acid catalysis may be brought about in this way because the positive charge in the oxonium salt increases still further the deficiency of electrons about the carbon atom.

$$R_2C{=}O \underset{B^-}{\overset{HB}{\rightleftharpoons}} \left\{R_2C{=}\overset{+}{O}H \longleftrightarrow R_2\overset{+}{C}{-}OH\right\} \overset{-:N}{\rightleftharpoons} R_2\underset{\underset{\displaystyle N}{|}}{C}{-}OH$$

Catalysis by bases operates chiefly in two ways. First, bases can convert compounds having reactive hydrogen into good nucleophiles.

$$N{:}H \underset{HB}{\overset{B^-}{\rightleftharpoons}} N{:}^-$$

Second, they can remove a proton from a carbon atom α to the carbonyl group, making the aldehyde or ketone itself a nucleophile.

$$RCH_2CH{=}O \underset{HB}{\overset{-:B}{\rightleftharpoons}} \left\{R\underset{\cdot\cdot}{\overset{-}{C}}HCH{=}O \longleftrightarrow RCH{=}CH{-}O{:}^-\right\}$$

This ion is identical with the ion that would be formed by removal of a proton from the hydroxyl group of an enol (p. 201) and is known as the *enolate ion.* The increased acidity of hydrogen atoms α to a carbonyl group compared to those of an alkane or an alkene may be ascribed to the electronegativity of the adjacent carbonyl group, or to resonance stabilization of the enolate ion (cf. p. 159), or to both. Resonance in the enolate ion requires that the four atoms attached to the two carbon atoms involved become coplanar (Fig. 6–*1*, p. 74). That resonance is the predominating factor is indicated by the fact that when this is not possible, the ion does not form readily. Moreover the stabilization of the anion by resonance makes the α hydrogen of aldehydes considerably more acidic than the aldehydic hydrogen, which is activated only by the inductive effect of the oxygen. Thus only the α hydrogens exchange with deuterium in deuterium oxide solution.

In general, aldehydes differ from ketones only in the relative rate of reaction and in the position of equilibrium. Since hydrogen is smaller than an alkyl group and since the electron deficiency on carbon caused by the oxygen atom is relieved by alkyl groups better than by hydrogen, the order of rate and extent of reaction when these factors are important is formaldehyde > other aldehydes > ketones.

Certain reactions, particularly of aldehydes, take place by free radical mechanisms. These mechanisms will be discussed as the occasion arises.

Because of the large number of reactions that aldehydes and ketones undergo, it is desirable to group them into four divisions according to the over-all effect of the reaction, namely simple addition, addition and loss of water, oxidation, and reduction, and to end with a group of miscellaneous tests and reactions. Mechanisms are postulated in accord with the foregoing discussion without detailed explanation.

SIMPLE ADDITION

1. ***Grignard Reagents.*** Addition of Grignard reagents to carbonyl groups undoubtedly involves initial coordination with magnesium, a solvent ether molecule being displaced. Reaction follows, probably by way of a four-membered cyclic transition state.

$$\rangle C{=}\ddot{O}: + RMgX \longrightarrow \underset{R-MgX}{\rangle C{=}\ddot{O}} \longrightarrow \underset{R\cdots MgX}{\rangle C{=}O} \longrightarrow \underset{R\quad MgX}{\rangle C{-}\ddot{O}:}$$

Cyclic six-membered transition states involving a second molecule of Grignard reagent or a molecule of magnesium halide also have been postulated. The alcohol is liberated from the magnesium alkoxide by the addition of water.

Formaldehyde yields primary alcohols, all other aldehydes yield secondary alcohols, and ketones yield tertiary alcohols. This procedure is one of the most important for the synthesis of complex alcohols, since by choosing the proper R groups in the aldehyde or ketone and in the Grignard reagent almost any desired alcohol may be synthesized, provided the R groups are not too highly branched to cause steric hindrance and prevent addition.

$$H{-}\overset{\overset{\displaystyle O}{\|}}{C}{-}H + RMgX \longrightarrow H{-}\underset{\underset{\displaystyle R}{|}}{\overset{\overset{\displaystyle OMgX}{|}}{C}}{-}H \xrightarrow{H_2O} RCH_2OH + HOMgX$$

$$RCHO + R'MgX \longrightarrow R{-}\underset{\underset{\displaystyle OMgX}{|}}{CH}{-}R' \xrightarrow{H_2O} R\underset{\underset{\displaystyle OH}{|}}{CH}R' + HOMgX$$

$$R_2CO + R'MgX \longrightarrow R_2\underset{\underset{\displaystyle OMgX}{|}}{C}{-}R' \xrightarrow{H_2O} R_2\underset{\underset{\displaystyle OH}{|}}{C}H' + HOMgX$$

Although the above reactions take place readily to give good yields of the products when simple aldehydes or ketones or simple Grignard reagents are involved, branching of the alkyl groups in either the carbonyl compound or the Grignard reagent or both may permit side reactions to predominate. These side reactions become so important with branched aldehydes, ketones, and Grignard reagents that it is not possible to synthesize many alcohols by the reaction of a Grignard reagent with an aldehyde or ketone. Frequently the desired alcohol can be synthesized if an alkyllithium (p. 111) is used in place of a Grignard reagent.

2. ***Hydrogen Cyanide.*** Anhydrous hydrogen cyanide adds to aldehydes and ketones to give α-hydroxy cyanides known as **cyanohydrins.** It was shown as early as 1903 by Lapworth[1] that the reaction is accelerated by bases, and he proposed that the initial step is the attack of the carbonyl carbon by cyanide ion.

$$HCN \underset{H_2O}{\overset{^-OH}{\rightleftarrows}} CN^- \overset{RCHO}{\rightleftarrows} R\underset{\underset{\displaystyle CN}{|}}{C}HO^- \underset{CN^-}{\overset{HCN}{\rightleftarrows}} R\underset{\underset{\displaystyle CN}{|}}{C}HOH$$

$$R_2CO + HCN \rightleftarrows R_2\underset{\underset{\displaystyle CN}{|}}{C}{-}OH$$

---

[1] Arthur Lapworth (1872–1941), professor at the University of Manchester who held in succession chairs in organic chemistry and in physical chemistry. He was a pioneer investigator of the mechanisms of organic reactions. His ideas were so advanced that his early work received little attention for many years. His investigations of the addition of hydrogen cyanide to carbonyl compounds, of the bromination of ketones, and of acid and base catalysis now are considered classical.

As expected, electron-withdrawing groups are favorable whereas electron-donating groups are unfavorable for cyanohydrin formation. *These compounds are named as addition products.* For example, that derived from acetaldehyde is known as *acetaldehyde cyanohydrin* and that from acetone as *acetone cyanohydrin.* As with any other nitrile, the cyanide group can be hydrolyzed to a carboxyl group (p. 239). Hence the cyanohydrins are intermediates for the synthesis of α-hydroxy acids (p. 615).

3. ***Sodium Bisulfite.*** When shaken with a saturated aqueous sodium bisulfite solution, most aldehydes and methyl ketones undergo nucleophilic attack by sulfur of the bisulfite anion to form a slightly soluble bisulfite addition compound in which hydrogen has added to oxygen and the sodium sulfonate group to carbon.

$$RCH{=}O + :SO_2(OH)^{-}\,{}^{+}Na \rightleftharpoons \left[RCH(O^{-}\,{}^{+}Na)(SO_2OH) \right] \rightleftharpoons RCH(OH)(SO_2O^{-}\,{}^{+}Na)$$

$$RCOCH_3 + NaHSO_3 \rightleftharpoons R{-}C(HO)(SO_3^{-}\,{}^{+}Na){-}CH_3$$

If both groups attached to the carbonyl group are larger than methyl, the addition compound does not form unless the groups are held out of the way of the carbonyl group as in cyclohexanone (p. 649). Even if one of the groups is methyl, the reaction may be very slow and the position of equilibrium unfavorable if the other group is branched, for example, a *t*-butyl group. As to nomenclature, the product from acetaldehyde is known as *acetaldehyde sodium bisulfite* or as the *bisulfite addition compound of acetaldehyde,* that from acetone is *acetone sodium bisulfite* or the *bisulfite addition compound of acetone.*

Since these compounds are salts, they are not soluble in organic solvents and may be freed from other organic compounds such as hydrocarbons or alcohols by filtering and washing with ether. The reactions are reversible, and hence the carbonyl compound is regenerated by any reagent that reacts irreversibly with bisulfite. Either alkali or acid can be used.

$$RCHOHSO_3Na + HCl \longrightarrow RCHO + NaCl + SO_2 + H_2O$$

$$R_2COHSO_3Na + Na_2CO_3 \longrightarrow R_2CO + Na_2SO_3 + NaHCO_3$$

Acids have the disadvantage that sulfur dioxide must be removed from the product. Alkalies have the disadvantage that they cause condensation reactions of aldehydes (p. 203). As a result alkalies usually are used to liberate ketones, and acids to liberate aldehydes. An alternative procedure consists of heating the bisulfite addition compound with a slight excess of an aqueous solution of formaldehyde.

$$RCHOHSO_3Na + HCHO \rightleftharpoons RCHO + H_2COHSO_3Na$$

$$R_2COHSO_3Na + HCHO \rightleftharpoons R_2CO + H_2COHSO_3Na$$

These exchange reactions take place because the position of equilibrium for the reaction of formaldehyde with bisulfite is farther to the right than it is with the other aldehydes and ketones. The chief importance of the bisulfite addition compounds is their use to separate carbonyl compounds from mixtures with other organic compounds.

4. ***Hydroxy Compounds.*** (*a*) WATER. Although compounds having two hydroxyl groups on the same carbon atom rarely can be isolated in the pure state, they may exist in water solution. The familiar example is carbonic acid, $O{=}C(OH)_2$, which behaves like a dibasic acid in aqueous solution but can be isolated only in the form of its salts or esters or of its anhydride, carbon dioxide. Similarly when aldehydes are dissolved in water, they may exist to a considerable extent in the hydrated form. This condition is true particularly for formaldehyde whose aqueous solutions contain almost exclusively methanediol.

$$\mathrm{HCHO + HOH \rightleftharpoons \left[\begin{array}{c} H_2C{-}O^- \\ | \\ {}^+OH_2 \end{array}\right] \rightleftharpoons \begin{array}{c} H_2C{-}OH \\ | \\ OH \end{array}}$$

Evidence for the almost complete reaction of formaldehyde with water is the absence of a characteristic carbonyl absorption band (p. 554) in the ultraviolet for aqueous solutions. At equilibrium in dilute aqueous solutions acetaldehyde is hydrated to the extent of 58 per cent, whereas hydration of acetone is negligible. The decrease in stability of the hydrates of acetaldehyde and acetone results from the electron-donating ability of the alkyl groups, which increases the electron density on the carbonyl carbon and makes it less capable of holding two electron-rich groups (p. 82). Electron-withdrawing groups, as expected, increase the stability of hydrates. Thus the hydrate of chloral (p. 215) and that of hexafluoroacetone (p. 596) can be isolated.

Not only is the position of equilibrium unfavorable for acetone but the rate of addition of water is slow. Thus the oxygen atoms of acetone do not exchange at a measurable rate with those of water when acetone is dissolved in water containing an increased concentration of the heavy oxygen isotope, O^{18}. Exchange takes place extremely rapidly in the presence of acids or hydroxide ion. In the acid catalysis, protonation of oxygen is followed by nucleophilic attack on carbon, whereas the reverse order holds for the base catalysis.

$$\mathrm{(CH_3)_2C{=}O \underset{B^-}{\overset{HB}{\rightleftharpoons}} (CH_3)_2C{=}\overset{+}{O}H \overset{HO^{18}H}{\rightleftharpoons} (CH_3)_2C(OH)(\overset{+}{O}{}^{18}H_2) \underset{HB}{\overset{B^-}{\rightleftharpoons}} (CH_3)_2C(OH)(O^{18}H)}$$

$$\mathrm{(CH_3)_2C{=}O \overset{O^{18}H^-}{\rightleftharpoons} (CH_3)_2C(O^-)(O^{18}H) \underset{OH^-}{\overset{HOH}{\rightleftharpoons}} (CH_3)_2C(OH)(O^{18}H)}$$

The reverse processes lead to the enrichment of the acetone in the O^{18} isotope, since H_2O^{16} or $^-O^{16}H$ can be eliminated as well as H_2O^{18} or $^-O^{18}H$.

(b) ALCOHOLS. In the presence of either acidic or basic catalysts, aldehydes add one mole of alcohol to form *hemiacetals*. In acid catalysis the effective point of attack is the carbonyl group.

$$\mathrm{RCHO \underset{B^-}{\overset{HB}{\rightleftharpoons}} RCH{=}\overset{+}{\ddot{O}}{:}H \overset{R'OH}{\rightleftharpoons} R{-}\underset{{}^+:\ddot{O}(H){-}R'}{\overset{H}{C}}{-}OH \underset{HB}{\overset{B^-}{\rightleftharpoons}} R{-}\underset{OR'}{\overset{H}{C}}{-}OH} \qquad (1)$$

A hemiacetal

The attack by the acid on the alcohol gives ROH_2^+, which can do nothing but transfer protons to the carbonyl group. In the base catalysis the alcohol must be the effective point of attack by the catalyst.

$$\mathrm{R'OH \underset{HOH}{\overset{OH^-}{\rightleftharpoons}} R'O^- \overset{RCHO}{\rightleftharpoons} \underset{OR'}{RCH}{-}\ddot{O}{:}^- \underset{OH^-}{\overset{HOH}{\rightleftharpoons}} \underset{OR'}{RCHOH}} \qquad (2)$$

Although the carbon atom of the carbonyl group of the aldehyde may be attacked by a base, this does not lead to the formation of a hemiacetal, but merely to the hydrated aldehyde.

$$\mathrm{RCHO \overset{OH^-}{\rightleftharpoons} \underset{OH}{RCH}{-}\ddot{O}{:}^- \underset{OR^-\ or\ OH^-}{\overset{HOR\ or\ HOH}{\rightleftharpoons}} \underset{OH}{RCH}{-}OH}$$

Like the hydrates, hemiacetals exist only in solution unless electron-attracting groups are present on the α carbon atom. Thus the hemiacetal of acetaldehyde cannot be isolated, but chloral alcoholate is a stable compound.

With an excess of alcohol and an acidic catalyst, water is eliminated and an *acetal* is formed.

$$\mathrm{R{-}\underset{OR}{\overset{H}{C}}{-}OH \underset{B^-}{\overset{HB}{\rightleftharpoons}} R{-}\underset{OR}{\overset{H}{C}}{-}\overset{+}{\ddot{O}}H_2 \underset{HOH}{\overset{HOR}{\rightleftharpoons}} R{-}\underset{OR}{\overset{H}{C}}{-}\overset{+}{\ddot{O}}HR \underset{HB}{\overset{B^-}{\rightleftharpoons}} R{-}\underset{OR}{\overset{H}{C}}{-}OR}$$

An acetal

Protonation of the alkoxyl group merely leads to the reversal of reaction *1*. Bases cannot catalyze acetal formation because removal of a proton from the hydroxyl group can lead only to the reversal of reaction *2*. Moreover although acids catalyze the hydrolysis of acetals, bases are unable to do so because the unprotonated alkoxyl is a poor leaving group.

Acetal formation can be used to "protect" an aldehyde group while reactions such as oxidation or reduction are being carried out on other functional groups. After the reaction is completed, the alkoxyl groups are removed by acid hydrolysis.

Acetal also is the specific name for the product from acetaldehyde and ethyl alcohol. The product from formaldehyde and methyl alcohol has the common name **methylal.**

In the reaction of ketones with alcohols the position of equilibrium at room temperature is so far to the left that ketals usually are not formed in any appreciable amount. The equilibrium is more favorable for the reaction of unhindered ketones with methyl alcohol at low temperature, and **2,2-dimethoxypropane** is made commercially from acetone and methyl alcohol using a strong-acid ion-exchange resin as catalyst.

$$\mathrm{CH_3COCH_3 + 2\,CH_3OH \underset{-27^\circ}{\overset{\text{Acid resin}}{\rightleftharpoons}} CH_3C(OCH_3)_2CH_3 + H_2O}$$

The easy reversibility of this reaction when catalyzed by acid makes 2,2-dimethoxypropane a useful reagent for the removal of water formed in other reactions such as esterification (p. 163). More general methods for the preparation of ketals involve an exchange reaction of ketones with alkyl orthoformates (p. 178) or sulfites (127).

$$\mathrm{R_2C{=}O + (R'O)_3CH \overset{H^+}{\rightleftharpoons} R_2C(OR')_2 + O{=}CHOR'}$$

$$\mathrm{R_2C{=}O + (R'O)_2SO \overset{H^+}{\rightleftharpoons} R_2C(OR')_2 + SO_2}$$

5. ***Acetylenes.*** The acetylide anion is a strong nucleophile and readily attacks a carbonyl group to give *alkynols*. It may be preformed as in the reaction of lithium or sodium acetylide (p. 150), or it may be produced catalytically.

$$\mathrm{RCH{=}O + Li^{+-}{:}C{\equiv}H \longrightarrow \underset{C\equiv H}{RCHO^-Li^+} \xrightarrow{H_2O} \underset{C\equiv CH}{RCHOH} + LiOH}$$

$$\mathrm{R_2C{=}O + Na^{+-}{:}C{\equiv}CR \xrightarrow{\text{Liq. } NH_3} \underset{C\equiv CR}{R_2CHO^{-+}Na} \xrightarrow{H_2O} \underset{C\equiv CH}{RCHOH} + NaOH}$$

Aqueous formaldehyde and acetylene at 100° and 90 p.s.i. in the presence of cuprous carbide (*copper acetylide*) gives **1,4-dihydroxy-2-butyne** (*butynediol*). Undoubtedly the carbide ion of the cuprous carbide reacts with the formaldehyde and is regenerated by the acetylene.

$$\mathrm{2\,H_2C{=}O \underset{2\,Cu^+}{\overset{Cu^{+-}C\equiv C^{-+}Cu}{\rightleftharpoons}} \underset{O^-}{H_2C}{-}C{\equiv}C{-}\underset{O^-}{CH_2} \underset{^-C\equiv C^-}{\overset{HC\equiv CH}{\rightleftharpoons}} \underset{OH}{H_2C}{-}C{\equiv}C{-}\underset{OH}{CH_2}}$$

Potassium hydroxide dissolved in ethers of ethylene glycol (p. 605), or a solution of sodium alkoxide or acetylide in methyl sulfoxide (p. 102) can be used as a catalyst for those carbonyl compounds not readily affected by alkali. Thus methyl ethyl ketone and acetylene at 400 p.s.i. give **3-hydroxy-3-methyl-1-pentyne** (*methylpentynol*), which is a useful soporific.

$$HC{\equiv}CH + {}^{-}{:}OH \rightleftharpoons HC{\equiv}C{:}^{-} + H_2O$$

$$C_2H_5\overset{CH_3}{\overset{|}{C}}{=}O \xrightarrow{{}^{-}{:}C\equiv CH} C_2H_5\underset{O^-}{\underset{|}{\overset{CH_3}{\overset{|}{C}}}}{-}C{\equiv}CH \xrightarrow{HC\equiv CH} C_2H_5\underset{OH}{\underset{|}{\overset{CH_3}{\overset{|}{C}}}}{-}C{\equiv}CH + {}^{-}{:}C{\equiv}CH$$

The HC≡C group is the *ethynyl group* and its introduction by means of acetylene is called *ethynylation.*

6. ***Cyclic Trimerization.*** Aliphatic aldehydes, but not ketones, undergo acid-catalyzed addition to give cyclic trimers. The reaction also takes place slowly in the absence of added catalyst.

$$R{-}\overset{H}{\overset{|}{C}}{=}\ddot{O}{:} \underset{B^-}{\overset{HB}{\rightleftharpoons}} R{-}\overset{H}{\overset{|}{C}}{=}\overset{+}{\ddot{O}}{:}H \xrightleftharpoons{R-\overset{H}{\overset{|}{C}}=\ddot{O}:} \text{R(H)C(:O}^{+}{=}\text{C(H)R)(OH)} \xrightleftharpoons{R-\overset{H}{\overset{|}{C}}=\ddot{O}:}$$

$$\text{[cyclic intermediate with } {:}\ddot{O}{:}H \text{ and } \overset{+}{O}\text{]} \rightleftharpoons \text{[cyclic intermediate with } {:}\overset{+}{O}{:}H\text{]} \underset{HB}{\overset{B^-}{\rightleftharpoons}} \text{2,4,6-trialkyl-1,3,5-trioxane (R, H on each C)}$$

These trimers resemble the acetals in that there is no effective point of attack for basic catalysts. Hence they are stable in neutral and alkaline solutions. The trimer can be reconverted to the monomer by heating with a nonvolatile acid catalyst and removing the lower-boiling aldehyde by distillation. Other types of polymerization are discussed under the individual aldehydes (pp. 214, 215).

7. ***Aldol Addition.***[2] In the presence of dilute aqueous alkalies or acids, aldehydes and ketones having at least one α hydrogen undergo also other self-addition reactions. These reactions may be repeated, and under certain conditions complex compounds are formed. In the presence of very dilute alkalies or acids, the product of reaction of two moles of aldehyde or ketone can be isolated.

In the acid-catalyzed reaction the carbonyl group is the first point of attack, and removal of a proton from the α carbon atom follows.

$$RCH_2\overset{H}{\overset{|}{C}}{=}O \underset{B^-}{\overset{HB}{\rightleftharpoons}} R\underset{H}{\underset{|}{C}}H{-}\overset{H}{\overset{|}{C}}{=}\overset{+}{O}H \underset{HB}{\overset{B^-}{\rightleftharpoons}} RCH{=}\overset{H}{\overset{|}{C}}OH$$

The product is the *enol form* of the carbonyl compound (p. 196), and the process is called *enolization*. In this process the equilibrium for simple aldehydes and ketones in the fluid state lies far to the left.[3] In the next stage of the reaction, the π electrons of the enol form

[2] This reaction frequently is called an aldol condensation. The term *condensation,* however, has been used very loosely by organic chemists. In this text it will be used only to mean the formation of a carbon-carbon bond with the elimination of some small molecule such as water, alcohol, or a metallic halide.

[3] Spontaneous reversible isomerization, especially of the type H—*X*—*Y*=*Z* ⇌ *X*=*Y*—*Z*—H, usually is referred to as *tautomerism,* and the isomers are called *tautomers* (p. 623). Tautomerism is not to be confused with resonance. Tautomerism involves a change in structure whereas resonance not only does not involve a change in the positions of the atoms in the molecule, but is a state of being and does not involve any change whatsoever (p. 159).

attack the electron-deficient carbon atom of a second molecule of the conjugate acid (p. 31) of the carbonyl compound.

$$RCH{=}C(H){-}\ddot{O}H \; + \; RCH_2C(H){=}\overset{+}{O}H \rightleftharpoons RCH(C(H){=}\overset{+}{O}H){-}CH(R)... \quad \text{i.e.}\quad \begin{matrix} RCH{-}CH{=}\overset{+}{O}H \\ | \\ RCH_2CH{-}OH \end{matrix} \underset{HB}{\overset{B^-}{\rightleftharpoons}} \begin{matrix} RCHCHO \\ | \\ RCH_2CHOH \end{matrix}$$

The first step in the base-catalyzed reaction is the removal of a proton from the α carbon atom (p. 196). The anion thus formed attacks the carbonyl carbon atom of a second molecule, and the product is stabilized by acquisition of a proton from a water molecule.

$$RCH_2CHO \underset{HOH}{\overset{OH^-}{\rightleftharpoons}} {}^{-}{:}CH(R)CHO \overset{R'CHO}{\rightleftharpoons} R'CH(:\ddot{O}:^-){-}CH(R)CHO \underset{OH^-}{\overset{HOH}{\rightleftharpoons}} R'CH(OH){-}CH(R)CHO$$

Since the product has both an alcohol and an aldehyde function, it is known as an *aldol.* **Acetaldol** was synthesized by Wurtz in 1872 (p. 145).

$$CH_3CHO + CH_3CHO \xrightleftharpoons{^{-}OH \text{ or } H^+} \underset{\text{Acetaldol or aldol}}{CH_3CHOHCH_2CHO}$$

With ketones the equilibrium is so far to the left that special means must be employed to obtain practical amounts of product.

$$CH_3COCH_3 + CH_3COCH_3 \xrightleftharpoons{^{-}OH \text{ or } H^+} \underset{\text{Diacetone alcohol}}{(CH_3)_2COHCH_2COCH_3}$$

For the production of **diacetone alcohol,** the acetone is converted to the equilibrium mixture by passing it over an insoluble catalyst such as calcium hydroxide or barium hydroxide. After the mixture is removed from the catalyst, the unchanged acetone is separated by distillation and recycled. Although diacetone alcohol is a good solvent, it is not useful in the formulation of lacquers because basic pigments catalyze the reversion to acetone.

Aldols lose water more readily than simple alcohols because the double bond in the resulting α,β-unsaturated aldehyde or ketone is conjugated with the carbonyl group. Conjugated systems usually are more stable thermodynamically than nonconjugated systems. The dehydration of either β-hydroxy aldehydes or ketones is catalyzed by dilute acids or bases. Even iodine catalyzes the reaction if the hydroxyl group is tertiary.

$$CH_3CHOHCH_2CHO \xrightarrow[\text{heat}]{H^+} \underset{\text{Crotonaldehyde}}{CH_3CH{=}CHCHO} + H_2O$$

$$(CH_3)_2COHCH_2COCH_3 \xrightarrow[\text{heat}]{I_2} \underset{\text{Mesityl oxide}}{(CH_3)_2C{=}CHCOCH_3} + H_2O$$

Iodine acts as an acidic catalyst because it is an electrophilic reagent. Familiar evidence for this statement is the greater solubility of iodine in potassium iodide solution than in water because of the formation of triiodide ions by the reaction $I_2 + I^- \longrightarrow I_3^-$. A related phenomenon is the brown color of solutions of iodine in alcohol in contrast to the violet color of solutions in hexane (cf. p. 387).

Although the dehydration of simple alcohols is not catalyzed by bases, the carbonyl group increases the acidity of the α hydrogen sufficiently to permit removal of a proton by a base with subsequent loss of hydroxide ion.

$$\text{RCHOHCH}_2\text{COR} \underset{\text{H}_2\text{O}}{\overset{^-\text{OH}}{\rightleftharpoons}} \text{R}\underset{\text{OH}}{\text{CH}}-\bar{\text{C}}\text{HCOR} \underset{^-\text{OH}}{\rightleftharpoons} \text{RCH=CHCOR}$$

Both the acid- and the base-catalyzed dehydrations are classed as β eliminations (p. 105).

Concentrated sodium hydroxide solutions convert aldehydes having at least two α hydrogen atoms to complex products known as **aldehyde resins.** Aldol addition, dehydration, and polymerization reactions probably play a part in their formation. The product from acetaldehyde is a sticky viscous orange-colored oil with a characteristic odor.

Ketones are not affected appreciably by concentrated sodium hydroxide solutions. The amide ion, however, is a stronger base than hydroxide ion, and under anhydrous conditions sodium amide converts acetone into a six-membered cyclic compound known as **isophorone,** which again is the result of a combination of aldol addition and dehydration.

$$3\,(\text{CH}_3)_2\text{C=O} \xrightarrow{\text{NaNH}_2} [\text{cyclic aldol intermediate}] \longrightarrow \text{Isophorone} + 2\,\text{H}_2\text{O}$$

Isophorone

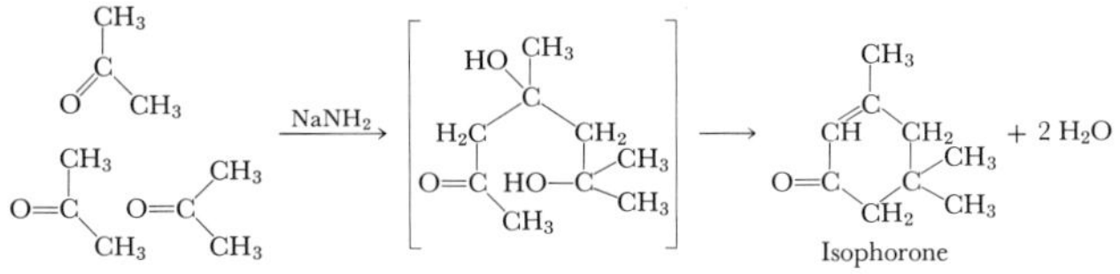

Strong acids favor the trimerization of aldehydes (p. 201) but lead to addition and dehydration of ketones. For example, acetone yields mesityl oxide directly in the acid-catalyzed reaction. Under more vigorous conditions of catalysis and dehydration, higher condensation products are formed.

$$(\text{CH}_3)_2\text{C=O} + \text{CH}_3\text{COCH}_3 + \text{O=C(CH}_3)_2 \xrightarrow[\text{or AlCl}_3]{\text{Dry HCl, ZnCl}_2,} (\text{CH}_3)_2\text{C=CH-CO-CH=C(CH}_3)_2 + 2\,\text{H}_2\text{O}$$

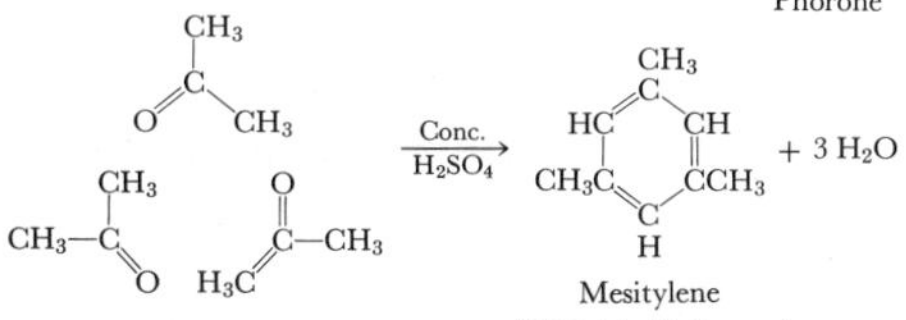

Phorone

$$3\,\text{CH}_3\text{COCH}_3 \xrightarrow[\text{H}_2\text{SO}_4]{\text{Conc.}} \text{C}_6\text{H}_3(\text{CH}_3)_3 + 3\,\text{H}_2\text{O}$$

Mesitylene
(1,3,5-trimethylbenzene)

The formation of mesitylene from acetone is an example of the synthesis of an aromatic-type compound (p. 93) from a member of the aliphatic series.

Addition and Loss of Water

1. ***Ammonia.*** Reaction products of aldehydes with ammonia have been isolated that appear to result from the addition of ammonia to the carbonyl group. Like the hydrates, however, the initial product is unstable and loses water to give an **aldimine,**[4] RCH=NH, which polymerizes to a cyclic trimer (cf. p. 201).

$$\text{RCH=O} + :\text{NH}_3 \longrightarrow \text{RCH(}^+\text{NH}_3\text{)-O}^- \longrightarrow \text{RCH(NH}_2\text{)OH} \longrightarrow \text{RCH=NH} + \text{H}_2\text{O}$$

$$3\,\text{RCH=NH} \longrightarrow (\text{RCHNH})_3 \text{ (cyclic trimer)}$$

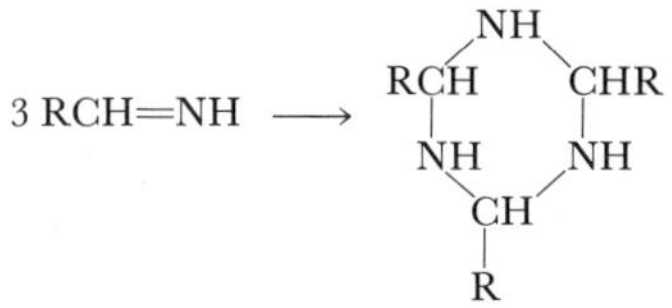

The product from acetaldehyde and ammonia, known as **aldehyde-ammonia,** is a crystalline trihydrate of the cyclic compound. Ketones do not yield analogous products, but

[4] The endings *ine* and *ime* are pronounced as if the *i* were an *e* and hence do not follow the usual rule for English pronunciation.

the initial addition of ammonia to the carbonyl group undoubtedly takes place in reactions such as the Strecker synthesis of α-amino acids (p. 357).

2. ***Hydroxylamine.*** Aldehydes and ketones add hydroxylamine, the hydroxy derivative of ammonia, H_2NOH, to give an unstable initial product analogous to that formed by addition of ammonia to the carbonyl group, but the subsequent loss of water gives a stable monomolecular product known as an **oxime.**[4] The reactions of aldehydes and ketones with hydroxylamine take place most readily at a hydrogen ion concentration such that the reagent is half converted into its salt. A satisfactory mechanism is one in which the reaction is catalyzed by acids, but the reagent reacts by way of the free base.

$$R_2C{=}O \underset{B^-}{\overset{HB}{\rightleftharpoons}} R_2C{=}\overset{+}{O}H \overset{H_2\ddot{N}OH}{\rightleftharpoons} \underset{H_2\overset{..}{\underset{+}{N}}OH}{R_2C{-}OH} \underset{HB}{\overset{B^-}{\rightleftharpoons}}$$

$$\underset{NHOH}{R_2C{-}OH} \underset{B^-}{\overset{HB}{\rightleftharpoons}} \underset{:NHOH}{R_2C{-}\overset{H}{\overset{..+}{O}}H} \underset{H_2O}{\rightleftharpoons} \underset{{}^+NHOH}{R_2C} \underset{HB}{\overset{B^-}{\rightleftharpoons}} \underset{NOH}{R_2C}$$

A ketoxime

$$RCHO + H_2NOH \longrightarrow RCH{=}NOH + H_2O$$

An aldoxime

The product from acetaldehyde, $CH_3CH{=}NOH$, is called **acetaldoxime,** that from acetone is **acetoxime,** and that from methyl ethyl ketone, $CH_3(C_2H_5)C{=}NOH$, is **methyl ethyl ketoxime.** Homologous compounds are named in the same way.

The aldehyde or ketone may be regenerated from its oxime by an acid-catalyzed exchange with an excess of a more reactive aldehyde or ketone such as formaldehyde, acetone, or levulinic acid (p. 330).

$$RCH{=}NOH + H_2O \overset{H^+}{\rightleftharpoons} RCHO + H_2NOH$$

$$HCH{=}O + H_2NOH \longrightarrow HCH{=}NOH + H_2O$$

The oximes are both weak bases and weak acids. They are more soluble in cold dilute acids or cold dilute alkali than they are in water.

$$RCH{=}NOH + HCl \longrightarrow [RCH{=}\overset{+}{N}HOH]Cl^-$$

$$RCH{=}NOH + NaOH \longrightarrow RCH{=}NO^{-+}Na + H_2O$$

They are, however, much weaker bases ($pK_a = 1.8$ for acetoxime) than hydroxylamine ($pK_a = 6.0$). Use is made of this fact in a procedure for the estimation of aldehydes and ketones. After reaction of the carbonyl compound with hydroxylamine hydrochloride, the hydrogen chloride liberated is titrated with standard alkali using a suitable indicator (*Bromophenol Blue*).

$$RCHO + H_3\overset{+}{N}OHCl^- \longrightarrow RCH{=}NOH + HCl + H_2O$$

Oximes frequently are crystalline solids and thus are useful derivatives for the identification of aldehydes and ketones. They are useful also for the synthesis of primary amines (p. 224) and alkyl cyanides (p. 237), and as intermediates for opening the rings of cyclic ketones (p. 649).

3. ***Substituted Hydrazines.*** Hydrazine has the formula H_2NNH_2, and although it reacts with aldehydes and ketones, the final product may not be typical. The substituted hydrazines having one free amino group (*NH_2 group*) behave in a regular fashion analogous to hydroxylamine. The hydrazines most used are *phenylhydrazine,* $C_6H_5NHNH_2$ (p. 432), and substituted phenylhydrazines, especially, *2,4-dinitrophenylhydrazine* (p. 400), the products

being known as **phenylhydrazones,** and *semicarbazide* (sometimes called *semicarbazine,* p. 287), $H_2NNHCONH_2$, the products being known as **semicarbazones.**

$$(CH_3)_2CO + H_2NNHC_6H_5 \longrightarrow \underset{\text{Acetone phenylhydrazone}}{(CH_3)_2C{=}NNHC_6H_5} + H_2O$$

$$CH_3CH{=}O + H_2NNHCONH_2 \longrightarrow \underset{\text{Acetaldehyde semicarbazone}}{CH_3CH{=}NNHCONH_2} + H_2O$$

Phenylhydrazones of aliphatic aldehydes and ketones tautomerize (p. 201) rapidly when dissolved in hexane or carbon tetrachloride to the **benzeneazoalkanes,** which are more stable because the double bond is conjugated with the benzene ring.

$$R_2C{=}N{-}NHC_6H_5 \longrightarrow R_2CH{-}N{=}NC_6H_5$$

Whereas the simpler aldehydes and ketones usually are liquids, many of the phenylhydrazones and semicarbazones are crystalline solids that can be purified readily and have definite melting points. Like the oximes they are useful derivatives for the identification of aldehydes and ketones.

OXIDATION

1. ***By Air*** (*Autoxidation*). Aldehydes, but not ketones, are among the most readily autoxidizable substances. They rapidly absorb oxygen from the air to give peroxy acids (p. 165). The chain reaction is initiated by adventitious free radicals.

$$RCHO + \cdot Z \longrightarrow R\dot{C}{=}O + HZ \qquad \textit{Initiation}$$

$$\left.\begin{array}{l} R\dot{C}{=}O + O_2 \longrightarrow RC({=}O){-}O{-}O\cdot \\ RC({=}O){-}O{-}O\cdot + RCH{=}O \longrightarrow RC({=}O){-}O{-}OH + R\dot{C}{=}O \end{array}\right\} \qquad \textit{Propagation}$$

2. ***By Other Oxidizing Agents.*** Aldehydes are oxidized easily to acids even by mild oxidizing agents. In the absence of other easily oxidized groups, however, the common oxidizing agents such as nitric acid, chromic acid, or potassium permanganate are used. The oxidation of aldehydes by chromic acid, $(HO)_2CrO_2$, or acid permanganate, $HOMnO_3$, probably takes place by way of the ester of the hydrate and is analogous to the oxidation of alcohols to aldehydes and ketones (p. 133).

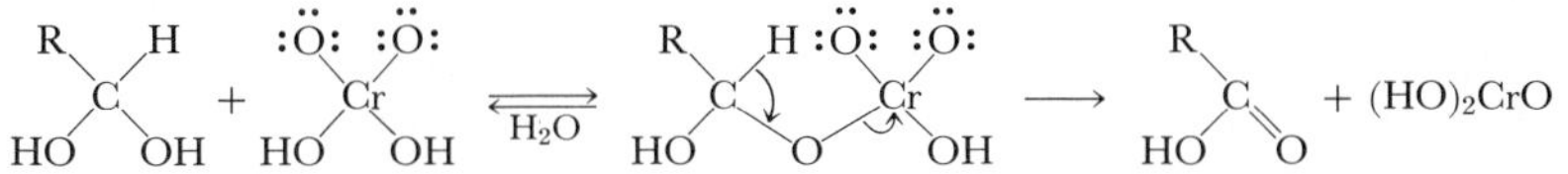

$$RCH(OH)_2 + (HO)_2Cr(=\ddot{O}{:})_2 \underset{H_2O}{\rightleftharpoons} RCH(OH){-}O{-}Cr(=\ddot{O}{:})_2{-}OH \longrightarrow RC({=}O)OH + (HO)_2CrO$$

The same intermediate chromate results from the addition of the hydrate to chromium trioxide, CrO_3. A product such as H_2CrO_3 disproportionates to Cr(V) and Cr(III). The Cr(V) can oxidize one mole of hydrate by the same mechanism as does Cr(VI).

$$2\,H_2CrO_3 \longrightarrow HCrO_3 + HCrO_2 + H_2O$$

$$RCH(OH)_2 + HCrO_3 \longrightarrow RCOOH + HCrO_2 + H_2O$$

Ketones on the other hand are fairly stable to oxidation. When oxidation is forced by using strong oxidizing agents under vigorous conditions, the carbon chain is broken and a mixture of acids having fewer carbon atoms is produced. Undoubtedly it is the enol form of the ketone that is oxidized.

$$RCH_2COCH_2R' \rightleftharpoons \begin{array}{c} RCH{=}C(OH)CH_2R' \\ \text{and} \\ RCH_2C(OH){=}CHR' \end{array} \xrightarrow[\text{oxidation}]{\text{Vigorous}} \begin{array}{c} RCOOH + HOOCCH_2R' \\ \text{and} \\ RCH_2COOH + HOOCR' \end{array}$$

Because of the difference between the ease of oxidation of aldehydes and ketones, it is possible to choose an oxidizing agent that attacks aldehydes but not ketones, and to use the reaction as a distinguishing test. One mild oxidizing agent used for this purpose is **Tollens reagent,** an ammoniacal solution of silver hydroxide. Reaction takes place with aldehydes at room temperature or on warming, and the reduction product, metallic silver, is readily visible.

$$RCHO + 2\,Ag(NH_3)_2OH \longrightarrow RCOO^{-+}NH_4 + 2\,Ag\downarrow + H_2O + 3\,NH_3$$

If the vessel in which the latter reaction takes place is clean and the rate of deposition slow enough, the silver deposits as a coherent silver mirror; otherwise it is a gray to black finely divided precipitate. Tollens reagent should be freshly prepared and discarded immediately after use. On standing, silver imide, Ag_2NH, is formed along with some silver amide, $AgNH_2$, and silver nitride, Ag_3N, all of which are violently explosive.

REDUCTION

1. ***Addition of Hydrogen.*** ***(a)*** BY CATALYTIC REDUCTION. Like the carbon-carbon double bond, the carbon-oxygen double bond adds hydrogen in the presence of hydrogenation catalysts such as finely divided platinum or palladium, or Raney nickel (p. 82). Aldehydes give primary alcohols, and ketones give secondary alcohols.

$$RCHO + H_2 \xrightarrow{\text{Pt, Pd, or Ni}} RCH_2OH$$

$$R_2CO + H_2 \longrightarrow R_2CHOH$$

(b) BY METAL HYDRIDE REDUCTION. The polarity of the double bond enables aldehydes and ketones to be reduced also by nucleophilic attack by hydride ion, using reagents such as lithium aluminum hydride in ether solution (p. 175), or sodium borohydride, $NaBH_4$, in water or ethanol.

$$4\,RCHO + LiAlH_4 \longrightarrow (RCH_2O^-)_4Li^+Al^{3+} \xrightarrow{H_2O} 4\,RCH_2OH + LiOH + Al(OH)_3$$

Reduction by sodium and alcohol or by sodium amalgam and water probably takes place at the surface of the metal (cf. p. 132), the hydride ion being formed from the proton donor (p. 175).

(c) PONNDORF REDUCTION. This method, also called the *Meerwein-Ponndorf-Verley reduction,* depends on the equilibrium that exists between carbonyl compounds and alcohols in the presence of aluminum alkoxides. Coordination of an unshared pair of electrons from the carbonyl group of the aldehyde or ketone with the metal facilitates the transfer of a hydride ion from the primary or secondary alkoxide group to the carbonyl group by means of a cyclic mechanism.

$$\begin{matrix} R'_2C{=}O \\ + \\ MOCHR_2 \end{matrix} \rightleftharpoons \begin{matrix} R' \quad\; R \\ R'{-}C\;\;H{-}C{-}R \\ :O\quad\;\; :O: \\ M \end{matrix} \rightleftharpoons \begin{matrix} R' \quad\; R \\ R'{-}C{-}H\;\;C{-}R \\ :O:\quad\; O: \\ M \end{matrix} \rightleftharpoons \begin{matrix} R'_2CHOM \\ + \\ OCR_2 \end{matrix}$$

An equivalent amount of metal alkoxide is not necessary because of the rapid equilibration of the alkoxides with the alcohols.

$$R'_2CHOM + R_2CHOH \rightleftharpoons R'_2CHOH + R_2CHOM$$

In addition to the cyclic transition state mechanism, a noncyclic mechanism may operate simultaneously in which two moles of metal alkoxide are involved in the transition state, one coordinating with the carbonyl oxygen and the other supplying the hydride ion. It is reported also that the actual reagent is the alkoxide trimer and that better yields are obtained when aluminum *i*-propoxide trimer is used as the reagent rather than as a catalyst.

If the alcohol used as the reducing agent is so chosen that the aldehyde or ketone formed boils at a lower temperature than any of the other reactants, it may be removed by slow distillation, and the reaction forced to completion. For example, when *i*-propyl alcohol and aluminum *i*-propoxide are used, the low-boiling acetone is removed. Ketones as well as aldehydes may be reduced. Not only are the yields excellent but other reducible groups such as carbon-carbon double bonds, halogen, carboxylic esters, or nitro groups are not affected. If the reaction is carried out in such a way that it is used to oxidize an alcohol by means of a ketone, usually cyclohexanone (p. 649), it is called an *Oppenauer oxidation.*

2. ***Reduction to Glycols and Pinacols.*** Adsorption of the carbonyl groups on the surface of amalgamated magnesium[5] followed by transfer of an electron to the carbon leads to bimolecular reduction to 1,2 glycols.

$$\text{Mg}\,\vdots\,\overset{\text{O}}{\underset{\text{CR}_2}{\|}}\;\; \text{Mg}\;\; \vdots\,\overset{\text{CR}_2}{\underset{\text{O}}{\|}}\;\text{Mg} \longrightarrow \text{Mg}^{2+}\; {}^{-}\!:\!\text{O}-\text{CR}_2-\text{CR}_2-\text{O}\!:^{-} \xrightarrow{2\,\text{H}^+} \text{Mg}^{2+} + \text{R}_2\underset{\text{HO}}{\text{C}}-\underset{\text{OH}}{\text{CR}_2}$$

Other reactive metals may be used such as amalgamated aluminum, or sodium in moist ether. Aldehydes yield a disecondary glycol.

$$2\,\text{RCHO} \longrightarrow \text{RCHOHCHOHR}$$

The product from acetone, $(CH_3)_2CHOHCHOH(CH_3)_2$, was called **pinacol** from the Greek word *pinax* meaning *plate* in reference to its crystalline structure. Symmetrically di- or tetrasubstituted 1,2 glycols in general are called *pinacols.*

The pinacols are freed from their salts with water or weak acids because strong acids cause dehydration with molecular rearrangement to a ketone.

$$\underset{\text{Pinacol}}{\text{CH}_3-\overset{\text{CH}_3}{\underset{\text{HO}}{\text{C}}}-\overset{\text{CH}_3}{\underset{\text{OH}}{\text{C}}}-\text{CH}_3} \underset{\text{B}^-}{\overset{\text{HB}}{\rightleftharpoons}} \text{CH}_3-\overset{\text{CH}_3}{\underset{\text{HO}:}{\text{C}}}-\overset{\text{(CH}_3)}{\underset{\overset{+}{\text{O}}\text{H}_2}{\text{C}}}-\text{CH}_3 \xrightarrow[\text{H}_2\text{O}]{} \text{CH}_3-\overset{\text{CH}_3}{\underset{\text{CH}_3}{\text{C}}}-\underset{{}^{+}\text{OH}}{\overset{}{\text{C}}}-\text{CH}_3 \underset{\text{HB}}{\overset{\text{B}^-}{\rightleftharpoons}} \underset{\text{Pinacolone}}{\text{CH}_3-\overset{\text{CH}_3}{\underset{\text{CH}_3}{\text{C}}}-\underset{\text{O}}{\overset{}{\text{C}}}-\text{CH}_3}$$

The product from pinacol is called **pinacolone** and the general reaction is known as the

[5] A pair of metals in contact, one of which is above hydrogen in the electromotive series and the other one below hydrogen in the series, is known as a *couple.* Common examples are amalgamated aluminum, amalgamated zinc, aluminum-copper couple, and zinc-copper couple. Usually the couples are more reactive as reducing agents than are the pure metals. The reducing action of metals depends on the tendency of the metal atoms to form positive ions, leaving electrons available at the metal surface to be supplied to other ions or molecules and bring about their reduction. The positive ions formed from the pure metals tend to be held by the negative charge remaining, and the molecules or ions being reduced are prevented from reaching the electrons left on the surface of the metal. If metals like aluminum and zinc are involved and the reaction is with water, the insoluble hydroxides produced form an adherent coating on the surface of the metal, which stops the reaction. With the couples, the electrons freed by solution of the active metal are transferred to the less active metal, and reduction takes place at the surface of the inactive metal, which is distant from the site of solution of the active metal.

Since sodium is one of the most active metals and sodium hydroxide is very soluble in water, reaction of sodium with water is violent. Here amalgamation decreases the reactivity by forming the much less reactive compound Na_4Hg. As a result, reductions with sodium amalgam may be carried out in aqueous and even slightly acidic solutions.

pinacol-pinacolone rearrangement. In these rearrangements, which are intramolecular, hydride ion migrates in preference to alkide ion.

$$RCHOHCH_2OH \xrightarrow{H^+} RCH_2CHO + H_2O$$

$$RCHOHCHOHR \xrightarrow{H^+} RCH_2COR + H_2O$$

3. ***Reduction of the Carbonyl Group to a Methylene Group.*** A carbonyl group may be converted to a methylene group by reducing it first to the alcohol and then converting the alcohol to the halide and reducing the halide to the hydrocarbon (p. 144); or the alcohol may be converted to the olefin and the double bond reduced catalytically (p. 145). More direct methods, however, are available.

(*a*) CLEMMENSEN REDUCTION. Reaction of an aldehyde or ketone with amalgamated zinc in the presence of concentrated hydrochloric acid replaces the oxygen by two hydrogen atoms. The best yield is obtained if the concentration of the ketone is kept low by dissolving it in a water-immiscible solvent. It would appear that the high acid concentration leads to adsorption of the oxonium salt on the metal surface. Satisfaction of the electron deficiency on the carbon atom by adsorption permits the oxygen atom to react with another proton, which permits elimination of the oxygen as a water molecule, solution of a zinc ion, and formation of an adsorbed methylene radical.

$$\begin{array}{llllllll}
 & :\overset{+}{O}:H & & :\ddot{O}:H & & H\,\overset{+}{:}\ddot{O}:H & & Zn^{2+} \quad + H_2O \\
Zn| & :+ \overset{\|}{C}R_2 & & \overset{+}{Zn}|:\overset{|}{C}R_2 & & \overset{+}{Zn}|:\overset{|}{C}R_2 & & Zn|: \\
Zn| & : & \longrightarrow & Zn|: & \xrightarrow{H^+} & Zn|: & \longrightarrow & Zn|:\ddot{C}R_2 \\
Zn| & : & & Zn|: & & Zn|: & & Zn|: \\
Zn| & : & & Zn|: & & Zn|: & &
\end{array}$$

The adsorbed methylene radical may react successively with two adsorbed hydrogen atoms, with the formation of the hydrocarbon and positive metallic ion. By keeping the concentration of the adsorbed ketone low, the formation of bimolecular reduction products is decreased.

$$\begin{array}{lllll}
Zn|: & & \overset{+}{Zn}|:H & & Zn^{2+} \quad H \\
Zn|:\ddot{C}R_2 & \xrightarrow{2\,H^+} & Zn|:\ddot{C}R_2 & \longrightarrow & Zn|: + \ddot{C}R_2 \\
Zn|: & & \overset{+}{Zn}|:H & & Zn|: \quad H
\end{array}$$

The over-all reaction is

$$RCOR + 2\,Zn + 4\,HCl \longrightarrow RCH_2R + H_2O + ZnCl_2$$

The Clemmensen reduction is limited to compounds that lack other reducible groups or groups affected by hot concentrated hydrochloric acid.

(*b*) WOLFF-KISHNER REDUCTION. Hydrazones or semicarbazones of aldehydes or ketones decompose at elevated temperatures, especially in the presence of bases, to yield nitrogen and the hydrocarbon.

$$R_2C{=}O \underset{H_2O}{\overset{H_2NNH_2}{\rightleftarrows}} R_2C{=}N{-}NH_2 \underset{HB}{\overset{B^-}{\rightleftarrows}} \{R_2C{=}N{-}\bar{N}H \longleftrightarrow R_2\bar{C}{-}N{=}NH\} \underset{B^-}{\overset{HB}{\rightleftarrows}}$$

$$R_2CH{-}N{=}NH \underset{HB}{\overset{B^-}{\rightleftarrows}} R_2CH{-}N{=}\bar{N} \longrightarrow N_2 + R_2\bar{C}H \underset{B^-}{\overset{HB}{\rightleftarrows}} R_2CH_2$$

The reaction usually is carried out by the *Huang-Minlon modification* of the earlier procedures.

A solution of the aldehyde or ketone, aqueous hydrazine, and sodium or potassium hydroxide in a high-boiling solvent (diethylene glycol, b.p. 245°, p. 605) is refluxed to form the hydrazone, and then the water is distilled until the decomposition temperature of around 200° is reached. The reaction can be carried out at room temperature if the preformed hydrazone is added to a solution of potassium *t*-butoxide in methyl sulfoxide (p. 270) because of the increased activity of the alkoxide ion in this solvent (p. 102). The Wolff-Kishner reduction is suitable only for compounds that do not contain other groups that react with alkali or hydrazine.

4. ***Intermolecular Oxidation and Reduction.*** Since an aldehyde can be either oxidized or reduced, it is possible for both oxidation and reduction to take place between two molecules, one being oxidized and the other reduced.

(*a*) CANNIZZARO REACTION. The reaction of two molecules of an aldehyde with one equivalent of a strong base can give one mole of the alcohol and one mole of the salt of the carboxylic acid.

$$RCH{=}O \overset{^{-}OH}{\rightleftharpoons} RCH(OH){-}O^{-}$$

$$RC(O^{-})(OH)H + O{=}CHR \longrightarrow RC(=O)OH + H{-}CHR(O^{-}) \longrightarrow RCOO^{-} + HOCH_2R$$

The negative charge on the first intermediate facilitates the transfer of a hydride ion to the second molecule of aldehyde. It is known that a direct transfer from carbon to carbon must take place because the alcohol does not contain carbon-bound deuterium when the reaction is carried out in heavy water. The over-all reaction is

$$2\,RCHO + NaOH \longrightarrow RCOO^{-+}Na + RCH_2OH$$

Since bases can bring about aldol condensation, the Cannizzaro reaction usually is limited to aldehydes such as formaldehyde or trimethylacetaldehyde (*pivalic aldehyde*) that lack an α hydrogen. Under certain conditions, however, even aldehydes having α hydrogen undergo a Cannizzaro reaction. For example, if *i*-butyraldehyde is heated with aqueous barium hydroxide in a sealed tube at 150°, a quantitative yield of *i*-butyl alcohol and barium *i*-butyrate is produced. Similarly *i*-valeraldehyde and *n*-heptaldehyde, when heated for a few hours with calcium oxide at 100°, give the corresponding alcohols and calcium salts as the chief products. With *i*-butyraldehyde an initial rapid reversible aldol addition takes place, which is followed by the slower irreversible Cannizzaro reaction.

(*b*) TISHCHENKO REACTION. Aldehydes are converted to alcohols under the conditions of the Ponndorf reduction (p. 206). In the absence of an alcohol, however, and in the presence of aluminum alkoxide activated by a small amount of anhydrous aluminum chloride or zinc chloride, aldehydes undergo an intermolecular oxidation and reduction to yield an ester. Nucleophilic attack by an alkoxide ion yields the aluminum salt of a hemiacetal, which can transfer a hydride ion to another molecule as in the Cannizzaro reaction, thus regenerating the alkoxide ion. A cyclic mechanism for the last step analogous to that for the Ponndorf reduction usually is postulated although other transition states are possible.

$$RCH{=}O + RCH_2\ddot{O}^{-+}Al(OCH_2R)_2 \longrightarrow [R(RCH_2O)C(H)O{:}Al(OCH_2R)_2 \cdots O{=}CHR] \longrightarrow RC(=O)OCH_2R + Al(OCH_2R)_3$$

Aluminum chloride catalyzes the reaction because it is a better Lewis acid than the aluminum alkoxide and can make the carbonyl group a better electrophile for reception of the alkoxide ion. The reaction is used commercially to prepare ethyl acetate from acetaldehyde.

MISCELLANEOUS TESTS AND REACTIONS

1. ***Schiff Fuchsin Aldehyde Reagent.*** Fuchsin is a magenta dye (p. 568) that can be decolorized in aqueous solution by sulfur dioxide. In the presence of aldehydes, but not ketones, a magenta color reappears. The reaction is not specific for aldehydes, since anything that removes sulfur dioxide, for example mild alkalies, amines, or even heating or exposure to air, regenerates the color, but in the absence of such interferences it serves to distinguish aldehydes from ketones.

The reaction with aldehydes is not merely a combination with sulfur dioxide and regeneration of the original fuchsin. The color is due to various reaction products of the aldehyde and the dye (p. 568). Thus the hue of the color produced differs with different aldehydes; for example, the color produced by formaldehyde is bluer than that produced by acetaldehyde. Moreover strong mineral acids destroy the color produced by acetaldehyde but not that produced by formaldehyde.

2. ***Replacement of Oxygen by Halogen.*** When aldehydes or ketones react with phosphorus pentachloride or phosphorus pentabromide, the oxygen of the carbonyl group is replaced by two halogen atoms. Haloalkenes often accompany the dihalo compounds. Both products may be explained by assuming decomposition of an initially formed addition compound to a halocarbonium ion followed by addition of halide ion or loss of a proton.

$$R{-}\underset{\underset{O}{\|}}{C}{-}CH_2R + PX_5 \longrightarrow R{-}C(X)(OPX_4){-}CH_2R \longrightarrow R{-}\overset{+}{C}(X){-}CH_2R + POX_3 + X^-$$

$$R{-}\overset{+}{C}(X){-}CH_2R \xrightarrow{X^-} R{-}CX_2{-}CH_2R$$

$$R{-}\overset{+}{C}(X){-}CH_2R \xrightarrow{X^-} R{-}\underset{\underset{X}{|}}{C}{=}CHR + HX$$

3. ***Salt Formation.*** The hydrogen atoms on a carbon atom α to a carbonyl group are sufficiently acidic (p. 196) to react with alkali metals and form salts. Thus acetone reacts readily with metallic sodium and hydrogen is evolved.

$$CH_3COCH_3 + Na \longrightarrow CH_3COCH_2^{-+}Na + \tfrac{1}{2}\,H_2$$

Because the enolate carbanion is a nucleophile, aldol addition and condensation products of acetone also are formed (pp. 202, 203). Equilibrium can be established between the enolate ion and other bases, their effectiveness increasing with their basicity. Thus the order of activity in removing a proton is sodium ethoxide $<$ *i*-propoxide $<$ *t*-butoxide $<$ amide $<$ hydride.

4. ***Halogenation and the Haloform Reaction.*** The halogenation of aldehydes and ketones is catalyzed by acids and by bases. The rate of reaction is dependent on the concentration of the ketone and on the concentration of the acid or base, but is independent of the concentration or the kind of halogen. Moreover the rate of halogenation is the same as the rate of deuterium exchange under the same conditions. Hence enolization or the formation of enolate ion is the rate-controlling step, the halogenation step being almost instantaneous.

In the acid-catalyzed reaction the carbonyl group is the initial point of attack and the essential steps may be represented by the following equilibria.

$$R\underset{\underset{O}{\|}}{C}CH_2R \underset{B^-}{\overset{HB}{\rightleftharpoons}} R\underset{\underset{:\overset{+}{O}:H}{\|}}{C}{-}\overset{\overset{H}{|}}{C}HR \underset{HB}{\overset{B^-}{\rightleftharpoons}} R\underset{\underset{OH}{|}}{C}{=}CHR \underset{HX}{\overset{X_2}{\rightleftharpoons}} R\underset{\underset{O}{\|}}{C}{-}\overset{\overset{X}{|}}{C}HR$$

The effect of adding the proton to the carbonyl group is to increase its attraction for electrons and hence make it easier for the α hydrogen to leave as a proton. Since halogen also is electron-attracting, a second hydrogen atom is replaced still more readily than the first hydrogen atom, and symmetrical polysubstitution results.

$$RCOCHXR + X_2 \longrightarrow RCOCX_2R + HX$$

In basic catalysis the initial step is removal of a proton from an α carbon atom.

$$\underset{\substack{\|\\O}}{RCCH_2R} \underset{HB}{\overset{B^-}{\rightleftharpoons}} \underset{\substack{\|\\O}}{RC\ddot{C}HR}^{-} \underset{X^-}{\overset{X_2}{\rightleftharpoons}} \underset{\substack{\|\\O}}{RC\overset{\overset{X}{|}}{C}HR}$$

Here again after the first hydrogen atom is substituted by halogen, the electron-attracting effect of the halogen makes the second hydrogen atom more readily removed than the first, and unsymmetrical substitution results.

If the carbonyl group is attached to a methyl group as in acetaldehyde or methyl ketones, a trihalogenated product is formed. In the presence of base, scission of the molecule analogous to the base-catalyzed hydrolysis of esters (p. 173) takes place to give a **haloform.** The trihalomethide ion is displaced readily because it is stabilized by the accumulation of electronegative halogen atoms.

$$X_3C{-}\underset{\substack{\|\\O}}{C}{-}R \overset{^-OH}{\rightleftharpoons} X_3C{-}\underset{\substack{|\\:O^-}}{\overset{\substack{OH\\|}}{C}}{-}R \rightleftharpoons X_3C{:}^- + \underset{\substack{\|\\O}}{\overset{\substack{OH\\|}}{C}}{-}R$$

$$X_3C{:}^- \underset{^-OH}{\overset{H_2O}{\rightleftharpoons}} X_3CH \qquad RCOOH \underset{H_2O}{\overset{^-OH}{\rightleftharpoons}} RCOO^-$$

The various steps in the reaction take place merely on mixing acetaldehyde or a methyl ketone with an alkaline solution of hypohalite. The over-all reaction is

$$RCOCH_3 + 3\,X_2 + 4\,NaOH \longrightarrow RCOO^{-+}Na + HCX_3 + 3\,NaX + 3\,H_2O$$

Trihalogenated methanes yield formic acid on hydrolysis, which led Dumas to give them the general name *haloforms* (chloroform, bromoform, and iodoform). To obtain a haloform from a carbonyl compound, at least one group united to the carbonyl group must be methyl. The other group may be hydrogen or any group linked through carbon, provided that any hydrogen atoms remaining on this carbon atom are not substituted more readily than those of the methyl group, and provided that other substituents in the group do not reduce the reactivity of the methyl group by steric hindrance. Acetic acid does not give a haloform because the carboxylate anion is formed, in which the effect of the carbonyl group on the α hydrogens is reduced by resonance (p. 159). Acetoacetic acid, CH_3CO-CH_2COOH, substitutes on the methylene rather than the methyl group to give CH_3-$COCX_2COOH$, which reacts with alkali to give the sodium salts of acetic acid and a dihaloacetic acid (cf. p. 662).

$$CH_3COCX_2COOH + 2\,NaOH \longrightarrow CH_3COONa + HCX_2COONa + H_2O$$

Pinacolone (p. 207) gives bromoform but does not give iodoform because the highly branched *t*-butyl group prevents the replacement of more than two hydrogens by the larger iodine.

$$(CH_3)_3CCOCH_3 + 2\,NaOI \longrightarrow (CH_3)_3CCOCHI_2 + 2\,NaOH$$

Not only do carbonyl compounds meeting the stated conditions undergo the haloform reaction, but all substances that oxidize to such compounds under the conditions of the reaction, for example properly constituted alcohols, also yield haloform.

$$RCHOHCH_3 + NaOX \longrightarrow RCOCH_3 + NaX + H_2O$$
$$RCOCH_3 + 3\,NaOX \longrightarrow HCX_3 + RCOONa + 2\,NaOH$$

The practical importance of the haloform reaction is its use to distinguish between different possible structures. For example, acetaldehyde is the only aldehyde, and ethyl alcohol is the only primary alcohol that gives a haloform. The two largest groups of compounds responding to the reaction are the methyl ketones and alkylmethylcarbinols. For conducting such a test, the reaction with alkali and iodine is used since iodoform is a yellow crystalline solid that is identified readily by its melting point.

Occasionally compounds not expected to react give iodoform. The reaction of acetoxime may be explained by the fact that it is a nitrogen analog of a ketone or that hydrolysis to acetone precedes formation of the haloform. The reaction of pulegone can be explained by its decomposition by alkali to give acetone. 2-Methyl-2-butene probably is converted first to 2-methyl-2,3-butanediol, which rearranges to methyl isopropyl ketone during the reaction (cf. p. 604). Compounds that have the structure $RCOCH_2COR$ or $RCHOHCH_2CHOHR$ give iodoform because 1,3 dicarbonyl compounds are cleaved by alkali.

$$RCOCH_2COR + NaOH \longrightarrow RCOCH_3 + NaOCOR$$

Isopropylamine, $(CH_3)_2CHNH_2$, and compounds of the type $CH_3CH(NHCOR)COR$ also give iodoform, presumably by oxidation to the imino compound, $>C{=}N{-}$, and hydrolysis (cf. oximes p. 204) to a methyl ketone.

5. ***Nitrosation.*** Nitrous acid reacts with ketones having an α hydrogen to give nitroso derivatives. Usually ethyl nitrite is passed into a solution of the ketone in hydrochloric acid. The reaction probably takes place by way of a nucleophilic attack of the enol on the protonated ester. If a methylene group is adjacent to the carbonyl group, the nitroso compound tautomerizes to the more stable oxime, which is known also as an *isonitroso* derivative.

$$RC(=O){-}CH_2R \underset{H^+}{\overset{H^+}{\rightleftharpoons}} RC(OH){=}CHR + H{-}\overset{+}{N}(=O){-}OC_2H_5 \xrightarrow{H^+} C_2H_5OH + RC(=O){-}CH(N{=}O)R \longrightarrow RC(=O){-}C(=NOH)R$$

Industrially Important Aldehydes and Ketones

Formaldehyde was prepared first by Butlerov[6] in 1859, more than fourteen years after the isolation of acetaldehyde by Liebig. In 1868 Hofmann (p. 222) obtained formaldehyde by oxidizing methanol with air in the presence of a platinum catalyst. Two methods of air oxidation are used commercially. In the older procedure a rich mixture of methanol and air (1 v.: 1 v.) is passed over a silver catalyst at 635°. Practically all of the oxygen is used, and the off-gases contain 18 to 20 per cent of hydrogen, indicating that the process is a combined dehydrogenation and oxidation. In the more recent process a lean mixture of air and 5 to 10 per cent methanol by volume is passed over an iron oxide-molybdenum oxide catalyst. Formaldehyde almost free of methanol is obtained and the off-gases contain oxygen and no hydrogen. The formaldehyde, or formaldehyde and excess methanol, are absorbed in water and the solution is concentrated by distillation to 37 per cent formaldehyde. The formaldehyde is present in solution as methylene glycol, $CH_2(OH)_2$, and its poly-

[6] Alexander Mikhailovich Butlerov (1828–1886), eminent Russian chemist and professor at the Universities of Kazan and St. Petersburg. He was the first to prepare aqueous formaldehyde, paraformaldehyde, hexamethylenetetramine, and the carbohydrate mixture formed by the action of dilute calcium hydroxide on formaldehyde. He was a strong advocate of the structural concept of organic chemistry and was the first to use the expression "the chemical structure of organic compounds." Butlerov prepared the first tertiary alcohol by the action of methylzinc on acetyl chloride. His studies of the reactions of *t*-butyl alcohol led him to the preparation of the isomeric butanes and butenes and to the discovery of the polymerization of isobutylene and the acid-catalyzed equilibrium between the two diisobutylenes (p. 87).

meric formaldehyde addition products, $HO(CH_2O)_xH$, where x has an average value of 3. This solution is called **formalin.** Ordinary formalin contains 7 to 10 per cent methanol in summer and 10 to 15 per cent in winter to prevent separation of higher polymers. Formalin free of methanol also is transported for industrial use but must be kept warm (*ca.* 30°) to prevent precipitation of polymer.

Formaldehyde is formed also during the controlled air-oxidation of natural gas or propane-butane mixtures in the vapor phase at 350° to 450°. Methane yields chiefly methanol and formaldehyde. Propane-butane mixtures give chiefly formaldehyde, acetaldehyde, and methanol, together with acetone, propyl and butyl alcohols, and organic acids.

Although formaldehyde can be liquefied readily (b.p. −21°), it cannot be handled safely in this form because it polymerizes explosively even at temperatures just above its freezing point (−118°), and no effective stabilizers are known. Besides the aqueous solution, formalin, formaldehyde is transported and stored as the solid polymer known as **paraformaldehyde,** which is prepared by concentrating formalin at reduced pressure. It is a linear polymer of the formula $HO(CH_2O)_xH$, where x has an average value of around 30. When x is less than 12, the product is soluble in water, acetone, or ether, but the higher polymers are insoluble. Slow solution of the higher polymers in water is accompanied by hydrolysis to fragments of lower molecular weight. Formaldehyde also gives solid polymers when the gas comes in contact with solid surfaces at temperatures below 137°. Gaseous formaldehyde containing about two per cent water is prepared most conveniently by heating paraformaldehyde above this temperature.

When a 60–65 per cent aqueous solution of formaldehyde is distilled with 2 per cent of sulfuric acid, the cyclic *trimer* (p. 201), **1,3,5-trioxane,** may be extracted from the distillate. It is a colorless highly refractive crystalline compound melting at 62° and boiling without decomposition or depolymerization at 115°. It has a pleasant odor resembling that of chloroform, in contrast to the sharp odor of formaldehyde, and is soluble in water and organic solvents. Strong acids initiate depolymerization, as with all compounds of this type, and it is useful as a source of formaldehyde in reactions carried out under anhydrous conditions.

Polymerization of extremely dry gaseous formaldehyde (0.1–0.01 per cent water) in a hydrocarbon solvent at −30° yields a **linear polyoxymethylene** of the formula $HO(CH_2O)_xH$, where $x = 600$–6000, depending on how much water was initially present. These products, because of the end hydroxyl groups and easy loss of a proton, are subject to depolymerization.

$$HO(-CH_2-O)_{x-2}-CH_2-O-CH_2-O-H \longrightarrow x\,CH_2O + H_2O$$

They can be stabilized, however, by replacing the end hydrogens by other groups, for example acetyl groups by acetylation.

$$HO(CH_2O)_xH + 2\,(CH_3CO)_2O \longrightarrow CH_3COO(CH_2O)_xCOCH_3 + 2\,CH_3COOH$$

The resulting product is a very strong, tough, high-melting thermoplastic known as *Delrin.* Dry formaldehyde may be obtained either by drying the gas produced by the depolymerization of paraformaldehyde or by depolymerizing pure 1,3,5-trioxane. *Celcon* is a copolymer of 1,3,5-trioxane with other monomers such as ethylene oxide (p. 605). These polymers belong to the class known as **polyoxymethylene plastics** or **acetal resins.**

Formaldehyde undergoes a number of reactions that do not take place with most other aldehydes because (*1*) it has only hydrogen atoms attached to the carbonyl group and is more reactive than other aldehydes, just as the aldehydes are more reactive than ketones, and (*2*) having only one carbon atom it does not have α hydrogen atoms and cannot undergo aldol-type additions with itself. For example, when it reacts with ammonia, the initially formed cyclic aldimine trimer reacts further and a compound is obtained that has the formula $(CH_2)_6N_4$ and is called **hexamethylenetetramine.**

$$3\,HCHO + 3\,NH_3 \underset{3\,H_2O}{\rightleftharpoons} (CH_2NH)_3 \xrightarrow[3\,H_2O]{3\,CH_2O + NH_3} (CH_2)_6N_4$$

This organic compound was the first whose structure, based on valence theory and chemical reactions, was confirmed by X-ray diffraction. When hexamethylenetetramine is taken internally, it is excreted in the urine and hydrolyzed to formaldehyde when the urine is acid. At one time it was used medicinally as a urinary antiseptic under the name *methenamine* or *urotropin,* but has been replaced largely by the sulfa drugs (p. 427) and mandelic acid (p. 477).

Hexamethylenetetramine assumed importance during World War II as an intermediate for the manufacture of the high explosive **trimethylenetrinitramine** (*cyclonite, hexogen, RDX*). Nitration is carried out in the presence of ammonium nitrate or ammonium sulfate, which permits utilization of 80–90 per cent of the methylene groups.

$$(CH_2)_6N_4 + 2\,NH_4NO_3 + 4\,HNO_3 \longrightarrow 2\,(CH_2NNO_2)_3 + 6\,H_2O$$

Trimethylenetrinitramine is one of the most powerful organic explosives and has a greater brisance than TNT (p. 417). Up to 30 per cent is mixed with TNT to insure complete detonation of large bombs and to increase the force of the explosion. The *plastic bombs* used by terrorists are a mixture of RDX and TNT.

Another product of the reaction of ammonia and formaldehyde, known as **pentamethylenetetramine,** is present in a freshly prepared solution of formalin and ammonia. Reaction of this solution with the proper amount of nitrous acid yields **dinitrosopentamethylenetetramine,** which is used as a foaming agent for rubber (p. 585).

$$5\,HCHO + 4\,NH_3 \longrightarrow 5\,H_2O + \underset{\text{Pentamethylenetetramine}}{HN(CH_2{-}N{-}CH_2)_2(CH_2)NH} \xrightarrow{HNO_2} \underset{\text{Dinitrosopentamethylenetetramine}}{ON{-}N(CH_2{-}N{-}CH_2)_2(CH_2)N{-}NO} + 2\,H_2O$$

Lacking α hydrogen atoms, formaldehyde does not give aldehyde resins with concentrated alkali, but undergoes the *Cannizzaro reaction* (p. 209). In aqueous solution in the presence of calcium hydroxide, a complex series of self-additions produces a mixture of sugars known as *formose.*

$$6\,HCHO \xrightarrow{\text{Dil. }Ca(OH_2)} C_6H_{12}O_6$$

A *specific test for formaldehyde* is the production of a violet color with casein of milk in the presence of ferric chloride and concentrated sulfuric acid. The test is sensitive to one part of formaldehyde in 200,000 parts of milk. This test was important in the enforcement of the law against the addition of formaldehyde as a preservative. It can be used as a general test for formaldehyde simply by adding to milk the material to be tested and then applying the test. Formaldehyde can be estimated quantitatively by measuring the amount of hydrogen evolved on reaction with alkaline hydrogen peroxide solution.

$$2\,CH_2O + H_2O_2 + 2\,NaOH \longrightarrow 2\,HCOO^{-+}Na + 2\,H_2O + H_2$$

Reaction of formaldehyde with zinc dithionite (*zinc hydrosulfite*) gives zinc hydroxyethylsulfinate (*zinc formaldehyde sulfoxylate*), which is converted to the sodium salt with sodium carbonate.

$$2\,ZnS_2O_4 + 2\,HCHO + 2\,H_2O \longrightarrow Zn(HSO_3)_2 + Zn(O_2SCH_2OH)_2 \xrightarrow{Na_2CO_3} ZnCO_3 + 2\,Na^{+-}O{-}\overset{\overset{O}{\uparrow}}{S}CH_2OH$$

The product is known as **sodium formaldehyde sulfoxylate** (*SFS, Rongalite*). It is a strong reducing agent used to destroy azo dyes (p. 416) in the discharge printing of cloth.

Formalin has some use as a disinfectant and for the preservation of biological specimens, but its chief use is for the manufacture of synthetic resins by condensation with urea (p. 284), melamine (p. 289), or phenol (p. 443). Frequently paraformaldehyde or hexamethylenetetramine may replace formalin for this purpose. Formaldehyde is used also to make *pentaerythritol* (p. 609), which is important for the manufacture of drying oils (p. 188) and the high explosive *pentaerythritol nitrate* (PETN) (p. 609). In the United States the production of formalin (37 per cent) in 1940 was 181 million pounds and increased to 2.8 billion pounds by 1964. The unit value is around 3 cents per pound.

Acetaldehyde is made commercially by the hydration of acetylene (p. 149) and by the oxidation of ethyl alcohol with air over a silver catalyst (p. 132). Since 1945 large amounts have been produced by the controlled air-oxidation of the propane-butane mixture from natural gas (p. 213). A new process permits its production by the direct oxidation of ethylene with air and a cupric chloride catalyst activated with palladium chloride. In the first stage the ethylene is simultaneously hydrated and oxidized at 100° and the acetaldehyde removed. The reactions believed to be taking place in the first stage are

$$C_2H_4 + PdCl_2 + H_2O \longrightarrow CH_3CHO + Pd + 2\,HCl$$

$$Pd + 2\,CuCl_2 \longrightarrow PdCl_2 + Cu_2Cl_2$$

In the second stage the cupric chloride is regenerated by air blown through the catalyst.

$$Cu_2Cl_2 + \tfrac{1}{2}\,O_2 + 2\,HCl \longrightarrow 2\,CuCl_2 + H_2O$$

The same process can be used to make acetone directly from propylene, and methyl ethyl ketone from the butenes. Some acetaldehyde is obtained as a by-product of the fermentation industries (p. 121). It boils at 20°, is miscible with water and organic solvents, and behaves typically in all of its reactions. The chief importance of acetaldehyde is its use as an intermediate for the synthesis of other organic compounds, especially acetic acid (p. 168) and acetic anhydride (p. 171) by air oxidation, ethyl acetate by the Tishchenko reaction (p. 209), and *n*-butyl alcohol by condensation to crotonaldehyde followed by catalytic reduction.

$$2\,CH_3CHO \xrightarrow[NaOH]{Dil.} CH_3CHOHCH_2CHO \xrightarrow{H^+} CH_3CH{=}CHCHO \xrightarrow[200°]{H_2/Cu} CH_3CH_2CH_2CH_2OH$$

It is converted into acetaldehyde cyanohydrin, which is used to make acrylonitrile (p. 619), lactic acid (p. 616), and acrylic esters (p. 619). Acetaldehyde is used also in the manufacture of rubber accelerators (p. 585), of the trimer, paraldehyde, and of the tetramer, metaldehyde. Production of acetaldehyde in 1964 was over 1 billion pounds valued at 7 cents per pound. **Paraldehyde** is a stable liquid, b.p. 125°, which is depolymerized readily by heating with acids and hence is a convenient source of acetaldehyde. Paraldehyde first was used medicinally as a sleep-producer (*hypnotic* or *soporific*) in 1882. It still is considered to be very efficient and one of the least toxic hypnotics. The chief objections to its use are that it is a liquid with a burning disagreeable taste, and that, because it is eliminated largely through the lungs, a patient's breath may smell of paraldehyde for as long as twenty-four hours after administration. **Metaldehyde,** a cyclic tetramer, is formed when acetaldehyde is treated with traces of acids or sulfur dioxide below 0°. One of the catalysts used commercially is a mixture of calcium nitrate and hydrogen bromide at −20°.

$$4\,CH_3CHO \xrightarrow[at\ -20°]{Ca(NO_3)_2,\ HBr} \begin{array}{ccc} & CH_3CH{-}O & \\ O & & CHCH_3 \\ | & & | \\ CH_3CH & & O \\ & O{-}CHCH_3 & \end{array}$$

Since it is a solid and has a fairly high vapor pressure despite its high melting point (m.p. 246°, sublimes below 150°), it is used as a solid fuel for heating liquids or foods when other fuels are not available or not convenient. Large quantities are used also in garden baits, the vapors being attractive and yet highly toxic to slugs and snails.

When chlorine is passed into ethanol, a series of reactions takes place. Oxidation of ethanol to acetaldehyde, chlorination to trichloroacetaldehyde, and acetal formation gives **trichloroacetal.**

$$CH_3CH_2OH + Cl_2 \longrightarrow 2\,HCl + CH_3CHO \xrightarrow{3\,Cl_2} 3\,HCl + Cl_3CCHO \xrightarrow{2\,C_2H_5OH,\ H^+} H_2O + Cl_3CCH(OC_2H_5)_2$$

Commercially the process is made continuous by running alcohol into the top on absorption column, chlorine into the bottom, and withdrawing hydrogen chloride from the top and trichloroacetal from the bottom. Treatment of the acetal with concentrated sulfuric acid gives **chloral** (*trichloroacetaldehyde*), used chiefly for the manufacture of DDT (p. 482).

$$Cl_3CCH(OC_2H_5)_2 + 2\,H_2SO_4 \longrightarrow Cl_3CCHO + 2\,C_2H_5OSO_3H + H_2O$$

Chloral is a liquid, b.p. 98°. Unlike most aldehydes, it reacts with water to give a stable crystalline hydrate, m.p. 52°. Whereas chloral gives a color with Schiff reagent, chloral hydrate does not. Hence water has added to the carbonyl group to give a 1,1-dihydroxy compound that is stabilized by the strongly electronegative trichloromethyl group.

$$Cl_3CCHO + H_2O \longrightarrow Cl_3CCH(OH)_2$$

Chloral hydrate is a quickly-acting soporific commonly known as *knock-out drops*. First prepared by Liebig in 1832 and introduced into medicine in 1869, it still is regarded highly by the medical profession and widely used as a sleep-producer.

***n*-Butyraldehyde** and ***i*-butyraldehyde** can be made by the dehydrogenation of the alcohols. A more important method, however, is the **hydroformylation** of alkenes (*oxo process*) based on an American observation in 1929 and developed in Germany during World War II. It consists of bringing the olefin and a slurry of cobalt salt, or a solution of preformed dicobalt octacarbonyl, $Co_2(CO)_8$, and cobalt tetracarbonyl hydride, $HCo(CO)_4$, in contact with carbon monoxide and hydrogen at 120° and 200 atmospheres. Propylene gives *n*-butyraldehyde and *i*-butyraldehyde in the ratio of 3:2. Patents claim that this ratio can be increased to 5:1.

$$CH_3CH{=}CH_2 + CO + H_2 \longrightarrow CH_3CH_2CH_2CHO \text{ and } \begin{array}{c} CH_3CHCH_3 \\ | \\ CHO \end{array}$$

In the commercial operation of this reaction, a mixture of octyl alcohols containing chiefly **2-ethylhexanol** also is produced. This alcohol probably arises by aldol addition of two moles of *n*-butyraldehyde, dehydration, and hydrogenation.

$$2\,CH_3CH_2CH_2CHO \longrightarrow CH_3CH_2CH_2CHOH\underset{\substack{|\\ C_2H_5}}{C}HCHO \longrightarrow CH_3CH_2CH_2CH{=}\underset{\substack{|\\ C_2H_5}}{C}{-}CHO \longrightarrow CH_3CH_2CH_2CH_2\underset{\substack{|\\ C_2H_5}}{C}HCH_2OH$$

Unsymmetrical dialkylethylenes, when subjected to the oxo reaction, give only one product. For example, isobutylene gives only isovaleraldehyde.

$$(CH_3)_2C{=}CH_2 + CO + H_2 \longrightarrow (CH_3)_2CHCH_2CHO$$

Moreover if the double bond is not terminal, isomerization to the terminal position takes place before reaction. Thus the mixture of diisobutylenes (p. 87) gives only **3,5,5-trimethylhexanal** (so-called "*nonyl aldehyde*").

$$\left.\begin{matrix}(CH_3)_3CCH_2C(CH_3){=}CH_2\\ \text{and}\\ (CH_3)_3CCH{=}C(CH_3)_2\end{matrix}\right. + CO + H_2 \longrightarrow (CH_3)_3CCH_2CH(CH_3)CH_2CHO$$

The aldehydes produced by the oxo synthesis can be reduced to alcohols (p. 206) or oxidized to acids (p. 205). Thus a mixture of primary octyl alcohols, known technically as "*isooctyl alcohol,*" is produced by the catalytic reduction of the aldehydes made from mixed heptenes. *n*-Butyraldehyde is used to manufacture rubber accelerators (p. 585) and poly(vinyl butyral) (p. 600). ***n*-Heptaldehyde** is one of the products of the destructive distillation of castor oil (p. 185).

Acetone is by far the most important ketone. It formerly was obtained by the destructive distillation of calcium acetate prepared from pyroligneous liquor (p. 168) but now is produced chiefly by the dehydrogenation of *i*-propyl alcohol (p. 132). Increasing amounts, however, are being obtained as a coproduct in the synthesis of other chemicals. Since 1954 the synthesis of phenol from cumene (p. 442) has yielded about three pounds of acetone for each five pounds of phenol. Acetone is formed also by the uncatalyzed oxidation of isopropyl alcohol with air at 420–460° and contact times of 80–30 seconds, the important product being **hydrogen peroxide.** When equal moles of alcohol and oxygen are used, conversion is 40 per cent per pass to give a 90 per cent yield of acetone and an 80 per cent yield of hydrogen peroxide. For each pound of hydrogen peroxide, two pounds of acetone are formed.

$$(CH_3)_2CHOH + O_2 \longrightarrow (CH_3)_2CO + H_2O_2$$

In the commercial production of allyl alcohol by a modified Ponndorf reduction of acrolein with isopropyl alcohol as the reducing agent, one pound of acetone is formed for each pound of allyl alcohol. The acrolein is made by the catalytic oxidation of propylene with air or oxygen.

$$CH_2{=}CHCH_3 + O_2 \xrightarrow[370^\circ]{Cu_2O} CH_2{=}CHCHO + H_2O$$

$$CH_2{=}CHCHO + (CH_3)_2CHOH \underset{400^\circ}{\overset{MgO-ZnO}{\rightleftharpoons}} CH_2{=}CHCH_2OH + (CH_3)_2CO$$

Acetone also is one of the products of the air oxidation of propane-butane (p. 213) and of the butyl alcohol fermentation of carbohydrates (p. 122). This process, in fact, was developed for the production of acetone during World War I because other sources were inadequate to supply the needs of the British for the manufacture of cordite, their standard smokeless powder (p. 608).[7]

About 54 per cent of the acetone consumed in the United States is used for the synthesis of other chemicals, especially methyl isobutyl ketone (p. 216) and methyl methacrylate (p. 620), 27 per cent is used as a solvent for cellulose acetate (p. 344), lacquers, resins, and acetylene (p. 151), and 19 per cent for miscellaneous uses. Production in 1964 amounted to over 1 billion pounds of which 72 per cent was obtained from *i*-propyl alcohol, 20 per cent from cumene, and 8 per cent from other sources. The unit value was 5 cents per pound.

Methyl ethyl ketone is made by the dehydrogenation of *s*-butyl alcohol and by the oxidation of butane (p. 168). It is known technically by the initials *MEK* and is used chiefly as a solvent for dewaxing lubricating oils. **Methyl *i*-butyl ketone** (*MIBK*) is made by the controlled catalytic reduction of mesityl oxide (p. 202).

PROBLEMS

13–*1*. Write skeleton formulas and give two names for each of the compounds belonging to the following groups: (*a*) aldehydes having five or less carbon atoms, (*b*) ketones having six or less carbon atoms.

---

[7] Scientists rarely have influenced directly the decisions of politicians and statesmen, but it has been stated that the Balfour Declaration establishing Palestine as a national home for the Jews was the result of Lloyd George's gratitude to Chaim Weizmann (1874–1952) for developing this process for the production of acetone. Weizmann later became the first president of the new state of Israel.

13–*2*. Pair each of the following types of compound with the appropriate formula: (*a*) an α,β-unsaturated aldehyde, (*b*) an acetal, (*c*) an ester, (*d*) a diether, (*e*) a pinacol; (*1*) $C_2H_5—O—COCH_3$, (*2*) $(CH_3)_2CHOHCHOH(CH_3)_2$, (*3*) $CH_3CH{=}CHCHO$, (*4*) $C_2H_5—O—CH_2CH_2—O—C_2H_5$, (*5*) $C_2H_5—O—CH_2—O—C_2H_5$.

13–*3*. Arrange the following compounds in the order of increasing dipole moment: (*a*) *n*-propyl alcohol, (*b*) propane, (*c*) propylene, (*d*) propionaldehyde.

13–*4*. Give balanced equations for the following oxidations: (*a*) ethyl alcohol to acetaldehyde with dichromate and sulfuric acid, (*b*) *n*-butyraldehyde to potassium *n*-butyrate with alkaline permanganate, (*c*) lauryl alcohol to dodecanal with *t*-butyl chromate, (*d*) 2-octanol to 2-octanone with chromium trioxide in acetic acid.

13–*5*. Give reactions for the synthesis of (*a*) nonanal from pelargonic acid, (*b*) hexanal from *n*-amyl bromide, (*c*) butanone from 2-butene, (*d*) isobutyrone from methylpropanal, (*e*) *i*-propyl *s*-butyl ketone from isobutyric acid.

13–*6*. (*a*) Which of the following reactions represents the best method for the synthesis of *n*-pentane? (*b*) Which procedure would not be expected to yield any pentane?

(*1*) $CH_3CH_2CH_2CH_2CH_2MgBr + H_2O \longrightarrow C_5H_{12} + Mg(OH)Br$

(*2*) $CH_3CH_2CH_2Br + 2\,Na + BrCH_2CH_3 \longrightarrow C_5H_{12} + 2\,NaBr$

(*3*) $CH_3CH_2CH_2CH_2CH_2Cl + Zn + HCl \longrightarrow C_5H_{12} + ZnCl_2$

(*4*) $CH_3CH_2CH_2CH_2CHO + H_2NNH_2 \longrightarrow C_5H_{12} + N_2 + H_2O$

(*5*) $CH_3CH_2CH_2CH_2CH_2OH + H_2 \xrightarrow[25^\circ]{Pt} C_5H_{12} + H_2O$

13–*7*. Which of the following reagents do not lead to a stable product when added to the carbonyl group of an aldehyde or ketone: (*a*) H_2/Pt, (*b*) HCN, (*c*) $NaHSO_3$, (*d*) Br_2, (*e*) C_2H_5MgBr, (*f*) H_2O, (*g*) NH_3, (*h*) HBr?

13–*8*. Which of the formulas given is the aldol addition product of (*a*) acetaldehyde, (*b*) propionaldehyde, (*c*) *n*-butyraldehyde, (*d*) isobutyraldehyde, (*e*) acetone?

(*1*) $CH_3CH_2CHOHCH(CH_3)CHO$ (*2*) $(CH_3)_2C(OH)CH_2COCH_3$

(*3*) $CH_3CH_2CH_2CHOHCH(C_2H_5)CHO$ (*4*) $CH_3CHOHCH_2CHO$

(*5*) $(CH_3)_2CHCHOHC(CH_3)_2CHO$

13–*9*. Starting with the proper aldehyde, give reactions for the synthesis of the following compounds: (*a*) diethylcarbinol, (*b*) 3-hydroxy-4-methyl-1-pentyne, (*c*) heptanaldoxime, (*d*) 2-ethyl-2-hexenal, (*e*) 1,3-dihydroxybutane.

13–*10*. Starting with the proper ketone, give reactions for the synthesis of the following compounds: (*a*) 3,6-dihydroxy-3,6-dimethyl-4-octyne, (*b*) *n*-butyrone semicarbazone, (*c*) hydroxydimethylacetic acid, (*d*) diethyl-*n*-propylcarbinol.

13–*11*. Give equations for the following reactions and name the organic products: (*a*) methyl ethyl ketone and sodium hypoiodite, (*b*) acetone and semicarbazide, (*c*) Clemmensen reduction of 5-nonanone, (*d*) 2-octanone and hydrogen cyanide, (*e*) acetaldehyde and *t*-butylmagnesium chloride, (*f*) ethyl ketone and amalgamated magnesium followed by mineral acid, (*g*) trimethylacetaldehyde and concentrated sodium hydroxide, (*h*) methyl ethyl ketone and nitrous acid.

13–*12*. Which of the following compounds do not give a yellow precipitate when allowed to react with iodine and sodium hydroxide: (*a*) methyl ethyl ketone, (*b*) ethyl ketone, (*c*) ethanol, (*d*) methanol, (*e*) formaldehyde, (*f*) acetaldehyde, (*g*) pinacolone, (*h*) *s*-butyl alcohol, (*i*) 2,4-hexanedione, (*j*) acetic acid?

13–*13*. Which olefin on ozonolysis gives two products, one of which reduces Tollens reagent and gives a negative iodoform reaction, whereas the other gives a positive iodoform reaction but a negative Tollens test: (*a*) $C_2H_5CH{=}CHCH_3$, (*b*) $(CH_3)_2C{=}CHC_2H_5$, (*c*) $(CH_3)_2C{=}C(CH_3)_2$, (*d*) $(CH_3)_2C{=}CHCH_3$, (*e*) $C_2H_5CH{=}CHC_2H_5$?

13–*14*. Which of the following compounds can be oxidized to a product that forms an oxime but does not give a color with Schiff reagent: (*a*) acetone, (*b*) 1-pentene, (*c*) *t*-butyl alcohol, (*d*) 2-propanol, (*e*) 1-propanol?

13–*15.* How many moles of hypochlorite is required to convert one mole of *s*-butyl alcohol to chloroform?

13–*16.* Describe a series of chemical tests that could be used to distinguish among the members of the following groups of compounds: (*a*) methyl-*n*-propylcarbinol, 3-pentanol, *n*-amyl alcohol, and dimethylethylcarbinol; (*b*) methyl alcohol, ethyl alcohol, 2-propanol, and *t*-butyl alcohol; (*c*) formaldehyde, acetaldehyde, acetone, and diethyl ketone.

13–*17.* Devise procedures for obtaining the components of the following mixtures in a reasonably pure state: (*a*) *n*-nonyl alcohol, pelargonic aldehyde, and nonane; (*b*) 2-octanone, 2-octanol, and capric acid; (*c*) 1-nonyne, dodecanal, and hexyl ether.

13–*18.* In the following diagrammatic chart, supply the structures and names of the compounds represented by the capital letters.

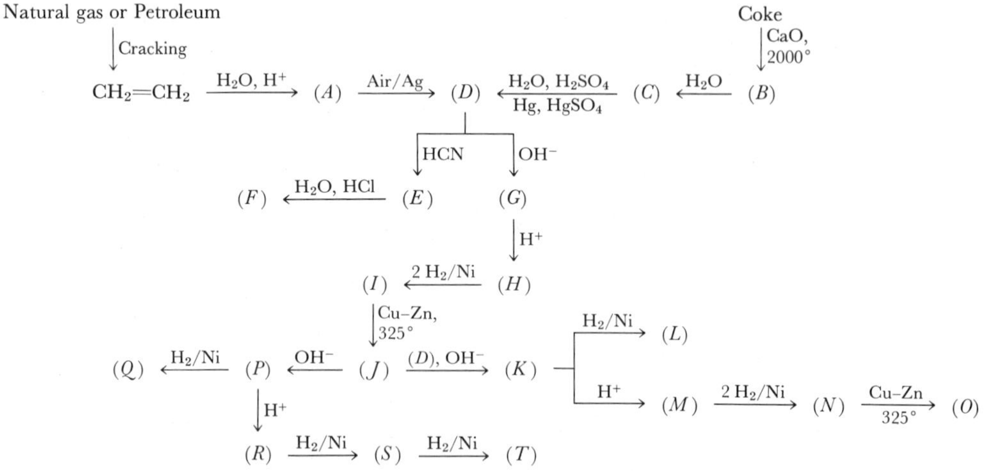

Natural gas or Petroleum $\xrightarrow{\text{Cracking}}$ $CH_2{=}CH_2$ $\xrightarrow{H_2O,\ H^+}$ (*A*) $\xrightarrow{\text{Air/Ag}}$ (*D*) $\xleftarrow[\text{Hg, HgSO}_4]{H_2O,\ H_2SO_4}$ (*C*) $\xleftarrow{H_2O}$ (*B*)

Coke $\xrightarrow{\text{CaO, } 2000^\circ}$ (*B*)

(*D*) $\xrightarrow{HCN}$ (*E*) $\xrightarrow{H_2O,\ HCl}$ (*F*)

(*D*) $\xrightarrow{OH^-}$ (*G*) $\xrightarrow{H^+}$ (*H*) $\xrightarrow{2\ H_2/Ni}$ (*I*) $\xrightarrow{\text{Cu–Zn, } 325^\circ}$ (*J*)

(*J*) $\xrightarrow{OH^-}$ (*P*) $\xrightarrow{H_2/Ni}$ (*Q*)

(*P*) $\xrightarrow{H^+}$ (*R*) $\xrightarrow{H_2/Ni}$ (*S*) $\xrightarrow{H_2/Ni}$ (*T*)

(*J*) $\xrightarrow{(D),\ OH^-}$ (*K*) $\xrightarrow{H_2/Ni}$ (*L*)

(*K*) $\xrightarrow{H^+}$ (*M*) $\xrightarrow{2\ H_2/Ni}$ (*N*) $\xrightarrow[325^\circ]{\text{Cu–Zn}}$ (*O*)

13–*19.* Arrange in the form of a diagrammatic chart indicating the successive steps and the conditions for the preparation of the following commercially important compounds starting with basic raw materials such as petroleum, natural gas, coke, or a plant product: (*A*) propylene, (*B*) *i*-propyl alcohol, (*C*) acetone, (*D*) acetone cyanohydrin, (*E*) methyl 2-methylpropenoate (*methyl methacrylate*), (*F*) diacetone alcohol, (*G*) mesityl oxide, (*H*) methyl *i*-butyl ketone, (*I*) methyl-*i*-butylcarbinol, (*J*) 2-methyl-2,4-pentanediol (*hexylene glycol*), (*K*) isophorone, (*L*) diisobutyl ketone.

13–*20.* Give a series of reactions for the following conversions: (*a*) *n*-butyraldehyde to *n*-hexane, (*b*) *i*-butyl bromide to *i*-valeraldehyde, (*c*) heptanoic acid to 3-nonanol, (*d*) *i*-propyl ether to *i*-butyl alcohol, (*e*) *i*-butyraldehyde to *i*-butyl iodide, (*f*) ethyl palmitate to hexadecanal, (*g*) propionic acid to methyldiethylcarbinol, (*h*) 1-pentene to 2-methyl-1-pentanol, (*i*) ethyl ketone to 2-methyl-3-ethyl-2-pentene, (*j*) propionaldehyde to *n*-propyl ether, (*k*) 3-pentanol to triethylcarbinol, (*l*) ethyl alcohol to acetal, (*m*) ethyl butyrate to butyrone, (*n*) methylacetylene to *i*-propyl alcohol, (*o*) *s*-butyl alcohol to 3,4-dimethyl-3,4-hexanediol, (*p*) acetylene to ethyl alcohol, (*q*) oleic acid to nonyl alcohol, (*r*) *n*-amyl bromide to *n*-hexyl bromide.

13–*21.* A neutral liquid, $C_4H_8O_3$, reacts with two moles of methylmagnesium bromide but evolves only one mole of methane. (*a*) What are the possible structures of the compound? (*b*) What are the possible structures if it had reacted with three moles of methylmagnesium bromide and evolved one mole of methane?

13–*22.* A compound *A* with the molecular formula C_8H_{16} absorbed bromine rapidly from carbon tetrachloride solution. After reaction with ozone and reduction of the ozonation products with zinc dust and acetic acid, two isomeric neutral products *B* and *C* were isolated; only *B* gave iodoform with iodine and sodium hydroxide solution. Give possible structures for the compound *A*.

13–*23*. A hydrocarbon having a molecular weight of 95 ± 5 decolorized aqueous permanganate with the formation of a brown precipitate. After the compound was allowed to react with an excess of permanganate, two products were isolated, an acid having a neutralization equivalent of 74 and a neutral compound that formed an oxime and gave a positive iodoform test. Give a structure for the hydrocarbon and balanced equations for the reactions involved.

13–*24*. An unknown compound formed a monoxime, reduced Fehling solution, gave iodoform on treatment with hypoiodite solution, and evolved methane on treatment with methylmagnesium iodide. Oxidation gave an acid that after purification had by titration an equivalent weight of 116. The acid still formed an oxime and gave a positive iodoform test but no longer reduced Fehling solution. Write a possible structural formula for the compound and show the reactions that it undergoes.

13–*25*. Compound *A* evolved hydrogen when warmed with metallic sodium. When 0.2 g. was treated with methylmagnesium iodide in a Zerevitinov apparatus, 34.5 cc. (S.T.P.) of methane was evolved. When compound *A* was heated with sulfuric acid, it gave chiefly a single lower-boiling compound, *B*, which decolorized bromine in carbon tetrachloride. Ozonolysis of *B* gave two products, *C* and *D*, neither of which produced a color with Schiff reagent and both of which gave iodoform when treated with sodium hypoiodite. Reduction of *D*, the higher boiling of the two compounds, with hydrogen and platinum gave an alcohol, which was dehydrated to an olefin *E*. Ozonolysis of *E* gave two products, one of which produced a color with Schiff reagent and the other did not. Give a structural formula for *A*, and the reactions that are described.

13–*26*. An unknown compound *A* reacted with hot hydrogen bromide to form two different halogen compounds *B* and *C*, which were separated by fractional distillation. Compound *B* on treatment with moist silver oxide gave a compound *D* that did not react at once with a cold mixture of concentrated hydrochloric acid and zinc chloride, but on standing with the reagent gave an insoluble liquid. Compound *C* likewise gave a new substance *E* on treatment with moist silver oxide, but *E* did not react at all with cold hydrochloric acid-zinc chloride mixture. Both *D* and *E* gave iodoform on treatment with iodine in alkaline solution. When *E* was oxidized under the proper conditions, an aldehyde *F* was obtained, which on treatment with methylmagnesium iodide solution and decomposition with water gave a compound identical with *D*. Write the structural formula for *A*, and illustrate by equations all of the reactions involved.

13–*27*. A compound having the molecular formula $C_6H_{12}O$ reacted with hydroxylamine but did not reduce Tollens reagent. Reduction with hydrogen and platinum gave an alcohol, which was dehydrated to give chiefly a single olefin. Ozonation and decomposition of the ozonide gave two liquid products, one of which reduced Tollens reagent but did not give a positive iodoform test whereas the other did not reduce Tollens reagent but did give a positive iodoform test. Give the structure of the original compound and show the reactions that it undergoes.

13–*28*. Compound *A*, after fusion with sodium and boiling with dilute nitric acid, gave a yellow precipitate with aqueous silver nitrate. Compound *A* was converted to the Grignard reagent, which was allowed to react with *i*-butyraldehyde and then decomposed with water. A compound *B* was obtained that reacted readily with hydrogen bromide to give an alkyl bromide *C*. Compound *C* likewise was converted to the Grignard reagent and decomposed with water to give a fourth compound *D*. When the original compound *A* was heated with metallic sodium, a compound identical with *D* was obtained. Write a structural formula for *A* and give all of the reactions involved.

13–*29*. A compound having the molecular formula $C_7H_{16}O$ did not react with sodium or phosphorus trichloride. When the compound was warmed with concentrated sulfuric acid, diluted, and distilled, a mixture was obtained that could not be separated satisfactorily by fractional distillation. After the mixture was oxidized by means of chromic acid, an acidic product was isolated having a neutralization equivalent of 74, along with a neutral product that reacted with semicarbazide but did not reduce Tollens reagent. Give the structure of the original compound and show the reactions that took place.

13–*30*. A water-insoluble compound, *A*, dissolved when boiled with aqueous sodium hydroxide. A distillate from the alkaline solution gave a positive iodoform reaction. When the distillate was heated with sodium dichromate and sulfuric acid and distilled again, the second distillate gave a positive Schiff reaction. When the alkaline solution that remained after the first distillation was acidified, compound *B* separated. When *B* was passed over thoria at 400°, compound *C* was obtained, which no longer was soluble in alkali but which formed a semicarbazone. When compound *C* was refluxed with amalgamated zinc and hydrochloric acid, compound *D* was formed, which was identical with the compound obtained by heating undecanal with hydrazine and sodium ethoxide. Give a formula for compound *A*, and equations for the reactions involved.

13–*31*. Postulate a likely series of steps for the following acid-catalyzed reactions: (*a*) the hydrolysis of acetals, (*b*) the conversion of a ketone to the ketal by means of ethyl orthoformate, (*c*) the hydrolysis of an oxime, (*d*) the formation of a semicarbazone, (*e*) the exchange reaction of formaldehyde with methyl ethyl ketoxime, (*f*) the trimerization of acetaldimine, (*g*) the phenylhydrazone-benzeneazoalkane tautomerism.

CHAPTER FOURTEEN

ALIPHATIC NITROGEN COMPOUNDS

Structurally the organic oxygen compounds have been considered as derivatives of the water molecule. Similarly many of the organic nitrogen compounds may be considered as ammonia molecules, the hydrogen atoms of which have been replaced by other groups. Since nitrogen has three replaceable hydrogens instead of two as in water, the number of possible combinations is increased.

H_2O	water	NH_3	ammonia
ROH	alcohols	RNH_2	primary amines
		R_2NH	secondary amines
ROR	ethers	R_3N	tertiary amines
RCOOH	acids	$RCONH_2$	amides
		$RC(NH)NH_2$	amidines
RCOOR	esters	RCONHR	*N*-alkyl amides
		$RCONR_2$	*N,N*-dialkyl amides
		RC(NH)OR	imidic esters
RCOOCOR	anhydrides	RCONHCOR	imides
RCH=O	aldehydes	RCH=NH	aldimines
R_2C=O	ketones	R_2C=NH	ketimines

This list is not complete, but contains the types of compounds most frequently encountered. Substituents on nitrogen other than hydrogen, alkyl, and acyl also may be present as for example in the cyanides (p. 237), the isocyanides (p. 240), the oximes (p. 204), the hydrazones (p. 205), and the nitro compounds (p. 242).

AMINES

Nomenclature

Aliphatic amines are alkyl substitution products of ammonia and are named as such, *ammonia* being contracted to *amine.*[1] Thus CH_3NH_2 is methylamine, $(CH_3)_2NH$ is dimethylamine, and $(CH_3)_3N$ is trimethylamine. With mixed amines the alkyl groups frequently are named in the order of increasing complexity, for example $CH_3(C_2H_5)NCH(CH_3)_2$ is methylethyl-*i*-propylamine. Compounds in which the nitrogen atom is united to one carbon atom, RNH_2, are called *primary amines;* to two carbon atoms, R_2NH, *secondary amines;* and to three carbon atoms, R_3N, *tertiary amines. The terms primary, secondary, and tertiary refer here to the condition of the nitrogen atom,* whereas when used with alcohols they refer to the carbon atom to which the hydroxyl group is attached. Thus although tertiary butyl alcohol, $(CH_3)_3COH$, is a tertiary alcohol, because the carbon atom is united to three other carbon atoms, tertiary butylamine, $(CH_3)_3CNH_2$, is a primary amine, because the nitrogen atom is united directly to only one carbon atom. When referring to the type of alkyl group in primary amines it is convenient to use the term *carbinamine* in analogy to the term *carbinol* despite the recommendation to discontinue the use of the latter term. Thus ethylamine is a

[1] Cf. footnote 4, p. 203.

primary carbinamine, isopropylamine is a secondary carbinamine, and *t*-butylamine is a tertiary carbinamine. All carbinamines are primary amines. The NH_2 group is called the *amino group,* and primary amines having other functional groups conveniently may be named as amino substitution products; for example, $CH_3CHNH_2CH_2CH_2OH$ could be called 3-amino-1-butanol (for order of precedence of suffixes, see footnote 1, page 435).

Preparation

Mixed Primary, Secondary, and Tertiary Amines

1. ***From Alkyl Halides and Ammonia.*** Wurtz prepared the first amines from alkyl isocyanates (p. 287) in 1849, but the most direct method for the preparation of amines is by the reaction of ammonia with an alkyl halide, reported by Hofmann[2] in 1850. Nucleophilic attack on carbon by the ammonia molecule displaces halide ion to give an alkyl-substituted ammonium halide.

$$\mathrm{H}:\underset{\mathrm{H}}{\overset{\mathrm{H}}{\ddot{\mathrm{N}}}}: + \mathrm{R}:\ddot{\mathrm{X}}: \longrightarrow \left[\mathrm{H}:\underset{\mathrm{H}}{\overset{\mathrm{H}}{\mathrm{N}}}{}^{+}:\mathrm{R}\right] :\ddot{\mathrm{X}}:^{-}$$

This initial reaction, however, is followed by a series of secondary reactions. The excess of ammonia present while the alkyl halide is reacting with ammonia competes with the primary amine for the hydrogen halide.

$$RNH_3X + NH_3 \rightleftharpoons RNH_2 + NH_4X$$

The primary amine thus formed also has an unshared pair of electrons and can react with a molecule of alkyl halide, giving rise to another pair of reactions to form a secondary amine.

$$RNH_2 + RX \longrightarrow R_2NH_2X$$
$$R_2NH_2X + NH_3 \rightleftharpoons R_2NH + NH_4X$$

Immediately a third pair of reactions is possible giving a tertiary amine.

$$R_2NH + RX \longrightarrow R_3NHX$$
$$R_3NHX + NH_3 \rightleftharpoons R_3N + NH_4X$$

Finally a single further reaction can take place giving a quaternary ammonium salt.

$$R_3N + RX \longrightarrow R_4NX$$

The reaction stops at this point, because there is no hydrogen attached to nitrogen in the quaternary ammonium salt (nitrogen united to four carbon atoms), and hence no proton can be transferred to another base.

Accordingly the reaction of alkyl halides with ammonia, known as the *Hofmann method for preparing amines,* gives rise to a mixture of primary, secondary, and tertiary amines, their salts, and the quaternary ammonium salt. Addition of strong alkali at the end of the reaction liberates a mixture of the free amines from their salts, but the quaternary salt is not affected. It is possible to control the reaction to a certain extent. If a very large excess of ammonia is used, the chance that the alkyl halide will react with ammonia molecules is greater than that it will react with amine molecules, and chiefly primary amine is produced. If increasing amounts of alkyl halide are used, more of the other products are formed.

---

[2] August Wilhelm Hofmann (1818–1895), German chemist who received his training under Liebig, and who was professor at the Royal College of Chemistry in London from 1845 to 1864, and at the University of Berlin from 1864 until his death. He is noted particulary for his work on amines and for his investigations of aromatic compounds. The latter work laid the basis for the coal tar chemical industry.

The order of reactivity of the alkyl halides with ammonia and amines is that expected for alkyl groups and for leaving groups in S_N2 reactions, namely methyl > primary > secondary > tertiary and iodides > bromides > chlorides (p. 103). Alkyl sulfates may replace alkyl halides (cf. p. 139). Because of the ease with which tertiary alkyl halides lose halogen acid by an $E1$ mechanism, reaction with ammonia yields only olefin and ammonium halide.

$$(CH_3)_3CCl + NH_3 \longrightarrow CH_2{=}C(CH_3)_2 + NH_4Cl$$

Tertiary carbinamines may be prepared by the Hofmann degradation of the proper amide (p. 224), which in turn is prepared readily by the hydrolysis of the nitrile (p. 239). They may be prepared also by the reaction of chloroamine with a tertiary alkyl Grignard reagent (p. 225) or by the hydrolysis of *N-t*-alkylformamides (p. 239).

2. ***Catalytic Reduction of Alkyl Cyanides.*** Addition of four atoms of hydrogen to the carbon-nitrogen triple bond gives a primary amine. Rhodium is the preferred catalyst, although platinum or Raney nickel also are active.

$$RC{\equiv}N + 2\,H_2 \xrightarrow{\text{Rh, Pt, or Ni}} RCH_2NH_2$$

Secondary and tertiary amines are formed at the same time by the addition of amine to the intermediate imine, followed by loss of ammonia and further reduction, or by reductive removal of the amino group.

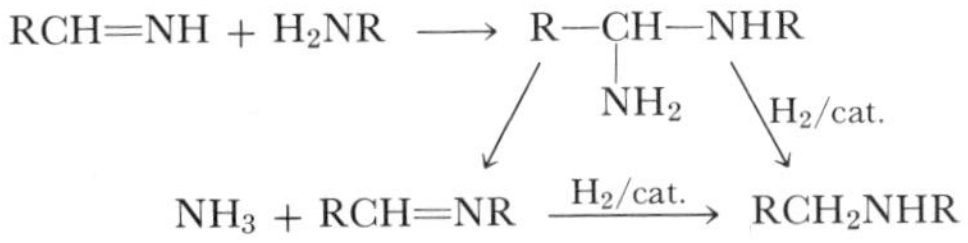

$$RCH{=}NH + H_2NR \longrightarrow R{-}\underset{\displaystyle NH_2}{\underset{|}{C}H}{-}NHR$$

$$R{-}CH(NH_2){-}NHR \longrightarrow NH_3 + RCH{=}NR \xrightarrow{H_2/\text{cat.}} RCH_2NHR$$

$$R{-}CH(NH_2){-}NHR \xrightarrow{H_2/\text{cat.}} RCH_2NHR$$

The secondary amine gives rise by a similar series of reactions to the tertiary amine. The formation of secondary and tertiary amines can be suppressed considerably by carrying out the reduction in the presence of a large excess of ammonia.

Pure Primary Amines

When only one type of amine is desired, reactions that do not give mixtures are of value. In the following reactions only the primary amine is formed.

1. ***From Amides* (Hofmann Rearrangement).** Electrophilic attack by halogen on nitrogen of ammonia leads to displacement of a proton and the formation of haloamine. Usually the reaction is carried out in alkaline solution to assist the removal of the proton.

$$X{:}X + {:}NH_3 + {}^{-}{:}OH \longrightarrow X{:}^{-} + X{:}NH_2 + H{:}OH$$

Hypohalite solutions made by adding halogen to aqueous alkali behave in the same way because free halogen always is present in equilibrium with halide and hypohalite ions.

$$X^{-} + {}^{-}OX + H_2O \rightleftharpoons X_2 + 2\,{}^{-}OH$$

All amino groups bearing hydrogen react as ammonia does. Thus when bromine is added to an alkaline solution of an amide, the *N*-halogenated amide is formed.

$$RCONH_2 + Br_2 + NaOH \longrightarrow RCONHBr + NaBr + H_2O$$

The electronegative halogen increases the acidity of the remaining hydrogen, and in the presence of excess base the conjugate base of the brominated amide is formed. A 1,2 shift of alkide ion with simultaneous loss of halide ion yields an isocyanate (p. 286), which immediately hydrolyzes to the carbamic acid. The latter is unstable and decomposes to the amine and carbon dioxide (p. 287). Several lines of evidence show that the migration of the alkyl group takes place intramolecularly (p. 311).

$$RCONHBr \underset{H_2O}{\overset{^-OH}{\rightleftharpoons}} R-\overset{\overset{O}{\|}}{C}-\ddot{N}:Br^- \xrightarrow{:Br^-} \underset{\text{An isocyanate}}{O{=}C{=}NR} \overset{H_2O}{\rightleftharpoons}$$

$$\underset{\text{A carbamic acid}}{\left[O{=}\underset{OH}{C}-NHR\right]} \rightleftharpoons O{=}\underset{O:^-}{C}-\overset{+}{N}H_2R \longrightarrow CO_2 + H_2NR$$

It is possible that the rearrangement takes place by way of an acylnitrene (p. 251) rather than by a concerted reaction. All of these steps take place spontaneously without the necessity of isolating any intermediates. The over-all reaction is

$$RCONH_2 + X_2 + 4\ NaOH \longrightarrow RNH_2 + Na_2CO_3 + 2\ NaX + 2\ H_2O$$

If the alkyl group is tertiary, the tertiary alkyl isocyanate resists hydrolysis in alkaline solution. The isocyanate can be isolated and subsequently hydrolyzed by 20 per cent hydrochloric acid.

2. ***From Carboxylic Acids*** **(Schmidt Reaction).** Carboxylic acids that are stable to concentrated sulfuric acid can be converted to primary amines having one less carbon atom by means of hydrazoic acid, which has the resonance structure

$$\{H\ddot{N}{=}\overset{+}{N}{=}\overset{-}{\ddot{N}}: \longleftrightarrow H\overset{-}{\ddot{N}}-\overset{+}{N}{\equiv}N:\}$$

Nucleophilic attack on the protonated carboxyl group by hydrazoic acid gives an intermediate that can lose nitrogen with simultaneous migration of the alkide ion. Loss of a proton gives the isocyanate, which is hydrolyzed to the amine and carbon dioxide under acidic conditions just as it is hydrolyzed in the Hofmann rearrangement under basic conditions.

$$R-\underset{OH}{C}{=}O \underset{HSO_4^-}{\overset{H_2SO_4}{\rightleftharpoons}} R-\underset{OH}{C}{=}\overset{+}{O}H \xrightarrow{H\ddot{N}{=}\overset{+}{N}{=}N} R-\overset{OH}{\underset{HN-\overset{+}{N}\equiv N}{C}}-OH \xrightarrow[H_2O]{} (R:)\underset{N-\overset{+}{N}\equiv N}{C}-OH \xrightarrow[N_2]{}$$

$$RN{=}\overset{+}{C}-\ddot{O}H \underset{H_2SO_4}{\overset{HSO_4^-}{\rightleftharpoons}} RN{=}C{=}O \xrightarrow{H_2O} RNH_2 + CO_2$$

Since hydrazoic acid is volatile, explosive, and very poisonous, it is more convenient to add sodium azide to a mixture of sulfuric acid and a solution of the compound in an inert solvent such as chloroform. The Schmidt reaction yields the same product as the Hofmann or the Curtius (p. 251) rearrangements but is preferred for small scale preparations because it is more direct, and the yields usually are better.

3. ***Chemical Reduction of Alkyl Cyanides.*** This reaction resembles the reduction of esters to alcohols (p. 175). The usual reagent is sodium and absolute alcohol.

$$RC{\equiv}N + 4\ Na + 4\ C_2H_5OH \longrightarrow RCH_2NH_2 + 4\ C_2H_5ONa$$

Lithium aluminum hydride is convenient for small scale work.

4. ***Reduction of Oximes.*** Oximes of aldehydes or ketones also may be reduced by sodium and alcohol to give good yields of primary amines.

$$RCH{=}NOH + 4\ Na + 3\ C_2H_5OH \longrightarrow RCH_2NH_2 + NaOH + 3\ C_2H_5ONa$$

This reaction provides a method for converting aldehydes and ketones to primary amines. Lithium aluminum hydride may be used as a reducing agent instead of sodium and alcohol. Ruthenium-catalyzed hydrogenation of oximes in methanol saturated with ammonia gives high yields of primary amine along with only a few per cent of secondary amine.

5. ***Other Methods.*** Pure primary amines can be prepared also by the hydrolysis of *N*-alkylformamides (p. 239) and by the reduction of amides (p. 235), nitro compounds (p. 245), and alkyl azides (p. 251).

PURE SECONDARY AMINES

Cyanamide ion from sodium cyanamide, $Na_2N{-}C{\equiv}N$ (p. 288), displaces halide ion from alkyl halides or alkyl sulfate ion from alkyl sulfates to give dialkylcyanamides.

$$N{\equiv}C{-}\ddot{\underset{..}{N}}{:}^{2-} + RX \underset{^{-}:X}{\rightleftharpoons} N{\equiv}C{-}\ddot{\underset{..}{N}}^{-}{-}R \underset{^{-}:X}{\overset{RX}{\rightleftharpoons}} N{\equiv}C{-}NR_2$$

The nitrile group can be hydrolyzed by aqueous acids or bases (p. 238) to give the *N,N*-dialkylcarbamic acid, which spontaneously loses carbon dioxide (p. 287).

$$R_2N{-}C{\equiv}N + 2\,H_2O \xrightarrow{H^+ \text{ or } ^-OH} NH_3 + [R_2NCOOH] \longrightarrow R_2NH + CO_2$$

In acid solution carbon dioxide is evolved and a mixture of the amine and ammonium salts is obtained, whereas in alkaline solution the free amine, ammonia, and alkali carbonate are formed.

PURE TERTIARY AMINES

Pure tertiary amines can be prepared from dialkylchloroamines and Grignard reagents.

$$R_2NH + Cl_2 + NaOH \longrightarrow R_2NCl + NaCl + H_2O$$
$$R_2NCl + R'MgCl \longrightarrow R_2NR' + MgCl_2$$

More frequently a secondary amine is allowed to react with an excess of alkyl halide, and the tertiary amine is separated from the accompanying quaternary salt (p. 226).

Physical Properties

Water, ammonia, and methane have very nearly the same polarizability, but their boiling points are 100°, −33°, and −161°, respectively. The abnormally high boiling point of water is explained by hydrogen bonding between hydrogen united to oxygen, and the unshared pair of electrons on the oxygen atom (p. 24). Ammonia also forms hydrogen bonds but its boiling point indicates that it is not so strongly associated as water. The nitrogen atom is nearer the center of the periodic table than oxygen and hence has less attraction for electrons (p. 21). Therefore the hydrogen atoms attached to nitrogen have less tendency to leave nitrogen as protons or to be attracted to the concentration of negative charge on other atoms than do those joined to oxygen. The decrease in tendency to form hydrogen bonds accounts for the smaller degree of association.

Primary and secondary amines also are associated. Methylamine ($Z = 18$) boils at −7° and dimethylamine ($Z = 26$) boils at +7°, but trimethylamine ($Z = 34$) boils at +4°. Thus even though its polarizability is greater than that of dimethylamine, trimethylamine has the lower boiling point. The explanation is that trimethylamine no longer has a hydrogen capable of forming hydrogen bonds, and hence it is not associated.

There is, however, nothing to prevent the unshared pair of electrons of tertiary amines from forming hydrogen bonds with water molecules, and all types of amines of low molecular weight are soluble in water. Since amines have a greater tendency to share their unshared pair of electrons than do alcohols, they form stronger hydrogen bonds with water and are somewhat more soluble. For example, *n*-butyl alcohol dissolves in water to the extent of about 8 per cent at room temperature, but *n*-butylamine is miscible with water; and although less than 1 per cent of *n*-amyl alcohol dissolves in water, this degree of insolubility is not reached in the amines until *n*-hexylamine. Like the alcohols and ethers, simple amines are soluble in most organic solvents. Like ammonia, the amines are pyramidal in shape (p. 15), and have dipole moments of 1.0 to 1.3 D. The N—H stretching absorption is in the 3500–3300 cm.$^{-1}$ region, which overlaps the O—H stretching absorption and makes differentiation of the two groups difficult. Primary amines can be identified by the presence of two bands in this region. The C—N stretching absorption is weak in the 1220–1020 cm.$^{-1}$ region.

The lower amines have an odor resembling that of ammonia; the odor of trimethylamine is described as "fishy." As the molecular weight increases, the odors become decidedly obnoxious, but they decrease again with increasing molecular weight and decreasing vapor pressure.

Reactions

1. ***Basic Properties.*** In aqueous solution, the basicity constant of ammonia ($pK_a = 9.2$) is approximately 10^{11} times greater than that of water ($pK_a = -1.7$). Since alkyl groups are electron-releasing, amines should be somewhat stronger bases than ammonia and generally this is true. Furthermore the order of basicity for the various amines should be ammonia $<$ primary $<$ secondary $<$ tertiary. This inductive effect is in accord with the ionization potentials of 9.41, 9.21, and 9.02 e.v. for methylamine, dimethylamine, and trimethylamine. The actual order of basicity, however, varies with the solvent and the acid. The inductive effect can be outweighed by the effect of solvation of the ions formed and by a steric effect.

The amine salts are named either as substituted ammonium salts or as acid addition products; for example $CH_3NH_3{}^{+-}Cl$ may be called methylammonium chloride or methylamine hydrochloride. In the older literature the amine salts frequently are represented by formulas such as $CH_3NH_2 \cdot HCl$, which is equivalent to writing ammonium chloride as $NH_3 \cdot HCl$.

Just as ammonia forms a water-insoluble salt with chloroplatinic acid, $(NH_4)_2PtCl_6$, so the amine chloroplatinates, for example $(RNH_3)_2PtCl_6$, are insoluble and are used for analytical purposes. To determine the equivalent weight of an amine, it is necessary only to ignite a weighed sample of its chloroplatinate and weigh the residue of platinum.

Calcium chloride may not be used for removing water from amines because it is solvated by them just as it is by water and ammonia. Amines usually are dried with potassium carbonate, potassium hydroxide, or barium oxide.

2. ***Alkylation.*** Since ammonia reacts with alkyl halides to give a mixture of primary, secondary, and tertiary amines, and quaternary ammonium salt (p. 222), amines also can react with primary alkyl halides to give secondary or tertiary amines, or quaternary salts. Because of steric hindrance, the rate of formation of quaternary salts from tertiary amines that have secondary or tertiary alkyl groups is very slow. Since amines such as ethyldiisopropylamine are strongly basic and practically nonalkylatable, they can be used as auxiliary bases to take up the hydrogen halide formed in the alkylation of other amines, and can be used as dehydrohalogenating agents in the conversion of alkyl halides to olefins.

3. ***Acylation.*** Acyl halides, acid anhydrides, and esters react with primary and secondary amines, just as they do with ammonia, to give amides (p. 233). Tertiary amines do not give amides because they do not contain a replaceable hydrogen atom.

$$2\,RNH_2 + R'COX \longrightarrow RNHCOR' + RNH_3X$$
$$RNH_2 + (R'CO)_2O \longrightarrow RNHCOR' + R'COOH$$
$$RNH_2 + R'COOR'' \longrightarrow RNHCOR' + R''OH$$

The reaction with acyl halides requires two moles of amine, only one of which is acylated because the second mole combines with the hydrogen halide. Although an acid is formed in the second reaction, it is a weak acid, and the salt of a weak base and a weak acid dissociates sufficiently to produce the acylated amine when heated with an excess of anhydride. Hence all of the amine can be converted into amide by this procedure.

The products formed are known as *N-substituted amides,* the *N* referring to the nitrogen atom of the amide. Thus the reaction product of methylamine with acetic anhydride, $CH_3CONHCH_3$, is called *N*-methylacetamide. Secondary amines in the above reactions yield *N,N*-disubstituted amides. *i*-Butyryl chloride and diethylamine give *N,N*-diethyl-*i*-butyramide, $(CH_3)_2CHCON(C_2H_5)_2$.

An important application of the acylation reaction is the separation of tertiary amines from a mixture with primary and secondary amines. After reaction of the mixture with acetic anhydride, the unchanged tertiary amine may be separated from the higher boiling amides by distillation, or by extraction with dilute acid. The tertiary amine is basic and forms water-soluble salts whereas the amides are neutral. Similarly acylation can be used to distinguish primary or secondary amines from tertiary amines.

4. ***Reactions with Aldehydes and Ketones.*** **(*a*)** ENAMINE FORMATION. Secondary amines react with aldehydes and ketones that have hydrogen on the α carbon atom to give **enamines.** The reaction is catalyzed by acids, and the initial product undoubtedly is the aldehyde-ammonia, which loses water. Azeotropic distillation with benzene removes the water and causes the reaction to go to completion.

$$R_2CHCHO + HNR_2 \rightleftharpoons \left[R_2CH\underset{\displaystyle OH}{\underset{|}{C}}H{-}NR_2\right] \rightleftharpoons H_2O + R_2C{=}CH{-}NR_2$$

The enamines are less reactive than aldehydes. Since the aldehyde can be regenerated by acid hydrolysis, formation of the enamine can protect the aldehyde group while reactions are carried out on other portions of the molecule. Reduction with formic acid yields the tertiary amine, and oxidation with chromic acid yields the carboxylic acid or ketone with one less carbon atom. Because of resonance, the carbon atom α to the original carbonyl group possesses anion character and can be alkylated or acylated. Hydrolysis regenerates the carbonyl group.

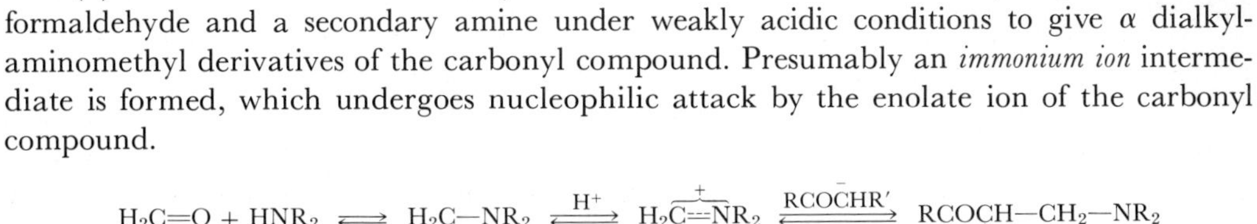

$$\left\{R_2C{=}CH{-}\ddot{N}R_2 \longleftrightarrow R_2\bar{\ddot{C}}{-}CH{=}\overset{+}{N}R_2\right\}$$

$$\xrightarrow{R'X} X^- + R_2\underset{\displaystyle R'}{\underset{|}{C}}{-}CH{=}\overset{+}{N}R_2 \xrightarrow{H_2O} R_2\underset{\displaystyle R'}{\underset{|}{C}}{-}CHO + H_2\overset{+}{N}R_2$$

$$\xrightarrow{RCOCl} Cl^- + R_2\underset{\displaystyle COR}{\underset{|}{C}}{-}CH{=}\overset{+}{N}R_2 \xrightarrow{H_2O} R_2\underset{\displaystyle COR}{\underset{|}{C}}{-}CHO + H_2\overset{+}{N}R$$

(*b*) MANNICH REACTIONS. Aldehydes and ketones that have an α hydrogen react with formaldehyde and a secondary amine under weakly acidic conditions to give α dialkylaminomethyl derivatives of the carbonyl compound. Presumably an *immonium ion* intermediate is formed, which undergoes nucleophilic attack by the enolate ion of the carbonyl compound.

$$H_2C{=}O + HNR_2 \rightleftharpoons H_2\underset{\displaystyle OH}{\underset{|}{C}}{-}NR_2 \underset{H_2O}{\overset{H^+}{\rightleftharpoons}} H_2C{=}\overset{+}{N}R_2 \overset{RCOCHR'}{\rightleftharpoons} RCO\underset{\displaystyle R'}{\underset{|}{C}}H{-}CH_2{-}NR_2$$

Mannich-type reactions are general also for other compounds that can lose a proton sufficiently readily to give a reactive nucleophile, such as acetylenes, nitroalkanes (p. 246), and thiophenes (p. 506).

5. ***Reactions with Nitrous Acid.*** Although nitrous acid is assigned the structural

formula HO—N=O, it is unstable and cannot be isolated. In aqueous solution it is in equilibrium with several species.

$$HONO + H_2O \rightleftharpoons H_3O^+ + \{^-O{-}\ddot{N}{=}O \longleftrightarrow O{=}\ddot{N}{-}O^-\}$$

Nitrite ion

$\updownarrow$ HONO

$$H_2\overset{+}{O}NO + {}^-ONO \xrightleftharpoons{0^\circ} H_2O + O{=}N{-}O{-}N{=}O \xrightarrow[\text{temp.}]{\text{Room}} O{=}N\cdot + O{-}\dot{N}{=}O$$

Nitrous acidium ion; Dinitrogen trioxide (blue); Nitric oxide; Nitrogen dioxide (brown)

Nitrous acidium ion $\xrightleftharpoons{X^-}$ $H_2O + X{-}N{=}O$ (Nitrosyl halide)

Nitrous acidium ion $\updownarrow$ $H_2O + {}^+NO$ (Nitrosonium ion)

Nitrogen dioxide $\downarrow$ $O_2N{-}NO_2$ Dinitrogen tetroxide (*colorless*)

As a weak acid ($pK_a = 3.35$) of about the same strength as simple carboxylic acids, it is ionized partially in aqueous solution. Proton transfer leads to the formation of the conjugate acid, nitrous acidium ion. Nucleophilic attack on the conjugate acid by nitrite ion yields dinitrogen trioxide, a blue liquid, which probably is responsible for the blue color formed when a cold solution of sodium nitrite is acidified. Dissociation of the nitrous acidium ion can lead to the nitrosonium ion. If a halogen acid is used to produce the nitrous acid from its salt, the halide ion can react by an S_N2 mechanism to give nitrosyl halide. Moreover dinitrogen trioxide is unstable at room temperature and decomposes into the free radicals, nitric oxide, NO, and brown nitrogen dioxide, NO_2. The latter is in equilibrium with its dimer, dinitrogen tetroxide, N_2O_4.

The behavior of amines toward nitrous acid depends on the conditions of the reaction and on the type of amine. In the absence of a strong acid, all types of amines merely yield amine nitrites.

$$RNH_2 + HONO \longrightarrow RNH_3{}^{+-}NO_2$$

The reaction may be carried out by passing carbon dioxide into a suspension of finely divided sodium nitrite in a solution of the amine in methyl alcohol. **Di-*i*-propylammonium nitrite** and **dicyclohexylammonium nitrite** (p. 649) are used as rust inhibitors. They volatilize slowly by dissociation into the free amine and are used in the packaging of machinery and parts.

In the presence of a strong acid, primary amines react with nitrous acid with the evolution of nitrogen. The various other products of the reaction depend on the conditions of the reaction and on the structure of the alkyl group. The initial step is the formation of the nitrosoamine, which appears to result from nucleophilic attack on dinitrogen trioxide with displacement of nitrite ion rather than by reaction with any of the other species present. The electronegativity of the nitroso group permits ready deprotonation of the intermediate ion.

$$RNH_2\!: + \overset{O}{\overset{\|}{N}}{-}ONO \longrightarrow ONO^- + RNH_2\overset{+}{-}NO \xrightleftharpoons[H_3O^+]{H_2O} RNH{-}N{=}O$$

Tautomerization of the nitrosoamine gives the diazohydroxide, which ionizes in the acid solution to the diazonium salt. Because of the great stability of the nitrogen molecule, the electron-donating power of the simple alkyl group is insufficient to stabilize the diazonium ion, and it decomposes to the carbonium ion and nitrogen.

$$RNH{-}N{=}O \xrightleftharpoons[H_2O]{H_3O^+} RNH{-}N{=}\overset{+}{O}H \xrightleftharpoons[H_3O^+]{H_2O} R{-}N{=}N{-}OH \xrightleftharpoons[2\,H_2O]{H_3O^+} R{-}\overset{+}{N}{\equiv}N\!: \longrightarrow R^+ + N_2$$

Alkyl diazohydroxide; Alkyldiazonium ion

The carbonium ion immediately undergoes a variety of reactions that depend on its structure and its environment, and a mixture of products usually is obtained. Thus it can rearrange (p. 89), and either the original carbonium ion or the rearranged carbonium ions can attack water molecules to give alcohols, attack alcohol molecules to give ethers, undergo elimination of protons to give alkenes, or combine with nitrite ion to give alkyl nitrites. If halogen acid is used to form the nitrous acid, reaction of the carbonium ion with halide ion yields alkyl halides. There is some stereochemical evidence that not all of the reactions take place by way of the free carbonium ion, but that some may result from a very rapid nucleophilic displacement of nitrogen from the diazonium ion.

Under comparable conditions, ethylamine gives 60 per cent ethyl alcohol and a trace of ether; *n*-propylamine gives 7 per cent *n*-propyl alcohol, 32 per cent *i*-propyl alcohol, and 28 per cent propylene; *n*-butylamine gives 25 per cent *n*-butyl alcohol, 13 per cent *s*-butyl alcohol, 8 per cent *n*- and *s*-butyl chlorides, and 36 per cent 1- and 2-butenes. Although the complexity of the reaction makes it of little value for preparative purposes, the evolution of nitrogen can be made quantitative, and the reaction is used to estimate primary amino groups from the volume of nitrogen evolved.

If a high concentration of halogen acid is used, fair yields of the halide may be obtained. Under these conditions it is possible that the active agent is nitrosyl halide.

$$NaNO_2 + 2\,HBr \longrightarrow NaBr + H_2O + NOBr$$

$$RNH_2 + NOBr \longrightarrow RBr + N_2 + H_2O$$

Secondary amines yield nitrosoamines, which are yellowish in color. Since they cannot tautomerize, they are relatively stable compounds. They are amides of nitrous acid and are nonbasic and insoluble in dilute acids (p. 234).

$$R_2NH + HONO \xrightarrow{H^+} R_2NNO + H_2O$$

Tertiary amines remain in solution as their salts, from which they can be recovered by the addition of alkali. If the reaction of tertiary amines with nitrous acid goes beyond salt formation, it is complex, the chief products being aldehydes and the nitroso derivatives of secondary amines.

6. ***Oxidation.*** The oxidation of primary and secondary amines can lead to a variety of products such as hydroxylamines, oximes, nitroso and nitro compounds, and aldehydes and ketones. The permanganate oxidation of *t*-alkyl primary amines to tertiary nitro compounds is of preparative value. Tertiary amines readily displace hydroxide ion from hydrogen peroxide with subsequent loss of a proton to give the class of compounds known as **amine oxides.**

$$R_3N\colon \underset{OH^-}{\overset{HO—OH}{\rightleftharpoons}} R_3\overset{+}{N}\colon OH \underset{H_2O}{\overset{OH^-}{\rightleftharpoons}} R_3\overset{+}{N}\colon\overset{-}{O}$$

A concerted mechanism analogous to that proposed for the oxidation of sulfides (p. 262) also is possible.

The amine oxides contain a semipolar bond (p. 16) between nitrogen and oxygen, and the formula may be written $R_3N\overset{+}{=}\!\!\!\overset{-}{}O$ or $R_3N\rightarrow O$. The presence of the semipolar bond gives rise to a high dipole moment ($\mu = 4.87$ D), which in turn causes the boiling point to be abnormally high. Trimethylamine oxide does not distill at temperatures up to 180°, where it decomposes, whereas dimethylethylamine, which would be expected to have about the same polarizability, boils at 38°.

The amine oxides are less basic than amines but form salts with strong acids.

$$R_3N\overset{+-}{—}\ddot{\underset{..}{O}}\colon + HX \longrightarrow \left[R_3\overset{+}{N}—\ddot{\underset{..}{O}}\colon H\right]X^-$$

Trimethylamine oxide has been isolated from the muscles of several varieties of marine animals such as the octopus and the spiny dogfish.

Individual Amines

Methylamine, dimethylamine, and **trimethylamine** occur in herring brine and are distributed widely in other natural products, probably as the result of the decomposition or metabolism of nitrogenous compounds. They are manufactured commercially by passing a mixture of methyl alcohol and ammonia over heated alumina.

$$CH_3OH + NH_3 \xrightarrow[400°]{Al_2O_3} \underset{H_2O}{\underset{+}{CH_3NH_2}} \xrightarrow{CH_3OH} \underset{H_2O}{\underset{+}{(CH_3)_2NH}} \xrightarrow{CH_3OH} \underset{H_2O}{\underset{+}{(CH_3)_3N}}$$

This procedure has not been practical for alcohols that can be dehydrated because they yield chiefly olefins. Methylamine hydrochloride is formed along with some di- and trimethylamine hydrochlorides when a solution of ammonium chloride in formalin is evaporated to dryness.

$$NH_4Cl + 2\,HCHO \longrightarrow CH_3NH_3^+Cl^- + HCOOH$$

If solid ammonium chloride and paraformaldehyde are mixed and heated, trimethylamine hydrochloride is formed.

$$2\,NH_4Cl + 9\,HCHO \longrightarrow 2\,(CH_3)_3NHCl + 3\,CO_2 + 3\,H_2O$$

Trimethylamine can be prepared also by the dry distillation of beet sugar residues, which contain betaine (p. 618).

$$2\,(CH_3)_3\overset{+}{N}CH_2COO^- \longrightarrow 2\,(CH_3)_3N + CH_2{=}CH_2 + 2\,CO_2$$

Although all three amines are obtained in the commercial synthesis from methanol and ammonia, dimethylamine is the most important and special catalysts have been developed to increase the yield at the expense of trimethylamine. United States production in 1964 was over 54 million pounds. The chief uses are for the preparation of dimethylformamide and dimethylacetamide (p. 235), of unsymmetrical dimethylhydrazine (p. 247), of dimethyllaurylamine oxide used as a foam stabilizer for liquid detergents, of rubber accelerators (p. 293), and of the dimethylamine salts of 2,4-D and 2,4,5-T (p. 444). Originally dimethylamine was used chiefly as a dehairing agent for kid hides. The total quantity used for this purpose has remained about constant, but it now accounts for only about 2 per cent of production. Of the 18 million pounds of methylamine produced in 1964, most of it was used for the manufacture of the insecticide Sevin (p. 491). Excess methylamine may be recycled. Production of trimethylamine was over 12 million pounds. The chief use is for the manufacture of choline chloride (p. 232) used in animal feeds. Excess trimethylamine may be utilized by conversion into dimethylamine and methyl chloride, or into ammonium chloride and methyl chloride.

$$(CH_3)_3NH^{+-}Cl + HCl \xrightarrow{Heat} (CH_3)_2NH_2^{+-}Cl + CH_3Cl$$

$$(CH_3)_3NH^{+-}Cl + 3\,HCl \xrightarrow{Heat} NH_4^{+-}Cl + 3\,CH_3Cl$$

***n*-Butylamines** are prepared commercially from butyl chloride and ammonia, and the **amyl amines** from amyl chlorides and ammonia. All have a wide variety of uses, for example as antioxidants and corrosion inhibitors, absorbents for acid gases, and in the manufacture of oil-soluble soaps. ***s*-Butylamine** applied to citrus fruits after harvesting controls spoilage caused by the mold *Pencillium digitatum.* ***t*-Butylamine** and **(1,1,3,3-tetramethylbutyl)amine** (so-called *t-octylamine*) are made commercially by the Ritter reaction (p. 239). **Higher normal amines** are made by the reduction of nitriles obtained from fat acids through the amides (cf. adiponitrile, p. 631).

QUATERNARY AMMONIUM SALTS

The final product of the reaction of ammonia or of an amine with an alkyl halide is a quaternary ammonium salt (p. 222). The properties of these salts are quite unlike those of the amine salts because the nitrogen no longer carries a hydrogen atom. They do not dissociate into amine and acid when heated, and strong alkalies have no effect on them. When a solution of a quaternary ammonium halide is shaken with silver hydroxide, or when a quaternary ammonium acid sulfate solution reacts with barium hydroxide, the insoluble silver halide or barium sulfate precipitates leaving the quaternary ammonium hydroxide in solution.

$$R_4N^{+-}X + AgOH \longrightarrow R_4N^{+-}OH + AgX$$
$$R_4N^{+-}SO_4H + Ba(OH)_2 \longrightarrow R_4N^{+-}OH + BaSO_4 + H_2O$$

The quaternary ammonium hydroxide dissociates completely into its ions in aqueous solution, and hence it has the same basic strength in water as sodium or potassium hydroxide. For example, glass is etched by solutions of quaternary ammonium hydroxides just as it is by solutions of sodium hydroxide.

When a quaternary ammonium salt is decomposed by heat, it dissociates into tertiary amine and alkyl halide. The dissociation merely is the reverse of the reaction for the formation of quaternary salts.

$$X{:}^- + R{-}\overset{+}{N}R_3 \rightleftharpoons X{:}R + {:}NR_3$$

An analogous reaction can be used for the esterification of sterically hindered acids. Reaction of the acid with the quaternary ammonium hydroxide yields the salt, which is decomposed by heat.

$$(CH_3)_4N^{+-}OOCCR_3 \longrightarrow (CH_3)_3N + CH_3OCOCR_3$$

When tetramethylammonium hydroxide is heated, the products are trimethylamine and methyl ether.

$$2\,(CH_3)_4N^{+-}OH \longrightarrow 2\,(CH_3)_3N + CH_3OCH_3 + H_2O$$

If one of the alkyl groups of a quaternary ammonium hydroxide carries hydrogen on a β carbon atom, the inductive effect of the positive charge on nitrogen increases the acidity of the β hydrogen. When such compounds are heated, the hydroxide ion removes the β hydrogen, tertiary amine is lost, and an alkene is formed.

$$[RCH_2CH_2\overset{+}{N}(CH_3)_3]^{-}OH \longrightarrow H_2O + (CH_3)_3N + RCH{=}CH_2$$

If all of the alkyl groups are primary, the alkyl group eliminated is that which yields the least-substituted olefin, a fact referred to as the *Hofmann rule* (kinetic control, p. 35).

Tertiary amine oxides are internal quaternary ammonium salts, and pyrolysis also yields an olefin, the other product being an *N,N*-dialkylhydroxylamine. The reaction appears to take place by way of a cyclic transition state in which the amine oxide group and the β hydrogen are essentially *cis* to each other.

$$\begin{array}{ll} R_2C{-}{-}CH_2 & \\ \;|\qquad\;\; | & \\ H \quad \overset{+}{N}(CH_3)_2 & \\ \quad\;\; {:}\ddot{O}{:}^- & \end{array} \longrightarrow R_2C{=}CH_2 + (CH_3)_2NOH$$

The relative rate of elimination of alkyl groups as olefin depends primarily on the number of β hydrogens available, provided steric factors or inductive effects are not important. The reaction is useful not only for the preparation of olefins but also for the preparation of *N,N*-dialkylhydroxylamines.

The positive charge on nitrogen in quaternary salts makes an α hydrogen even more acidic than a β hydrogen, and in the absence of a β hydrogen, whose removal permits elimination of a stable alkene, strong bases remove the α hydrogen. The resulting products are another type of internal quaternary salt known as **ylides.**[3] Thus reaction of tetramethylammonium iodide with an alkyllithium yields **trimethylammonium methylide.**

---

[3] Pronounced "illides," not "eyelids."

$$(CH_3)_3\overset{+}{N}CH_3\ Br^- + \overset{\delta-}{R}-\overset{\delta+}{Li} \longrightarrow RH + (CH_3)_3N^{\pm}{=}CH_2{:} + Li^+ + Br^-$$

The simple ylides cannot be isolated free of lithium bromide and may be considered as organolithium compounds of the type $[R_3\overset{+}{N}-\overset{\delta-}{C}H_2-\overset{\delta+}{Li}]Br^-$. Stabilization of the methylide can be brought about by resonance. Thus trimethylammonium dicyanomethylide has been isolated as a pure compound, m.p. 153°.

$$\left\{ {}^{-}{:}\ddot{N}{=}C{=}\underset{{}^{+}N(CH_3)_3}{C}{-}C{\equiv}N{:} \longleftrightarrow {:}N{\equiv}C{-}\underset{{}^{+}N(CH_3)_3}{\overset{\ddot{-}}{C}}{-}C{\equiv}N{:} \longleftrightarrow {:}N{\equiv}C{-}\underset{{}^{+}N(CH_3)_3}{C}{=}C{=}\ddot{N}{:}^{-} \right\} \quad \text{or} \quad N{\equiv}C{-}\underset{{}^{+}N(CH_3)_3}{\overset{-}{C}}{-}C{\equiv}N$$

Ylides have been isolated also in which the negative charge is distributed over aromatic nuclei. The ionic nature of the methylene group is shown by its behavior as a nucleophile in reaction with methyl iodide.

$$(CH_3)_3N^{\pm}{=}CH_2{:} + CH_3I \longrightarrow (CH_3)_3\overset{+}{N}-CH_2-CH_3 + I^-$$

Choline chloride is trimethyl-(2-hydroxyethyl)-ammonium chloride, $[(CH_3)_3\overset{+}{N}CH_2CH_2OH]Cl^-$. Its quaternary ammonium ion, commonly called **choline,** is an extremely important factor in biological processes and must be supplied by the diet of the animal organism. It is a component of the lecithins (p. 610) and has been shown to be a factor which (*1*) is necessary for growth, (*2*) affects fat transport and carbohydrate metabolism, (*3*) is involved in protein metabolism, (*4*) prevents hemorrhagic kidney disintegration, and (*5*) prevents the development of fatty livers in depancreatized dogs. Over 25 million pounds of choline chloride was produced in 1964 for use in animal and poultry feeds to stimulate growth.

Choline ion also is the precursor in the animal organism of acetylcholine ion, $(CH_3)_3\overset{+}{N}CH_2CH_2OCOCH_3$, commonly called **acetylcholine,** which is extremely important in controlling the functions of the body. Nerve impulses arriving at a junction between nerve cells or nerve fibers liberate chemicals that effect the transfer of the impulse to the next cell or fiber. One of these chemicals is acetylcholine. Its action is confined to the site where it is liberated because the blood and neighboring tissues contain an esterase, an enzyme that catalyzes the hydrolysis of esters. When acetylcholine diffuses from the junction of the nerve fibers, it is hydrolyzed immediately to the practically inactive choline thus permitting transmission of single rather than continuous nerve impulses. Some of the most toxic substances known, such as curare, are quaternary salts that can displace acetylcholine at the nerve junctions. By preventing the transmission of nerve impulses by acetylcholine, they lead to total muscular relaxation and paralysis. The highly toxic esters of phosphorus acids (p. 272) and carbamic acid (p. 491), which are used widely as insecticides, exert their action by phosphorylating or acylating acetylcholine esterase, thereby rendering it inactive (p. 272).

Excessive activity of the sympathetic nervous system can lead to high blood pressure and nervous irritability. It can be relieved by blocking the function of the nerve ganglia by the administration of quaternary ammonium salts. **Tetraethylammonium bromide** was used first for this purpose but has been replaced by **hexamethonium** (*hexamethylene-bis-triethylammonium bromide*), $(C_2H_5)_3\overset{+}{N}(CH_2)_6\overset{+}{N}(C_2H_5)_3(Br^-)_2$. **Succinylcholine chloride** (*Sucostrin chloride*), $[(CH_3)_3\overset{+}{N}CH_2CH_2$-$OCOCH_2$-$]_2[Cl^-]_2$, is used as an adjuvant in ether anesthesia. Administration makes it possible to produce complete relaxation with very light ether anesthesia, thus avoiding the undesirable effects of deep anesthesia. **Chlorocholine chloride,** (*2-chloroethyltrimethylammonium chloride*) stunts the growth of plants to as much as 60 per cent of normal and enables florists to regulate plant growth to desired size. Many quaternary ammonium compounds, such as betaine (p. 618), muscarine (p. 514), thiamine (p. 526), and nicotinamide adenine dinucleotide (p. 517), have been isolated from biological sources.

Quaternary ammonium salts in which one of the alkyl groups attached to nitrogen is a long chain hydrocarbon group, such as **cetyltrimethylammonium chloride,** $C_{16}H_{33}(CH_3)_3N^{+-}Cl$, have properties similar to soaps and are known as *invert soaps* or *cationic detergents* because the detergent action resides in a positive ion rather than in a negative ion as is the case with ordinary soaps. Many of these invert soaps have high germicidal action. **Dioctadecyldimethylammonium chloride,** $(C_{18}H_{37})_2(CH_3)_2N^{+-}Cl$, is strongly adsorbed as monomolecular films on textile fibers, and the lubricating action of the long hydrocarbon chains reduces interfiber friction and imparts softness and flexibility and increased strength to the fabric.

AMIDES

The monoacyl derivatives of ammonia, primary amines, and secondary amines, having the general formula $RCONH_2$, $RCONHR$, or $RCONR_2$, are known as *amides*. Diacyl derivatives of ammonia, $(RCO)_2NH$, and of primary amines, $(RCO)_2NR$, are called *imides*. Except for cyclic imides (pp. 475, 525), their preparation usually is more difficult because the first acyl group greatly reduces the basicity of the nitrogen. The direct preparation of triacylamines is still more difficult. From structural theory, isomeric forms of amides that have hydrogen on nitrogen should be possible, but such isomers are tautomeric, and, as with keto-enol tautomers (p. 201), the keto form is the more stable.

$$R{-}\overset{\overset{\displaystyle O}{\|}}{C}{-}NH_2 \rightleftarrows R{-}\overset{\overset{\displaystyle OH}{|}}{C}{=}NH$$

Preparation

The more common methods for the preparation of amides have been described, namely the reaction of ammonia or primary or secondary amines with acyl halides (p. 169), acid anhydrides (p. 171), or esters (p. 174), and the pyrolysis of ammonium or substituted ammonium salts (p. 166).

When amides are heated with carboxylic acids, an exchange reaction takes place leading to an equilibrium mixture of two amides and two acids.

$$RCONH_2 + HOOCR' \rightleftharpoons \begin{matrix} & O & \\ & \| & \\ R{-} & C & \cdots NH_2 \\ & \vdots & \vdots \\ HO & \cdots & C{=}O \\ & & | \\ & & R' \end{matrix} \rightleftharpoons RCOOH + H_2NCOR'$$

If urea, the amide of carbonic acid (p. 282), is heated with a carboxylic acid, one product is the unstable carbamic acid, which decomposes, and the reaction goes to completion.

$$CO(NH_2)_2 + RCOOH \rightleftharpoons RCONH_2 + [H_2NCOOH] \longrightarrow NH_3 + CO_2$$

Amides can be prepared also by the partial hydrolysis of nitriles (p. 238) and by the addition of olefins to nitriles (p. 239).

Nomenclature

The simple amides are named by replacing *ic acid* or *oic acid* in the name of the acid by *amide*. Thus $HCONH_2$ is formamide or methanamide, and $(CH_3)_2CHCONH_2$ is *i*-butyramide or methylpropanamide. Amides derived from amines are named as nitrogen substitution products; for example $CH_3CONHCH_3$ is *N*-methylacetamide, or *N*-methylethanamide.

Physical Properties

Most amides containing the $CONH_2$ group are solids. Formamide melts at 2°. The *N*-alkyl substituted amides of the aliphatic acids usually are liquids. Amides that have hydrogen on nitrogen are associated liquids and have high boiling points. Formamide, *N*-methylformamide, and *N,N*-dimethylformamide boil at 193°, 180°, and 155°. As expected, although the polarizability increases with increasing molecular weight, the boiling point decreases as the ability to form hydrogen bonds decreases. Nevertheless the boiling point of *N,N*-dimethylformamide is abnormally high even though it cannot form hydrogen bonds. It would be expected to have about the same polarizability as ethyl formate, b.p. 54°, but because of its high dipole moment of 3.84 D compared with 1.92 D for ethyl formate, it boils at 155°.

Because amides that have hydrogen on nitrogen form hydrogen bonds with other compounds of oxygen and nitrogen, they are solvents for and are soluble in such compounds, but insoluble in hydrocarbon solvents. Unsubstituted amides of monocarboxylic acids containing five atoms or less are soluble in water. Amides have high dielectric constants ($\epsilon = 37.6$ for *N,N*-dimethylformamide) and are excellent solvents not only for covalent compounds but for many salts as well. Nucleophilic displacement reactions take place much more rapidly in *N,N*-dimethylformamide or *N,N*-dimethylacetamide, which cannot form hydrogen bonds, than in water or the alcohols because the nucleophile is much less strongly solvated in the aprotic solvent than in the protic solvent (p. 102).

The position of the infrared absorption bands of amides varies considerably depending on the physical state of the sample, on the possibility of hydrogen bonding, and on whether the nitrogen is primary, secondary, or tertiary. The C=O and N—H stretching absorptions lie in the regions 1700–1630 cm.$^{-1}$ and 3500–3070 cm.$^{-1}$ respectively.

Reactions

1. ***Basic and Acidic Properties.*** In contrast to ammonia and the amines ($pK_a = 9$ to 11), the simple amides are such weak bases ($pK_a = -1$ to -2) that they do not form salts that are stable in aqueous solution; that is, the replacement of a hydrogen atom of the ammonia molecule by an acyl group gives a compound that has about the same basicity as water ($pK_a = -1.74$). This effect is in harmony with the effect of replacement of a hydrogen atom of a water molecule by an acyl group. The carbonyl group exerts an electron-attracting effect, which increases the ease of ionization of the remaining proton of carboxylic acids and decreases the ability of the nitrogen atom of amides to share its unshared pair. The reduced basicity of the nitrogen atom makes the amide a poorer nucleophile and accounts for the fact that only strong electrophiles such as acyl chlorides are capable of introducing a second acyl group to form imides. When two hydrogen atoms of ammonia are replaced by acyl groups, the attraction for electrons is sufficient to permit the remaining hydrogen to be removed as a proton by strong bases in aqueous solution. In other words the imides are weak acids.

$$(RCO)_2NH + NaOH \longrightarrow (RCO)_2N^{-+}Na + H_2O$$

Unlike the uncharged imide, the negative imide ion is a sufficiently good nucleophile to react with acid anhydrides and yield the triacylated ammonia.

$$(RCO)_2N^{-+}Na + R'COCl \longrightarrow (RCO)_2NCOR' + NaCl$$

2. ***Hydrolysis.*** The hydrolysis of amides produces the acid and the amine. Like the hydrolysis of esters, the reaction is catalyzed by both acids and bases. With esters, however, only the base-catalyzed reaction goes to completion, whereas the hydrolysis of amides by either acid or base involves an irreversible step.

Catalysis by acid

$$\mathrm{R\overset{O}{\overset{\|}{C}}{-}NH_2} \underset{B^-}{\overset{HB}{\rightleftharpoons}} \left\{ \mathrm{R\overset{^+OH}{\overset{\|}{C}}{-}NH_2} \longleftrightarrow \mathrm{R\overset{OH}{\overset{|}{C}}{=}\overset{+}{N}H_2} \right\} \overset{H_2O}{\rightleftharpoons}$$

$$\mathrm{R\underset{^+OH_2}{\underset{|}{\overset{OH}{\overset{|}{C}}}}{-}NH_2} \underset{HB}{\overset{B^-}{\rightleftharpoons}} \mathrm{R\underset{OH}{\underset{|}{\overset{OH}{\overset{|}{C}}}}{-}NH_2} \underset{B^-}{\overset{HB}{\rightleftharpoons}} \mathrm{R\underset{OH}{\underset{|}{\overset{O-H}{\overset{|}{C}}}}{-}\overset{+}{N}H_3} \longrightarrow RCOOH + {}^+NH_4$$

Catalysis by base

$$\mathrm{R\overset{O}{\overset{\|}{C}}{-}NH_2} \overset{^-OH}{\rightleftharpoons} \mathrm{R\underset{O-H}{\underset{|}{\overset{O^-}{\overset{|}{C}}}}{-}NH_2} \longrightarrow RCOO^- + NH_3$$

3. ***Dehydration of Primary Amides.*** Distillation of an unsubstituted amide with an acid anhydride or an acid halide such as phosphorus pentoxide or thionyl chloride removes a molecule of water to yield an alkyl cyanide. The reaction may be carried out under milder conditions by heating the amide with phosphorus oxychloride and a tertiary amine such as pyridine. Presumably nucleophilic attack on the acid chloride is followed by removal of protons in a series of steps.

$$RC(=O)NH_2 \underset{Cl^-}{\overset{POCl_3}{\rightleftharpoons}} RC(=\overset{+}{O}{:}POCl_2)(HN{-}H) \underset{R_3NH^+}{\overset{R_3N}{\rightleftharpoons}} RC(-OPOCl_2)(=N{-}H) \underset{R_3NH^+}{\overset{R_3N}{\rightleftharpoons}} RC{\equiv}N + {}^-OPOCl_2$$

The reactions of phosphorus pentoxide or thionyl chloride in the absence of pyridine may involve cyclic transition states. Nitriles are made commercially by the vapor phase dehydration of amides over a solid, acid catalyst such as alumina or boron phosphate at 350–400° (p. 631).

4. ***Conversion to Amines.*** ***(a)*** WITHOUT LOSS OF CARBON. Lithium aluminum hydride reduces primary amides to primary amines.

$$RC(=O){-}NH_2 \overset{H:^-}{\rightleftharpoons} RC(O^-)(H){-}NH_2 \rightleftharpoons RC(OH)(H){-}\ddot{N}H^- \rightleftharpoons HO^- + RC(H){=}NH \xrightarrow{H:^-} RCH_2NH^- \underset{^-OH}{\overset{H_2O}{\rightleftharpoons}} RCH_2NH_2$$

Amines are formed instead of alcohols because the more electronegative oxygen enables the hydroxyl group to leave as hydroxide ion more readily than the amino group as amide ion. *N*-Alkyl amides are converted to secondary amines and *N,N*-dialkyl amides to tertiary amines. The theoretical amounts of lithium aluminum hydride required per mole of amide are respectively 1, 0.75, and 0.5 mole. Borane in tetrahydrofuran appears to be preferable to lithium aluminum hydride as a reducing agent in that the reaction is faster and yet more selective. Thus amide groups can be reduced without affecting other reducible groups such as the nitro group.

(b) WITH LOSS OF CARBON. The conversion of amides to primary amines with one less carbon atom by the Hofmann rearrangement has been discussed on page 223.

5. ***Conversion to Aldehydes or Ketones.*** Lithium aluminum triethoxide hydride is a less powerful reducing agent than lithium aluminum hydride (p. 170). It reduces amides only to the hydroxy amino stage, which loses amine to give the aldehyde. *N,N*-Dialkyl amides use only one third the amount of reagent required by unsubstituted amides.

$$RC(=O){-}N(CH_3)_2 \overset{H:^-}{\rightleftharpoons} RCH(O^-){-}N(CH_3)_2 \longrightarrow RCHO + {}^-{:}N(CH_3)_2$$

Alkide ion from reactive organometallic compounds displaces amide ion from *N,N*-dialkyl amides just as hydride ion does to give either aldehydes or ketones.

$$HCON(CH_3)_2 + R'MgX \longrightarrow R'CHO + (CH_3)_2N^{-+}Mg^+X^-$$

$$RCON(CH_3)_2 + R'CH_2Li \longrightarrow RCOCH_2R' + Li^{+-}N(CH_3)_2$$

Organolithium compounds are limited practically to only a few primary alkyl groups, but they give better yields of ketones than do Grignard reagents.

Uses

Of the simple amides, only a few are of commercial importance. ***N,N*-Dimethylformamide** and ***N,N*-dimethylacetamide** are made by the aminolysis of methyl formate or methyl acetate with dimethylamine and are used as solvents for the spinning of acrylic fibers. Dimethylformamide is used widely as a reaction medium (p. 102). **Oleamide** is used to improve the surface characteristics of polyethylene and polypropylene film. Since many amides are solids that can be crystallized readily and have characteristic melting points, they are used as derivatives for the identification of carboxylic acids, esters, acid halides, and nitriles. Of the more complex amides, urea and its derivatives (p. 282), the nylons (pp. 631, 649), and the proteins (p. 359) are among the most important.

DERIVATIVES OF IMIDIC ACIDS

The tautomeric form of an amide, $\mathrm{RC({=}NH){-}OH}$, is a nitrogen analog of a carboxylic acid and is known as an **imidic acid.** Although the imidic acids cannot be isolated because of the greater stability of the amide form, their derivatives are easily prepared. Their acid chlorides are known as **imide chlorides** and are obtained when an *N*-alkyl amide reacts with phosphorus pentachloride.

$$\mathrm{RC(NHR'){=}O \rightleftharpoons RC({=}NR'){-}OH \xrightarrow{PCl_5} RC({=}NR'){-}Cl + POCl_3 + HCl}$$

When dry hydrogen chloride is passed into a mixture of an alkyl cyanide and an alcohol in ether solution, the hydrochloride of an **alkyl imidate** (also called *imido ester, imino ester,* or *imino ether*) precipitates (*Pinner synthesis*). Undoubtedly the imide chloride is an intermediate.

$$\mathrm{RC{\equiv}N + HCl \longrightarrow RC({=}NH){-}Cl \xrightarrow{R'OH} RC({=}NH)(\overset{+}{H}OR') + Cl^- \longrightarrow [RC({=}\overset{+}{N}H_2){-}OR']\,Cl^-}$$

If sodium carbonate is added in the presence of ether, the free alkyl imidate is formed and goes into the ether solution.

$$\mathrm{[RC({=}\overset{+}{N}H_2){-}OR']\,Cl^- + Na_2CO_3 \longrightarrow RC({=}NH){-}OR' + NaCl + NaHCO_3}$$

Addition of water hydrolyzes the alkyl imidate hydrochloride to an ester.

$$\mathrm{[RC({=}\overset{+}{N}H_2){-}OR']\,Cl^- + H_2O \longrightarrow RC({=}O){-}OR' + NH_4Cl}$$

When the hydrochloride is warmed with an excess of alcohol, an ortho ester (p. 177) is formed.

$$\mathrm{[RC({=}\overset{+}{N}H_2){-}OR']\,Cl^- + 2\,HOR' \longrightarrow RC(OR')_2{-}OR' + NH_4Cl}$$

Reaction of either an imide chloride or an alkyl imidate with ammonia yields the imidic amide, which is known as an **amidine.**

$$\mathrm{RC({=}NR'){-}Cl + 2\,NH_3 \longrightarrow RC({=}NR'){-}NH_2 + NH_4Cl}$$

$$\mathrm{RC({=}NH){-}OR' + NH_3 \longrightarrow R{-}C({=}NH){-}NH_2 + HOR'}$$

In contrast to the amides, which are very weak bases ($pK_a = \sim 1$), the amidines are strong bases ($pK_a = \sim 12$). The stability of the salt is explained by its high resonance energy, the contributing resonance forms being identical. In the amide salt, resonance contributes very little to the stability of the cation relative to the free amide because the possible forms are not identical, resonance being inhibited by the greater electronegativity of oxygen compared to nitrogen (cf. p. 161).

$$\left\{ \begin{matrix} \overset{+}{N}H_2 \\ \| \\ RC{-}NH_2 \end{matrix} \longleftrightarrow \begin{matrix} NH_2 \\ | \\ RC{=}\overset{+}{N}H_2 \end{matrix} \right\} \qquad \left\{ \begin{matrix} \overset{+}{O}H \\ \| \\ RC{-}NH_2 \end{matrix} \longleftrightarrow \begin{matrix} OH \\ | \\ RC{=}\overset{+}{N}H_2 \end{matrix} \right\}$$

In neither the amides nor the amidines is there any tendency for a proton to combine with an NH_2 group rather than oxygen or NH, since the slight amount of resonance in the amide or the amidine group would be destroyed without producing a more highly stabilized ion (p. 283).

ALKYL CYANIDES (NITRILES)

Nomenclature

The alkyl cyanides, $RC{\equiv}N$, are alkyl derivatives of hydrogen cyanide and generally are named as if they were salts; for example CH_3CN is called methyl cyanide. Organic cyanides also are called *nitriles*.[4] The common names are derived from the common names of the corresponding carboxylic acids. They are formed by dropping *ic acid* and adding *nitrile* with a connective *o;* for example methyl cyanide, which yields acetic acid on hydrolysis, is called acetonitrile. In the international system the longest chain having the nitrile group as a terminal group determines the parent name. Thus $CH_3(CH_2)_{11}CN$ is tridecanenitrile. When it is necessary to express the $C{\equiv}N$ group as a substituent, it is called *cyano;* for example, $N{\equiv}CCH_2COOH$ is cyanoacetic acid.

Preparation

Alkyl cyanides usually are prepared by the reaction of alkyl halides, sulfates, or toluenesulfonates (p. 407) with sodium or potassium cyanide in aqueous alcoholic solution, or better in dimethylformamide or methyl sulfoxide (p. 102). Small amounts of isocyanides (p. 240) also are formed because the cyanide ion has an unshared pair of electrons on nitrogen as well as on carbon.

$$:C{\equiv}N:^- + RX \longrightarrow :C{\equiv}N:R + X^-$$

Ions or molecules that have two nucleophilic sites are said to be *ambident* (L. *ambo* both, *dens* tooth). The cyanide is the chief product, however, and formation of the more covalent bond is the usual course of the reactions of ambident anions (pp. 242, 267). The isocyanide can be removed from the mixture by partial hydrolysis because it is hydrolyzed more rapidly than the nitrile. Alkyl cyanides can be prepared also by the dehydration of amides (p. 235) or of aldoximes.

$$RCH{=}NOH + P_2O_5 \longrightarrow RC{\equiv}N + 2\,HPO_3$$

The latter methods are indirect procedures for converting acids or aldehydes into nitriles.

Hydrogen cyanide adds to olefins in the gas phase at 350° in the presence of activated alumina, or homogeneously at 130° in the presence of dicobalt octacarbonyl (p. 215). Thus isobutylene gives *t*-butyl cyanide (*pivalonitrile*).

---

[4] Fehling first prepared the compound C_7H_5N in 1844 by pyrolysis of ammonium benzoate. Since the names *nitrobenzoyl* and *azobenzoyl* already had been assigned to other compounds, he called the new compound *benzonitril,* merely to indicate that it also contained nitrogen and was derived from benzoic acid.

$$(CH_3)_2C{=}CH_2 + HCN \longrightarrow (CH_3)_3CCN$$

The yields are best from olefins having a terminal double bond. Secondary and tertiary alkyl cyanides can be prepared also by the alkylation of primary or secondary alkyl cyanides (p. 239).

Physical Properties

Nitriles have abnormally high boiling points because of their high dipole moments. Thus propane, methyl ether, and methyl cyanide, all of about the same polarizability, equally unbranched, and incapable of forming hydrogen bonds, have dipole moments of 0, 1.3, and 4.0 D and boil at −42°, −24°, and 82° respectively.

Cyanides are less soluble in water than amines having the same number of carbon atoms. Hydrogen cyanide and acetonitrile are miscible with water, and propionitrile is fairly soluble, but the higher nitriles are only slightly soluble. The decreased tendency to form hydrogen bonds with water can be ascribed to the reduced basicity of the nitrogen atom (reduced tendency to share the unshared pair of electrons). The cyanides do not dissolve in aqueous acids because they are too weakly basic to form salts. The weak basic properties of nitriles, despite the presence of an unshared pair of electrons on nitrogen, may be explained in the same way as the acidity of acetylenic hydrogen (p. 148). In the conjugate acid of the nitrile, $RC{\equiv}\overset{+}{N}{-}H$, the σ bonds to nitrogen are *sp* hybrids. The pair of electrons bonding hydrogen is much closer to the nitrogen nucleus than that in the conjugate acid of an amine, $R_3\overset{+}{N}{:}H$, where the bonds are sp^3 hybrids and have much less *s* character. Hence the conjugate acid of a nitrile is a much stronger acid ($pK_a = -10$) than that of an amine ($pK_a = 10$), and the nitrile is a much weaker base than the amine.

Just as the carbon-carbon triple bond is stronger and absorbs in the infrared at higher frequencies than the carbon-carbon double bond (p. 147), so the carbon-nitrogen triple bond is stronger and absorbs at higher frequencies than the carbon-nitrogen double bond. The range for the C≡N stretching absorption is 2260–2240 $cm.^{-1}$ (cf. Fig. 20–6, p. 393) compared to 2140–2100 $cm.^{-1}$ for the monosubstituted acetylenes and 1690–1640 $cm.^{-1}$ for the carbon-nitrogen double bond.

Physiological Action

Pure alkyl cyanides have a pleasant odor and are only moderately toxic. Usually, however, they are contaminated with the disagreeably odorous and highly toxic isocyanides. α-Hydroxy and α-amino nitriles are toxic, probably because of easy loss of hydrogen cyanide.

Reactions

The triple bond of nitriles is strongly polarized as shown by the high dipole moment of 4.0 D. Because of the dipole they are subject both to electrophilic attack on nitrogen and to nucleophilic attack on carbon. Hence addition reactions are catalyzed by both acids and bases. The strongly electronegative character of the nitrile group that results from the *sp* hybridization and the electronegative nitrogen makes the α hydrogen readily removable as a proton. Hence bases can also convert nitriles into good nucleophiles.

1. ***Hydrolysis to Amides and Acids.*** The acid- or base-catalyzed hydrolysis of nitriles is analogous to the hydrolysis of esters. The first stable product is an amide.

Catalysis by acid

$$RC{\equiv}N{:} \underset{B^-}{\overset{HB}{\rightleftharpoons}} \overbrace{RC{\equiv}NH}^{+} \xrightarrow{H_2O} \underset{{}^+OH_2}{RC{=}NH} \underset{HB}{\overset{B^-}{\rightleftharpoons}} \underset{OH}{RC{=}NH} \rightleftharpoons \underset{O}{\overset{}{RC{-}NH_2}}$$

Catalysis by base

$$\mathrm{RC{\equiv}N\colon} \underset{}{\overset{^{-}OH}{\rightleftharpoons}} \underset{\displaystyle\mathrm{OH}}{\mathrm{RC{=}N\colon^{-}}} \underset{OH^{-}}{\overset{H_2O}{\rightleftharpoons}} \underset{\displaystyle\mathrm{OH}}{\mathrm{RC{=}NH}} \rightleftharpoons \underset{\displaystyle\mathrm{O}}{\mathrm{RC{-}NH_2}}$$

The amide, however, is irreversibly hydrolyzed further under the same conditions (p. 234). The result is that the nitrile ends up as the carboxylic acid or its salt. Hydrochloric acid or aqueous sodium hydroxide are the usual reagents, and the over-all reactions are

$$\mathrm{RC{\equiv}N + 2\,H_2O + HCl \longrightarrow RCOOH + NH_4{}^{+-}Cl}$$
$$\mathrm{RC{\equiv}N + H_2O + NaOH \longrightarrow RCOO^{-+}Na + NH_3}$$

If the further hydrolysis of the amide is sterically hindered by the presence of a tertiary alkyl group, the intermediate amide can be obtained in good yield. Thus trimethylacetamide results when *t*-butyl cyanide is boiled with 80 per cent sulfuric acid.

$$\mathrm{(CH_3)_3CCN + H_2O \xrightarrow{H_2SO_4,\ heat} (CH_3)_3CCONH_2}$$

To convert primary alkyl cyanides to amides, the best reagent is alkaline hydrogen peroxide. The rate of nucleophilic attack by hydroperoxide ion is 10,000 times that of attack by hydroxide ion, with the result that amide formation is very much faster than hydrolysis of the amide to acid.

$$\mathrm{H_2O_2} \underset{H_2O}{\overset{OH^{-}}{\rightleftharpoons}} \mathrm{HO_2{}^{-}} \xrightarrow{RC\equiv N} \underset{\displaystyle\mathrm{O{-}OH}}{\mathrm{RC{=}N^{-}}} \underset{OH^{-}}{\overset{H_2O}{\rightleftharpoons}} \underset{\displaystyle\mathrm{O{-}OH}}{\mathrm{RC{=}NH}}$$

$$\underset{\displaystyle\mathrm{O{-}OH}}{\mathrm{RC{=}NH}} + \mathrm{H{-}O{-}O{-}H} \longrightarrow \underset{\displaystyle\mathrm{O}}{\mathrm{RC{-}NH_2}} + \mathrm{O_2 + H^{+}} + {}^{-}\mathrm{OH}$$

2. ***Alkylation.*** Primary and secondary alkyl cyanides are alkylated readily when heated with an alkyl halide and finely powdered sodium amide.

$$\mathrm{RCH_2CN} \underset{NH_3}{\overset{^{-}NH_2}{\rightleftharpoons}} \mathrm{R\bar{C}HCN} \underset{X^{-}}{\overset{R'X}{\rightleftharpoons}} \underset{\displaystyle\mathrm{R'}}{\mathrm{RCHCN}}$$

$$\mathrm{R_2CHCN} \underset{NH_3}{\overset{^{-}NH_2}{\rightleftharpoons}} \mathrm{R_2\bar{C}CN} \underset{X^{-}}{\overset{R'X}{\rightleftharpoons}} \underset{\displaystyle\mathrm{R'}}{\mathrm{R_2CCN}}$$

This reaction provides the most satisfactory general method for preparing tertiary alkyl cyanides.

3. ***Addition of Olefins.*** Tertiary carbonium ions formed from either alkenes or alcohols add to nitriles to give **nitrilium salts.** Addition of water gives an ***N-t*-alkyl amide** (*Ritter reaction*). The reaction is carried out by adding sulfuric acid to an acetic acid solution of the nitrile and of the alkene or alcohol.

$$\mathrm{R_2C{=}CH_2} \underset{^{-}SO_4H}{\overset{H_2SO_4}{\rightleftharpoons}} \underset{\displaystyle\mathrm{CH_3}}{\mathrm{R_2C^{+}}} \xrightarrow{R'C\equiv N} \underset{\displaystyle\mathrm{CH_3}}{\mathrm{R_2C{-}\overset{+}{N}{\equiv}CR'}} \underset{H^{+}}{\overset{H_2O}{\rightleftharpoons}} \underset{\displaystyle\mathrm{CH_3}\quad\mathrm{OH}}{\mathrm{R_2C{-}N{=}CR'}} \rightleftharpoons \underset{\displaystyle\mathrm{CH_3}\quad\mathrm{O}}{\mathrm{R_2C{-}NHCR'}}$$

The amides derived from alkyl cyanides are difficult to hydrolyze in alkaline solution because of steric hindrance, and acid hydrolysis regenerates the olefin. The formamides derived from hydrogen cyanide, however, can be hydrolyzed by bases, and the reactions provide a convenient synthesis of **tertiary carbinamines.** Isobutylene, for example, gives *N-t*-butylformamide, which can be hydrolyzed to *t*-butylamine.

$$(CH_3)_2C{=}CH_2 + HCN \underset{}{\overset{H^+}{\rightleftharpoons}} (CH_3)_3C{-}\overset{+}{N}{\equiv}CH \underset{H^+}{\overset{H_2O}{\rightleftharpoons}} (CH_3)_3C{-}NHCHO \xrightarrow{NaOH} (CH_3)_3CNH_2 + Na^{+-}OCHO$$

4. ***Other Reactions.*** The conversion of nitriles to esters and to ortho esters by way of the alkyl imidate hydrochlorides (p. 236), and the reduction of nitriles to amines (p. 223) have been discussed. By the proper choice of reducing agent, reduction can be stopped at the imine stage and subsequent hydrolysis yields aldehydes. Suitable reagents for this purpose are lithium aluminum triethoxide hydride (p. 170), the sodium analog made from sodium hydride and aluminum ethoxide, or di-*i*-butylaluminum hydride (p. 113).

$$RC{\equiv}N \xrightarrow{LiAl(OC_2H_5)_3H} \underset{\displaystyle H}{RC}{=}N{:}^{-+}LiAl(OC_2H_5)_3 \xrightarrow{5\,H_2O} RCHO + NH_3 + LiOH + Al(OH)_3 + 3\,C_2H_5OH$$

Reduction with sodium hypophosphite in the presence of Raney nickel also yields aldehydes.

$$RC{\equiv}N + Na^{+-}O\overset{O^-}{\overset{|+}{P}}H_2 + 2\,H_2O \xrightarrow{Ni} RCHO + Na^{+-}O\overset{O^-}{\overset{|+}{P}}H(O^{-+}NH_4)$$

Uses

Acetonitrile can be prepared from a mixture of acetic acid vapor and ammonia over a dehydrating catalyst such as alumina at around 400°. It is a coproduct in the production of acrylonitrile from propylene and ammonia (p. 619). Most of the acetonitrile is used in the separation and purification of butadiene (p. 587) and isoprene (p. 588) by extractive distillation. It is used also as a solvent for polymers, for extraction and crystallization, and as a reaction medium. **Adiponitrile** (p. 631) is an intermediate in the synthesis of nylon 6-6. Other nitriles have not found large scale commercial use, but they are important intermediates in the laboratory synthesis of acids, esters, amides, and amines.

ALKYL ISOCYANIDES (ISONITRILES)

As the name implies, the isocyanides or isonitriles are isomeric with the cyanides or nitriles. One of the classical methods of synthesis is that from primary amines, chloroform, and alcoholic potassium hydroxide reported by Hofmann in 1867. The strong base is believed to convert the chloroform into the very reactive intermediate, dichlorocarbene (p. 178), which is a strong electrophile and reacts immediately with the amine. Loss of two molecules of hydrogen chloride yields the isocyanide.

$$HCCl_3 \xrightarrow[H_2O\,+\,Cl^-]{^-OH} \ddot{C}Cl_2 \xrightarrow{RNH_2} R\overset{H}{\underset{H}{N}}{}^{+}{:}^{-}\underset{Cl}{C}{-}Cl \xrightarrow[H_2O\,+\,Cl^-]{^-OH} R\underset{H}{N}\overset{+-}{=}\underset{Cl}{C}{:} \xrightarrow[H_2O\,+\,Cl^-]{^-OH} \underset{\text{An isocyanide}}{RN\overset{+-}{\equiv}C{:}}$$

The other method, published by Gautier in 1868, is by the reaction of an alkyl halide with silver cyanide.

$$RX + AgCN \longrightarrow R\overset{+}{N}{\equiv}\overset{-}{C}{:} + AgX$$

In 1958 a more general and convenient procedure than the classical methods for the production of isonitriles led to a revival of interest in them. Whereas dehydration of unsubstituted amides yields nitriles (p. 235), the dehydration of *N*-alkylformamides gives isonitriles. The preferred reagent is phosphorus oxychloride in the presence of pyridine or potassium *t*-butoxide. The initial step is analogous to that in the dehydration of unsubstituted amides (p. 235).

$$\underset{\text{H}\ \ \text{H}}{\text{RN—C=O}} \underset{\text{Cl}^-}{\overset{\text{POCl}_3}{\rightleftharpoons}} \underset{\text{H}\ \ \text{H}}{\text{RN—C=}\overset{+}{\text{O}}\text{:POCl}_2} \underset{\text{C}_5\text{H}_5\text{NH}^+}{\overset{\text{C}_5\text{H}_5\text{N:}}{\rightleftharpoons}} \underset{\text{H}}{\text{R}\ddot{\text{N}}\text{=C—OPOCl}_2} \underset{\text{C}_5\text{H}_5\text{NH}^+}{\overset{\text{C}_5\text{H}_5\text{N:}}{\rightleftharpoons}} \text{RN}\equiv\text{C: + }^-\text{OPOCl}_2$$

Physical Properties

The isocyanides boil about 20° lower than the cyanides. Methyl cyanide boils at 81° and methyl isocyanide at 60°. The corresponding ethyl derivatives boil at 97° and 78°. The lower boiling points of the isocyanides presumably are due to their lower dipole moment (~3 D) compared to that of the cyanides (~4 D). Because the isocyanide has a semipolar bond (p. 240), its dipole moment might be expected to be higher than that of the cyanide. However, since the moment due to the semipolar bond is in the opposite direction to that due to the difference in electronegativity of carbon and nitrogen, the resulting moment is smaller for isocyanides than for nitriles, and the $\overset{+}{\text{N}}\equiv\overset{-}{\text{C}}$ stretching absorption is at lower frequencies (2183–2144 cm.$^{-1}$).

The isonitriles have an obnoxious odor that can be detected in minute amounts. They are highly toxic.

Reactions

When isonitriles are heated with metal cyanides at 100°, nucleophilic displacement brings about rearrangement to the more stable nitrile.

$$\text{:}\overset{-}{\text{N}\equiv\text{C}}\text{: + R:N}\equiv\text{C:} \rightleftharpoons \text{:N}\equiv\text{C:R + :}\overset{-}{\text{N}\equiv\text{C}}\text{:}$$

Because of their high polarity and the unshared pair of electrons on carbon, the isocyanides are highly reactive. The reactions usually involve nucleophilic attack on the reagent followed by nucleophilic attack on carbon, the result being the addition of both portions of the reagent to the isonitrile carbon atom. Conceivably the reaction may be concerted.

$$\text{R}\overset{+}{\text{N}}\equiv\overset{-}{\text{C}}\text{: + }A\text{:}B \longrightarrow \text{R}\overset{+}{\text{N}}\equiv\text{C:}A\text{ + }^-\text{:}B \longrightarrow \text{RN=C}\begin{matrix}A\\B\end{matrix}$$

Halogen, Grignard reagents, amines, water, alcohols, hydrogen sulfide, mercaptans, acyl chlorides, and mild oxidizing agents are among the many substances that react. Isonitriles are stable to bases but are hydrolyzed more readily by acids than nitriles, yielding an amine salt and formic acid.

$$\text{RN}\equiv\text{C + H}_2\text{O} \xrightarrow{\text{H}^+} \text{RN=C}\begin{matrix}\text{OH}\\\text{H}\end{matrix} \rightleftharpoons \text{RNH—}\overset{\text{O}}{\overset{\|}{\text{C}}}\text{—H} \xrightarrow{\text{H}_2\text{O, H}^+} \text{RNH}_3^+ + \text{HO}\overset{\text{O}}{\overset{\|}{\text{C}}}\text{H}$$

Reduction by sodium in alcohol, by metal hydrides, or by catalytic hydrogenation yields the secondary amine.

$$\text{R}\overset{+}{\text{N}}\equiv\overset{-}{\text{C}}\text{: + 2 H}_2 \xrightarrow{\text{Pt}} \text{RNHCH}_3$$

This reaction was an early proof that the nitrogen is bound to an alkyl group and a single carbon atom.

Few of the simple reactions of isonitriles are of practical value. Recent studies, however, have revealed numerous reactions that are more involved and lead to new types of complex nitrogen compounds that may prove useful.

NITROALKANES

Nitro compounds have the general formula RNO_2 in which the nitro (NO_2) group is linked through nitrogen to carbon. Like the carboxylate ion (p. 159), a nitroalkane is a resonance hybrid.

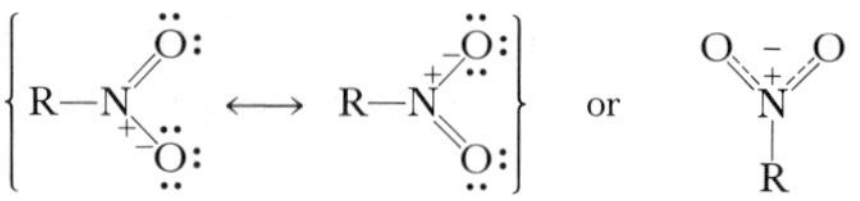

$$\left\{ R-\overset{+}{N}\begin{matrix} \nearrow \ddot{O}: \\ \searrow :\ddot{O}:^- \end{matrix} \longleftrightarrow R-\overset{+}{N}\begin{matrix} \nearrow :\ddot{O}:^- \\ \searrow \ddot{O}: \end{matrix} \right\} \quad \text{or} \quad \overset{O \; - \; O}{\underset{R}{\overset{+}{N}}}$$

The barrier to free rotation about the carbon-carbon bond in ethane is about 3000 cal. per mole because the hydrogen atoms prefer to occupy staggered positions rather than opposed positions (p. 56). In contrast the barrier to free rotation about the carbon-nitrogen bond in nitromethane is only about 6 cal. per mole as determined by microwave spectroscopy (p. 542). This result confirms the view that trigonal hybridization leads to a planar distribution of the nitrogen bonds in nitro compounds, since there is no favored position for the oxygen atoms in nitromethane. When one oxygen is between two hydrogens of the methyl group, the other is opposed to the third hydrogen.

Prior to 1940 the nitroalkanes were largely of theoretical interest. With the advent of the commercial development of vapor phase nitration of propane, however, the lower nitroalkanes were made available, and their numerous reactions became of considerable importance. Nitro compounds are named as substitution products. Thus CH_3NO_2 is nitromethane and $CH_3CH(NO_2)CH_3$ is 2-nitropropane.

Preparation

1. ***From Alkyl Halides and Nitrites.*** The nitrite ion, like the cyanide ion, is an ambident anion (p. 237), and nucleophilic attack on an alkyl halide can yield either an alkyl nitrite or a nitroalkane.

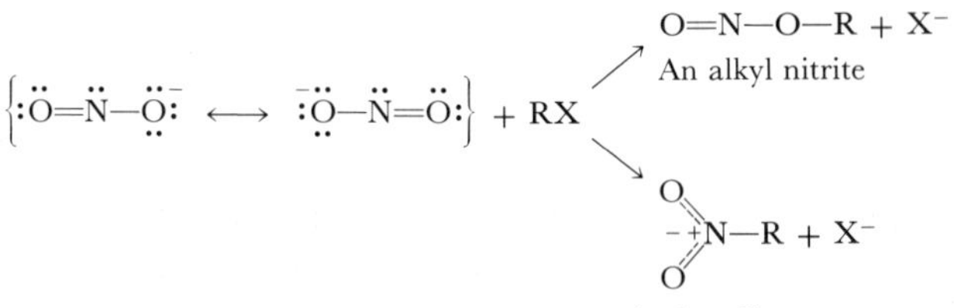

$$\left\{ :\ddot{O}=\ddot{N}-\ddot{O}:^- \longleftrightarrow {}^-:\ddot{O}-\ddot{N}=\ddot{O}: \right\} + RX \begin{matrix} \nearrow O=N-O-R + X^- \quad \text{An alkyl nitrite} \\ \searrow O_2\overset{+}{N}-R + X^- \quad \text{A nitroalkane} \end{matrix}$$

Sodium nitrite in methyl sulfoxide or in dimethylformamide cantaining some urea reacts with primary or secondary alkyl bromides or iodides to give 50–60 per cent yields of nitroalkane along with 25–35 per cent of alkyl nitrite. Tertiary halides yield chiefly alkyl nitrite or alkene. Silver nitrite, first used by Victor Meyer[5] in 1874, gives better yields of primary nitro compounds (80 per cent) but is a more costly reagent.

Nitromethane is prepared readily in fair yield by the reaction of an aqueous solution of sodium chloroacetate with sodium nitrite and distillation of the mixture. The intermediate sodium nitroacetate decomposes to nitromethane and sodium bicarbonate (cf. *haloform reaction*, p. 211).

---

[5] Victor Meyer (1848–1897), professor of chemistry successively at Zuerich, Goettingen, and Heidelberg. He is noted for his discovery and investigation of nitro compounds and of thiophene (p. 504), and for his work on steric hindrance (p. 469). He was the first to use the term *stereochemistry*. A once widely used apparatus for the determination of molecular weights bears his name.

$$ClCH_2COONa + NaNO_2 \longrightarrow NO_2CH_2COO^{-+}Na + NaCl$$

$$O_2NCH_2{-}\overset{O}{\overset{\|}{C}}{-}O{:}^{-} \longrightarrow CO_2 + O_2NCH_2^{-} \xrightarrow{H_2O} O_2NCH_3 + HOCO_2^{-}$$

2. ***Vapor Phase Nitration of Hydrocarbons.*** The lower nitroalkanes are made commercially by the reaction of propane with nitric acid at 420°.

$$RH + HONO_2 \xrightarrow{420^\circ} RNO_2 + H_2O$$

Attack by free nitro radicals at this temperature is nonselective and carbon-carbon as well as carbon-hydrogen bonds are broken. Propane yields not only 1- and 2-nitropropane but also nitroethane and nitromethane. These products are separated by distillation.

3. ***Oxidation of Oximes.*** Primary and secondary nitro compounds can be prepared by the oxidation of oximes with trifluoroperoxyacetic acid (*trifluoroacetic acid and hydrogen peroxide*).

$$RCH{=}NOH + CF_3\overset{O}{\overset{\|}{C}}{-}O{-}OH \longrightarrow CF_3COOH + RCH{=}\overset{O^-}{\overset{|}{N^+}}{-}OH \rightleftharpoons RCH_2NO_2$$

$$R_2C{=}NOH + CF_3\overset{O}{\overset{\|}{C}}{-}O{-}OH \longrightarrow CF_3COOH + R_2C{=}\overset{O^-}{\overset{|}{N^+}}{-}OH \rightleftharpoons R_2CHNO_2$$

The reaction is carried out best in solution in acetonitrile and in the presence of sodium bicarbonate or disodium acid phosphate to neutralize the strong trifluoroacetic acid.

4. ***Oxidation of Carbinamines.*** The oxidation of tertiary carbinamines with an aqueous or aqueous-acetone solution of potassium permanganate is the best general method for preparing tertiary nitroalkanes.

$$R_3CNH_2 + 2\,KMnO_4 \longrightarrow R_3CNO_2 + 2\,KOH + 2\,MnO_2$$

Peracetic acid also oxidizes tertiary carbinamines to nitroalkanes and appears to be generally suitable for the conversion of secondary carbinamines to secondary nitroalkanes.

Physical Properties

Because their semipolar character reinforces the inductive effect of the oxygen atoms, the nitroalkanes have a high dipole moment (~3.6 D) and abnormally high boiling points. Thus nitromethane and nitroethane boil at 101° and 115° compared to −12° and 28° for isobutane and isopentane. Although the propyl alcohols are miscible with water, nitromethane is soluble to about the same extent as *n*- or *i*-butyl alcohol, and the higher nitroalkanes are practically insoluble. Hence they are less able than alcohols to form hydrogen bonds. Because of their high dielectric constants ($\epsilon = 25$–35), they usually are good solvents for dipolar compounds. Primary and secondary nitro compounds absorb infrared radiation in the 1565–1545 cm.$^{-1}$ and the 1385–1360 cm.$^{-1}$ regions, which have been assigned to the asymmetric and the symmetric vibrations of the C—N bonds. Tertiary nitro compounds absorb in the ranges 1545–1530 cm.$^{-1}$ and 1360–1340 cm.$^{-1}$ and conjugated nitro alkenes in the ranges 1530–1515 cm.$^{-1}$ and 1360–1330 cm.$^{-1}$

Reactions

1. ***Salt Formation.*** Just as hydrogen on the α carbon atom of an aldehyde or ketone is removable by bases because of resonance stabilization of the resulting enolate anion (p. 196), so hydrogen on the α carbon of a nitroalkane has acidic properties because the anion is stabilized by resonance.

$$R_2CHNO_2 \underset{H_2O}{\overset{^-OH}{\rightleftharpoons}} \left\{ R_2\ddot{C}{-}\overset{+}{N}(=O)O^- \longleftrightarrow R_2C{=}\overset{+}{N}(O^-)O^- \right\}$$

The nitroalkanes, however, are much stronger acids ($pk_a \sim 9$) than the aldehydes or ketones ($pk_a \sim 20$). When the sodium salt of a ketone reacts with water, the ketone is regenerated at once. The primary and secondary nitroalkanes dissolve in aqueous alkali and are regenerated from their salts only by reaction with a stronger acid. Moreover, on acidification the proton adds much more rapidly to the oxygen atoms of the anion than to the α carbon atom with the result that the initial product is an isomer of the original nitroalkane. This isomer is less stable thermodynamically and reverts through the anion to the nitroalkane.

$$\left\{ R_2\ddot{C}{-}\overset{+}{N}(=O)O^- \longleftrightarrow R_2C{=}\overset{+}{N}(O^-)O^- \right\} \underset{\text{Rapid}}{\overset{H^+}{\rightleftharpoons}} \underset{\textit{aci}\text{ Form}}{R_2C{=}\overset{+}{N}(OH)O^-} \underset{\text{Slow}}{\rightleftharpoons} \underset{\textit{nitro}\text{ Form}}{R_2CHNO_2}$$

The equilibrium between the *nitro* form and the *aci* form is another case of tautomerism (p. 201) but here, with certain compounds (p. 623), both tautomers can be isolated. The acidity of the *aci* form is as great as that of the carboxylic acids ($pK_a = 4$ to 6). 2-Nitropropane, although practically insoluble in water, will maintain the acidity of water at pH 4.3 presumably because of the small amount of *aci* form in equilibrium with the *nitro* form. The *aci* forms are called *nitronic acids.* Tertiary nitro compounds, R_3CNO_2, in which no hydrogen is present on the α carbon, do not form salts.

2. ***Bromination.*** Primary and secondary nitro compounds, like aldehydes and ketones (p. 210), brominate easily in alkaline solution.

$$[R\overset{-}{C}HNO_2]\overset{+}{N}a + Br_2 \longrightarrow RCH(Br)NO_2 + NaBr$$

$$[R_2\overset{-}{C}NO_2]\overset{+}{N}a + Br_2 \longrightarrow R_2C(Br)NO_2 + NaBr$$

Only hydrogen on the carbon atom adjacent to the nitro group is acidic. Hence tertiary nitro compounds are not brominated.

3. ***Reaction with Nitrous Acid.*** Like ketones (p. 212), primary nitro compounds react with nitrous acid to give nitroso derivatives. The products, known as *nitrolic acids,* dissolve in alkali to form red salts.

$$RCH_2NO_2 + HONO \longrightarrow RCH(NO){-}NO_2 + H_2O$$

$$RCH(NO)NO_2 + NaOH \longrightarrow \underset{\text{Red solution}}{\left\{ R\ddot{C}(NO){-}\overset{+}{N}(=O)O^- \longleftrightarrow RC(NO){=}\overset{+}{N}(O^-)O^- \right\} Na^+}$$

Secondary nitro compounds give blue nitroso derivatives, which are insoluble in alkali.

$$R_2CHNO_2 + HONO \longrightarrow \underset{\text{Blue, insoluble in water and dilute alkali}}{R_2C(NO){-}NO_2} + H_2O$$

These nitroso derivatives are blue only in the liquid monomolecular state. They solidify to white crystalline dimers, which probably have the structure

$$R_2C(NO_2)-\overset{\overset{\large O^-}{|}}{N^+}=\overset{\overset{\large O^-}{|}}{N^+}-C(NO_2)R_2$$

Tertiary nitro compounds do not react with nitrous acid. The difference in behavior of primary, secondary, and tertiary nitro compounds to nitrous acid and alkali may be used to distinguish between the three types and sometimes is referred to as the "red, white, and blue reaction."

4. ***Reduction to Primary Amines.*** The reduction of nitroalkanes may be brought about by hydrogen and a platinum or nickel catalyst, or by an active metal, for example iron, zinc, or tin, and hydrochloric acid.

$$RNO_2 + 6\,[H] \longrightarrow RNH_2 + 2\,H_2O$$

They are reduced also by lithium aluminum hydride but not by sodium borohydride.

5. ***Acid Hydrolysis.*** ***(a)*** HYDROLYSIS OF THE *nitro* FORM OF A PRIMARY NITRO COMPOUND. When a primary nitro compound is boiled with concentrated aqueous hydrochloric acid or 85 per cent sulfuric acid, a carboxylic acid and a salt of hydroxylamine are formed. The reaction involves an oxidation of the methylene group and a reduction of the nitro group to give a hydroxamic acid, which then hydrolyzes to the final products. This hydrolysis is analogous to that of an oxime to the carbonyl compound (p. 204).

$$RCH_2NO_2 \longrightarrow \underset{\underset{\large OH}{|}}{RCHN}{=}O \rightleftharpoons \underset{\underset{\large OH}{|}}{RC}{=}NOH \xrightarrow{HCl,\ H_2O} RCOOH + HONH_3Cl$$

A hydroxamic acid

The price of hydroxylamine, which formerly was produced by the reduction of nitrous acid and isolated by way of acetoxime (p. 204), has been reduced greatly because of the above process.

(b) HYDROLYSIS OF THE *aci* FORM OF A PRIMARY OR SECONDARY NITRO COMPOUND. If a primary or secondary nitro compound first is converted to the salt of the *aci* form by alkali and then hydrolyzed by 25 per cent sulfuric acid, aldehydes and ketones are produced with the evolution of nitrous oxide (*Nef*[6] *reaction*).

$$2\,RCH{=}NOONa + 2\,H_2SO_4 \longrightarrow 2\,RCHO + N_2O + 2\,NaHSO_4 + H_2O$$

$$2\,R_2C{=}NOONa + 2\,H_2SO_4 \longrightarrow 2\,R_2CO + N_2O + 2\,NaHSO_4 + H_2O$$

The Nef reaction is used not only for the synthesis of simple aldehydes and ketones but also for more complex compounds such as the γ-keto acids and the sugars.

6. ***Addition to Aldehydes and Ketones.*** Primary and secondary nitroalkanes add to aldehydes or ketones in the presence of dilute alkali by a nucleophilic attack on the carbonyl group analogous to aldol addition.

$$R_2CHNO_2 \underset{H_2O}{\overset{^-OH}{\rightleftharpoons}} R_2\ddot{\bar{C}}NO_2 \xrightarrow{HCHO} \underset{\underset{\large NO_2}{|}}{R_2C}{-}CH_2O^- \underset{^-OH}{\overset{H_2O}{\rightleftharpoons}} \underset{\underset{\large NO_2}{|}}{R_2C}CH_2OH$$

The product from reaction with a primary nitroalkane can react with a second mole

[6] John Ulric Nef (1862–1915), American chemist, professor at the University of Chicago. He is noted chiefly for his studies of the products formed by the action of alkali on carbohydrates (p. 320) and for his ideas concerning the nature of compounds containing so-called *divalent carbon,* such as the isocyanides and the fulminates.

of formaldehyde, but steric hindrance prevents reaction with more than one mole of ketone or a higher aldehyde.

$$RCH_2NO_2 + 2\,HCHO \xrightarrow{^-OH} \underset{\displaystyle NO_2}{RC}(CH_2OH)_2$$

$$RCH_2NO_2 + R'CHO \xrightarrow{^-OH} \underset{\displaystyle NO_2}{RCH}-\underset{\displaystyle R'}{CHOH}$$

These nitro alcohols can be reduced readily to amino alcohols (p. 245). They are dehydrated by acids to nitroalkenes.

$$\underset{\displaystyle NO_2}{RCH}-\underset{\displaystyle R'}{CHOH} \overset{H^+}{\rightleftharpoons} \underset{\displaystyle NO_2}{RC}=CHR' + H_2O$$

A Mannich reaction (p. 227) takes place between a nitroalkane, formaldehyde, and a secondary amine to give a dialkylaminomethyl derivative.

$$H_2C=O + HNR_2 \longrightarrow H_2\underset{\displaystyle OH}{C}-NR_2 \underset{H_2O}{\overset{H^+}{\rightleftharpoons}} H_2\overset{+}{C}\text{---}NR_2 \underset{H^+}{\overset{R_2'CHNO_2}{\rightleftharpoons}} H_2\underset{\displaystyle R_2'CNO_2}{C}-NR_2$$

7. ***Carboxylation.*** Reaction of nitro compounds that have two α hydrogens with magnesium methyl carbonate (p. 281) gives the magnesium salts of α-nitro carboxylic acids. Acidification gives the free acid.

$$\underset{\displaystyle NO_2}{RCH_2} + Mg^{2+}(^-OCOOCH_3)_2 \longrightarrow CO_2 + 2\,CH_3OH + RC\begin{matrix}\diagup CO-O^- \\ \diagdown \overset{+}{N}(-O^-)=O\end{matrix}\;Mg^{2+} \xrightarrow{2\,H^+} \underset{\displaystyle NO_2}{RCHCOOH}$$

Reduction of the nitro group gives α-amino acids (p. 245).

Uses

The aliphatic nitro compounds are excellent solvents, but their chief use has been as intermediates for the synthesis of other organic compounds, especially the amino alcohols (p. 609). **Tetranitromethane** is prepared by the action of nitric acid on acetic anhydride or ketene.

$$(CH_3CO)_2O + 4\,HONO_2 \longrightarrow C(NO_2)_4 + CO_2 + CH_3COOH + 3\,H_2O$$

$$CH_2=C=O + 4\,HONO_2 \longrightarrow C(NO_2)_4 + CO_2 + 3\,H_2O$$

It is a valuable reagent for the detection of carbon-carbon unsaturation. When a dilute solution in chloroform is added to a solution of an unsaturated compound, a yellow to red color is produced. The simpler olefins and acetylenes give a yellow color, the tetraalkyl-substituted ethylenes and simple conjugated dienes (p. 582) give orange to light red colors, and the alkyl-substituted dienes give a deep red color. An advantage of the test is that even unreactive double bonds that do not react with bromine or undergo catalytic reduction give a color with tetranitromethane. α,β-Unsaturated carbonyl compounds (p. 618) do not respond to the test. The colored complexes probably are of the charge-transfer type (p. 387). Mixtures of tetranitromethane with organic compounds are violent explosives, and fatal accidents have occurred during its use. Only very dilute solutions should be used, and the solutions should not be heated.

Tetranitromethane reacts with alkaline hydrogen peroxide to give the salt of **nitroform,** $HC(NO_2)_3$.

$$C(NO_2)_4 + 2\,KOH + H_2O_2 \longrightarrow K^{+-}C(NO_2)_3 + \tfrac{1}{2}\,O_2 + KNO_3 + 2\,H_2O$$

Because of the presence of three electronegative nitro groups, nitroform is as strong an acid as phosphoric acid. Nitroform is prepared commercially by the reaction of acetylene with nitric acid and is used as an oxidizing agent in explosives.

Chloropicrin, Cl_3CNO_2, is made by the chlorination of nitromethane in the presence of calcium carbonate, or by the reaction of picric acid with sodium hypochlorite (p. 445). It is a powerful lachrymator and is used as a tear gas for dispelling mobs and as a warning agent for toxic gases. Its principal use, however, is as a disinfectant of soil for agricultural purposes.

1,1-Dinitroethane is made by the action of nitrite ion and silver nitrate on the sodium salt of nitroethane, a process that has been termed *oxidative nitration.*

$$CH_3\bar{C}HNO_2 + NO_2^- + 2\,Ag^+ \longrightarrow CH_3CH(NO_2)_2 + 2\,Ag$$

The silver can be recovered quantitatively. Addition of 1,1-dinitroethane to formaldehyde gives **2,2-dinitro-1-propanol,** which is converted to **bis(2,2-dinitropropyl)acetal** by reaction with paraldehyde using boron fluoride as catalyst.

$$CH_3CH(NO_2)_2 + HCHO \longrightarrow CH_3CH(NO_2)_2CH_2OH \xrightarrow{(CH_3CHO)_3,\ BF_3} [CH_3CH(NO_2)_2CH_2O]_2CHCH_3$$

The acetal is used as an oxidizing plasticizer for rocket fuels.

Until 1949, no natural product had been known to contain a nitro group. In that year, however, **β-nitropropionic acid,** $O_2NCH_2CH_2COOH$, was shown to be a hydrolytic product of the glycoside (p. 324) **hiptagin** present in the bark of *Hiptage mandoblata* and in the kernels of *Corynocarpus laevigata.* Since then it has been isolated from cultures of *Aspergillus* and *Penicillium* molds.

HYDRAZINES AND HYDRAZIDES

Monoalkylhydrazines ordinarily are prepared by the reaction of chloroamine with a primary amine.

$$NH_3 + Cl_2 + NaOH \longrightarrow NH_2Cl + NaCl + H_2O$$

$$RNH_2 + ClNH_2 \longrightarrow \left[R\overset{+}{N}H_2{-}NH_2\right]Cl^- \xrightarrow{NaOH} RNH{-}NH_2$$

Monomethylhydrazine (*MMH*, m.p. $< -80°$, b.p. $87°$) reacts spontaneously with dinitrogen tetroxide to give only gaseous products with the evolution of heat and is used as fuel in the steering system of large rockets.

$$4\,CH_3NHNH_2 + 5\,N_2O_4 \longrightarrow 4\,CO_2 + 12\,H_2O + 9\,N_2$$

Both fuel and oxidizing agent can be stored in the missile as liquids without refrigeration. **Unsymmetrical dimethylhydrazine** (*UDMH*), made from chloroamine and dimethylamine or by the reduction of nitrosodimethylamine, has been used in combination with hydrazine as a rocket missile fuel.

Acid hydrazides, the monoacyl derivatives of hydrazine, are prepared by the reaction of acyl halides or of esters with hydrazine. They are used for the preparation of derivatives of aldehydes and ketones (cf. hydrazones, p. 205), and some have therapeutic properties (p. 518).

ALIPHATIC AZO AND DIAZO COMPOUNDS

When a symmetrically disubstituted hydrazine is oxidized with sodium dichromate and sulfuric acid, an azo compound (Fr. *azote* nitrogen) is formed in which the two nitrogen atoms are joined by a double bond.

$$RNHNHR \xrightarrow{Na_2Cr_2O_7,\ H_2SO_4} RN{=}NR + H_2O$$

Azomethane, $CH_3N{=}NCH_3$, prepared from *sym*-dimethylhydrazine, is a yellow gas (p. 555). Although the nitrogens each have an unshared pair of electrons, they are not basic in aqueous solution. As with the oximes, and to a greater extent the nitriles and the acetylide ion, the presence of a multiple bond reduces the availability of unshared pairs of electrons (p. 238). The azo compounds are reduced readily to the hydrazines and further to the primary amines. Azomethane is hydrolyzed, when boiled with aqueous hydrochloric acid, to methylhydrazine and formaldehyde, presumably because it can tautomerize to the easily hydrolyzed hydrazone of formaldehyde.

$$CH_3\ddot{N}{=}\ddot{N}CH_3 \underset{B^-}{\overset{HB}{\rightleftharpoons}} CH_3\overset{+}{N}(H){=}\ddot{N}{-}CH_2(H) \underset{HB}{\overset{B^-}{\rightleftharpoons}} \underset{\text{Formaldehyde methylhydrazone}}{CH_3NHN{=}CH_2} \xrightarrow{H_2O,\ HCl} \underset{\text{Methylhydrazine hydrochloride}}{CH_3\overset{+}{N}H_2NH_2\,{}^-Cl} + HCHO$$

Frequently the azo compound cannot be isolated, because it isomerizes spontaneously to the hydrazone.

Azomethane is unstable above 200° and yields free methyl radicals (p. 38) and nitrogen.

$$CH_3N{=}NCH_3 \longrightarrow 2\,CH_3\cdot + N_2$$

The methyl radicals may combine to give ethane, but in the presence of other molecules, they may abstract hydrogen or initiate free radical chain reactions. That free methyl radicals actually can have an independent existence in the gas phase was demonstrated first by Paneth in 1929. When a stream of inert gas saturated with tetramethyllead vapor is passed through a silica tube heated at one spot, a lead mirror forms at this spot because of the decomposition of the tetramethyllead. If then the tube is heated some distance nearer the source of the gas stream, a new mirror is formed, and at the same time the first mirror disappears. Evidently the decomposition of tetramethyllead yields free methyl radicals that can exist for a sufficient length of time to travel the distance to the first lead mirror and convert it to volatile tetramethyllead.

$$Pb(CH_3)_4 \xrightarrow{\text{Hot}} Pb + 4\,CH_3\cdot$$

$$Pb + 4\,CH_3\cdot \xrightarrow{\text{Cold}} Pb(CH_3)_4$$

From the rate of streaming of the gas and the distance between the two mirrors, the average life of the methyl radical under the conditions of the experiment was calculated to be 8.4×10^{-3} second. Thus even in the gaseous state at 1 to 2 mm. pressure, the methyl radicals are destroyed rapidly by collision with gaseous particles or with the walls of the tube. The same technique has been used to demonstrate the formation of methyl radicals by the pyrolysis of azomethane at 200°, of butane at 600°, and of acetone at 700°.

2,2′Dicyano-2,2′-azopropane (*azobisisobutyronitrile or Vazo*) was prepared first by Thiele in 1896 by the reaction of acetone with potassium cyanide and hydrazine hydrochloride. The cyanide frees the hydrazine and is converted to hydrogen cyanide.

$$2\,KCN + Cl^{-+}H_3N{-}NH_3^{+-}Cl \longrightarrow 2\,KCl + 2\,HCN + H_2N{-}NH_2$$

The acetone presumably forms the azine (*a dihydrazone,* p. 205), which adds hydrogen cyanide.

$$2\,(CH_3)_2CO + H_2N{-}NH_2 \longrightarrow (CH_3)_2C{=}N{-}N{=}C(CH_3)_2 \xrightarrow{2\,HCN} (CH_3)_2\underset{CN}{C}{-}NH{-}NH{-}\underset{CN}{C}(CH_3)_2$$

The substituted hydrazine is oxidized to the azo compound with sodium hypochlorite.

$$(CH_3)_2\underset{CN}{C}{-}NHNH{-}\underset{CN}{C}(CH_3)_2 + NaOCl \longrightarrow (CH_3)_2\underset{CN}{C}{-}N{=}N{-}\underset{CN}{C}(CH_3)_2 + H_2O + NaCl$$

This compound decomposes when heated to 100° into nitrogen and free radicals, which can act as chain reaction initiators or combine to form tetramethylsuccinonitrile (p. 629).

$$(CH_3)_2\underset{CN}{C}{-}N{=}N{-}\underset{CN}{C}(CH_3)_2 \xrightarrow{100^\circ} N_2 + 2\left[(CH_3)_2\underset{CN}{C}\cdot\right] \longrightarrow (CH_3)_2\underset{CN}{C}{-}\underset{CN}{C}(CH_3)_2$$

Since 1948 it has been used widely to initiate free radical polymerizations and as a foaming agent in the production of foam rubber and expanded plastics.

In azo compounds nitrogen is bound to nitrogen and there is one carbon-nitrogen bond for each nitrogen. Another type of compound also has nitrogen bound to nitrogen but both nitrogens are bound to a single carbon atom. Hence these compounds are called

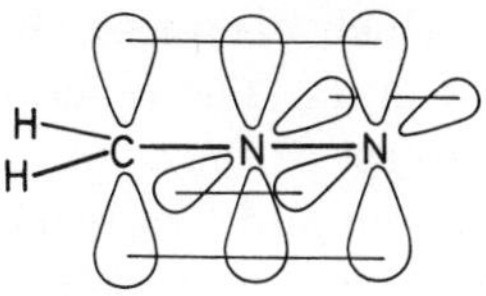

$$\{\mathrm{CH_2{=}\overset{+}{N}{=}\ddot{\overset{-}{N}}:} \longleftrightarrow \mathrm{\ddot{\overset{-}{C}}H_2{-}\overset{+}{N}{\equiv}N:}\}$$

Figure 14–*1*. Resonance in diazomethane.

diazo compounds. The simplest aliphatic diazo compound has the molecular formula, CH_2N_2, and is called *diazomethane.* From dipole moment and electron diffraction measurements, it has been concluded that diazomethane has a linear structure and is a resonance hybrid. Thus the two extreme electronic structures possible for a linear molecule coalesce because of the overlapping of *p* atomic orbitals of all three atoms, giving rise to π-type molecular orbitals as indicated in Fig. 14–*1*.

Theoretically diazomethane should be formed by the action of a base on methyldiazonium ion (p. 228).

$$\mathrm{CH_3{-}\overset{+}{N}{\equiv}N + B^-} \longrightarrow \{\mathrm{\overset{-}{C}H_2{-}\overset{+}{N}{\equiv}N:} \longleftrightarrow \mathrm{CH_2{=}\overset{+}{N}{=}\ddot{\overset{-}{N}}:}\} + \mathrm{HB}$$

No base that is present under the conditions for the formation of the diazonium ion is strong enough to remove the proton, and the ion decomposes with evolution of nitrogen and the formation of other products (p. 229). *N*-Methyl-*N*-nitroso amides, however, are reasonably stable compounds and react readily with aqueous potassium hydroxide to give diazomethane. The nitroso group greatly increases the acidity of the hydrogen on the methyl group, and a mechanism is available for eliminating the rest of the molecule.

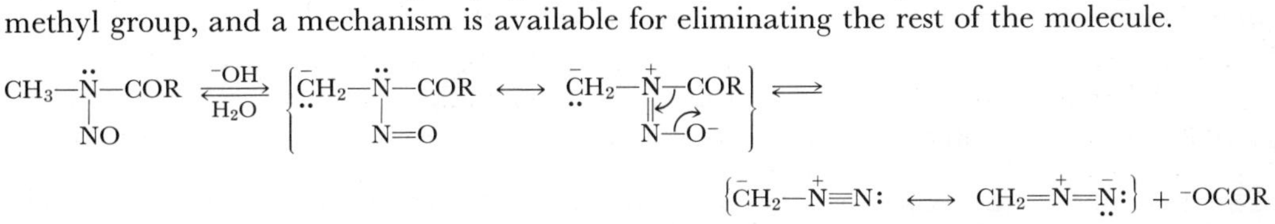

$$\mathrm{CH_3{-}\ddot{N}(NO){-}COR} \underset{\mathrm{H_2O}}{\overset{^-\mathrm{OH}}{\rightleftharpoons}} \{\mathrm{\ddot{\overset{-}{C}}H_2{-}\ddot{N}(N{=}O){-}COR} \longleftrightarrow \mathrm{\ddot{C}H_2{-}\overset{+}{N}(N{-}O^-){-}COR}\} \rightleftharpoons$$

$$\{\mathrm{\overset{-}{C}H_2{-}\overset{+}{N}{\equiv}N:} \longleftrightarrow \mathrm{CH_2{=}\overset{+}{N}{=}\ddot{\overset{-}{N}}:}\} + {}^-\mathrm{OCOR}$$

Of the various *N*-methyl-*N*-nitroso amides that have been used, *N,N'*-dinitroso-*N,N'*-dimethylterephthalamide (p. 476) is most convenient because it is available commercially in a stabilized form.

$$\mathrm{CH_3N(NO){-}COC_6H_4CO{-}N(NO)CH_3 + 2\,KOH} \longrightarrow \mathrm{2\,CH_2N_2 + 2\,H_2O + K^{+\,-}OCOC_6H_4COO^{-\,+}K}$$

The ethoxycarbonyl group in ethyl aminoacetate increases the acidity of the α hydrogen sufficiently to permit the formation of **ethyl diazoacetate** (*diazoacetic ester*) by the direct action of nitrous acid.

$$\mathrm{C_2H_5OCOCH_2NH_2 + HONO} \longrightarrow \mathrm{C_2H_5OCOCHN_2 + 2\,H_2O}$$

Diazomethane is a highly toxic, yellow gas that sometimes explodes even in the gaseous state at room temperature. It usually is handled in ether solution under a hood and behind safety glass. Nevertheless it is a very valuable reagent. In general it acts as a good nucleophile. Its use to prepare methyl esters has been discussed (p. 164).

Nucleophilic attack by diazomethane converts aldehydes into methyl ketones and converts ketones into higher homologs.

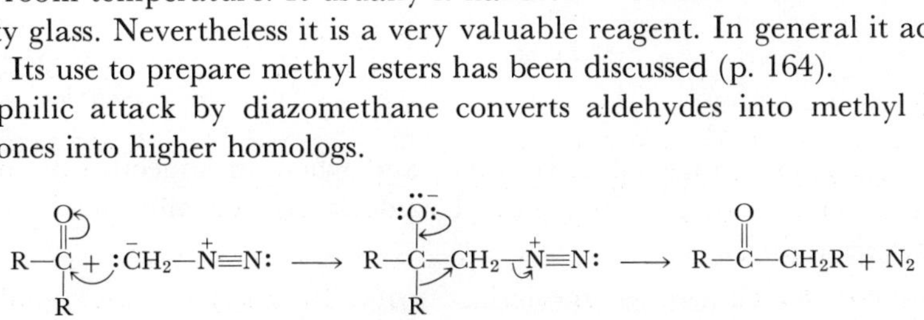

$$\mathrm{R{-}\overset{O}{\overset{\|}{C}}{-}R + :\overset{-}{C}H_2{-}\overset{+}{N}{\equiv}N:} \longrightarrow \mathrm{R{-}\overset{:\ddot{O}:^-}{\overset{|}{C}}(R){-}CH_2{-}\overset{+}{N}{\equiv}N:} \longrightarrow \mathrm{R{-}\overset{O}{\overset{\|}{C}}{-}CH_2R + N_2}$$

This reaction may be catalyzed by boron fluoride, which complexes with the carbonyl group and increases the electrophilicity of the carbon atom. The reaction is particularly useful for increasing the size of the ring of cyclic ketones (p. 652).

Pyrolysis or photolysis of diazomethane gives nitrogen and two reactive species of the composition CH_2, one of which, designated as *methylene,* has two unpaired electrons (*triplet state*), whereas the other, designated as *carbene,* has a pair of electrons (*singlet state*) and an empty orbital (p. 178). Carbene in an excited state appears to be the initial product, which decays to the more stable methylene.

$$^{-}\!:CH_2{-}\overset{+}{N}{\equiv}N \xrightarrow{\text{Heat or } h\nu} N_2 + \underset{\text{Carbene}}{H{-}\ddot{C}{-}H} \longrightarrow \underset{\text{Methylene}}{H{-}\dot{C}{-}H}$$

These species cannot be isolated but have been observed spectroscopically and are considered to be intermediates in reactions of diazomethane induced by heat or light.

Carbene produced photochemically has a large excess of energy and is a highly reactive species. It inserts itself almost randomly between hydrogen and the element to which hydrogen is bonded. Thus *n*-pentane gives *n*-hexane and the two methyl pentanes, *i*-propyl alcohol gives *s*-butyl and *t*-butyl alcohols and *i*-propyl methyl ether. This behavior is referred to as the *insertion reaction.*

Decomposition of diazomethane in a pentane solution containing a catalytic amount of cuprous chloride yields a less energetic form of carbene. In the presence of an alkene it adds entirely *cis* to a double bond to give a *cis*-substituted cyclopropane.

$$\text{cis-}H_3C(H)C{=}C(H)CH_3 + CH_2N_2 \xrightarrow{CuCl} \text{cis-1,2-dimethylcyclopropane} + N_2$$

The diazoalkanes themselves add to polarized unsaturated compounds to give heterocyclic compounds. The reaction is believed to take place by way of the dipolar form of the diazo compound and is classed as a **1,3-dipolar cycloaddition.**

$$C_2H_5OOCC{\equiv}CCOOC_2H_5 + CHCOOC_2H_5{=}\overset{+}{N}{=}\ddot{N}{:}^{-} \longrightarrow \text{(pyrazoline)} \rightleftharpoons \text{(pyrazole, N–H)}$$

Ethyl acetylene dicarboxylate; Ethyl diazo acetate; Ethyl pyrazole-3,4,5-tricarboxylate

AZIDES

The azides are derivatives of hydrazoic acid, HN_3. Sodium azide is prepared by passing nitrous oxide into molten sodium amide.

$$^{-}\!:\ddot{N}{=}\overset{+}{N}{=}O + {}^{-}\!:NH_2 \longrightarrow {}^{-}\!:\ddot{N}{=}\overset{+}{N}(NH_2){-}O^{-} \rightleftharpoons {}^{-}\!:N{=}\overset{+}{N}(\ddot{N}H){-}OH \longrightarrow \underset{\text{Azide ion}}{:\overset{-}{\ddot{N}}{=}\overset{+}{N}{=}\overset{-}{\ddot{N}}:} + H_2O$$

The azide ion is a good nucleophile (p. 103), and when an aqueous solution of sodium azide is treated with methyl sulfate, **methyl azide** is evolved and can be condensed to a liquid that boils at 20°.

$$(CH_3)_2SO_4 + N_3^- \longrightarrow \underset{\text{Methyl azide}}{CH_3N_3} + CH_3SO_4^-$$

Ethyl azide can be prepared by a similar process. The azide group has a linear structure analogous to that of the diazo group (p. 249) and may be represented as a resonance hybrid.

$$\left\{ R\ddot{N}=\overset{+}{N}=\overset{-}{\ddot{N}}: \longleftrightarrow R\overset{\cdot\cdot}{\underset{\cdot\cdot}{N}}-\overset{+}{N}\equiv N: \right\}$$

Just as diazoalkanes yield carbene on photolysis (p. 250), so photolysis of hydrazoic acid yields an analogous reactive intermediate that usually is called *nitrene.*

$$H-\overset{\cdot\cdot}{\underset{\cdot\cdot}{N}}-\overset{+}{N}\equiv N: \xrightarrow{h\nu,\ Hg} H-\overset{\cdot\cdot}{\underset{\cdot\cdot}{N}} + N_2$$

Nitrenes have been postulated as intermediates in the reactions of many nitrogen compounds.

Reduction of alkyl azides either catalytically or with lithium aluminum hydride gives the primary amine.

$$RN_3 \xrightarrow[\text{or } LiAlH_4]{H_2/Pt} RNH_2 + N_2$$

The series of reactions alkyl halide → azide → amine has the advantage over direct ammonolysis of alkyl halide (p. 222) in that no secondary or tertiary amine is formed.

The most important azides are the acyl derivatives. They are more stable than the alkyl azides because of the additional resonance stabilization by the carbonyl group. Usually acyl azides are prepared by the reaction of an acyl chloride with sodium azide.

$$RCOCl + NaN_3 \longrightarrow RCON_3 + NaCl$$

When heated, they undergo the *Curtius rearrangement*[7] with the formation of the isocyanate (p. 286). The reaction is analogous to the Hofmann rearrangement of amides (p. 223) except that the rearrangement of the azides can be brought about in anhydrous media, and the isocyanate can be isolated without difficulty.

$$R-\overset{O}{\overset{\|}{C}}-\overset{-}{\underset{\cdot\cdot}{N}}-\overset{+}{N}\equiv N: \longrightarrow O=C=\underset{\cdot\cdot}{N}-R + N_2$$

There is some evidence that an acyl nitrene is an intermediate in the rearrangement.

PROBLEMS

Amines

14–*1.* Compare the number of possible alcohols having the molecular formula $C_4H_{10}O$ with the number of possible amines having the formula $C_4H_{11}N$.

14–*2.* Give balanced equations for the following preparations: (*a*) ethylamine from an amide, (*b*) *n*-butylamine from an oxime, (*c*) *i*-propyl-di-*n*-butylamine from a chloroamine, (*d*) *n*-pentadecylamine from an acid, (*e*) *i*-butylamine from a cyanide, (*f*) diethylamine from sodium cyanamide,

14–*3.* Write electronic formulas for each of the following compounds and indicate from which atom the electrons were derived: (*a*) ethylamine, (*b*) tetramethylammonium iodide, (*c*) trimethylamine oxide, (*d*) triethylamine-boron fluoride complex, (*e*) nitrosodimethylamine, (*f*) methylamine sulfate.

14–*4.* Arrange the following types of compounds in the order of decreasing basicity: (*a*) alcohols, (*b*) trialkylammonium chlorides, (*c*) ammonia, (*d*) aliphatic amines, (*e*) alkenes, (*f*) quaternary ammonium hydroxides.

[7] Theodor Curtius (1857–1928), professor at the University of Heidelberg. He is noted for his discovery of the aliphatic diazo compounds in 1883 and of the rearrangement of acid azides, for his work on hydrazines and heterocyclic compounds, and for his syntheses of amino acids and polypeptides.

14–*5*. Arrange the following compounds or ions in the order of decreasing basicity: (*a*) acetate ion, (*b*) *n*-butylamine, (*c*) amide ion, (*d*) chloride ion, (*e*) acetic acid, (*f*) hydroxide ion, (*g*) water.

14–*6*. Give a series of reactions for the following conversions: (*a*) methyl iodide to trimethylamine oxide, (*b*) ethyl palmitate to *n*-pentadecylamine, (*c*) ethyl cyanide to *N*-acetyl-*n*-propylamine, (*d*) 2-octanone to 2-aminooctane, (*e*) *i*-propyl ether to *i*-propylamine, (*f*) 1-butene to *s*-butyl isocyanide, (*g*) ethyl laurate to *n*-undecylamine, (*h*) *i*-amyl bromide to *i*-hexylamine.

14–*7*. Pair the following names with the appropriate reaction: (*a*) Markovnikov, (*b*) Wurtz, (*c*) Williamson, (*d*) Hofmann, (*e*) Baeyer, (*f*) Walden; (*1*) synthesis of alkanes, (*2*) S_N2 reactions, (*3*) synthesis of amines, (*4*) synthesis of ethers, (*5*) mode of addition of unsymmetric reagents to unsymmetric alkenes, (*6*) test for alkenes.

14–*8*. Give an equation for the reaction of dimethylamine and formaldehyde with methylacetylene. Name the product formed.

14–*9*. How can one distinguish readily among the members of each of the following groups of compounds: (*a*) *n*-butyl alcohol, *n*-butylamine, and diethylamine; (*b*) propionic acid, *i*-butylamine, and ethyl ether; (*c*) *n*-octane, hexylamine, and triethylamine; (*d*) tetraethylammonium bromide, ammonium bromide, and sodium bromide; (*e*) *n*-hexyl bromide, *n*-hexylammonium bromide, and ammonium bromide; (*f*) *n*-butyl nitrite, di-*i*-propylammonium nitrite, and trimethylammonium nitrite; (*g*) diethyl ketoxime, *i*-octylamine, and caproamide.

14–*10*. Which of the following reagents, (*a*) dilute HCl, (*b*) dilute NaOH, (*c*) saturated $NaHSO_3$, and (*d*) cold concentrated H_2SO_4, can be used to separate a mixture of (*1*) caproic acid and 3-heptanone; (*2*) hexane and ethyl ether; (*3*) methyl ethyl ketone and *n*-butyrone; (*4*) triethylamine and 2-octanol?

14–*11*. Devise a procedure for obtaining in a relatively pure state each component of the following mixtures: (*a*) 2-octanol, *n*-octane, tri-*n*-butylamine, and methyl *i*-butyl ketone; (*b*) methylamine, acetic acid, methanol, and *n*-hexane; (*c*) ethyl alcohol, dimethylamine, trimethylamine, and tetramethylammonium bromide.

14–*12*. Explain how a large excess of ammonia can suppress the formation of secondary and tertiary amines during the catalytic reduction of alkyl cyanides.

14–*13*. Show how the reaction of *n*-butylamine with nitrous acid in the presence of hydrochloric acid can lead to the formation of the various products reported on page 228.

14–*14*. Give a probable mechanism for the conversion of a primary amine to the alkyl halide by means of nitrosyl bromide.

Amides and Imidic Acids

14–*15*. Give equations for the following syntheses: (*a*) *N*-ethylpropionamide from an acyl chloride, (*b*) valeramide from an ester, (*c*) *N,N*-dimethyl-*n*-butyramide from an anhydride, (*d*) lauramide from an acid.

14–*16*. Which of the following compounds react with ammonia to give the same organic nitrogen compound: (*a*) acetone, (*b*) acetyl chloride, (*c*) acetaldehyde, (*d*) ethyl formate, (*e*) methyl acetate, (*f*) ethyl alcohol, (*g*) ethyl chloride, (*h*) acetic anhydride?

14–*17*. Give a series of reactions for carrying out the following conversions: (*a*) *n*-butyramide to *N*-*n*-propylacetamide, (*b*) *N*-methylcaproamide to ethyl caproate, (*c*) *N,N*-dimethylformamide to nitrosodimethylamine, (*d*) caprylamide to *n*-octylamine, (*e*) *N,N*-di-*i*-propylacetamide to *N,N*-di-*i*-propyl-*n*-valeramide, (*f*) *i*-amyl alcohol to *N*-*s*-butyl-*i*-valeramide, (*g*) *n*-butyl bromide to *n*-valeraldehyde, (*h*) acetonitrile to ethyl orthoacetate, (*i*) *N*-methylpropionamide to *N*-methylpropionamidine.

14–*18*. What can be said about the structures of the members of the following pairs of compounds: (*a*) Two compounds have the molecular formula C_3H_7NO. One reduces Tollens reagent but evolves no nitrogen when treated with nitrous acid, whereas the other evolves nitrogen but does not reduce Tollens reagent. When boiled with aqueous sodium hydroxide, each evolves a product that turns red litmus blue. (*b*) Two compounds have the molecular formula $C_6H_{13}NO$. After each compound is boiled for several hours with hydrochloric acid, the nonaqueous layer from one compound gives the iodoform reaction and that from the other dissolves in dilute aqueous alkali. (*c*) Two compounds have the molecular formula C_3H_9NO. One evolves methane when treated with methylmagnesium iodide whereas the other does not. Neither evolves nitrogen when treated with nitrous acid.

14–*19*. Suggest a likely mechanism involving cyclic transition states for the dehydration of an amide by thionyl chloride.

Alkyl Cyanides and Isocyanides

14–*20*. Give equations for the following preparations: (*a*) propionitrile from an alkyl iodide, (*b*) 2-methylbutanoic acid from an alkyl cyanide, (*c*) *i*-propyl isocyanide from an amine, (*d*) (4-methylpentyl)amine from an alkyl cyanide, (*e*) 2-cyano-3-methylbutane from an olefin, (*f*) acetamide from an alkyl cyanide, (*g*) *i*-butyl cyanide from an amide, (*h*) methyl trimethylacetate from an alkyl cyanide, (*i*) *n*-hexyl cyanide from an aldoxime, (*j*) methyl-*n*-butylamine from an alkyl isocyanide, (*k*) *i*-butyl isocyanide from an *N*-alkylformamide.

14–*21*. Give a series of reactions for the following conversions: (*a*) caprylamide to *n*-heptyl isocyanide, (*b*) *i*-butyraldehyde to *i*-propyl cyanide, (*c*) *t*-butyl alcohol to *t*-butyl cyanide, (*d*) 2-methyl-2-butene to *t*-amylamine, (*e*) methyl palmitate to hexadecyl cyanide, (*f*) *n*-butyl ether to *n*-amylamine, (*g*) propionic acid to *i*-propyl cyanide.

14–*22*. Give the structure and name of the product formed by the reaction of ethyl isocyanide with (*a*) water; (*b*) methanol; (*c*) diethylamine; (*d*) methylmagnesium bromide followed by water.

14–*23*. (*a*) Which of the following compounds would react rapidly with cold aqueous sodium hydroxide: (*1*) $(CH_3)_3NH^+Cl^-$, (*2*) $CH_3C{\equiv}N$, (*3*) $CH_3CH{=}NOH$, (*4*) $(CH_3)_4N^+Cl^-$, (*5*) CH_3CONH_2, (*6*) $(C_2H_5)_3N$, (*7*) $CH_3COO^{-+}NH_4$? (*b*) Which would not react with hot aqueous sodium hydroxide?

14–*24*. How can one distinguish readily among the members of the following groups of compounds: (*a*) *n*-butyl ether, *n*-butyl cyanide, and *n*-butylamine; (*b*) *i*-octyl cyanide, *i*-octyl isocyanide, and *i*-capramide; (*c*) butyronitrile, butyric acid, and butyrone; (*d*) ammonium acetate, acetamide, and acetonitrile?

14–*25*. Which of the following compounds have a *carbon* atom with (*a*) an *sp*-hybridized orbital, (*b*) an sp^2-hybridized orbital, (*c*) an sp^3-hybridized orbital, (*d*) a semipolar bond: (*1*) formaldehyde, (*2*) methyl isocyanide, (*3*) acetaldoxime, (*4*) acetamide, (*5*) acetone, (*6*) acetonitrile, (*7*) *s*-butyl alcohol, (*8*) trimethylamine, (*9*) dimethylacetylene, (*10*) trimethylamine oxide?

Nitroalkanes

14–*26*. Give equations for the following preparations: (*a*) 2-nitro-2-methylpropane from an amine, (*b*) *i*-propylamine from a nitro compound, (*c*) propionaldehyde from a nitro compound, (*d*) 1-nitroheptane from an oxime, (*e*) acetic acid from a nitro compound, (*f*) 2-nitro-2-ethyl-1,3-propanediol from a nitro compound.

14–*27*. Give a series of reactions for each of the following conversions: (*a*) *i*-amyl ether to 1-nitro-3-methylbutane, (*b*) 2-octanone to 2-nitrooctane, (*c*) trimethylacetamide to 2-nitro-2-methylpropane, (*d*) 1-nitropropane to α-aminopropionic acid, (*e*) nitroethane to 2-amino-1-propanol.

14–*28*. What tests may be used to distinguish among the members of the following groups of compounds: (*a*) 1-nitropropane, butyronitrile, and *n*-hexylamine; (*b*) nitromethane, 2-nitropropane, and heptanoic acid; (*c*) 2-nitro-2-methylpropane, *n*-propyl ether, and *n*-octane; (*d*) 1-nonene, 1-nitropropane, and 1-nonanol; (*e*) *s*-butyl nitrite, 2-nitrobutane, and butyronitrile.

14–*29*. Write electronic formulas for methyl nitrite, nitromethane, the *aci* form of nitromethane, methyl nitrate, and nitrosonitromethane.

14–*30*. Analysis indicated that two compounds had the molecular formula $C_5H_{11}NO_2$. One was fairly soluble in water and the other was insoluble but dissolved in dilute alkali. The water-soluble compound evolved ammonia when boiled with alkali, and vigorous oxidation of the hydrolysis product gave an acid having a neutralization equivalent of 66. When the alkaline solution of the water-insoluble compound was poured into hot 25 per cent sulfuric acid, a product was obtained that gave an iodoform test. What can be said about the structures of the two compounds? Give equations for the reactions involved.

Other Nitrogen Compounds

14–*31*. Indicate the probable steps in the formation of ethyl diazoacetate when sodium nitrite is added to a solution of ethyl aminoacetate in hydrochloric acid.

14–*32*. Give reactions for the following conversions: (*a*) methyl ethyl ketone to ethyl ketone, (*b*) *n*-amyl bromide to pure *n*-amylamine, (*c*) octadecanohydrazide to heptadecylamine.

14–*33*. A pure neutral organic compound, *A*, gives a positive test for nitrogen but negative tests for halogen, phosphorus, and sulfur, and leaves no residue when burned. A 1.26-g. sample of this substance depresses the freezing point of a definite quantity of benzene to the same amount as 0.84 g. of *n*-butyl alcohol. Combustion analysis gives C, 41.35 per cent; H, 6.89 per cent; N, 24.12 per cent. When *A* is refluxed with dilute sulfuric acid, a volatile acid is isolated from the distillate as the sodium salt, which contains 28.05 per cent sodium. When a large excess of sodium hydroxide is added to the sulfuric acid residue and the mixture is boiled, a basic liquid distills. From these data, what can be said concerning the structure of the compound? Tell which of the possible products is the more likely and give your reasoning.

CHAPTER FIFTEEN

ALIPHATIC SULFUR, PHOSPHORUS, AND SILICON COMPOUNDS

Elements of the second period are characterized by an electron shell capable of holding a maximum of eight electrons, which are closer to the nucleus than the valence electrons of higher period elements and are shielded from the nucleus only by the two 1*s* electrons. Furthermore there are no orbitals of quantum number 2 besides the 2*s* and 2*p* orbitals, and the orbitals of quantum number 3 are of considerably higher energy than the 2*p* orbitals (Fig. 2–*1*, p. 8). The valence electrons of elements of the third period are not only farther from the positive nucleus, but are shielded from it by a core of ten negative electrons. Hence the valence electrons of third period elements are held much more loosely than those of their second period congeners. This difference is evident from the lower ionization potentials (Table 2–*1*, p. 7) of the third period elements; that is, third period elements are less electronegative than corresponding elements of the second period.

For reactions of compounds analogous to those of second period elements, the change in electronegativity can alter the position of equilibrium either favorably or unfavorably depending on whether the bonds are stronger or weaker. Thus sulfur-sulfur bonds are stronger than oxygen-oxygen bonds, and alkyl disulfides, RS—SR, are more stable than alkyl peroxides, RO—OR. On the other hand, carbon-sulfur bonds are weaker than carbon-oxygen bonds, and the equilibrium position for the formation of a thio ester, RCOSR, from an acid and a thiol, RSH, is less favorable than for the formation of an ester, RCOOR, from an acid and an alcohol.

The decreased electronegativity also can alter the rates of reactions because of a change in polarity of bonds. Even though the carbon-oxygen bond is the stronger, the decreased electronegativity of sulfur makes the carbon-sulfur bond less polar and less reactive than the carbon-oxygen bond. The opposite effect also is possible. Just as sulfur is more electropositive than oxygen, so silicon is more electropositive than carbon. Hence the silicon-chlorine bond is more polar and more reactive than the carbon-chlorine bond.

Three other facts should be recalled. With increasing numbers of electrons and increasing distance and greater shielding of valence electrons from the nucleus, atoms become more polarizable with consequent changes in their nucleophilicity and in their ability to leave a molecule (p. 103). Although the electronegativity of the third period elements is less than that of their second period congeners, the greater dispersion of the valence electrons of third period elements makes a pair of electrons less able to bond with the small proton; that is, the overlap of a 3*p* orbital with the 1*s* orbital of hydrogen is poorer than that of a 2*p* orbital. The result is that a σ-bonded hydrogen is more acidic, and an unshared pair of electrons has less tendency to hydrogen-bond with other molecules (p. 32). Secondly, the greater size of the valence shell allows more room for bonded groups. Steric effects are less, both as to ease of nucleophilic attack and as to the tendency to to assume the higher coordination numbers. Finally, elements of the third period have unfilled $3d$ and $4s$ orbitals that are sufficiently close in energy to the $3p$ orbitals to hybridize with them for bonding purposes (Fig. 2–*1*, p. 8). Frequently it is stated that elements such as sulfur, phosphorus, and silicon can expand their valence shells from eight to ten or twelve electrons. What this means is that two of the five d orbitals can be used for bonding in addition to the s and p orbitals. Actually there is no more an expansion of the valence shell when this happens than when, for example, a tricovalent boron compound makes use of its empty p orbital to form a tetracovalent complex (p. 13).

SULFUR

Since sulfur is the element below oxygen in the periodic table, a series of organic compounds analogous to the oxygen compounds is known.

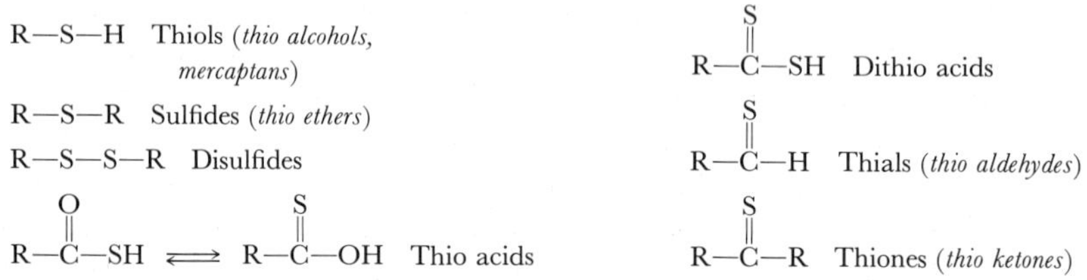

R—S—H Thiols (*thio alcohols, mercaptans*)

R—S—R Sulfides (*thio ethers*)

R—S—S—R Disulfides

R—C(=O)—SH ⇌ R—C(=S)—OH Thio acids

R—C(=S)—SH Dithio acids

R—C(=S)—H Thials (*thio aldehydes*)

R—C(=S)—R Thiones (*thio ketones*)

It should be noted here that whereas the carbon-oxygen double bond results from a $2p$-$2p$ overlap of orbitals, that of a carbon-sulfur double bond results from a less favorable $2p$-$3p$ overlap. Hence carbon-sulfur double bonds are encountered less frequently and are more reactive than carbon-oxygen double bonds.

In addition to the oxygen analogs, several types of compounds are known in which sulfur is linked to three or four atoms or groups. In these compounds use is made of the unshared pairs of electrons on sulfur for bonding purposes. When the third and fourth atoms are oxygen, the nature of the bond cannot be described in a simple way. For most purposes it is satisfactory to extend the simple octet rule to the third period and assume that the bonds to oxygen are semipolar (p. 16), both electrons being supplied by the sulfur atom.

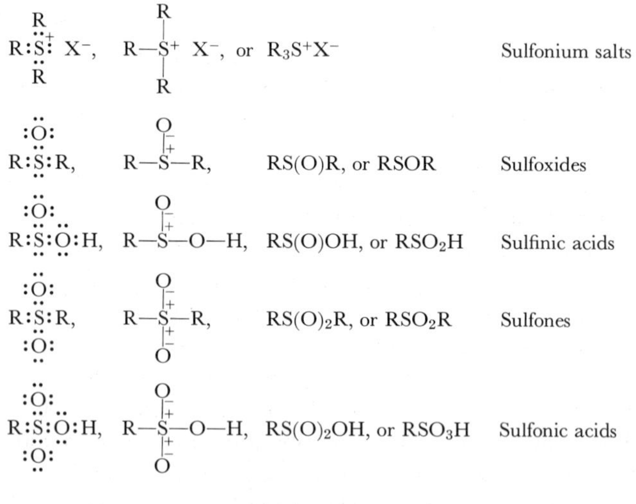

Electron-dot formula	Bond formula	Condensed formula	Name
$R{:}\ddot{S}{:}^{+}R$ (R above) X^-	R—S^+(—R)(—R) X^-	or $R_3S^+X^-$	Sulfonium salts
$R{:}\ddot{S}(:\ddot{O}:){:}R$	R—$\overset{+}{S}$(—$\overset{-}{O}$)—R	RS(O)R, or RSOR	Sulfoxides
$R{:}\ddot{S}(:\ddot{O}:){:}\ddot{O}{:}H$	R—$\overset{+}{S}$(—$\overset{-}{O}$)—O—H	RS(O)OH, or RSO_2H	Sulfinic acids
$R{:}S(:\ddot{O}:)_2{:}R$	R—S^{2+}(—$\overset{-}{O}$)₂—R	$RS(O)_2R$, or RSO_2R	Sulfones
$R{:}S(:\ddot{O}:)_2{:}\ddot{O}{:}H$	R—S^{2+}(—$\overset{-}{O}$)₂—O—H	$RS(O)_2OH$, or RSO_3H	Sulfonic acids

Previous to the publication in 1916 of the Lewis theory of electronic structure, compounds such as the sulfoxides and sulfinic acids were considered to contain tetracovalent sulfur, one oxygen being bound to sulfur by a double bond. Similarly sulfones and sulfonic acids were considered to contain hexacovalent sulfur and two doubly bound oxygen atoms.

$R-\overset{\overset{O}{\|}}{S}-R$	$R-\overset{\overset{O}{\|}}{S}-OH$	$R-\underset{\underset{O}{\|}}{\overset{\overset{O}{\|}}{S}}-R$	$R-\underset{\underset{O}{\|}}{\overset{\overset{O}{\|}}{S}}-OH$
Sulfoxides	Sulfinic acids	Sulfones	Sulfonic acids

The nitrogen atom in nitric acid, nitro compounds, and amine oxides was thought to be pentacovalent.

$$HO-\overset{\overset{O}{\|}}{N}=O \qquad R-\overset{\overset{O}{\|}}{N}=O \qquad R_3N=O$$

Postulation by Lewis and by Langmuir of the octet rule required the nitrogen atom in the nitro group and in amine oxides to be tetracovalent, just as it is in the ammonium ion, which in turn required that the bond to one of the oxygen atoms be semipolar (p. 16). This postulation was applied also to the oxygenated compounds of sulfur and phosphorus, and compounds such as sulfur hexafluoride and phosphorus pentachloride were considered to be anomalous exceptions to the general rule.

During the next twenty years, several lines of investigation seemed to confirm the newer concept, but in 1937 experimental results on bond distances indicated that these sulfur-oxygen bonds were shorter, and hence stronger, than would be expected for a semipolar bond. Yet it was not until 1944 that the semipolar viewpoint was questioned seriously. Since then, considerable evidence has been produced that indicates that some additional bonding results from the interaction of the unshared pairs of electrons on oxygen with the sulfur atom.

When oxygen bonds with another atom by means of a coordinate covalence as in the amine oxides (p. 229), the donor atom becomes deficient in electrons and positively charged, while the oxygen atom acquires an excess of negative charge. This imbalance is at a maximum with elements of the second period because only $2s$ and $2p$ orbitals are available to accommodate the electrons, and the imbalance cannot be relieved in any way; in other words, elements of the second period obey the octet rule. The valence shell of the elements of the third period has five $3d$ orbitals available (p. 8), three of which have shapes and orientations that permit them, in the case of the sulfoxides, R_2SO, to overlap a $2p$ orbital of oxygen as shown in Fig. 15–*1*. Accordingly the imbalance can be relieved by a type of $p—d$ π bonding. Although two such π bonds can be formed from two $2p$ orbitals of oxygen and two $3d$ orbitals of sulfur, they are weak because of poor overlap, and the strength of the σ and two π bonds in the sulfur-oxygen linkage (125 kcal.) is about 25 per cent less than that of the carbon-oxygen double bond in formaldehyde (166 kcal.), although the latter has only a σ bond and one π bond. Similar considerations apply to the so-called pentavalent oxygen compounds of phosphorus (p. 271). In the sulfones the situation is still more complicated, it being necessary to hybridize $3d$ and $4p$ orbitals to form orbitals suitable for the π bonding.

Recent work shows that the hydrogens of methyl sulfide are much more acidic than those of tetramethylsilane. Since both silicon and sulfur have empty $3d$ orbitals, it appears that resonance stabilization of the carbanion from methyl sulfide does not result from direct interaction of the unshared $2p$ electrons on carbon with $3d$ orbitals of sulfur. Instead it is suggested that the interaction is with an empty $3p$ orbital of sulfur formed by promotion of a pair of $3p$ electrons to an empty $3d$ orbital. Accordingly a similar interpretation may be applied to the interaction of unshared $2p$ electrons on oxygen with $3p$ orbitals on sulfur in sulfoxides and sulfones.

In current literature the tendency has been to revert to the old symbolism, $>S=O$ and $>S\begin{smallmatrix}\nearrow\!\!\!\!\!/\, O \\ \searrow\!\!\!\!\!\backslash\, O\end{smallmatrix}$, to indicate the multiple-bond character of the sulfur-oxygen linkages. No distinction is made between a carbon-oxygen double bond and a sulfur-oxygen multiple bond. Yet the differences are far greater than the similarities. In a carbon-oxygen double bond, such as that in a ketone, $R_2C=O$, the σ bond is the result of the overlapping of an sp^2 hybrid orbital and a p orbital, and the carbon valences are planar. The π bond results from the overlapping of two p orbitals (cf. p. 159). Reagents, such as hydrogen, can add to the double bond with the formation of new strong

Figure 15–*1*. Bond formation resulting from overlapping of a $3d$ orbital with a $2p$ orbital.

single σ bonds. Addition of reagents to the sulfur-oxygen type of double bond does not take place because the overlap of the orbitals of the addenda with $3sp^3d$ or $3sp^3d^2$ hybrid orbitals of sulfur is too poor to form strong σ bonds. Only with highly electronegative fluorine does sulfur yield a neutral molecule, SF_6, in which it is joined to more than two groups by single covalent bonds. The use of the double bond symbol is particularly undesirable because even electronically the sulfoxide bond resembles a triple bond more closely than a double bond. Since in its chemical behavior the bond resembles the nitrogen-oxygen semipolar bond more closely than the carbon-oxygen double bond, it seems preferable to retain the semipolar symbolism with the understanding that the difference in charge and the bond distance are less than in the nitrogen-oxygen semipolar bond. When a formula is printed in a linear style, the oxygen involved may be placed in parentheses, for example RS(O)R, to avoid confusion with the hypothetical compound R—S—O—R.

THIOLS (THIO ALCOHOLS, MERCAPTANS)

Preparation

The hydrosulfide ion is a strong nucleophile (p. 103), and thiols (*mercaptans*) are formed when an alkyl halide is refluxed with an alcoholic solution of sodium hydrosulfide, NaSH.

$$HS^- + RX \longrightarrow HSR + X^-$$

A more convenient procedure involves the reaction of hydrosulfide ion with alkyl sulfate ion.

$$HS^- + ROSO_3^- \longrightarrow HSR + SO_4^{2-}$$

The alkyl sulfate ion is prepared by allowing an alcohol to react with sulfuric acid and neutralizing with sodium carbonate. The hydrosulfide ion is formed by saturating an aqueous solution of sodium hydroxide with hydrogen sulfide. Whether alkyl halides or alkyl sulfates are used, some alkyl sulfide always is formed because of the equilibrium.

$$RSH + {}^-SH \rightleftharpoons RS^- + H_2S$$

The thioalkoxide ion then reacts with a second mole of alkyl halide or sulfate.

$$RS^- + RX \longrightarrow RSR + X^-$$

The sodium hydrosulfide ordinarily is used in large excess. Because of the basicity of the hydrosulfide ion, another competing reaction is present, namely loss of mineral acid to give olefins and hydrogen sulfide. Hence yields of mercaptans or sulfides are best if the alkyl group is primary, whereas only olefin is formed if it is tertiary. Thiols can be prepared also by the reduction of disulfides (p. 263).

Nomenclature

The following examples indicate the usual methods of nomenclature: CH_3SH, methyl mercaptan or methanethiol; $CH_3CH_2CHSHCH_3$, *s*-butyl mercaptan or 2-butanethiol. The —*SH* group is known as the *sulfhydryl* or *thiol* group, or more commonly as the *mercapto* group. The last term may be used as a prefix for polyfunctional compounds, for example $HSCH_2COOH$, mercaptoacetic acid.

Physical Properties

If the boiling points of dicovalent sulfur compounds are compared with those of analogous oxygen, nitrogen, and carbon compounds of approximately the same polarizability (Table 15–*1*), it is evident that the boiling points of mercaptans are much more nearly normal than those of alcohols or of primary or secondary amines. Although the mercaptans boil somewhat higher than hydrocarbons with the same number of electrons, the higher boiling

TABLE 15–1. BOILING POINTS OF WATER, AMMONIA, HYDROGEN SULFIDE, AND METHANE, AND OF THEIR ALKYL DERIVATIVES

OXYGEN COMPOUNDS		NITROGEN COMPOUNDS		SULFUR COMPOUNDS		CARBON COMPOUNDS	
COMPD.	B.P.	COMPD.	B.P.	COMPD.	B.P.	COMPD.	B.P.
H_2O	+100	NH_3	−33			CH_4	−161
CH_3OH	+65	CH_3NH_2	−7	H_2S	−61	C_2H_6	−89
C_2H_5OH	+78	$C_2H_5NH_2$	+17	CH_3SH	+6	C_3H_8	−42
$(CH_3)_2O$	−24	$(CH_3)_2NH$	+7				
n-C_3H_7OH	+98	n-$C_3H_7NH_2$	+49	C_2H_5SH	+37	n-C_4H_{10}	−1
i-C_3H_7OH	+82	$CH_3NHC_2H_5$	+32			i-C_4H_{10}	−10
$CH_3OC_2H_5$	+11	$(CH_3)_3N$	+4	$(CH_3)_2S$	+38		

point cannot be ascribed to hydrogen bonding, since ethyl mercaptan and methyl sulfide boil at almost the same temperature. This behavior is in marked contrast to that of the propyl alcohols and methyl ethyl ether, or of methylethylamine and trimethylamine. The rise in boiling points of the series *n*-butane, trimethylamine, methyl ethyl ether, and methyl sulfide can be ascribed to the rise in dipole moment, the measured values of μ being respectively 0, 0.6, 1.2, and 1.5 D.

Mercaptans are much less soluble in water than the corresponding alcohols, only 1.5 g. of ethyl mercaptan being soluble in 100 cc. of water at room temperature. This low water solubility can be ascribed to the inability of sulfur to form strong hydrogen bonds with hydrogen attached to oxygen, just as it does not with hydrogen attached to sulfur (p. 255). The S—H bond strength is weaker than that of C—H and leads to the lower stretching absorption frequency of 2600–2500 cm.$^{-1}$ compared to 2960–2840 cm.$^{-1}$ for the C—H bond.

Physiological Properties

The volatile mercaptans have an extremely disagreeable odor. E. Fischer found that the nose can detect one volume of ethyl mercaptan in 50 billion volumes of air. This concentration expressed in terms of weight per cubic centimeter of air is 1/250 of the concentration of sodium that Kirchhoff and Bunsen were able to detect by means of the spectroscope. The obnoxious odor of mercaptans decreases with increasing molecular weight, and the odor becomes pleasant above nine carbon atoms. Like hydrogen sulfide, the lower mercaptans are toxic.

Reactions

Although dicovalent organic sulfur compounds are analogous structurally to organic oxygen compounds, the reactions of the two classes differ considerably. These differences arise because sulfur is less electronegative than oxygen (p. 21) and because the valence shell of sulfur is farther away from the nucleus and the electrons are more dispersed than those of oxygen. The ionization potentials of only a few sulfur compounds are known, but all are lower than those of the oxygen analogs. Thus the ionization potential of ethanethiol is 9.21 e.v. and that of ethanol is 10.65 e.v., which indicates that the unshared electrons of sulfur are not held so tightly as those of oxygen. Nevertheless the affinity of a mercaptide ion for a proton is less than that of an alkoxide ion because the more diffuse negative charge on sulfur makes it more difficult for the small proton to share a pair of electrons; that is, the overlap of a $3p$ orbital with a $1s$ orbital is poor. Hence thiols are more acidic than alcohols, just as hydrogen sulfide is a stronger acid than water (p. 32). By the same token thiols are less basic than alcohols.

Because sulfur is less electronegative than oxygen, the carbon-sulfur bond and the sulfur-hydrogen bond are less polar than the carbon-oxygen and the oxygen-hydrogen bonds. The decreased polarity is shown by lower dipole moments, which are 0.9 D and 1.8 D for hydrogen sulfide and water respectively, and 1.6 D and 1.7 D for ethanethiol and ethanol. This reduced polarity of the carbon-sulfur bond, together with the reduced basicity, makes scission of the bond, for example by halogen acids, more difficult than scission of the carbon-oxygen bond.

On the other hand, the decreased electronegativity of sulfur makes it easier for other atoms, which have larger valence shells than hydrogen, to share sulfur electrons. Hence, although the mercaptide ion is a weaker base than alkoxide ion, it is a better nucleophile (p. 103).

1. ***Salt Formation.*** Salts of thiols, commonly called *mercaptides,* are formed readily in aqueous solutions of strong bases.

$$RSH + HO^{-+}Na \longrightarrow RS^{-+}Na + H_2O$$

Like sodium sulfide, the soluble mercaptides are hydrolyzed markedly when dissolved in water. The heavy metal salts, such as those of lead, mercury, copper, cadmium, and silver, are insoluble in water. The ease of formation of insoluble mercury salts gave rise to the name mercaptan (L. *mercurium captans* seizing mercury). Standardized aqueous silver nitrate can be used for the estimation of mercaptans. The mercaptide ion, like the sulfide ion, reacts with aqueous sodium nitroprusside to give a transient purple color, which is used as a test for the thiol group.

$$[Fe(CN)_5(N{\equiv}O)]^{2-} + {}^{-}{:}SR \longrightarrow [Fe(CN)_5(\underset{\displaystyle SR}{\underset{|}{N}}{=}O)]^{3-}$$

2. ***Oxidation to Disulfides.*** The sulfur-sulfur bond is much stronger than the oxygen-oxygen bond and the sulfur-hydrogen bond is weaker than the oxygen-hydrogen bond. The bond dissociation energy for methyl disulfide, $CH_3S{-}SCH_3$, is 76 kcal., compared to 37 kcal. for methyl peroxide, $CH_3O{-}OCH_3$, and 90 kcal. for $CH_3S{-}H$ compared to 100 kcal. for $CH_3O{-}H$. Hence ready oxidation of thiols to disulfides is not surprising. Mild oxidizing agents suffice, and strong oxidizing agents are avoided since they oxidize disulfides further to sulfonic acids (p. 263). The usual reagent is iodine, which is added to an alcoholic or acetic acid solution of the thiol.

$$RSH + I_2 \longrightarrow RS^+ + HI + I^-$$
$$RS^+ + \underset{\displaystyle H}{\underset{|}{S}}R \longrightarrow RS{-}SR + H^+$$

If stronger oxidizing conditions are desired, the reaction is carried out in the presence of a base to remove the hydrogen iodide, or the preformed sodium or lead salt of the thiol is used.

The *doctor process* for sweetening gasoline depends on the oxidation of mercaptans to less odorous disulfides by means of sodium plumbite solutions and a small amount of free sulfur.

$$2\,RSH + Na_2PbO_2 \longrightarrow Pb(SR)_2 + 2\,NaOH$$
$$Pb(SR)_2 + S \longrightarrow PbS + (RS)_2$$

The doctor solution is regenerated by passing air through the hot solution.

$$PbS + 4\,NaOH + 2\,O_2 \longrightarrow Na_2PbO_2 + Na_2SO_4 + 2\,H_2O$$

Doctor solution can be used also to test for the presence of the sulfhydryl group. The formation of black lead sulfide indicates a positive reaction. The oxidation of sulfhydryl groups from the cysteine portion of protein molecules (pp. 359, 361) to disulfide groups and reduction of disulfide to sulfhydryl plays an important part in biological processes.

3. ***Acylation.*** Although thioesters are formed by the reaction of thiols with carboxylic acids, the position of equilibrium is much less favorable than for the esterification of alcohols, the respective equilibrium constants being < 0.06 and > 4.0. Hence acylation by acid anhydrides or acyl chlorides is the usual procedure. These reactions go very easily because of the good nucleophilicity of thiols, the mechanisms being the same as those for the acylation of alcohols (pp. 169, 170).

$$RSH + (R'CO)_2O \longrightarrow \underset{\text{A thiol ester}}{RS{-}\overset{\overset{\displaystyle O}{\|}}{C}{-}R'} + R'COOH$$

$$RSH + R'COCl \longrightarrow RSCOR' + HCl$$

4. ***Thio Acetal Formation.*** Good nucleophilicity, the low polarity of the carbon-sulfur bond, and the low basicity of sulfur all contribute to make the acid-catalyzed reaction of thiols with a carbonyl group to give acetals faster and the equilibrium more favorable than for alcohols (p. 200). Thus even ketones in the presence of hydrogen chloride or zinc chloride can be converted directly to thioketals.

$$RCHO + 2\,R'SH \longrightarrow RCH(SR')_2 + H_2O$$
$$R_2CO + 2\,R'SH \longrightarrow R_2C(SR')_2 + H_2O$$

As expected, thioacetals and thioketals are much more stable to acid hydrolysis than the oxygen analogs. The aldehyde or ketone can be regenerated, however, by hydrolysis in the presence of mercuric oxide or cadmium chloride, which shift the equilibrium by forming the insoluble mercury or cadmium mercaptides.

When thioacetals are heated with Raney nickel, which essentially has a nickel hydride surface, a methyl or methylene group and nickel mercaptide are formed by reduction.

$$R_2C(SR')_2 + 2\,H_2 + Ni \longrightarrow R_2CH_2 + NiS + 2\,R'H$$

The thioacetal group can be reduced to a methylene group also merely by heating with hydrazine in diethylene glycol or triethylene glycol (p. 605) until gas no longer is evolved, the other products being nitrogen, hydrogen sulfide, and alkane.

$$R_2C(SR')_2 + 2\,H_2NNH_2 \longrightarrow R_2CH_2 + 2\,N_2 + 2\,H_2S + 2\,R'H$$

Thus two series of reactions are available for converting a carbonyl group to a methylene group in addition to the Wolff-Kishner or thc Clemmensen reductions (p. 208).

SULFIDES

Preparation

Sulfides ordinarily are prepared by the reaction of alkyl halides or sulfates with sodium sulfide or sodium mercaptides in alcoholic solution.

$$2\,RX + Na_2S \longrightarrow R_2S + 2\,NaX$$
$$RX + NaSR' \longrightarrow RSR' + NaX$$

The first reaction gives simple sulfides, and the second may be used to prepare mixed sulfides.

Nomenclature

The nomenclature of sulfides is evident from the following examples: $C_2H_5SC_2H_5$, ethyl sulfide or ethylthioethane; $CH_3SCH_2CH(CH_3)_2$, methyl *i*-butyl sulfide or 1-methylthio-2-methylpropane. The international names rarely are used except for polyfunctional compounds. Although names such as diethyl sulfide instead of ethyl sulfide are used, the prefix *di* is unnecessary.

Reactions

Just as thiols are more acidic than alcohols (p. 259), sulfides are less basic than ethers. With elements other than hydrogen, however, the unshared electrons of sulfur are more readily shared (p. 255). As a result sulfur behaves more like nitrogen than like oxygen and forms stable tri- and tetracovalent compounds that have no oxygen analogs.

1. ***Oxidation to Sulfoxides and Sulfones.*** Most oxidizing agents that convert sulfides to sulfoxides also convert sulfoxides to sulfones. As expected from steric and electronic considerations, however, the first reaction is much faster than the second. By the proper choice of reagent and conditions, either sulfoxide or sulfone may be obtained. At room temperature or below, the calculated amount of 30 per cent hydrogen peroxide or an equivalent amount of a peroxy acid usually yields only sulfoxides, whereas an excess of oxidizing agent at 90–100° converts the sulfoxides to sulfones. The sulfones are very resistant to further oxidation.

$$R_2S + H_2O_2 \xrightarrow{25^\circ} R_2\overset{+}{S}\!-\!\overset{-}{O} + H_2O$$

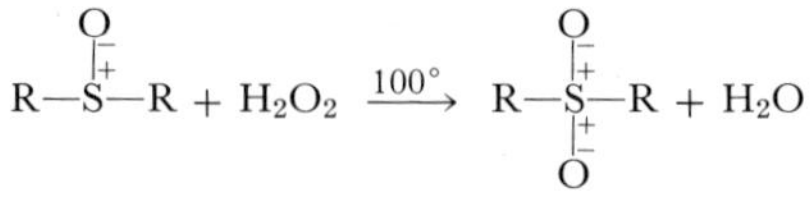

$$R-\overset{\overset{\large O^-}{|}}{\underset{}{S^+}}-R + H_2O_2 \xrightarrow{100^\circ} R-\overset{\overset{\large O^-}{|}}{\underset{\underset{\large O^-}{|}}{S^{2+}}}-R + H_2O$$

The oxidation by hydrogen peroxide is catalyzed by acids. It has been concluded that a one-step concerted mechanism involving the hydrogen-bonded peroxide-acid complex and the sulfide best explains the effects of substituents on the rate of reaction.

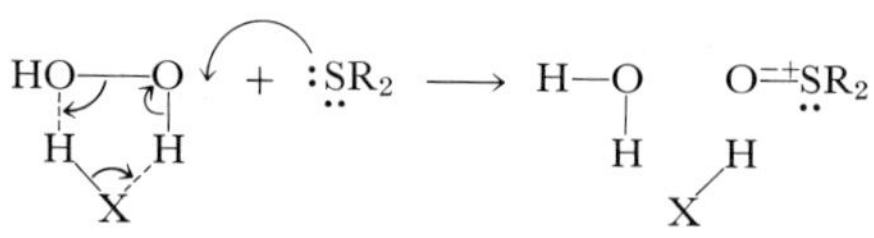

$$HO-O(H)\cdots H\cdots X + :\!SR_2 \longrightarrow H-O-H \quad \overset{-}{O}-\overset{+}{S}R_2 \quad H-X$$

The reaction with peroxy acids is not catalyzed by acids because no additional hydrogen bonding is needed.

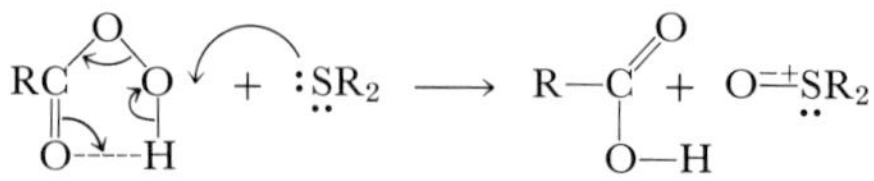

$$RC(=O)(-O-O-H) + :\!SR_2 \longrightarrow R-C(=O)-O-H + \overset{-}{O}-\overset{+}{S}R_2$$

Oxidation of the sulfoxide to the sulfone can proceed by analogous mechanisms but the blocking effect and the electron-withdrawing effect of the oxygen atom in the sulfoxide makes the latter a much less effective nucleophile, and the rate of reaction is much slower.

2. ***Formation of Sulfonium Salts.*** Sulfides react with alkyl halides to give ternary sulfonium salts analogously to the reaction of tertiary amines to give quaternary ammonium salts.

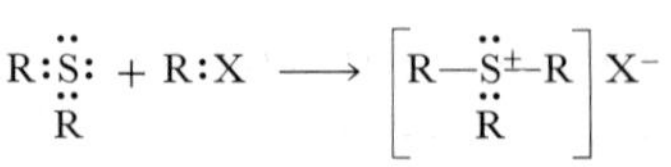

$$R:\underset{R}{\ddot{S}}: + R:X \longrightarrow \left[R-\underset{R}{\overset{+}{\ddot{S}}}-R\right]X^-$$

If the R groups are not all alike, this reaction is complicated by the fact that the sulfonium salts dissociate more readily than quaternary ammonium salts with the result that a mixture of all of the possible sulfonium salts may be formed.

DISULFIDES

Preparation

Disulfides may be prepared by the oxidation of thiols (p. 260) or by the reaction of sodium disulfide with an alkyl halide or sulfate.

$$Na^{+-}S{-}S^{-+}Na + RX \longrightarrow NaX + Na^{+-}S{-}S{-}R \xrightarrow{RX} NaX + R{-}S{-}S{-}R$$

Sodium disulfide is prepared by dissolving an equivalent amount of sulfur in a concentrated aqueous solution of sodium sulfide. The aqueous solution is diluted with alcohol, heated to boiling, and the alkyl halide added.

Nomenclature

Disulfides are named as such; for example, $(CH_3)_2S_2$ is methyl disulfide, and $CH_3SSC_2H_5$ is methyl ethyl disulfide.

Reactions

The only reactions expected of disulfides are scission of the sulfur-sulfur bond, formation of coordination compounds, or both. Although numerous reactions of these types are known, those of most interest are reduction to thiols and oxidation to sulfonic acids.

1. ***Reduction to Thiols.*** When zinc dust is added to a boiling mixture of a disulfide in 50 per cent sulfuric acid, the disulfide is reduced to the mercaptan.

$$RS{-}SR + Zn + H_2SO_4 \longrightarrow 2\,RSH + ZnSO_4$$

The reaction of disulfides with metallic sodium probably involves homolysis of the sulfur-sulfur bond.

$$RS{:}SR + 2\,Na\cdot \longrightarrow 2\,RS{:}^{-+}Na$$

The rate of this reaction increases rapidly with increasing molecular weight of the alkyl group. Methyl and ethyl disulfides are unreactive, whereas the reaction of *n*-butyl disulfide must be controlled by cooling. The chemical reactions involved in the permanent waving of hair consist of the reduction of disulfide linkages followed by regeneration of disulfide bonds by oxidation (p. 364).

2. ***Oxidation to Sulfonic Acids.*** Strong oxidizing agents such as permanganate, chromic acid, or nitric acid oxidize disulfides with cleavage of the sulfur-sulfur bond to give sulfonic acids.

$$RS{-}SR + 10\,HNO_3 \longrightarrow 2\,R\overset{\overset{\displaystyle O^-}{|+}}{\underset{\underset{\displaystyle O^-}{|+}}{S}}{-}OH + 10\,NO_2 + 4\,H_2O$$

Often it is more convenient to oxidize mercaptans or mercaptides but undoubtedly the first step is formation of the disulfide.

POLYSULFIDES

The ability of sulfur atoms to bond with each other and form long chains and large rings is reflected in the existence of organic polysulfides. Thus the reaction of alkyl halides with sodium polysulfides leads to the formation of mixtures of alkyl polysulfides in which x varies from 3 to 8.

$$2\,RX + Na_2S_x \longrightarrow R{-}S_x{-}R + 2\,NaX$$

Physical evidence indicates that the sulfur atoms in these compounds are linked linearly even though some of the sulfur can be removed by boiling with aqueous sodium hydroxide.

THIO CARBOXYLIC ACIDS AND DITHIO CARBOXYLIC ACIDS

Thio carboxylic acids may be prepared by the action of phosphorus pentasulfide on a carboxylic acid.

$$RCOOH + P_2S_5 \longrightarrow RCOSH + P_2OS_4$$

Structural theory calls for the existence of a second isomer, RCSOH. Only one substance has been isolated, however, because the two forms readily interconvert; that is, they are tautomeric (pp. 201, 623).

$$R{-}\overset{\overset{\displaystyle O}{\|}}{C}{-}SH \rightleftharpoons R{-}\overset{\overset{\displaystyle OH}{|}}{C}{=}S$$

The **esters** of the two structures are stable isomers. One isomer can be made from an acid chloride and a mercaptan and the other from an alkyl imidate hydrochloride (p. 236) and hydrogen sulfide.

$$RCOCl + HSR \longrightarrow R\overset{\overset{\displaystyle O}{\|}}{C}{-}SR + HCl$$

$$\left[R\overset{\overset{\displaystyle {}^{+}NH_2}{\|}}{C}{-}OR\right]Cl^- + H_2S \longrightarrow R\overset{\overset{\displaystyle S}{\|}}{C}{-}OR + NH_4Cl$$

To distinguish between the isomeric derivatives of the tautomeric acids, the acid having the thiol group is termed a *thiolic acid* and that having a thiocarbonyl group is termed a *thionic acid.* Thus $CH_3CO{-}SC_2H_5$ is ethyl thiolacetate, whereas $CH_3CS{-}OC_2H_5$ is ethyl thionacetate. The direct esterification of a thio acid with an alcohol leads to the formation of an ester and hydrogen sulfide.

$$RCOSH + HOR' \overset{H^+}{\rightleftharpoons} RCOOR' + H_2S$$

This reaction was the original basis for the assumption that in the esterification of carboxylic acids with primary alcohols, the hydroxyl group is removed from the acid rather than from the alcohol, an assumption that has been proved by the use of the O^{18} isotope (p. 163).

Thio amides that have hydrogen on nitrogen are again tautomeric.

$$R{-}\overset{\overset{\displaystyle S}{\|}}{C}{-}NH_2 \rightleftharpoons R{-}\overset{\overset{\displaystyle SH}{|}}{C}{=}NH$$

They are prepared by the reaction of an amide with phosphorus pentasulfide.

$$RCONH_2 + P_2S_5 \longrightarrow RCSNH_2 + P_2OS_4$$

The **dithio acids** are made by the reaction of carbon disulfide with Grignard reagents.

$$S{=}C{=}S + RMgX \longrightarrow R{-}\overset{\overset{\displaystyle S}{\|}}{C}{-}SMgX \xrightarrow{H^+} R{-}\overset{\overset{\displaystyle S}{\|}}{C}{-}SH$$

They are colored oils having an unbearable odor and are oxidized by air.

$$2\ RCSSH + [O]\ (air) \longrightarrow R\overset{S}{\overset{\|}{C}}{-}S{-}S{-}\overset{S}{\overset{\|}{C}}{-}R + H_2O$$

SULFONIUM SALTS

Sulfonium salts, $R_3S^+X^-$, ordinarily formed by the reaction of sulfides with alkyl iodides (p. 262), are strongly ionized in aqueous solution. They react with silver hydroxide to give sulfonium hydroxides.

$$R_3S^+X^- + AgOH \longrightarrow R_3S^+OH^- + AgX$$

Sulfonium hydroxides are strong bases analogous to the quaternary ammonium hydroxides (p. 231). They also pyrolyze in a similar way to give sulfides and olefins.

$$(C_2H_5)_3S^{+-}OH \xrightarrow{\text{Heat}} (C_2H_5)_2S + C_2H_4 + H_2O$$

Sulfonium salts dissociate more easily than quaternary ammonium salts (cf. p. 231).

$$R_3S^+X^- \rightleftharpoons R_2S + RX$$

Although the sulfur atom in the sulfonium ion still has an unshared pair of electrons, the positive charge holds the pair too strongly to permit reaction with a second molecule of alkyl halide to give a tetraalkylsulfonium ion with two positive charges of the type $[R_4S^{2+}][X^-]_2$. Sulfonium salts react with halogen and with some metallic salts to give stable complexes, which probably have this type of structure.

$$R_3S^+X^- + X_2 \longrightarrow [R_3SX^{2+}][X^-]_2$$

Selenium, in which the valence electrons are farther from the nucleus, is able to form tetraalkylselenonium dihalides of the type $[R_4Se^{2+}]\,[X^-]_2$.

SULFOXIDES

Sulfoxides usually are prepared by the oxidation of sulfides with the theoretical amount of 30 per cent hydrogen peroxide in acetone or acetic acid solution at room temperature (p. 262). Ordinarily sulfoxides are named as such. In the international system the *SO group* is called a *sulfinyl group*. Thus $C_4H_9SOC_4H_9$ is *n*-butyl sulfoxide or 1-butylsulfinylbutane, and $CH_3SOCH_2CH(CH_3)_2$ is methyl *i*-butyl sulfoxide or 1-methylsulfinyl-2-methylpropane.

The higher sulfoxides are solids or thick viscous oils that supercool and crystallize with difficulty. Methyl sulfoxide is a mobile liquid with about the same viscosity as *n*-propyl alcohol. Sulfoxides are soluble in organic solvents, and the lower members are very soluble in water because of hydrogen bonding of water molecules to the oxygen atom. Like amine oxides, they have a high dipole moment (~4 D) and a high boiling point. The stretching modes of the sulfur-oxygen semipolar bond usually can be recognized by the high intensity absorption and its complex appearance in the 1060–1040 $cm.^{-1}$ region.

The oxygen of sulfoxides is sufficiently basic to form salts with strong acids and to form complexes with Lewis acids and with compounds of transition metals that have vacant orbitals. Although nitrogen is more electronegative than sulfur, the sulfoxides are less basic than the amine oxides. The electron density on oxygen that results from the formation of the semipolar bond is reduced in the sulfoxides by the use of $3d$ orbitals to form π bonds, whereas this is not possible in the amine oxides (p. 229).

The positive charge on sulfur, which results from the semipolar bond, makes a hydrogen on an adjacent carbon sufficiently acidic to permit the formation of a sodium salt by reaction with sodium hydride. Thus methyl sulfoxide yields sodium methylsulfinylmethide, which whimsically has been called *dimsyl sodium* or *sodium dimsyl.*

$$CH_3SOCH_3 + NaH \longrightarrow CH_3SOCH_2^{-+}Na + H_2$$

Its solution in methyl sulfoxide or in tetrahydrofuran has been used to titrate weak acids such as alcohols, amines, acetylenes, and ketones, the endpoint being determined with triphenylmethane (p. 480) as indicator. The anions thus formed in methyl sulfoxide are highly reactive. For example, they react readily with carbon dioxide to give carboxylic acids.

$$RCOCH_3 + {}^-CH_2SOCH_3 \longrightarrow (CH_3)_2SO + RCOCH_2^- \xrightarrow{CO_2} RCOCH_2COO^- \xrightarrow{H^+} RCOCH_2COOH$$

$$RC{\equiv}CH + {}^-CH_2SOCH_3 \longrightarrow (CH_3)_2SO + RC{\equiv}C^- \xrightarrow{CO_2} RC{\equiv}CCOO^- \xrightarrow{H^+} RC{\equiv}CCOOH$$

The sulfoxides have an unshared electron pair on sulfur also and can form quaternary salts. When methyl sulfoxide is refluxed with methyl iodide for several days, the crystalline salt separates.

$$(CH_3)_2\overset{\overset{O^-}{|}}{\underset{+}{S}}: + CH_3I \longrightarrow \left[(CH_3)_2\overset{\overset{O^-}{|}}{\underset{+}{S}}{-}CH_3\right]I^-$$

Removal of hydrogen iodide from the quaternary salt gives the ylide (p. 231).

$$\left[(CH_3)_2\overset{\overset{O^-}{|}}{\underset{+}{S}}{-}CH_3\right]I^- + NaH \longrightarrow (CH_3)_2\overset{\overset{O^-}{|+}}{S}{\overset{+-}{=}}CH_2 + H_2 + NaI$$

The ylide reacts with alkenes or ketones to give three-membered ring compounds, presumably by addition of carbene (p. 250) to the double bond.

$$CH_3CH{=}CHCH_3 + (CH_3)_2\overset{\overset{O^-}{|+}}{S}{\overset{+-}{=}}CH_2 \longrightarrow CH_3\underset{\diagdown\ \diagup}{CH{-}CH}CH_3 \ (\text{ring } CH_2) + (CH_3)_2SO$$

$$R\overset{\overset{O}{\|}}{C}R + (CH_3)_2\overset{\overset{O^-}{|+}}{S}{\overset{+-}{=}}CH_2 \longrightarrow R{-}\overset{O{-}CH_2\ (\text{ring})}{C}{-}R + (CH_3)_2SO$$

Sulfoxides can be reduced to sulfides with zinc and acetic acid. They can be oxidized to sulfones with excess hydrogen peroxide in glacial acetic acid at 100° (p. 262), or with other oxidizing agents such as nitric acid at 120–180°, alkaline permanganate, or chromic acid.

SULFINIC ACIDS

Sulfinic acids can be obtained by passing sulfur dioxide into a solution of Grignard reagent and liberating the free acid from the halomagnesium salt by adding mineral acid.

$$RMgX + SO_2 \longrightarrow R\overset{\overset{O^-}{|+}}{S}{-}OMgX \xrightarrow{HX} R\overset{\overset{O^-}{|+}}{S}{-}OH + MgX_2$$

The sulfinic acids can be oxidized to sulfonic acids

$$RSOOH + [O](air) \longrightarrow RSO_2OH$$

and reduced to mercaptans.

$$RSO_2H + 4\,[H]\,(Zn + H_2SO_4) \longrightarrow RSH + 2\,H_2O$$

The last reaction affords proof that the sulfur atom in sulfinic and sulfonic acids is united directly to carbon.

SULFONES

Sulfones usually are prepared by oxidizing sulfides or sulfoxides at elevated temperature with an excess of hydrogen peroxide in glacial acetic acid, with chromic or nitric acid, or with potassium permanganate (p. 266). They usually are named as such. In the international system the *SO_2 group* is called *sulfonyl,* and sulfones are named as sulfonyl derivatives of saturated hydrocarbons. Thus $CH_3SO_2CH_3$ is methyl sulfone or methylsulfonylmethane, and $CH_3(CH_2)_3SO_2CH(CH_3)_2$ is *i*-propyl *n*-butyl sulfone or 1-*i*-propylsulfonylbutane.

Sulfones are colorless stable solids. The lower members are soluble in water because the oxygens form hydrogen bonds with water molecules. Although high-boiling, they are distillable without decomposition. Sulfones show two intense stretching absorption bands in the 1350–1300 cm.$^{-1}$ and 1160–1140 cm.$^{-1}$ regions.

In contrast to the sulfoxides, sulfones are not readily reduced to sulfides. If heated with Raney nickel containing absorbed hydrogen, the sulfur is removed completely to yield saturated hydrocarbon.

$$R_2SO_2 + 3\,H_2 + Ni \longrightarrow 2\,RH + 2\,H_2O + NiS$$

The same type of reaction takes place with sulfoxides, sulfides, mercaptans, and other types of sulfur compounds (p. 261).

Because of the positive charge on sulfur, the sulfone group is strongly electron-attracting and increases the acidity of hydrogen on α carbon atoms. The acid pK_a's of methyl sulfone, bis(methylsulfonyl)methane, and tris(methylsulfonyl)methane are approximately 23, 14, and 0, compared with approximately 20 for acetone. Like other compounds that readily form carbanions (pp. 621, 632), sulfones undergo reactions such as carbon alkylation, the Claisen ester condensation (p. 620), and the Knoevenagel reaction (p. 622). The sulfone group, like the carbonyl group, also has a strong polarizing effect on electrons in an adjacent carbon-carbon double bond, which permits the double bond to undergo addition reactions characteristic of α,β-unsaturated carbonyl compounds (p. 618).

SULFONIC ACIDS AND DERIVATIVES

Preparation of Sulfonic Acids

1. ***By the Oxidation of Mercaptans or Disulfides.*** This procedure is that most generally used for the preparation of alkanesulfonic acids (p. 263). Frequently mercaptans are purified as the lead mercaptides, and the latter are oxidized directly to lead salts of sulfonic acids. The free acid may be obtained by passing hydrogen chloride into an alcoholic suspension of the lead sulfonate.

2. ***From Alkyl Halides and Alkali Sulfite.*** Primary alkyl halides react with aqueous solutions of sodium or ammonium sulfite at elevated temperatures to give alkanesulfonates.

$$RX + Na_2SO_3 \xrightarrow{200^\circ} RSO_3^-Na^+ + NaX$$

As in the reaction of primary alkyl halides with other ambident anions (p. 237), the alkyl group combines with sulfur rather than oxygen to form the more covalent bond and give a sulfonate rather than an alkyl sulfite (cf. p. 242). The free sulfonic acids can be obtained by passing dry hydrogen chloride into an alcoholic solution of the sodium salt, the sulfonic acid being soluble and the sodium chloride insoluble in alcohol.

3. ***From Sulfonyl Chlorides.*** Sulfonyl chlorides (p. 268) are purified readily by distillation and can be hydrolyzed to the sulfonic acid by boiling with water.

$$RSO_2Cl + H_2O \longrightarrow RSO_3H + HCl$$

Since the sulfonic acids are nonvolatile, the hydrochloric acid and water can be removed by vacuum distillation leaving the pure acid as a residue.

4. ***By the Addition of Bisulfite to Double Bonds.*** Sodium bisulfite adds to simple olefins only in the presence of oxygen or oxides of nitrogen. The addition takes place by a free radical mechanism and does not follow the Markovnikov rule (p. 85).

$$^{-}OSOOH + \cdot O{-}O\cdot \text{ (or } \cdot NO) \longrightarrow {}^{-}OSOO\cdot + HOO\cdot \text{ (or HNO)} \qquad \textit{Initiation}$$

$$\left.\begin{array}{l} RCH{=}CH_2 + \cdot OSO_2^- \longrightarrow R\dot{C}H{-}CH_2SO_2^- \\ R\dot{C}H{-}CH_2SO_2^- + HOSO_2^- \longrightarrow RCH_2CH_2SO_2^- + \cdot OSO_2^- \end{array}\right\} \quad \textit{Propagation}$$

Double bonds conjugated with a carbonyl group add bisulfite by an ionic mechanism (p. 635). Aldehydes and methyl ketones usually add bisulfite to give bisulfite addition compounds, which are the sodium salts of hydroxy sulfonic acids (p. 198).

Nomenclature

The international system commonly is used in which the suffix *sulfonic acid* is added to the name of the hydrocarbon; for example CH_3SO_3H is methanesulfonic acid. For higher molecular weight compounds the alkyl group frequently is named, for example, $C_{12}H_{25}SO_3H$, laurylsulfonic acid, and $C_{16}H_{33}SO_3H$, cetylsulfonic acid, but this practice should be discontinued. Unobjectionable names are dodecanesulfonic acid and hexadecanesulfonic acid.

Reactions of Sulfonic Acids and Their Derivatives

Owing to the high electronegativity of the sulfonyl group, the sulfonic acids, in contrast to carboxylic acids, are strong acids. Because of almost complete ionization, they are very soluble in water, insoluble in saturated hydrocarbons, and relatively nonvolatile.

Sulfonyl chlorides may be prepared by the action of chlorine and water on a variety of sulfur compounds. One of the best methods starts with disulfides.

$$RSSR + 5\,Cl_2 + 4\,H_2O \longrightarrow 2\,RSO_2Cl + 8\,HCl$$

The oxidation is carried out in the presence of water despite the fact that an acid chloride is produced. The sulfonyl chlorides are very insoluble in water, and their rate of reaction with water is so slow that they can be removed before any appreciable hydrolysis has taken place.

Methanesulfonic acid reacts with thionyl chloride or phosphorus trichloride to give good yields of methanesulfonyl chloride, and the reactions should be general.

$$CH_3SO_2OH + SOCl_2 \longrightarrow CH_3SO_2Cl + HCl + SO_2$$

$$3\,CH_3SO_2OH + PCl_3 \longrightarrow 3\,CH_3SO_2Cl + P(OH)_3$$

Sulfonic acids do not give satisfactory yields of esters by direct esterification, or of amides by heating ammonium salts. Hence these derivatives always are made from the sulfonyl chlorides. Because they are insoluble in water, sulfonyl chlorides hydrolyze slowly, and their reactions may be carried out in the presence of water. Usually sodium hydroxide is added to increase the nucleophilicity of the alcohol and to remove the hydrogen chloride (cf. *Schotten-Baumann reaction,* p. 469).

$$RSO_2Cl + HOR' + NaOH \longrightarrow RSO_2OR' + NaCl + H_2O$$

$$RSO_2Cl + NH_3 + NaOH \longrightarrow RSO_2NH_2 + NaCl + H_2O$$

Sulfonamides are characterized by the relatively high acidity of the hydrogen atoms attached to nitrogen, which results from the strong electronegativity of the sulfonyl group. They are sufficiently acidic to form stable salts with strong alkalies in aqueous solutions.

$$RSO_2NH_2 + NaOH \longrightarrow [RSO_2\bar{N}H]Na^+ + H_2O$$

Therefore sulfonamides derived from ammonia or primary amines are soluble in dilute sodium hydroxide solutions.

Sulfonic esters behave more like sulfuric esters than like carboxylic esters. The sulfonate ion, like the alkyl sulfate ion (p. 139), is a good leaving group. Hence the esters have an alkylating action. Thus ammonia yields amine salts rather than sulfonamides.

$$RSO_2OR' + NH_3 \longrightarrow RSO_2O^{-+}NH_3R'$$

When sulfonic esters are heated with alcohols, ethers are formed.

$$RSO_2OR' + HOR'' \longrightarrow RSO_2OH + R'OR''$$

Reaction with Grignard reagents gives hydrocarbons.

$$RSO_2OR' + R''MgX \longrightarrow R'R'' + RSO_2O^{-+}MgX$$

When sulfonic acids are heated with phosphorus pentoxide, the **anhydrides** are formed.

$$RSO_2OH + P_2O_5 \longrightarrow (RSO_2)_2O + 2\,HPO_3$$

Anhydrides may be prepared also from the acid and carbodiimides (p. 291). The reactions of the sulfonic anhydrides are analogous to those of the sulfonyl halides but are more rapid.

INDIVIDUAL SULFUR COMPOUNDS

Methyl mercaptan is made commercially by passing a mixture of methyl alcohol and hydrogen sulfide over alumina-potassium tungstate (9:1) at about 400°.

$$CH_3OH + H_2S \longrightarrow CH_3SH + H_2O$$

Production capacity of one plant in the United States is over 5 million pounds per year. It is used chiefly for the synthesis of methionine (p. 358). ***n*-Propyl mercaptan** has been identified in the volatile products of freshly crushed onion. **Trichloromethanesulfenyl chloride** (*perchloromethyl mercaptan*), made by the reaction of chlorine with carbon disulfide, is the only important derivative of the unknown aliphatic sulfenic acids, R—S—OH.

$$2\,CS_2 + 5\,Cl_2 \longrightarrow 2\,Cl_3CSCl + S_2Cl_2$$

It is used in the synthesis of the powerful fungicides, Phaltan (p. 475) and Captan (p. 649). ***n*-Butyl mercaptan** has been isolated as a component of the malodorous secretion of skunks. A ***t*-dodecyl mercaptan** is used to regulate the chain length in the manufacture of synthetic rubbers (p. 597).

Methyl sulfide is derived commercially from the methoxyl groups of lignin by heating the spent liquor from the manufacture of kraft or sulfite pulp (p. 342) with sodium sulfide. It is used as an odorant for natural gas and to make methyl sulfoxide (p. 270). **Mustard gas** (*2-chloroethyl sulfide*) is one of the more powerful vesicants used in chemical warfare. It is not a gas but a heavy oily liquid boiling at 217°. It is made by reaction of ethylene and sulfur monochloride.

$$2\,CH_2{=}CH_2 + S_2Cl_2 \longrightarrow ClCH_2CH_2SCH_2CH_2Cl + S$$

It may be made also by a process that avoids the presence of sulfur in the final product.

$$CH_2{=}CH_2 \xrightarrow[\text{Ag catalyst}]{O_2\text{ (air)}} \underset{\text{Ethylene oxide}}{CH_2{-}CH_2 \text{ (}{-}O{-}\text{)}} \xrightarrow{H_2S} \underset{\text{2-Hydroxyethyl sulfide}}{(HOCH_2CH_2)_2S} \xrightarrow{HCl} \underset{\text{Mustard gas}}{(ClCH_2CH_2)_2S}$$

Although not used during World War II, enormous quantities of mustard gas were manufactured by both sides, which possibly acted as a deterrent to its use.

Oil of garlic is a complex mixture of which the chief component is allyl disulfide, $(CH_2{=}CHCH_2)_2S_2$. **Allicin,** the monosulfoxide, appears to be the immediate precursor. **Linear polymeric disulfides** and **polysulfides** of high molecular weight have rubber-like properties (p. 589).

Methyl sulfoxide (*dimethyl sulfoxide, DMSO*), first prepared by Zaĭtsev in 1866, has been made commercially only since 1960 by the vapor or liquid phase air oxidation of methyl sulfide with nitrogen dioxide as catalyst.

$$(CH_3)_2S + NO_2 \longrightarrow (CH_3)_2SO + NO$$
$$NO + \tfrac{1}{2}\,O_2\,(\text{air}) \longrightarrow NO_2$$

The methyl sulfoxide, b.p. 189°, is separated and the unreacted materials are recycled. It is miscible with water and is highly hygroscopic. Methyl sulfoxide is an excellent aprotic solvent for dipolar compounds and even for many salts because of its high dielectric constant ($\epsilon = 45$). Actually it is not, strictly speaking, aprotic, since the hydrogens are sufficiently acidic to exchange with deuterium in the presence of base. It does not, however, form strong hydrogen bonds with unshared electron pairs, and hence the rate of many reactions that take place by a polar mechanism are much faster in methyl sulfoxide than in hydroxylic solvents (p. 102). Now that it is available commercially, it has become widely used as a reaction medium. It is used also as a selective solvent in the separation of mixtures of gases and liquids, for example aromatic hydrocarbons from paraffins and olefins. Methyl sulfoxide, however, is a much more reactive compound than dimethylformamide, which limits its use as a solvent but also makes it a useful reagent (p. 266). Methyl sulfoxide has been reported to have most unusual physiological properties and a remarkable ability to penetrate animal tissues.

Thioacetamide is being used instead of gaseous hydrogen sulfide in qualitative and quantitative inorganic analysis. It is reasonably stable in neutral aqueous solution, but in the presence of acids or bases, especially on warming, it hydrolyzes to give hydrogen sulfide or sulfide ion.

$$CH_3CSNH_2 + H_2O + HCl \longrightarrow H_2S + CH_3COOH + NH_4Cl$$
$$CH_3CSNH_2 + 3\,NaOH \longrightarrow Na_2S + CH_3COONa + NH_3 + H_2O$$

Methanesulfonyl chloride (*mesyl chloride*) is used in research work for the preparation of methanesulfonates, particularly in the field of carbohydrates (p. 328). The salts of many sulfonic acids that have long hydrocarbon chains are important wetting agents and detergents (pp. 635, 406).

PHOSPHORUS

Phosphorus forms stable compounds in which it is bound to three, four, five, or six other atoms or groups. From its position in the periodic table, it would be expected to resemble nitrogen, and several thousand tricovalent phosphorus compounds are known. As noted earlier (p. 15), these compounds make use of more nearly pure *p* orbitals for bonding (Fig. 15–2*a*) than does nitrogen. Several thousand compounds having a coordination number of four, making use of sp^3 hybrid orbitals, also are known (Fig. 15–2*b*). In this respect phosphorus resembles carbon more nearly than does nitrogen. Unlike second period elements, however, third period elements have *d* orbitals available in the valence shell. These 3*d* orbitals are sufficiently close in energy to the 3*p* and 3*s* orbitals for hybridization to take place. Thus sp^3d hybridization leads to five bonds directed to the corners of a triangular bipyramid and accounts for a few compounds such as PCl_5 in the vapor and

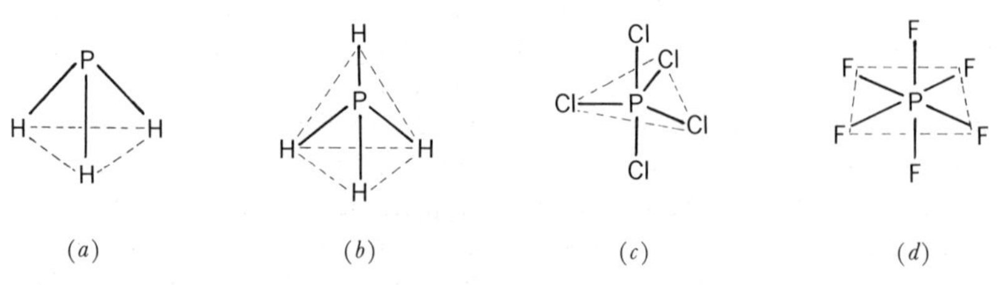

Figure 15–2. Compounds of phosphorus using (*a*) *p* orbitals, (*b*) sp^3 hybrid orbitals, (*c*) sp^3d hybrid orbitals, and (*d*) sp^3d^2 hybrid orbitals.

$(HO)_3P{=}O$	$Br_3P{=}S$	$R_3P{=}CR_2$
(*a*)	(*b*)	(*c*)
$(HO)_3P\overset{\pm}{-}\ddot{\underset{..}{O}}:$	$Br_3P\overset{\pm}{-}\ddot{\underset{..}{S}}:$	$R_3P\overset{\pm}{-}\ddot{C}R_2$
(*d*)	(*e*)	(*f*)

Figure 15–*3*. Representations of multiple-bonded phosphorus compounds.

liquid states and $P(C_6H_5)_5$ (Fig. 15–*2c*). The hexafluorophosphate ion PF_6^- exists because of sp^3d^2 hybridization, which leads to six bonds directed to the corners of an octahedron (Fig. 15–*2d*). Such compounds are the only ones in which phosphorus can, strictly speaking, be called pentavalent or hexavalent.

Just as sulfoxides and sulfones frequently are represented as having double bonds between sulfur and oxygen (p. 257), the structures of compounds in which phosphorus is united to four other atoms or groups frequently are represented with double bonds between phosphorus and oxygen, nitrogen, sulfur, or carbon as in Fig. 15–*3a,b,c*. It is preferable, however, to represent these compounds as having a semipolar bond as in Fig. 15–*3d,e,f*, with the understanding that the separation of charge is relieved by a type of π bonding resulting from some sharing of unshared pairs with phosphorus, which is permitted by overlap of *p* orbitals of the second element with *d* orbitals of phosphorus. The extent of this multiple bonding depends on the deficiency in electron density on phosphorus. Thus in phosphorus oxyfluoride, F_3PO, the electronegative fluorine atoms further decrease the electron density on phosphorus and considerable multiple bonding between oxygen and phosphorus undoubtedly occurs. On the other hand, practically none may be present in trimethylphosphine oxide, $(CH_3)_3PO$, where the electron deficiency on phosphorus is relieved by the electron-donating methyl groups.

As already noted, phosphorus has a greater tendency than nitrogen to exhibit a coordination number of four. Thus hydroxylamine is stable whereas phosphine oxide is the stable phosphorus analog.

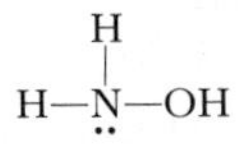

```
     H                 H
     |                 |
  H—N—OH            H—P±—O
     ··                |
                       H
 Hydroxylamine     Phosphine oxide
```

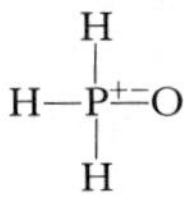

Similarly, although phosphoric acid, H_3PO_4, is tribasic, phosphorous acid, H_3PO_3, and hypophosphorous acid, H_3PO_2, are dibasic and monobasic respectively.

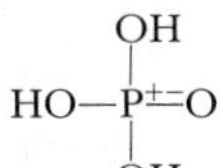

```
      OH                OH                  H
      |                 |                   |
  HO—P±—O           HO—P±—O            HO—P±—O
      |                 |                   |
      OH                H                   H
 Phosphoric acid   Phosphorous acid   Hypophosphorous acid
```

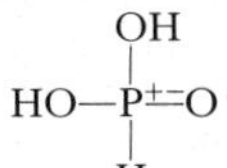

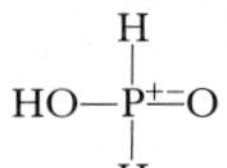

Because the valence shell of phosphorus is farther from the nucleus than that of nitrogen, there is more room for four groups on phosphorus than on nitrogen.

It should be noted that although phosphorous acid and hypophosphorous acid have only two and one titratable hydrogen atoms, trialkyl phosphites and dialkyl hypophosphites are stable compounds. Hence for the purpose of naming derivatives that have an organic group in place of hydrogen, the acids are given two names, and the derivatives are named according to whether the phosphorus atom in the compound has a coordination number of three or four.

$$\underset{\text{Phosphorous acid}}{\text{HO—}\overset{\text{OH}}{\underset{..}{\text{P}}}\text{—OH}} \qquad \underset{\substack{\text{Phosphonous acid}\\(\textit{Hypophosphorous acid})}}{\text{HO—}\overset{\text{OH}}{\underset{..}{\text{P}}}\text{—H}} \qquad \underset{\text{Phosphinous acid}}{\text{HO—}\overset{\text{H}}{\underset{..}{\text{P}}}\text{—H}}$$

$$\underset{\text{Phosphonic acid}}{\text{HO—}\overset{\text{OH}}{\underset{\text{H}}{\text{P}}}{}^{\pm}\text{=O}} \qquad \underset{\text{Phosphinic acid}}{\text{HO—}\overset{\text{H}}{\underset{\text{H}}{\text{P}}}{}^{\pm}\text{=O}} \qquad \underset{\text{Phosphine oxide}}{\text{H—}\overset{\text{H}}{\underset{\text{H}}{\text{P}}}{}^{\pm}\text{=O}}$$

Esters of Acids of Phosphorus. The esters of the acids of phosphorus are by far the most important of the organic compounds containing phosphorus. They are used to make true organophosphorus compounds having a carbon-phosphorus bond, they are valuable as plasticizers, nonflammable hydraulic fluids (p. 445), and insecticides (p. 444), and they play an important part in biological processes (pp. 533, 667).

Alkyl phosphites usually are made from primary alcohols and phosphorus trichloride in the presence of a tertiary amine, preferably *N,N*-dimethylaniline, $C_6H_5N(CH_3)_2$ (p. 426), to remove the hydrogen chloride.

$$3\,ROH + PCl_3 + 3\,C_6H_5N(CH_3)_2 \longrightarrow (RO)_3P + 3\,[C_6H_5\overset{+}{N}H(CH_3)_2]Cl^-$$

In the absence of the base, the hydrogen chloride cleaves the phosphite to give alkyl chloride and the **alkyl phosphonate,** which commonly is called the *dialkyl phosphite.*

$$(RO)_3P + HCl \longrightarrow RCl + (RO)_2POH \rightleftharpoons (RO)_2\underset{\text{H}}{\text{P}}{}^{\pm}\text{=O}$$

The **alkyl phosphates** are made from primary alcohols and phosphorus oxychloride, usually in the presence of a tertiary amine.

$$3\,ROH + POCl_3 + 3\,R_3N \longrightarrow (RO)_3PO + 3\,R_3NH^{+-}Cl$$

For the preparation of either phosphites or phosphates, an equivalent amount of the sodium alkoxide may be used instead of the alcohol and a tertiary amine. The inability of phosphorus to form ordinary double bonds is reflected in the nonexistence of alkyl metaphosphates, R—O—PO_2, the analogs of alkyl nitrates, R—O—NO_2.

The alkyl phosphonates (*dialkyl hydrogen phosphites* or *dialkyl phosphites*), trialkyl phosphites, and trialkyl phosphates in which the alkyl groups are methyl, ethyl, *n*-butyl, and 2-ethylhexyl, are manufactured commercially on a large scale, as are the phosphates from phenols (p. 445). **Methyl phosphate** is used as a gasoline additive to reduce preignition. **β-Diethylaminoethyl phosphate,** $[(C_2H_5)_2NCH_2CH_2O]_3PO$, when applied to the leaves of plants, inhibits flowering. When **ethyl phosphate** is heated with phosphorus oxychloride or phosphorus pentoxide, a product is obtained that is an effective insecticide. This material, commonly called **tetraethyl pyrophosphate** (*TEPP*), is a mixture of esters, the active component of which appears to be ethyl pyrophosphate, $(C_2H_5O)_2P(O)$—O—$P(O)(OC_2H_5)_2$. The so-called *nerve gases* developed in Germany during World War II are highly toxic esters of phosphorus acids. **Isopropyl methylfluorophosphonate,** $CH_3PF(O)OC_3H_7$-*i*, called *Sarin,* is over ten times more toxic than hydrogen cyanide. The high toxicity of the esters results from their inhibiting action on acetylcholine esterase (p. 232). The hydrolysis of acetylcholine involves acetylation of the enzyme followed by hydrolysis of the acetylated enzyme. This process is exceedingly rapid, the cycle taking place at least 5000 times per second. The phosphorus esters also phosphorylate the enzyme very rapidly, but regeneration of the enzyme by hydrolysis is very slow. The phosphorylated enzymes have a life of from hours to days or weeks and in effect destroy the enzyme, leading to general muscular contraction and death of the organism.

The *O,O*-dialkyl hydrogen dithiophosphates are made by the reaction of an alcohol with phosphorus pentasulfide.

$$P_2S_5 + 5\,ROH \longrightarrow 2\,(RO)_2P(S)SH + H_2O + RSH$$

Their salts are used as additives for lubricating oils and as mineral flotation agents. **O,O-Dimethyl hydrogen dithiophosphate** is an intermediate in the preparation of the insecticide, malathion (p. 635), which is relatively nontoxic to mammals. Reaction with *N*-methylchloroacetamide gives a product, called *Dimethoate,* that is translocated by plants and makes the plant as a whole insecticidal.

$$(CH_3O)_2P(S)S^{-+}Na + ClCH_2CONHCH_3 \longrightarrow (CH_3O)_2P(S)SCH_2CONHCH_3 + NaCl$$

The boiling points of alkyl phosphates are high because of the polarity of the phosphorus-oxygen bond. The lowest member, methyl phosphate, boils at 197° compared to 111° for methyl phosphite. The phosphites, like most esters of inorganic acids, are insoluble in water, but the high electron density on the coordinately bound oxygen of phosphates causes them to be much more capable of forming strong hydrogen bonds with water molecules, and methyl and ethyl phosphates are miscible with water.

Trivalent Organophosphorus Compounds. Phosphine, PH_3, the starting point for the synthesis of certain organophosphorus compounds, can be made conveniently by heating anhydrous phosphorous acid or by hydrolyzing aluminum phosphide with dilute acids. It results also from the reaction of white phosphorus with concentrated aqueous sodium hydroxide and from the direct combination of hydrogen with white phosphorus.

$$3\,H_3PO_3 \xrightarrow{210^\circ} 3\,H_3PO_4 + PH_3$$

$$AlP + 3\,HCl \longrightarrow AlCl_3 + PH_3$$

$$4\,P + 3\,NaOH + 3\,H_2O \longrightarrow 3\,NaH_2PO_2 + PH_3$$

The boiling point of phosphine, −88°, is practically identical with that of ethane, which has the same number of electrons, indicating that, unlike ammonia, phosphine molecules show no tendency to hydrogen-bond with each other. Neither does phosphine form hydrogen bonds with water; its solubility in 100 g. of water at 20° is only 0.02 g. compared to 53 g. for ammonia. As with divalent sulfur compounds (p. 259), the absence of hydrogen bonding results from the diffuseness of the negative charge of the unshared electrons in the third shell (p. 255).

Phosphine is such a weak base ($pK_a = -13.5$) that it forms a stable salt only with the strongest halogen acid, hydrogen iodide, and has practically no nucleophilicity. Hence alkylphosphines are not made directly from phosphine and alkyl halides, analogously to the Hofmann preparation of amines (p. 222). **Trialkylphosphines** are obtained readily by reaction of Grignard reagents or alkyllithiums with phosphorus trichloride.

$$PCl_3 + 3\,RMgCl \longrightarrow R_3P + 3\,MgCl_2$$

Like phosphine, the alkylphosphines do not form hydrogen bonds and are insoluble in water. Replacement of the hydrogen of phosphine by alkyl groups markedly increases the nucleophilicity and the basicity, although the alkylphosphines still are much less basic than the amines. For example, the pK_a's of ethyl-, diethyl-, and triethylphosphine are 2.69, 2.73, and 6.68, whereas those of the corresponding amines are 10.63, 10.98, and 10.65. Like phosphine, the alkylphosphines are highly toxic.

The phosphines are strong reducing agents and the lower members are oxidized by air, some being spontaneously flammable. Phosphine is a very poor nucleophile compared to ammonia, but trialkylphosphines are even better nucleophiles than trialkylamines. Like the latter, they react with alkyl iodides to form quaternary salts. Alkyl groups attached to phosphorus are farther from the nucleus and less crowded than when attached to nitrogen, and steric effects are less.

Although reactive organometallic compounds convert phosphorus trichloride to trialkylphosphines, less reactive compounds such as those of mercury, cadmium, zinc, or lead give **alkyldichlorophosphines** and **dialkylchlorophosphines.**

$$2\,PCl_3 + R_2Hg \longrightarrow 2\,RPCl_2 + HgCl_2$$
$$PCl_3 + R_2Hg \longrightarrow R_2PCl + HgCl_2$$

Reaction of these alkylchlorophosphines with alkoxides in alcohol solution yields **esters of phosphonous acids** and of **phosphinous acids.**

$$RPCl_2 + 2\,^{-}OR \longrightarrow \underset{\text{An alkyl alkylphosphonite}}{RP(OR)_2} + 2\,Cl^-$$

$$R_2PCl + {}^{-}OR \longrightarrow \underset{\text{An alkyl dialkylphosphinite}}{R_2POR} + Cl^-$$

Organophosphorus Compounds with a Coordination Number of Four (*"Pentavalent" Phosphorus Compounds*). The most widely studied compounds of this type have been the **esters of phosphonic acids** because they can be made easily from readily available starting materials. In the *Arbusov rearrangement* an alkyl phosphite is heated with a small amount of the corresponding alkyl halide. The initial product formed by an S_N2 reaction is a phosphonium salt, which decomposes to a dialkyl alkylphosphonate.

$$(RO)_3P + RX \longrightarrow (RO)_3\underset{\displaystyle R}{\overset{}{P}}{}^{+}\,{}^{-}X \longrightarrow (RO)_2\underset{\displaystyle R}{P}{}^{\pm}\!\!=\!O + RX$$

The formation of the phosphonium salt is so fast compared to its degradation that if an equivalent amount of a different halide is used, up to 95 per cent yield of the product having the new alkyl group attached to phosphorus is obtained. This behavior is called the *Arbusov reaction.*

$$(RO)_3P + R'X \longrightarrow (RO)_3\underset{\displaystyle R'}{P}{}^{+}\,{}^{-}X \longrightarrow (RO)_2\underset{\displaystyle R'}{P}{}^{\pm}\!\!=\!O + RX$$

Phosphonic esters can be prepared equally readily by the *Michaelis reaction* in which a sodium derivative of an alkyl phosphonate (*dialkyl phosphite*) reacts with an alkyl halide.

$$(RO)_2\underset{\displaystyle H}{P}{}^{\pm}\!\!=\!O + \cdot Na \longrightarrow \tfrac{1}{2}\,H_2 + (RO)_2\ddot{P}\!-\!O^{-}\,{}^{+}Na \xrightarrow{R'X} (RO)_2\underset{\displaystyle R'}{P}{}^{\pm}\!\!=\!O + NaX$$

Thus the more stable phosphonate is formed rather than a trialkyl phosphite. Ordinarily butyl phosphonate (*dibutyl hydrogen phosphite*) is used because the sodium derivative is soluble in hydrocarbon solvents.

Phosphine adds to aldehydes in the presence of aqueous hydrochloric acid to give tetrakis(α-hydroxyalkyl)phosphonium chlorides, $(RCHOH)_4P^{+}\,{}^{-}Cl$. The reaction undoubtedly results from electrophilic attack by the protonated aldehyde on the phosphine.

$$RCH\!=\!O \xrightarrow{H^+} \left\{RCH\!=\!\overset{+}{O}H \longleftrightarrow R\overset{+}{C}H\!-\!OH\right\} \xrightarrow{:PH_3} RCHOH\overset{+}{P}H_3 \longrightarrow RCHOHPH_2 + H^+$$

Repetition of these steps leads to the final product. **Tetrakis(hydroxymethyl)phosphonium chloride** (*THPC*) from phosphine and formaldehyde is used in flame-retardant finishes for cotton.

SILICON

Two aspects of the chemistry of silicon are of interest: (*1*) the preparation and properties of silicon analogs of carbon compounds, and (*2*) the preparation and properties of organosilicon compounds in which carbon is bound to silicon. The position of silicon just below carbon in the periodic table leads to the expectation that it would form tetracovalent compounds with tetrahedral bond angles, and most silicon compounds are of this type. Unlike carbon, however, the valence shell has *d* orbitals available, and sp^3d^2 hybridization leads to a few compounds in which the silicon has a coordination number of six. Thus the fluosilicate ions have the composition SiF_6^{2-}, and the bonds are directed to the corners of an octahedron (cf. PF_6^-, p. 270). Tetracovalent silicon compounds are more vulnerable to nucleophilic attack than carbon compounds because empty *d* orbitals are available for receiving an unshared pair of electrons (cf. Fig. 15–*4*).

The differences between the properties of silicon and carbon compounds arise not only from the availability of *d* orbitals, but also because the valence electrons are farther from the nucleus in silicon than in carbon. This leads to three important consequences: (*1*) the valence electrons are held less strongly by the nucleus and hence silicon is less electronegative than carbon; (*2*) there is more room for atoms or groups to bond to silicon; (*3*) overlap of *p* orbitals with *p* orbitals of other elements is less, with the result that ordinary double bonds do not form.

The **silicon hydrides** are analogs of alkanes. When crude magnesium silicide, prepared by heating silica with magnesium, reacts with mineral acids, a mixture is obtained consisting of 40 per cent silane, SiH_4; 30 per cent disilane, Si_2H_6; 15 per cent trisilane, Si_3H_8; and 10 per cent tetrasilane, Si_4H_{10}. The remaining 5 per cent is a mixture of higher silanes, the highest identified being Si_6H_{14}. **Silane** boils at $-112°$ and decomposes at 400° to silicon and hydrogen. The higher silanes decompose to mixtures of lower silanes and silicon at progressively lower temperatures.

Silicon-hydrogen bonds are much more reactive than carbon-hydrogen bonds. Thus, although stable to aqueous acids, they are hydrolyzed with the evolution of hydrogen when boiled with strong bases.

$$SiH_4 + H_2O + 2\,NaOH \longrightarrow Na_2SiO_3 + 4\,H_2$$

All of the factors enumerated previously probably contribute to the increased reactivity. First, the silicon-hydrogen bond is more polar than the carbon-hydrogen bond. Since the carbon-hydrogen bond is practically nonpolar, the silicon atom is positive with respect to

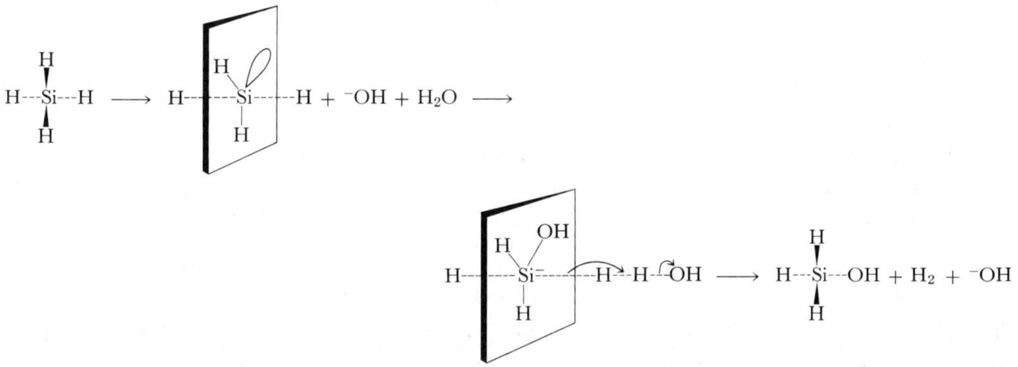

$$SiH_4 \longrightarrow SiH_4 + {}^-OH + H_2O \longrightarrow$$

$$[H_3Si(OH)\text{---}H\text{---}H\text{---}OH] \longrightarrow H_3Si\text{---}OH + H_2 + {}^-OH$$

Figure 15–*4*. Displacement of hydride ion by a flank nucleophilic attack on silicon.

hydrogen, and hydrogen can leave more readily as hydride ion. Second, there is more room for the approach of the hydroxide ion. Third, it is possible that a metastable intermediate involving *sp*³*d* hybridization can form that permits attack from the side as illustrated in Fig. 15–*4* rather than from the rear of the leaving group. The energy profile for this type of reaction would resemble one or the other of those illustrated in Fig. 3–*3*, page 36.

These same considerations apply to the nucleophilic displacement of other atoms or groups. The presence of electron-releasing groups always leads to a decrease in reactivity. This effect is in accord with the formulation of an intermediate in which silicon bears a negative formal charge. Despite the fact that silicon is more electropositive than carbon, there is no evidence that silonium ions, R_3Si^+, analogous to carbonium ions, R_3C^+, are intermediates in the reactions of silicon compounds, and there is no kinetic evidence for reactions of the S_N1 type.

Silicon tetrachloride, $SiCl_4$, was discovered by Berzelius in 1823. It is prepared by the reaction of chlorine with silicon, some Si_2Cl_6 and Si_3Cl_8 being formed at the same time.

$$Si + 2\,Cl_2 \longrightarrow SiCl_4 \text{ (and } Si_2Cl_6 \text{ and } Si_3Cl_8)$$

The boiling point is 58°, or 19° lower than carbon tetrachloride. It is hydrolyzed readily by water, although it may be distilled from sodium, with which it does not react below 200°.

$$SiCl_4 + 3\,H_2O \longrightarrow H_2SiO_3 + 4\,HCl$$

Reaction with alcohols gives the orthosilicates (p. 127).

$$SiCl_4 + 4\,ROH \longrightarrow (RO)_4Si + 4\,HCl$$

Reduction with lithium aluminum hydride gives almost quantitative yields of silane.

$$SiCl_4 + LiAlH_4 \longrightarrow SiH_4 + LiAlCl_4$$

Trichlorosilane (*silicochloroform*), first prepared by Woehler in 1857, results along with silicon tetrachloride from the action of dry hydrogen chloride on silicon.

$$Si + 3\,HCl \longrightarrow SiHCl_3 + H_2 \text{ (and } SiCl_4)$$

It boils at 32°, or 29° below chloroform. It fumes in moist air because of its ready hydrolysis.

$$SiHCl_3 + 3\,H_2O \longrightarrow H_2SiO_3 + 3\,HCl + H_2$$

The trisubstituted silanes add to carbon-carbon double and triple bonds in the presence of platinum catalyst or free radical initiators. As in hydrogenation of acetylenes by metal catalysts (p. 148), addition is *cis* but here the product that results is *trans* (Fig. 15–*5a*). Free radical addition, on the other hand, is *trans* to yield a *cis* product (Fig. 15–*5b*).

The first compound to be prepared that contained a carbon-silicon bond was **tetraethylsilane,** b.p. 153°. It was made by Friedel and Crafts (p. 377) in 1863 by the reaction of ethylzinc and silicon tetrachloride.

$$SiCl_4 + 2\,Zn(C_2H_5)_2 \xrightarrow{160^\circ} Si(C_2H_5)_4 + 2\,ZnCl_2$$

Kipping[1] found in 1904 that Grignard reagents react with silicon halides, and this reaction has been one of the most fruitful in the synthesis of organosilicon compounds. Silicon tetra-

[1] Frederick Stanley Kipping (1863–1949), professor of chemistry at the University of Nottingham. He worked under W. H. Perkin, Jr. (p. 638) in Baeyer's laboratory in Munich, and had as a student Arthur Lapworth (p. 197) in the laboratory of H. E. Armstrong (p. 558). Since Perkin and Lapworth later married sisters of Kipping's wife, all three were associated closely throughout most of their lives. "Organic Chemistry" by Perkin and Kipping was the standard text in England for many years.

$$R{-}C{\equiv}C{-}H + H{-}SiCl_3 \longrightarrow \underset{H}{\overset{R}{>}}C{=}C\underset{SiCl_3}{\overset{H}{<}} \quad (a)$$

$$R{-}\overset{H}{C}{\equiv}C{-}H \; (SiCl_3) \longrightarrow \underset{R}{\overset{H}{>}}C{=}C\underset{SiCl_3}{\overset{H}{<}} \quad (b)$$

Figure 15–5. Addition of trichlorosilane to acetylenes: (*a*) platinum-catalyzed; (*b*) free radical–initiated.

chloride reacts with Grignard reagents to give a mixture of the compounds having from one to four chlorine atoms replaced by hydrocarbon groups.

$$SiCl_4 \xrightarrow{RMgX} RSiCl_3 \xrightarrow{RMgX} R_2SiCl_2 \xrightarrow{RMgX} R_3SiCl \xrightarrow{RMgX} R_4Si$$

By regulating conditions it is possible to obtain good yields of the desired products. Trichlorosilane yields in the same way alkyldichloro-, dialkylchloro-, and trialkylsilanes.

Trialkylchlorosilanes, R_3SiCl, are silicon analogs of tertiary alkyl chlorides; yet it is not possible to eliminate hydrogen chloride to form an unsaturated compound with a carbon-silicon double bond. Reaction with anhydrous ammonia or amines gives the trialkylsilylamine.

$$R_3SiCl + 2\,NH_3 \longrightarrow R_3SiNH_2 + NH_4Cl$$

Further reaction of the silylamine with trialkylsilyl chloride gives the disilylamine (*disilazane*).

$$R_3SiCl + H_2NSiR + NH_3 \longrightarrow R_3SiNHSiR_3 + NH_4Cl$$

The amino and imino groups, as well as halogen, are easily displaced by nucleophiles. This property has led to the use of trimethylsilylamine and bis(trimethylsilyl)amine (*1,1,1,3,3,3-hexamethyldisilazane*) for the replacement of hydrogen in all types of hydroxyl, amino, and imino groups, as well as other types of reactive hydrogen, by trimethylsilyl groups. Usually a mixture of hexamethyldisilazane, trimethylsilyl chloride, and a tertiary amine are allowed to react with the compound.

$$(CH_3)_3SiCl + [(CH_3)_3Si]_2NH + 3\,ROH \longrightarrow 3\,ROSi(CH_3)_3 + NH_4Cl$$

The resulting derivatives are remarkably volatile and stable to heat. They have found wide use for converting nonvolatile compounds such as amino acids, peptides, and carbohydrates into derivatives sufficiently volatile to be separated by gas chromatography (p. 45).

Hydrolysis of trialkylsilylamines or of trialkylchlorosilanes gives the silanol.

$$R_3SiNH_2 + H_2O \longrightarrow R_3SiOH + NH_3$$

$$R_3SiCl + H_2O + NH_3 \longrightarrow R_3SiOH + NH_4Cl$$

These silanols lose water easily and form the oxides known as *hexaalkyldisiloxanes.*

$$2\,R_3SiOH \longrightarrow R_3SiOSiR_3 + H_2O$$

Often this reaction takes place spontaneously, but the more stable compounds require the presence of alkali. The silanols do not esterify with acids, but acyl chlorides give the chloro derivatives as occurs with tertiary alcohols.

$$R_3SiOH + CH_3COCl \longrightarrow R_3SiCl + CH_3COOH$$

It is impossible to obtain unsaturated compounds from silanols by the loss of water.

Hydrolysis of dialkyldichlorosilanes, R_2SiCl_2, gives diols, which are insoluble in water but soluble in aqueous alkali.

$$R_2SiCl_2 + 2\,H_2O \longrightarrow 2\,HCl + R_2Si(OH)_2 \xrightarrow{NaOH} R_2Si(ONa)_2 + 2\,H_2O$$

The diols lose water readily to give open-chain and cyclic polymers, which are called **silicones,** a name that was given to them before their structure was known in the belief that they corresponded to ketones. No compounds are known, however, in which silicon is united to another element by a double bond. The first polymeric product was obtained in 1872 by Ladenburg (p. 651), who hydrolyzed diethyldiethoxysilane, $(C_2H_5)_2Si(OC_2H_5)_2$, and obtained a viscous oil that did not freeze at $-15°$ and decomposed only at a very high temperature. Polymers of this type sometimes are called *polysiloxanes* although the term *silicone* is used more frequently.

The products of hydrolysis of alkyltrichlorosilanes, $RSiCl_3$, were believed at first to be analogous to the carboxylic acids and were assigned the formula RSiOOH, but Kipping found them to be complex compounds of high molecular weight. These compounds also now are grouped with the silicones.

In a summary of his work in 1937, Kipping pointed out that the number of types of silicon compounds was small compared to carbon, and because of the limited number of reactions that they undergo, he concluded that "the prospect of any immediate and important advance in this section of organic chemistry does not seem to be very hopeful." At that very time, however, industrial chemists were trying to make use of Kipping's polymeric compounds because of their insolubility, nonreactivity, and stability to heat. A Russian publication in 1939 indicated that certain polymers were suitable as dielectrics and insulating materials at elevated temperatures. American publications followed in 1941. Commercial production in the United States of the silicones in the form of oils, greases, resins, and elastomers was announced in 1944. In general the silicones are made by hydrolyzing the dimethyldichloro- or methyltrichlorosilanes.

$$x\,(CH_3)_2SiCl_2 + (x+1)\,H_2O \longrightarrow HO\left[-\underset{CH_3}{\overset{CH_3}{Si}}-O\right]_{x-1}-\underset{CH_3}{\overset{CH_3}{Si}}OH + 2x\,HCl$$

The products from dimethyldichlorosilane are oils, the molecular weight of which can be controlled by the amount of trimethylchlorosilane added, which hydrolyzes to trimethylsilanol. Reaction of this compound with a terminal hydroxyl group stops the chain. Greases are made by thickening the oils with soaps.

Hydrolysis of methyltrichlorosilane permits cross linking of the chains and yields a three dimensional polymer of low molecular weight.

$$CH_3SiCl_3 \xrightarrow{H_2O} \begin{array}{c} -O-\underset{OH}{\overset{CH_3}{Si}}-O-\underset{CH_3}{\overset{OH}{Si}}-O-\underset{CH_3}{\overset{|}{Si}}-OH \\ -O-\underset{CH_3}{\overset{OH}{Si}}-O-\underset{OH}{\overset{CH_3}{Si}}-O-\underset{|}{\overset{CH_3}{Si}}-OH \end{array} \xrightarrow{\text{Dehydration}} \begin{array}{c} -O-\underset{|}{\overset{CH_3}{Si}}-O-\underset{CH_3}{\overset{O-}{Si}}-O-\underset{CH_3}{\overset{|}{Si}}-O- \\ O \\ -O-\underset{CH_3}{\overset{|}{Si}}-O-\underset{O-}{\overset{CH_3}{Si}}-O-\underset{|}{\overset{CH_3}{Si}}-O- \end{array}$$

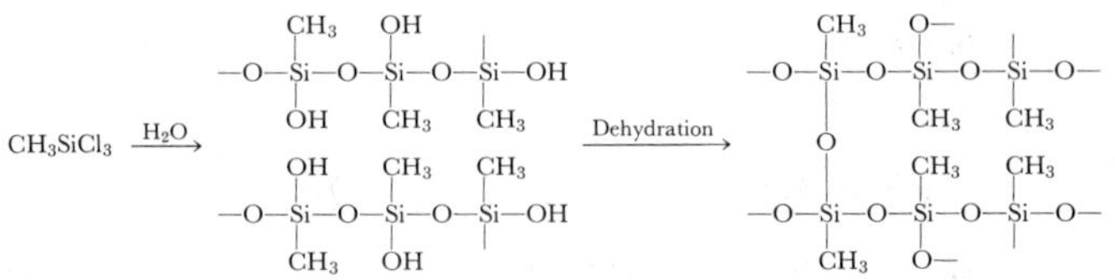

Usually some phenyltrichlorosilane, $C_6H_5SiCl_3$, is used along with the methyltrichlorosilane to reduce the brittleness of the final product. After the hydrogen chloride has been removed by washing, the polymer of low molecular weight can be polymerized further by heat to give solid resins.

All of the silicones have to a marked extent the properties of not being wet by water and of resistance to a relatively high temperature. Films and coatings are used to waterproof materials and machines, and the resins are used to fill the voids and act as a binder for glass fiber insulation of magnet wire, thus permitting motors to operate at a higher temperature than if the usual organic resins were used. The oils have the unique property of remaining fluid at low temperatures. Silicone fluids are used to control the cell size and structure of rigid polyurethan foams (p. 605). Silicone rubber in contrast to other rubbers is not attacked by ozone, does not become set under compression at high temperatures, and retains its flexibility at low temperatures (p. 589).

The **methylchlorosilanes** are made commercially by passing methyl chloride over silicon and copper at 300°. The process is operated to give chiefly dimethyldichlorosilane, although other products such as methyltrichlorosilane, trimethylchlorosilane, silicon tetrachloride, and hydrocarbons also are formed.

$$2\,CH_3Cl + Si(Cu) \longrightarrow (CH_3)_2SiCl_2$$

$$4\,CH_3Cl + 2\,Si(Cu) \longrightarrow CH_3SiCl_3 + (CH_3)_3SiCl$$

Pure **methylphenyldichlorosilane,** required for the production of one type of silicone rubber, is made by the reaction of methyltrichlorosilane with phenylmagnesium chloride in tetrahydrofuran.

$$CH_3SiCl_3 + C_6H_5MgCl \longrightarrow CH_3(C_6H_5)SiCl_2 + MgCl_2$$

Methylvinyldichlorosilane, used for rubbers of low compression set, is made by the addition of methyldichlorosilane to acetylene in the presence of either peroxide or platinum catalyst (p. 276).

$$HC{\equiv}CH + HSiCl_2CH_3 \longrightarrow H_2C{=}CH{-}SiCl_2CH_3$$

(3-Cyanopropyl)methyldichlorosilane, used for oil-resistant rubbers, is made in the same way from allyl cyanide, $CH_2{=}CHCH_2CN$.

$$(CH_3)Cl_2SiH + CH_2{=}CHCH_2CN \longrightarrow (CH_3)Cl_2SiCH_2CH_2CH_2CN$$

Thousands of other organosilicon compounds with various functional groups have been synthesized, chiefly in the search for compounds that can be used to impart specific properties to silicones. Total commercial production, however, is low compared to that of other resins and elastomers, chiefly because of the difficulty of manufacture and the consequent high price. In 1964 production of silicone resins was about 11 million pounds, and they sold for $2.28 per pound. Production of silicone elastomers was 8.3 million pounds, and the selling price was $3.77 per pound.

PROBLEMS

Sulfur Compounds

15–*1*. Give balanced equations for the following oxidations using an excess of oxidizing agent: (*a*) *n*-butyl mercaptan with dichromate and sulfuric acid, (*b*) *i*-propyl disulfide with nitric acid, (*c*) ethyl sulfide with aqueous permanganate.

15–*2*. Give equations for the synthesis of the following compounds: (*a*) *i*-propyl sulfide, (*b*) ethyl *n*-butyl sulfide, (*c*) methyl sulfone, (*d*) dithioacetal, (*e*) ethyl mercaptan from ethyl disulfide, (*f*) *n*-amyl disulfide from *n*-amyl chloride, (*g*) *i*-amyl mercaptan from sodium *i*-amyl sulfate, (*h*) *n*-butyl disulfide from *n*-butyl mercaptan, (*i*) *n*-hexane from *n*-hexyl sulfone.

15–*3*. Give equations for the preparation of the following compounds: (*a*) thiopropionic acid, (*b*) thio-*n*-butyramide, (*c*) methyl thiolacetate, (*d*) ethyl methanesulfonate, (*e*) tri-*n*-propylsulfonium iodide, (*f*) propanesulfonamide, (*g*) 2-methylpropanesulfonic acid from 2-methyl-1-propanethiol, (*h*) butanesulfinic acid from *n*-butyl bromide, (*i*) ethanesulfonyl chloride from ethyl disulfide, (*j*) dithioacetic acid from methyl iodide, (*k*) dodecanesulfonic acid from lauryl disulfide, (*l*) 1-pentanethiol from pentanesulfonic acid, (*m*) hexadecanesulfonic acid from cetyl iodide, (*n*) neohexane from ethyl methanesulfonate.

15–*4*. Write electronic formulas for each of the following compounds: (*a*) ethyl sulfoxide, (*b*) triethylsulfonium iodide, (*c*) potassium methyl sulfate, (*d*) ethyl *i*-propyl sulfone, (*e*) *n*-butyl sulfite, (*f*) ethyl butanesulfonate, (*g*) dithioacetic acid, (*h*) sodium ethanesulfinate, (*i*) thioacetamide.

15–*5*. Give equations for the following reactions: (*a*) thermal decomposition of dimethyl-*n*-butyl sulfonium hydroxide, (*b*) air oxidation of dithiopropionic acid, (*c*) *n*-propyl mercaptan with sodium plumbite in the presence of sulfur, (*d*) the action of chlorine on *n*-butyl sulfide in aqueous acetic acid.

15–*6*. What chemical reactions may be used to distinguish among the members of each of the following groups: (*a*) methyl sulfite, methyl sulfate, and methyl methanesulfonate; (*b*) ethyl mercaptan, ethyl sulfide, and ethyl disulfide; (*c*) *n*-propyl sulfoxide, *n*-propyl sulfide, and *n*-propyl sulfone; (*d*) *i*-butyryl chloride, methanesulfonyl chloride, and trimethylsulfonium chloride; (*e*) ethyl hydrogen sulfate, ethanesulfonic acid, and propionic acid; (*f*) ammonium hexanesulfonate, hexanesulfonamide, and ethyl hexanesulfonate?

15–*7*. How can the following mixtures be separated into their components: (*a*) ethyl alcohol, acetic acid, and methanesulfonic acid; (*b*) decanesulfonamide, decanoamide, and ammonium decanoate; (*c*) *n*-amyl alcohol, *n*-amyl sulfide, and *n*-amyl mercaptan; (*d*) *n*-butyl ether, *n*-butyl bromide, and *n*-butyl mercaptan; (*e*) *n*-propyl sulfide, pentanesulfonic acid, and pentanesulfonamide?

15–*8*. Give a series of reactions for the following conversions: (*a*) ethyl mercaptan to ethanesulfonic anhydride, (*b*) *n*-butyl bromide to *n*-butyl sulfoxide, (*c*) *i*-propyl disulfide to methyl *i*-propyl sulfide, (*d*) methyl laurate to dodecyl mercaptan, (*e*) ethyl cyanide to thiopropionamide, (*f*) pentanesulfonic acid to *n*-amyl mercaptan, (*g*) methyl ethyl ketone to *s*-butyl disulfide.

15–*9*. Indicate the likely mechanistic steps for (*a*) the acylation of a mercaptan by an acid anhydride, (*b*) the reaction of an alkyl sulfonate with an alcohol to give an ether.

Phosphorus, Silicon, and Tin Compounds

15–*10*. Give reactions for the preparation of the following compounds: (*a*) *n*-butyl phosphate, (*b*) ethyl phosphonate, (*c*) ethyl pyrophosphate with use of phosphorus oxychloride (C_2H_5Cl is one of the products), (*d*) *O,O*-dimethyl hydrogen dithiophosphate, (*e*) triethylphosphine.

15–*11*. Give a series of reactions for each of the following conversions: (*a*) phosphorus trichloride to ethyl di-*n*-propylphosphinite, (*b*) *n*-butyl phosphonate to *n*-butyl ethylphosphonate, (*c*) *n*-propyl alcohol to *n*-propyl *n*-propylphosphonate, (*d*) phosphine to tetrakis(hydroxymethyl)phosphonium chloride.

15–*12*. The free energy change for the reaction

$$(CH_3)_2N{-}P(CH_3)_2 + (CH_3)_2PH \rightleftharpoons (CH_3)_2P{-}P(CH_3)_2 + (CH_3)_2NH$$

has been reported to be -1.12 kcal. at 25°. The empirical bond energy for the P—H bond is 76 kcal., for the P—P bond, 55 kcal., and for the N—H bond, 93 kcal. Assuming that for the given reaction $\Delta S = 0$, calculate the energy for the N—P bond.

15–*13*. Give a series of reactions for the conversion of (*a*) acetylene to *trans*-2-ethylvinyltrichlorosilane, (*b*) tri-*n*-propylchlorosilane to hexa-*n*-propyldisiloxane.

CHAPTER SIXTEEN

DERIVATIVES OF CARBONIC ACID AND OF THIOCARBONIC ACID

DERIVATIVES OF CARBONIC ACID

From the theory of structure, the compound $C(OH)_4$ should exist, but the presence of more than one hydroxyl group on the same carbon atom usually results in an unstable molecule that loses water (p. 198). It is not surprising therefore that the hypothetical orthocarbonic acid has not been isolated as a stable compound. Loss of one molecule of water should give ordinary carbonic acid $O{=}C(OH)_2$. Although it is believed that this compound exists in aqueous solutions of carbon dioxide, all attempts to isolate it result in its decomposition to carbon dioxide and water.

$$\underset{\text{Orthocarbonic acid (hypothetical)}}{[C(OH)_4]} \underset{H_2O}{\rightleftarrows} \underset{\text{Carbonic acid (hypothetical)}}{[O{=}C(OH)_2]} \underset{H_2O}{\rightleftarrows} \underset{\text{Carbon dioxide}}{O{=}C{=}O}$$

Despite the instability of orthocarbonic acid and carbonic acid, numerous derivatives of these compounds are known, some of which are of considerable importance.

The **derivatives of orthocarbonic acid** resemble the derivatives of orthocarboxylic acids (p. 177) in that, lacking a carbonyl group, they do not possess the properties of carboxylic acid derivatives. Carbon tetrachloride, for example, does not react like an acyl halide. It does not undergo hydrolysis, alcoholysis, or ammonolysis readily. Similarly ethyl orthocarbonate resembles the acetals rather than esters in that it is stable to alkali but hydrolyzes easily with aqueous acids. The **orthocarbonates** are prepared by the reaction of chloropicrin (*trichloronitromethane,* p. 445) with sodium alkoxides.

$$Cl_3CNO_2 + 4\,RONa \longrightarrow C(OR)_4 + 3\,NaCl + NaNO_2$$

Derivatives of carbonic acid, on the other hand, retain a carbonyl group or its equivalent. Their reactions therefore are those of typical carboxylic acid derivatives. Since carbonic acid contains two hydroxyl groups united to a carbonyl group, each hydroxyl group can behave like the hydroxyl of a carboxylic acid. The products of addition of alcohols to carbon dioxide, that is, the alkyl hydrogen carbonates, are no more stable than carbonic acid, the product from the addition of water. Metallic salts, however, are known. Thus **magnesium methyl carbonate** can be prepared readily by passing dry carbon dioxide into a solution of magnesium methoxide in dimethylformamide.

$$(CH_3O^-)_2Mg^{2+} + CO_2 \longrightarrow (CH_3OCO_2^-)_2Mg^{2+}$$

It is a useful reagent for the preparation of α-nitro carboxylic acids (p. 246).

Phosgene, $COCl_2$, the acid chloride of carbonic acid, first was made by the action of light on a mixture of carbon monoxide and chlorine (Gr. *phos* light, *genes* born). In the technical process activated carbon is used as a catalyst.

$$CO + Cl_2 \xrightarrow[\text{carbon}]{\text{Activated}} \underset{\text{Phosgene}}{COCl_2}$$

Phosgene is a sweet-smelling gas (b.p. 8°) that is ten times as toxic as chlorine and was the principal offensive battle gas of World War I. The toxic action apparently is due to its ready hydrolysis in the lungs with the liberation of hydrogen chloride.

When phosgene reacts with alcohols, the **alkyl chloroformate** is formed first, and further reaction gives the **alkyl carbonate.**

$$\underset{\text{Phosgene}}{COCl_2} + C_2H_5OH \longrightarrow \underset{\text{Ethyl chloroformate}}{ClCOOC_2H_5} + HCl$$

$$ClCOOC_2H_5 + C_2H_5OH \longrightarrow \underset{\text{Ethyl carbonate}}{OC(OC_2H_5)_2} + HCl$$

Phosgene is used for the manufacture of certain dye intermediates (p. 464), for the preparation of **ethyl carbonate,** a useful solvent, and for the synthesis of isocyanates used to make polyurethan rubbers and plastics (p. 423).

Urea, $CO(NH_2)_2$, is the most important derivative of carbonic acid. It may be considered as the diamide of carbonic acid, the monoamide being the unstable carbamic acid, H_2NCOOH. It is because of its relation to carbamic acid, however, that urea frequently is called *carbamide*.

Prior to the development in recent times of technical methods of synthesis and uses, urea was of interest primarily because it is the chief final product of nitrogen metabolism in mammals, being eliminated in the urine. Adult man excretes about 30 g. of urea in 24 hours. It is produced in the liver from ammonia and carbon dioxide by the ornithine cycle (p. 365). Urea was isolated from urine in 1773, although it was not characterized fully and named urea until 1799. It probably was synthesized first by John Davy, a brother of Sir Humphrey Davy, who in 1811 prepared phosgene by the action of sunlight on a mixture of chlorine and carbon monoxide and in 1812 reported that the product reacted with dry ammonia to give a solid that did not evolve carbon dioxide on treatment with acetic acid and hence was not ammonium carbonate. He did not identify his product with urea, however, and credit for the synthesis of urea has been given to Woehler, who in 1828 recognized that the product obtained by boiling a solution of ammonium cyanate with water is identical with urea isolated from urine (p. 365). This reaction is the result of the dissociation of ammonium cyanate into ammonia and cyanic acid (p. 285), followed by addition.

$$NH_4NCO \rightleftharpoons NH_3 + HN{=}C{=}O \longrightarrow H_2NCONH_2$$

Since its discovery, urea has been isolated as a product of over fifty reactions, two of which have been utilized for its synthesis on a large scale. About the time of World War I, it was prepared by the hydrolysis of calcium cyanamide (p. 288), and sold for 60 cents per pound in 1920. The present commercial synthesis is from ammonia and carbon dioxide. The preheated gases in a ratio of 2 or 3 to 1 are forced at around 2500 p.s.i. into an autoclave through which an aqueous solution or oil-water slurry of ammonia and carbon dioxide is being circulated at 170–180°. Nucleophilic attack by ammonia on carbon dioxide gives an intermediate that transfers a proton to a second molecule of ammonia to yield ammonium carbamate, which decomposes at the elevated temperature to urea (p. 166).

$$CO_2 + NH_3 \rightleftharpoons \left[O{=}\overset{O^-}{\underset{|}{C}}{-}\overset{+}{N}H_3\right] \xrightleftharpoons{NH_3} \underset{\text{Ammonium carbamate}}{\left[O{=}\overset{O^-}{\underset{|}{C}}{-}NH_2\right]NH_4^+} \xrightleftharpoons{175°} \underset{\text{Urea}}{O{=}\overset{NH_2}{\underset{|}{C}}{-}NH_2} + H_2O$$

The aqueous solution, after leaving the reactor, is distilled to remove ammonium carbonate and undecomposed ammonium carbamate, which dissociate to ammonia and carbon dioxide. The latter are dissolved in water and recirculated. The urea is separated from the concentrated aqueous still-residue by crystallization or by a form of spray-drying known as *prilling*. It is desirable to conduct the isolation of the urea in the solid form at as low a temperature as possible to prevent the formation of biuret (p. 285), particularly if the urea is to be used as fertilizer since biuret is toxic to plants. Commercial crystalline urea contains from 0.1 to 0.9 per cent biuret, whereas prilled urea contains 1 to 8 per cent. Since the development of the carbon dioxide–ammonia process, the price of urea has dropped remarkably and production has greatly increased. Production in the United States in 1964 was over 2.4 billion pounds, and the price was around four cents per pound. About 80 per cent is used as a fertilizer and 8 per cent for the manufacture of urea-formaldehyde plastics (p. 285). A part of the nitrogen required by ruminants for the synthesis of proteins can be supplied by urea, and most of the remaining production is added to commercial cattle feeds. A small amount is used in the manufacture of pharmaceuticals (p. 525).

Urea has the useful property of forming crystalline **inclusion compounds** with many straight-chain organic compounds having more than four carbon atoms, for example hydrocarbons, alcohols, mercaptans, alkyl halides, ketones, acids, and esters, but not with most branched-chain or cyclic compounds. The procedure for preparing these inclusion compounds merely involves mixing saturated methanol solutions of urea and of the compound. A crystalline precipitate forms from which the components can be separated either by extracting the urea with water or the organic compound with ether. The process can be used to remove normal hydrocarbons and thus improve the octane number of gasoline, lower the freezing point of fuel for jet planes, and lower the pour point of lubricating oil. In the laboratory it has been used to isolate and purify compounds and to resolve alkyl halides into optically active components (p. 309). Thiourea (p. 293) forms inclusion compounds with many branched and cyclic compounds.

There is no evidence that the straight-chain compounds are bound to the urea molecules by other than adsorption forces at the solid surface. The chains merely occupy channels in the urea crystal lattice. The melting point of each compound is that of urea, and the heat of formation is even less than the usual heats of adsorption on solid surfaces. Although each compound has a definite composition, the ratio of urea molecules to straight-chain molecules is not stoichiometrical but is proportional to the number of carbon atoms in the chain. Approximately two-thirds mole of urea is combined for each angstrom of chain length. It is of interest that unsaturated compounds included in urea crystals are not subject to autoxidation. Numerous other compounds also form crystalline clathrates that can hold small molecules of gases (p. 447) or large molecules of great variety depending on the size and shape of the cavities in the crystal lattice.

The reactions of urea are those of an amide. It is a very weak base, ($pK_a = 0.18$), although it appears to be somewhat stronger than acetamide ($pK_a = -1.49$). Despite the low basicity of urea, the addition of concentrated nitric acid to a concentrated aqueous solution gives a precipitate of **urea nitrate** because of the insolubility of the latter in concentrated nitric acid solution. Although the nitrogen atom of an amino group ordinarily is more basic than the oxygen atom of a carbonyl group, salt formation of an amide group undoubtedly results from addition of a proton to the oxygen rather than to the nitrogen. If the proton added to nitrogen, the resonance stabilization of the ion would be very low, whereas if it adds to oxygen, considerable stabilization by resonance results. Hence the cation of urea nitrate is best represented as a resonance hybrid (p. 160).

$$\left\{ H_2N{-}\overset{\overset{+}{O}H}{\overset{\|}{C}}{-}NH_2 \longleftrightarrow H_2\overset{+}{N}{=}\overset{OH}{\overset{|}{C}}{-}NH_2 \longleftrightarrow H_2N{-}\overset{OH}{\overset{|}{C}}{=}\overset{+}{N}H_2 \right\}$$

Hydrolysis of urea yields ammonium carbonate.

$$CO(NH_2)_2 + 2\,H_2O \longrightarrow [CO(OH)_2] + 2\,NH_3 \longrightarrow (NH_4)_2CO_3$$

If the reaction is catalyzed by alkali, the products are the alkali carbonate and ammonia, whereas acid catalysis yields carbon dioxide and the ammonium salt. The hydrolysis also is catalyzed rapidly at room temperature by the enzyme *urease,* which is present in soybean and jack bean, and is elaborated by certain bacteria. The hydrolysis of urea in the soil liberates the nitrogen as ammonia, a part of the nitrogen cycle (Fig. 19–*1*, p. 352). Use is made of the urease catalysis for the estimation of urea in biological fluids. After hydrolysis in a suitably buffered solution, the ammonia is removed and estimated colorimetrically or by titration with standard acid.

Since urea contains primary amino groups, nitrogen is evolved on reaction with nitrous acid (p. 228).

$$CO(NH_2)_2 + 2\,HONO \longrightarrow CO_2 + 2\,N_2 + 3\,H_2O$$

This reaction is useful as a means of destroying nitrous acid and oxides of nitrogen.

As an amide, urea undergoes the Hofmann rearrangement (p. 223), but the product, hydrazine, is oxidized by hypobromite to nitrogen and water.

$$H_2NCONH_2 + NaOBr + 2\,NaOH \longrightarrow \underset{\text{Hydrazine}}{H_2NNH_2} + Na_2CO_3 + NaBr + H_2O$$

$$H_2NNH_2 \xrightarrow{2\,NaOBr} N_2 + 2\,H_2O + 2\,NaBr$$

Amides can be acylated to diacyl derivatives of ammonia, which are known as *imides.* The imides are more acidic than the amides and react with strong bases in aqueous solution to form salts (p. 234). Urea can undergo diacylation, the products being known as **ureides.**

$$CO(NH_2)_2 + 2\,(CH_3CO)_2O \longrightarrow \underset{\text{Diacetylurea}}{CO(NHCOCH_3)_2} + 2\,CH_3COOH$$

Many of the cyclic ureides such as the barbituric acids (p. 525) and alloxan (p. 526) have important physiological properties.

Urea adds to formaldehyde in aqueous solution in the presence of acidic or basic catalysts to give *N*-(hydroxymethyl)urea and bis[*N*-(hydroxymethyl)]urea (cf. p. 227) commonly known as **methylolurea** and **dimethylolurea.**

$$H_2NCONH_2 \xrightarrow{HCH=O} \underset{\substack{N\text{-(Hydroxymethyl)urea} \\ (\textit{methylolurea})}}{H_2NCONHCH_2OH} \xrightarrow{HCHO} \underset{\substack{\text{Bis[}N\text{-(hydroxymethyl)]urea} \\ (\textit{dimethylolurea})}}{HOCH_2NHCONHCH_2OH}$$

As with the aldehyde-amine addition products, the hydroxymethyl derivatives lose water readily to give the aldimino derivative, which can add another mole of amino compound. Since urea has two amino groups, a linear polymer can result. Polymers formed by the elimination of a small molecule such as water are called *condensation polymers.*

$$H_2NCONHCH_2OH \longrightarrow H_2O + H_2NCON{=}CH_2 \xrightarrow{H_2NCONH_2} H_2NCONHCH_2NHCONH_2$$

$$\xrightarrow{x\,H_2CO\,+\,x\,H_2NCONH_2} H_2NCONH(CH_2NHCONH)_xCH_2NHCONH_2 + x\,H_2O$$

These linear polymers, which still have hydrogen on nitrogen, can cross link by condensation with aldimino derivatives of low molecular weight to give solid polymers.

$$\begin{matrix} (\text{—}NHCONHCH_2\text{—})_x \\ + \\ CH_2{=}NCON{=}CH_2 \\ + \\ (\text{—}NHCONHCH_2\text{—})_y \end{matrix} \longrightarrow \begin{matrix} (\text{—}NHCONCH_2\text{—})_x \\ | \\ CH_2NHCONHCH_2 \\ | \\ (\text{—}NHCONCH_2\text{—})_y \end{matrix}$$

In addition to linear condensation, cyclic trimerization of the aldimino compound can take place (p. 203), which can lead to still more complex polymers.

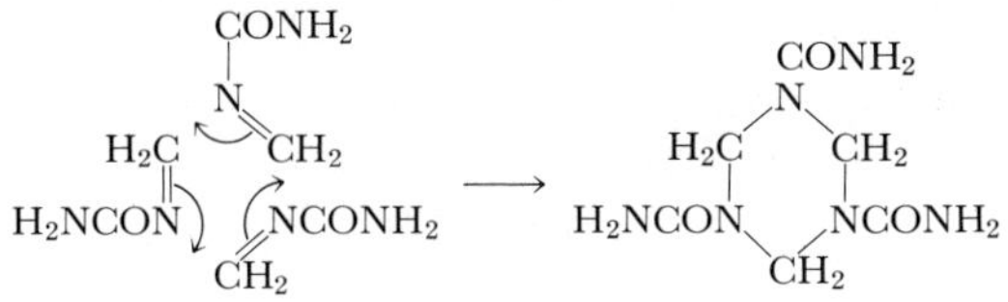

$$\text{cyclic trimerization: } 3\ H_2NCON{=}CH_2 \longrightarrow \text{ring of } N(CONH_2),\ CH_2,\ N(CONH_2),\ CH_2,\ N(CONH_2),\ CH_2$$

These polycondensation products constitute the commercially important group of plastics known as the **urea-formaldehyde resins.** Two to three moles of formaldehyde and one of urea are condensed in aqueous solution in the presence of ammonia as an alkaline catalyst. The reaction is stopped at the syrupy linear stage and mixed with a filler, usually high-grade wood pulp. The mixture is dried and ground and constitutes the *thermosetting* molding powder. To form an object, the powder is subjected to heat and pressure in a mold. It first flows to fill the mold and then sets to an infusible solid because of completion of the reaction in which cross linking of the chains takes place. Because the urea-formaldehyde plastics are colorless, color can be added to produce any desired shade. For molding purposes, however, they have been replaced largely by the melamine-formaldehyde polymers (p. 289). The intermediate condensation products are used widely as water-proof adhesives in the manufacture of plywood. Some of the products of lower molecular weight are used as fertilizer when a slow release of ammonia is desired. Production of urea-formaldehyde resins in the United States amounted to 418 million pounds in 1964.

When urea is heated strongly above its melting point, it decomposes into ammonia and **isocyanic acid,** which is tautomeric with **cyanic acid.** The isocyanic acid polymerizes at once to a mixture of about 30 per cent of the linear polymer, **cyamelide,** and 70 per cent of the trimer, **cyanuric acid.**

$$H_2NCONH_2 \longrightarrow NH_3 + \underset{\text{Isocyanic acid}}{HN{=}C{=}O} \rightleftharpoons \underset{\text{Cyanic acid}}{N{\equiv}COH}$$

$$x\ HNCO \longrightarrow \underset{\text{Cyamelide}}{\left[-NH-\overset{\overset{O}{\|}}{C}-\right]_x}$$

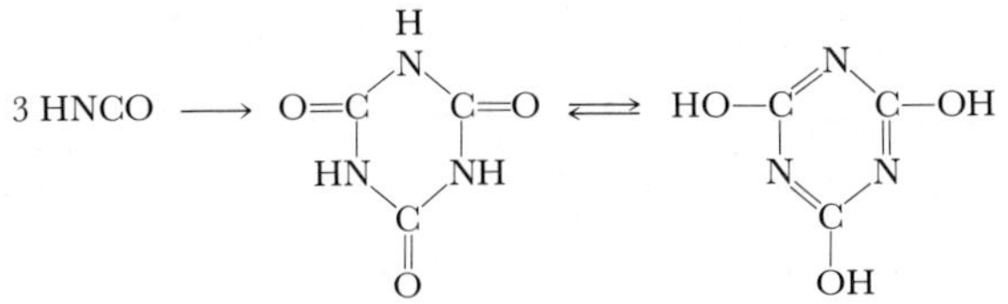

$$3\ HNCO \longrightarrow (-NH-CO-)_3 \rightleftharpoons (-N{=}C(OH)-)_3$$

Cyanuric acid

If urea is heated gently, the isocyanic acid first produced adds a molecule of urea to form **biuret.**

$$HN{=}C{=}O + H_2NCONH_2 \longrightarrow \underset{\text{Biuret}}{H_2NCONHCONH_2}$$

An alkaline solution of biuret gives a violet-pink color when copper sulfate solution is added. The color is due to the presence of a coordination complex with cupric ion in which the four water molecules normally coordinated with the cupric ion are displaced by the amino groups. The alkali removes two protons from the coordinated amino groups to give the neutral insoluble complex, and then two more protons to give the water-soluble salt.

$$[H_2O{:}Cu{:}OH_2(OH_2)_2]^{2+}SO_4^{2-} + 2\ O{=}C(NH_2)\text{–}NH\text{–}C(NH_2){=}O \longrightarrow 4\ H_2O + [Cu(H_2NCONHCONH_2)_2]^{2+}SO_4^{2-} \xrightarrow{2\ NaOH}$$

$$\begin{matrix}Na_2SO_4 \\ + \\ 2\ H_2O\end{matrix} + [Cu(H_2NCONHCONH)(HNCONHCONH_2)] \xrightarrow{2\ NaOH} 2\ H_2O + [Cu(HNCONHCONH)_2]^{2-}\ 2\ Na^+$$

Insoluble complex — Soluble complex

The reaction takes place because of the increased stability that arises when rings are formed. Since only five- and six-membered rings can be formed readily because of the limitations imposed by bond angles, complexes of this type are formed only when the electron-donating groups, such as the amino groups, are spaced properly in the molecule. Three consecutive peptide linkages in proteins or peptides can lead to a stable complex with copper in which three five-membered rings are formed, and proteins and tetra- or higher peptides give the biuret test (p. 359).

Alcohols add to isocyanic acid to give the **alkyl carbamates,** which commonly are called *urethans.*

$$HN{=}C{=}O + HOR \longrightarrow H_2NCOOR$$

They can be prepared also by the reaction of alkyl chloroformates with ammonia.

$$2\ NH_3 + ClCOOR \longrightarrow H_2NCOOR + NH_4Cl$$

Amines give *N*-substituted urethans.

$$2\ R'NH_2 + ClCOOR \longrightarrow R'NHCOOR + R'NH_3Cl$$
$$2\ R'_2NH + ClCOOR \longrightarrow R'_2NCOOR + R'_2NH_2Cl$$

The simple urethans, such as ethyl carbamate, are mild soporifics. The dicarbamate derived from 2-methyl-2-*n*-propyl-1,3-propanediol (Problem 35–6) is known as **meprobamate** (*Equanil* or *Miltown*) and is one of the more widely used tranquilizing (*ataractic*) drugs. Production in 1964 was 1.2 million pounds valued at $3.2 million.

The mixed anhydride of carbamic acid and phosphoric acid, $H_2NCOOPO_3H_2$, commonly called **carbamyl phosphate,** is synthesized enzymatically from ammonia, carbon dioxide, and pyrophosphate by both higher animals and microorganisms. It enters the ornithine cycle in the production of urea (p. 365) and also is a precursor of pyrimidines (p. 525) and purines (p. 528).

Although isocyanic acid is tautomeric, two series of stable alkyl derivatives (*esters*) should be possible, the *alkyl cyanates,* $ROC{\equiv}N$, and the *alkyl isocyanates,* $O{=}C{=}NR$. The former have been isolated only when the alkyl group is sufficiently bulky to prevent trimerization. The **alkyl isocyanates** on the other hand are stable compounds. They are formed when potassium cyanate is heated with an alkyl sulfate in the presence of dry sodium carbonate. The ambident cyanate ion thus forms the more covalent carbon-nitrogen bond (p. 237).

$$R_2SO_4 + KOCN \xrightarrow{Na_2CO_3} RN{=}C{=}O + RSO_4K$$

Phosgene reacts with primary amines to give alkylcarbamyl chlorides, which can be pyrolyzed to alkyl isocyanates.

$$2\ RNH_2 + COCl_2 \longrightarrow RNH_3Cl + RNHCOCl \xrightarrow{Heat} RN{=}C{=}O + HCl$$

Now that isocyanides are readily available (p. 240), their oxidation by ozone becomes a practical method for the preparation of the higher alkyl isocyanates.

$$RN{\equiv}C\colon + O_3 \longrightarrow RN{=}C{=}O + O_2$$

The alkyl isocyanates are intermediates in the Hofmann conversion of amides to amines (p. 223) and in the Curtius (p. 251) rearrangement.

Like isocyanic acid, the isocyanates readily undergo nucleophilic attack on carbon. Hydrolysis yields the carbamic acid, which is unstable and decomposes spontaneously to the amine and carbon dioxide.

$$RN{=}C{=}O + H_2O \longrightarrow [RNHCOOH] \longrightarrow RNH_2 + CO_2$$

It is of interest that this reaction led to the discovery of amines by Wurtz (p. 145), who first prepared methylamine and ethylamine in 1849 by the hydrolysis of methyl and ethyl isocyanates obtained from potassium cyanate and the alkyl iodides.

Isocyanates add alcohols or amines to yield urethans or ***N*-alkylureas.**

$$\begin{aligned} RN{=}C{=}O + HOR' &\longrightarrow RNHCOOR' \\ + HNH_2 &\longrightarrow RNHCONH_2 \\ + HNHR' &\longrightarrow RNHCONHR' \\ + HNR'_2 &\longrightarrow RNHCONR'_2 \end{aligned}$$

Because of the ease of reaction and the formation of solid products, the isocyanates, especially **phenyl isocyanate,** $C_6H_5N{=}C{=}O$ (p. 423), frequently are used to prepare derivatives of alcohols and amines for identification purposes. Other isocyanates are technically important for the synthesis of polyurethans (p. 605).

Reaction of urea with fuming sulfuric acid gives **sulfamic acid.**

$$H_2NCONH_2 + SO_3 + H_2SO_4 \longrightarrow 2\,H_2NSO_3H + CO_2$$

Ammonium sulfamate has important uses as a flame-proofing agent and as an herbicide.

When urea nitrate is dissolved in concentrated sulfuric acid, **nitrourea** is formed.

$$[H_2NC(OH)NH_2]^+NO_3^- \xrightarrow[H_2SO_4]{\text{Conc.}} \underset{\text{Nitrourea}}{H_2NCONHNO_2} + H_2O$$

Electrolytic reduction of nitrourea yields **semicarbazide** (*semicarbazine*), a valuable reagent for aldehydes and ketones (p. 205).

$$H_2NCONHNO_2 + 6\,[H]\ (\text{electrolysis}) \longrightarrow \underset{\text{Semicarbazide}}{H_2NCONHNH_2} + 2\,H_2O$$

Semicarbazide solutions decompose slowly to form **hydrazinedicarbonamide.**

$$H_2NCONHNH_2 + H_2NNHCONH_2 \longrightarrow H_2NCONHNHCONH_2 + H_2NNH_2$$

This sparingly soluble compound, which melts at 245°, precipitates when solutions have been boiled for several hours and may be mistaken for a semicarbazone.

Various methods are available for the preparation of *N*-alkylureas (p. 287). Direct alkylation of urea in the presence of a base, however, gives ***O*-alkylureas.**

$$O{=}C(NH_2)_2 \underset{H_2O}{\overset{^-OH}{\rightleftarrows}} \left\{ O{=}C(\bar{N}H)(NH_2) \longleftrightarrow {}^-O{-}C({=}NH)(NH_2) \right\} \xrightarrow{(CH_3)_2SO_4} CH_3SO_4^- + CH_3{-}O{-}C({=}NH)(NH_2)$$

O-Alkylureas can be prepared also by the addition of alcohols to carbodiimides (p. 290).

The **cyanogen halides** may be regarded as the acid halides of cyanic acid. They are prepared by the reaction of halogens with metallic cyanides.

$$MCN + X_2 \longrightarrow XCN + MX$$

They are highly toxic lachrymators. **Cyanogen chloride** melts at −6° and boils at 15.5°. **Cyanogen bromide** melts at 52° and boils at 61°. **Cyanogen iodide** sublimes at atmospheric pressure. The cyanogen halides are stable when pure but polymerize readily in the presence of free halogen to give the cyanuryl halides, the acid halides of cyanuric acid (p. 285).

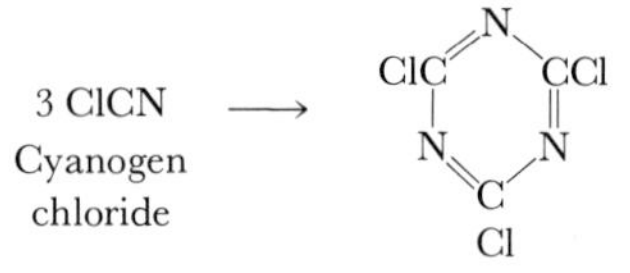

$$3\ ClCN \longrightarrow (ClC{=}N)_3$$

3 ClCN (Cyanogen chloride) → Cyanuryl chloride

Cyanuryl chloride (*cyanuric chloride*) is prepared best by the reaction of chlorine with hydrogen cyanide, using chloroform containing 1 per cent of ethanol as a solvent. The ease of displacement of chlorine in cyanuryl chloride lies between that of an alkyl chloride and an acyl chloride. Thus cyanuryl chloride reacts with alcohols in the presence of base to give alkyl cyanurates and with amines to give "amides." It is the starting point for the manufacture of triazine herbicides. *Simazine* is used as a preemergence herbicide. It results from the reaction of one mole of cyanuryl chloride with two moles of ethylamine.

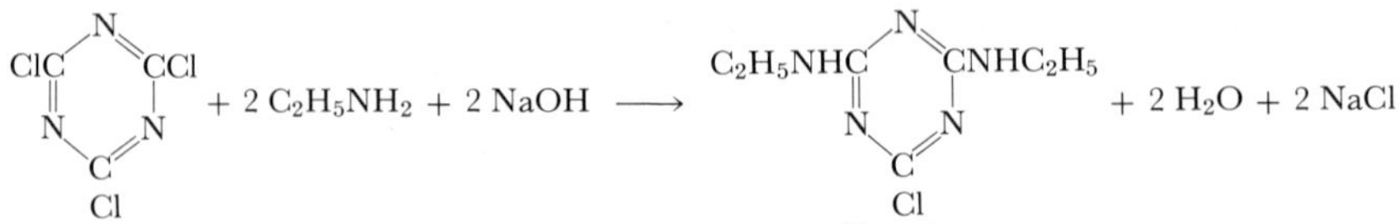

$$(ClC{=}N)_3 + 2\ C_2H_5NH_2 + 2\ NaOH \longrightarrow \text{Simazine} + 2\ H_2O + 2\ NaCl$$

Simazine

Cyanuryl chloride is used also in the synthesis of "reactive" dyes for cotton (p. 565) and of optical bleaches (p. 567).

Cyanamide, $H_2NC{\equiv}N$, may be considered as the amide of cyanic acid. Its most important derivative is the calcium salt, **calcium cyanamide,** which results when nitrogen is passed through a mixture of calcium carbide with 10 per cent of calcium oxide at about 1100°.

$$CaC_2 + N_2 \xrightarrow{CaO} CaNCN + C$$

The mixture is brought to reaction temperature by an electrically heated carbon rod, but after the exothermic reaction has been started, the carbon rod may be removed. The cyanamide process was the first important method for the fixation of atmospheric nitrogen, and calcium cyanamide has continued to be an important nitrogen fertilizer. It is used also as a soil fumigant and as a defoliant.

Cyanamide is stable in aqueous solutions of $pH < 5$ but dimerizes readily to **dicyandiamide** at pH 7–12. Hence dicyandiamide is the product formed when calcium cyanamide is heated with water.

$$CaNCN + 2\ H_2O \longrightarrow Ca(OH)_2 + H_2NCN$$

$$H_2NC{\equiv}N + H_2NC{\equiv}N \longrightarrow H_2NC({=}NH)NHC{\equiv}N$$

Dicyandiamide

When dicyandiamide is heated in the presence of anhydrous ammonia and methyl alcohol, **melamine,**[1] the cyclic trimer of cyanamide, is formed.

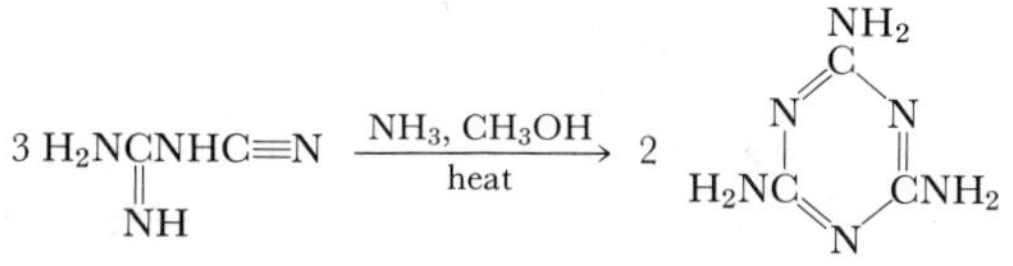

$$3\ H_2N\underset{\underset{NH}{\|}}{C}NHC{\equiv}N \xrightarrow[\text{heat}]{NH_3,\ CH_3OH} 2\ \text{Melamine (cyclic } C_3N_3(NH_2)_3\text{: ring of alternating C and N, each C bearing } NH_2)$$

Melamine

The amino groups of melamine, like the amino groups of urea, condense with formaldehyde to give products of high molecular weight known as **melamine resins.** They are superior to the urea-formaldehyde resins in resistance to heat and water, and are used in the manufacture of plastic dinnerware.

Guanidine, $HN{=}C(NH_2)_2$, the amidine (p. 236) of carbamic acid, is formed when dicyandiamide is heated with an excess of ammonia. If dicyandiamide is heated with ammonium chloride, guanidine hydrochloride is formed.

$$H_2N\underset{\underset{NH}{\|}}{C}NHCN + 2\ NH_3 \longrightarrow 2\ HN{=}C(NH_2)_2$$

Guanidine

Except for the quaternary ammonium hydroxides and certain carbanions, guanidine is the strongest organic base known ($pK_a = 13.6$). The high basicity is believed to result from the large resonance stabilization of the guanidinium ion relative to free guanidine. A large resonance stabilization is to be expected because the contributing structures are identical (cf. urea, p. 283).

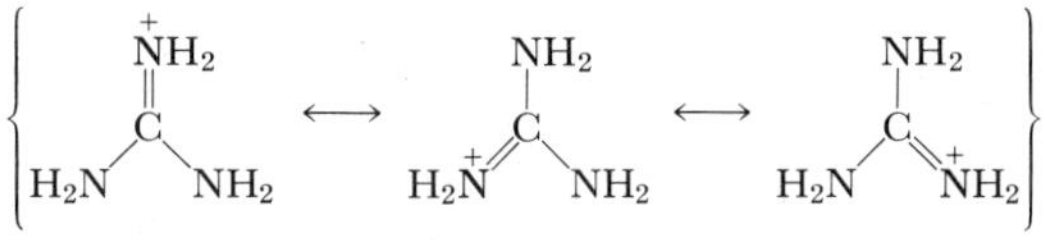

$$\left\{ (H_2N)_2C{=}\overset{+}{N}H_2 \longleftrightarrow H_2\overset{+}{N}{=}C(NH_2)NH_2 \longleftrightarrow H_2N(NH_2)C{=}\overset{+}{N}H_2 \right\}$$

When guanidinium nitrate, prepared by heating dicyandiamide with ammonium nitrate, is mixed with concentrated sulfuric acid, **nitroguanidine** is formed.

$$[H_2\overset{+}{N}{=}C(NH_2)_2]NO_3^- \xrightarrow{H_2SO_4} H_2N\underset{\underset{NH_2}{|}}{C}{=}NNO_2 + H_2O$$

Nitroguanidine

It is used as a component of some explosives. It is about as powerful as TNT and explodes without producing a flash. When mixed with colloided cellulose nitrate (p. 343), it gives a flashless propellant powder. Reduction of nitroguanidine with zinc and acetic acid gives **aminoguanidine.** It is made commercially by the addition of hydrazine to cyanamide.

$$H_2NC{\equiv}N + H_2NNH_2 \longrightarrow H_2N\underset{\underset{NH}{\|}}{C}NHNH_2$$

Aminoguanidine is an intermediate for the synthesis of numerous heterocyclic compounds used as herbicides or medicinals.

Several derivatives of guanidine are of importance in biological processes. **Creatine,**

[1] Liebig (p. 3) heated ammonium thiocyanate and obtained a product that he called *melam,* a name that he says he "grabbed out of the air, which is just as good for the purpose as if it were derived from the color or some other property." Hydrolysis of melam gave as one of the products a basic compound, which was called *melamine.*

isolated by Chevreul in 1832, is methylguanidinoacetic acid. Creatine is dehydrated readily to **creatinine,** which is excreted in the urine.

$$H_2\overset{+}{N}{=}C(NH_2){-}N(CH_3){-}CH_2COO^- \text{ (or } HN{=}C(NH_2){-}N(CH_3){-}CH_2COOH) \rightleftharpoons HN{=}C\langle{}^{NH-CO}_{N(CH_3)-CH_2} + H_2O$$

Creatine — Creatinine

Creatine is synthesized commercially for use as a nontoxic buffering agent for certain pharmaceuticals. The amino acid **arginine** (p. 353) is α-amino-δ-guanidino-*n*-valeric acid and is involved in nitrogen metabolism in mammals. **Phosphoarginine** and **phosphocreatine,** the phosphoric acid amides, hydrolyze rapidly to the base and phosphoric acid and are known as *phosphagens.* They play an important part in muscular processes of invertebrates and vertebrates respectively. Numerous other guanidine derivatives have been isolated from both plants and animals.

The alkyl isocyanates, O=C=NR, are mononitrogen analogs of carbon dioxide. The dinitrogen analogs, RN=C=NR, are known as **carbodiimides.** They can be made from *N,N*-dialkylureas (p. 287) by dehydration with *p*-toluenesulfonyl chloride (p. 406) in the presence of triethylamine.

$$RNHC({=}O)NHR' + CH_3C_6H_4SO_2Cl + 2\,(C_2H_5)_3N \longrightarrow RN{=}C{=}NR' + CH_3C_6H_4SO_2^- + Cl^- + 2\,(C_2H_5)_3\overset{+}{N}H$$

Symmetrical carbodiimides can be made directly by a phosphine oxide-catalyzed loss of carbon dioxide from two molecules of isocyanate. It appears that the reaction takes place through the phosphine imide by way of four-membered cyclic transition states.

$$R_3\overset{+}{P}{-}\overset{-}{O} + RN{=}C{=}O \rightleftharpoons \begin{matrix} R_3P\text{---}O \\ R'N\text{---}C{=}O \end{matrix} \rightleftharpoons R_3\overset{+}{P}{-}\overset{-}{N}R' + CO_2$$

$$R_3\overset{+}{P}{-}\overset{-}{N}R' + O{=}C{=}NR' \rightleftharpoons \begin{matrix} R_3P\text{---}NR' \\ O\text{---}C{=}NR' \end{matrix} \longrightarrow R_3\overset{+}{P}{-}\overset{-}{O} + R'N{=}C{=}NR'$$

Like isocyanic acid and the alkyl isocyanates, the carbodiimides contain a twin double bond and readily undergo addition of compounds containing reactive hydrogen, the initial step being a nucleophilic attack on carbon. Water yields the disubstituted ureas.

$$RN{=}C{=}NR + H_2O \longrightarrow \left[RNHC(OH){=}NR\right] \longrightarrow RNHCONHR$$

Alcohols give *O*-alkylureas, mercaptans give *S*-alkylthioureas, and amines give substituted guanidines.

$$RN{=}C{=}NR + HOR' \longrightarrow RNHC(OR'){=}NR$$

$$+ HSR' \longrightarrow RNHC(SR'){=}NR$$

$$+ HNHR' \longrightarrow RNHC(NHR'){=}NR$$

Carboxylic acids give either *N*-acyl ureas or the urea and the carboxylic anhydride. The *N*-acylurea arises from rearrangement of the intermediate addition product.

$$RN{=}C{=}NR + HOCOR \longrightarrow RNH{-}\underset{OCOR}{\underset{|}{C}}{=}NR \longrightarrow RNHC{=}NR \cdots O{-}COR \longrightarrow RNH\underset{\overset{\|}{O}}{C}{-}\underset{COR}{\underset{|}{N}}R$$

Attack on the intermediate by a second molecule of acid yields the anhydride and the urea.

$$\begin{array}{c} RNHC{=}N{-}R \\ O \quad\quad H \\ C \;\; O \\ R \;\; O \;\; COR \end{array} \longrightarrow RNH\underset{\overset{\|}{O}}{C}{-}NHR + RCO{-}O{-}COR$$

If R is an aromatic group, the acyl urea is the chief product, whereas if R is an aliphatic group, the urea and the anhydride are formed. Anhydrides are produced not only from carboxylic acids but also from sulfonic acids and dialkyl esters of phosphoric acid.

$$RN{=}C{=}NR + 2\,R'SO_3H \longrightarrow (R'SO_2)_2O + RNHCONHR$$

$$+ 2\,(R'O)_2\underset{O^-}{\underset{|+}{P}}OH \longrightarrow (R'O)_2\underset{O^-}{\underset{|+}{P}}{-}O{-}\underset{O^-}{\underset{|+}{P}}(OR')_2 + RNHCONHR$$

The last reaction is useful for the synthesis of derivatives of polyphosphoric esters of biological importance (p. 533).

When an aliphatic carbodiimide reacts with a mixture of a carboxylic acid and an alcohol, an ester is formed. The initial product of addition of the acid to the diimide acts as a strong acylating agent.

$$\begin{array}{c} RNHC{=}N{-}R \\ O \quad\quad H \\ C \;\; O \\ R \;\; O \;\; R' \end{array} \longrightarrow RNHCONHR + R'OCOR$$

If an amine is used instead of an alcohol, an amide is formed.

$$RN{=}C{=}NR + RCOOH + R'NH_2 \longrightarrow RNHCONHR + RCONH_2$$

This reaction is important for the synthesis of peptides (p. 366). **Dicyclohexylcarbodiimide,** $C_6H_{11}N{=}C{=}NC_6H_{11}$, and **di-*i*-propylcarbodiimide** have been widely used for this purpose. The simultaneous formation of the urea, however, frequently makes isolation of the desired product difficult. **Ethyl-(3-dimethylaminopropyl)carbodiimide** yields *N,N'*-ethyl-(3-dimethylaminopropyl)urea, which is soluble in dilute acids and thus readily separated from the amide.

DERIVATIVES OF THIOCARBONIC ACID

Many of the derivatives of carbonic acid have sulfur analogs, some of which are of general interest. Carbon oxysulfide and carbon disulfide are analogs of carbon dioxide. **Carbon oxysulfide** may be prepared by passing a mixture of carbon monoxide and sulfur vapor through a hot iron tube at 500°.

$$CO + S \xrightarrow{500^\circ} COS$$

It is an odorless, toxic gas, boiling at −47.5°. **Carbon disulfide** is manufactured on a large scale by reaction of sulfur with charcoal at a high temperature, either in direct-fired retorts or in a continuous-type furnace in which the charcoal is heated by the resistance it offers to the electric current.

$$C + 2\,S \longrightarrow CS_2$$

It can be prepared also by the reaction of methane and sulfur at 700° over an alumina catalyst, but this process requires recovery of sulfur by the controlled air-oxidation of the hydrogen sulfide.

$$CH_4 + 4\,S \xrightarrow[Al_2O_3]{700^\circ} CS_2 + 2\,H_2S$$

Carbon disulfide is a toxic, low-boiling, highly flammable liquid that is used to some extent as a solvent, as a toxic agent for rodents, and as an intermediate for the manufacture of carbon tetrachloride (p. 591). The principal uses are in the manufacture of viscose rayon (p. 344), rubber accelerators (p. 293), and fungicides (p. 293).

When solutions of alkali hydroxide or alkoxides in alcohols are mixed with carbon disulfide, nucleophilic attack on carbon by alkoxide ion occurs, and the *alkali O-alkyl dithiocarbonates* are formed. These ester salts commonly are known as **xanthates.**[2]

$$C_2H_5OH + CS_2 + NaOH \longrightarrow \underset{\text{Sodium ethyl xanthate}}{C_2H_5O\overset{\overset{\large S}{\|}}{C}S^{-+}Na} + H_2O$$

The alkyl hydrogen dithiocarbonates decompose at room temperature into carbon disulfide and alcohol. Hence the above reaction is reversed by the addition of acid.

$$C_2H_5O\overset{\overset{S}{\|}}{C}S^-Na^+ + HCl \longrightarrow NaCl + \left[C_2H_5O\overset{\overset{S}{\|}}{C}SH\right] \longrightarrow C_2H_5OH + CS_2$$

Reaction of the sodium alkyl xanthates with methyl iodide gives the **S-methyl *O*-alkyl xanthates.** If the *O*-alkyl group has a β hydrogen, pyrolysis of these esters takes place in the same way as the pyrolysis of carboxylic acid esters (p. 176) to yield olefins without rearrangement of the carbon skeleton. The pyrolysis of xanthates is known as the *Chugaev reaction.* As with the carboxylic esters it is believed to take place by way of a cyclic transition state. The *S*-methyl dithiocarbonate decomposes to mercaptan and carbon oxysulfide.

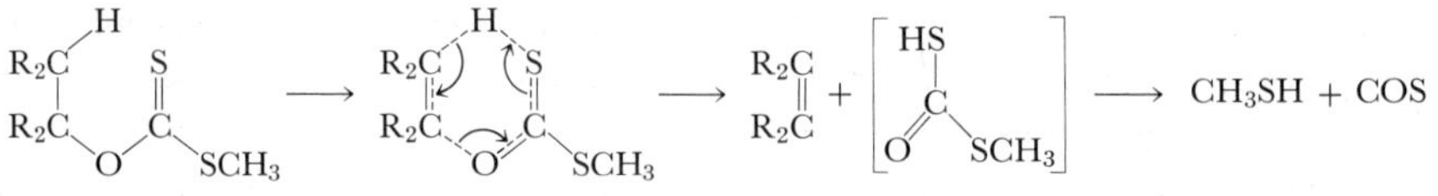

$$\text{(cyclic: } R_2C(H)\text{–}R_2C\text{–O–C(=S)SCH}_3) \longrightarrow \text{(cyclic transition state)} \longrightarrow \begin{matrix}R_2C\\ \|\\ R_2C\end{matrix} + \left[HS\text{–}\overset{}{C}(=O)\text{–}SCH_3\right] \longrightarrow CH_3SH + COS$$

The **sodium alkyl xanthates** are used as collecting agents in the flotation process for the concentration of ores. The most important use of the reaction of carbon disulfide with hydroxyl groups is for the production of viscose solutions from cellulose (p. 344).

Analogous to the reaction of alcohols with carbon disulfide is the reaction of ammonia and of primary and secondary amines to give the amine salts of **dithiocarbamic acids.**

$$S{=}C{=}S + 2\,HNH_2 \longrightarrow H_2N\overset{\overset{S}{\|}}{C}S^{-+}NH_4$$

$$+ 2\,HNHR \longrightarrow RNH\overset{\overset{S}{\|}}{C}S^{-+}NH_3R$$

$$+ 2\,HNR_2 \longrightarrow R_2N\overset{\overset{S}{\|}}{C}S^{-+}NH_2R_2$$

[2] Originally, the reaction product of carbon disulfide, ethyl alcohol, and potassium hydroxide was known as potassium xanthate because it gave a yellow precipitate with copper sulfate (Gr. *xanthos* yellow). Accordingly xanthic acid should be C_2H_5OCSSH. Since the chief variation in the xanthates is in the nature of the alkyl group, it is preferable to give the common name xanthic acid to the hypothetical dithiocarbonic acid, HOCSSH.

Metallic or amine salts of a variety of *N*-substituted dithiocarbamic acids are important agricultural fungicides and soil sterilants, and are used along with zinc oxide to accelerate the vulcanization of rubber (p. 584).

Like all sulfhydryl compounds, the dithiocarbamates are oxidized readily to disulfides. The products are known as *thiuram disulfides.* **Tetramethylthiuram disulfide,** $(CH_3)_2NCSS—SCSN(CH_3)_2$, is a valuable rubber accelerator known as *Tuads* (p. 584). It is used also in fungicidal preparations for disinfecting seeds and turf. Although it had been used as a rubber accelerator since 1918, its fungicidal properties were not recognized until 1931. It may cause severe allergic skin rashes in susceptible individuals. **Tetraethylthiuram disulfide** (*Antabuse*) has been used for the treatment of chronic alcoholism. A patient who has been given the compound orally becomes violently ill on drinking alcoholic beverages because of the increase in the concentration of acetaldehyde in the blood.

The reaction of alkali thiocyanates with alkyl halides or sulfates gives **alkyl thiocyanates** in contrast to the behavior of alkali cyanates, which yield alkyl isocyanates (p. 286).

$$RX + NaSCN \longrightarrow RSCN + NaX$$

In the reaction of both the cyanate and the thiocyanate ion, the product is that in which the more covalent bond has been formed, this being the usual course of reaction of a primary alkyl halide with an ambident ion (p. 237). Some of the alkyl thiocyanates are useful insecticides (*Lethanes*).

The **isothiocyanates,** or **mustard oils,** are obtained by the removal of hydrogen sulfide from *N*-alkyldithiocarbamates by means of a reagent, such as lead nitrate, that can give an insoluble sulfide.

$$RNHCSS^{-+}NH_3R + Pb(NO_3)_2 \longrightarrow RN{=}C{=}S + PbS + RNH_3NO_3 + HNO_3$$

All of the volatile isothiocyanates have pungent odors and tastes. They are called *mustard oils* because **allyl isothiocyanate,** $CH_2{=}CHCH_2NCS$, is one of the hydrolytic products of a glycoside occurring in black mustard (*Brassica nigra*). Many plants, especially members of the *Cruciferae,* contain glycosides that yield isothiocyanates when the plant cells are crushed, which liberates a hydrolytic enzyme.

Like the alkyl isocyanates, the alkyl isothiocyanates readily add alcohols and amines, the products being **alkyl thiocarbamates** (*thiourethans*) and **thioureas.**

$$\begin{aligned} RNCS + HOR' &\longrightarrow RNHCSOR' \\ + HNHR' &\longrightarrow RNHCSNHR' \\ + HNR'_2 &\longrightarrow RNHCSNR'_2 \end{aligned}$$

Solid substituted thioureas formed from **phenyl isothiocyanate,** $C_6H_5N{=}C{=}S$ (p. 425), are useful derivatives for amines. The rate of reaction with amines is so much faster than with water that the reaction can be carried out on aqueous solutions of amines.

Thiourea is made by the action of hydrogen sulfide on calcium cyanamide.

$$CaNCN + 2\,H_2S \xrightarrow{150-180^\circ} CaS + H_2NCSNH_2$$

Like urea, it forms inclusion compounds with other organic compounds (p. 283). Unlike urea, it occludes compounds with branched chains and cyclic compounds, and only with those straight-chain compounds having more than fourteen carbon atoms. Solutions of thiourea applied to green hemlock lumber prevent "brown stain."

PROBLEMS

16–*1*. Give a general formula for each of the products represented by italic capital letters in the following diagrammatic chart.

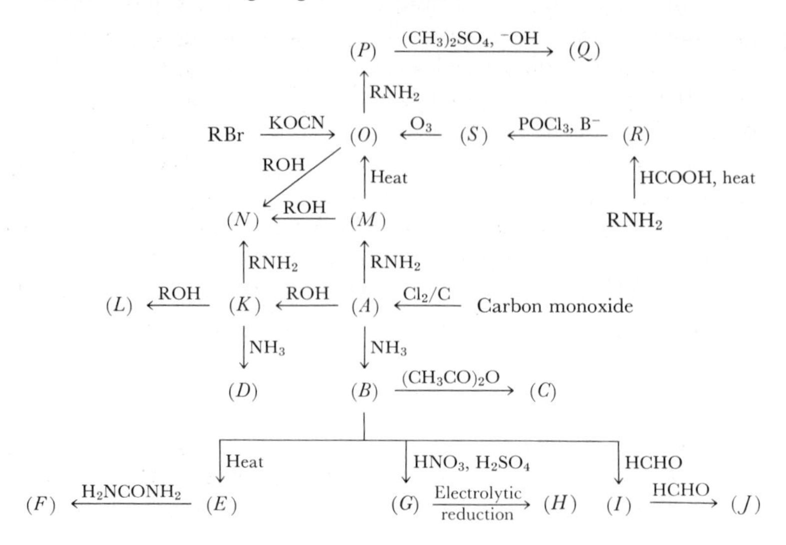

16–*2*. Which of the following compounds cannot be converted readily into urea: (*a*) cyanic acid, (*b*) carbon dioxide, (*c*) cyanogen, (*d*) phosgene, (*e*) ethyl carbamate, (*f*) hydrogen cyanide, (*g*) cyanamide, (*h*) acetamide, (*i*) ethyl carbonate, (*j*) methylal, (*k*) cyanuryl chloride?

16–*3*. Which of the following structures represent compounds that can be isolated in a pure state: (*a*) $H_2C(OH)_2$, (*b*) $H_2C(OCH_3)_2$, (*c*) H_2NCOOH, (*d*) HOCOOH, (*e*) ClCOOH, (*f*) $HOCH_2NH_2$, (*g*) $C_2H_5NHCOOH$, (*h*) $H_2NCOOC_2H_5$, (*i*) $H_2C{=}CHOH$, (*j*) $CH_3CH(OH)CN$, (*k*) $(CH_3)_2C(OH)OC_2H_5$, (*l*) $CH_2(OC_2H_5)Cl$?

16–*4*. Give equations for the preparation of the following compounds: (*a*) methyl chloroformate, (*b*) *n*-butyl carbonate, (*c*) *O*-ethylurea, (*d*) *S*-*i*-butylthiourea, (*e*) *i*-propyl thiocyanate, (*f*) *S,N,N'*-triethylthiourea, (*g*) potassium *i*-amyl xanthate, (*h*) sodium *N*-methyldithiocarbamate, (*i*) di-*n*-propylcarbodiimide, (*j*) tetra-*N*-methylurea.

16–*5*. Give a series of reactions for the following syntheses: (*a*) *N*-methyl-*N'*-ethylurea from potassium cyanate, (*b*) *n*-propyl isothiocyanate from carbon disulfide, (*c*) tetraethylthiuram disulfide from diethylamine, (*d*) *N,N'*-dimethylthiourea from methylamine, (*e*) ethyl *N*-*s*-butylcarbamate from *s*-butylamine.

16–*6*. Tell how one can distinguish between the members of the following pairs of compounds, and give the reactions involved: (*a*) urea and acetamide, (*b*) ethyl isocyanate and ethyl isocyanide, (*c*) *N,N'*-dimethylthiourea and *N,S*-dimethylthiourea, (*d*) ethyl chloroformate and acetyl chloride, (*e*) urea and *N,N'*-dimethylurea, (*f*) ethyl carbonate and ethyl orthocarbonate, (*g*) *i*-propyl thiocyanate and *i*-propyl isothiocyanate, (*h*) diethylcarbodiimide and *N,N'*-dimethylurea, (*i*) phosgene and ethyl chloride.

16–*7*. (*a*) Predict the number of moles of ammonia that should be evolved when arginine is boiled with an aqueous solution of sodium hydroxide. Give equations illustrating the stepwise course of the reaction. (*b*) Predict the number of moles of nitrogen that should be evolved when arginine is treated with aqueous nitrous acid. Give a series of equations illustrating the stepwise course of the reaction.

16–*8*. Give a likely mechanism for the acid-catalyzed hydrolysis of an isocyanate.

16–*9*. Compound *A* has the molecular formula $C_4H_7ClO_2$. It reacts with ammonia to give compound *B* having the molecular formula $C_4H_9NO_2$. When *B* is heated with dilute sulfuric acid, a gas is evolved that gives a precipitate when passed into a solution of barium hydroxide. When the acid solution is distilled and the distillate saturated with sodium hydroxide, an oily liquid separates. This liquid when shaken with an aqueous solution of sodium hypoiodite gives a yellow precipitate. When the acid solution is made alkaline with sodium hydroxide, a gas is evolved that turns red litmus blue. Write a structural formula for *A*, and give equations for the reactions that it undergoes.

16–*10*. A colorless compound contains C, H, N, and O. Its aqueous solution is acidic and titration gives an equivalent weight of 123. Treatment with sodium nitrite and hydrochloric acid gives a gas, a portion of which is soluble in potassium hydroxide solution. A quantitative determination shows that two moles of alkali-insoluble gas is produced from one equivalent weight of the compound. Evaporation of the solution from the nitrous acid treatment to dryness leaves a residue of sodium chloride, sodium nitrite, and sodium nitrate. When the original compound is heated with sodium hydroxide solution, ammonia is evolved, and the solution when acidified evolves a gas. Evaporation of the solution leaves only inorganic salts. Give the name for the original compound and the reactions that it undergoes.

CHAPTER SEVENTEEN

OPTICAL ISOMERISM

The existence of two or more compounds having the same number and kinds of atoms and the same molecular weight is called *isomerism* (p. 54). Isomers have the same *composition* and are represented by the same molecular formulas. Two main types of isomerism exist. The most common type is known as **structural isomerism** because it is assumed that the differences among the isomers result from the different order in which the atoms are attached to each other. For example, butane and isobutane differ in that butane has the carbon atoms linked in a chain, and isobutane has a branched carbon skeleton. Such structural isomers may be called *skeletal isomers*. Structural isomers that arise from the possibility of more than one position for some other element or group as with 1-chloropropane and 2-chloropropane are called *position isomers*. Structural isomers such as methyl ether and ethyl alcohol have different functional groups in the molecule and may be called *functional isomers*. The order in which atoms are joined together is spoken of as the *constitution* of the compound and is represented by a structural formula.

In the second main type of isomerism, however, the isomers have the same structural formulas. To explain this type, it is necessary to postulate a different distribution of the atoms in space, and the phenomenon is known as **stereoisomerism** (Gr. *stereos* solid). The subject of stereoisomerism usually is divided into two parts, *geometric isomerism*, which already has been discussed (p. 76), and *optical isomerism*, the subject of this chapter. The space arrangement of the atoms is referred to as the *configuration* of the molecule, and three-dimensional models, or perspective drawings, or projections of the space models must be used to illustrate the difference between stereoisomers.

Because of essentially free rotation about single bonds and a certain flexibility of bond angles, the same kinds of molecules, that is, molecules having the same structure and configuration, may assume different shapes in space. The particular shape that a molecule assumes is referred to as its *conformation* (p. 56). The four terms *composition, constitution, configuration,* and *conformation* have definite and distinct meanings. They should not be used interchangeably.

Polarized Light

Wave motion may be caused by longitudinal vibrations or transverse vibrations. In a longitudinal vibration, such as a sound wave, the vibrations are parallel to the direction of propagation and symmetrical about the line of propagation. In a transverse vibration, such as an ocean wave, the vibrations are perpendicular to the direction of propagation and there is a lack of symmetry about the line of propagation. The propagation of such a wave may be represented by Fig. 17–*1*, which shows the instantaneous magnitude of the vibrations over a given distance. The behavior of the vibrators during propagation of the wave may be visualized by moving the boundary of the wave along the direction of propagation. Each vector maintains a fixed position and direction but varies continuously in magnitude from zero to $+1$ to zero to -1 to zero.

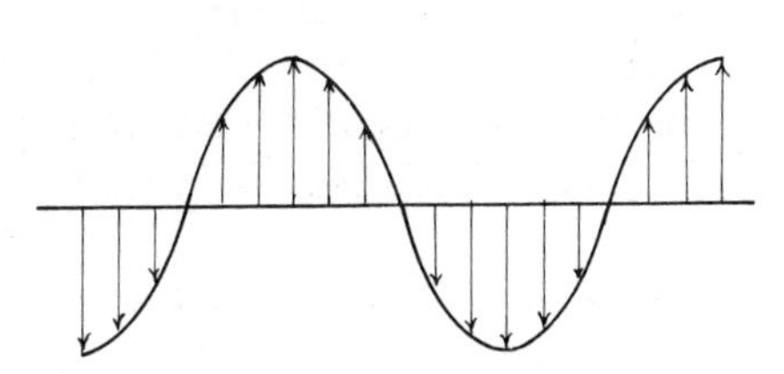

Figure 17–*1*. Propagation of a wave by transverse vibrations.

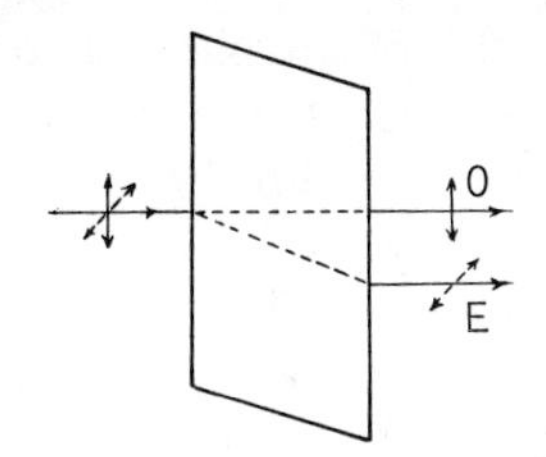

Figure 17–*2*. Double refraction by a calcite crystal.

Ordinary light does not show a lack of symmetry, but in 1669 Erasmus Bartholinus[1] discovered that a properly oriented crystal of Iceland spar (*calcite,* a crystalline calcium carbonate) divides a single ray of ordinary light into two rays. Thus a single line viewed through a properly oriented crystal appears as two lines. This phenomenon is known as *double refraction.* Eight years later Huygens[2] found that each of the rays formed by double refraction is vibrating in a single plane and that the plane of vibration of one ray is perpendicular to the plane of vibration of the other ray (Fig. 17–*2*).

Thus the symmetry of a ray of ordinary light about the direction of propagation is caused by transverse vibrations in all directions perpendicular to the direction of propagation. If two mutually perpendicular planes are passed through the ray with the intersection of the planes coinciding with the axis of the ray, each vector will have a component in each plane as indicated in Fig. 17–*3*. The action of the calcite crystal is to separate the vectors into their components. The emergent rays, each of which is vibrating in a single plane, are said to be *plane-polarized.*

The Nicol prism,[3] invented in 1828, is a device for separating one plane-polarized ray from the other. The calcite crystal is a rectangular rhombohedron, the acute angles measuring 71 degrees. In the Nicol prism the two end faces are cut away until these angles are reduced to 68 degrees, and the crystal is cut in a plane perpendicular to the two end faces and diagonally through the corners of the obtuse angles. The surfaces are polished and the two halves cemented together with Canada balsam, which has an index of refraction less than that of calcite for one of the polarized rays and greater than that of calcite for the other. The action of the Nicol prism is illustrated in Fig. 17–*4*. A light ray

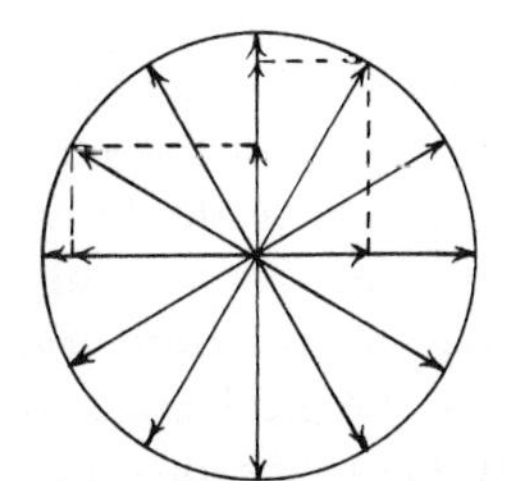

Figure 17–*3*. Vibrating vectors of light ray.

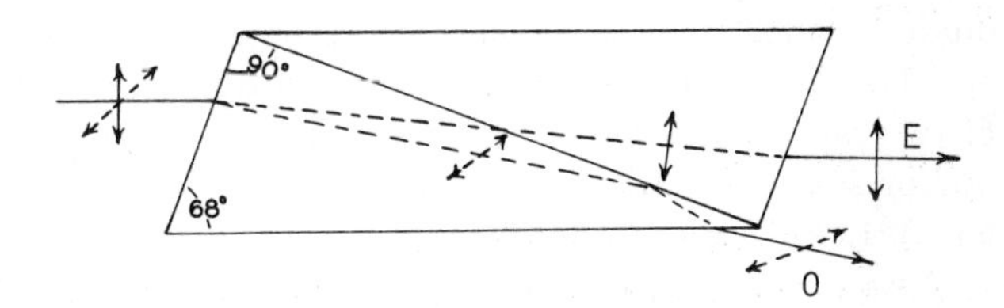

Figure 17–*4*. Production of plane-polarized light by a Nicol prism.

[1] Erasmus Bartholinus (1625–1698), Danish professor of mathematics and medicine at the University of Copenhagen.

[2] Christiaan Huygens (1629–1695), Dutch mathematician, astronomer, and physicist. He is known best for his contributions to physical optics.

[3] William Nicol (1768–1851), Scottish physicist who pursued his investigations privately. He devoted himself chiefly to the examination of fluid-filled cavities in crystals, to the manufacture of microscope lenses, and to the microscopic examination of fossil wood.

entering the prism parallel to the long axis is doubly refracted. The ordinary ray, *O*, is totally reflected from the surface of the Canada balsam. The extraordinary ray, *E*, is transmitted through the crystal. The reduction in the acute angles of the original calcite crystal from 71 degrees to 68 degrees is for the purpose of securing the proper angle of incidence on the balsam to produce this effect.

If a similarly oriented second Nicol prism is placed in the path of the emergent plane-polarized ray, the ray will pass through the second prism without being affected. If, however, the second prism is rotated about its long axis through 90 degrees, the effect is the same as if the ray were vibrating at right angles to its original direction, and it is totally reflected from the Canada balsam layer of the second prism. Two prisms so placed that the plane-polarized ray transmitted by one is not transmitted by the other are spoken of as *crossed Nicols.*

Light may be polarized by processes other than double refraction. In 1808 Malus[4] discovered that light reflected from a glass surface at a particular angle is plane-polarized. When ordinary light passes through a crystal of the mineral tourmaline, the component vibrating in one plane is absorbed much more strongly than that vibrating perpendicular to this plane. This phenomenon is known as *dichroism.* If a dichroic crystal is of the proper thickness, the more strongly absorbed component will be practically extinguished, whereas the other is transmitted in appreciable amount as plane-polarized light. The modern Polaroid operates on the same principle, the absorbing medium being a film containing properly oriented microscopic crystals of a dichroic substance such as the periodide sulfate of quinine. The transmitted light is slightly colored and not completely polarized, but it is possible by this method to make polarizing plates of large area at reasonable cost.

Optical Activity

In 1811 Arago,[5] a pupil of Malus, found that a quartz plate obtained by cutting a quartz crystal perpendicular to the crystal axis causes the rotation of the plane of polarization of plane-polarized light. This phenomenon can be observed best by placing a plate of quartz between crossed Nicols, the face of one of the Nicol prisms being illuminated from a light source. Before the quartz plate is placed between the two Nicol prisms, no light passes through the second prism. With the quartz plate between them some light passes through the second prism, which now must be rotated through a definite angle to become dark again. *The ability to rotate the plane of polarization of plane-polarized light* is called **optical activity.** Substances possessing the ability are said to be *optically active.* The number of degrees of arc through which the second crystal must be rotated to restore the original condition is called the *optical rotation* of the optically active substance and is given the symbol α.

Häuy[6] had discovered two kinds of quartz, the crystals differing only in the location of two facets that caused the crystals to be nonidentical mirror images. Because of the mirror image relationships, they were called *enantiomorphs* (Gr. *enantios* opposite, *morph* form). In 1815 Biot,[7] another pupil of Malus, found that plates of the same thickness from the two kinds of quartz rotate plane-polarized light the same amount but in opposite

---

[4] Etienne Louis Malus (1775–1812), French military engineer. He left the army in 1801 and died of tuberculosis in Paris at the age of 37.

[5] Dominique François Jean Arago (1786–1853), French physicist who, after successfully completing a geodetic survey through Spain, was appointed an astronomer of the French Royal Observatory, a post that he held until his death. He was active in French politics and did much to enhance the prestige of French science.

[6] Rene Just Häuy (1743–1822), French mineralogist who is regarded as one of the founders of the science of crystallography.

[7] Jean Baptiste Biot (1774–1862), French physicist who was associated with Arago in various geodetic surveys. His most important work dealt with optics, especially the polarization of light.

directions. The form that rotates the plane of polarization to the right when facing the light source is called *dextrorotatory* and that which rotates the plane of polarization to the left is called *levorotatory*. Biot found also that other substances such as sugar solutions and turpentine are optically active, the latter even in the vapor phase.

Measurement of Optical Rotation

The instrument used to measure the extent of rotation of plane-polarized light is called a *polarimeter* or *polariscope* (Fig. 17–5). It consists of a fixed Nicol prism, *A*, known as the *polarizer*, for polarizing the monochromatic light from the light source, *B*. A second Nicol prism, *C*, known as the *analyzer*, is attached to a disk, *D*, graduated in degrees and fractions of a degree, that can be rotated. The container for the sample is a tube, *E*, with clear glass ends, known as a *polarimeter tube*. The polarizer and analyzer are mounted on a suitable stand with a trough between them to hold the polarimeter tube in the path of the polarized light. Because it is easier for the eye to match two adjacent areas to the same degree of brightness than to determine a point of maximum darkness or brightness, a third smaller Nicol prism, *F*, is placed behind the polarizer and rotated through a small angle. In this way the field is divided into two halves of unequal brightness. An eyepiece, *G*, focuses on this field. By rotating the analyzer the fields may be brought to equal brightness, which provides a zero point. When an optically active substance is placed in the path of the light, the fields become unequally bright. Rotation of the analyzer returns the two fields to equal intensity. The number of degrees through which the analyzer is rotated measures the activity of the sample.

The amount of rotation is directly proportional to the length of the path through the active material, and this distance must be accurately known. For solutions the extent of rotation depends on the concentration, that is, the weight per unit volume of the substance in the solution. These statements are summarized in the equation

$$\alpha = [\alpha]cl \quad \text{or} \quad [\alpha] = \frac{\alpha}{cl}$$

where α = the observed rotation
c = the concentration in grams per cubic centimeter of solution[8] = the density of a pure liquid
l = length of tube in decimeters
$[\alpha]$ = a constant that is characteristic of the compound and is called the *specific rotation*.

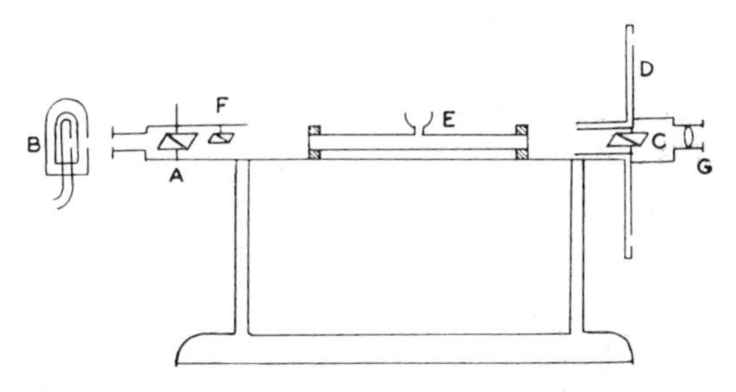

Figure 17–5. Cross section of a polarimeter.

[8] Frequently the concentration of solutions is expressed as grams per 100 cc. of solution, whence $[\alpha] = \dfrac{100\alpha}{cl}$.

The *molecular rotation*, $[M]$, is the specific rotation multiplied by the molecular weight, M, and divided by 100 to reduce the size of the figure.[9]

$$[M] = \frac{M[\alpha]}{100}$$

The extent of rotation varies with the wavelength of the light. In a region of the spectrum that is distant from an absorption band, it varies approximately inversely with the square of the wavelength. This phenomenon is known as *rotatory dispersion.* If white light were used as a source of light, each wavelength would be rotated a different amount while passing through the solution. Accordingly it is necessary to use monochromatic light when measuring optical activity. Usually the D line of sodium is used, although frequently it is preferable to use the green line of the mercury arc or the red line of the cadmium arc. The rotation varies somewhat also with the temperature. For accurate work the polarimeter tube is maintained at some fixed temperature, usually 25°. The wavelength used and the temperature of the solution are designated by subscript and superscript. For example, $[\alpha]_D^{25}$ indicates that the rotation was determined at 25° using the D line of sodium. Usually there is more or less electrical effect between solute molecules and between solvent and solute molecules, which may cause the specific rotation to vary considerably with different concentrations and with different solvents. Hence it is necessary to indicate both the concentration and the solvent used. Thus a proper description of a specific rotation would be

$$[\alpha]_D^{25} = +95.01° \text{ in methyl alcohol } (c = 0.105 \text{ g./cc.})$$

Optical Isomerism

By 1848 two isomeric acids had been isolated from the tartar of grapes. The common acid, called *tartaric acid,* first was isolated by Scheele in 1769 and found by Biot to be dextrorotatory. Its isomer, which was isolated by Kestner[10] some time previous to 1819 and named *racemic acid* (L. *racemes* a bunch of berries or grapes) by Gay-Lussac, was optically inactive. In the spring of 1848, Pasteur (p. 120) was studying the crystal structure of sodium ammonium tartrate. He noticed that the crystals were characterized by facets that eliminated certain elements of symmetry from the crystal (cf. p. 302). Such facets are known as *hemihedral facets* because they occur in only half the number required for complete symmetry. As a result of the occurrence of the hemihedral facets, the crystals were not identical with their mirror images; that is, the mirror image could not be superposed on the crystal with coincidence at all points. Figure 17–6 shows hypothetical enantiomorphic crystals with hemihedral facets. These models have fewer faces than crystals of the sodium

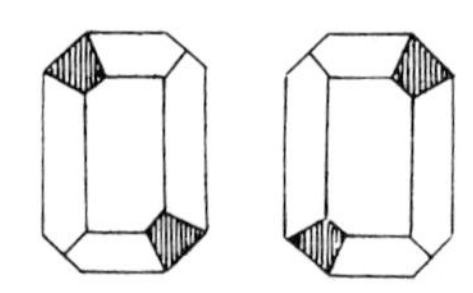

Figure 17–6. Models of hypothetical enantiomorphic crystals with hemihedral facets.

[9] Unfortunately a second symbol, $[\phi]$, for molecular rotation has been introduced recently.

[10] Kestner, the owner of a chemical plant, had obtained as a by-product of tartaric acid manufacture an acid that he thought to be oxalic acid and that he sold as such. It was pointed out in a handbook in 1819 that this compound was neither oxalic nor tartaric acid. Gay-Lussac obtained a sample from Kestner and found that the analytical values were the same as those for tartaric acid. He called the new compound racemic acid. It was not until 1830 that Berzelius convinced himself that both compounds had the same composition, and coined the term *isomerism* for the phenomenon.

ammonium tartrates, and the hemihedral facets can be seen readily. If the hemihedral facets were lacking or if the crystals were holohedral, that is, if the facets appeared on all corners, the mirror images would be identical.

Recalling that the active quartz crystals had hemihedral facets and that Herschel[11] in 1820 had suggested that there may be a connection between the hemihedralism of quartz and its optical activity, Pasteur proceeded to examine crystals of the inactive sodium ammonium racemate, expecting to find them holohedral. He found instead that all of the crystals obtained had hemihedral facets, but that two kinds of crystals were present. One kind was identical with the crystals of sodium ammonium tartrate, and the other kind consisted of mirror images of the tartrate crystals. Pasteur separated the two types of crystals under the microscope and found that the type that looked like sodium ammonium tartrate was indeed dextrorotatory and identical with it, but that the mirror image crystals, when dissolved in water, rotated plane-polarized light exactly the same amount in the opposite direction. When equal weights of the two crystals were dissolved and mixed, the solution was optically inactive. In other words, the reason that racemic acid is inactive is because it is composed of equal quantities of two different kinds of molecules, one dextrorotatory and the other levorotatory.

Quartz and other optically active crystals such as those of sodium chlorate and magnesium sulfate give inactive solutions. Similarly amorphous silica is optically inactive. Hence the cause of the activity in the *crystal* lies in the arrangement of the atoms in the crystal. Tartaric acid, on the other hand, is active in solution. Moreover pinene from oil of turpentine is active in both the liquid and gaseous states. In these compounds the activity must be due to the arrangement of the atoms in the individual molecules. Pasteur himself came to this conclusion, but since the theories of structural organic chemistry were not developed until around 1860 (p. 4), he did not recognize the principle necessary to relate optical activity to molecular structure.

By 1874 the constitutions of several active compounds were known. In September and November of that year two papers appeared, one by van't Hoff[12] and the other by Le Bel,[13] in which each pointed out that every optically active compound whose structure was known had at least one carbon atom that was combined with four different groups. The following examples may be cited, the carbon atoms under discussion being marked by an asterisk.

$$CH_3CH_2\underset{\underset{\displaystyle CH_3}{|}}{\overset{*}{C}}HCH_2OH$$

Active amyl alcohol

$$CH_3\underset{\underset{\displaystyle OH}{|}}{\overset{*}{C}}HCOOH$$

Lactic acid

$$HOOCCH_2\underset{\underset{\displaystyle OH}{|}}{\overset{*}{C}}HCOOH$$

Malic acid

$$HOOCCH_2\underset{\underset{\displaystyle NH_2}{|}}{\overset{*}{C}}HCOOH$$

Aspartic acid

[11] John Frederick William Herschel (1792–1871), noted English astronomer who by inclination was more interested in chemistry and the properties of light and made many valuable contributions in these fields.

[12] Jacobus Hendricus van't Hoff (1852–1911), Dutch physical chemist, professor at Amsterdam University and after 1896 at the Prussian Academy of Sciences and the University of Berlin. He is noted not only for his theoretical contributions to stereoisomerism, but also for his contributions to the theories of solutions and of chemical equilibria. He was the first recipient of the Nobel Prize in Chemistry in 1901. When criticized by the Dutch for accepting the essentially research post at Berlin, he replied that one gets tired of teaching that potassium permanganate is an oxidizing agent.

[13] Jules Achille Le Bel (1847–1930), French chemist who was financially independent and conducted his investigations privately. His experimental work dealt largely with the verification of predictions based on his stereochemical theories.

Wherever a pair of isomers differing in sign of rotation was known, the members of the pair had identical chemical and physical properties with the exception of their action on plane-polarized light, and even here they differed only in the sign of rotation and not in the magnitude of rotation. Accordingly the space relationship between atoms of one isomer must be the same as that between the atoms of the other isomer. van't Hoff and Le Bel pointed out that if the four different groups about the carbon atom are placed at the four corners of a tetrahedron, two arrangements are possible. Two molecules result that are mirror images of each other, but that are not superposable and hence not identical (Fig. 17–7). The asymmetry of such arrangements is of the same type as the asymmetry of the quartz crystals or of the sodium ammonium tartrate crystals; that is, *the condition necessary for the existence of optical activity is that the arrangement of the atoms be such that a crystal or molecule and its mirror image are not superposable.* Objects that are not superposable on their mirror images are said to be *dissymmetric.* It should be noted that they need not be lacking in all elements of symmetry and that although all asymmetric objects are also dissymmetric, dissymmetric objects are not necessarily asymmetric. A carbon atom joined to four different atoms or groups of atoms lacks symmetry and is known as an *asymmetric carbon atom.* Two molecules that are nonsuperposable mirror images of each other are known as *optical antipodes, enantiomorphs,* or *enantiomers.* The term *mirror images* also frequently is used, *nonsuperposable* being implied.

Actually the presence of an asymmetric carbon atom is not necessary for optical activity. Its presence is merely the most frequently encountered condition that removes the elements of symmetry that make mirror images identical. These elements of symmetry are (*1*) a plane of symmetry, (*2*) a center of symmetry, and (*3*) a fourfold alternating (or mirror) axis of symmetry. A *plane of symmetry* is a plane that divides an object into two parts that are mirror images of each other. The compound C*abda* (Fig. 17–*8a*), for example, has a plane of symmetry, namely that which divides the carbon atom and the groups *b* and *d* into like halves, whereas the compound C*abdc* (Fig. 17–8*b*), does not have a plane of

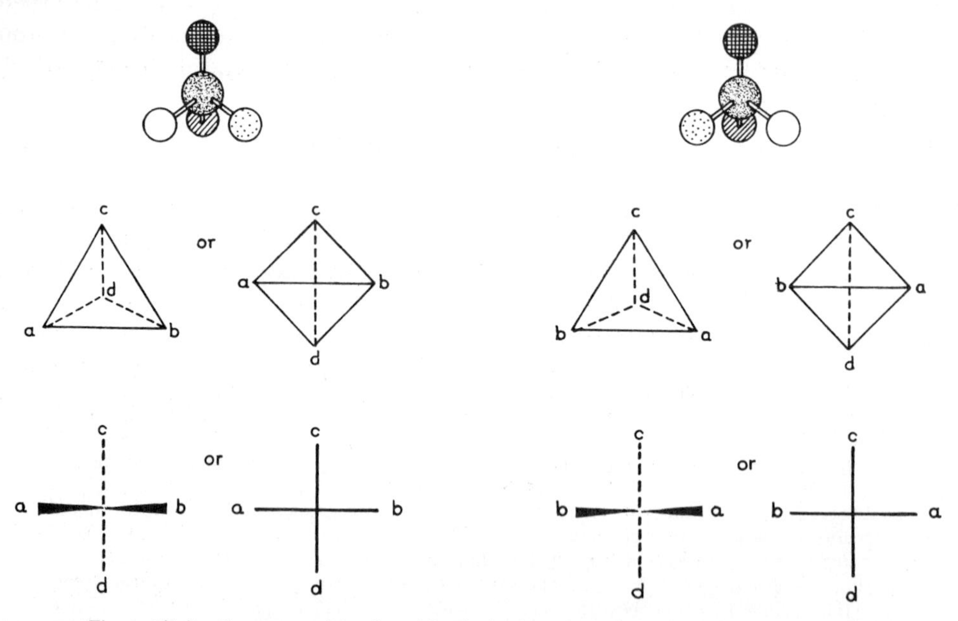

Figure 17–7. Representation of nonidentical mirror images by molecular models, by tetrahedra, and by projection formulas.

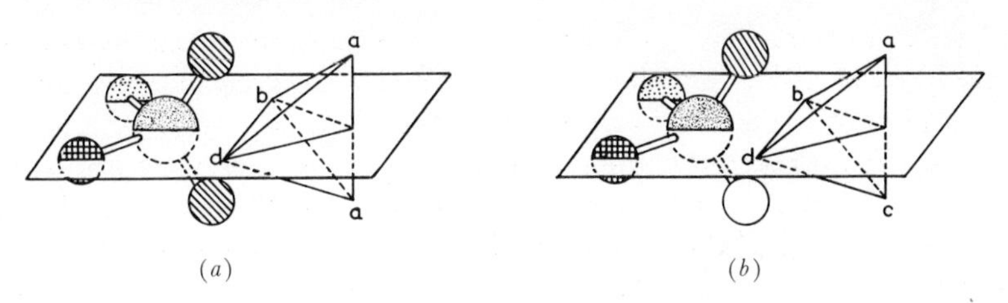

Figure 17–8. (*a*) Plane of symmetry in compound *Cabda;* (*b*) lack of a plane of symmetry in compound *Cabdc.*

symmetry. If an object possesses a plane of symmetry, the object and its mirror image are identical. The same statement holds for the other two elements of symmetry, but since they need to be considered only infrequently, discussion of them is omitted.

Expected Number of Optical Isomers

Molecules containing a single asymmetric carbon atom exist in only two forms. The asymmetric atoms may be designated as $A+$ and $A-$ for the dextrorotatory and levorotatory forms respectively. If a second asymmetric carbon atom is present, the configuration of this atom may be designated as $B+$ or $B-$. Therefore a total of four isomers is possible, namely those in which the four configurations are $A+B+$, $A+B-$, $A-B+$, and $A-B-$. Since $A+$ is the mirror image of $A-$, and $B+$ is the mirror image of $B-$, $A+B+$ is the mirror image of $A-B-$, and the two constitute an enantiomorphic pair and have identical properties except for the sign of rotation. Similarly $A+B-$ and $A-B+$ are enantiomorphic. $A+B+$, however, is not a mirror image of either $A+B-$ or $A-B+$ and has different chemical and physical properties. Similarly any active compound has only one mirror image and all of its other optical isomers differ from it in chemical and physical properties. Optical isomers that are not mirror images are called *diastereoisomers.*

If a third asymmetric atom is present, it also may exist in two forms, $C+$ and $C-$, and similarly for a fourth atom, $D+$ and $D-$. Hence two active forms exist if a single asymmetric carbon atom is present, and the number of active forms is doubled each time a new asymmetric atom is added. The total number of active forms, therefore, is 2^n, where n is the number of different asymmetric carbon atoms.

To represent space models on plane paper, perspective drawings of tetrahedra may be used. It is more convenient, however, to use projection formulas. The convention has been adopted that the two groups at the top and bottom of the projection formulas always are directly over each other and behind the plane of the paper, and the two groups joined by the horizontal line are in front of the plane of the paper. Although dotted lines and wedges frequently are used (Fig. 17–7), solid lines should be interpreted in the same way. In any comparison of projection formulas with each other, neither the formula as a whole, nor any portion of the formula may be rotated out of the plane of the paper. Otherwise the top and bottom groups would not bear the same relation to the other formulas as was assumed when the formulas were projected. Any rotation within the plane of the paper must be through 180°. If a projection formula has been rotated through 90°, it must be rotated clockwise or counterclockwise through 90° before an interpretation of the configuration can be made. Figure 17–9 illustrates several methods for representing the active forms of a compound that has two unlike asymmetric carbon atoms.

In the formulation of the 2^n rule the assumption was made that all of the asymmetric carbon atoms were structurally different, that is, that no two of them were attached to the same four kinds of groups. If any of the asymmetric carbon atoms are alike, the number of

Perspective formulas

Projection formulas

Active group representation

Figure 17–*9*. Methods of representing active forms having two different asymmetric carbon atoms.

possible isomers is decreased. Thus if a compound contains two like asymmetric carbon atoms, the possible configurations are $A+A+$, $A+A-$, $A-A+$, and $A-A-$. However, $A+A-$ is identical with $A-A+$ and only three optical isomers exist. The tartaric acids are examples of this situation. They have the constitution HOOCCHOHCHOHCOOH. Each of the two asymmetric carbon atoms bears the groups H, OH, COOH, and CHOHCOOH, and hence they are structurally alike. The four possible combinations of active groups are shown by both perspective and projection formulas in Fig. 17–*10*.

The first two forms are nonsuperposable. The second two arrangements, however, have

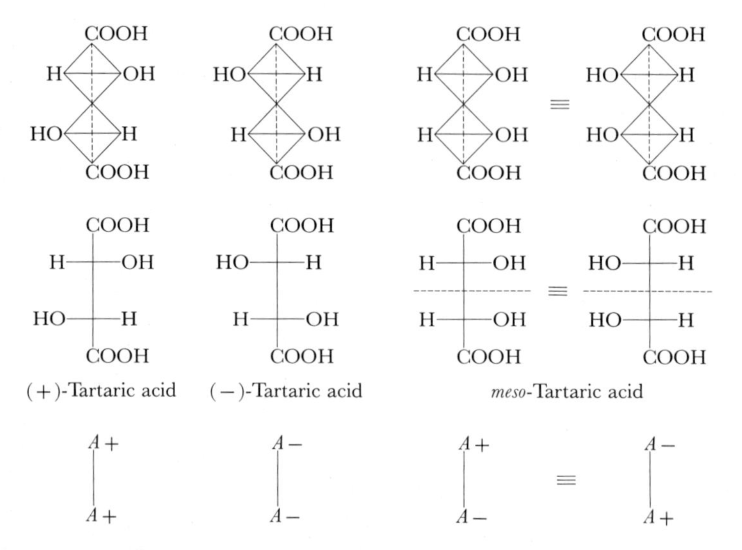

Figure 17–*10*. Methods of representing the optical isomers of a compound having two like asymmetric carbon atoms.

a plane of symmetry. These structures are not superposable in the positions shown, but if one or the other is rotated 180° in the plane of the paper, they become superposable. Similarly in the projection formulas the dotted line indicates a plane of symmetry, and rotation of one or the other formula in the plane of the paper through 180° causes the two to coincide. Although it may appear that rotation of a projection formula about a vertical axis also makes it coincide with the other formula, such is not the case because the carboxyl groups are not in the plane of the paper.

Not only is the number of optical isomers reduced from four to three when two like asymmetric carbon atoms are present, but the third form, having a plane of symmetry, is optically inactive. It is known as a *meso form* (Gr. *mesos* middle). Therefore a compound having two like asymmetric carbon atoms has three optical isomers. Two are active enantiomorphs with identical chemical and physical properties except for their opposite effect on plane-polarized light. The third is an inactive *meso* form, which is a diastereoisomer of the other two and hence differs from them in chemical and physical properties. In addition a fourth form, known as the *racemic form* (after racemic acid), consists of a mixture of equal amounts of the two active forms and hence is optically inactive. It differs from the *meso* form, however, in that it can be separated into active forms whereas the *meso* form cannot. The solid racemic form usually differs in physical properties from the other forms, but in solution it dissociates into the two active forms, and its properties in solution are identical with the properties of the active forms except that the solution is not optically active.

Geometric Isomers and Optical Isomers

Although both geometric isomerism and optical isomerism have been discussed, no definitions have been given for geometric or optical isomers. Because similar spatial features may give rise to either type of stereoisomerism, a convention is necessary to avoid confusion. The convention adopted in this text is that ***geometric isomers*** *comprise a set of stereoisomers, no pair of which are nonsuperposable mirror images.* Conversely, ***optical isomers*** *are a set of stereoisomers, at least two members of which are nonsuperposable mirror images.* Although geometric isomers never are optically active, it is not necessary that all members of a set of optical isomers be active. Thus *meso*-tartaric acid and the *dextro*- and *levo*-tartaric acids all are optical isomers of each other, although *meso*-tartaric acid is optically inactive. Moreover it is possible for the rotations of enantiomorphs to be so low as to be nondetectable. Usually, however, a nondetectable rotation in the visible will become detectable at shorter wavelengths, or the compound can be converted into a derivative that has a detectable rotation.

Cis-trans isomerism can give rise to either geometric or optical isomers. Thus *cis*- and *trans*-1,4-cyclohexanediol are geometric isomers because neither has a nonsuperposable mirror image, whereas the *cis*- and the two *trans*-1,3-cyclohexanediols are optical isomers, the *cis* compound being a *meso* form.

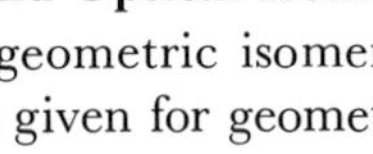

cis-1,4-Cyclohexanediol

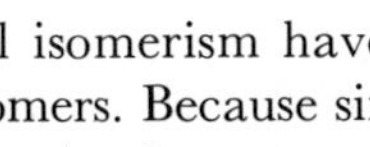

trans-1,4-Cyclohexanediol

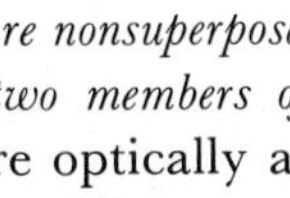

cis(= *meso*)-1,3-Cyclohexanediol

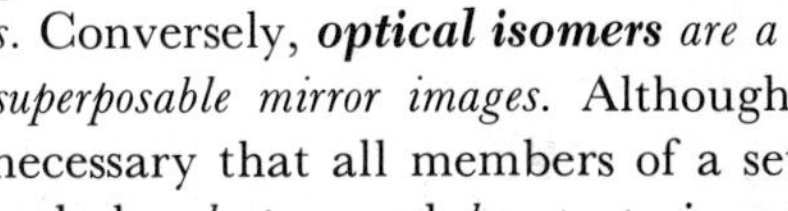

Active *trans*-1,3-Cyclohexanediols

Similarly a *meso* and two active 1,2-cyclohexanediols are known. If a double bond capable

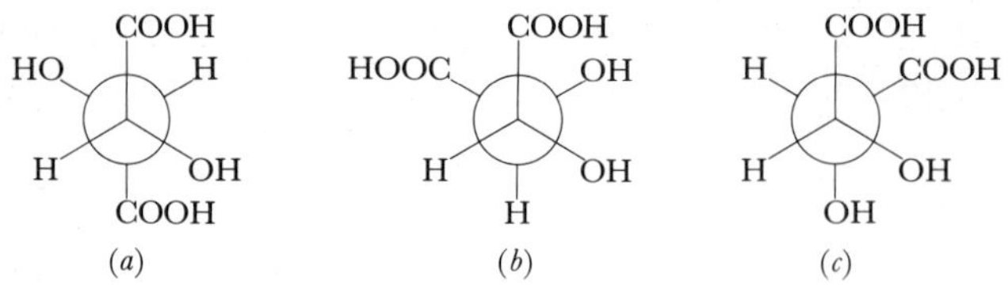

Figure 17–*11*. Conformers of *meso*-tartaric acid.

of *cis-trans* isomerism is present in a molecule along with an asymmetric carbon atom, all four isomers are optically active and give rise to two racemic mixtures.

As with all compounds that are classed as isomers, the term implies that the energy barrier to isomerization is sufficient to permit the isolation of the individual species at ordinary temperature. If the barrier to free rotation about the central carbon atoms were sufficiently great, that is, $> \sim 20$ kcal., *meso*-tartaric acid should be separable into the three forms shown in Fig. 17–*11*. A comparison of models readily shows that because of the center of symmetry the mirror image of (*a*) is identical with (*a*), whereas (*b*) and (*c*) are nonsuperposable mirror images of each other and should be optically active. Since the barrier to free rotation is too low for the isolation of (*a*), (*b*), and (*c*), they are called *conformers* and are said to differ in *conformation.* If the barrier were sufficiently great to permit isolation, they would be called *stereoisomers* and would be said to differ in *configuration.* As a corollary it follows that ordinary *meso*-tartaric acid is optically inactive and nonresolvable, not because the two asymmetric carbon atoms have opposite configurations, but because the barrier to free rotation about the bond joining them is low.

Separation of Racemic Forms into Active Components (*Resolution*)

Many naturally occurring organic compounds that contain an asymmetric carbon atom are optically active. For example, the lactic acid, $CH_3CHOHCOOH$, isolated from muscle is the (+) isomer. Lactic acid produced by fermentation may be (+), or (−), or racemic, depending on the fermenting organism. On the other hand, lactic acid produced by a synthesis that does not involve an optically active compound always is racemic. The reason is that when the asymmetric carbon atom is formed by synthesis, there always is an equal chance of producing the (+) or the (−) form, and since the number of molecules involved is very large, equal quantities of both forms are produced. Suppose, for example, that lactic acid is being synthesized by brominating propionic acid to α-bromopropionic acid and hydrolyzing to the hydroxy acid. In the bromination step either of the two α hydrogen atoms may be replaced. Replacement of one gives rise to the (+) form and of the other to the (−) form.

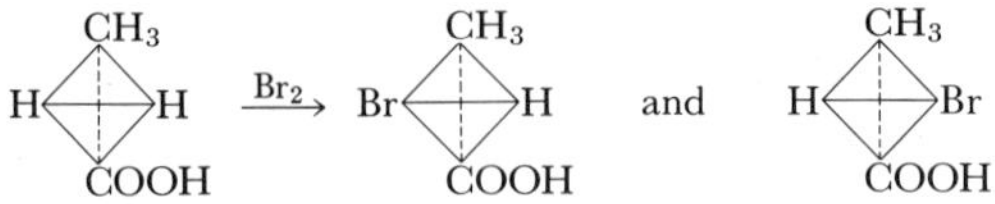

Or if acetaldehyde is converted to the cyanohydrin, a racemic mixture is formed because the carbon atom of the carbonyl group may be attacked from either side with equal ease during the addition of hydrogen cyanide.

Because synthesis always produces racemic mixtures, the separation of racemic mixtures into their active components is of considerable importance. This process is known as *resolution,* and the racemate is said to be *resolved* into its active components. Pasteur's original separation of sodium ammonium racemate into the (+)-tartrate and the (−)-tartrate is largely of historical interest only, since not all racemates separate into enantiomorphic crystals. Even sodium ammonium tartrate does so only below 27.7°. Furthermore the mechanical separation is tedious.

All chemical methods for the resolution of racemates or for the production of active compounds depend on the fact that diastereoisomers, unlike enantiomorphs, have different chemical and physical properties. If a racemic mixture undergoes a reaction involving dissymmetric transition states, which it must if one of the reagents or a catalyst is a single dissymmetric form, that is, if the reagent or catalyst is optically active, the two transition states are diastereoisomers, and the two enantiomorphs can react at different rates. Enzymes, for example, consist of a single form of asymmetric protein molecules and can lead to the decomposition of one form of a pair of enantiomorphs and not the other. Pasteur found that the mold *Penicillium glaucum,* when allowed to grow in the presence of ammonium racemate, metabolized and hence destroyed the (+)-tartrate and left the (−)-form unchanged. The general procedure for enzymatic resolution is to make a solution of the racemate, add nutrient salts, and inoculate with the desired organism. This method has the serious disadvantage that one form is lost, and the other form usually is obtained in poor yield. Moreover it is necessary to work with dilute solutions. The method can be applied, however, to a fairly large variety of compounds and frequently provides a simple method for determining whether a compound is capable of resolution.

The differences in behavior of enantiomorphs in living matter are explainable by the fact that reactions in living matter are catalyzed largely by optically active enzymes. A few examples may be given: only one form of an amino acid commonly is present in proteins and is utilizable by the organism in the synthesis of proteins (p. 354); (+)-glucose (p. 317) is the form synthesized by plants and the only form fermentable by yeast or utilizable by living matter; (+)-leucine (p. 353) is sweet whereas (−)-leucine is faintly bitter; (−)-tartaric acid is more toxic than (+)-tartaric acid; (−)-epinephrine has twenty times the activity of (+)-epinephrine in raising blood pressure (p. 459). These differences merely emphasize the fact that enantiomorphs have identical properties only if they are reacting with optically nonactive reagents.

Enzymatic decomposition or synthesis is a complicated process, but the principles underlying the production of active compounds by means of enzymes apply also to simpler reactions. For example, if racemic lactic acid is partially esterified with an active alcohol such as (+)-amyl alcohol, the (+) and (−) forms of the lactic acid react at different rates, and the unreacted portion contains an excess of one of the forms and has some optical activity. Even partial dehydration of racemic *s*-butyl alcohol over one of the dissymmetric forms of quartz leads to an observable optical activity in the undecomposed portion.

By far the most generally practical procedure for the separation of racemic mixtures, also developed by Pasteur, involves the conversion of the enantiomorphs into intermediate compounds that are diastereoisomers. Since diastereoisomers are not mirror images, they do not have the same physical properties and may be separated by ordinary physical methods such as fractional crystallization. After the separation of the diastereoisomers, they are converted into the original reactants. If (A+) and (A−) represent the two active forms present in the racemic mixture and (B+) represents a single active form of another compound that will combine with the racemic mixture, the process of separation may be illustrated schematically as follows:

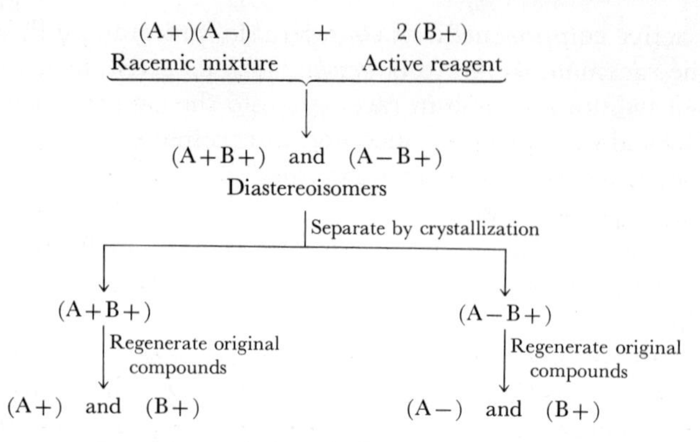

In practice the type of compound that is formed most readily and from which the original reactants can be regenerated most easily is a salt. Thus a racemic acid can be resolved after salt formation with an active base.

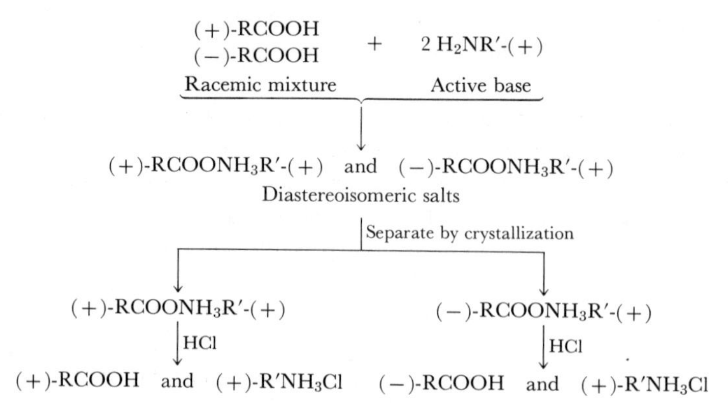

Likewise a racemic base can be resolved after salt formation with an active acid.

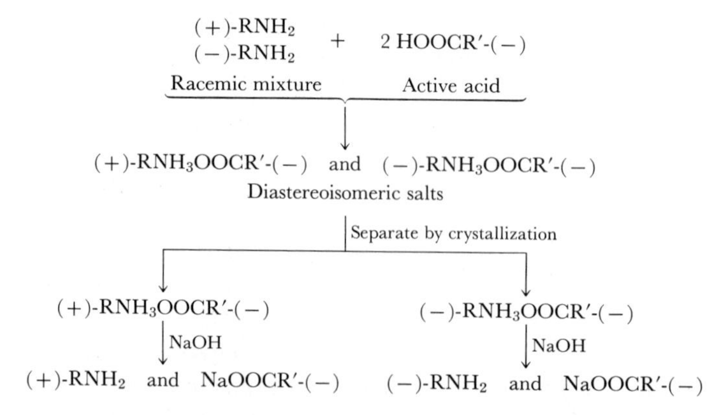

The naturally occurring alkaloids (p. 534) are active bases that give readily crystallizable salts. Those most commonly used for effecting resolution of racemic acids are (−)-brucine, (−)-strychnine, (−)-quinine, (+)-cinchonine, (+)-quinidine, and (−)-morphine. The easily available naturally occurring acids suitable for the resolution of racemic bases are (+)-tartaric acid and (−)-malic acid. Once a synthetic racemic mixture has been resolved, the active components can be used to resolve other racemic mixtures.

Indirect methods and reactions other than salt formation have been used for the resolution of compounds that do not contain acidic or basic groups. Selective adsorption on optically active adsorbents and ion exchange resins (p. 484), vapor phase chromatography on an active immobile phase, and extraction with or diffusion through optically active solvents have been reported to bring about resolution of racemic mixtures. In 1952 it was found that certain compounds, for example urea, crystallize spontaneously into dissymmetric forms that are capable of occluding other compounds (p. 283). If crystallization takes place in the presence of a racemic mixture, one form can be preferentially occluded and a resolution effected. This procedure is particularly promising for resolving racemic hydrocarbons and alkyl halides, which, because of the absence of reactive groups, are difficult to resolve by the usual methods.

The question arises as to how different groups must be to permit the detection of optical activity. It has been demonstrated in a number of compounds that even the difference between the isotopes hydrogen and deuterium is sufficient to give rise to observable activity. The enantiomorphs of 1-deuteriobutyl acetate $CH_3CH_2CH_2CHD(OCOCH_3)$, for example, have been prepared with specific rotations of $[\alpha]_D^{25} + 0.094 \pm 0.001°$ and $[\alpha]_D^{25} - 0.090 \pm 0.007°$. They have been used to study the mechanisms of reactions at primary carbon atoms. Others have been used in the study of asymmetric synthesis.

Asymmetric Synthesis

If a new asymmetric carbon atom is produced by a reaction involving a dissymmetric reagent, diastereoisomeric transition states also result, and the two forms of the new asymmetric carbon atom are produced at different rates and hence in unequal amounts. Such a process is known as an *asymmetric synthesis*. Thus although catalytic reduction of pyruvic acid yields racemic lactic acid, reductase of yeast yields (−)-lactic acid.

$CH_3COCOOH$ (Pyruvic acid) $\xrightarrow{H_2/Pt}$ HO–C(COOH)(CH₃)–H (+) + H–C(COOH)(CH₃)–OH (−)

Racemic lactic acid

$CH_3COCOOH$ (Pyruvic acid) $\xrightarrow{\text{Reductase}}$ H–C(COOH)(CH₃)–OH

(−)-Lactic acid

Moreover if (+)-amyl pyruvate is reduced catalytically and the (+)-amyl lactates hydrolyzed, the lactic acid obtained is optically active to a small extent because of an excess of one of the active forms.

$$CH_3COCOOC_5H_{11}(+) \xrightarrow{H_2/Pt} CH_3CHOHCOOC_5H_{11}(+)$$

$$\downarrow H_2O$$

$$CH_3CHOHCOOH + C_5H_{11}OH(+)$$

Active

Even the presence of an optically active solvent can lead to an asymmetric synthesis.

Racemization

Two methods for producing racemic mixtures have been discussed, one by mixing equal amounts of the (+) and (−) forms (p. 301) and the other by producing an asymmetric carbon atom by a synthesis in which none of the reagents is optically active (p. 306). A third method consists in converting either a (+) or (−) form into a mixture of equal amounts of both forms. This process is known as *racemization*.

Since conversion of a (+) form to a (−) form or of a (−) form to a (+) form involves the change in position of at least two groups on the asymmetric atom, bonds must be broken during the racemization, and the molecule must pass through a nondissymmetric intermediate. In the subsequent reformation of the compound equal amounts of the two forms are produced. The most easily racemized compounds, and those for which the racemization is most readily explainable, have a carbonyl group α to an asymmetric atom that carries a hydrogen atom. The racemization undoubtedly involves enolization (p. 201). The enol form does not contain an asymmetric atom, and when ketonization takes place, the double bond may be attacked from either side with equal ease, and equal quantities of both (+) and (−) forms are produced. Eventually racemization becomes complete. Thus the racemization of active lactic acid may be represented by the following equilibria.

$$\underset{\text{(−)-Lactic acid}}{\text{H—C(COOH)(OH)—}CH_3} \rightleftharpoons \underset{\text{Enol form}}{(HO)(HO)C{=}C(CH_3)(OH)} \rightleftharpoons \underset{\text{(+)-Lactic acid}}{\text{HO—C(COOH)(H)—}CH_3}$$

If more than one asymmetric carbon atom is present, only those having enolizable hydrogen are isomerized. For example, an active form of 2,3-dichlorobutyric acid isomerizes only to its diastereoisomer and not to the mixture of racemates.

$$\begin{array}{c} COOH \\ H{-}C{-}Cl \\ Cl{-}C{-}H \\ CH_3 \end{array} \rightleftharpoons \begin{array}{c} COOH \\ Cl{-}C{-}H \\ Cl{-}C{-}H \\ CH_3 \end{array}$$

Optical isomers that differ only in the configuration of one carbon atom are called *epimers,* and the conversion of an active compound into its epimer is known as *epimerization.*

Absolute and Relative Configuration

Absolute configuration is the actual configuration of the molecule in space, that is, it is the answer to the question, "Which of two enantiomorphic models represents the dextrorotatory form of a compound and which the levorotatory form?" Prior to 1951, no procedure was known by which this question could be answered. If, however, the configuration of some reference compound is assumed, it is possible to relate the configuration of other active compounds to this substance and to each other. *The configuration of a compound with reference to the arbitrarily assigned configuration of the reference substance is known as its* **relative configuration.**

The substance used for reference is glyceraldehyde, the (+) form of which was assigned arbitrarily the configuration represented by the perspective formulas *I* or *II* (cf. p. 318). In these formulas the asymmetric carbon atom is in the plane of the paper, the aldehyde and the hydroxymethyl groups behind the plane of the paper, and the hydrogen atom and the hydroxyl group in front of the plane of the paper. Formula *III* is a projection of this space formula in the plane of the paper.

$$\underset{I}{\begin{array}{c} CHO \\ H \lozenge OH \\ CH_2OH \end{array}} \qquad \underset{II}{\begin{array}{c} CHO \\ H \blacktriangleright\!\!\blacktriangleleft OH \\ CH_2OH \end{array}} \qquad \underset{III}{\begin{array}{c} CHO \\ H{-}C{-}OH \\ CH_2OH \end{array}}$$

(+)-Glyceraldehyde

The classical chemical methods for the assignment of relative configuration depend on the assumption that no change in configuration occurs so long as bonds to the asymmetric atom are not involved. The following series of reactions illustrates how the configurations of several compounds have been interrelated. The carbon atom corresponding to that in the reference substance is indicated by an asterisk.

```
        CHO                                      COOH                     COOH
         |              Oxidation                 |          HNO2          |
     H—C*—OH           ---------->           H—C*—OH      <--------    H—C*—OH
         |                                        |                        |
       CH2OH                                    CH2OH                    CH2NH2
  (+)-Glyceraldehyde                         (−)-Glyceric             (+)-Isoserine
         |                                      acid                       |
         | HCN                                                             | NaNO2
         v                                       COOH                      |  + 2 HBr
   CN            CN                               |                        v
   |             |          Hydrolysis       HO—C—H                      COOH
H—C—OH   +  HO—C—H             and           ---->  |                    |
H—C*—OH     H—C*—OH         oxidation         H—C*—OH                 H—C*—OH
   |             |                                |                        |
 CH2OH         CH2OH                            COOH                     CH2Br
   |                                         (−)-Tartaric                  |
   | Hydrolysis                                 acid                       | Zn-HCl
   |  and oxidation                               | HI                     |
   v                                              v                        v
  COOH                          CONH2            COOH                    COOH
   |          CH2NH2              |               |                        |
H—C—OH           |     NaOCl     CH2             CH2                   H—C*—OH
H—C*—OH      H—C*—OH  <-----  H—C*—OH  <----  H—C*—OH                     |
   |             |                |               |                       CH3
  COOH          COOH             COOH            COOH              (−)-Lactic acid
meso-Tartaric (−)-Isoserine                  (−)-Malic acid
   acid
```

In 1951 the *absolute* configuration of (+)-tartaric acid, as determined by a special type of X-ray analysis of its sodium rubidium salt, was found to be the same as that based on the arbitrarily assigned configuration of (+)-glyceraldehyde. Hence all configurations that have been assigned relative to (+)-glyceraldehyde are, conveniently, absolute configurations.

Various less certain procedures have been developed for arriving at the relative configuration of dissymmetric compounds that cannot be related by chemical interconversions because they would require the breaking of a bond attached to the asymmetric center. Among these may be mentioned (*1*) the **rule of shift,** which states that active compounds that undergo similar changes in structure show similar changes in optical activity; (*2*) **quasiracemic compound formation,** which depends on the fact that compounds with similarly constituted asymmetric centers of opposite configuration often undergo compound formation, analogous to compound formation in true racemates, which can be detected from melting point diagrams of mixtures; (*3*) **asymmetric syntheses** that show, when a reagent is added to a carbonyl group near an asymmetric center, a bias that depends on a steric effect owing to differences in size of the groups attached to the asymmetric center; and (*4*) **optical rotatory dispersion** (p. 300), which depends on the fact that if the close environment of an absorbing group in two different compounds is the same in conformation and configuration, the rotatory dispersion curves of the compounds will be very similar, and if the environment of one is the mirror image of that of the other, the curves will be of opposite sign. In addition, correlations can be made even though bonds to the asymmetric center are broken, provided it can be demonstrated that such steps used to establish the correlation take place with complete inversion as in S_N2 reactions (p. 100), or complete retention of configuration as in the Hofmann rearrangement of amides (p. 224).

Configurational Nomenclature

The examples given on page 311 show that there is no obvious relation between the configuration of a compound and its sign of rotation. Thus (+)-glyceraldehyde and (−)-glyceric acid have the same configuration but rotate polarized light in opposite directions. The sign of rotation for certain compounds is known to change with a change in temperature, concentration, solvent, acidity of the solution, or the presence of neutral salts. Accordingly some method for designating configuration must be used. Several systems of limited application have been devised, and they still are used in certain fields such as carbohydrates (pp. 318, 331), amino acids (p. 354), and steroids (p. 663). A general system, however, is desirable, and although none has been adopted officially, the Cahn-Ingold-Prelog system[14] has gained wide acceptance. The basis of the system is the following **sequence rule,** which establishes the priority of the groups attached to any asymmetric atom.

(*1*) Groups are arranged in the order of decreasing atomic number of the atoms bound to the asymmetric atom. For chlorobromoacetic acid, the resulting order is Br, Cl, COOH, H.

(*2*) If the relative priority of two groups cannot be decided by the first atoms, the atomic numbers of the next atoms to which they are attached are considered. Thus $—CH_2CH_3$ precedes $—CH_3$ and the order for 2-butanol is HO, C_2H_5, CH_3, H. Similarly the order for 2,3-dimethyl-3-ethylhexane is $CH(CH_3)_2$, $CH_2CH_2CH_3$, CH_2CH_3, CH_3, and for 1-bromo-3,3-dichloro-2-methylpropane it is CH_2Br, $CHCl_2$, CH_3, H.

(*3*) When a double or triple bond is encountered, the atom remote from the asymmetric center is counted twice or three times. Hence the aldehyde group, which may be considered as —C(H)(O)—O, precedes the hydroxymethyl group, —C(H)(O)—H, and the sequence for glyceraldehyde is HO, CHO, CH_2OH, H.[15]

If any two groups attached to an asymmetric carbon atom are interchanged, the enantiomorph results. If a second interchange is made, the original form is regained. To designate the configuration of an asymmetric carbon atom, the necessary double interchange is made to place the group of lowest priority at the bottom of the perspective or projection formula as illustrated for (+)-glyceraldehyde,

CHO / H◁ ▷OH / CH_2OH ≡ CHO / HO◁ ▷CH_2OH / H (*R*) or CHO / H—+—OH / CH_2OH ≡ CHO / HO—+—CH_2OH / H (*R*)

If now in following the sequence of groups from highest to lowest priority one moves in a clockwise direction, the configuration is designated as *R* (L. *rectus* right), whereas if one moves in a counterclockwise direction, the configuration is designated as *S* (L. *sinister* left). Thus (+)-glyceraldehyde is *R*-glyceraldehyde. For glyceric acid the sequence is HO, COOH, CH_2OH, H. Hence (+)-glyceric acid has the *S* configuration.

COOH / HO—+—H / CH_2OH ≡ COOH / $HOCH_2$—+—OH / H

S-Glyceric acid

---

[14] R. S. Cahn, C. K. Ingold, V. Prelog, *Experientia,* **12,** 81 (1956).

[15] Where these rules based on atomic number do not suffice to establish priority, easily applied subrules are available; for example, *cis* precedes *trans,* *R* precedes *S,* and the higher mass number precedes the lower. For compounds having molecular asymmetry owing to factors other than asymmetric atoms, the original article[14] should be consulted.

For (−)-tartaric acid the sequence of groups attached to each asymmetric carbon atom is HO, COOH, CHOHCOOH, H. Hence (−)-tartaric acid is *S,S*-tartaric acid.

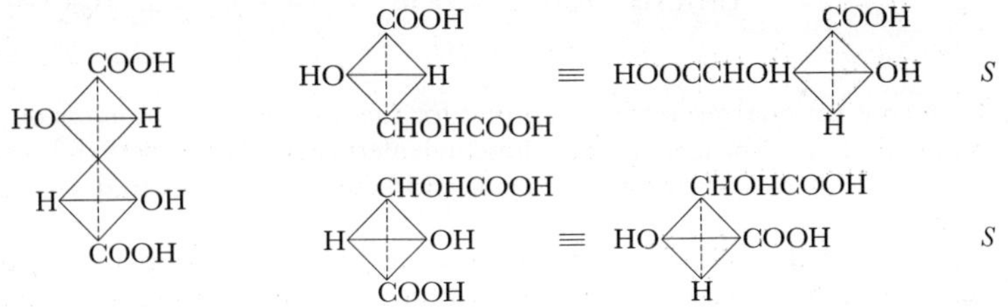

COOH
HO—C—H
H—C—OH
COOH

COOH
HO—C—H
CHOHCOOH
≡
COOH
HOOCCHOH—C—OH
H
S

CHOHCOOH
H—C—OH
COOH
≡
CHOHCOOH
HO—C—COOH
H
S

If a compound has two or more different asymmetric carbon atoms, the position of the asymmetric atom is indicated by a numeral preceding the configurational designation, as is demonstrated in the following example. Here *Y* is used to represent the group that constitutes the rest of the molecule attached to each asymmetric carbon atom.

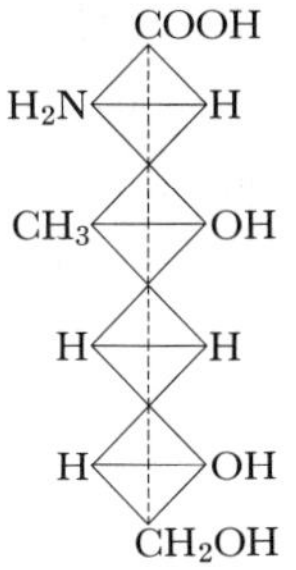

COOH
H_2N—C—H
CH_3—C—OH
H—C—H
H—C—OH
CH_2OH

H_2N, COOH, *Y*, H — COOH / *Y*—C—NH_2 / H — *S*

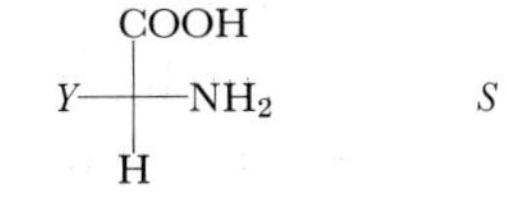

HO, $CHNH_2COOH$, *Y*, CH_3 — $CHNH_2COOH$ / HO—C—*Y* / CH_3 — *R*

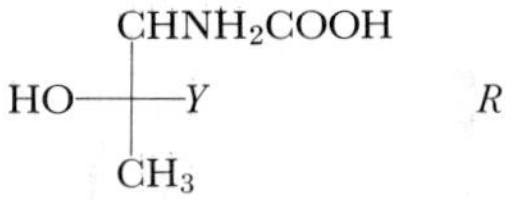

HO, CH_2OH, *Y*, H — *Y* / HO—C—CH_2OH / H — *S*

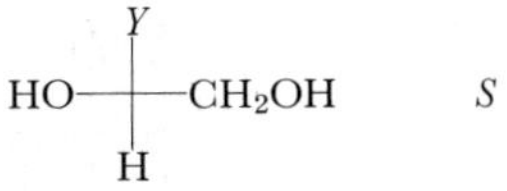

2*S*,3*R*,5*S*-2-Amino-3-methyl-3,5,6-trihydroxycaproic acid

PROBLEMS

17–*1*. Pair each of the terms (*a*) composition, (*b*) constitution, and (*c*) configuration, with the proper counterpart among the terms (*1*) structural formula, (*2*) space formula, and (*3*) molecular formula.

17–*2*. Pair each of the terms (*a*) optical activity, (*b*) double refraction, and (*c*) differential absorption of light vibrating in two perpendicular planes, with (*1*) calcite crystal, (*2*) rotation of the plane of plane-polarized light, or (*3*) Polaroid.

17–*3*. Predict the number and kind of stereoisomers (active, *meso,* or geometric) theoretically possible for each of the following compounds:
(*a*) $(CH_3)_2CHCH(NH_2)COOH$, (*b*) $C_2H_5CH(CH_3)CH_2OH$,
(*c*) $CH_3CHBrCH_2CHBrCH_3$, (*d*) $CH_2OH(CHOH)_4CHO$,
(*e*) $CH_2(NH_2)CH_2COOH$, (*f*) $CH_2ClCHClCH_2COOH$,
(*g*) $CH_3CH_2CHBrCHBrCH_3$, (*h*) $CH_3(CHOH)_2CH_3$,
(*i*) serine, (*j*) threonine, (*k*) methionine, (*l*) lysine, (*m*) cystine, (*n*) arginine, (*o*) hydroxyproline, (*p*) isoleucine, (*q*) $CH_2{=}CHCH_2COOH$,
(*r*) $CH_3(CH{=}CCH_3)_2(CH{=}CH)_2CH_3$, (*s*) linolenic acid, (*t*) ricinoleic acid. (See index to locate formulas of unfamiliar compounds.)

17–*4*. For each compound in the following series of reactions, predict the number and kind of stereoisomers possible.

(*a*) $CH_3OOCCH(CH_3)CHOHCH(CH_3)COOH \longrightarrow CH_3OOCCH(CH_3)CH{=}C(CH_3)COOH \longrightarrow$
$CH_3OOCCH(CH_3)CH_2CH(CH_3)COOH \longrightarrow CH_3OOCCH(CH_3)CH_2CH(CH_3)COOCH_3$

(*b*) $C_2H_5CH(CH_3)CH_2SCH_2CH_2COOH \longrightarrow C_2H_5CH(CH_3)CH_2{-}SO{-}CH_2CH_2COOH \longrightarrow$
$C_2H_5CH(CH_3)CH_2{-}SO{-}CH_2CHBrCOOH \longrightarrow C_2H_5CH(CH_3)CH_2{-}SO{-}CH{=}CHCOOH$

(*c*) $CH_3CHOHCH(CH_3)CHO \longrightarrow CH_3CHOHCH(CH_3)—CHOH(CN) \longrightarrow CH_3CHOHCH(CH_3)CH(CN)(NC_2H_5C_3H_7) \longrightarrow$

$CH_3CH{=}C(CH_3)CH(COOH)(NC_2H_5C_3H_7) \longrightarrow CH_3CH{=}C(CH_3)CH(COO^-)—N^+(CH_3)(C_2H_5)(C_3H_7)$

17–5. Give conventional projection formulas for the possible stereoisomers of (*a*) 2,3-butanediol, (*b*) 3-hexene, (*c*) α,β-dimethylvaleric acid, (*d*) 2-methyl-3-pentenoic acid, and (*e*) 2-methylpropionic acid, and label those forms that are optically active.

17–6. Pair each of the following terms with the appropriate example: (*a*) geometric isomerism, (*b*) enolization, (*c*) epimerization, (*d*) enantiomorphism, (*e*) racemization, (*f*) asymmetric synthesis, (*g*) diastereoisomers, (*h*) *meso* forms, (*i*) asymmetric decomposition;

(*1*) $CH_3CH_2CH_2CHO \rightleftharpoons CH_3CH_2CH{=}CHOH$

(*2*) $CH_3—C(C_2H_5)(H)—OH$ and $HO—C(C_2H_5)(H)—CH_3$ (*3*) $CH_3—C(C_2H_5)(H)—I \overset{I^-}{\rightleftharpoons} I—C(C_2H_5)(H)—CH_3$

(*4*) $CH_3(H)C{=}C(CH_3)(H)$ (cis) and $H(H_3C)C{=}C(CH_3)(H)$ (trans) (*5*) CH_3---C(Cl)(H)---C(H)(Cl)---COOH and CH_3---C(Cl)(H)---C(Cl)(H)---COOH

(*6*) CH_3---C(Cl)(H)---C(H)(Cl)---COOH $\overset{^-OH}{\rightleftharpoons}$ CH_3---C(Cl)(H)---C(Cl)(H)---COOH

(*7*) CH_3---C(Cl)(H)---C(Cl)(H)---C(Cl)(H)---CH_3 and CH_3---C(Cl)(H)---C(H)(Cl)---C(Cl)(H)---CH_3

(*8*) Ammonium racemate $\xrightarrow{\text{Penicillium}}$ (−)-Tartrate

(*9*) $CH_3COCOOH \xrightarrow{\text{Reductase of yeast}}$ (−)-Lactic acid

17–7. Which of the following stereochemical results, (*a*) racemization, (*b*) formation of optically active diastereoisomers, (*c*) optical resolution, (*d*) inversion of configuration, (*e*) retention of configuration, (*f*) formation of a racemic mixture, and (*g*) formation of a *meso* compound, occurs in the following operations: (*1*) methylmagnesium bromide reacts with optically active ethyl *s*-butyl ketone, (*2*) optically active ethyl *s*-butyl ketone is treated with alcoholic sodium hydroxide, (*3*) optically active *s*-butyl alcohol reacts with acetic anhydride to give an optically active acetate, (*4*) bromine addition to *cis*-3-hexene, (*5*) optically active 2-iodooctane reacts with sodium ethoxide, (*6*) a solution of optically inactive 2-chlorooctane and urea in methanol is allowed to crystallize, (*7*) permanganate oxidation of *cis*-3-hexene.

17–8. (*a*) When 5.678 g. of cane sugar is dissolved in water and brought to a total volume of 20 cc. at 20°, the rotation of the solution in a 10-cm. tube is 18.88 degrees. What is the specific rotation of cane sugar? (*b*) The observed rotation of an aqueous solution of cane sugar in a 2-dcm. tube is 10.75 degrees. What is the concentration of the sugar solution?

17–9. (*a*) A sample of pure active amyl alcohol having a density of 0.8 g. per cc. at 20° gives a rotation of 9.44 degrees in a 20-cm. tube. Calculate the specific rotation of the alcohol. (*b*) A fraction of fusel oil boiling between 125° and 135° and having a density of 0.8 g. per cc. at 20° gives a rotation of 3.56 degrees in a 4-dcm. tube. What per cent of active amyl alcohol is present?

17–*10*. Specify as *R* or *S* the configuration of each asymmetric carbon atom in the following compounds.

(*a*) COOH; $ClCH_2$—C—CH_3; $CONH_2$

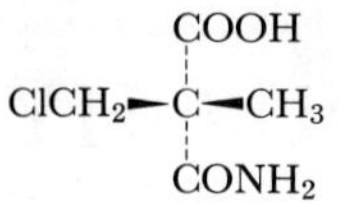

(*b*) OH; CH_3, C_2H_5; H

(*c*) CH_3 OH Cl COOH
HOOC----C----C----C----C----$COCH_3$
Br H Br OH
(*1*) (*2*) (*3*) (*4*)

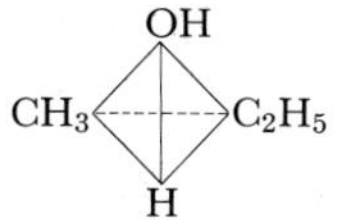

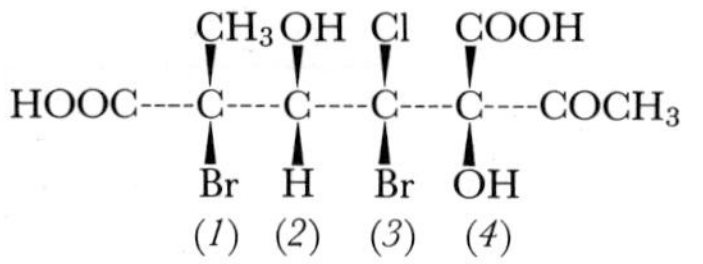

(*d*) C≡N
H, $CONH_2$ (*1*)
CH_3, OH (*2*)
C≡N

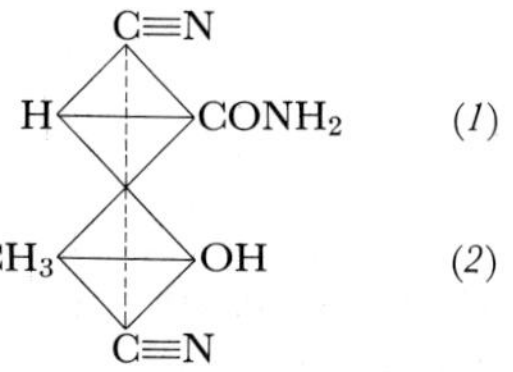

17–*11*. Give formulas showing the configurations of (*a*) 12*R*,12-hydroxy-*cis*-9-octadecenoic acid, (*b*) *S*-α-azidopropionic acid, (*c*) 2*R*,3*S*,4*R*,5*R*-2,3,4,5,6,-pentahydroxyhexanal.

CHAPTER EIGHTEEN

CARBOHYDRATES

Carbohydrates are polyhydroxy aldehydes or polyhydroxy ketones, or substances that yield such compounds on hydrolysis. They are distributed universally in plants and animals, and make up one of the three important classes of animal foods. The combustion of carbohydrates to carbon dioxide and water yields about 4 kcal. of energy per gram. The term *carbohydrate* came into use for these compounds because ordinarily the ratio of hydrogen to oxygen is 2 to 1; for example starch has the empirical formula $C_6H_{10}O_5$, glucose $C_6H_{12}O_6$, and maltose $C_{12}H_{22}O_{11}$. For some carbohydrates, however, this ratio does not hold; thus rhamnose has the molecular formula $C_6H_{12}O_5$, and deoxyribose $C_5H_{10}O_4$.

Nomenclature and Classification of Carbohydrates

The simpler carbohydrates commonly are called *sugars* or *saccharides* (L. *saccharum* sugar). The ending for sugars is *ose,* for example, arabinose, glucose, maltose. Frequently the generic term *glycose* (Gr. *glykys* sweet) is used from which is derived the prefix *glyco.* The generic terms are used when it is not desired to designate a particular sugar or derivative. The number of carbon atoms may be indicated by a prefix; thus a sugar having five carbon atoms is called a *pentose,* and one with six carbon atoms is called a *hexose.* Similarly the prefix may indicate whether the sugar contains an aldehyde group or a ketone group, giving rise to the terms *aldose* and *ketose.* Both the number of carbon atoms and the type of carbonyl group may be indicated by terms such as *aldopentose* and *ketohexose.*

Carbohydrates may be subdivided according to the following classification.

A. Monosaccharides. Carbohydrates that do not hydrolyze.

B. Oligosaccharides. Carbohydrates that yield a few molecules of monosaccharide on hydrolysis.

- *1.* Disaccharides. One molecule yields two molecules of monosaccharide on hydrolysis.
  - (*a*) Reducing disaccharides. Disaccharides that reduce Fehling solution.
  - (*b*) Nonreducing disaccharides. Disaccharides that do not reduce Fehling solution.
- *2.* Trisaccharides. One molecule yields three molecules of monosaccharides on hydrolysis.
- *3.* Tetra-, penta-, and hexasaccharides.

C. Polysaccharides. Carbohydrates that yield a large number of molecules of monosaccharides on hydrolysis.

- *1.* Homopolysaccharides. Polysaccharides that yield only one kind of sugar on hydrolysis.
- *2.* Heteropolysaccharides. Polysaccharides that yield more than one kind of sugar on hydrolysis.

MONOSACCHARIDES

Aldoses

The most important carbohydrate is **(+)-glucose** (Gr. *gleukos* must, sweet wine). It is called also *dextrose* or *grape sugar*. The dextrorotatory form is the only one that occurs naturally, and the word *glucose* without indication of sign of rotation always means (+)-glucose. Glucose is obtained most readily by the hydrolysis of starch or cellulose. It also may be isolated as one of the hydrolytic products of most oligosaccharides and of many other plant products known as glucosides. It occurs free, along with fructose and sucrose, in plant juices. Honey contains chiefly glucose and fructose. Glucose is present to the extent of about 0.1 per cent in the blood of normal mammals. Glucose in free or combined form probably is the most abundant organic compound.

Because glucose is typical of the aldoses, it is discussed here in considerable detail. Only the characteristic differences of other aldoses are presented later.

Constitution. Glucose has the molecular formula $C_6H_{12}O_6$. Its constitution may be arrived at from the following chemical behavior.

(*a*) Reduction with hydrogen iodide and phosphorus yields *n*-hexane. Therefore all of the carbon atoms must be linked consecutively without branching.

(*b*) Glucose reacts with hydroxylamine to form a monoxime, or adds one mole of hydrogen cyanide to form a cyanohydrin, indicating the presence of one carbonyl group.

(*c*) On mild oxidation, for example with bromine in an aqueous buffered solution, glucose yields the monobasic gluconic acid, $C_5H_{11}O_5COOH$ (p. 346). Since no carbon atoms are lost, the carbonyl group must be present as an aldehyde group, which can occupy only an end position of the chain.

(*d*) On reduction with sodium amalgam, two hydrogen atoms are added to give $C_6H_{14}O_6$, a compound known as *sorbitol* (p. 349). Sorbitol reacts with acetic anhydride to give a hexa-acetate. Accordingly the six oxygen atoms of sorbitol must be present as six hydroxyl groups. Compounds containing two hydroxyl groups on the same carbon atom are rare, and those that are known readily lose water. Sorbitol, however, does not dehydrate easily. Hence one hydroxyl group must be located on each of the six carbon atoms. Since one hydroxyl group was formed by reduction of the aldehyde group of glucose, the constitution of glucose can be represented by $HOCH_2(CHOH)_4CHO$. Most of the isomeric aldohexoses undergo the same reactions, which indicates the same constitution. Therefore they are stereoisomers of glucose.

It has been shown by similar methods that most aldopentoses, $C_5H_{10}O_5$, have the structure $HOCH_2(CHOH)_3CHO$. Similarly the aldotetroses have the structure $HOCH_2(CHOH)_2$-CHO, and the simplest compound grouped with the aldoses is the triose named *glycerose* or *glyceraldehyde,* $HOCH_2CHOHCHO$.

Configuration. The unbranched aldohexoses have four different asymmetric carbon atoms, and hence 16 optical isomers are possible, all of which are known. Of these 16 isomers, only two besides (+)-glucose are common, namely (+)-mannose and (+)-galactose. **(+)-Mannose** is one of the products of hydrolysis of a number of polysaccharides. It is obtained most readily by the hydrolysis of the vegetable ivory nut, which is the hard endosperm of the seed of the Taguà palm, *Phyletephas macrocarpa.* Vegetable ivory is used for the manufacture of buttons. **(+)-Galactose** is formed along with (+)-glucose by the hydrolysis of the disaccharide *lactose* or *milk sugar* and is one of the products of hydrolysis of several polysaccharides. The following configurations are assigned to the three common aldohexoses.

```
    CHO                 ¹CHO
H       OH          H—²C—OH
HO      H    or    HO—³C—H
H       OH          H—⁴C—OH
H       OH          H—⁵C—OH
    CH2OH               ⁶CH2OH
```

(+)-Glucose

```
    CHO                  CHO
HO      H           HO—C—H
HO      H    or     HO—C—H
H       OH           H—C—OH
H       OH           H—C—OH
    CH2OH                CH2OH
```

(+)-Mannose

```
    CHO                  CHO
H       OH           H—C—OH
HO      H    or     HO—C—H
HO      H           HO—C—H
H       OH           H—C—OH
    CH2OH                CH2OH
```

(+)-Galactose

Eight aldopentoses that have an unbranched carbon chain are possible and all are known. **(+)-Arabinose** and **(+)-xylose** may be obtained by the hydrolysis of a wide variety of plant polysaccharides. Corn cobs, straw, oat hulls, or cottonseed hulls yield 8 to 12 per cent of (+)-xylose. (+)-Arabinose is obtained by the hydrolysis of many plant gums (p. 348). **(−)-Ribose** is important because it is one of the products of hydrolysis of one class of nucleic acids (p. 530). **(+)-Apiose** (L. *apium* parsley), a branched-chain sugar, is one of the hydrolytic products of *apiin,* a glycoside that occurs in parsley, and also is a product of the hydrolysis of the leaves and fibers of Australian pond weed (*Posidonia australis*).

```
    CHO            CHO            CHO                CHO
 H—C—OH         H—C—OH         H—C—OH             H—C—OH
HO—C—H         HO—C—H          H—C—OH      HOCH2—C—OH
HO—C—H          H—C—OH         H—C—OH               CH2OH
   CH2OH          CH2OH          CH2OH
```

(+)-Arabinose   (+)-Xylose   (−)-Ribose   (+)-Apiose

With the application of paper and column chromatography, many new sugars have been and are being identified and isolated from natural sources. Aldoheptoses have been obtained by the hydrolysis of bacterial polysaccharides, and other sources have yielded numerous branched-chain sugars, deoxy sugars (p. 319), higher ketoses (p. 333), and amino sugars (p. 346). Aldoheptoses, octoses, nonoses, and decoses have been synthesized.

Configurational Nomenclature of Carbohydrates. It has been noted (p. 312) that several systems for the designation of configuration were used in certain fields long before a general system was devised. These limited systems are strongly entrenched and have certain advantages for their particular purpose. Hence they continue to be used. One of the first of these systems was that originated by Emil Fischer[1] for carbohydrates. Figure 18–*1* indicates that half of the aldoses may be considered as having D(+)-glyceraldehyde (p. 310) as a parent compound. Similarly their enantiomorphs may be considered as having L(−)-glyceraldehyde as a parent compound. Sugars for which the highest-numbered asymmetric carbon atom has the same configuration as D(+)-glyceraldehyde (hydroxyl group on the right in the Fischer projection formula) are said to belong to the **D** (*dee*) **family;** those having the enantiomorphic configuration belong to the **L** (*ell*) **family.** This system specifies the configuration of only the highest-numbered asymmetric carbon atom. The common name specifies the configuration of the remaining carbon atoms, but either the configuration of

[1] Emil Fischer (1852–1919), professor of organic chemistry at the University of Berlin, and the outstanding director of organic chemical research. He and his co-workers did monumental work in the fields of carbohydrates, amino acids and proteins, and purines, and made important contributions in numerous other fields such as enzymes, stereochemistry, triphenylmethane dyes, hydrazines, and indoles. He was the second recipient of the Nobel Prize in Chemistry in 1902, the first being van't Hoff in 1901.

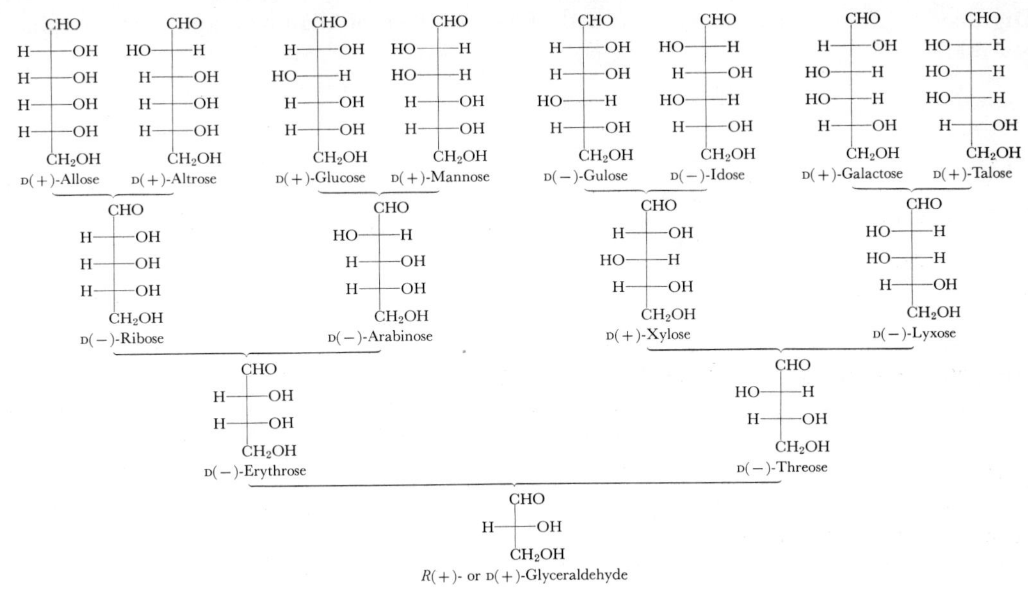

Figure 18–*1*. Configurations of the D family of aldoses.

each sugar must be memorized or one must refer to a chart such as that in Fig. 18–*1*. According to the general system (p. 312), the configurations of the individual carbon atoms turn out to be *R* if in the projection formula the hydroxyl group is on the right, and *S* if it is on the left. Thus (+)-glucose may be called 2*R*,3*S*,4*R*,5*R*-2,3,4,5,6-pentahydroxyhexanal or 2*R*,3*S*,4*R*,5*R*-aldohexose. Clearly it is simpler to say D-glucose and for this reason the older system has been retained by carbohydrate chemists.

Deoxyaldoses

Some naturally occurring sugars have a hydrogen atom in place of one or more of the hydroxyl groups. In naming these compounds the prefix *deoxy* is used with the name of the oxygen analog to indicate the lack of oxygen, and the class as a whole is known as the **deoxy sugars. L(+)-Rhamnose** (*6-deoxy-L-mannose*) is a hydrolytic product of many glycosides and the most common naturally occurring deoxy sugar. **L(−)-Fucose** (*6-deoxy-L-galactose*) is one of the products of the hydrolysis of the cell walls of marine algae, of gum tragacanth (p. 348), and of certain carbohydrates of animal origin (p. 365).

L(+)-Rhamnose	L(−)-Fucose
CHO	CHO
H—C—OH	HO—C—H
H—C—OH	H—C—OH
HO—C—H	H—C—OH
HO—C—H	HO—C—H
CH_3	CH_3

Rhamnose and fucose sometimes are called *methylpentoses,* but this term may cause confusion with the methylated sugars (p. 329).

2-Deoxy-D-ribose is a hydrolytic product of deoxyribonucleic acids (*DNA*), which are present in chromosomes and are the probable carriers of genetic information (p. 532).

Digitoxose (*2,6-dideoxy-D-allose*) is a hydrolytic product of digitoxin and other cardiac glycosides.

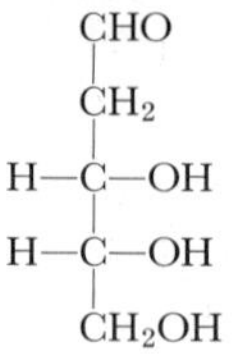

```
     CHO                      CHO
     |                        |
     CH2                      CH2
     |                        |
 H—C—OH                   H—C—OH
     |                        |
 H—C—OH                   H—C—OH
     |                        |
     CH2OH                H—C—OH
                              |
                              CH3
 2-Deoxy-D-ribose         2,6-Dideoxy-D-allose
                          (digitoxose)
```

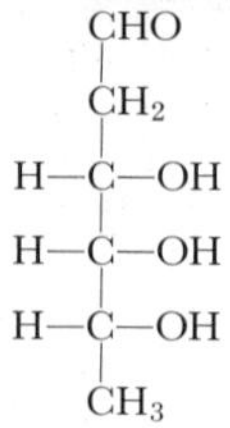

Some Special Reactions of the Aldoses

The Action of Alkalies. Sugars are fairly stable to acids at room temperature, but in aqueous alkaline solution they undergo numerous reactions under very mild conditions to give complex mixtures of products. These reactions can for convenience be divided into (*1*) isomerizations, (*2*) fragmentations, and (*3*) intramolecular oxidations and reductions. The various products thus formed also can undergo further condensation reactions.

Isomerizations result from the enolizing action of dilute alkali at room temperature (p. 196). Thus if glucose dissolved in aqueous calcium hydroxide solution is allowed to stand for several days, a mixture is formed from which has been isolated 63.5 per cent glucose, 21 per cent fructose, 2.5 per cent mannose, 10 per cent unfermentable ketoses, and 3 per cent of other substances. Other aldoses and ketoses behave similarly. The reaction is called the *Lobry de Bruyn-Alberda van Ekenstein* transformation after its discoverers and has been used for the preparation of ketoses.

Sugars have hydroxyl groups β to a carbonyl group, and since aldol additions are reversible (p. 201), sugars in the presence of alkali are subject to **fragmentation reactions** resulting from scission between the α and β carbon atoms. This behavior frequently is referred to as *degradation.*

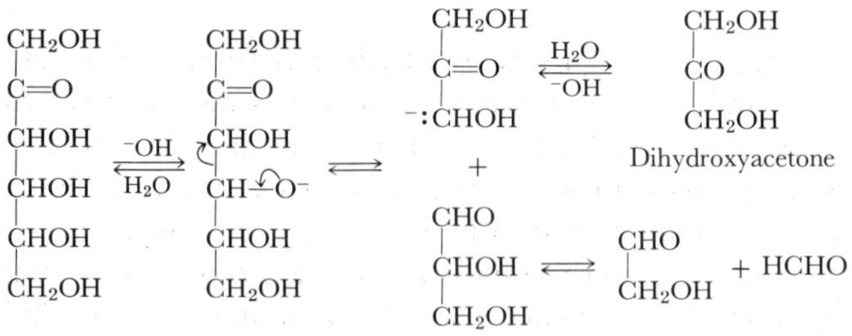

```
                                       CH2OH                CH2OH
CH2OH          CH2OH                   |        H2O         |
|              |                       C=O     <====>       CO
C=O            C=O                     |        -OH         |
|              |                     -:CHOH                 CH2OH
CHOH    -OH    CHOH                                     Dihydroxyacetone
|      <====>  |             <==>        +
CHOH    H2O    CH—O-
|              |                       CHO
CHOH           CHOH                    |                CHO
|              |                       CHOH    <==>     |        + HCHO
CH2OH          CH2OH                   |                CH2OH
                                       CH2OH
                                   Glyceraldehyde
```

Clearly these fragmentation products can recombine by aldol additions to give a complex mixture of products.

Lactic acid, in the form of lactate ion, is a common product of the action of alkali on carbohydrates. Presumably it arises by a rearrangement of the glyceraldehyde formed by fragmentation.

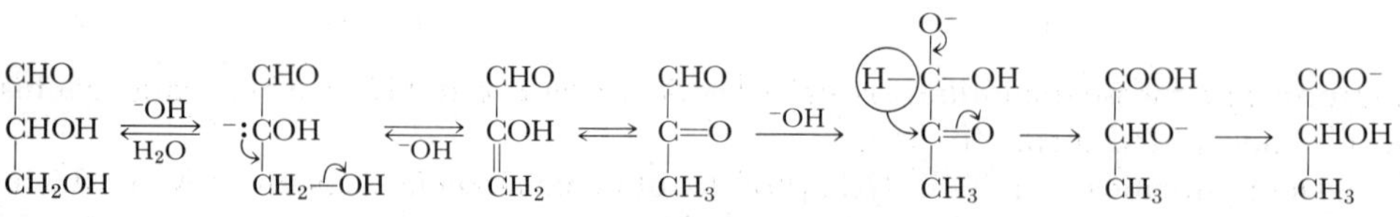

```
                                                             O-
                                                             |
CHO            CHO              CHO        CHO          (H)—C—OH     COOH        COO-
|       -OH    |                |          |      -OH        |        |           |
CHOH   <====> -:COH    <====>   COH  <==>  C=O   ---->       C=O ---> CHO-  --->  CHOH
|       H2O    |         -OH    ||         |                 |        |           |
CH2OH          CH2—OH           CH2        CH3               CH3      CH3         CH3
```

The first half of the reaction is the β elimination of a hydroxyl group (p. 203), and the last half is an **intramolecular oxidation and reduction.** It is a type of Cannizzaro reaction involving a hydride ion transfer (p. 209), which commonly is referred to as a benzilic acid rearrangement (p. 465). By similar mechanisms, acidic products known as *saccharinic acids* are formed from sugars in the presence of alkali.

Oxidation by Cupric Ion in Alkaline Solution. Although the salts of many heavy metals such as copper, silver, mercury, and bismuth are reduced by alkaline sugar solutions, reagents containing copper usually are used for analytical purposes. Cupric hydroxide oxidizes carbohydrates, but its slight solubility in dilute alkali permits only a slow reaction. The rate of oxidation is increased greatly if the copper is kept in solution by the formation of a complex salt with tartrate ion (*Fehling solution*) or citrate ion (*Benedict solution*) (p. 636). The former uses sodium hydroxide for alkaline reaction, whereas the latter uses sodium carbonate. Benedict solution is the more stable and is not affected by substances such as creatine (p. 290) and uric acid (p. 528). Hence it is preferred for detecting and estimating glucose in urine.

Evidence for reduction is the formation of red cuprous oxide, which precipitates because the cuprous ion does not form complexes with tartrates or citrates. If protective colloids are present, as in urine samples, the color of the precipitate may vary from yellow to red depending on the state of subdivision of the particles. In the presence of an excess of carbohydrate, some of the cuprous oxide may be reduced to metallic copper.

The equation for the oxidation of an aldehyde group to a carboxyl group requires a ratio of two moles of cupric salt per mole of aldehyde.

$$\mathrm{RCHO} + 2\,\mathrm{Cu(OH)_2} \longrightarrow \mathrm{RCOOH} + \mathrm{Cu_2O} + 2\,\mathrm{H_2O}$$

Actually one mole of glucose reduces between five and six moles of cupric salt depending on the conditions of the reaction. Moreover ketoses reduce just as well as aldoses, although simple ketones do not reduce Fehling solution. The explanation lies in the reaction of the sugars with alkali, which not only interconverts aldoses and ketoses but gives degradation products that have reducing properties (p. 320). Furthermore, some of the excess reduction may be caused by the CHOH groups adjacent to the carbonyl group, since even simple compounds containing this grouping are oxidized by alkaline cupric solutions to give dicarbonyl compounds (p. 464).

$$\mathrm{R{-}\underset{\underset{\displaystyle O}{\|}}{C}{-}\underset{\underset{\displaystyle OH}{|}}{CH}{-}R} + 2\,\mathrm{Cu(OH)_2} \longrightarrow \mathrm{R{-}\underset{\underset{\displaystyle O}{\|}}{C}{-}\underset{\underset{\displaystyle O}{\|}}{C}{-}R} + \mathrm{Cu_2O} + 3\,\mathrm{H_2O}$$

Despite the complexity of the reaction, oxidation can be used successfully for the quantitative estimation of sugars by standardizing conditions rigidly and by using empirically determined tables relating the amount of sugar to the amount of cupric ion reduced.

Oxidation to Glycaric Acids. When aldoses are oxidized by strong nitric acid, both the aldehyde group and the primary alcohol group are converted to carboxyl groups. The dicarboxylic acids are known as **glycaric acids** (older name *saccharic acids*). By thus making both ends of the sugar molecule alike, it is possible to determine in which aldoses the configurations of the top and bottom pair of asymmetric atoms are enantiomorphic, since the glycaric acids from such molecules have a plane of symmetry and are *meso* forms and inactive. Thus (+)- or (−)-galactose gives the inactive galactaric acid (*mucic acid*), whereas (+)- and (−)-mannose give active mannaric acids (*mannosaccharic acids*).

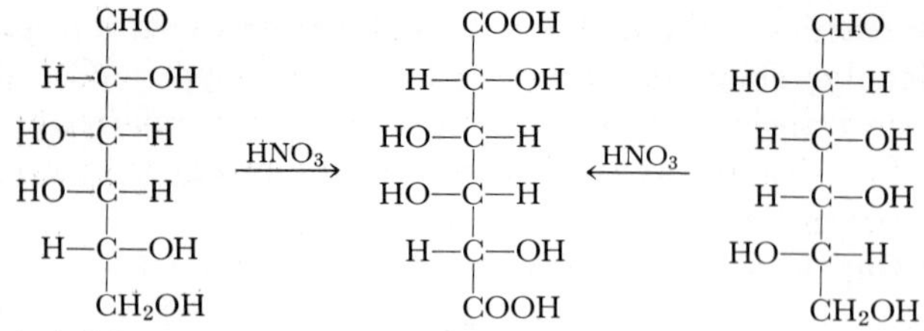

$$\begin{array}{c} CHO \\ H-C-OH \\ HO-C-H \\ HO-C-H \\ H-C-OH \\ CH_2OH \\ (+)\text{-Galactose} \end{array} \xrightarrow{HNO_3} \begin{array}{c} COOH \\ H-C-OH \\ HO-C-H \\ HO-C-H \\ H-C-OH \\ COOH \\ \text{Galactaric acid} \\ (\textit{mucic acid})\ \text{Inactive} \end{array} \xleftarrow{HNO_3} \begin{array}{c} CHO \\ HO-C-H \\ H-C-OH \\ H-C-OH \\ HO-C-H \\ CH_2OH \\ (-)\text{-Galactose} \end{array}$$

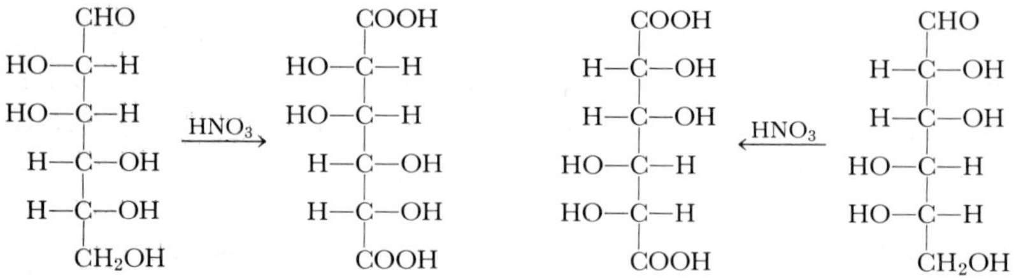

$$\begin{array}{c} CHO \\ HO-C-H \\ HO-C-H \\ H-C-OH \\ H-C-OH \\ CH_2OH \\ (+)\text{-Mannose} \end{array} \xrightarrow{HNO_3} \begin{array}{c} COOH \\ HO-C-H \\ HO-C-H \\ H-C-OH \\ H-C-OH \\ COOH \end{array} \quad \begin{array}{c} COOH \\ H-C-OH \\ H-C-OH \\ HO-C-H \\ HO-C-H \\ COOH \end{array} \xleftarrow{HNO_3} \begin{array}{c} CHO \\ H-C-OH \\ H-C-OH \\ HO-C-H \\ HO-C-H \\ CH_2OH \\ (-)\text{-Mannose} \end{array}$$

Mannaric acids
(*mannosaccharic acids*) Active

Formation of Osazones. Fischer, in his study of the reaction of phenylhydrazine with aldehydes and ketones, found that reducing sugars yield derivatives containing two phenylhydrazine residues instead of one. A yellow crystalline product known as a **phenylosazone** separates from the hot aqueous solution. The other products of the reaction are aniline (*phenylamine*) and ammonia. This reaction is characteristic of the grouping RCOCHOHR.

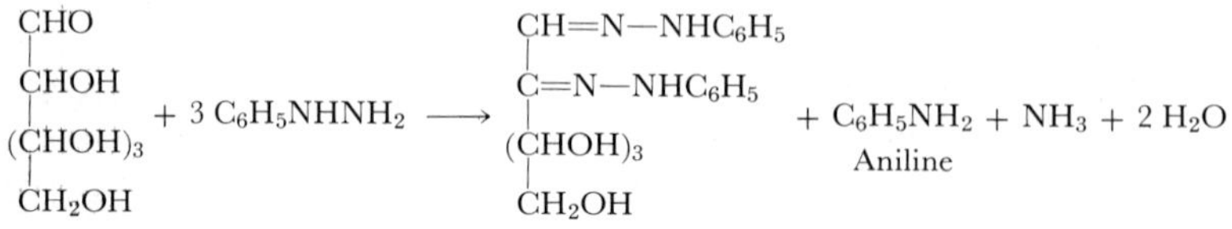

$$\begin{array}{l} CHO \\ CHOH \\ (CHOH)_3 \\ CH_2OH \end{array} + 3\ C_6H_5NHNH_2 \longrightarrow \begin{array}{l} CH=N-NHC_6H_5 \\ C=N-NHC_6H_5 \\ (CHOH)_3 \\ CH_2OH \\ \text{A phenylosazone} \end{array} + \underset{\text{Aniline}}{C_6H_5NH_2} + NH_3 + 2\ H_2O$$

Because sugars are highly soluble in water and tend to form viscous syrups, many crystallize with difficulty. Hence the easily formed crystalline osazones were invaluable in the early work on the sugars.

The course of osazone formation is not known with certainty. Of the various mechanisms that have been proposed, the following involving an enamine and an imino keto intermediate appears at present to be the most satisfactory.

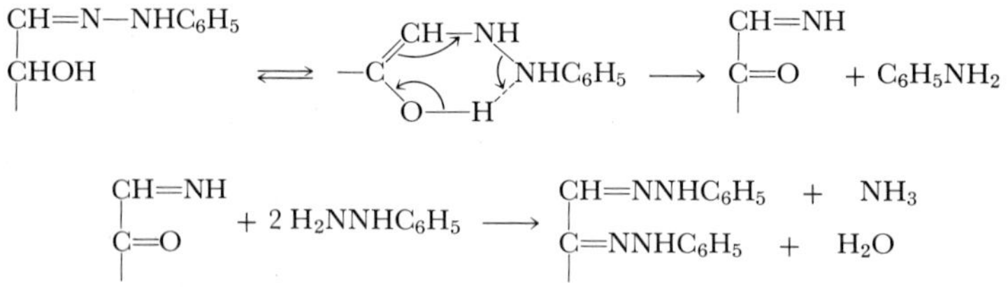

$$\begin{array}{l} CH=N-NHC_6H_5 \\ CHOH \\ | \end{array} \rightleftharpoons \begin{array}{c} CH-NH \\ -C \quad\quad NHC_6H_5 \\ O-H \end{array} \longrightarrow \begin{array}{l} CH=NH \\ C=O \\ | \end{array} + C_6H_5NH_2$$

$$\begin{array}{l} CH=NH \\ C=O \\ | \end{array} + 2\ H_2NNHC_6H_5 \longrightarrow \begin{array}{l} CH=NNHC_6H_5 \quad + \quad NH_3 \\ C=NNHC_6H_5 \quad + \quad H_2O \\ | \end{array}$$

It accommodates the known facts that glucose containing tritium in the 1-position forms an osazone without loss of tritium, and that when the hydrazone having N^{15} bound to C-1 is converted to the osazone with ordinary phenylhydrazine, most of the N^{15} appears in the ammonia.

The reason that α-hydroxy carbonyl compounds or their imino derivatives are oxidized more easily than simple alcohols, that is, that they are strong enough reducing agents to reduce cupric ion or the N—N bond, is that conjugation in the 1,2 diketone or its imino derivatives makes them more stable than simple carbonyl compounds. Reaction with phenylhydrazine stops at the osazone stage, apparently because cyclic hydrogen bonding between the phenylhydrazine residues permits resonance stabilization in addition to that provided by simple conjugation.

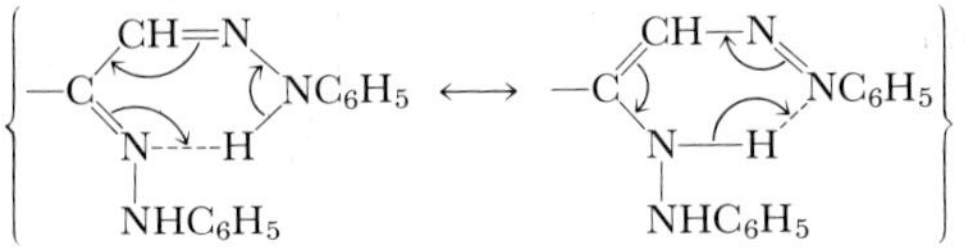

$$\left\{ -C\begin{matrix} CH{=}N \\ \\ N{\cdots}H \\ | \\ NHC_6H_5 \end{matrix} NC_6H_5 \longleftrightarrow -C\begin{matrix} CH{-}N \\ \\ N{-}H \\ | \\ NHC_6H_5 \end{matrix} NC_6H_5 \right\}$$

Osazone stabilization

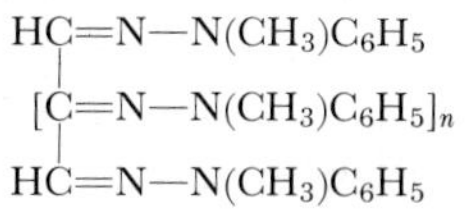

$$\begin{array}{l} HC{=}N{-}N(CH_3)C_6H_5 \\ [C{=}N{-}N(CH_3)C_6H_5]_n \\ HC{=}N{-}N(CH_3)C_6H_5 \end{array}$$

A methylphenylalkazone

This view is supported by the behavior of 1-methyl-1-phenylhydrazine, $C_6H_5N(CH_3)NH_2$, which cannot form a cyclic hydrogen-bonded structure and which oxidizes sugars beyond the osazone stage. Thus all of the hydroxyl groups of pentoses are oxidized with the formation of a compound that has been called an *alkazone.*

When osazones are formed, the asymmetry of carbon atom 2 is destroyed. Therefore sugars that differ only in the configuration of that carbon atom, such as glucose and mannose, yield the same osazone.

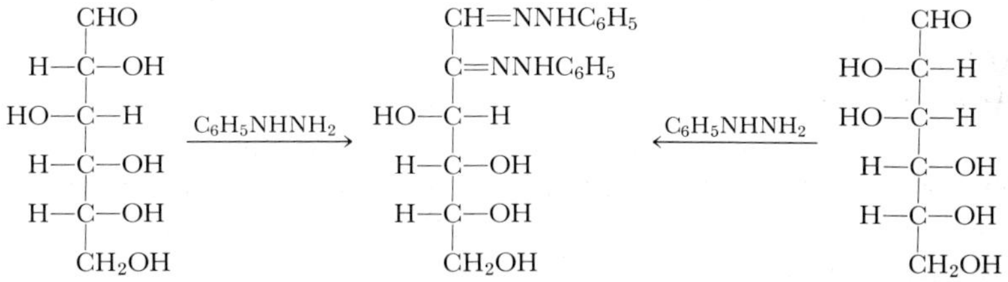

$$\begin{array}{c} CHO \\ H{-}C{-}OH \\ HO{-}C{-}H \\ H{-}C{-}OH \\ H{-}C{-}OH \\ CH_2OH \end{array} \xrightarrow{C_6H_5NHNH_2} \begin{array}{c} CH{=}NNHC_6H_5 \\ C{=}NNHC_6H_5 \\ HO{-}C{-}H \\ H{-}C{-}OH \\ H{-}C{-}OH \\ CH_2OH \end{array} \xleftarrow{C_6H_5NHNH_2} \begin{array}{c} CHO \\ HO{-}C{-}H \\ HO{-}C{-}H \\ H{-}C{-}OH \\ H{-}C{-}OH \\ CH_2OH \end{array}$$

D-Glucose — D-Glucose phenylosazone (yellow precipitate) — D-Mannose

An intermediate in this reaction is the phenylhydrazone in which only the carbonyl group has reacted. Although glucose phenylhydrazone is soluble in hot water, mannose phenylhydrazone is insoluble and precipitates as a white solid from aqueous solution even at room temperature. This difference in behavior is used to distinguish between glucose and mannose.

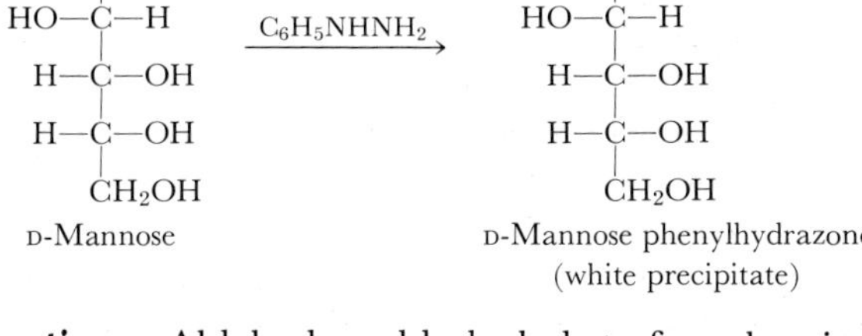

$$\begin{array}{c} CHO \\ HO{-}C{-}H \\ HO{-}C{-}H \\ H{-}C{-}OH \\ H{-}C{-}OH \\ CH_2OH \end{array} \xrightarrow{C_6H_5NHNH_2} \begin{array}{c} CH{=}NNHC_6H_5 \\ HO{-}C{-}H \\ HO{-}C{-}H \\ H{-}C{-}OH \\ H{-}C{-}OH \\ CH_2OH \end{array}$$

D-Mannose — D-Mannose phenylhydrazone (white precipitate)

Glycoside Formation. Aldehydes add alcohols to form hemiacetals (p. 199). In the presence of acid catalysts, hemiacetals react with alcohols to give acetals.

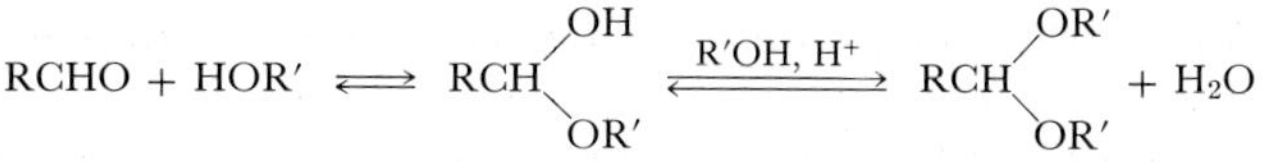

$$RCHO + HOR' \rightleftharpoons RCH\begin{matrix} OH \\ OR' \end{matrix} \xrightleftharpoons{R'OH,\ H^+} RCH\begin{matrix} OR' \\ OR' \end{matrix} + H_2O$$

Hemiacetal — Acetal

When Fischer attempted to prepare an acetal from glucose, methanol, and hydrogen chloride, definite crystalline products were obtained, but analysis showed that although a molecule of water had been eliminated, only one methyl group had been introduced. Furthermore two isomeric products were obtained. These compounds no longer reduced Fehling solution nor formed osazones. Moreover they behaved like acetals in that they were hydrolyzed readily in acid solutions but were stable to alkali. This behavior is explainable if it is assumed that the aldehyde group of the sugar first reacts with a hydroxyl group in the chain to give an internal cyclic hemiacetal. The hemiacetal then reacts with methanol to form the acetal. Since a new asymmetric carbon atom is produced in this process, two diastereoisomers are formed. The crystalline methyl glycosides have been shown to have a six-membered ring, indicating that hemiacetal formation takes place between the aldehyde group and the hydroxyl group on C-5.

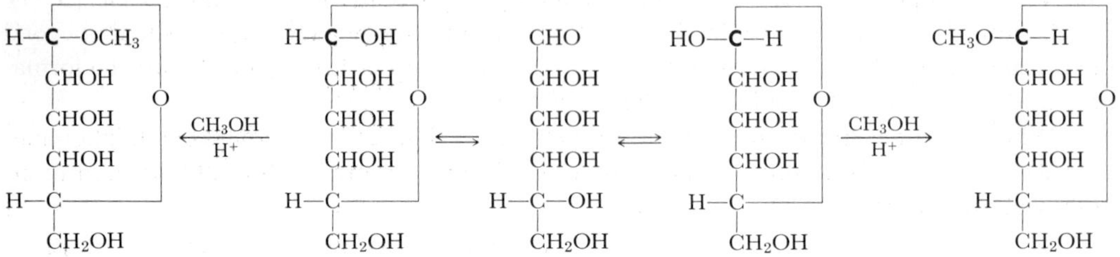

Methyl α-D-glycoside								Methyl β-D-glycoside
H—**C**—OCH_3 (ring O)		H—**C**—OH (ring O)		CHO		HO—**C**—H (ring O)		CH_3O—**C**—H (ring O)
CHOH		CHOH		CHOH		CHOH		CHOH
CHOH	$\xleftarrow[H^+]{CH_3OH}$	CHOH	⇌	CHOH	⇌	CHOH	$\xrightarrow[H^+]{CH_3OH}$	CHOH
CHOH		CHOH		CHOH		CHOH		CHOH
H—C—		H—C—		H—C—OH		H—C—		H—C—
CH_2OH		CH_2OH		CH_2OH		CH_2OH		CH_2OH

These acetals are called *glycosides,* and the two forms are designated α and β. Isomers of this type are known as *anomers,* and the carbon atom responsible for the existence of anomers is known as the *anomeric* (Gr. *ano* above) carbon atom. Hereafter the anomeric carbon atom is indicated by boldface *sans serif* type. It can be distinguished readily from the other carbon atoms by the fact that it is united to two oxygen atoms. For glycosides the anomeric carbon atom also is known as the *glycosidic* carbon atom. The designation α is given to the form in which the hydroxyl or substituted hydroxyl group is on the right in the projection formulas for members of the D family and on the left for members of the L family. Hence when indicating the configuration of the anomeric carbon atom, it is necessary to indicate the family of the sugar as well.

Mutarotation. Many optically active compounds give solutions, the rotation of which changes with time. This phenomenon, which is called *mutarotation* (L. *mutare* to change), must result from some structural change in the molecule, which may or may not involve an asymmetric center. The mutarotation of glucose solutions was reported first in 1846. By 1895 readily interconvertible isomeric modifications of lactose and of glucose had been isolated. Three such forms are obtainable from glucose. The form designated α crystallizes as the hydrate from 70 per cent alcohol below 30°. Its freshly prepared aqueous solutions have a specific rotation of +112°. The β form crystallizes from aqueous solutions evaporated at temperatures above 98°. It has a specific rotation of +18.7°. The third form is obtained by adding alcohol to a concentrated aqueous solution and has a rotation of +52.7°. Only α- and β-glucose mutarotate, however, and each finally attains the value +52.7°. The third form therefore is nothing more than the equilibrium mixture of the α and β forms.

The existence of the two forms of glucose and their mutarotation is explainable if it is assumed that they have cyclic hemiacetal structures analogous to the acetal structures postulated for the methyl α- and β-glycosides. It usually is assumed that their interconversion takes place by way of the open-chain form.

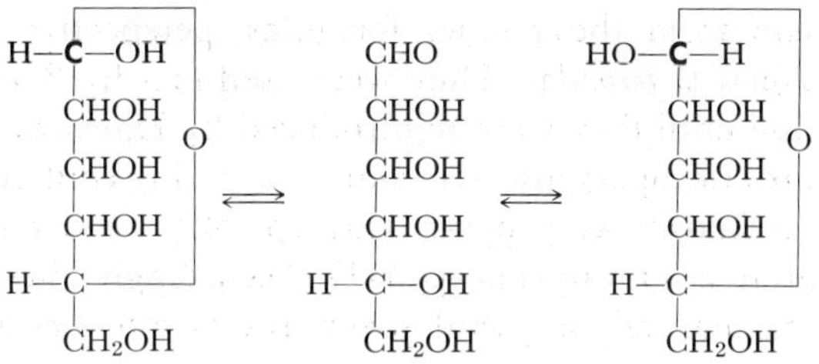

H—C—OH ⇌ CHO ⇌ HO—C—H
CHOH CHOH CHOH
CHOH CHOH CHOH
CHOH CHOH CHOH
H—C— H—C—OH H—C—
CH_2OH CH_2OH CH_2OH

α-D-Glycose β-D-Glycose

Since the mutarotation of sugars is caused merely by an aldehyde-hemiacetal equilibrium, it is catalyzed by both acids and bases. The detailed mechanism of this type of reaction has been discussed (p. 199). That the equilibrium rotation of glucose is not the average of the rotations of α- and β-glucose is not surprising, since they are diastereoisomers and not enantiomorphs, and the equilibrium composition need not be a 50–50 mixture. The anomeric carbon atom of the hemiacetal structure frequently is called the *reducing* carbon atom, because it is involved in the reduction of Fehling solution. The anomeric hydroxyl group frequently is called the hemiacetal hydroxyl group.

Representation of Ring Structures and Conformations

The Fischer projection formulas used thus far are convenient for indicating the structure of the sugars and the relative configurations of the asymmetric carbon atoms but not the space relationships within the molecules. With a model, the open-chain formula for glucose (*I*) can be made to assume a coiled arrangement (*II*).

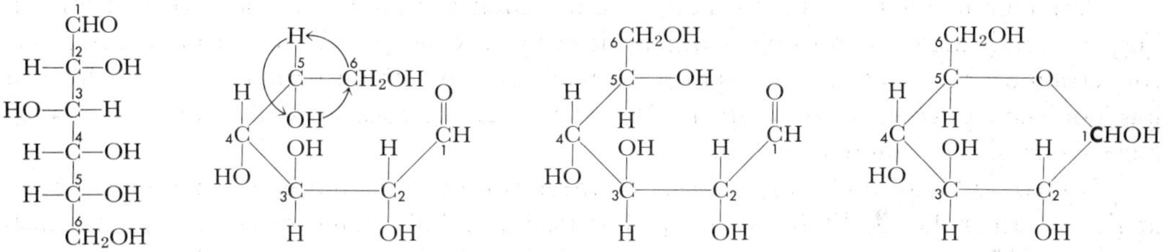

CHO / H—C—OH / HO—C—H / H—C—OH / H—C—OH / CH_2OH

I *II* *III* *IV*

For the oxygen atom of the 5-hydroxyl group to become a member of the ring in the cyclic hemiacetal form (*IV*), however, rotation must take place about the bond between the fourth and fifth carbon atoms (*III*), with the result that the hydrogen atoms on the fourth and fifth carbon atoms are on opposite sides of the ring instead of on the same side as in the projection representation (*I*). Formula *IV* is not a projection formula but is a **perspective formula.** The ring may be considered as perpendicular to the plane of the paper with the lower portion in front and the groups extending above and below the plane of the ring.

Formula *IV* does not indicate the configuration of the reducing carbon atom. In the following formulas these configurations are given. Except for the anomeric carbon atoms the ring carbon atoms are omitted. The heavy bonds are used as an aid in visualizing the ring as perpendicular to the plane of the paper with the heavy bonds in front.

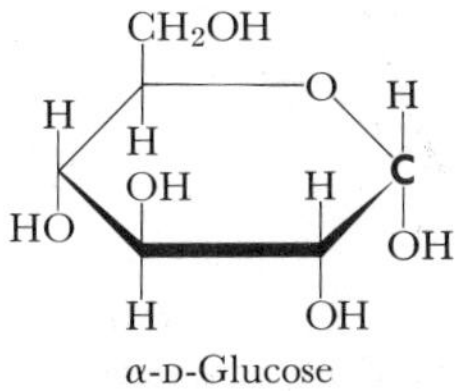

α-D-Glucose 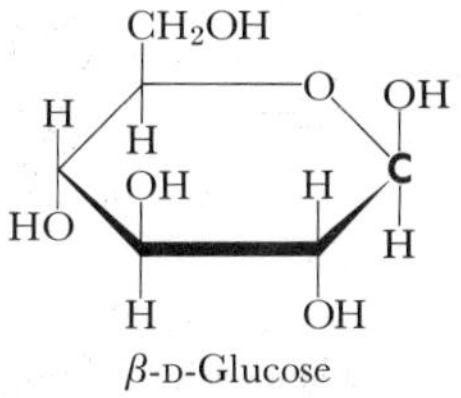 β-D-Glucose

To distinguish them from the Fischer formulas, perspective formulas of this type frequently are called *Haworth formulas.* They were used first by Boeseken in 1913 but did not come into general use until they were reintroduced by Haworth (p. 329) in 1926. Five-membered ring structures in sugars also are common. A ring containing five carbon atoms and an oxygen atom is known as a pyran ring (p. 521), and one having four carbon atoms and an oxygen atom is a furan ring (p. 511). Accordingly Haworth designated sugars having six-membered hemiacetal or acetal structures as *pyranoses* and those having five-membered rings as *furanoses*.

Since about 1950 the importance of **conformation** in the ring structures has been emphasized. If the pyranose ring were a regular planar hexagon, the internal bond angles would be 120°, which is considerably larger than the natural bond angle of 109.5°, and the ring structure would be less stable than the open-chain structure. It is not necessary for the ring atoms to be planar, however. If the ring is allowed to buckle, two forms can be constructed from models in which all of the atoms can have their normal bond angles. In one form (Fig. 18–2*a,b*), alternate atoms occupy two separate planes to give a rigid structure that commonly is called the *chair form* because of the fancied resemblance to a chair. In the extreme conformation of the other form, two opposite corners of the hexagon are bent up, leaving the other four atoms in a plane. It is called the *boat form* because of the resemblance to a boat (Fig. 18–2*c,d*). This form is flexible, however, and it may pass through an infinite number of conformations with any opposite points of the hexagon being turned up or down in the extreme forms. During these changes only rotation about the single bonds takes place with no change in the bond angles occurring. The stage halfway between one boat conformation and another is called the *twist* or *skew-boat* form (Fig.18–*2e*).

Although it is not possible to change from a chair form to a boat form without spreading the bond angles, bond angles are sufficiently flexible (p. 56) to permit ready interconversion of chair and boat forms at room temperature without breaking bonds. Hence it has not been possible to separate the two forms as stereoisomers, the difference being merely one of conformation (p. 56).

Insofar as bond angles are concerned there is no strain, that is, no changes in bond angles, in either the rigid form or the flexible forms, and all should be equally stable. Such is found not to be the case, however. Most of the quantitative work has been done with

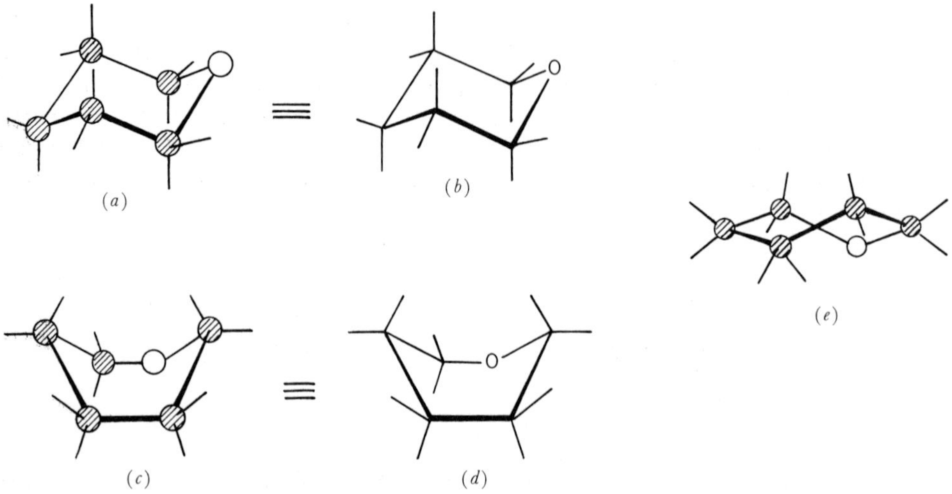

Figure 18–*2*. Conformations of pyranose rings.

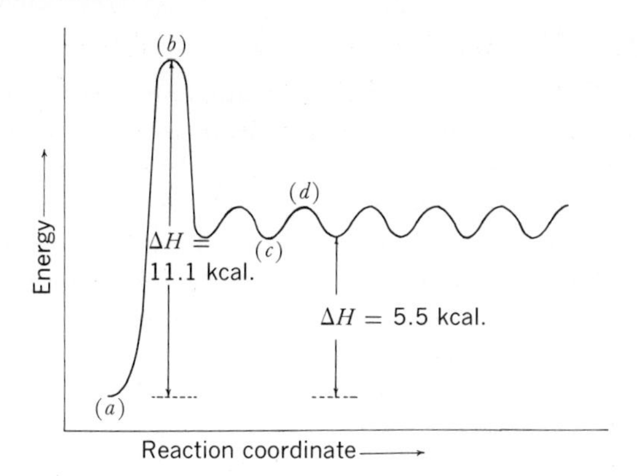

Figure 18–3. Energy profile for the interconversion of conformers of cyclohexane.

cyclohexane (p. 640), but since the C—O bond length and C—O—C bond angle are similar to the C—C bond length and C—C—C bond angle, the same considerations should apply to both the cyclohexane ring and the tetrahydropyran ring. For cyclohexane, the rigid chair form is more stable than the most stable conformation of the flexible form by about 5.5 kcal. The reason for the difference is that in the chair form the hydrogen atoms are completely staggered as in ethane (p. 56), whereas this is not true in any of the flexible forms. This difference in stability is sufficient to ensure that at any instant only one molecule will exist in the flexible form of cyclohexane for approximately a thousand molecules in the rigid chair form.

The energy relationships between the chair form and the various flexible forms are illustrated in Fig. 18–3. The barrier to the conversion of the chair form to the flexible forms is 11.1 kcal., (*a*) being the energy level of the chair form, (*b*) that of the transition state for interconversion of the chair form and the flexible forms, (*c*) that of one of the six most stable conformations of the flexible forms, and (*d*) that of the transition state for the interconversion of the flexible forms.

The considerations thus far relate to cyclohexane or tetrahydropyran, which have only hydrogen attached to carbon. If other groups are present, their greater size alters the relative stability of different conformations. Figure 18–2*a* and *b* shows that half of the bonds to carbon are directed perpendicular to the plane of the ring and above and below the plane of the ring, whereas the other half are directed outward peripherally and more or less parallel to the plane of the ring. The former are designated as **axial bonds, *a*,** and the latter as **equatorial bonds, *e*.** Clearly, groups attached to axial bonds are closer together than groups attached to equatorial bonds. If hydrogen atoms are replaced by other groups, that conformation normally predominates that has the maximum number of such groups in the equatorial position. Thus, of the two possible chair conformations *I* and *II* for β-D-glucose, *I* is expected to be the more stable by from 6 to 9 kcal. (p. 349).

I *II*

The barrier to interconversion is sufficiently low, however, that if the occasion requires it, the aldose can react readily in the less stable conformation.

Some Other Reactions of Aldoses

Ester Formation. The hydroxyl groups of sugars can be converted readily to ester groups. The aldoses acetylate with acetic anhydride in the presence of acidic or basic catalysts. The hexoses yield pentaacetates and the pentoses tetraacetates, thus indicating the presence of five and four hydroxyl groups. The acetates, however, do not reduce Fehling solution and therefore do not contain a free aldehyde group. Moreover two isomeric α and β forms of each acetate are obtained. Hence it is the cyclic form of the sugar that is acetylated. Sulfuric acid or zinc chloride commonly are used as acidic catalysts, and pyridine (p. 515) or sodium acetate (acetate ion)[2] are used as basic catalysts.

It is possible by choosing the proper catalyst and conditions to obtain chiefly α- or β-glucose pentaacetate as desired. Acids above room temperature rapidly catalyze the interconversion of α and β forms of the acetates, and the equilibrium mixture of 90 per cent α and 10 per cent β is obtained. In the presence of bases at temperatures below 0°, the rate of anomerization of the acetate is much slower than acylation, and since the less hindered equatorial β hydroxyl group reacts much faster than the axial α hydroxyl group, the base-catalyzed acetylation yields chiefly the β acetate.

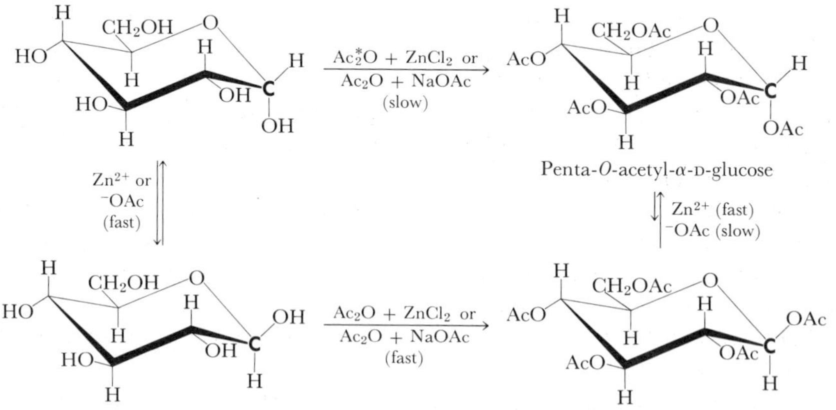

Penta-*O*-acetyl-α-D-glucose

Penta-*O*-acetyl-β-D-glucose

* The abbreviation *Ac* commonly is used to designate the acetyl group, CH_3CO. Thus HOAc is acetic acid, Ac_2O is acetic anhydride, and ROAc is an acetate. *Ac* should not be used for the acetate group, CH_3CO_2.

Many other esters of the sugars have been prepared. Among those of importance are the **benzoates** (p. 470), the **methanesulfonates** (p. 270), and the ***p*-toluenesulfonates** (p. 407). In ribonucleic acids (*RNA*) and deoxyribonucleic acids (*DNA*), phosphoric acid ester linkages join the nucleotides through the 3- and 5-hydroxyl groups of the ribose or deoxyribose portions of the molecule (p. 533).

Ether Formation. Once a sugar is converted to the methyl glycoside, it is stable to alkali because the hydrolysis of acetals is not catalyzed by base (p. 200) and because the acetal grouping does not activate α hydrogens (p. 196). Hence the remaining hydroxyl groups can be converted to ethers by a modified Williamson synthesis (p. 139). The **methyl ethers** are the most important. The original procedure, discovered by Purdie[3] and widely

---

[2] Zinc chloride is a Lewis acid (p. 34) because the zinc atom lacks electrons in its valence and like the proton can combine with an unshared pair of electrons. Acetate ion in acetic acid solution is a base analogous to the hydroxide ion in water solution (p. 31).

[3] Thomas Purdie (1843–1916), professor of chemistry at the University of St. Andrews, Scotland. He was interested primarily in the relation of optical rotation to structure. During this work, he observed that ethyl lactate, $CH_3CHOHCOOC_2H_5$, made from the silver salt and ethyl iodide

used by Purdic and Irvine,[4] treated the hydroxy compound with methyl iodide in the presence of silver oxide.

$$2\,ROH + 2\,CH_3I + Ag_2O \longrightarrow 2\,ROCH_3 + 2\,AgI + H_2O$$

Later Haworth[5] developed the use of methyl sulfate in the presence of sodium hydroxide. Frequently the latter reagent brings about only partial methylation, which is completed by the Purdie procedure.

$$\begin{array}{l} \text{CHOMe*} \\ \text{CHOH} \\ \text{CHOH} \\ \text{CHOH} \\ \text{CH—} \\ \text{CH}_2\text{OH} \end{array} \;(\text{O ring}) + 4\,Me^*_2SO_4 + 4\,NaOH \longrightarrow \begin{array}{l} \text{CHOMe} \\ \text{CHOMe} \\ \text{CHOMe} \\ \text{CHOMe} \\ \text{CH—} \\ \text{CH}_2\text{OMe} \end{array} \;(\text{O ring}) + 4\,NaMeSO_4 + 4\,H_2O$$

Methyl glycoside — Methyl tetra-*O*-methyl glycoside

* The abbreviation *Me* commonly is used to designate a methyl group, CH_3. Thus Me_2SO_4 is methyl sulfate, MeOH is methanol, and MeOR is a methyl ether.

In the Haworth procedure, the sugar first is converted to the methyl glycoside by adding the sodium hydroxide to the sugar and methyl sulfate at 30–40° and at such a rate that the mixture never becomes alkaline. After the mixture no longer reduces Fehling solution, the methylation is completed at a higher temperature.

The methoxyl on C-1 differs markedly from the other four in that it is an acetal methoxyl whereas the others are ether methoxyls. The acetal methoxyl can be removed readily by acid hydrolysis to give the 2,3,4,6-tetra-*O*-methylglycose, which, because the ring can open easily, again reduces Fehling solution.

$$\begin{array}{l} \text{CHOMe} \\ \text{CHOMe} \\ \text{CHOMe} \\ \text{CHOMe} \\ \text{CH—} \\ \text{CH}_2\text{OMe} \end{array} \;(\text{O ring}) + H_2O \xrightarrow{H^+} CH_3OH + \begin{array}{l} \text{CHOH} \\ \text{CHOMe} \\ \text{CHOMe} \\ \text{CHOMe} \\ \text{CH—} \\ \text{CH}_2\text{OMe} \end{array} \;(\text{O ring}) \rightleftharpoons \begin{array}{l} \text{CHO} \\ \text{CHOMe} \\ \text{CHOMe} \\ \text{CHOMe} \\ \text{CHOH} \\ \text{CH}_2\text{OMe} \end{array}$$

The 2,3,4,6-tetra-*O*-methylglycose, however, can form only a hydrazone and not an osazone

---

always had a higher rotation than that made by direct esterification. He found that the former was contaminated with the ethyl ether, which led to the discovery of alkylation by an alkyl iodide and silver oxide. According to Irvine, Purdie was quick to realize the application of the reaction to a study of the structure of sugars and soon found that methyl α- and β-tetra-*O*-methylglucosides hydrolyzed to the same tetra-*O*-methylglucose and hence had the same ring structure.

[4] James Colquhoun Irvine (1877–1952), student of Purdie and his successor as professor of chemistry at St. Andrews. Later he became Principal and Vice-Chancellor of St. Andrews. He made a systematic study of the methylation of sugars and laid the groundwork for the determination of the structure of di- and polysaccharides.

[5] Walter Norman Haworth (1883–1950), professor of chemistry at the University of Birmingham. He began the study of chemistry under W. H. Perkin, Jr. (p. 638) and later worked in Wallach's laboratory at Goettingen. A lectureship at St. Andrews brought him into contact with the work of Purdie and Irvine, which changed his interest from the field of terpenes to that of carbohydrates. He developed the methylation procedure using methyl sulfate and sodium hydroxide, and largely as the result of his work, the ring structures and constitutions of most of the mono- and disaccharides became established, as well as the basic structure of many polysaccharides. He was awarded the Nobel Prize in Chemistry in 1937.

because the hydroxyl group on C-2 is methylated. The methylated sugars played an important part in the determination of ring structures and of the constitution of oligo- and polysaccharides (p. 340). Some *O*-methyl ethers have been isolated from natural sources.

Action of Mineral Acids. Although dilute acids at room temperature have little effect on aldoses other than catalyzing α,β interconversion, hot mineral acids produce complex changes that involve dehydration. Thus all pentoses when distilled with 12 per cent hydrochloric acid give approximately theoretical yields of **furfural.** The reaction is used both as a qualitative test for pentoses or substances yielding pentoses on hydrolysis and for their quantitative estimation, by detecting or estimating the furfural that distills (p. 512). The 6-deoxyaldohexoses are converted into **5-methylfurfural.** Dehydration is believed to take place by a series of β eliminations followed by cyclic hemiacetal formation and a final dehydration.

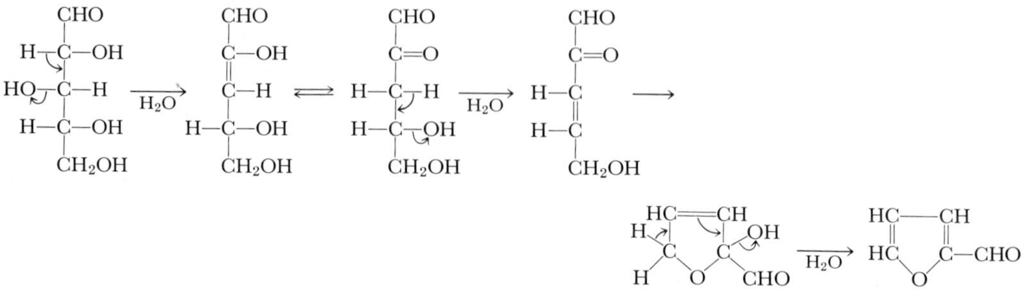

CHO–CHOH–CHOH–CHOH–CH_2OH $\xrightarrow{-H_2O}$ CHO–C(OH)=CH–CHOH–CH_2OH ⇌ CHO–C(=O)–CH_2–CHOH–CH_2OH $\xrightarrow{-H_2O}$ CHO–C(=O)–CH=CH–CH_2OH ⟶ (cyclic hemiacetal: HC=CH, C(H), C(OH)(CHO), O) $\xrightarrow{-H_2O}$ HC=CH–CH=C(CHO)–O (ring)

Furfural

Hexoses yield **5-(hydroxymethyl)furfural,** but this compound is more soluble in the acid than furfural and is not volatile with steam. Hence it is acted upon further by the hot acid, giving **levulinic acid** and considerable amounts of dark insoluble condensation products known as *humins.*

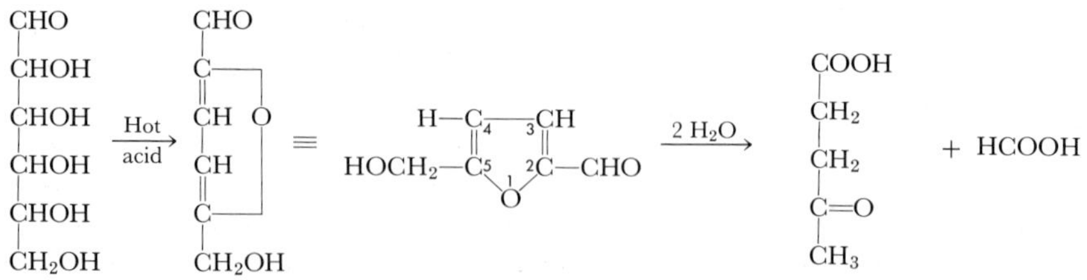

CHO–(CHOH)$_4$–CH_2OH $\xrightarrow{\text{Hot acid}}$ $HOCH_2$–C_5=C_4H–C_3H=C_2(CHO)–O_1 (ring) $\xrightarrow{2\,H_2O}$ COOH–CH_2–CH_2–C(=O)–CH_3 + HCOOH

5-(Hydroxymethyl)furfural | Levulinic acid

For the preparation of hydroxymethylfurfural, the best yields are obtained when phosphoric acid buffered with ammonium phosphate to *p*H 4 is used. The different furfurals give characteristic color reactions with polyhydric phenols such as resorcinol (p. 446), orcinol (p. 447), and phloroglucinol (p. 448), which may be used to distinguish among pentoses, 6-deoxyhexoses, and hexoses.

Ketoses

Ketoses are isomeric with aldoses that have the same number of carbon atoms and differ in structure only in that the carbonyl group contains a nonterminal carbon atom. The common ketoses all have the carbonyl oxygen on the second carbon atom and may be represented by the general formula, $HOCH_2(CHOH)_nCOCH_2OH$.

CH_2OH	CH_2OH	CH_2OH	CH_2OH
C=O	C=O	C=O	C=O
H—C—OH	HO—C—H	H—C—OH	HO—C—H
H—C—OH	H—C—OH	HO—C—H	HO—C—H
H—C—OH	H—C—OH	H—C—OH	H—C—OH
CH_2OH	CH_2OH	CH_2OH	CH_2OH
D-Psicose (D-*ribo*hexulose)	D-Fructose (D-*arabo*hexulose)	D-Sorbose (D-*xylo*hexulose)	D-Tagatose (D-*lyxo*hexulose)

CH_2OH	CH_2OH
C=O	C=O
H—C—OH	HO—C—H
H—C—OH	H—C—OH
CH_2OH	CH_2OH
D-Ribulose (D-*erythro*pentulose)	D-Xylulose (D-*threo*pentulose)

CH_2OH	CH_2OH
C=O	C=O
	H—C—OH
CH_2OH	CH_2OH
Dihydroxyacetone	D-Erythrulose (D-*glycero*tetrulose)

Figure 18–*4*. Configurations of the D family of ketoses.

Just as glyceraldehyde may be considered as the first member of the aldoses (Fig. 18–*1*, p. 319), dihydroxyacetone may be considered as the first member of the ketoses, although D-*glycero*tetrulose is the first member of the D family of ketoses (Fig. 18–*4*). In systematic nomenclature, the ending for ketoses is *ulose,* an ending derived from *levulose,* an old name for fructose. A prefix is used to indicate the number of carbon atoms and another, in italics, to indicate the relation of the configuration of the ketose to that of the aldose having one less carbon atom. More frequently common names for the known ketoses have been used.

D(−)-Fructose is the most abundant ketose. It occurs free, along with glucose and sucrose, in fruit juices and honey, and combined with other sugars in oligosaccharides. It is the chief product of the hydrolysis of the polysaccharide inulin (p. 346). Naturally occurring fructose has a high negative rotation, which gave rise to the earlier name *levulose.* It is the sweetest of the sugars.

Since individuals vary greatly in their sensory perceptions, it is possible to state only average opinions regarding taste. The results of early workers based on threshold methods, that is, the highest dilution that can be tasted, are valueless. When concentrations of equal sweetness are compared, the relative sweetness varies with the concentration. Moreover since the sweetness of an α form differs from that of a β form, solutions that have reached equilibrium must be used. When compared at 10 per cent concentration with glucose, the relative sweetness of some of the common sugars is lactose 0.55, galactose 0.95, glucose 1.00, sucrose 1.45, and fructose 1.65. Many synthetic compounds are known that are from several hundred to several thousand times sweeter than the sugars (pp. 425, 477).

Fructose gives the usual addition reactions of carbonyl compounds, but it is not oxidized by aqueous bromine to give a monocarboxylic acid. Hence it is a ketose and not an aldose. It is oxidized by Fehling solution and has a reducing power comparable to that of glucose. As a ketone it would not be expected to reduce Fehling solution. In the presence of alkali, however, it is converted to glucose and mannose (p. 318), which are oxidized readily. Moreover alkali causes fragmentation to reducing compounds of lower molecular weight, and CHOH groups adjacent to a carbonyl group also are oxidized by Fehling solution (p. 464).

When fructose reacts with phenylhydrazine, a phenylosazone is formed that is identical with that obtained from (+)-glucose. Hence the carbonyl group must contain C-2, the configuration of the rest of the molecule must be identical with that of (+)-glucose, and (−)-fructose must belong to the D family.

CHO		$CH{=}NNHC_6H_5$		CH_2OH
H—C—OH		$C{=}NNHC_6H_5$		C=O
HO—C—H	$\xrightarrow{C_6H_5NHNH_2}$	HO—C—H	$\xleftarrow{C_6H_5NHNH_2}$	HO—C—H
H—C—OH		H—C—OH		H—C—OH
H—C—OH		H—C—OH		H—C—OH
CH_2OH		CH_2OH		CH_2OH
D(+)-Glucose		D-Glucose phenylosazone		D(−)-Fructose

It was the discovery that glucose, mannose (p. 323), and fructose give the same osazone, and the realization of its significance, that led Fischer to undertake the unraveling of the configurations of the sugars. α-Methylphenylhydrazine, $C_6H_5N(CH_3)NH_2$, can be used to distinguish fructose from glucose or mannose, since conditions can be chosen such that fructose gives an osazone, whereas glucose and mannose do not (see, however, p. 323).

L(−)-Sorbose is the only other ketohexose that readily is available. It does not occur naturally but is produced by the action of sorbose bacteria (*Acetobacter xylinum* or, better, *Acetobacter suboxydans*) on sorbitol (p. 349).

CH_2OH		CH_2OH
HO—C—H		C=O
HO—C—H	$\xrightarrow[\text{Acetobacter}]{O_2}$	HO—C—H
H—C—OH		H—C—OH
HO—C—H		HO—C—H
CH_2OH		CH_2OH
Sorbitol		L(−)-Sorbose

L-Sorbose is important as an intermediate for the commercial synthesis of **vitamin C** (*ascorbic acid*), which is a necessary dietary factor for the prevention of scurvy. Sorbose can be oxidized with nitric acid to 2-deoxy-2-oxo-L-gulonic acid, which readily undergoes an internal esterification to give the cyclic 1,4 lactone (cf. p. 616). Because of the acidic properties of ascorbic acid and its ready oxidation, it is believed to exist in the enediol form. Production in 1964 was 7.4 million pounds valued at $14.3 million.

CH_2OH		COOH		C(=O)— (ring O to C-4)		C(=O)— (ring O to C-4)
C=O		C=O		C=O		HO—C
HO—C—H	$\xrightarrow{HNO_3}$	HO—C—H	$\xrightarrow{\text{Heat}}$	HO—C—H	⇌	HO—C (double bond C2=C3)
H—C—OH		H—C—OH		H—C— (ring O)		H—C— (ring O)
HO—C—H		HO—C—H		HO—C—H		HO—C—H
CH_2OH		CH_2OH		CH_2OH		CH_2OH
L-Sorbose		2-Deoxy-2-oxo-L-gulonic acid		Vitamin C (*ascorbic acid*)		

D-Ribulose (D-*erythro*pentulose) and **D-sedoheptulose** (D-*altro*heptulose) play a part in the cycle whereby plants convert carbon dioxide and water to carbohydrates (*photosynthetic cycle*). D-Sedoheptulose was isolated originally from *Sedum spectabile,* a common herbaceous garden plant. D-*Manno*heptulose and D-*glycero*-D-*manno*octulose are present in the fruit of the avocado tree (*Persea gratissima*). They are not metabolized but are eliminated in the urine and may give rise to a positive test for glycosuria, although no glucose is present.

```
      CH2OH                 CH2OH                 CH2OH
       |                     |                     |
       C=O                   C=O                   C=O
       |                     |               ------+------
   HO—C—H                HO—C—H              HO—C—H    \
       |                     |                   |      |
    H—C—OH               HO—C—H              HO—C—H     |
       |                     |                   |      } manno
    H—C—OH                H—C—OH              H—COH     |
       |                     |                   |      |
    H—C—OH                H—C—OH              H—C—OH   /
       |                     |               ------+------
      CH2OH                 CH2OH             H—C—OH   \
                                                 |      } glycero
                                                CH2OH  /
                                             -------------
  D-Sedoheptulose      D-Mannoheptulose
 (D-altroheptulose)                       D-Glycero-D-mannooctulose
```

It will be noted that the configuration of sugars having more than four asymmetric carbon atoms in a chain is designated also according to the relation to the aldoses having six or less carbon atoms. A second configurational prefix merely is added for the additional asymmetric carbon atoms. The prefix for the higher numbered carbon atoms is given first.

OLIGOSACCHARIDES

Disaccharides

Reducing Disaccharides. **Maltose,** $C_{12}H_{22}O_{11}$, is formed by the enzyme-catalyzed hydrolysis of starch.

$$\underset{\text{Starch}}{(C_6H_{10}O_5)_x} + \frac{x}{2}\,H_2O \xrightarrow{\text{Diastase}} \frac{x}{2}\,\underset{\text{Maltose}}{C_{12}H_{22}O_{11}}$$

Acid- or enzyme-catalyzed hydrolysis of maltose yields two molecules of glucose.

$$\underset{\text{Maltose}}{C_{12}H_{22}O_{11}} + H_2O \xrightarrow[\text{Maltase}\ (\alpha\text{-glucosidase})]{H^+\ \text{or}} \underset{\text{Glucose}}{2\,C_6H_{12}O_6}$$

Maltose reduces Fehling solution, forms an osazone, and undergoes mutarotation. Hence it contains a potential aldehyde group and a hemiacetal ring structure. The formation of an octaacetate and octamethyl derivative indicates the presence of eight hydroxyl groups.

Hydrolysis to two molecules of glucose points to a linkage through oxygen rather than directly between two carbon atoms, and the ease of hydrolysis by enzymes and acids indicates an acetal linkage rather than an ether linkage. Degradation of methylated derivatives has shown that the two glucose units are linked through the 1 position of the nonreducing half of the molecule and the 4 position of the reducing half. Maltose is hydrolyzed by *maltase,* a glucosidase that hydrolyzes methyl α-glucoside (p. 324), and not by *emulsin,* the glucosidase that hydrolyzes methyl β-glucoside. Hence the configuration of C-1 of the nonreducing portion is believed to be α. These conclusions are embodied in the following perspective formulas.

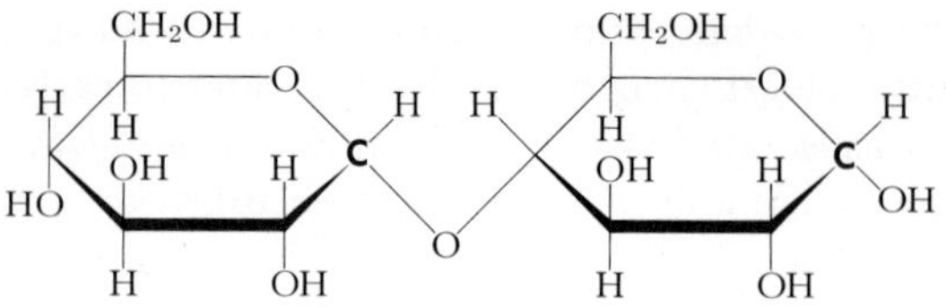

α-Maltose
4-*O*-(α-D-Glucopyranosyl)-α-D-glucopyranose

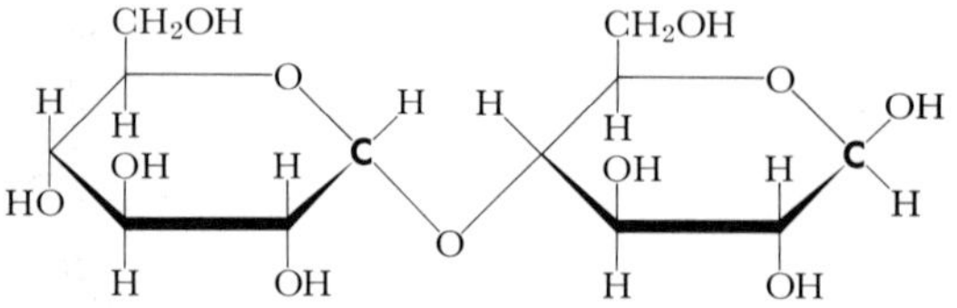

β-Maltose
4-*O*-(α-D-Glucopyranosyl)-β-D-glucopyranose

α- and β-Maltose differ only in the configuration of the anomeric carbon atom of the reducing half of the molecule. The second name given under the formulas is more descriptive in that it indicates both the point at which the nonreducing half of the molecule is attached to the reducing half and the structure and configuration of each half.

Cellobiose is a disaccharide formed by the hydrolysis of its octaacetate. The latter compound is obtained in 40 per cent yield when cellulose (cotton or paper) is dissolved in a mixture of acetic anhydride and sulfuric acid, and allowed to stand for one week at 35°. Cellobiose like maltose yields two molecules of glucose on hydrolysis and undergoes all of the reactions of maltose. The results of methylation and degradation experiments indicate that it has the same structure as maltose. The only difference in behavior of the two sugars is that cellobiose is hydrolyzed in the presence of emulsin (*β-glucosidase*) and not in the presence of maltase (*α-glucosidase*). Hence cellobiose differs from maltose in that the linkage between the two glucose units has the β configuration instead of the α configuration.

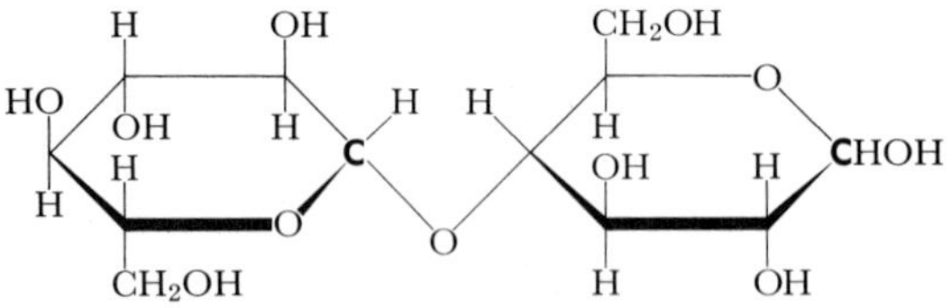

Cellobiose
4-*O*-(β-D-Glucopyranosyl)-D-glucopyranose

Cellobiose differs also in that in the chair conformations of the rings all of the hydroxyl and hydroxymethyl groups in both rings can be equatorial, whereas the α linkage in maltose does not permit this in the nonreducing ring. X-ray analysis of crystalline cellobiose shows that both rings are in the chair form but twisted 26° with respect to each other. All of the oxygen atoms except that joining the two rings are hydrogen bonded. *O*-3′ is intramolecularly bonded to *O*-5, but all of the other hydrogen bonds are intermolecular. These relationships have a significant bearing on the structure of cellulose (p. 341).

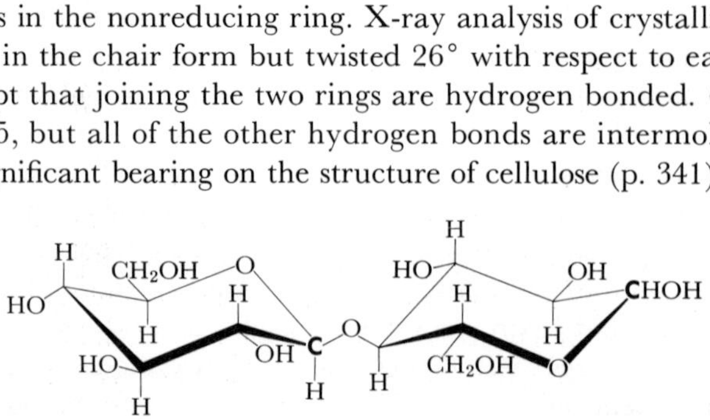

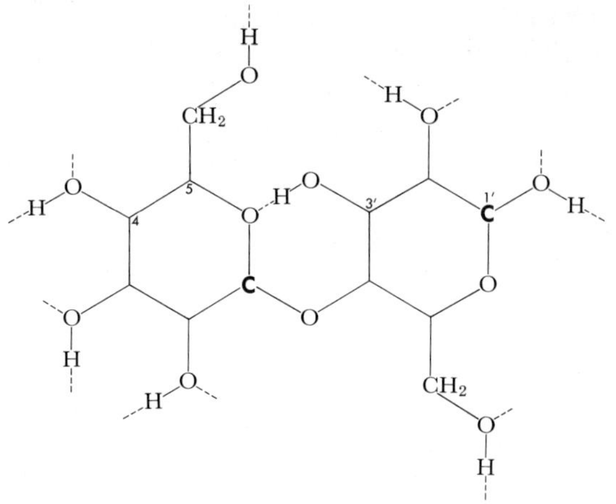

Neither of these representations gives a true picture of the shape of the cellobiose molecule. In the crystal, the rings not only are in the chair form but are angled and twisted with respect to each other in order best to form hydrogen bonds and to fit into the crystal lattice. In solution solvation greatly changes the attractive forces between the molecules, and thermal agitation causes the molecules to assume varying shapes and conformations.

Lactose (*milk sugar*) is present in about 5 per cent concentration in all mammalian milk investigated with the exception of that of the California sea lion (*Zalophus californianus*), whose milk appears to contain glucose but no lactose. It has been isolated also from the ripe fruit of the sapodilla tree (*Achras zapota*) from which chicle is obtained. Lactose is prepared commercially from whey, the aqueous solution left after the coagulation of the proteins of milk in the manufacture of cheese. Lactose is a reducing disaccharide, and on acid or enzyme hydrolysis it yields one molecule of galactose and one molecule of glucose. If it first is oxidized to lactonic acid and then hydrolyzed, the products are galactose and gluconic acid. Hence the glucose unit contains the reducing portion of the molecule. Lactose is hydrolyzed by β-galactosidase, and hence the linkage between the galactose and glucose portions is believed to be β. Methylation and degradation experiments have established its structure as 4-*O*-(β-D-galactopyranosyl)-D-glucopyranose.

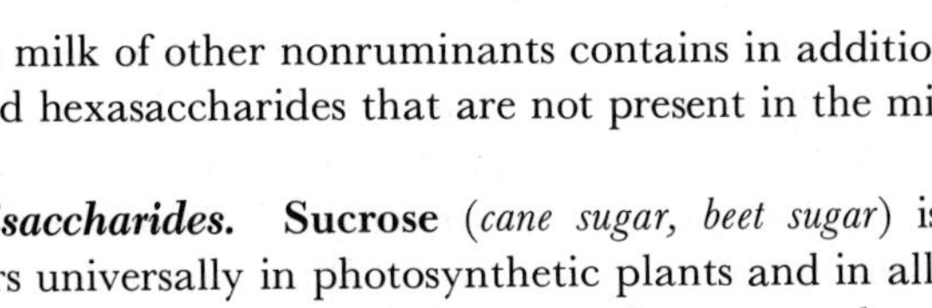

Lactose
4-*O*-(β-D-Galactopyranosyl)-D-glucopyranose

Human milk and the milk of other nonruminants contains in addition to lactose numerous tri-, tetra-, penta-, and hexasaccharides that are not present in the milk of cows, sheep, and goats.

Nonreducing Disaccharides. **Sucrose** (*cane sugar, beet sugar*) is the most important disaccharide. It occurs universally in photosynthetic plants and in all portions of the plant. It appears to serve as an easily transported source of energy and as a basic unit to which molecules of glucose, fructose, or galactose may be attached in progressively larger numbers until polysaccharides are formed (p. 347). Although present in honey to the extent of less than one per cent, and occasionally in human urine, sucrose has not been found in animals.

The principal sources are sugar cane, sugar beets, and the sugar maple tree. The great ease with which sucrose crystallizes probably accounts for its isolation in a pure state as early as 300 A.D. (p. 2). Sugar cane (*Saccharum officinarum*) belongs to the grass family and probably originated in northeastern India (Skr. *sakara* gravel or sugar). From there it was introduced into China about 400 A.D. and into Egypt by the Arabians in 640 A.D. It is believed that it was introduced into Santo Domingo by Columbus on his second voyage to America in 1494. Although sucrose was discovered in beet juice in 1747, it was not produced from this source until the Napoleonic Wars (1796–1814) made the price of cane sugar prohibitive in Europe.

In the extraction of sucrose from cane, the stalk is ground and the juice expressed with rollers. The juice, which contains about 15 per cent sucrose, is made slightly alkaline with lime to prevent hydrolysis. When heated, most of the impurities separate as a heavy scum and a precipitate, which are removed. This process is known as *defecation.* The clear juice is concentrated under reduced pressure and allowed to crystallize. The raw sugar is removed by centrifuging and is washed in the centrifuge. The centrifugate is reconcentrated, and the operation is repeated until no more crystals can be obtained economically. The final mother liquor is a dark viscous liquid that contains about 50 per cent fermentable sugars and is known as *blackstrap.* It is used for preparing cattle feeds and for the production of alcohol (p. 121).

The raw sugar is shipped to refineries where the brown color is removed. The sugar is dissolved in water and passed through decolorizing carbon. Concentration and crystallization yield the pure sugar of commerce. *Brown sugars* are mixtures of the purified sugar with various amounts of molasses. *Demerara* is a crude sugar from British Guiana that is popular in Great Britain. If the white crystalline sugar is ground with 3 per cent of starch to prevent caking, the product is known as *powdered sugar.*

The white sugar beet, a cultivated variety of *Beta maritima,* has been bred to contain up to 18 per cent sucrose. The beet root, free of leaves, is washed and sliced into V-shaped pieces about 1 cm. thick. These pieces are extracted countercurrently by hot water to yield a dark solution containing about 12 per cent sucrose. This diffusion of the sugar from the plant cells extracts much less protein and impurities of high molecular weight than if the beet were crushed and pressed. The warm extract is agitated with 2 to 3 per cent of lime and the mixture then saturated with carbon dioxide. The precipitate carries down most of the impurities, and the yellow filtrate is decolorized with sulfur dioxide. After it is concentrated, crystallized, and washed, the product is ready for market. The molasses may be sold for cattle feed or the sucrose may be recovered. In the *Steffen process,* the molasses is diluted to about 7 per cent sugar, and lime is added to precipitate the slightly soluble tricalcium saccharate, which is removed by filtration and added instead of lime to the raw juice in the defecation step. The Steffen filtrate is concentrated and sold for the production of monosodium glutamate (p. 358).

Production of sucrose from the sugar maple (*Acer saccharinum*) is relatively unimportant. Small amounts are prepared also from other sources such as sorghum and palm sap, but sucrose from these sources does not enter the world market.

Sucrose is produced in larger amount than any other pure organic chemical. World production in 1963 was 51.9 million metric tons of which about one third was from beets and two thirds from cane. Production in the United States mainland was 3.4 million tons of which only 30 per cent was from cane. Consumption was 9 million tons or 98 pounds per capita.

Hydrolysis of sucrose in the presence of acids or of the enzyme *sucrase* (invertase) yields one molecule of fructose and one molecule of glucose. Sucrose has a positive rotation, $[\alpha]_D^{20} = +66.53$, but during hydrolysis the sign of rotation changes because the high negative rotation of fructose, $[\alpha]_D^{20} = -92.4$ for the equilibrium mixture, more than balances the positive rotation of glucose, $[\alpha]_D^{20} = +52.7$ for the equilibrium mixture. Because of the change in sign of rotation during hydrolysis the process is known as *inversion,* and the mixture of fructose and glucose is known as *invert sugar.* It has about the same sweetening power as sucrose (p. 331) but has much less tendency to crystallize and hence is used in the manufacture of candies and syrups.

Sucrose does not reduce Fehling solution, form an osazone, undergo mutarotation, or form methyl glycosides. Accordingly the fructose and glucose portions are linked through the two anomeric carbon atoms. Methylation gives octamethylsucrose, which can be hydrolyzed to 2,3,4,6-tetra-*O*-methylglucose and 1,3,4,6-tetra-*O*-methylfructose. The first product proves the presence of a six-membered ring in the glucose portion of the molecule, and the second product proves that a five-membered ring is present in the fructose portion of the molecule. Sucrose is hydrolyzed by yeast α-glucosidase and not by the β-glucosidase of emulsin. It is hydrolyzed also by sucrase, an enzyme that hydrolyzes β- but not α-fructofuranosides. Accordingly the anomeric carbon of the glucose portion is assigned the α configuration, and that of the fructose portion the β configuration. This structure has been confirmed by X-ray analysis.

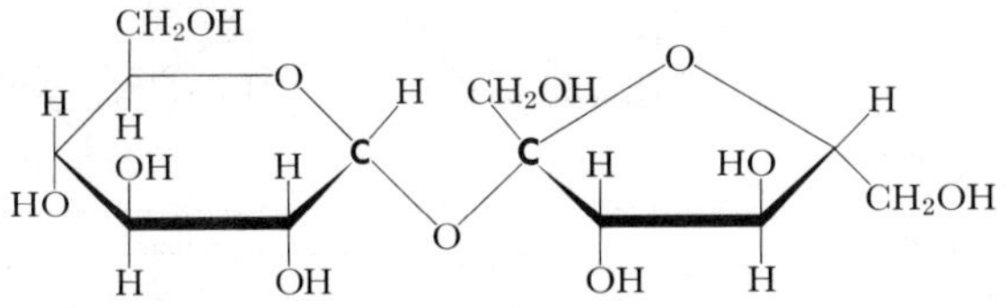

CH_2OH — O — H — CH_2OH — O — H
H — H — OH — H — C — C — H — HO — CH_2OH
HO — O
H — OH — OH — H

Sucrose
α-D-Glucopyranosyl β-D-fructofuranoside

The synthesis of sucrose from D-glucopyranose-1-phosphate and fructose by means of an enzyme from the microorganism *Pseudomonas saccharophila* was accomplished in 1943. Many early attempts to synthesize sucrose by chemical methods failed because of the difficulty of obtaining the correct configuration of the glycosidic linkage. In 1953 sucrose octaacetate was obtained in 5.5 per cent yield by heating 1,2-anhydro-3,4,6-tri-*O*-acetyl-α-D-glucopyranose with 1,3,4,6-tetra-*O*-acetyl-D-fructofuranose. Purification and deacetylation gave sucrose identical with the natural product.

Sucrose can be converted to esters either by acylation with acid anhydrides or with esters of the lower alcohols (alcoholysis, p. 173). **Sucrose octaacetate** is extremely bitter and is used in certain formulas of denatured alcohol. **Sucrose acetate isobutyrate** is a mixed ester containing acetyl and isobutyryl groups in the ratio of about 1:3. It is made by the reaction of sucrose with a mixture of acetic and isobutyric anhydrides and is used to modify the properties of plastics and protective coatings. Alcoholysis of the methyl esters of long-chain fat acids in the presence of potassium carbonate and in *N,N*-dimethylformamide or methyl sulfoxide as solvent yields chiefly mono esters that are **nonionic detergents.** They are cheaper than nonionic detergents derived from ethylene oxide, they are nontoxic solids, and they are easily destroyed by sewage bacteria. **Sucrose monostearate** acts as an anti-staling agent in bread. **Octa-*O*-(2-hydroxypropyl)-sucrose,** prepared by the reaction of sucrose with propylene oxide (p. 606), is reported to be useful as a plasticizer for various products and as a cross-linking agent for polyurethan foams (p. 605).

Trisaccharides

Raffinose is present in small amount in many plants and is the most readily available trisaccharide. It accumulates in the mother liquor during the preparation of beet sugar and can be prepared from beet sugar molasses. It may be obtained also by extracting defatted ground cottonseed with water. Complete hydrolysis of raffinose yields one mole each of glucose, fructose, and galactose. Since raffinose is nonreducing, all of the anomeric carbon atoms must be involved in glycosidic links. When raffinose is hydrolyzed in the presence of the α-galactosidase of emulsin, the products are galactose and sucrose, whereas if it is hydrolyzed by sucrase (*β-fructosidase*) the products are fructose and melibiose. Since the structures of sucrose and melibiose are known, the structure of raffinose is established.

Galactose$^{1\alpha}$———^{6}glucose$^{1\alpha}$———$^{2\beta}$fructose

Hydrolyzed here by α-galactosidase → Galactose + sucrose

Hydrolyzed here by β-fructosidase → Melibiose + fructose

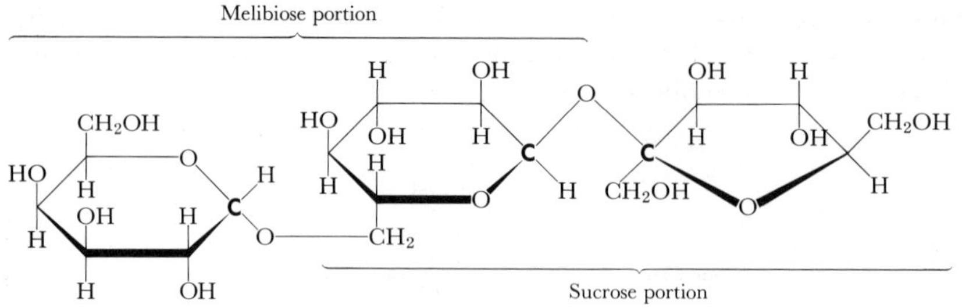

Melibiose portion

Sucrose portion

Raffinose

POLYSACCHARIDES

As the name implies, polysaccharides are carbohydrates of high molecular weight (30,000 to 400,000,000). They are insoluble in liquids and are altered readily by the acids and alkalies required to catalyze their conversion into soluble derivatives. Hence their purification is extremely difficult. Moreover even the purified products are not molecularly homogeneous, since the same substance may consist of polymeric molecules of a range of molecular weights. Those polysaccharides made up of only a single type of building unit are known as *homopolysaccharides*. Other polysaccharides are derived from several different types of building units and are known as *heteropolysaccharides*. In general the nomenclature of polysaccharides is unsatisfactory. Those substances without well established common names are given names ending in *an* and generically are called *glycans*. The terms *hexosan, pentosan,* and *glycosan* should be dropped, since the first two refer to polysaccharides and the last refers to glycosidic anhydro sugars.

The biological function of the polysaccharides varies considerably. Starch, glycogen, and inulin are reserve foodstuffs for the plant or animal; the cellulose of plants and the chitin of crustacea have a structural function; for others, such as gums and mucilages, the function is unknown.

Homopolysaccharides

Starch. During the growth of a plant, carbohydrate is stored in various parts, such as the roots, seeds, and tubers, in the form of microscopic granules of starch. Seed may contain up to 70 per cent starch, and roots and tubers up to 30 per cent. Corn, potatoes, wheat, rice, tapioca, sago, and sweet potatoes constitute the chief commercial sources. Starch granules from different sources vary in shape, size, and general appearance.

In the United States most starch is derived from corn (*maize*). The corn grains are soaked in warm water containing sulfur dioxide and then are shredded to free the germ. When the mass is mixed with water, the germ floats because of the high oil content and is collected, dried, and pressed to produce corn oil (p. 185). The remainder of the grain is ground as finely as possible without rupturing the starch granules, and the mixture is washed through screens to remove most of the hull. The starch is allowed to settle along inclined troughs or is removed from the aqueous slurry by continuous centrifuges.

The protein constituents of the grain (*gluten*) are recovered by evaporation of the steep water and the aqueous centrifugates. The concentrate finds an important outlet as a nutrient for the mold in the production of penicillin (p. 524). It also is mixed with the dried hulls and germ press cake for use in animal feeds. Extraction of the gluten fraction with isopropyl alcohol yields zein (p. 364), and hydrolysis of the gluten yields glutamic acid (p. 358). The separation of starch from potatoes is somewhat simpler in that steeping and germ removal are not required. Annual production of cornstarch in the United States is about 3 billion pounds, and of potato starch (*farina*), about 100 million pounds.

When starch is heated with water, the granules swell greatly and a colloidal suspension is produced. This dispersion has been separated into two components. When it is saturated with a slightly soluble alcohol such as butanol or the commercial mixture of amyl alcohols known as Pentasol (p. 107), or with thymol (p. 445), a microcrystalline precipitate forms, which is known as *A-fraction* or **amylose.** Addition to the mother liquors of a water-miscible alcohol such as methanol, or coprecipitation with aluminum hydroxide, gives an amorphous material known as *B-fraction* or **amylopectin.** A commercial process for the separation of amylose and amylopectin uses fractional precipitation from a solution in hot aqueous magnesium sulfate. Most starches contain amylose and amylopectin in the ratio of from 1:3 to 1:4, together with 4 to 9 per cent of a fraction having intermediate properties. Some genetically pure mutant varieties of grains, known as waxy corn, sorghum, millet, rice, and barley, contain only amylopectin, whereas another recessive mutant strain of wrinkled pea has been reported to contain from 70 to 98 per cent of amylose.

The two fractions differ markedly in their physical and chemical properties. Purified amylose is only sparingly soluble in hot water, and the amylose reprecipitates when the solutions are allowed to stand (*retrogradation*). It disperses readily in dilute sodium hydroxide but is degraded by the alkali. Amylopectin dissolves readily in water, and the solution does not gel or precipitate on standing. Solutions of native starches gel rapidly and precipitate slowly. Apparently the amylopectin acts as a protective colloid for the amylose.

The typical blue color reaction of starch with iodine is due to the amylose fraction, which absorbs 18 to 20 per cent of its weight of iodine as determined by potentiometric titration. The blue product is an inclusion compound (p. 283) in which iodine molecules fit into the central open spaces of a helix of the C_6 units that constitute the amylose molecule. Amylopectin gives a red to purple color and absorbs only 0.5 to 0.8 per cent of iodine. Iodine titration of whole starch may be used to determine the relative amounts of the two components.

Both fractions give only glucose on complete hydrolysis by acids. More information can be obtained from partial hydrolysis by means of enzymes. Two groups of starch-splitting enzymes, the α- and *β-amylases,* have been recognized for some time. Here the α and β merely differentiate the two groups of enzymes and do not refer to the configuration of the glycosidic linkage that they attack, since both are 1,4-α glucosidases. They differ in that α-amylase hydrolyzes the linkages randomly, whereas β-amylase attacks only the second link from the nonreducing end of a chain and liberates maltose. Amylose that has been prepared in the usual way is hydrolyzed to maltose to the extent of about 70 per cent by pure β-maltase. In 1952 it was found that if another enzyme called *Z-enzyme* is present, hydrolysis to maltose proceeds almost to completion. Z-enzyme is a β-glucosidase. Hence it is believed that amylose consists of glucose units joined mainly by α-glycosidic links through the 1,4 positions to give an essentially linear polymer coiled into a spiral with a period of six units per turn, but that there are a few single β-glucosyl units as branches. These branches prevent the β-amylase from converting the amylose completely to maltose unless the branches first are removed by the Z-enzyme. Amylose prepared by leaching starch with hot water in the absence of oxygen is more than 95 per cent hydrolyzed by β-amylase alone. When amylose is completely methylated and then hydrolyzed, the chief product is 2,3,6-tri-*O*-methylglucose, but about 0.3 per cent of 2,3,4,6-tetra-*O*-methylglucose is obtained, which results from the terminal nonreducing glucose units. Ultracentrifuge studies indicate that amylose is heterogeneous, the molecules having molecular weights between 17,000 and 225,000. Osmotic pressure measurements indicate that amyloses from different sources or prepared in different ways may have average molecular weights ranging from 100,000 to 2,000,000.

Amylopectin is hydrolyzed to the extent of only about 50 per cent by β-amylase, with or without the presence of Z-enzyme, the remainder being a product of high molecular weight known as *limit dextrin.* β-Amylase in conjunction with another enzyme known as *R-enzyme*, a 1,6-α glucosidase, brings about complete hydrolysis. Methylation of amylopectin, followed by hydrolysis, gives as much as 4 per cent of 2,3,4,6-tetra-*O*-methylglucose and an equal amount of 2,3-di-*O*-methylglucose. The tetra-*O*-methylglucose corresponds to about one nonreducing end for each 25 glucose units. Molecular weight determinations on amylopectin indicate an average value of from one to six million (6000–37,000 C_6 units) from osmotic pressure measurements and 400 million (2.5×10^6 C_6 units) from light scattering. Hence amylopectin is a highly branched molecule. The branches consist of 20 to 25 glucose units joined by 1,4-α linkages and are in turn joined to other branches by 1,6-α linkages. Branching is believed to be random rather than to follow any regular pattern.

The chief use of starch is as food. It is hydrolyzed stepwise by the *salivary amylases,* the final product being maltose if the reaction goes to completion. Because these enzymes are destroyed by strong acid, starch digestion stops as soon as the food is mixed completely with the hydrochloric acid of the stomach. Digestion continues, however, in the intestine, where the hydrochloric acid is neutralized and *pancreatic amylase* completes the hydrolysis to maltose. Before passage through the mucous membrane of the intestine, disaccharides such as maltose, sucrose, and lactose are hydrolyzed by specific intestinal enzymes to the monosaccharides. There is some evidence that glucose, fructose, and galactose may undergo phosphorylation and dephosphorylation during the process of passing through the walls of the intestine into the blood stream. These sugars are removed rapidly from the blood by the liver and muscle tissue and converted into glycogen.

Industrially starch is converted by acid hydrolysis into **corn syrup,** a mixture of saccharides of various molecular weights, or into **crystalline glucose,** both of which are used as food. Large quantities of starch are used as stiffening agents and adhesives, often after modification by heat to **dextrins.** Starch can be converted to esters and ethers, but so far these functional derivatives have not found large-scale uses. **Starch nitrates** containing 11 to 13 per cent nitrogen (so-called *nitrostarch*) have been used as a demolition explosive to replace TNT.

Glycogen. This reserve carbohydrate of animals does not occur as granules but is distributed throughout the protoplasm. Some of it is water-soluble, but the remainder is bound to protein. The best source is mussels. Glycogen is isolated by hydrolyzing the tissue with hot aqueous alkali, neutralizing, and precipitating with alcohol. Glycogen gives a violet to brown color with iodine. Acid hydrolysis yields only glucose, but β-amylase gives about 45 per cent maltose. Methylation and hydrolysis indicates one end group for each 10 to 12 C_6 units. Glycogen preparations are heterogeneous and average molecular weights vary from 3 to 15 million (osmotic pressure). Glycogens isolated without the use of alkali have molecular weights up to 100 million. The shorter chain length and higher molecular weight indicate that glycogen is even more highly branched than amylopectin, but that otherwise it has a similar structure.

Cellulose. This polysaccharide constitutes the cell membranes of the higher plants. It makes up about 10 per cent of the dry weight of leaves, about 50 per cent of the woody structures of plants, and 98 per cent of cotton fiber. A membrane of cellulose, indistinguishable from plant cellulose, is produced when *Acetobacter xylinum* is grown in a nutrient medium containing glucose, fructose, mannitol, or glycerol. Cellulose has been produced in the absence of living cells from uridine diphosphate, D-glucose, and extracts of *Acetobacter xylinum,* and from guanosine diphosphate, D-glucose, and an enzyme preparation isolated from mung beans (*Phaseolus aureus*). An animal cellulose known as *tunicin* forms a leathery mantle on tunicates. Pure cellulose is obtained best from cotton by dewaxing with an organic solvent, and removing pectic substances (p. 345) by extracting with hot 1 per cent sodium hydroxide solution in the absence of oxygen.

As with starch, glucose is the only monosaccharide found in the hydrolysis products of cellulose, and 90–95 per cent yields of 2,3,6-tri-*O*-methylglucose have been obtained from methylated cellulose. Unlike starch, no tetra-*O*-methylglucose is detected provided the methylation has been carried out in the absence of air. Apparently the molecule is so large and unbranched that the amount of tetra-*O*-methylglucose produced cannot be detected. By the ultracentrifugal method molecular weights of one to two million (6000 to 12,000 C_6 units) are indicated.

Acetolysis yields 40–50 per cent of cellobiose octaacetate, which has a β-glycosidic linkage (p. 334). Hence it is assumed that cellulose is a continuous chain of cellobiose units, all β-linked through the 1′–4 positions.

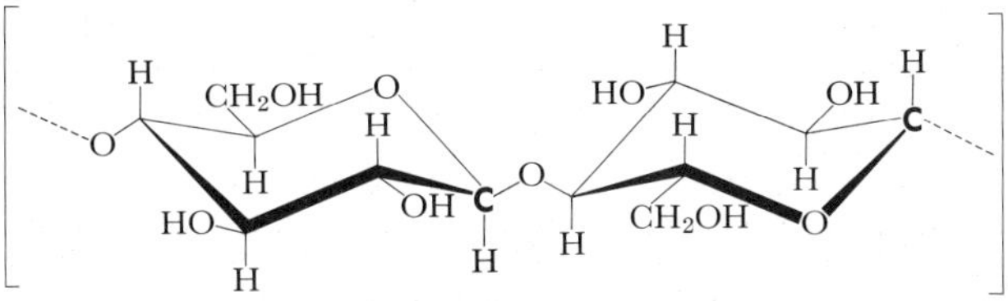

Cellulose

The oxygen-oxygen distance of 10.3 A found from X-ray diagrams of cellulose is the same as the 0–1′ to 0–4 distance found in cellobiose (p. 334). To account for this distance, it is necessary to assume that the rings are angled and twisted with respect to each other. The insolubility of cellulose in water and alcohols despite the presence of five oxygen atoms for each six carbon atoms results from the strong hydrogen bonding between the chains, while the long unbranched molecules account for its fibrous nature and high tensile strength.

Cellulose is soluble in sulfuric acid and concentrated solutions of zinc chloride, but since a scission of glycosidic links takes place in the presence of acids, the product regenerated by pouring the solution into water has a much lower molecular weight than the original cellulose. Such products are said to be *degraded.* Degradation also takes place on exposure to air in the presence of alkali.

The best solvent for cellulose is an aqueous solution of cupric ammonium hydroxide, $Cu(NH_3)_4(OH)_2$, which is known as *cuprammonium solution* or *Schweitzer reagent.* A more convenient reagent currently used is *cupriethylenediamine,* $Cu(H_2NCH_2CH_2NH_2)_2(OH)_2$, made by dissolving cupric hydroxide in ethylenediamine (p. 611) and diluting with water to the desired concentration. Since the viscosity of solutions decreases with decreasing chain length of a molecule, the extent of degradation of cellulose can be estimated by determining the viscosity of solutions in these reagents. Cellulose regenerated by acidifying cuprammine solutions gives, when redissolved, a solution having a lower viscosity than the original cellulose. Hence even these reagents bring about some degradation. The degradation is greater if the solutions are exposed to air.

Although considerable study of these solutions has been made, little is known about the constitution of the solute. In concentrated copper solutions, one atom of copper combines with each C_6 unit. It seems likely that complexes are formed analogous to those existing in Fehling solution (p. 636). The large size of the copper atom increases the distance between the chains and prevents hydrogen bonding between the chains, thus permitting hydration and solution (cf. p. 344). The solubility of cellulose in alkaline iron tartrate solution must be due also to complexing of the hydroxyl groups with the metal.

Commercial Sources of Cellulose. The technical uses of cellulose are dependent chiefly on its fibrous form and on the strength and flexibility of products prepared from it. Cotton fibers have an average tensile strength of 80,000 p.s.i. compared with 65,000 p.s.i. for medium steel. Fiber length varies with the source. The bast fibers, such as those from flax, jute, sisal, and hemp, vary from 20 to 350 cm. in length. Cotton, which is a seed hair, has fibers 1.5 to 5.5 cm. long. Wood fibers vary from 0.2 to 5 mm. in length. The long fibers are used chiefly for making textiles, and cotton is by far the most important for this purpose.

After removal of moisture, wood consists of 40 to 50 per cent cellulose, 15 to 25 per cent of other polysaccharides known as *hemicelluloses* (p. 348), 30 per cent lignin (p. 457), which acts as a matrix for the cellulose fibers, and 5 per cent of other substances such as mineral salts, sugars, fat, resin, and protein. Wood pulp is produced by dissolving the lignin with hot solutions of (*1*) sodium hydroxide, (*2*) sodium, calcium, magnesium, or ammonium bisulfite, or (*3*) a mixture of sodium hydroxide and sodium sulfide made from lime and reduced sodium sulfate. The products, known as soda pulp, sulfite pulp, and sulfate (*kraft*) pulp, respectively, consist of impure cellulose that has been more or less degraded.

Most wood pulp is used in the manufacture of paper, where the length of the fibers and the strength are most important. Highly purified pulp is used to make rayon (p. 344) and for conversion to esters and ethers used for fibers, plastics, and thickening agents (p. 344). Fermentable sugars in sulfite liquors can be converted to alcohol. The sulfate process yields for each ton of paper about 50 pounds of **tall oil** (Sw. *tallolja* pine oil), which consists of about 50 per cent unsaturated fatty acids, chiefly oleic and linoleic acids, and 50 per cent resin acids. It is used in the manufacture of soap (p. 190) and synthetic detergents (p. 606).

Mercerized Cotton. John Mercer, an English chemist and calico printer, attempted in 1814 to filter a concentrated solution of sodium hydroxide through cotton cloth and found that the fibers swelled and stopped the filtration. After the cloth was washed free of sodium hydroxide and dried, it dyed more readily than the original cloth. He patented his process in 1850. This treatment, however, shrinks the cloth considerably. Around 1889 Horace Lowe, another English technical chemist, attempted to prevent this shrinkage mechanically by carrying out the whole process with the cloth or yarn under tension. Not only was shrinkage prevented, but the surface irregularities of the fiber were removed, giving it a high luster like silk. Moreover the strength of the fiber increased about 20 per cent. This process now is known as *mercerization* and the product is called *mercerized cotton*. It is by far the most widely used chemical finish for cotton in which the fiber is modified.

Parchment Paper. When paper is treated for a few seconds with about 80 per cent sulfuric acid, an alteration of the surface of the fiber takes place. After washing and drying, the paper is stiffer and tougher and does not disintegrate in water. Because of its resemblance to parchment, it is called *parchment paper*.

Glucose. Cellulose can be hydrolyzed to glucose and in this way becomes usable for food or for the production of alcohol or yeast by fermentation. In the technical processes the wood chips are extracted with 40 per cent hydrochloric acid at 20° (*Bergius process*) or 0.5 per cent sulfuric acid at 130° (*Scholler process*). The cellulose dissolves leaving the lignin as a solid. Concentration of the extract yields the sugars. Acetic acid also may be recovered from the extracts. The residual lignin can be briquetted and distilled to give charcoal and methyl alcohol. Wood saccharification has been operated in countries with a controlled economy, particularly in Germany before World War II, although even there the operation was not so extensive as generally is believed. Attempts have been made periodically since 1900 to operate plants in the United States, but they never have been feasible economically.

Cellulose Esters. Many important products are derived from cellulose by esterifying the hydroxyl groups. Because hydrogen bonding is not possible, these derivatives are soluble in organic solvents and thus permit the formation of films or fibers. Formerly the cellulose used for this purpose was chiefly purified cotton linters, the short cotton fibers cut from the cotton seed after the removal of the long fibers by the cotton gin. More recently wood pulp having a high *α-cellulose* content has become the dominant raw material. α-Cellulose is a technical term for the portion of wood pulp that is insoluble in alkali of mercerizing strength (17.5 per cent). The soluble portion consists of the so-called *hemicelluloses*, which are chiefly xylans and mannans (p. 348).

The **cellulose nitrates,** commonly called *nitrocelluloses*, were the first esters to be of technical importance. Dry purified cotton linters are soaked in a mixture of nitric and sulfuric acid. It is desirable to carry out the reaction for as short a time (15–30 minutes), and at as low a temperature (30–40°) as possible and to keep the water concentration low to reduce degradation. After centrifuging and washing, the nitrated cotton is boiled with water to hydrolyze sulfates and thus produce good stability. It is stored wet and dehydrated before use by washing with ethanol or butanol. Three chief types of cellulose nitrate are produced: (*1*) *celluloid pyroxylin* containing 10.5–11 per cent nitrogen, soluble in ethanol-ether mixture and in absolute ethanol; (*2*) *soluble pyroxylin* (*collodion* *cotton,* or *dynamite cotton*) containing 11.5–12.3 per cent nitrogen and soluble in absolute ethanol; and (*3*) *guncotton,* containing 12.5–13.5 per cent nitrogen. Guncotton containing over 13 per cent nitrogen is insoluble in ethanol-ether and in absolute ethanol. All of the commercial nitrates dissolve in acetone and in esters of lower molecular weight. The dinitrates of cellulose should contain 11.11 per cent nitrogen and the trinitrate 14.15 per cent nitrogen. Hence the commercial products are mixtures rather than pure compounds.

Guncotton is the least degraded, having a chain length of about 3000 C_6 units. It gives solutions of extremely high viscosity. It is used for making smokeless powder, the chief propellant explosive for guns. The guncotton is *gelatinized* or *colloided* to a dough by mixing with a solvent, and expressed and cut into cylindrical perforated pellets called *grains.* The perforations permit a progressive increase in burning area and hence an increasing rate of combustion as the projectile moves through the barrel of the gun. The equation for the decomposition of cellulose trinitrate may be written as

$$2\left[\begin{array}{l}-CH(CHONO_2)_2-CH-O- \\ \;\; CH\text{———————}O \\ \;\; CH_2ONO_2\end{array}\right]_x \longrightarrow 9x\,CO + 3x\,CO_2 + 7x\,H_2O + 3x\,N_2$$

Sufficient oxygen is present to convert the compound completely to gaseous products, and the large amount of energy liberated in transferring oxygen from nitrogen to hydrogen and carbon raises the gases to a high temperature and produces the pressure necessary to expel the projectile at high velocity. With the advent of rockets that use as propellant a mixture of an oxidizing agent with an organic plastic (p. 589), the military use of smokeless powder has greatly decreased.

In 1865 Alexander Parkes, an Englishman, obtained a horn-like mass by mixing cellulose nitrate, alcohol, and camphor. In 1869 the American inventor John Wesley Hyatt obtained a patent for a similar product, which he had developed as a substitute for ivory for billiard balls, and for which he coined the name **celluloid.** It is made by mixing two parts of celluloid pyroxylin with one part of powdered camphor (p. 661) and kneading in a mixer with enough alcohol to form a dough. The soft mass is pressed into various forms or rolled into sheets. In the curing the alcohol evaporates and the articles solidify. Celluloid has many disadvantages, such as undesirable odor and taste, nonresistance to acid and alkali, discoloration in light, and high flammability. Yet for about fifty years it was the only important plastic.

Celluloid pyroxylin and soluble pyroxylin are more degraded than guncotton, having chain lengths of only 500 to 600 C_6 units. It is this smaller molecular size that permits them to give free-flowing solutions at fairly high concentration. Solutions of sufficient concentration to leave a film thick enough to act as a binding agent for pigments and to serve as a protective coating are, however, too viscous to be applied with a brush or spray gun. Further degradation can be brought about by heating the pyroxylin with water in pressure digestors. Pyroxylin giving solutions of almost any desired viscosity thus can be obtained, but the lower the viscosity the shorter the chain length, and hence the lower the film strength and the greater the brittleness of the film. The material commonly used in lacquers is known as **half-second cotton,** the viscosity being measured by the time required for a ball of specified size and material to drop through a solution of specified composition under standard conditions. It has a chain length of 150 to 200 C_6 units. **Nitrocellulose lacquers** consist of pigment, half-second cotton, and a mixture of solvents. The solvents are grouped as *low boilers, medium boilers,* and *high boilers.* The low boilers evaporate rapidly leaving a thick film that still can flow enough to remove brush or spray marks. Evaporation of the medium boilers leaves a dry film. The high boilers remain in the film to plasticize it, that is, to keep it flexible. Addition of pigments gives colored lacquers. Lacquers were introduced as finishes for automobiles in 1924, to replace varnish, and the ease of application, rapid drying, and durability revolutionized the industry. It is said that the development of chromium plating received its chief impetus because the new body finishes outlasted the nickel-plated trim. Nitrocellulose lacquers have been supplanted largely by synthetic alkyd baking enamels (p. 474) or coatings based on poly(methyl methacrylate) (p. 620) for automobile finishes but still are used extensively for refinishing automobiles and for household decoration.

The **cellulose acetates** are the most important esters. The maximum amount of acetyl that can be introduced by direct esterification corresponds to the acetylation of one of the three hydroxyl groups of each C_6 unit, and the product is not soluble in organic solvents and has no commercial use. When cellulose is pretreated by soaking under carefully controlled conditions in a dilute solution of sulfuric acid in glacial acetic acid and then is added to a mixture of acetic anhydride and acetic acid, acetylation takes place and a viscous solution results. Dilution with water precipitates a product having 2.9 hydroxyl groups esterified with acetic acid to 0.1 esterified with sulfuric acid. Treatment with water and magnesium hydroxide hydrolyzes the sulfuric ester and a slight amount of the acetyl groups to give 2.85 acetyl groups to 0.15 hydroxyl groups. This product is insoluble in acetone but is soluble in methylene chloride containing a small amount of ethanol and is the **"triacetate"** used to manufacture photographic film base. A further slight hydrolysis with water gives a product having 2.4 acetyl groups to 0.6 hydroxyl groups, which is the "triacetate" used for fiber or plastics.

The older commercial material is the so-called **cellulose diacetate.** After acetylation to the triacetate stage, some water is added to the acetylation mixture, and hydrolysis is permitted to take place until approximately the diacetate is formed. The product is precipitated with water, washed, and dried. It is soluble in acetone, gives flexible films and fibers, and is thermoplastic. This material is not molecularly homogeneous but is a mixture having approximately the composition of the diacetate.

Acetate-propionates and **acetate-butyrates** are prepared by using propionic or butyric acid along with acetic anhydride during the acetylation. Because of the longer hydrocarbon chains, these products are thermoplastic without being partially hydrolyzed, and since they contain fewer free hydroxyl groups, they are less permeable to water, weather better, and are more compatible with gums and plasticizers.

Cellulose Ethers. A number of cellulose ethers are of technical importance. **Ethylcellulose** is prepared by soaking high α-cellulose wood pulp or cotton linters with 17.5 per cent sodium hydroxide solution and removing the excess solution in a hydraulic press. The product is known as *alkali cellulose* and has the approximate composition $(C_6H_{10}O_5 \cdot 2\ NaOH)_x$, although it does not seem to be a definite compound. The alkali cellulose is shredded to crumbs and treated in an autoclave with ethyl chloride at 125° for 8 to 12 hours. The ethoxyl content obtained depends on the concentration of the alkali and on the time and number of treatments. The ethylcelluloses differ in solubility characteristics depending on the extent of reaction. Products containing 0.8 to 1.4 ethoxyl groups per C_6 unit are soluble in water, whereas those containing more than 1.5 ethoxyl groups are soluble in most organic solvents. The most widely used commercial product contains 2.2 to 2.6 ethoxyl groups. It is highly compatible with resins and plasticizers and is soluble in mixtures of alcohol and hydrocarbons. It is used in the manufacture of coatings, films, and plastics. Unlike the esters, it is resistant to the action of alkalies.

The **methylcelluloses** of low methoxyl content are soluble in cold water and precipitate when the solution is heated. They are used chiefly as thickeners for textile printing and cosmetics, as sizing, and as finish for textile fibers.

The water solubility of the lower acetates and ethers seems anomalous because acylation or etherification of a hydroxyl group ordinarly decreases the solubility in water. The insolubility of cellulose in water, however, may be explained by hydrogen bonding between chains (p. 117). The replacement of a few hydrogen atoms by hydrocarbon or acyl groups separates the chains and renders these forces inoperative. Hydration of the hydroxyl groups and solution then can take place. In accord with this view, the solubility in water decreases at higher temperatures because the hydrates become less stable.

Carboxymethylcellulose (*CMC*) is prepared by the reaction of alkali cellulose with sodium chloroacetate.

$$ROH + ClCH_2COONa + NaOH \longrightarrow R{-}O{-}CH_2COONa + NaCl + H_2O$$

The commercial product contains about 0.5 carboxymethyl groups per C_6 unit, and the sodium salt is soluble in both cold and hot water. Metallic salts other than those of the alkali metals are insoluble in water. The *sodium salt* is used on a large scale as a thickening agent and protective colloid, as an additive for synthetic detergents, and as a size for textiles. It was used widely in Germany during World War II to extend detergents and as a replacement for starch. Significant commercial production began in the United States in 1946 and was over 45 million pounds in 1964. A purified product called *cellulose gum* is used as a thickening agent in foods, especially ice cream. **Hydroxyethylcellulose** is manufactured commercially by the reaction of cellulose with ethylene oxide (p. 605). It is used as a water-soluble thickening agent.

Synthetic Fibers from Cellulose. Synthetic fibers are textile fibers produced from a homogeneous solution or melt, whether the raw material is a synthetic or a natural product. Thus the source of carbon for some synthetic fibers is coke, but that for others is essentially unmodified cellulose or protein. Cellulose is the raw material for both rayon and acetate fibers. The general process of manufacture is the same for both, namely dissolving the cellulose derivative, forcing the solution through the fine holes of a die called a *spinneret,* and precipitating the cellulose or evaporating the solvent to give a thread. The threads have smooth cylindrical surfaces, which reflect light and give them a high luster. The strength of the threads can be increased greatly by stretching them while in a plastic state, which brings about a linear orientation of the molecules and hence an increase in the attractive forces between the molecules.

Rayon manufacture is based on the reaction of the hydroxyl groups of cellulose with carbon disulfide in the presence of sodium hydroxide to give *xanthates* (p. 292), which are soluble in water. The reaction can be reversed by acidification.

$$ROH + CS_2 + NaOH \longrightarrow \left[ROC(=S){-}S^-\right]Na^+ + H_2O$$

Sodium alkyl xanthate

$$\left[ROC(=S)-S^-\right]Na^+ + NaHSO_4 \longrightarrow ROH + CS_2 + Na_2SO_4$$

Alkali cellulose (p. 344) from high α-cellulose sulfite pulp or from cotton linters is treated in rotating drums with an amount of carbon disulfide equal to 30 to 40 per cent of the weight of dry cellulose (calculated amount for 1 mole of carbon disulfide per C_6 unit is 47 per cent). An orange-colored crumbly product is formed, which is dissolved in 3 per cent sodium hydroxide solution to give a viscous solution known as *viscose,* in which the cellulose molecule has been degraded to an average chain length of 400 to 500 C_6 units. The solution initially is unstable, and the viscosity drops rapidly during the first day of standing. A slow hydrolysis of xanthate radicals then takes place with a gradual rise in viscosity for about eight days when the solution rapidly sets to a gel. Spinning of the solution is carried out after it has ripened four to five days, when the viscosity is not changing rapidly but after considerable hydrolysis has taken place. The filaments from the spinneret are passed through a bath of sodium bisulfate and additives, where the hydrolysis is completed to give a regenerated cellulose fiber. After coagulation and twisting of the filaments into a thread, the product is thoroughly washed and sometimes bleached. The number of holes in the spinneret depends on the number of filaments desired in the thread and may vary from 15 to 500. The size of the filaments or thread is measured in *denier,* which is the weight in grams of 9000 meters of thread. The common size of the filaments is 2 to 3 denier and of the thread 150 denier, but rayon thread is made as fine as 50 and as coarse as 1000 denier. Ordinary viscose rayon has 50 to 80 per cent of the strength of silk and loses 50 to 60 per cent of its strength when wet. If the spinning is carried out under tension while the filaments still are plastic, orientation of the molecules takes place and the product has a strength equal to that of silk, wet or dry. Since around 1930, filaments of 1 to 20 denier have been cut to lengths of 1 to 6 inches to give **staple fiber.** Here the number of filaments from a single spinneret may be as high as 10,000. Staple fiber is combined with wool, cotton, or linen fibers or with other synthetic fibers in spun yarn, or is converted to **spun rayon** by the usual spinning process for cotton or wool.

Viscose solution is converted into **cellophane** by extruding through a slot into the coagulating bath. The addition of glycerol improves the flexibility, and a coating of wax and nitrocellulose lacquer makes it impervious to water vapor. Cellophane once was the dominant transparent film, but polyethylene film now is cheaper per square foot. Yearly production of cellophane still is increasing, but it now accounts for less than 50 per cent of the market. Modification of the extruding die gives sausage casings, artificial straw, or filaments containing bubbles. Sealing caps of cellophane are kept moist and pliable in dilute glycerol. When placed over a bottle top, they dry and shrink to a tight fit. Synthetic sponges are made by incorporating crystals of Glauber's salt ($Na_2SO_4 \cdot 10\,H_2O$) of all sizes in the viscose solution and coagulating in blocks. Leaching with warm water dissolves the crystals and leaves the sponge-like mass.

Acetate fiber is made by dissolving cellulose diacetate in acetone or triacetate in methylene chloride and forcing the solution through spinnerets into warm air, where the solvent evaporates. The thread does not require a finishing treatment, and since two thirds or more of the hydroxyl groups are esterified, it has a higher wet strength than ordinary viscose rayon. The smaller number of hydroxyl groups, however, makes direct dyeing more difficult and special dyes are required (p. 562). Alternatively the dye may be incorporated in the spinning solution to give a colored thread. Acetate fiber can be distinguished easily from rayon by the fact that it still is soluble in acetone or methylene chloride. Rayon is regenerated cellulose and hence insoluble in organic solvents.

World production of rayon and acetate was 3 million metric tons in 1963, of which about 40 per cent was filament yarn and 60 per cent staple fiber. Of this amount about three fourths was made by the viscose process. In the United States viscose rayon and acetate each account for about half of production. World production of noncellulosic synthetic fibers such as the nylon (p. 631), polyester (p. 476), and acrylic (p. 619) fibers was over 1.3 million tons.

Dextrans. These highly branched glucosans are produced by the fermentation of sucrose solutions by certain bacteria (*Leuconostoc mesenteroides, Betacoccus arabinosaceus*). Dextrans having 90 per cent of the molecules in the molecular weight range of 50,000 to 100,000 are prepared by partial hydrolysis of native dextran, and have been used as blood plasma extenders in the treatment of shock caused by loss of body fluids. Industrial uses for nonclinical dextrans also are being developed.

Pectins and Pectic Acids. Fruits and berries contain water-insoluble complex carbohydrates known as *protopectins,* which yield on partial hydrolysis **pectins** or **pectinic acids.** Pectins have the property of forming gels with sugar and acid under proper conditions and are the agents necessary for the production of jellies from fruit juices. The pectins or pectinic acids contain both carboxyl groups and methoxycarbonyl groups. Removal of the remainder of the methoxy groups by hydrolysis yields the **pectic acids.** Enzymic hydrolysis of purified pectic acids gives up to 85 per cent of D-galacturonic acid and no other carbohydrates have been identified. The pectic acids contain one free carboxyl group per C_6 unit. These and other experimental data have led to the view that the pectic acids essentially are linear polygalacturonides with α linkages. In the pectins the carboxyl groups are esterified partially with methyl groups.

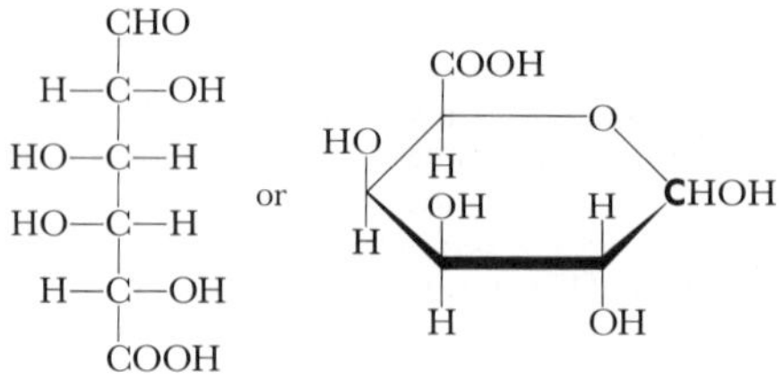

D-Galacturonic acid

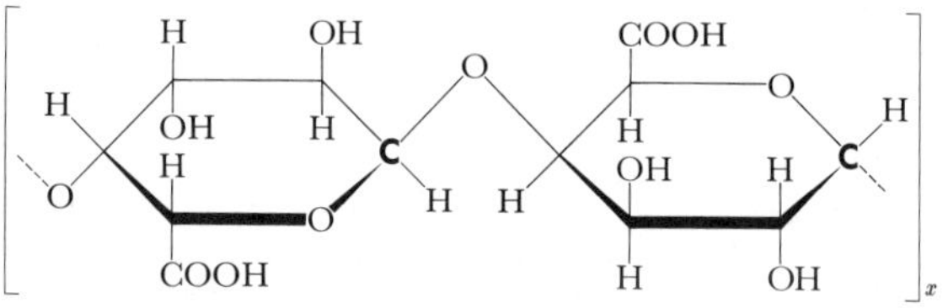

Pectic acid

Galacturonic acid belongs to the group of substances known as **uronic** or **glycuronic acids** in which the hydroxymethyl group of the aldose is replaced by a carboxyl group, the aldehyde group being unchanged. In the *glyconic acids* (cf. p. 317) the aldehyde group is replaced by the carboxyl group, and in the *glycaric acids* (p. 321) both terminal carbon atoms are in carboxyl groups. The uronic acids and their polymers, when heated with 12 per cent hydrochloric acid, lose one molecule of carbon dioxide for each carboxyl group. The other products are furfural and water.

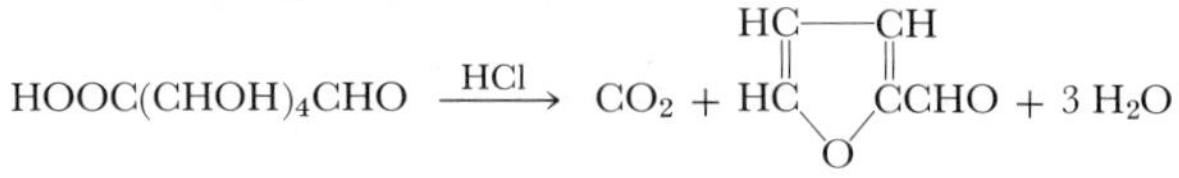

The reaction is used for the quantitative estimation of uronic acid and polyuronides. The carbon dioxide is absorbed in barium hydroxide solution and the amount of barium carbonate that precipitates is weighed.

Chitin. The shells of crustacea and the structural substance of insects and fungi consist of the polysaccharide **chitin,** which contains nitrogen. Enzymic hydrolysis yields *N*-acetyl-2-amino-2-deoxyglucose (*N-acetyl-2-glucosamine*). The structure of chitin appears to be identical with that of cellulose except that the acetylamino group, $CH_3CONH—$, replaces the hydroxyl at C-2.

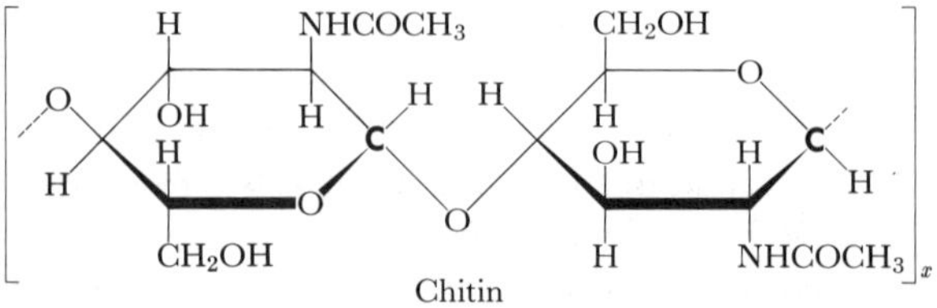

Chitin

Heteropolysaccharides

Inulin. This starch-like substance occurs in many plants, particularly members of the *Compositae,* for example inula, Jerusalem artichoke, goldenrod, dandelion, dahlia, and chicory. It is obtained readily from dahlia tubers or chicory roots. Hydrolysis yields fructose and a small amount of glucose. Methylation followed by hydrolysis yields 91 per cent of 3,4,6-tri-*O*-methylfructose, 3 per cent of a tetra-*O*-methylfructose, 2 per cent of a tetra-*O*-methylglucose and smaller amounts of a mixture of tri-*O*-methylglucoses. The amount of tetra-*O*-methylfructose indicates a chain length of about 33 C_6 units. Osmotic pressure measurements on the acetate and methyl ether indicate a molecular weight corresponding to about 30 C_6 units. Hydrolysis by sucrase and the levo rotation of solutions point to a β linkage. It has been postulated that the structure is a chain of β-fructofuranose units joined at the 1 and 2 positions and terminated with a sucrose-type linkage to glucose. Possibly a second glucose unit also forms a part of the molecule.

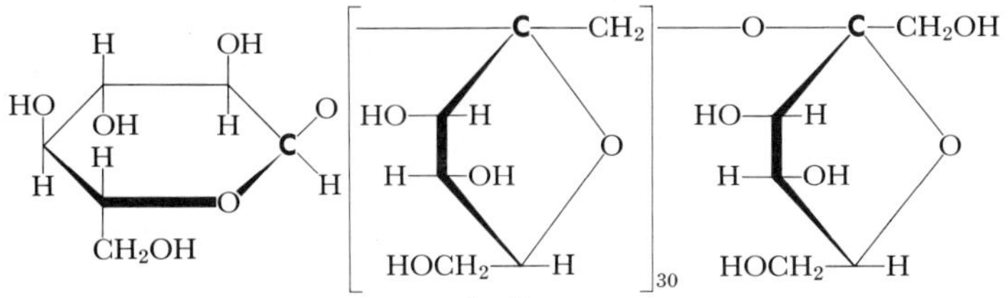

Inulin

Alginic Acid. The gelatinous material called *algin* that is present in the cell walls of most brown algae (*Phaeophyceae*) consists of salts of the polysaccharide alginic acid. It is extracted commercially from sea weeds, chiefly *Macrocystis pyrifera,* the giant kelp that grows off the coast of southern California. It is obtained commercially also from other species such as *Laminaria digitata* and *Ascophyllum nodosum.* The various commercial algins are sodium, ammonium, potassium, or calcium salts. They are used as thickening agents and emulsifiers for foods, especially ice cream, and for pharmaceuticals and cosmetics, as textile sizes, and as modifying agents for numerous industrial products. Alginic acid yields on hydrolysis roughly two moles of D-mannuronic acid to one mole of L-guluronic acid and is believed to be a poluronide with β linkages between the 1,4 positions.

Agar. Hot water extracts of certain East Indian seaweeds (various species of *Gelidium*) set to a gel on cooling. Freezing of the gel and subsequent thawing precipitates a material that can be removed by filtration. Drying of the precipitate to about 35 per cent moisture content gives translucent flakes known as **agar.** Solutions of agar in hot water again set to a gel when cooled. Its chief use is to prepare a support for nutrient media in the growing of microorganisms. The dry product swells in warm water without dissolving and is indigestible, making it useful also as a bulk cathartic.

Two components have been isolated from agar, one of which has been named *agarose* and the other *agaropectin.* **Agarose,** the main component, is believed to consist of a chain of alternating D-galactose and 3,6-anhydro-L-galactose units. The D-galactose units are linked β to the 4 positions of the anhydro-L-galactose units, and the latter are linked α to the 3 positions of the D-galactose units.

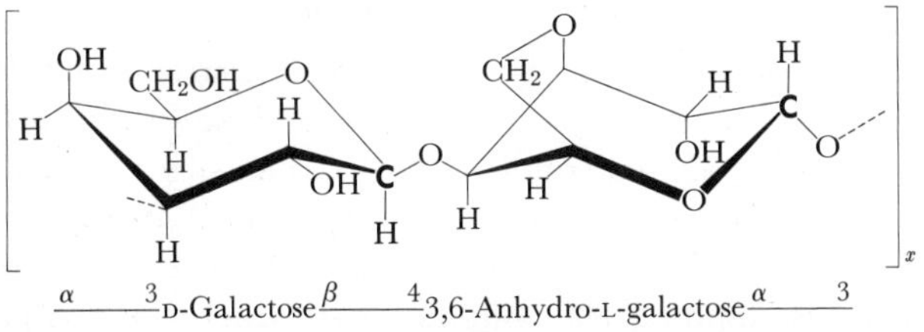

α — 3 D-Galactose β — 4 3,6-Anhydro-L-galactose α — 3

Agarose

Agaropectin is more complicated and less well characterized. It yields on hydrolysis some D-glucuronic acid and sulfuric acid in addition to D-galactose and 3,6-anhydro-L-galactose.

Hemicelluloses. This material formerly was defined as water-insoluble, alkali-soluble material closely associated with cellulose. More recently the class has been redefined as noncellulosic, nonpectin polysaccharides present in the cell walls of higher land plants. It makes up 15 to 25 per cent of woods and 25 to 40 per cent of agricultural crop residues such as straw, cornstalks, corncobs, and bean and grain hulls. The main classes are grouped as *xylans, mannans,* and *galactans* according to the principal monose unit. The molecular sizes range from 50 to 200 sugar units.

Xylans are the most abundant and most widely distributed. Xylan from asparto grass appears to be a linear homopolysaccharide consisting of 1,4-linked β-D-xylopyranose units. This structure is like that of cellulose, except that hydrogen replaces the hydroxymethyl group. Like cellulose it exists in the chair form with all hydroxyl groups equatorial. Many other xylans have the same basic structure but have other sugars as end groups or attached as side chains. L-Arabinose is especially common and usually is present as the nonreducing furanose form. D-Glucuronic acid and its 4-*O*-methyl ether also frequently are present as side chains.

Ivory nuts yield a **mannan** having 1,4-linked β-D-mannopyranose units as a backbone. Some nonreducing D-galactose units are attached as end groups or as single-unit side chains. **Glucomannans** consisting of D-glucose and D-mannose joined 1,4 by β linkages make up half of the hemicellulose fraction of the wood of conifers and are present also in hardwoods. The ratio of glucose to mannose varies, and it is not known if they are linked in a regular or a random fashion. Most are essentially linear in structure but some may be branched.

Galactans and **arabinogalactans** are especially abundant in larches. They are all highly branched with predominantly 1,6- and 1,3-linked galactose units. The principal heterosugar is L-arabinose, which is present mostly as nonreducing furanose side chains attached 1,6, but β-L-arabinopyranose units also may be present. An arabinogalactan is extracted commercially from western larch (*Larix occidentalis*) and is competitive with gum arabic (p. 348).

Plant Gums and Mucilages. These substances resemble the hemicelluloses chemically and sometimes are classed with them. They are much more complex, however, have different physical properties, and frequently are not associated with cellulose. They dissolve in water to give mucilaginous solutions or gels that are used widely as adhesives or thickening agents. **Gum arabic** is the most important polysaccharide gum of commerce, the annual production being around 50 million pounds. It is obtained as an artificially induced exudation from various species of *Acacia,* particularly *Acacia verek.* Other gums of commercial importance are **gum tragacanth,** an exudation of shrubs of the genus *Astragalus,* **locust bean** or **carob bean flour** from the endosperm of the seed of the Mediterranean locust tree (*Ceratonia siliqua*), and **guar flour** from the endosperm of the guar bean (*Cyamopsis proraloides* or *C. tetragona*), a native of India, which now is grown commercially in southern United States.

SUGAR ALCOHOLS

Although sugar alcohols are not carbohydrates in the strict sense, the naturally occurring sugar alcohols are so closely related to the carbohydrates that they are discussed here rather than with other polyhydric alcohols (p. 302). The sugar alcohols most widely distributed in nature are sorbitol (identical with D-glucitol), D-mannitol, and galactitol (*dulcitol*). They correspond to the reduction products of glucose, mannose, and galactose. Since the reduction of the sugars gives compounds in which the end groups are identical, the reduction products of two different sugars frequently are identical. Thus L-arabitol is identical with L-lyxitol. Similarly D-glucose on reduction gives a sugar alcohol identical with sorbitol, one of the two reduction products of L-sorbose. Sorbitol was the name first assigned to the naturally occurring compound and commonly is used rather than D-glucitol. The designation of family for *meso* forms such as galactitol is meaningless.

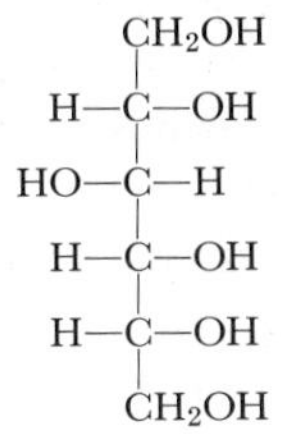

Sorbitol (D-*glucitol*)	D-Mannitol	Galactitol (*dulcitol*)
CH_2OH	CH_2OH	CH_2OH
H—C—OH	HO—C—H	H—C—OH
HO—C—H	HO—C—H	HO—C—H
H—C—OH	H—C—OH	HO—C—H
H—C—OH	H—C—OH	H—C—OH
CH_2OH	CH_2OH	CH_2OH

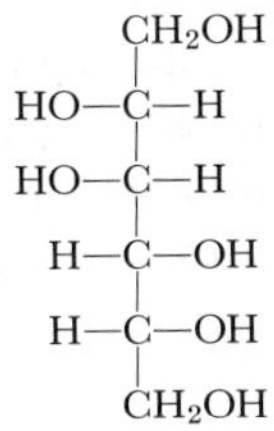

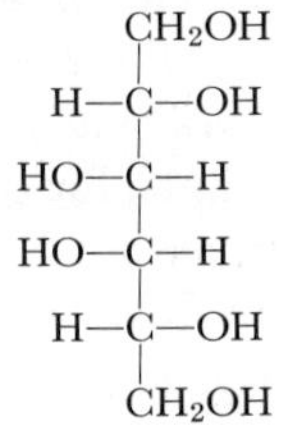

Sorbitol first was isolated from the berries of the mountain ash (*Sorbus aucuparia*) in 1872. The red seaweed, *Bostrychia scorpoides,* contains almost 14 per cent sorbitol. It has been isolated in appreciable quantities from many other plants ranging from the algae to the higher orders. Sorbitol is the most important of the sugar alcohols and is manufactured by the catalytic hydrogenation of glucose. United States production is over 50 million pounds per year. The largest amount (25 per cent) is used for the synthesis of vitamin C (p. 332). Surface-active agents, foods, cosmetics, pharmaceuticals, and miscellaneous industrial uses consume the rest.

D-Mannitol occurs in many land and marine plants and, in contrast to sorbitol, is present frequently in the plant exudates known as *mannas,* for example in the exudates of the manna ash (*Fraxinus ornus*), of the olive, and of the plane trees.[6] The amount of mannitol in the seaweed *Laminaria digitata,* from which alginic acid is obtained (p. 347), varies from as low as 3 per cent (dry weight) in the winter months to as high as 37 per cent in the summer. **Galactitol** (*dulcitol*) also is present in many plants and plant exudates and can be made by the catalytic hydrogenation of galactose.

Inositols. Related to the sugar alcohols are the polyhydroxycyclohexanes or **cyclitols,** of which the hexahydroxy derivatives, known as the **inositols,** are of most interest. Nine stereoisomeric forms are possible, two of which are optically active and seven of which are *meso* forms. The all-*trans* cyclitol is the most stable because all of the hydroxyl groups can be equatorial in one of the chair forms. It has been estimated that an isomer that must have one axial hydroxyl is less stable than the isomer with all hydroxyl groups equatorial by 0.9 kcal., one with two on the same side of the ring by 2.8 kcal., and one with three on the same side of the ring by 5.7 kcal. It is of interest that none of the four naturally occurring inositols need have more than one axial hydroxyl on the same side of the ring and that where a methyl ether occurs naturally, its structure and configuration are such that in the more stable conformation the methoxyl group is equatorial.

***scyllo*-Inositol** (*scyllitol*) has been isolated from the organs of elasmobranch fish, and from acorns, coconut palm leaves, and the flowers and bracts of the flowering dogwood.

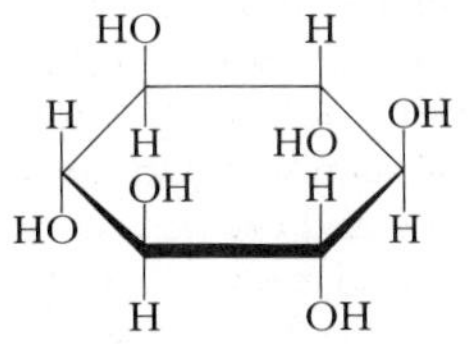

scyllo-Inositol

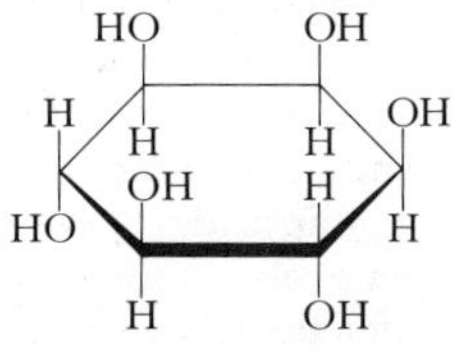

myo-Inositol

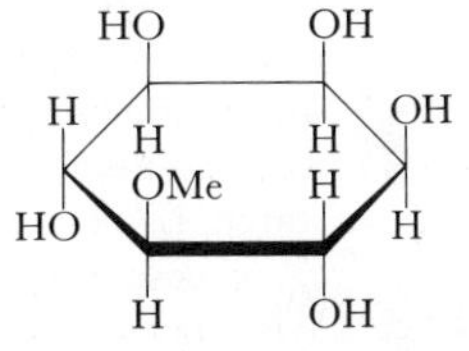

Sequoyitol

[6] The nature of the manna eaten by the Israelites in the wilderness (*Exodus XVI, 14*) is not known. The best evidence indicates that it was a honeydew secreted by insects.

***myo*-Inositol,** formerly called *meso*-inositol, probably is universally present in all living cells. The hexaphosphoric acid ester is called **phytic acid,** and its calcium magnesium salt, isolated commercially from corn steep water (p. 339), has the trade name **Phytin.** *myo*-Inositol (*bios I*) is a component of the vitamin B complex and at one time was believed to be required in the diet. It is necessary for the synthesis of certain phospholipids, but the animal organism can synthesize it from other compounds. Various methyl ethers of *myo*-inositol occur in plants, for example **sequoyitol** from the heartwood of *Sequoia sempervirens.*

Streptomycin, an antibiotic from *Streptomyces griseus* that is more active against gram-negative organisms than penicillin (p. 524), is used along with *p*-aminosalicylic acid (p. 473) and isoniazide (p. 518) in the treatment of tuberculosis. It yields **streptamine** as one of the products of hydrolysis. Streptamine has two amino groups instead of two of the hydroxyl groups of scyllitol. The amino groups are derived from guanidyl groups in the streptomycin molecule. The other constituents are a dialdo deoxy sugar named **streptone,** and *N*-methyl-L-glucosamine.

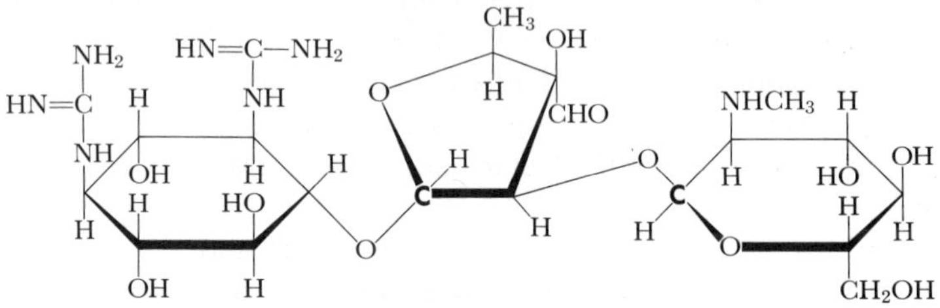

Streptomycin

PROBLEMS

18–*1*. How many straight-chain aldoheptoses are possible (*a*) excluding α and β forms; (*b*) including α and β forms of both furanoses and pyranoses?

18–*2*. To which of the given materials do the following descriptive phrases apply: (*a*) a pentose, (*b*) has a furanose ring, (*c*) is highly branched, (*d*) regenerated cellulose, (*e*) a ketose, (*f*) contains a β linkage, (*g*) a linear polymer, (*h*) exists in α and β forms, (*i*) a polyfructoside, (*j*) an explosive, (*k*) a component of starch, and (*l*) an acetal; (*1*) amylose, (*2*) sucrose, (*3*) viscose rayon, (*4*) xylose, (*5*) maltose, (*6*) cellulose, (*7*) fructose, (*8*) glycogen, (*9*) methyl α-glycoside, (*10*) inulin, and (*11*) cellulose nitrate?

18–*3*. Pair the given structural features of monosaccharides with the proper given chemical behavior: (*a*) the presence of hydroxyl β to a carbonyl group, (*b*) the effect of a carbonyl group on an adjacent CHOH group, (*c*) the presence of hydrogen on carbon α to a carbonyl group, and (*d*) hemiacetal ring formation; (*1*) mutarotation, (*2*) easy isomerization by alkali, (*3*) oxidation by Fehling solution, and (*4*) fragmentation by alkali.

18–*4*. How may the members of the following pairs of compounds be distinguished easily from each other: (*a*) arabinose and glucose, (*b*) ribose and arabinose, (*c*) glucose and mannose, (*d*) rhamnose and arabinose, (*e*) xylose and galactose, (*f*) glucose and maltose, (*g*) maltose and lactose, (*h*) cellobiose and sucrose, (*i*) inulin and starch, (*j*) pectic acid and alginic acid, (*k*) cellulose acetate and cellulose nitrate, (*l*) viscose rayon and acetate fiber, (*m*) ethylcellulose and cellulose acetate?

18–*5*. Write perspective formulas for the following disaccharides: (*a*) 4-*O*-(α-L-arabinopyranosyl)-β-D-galactopyranose, (*b*) 6-*O*-(β-D-ribofuranosyl)-α-D-glucopyranose, (*c*) β-D-mannopyranosyl α-L-rhamnopyranoside.

18–*6*. Write the Haworth representation for β-L-talopyranose, and the conformational representation for the more stable form.

18–7. A pentose of the D-family gave an optically inactive glycaric acid on oxidation. When it was degraded to the tetrose and the tetrose was oxidized, *meso*-tartaric acid was formed. What is the configuration and name of the pentose?

18–8. An aldoheptose belonging to the D-family was oxidized with nitric acid and gave a *meso* pentahydroxy dibasic acid. When the heptose was degraded to the hexose and oxidized to the dibasic acid, an active form was obtained. If the hexose first was made to undergo epimeric change and then oxidized, the dibasic acid was inactive. What is the configuration of the heptose?

CHAPTER NINETEEN

α-AMINO ACIDS, PROTEINS, AND PEPTIDES

α-Amino acids, $RCH(NH_2)COOH$, are the final products of hydrolysis of proteins (Gr. *proteios* of first importance). The latter are complex polyamides of high molecular or particle weight (6000 to 322,000,000). Proteins are present in all living tissue, but certain tissues, such as seeds and flesh, contain larger amounts than others, such as fat and bone. Other biologically indispensable substances that are proteins are the enzymes, which catalyze most biochemical reactions; some of the hormones, which regulate metabolic processes; respiratory proteins such as hemoglobin; antibodies, which are produced to counteract the invasion of harmful agents; and viruses.

Plants and bacteria are the ultimate source of all proteins because the animal organism is unable to synthesize certain essential amino acids (p. 365) from inorganic nitrogen compounds. In the synthesis of amino acids, plants are aided by soil microorganisms. The nitrite bacteria change ammonia to nitrites, and nitrate bacteria change nitrites to nitrates. The nitrates and ammonia are converted by the plant first into α-amino acids and then into proteins. Other soil bacteria are able to change organic nitrogen into ammonia, thus completing the cycle. In addition, certain soil bacteria in conjunction with the plants on whose roots they grow, namely the legumes, are capable of converting atmospheric nitrogen into amino acids. Some soil organisms can convert nitrogen into ammonium ion nonsymbiotically, and still others are able to carry out the reverse processes of reduction of nitrate and nitrite to nitrogen and ammonia. These various processes, known as the **nitrogen cycle,** are illustrated diagrammatically in Fig. 19–*1*. The animal organism ingests proteins from plants or other animals, hydrolyzes them to α-amino acids, and then uses the α-amino acids to synthesize its own body proteins (p. 365).

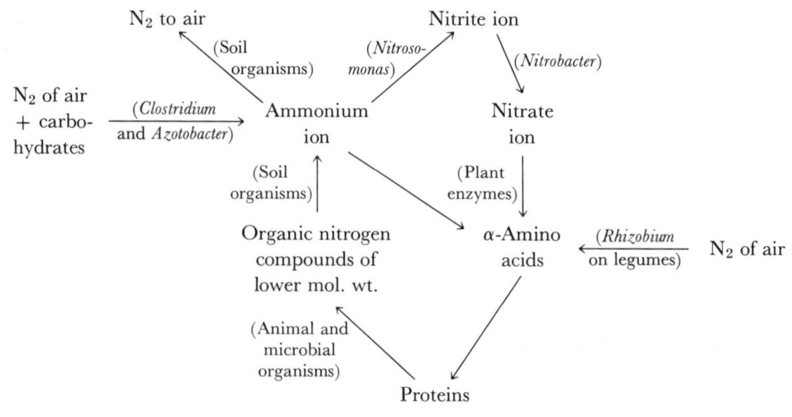

Figure 19–*1*. The nitrogen cycle.

α-Amino Acids

Natural Occurrence. The amino acids commonly obtained by the hydrolysis of proteins are listed in Table 19–*1*. They are divided into *neutral, basic,* and *acidic* amino acids according to whether the number of basic groups is equal to the number of acidic groups,

TABLE 19–*1*. COMMON AMINO ACIDS DERIVED FROM PROTEINS

Neutral Amino Acids (equal number of amino and carboxyl groups)

1. Glycine or aminoacetic acid, $CH_2(NH_2)COOH$
2. Alanine or α-aminopropionic acid, $CH_3CH(NH_2)COOH$
3. Valine or α-aminoisovaleric acid, $(CH_3)_2CHCH(NH_2)COOH$
4. Leucine or α-aminoisocaproic acid, $(CH_3)_2CHCH_2CH(NH_2)COOH$
5. Isoleucine or α-amino-β-methylvaleric acid, $CH_3CH_2CH(CH_3)CH(NH_2)COOH$
6. Serine or α-amino-β-hydroxypropionic acid, $HOCH_2CH(NH_2)COOH$
7. Threonine or α-amino-β-hydroxybutyric acid, $CH_3CH(OH)CH(NH_2)COOH$
8. Cysteine or α-amino-β-mercaptopropionic acid, $HSCH_2CH(NH_2)COOH$
9. Cystine or bis-(2-amino-2-carboxyethyl)disulfide,
   $HOOCCH(NH_2)CH_2SSCH_2CH(NH_2)COOH$
10. Methionine or α-amino-γ-(methylthio)butyric acid, $CH_3SCH_2CH_2CH(NH_2)COOH$
11. Phenylalanine or α-amino-β-phenylpropionic acid, $C_6H_5CH_2CH(NH_2)COOH$
12. Tyrosine or α-amino-β-(4-hydroxyphenyl)propionic acid,

```
          H       H
           \     /
            C5—C6
           //     \\
   HO—C4           1C—CH2CH(NH2)COOH
           \      /
            C3=C2
           /     \
          H       H
```

```
                                            H2C——CH2
                                             |    |
13. Proline or 2-pyrrolidinecarboxylic acid,* H2C    CHCOOH
                                              \  /
                                               N
                                               H
```

14. Tryptophan or α-amino-β-(3-indolyl)propionic acid,*

```
          H
          |
     H    C
      \ //  \
       C     C——————3C—CH2CH(NH2)COOH
       |     ||     ||
       C     C  1  2C
      /  \\ /  \  /  \
     H    C     N     H
          |     H
          H
```

Basic Amino Acids (more basic groups than carboxyl groups)

15. Lysine or α,ε-diaminocaproic acid, $H_2NCH_2CH_2CH_2CH_2CH(NH_2)COOH$
16. Arginine or α-amino-δ-guanidinovaleric acid,

```
      NH
      ||
  H2NCNHCH2CH2CH2CH(NH2)COOH
```

17. Histidine or α-amino-β-(5-imidazolyl)propionic acid,*

```
     N3——4CH
     ||    ||
    HC2   5C—CH2CH(NH2)COOH
      \ 1 /
        N
        H
```

Acidic Amino Acids (more carboxyl groups than amino groups)

18. Aspartic acid or aminosuccinic acid, $HOOCCH_2CH(NH_2)COOH$
19. Glutamic acid or α-aminoglutaric acid, $HOOCCH_2CH_2CH(NH_2)COOH$

* Heterocyclic compounds are discussed in Chapter 30.

greater than the number of acidic groups, or less than the number of acidic groups. In addition to the amino acids generally obtained, a few have been derived from specific proteins. **Hydroxyproline** (*4-hydroxy-2-pyrrolidinecarboxylic acid*) and **hydroxylysine** (*α,ε-diamino-δ-hydroxycaproic acid*) have been obtained only from collagen. **Iodogorgoic acid** (*3,5-diiodotyrosine*) and **thyroxine** (*3,5,3′,5′-tetraiodothyronine*) are hydrolytic products of thyroglobulin.

$$\text{HO}-\underset{\text{(3′,5′-diiodo)}}{C_6H_2I_2}-\text{O}-\underset{\text{(3,5-diiodo)}}{C_6H_2I_2}-\text{CH}_2\text{CH(NH}_2\text{)COOH}$$

Thyroxine

Bromotyrosine and other iodotyrosines have been obtained from marine organisms. Both cysteine and cystine are listed in the table although they readily can be interconverted by oxidation of the thiol group or reduction of the disulfide. On the other hand *asparagine*, $HOOCCH(NH_2)CH_2CONH_2$, and *glutamine*, $HOOCCH(NH_2)CH_2CH_2CONH_2$, the mono amides of aspartic and glutamic acids, which also occur in proteins, are not considered as additional amino acids because ammonia is evolved on hydrolysis and only the parent acids are isolated. Numerous other α-amino acids such as *citrulline*, $H_2NCONH(CH_2)_3CH(NH_2)COOH$, *ornithine*, $H_2N(CH_2)_3CH(NH_2)COOH$, and *α-aminoadipic acid*, $HOOC(CH_2)_3CH(NH_2)$-$COOH$, have been isolated from natural sources but are not building units of proteins. In fact by 1960 some 70 nonprotein amino acids, the majority of which are α-amino acids, had been isolated, chiefly from plants.

For all α-amino acids except glycine, the α carbon is asymmetric. Chemical interrelations have shown that for all amino acids derived from plants or animals, the configuration of the α carbon is the same. Physical and chemical evidence that this configuration is *S* (p. 312) was confirmed in 1954 by X-ray analysis of natural isoleucine hydrobromide.

$$\begin{array}{c} COOH \\ H_2N \diamond H \\ R \end{array} \equiv \begin{array}{c} COOH \\ H_2N-\!\!\!+\!\!\!-H \\ R \end{array} \equiv \begin{array}{c} COOH \\ R-\!\!\!+\!\!\!-NH_2 \\ H \end{array}$$

S- (or L-)α-Amino acids

If in the formula $RCH(NH_2)COOH$, the carboxyl, amino, hydrogen, and R groups are considered to correspond respectively to the aldehyde, hydroxyl, hydrogen, and hydroxymethyl groups of glyceraldehyde, then the natural α-amino acids have the same configuration as (−)-glyceraldehyde. By analogy to the usual nomenclature of the carbohydrates, they commonly are said to belong to the L family. Isoleucine and threonine have two asymmetric β carbon atoms, which have the configurations *R* and *S* respectively, and hence the systematic names are 2*S*,3*S*-2-amino-3-methylvaleric acid, and 2*S*,3*R*-2-amino-3-hydroxybutyric acid.

$$\begin{array}{cc} COOH & \\ H_2N-\!\!\!+\!\!\!-H & S \\ CH_3-\!\!\!+\!\!\!-H & S \\ C_2H_5 & \end{array} \qquad \begin{array}{cc} COOH & \\ H_2N-\!\!\!+\!\!\!-H & S \\ H-\!\!\!+\!\!\!-OH & R \\ CH_3 & \end{array}$$

L-Isoleucine L-Threonine

It will be noted that for the α-amino acids the lowest-numbered asymmetric carbon atom determines the family whereas for carbohydrates it is the highest-numbered asymmetric carbon.

Although amino acids having the D configuration have not been obtained from animal or vegetable sources, numerous products isolated from microorganisms have yielded on hydrolysis the D form of valine, leucine, serine, cysteine, phenylalanine, aspartic acid, and glutamic acid. Microorganisms account also for around a dozen other D amino acids that have structures different from those of amino acids derived from proteins.

Some Properties of Amino Acids. Since amino acids contain potentially free amino and carboxyl groups, they are amphoteric electrolytes. The solid dry amino acid is an *inner salt* or **dipolar ion,** the acidic carboxyl group having transferred its proton to the basic amino group. In aqueous solution, however, equilibrium is established among this form and the anion and cation forms of the amino acid.

$$\underset{\text{Anion form}}{H_2N{-}\overset{\overset{\displaystyle R}{|}}{C}H{-}COO^-} \underset{H_2O}{\overset{^-OH}{\rightleftharpoons}} \underset{\text{Dipolar ion}}{H_3\overset{+}{N}{-}\overset{\overset{\displaystyle R}{|}}{C}H{-}COO^-} \underset{H_2O}{\overset{H_3O^+}{\rightleftharpoons}} \underset{\text{Cation form}}{H_3\overset{+}{N}{-}\overset{\overset{\displaystyle R}{|}}{C}H{-}COOH}$$

The position of equilibrium varies with the amino acid and with the acidity of the solution. In strongly acidic solution the amino acid is largely in the cation form, whereas in strongly basic solution it is largely in the anion form. The acidity of the solution, expressed as *p*H (the logarithm of the reciprocal of the hydrogen ion concentration $= \log 1/[H^+]$), at which the concentration of the dipolar form is at a maximum, is known as the **isoelectric point** of the amino acid and frequently is designated as the *p*I of the amino acid. This point usually is not the neutral point (*p*H 7). The isoelectric point depends on the basicity of the amino group and on the acidity of the carboxyl group, which vary with the nature of the R group. Moreover an excess of amino groups over carboxyl groups or of carboxyl over amino groups, as well as the presence of other functional groups such as the phenolic hydroxyl in tyrosine or the thiol group of cysteine, also affect the isoelectric point. Hence it is characteristic of each amino acid. Since proteins and peptides also contain carboxyl and amino groups (p. 359), the same considerations apply to them as apply to the amino acids.

Differences in isoelectric point are useful for identification purposes, but they are particularly important in the isolation and purification of amino acids and proteins. Usually minimum solubility occurs at the isoelectric point. Hence the order of precipitation or crystallization can be varied by changing the *p*H of a solution. If an electromotive force is applied to two electrodes placed in a solution of an amino acid at its isoelectric point, no migration takes place because the negative charge balances the positive charge. At a *p*H greater than the isoelectric point, the amphoteric electrolyte has an excess of negative charge and moves towards the anode; at a *p*H below the isoelectric point, it has an excess of positive charge and moves towards the cathode. Clearly a mixture of amino acids, peptides, or proteins can be separated in this way, since they will migrate in one direction or the other at different rates that depend on their isoelectric point and the *p*H of the solution. This process is called **electrophoresis** (Gr. *phoresis* being borne). For micro separations, a potential difference may be applied to the two ends of a horizontal strip of paper or a layer of inert powder impregnated with the solution (p. 45). Electrophoresis also may be combined with elution paper chromatography (p. 45) to produce a more efficient separation.

Certain reactions are useful for the **analysis** of amino acids. Free amino groups in either amino acids or proteins can be estimated quantitatively by measuring the volume of nitrogen liberated on reaction with nitrous acid (*Van Slyke method*).

$$RCH_2CHNH_2COOH + HONO \longrightarrow RCH{=}CHCOOH + N_2 + 2\,H_2O$$

Although the reaction as usually written indicates the replacement of the amino group by a hydroxyl group, unsaturated and rearranged compounds also may be formed (p. 229). The important fact is that one mole of nitrogen is eliminated for each free amino group in the molecule. Cupric ion in alkaline solution gives with α-amino acids deep-blue copper chelate compounds (p. 604) in which one atom of copper is combined with two molecules of amino acid. They can be used for the colorimetric estimation of α-amino acids or for their isolation and purification.

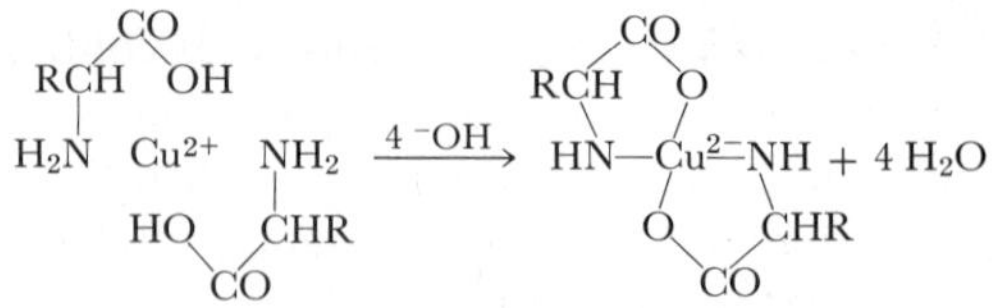

$$\begin{array}{ccc} & CO & \\ RCH & & OH \\ | & & \\ H_2N & Cu^{2+} & NH_2 \\ & & | \\ HO & & CHR \\ & CO & \end{array} \xrightarrow{4\ ^-OH} \begin{array}{ccc} & CO & \\ RCH & & O \\ HN & -Cu^{2-}- & NH \\ O & & CHR \\ & CO & \end{array} + 4\,H_2O$$

A more widely used colorimetric procedure involves the **ninhydrin reaction.** Most α-amino carboxylic acids when treated with a dilute solution of triketohydrindene hydrate (*ninhydrin,* p. 646) give a blue to purple product with intense absorption in the region 550–570 mμ. Proline, hydroxyproline, and asparagine give yellow to orange-brown solutions with a broad absorption band around 440 mμ. The reaction is used not only to locate amino acids on a paper chromatogram but also for colorimetric quantitative estimation. Measurement of the carbon dioxide evolved in the reaction (p. 647) also can be used for quantitative analysis.

Analysis of Amino Acid Mixtures. Numerous specific reagents have been used for the isolation and estimation of individual amino acids from protein hydrolysates. Since about 1950, however, machines developed at the Rockefeller Institute have become available for the automatic analysis of complex mixtures of amino acids. If a solution of a base is passed through a column packed with an insoluble resin containing strongly acidic sulfonic acid groups (a cation-exchange resin, p. 484), the resin transfers a proton to the base, and the base is adsorbed because of the attraction between the negative charge on the resin and the positive charge on the substituted ammonium ion. The attractive force varies with the acidity constant of the base, the strongest bases being held most strongly. Hence if a solution of a mixture of amino acids of varying basicity is placed on the column and eluted by means of a buffer of the proper *p*H, the various amino acids pass through the column at different rates and thus are separated. In the automatic amino acid analyzer, the hydrolysate from about 2 mg. of protein is placed on the column, and a buffer of the desired *p*H is pumped through at a constant rate. The effluent is met by a capillary stream of ninhydrin reagent, which produces a color that has an intensity proportional to the concentration of the amino acid. The absorbance of the resulting solution is measured continuously at 570 mμ and 440 mμ as it flows through a cylindrical cell of 2 mm. bore and the absorbance is recorded on a chart as a function of the volume of effluent (Fig. 19–*2*). The positions of the peaks are characteristic of the amino acids, and the areas under the peaks correspond to the relative amounts. In the example given, the acids have been eluted in three stages using buffers of *p*H 3.25 and 4.25 on a 150-cm. column to elute the acidic and neutral amino acids, and one of *p*H 5.28 on a 15-cm. column to elute the basic amino acids. Once the process is started, it is completely automatic and is completed in about 24 hours. The accuracy is 100 ± 3 per cent.

Preparation of Amino Acids. α-Amino acids are important to the biochemist for use in research, to the physician for the treatment of deficiency diseases and for intravenous feeding, and to the public at large because of their use as supplements to foods for man and animals. Many of the L amino acids still are obtained best by the acid or enzymatic hydrolysis of proteins. Alkaline hydrolysis is avoided because the presence of hydrogen on the asymmetric carbon atom adjacent to the amide carbonyl group permits rapid racemization in the presence of base (p. 310). Racemization is not so rapid in acidic solutions but tryptophan is destroyed by acids because of the presence of the pyrrole nucleus (p. 508).

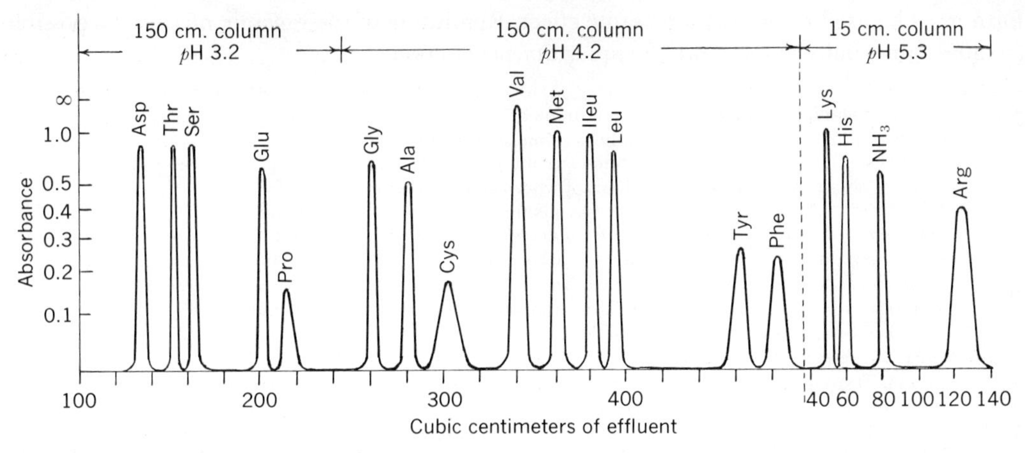

Figure 19–2. Typical recording from an automatic amino acid analyzer.

All of the natural amino acids have been synthesized in the laboratory. One of the more general methods is by the reaction of α-halogen acids with ammonia.

$$\underset{\displaystyle X}{\text{RCHCOOH}} + 2\,NH_3 \longrightarrow RCH(NH_2)COOH + NH_4X$$

A large excess of ammonia is used to decrease the amount of by-product formed by the reaction of two moles of halogen acid with one mole of ammonia. These by-products can be avoided by converting the α-halogen acid to the α-azido derivative with sodium azide and reducing the azido group catalytically (p. 251).

$$\underset{\displaystyle X}{\text{RCHCOOH}} + NaN_3 \longrightarrow NaX + \underset{\displaystyle N_3}{\text{RCHCOOH}} \xrightarrow{H_2/Pt} \underset{\displaystyle NH_2}{\text{RCHCOOH}} + N_2$$

Bromo acids most commonly are used in both of these procedures, since they are prepared readily by the Hell-Volhard-Zelinsky reaction (p. 169).

A second method is known as the *Strecker synthesis,* which consists of the reaction of an aldehyde or ketone with a mixture of ammonium chloride and sodium cyanide, followed by acid hydrolysis of the amino nitrile. The first stage of the reaction involves the formation of ammonium cyanide, which dissociates into ammonia and hydrogen cyanide. Reaction of ammonia with the aldehyde or ketone gives the ammonia-addition product, which reacts with hydrogen cyanide to give the amino nitrile.

$$NH_4Cl + NaCN \longrightarrow NH_4CN + NaCl$$

$$NH_4CN \rightleftharpoons NH_3 + HCN$$

$$RCHO \xrightleftharpoons{NH_3} \underset{\displaystyle NH_2}{\text{RCHOH}} \xrightleftharpoons[H_2O]{} RCH{=}NH \xrightarrow{HCN} \underset{\displaystyle NH_2}{\text{RCHCN}} \xrightarrow{H_2O,\ H^+} \underset{\displaystyle NH_2}{\text{RCHCOOH}}$$

Numerous other syntheses have been devised.

The synthetic product always is racemic (p. 306), but only the L form of amino acids can be incorporated into proteins. Of the essential amino acids (p. 365), animal organisms can convert the D form of tryptophan, phenylalanine, methionine, and histidine into the L form. Hence the DL mixture is as useful as the pure L form. For the other essential amino acids, only half of the synthetic mixture can be utilized, and an amount twice that of the L form must be used to produce the same effect. Resolution of the racemic mixtures is possible (p. 306) but usually is too costly for quantity production.

Commercial Production. Several amino acids are prepared in large amounts commercially either from natural sources or by synthesis. **Monosodium glutamate** (*MSG, Accent, Ajinomoto*) is used widely for enhancing the flavor of protein-containing foods. About half of that produced in the United States is obtained by the acid hydrolysis of wheat gluten, which yields up to 35 per cent glutamic acid. Most of the other half is prepared by alkaline hydrolysis of the residual liquor (*Steffen waste*) obtained in the production of sucrose from beets (p. 336). Recently a plant has been built to produce L-glutamic acid by fermentation of carbohydrate in a nutrient medium by means of *Micrococcus glutamicus*. Synthetic processes have been developed. Resolution is by seeding supersaturated solutions with the active forms. Total U.S. production of monosodium glutamate in 1964 amounted to 38.6 million pounds valued at 24.8 million dollars. Japanese production was over 114 million pounds, chiefly by fermentation and synthesis.

DL-Tryptophan is synthesized commercially because most of that occurring in proteins is destroyed during acid hydrolysis. Hence as an indispensable amino acid, it must be added to the protein hydrolysates used to treat cases of serious malnutrition. Synthetic **DL-methionine** or its α-hydroxy analog, which is equally effective, is used to fortify protein hydrolysates and to enrich foodstuffs. Small amounts greatly increase the rate of growth of young animals. The rate of consumption in the formulation of poultry feeds in 1964 was about 6 million pounds annually. **L-Lysine** is made commercially by synthesis and resolution, and by two fermentation processes. Total U.S. production of amino acids for use as therapeutic nutrients was over 1.7 million pounds in 1964.

PROTEINS

Structure of Proteins

The proteins are classified into **simple proteins,** yielding only α-amino acids on hydrolysis, and **conjugated proteins,** which yield α-amino acids and one or more groups of a nonprotein nature. The latter are known as **prosthetic groups** (Gr. *prosthesis* an addition). Because of the high molecular weight of the proteins and their similarity in being composed of amino acid residues, it is not possible to determine the empirical formulas of proteins from analysis for the elements. The elemental composition of all proteins is approximately 50 per cent carbon, 7 per cent hydrogen, 16 per cent nitrogen, 25 per cent oxygen, and 0–2 per cent sulfur. Phosphorus, iron, copper, and other elements also may be present. Proteins, when mixed with non-nitrogenous substances as in foodstuffs, usually are estimated by determining the total per cent nitrogen by the Kjeldahl method (p. 48) and dividing by 0.16. Since the nitrogen content for different proteins varies from 12 to 19 per cent and nitrogenous compounds other than protein may be present, this procedure is only approximate. Where the empirical formula of a protein is known, it has been arrived at only after determination of the exact proportion of each of the amino acid residues (p. 362).

To the organic chemist, structure refers only to the order of linkage of atoms by covalent bonds. For linear molecules of high molecular weight, however, the molecules may assume various conformations and shapes, which are important in determining the physical properties of proteins, properties that are as important to the biochemist as covalent structure. Hence several classes of structure are recognized. Covalent bonding is referred to as *primary structure;* the conformation of the molecule that results from hydrogen bonding between the imino and carbonyl groups is called *secondary structure;* and the folding and twisting of the chains owing to attractive forces between different parts of the chain is *tertiary structure.* Not all workers in the field are happy with these terms or interpret them in exactly the same way, but they are in common use.

The Peptide Linkage. Since proteins are hydrolyzed readily, and since the amino group and the carboxyl group are the only reactive functional groups present in all but four of the amino acids, an amide linkage is the only logical mode of joining them. A linkage between two amino groups would not hydrolyze, and an anhydride linkage between two carboxyl groups would not be stablc in an aqueous medium.

If the α-amino acids are represented by the structure, $RCH(NH_2)COOH$, the protein may be represented by the partial structure *I*.

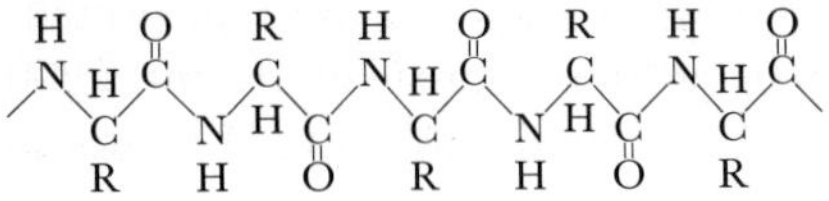

H O R H O R H O
N H C C N H C C N H C
C N H C C N H C C
R H O R H O R

I

HOOC Lysine Glutamine
H O CH₂ H O R H O
N H C C N H C C N H C
C N H C C N H C C
R H O $(CH_2)_4$ H O $(CH_2)_2$
Aspartic acid NH_2 $CONH_2$

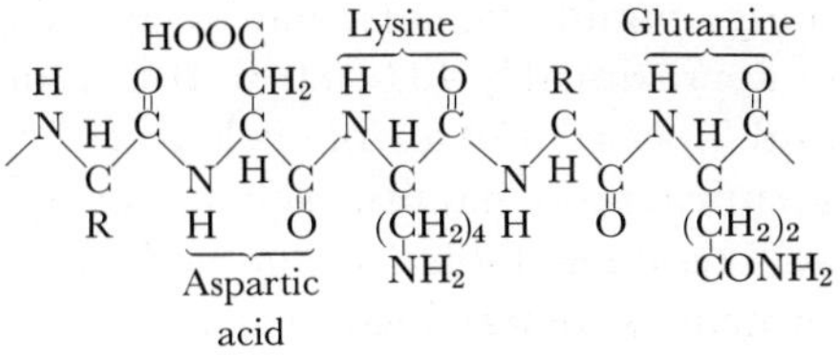

II

A few of the amino acids, however, have more than one carboxyl group, and a few have more than one amino group. Therefore in a molecule such as *I* some free amino groups and some free carboxyl groups may be expected to be present. Since ammonia also is a product of protein hydrolysis, some of the carboxyl groups must be combined with ammonia as simple amide groups, $CONH_2$. Thus a section of a protein molecule containing lysine, aspartic acid, and glutamic acid residues conceivably could have the structure *II*. Moreover it is known that chains of amino acids may be cross linked by means of the disulfide linkage in the cystine portions of the molecule as represented in structure *III*.

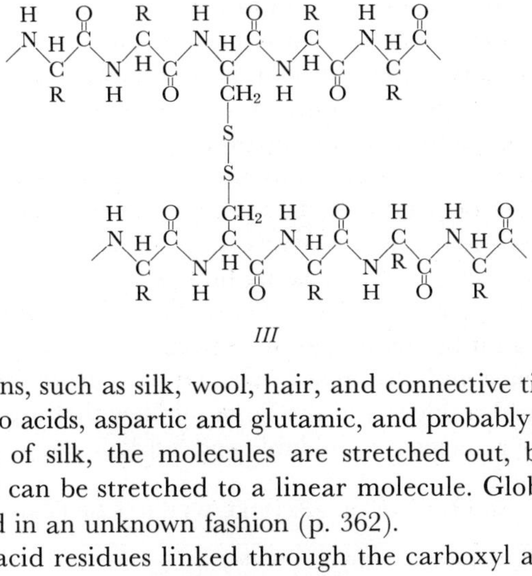

H O R H O R H O
N H C C N H C C N H C
C N H C C N H C C
R H O CH_2 H O R
S
S
H O CH_2 H O H H O
N H C C N H C C N H C
C N H C C N R C C
R H O R H O R

III

The fibrous proteins, such as silk, wool, hair, and connective tissues, are notably lacking in the dibasic amino acids, aspartic and glutamic, and probably are best represented by structure I. In fibroin of silk, the molecules are stretched out, but keratin of hair has a folded structure that can be stretched to a linear molecule. Globular proteins also may have their chains folded in an unknown fashion (p. 362).

Chains of amino acid residues linked through the carboxyl and α-amino groups by amide linkages frequently are called **peptides,** and this particular type of amide linkage is called a **peptide bond.** The units making up the peptide are amino acids less the elements of water. They are referred to as *amino acid residues* or simply as *residues.*

Molecular or Particle Weights. The molecular weights of proteins are so high that methods of determination such as those dependent on the lowering of the freezing point or elevation of the boiling point cannot be used. If the protein contains a characteristic group or atom, the equivalent weight can be found by determining the weight of the group or atom in a given weight of protein and calculating the weight of protein associated with one gram molecular weight of the group, or one atomic weight of the atom. For example, an equivalent weight can be calculated for hemoglobin from the amount of iron present. So-called molecular weights have been determined by the measurement of osmotic pressure, rates of diffusion, sedimentation in the ultracentrifuge, and the scattering of light. These values are particle weights and do not preclude the possibility of subdivision into smaller units without breaking covalent bonds. A particle weight as high as 48,000 has been reported for the insulin molecule as determined by sedimentation, but osmotic pressure measurements indicate a value of 12,000 under conditions of maximum dissociation. Still other work supports a minimum molecular weight of about 6000. Tobacco mosaic virus has a particle weight of approximately 40,000,000. It is a nucleoprotein that consists of about 95 per cent protein and 5 per cent of a ribonucleic acid (p. 530). The protein portion consists of around 2200 identical protein units, each having a molecular weight of 18,000 and made up of 158 amino acid residues. Table 19–*2* gives the particle weights and equivalent weights for a number of proteins as determined by several methods. There is fair agreement among the values obtained by the different procedures. The fact that the particle weights are not integral multiples of the equivalent weights must be ascribed to inexact methods of analysis. The only proteins whose molecular weights are known exactly are those whose constitution has been fully elucidated, such as insulin and ribonuclease (p. 362).

Some proteins have been shown by sedimentation in the ultracentrifuge to be definitely inhomogeneous; that is, they are composed of mixtures of particles having different weights. For example, the particle weights of lactalbumin vary from 12,000 to 25,000; of gelatin, from 10,000 to 100,000; and of casein, from 75,000 to 375,000.

Purification of Proteins. Because of the complexity of the proteins, their high molecular weights, the complex mixtures that occur naturally, and the similarity in their chemical and physical properties, it is difficult to be certain whether a protein is pure, that is, a single chemical species. Several proteins have been obtained in a crystalline state but even crystallinity does not guarantee that a protein is homomolecular. Proteins that once were thought to be pure have been separated into a number of fractions by more discriminating methods of separation. Counter-current distribution (p. 44) and the various forms of chromatography (p. 45) have been especially useful. Advantage also may be taken of the fact that proteins have free or potentially free amino groups and carboxyl groups (formula *II*, p. 359) and characteristic isoelectric points (p. 355). They can be separated and purified by electrophoresis (p. 355) and by passing buffered solutions through columns of ion-exchange resins. Only if a protein behaves as a single entity when subjected to all known methods of separation can it be considered to be pure.

Amino Acid Sequence and Disulfide Links (Primary Structure). Because of the different kinds of amino acid residues and the large number that are present, the task of unravelling the structure of proteins is formidable and not long ago seemed almost impossible.

TABLE 19–*2*. MOLECULAR OR PARTICLE WEIGHTS OF PURIFIED PROTEINS

PROTEIN \ PROCEDURE	OSMOTIC PRESSURE	SEDIMENTATION-DIFFUSION	LIGHT SCATTERING	EQUIV. WT. BY CHEM. ANAL.
Insulin (beef)	12,000	6, 12, 24, 36, and 48 thousand		5734 (complete structure)
Ribonuclease (beef)		14,000		13,683 (complete structure)
Lysozyme	17,500	14,000–17,000	15,000	
Pepsin	36,000	35,000		
β-Lactoglobulin	35,000	41,000	36,000	
Ovalbumin	45,000	44,000	46,000	35,000 (tryptophan)
Hemoglobin (human)	67,000	63,000		16,700 (iron) 64,450 (complete structure)
Serum albumin (horse)	73,000	70,000	76,000	8,000 (cystine)
Hemocyanin (lobster)		760,000	625,000	
Hemocyanin (snail)		8,900,000	6,300,000	25,500 (copper)
Tobacco mosaic virus		59,000,000	40,000,000	
Influenza virus			322,000,000	

Remarkable progress has been made since 1937, however, in the development of methods for the separation and estimation of amino acids and for arriving at the sequence of amino acid residues in peptides and polypeptides. In 1954 the first complete elucidation of the structure of a protein, namely the subunit of the **insulin** molecule of molecular weight around 6,000, was reported.

The principle of the method for determining the constitution of a linear protein molecule is relatively simple. First it is necessary to know the amino acid composition of the protein. At one end of the chain an amino acid residue must have a free amino group (*terminal* amino group), and at the other end an amino acid residue must have a free carboxyl group. Moreover regardless of the number of amide bonds broken by hydrolysis, each fragment will have the same two features. It has been found possible to "tag" the terminal amino groups in several ways. One important method is by the use of dinitrophenyl fluoride, DNP-F, (*2,4-dinitrofluorobenzene,* p. 400), which reacts with the terminal amino group as would an alkyl halide.

$$\mathrm{H_2N}\underset{\mathrm{R}}{\underset{|}{\mathrm{CH}}}-\underset{\mathrm{O}}{\underset{\|}{\mathrm{C}}}-(\mathrm{NH}\underset{\mathrm{R}}{\underset{|}{\mathrm{CH}}}-\underset{\mathrm{O}}{\underset{\|}{\mathrm{C}}}-)_x\mathrm{NH}\underset{\mathrm{R}}{\underset{|}{\mathrm{CH}}}\mathrm{COOH} + \text{DNP-F} + \mathrm{NaHCO_3} \longrightarrow$$

$$\text{DNP-NH}\underset{\mathrm{R}}{\underset{|}{\mathrm{CH}}}-\underset{\mathrm{O}}{\underset{\|}{\mathrm{C}}}-(\mathrm{NH}\underset{\mathrm{R}}{\underset{|}{\mathrm{CH}}}-\underset{\mathrm{O}}{\underset{\|}{\mathrm{C}}}-)_x\mathrm{NH}\underset{\mathrm{R}}{\underset{|}{\mathrm{CH}}}\mathrm{COOH} + \mathrm{NaF} + \mathrm{CO_2} + \mathrm{H_2O}$$

Hydrolysis of the product and separation and identification of the yellow DNP derivative identifies the terminal amino acid. Next the protein is hydrolyzed partially under a variety of conditions to obtain as many different fragments as possible having two or more amino acid residues in the molecule. All of these fragments are converted into their DNP derivatives, the derivatives are separated, and the amino acid associated with DNP is determined. With the amino acid at the amino end of each fragment known, it is possible to fit the amino acids together in the proper sequence. A chain of four amino acids, a, b, c, and d, may be used as a simple example. If it gives DNP-b(a,c,d) and after hydrolysis gives DNP-a(c,d), DNP-b(a,d), DNP-b-a, and DNP-d-c, then the structure of DNP-(a,b,c,d) is DNP-b-a-d-c.

The insulin subunit gives on hydrolysis 48 molecules of amino acids of sixteen different kinds and has the molecular formula

$$\mathrm{Glu_7Leu_6Val_5Gly_4Tyr_4Ala_3Asp_3(CysSSCys)_3Phe_3Ser_3His_2Arg_1Ileu_1Lys_1Pro_1Thr_1}.$$ [1]

Oxidation of the protein molecule with peroxyformic acid (p. 165) converts the disulfide linkages of the cystine units into sulfonic acid groups (p. 263) and splits the molecule into two chains. One chain, on hydrolysis, yields twenty-one molecules of amino acids, including four molecules of cysteic acid, $\mathrm{HO_3SCH_2CH(NH_2)COOH}$, and the other yields thirty molecules of amino acids including two molecules of cysteic acid. The structure of each chain has been established, chiefly by the DNP procedure. Finally the mode of union of the two chains by means of disulfide linkages has been determined by isolating fragments containing the disulfide linkages and determining their constitution. The complete structure of the protein molecule is indicated schematically in the following formula.

```
                           H2NCO   ┌─────────S─S─────┐                    CONH2        CONH2      CONH2
                               |   |                 |                      |            |          |
H2N-Gly-Ileu-Val-Glu-Glu-Cys-Cys[-Ala-Ser-Val-]Cys-Ser-Leu-Tyr-Glu-Leu-Glu-Asp-Tyr-Cys-Asp-COOH
                                 |                                                     |
                                 S                                                     S
                                 |                                                     |
          H2NCO CONH2            S                                                     S
              |     |            |                                                     |
H2N-Phe-Val-Asp-Glu-His-Leu-Cys-Gly-Ser-His-Leu-Val-Glu-Ala-Leu-Tyr-Leu-Val-Cys-Gly-Glu-Arg
                                                                                         |
                                                     HOOC-Ala-Lys-Pro-Thr-Tyr-Phe-Phe-Gly
```

Beef insulin

---

[1] The amino acid residues in proteins are designated by the first three letters of the common names of the amino acids except for isoleucine (*Ileu*).

A total of twelve carboxyl groups, three from aspartic acid, seven from glutamic and two from the ends of the chains, are not used in peptide linkages. Since six molecules of ammonia are formed on hydrolysis, six of these carboxyl groups occur as simple amide groups. Hence the exact molecular weight of the insulin subunit is 5734. It is of interest that insulins from different species are not identical. The above formula is for insulin from beef pancreas. Other insulins that have been investigated differ in structure only in the portion of the molecule enclosed in the dotted rectangle. In sheep insulin this portion is alanine-glycine-valine, in horse insulin it is threonine-glycine-isoleucine, and in hog or whale insulin it is threonine-serine-isoleucine.

More recently 5-dimethylamino-1-naphthalenesulfonyl chloride (*dansyl chloride*) (p. 493) has been used to convert end amino groups to sulfonamide groups (p. 408), which are very resistant to hydrolysis. Although the derivative is colorless, it fluoresces intensely yellow in ultraviolet light and can be detected at one hundredth the concentration necessary to detect DNP derivatives. Another method for determining the amino end residue is by reaction with phenylisothiocyanate (p. 425). Treatment of the resulting phenylthiourea with dry hydrogen chloride in nitromethane removes the end residue as the cyclic 3-phenyl-2-thiohydantoin (p. 523).

$$C_6H_5N{=}C{=}S + H_2N\underset{\displaystyle R'}{\underset{|}{C}}HCO(NH\underset{\displaystyle R}{\underset{|}{C}}HCO)_xNH\underset{\displaystyle R}{\underset{|}{C}}HCOOH \longrightarrow$$

$$C_6H_5NH{-}\underset{\displaystyle S}{\underset{\|}{C}}{-}NH\underset{\displaystyle R'}{\underset{|}{C}}HCO(NH\underset{\displaystyle R}{\underset{|}{C}}HCO)_xNH\underset{\displaystyle R}{\underset{|}{C}}HCOOH \xrightarrow{H^+}$$

$$\text{(3-phenyl-2-thiohydantoin: } C_6H_5N{-}C{=}S,\ O{=}C,\ NH,\ CH{-}R' \text{ ring)} + H_3\overset{+}{N}\underset{\displaystyle R}{\underset{|}{C}}HCO(NH\underset{\displaystyle R}{\underset{|}{C}}HCO)_{x-1}NH\underset{\displaystyle R}{\underset{|}{C}}HCOOH$$

The thiohydantoin can be identified by paper chromatography. Alternatively the end residue may be identified by quantitative amino acid analysis (p. 357) of the peptide that remains. Because of the advances in the methods of analysis and improved methods for the separation and purification of peptides, the full structure of twelve natural peptides or proteins having more than 18 amino acid residues had been determined by the end of 1961, including **β-ACTH** (*adrenocorticotropic hormone*) made up of 39 amino acid residues, the enzyme **ribonuclease** with 124 residues, the α and β chains of the protein portion of **human hemoglobin** with 141 and 146 residues respectively, and the protein subunit of **tobacco mosaic virus** with 158 residues.

Secondary and Tertiary Structures. The long chains of hydrocarbons, alcohols, and carboxylic acids have in the solid crystalline state an extended zig-zag conformation (p. 64) because there are no strong forces between the methylene groups that prevent them from assuming the arrangement of closest packing. The hydrogen of the amide nitrogen, however, can form hydrogen bonds with the oxygen of the carbonyl group, provided they can approach each other at the proper distance. If the protein chain of L amino acid residues assumes a helical arrangement with 3.7 residues per turn, each hydrogen attached to nitrogen can form a hydrogen bond with the carbonyl oxygen of the fourth subsequent residue to give a completely hydrogen-bonded, coiled **secondary structure** known as the **α-helix** (Fig. 19–3). It appears to be the predominant arrangement for fibrous proteins and also makes up portions of globular proteins. Some fibrous proteins, such as silk and keratin, have a sheeted structure that results from hydrogen bonds between adjacent extended chains.

The shape that the helix takes, known as the **tertiary structure,** depends on the interactions between groups in the side chains of the amino acid residues and may result

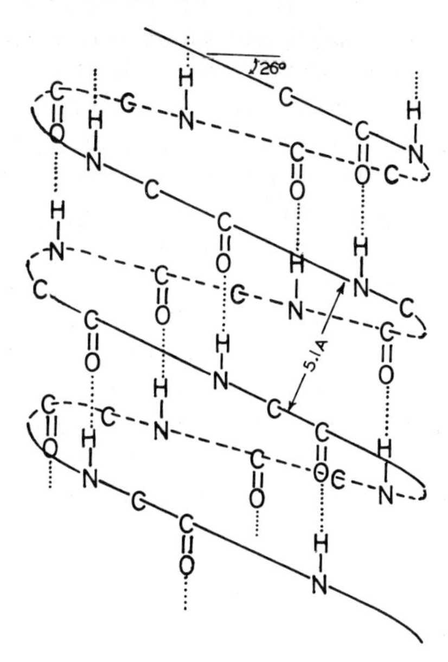

Figure 19–*3*. Hydrogen bonding in the α-helix secondary structure of proteins.

from hydrogen bonding and other van der Waals forces (p. 21) or from cross linking by disulfide bonds. For fibrous proteins, the coiled chains are aligned to give the fiber bundle. Most proteins are globular rather than fibrous. Here the coiled chains are tightly folded and looped to give a compact molecule. The helical secondary structure may be partially or almost completely disrupted. Figure 19–*4* indicates the tertiary structure, as determined by X-ray diffraction, of crystalline **myoglobin,** an oxygen-carrying chromoprotein isolated from skeletal muscle. It consists of eight segments of right-handed, single-strand α-helices, each having 7 to 20 residues. The bends require 2 to 3 residues each, and the section on the upper left consists of 13 to 18 irregularly arranged residues. The prosthetic group, an iron-containing porphyrin (p. 510) represented by the disc, is joined to the protein by a bond between the iron atom and the nitrogen of a histidine residue. The **oxyhemoglobin** of blood appears to consist of two pairs of subunits whose amino acid sequences have been determined (p. 362). Each subunit has a tertiary structure similar to that of myoglobin. The four subunits are arranged tetrahedrally in crystalline hemoglobin (Fig. 19–*5*).

The solution of the structure of a protein crystal by X-ray analysis requires an enormous amount of labor. To determine the three-dimensional structure of myoglobin, the recording and measurement of about a quarter of a million spots was necessary. In the final

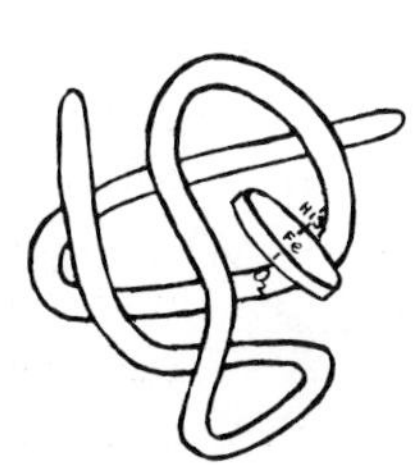

Figure 19–*4*. Tertiary structure of myoglobin.

Figure 19–*5*. Tertiary structure of oxyhemoglobin. Two subunits face the observer. Rotation through 180° about the pin gives the position of the other two units.

calculation 5×10^9 figures were added and subtracted. Work on the structure of horse oxyhemoglobin begun in 1937 was not completed until 1959.

Globular proteins are rather labile substances and their properties may be changed markedly by physical or chemical agents without the scission of peptide bonds. Such changes are called **denaturation** because they usually destroy the characteristic biological activity of enzymes, hormones, or viruses. Denaturation may be reversible or irreversible. Agents that bring about denaturation may be heat, ultraviolet light, acids, alkalies, certain anions, organic solvents, or concentrated aqueous solutions of urea. Usually the protein becomes less soluble at its isoelectric point, the optical activity becomes more levorotatory, the functional groups in side chains become more reactive, and the molecule is more susceptible to attack by proteolytic enzymes. These effects are believed to be due to unfolding of the tightly coiled chains, which results from the breaking of hydrogen bonds or disulfide bonds that hold the native protein in its characteristic tertiary structure.

Practical use is made of denaturation. It is possible to convert globular proteins such as zein, casein, albumin, or soybean protein into a fibrous form by dissolving in aqueous sodium hydroxide, forcing the solution through fine holes, and coagulating the filament by passing it into an acidified aqueous formaldehyde bath. The molecules become uncoiled on solution and are stretched out and oriented parallel to each other during the spinning process. The product is known as *synthetic wool.* In another process the disulfide bonds in the keratin of chicken feathers are reduced to sulfhydryl groups. The keratin then dissolves in aqueous urea and can be extruded into fibers to give artificial bristles. The permanent waving of hair is based on the reduction of the disulfide linkages with ammonium thioglycolate (p. 616) to permit shaping of the hair.

$$RSSR + 2\,HSCH_2COONH_4 \longrightarrow 2\,RSH + (NH_4OOCCH_2S)_2$$

Subsequent reformation of the disulfide linkage by oxidation with potassium bromate sets the wave.

$$6\,RSH + KBrO_3 \longrightarrow 3\,RSSR + KBr + 3\,H_2O$$

Glycoproteins. Numerous proteins have carbohydrates as prosthetic groups. **Ovalbumin** from egg white, after enzymatic removal of protein, yields a polysaccharide having a molecular weight of around 2000. Hydrolysis gives D-mannose and *N*-acetyl-D-glucosamine in the molar ratio of about 5 to 3.

The **mucins** and **mucoids** (*mucus-like substances*) confer the viscid, slippery properties to fish eggs, slugs, saliva, and various other tissue and glandular secretions. Substances responsible for blood group specificity are glycoproteins that appear to be related closely in structure to the mucins. Hydrolysis of the carbohydrate fraction gives hexoses, hexosamines, and **sialic acids.** The last are *N*- or *O*-acyl derivatives of an unusual carbohydrate, **neuraminic acid.**

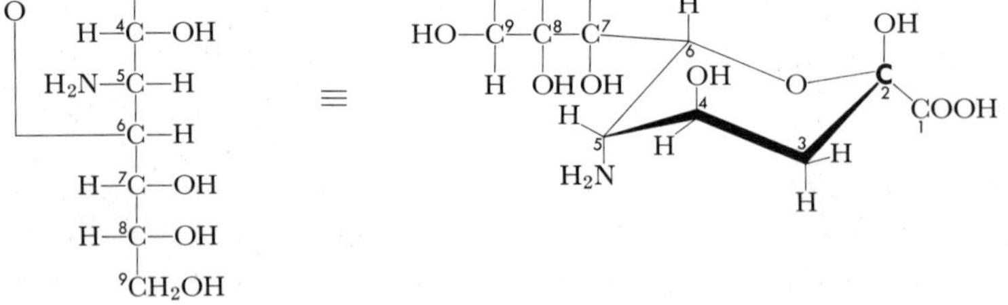

HO—²C—¹COOH
³CH₂
H—⁴C—OH
H₂N—⁵C—H
⁶C—H
H—⁷C—OH
H—⁸C—OH
⁹CH₂OH

≡

HO—C⁹(H)(H)—C⁸(H)(OH)—C⁷(H)(OH)— [ring: C6(H), O, C2(OH)(COOH), C3(H)(H), C4(H)(OH), C5(H)(H₂N)]

β-D-Neuraminic acid

Sialic acids are widely distributed. They have been isolated from milk, urine, blood fluid, seminal fluids, erythrocytes, brain phospholipids (*gangliosides,* p. 610), and *Escherichia coli.* Much interest has attached to the fact that influenza virus is able to remove sialic acid from erythrocytes. The sialic acids probably always are part of a polysaccharide structure. The mucin from egg white is reported to yield 4.4 per cent galactose, 2.2 mannose, 4.7 glucosamine, 2.4 galactosamine, and 6.0 sialic acid; that from the seminal fluid of hogs, 7.5 galactose, 5.0 fucose, 6.6 glucosamine, 3.9 galactosamine, and 7.4 sialic acid. It is questionable, however, whether any glycoprotein yet has been isolated in a pure state.

Metabolism of Proteins

The animal organism normally obtains its supply of nitrogen predominantly by ingesting proteins from plants and other animals. These proteins are hydrolyzed progressively to peptides (p. 359) and amino acids. The hydrolyses are catalyzed by the enzyme *pepsin* under acid conditions (*p*H 1–2) in the stomach and by *trypsin, chymotrypsins,* and *peptidases* under slightly acid to slightly alkaline conditions (*p*H 6–8) in the intestines. The amino acids pass through the walls of the intestine into the portal blood stream, which carries them to the liver and other tissues of the body. Under the influence of specific cellular enzymes, the amino acids are reconverted into proteins characteristic of the particular tissue. The liver converts the amino acids also into plasma proteins, which are carried to peripheral tissues where they can be used as a source of amino acids for the synthesis of tissue proteins. That an organism can synthesize a specific protein for each particular purpose by picking out the desired amino acids from the blood stream and putting them together in the proper order is a striking example of the exactness with which life processes are regulated.

The organism also is able to convert some amino acids into others and to synthesize them from ammonium salts and α-keto acids derived from carbohydrates. Amino acids that can be so formed need not be ingested. On the other hand it is known that the rat cannot produce valine, leucine, isoleucine, phenylalanine, threonine, tryptophan, methionine, lysine, arginine,[2] or histidine, and hence they have been termed *indispensable* or *essential amino acids.* The term "indispensable" should not be taken to mean that these amino acids are more important than the "dispensable" amino acids, since all natural amino acids undoubtedly are necessary for the development and maintenance of the organism. Indispensable merely means that these amino acids must be supplied in the proteins of the diet, and cannot be synthesized from other amino acids or from ammonia nitrogen. Approximately 6 per cent of the protein intake must consist of these amino acids. Studies made so far indicate that the requirements of other species appear to be similar to those of the rat. Arginine, however, is dispensable in dogs and histidine is dispensable in man. For growing chicks glycine is an indispensable amino acid, although it is dispensable in rats. One of the more interesting facts discovered through the aid of isotopes is that the proteins in the body are being actively built up and torn down continuously, and that there is a fairly rapid turnover of amino acids. It has been calculated that the average "half-life" of proteins, that is, the time required for half of the original amino acids in the body protein to be replaced by other amino acids, is about 17 days for the rat and 80 days for the adult human.

Most of the ingested amino acids finally are deaminated and burned to carbon dioxide and water. The ammonia is converted to urea by way of carbamyl phosphate (p. 286), ornithine, citrulline (p. 354), and arginine (*ornithine cycle*) and eliminated in the urine.

[2] Arginine can be produced but not at a rate sufficient for normal growth.

Synthesis of Peptides and Polypeptides

Ever since the postulation by Hofmeister[3] that proteins are made up of α-amino acid residues joined by amide linkages, there has been continued interest in the synthesis of peptides and polypeptides. At first the purpose was to confirm the theory. Then it became desirable to prepare simple compounds of known constitution for the investigation of the relation between constitution and enzyme activity. With the establishment of the structure of proteins, world-wide efforts are under way to synthesize them. The task is a formidable one because it requires the use of special reactions for protecting groups and for bringing about stepwise condensations that join each amino acid to the chain in the proper way and in the proper order. Great progress has been made, however, and syntheses of *β-ACTH* and the *A* and *B* chains of insulin (p. 361) were reported in 1963. An apparatus developed at the Rockefeller Institute for the automated synthesis of peptide chains in the solid phase was described in 1965. It promises to simplify tremendously the problem of synthesizing proteins of still higher molecular weight.

PROBLEMS

19–*1*. (*a*) Which of the following types of linkage is most numerous in the protein molecule: (*1*) amide, (*2*) ether, (*3*) anhydride, (*4*) proton bonding, (*5*) disulfide, or (*6*) imino? (*b*) Which is responsible for cross linking of protein chains? (*c*) Which is responsible for the helical conformation?

19–*2*. Pair each member of the first group of items with the appropriate member of the second group: (*a*) prosthetic group, (*b*) secondary structure, (*c*) electrophoresis, (*d*) isoelectric point, (*e*) basic amino acid, (*f*) cystine, (*g*) indispensable amino acid, (*h*) ninhydrin, (*i*) 2,4-dinitrofluorobenzene; (*1*) disulfide cross link, (*2*) conjugated protein, (*3*) minimum solubility, (*4*) α helix, (*5*) reagent for amino end residue, (*6*) tryptophan, (*7*) fractionation of proteins, (*8*) reagent for α-amino acids, (*9*) more amino than carboxyl groups.

19–*3*. Give equations for the following syntheses: (*a*) leucine from *i*-caproic acid, (*b*) valine from *i*-butyraldehyde, (*c*) alanine from nitroethane, (*d*) isoleucine from *s*-butyl bromide.

19–*4*. The following pentapeptides were converted to their DNP derivatives, hydrolyzed, and the products determined qualitatively and quantitatively. The peptides then were partially hydrolyzed, the products were isolated as pure DNP derivatives, and these derivatives were hydrolyzed and the products determined qualitatively and quantitatively. From the data given derive the structural formula of the peptide.

Hydrolysis Products of DNP-Pentapeptide	*Hydrolysis Products of DNP Derivatives of the Hydrolysis Products of Peptide*
A. DNP-glycine and two moles each of glycine and serine	DNP-serine, one mole of serine, and two moles of glycine
	DNP-glycine and one mole each of serine and glycine
	DNP-serine and one mole each of serine and glycine
B. DNP-cysteine and two moles each of cysteine and leucine	DNP-leucine, one mole of leucine, and two moles of cysteine
	DNP-cysteine and one mole each of cysteine and leucine
	DNP-cysteine and one mole of cysteine

[3] Franz Hofmeister (1850–1922), professor of biochemistry and experimental pharmacology at the University of Prague, and later at the University of Strassburg.

C. DNP-isoleucine and two moles each of isoleucine and aspartic acid	DNP-isoleucine and one mole each of isoleucine and aspartic acid
	DNP-aspartic acid and one mole of aspartic acid
	DNP-aspartic acid and one mole of isoleucine
D. DNP-methionine, two moles of methionine, and one mole each of serine and glycine	DNP-methionine and one mole each of methionine and glycine
	DNP-methionine and one mole of methionine
	DNP-serine and one mole of meth-ionine
	DNP-methionine and one mole each of methionine and serine

CHAPTER TWENTY

AROMATIC HYDROCARBONS. SOURCES OF AROMATIC COMPOUNDS. INFRARED SPECTRA

Most of the compounds considered thus far are classed as aliphatic because their composition is related to that of the fats insofar as the hydrocarbon portion of the molecule is concerned (Gr. *aliphatos* fat). It was recognized at an early date that many organic compounds have a hydrocarbon portion with a higher ratio of carbon to hydrogen than the alkanes and with distinctly different chemical properties. These substances frequently are pleasantly odorous or derivable from aromatic substances. For example, the essential components of the volatile oils of cloves, cinnamon, sassafras, anise, bitter almonds, wintergreen, and vanilla exhibit these properties. The hydrocarbon *benzene* received its name because it was obtained by the decarboxylation of benzoic acid isolated from the aromatic substance *gum benzoin,* and the name *toluene* was assigned to another hydrocarbon because it had been obtained by heating the fragrant *tolu balsam.* Loschmidt[1] in 1861 was the first to state that most of the aromatic compounds could be considered as derivatives of the hydrocarbon benzene, C_6H_6, just as the aliphatic compounds were considered as derivatives of methane, CH_4. Since then the term *aromatic compounds* has been applied to those compounds having the characteristic chemical properties of benzene. *Arene* is a general term for an aromatic hydrocarbon. Under the designation *nonbenzenoid aromatic compounds,* a still different application of the term *aromatic* is discussed in Chapter 38.

BENZENE AND ITS HOMOLOGS

Isolation and Structure of Benzene

During the latter part of the eighteenth century an illuminating gas was manufactured in England by the thermal decomposition of whale oil and other fat oils. When this gas was compressed for distribution in tanks, a light mobile liquid separated. This liquid was brought to the attention of Michael Faraday (p. 121) in 1820, and in 1825 he reported the isolation, by distillation and crystallization, of a compound that he called *bicarburet of hydrogen.* In 1833 Mitscherlich[2] reported the isolation of the same hydrocarbon by distilling

[1] Joseph Loschmidt (1821–1895), Austrian physicist and professor at the University of Vienna. He originally was a chemist and published privately in 1861 his views on the use of graphic constitutional formulas in organic chemistry. In this work he proposed formulas for 368 compounds of which 21 belonged to the aromatic series.

[2] Eilhard Mitscherlich (1794–1855), professor of chemistry at the University of Berlin. He first specialized in Oriental languages, particularly Persian, and began the study of medicine because as a physician he would have more freedom than other Europeans to travel in Persia. He soon became so interested in chemical subjects that he gave up his aspirations in the other fields. He is noted chiefly for his work on isomorphism.

benzoic acid with lime. He named his product *benzin,* but Liebig, as editor of the Annalen der Chemie, added a note to Mitscherlich's paper stating that a more suitable name would be *benzol,* the ending *ol* indicating that it was a liquid (L. *oleum* oil) obtained from the solid benzoic acid. Benzol still is the common term in the German chemical literature and is used in all countries by technicians in the coal tar industry. It has been replaced in the English and French scientific literature by the term *benzene,* the ending *ol* being reserved for alcohols. Although the separation of benzene from the products of the destructive distillation of coal (p. 389) usually is credited to Leigh (1842) and Hofmann (1845), Liebig's note to Mitscherlich's work indicates that Faraday's compound was recognized then as one of the products of the distillation of coal.

The data of both Faraday and Mitscherlich indicated that benzene has the molecular formula C_6H_6. Since the corresponding alkane has the molecular formula C_6H_{14}, benzene would be expected to be highly unsaturated. On the contrary, it is almost as stable to oxidation and to the usual addition reactions as the saturated hydrocarbons. Hydrogen adds catalytically and chlorine or bromine adds in the presence of sunlight, but only six atoms add and not eight (p. 374). Moreover the number of isomers formed on replacing hydrogen atoms by other elements or groups does not correspond to that expected for the aliphatic hydrocarbons. Thus only one monosubstitution product, for example one chlorobenzene, one hydroxybenzene, or one aminobenzene, is known. Hence each hydrogen atom bears the same relationship to the molecule as a whole as every other hydrogen atom; that is, the molecule is symmetrically constituted.

Kekulé, who originated or consolidated most of the views concerning the structure of organic compounds (p. 4), assigned the first definite structural formula to benzene in 1865. He proposed that the six carbon atoms were at the corners of a regular hexagon and that one hydrogen atom was joined to each.

```
        H
        |
   H    C    H
     \ / \ /
      C   C
      |   |
      C   C
     / \ / \
   H    C    H
        |
        H
```

Four years earlier Loschmidt (p. 368) had represented the benzene nucleus by a circle but had made no attempt to indicate the arrangement of the carbon atoms in the nucleus. Kekulé's formula accounts also for the fact that if two hydrogen atoms are replaced by other groups, *Y*, three and only three isomers are known. If two adjacent hydrogen atoms are replaced, the resulting compound is known as the *ortho* isomer; if two alternate hydrogen atoms are replaced, the compound is known as the *meta* isomer; and if two opposite hydrogen atoms are replaced, it is known as the *para* isomer.

```
        Y                    Y                    Y
        |                    |                    |
   H    C    Y          H    C    H          H    C    H
     \ /1\ /              \ / \ /              \ / \ /
     C6   2C               C   C                C   C
      |   |                |   |                |   |
     C5   3C               C   C                C   C
     / \4/ \              / \ / \              / \ / \
   H    C    H          H    C    Y          H    C    H
        |                    |                    |
        H                    H                    Y
```

ortho Isomer   *meta* Isomer   *para* Isomer

Regardless of which two adjacent hydrogen atoms are replaced, the same compound results. For example, if hydrogen atoms at positions 2 and 3 are replaced instead of those at 1 and 2, rotation through 60 degrees in the plane of the paper brings the two molecules into coincidence. Similar considerations hold for the *meta* and *para* isomer. Moreover the two groups, *Y*, may be alike or different without changing the number of isomers.

Objections to the simple hexagon formula for benzene are that it violates Kekulé's own advocacy of the tetravalence of carbon, and that it does not explain the addition of six atoms of halogen or six atoms of hydrogen. Both difficulties are overcome by the introduction of three double bonds in continuous conjugation.

(*a*) (*b*)

Now, however, the objection arises that two *ortho* substitution products would be expected, one (*a*) in which the two carbon atoms carrying the *Y* groups are linked by a double bond, and another (*b*) in which they are linked by a single bond. To overcome this difficulty, Kekulé proposed in 1872 that the positions of the double bonds are not fixed, but that an equilibrium exists between two structures that is so mobile that individual isomers such as (*a*) and (*b*) cannot be isolated.

(*a*) (*b*)

The current view is that in neither benzene nor its derivatives do molecules of two structures exist in mobile equilibrium with each other, but that only one kind of molecule is present that is a hybrid of the two structures. The most direct evidence for this view comes from the measurement of the distances between carbon atoms by means of electron diffraction. In ethane and other saturated compounds, the distance between adjacent carbon atoms is 1.53 A, in ethylene the carbon-carbon distance is 1.33 A, and in acetylene it is 1.20 A. In other words the distance is shorter for a double bond than for a single bond and shorter for a triple bond than for a double bond (p. 147). This result is to be expected because the attraction between atoms increases with the formation of each bond. In compounds containing both single bonds and double bonds, both distances are observed (cf. p. 161), but for benzene only a single carbon-carbon distance can be detected, namely 1.39 A, which lies between the single bond and double bond distances. Therefore all of the carbon-carbon bonds in benzene are alike.

The problem of benzene structure is analogous to that for the carboxylate ion (p. 160) and the nitro group (p. 242). The two classical Kekulé structures correspond to two equivalent electronic structures. The so-called *resonance hybrid* of the two electronic structures is more stable than either single structure by the resonance energy, which for benzene is about 36 kcal. per mole. A better term would be stabilization energy, since the effect arises from several factors of which delocalization is only one. The nature of the resonance hybrid can be made clear most readily by the use of molecular orbitals. Each carbon atom is bonded to three other atoms. Hence sp^2 hybridization of the atomic orbitals of the carbon atoms takes place on bonding, and three σ-type molecular orbitals are formed that make angles of

120 degrees with each other and cause the four atoms to lie in a plane (p. 14). Therefore all of the atoms of benzene lie in a plane and all bond angles are 120 degrees. These σ-type bonds are indicated by the usual dash. One *p* orbital containing an unpaired electron remains at each carbon atom as illustrated in Fig. 20–*1a*.

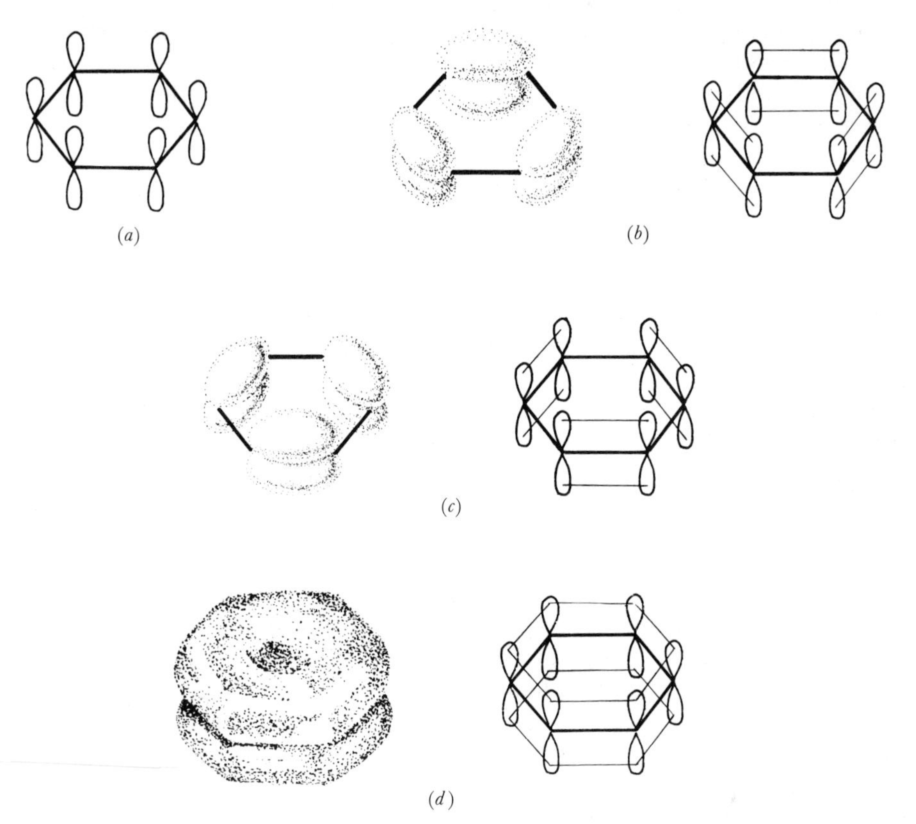

Figure 20–*1*. Resonance in benzene.

If the *p* orbitals overlapped as three pairs as indicated in Fig. 20–*1b*, one of the Kekulé structures with three double bonds would be obtained, whereas if they overlapped as in Fig. 20–*1c*, the other Kekulé structure would be obtained. Since any *p* orbital overlaps both *p* orbitals on either side of it equally well, however, all six orbitals overlap each other giving a hexagonal π orbital above and below the plane of the carbon atoms (Fig. 20–*1d*). These structures are shown schematically by drawing light lines connecting *p* orbitals to indicate overlapping. The heavier lines indicate the σ-type bonds between the atoms (p. 159). The movement of the π electrons about six positive nuclei instead of only two accounts for the greater stability of this system and the large resonance energy of the aromatic ring. Because of the Pauli exclusion principle (p. 7), the molecular π orbital indicated can accommodate only two electrons. Actually three additional bonding π orbitals are available to accommodate the remaining four electrons. These π orbitals also encompass all six nuclei, but each has a nodal plane perpendicular to the plane of the ring in addition to that coinciding with the plane of the ring.

Resonance in the aromatic nucleus is represented in various ways less cumbersome than those illustrated in Fig. 20–*1*. Thus a double-headed arrow between two Kekulé structures enclosed in braces may be used as in (*a*). Frequently the π electrons are indicated by

(a) (b) (c) (d)

a full or a dotted circle as in (*b*) or (*c*). Because electrons repel each other they tend to stay away from each other as far as possible, a behavior known as *electron correlation.* This behavior is indicated by (*d*) where the electrons are centered between the ring carbon atoms and alternate above and below the plane of the ring. None of these representations is entirely satisfactory, and in this text the simple single Kekulé structure ordinarily is used with the understanding that the three pairs of electrons are delocalized. Various other types of symbolism will be used where they are deemed helpful in clarifying the mechanism of reactions.

Nomenclature

Aromatic hydrocarbons frequently are called *arenes.* Their substitution products may be named as derivatives of benzene or of some other aromatic compound. Thus the compound in which a chlorine atom has replaced a hydrogen atom in benzene is known as *chlorobenzene.* When more than one substituent is present, the positions occupied are indicated by numbers. Usually a formula is written with one group at the top carbon atom of the hexagon, which then becomes the number 1 carbon atom. The six carbon atoms of benzene are numbered from 1 to 6 in the direction around the ring that gives the substituents the smaller numbers. It is preferable to name the substituent groups in alphabetical order. When only two like substituents are present, the symbols *o*-, *m*-, or *p*- for *ortho, meta,* or *para* may be used. These terms may be combined with numerals as in 4-chloro-*m*-xylene (*not* 2-chloro-) where it does not lead to ambiguity.

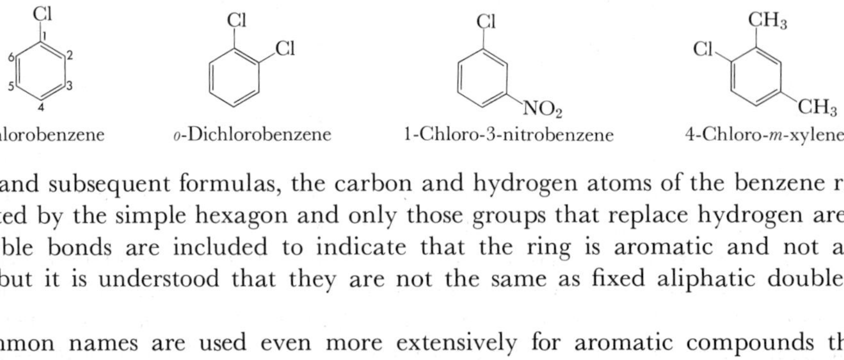

Chlorobenzene　*o*-Dichlorobenzene　1-Chloro-3-nitrobenzene　4-Chloro-*m*-xylene

In these and subsequent formulas, the carbon and hydrogen atoms of the benzene ring are represented by the simple hexagon and only those groups that replace hydrogen are given. The double bonds are included to indicate that the ring is aromatic and not alicyclic (p. 91), but it is understood that they are not the same as fixed aliphatic double bonds (p. 370).

Common names are used even more extensively for aromatic compounds than for aliphatic compounds. Thus many of the hydrocarbons have common names. Methylbenzene is called *toluene,* and 1,2-, 1,3-, and 1,4-dimethylbenzene are known as *ortho-, meta-,* and *para-xylene.*

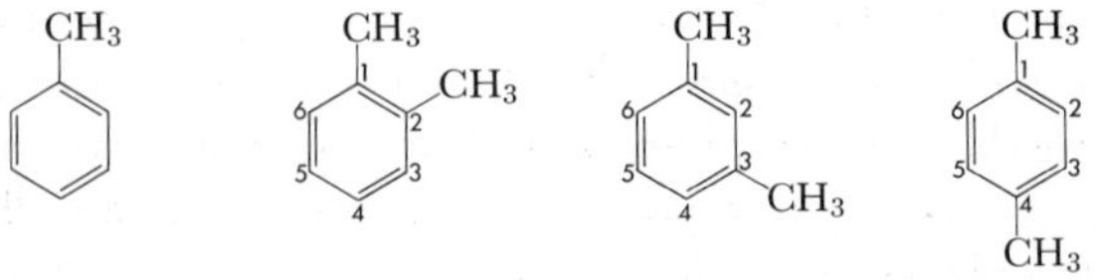

Toluene　*o*-Xylene　*m*-Xylene　*p*-Xylene

1,3,5-Trimethylbenzene is called *mesitylene,* isopropylbenzene is *cumene,* 1,2,4,5-tetramethylbenzene is *durene,* and 4-isopropyltoluene is *p-cymene.*

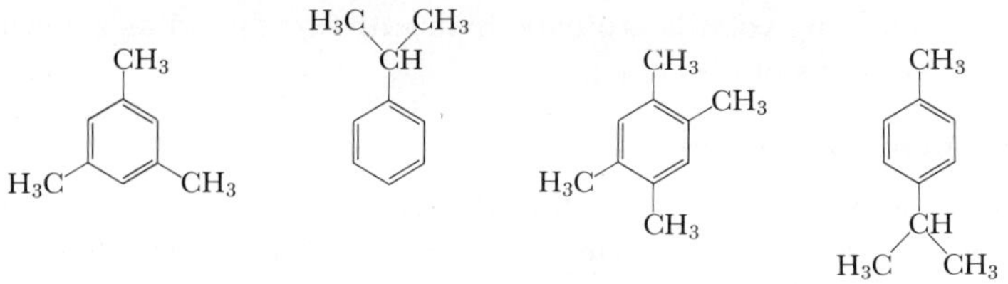

Mesitylene   Cumene   Durene   *p*-Cymene

Kekulé introduced the terms *ring* and *nucleus* to designate the characteristic portion of aromatic compounds. He called the aliphatic portions *side chains.* Groups obtained by dropping a hydrogen atom from the nucleus are called *aryl* groups. Frequently they are indicated by the symbol Ar, just as R is used to designate an alkyl group. The group C_6H_5 derived from benzene is known as *phenyl.* This name is derived from *phene,* a name that was proposed by Laurent for benzene, because it is present in illuminating gas (Gr. *phainein* to bring to light). The phenyl group frequently is indicated by the symbol *Ph* or the Greek letter ϕ (*phi*); for example PhBr or ϕBr represents phenyl bromide, that is, bromobenzene. If two hydrogen atoms are dropped, the residue, C_6H_4, is known as a *phenylene* group. Since the two hydrogen atoms may be removed in three ways, there are *o*-, *m*-, and *p*-phenylene groups. If a hydrogen atom is dropped from the methyl group of toluene, the residue is known as a *benzyl* group, but if from the nucleus, three *tolyl* groups, *ortho, meta,* or *para,* result. If two hydrogens are dropped from the methyl group of toluene, the residue is called a *benzylidene* (formerly *benzal*) group.

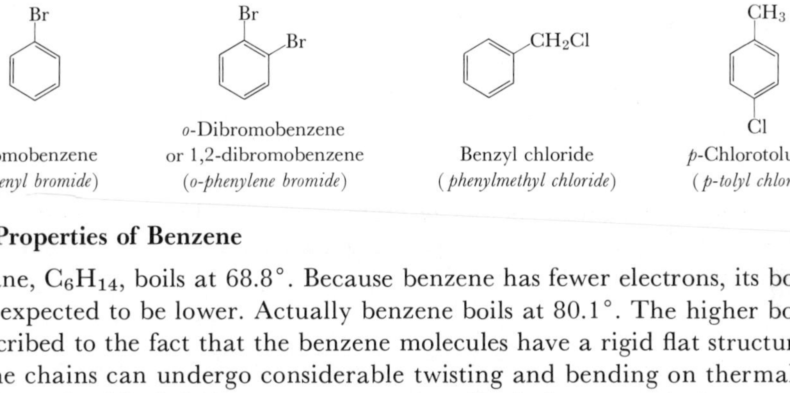

Bromobenzene (*phenyl bromide*)   *o*-Dibromobenzene or 1,2-dibromobenzene (*o-phenylene bromide*)   Benzyl chloride (*phenylmethyl chloride*)   *p*-Chlorotoluene (*p-tolyl chloride*)

Physical Properties of Benzene

Hexane, C_6H_{14}, boils at 68.8°. Because benzene has fewer electrons, its boiling point might be expected to be lower. Actually benzene boils at 80.1°. The higher boiling point can be ascribed to the fact that the benzene molecules have a rigid flat structure, whereas the hexane chains can undergo considerable twisting and bending on thermal agitation. Hence the van der Waals forces can operate more effectively among the benzene molecules. The more symmetrical structure for benzene accounts also for its relatively high melting point of +5.5° compared with −95° for hexane or for toluene.

Being a hydrocarbon, benzene is practically insoluble in water, but the solubility at 15° of 0.18 g. per 100 g. of water is over ten times that of the 0.014 g. per 100 g. of water for *n*-hexane. The greater attraction of benzene for water molecules would be expected because of the greater polarizability of the unsaturation electrons. The solubility of water in benzene is 0.06 g. per 100 g. of benzene. In general the solubility of highly associated liquids and solids in aromatic hydrocarbons and of aromatic hydrocarbons in associated liquids is greater than the corresponding solubilities of the saturated hydrocarbons. It is for this reason that aromatic hydrocarbons are more useful as solvents and diluents for paints, lacquers, and synthetic enamels than are alkanes. Aromatic hydrocarbons should be used only with proper precautions, however, because of their high toxicity. They cause the destruction of red blood corpuscles, and even very low concentrations, especially of benzene, are dangerous on prolonged exposure. All workers who use for long periods materials containing volatile aromatic hydrocarbons should be subjected to frequent blood counts to detect signs of poisoning.

Reactions of the Benzene Nucleus

Addition Reactions. Unlike the olefins, the benzene nucleus does not readily undergo reactions such as the addition of halogen or ozone, catalytic hydrogenation, or oxidation by permanganate, although all of these reactions can be brought about under the proper conditions. As a measure of the relative reactivity of benzene and olefins to addition, the heat liberated on catalytic hydrogenation may be cited. A doubly substituted olefin such as *cis*-butene liberates 28.6 kcal. per mole on catalytic hydrogenation. If three ordinary double bonds were present in benzene, the heat liberated should be $3 \times 28.6 = 85.8$ kcal. The actual amount of heat liberated is 49.8 kcal. The difference of 36 kcal. is called the *resonance energy* of benzene; that is, of the 85.8 kcal. that is liberated on the catalytic hydrogenation of three double bonds, 36 kcal. is used up overcoming the resonance energy of the benzene molecule. Since the heat liberated on hydrogenating 1,3-cyclohexadiene, the cyclic compound with two conjugated double bonds, is 57.2 kcal., the hydrogenation of benzene to cyclohexadiene actually is endothermic by 7.4 kcal. Similarly in other addition reactions, except for addition of fluorine (cf. p. 67), energy must be supplied to get the first mole of reagent to react. Once one mole of reagent is added, the remaining two pairs of electrons have the same high reactivity as the ordinary π electrons of alkenes. Hence the product obtained is the result of the addition of three moles of reagent. Catalytic hydrogenation of benzene yields cyclohexane. In the presence of a high concentration of ozone, benzene slowly adds three moles to give a triozonide, which can be hydrolyzed to glyoxal.

$$C_6H_6 + 3\,O_3 \longrightarrow C_6H_6(O_3)_3 \xrightarrow{3\,H_2O} 3\,O{=}CH{-}CH{=}O + 3\,H_2O_2$$

Glyoxal

Benzene does not react with chlorine or bromine under ordinary conditions, but in the presence of light of the proper energy to yield free halogen atoms (360–400 mμ), addition of six atoms of halogen takes place by a free radical chain reaction.

$$C_6H_6 + 3\,Cl_2 \xrightarrow[(360\text{–}400\ m\mu)]{\text{Light}} C_6H_6Cl_6$$

γ-Benzene hexachloride

The production of the hexachlorides was reported first by Faraday in 1825 in his paper on the isolation of benzene. Nine stereoisomers of benzene hexachloride are possible, corresponding to the nine inositols (p. 349). Two of the nine are enantiomorphs and are called the *dextro* and *levo* α forms. The remaining known forms are designated as β, γ, δ, ϵ, η, and θ. Their configurations have been assigned on the basis of their electron diffraction patterns. Only the all-*cis* ι isomer, whose chair form would require three axial chlorines on the same side of the ring, is unknown.

The mixture of isomers has become of considerable commercial importance because of the discovery in 1943 of its insecticidal properties. It is called **BHC** or **666,** abbreviations of its name or of the formula $C_6H_6Cl_6$. It must be used at the proper dilution because of its persistent musty odor at higher concentrations. The insecticidal property is due solely to the γ form, which makes up only

13 to 18 per cent of the mixed halides. The insecticidal products are sold on the basis of the content of the γ isomer, which has been given the names *gammexane* and *lindane* (after van der Linden, who established the existence of the first four isomers in 1912). Pure γ-isomer is reported to be free of the musty odor. United States sales of benzene hexachloride in 1964 were equivalent to 3.2 million pounds of the γ isomer, valued at almost $2 million.

Electrophilic Substitution. The most characteristic behavior of aromatic compounds is the ready displacement of nuclear hydrogen by other groups. Thus whereas the direct nitration of alkanes takes place only at high temperature (p. 243), and halogenation results only from a free radical chain reaction initiated by light or high temperature (p. 68), the nitration of benzene takes place with a mixture of nitric and sulfuric acids at 60°, and halogenation occurs readily with chlorine or bromine in the presence of a Lewis-acid catalyst. Moreover numerous other groups such as sulfonic acid, alkyl, acyl, phenylazo, and acetoxymercuri can be introduced directly into aromatic nuclei. Various lines of evidence indicate that the active agent is a positive ion or an electron-deficient group, and hence these reactions are **electrophilic displacement reactions.**

(*a*) NITRATION. Nitric acid reacts with concentrated sulfuric acid to give nitronium ion, hydronium ion, and bisulfate ion. Thus the lowering of the freezing point of anhydrous sulfuric acid by nitric acid corresponds to the formation of four entities (van't Hoff factor = 4) according to the equation

$$HONO_2 + 2\,H_2SO_4 \longrightarrow {}^{+}NO_2 + H_3O^+ + 2\,HSO_4^-$$

Attack of the π electron system of the benzene ring by the nitronium ion leads to an intermediate *benzenonium ion,* commonly referred to as a σ *complex* because the new group is held by a σ bond. The intermediate ion is converted to the substituted benzene by loss of a proton to a bisulfate anion. The over-all result is **nitration** of the nucleus.

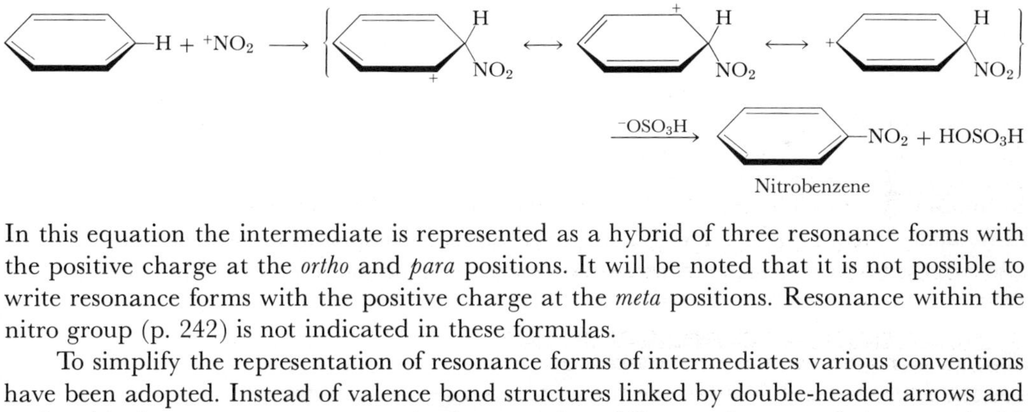

$$C_6H_5\text{—}H + {}^{+}NO_2 \longrightarrow \left\{ C_6H_5(H)(NO_2)^+ \longleftrightarrow C_6H_5(H)(NO_2)^+ \longleftrightarrow C_6H_5(H)(NO_2)^+ \right\} \xrightarrow{^-OSO_3H} C_6H_5\text{—}NO_2 + HOSO_3H$$

Nitrobenzene

In this equation the intermediate is represented as a hybrid of three resonance forms with the positive charge at the *ortho* and *para* positions. It will be noted that it is not possible to write resonance forms with the positive charge at the *meta* positions. Resonance within the nitro group (p. 242) is not indicated in these formulas.

To simplify the representation of resonance forms of intermediates various conventions have been adopted. Instead of valence bond structures linked by double-headed arrows and enclosed in braces, a common practice is to use dotted lines and a central charge as in (*a*) to indicate that the remaining electrons and the positive charge are distributed over the remaining five sp^2-hybridized carbon atoms.

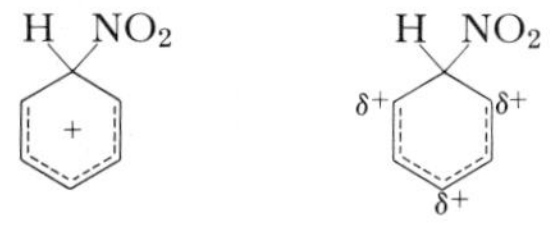

H NO₂ (ring with central +) (*a*)

H NO₂ (ring with $\delta+$ at ortho and para positions) (*b*)

Actually the positive charge is not evenly distributed but is more concentrated at the *ortho* and *para* positions than at the *meta* positions. Hence from a pedagogical standpoint a representation such as (*b*) is to be preferred.

There are various reasons why the aromatic nucleus reacts differently from the alkenes and the alkanes. The resistance to addition to the π system in contrast to the easy addition to a simple carbon-carbon double bond has been discussed (p. 374). The relatively easy displacement of aromatic hydrogen by an electrophile compared to the inactivity of an alkane is the result of several factors. In the first place, attack of the π system from either side of the benzene ring is easy, steric hindrance being entirely lacking except for that resulting from solvation. Electrophilic displacement of a proton from an alkane would require attack from the back side directly in line with the group being displaced. Of more importance is the fact that the C—H bond of alkanes must be broken at the same time as bond formation by the electrophilic reagent. With the arenes, the availability of the π electrons permits bond formation with the entering group and the production of the arenonium intermediate without the necessity of simultaneous breaking of the C—H bond. Bond-breaking is a separate step in which the proton is removed by a base. Although the formation of the intermediate undoubtedly is somewhat endothermic because of some loss of resonance energy in the intermediate, this resonance energy is regained on loss of the proton. The over-all reaction is exothermic. The bond-making process probably is preceded and aided by the formation of what has come to be known as a **π complex.** Because the π electrons are not held too strongly, they can interact with empty atomic or molecular orbitals to give 1:1 complexes of varying stability. Thus aromatic hydrocarbons form complexes with silver ions, with halogen acids, and with iodine. The energy profile for the over-all behavior may be represented as in Figure 20–*2*.

(*b*) HALOGENATION. Chlorine or bromine in the presence of a Lewis acid readily bring about the **chlorination** or **bromination** of arenes. Ferric chloride or bromide generated by the reaction of halogen with iron powder is the usual catalyst, although other strong Lewis acids such as the anhydrous aluminum halides also are effective.

$$C_6H_6 + :\ddot{X}:\ddot{X}: + FeX_3 \longrightarrow \left[C_6H_6X\right]^{\delta+}\ {}^{-}FeX_4 \longrightarrow C_6H_5X + HX + FeX_3$$

Iodination is reversible and usually is brought about in the presence of mercuric oxide or nitric acid. These reagents not only act as catalysts to make iodine more electrophilic but also react with the hydrogen iodide and enable the iodination to go to completion.

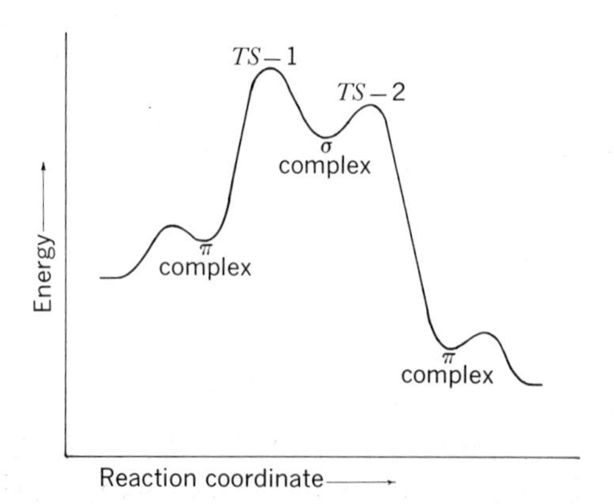

Figure 20–*2*. Energy profile for nitration.

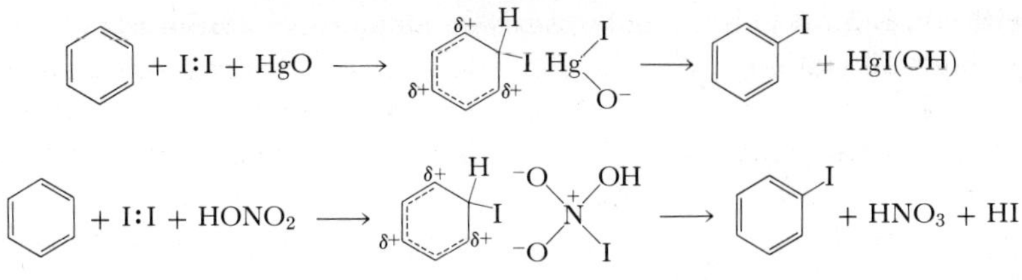

$$C_6H_6 + I{:}I + HgO \longrightarrow [C_6H_6 \cdot \text{I}]^{\delta+}\ \text{Hg(I)O}^- \longrightarrow C_6H_5I + HgI(OH)$$

$$C_6H_6 + I{:}I + HONO_2 \longrightarrow [C_6H_6 \cdot \text{I}]^{\delta+}\ {}^-O{-}\overset{+}{N}(OH)(O^-)I \longrightarrow C_6H_5I + HNO_3 + HI$$

$$4\ HI + 2\ HNO_3 \longrightarrow 2\ I_2 + N_2O_3 + 3\ H_2O$$

(*c*) ACYLATION. Reaction of the aromatic nucleus with acyl chlorides or acid anhydrides in the presence of anhydrous aluminum chloride yields ketones. Aromatic substitutions catalyzed by anhydrous aluminum chloride usually are called **Friedel-Crafts reactions.**[3,4]

Since the acyl chloride contains oxygen, aluminum chloride reacts with it to form a stable complex that is devoid of catalytic activity. Hence an excess of aluminum chloride over that necessary to form the complex is required to catalyze the reaction. Decomposition of the product with water yields the ketone.

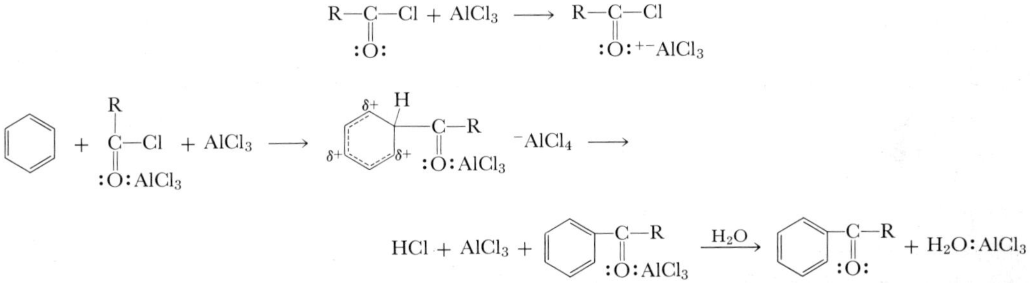

$$R{-}\underset{\underset{:O:}{\|}}{C}{-}Cl + AlCl_3 \longrightarrow R{-}\underset{\underset{:O:^{+-}AlCl_3}{\|}}{C}{-}Cl$$

$$C_6H_6 + R{-}\underset{\underset{:O:AlCl_3}{\|}}{C}{-}Cl + AlCl_3 \longrightarrow [C_6H_6{-}\underset{\underset{:O:AlCl_3}{\|}}{C}{-}R]^{\delta+}\ {}^-AlCl_4 \longrightarrow$$

$$HCl + AlCl_3 + C_6H_5{-}\underset{\underset{:O:AlCl_3}{\|}}{C}{-}R \xrightarrow{H_2O} C_6H_5{-}\underset{\underset{:O:}{\|}}{C}{-}R + H_2O{:}AlCl_3$$

In a solvent of high dielectric constant such as nitrobenzene, the acyl chloride is partially ionized.

$$RCOCl + AlCl_3 \longrightarrow RCO^{+-}AlCl_4$$

Here the uncomplexed ion may take part in the reaction with subsequent complexing of the product.

Organic acid anhydrides behave like acyl halides and frequently give better yields of ketone. They require over two moles of aluminum chloride, however, because the organic acid produced also reacts irreversibly with aluminum chloride.

$$C_6H_6 + (RCO)_2O + AlCl_3 \longrightarrow C_6H_5\underset{\underset{:O:AlCl_3}{\|}}{C}R + RCOOH$$

$$RCOOH + AlCl_3 \longrightarrow RCOO^{-+}AlCl_2 + HCl$$

(*d*) ALKYLATION. Displacement of hydrogen by alkyl groups can be brought about by using alkyl halides and anhydrous aluminum halides. Since alkyl halides do not form stable

---

[3] Charles Friedel (1832–1899), successor to Wurtz at the Sorbonne. He was the first to produce isopropyl alcohol by the reduction of acetone, and to report, with Crafts, transesterification (p. 174). He investigated the preparation of organosilicon compounds (p. 276) and in his later years devoted himself to the chemical aspects of mineralogy.

[4] James Mason Crafts (1839–1917), graduate of Harvard University and student of Bunsen and Wurtz. He became professor of chemistry first at Cornell and then at Massachusetts Institute of Technology, but resigned his post in 1874 because of poor health. He revisited the laboratory of Wurtz, planning to stay about a year, and remained there for seventeen years. Much of this time he worked in collaboration with Friedel. He returned to the United States in 1891 and resumed teaching at the Institute. He became head of the chemistry department in 1895 and was president of the Institute from 1897 until he resigned in 1900 to continue his scientific investigations.

complexes with aluminum halides, only catalytic amounts of aluminum halide are necessary to bring about **alkylation.**

$$C_6H_6 + R:X: + AlX_3 \longrightarrow [\delta+\ C_6H_6R\ \delta+\ \delta+]\ {}^-AlX_4 \longrightarrow C_6H_5R + HX + AlX_3$$

Alkenes and alcohols also can be used to alkylate arenes. Aluminum chloride catalyzes the reaction of alkenes with aromatic compounds only in the presence of hydrogen chloride, but hydrogen fluoride, concentrated sulfuric, or phosphoric acid alone suffice for the more reactive olefins.

$$C_6H_6 + CH_2{=}CH_2 + HCl + AlCl_3 \longrightarrow [\delta+\ C_6H_6CH_2CH_3\ \delta+\ \delta+]\ {}^-AlCl_4 \longrightarrow C_6H_5CH_2CH_3 + HCl + AlCl_3$$

$$C_6H_6 + CH_3CH{=}CH_2 + H_3PO_4 \longrightarrow [\delta+\ C_6H_6CH(CH_3)CH_3\ \delta+\ \delta+]\ {}^-OPO_3H_2 \longrightarrow C_6H_5CH(CH_3)CH_3 + H_3PO_4$$

Alcohols undoubtedly first form oxonium salts.

$$C_6H_6 + R\overset{+}{O}H_2\,{}^-SO_4H + H_2SO_4 \longrightarrow H_3O^{+-}SO_4H + [\delta+\ C_6H_6R\ \delta+\ \delta+]\ {}^-SO_4H \longrightarrow C_6H_5R + H_2SO_4$$

(*e*) SULFONATION. Reaction of arenes with concentrated sulfuric acid brings about displacement of hydrogen by the sulfonic acid group, a process known as **sulfonation.** Sulfonation follows a different course from nitration in that the reaction is reversible, and the active agent is sulfur trioxide rather than the bisulfonium ion, $^+SO_3H$. Sulfur trioxide is one of the components of sulfuric acid because of the equilibrium

$$2\,H_2SO_4 \rightleftharpoons SO_3 + H_3O^+ + HSO_4^-$$

The three oxygen atoms in sulfur trioxide and the semipolar bonds greatly decrease the electron density on sulfur, making sulfur trioxide a strongly electron-seeking reagent.

$$C_6H_6 + SO_3 \rightleftharpoons \left[\delta+\ C_6H_6SO_2O^-\ \delta+\ \delta+\right] \underset{HB}{\overset{B^-}{\rightleftharpoons}} C_6H_5SO_2O^- \underset{B^-}{\overset{HB}{\rightleftharpoons}} C_6H_5SO_2OH$$

Benzenesulfonic acid

It is the ability to undergo these electrophilic substitution reactions that is one of the important characteristics of aromatic compounds. Numerous other electrophilic substitution reactions will be discussed from time to time. Under certain conditions aromatic compounds are capable also of undergoing nucleophilic substitution (p. 412) and attack by free radicals.

Effect of Nuclear Groups Other Than Hydrogen on Electrophilic Substitution

Activation and Deactivation of the Benzene Nucleus. Thus far the discussion has been limited primarily to the behavior of benzene, where all of the hydrogens have identical properties. If one hydrogen has been replaced by another group, the behavior of the remaining hydrogens is affected in various ways. The first noticeable effect is that some groups increase the rate of substitution compared to benzene whereas other groups decrease the rate of substitution. The former groups are said to *activate* the ring and the latter to *deactivate* the ring. Activating groups are those that can make the π electrons more readily available, that is, increase the nucleophilicity of the ring, whereas deactivating groups are electron-withdrawing groups. Experimentally the most strongly activating group is the oxide ion, $—O^-$, which carries a full negative charge of one electron, whereas the most strongly deactivating group is the trialkylammonium group, $—NR_3^+$, which carries a full positive charge equivalent to one electron. For groups that do not carry a full negative or positive charge, the relative effectiveness for increasing or decreasing the rate of substitution should depend on the size and direction of the electric dipole of the monosubstituted benzene.

Electric Dipole Moments and Their Direction. The methyl group is believed to increase the electron density at an unsaturated carbon atom (p. 82). If a change in electron distribution takes place when one group replaces another, it should lead to a change in the electric dipole moment (p. 18). The dipole moment of benzene is zero, as is that of methane, *p*-xylene, and all other symmetrically constituted compounds (p. 20). If there were no difference between the phenyl group and the methyl group in their ability to attract or repel electrons, the moment of toluene would be zero, as is true for ethane or biphenyl. Toluene, however, has a moment of 0.4 D. Hence there is a shift of the electron density either towards the phenyl group or towards the methyl group.

The direction of the polarization has been determined by comparing the moment of nitrobenzene with that of *p*-nitrotoluene. For nitrobenzene the moment is 4.3 D. The high dipole moment of nitro compounds is explained by assuming the presence of a semipolar bond in which the nitrogen is positive with respect to the oxygen atoms (p. 242). Hence the effect of the group must be to decrease the electron density of the ring, which agrees with the fact that experimentally it is deactivating. Since *p*-dinitrobenzene has zero moment, the moments caused by the nitro groups must be opposed to each other and lie on a line passing through the *para* positions.[5] If now any other group is *para* to the nitro group and the vector representing the electric moment that it induces lies on the line passing through the *para* positions, the moment of the disubstituted benzene will be greater or less than that of nitrobenzene depending on whether the moment is in the same direction as that caused by the nitro group or is opposed to it. Thus if the methyl group has a greater attraction for electrons than a hydrogen atom, the moment of *p*-nitrotoluene should be less than that of nitrobenzene and should approximate $4.3 - 0.4 = 3.9$ D, but if the

---

[5] Here is direct evidence that the nitro group is a resonance hybrid $\left\{—\overset{+}{N}(=O)—O^- \longleftrightarrow —\overset{+}{N}(—O^-)=O\right\}$. If it were unsymmetrically constituted as represented by one of these structures, *p*-dinitrobenzene could have zero moment only when the N $\leftrightarrow$ O vectors are opposed to each other as in (*a*). In all other positions the molecule would have a resultant moment as in (*b*). Since isomeric forms cannot be isolated, essentially free rotation about the nitrogen-carbon bond must be assumed. Therefore the moments of unsymmetrical nitro groups would not always be opposed, and the molecule would have a resultant electric moment. Hence the zero moment must be ascribed to a symmetrical structure for the nitro group (*c*).

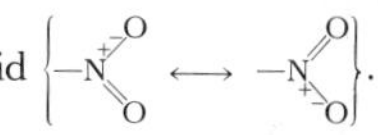

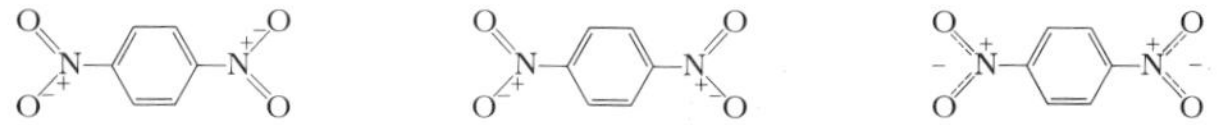

$\mu = 0$ (*a*) $\mu =$ maximum (*b*) $\mu = 0$ (*c*)

Groups known to have unsymmetrical structures, such as the hydroxyl group and the amino group, give rise to electric moments in the *para*-disubstituted compounds.

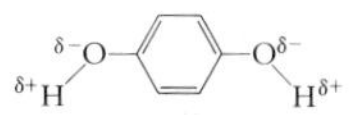

$\mu = 2.5$ D 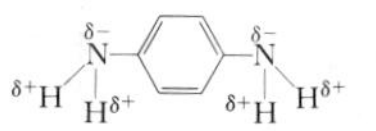 $\mu = 1.5$ D

methyl group has less attraction for electrons than hydrogen (*electron-repelling* or *electron-donating*), the moment of *p*-nitrotoluene should be larger than that of nitrobenzene and should approximate 4.3 + 0.4 = 4.7 D. The observed moment for *p*-nitrotoluene is 4.5 D, and hence the methyl group increases the electron density of the benzene nucleus, which agrees with the fact that it is activating. It is convenient to indicate the direction of the induced moments by the sign $+\!\!\longrightarrow$. The arrow symbolizes the direction in which electron density is increased, the tail being the positive end of the dipole.

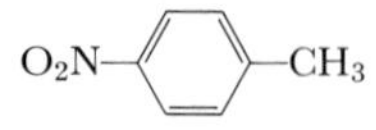

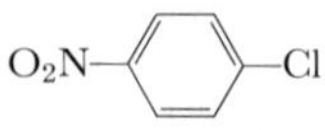

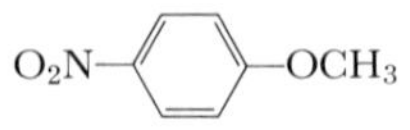

O_2N–C_6H_4–CH_3		O_2N–C_6H_4–Cl		O_2N–C_6H_4–OCH_3	
←+ 4.3 D	←+ 0.4 D	←+ 4.3 D	+→ 1.7 D	←+ 4.3 D	←+ 1.4 D
Difference	= 3.9 D	Difference	= 2.6 D	Difference	= 2.9 D
Sum	= 4.7 D	Sum	= 6.0 D	Sum	= 5.7 D
Experimental	= 4.5 D	Experimental	= 2.8 D	Experimental	= 5.3 D

Chlorobenzene has a moment of 1.7 D and *p*-chloronitrobenzene 2.8 D. Hence the chlorine atom has a greater attraction for electrons than hydrogen, in agreement with the deactivating effect. The directions of the moments for other groups have been determined by similar methods and are listed in Table 20–*1*. Groups above hydrogen in Table 20–*1* are activating, whereas those below hydrogen are deactivating.

The behavior of the groups above hydrogen in which nitrogen or oxygen is bound to a ring carbon appear to be anomalous. Both nitrogen and oxygen are more electronegative than hydrogen (p. 21) and should be electron-withdrawing. As is demonstrated for methoxybenzene, however, the dipole moments of the *p*-nitro derivatives shows that they are electron-donating. The explanation lies in the fact that these groups have unshared pairs of electrons in a 2*p* orbital, which can interact with the π electrons of the benzene ring thus increasing the electron density of the nucleus. The effect for aniline (*aminobenzene*)

TABLE 20–*1*. MAGNITUDE AND DIRECTION OF ELECTRIC MOMENTS OF BENZENE DERIVATIVES

GROUP *A* IN C_6H_5-*A*	ELECTRIC MOMENT OF C_6H_5-*A*	DIRECTION OF MOMENT C_6H_5-*A* ←+ or C_6H_5-*A* +→
$N(CH_3)_2$	1.68	←+
NH_2	1.53	
OH	1.45	
OCH_3	1.38	
$C(CH_3)_3$	0.83	
$CH(CH_3)_2$	0.79	
CH_2CH_3	0.59	
CH_3	0.36	
H	0.0	
COOH	1.6	+→
F	1.6	
Cl	1.7	
Br	1.8	
CH_2Cl	1.8	
$COOC_2H_5$	1.9	
$CHCl_2$	2.0	
CCl_3	2.1	
CHO	2.8	
$COCH_3$	3.0	
NO	3.1	
SO_3H	3.8	
NO_2	4.3	
CN	4.4	

may be represented either by overlap of the p orbitals as in Fig. 20–3a to give a new molecular orbital, or by resonance among valence-bond structures as in Fig. 20–3b.

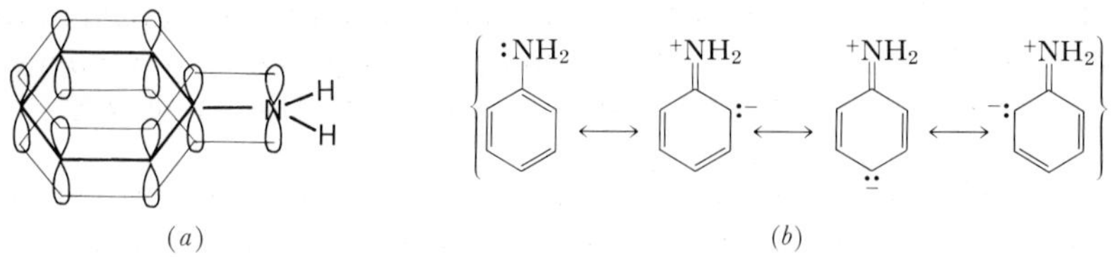

(a) (b)

Figure 20–3. Interaction of an unshared pair of electrons on nitrogen with the π electrons of the nucleus: (a) molecular orbital representation; (b) valence-bond representation.

This behavior is known as the **resonance effect,** and predominates over the usual inductive effect of nitrogen and oxygen. Resonance should occur also with halogen and there is evidence that it does. Thus the interatomic carbon-chlorine distances for t-butyl chloride and chlorobenzene are 1.76 A and 1.69 A respectively, and their dipole moments are 2.14 D and 1.56 D. Both the shortening of the bond and the lower dipole moment for the aromatic compound are understandable if there is interaction of the unshared electrons of chlorine with the π electrons of the nucleus. Because the halogens are much more electronegative than nitrogen or oxygen, however, the inductive effect here predominates and the over-all effect is electron-withdrawing.

The order within the halogen group also appears anomalous since it is the reverse of their electronegativities, and it can mean only that a decrease in resonance from fluorine to iodine more than offsets the effect of decreasing electronegativity. The explanation is the same as that for increase in acidity of the halogen acids from hydrogen fluoride to hydrogen iodide; that is, the increasingly poorer overlap of 3, 4, and 5p orbitals with the π orbitals of the nucleus leads to progressively lower resonance energy just as poorer overlap with the 1s orbital of hydrogen leads to progressively lower H—X bond energy.

Table 20–1 explains why nitration, sulfonation, halogenation, and acylation by the Friedel-Crafts reaction are practical preparative procedures. All of the groups introduced are deactivating. Hence in contrast to the free radical nitrations and halogenations of alkanes, *the reactions are controllable*. The introduction of one group makes it more difficult to introduce a second, and the introduction of a second makes it still more difficult to introduce a third. Thus conditions of temperature and concentrations can be chosen to give chiefly the desired product. An exception is alkylation by the Friedel-Crafts reaction. Since the alkyl groups are somewhat activating, the ease of introduction of alkyl groups increases with increasing substitution by alkyl groups, and mixtures of polyalkylated benzenes result. Although it is not possible to replace more than three hydrogens of benzene by nitro groups, all six hydrogens can be replaced by methyl groups. The resulting product is so basic (p. 387) that it forms with methyl chloride a stable σ complex that in effect is a quaternary salt.

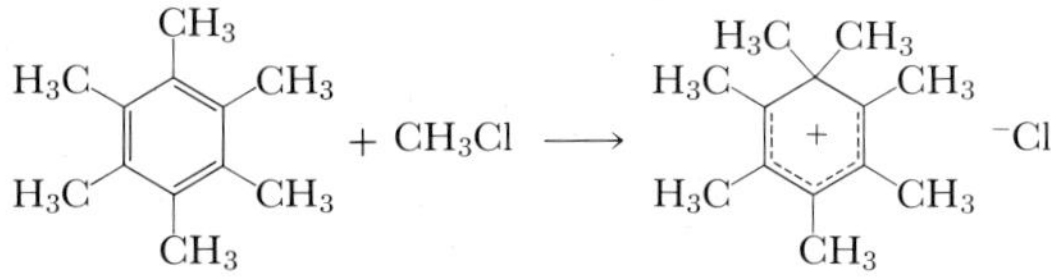

$$C_6(CH_3)_6 + CH_3Cl \longrightarrow [C_6(CH_3)_7]^+ \; ^-Cl$$

Directive Influence of Substituents. All six hydrogens of benzene are equivalent. Hence displacement of any hydrogen takes place with equal ease and yields the same product. If one hydrogen has been replaced by another group, however, three kinds of nuclear hydrogen remain, depending on whether they are in the *ortho, meta,* or *para* positions. If all hydrogens could be displaced with equal ease, the ratios of *ortho* to *meta* to *para* substitution should be the statistical one of 2:2:1. Rarely is this ratio found experimentally. Table 20–2 lists the relative amounts of *ortho, para,* and *meta* isomers formed in the nitration of monosubstituted benzenes. The values for the different substituents are not strictly comparable because the conditions of nitration were not always the same. They reflect, however, the general behavior.

It will be noted that the amount of *meta* isomer formed varies from zero to 100 per cent and follows approximately the direction and magnitude of the dipole moment of the monosubstituted benzene as listed in Table 20–1; that is, electron-donating groups favor *ortho,para* substitution, and electron-withdrawing groups favor *meta* substitution.

The substitution reactions for the most part are irreversible, and the relative amounts of isomers formed depends on the relative rates of substitution at the various positions, which in turn depend on the difference in free energy between the ground state of the reactants and the transition state for the reaction. Since the rate-determining step is the formation of the benzenonium ion (p. 375), factors that stabilize the ion should stabilize the transition state and increase the rate of substitution. Hence the σ complex may be used as a model for the transition state.

If A is the directing group and E the entering electrophile, the more stable resonance forms will be those in which the positive charge is located at the positions *ortho* and *para* to

TABLE 20–2. RELATIVE AMOUNTS OF *ortho, para,* AND *meta* ISOMERS FORMED IN THE NITRATION OF MONOSUBSTITUTED BENZENES

GROUP PRESENT IN RING	ISOMERS FORMED ON NITRATION (PER CENT)			
	ortho	*para*	$o + p$	*meta*
OH*	40	60	100	0
$CH(CH_3)_2$	14	86	100	0
CH_2CH_3	55	45	100	0
F	12	88	100	trace
Cl	29.6	69.5	99.1	0.9
Br	36.5	62.4	98.9	1.1
I	38.4	59.8	98.2	1.8
$NHCOCH_3$	19	79	98	2
CH_3	59	37	96	4
$C(CH_3)_3$	12	80	92	8
$CH_2COOC_2H_5$	42	47	89	11
$CH_2CH_2NO_2$	35	52	87	13
CH_2Cl	32	52	84	16
CH_2NO_2			67	33
$CHCl_2$	23	43	66	34
$[NH_3]^+$	1	52	53	47
$COCH_3$	45	0	45	55
CCl_3	7	29	36	64
$CONH_2$	27	3	30	70
$COOC_2H_5$	28	4	32	68
SO_3H	21	7	28	72
CHO	19	9	28	72
COOH	19	1	20	80
CN	17	2	19	81
NO_2	7	trace	7	93
SO_2CH_3	trace	trace	0	100
$[N(CH_3)_3]^+$	0	0	0	100

* In the absence of nitrous acid.

E. Resonance forms in which the positive charge is located *meta* to *E* are not important because they would require the unpairing of electrons or a separation of charge in the molecule with like or opposite charges on adjacent carbon atoms. When *E* is *ortho* or *para* to *A*, a positive charge is located at the ring carbon atom to which *A* is attached in (*a*) and (*d*) of intermediates *I* and *II*, whereas if *E* is *meta* to *A* as in intermediate *III*, this carbon does not bear a positive charge in any of the resonance forms.

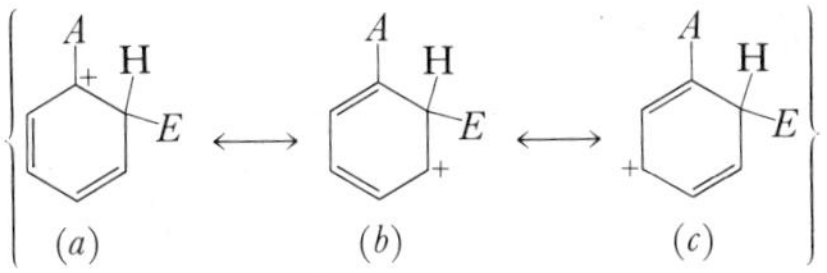

If *A* is an electron-donating group, it will stabilize a positive charge adjacent to it, and hence favor the formation of intermediates *I* and *II* in which *E* is the in *ortho* and *para* positions, over *III* in which *E* is in the *meta* position. On the other hand, if *A* is electron-attracting, it is more difficult to withdraw electrons from the carbon atom to which *A* is attached, and intermediates *I* and *II* can be formed less readily than intermediate *III*. Thus electron-donating groups favor *ortho,para*-substitution, whereas electron-attracting groups favor *meta* substitution.

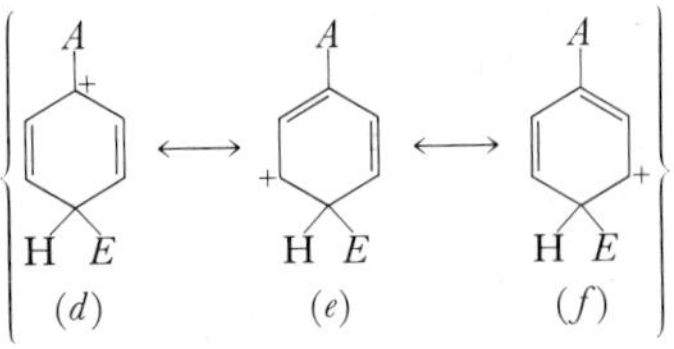

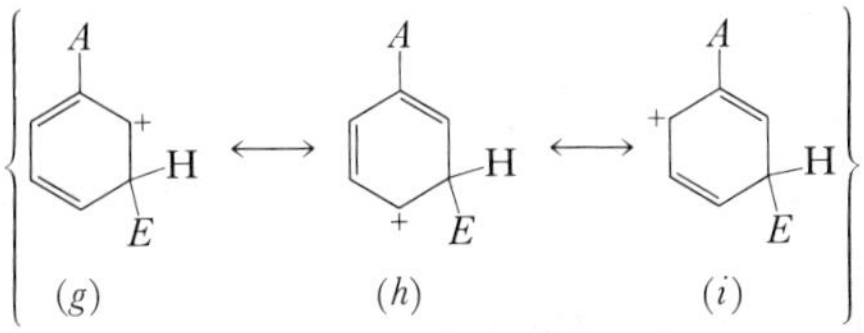

If *A* is a group that is not complicated by having unshared electrons or multiple bonds on the atom joined to the ring, such as a methyl group, CH_3, or a trichloromethyl group, CCl_3, these considerations hold. Because of its greater polarizability, a methyl group is better able than hydrogen to supply electrons to a more electronegative group such as phenyl (Table 20–*1*, p. 380). Hence the rate of substitution at the *ortho* and *para* positions of toluene is increased over that expected statistically. This type of directive influence of the group *A* is known as an **inductive effect.** According to Table 20–*1* the trichloromethyl group attracts electrons more strongly than a hydrogen atom; that is, it has an inductive effect opposite to that of a methyl group, and, by decreasing the electron density, tends to prevent the formation of intermediates *I* and *II*. Hence the rate of substitution at the *ortho* and *para* positions is decreased over that expected statistically, which results in a higher proportion of *meta* substitution.

The series NO_2, CH_2NO_2, and $CH_2CH_2NO_2$ of Table 20–*2* indicates that the effect of a group rapidly diminishes as saturated hydrocarbon groups are interposed between it and the ring. The series CH_3, CH_2Cl, $CHCl_2$, CCl_3 illustrates the effect of replacing a less strongly electron-attracting group by a more strongly electron-attracting group.

The *ortho,para*-directing effect of halogen appears to be anomalous. Since halogen is more electron-attracting than hydrogen, its inductive effect deactivates the nucleus. When called upon to do so by an electrophilic reagent, however, it can supply electrons by its resonance effect, and a resonance form in addition to those illustrated by *I* or *II* is possible when a group enters the *ortho* or *para* positions.

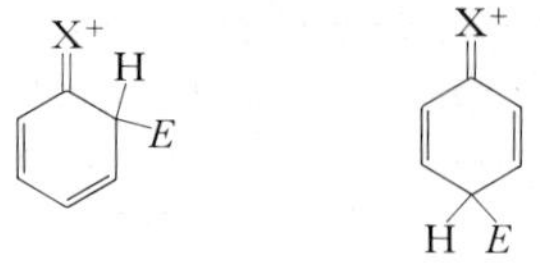

No such additional resonance form is possible to aid the entrance of E to a *meta* position. This resonance effect accounts also for the strongly *ortho,para*-directing effect of hydroxy or amino groups, since the inductive effects of oxygen and nitrogen are less than those of the halogens.

The hydroxyl group is more strongly activating in alkaline solution than in neutral or acid solution because the negative charge on the oxygen atom in the phenoxide ion reverses the inductive effect; moreover, the resonance effect is increased because no separation of charge is present in the resonance structures of the phenoxide ion.

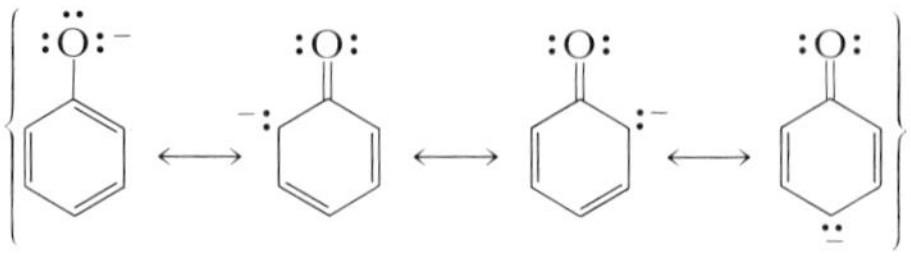

Since the inductive and resonance effects now reinforce each other, the negatively-charged oxygen atom is the strongest activating and strongest *ortho,para*-directing group.

The amino group is less strongly electron-attracting than the hydroxyl group. Hence the effect of the amino group lies between that of the undissociated hydroxyl group and the negatively-charged oxygen atom. In acid solution the unshared pair of electrons of the amino group acquires a proton and the group a positive charge, thereby removing the resonance effect and increasing the electron-attracting power of the group, with the result that it becomes deactivating and *meta*-directing. In the nitration of aniline in acid solution most of the *ortho* and *para* isomers arise from the reaction of the free amine that is in equilibrium with its salt. Thus the proportion of the *meta* isomer increases with increasing concentration of acid. Recent work shows, however, that the increase is not as great as expected, which indicates that even the cation gives some *para* nitro derivative. Moreover the trimethylanilinium ion undergoes up to 11 per cent *para* substitution.

Directive influence frequently is ascribed to differences in electron densities at the various positions of the benzene ring as the result of resonance. Thus halogen is said to lead to a greater electron density at the *ortho* and *para* positions than at the *meta* positions, whereas a nitro group leads to a relatively higher electron density at the *meta* positions.

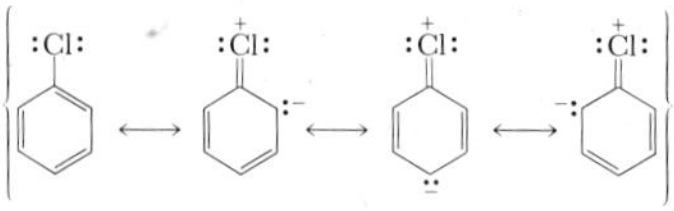

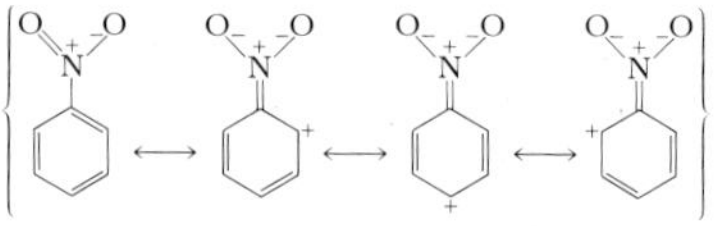

This explanation does not account for the strong *meta*-directing effect of the trimethylammonium group. Moreover relative reactivity depends not on the ground state of a particular reactant but on the difference in free energy between the reactants and the transition state. Hence it is the relative stability of the various possible transition states, which are believed to be approximated by those of the corresponding σ complexes, that accounts for the directive influence of different groups.

It will be noted from Table 20–2, p. 382, that where the atom bound to the ring is not joined to another atom by a multiple bond, the amount of *ortho* isomer is never twice that of the *para* as expected statistically. Although electronic effects have some influence, the chief reason that the *para* position is favored is that it is less sterically hindered. Similarly very little substitution takes place between two groups that are *meta* to each other (p. 386). A steric effect is observable also as regards the entering group. Thus the ratio of *ortho* to *para* substitution decreases in the order chlorination, nitration, bromination, sulfonation. This order is the same as that of increasing size of the entering group. When the atom bound to the ring is attached to another atom by a multiple bond, the *ortho/para* ratio is greater than two. Here the *ortho* σ complex is stabilized relative to the *para* by an additional resonance form that is not available to the *para* complex without placing like charges on adjacent carbon atoms.

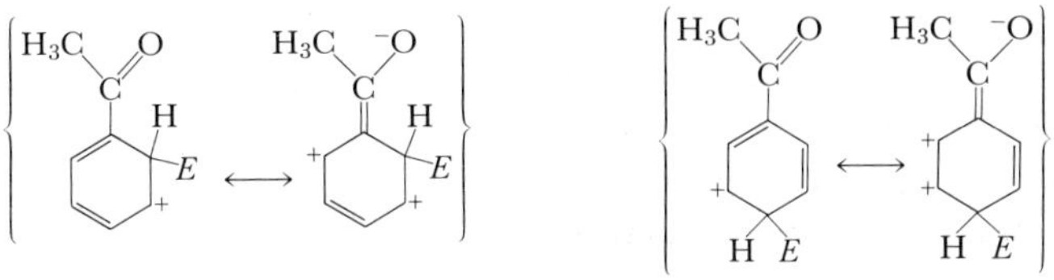

The behavior of aromatic compounds on electrophilic substitution may be summarized as follows:

(*1*) A number of groups, such as halogen (X), nitro (NO_2), sulfonic acid (SO_3H), alkyl (R), and acyl (RCO or ArCO), may be introduced directly into the benzene nucleus, hydrogen being displaced.

(*2*) When a second substituent is introduced into a benzene nucleus, the relative amounts of the *ortho, meta,* and *para* isomers should be 40, 40, and 20 per cent respectively on a statistical basis. Usually this ratio is not obtained. The relative amounts of the isomers depends primarily on the nature of the group already present in the ring, although the nature of the entering group and the conditions of the reaction, such as temperature and concentrations, have some influence.

(*3*) Groups vary greatly in their directing power, from the $^+NR_3$ group, which causes substitution almost exclusively in the *meta* position, to the O^- group, which causes substitution almost exclusively in the *ortho* and *para* positions. Other groups fall between these extremes in directing power (Table 20–*2*). Groups causing the production of more than 60 per cent of the *ortho* and *para* isomers combined are called *ortho,para-directing groups,* and those causing the production of more than 40 per cent of the *meta* isomer are called *meta-directing groups.*

(*4*) *ortho,para*-Directing groups, *with the exception of halogen,* increase the ease with which a second hydrogen can be displaced. Such groups are said to *activate* the ring. *meta*-Directing groups *and halogen* decrease the ease with which a second hydrogen can be displaced; that is, they are *deactivating* groups.

Activation and deactivation are reflections of the rates of substitution compared with benzene. The effect of a group on the relative amounts of the three isomers formed reflects the relative rates of substitution at the unsubstituted positions. When the group already present is activating and *ortho,para*-directing, it increases the rate of substitution over that of benzene, and at the *ortho* and *para* positions more than at the *meta* positions. When a group is deactivating and *meta*-directing, it decreases the rate of substitution compared with that for benzene, but decreases the rate less at the *meta* positions than at the *ortho* and *para* positions. Halogen, which is deactivating and *ortho,para*-directing, decreases the rate of substitution over that for benzene, but decreases the rate less at the *ortho* and *para* positions than at the *meta* positions.

Table 20–*2*, p. 382, does not list the free NH_2, NHR, or NR_2 groups nor the O^- group, because they cannot exist as such in acid solution. These groups are known to direct strongly to the *ortho* and *para* positions from other reactions such as halogenation.

If several groups are all strongly *ortho,para*-directing, it is difficult to determine their relative directing power because so little of the *meta* isomer is formed. An answer has been obtained by determining which isomer is produced in larger amount when a third group enters a *para*-disubstituted compound. For example, chlorination of *p*-hydroxytoluene gives chiefly 3-chloro-4-hydroxytoluene, indicating that the hydroxyl group is more strongly *ortho,para*-directing than the methyl group.

$$CH_3C_6H_4OH + Cl_2 \longrightarrow CH_3C_6H_3(Cl)OH + HCl$$

By such methods the relative order of directive influence has been shown to be $O^- > NH_2 > OH > X > CH_3$.

When two substituents are present in the benzene ring, the position taken by the entering group is influenced by both groups already present. If these groups direct to the same position, a third group enters almost exclusively at this position. For example, further nitration of 2-nitrotoluene gives a mixture of 2,4- and 2,6-dinitrotoluene, and 4-nitrotoluene gives 2,4-dinitrotoluene (p. 411). In 2-nitrotoluene the 4 and 6 positions are *para* and *ortho* to the methyl group and *meta* to the nitro group. In 4-nitrotoluene the 2 position is *ortho* to the methyl group and *meta* to the nitro group. Similarly two *ortho,para*-directing groups or two *meta*-directing groups in the *meta* position to each other direct to a single position.

$$C_6H_4(CH_3)_2 \xrightarrow[(+\ H_2SO_4)]{HONO_2} C_6H_3(CH_3)_2NO_2 + H_2O$$

4-Nitro-*m*-xylene

$$C_6H_4(NO_2)_2 \xrightarrow[(+\ \text{fuming}\ H_2SO_4)]{\text{Fuming}\ HONO_2} C_6H_3(NO_2)_3 + H_2O$$

1,3,5-Trinitrobenzene

When two *ortho,para*-directing groups or two *meta*-directing groups are *ortho* or *para* to each other, that with the stronger directive power determines the predominant isomer. Thus in the chlorination of *p*-hydroxytoluene, the chief product is 3-chloro-4-hydroxytoluene.

If an *ortho,para*-directing group is *meta* to a *meta*-directing group, the influence of the *ortho,para*-directing group usually predominates over that of the *meta*-directing group. For example, in the nitration of *m*-nitrotoluene, a third nitro group enters the 2, 4, and 6 positions but not the 5 position.

$$CH_3C_6H_4NO_2 \xrightarrow[(+\ H_2SO_4)]{HONO_2} CH_3C_6H_3(NO_2)_2 \text{ and } CH_3C_6H_3(NO_2)_2 \text{ and } CH_3C_6H_3(NO_2)_2$$

Many rules have been formulated for remembering whether a group is predominantly *ortho,para*-directing or predominantly *meta*-directing. Some are purely empirical and others have some theoretical basis, but none is quantitative. Probably the simplest empirical rule is that halogen and groups in which the atom joined to the ring is united to other elements by single homopolar bonds or carries a negative charge are chiefly *ortho,para*-directing, whereas groups in which the atom joined to the ring is united to other elements by multiple bonds or semipolar bonds or carries a positive charge are chiefly *meta*-directing. The principal exceptions to this rule are the trichloromethyl group (CCl_3), which directs chiefly to the *meta* position, and the vinyl group ($CH{=}CH_2$), which directs chiefly to the *ortho* and *para* positions.

Some Other Chemical Properties

Basicity of Aromatic Hydrocarbons. The fact that arenes form σ complexes may be thought of as an exhibition of basicity. The basicity of benzene is increased if electron-donating substituents are present, especially if they are *meta* to each other. Thus the relative basicities of some hydrocarbons are: benzene, 1; toluene, 10; *p*-xylene, 100; *m*-xylene, 2000; mesitylene, 2.8×10^6; hexamethylbenzene, 8.9×10^7 (cf. p. 381).

The variation in basicity can be observed visually by the differences in color of the complexes with iodine. Iodine in the vapor state is violet as are solutions in hexane. Both have an absorption maximum at 520 mμ. Solutions in electron-donor solvents such as ether are brown and absorb around 450 mμ with no appreciable change in the intensity or shape of the absorption band. This shift in wavelength is ascribed to the formation of a 1:1 complex resulting from the interaction of the unshared electrons on oxygen with an unoccupied molecular orbital of the iodine molecule. The complex may be represented as a resonance hybrid of the type $\{DA \longleftrightarrow D^+A^-\}$, where DA is a close association of the electron-donor molecule, D, and the electron-acceptor molecule, A, whereas in D^+A^- an electron has been transferred from D to A. Hence they are called **charge-transfer complexes.** The intense absorption band often associated with these complexes may be ascribed to a transition from a ground state, in which the amount of charge transfer is small, to an excited state, in which the amount of charge transfer is large. The colors of dilute solutions of iodine in hexane, benzene, *m*-xylene, and mesitylene vary progressively from violet to reddish brown and indicate progressively increasing electron-donor properties. The respective ionization potentials of the hydrocarbons are 10.4, 9.5, 9.0, and 8.7 e.v. and are proportional to the wavelength of the absorption maximum.

Disproportionation and Isomerization. As would be predicted from bond energies, the alkylation of benzene is reversible. When toluene is treated with catalytic amounts of aluminum chloride and hydrogen chloride, a mixture of benzene, toluene, xylenes, and higher alkylated benzenes results. If higher alkyl groups are present, rearrangement of the side chain takes place, as well as degradation caused by catalytic cracking of the carbon chain. The interconversion of xylenes is of some practical importance (p. 392). This reaction might be expected to result from reversible demethylation and remethylation to give the thermodynamic equilibrium mixture. It has been found, however, that the first product of the isomerization of *o*-xylene is *m*-xylene, which then rearranges to *p*-xylene. Since methylation of toluene initially yields *o*- and *p*-xylene, the isomerization reaction is believed to take place by a methide ion 1,2 shift.

$$C_6H_4(CH_3)_2\ (o) + HF + BF_3 \rightleftharpoons \left[C_6H_5(CH_3)_2^+\right] BF_4^- \rightleftharpoons \left[C_6H_5(CH_3)_2^+\right] BF_4^- \rightleftharpoons C_6H_4(CH_3)_2\ (m) + HF + BF_3$$

Similarly a methyl group can migrate from the *para* position to either *meta* position and from the *meta* position to an *ortho* or *para* position. If the molar concentrations of hydrogen fluoride and boron fluoride are at least as high as that of the xylene, the product is chiefly *m*-xylene rather than the equilibrium mixture because of the greater basicity of the *meta* isomer. Isomerization induced by ultraviolet radiation involves ring carbon transpositions.

Oxidation of Side Chains. Although the oxidation of side chains does not involve the aromatic nucleus, it usually is considered to be one of the typical reactions of aromatic compounds. When the alkylbenzenes are subjected to vigorous oxidation with chromic acid (*sodium dichromate and sulfuric acid*) or potassium permanganate, or to catalytic air-oxidation, the alkyl group is converted to a carboxyl group. The carbon atom adjacent to the ring is more susceptible to oxidation than the remaining carbon atoms of the alkyl group and is the initial point of attack. For chains having two or more carbon atoms, the probable course of the reaction is by way of the alcohol, ketone, and enol (p. 205). Since all of the intermediates are oxidized more readily than the hydrocarbon, the only isolable aromatic compound is the carboxylic acid.

$$ArCH_2CH_2R \longrightarrow ArCHOHCH_2R \longrightarrow Ar\underset{\underset{O}{\|}}{C}CH_2R \longrightarrow Ar\underset{\underset{OH}{|}}{C}{=}CHR \longrightarrow ArCOOH + HOOCR$$

Clearly any substituted carbon chain also can be oxidized to a carboxyl group.

Orientation of Substituents in the Benzene Nucleus

Table 20–2, p. 382, lists the ratios of *ortho, meta,* and *para* isomers formed on electrophilic substitution of a monosubstituted benzene. These isomers have different physical and chemical properties, but nothing has been said about how the structure is assigned to any particular isomer. For example, three xylenes are known (p. 372). One boils at 138.4° and melts at +13.4°, another boils at 139.3° and melts at −47.4°, and the third boils at 144.1° and melts at −25°. The question is, "Which isomer has the methyl groups in the *ortho* position, which in the *meta* position, and which in the *para* position?" The procedure for assigning structures to isomeric benzene derivatives is known as *orientation.* Two methods have found general use. One is known as the Koerner absolute method, and the other may be called the interconversion method.

Koerner*[6] *Absolute Method. If a third group is introduced into each member of any set of three isomeric disubstituted benzenes in which the two original substituents are alike, the *ortho* isomer should yield two trisubstitution products, the *meta* three trisubstitution products, and the *para* only one trisubstitution product. Thus *o*-xylene on mononitration should yield two nitro-*o*-xylenes, *m*-xylene should yield three nitro-*m*-xylenes, and *p*-xylene should yield one nitro-*p*-xylene.

The nitro group has, of course, a larger total number of positions to enter, but more than one position may yield the same compound. For example, in *p*-xylene four hydrogen atoms may be replaced by a nitro group but replacement of any one of them yields the same product.

[6] Wilhelm Koerner (1839–1925), assistant to Kekulé and Cannizzaro and later professor of chemistry and director of the School of Technology and Agriculture in Milan. He devised one of the chemical proofs of the equivalence of the six hydrogen atoms of benzene, and first proposed the accepted structural formula for pyridine (p. 514).

When the three known xylenes are nitrated, that boiling at 138.4° yields only one mononitro compound and hence is the *para* isomer; that boiling at 139.3° yields three mononitro compounds and hence is the *meta* isomer; that boiling at 144.1° yields only two mononitro compounds and hence is the *ortho* isomer. The constitutions of a number of disubstituted benzenes have been determined by this procedure, and the principle can be extended to other types of substitution.

Interconversion Method. Once the orientation of the groups in a substituted benzene has been determined, the orientation is known for all compounds that can be derived from it by converting one or more substituents into other groups. For example, three benzenedicarboxylic acids are known. One called *phthalic acid* melts with decomposition at about 200°, another, *isophthalic acid,* melts at 348°, and a third, *terephthalic acid,* sublimes without melting at 300°. These acids can be formed respectively by the vigorous oxidation of *o*-, *m*-, and *p*-xylene and hence are the *o*-, *m*-, and *p*-dicarboxylic acids.

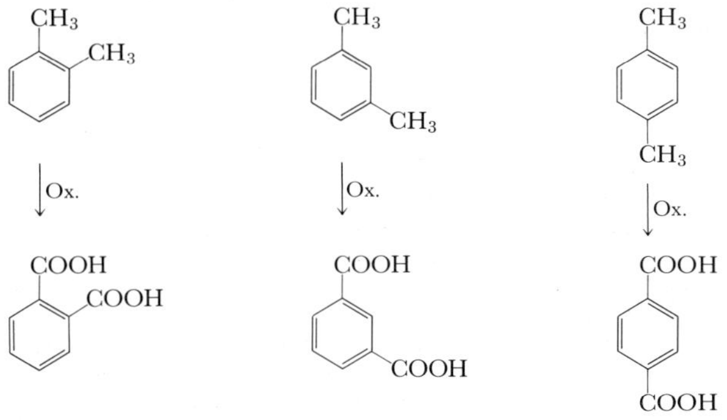

CH_3 CH_3 CH_3 CH_3 CH_3 CH_3

Ox. Ox. Ox.

COOH COOH COOH COOH COOH COOH

Phthalic acid — Isophthalic acid — Terephthalic acid

Since the three methylbenzenecarboxylic acids (*toluic acids*) also yield the three phthalic acids on oxidation, their constitution likewise is established.

SOURCES OF AROMATIC COMPOUNDS

Carbonization of Coal

Coal is a compact stratified mass derived from plants that have suffered partial decay and have been subjected to various degrees of heat and pressure. Most normal banded coals are believed to have originated in peat swamps. The substances peat, lignite, soft or bituminous coal, and anthracite or hard coal are progressive stages of metamorphosis in which the ratio of the amount of carbon to the amount of other elements increases. When bituminous coal is heated to a sufficiently high temperature (350–1000°) in the absence of air, volatile products are formed, and a residue of impure carbon remains, which is called *coke.* The process is known as the *destructive distillation* or *carbonization* of coal. When the volatile products cool to ordinary temperature, a portion condenses to a black viscous liquid known as *coal tar.* The noncondensable gases are known as *coal gas.* One ton of coal yields about 1500 pounds of coke, 8 gallons of tar, and 10,000 cubic feet of coal gas. About 20 pounds of ammonium sulfate is obtained by washing the gas with sulfuric acid to remove the ammonia. Production of coal tar in the United States in 1964 was 762 million gallons. The principal reason for the commercial carbonization of coal is the production of **coke,** which is used for the reduction of ores in blast furnaces. Coke is used also as a smokeless industrial and household fuel.

Coal gas varies in composition during the course of the distillation but consists chiefly of hydrogen and methane in about equal volumes, along with some carbon monoxide, ethane, ethylene, benzene, carbon dioxide, oxygen, and nitrogen, and smaller amounts of cyclopentadiene (p. 643), toluene, naphthalene (p. 487), water vapor, ammonia, hydrogen sulfide, hydrogen cyanide, cyanogen, and nitric oxide. After removal of the noxious components, the gas is run into mains for use as illuminating gas and as a domestic fuel. When economically profitable, the benzene, toluene, and other less volatile hydrocarbons also are extracted from the gas by washing (*scrubbing*) with a high-boiling petroleum fraction (b.p. 285–350°) known as *straw oil.* The hydrocarbons are recovered by heating (*stripping*) the oil and condensing the vapors. The condensate contains chiefly benzene and toluene and is known as *light oil* because of its low specific gravity. Although the benzene and toluene are liquids at room temperature, the coal gas is saturated with them, and a larger amount can be obtained by washing the coal gas than can be obtained by distilling the coal tar. About three gallons of light oil can be recovered per ton of coal carbonized.

The composition of **coal tar** varies with the process used for carbonization. Tar obtained from high temperature distillation is the most useful for chemical purposes. The first step in the separation of the black foul-smelling liquid into its components is distillation. The fractions obtained in a typical procedure are (*1*) *light oil* (so-called because it floats on water) distilling up to 200°, 5 per cent; (*2*) *middle oil* (carbolic oil), 200–250°, 17 per cent; (*3*) *heavy oil* (dead oil, creosote oil), 250–300°, 7 per cent; (*4*) *anthracene oil* (green oil), 300–350°, 9 per cent; and (*5*) *pitch,* the residue, 62 per cent.

The further separation of the fractions depends on a combination of chemical and physical methods. Three main groups of substances are present: (*1*) *neutral compounds,* chiefly hydrocarbons; (*2*) *tar acids,* very weakly acidic substances soluble in sodium hydroxide solution; (*3*) *tar bases,* weakly basic substances soluble in dilute sulfuric acid.

The **light oil** fraction from coal tar is rather small, but about eight times as much can be recovered by scrubbing coal gas. It usually is separated directly into its components by fractional distillation. The so-called *crude 90 per cent benzol* is a fraction, 90 per cent of which distills between 80° and 100°; *crude 90 per cent toluol, solvent naphtha,* and *heavy naphtha* are the fractions, 90 per cent of which distill between the ranges 100–120°, 130–160°, and 160–210° respectively. They consist chiefly of aromatic hydrocarbons and olefins. Most of the tar bases remain behind with the tar acids. The distilled fractions are treated with concentrated sulfuric acid to polymerize the olefins and to remove any nitrogen or oxygen compounds. They then are washed with 10 per cent aqueous sodium hydroxide and redistilled to give benzene, toluene, and xylenes.

The **middle** or **carbolic oil** is combined with the high-boiling fraction from the light oil and cooled in large shallow pans. The solid hydrocarbon *naphthalene* (p. 487) crystallizes and is removed by centrifuging. The crude naphthalene is distilled, washed while molten with sulfuric acid, water, and aqueous alkali, and distilled again to give refined naphthalene. The oil that is separated from the crude naphthalene during the initial crystallization is washed with aqueous sodium hydroxide to remove the tar acids. Steam is passed into the aqueous extract to remove volatile materials, and then the solution is cooled and saturated with carbon dioxide, which liberates the weakly acidic tar acids from their salts. The tar acid layer is separated and fractionally distilled to give *phenol* (p. 441), which is purified further by crystallization. The aqueous sodium carbonate layer is reconverted to sodium hydroxide solution (*causticized*) by the addition of lime, and the precipitate of calcium carbonate is reconverted to lime and carbon dioxide by heating in lime kilns. The oil remaining after the extraction of acidic substances with alkali is washed with dilute sulfuric acid, which forms salts with the tar bases and takes them into solution. The tar bases are liberated by the addition of sodium hydroxide and distilled to obtain the *pyridine* (p. 514). The oil remaining after extraction of the tar bases is treated with concentrated sulfuric acid to polymerize the benzofurans (p. 514) and indenes (p. 646). Distillation gives solvent naphtha and a solid residue of polymers known as *cumar* or *coumarone resin.*

The **heavy oil** may be treated in much the same way to yield naphthalene, the higher tar acids (*cresols,* p. 435), and the higher tar bases (*quinolines,* p. 519). The **anthracene oil** is run into tanks and allowed to crystallize over a period of one to two weeks. After filtration, the nearly dry cakes are subjected to about 60,000 pounds pressure in a warm hydraulic press to remove more liquid impurities and then ground and washed with solvent naphtha to remove most of the hydrocarbon *phenanthrene* (p. 500), and with pyridine to remove most of a nitrogen-containing compound, *carbazole* (p. 511). The residue is sublimed to give the hydrocarbon *anthracene* (p. 496) of 85 to 90 per cent purity. With the development of commercial methods of synthesis for pyridine, pyridine derivatives, anthracene derivatives, and carbazole, the tendency has been to use the higher fractions as wood preservatives under the name *creosote oil,* rather than separate them into their components.

Over 215 individual compounds have been isolated from coal tar, the first being naphthalene, described in 1820. Anthracene was isolated in 1832, phenol, aniline (p. 418), quinoline, and pyrrole (p. 507) in 1834, and chrysene in 1837. Forty-six components were isolated during the thirty-year period 1861–1890 and over 100 during the 15-year period 1931–1946. The remainder are scattered over the intervening years. The period 1861–1890 was that of the most intensive development of the chemistry of coal tar. The large number of compounds isolated during the period 1931–1946 reflected new techniques for separation and an exhaustive effort to identify all of the components of coal tar. The more important components are listed in Table 20–*3*. The percentages of benzene, toluene, and xylenes given do not include the amounts present in the light-oil fraction extracted from coal gas by scrubbing (p. 389).

Aromatic Hydrocarbons from Petroleum and Other Sources

With increased demand for aromatic compounds and the development of newer processes, the production of aromatic compounds from petroleum has become economically feasible. Reformate

TABLE 20–3. CHIEF COMPONENTS OF COAL TAR

COMPOUND	PER CENT	COMPOUND	PER CENT
Benzene	0.1	Phenanthrene	4.0
Toluene	0.2	Anthracene	1.1
Xylenes	1.0	Carbazole	1.1
Naphthalene	10.9	Crude tar bases	2.0
α- and β-Methylnaphthalenes	2.5	(Pyridine 0.1)	
Dimethylnaphthalenes	3.4	Crude tar acids	2.5
Acenaphthene	1.4	(Phenol, 0.7, cresols, 1.1,	
Fluorene	1.6	xylenols, 0.2)	

(reformed gasoline fractions, p. 94) contains large amounts of the lower aromatic hydrocarbons. They can be removed from the alkanes by selective extraction with ethylene glycol ($HOCH_2CH_2OH$, p. 604, *Udex process*) in which the aromatics are more soluble than the alkanes. Fractional distillation of the extract yields **benzene, toluene,** and a mixture of the **xylenes** and **ethylbenzene.** A typical composition of the last mixture is one fifth each of ethylbenzene, *p*-xylene, and *o*-xylene, and two fifths *m*-xylene. Their boiling points are 277°, 281°, 292°, and 282° respectively, but by the use of three 200-ft. columns it is possible to separate the mixture into ethylbenzene, the mixture of *p*- and *m*-xylene, and the *o*-xylene. The high melting point of the *p*-xylene (13.4°) compared to that of *m*-xylene (−47.4°) permits the separation of pure *p*-xylene by low temperature crystallization. Some of the higher methylated benzenes such as **pseudocumene** (*1,2,4-trimethylbenzene*) and **durene** (*1,2,4,5-tetramethylbenzene*) also are being produced commercially from reformed petroleum fractions.

Most of the production of **ethylbenzene** is by synthesis from benzene and ethylene.

$$C_6H_6 + CH_2{=}CH_2 \xrightarrow[\text{5 p.s.i.}]{\text{AlCl}_3\text{-HCl, 95°}} C_6H_5CH_2CH_3$$

Ethylbenzene

The catalyst is a liquid aluminum chloride-hydrogen chloride-hydrocarbon complex through which the mixture of ethylene and an excess of benzene vapor is passed. The aromatic hydrocarbons are insoluble in the complex and are separated readily and the ethylbenzene is purified by distillation. Unreacted benzene and the higher-alkylated benzenes are recycled to give over-all yields of better than 95 per cent based on either the ethylene or benzene.

Cumene (*isopropylbenzene*) similarly is synthesized from benzene and propylene. Here the preferred catalyst is a supported phosphoric acid similar to that used for the polymerization of olefins (p. 87). The reaction is carried out at higher temperatures and pressures.

$$C_6H_6 + CH_2{=}CHCH_3 \xrightarrow[\text{400 p.s.i.}]{\text{H}_3\text{PO}_4\text{, 250°}} C_6H_5CH(CH_3)_2$$

***p*-Cymene** (*p-isopropyltoluene*) is one of the minor products of the sulfite process for making wood pulp (p. 342). The crude material is called *spruce turpentine.* **Mesitylene** can be made by the condensation of three moles of acetone in the presence of concentrated sulfuric acid (p. 203), but any large-scale commercial demand could be met by isolation from the C_9 fraction of petroleum reformates.

Production and Uses

In 1964 the total United States production of benzene was 730 million gallons of which 84 per cent came from petroleum. Of the 495 million gallons of toluene produced, 95 per cent came from petroleum as did 98 per cent of the 343 million gallons of xylenes. Most of the 3 billion pounds of ethylbenzene was produced by synthesis from benzene. This production was greater than that of any other synthetic organic compound.

Approximately 38 per cent of the benzene is used to make ethylbenzene (p. 391), 20 per cent to synthesize phenol (p. 441), 14 per cent for cyclohexane (p. 647), 4 per cent for synthetic detergents (p. 406), 4 per cent for maleic anhydride (p. 634), and 2 per cent each for DDT (p. 482), aniline (p. 425), and chlorobenzenes (p. 397). The chief use for toluene is as a solvent. Minor amounts are used to synthesize other chemicals. The greater demand for benzene has led to some production from toluene by catalytic hydrodemethylation.

$$C_6H_5CH_3 + H_2 \xrightarrow[\text{600°, 800 p.s.i.}]{\text{Cr}_2\text{O}_3\text{-Al}_2\text{O}_3\text{-NaOH}} C_6H_6 + CH_4$$

o-Xylene on oxidation yields phthalic acid (p. 474) and *p*-xylene yields terephthalic acid (p. 475). Some *m*-xylene is used to make isophthalic acid (p. 475) but most is used as a solvent. It may be isomerized also to the equilibrium mixture of 25 per cent *ortho,* 52 per cent *meta,* and 23 per cent *para* (p. 387). The sole use for ethylbenzene is in the production of styrene (p. 483). Cumene is used as a high-octane component of aviation fuel for piston-type engines and as an intermediate for the synthesis of phenol (p. 442).

INFRARED SPECTRA OF AROMATIC COMPOUNDS

Aromatic structures are recognized readily by characteristic infrared absorption resulting from C—H stretching and out-of-plane bending of hydrogen attached to the aromatic ring, and from the C—C stretching of the ring carbon atoms. Except for strongly electronegative groups such as nitro and fluorine, the nature of substituents has little effect on the frequencies of these absorptions.

The C—H stretching absorption usually consists of three or more bands between 3100 and 3000 cm.$^{-1}$. One band usually is more intense than the others (Figs. 20–*4* to 20–*10*). Thus the greater electronegativity of sp^2 hybridized carbon compared to sp^3 hybridized carbon causes the C—H stretching absorptions of aromatic compounds to be at somewhat higher frequencies than those of methyl and methylene groups, which lie below 3000 cm.$^{-1}$ (Figs. 20–*4*, 20–*8*, 20–*9*, 20–*10*).

Four bands in the double bond region in the neighborhood of 1600, 1580, 1500, and 1450 cm.$^{-1}$ are assigned to the C—C stretching vibrations of the ring carbon atoms. They vary widely in intensity. That near 1580 cm.$^{-1}$ often is very weak. These bands confirm the presence of an aromatic ring indicated by the C—H stretching absorption.

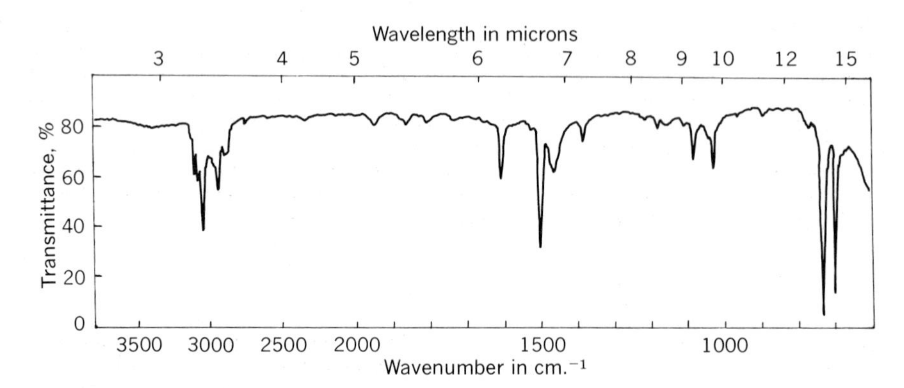

Figure 20–*4*. Infrared spectrum of toluene.

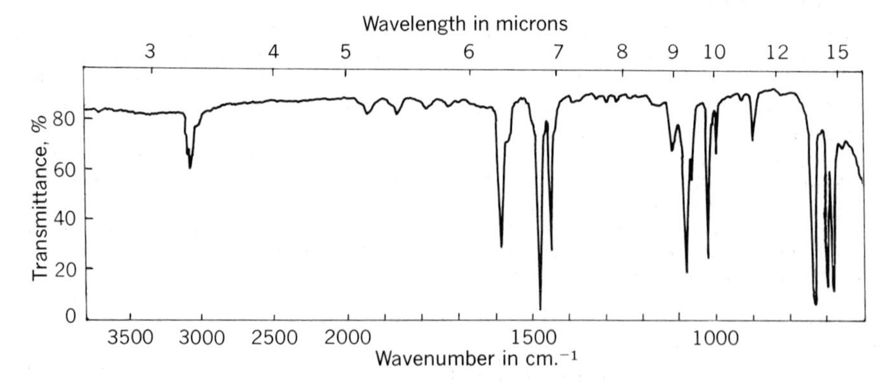

Figure 20–*5*. Infrared spectrum of chlorobenzene.

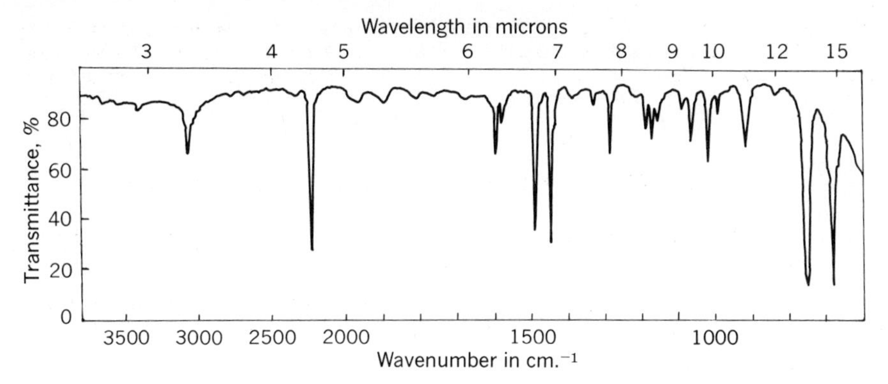

Figure 20–6. Infrared spectrum of benzonitrile.

The bands in the 2000–1660 cm.$^{-1}$, 1250-1000 cm.$^{-1}$, and 1000–650 cm.$^{-1}$ regions result from out-of-plane C—H vibrations. The number and shapes of these bands are characteristic of the number of hydrogens that are adjacent to each other on the ring and hence of the number and position of substituents on the aromatic ring. The bands in the low frequency group result from the fundamental vibrations of the C—H bonds. Their high intensity makes them particularly useful for quantitative estimation of relative amounts in mixtures of compounds. For compounds having five adjacent ring hydrogens, that is, mono-substituted benzenes, one band is present in each of the ranges 770–730 cm.$^{-1}$ and 710–690 cm.$^{-1}$ (Figs. 20–*4* to 20–*6*). Chlorobenzene (Fig. 20–*5*) has an additional band at 680 cm.$^{-1}$ resulting from the C—Cl stretching absorption. In the spectrum for benzonitrile (Fig. 20–*6*), the band at 2220 cm.$^{-1}$ is due to the C≡N stretching absorption, which is at slightly lower frequency than that for aliphatic nitriles (p. 238). Four weak bands are present in the 2000–1700 cm.$^{-1}$ range. They usually decrease in intensity with decreasing frequency (Fig. 20–*4*) but for chloro- and bromobenzene the second band is slightly more intense than the first (Fig. 20–*5*), and in phenol the fourth band is stronger than the second and third. Three bands of variable intensity are present in the ranges 1175–1125, 1110–1070, and 1070–1000 cm.$^{-1}$. The 2000–1700 cm.$^{-1}$ and 1175–1000 cm.$^{-1}$ regions are due to overtone and combination bands of the fundamentals in the 1000–650 cm.$^{-1}$ region.

The infrared spectrum for nitrobenzene (Fig. 20–*7*) illustrates the effect of a strongly electronegative group in increasing the complexity of the absorption and displacing bands from their normal positions. The bands of very high intensity at 1515 and 1345 cm.$^{-1}$ are due to the C—NO_2 stretching absorption.

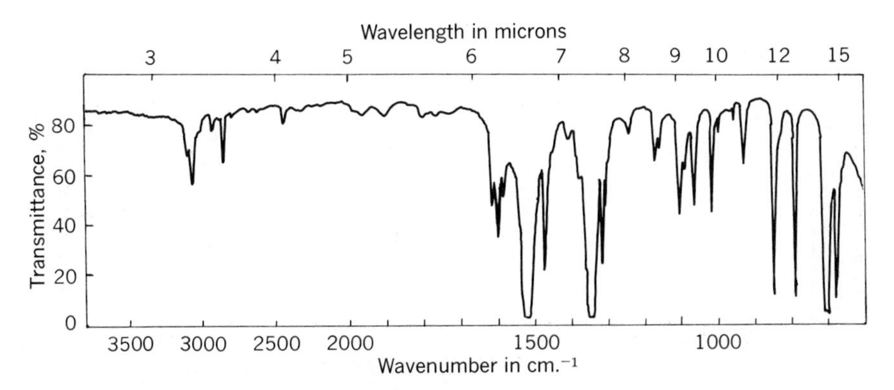

Figure 20–7. Infrared spectrum of nitrobenzene.

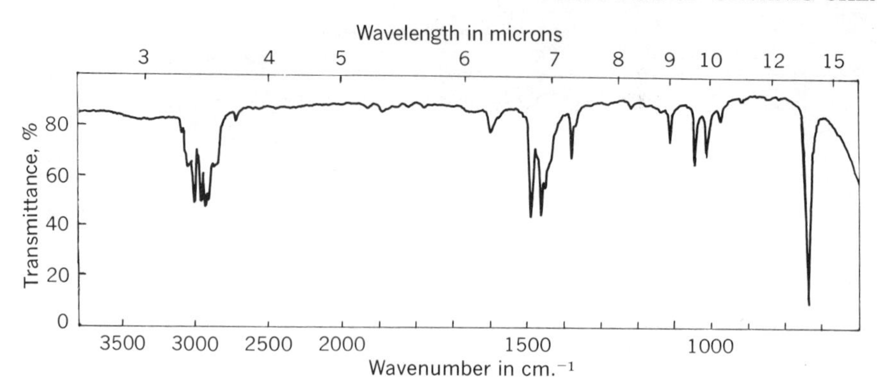

Figure 20–*8*. Infrared spectrum of *o*-xylene.

ortho-Disubstituted benzenes (Fig. 20–*8*) have four adjacent ring hydrogens and show a single intense absorption band in the range 770–735 cm.$^{-1}$. They show one weak to medium-strength absorption band in each of the ranges 1225–1175 cm.$^{-1}$ and 1125–1090 cm.$^{-1}$, and two bands in the range 1070–1000 cm.$^{-1}$.

meta-Disubstituted benzenes (Fig. 20–*9*) have three adjacent ring hydrogens and a hydrogen with no adjacent partner. The three adjacent hydrogens give rise to two bands in the ranges 810–750 cm.$^{-1}$ and 725–680 cm.$^{-1}$ and the single hydrogen to a medium-strength band in the 900–860 cm.$^{-1}$ range. Three medium-strength bands lie in the ranges 1175–1125, 1110–1070, and 1070–1000 cm.$^{-1}$.

The absorption of *para*-disubstituted benzenes (Fig. 20–*10*) with two adjacent ring hydrogens closely resembles that of *ortho*-disubstituted benzenes in the 1225–1000 cm.$^{-1}$ region, and they also have a single high-intensity band in the low frequency region. This band, however, is in a higher frequency range (860–800 cm.$^{-1}$) than that for *ortho*-disubstituted benzenes (770–735 cm.$^{-1}$). Moreover frequencies of both the 1600 cm.$^{-1}$ and the 1500 cm.$^{-1}$ bands for the C—C stretching absorption are about 20 cm.$^{-1}$ higher than those for *ortho* compounds.

Ortho, meta, and *para* isomers can be distinguished also by the absorption bands in the 2000–1700 cm.$^{-1}$ range. Here the intensities and numbers of bands are more significant than the frequencies at which they occur, and reference spectra of known compounds are used for comparison.

In addition to *meta*-disubstituted compounds, 1,2,4- and 1,3,5-trisubstituted benzenes,

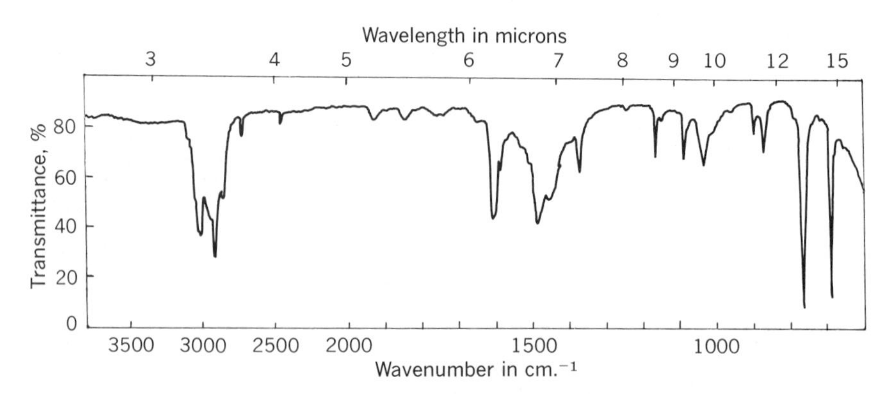

Figure 20–*9*. Infrared spectrum of *m*-xylene.

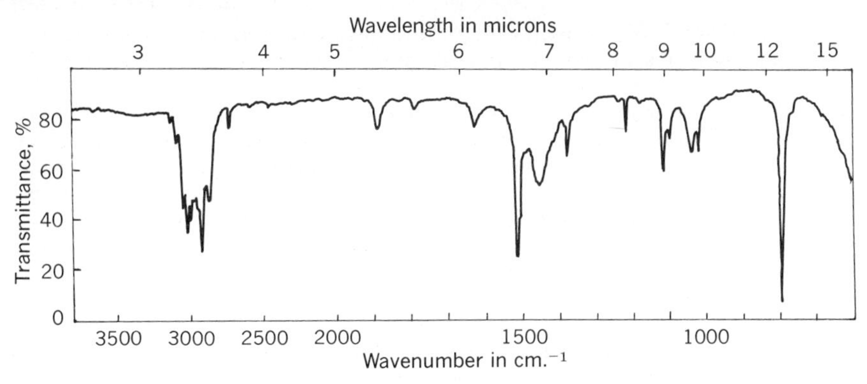

Figure 20–*10*. Infrared spectrum of *p*-xylene.

1,2,3,5- and 1,2,4,5-tetrasubstituted benzenes, and pentasubstituted benzenes have hydrogen with no adjacent neighbor and show medium absorption at 900–860 cm.$^{-1}$. Like the *para*-disubstituted benzenes, 1,2,4-trisubstituted and 1,2,3,4-tetrasubstituted benzenes have two adjacent ring hydrogens and show strong absorption in the range 860–800 cm.$^{-1}$. 1,2,3-Trisubstituted benzenes resemble *meta*-disubstituted benzenes in that they have three adjacent hydrogens responsible for strong absorption at 800–710 cm.$^{-1}$ and medium absorption at 720–685 cm.$^{-1}$. The 1,3,5-trisubstituted benzenes not only show the medium absorption in the 900–860 cm.$^{-1}$ range characteristic of an isolated hydrogen but also absorb strongly in the ranges 865–810 and 730–675 cm.$^{-1}$.

PROBLEMS

20–*1*. Pair the following names with the appropriate phrase: (*a*) Mitscherlich, (*b*) Faraday, (*c*) Loschmidt, (*d*) Kekulé, (*e*) Friedel and Crafts; (*1*) aluminum chloride-catalyzed electrophilic substitution, (*2*) the name *benzene,* (*3*) the term *aromatic compound,* (*4*) the first isolation of benzene, (*5*) the structure of benzene.

20–*2*. For which of the following entities does resonance make an important contribution to thermodynamic stability: (*a*) carboxylic acids, (*b*) nitro compounds, (*c*) ketones, (*d*) cyclohexane, (*e*) carboxylate anions, (*f*) cyclohexene, (*g*) benzene, (*h*) enolate anions, and (*i*) trialkylcarbonium ions?

20–*3*. Place the following compounds in the order of increasing basicity: (*a*) methanol, (*b*) hexane, (*c*) benzene, (*d*) mesitylene, (*e*) acetic acid, and (*f*) *m*-xylene.

20–*4*. Arrange the following compounds in the order of increasing dipole moment: (*a*) nitrobenzene, (*b*) toluene, (*c*) benzotrichloride, (*d*) benzyl chloride, and (*e*) *p*-dichlorobenzene.

20–*5*. (*a*) Give perspective formulas for the optically active benzene hexachlorides. (*b*) Designate the consecutive axial, *a*, or equatorial, *e*, positions of the halogens in the more stable conformation of the isomeric benzene hexachlorides. (For example, γ-benzene hexachloride is *aaaeee*. Remember that the compound is a cycle and that *eaaaee*, *aaeeea*, etc., are all the same.)

20–*6*. Give structural formulas and name all of the possible compounds in each of the following groups: (*a*) the trimethylbenzenes, (*b*) the monochloromononitrotoluenes, (*c*) the monochlorodinitrobenzenes, (*d*) the monobromotrimethylbenzenes.

20–*7*. Show why it is not possible to use the simple Koerner method to distinguish among *ortho-*, *meta-*, and *para*-disubstituted benzenes when the two groups are different.

20–*8*. What are the possible structural formulas of aromatic hydrocarbons, $C_{10}H_{14}$, that meet the following specifications: (*a*) oxidation gives a monocarboxylic acid; (*b*) nitration can give three mononitro derivatives and oxidation a dicarboxylic acid; (*c*) nitration can give two mononitro derivatives and oxidation a dicarboxylic acid; (*d*) nitration can give two mononitro derivatives and oxidation a tricarboxylic acid?

20–*9*. Three tribromobenzenes are known, which melt at 44°, 87.4°, and 119°. When each is nitrated, they give respectively three, two, and one mononitrotribromobenzenes. Give the structures of the tribromobenzenes and of their nitration products.

20–*10*. Six dibromonitrobenzenes melt respectively at 57.8°, 61.8°, 82.6°, 83.6°, 85.2°, and 104.5°. When the nitro group is replaced by hydrogen by a series of reactions, the compound melting at 83.6° gives a dibromobenzene melting at 87°. The two dibromonitrobenzenes melting at 57.8° and 85.2° give a liquid dibromobenzene boiling at 225°. The remaining three dibromonitrobenzenes give a liquid dibromobenzene boiling at 219°. Without consulting the literature, assign structures to the three dibromobenzenes.

20–*11*. Three chloronitrobenzenes, *A*, *B*, and *C*, when nitrated further give two, four, and four dinitro derivatives respectively. When *o*-dinitrobenzene is monochlorinated, one of the products is identical with one of the four products obtained from *B* and with one of the four products obtained from *C*. The other isomer also is obtained from *C* but not from *B*. Assign structures to the three chloronitrobenzenes.

20–*12*. Give the formulas and names of the chief product or products that would be expected when one more group is introduced by the following substitution reactions: (*a*) nitration of bromobenzene, (*b*) chlorination of nitrobenzene, (*c*) sulfonation of toluene, (*d*) bromination of benzoic acid, (*e*) nitration of benzenesulfonic acid, (*f*) chlorination of *m*-xylene in the presence of ferric chloride, (*g*) nitration of *m*-dinitrobenzene, (*h*) sulfonation of *o*-nitrocumene, (*i*) bromination of chlorobenzene.

20–*13*. Give a series of reactions for the synthesis of the following compounds from an aromatic hydrocarbon: (*a*) *m*-bromonitrobenzene, (*b*) 5-chloro-2,4-dimethyl-1-nitrobenzene, (*c*) 4-bromo-2-nitroethylbenzene, (*d*) *p*-chlorobenzenesulfonic acid, (*e*) 2-chloro-4-nitrotoluene, (*f*) *m*-nitrobenzoic acid, (*g*) *p*-bromobenzoic acid, and (*h*) 2,4-dinitrochlorobenzene.

CHAPTER TWENTY-ONE

HALOGEN DERIVATIVES OF AROMATIC HYDROCARBONS

In a strict sense aromatic halides are halogen compounds in which the halogen is attached directly to an aromatic nucleus. It is convenient, however, to consider also at this point those halogen compounds in which the halogen is present in a side chain. The former are called *aryl halides* and the latter *aralkyl halides.*

ARYL HALIDES

Preparation

The aryl halides are prepared either by the previously discussed direct halogenation of benzene or its derivatives (p. 376), or by replacement of an amino group by halogen by way of the diazonium halide, a reaction to be described later (p. 432). The replacement of a hydroxyl group by halogen, which is of primary importance in the preparation of alkyl halides, seldom is used because it is difficult to replace a hydroxyl group attached to a benzene ring (p. 437). Halogenation follows the previously discussed rules for electrophilic substitution in aromatic compounds. Thus further chlorination of chlorobenzene yields chiefly *o*- and *p*-dichlorobenzenes.

$$C_6H_5Cl + Cl_2 \xrightarrow{FeCl_3} o\text{-}C_6H_4Cl_2 \text{ and } p\text{-}C_6H_4Cl_2 + HCl$$

o-Dichlorobenzene *p*-Dichlorobenzene

In the above formulas and subsequently, bonds usually are not indicated between the substituent and the aromatic ring.

Chlorobenzene is an intermediate in the manufacture of phenol, aniline, DDT, and dyes. *p*-Dichlorobenzene is used as a larvicide against the clothes moth and the peach-tree borer. *o*-Dichlorobenzene containing 4 per cent of the *para* isomer is used as a heat transfer medium in the 150–250° range. Its biggest use is as a solvent for the preparation of tolylene diisocyanate (p. 423). The relative technical importance of chlorobenzene, *p*-dichlorobenzene, and *o*-dichlorobenzene is indicated by the production in the United States in 1964 of 538, 63, and 52 million pounds respectively.

Other aromatic compounds halogenate in a similar manner. For example, toluene in the presence of ferric halide chlorinates or brominates more easily than benzene and yields a mixture of *o*- and *p*-chloro- or bromotoluenes.

$$C_6H_5CH_3 + X_2 \xrightarrow{FeX_3} o\text{-}XC_6H_4CH_3 \text{ and } p\text{-}XC_6H_4CH_3 + HX$$

Iodination is discussed on page 377.

Physical Properties

Chloro-, bromo-, and iodobenzene are liquids at room temperature, and boil at 132°, 156°, and 189°, within ±10° of the corresponding *n*-hexyl or cyclohexyl halides. The *para* dihalogenated benzenes melt higher than the *ortho* or *meta* isomers because the symmetric structure permits the molecules to pack more closely in the crystal.

Reactions

In the absence of other reactive functional groups, aryl halides react readily with the more active metals to give organometallic compounds. Thus chlorobenzene and a dispersion of metallic sodium (particle size $< 25\mu$) in benzene gives phenylsodium.

$$C_6H_5Cl + 2\ Na \xrightarrow{C_6H_6} C_6H_5Na + NaCl$$

Similarly bromobenzene in ether solution reacts with lithium to give phenyllithium (cf. p. 111).

$$C_6H_5Br + 2\ Li \xrightarrow{Ether} C_6H_5Li + LiBr$$

Aryl bromides that do not react directly with lithium can be converted to aryllithiums by an exchange reaction with butyllithium (p. 111).

$$ArBr + C_4H_9Li \longrightarrow ArLi + C_4H_9Br$$

Chloro- or bromoarenes in tetrahydrofuran react with magnesium to give Grignard reagents.

$$ArX + Mg \xrightarrow{THF} ArMgX$$

All of these products undergo the typical reactions of reactive organometallic compounds (pp. 110, 176, 178, 193, 194, 197, 235, 264, 266) and are useful for the preparation of substituted aromatic compounds such as carboxylic acids and alcohols. The Grignard reagents and arylsodium compounds are the most important. The chief use for phenyllithium is for the preparation of other organolithium compounds (p. 519). Compounds having hydrogen more acidic than that in benzene undergo an acid-base exchange, a process known as *metalation*. Toluene, for example, is metalated by phenyllithium.

$$C_6H_5CH_3 + C_6H_5Li \longrightarrow C_6H_5CH_2Li + C_6H_5$$

When aryl iodides are heated with activated copper powder, coupling of two aromatic nuclei takes place (*Ullmann reaction*).

$$C_6H_5I + 2\ Cu + IC_6H_5 \longrightarrow C_6H_5\text{-}C_6H_5 + 2\ CuI$$

Biphenyl

When a mixture of aryl halide and alkyl halide is allowed to react with metallic sodium, a mixed coupling results (*Wurtz-Fittig*[1] *reaction*).

[1] Rudolph Fittig (1835–1910), professor at the University of Strassburg. He discovered the pinacol reaction in 1858, and applied the Wurtz reaction to mixed alkyl and aryl halides in 1863. He is known also for his synthesis of β-naphthol from benzaldehyde (1883) and his discovery of the interconversion of α,β- and β,γ-unsaturated acids by alkali (1894).

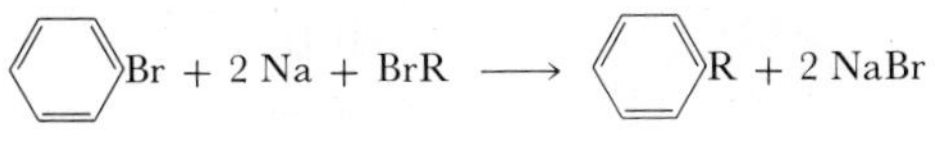

$$C_6H_5Br + 2\,Na + BrR \longrightarrow C_6H_5R + 2\,NaBr$$

The satisfactory yield of arylalkane may be ascribed to the fact that both the aryl and the alkyl anion can be formed (p. 144), but that only the alkyl halide undergoes ready nucleophilic displacement of halogen.

Reactions of the S_N2 or S_N1 type characteristic of the alkyl halides do not take place with aryl halides. Because the halogen lies in the plane of the ring, a nucleophile cannot approach from the back side, and an S_N2 mechanism cannot operate. Because of resonance with the nucleus (p. 384), halogen has little tendency to ionize and permit an S_N1 mechanism. Hence simple aryl halides do not react with nucleophiles such as hydroxide, alkoxide, sulfide, or cyanide ion, or with ammonia under conditions that bring about displacement reactions with alkyl halides. Nevertheless displacement of "unactivated" halogen can be brought about by nucleophiles in the presence of strong bases. If the halogen is "activated" by the presence of electronegative groups in the *ortho* or *para* positions, nucleophilic displacement takes place under mild conditions (p. 400). For each type, a new mechanism is involved.

A large amount of evidence supports the view that reaction in the presence of strong bases takes place by elimination of halogen acid to give a neutral **dehydrobenzene** as a highly reactive intermediate, which then adds a molecule of the solvent. The reactive intermediate can be visualized as having a structure analogous to that of an olefin. When a double bond is formed by elimination of halogen acid from an alkyl halide, the remaining unshared pair of electrons occupies a π orbital that results from the overlap of two p orbitals (p. 75). When halogen acid is removed from an aryl halide, the remaining pair of electrons occupies a π-type molecular orbital formed by the overlap of the two sp^2 hybrid orbitals formerly used to bond with the hydrogen and the halogen.

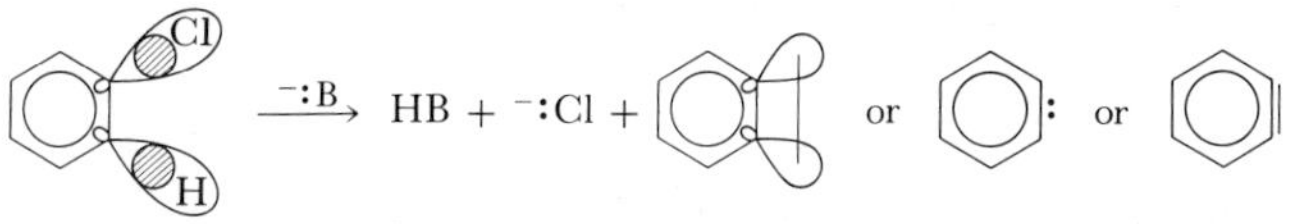

$$\xrightarrow{^-:B} HB + {^-}:Cl + \quad \text{or} \quad \text{or}$$

Addition to the intermediate, usually of a molecule of solvent, completes the reaction.

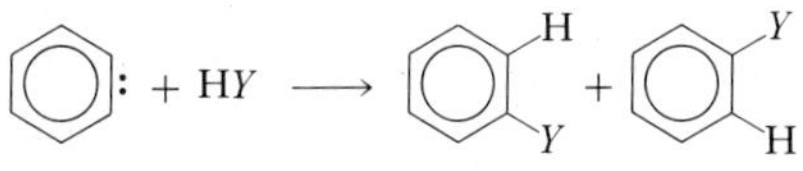

$$C_6H_4: + HY \longrightarrow C_6H_5Y + C_6H_5Y$$

It will be noted that the reactive pair of electrons of the intermediate occupies a molecular orbital that lies in the plane of the benzene ring and is perpendicular to the π system of the ring. Hence it does not interact with this system and is associated with only one of the sp^2—sp^2 σ bonds of the ring. It differs from the π bond of an olefin in that (*1*) the electron density is concentrated on one side of the σ bond instead of being distributed on both sides and (*2*) the overlap of the two sp^2 orbitals is poorer than that of the two p orbitals of an olefin. Both factors account for the high reactivity of the intermediate and the fact that it has not been isolated but has been detected in solution only by its reactions. Spectroscopic evidence indicates that it is present in the gas phase as a transient intermediate in the formation of biphenylene by the flash photolysis of diazotized anthranilic acid (p. 471). Moreover decomposition of this compound in the mass spectrograph (p. 543) gives a particle of mass 76.

Although the reactive intermediates are dehydrobenzenes, they more commonly are called **"benzynes"** and are represented as if the nucleus contained a triple bond as in *I*.

$$\underset{I}{\text{(benzyne)}} \qquad \underset{II}{H_2C{-}C{\equiv}C{-}CH_2 \text{ (ring closed by } (CH_2)_4)}$$

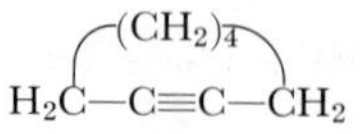

The general subject involving such intermediates is referred to as *benzyne chemistry*. Because the *sp* bonding orbitals of acetylenes are linear, however, a structure such as *I* would be highly strained. The smallest ring known to contain a triple bond contains eight members (*II*) and is nonplanar. Moreover the triple bond in a structure such as *I* would be different from a normal triple bond in that one pair of π electrons would be localized while the other pair is delocalized.

If halogen is activated, reactions take place by way of a σ intermediate analogous to the intermediate in electrophilic substitution (p. 375).

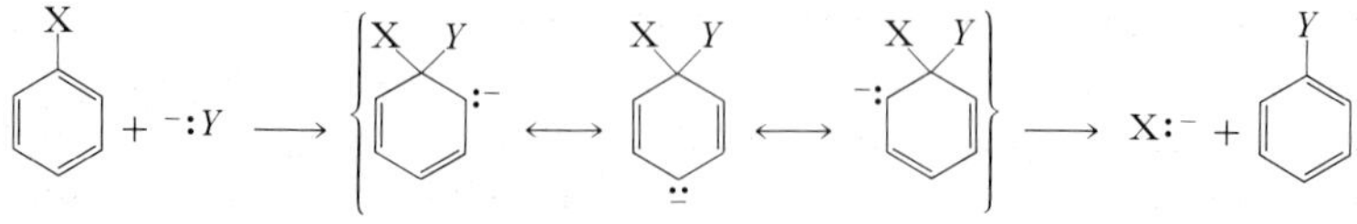

$$C_6H_5X + {}^-{:}Y \longrightarrow \left\{ [C_6H_5XY]^- \longleftrightarrow [C_6H_5XY]^- \longleftrightarrow [C_6H_5XY]^- \right\} \longrightarrow X{:}^- + C_6H_5Y$$

Activating electron-attracting groups in the *ortho* or *para* position assist the formation of the transition state by relieving the increased electron density at these positions. The nitro and carboxyl groups are particularly effective because either can provide a favorable site for the location of the negative charge.

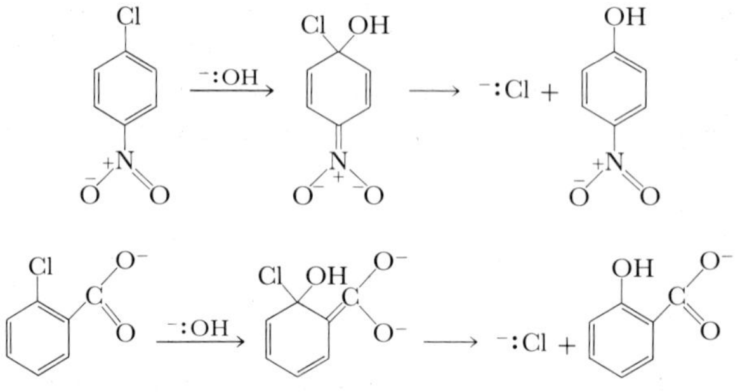

$$p\text{-}ClC_6H_4NO_2 \xrightarrow{^-{:}OH} [Cl(OH)C_6H_4{=}NO_2]^- \longrightarrow {}^-{:}Cl + p\text{-}HOC_6H_4NO_2$$

$$o\text{-}ClC_6H_4CO_2^- \xrightarrow{^-{:}OH} [Cl(OH)C_6H_4{=}C(O^-)_2]^- \longrightarrow {}^-{:}Cl + o\text{-}HOC_6H_4CO_2^-$$

Thus although chlorobenzene reacts with concentrated alkali only at 350°, 2,4-dinitrochlorobenzene is hydrolyzed by boiling with an aqueous solution of sodium carbonate. Similarly aqueous sodium hydrogen sulfide gives 2,4-dinitrothiophenol, sodium sulfide gives 2,4-dinitrophenyl sulfide, sodium disulfide gives 2,4-dinitrophenyl disulfide, ammonia gives 2,4-dinitroaniline, and hydrazine gives 2,4-dinitrophenylhydrazine. The last compound is a valuable reagent for the preparation of derivatives of aldehydes and ketones (p. 204). 2,4-Dinitrofluorobenzene has found important use in determining the structure of polypeptides (p. 361).

These bimolecular nucleophilic displacement reactions resemble S_N2 reactions but it will be noted that the carbon atom of the ring is never bonded to more than four other atoms. To distinguish it from bimolecular nucleophilic displacement at a saturated carbon atom, the mechanism for aromatic compounds frequently is designated as $S_{N_{Ar}}2$.

Although nonactivated aromatic halides are believed to react usually by way of dehydrobenzene intermediates, examples are known in which reaction takes place also by the $S_{N_{Ar}}2$ mechanism. Nonactivated aromatic halides react readily with cuprous salts, or with a nucleophile in the presence of cuprous oxide, when a high dielectric aprotic solvent (p. 102) is used. Thus chloroarenes and cuprous iodide or cuprous cyanide in dimethylformamide, *N*-methylpyrrolidone (p. 617), or methyl sulfoxide give good yields of iodoarenes or aryl cyanides.

$$ArCl + CuCN \xrightarrow{MSO} ArCN + CuCl$$

Thiols in the presence of cuprous oxide give sulfides.

$$2\,ArCl + Cu_2O + 2\,RSH \longrightarrow 2\,ArSR + 2\,CuCl + H_2O$$

Carboxylic acids lead to reduction.

$$ArCl + 2\,Cu_2O + RCOOH \longrightarrow ArH + CuCl + RCOOCu + 2\,CuO$$

The catalytic activity of cuprous ion is believed to be due to its ability to form solvated complexes with the reactants.

ARALKYL HALIDES

Aralkyl halides are aryl-substituted alkyl halides. Only those having halogen on a carbon atom adjacent to the nucleus are of special interest. They usually are referred to as *benzyl halides.*

Preparation of Benzyl Halides

The chemical properties of alkyl side chains are for the most part those of aliphatic compounds. Hence if a mixture of toluene vapor and halogen is exposed to light of short wavelength, halogenation takes place in the side chain more rapidly than substitution in the nucleus or than addition to the double bonds. The rates of replacement of a second and third hydrogen of a methyl group by halogen differ little from the rate of replacement of the first hydrogen, and mono-, di-, and trisubstitution products are formed (p. 68).

$$C_6H_5CH_3 \xrightarrow[+\text{ light}]{X_2} HX + C_6H_5CH_2X \xrightarrow[+\text{ light}]{X_2} HX + C_6H_5CHX_2 \xrightarrow[+\text{ light}]{X_2} HX + C_6H_5CX_3$$

Benzyl halide; Benzylidene halide (*benzal halide*); Benzylidyne halide (*benzotrihalide*)

A certain amount of regulation of the ratio of mono- to di- to trisubstitution can be obtained by varying the hydrocarbon-halogen ratio. In the technical preparation of benzyl chloride, reaction takes place at 130–140°. This temperature is above the boiling point of toluene (111°) but below the boiling point of benzyl chloride (179°). Hence the benzyl chloride condenses as it is formed and further chlorination is prevented. The reaction is carried out in a lead-lined vessel illuminated internally with quartz mercury-vapor lamps. At higher temperatures the reaction takes place in the absence of light.

Benzyl chlorides can be prepared also by the reaction of an aromatic nucleus with formaldehyde and hydrogen chloride in the presence of zinc chloride or phosphoric acid.

$$C_6H_6 + HCHO + HCl \xrightarrow[\text{or } H_3PO_4]{ZnCl_2} C_6H_5CH_2Cl + H_2O$$

The process is known as **chloromethylation** and is another substitution reaction that is applicable to aromatic compounds in general. Either the formaldehyde condenses with the aromatic compound to give benzyl alcohol, which then reacts with hydrogen chloride, or the hydrogen chloride adds to the formaldehyde to give chloromethyl alcohol, which reacts with the aromatic compound to give the benzyl chloride.

Reactions of Aralkyl Halides

Halogen in the side chain undergoes the same reactions as that in simple alkyl halides (p. 100). Even the rates of reaction are about the same if the halogen is located beyond the carbon atom attached directly to the ring. If the halogen is located on the carbon atom attached to the ring, the reactivity is greatly increased. For example, although benzyl chloride is a primary halide, it is hydrolyzed readily by aqueous solutions of sodium carbonate, and benzyl bromide reacts with tertiary amines to form quaternary salts about 300 times faster than *n*-propyl bromide.

In the transition state for an S_N2 reaction, the carbon atom involved may have either a higher electron density or a lower electron density than in the halide, depending on whether bond formation with the nucleophile takes place to a greater or lesser extent than bond-breaking by the leaving group. Since a phenyl group can either absorb a negative charge or relieve a positive charge on an adjacent carbon atom by resonance (p. 454), it accelerates S_N2 reactions. Moreover both electron-withdrawing groups and electron-donating groups on the nucleus accelerate reaction. The former decrease the electron density on the carbon atom and make bond formation easier for the nucleophile, whereas the latter increase the electron density, and bond-breaking by the leaving group is facilitated. In S_N1 reactions bond-breaking always is more important than bond-making, and the phenyl group increases the reactivity. As would be expected, however, only electron-donating substituents give a further increase in reactivity, whereas electron-withdrawing substituents decrease the reactivity.

PROBLEMS

21–*1*. Give reactions for the preparation of the following compounds: (*a*) benzene hexabromide, (*b*) *o*- and *p*-chlorocumene, (*c*) *o*-methylbenzyl bromide, (*d*) *o*- and *p*-bromochlorobenzene, (*e*) 4-methylbiphenyl, (*f*) *p*-bis(chloromethyl)benzene, (*g*) *p*-(chloromethyl)ethylbenzene, (*h*) 2,4-dinitrophenylhydrazine, (*i*) *p*-cyano-*i*-propylbenzene, and (*j*) ethyl phenyl sulfide.

21–*2*. Place the following compounds in the order of increasing reactivity towards nucleophiles: (*a*) 2,4-dinitrobenzyl chloride, (*b*) *n*-butyl chloride, (*c*) chlorobenzene, (*d*) benzyl chloride, and (*e*) iodobenzene.

21–*3*. Give a series of reactions for the following conversions: (*a*) toluene to phenylacetic acid, (*b*) benzene to *p*-*i*-propylbenzyl chloride, (*c*) mesitylene to bromobenzene-2,4,6-tricarboxylic acid, (*d*) *p*-xylene to 1,4-dimethyl-2-*n*-propylbenzene, (*e*) acetone to iodomesitylene.

21–*4*. Give reactions that may be used to distinguish between the members of each of the following pairs of compounds: (*a*) benzene and *n*-hexane, (*b*) bromobenzene and benzyl bromide, (*c*) chlorobenzene and *m*-xylene, (*d*) benzyl chloride and benzylidene chloride, (*e*) benzyl chloride and *t*-amyl chloride, (*f*) iodobenzene and bromobenzene.

21–*5*. Analysis of an organic compound showed that it had the composition C_8H_8BrCl. Oxidation yielded *p*-bromobenzoic acid. Write possible structures for the compound.

CHAPTER TWENTY-TWO

AROMATIC SULFONIC ACIDS AND THEIR DERIVATIVES

SULFONIC ACIDS

Since aromatic sulfonic acids can be obtained by direct sulfonation, they are more readily available than aliphatic sulfonic acids. The aromatic sulfonic acids are used as intermediates for the introduction of other groups and to confer water-solubility on aromatic compounds, particularly dyes.

Nomenclature

Sulfonic acids are named by attaching the ending *sulfonic* to the name of the compound that has been substituted and adding *acid*. The positions of the sulfonic acid groups are indicated by numbers or letters.

SO_3H Cl, SO_3H NO_2, HO_3S, SO_3H

Benzenesulfonic acid   2-Chlorobenzenesulfonic acid (*o-chlorobenzenesulfonic acid*)   5-Nitro-1,3-benzenedisulfonic acid

Physical Properties

Sulfonic acids are very soluble in water because they are strong acids and are completely ionized in aqueous solution. They are insoluble or only slightly soluble in nonoxygenated solvents. The pure acids are hygroscopic and difficult to obtain anhydrous. They usually crystallize from aqueous solutions with water of hydration.

Sulfonic acids are relatively nonvolatile, although some can be distilled at low pressures without decomposition. Thus benzenesulfonic acid boils at 135–137° at a pressure below 0.01 mm. of mercury. The melting points of the anhydrous sulfonic acids are lower than those of the corresponding carboxylic acids.

Preparation

Aromatic sulfonic acids generally are prepared by direct sulfonation (p. 378). The ease of sulfonation depends on the substituents already present in the ring. Thus if activating groups are present, concentrated sulfuric acid at room temperature may suffice to bring about sulfonation. When deactivating groups are present, the use of elevated temperatures and fuming sulfuric acid (oleum) containing varying amounts of dissolved sulfur trioxide may be necessary. Benzene can be converted to the monosulfonic acid using 10 per cent fuming sulfuric acid (100 per cent sulfuric acid containing 10 per cent sulfur trioxide) at room temperature. To convert the monosulfonic acid to the *m*-disulfonic acid, a temperature of 200–245° is used, and to convert the *m*-disulfonic acid to the 1,3,5-trisulfonic acid, a temperature of 280–300° is required.

$$C_6H_6 \xrightarrow[25^\circ]{\text{Fuming } H_2SO_4} C_6H_5SO_3H \xrightarrow[200\text{–}245^\circ]{\text{Fuming } H_2SO_4} C_6H_4(SO_3H)_2 \xrightarrow[280\text{–}300^\circ]{\text{Fuming } H_2SO_4} C_6H_3(SO_3H)_3$$

Benzenesulfonic acid; *m*-Benzenedisulfonic acid; 1,3,5-Benzenetrisulfonic acid

Since 1947 a stabilized form of liquid sulfur trioxide has been available commercially. Because it is much more reactive than sulfuric acid, it can be used in stoichiometric amounts, and the product is free of excess sulfuric acid. When fuming sulfuric acid or sulfur trioxide is used some *sulfone* (p. 267) may be formed as a by-product. Since the sulfone is insoluble in water, it can be removed after dilution of the sulfonation mixture.

$$C_6H_5SO_2OH + C_6H_6 + SO_3 \longrightarrow \underset{\text{Phenyl sulfone}}{C_6H_5SO_2C_6H_5} + H_2SO_4$$

In general sulfonation follows the rules for substitution outlined for nitration (p. 385). It appears, however, to be less predictable and more influenced by conditions. Phenol, for example, gives chiefly *o*-phenolsulfonic acid, whereas chlorobenzene gives chiefly *p*-chlorobenzenesulfonic acid. Temperature and catalysts such as mercuric sulfate frequently have a pronounced influence on the position taken by the sulfonic acid group (p. 498). Sulfonation is more readily reversible than other electrophilic aromatic substitutions. Hence the initial product composition depends on relative rates, but the equilibrium product composition depends on relative thermodynamic stability (p. 35). Sulfonic acids may be prepared also by the oxidation of disulfides or by the hydrolysis of sulfonyl chlorides (pp. 263, 407).

Most sulfonic acids are used in the form of their salts. Fortunately the salts can be isolated much more readily than the sulfonic acid. The sodium salts usually are less soluble than the sulfonic acid, particularly in a solution saturated with sodium chloride. Hence they can be isolated by *salting out*. The reaction mixture is poured into water and a brine solution is added.

$$C_6H_5SO_3H + NaCl \longrightarrow \underset{\text{Sodium benzenesulfonate}}{C_6H_5SO_3Na} + HCl$$

The *calcium, barium,* and *lead sulfonates* are soluble in water in contrast to the sulfates. After dilution of the sulfonation mixture with water, the hydroxide, oxide, or carbonate of calcium, barium, or lead can be added to precipitate the sulfate present and leave the sulfonate in solution. Evaporation of the filtrate gives the calcium, barium, or lead sulfonate.

Commercially the **sodium salts** are most useful and are prepared by the *liming out process*. After dilution of the sulfonation mixture, slaked lime is added, and the precipitate of calcium sulfate is removed by filtration. Addition of sodium carbonate to the filtrate precipitates calcium carbonate and leaves the sodium sulfonate in solution. Removal of the calcium carbonate and evaporation give the sodium salt.

Free sulfonic acids can be obtained by adding sufficient sulfuric acid to an aqueous solution of a calcium, barium, or lead salt to precipitate the metallic sulfate, and evaporating the filtrate. Usually it is more convenient to hydrolyze the pure sulfonyl chloride (p. 407).

Reactions of Aromatic Sulfonic Acids and Their Salts

Substitution of the Nucleus

The aromatic ring can be halogenated or nitrated, the SO_3H group being deactivating and *meta*-directing.

$$C_6H_5SO_3H + Br_2 \xrightarrow{Fe} C_6H_4(SO_3H)Br + HBr$$

$$C_6H_5SO_3H + HNO_3 \xrightarrow{H_2SO_4} C_6H_4(SO_3H)NO_2 + H_2O$$

Frequently the sulfonic acid group is displaced, especially when the reaction is carried out in the presence of water.

$$C_6H_3(CH_3)_2SO_3H + 2\ Br_2 + H_2O \longrightarrow C_6H_2(CH_3)_2Br_2 + 2\ HBr + H_2SO_4$$

$$C_6H_3(OH)(Cl)SO_3H + 2\ HNO_3 \longrightarrow C_6H_2(OH)(Cl)(NO_2)_2 + H_2SO_4 + H_2O$$

These reactions appear to involve electrophilic displacement of sulfur trioxide from the sulfonate ion, followed by reaction with water.

$$C_6H_5SO_3^- + {}^+E \longrightarrow [C_6H_5(E)SO_3^-]^+ \longrightarrow C_6H_5E + SO_3 \xrightarrow{H_2O} H_2SO_4$$

Reactions of Free Sulfonic Acids

1. ***Salt Formation.*** Sulfonic acids are considerably stronger acids than carboxylic acids. Hence they react with bases to form stable salts in aqueous solution. For benzenesulfonic acid, pK_a is 0.8, compared to 4.2 for benzoic acid (C_6H_5COOH) and 2.1 for phosphoric acid. Electronegative groups increase the acidity, especially if they are in the *ortho* or *para* positions.

2. ***Hydrolysis.*** The sulfonation reaction is reversible. Hence if sulfonic acids are boiled with an excess of water, the sulfonic acid group slowly is removed. Since the reaction involves an electrophilic displacement by a proton, a high concentration of an added mineral acid greatly increases the rate of hydrolysis. The reaction is fairly rapid if carried out in a sealed tube at 150–170°.

$$C_6H_5SO_3H + H_2O \xrightleftharpoons{\text{Heat (25\% HCl)}} C_6H_6 + H_2SO_4$$

Reactions of Sodium Sulfonates

1. ***Replacement by Hydroxyl.*** When a sodium sulfonate is heated with molten sodium hydroxide, the sodium salt of a phenol and sodium sulfite are formed.

$$C_6H_5SO_3Na + 2\ NaOH \xrightarrow{\text{Fusion}} \underset{\text{Sodium phenoxide}}{C_6H_5ONa} + Na_2SO_3 + H_2O$$

Acidification yields the phenol (p. 441).

$$C_6H_5ONa + H_2SO_4 \longrightarrow C_6H_5OH + NaHSO_4$$

2. ***Replacement by the Nitrile Group.*** Fusion of a sodium sulfonate with sodium cyanide yields a nitrile and sodium sulfite.

$$C_6H_5SO_3Na + NaCN \xrightarrow{\text{Fusion}} \underset{\substack{\text{Phenyl cyanide} \\ (\textit{benzonitrile})}}{C_6H_5CN} + Na_2SO_3$$

3. ***Replacement by Carboxyl.*** Heating a sodium sulfonate with sodium formate yields the sodium salt of a carboxylic acid.

$$C_6H_5SO_3Na + HCOONa \longrightarrow \underset{\text{Sodium benzoate}}{C_6H_5COONa} + NaHSO_3$$

The first two reactions probably are $S_{N_{Ar}}2$ displacements, and the third may involve a cyclic mechanism, but the mechanisms have not been established.

Uses of Sulfonic Acids and Their Salts

The free sulfonic acids find very little use. Because they are strong acids and have a much weaker oxidizing action than sulfuric acid, they frequently are used as acid catalysts.

An important use of the sodium sulfonates is for the manufacture of phenols by fusion with sodium hydroxide (p. 441). The sodium sulfonate group usually is present in direct or substantive dyes (p. 561), its function being to confer water solubility on the dyestuff. Sodium salts of alkylated aromatic sulfonic acids are important synthetic detergents. Since their calcium, magnesium, and iron salts are soluble in water, they are as effective in hard water as in soft water (cf. p. 190). Previous to 1966, the most widely used synthetic detergent was that obtained by the alkylation of benzene with tetrapropylene (p. 87), followed by sulfonation and conversion to the sodium salt.

$$CH_3CH(CH_3)CH_2CH(CH_3)CH_2CH(CH_3)CH_2CH{=}CH_2 + C_6H_6 \xrightarrow[\text{AlCl}_3\text{-HCl}]{\text{HF or}} C_{10}H_{21}CH(CH_3)C_6H_5 \xrightarrow[\text{or SO}_3]{\text{H}_2\text{SO}_4}$$

$$p\text{-}C_{10}H_{21}CH(CH_3)C_6H_4SO_3H \xrightarrow{\text{Na}_2\text{CO}_3} CH_3CH(CH_3)CH_2CH(CH_3)CH_2CH(CH_3)CH_2CH(CH_3)C_6H_4SO_3Na$$

Unfortunately the highly branched chain is not destroyed readily by microorganisms, and the wide use of these detergents resulted in their appearance in ground waters and rivers. They also cause undesirable foaming in sewage disposal plants. Unbranched hydrocarbon chains are attacked by microorganisms, which convert the end methyl group to a carboxyl group and then decrease the length of the chain two carbon atoms at a time by β oxidation. Thus detergents with straight side chains are converted to products that give no visible evidence of their presence, namely foaming, and are said to be *biodegradable,* or *biologically soft.* Major manufacturers planned to convert completely to soft detergents by the end of 1965, replacing propylene polymers by 1-alkenes or 1-alkanols (p. 122).

DERIVATIVES OF SULFONIC ACIDS

Derivatives of the aliphatic sulfonic acids and of the aromatic sulfonic acids, such as the amides and esters (p. 269), are prepared from the acid chlorides. All of the methods used for the preparation of alkanesulfonyl chlorides (p. 268) can be used for the preparation of aromatic sulfonyl chlorides. The latter, however, usually are prepared by a method applicable only to aromatic compounds, namely direct **chlorosulfonation** by means of chlorosulfonic acid. Two moles of chlorosulfonic acid are required per mole of aromatic compound. Toluene yields a mixture of *o*- and *p*-toluenesulfonyl chlorides.

$$C_6H_5CH_3 + 2\,HOSO_2Cl \longrightarrow \underset{o\text{-Toluenesulfonyl chloride}}{o\text{-}CH_3C_6H_4SO_2Cl} \text{ and } \underset{p\text{-Toluenesulfonyl chloride}}{p\text{-}CH_3C_6H_4SO_2Cl} + H_2SO_4 + HCl$$

Undoubtedly sulfonation is brought about by sulfur trioxide formed by dissociation of the chlorosulfonic acid and is followed by conversion of the sulfonic acid to the acid chloride.

$$HOSO_2Cl \rightleftharpoons SO_3 + HCl$$

$$C_6H_6 + SO_3 \rightleftharpoons C_6H_5SO_2OH$$

$$C_6H_5SO_2OH + ClSO_3H \rightleftharpoons \underset{\text{Benzenesulfonyl chloride}}{C_6H_5SO_2Cl} + H_2SO_4$$

Readily available aryl disulfides, such as 2,4-dinitrophenyl disulfide (p. 400), can be converted to the sulfonyl chlorides by the action of chlorine and water (p. 268).

The sulfonyl chlorides boil much lower than the sulfonic acids. They are insoluble in water and soluble in organic liquids. Hence they can be isolated more readily than sulfonic acids and can be purified by distillation or crystallization.

Because of the insolubility of sulfonyl chlorides in water, they react only slowly with it. When boiled with water, the sulfonyl chloride hydrolyzes to the sulfonic acid and hydrogen chloride, and evaporation at reduced pressure gives the free sulfonic acid.

$$p\text{-}CH_3C_6H_4SO_2Cl + H_2O \xrightarrow{\text{Heat}} \underset{p\text{-Toluenesulfonic acid}}{p\text{-}CH_3C_6H_4SO_3H} + HCl$$

Alcohols and amines react more rapidly than water. Hence the reactions may be carried out in the presence of water, usually with the addition of alkali to remove the hydrogen chloride.

$$ArSO_2Cl + HOR + NaOH \longrightarrow ArSO_2OR + NaCl + H_2O$$

In the absence of alkali, the ester acts as an alkylating agent and undergoes side reactions with the hydrogen chloride and excess alcohol.

$$ArSO_2OR + HCl \longrightarrow ArSO_3H + RCl$$

$$ArSO_2OR + HOR \xrightarrow{H^+} ArSO_3H + ROR$$

Pyridine, a tertiary amine (p. 514), frequently is used instead of aqueous alkali in the preparation of sulfonic acid esters. Not only does it act as a basic catalyst and combine with the hydrogen chloride formed in the reaction, but it is a solvent for the sulfonyl chlorides and the alcohol. The reaction of *p*-toluenesulfonyl chloride (*tosyl chloride*) with alcohols frequently is called *tosylation*. *p*-Bromobenzenesulfonyl chloride has been contracted to *brosyl chloride*. *p*-Toluenesulfonates and *p*-bromobenzenesulfonates frequently are called *tosylates* and *brosylates*, but there is no justification for these terms since no "tosylic" or "brosylic" acids exist. The general behavior of sulfonic esters has been discussed (p. 269). Because of their alkylating action and ready availability, the *p*-toluenesulfonates have been used frequently for the preparation of hydrocarbons, quaternary salts, and esters of carboxylic acids.

$$p\text{-}CH_3C_6H_4SO_2OR + R'MgX \longrightarrow R{-}R' + p\text{-}CH_3C_6H_4SO_2OMgX$$

$$+ 2\,NR_3 \longrightarrow p\text{-}CH_3C_6H_4SO_3{}^{-+}NR_4$$

$$+ Na^{+-}OCOR' \longrightarrow ROCOR' + p\text{-}CH_3C_6H_4SO_3{}^{-+}Na$$

As with other S_N2 displacement reactions, the configuration of the carbon atom in group R that is attached to oxygen is inverted (p. 100).

Ammonia and primary and secondary amines, like the alcohols, react with sulfonyl chlorides in the presence of water. Here also an equivalent quantity of a strong alkali is used but for a different reason. The hydrogen chloride produced in the reaction combines at once with unreacted amine. Since the amine salt does not react with the sulfonyl chloride, only half of the amine can be converted to the sulfonamide.

$$ArSO_2Cl + 2\,HNHR \longrightarrow ArSO_2NHR + RNH_3Cl$$

If an equivalent quantity of strong base is present, it reacts with the hydrogen chloride, and all of the amine is convertible into sulfonamide.

$$ArSO_2Cl + HNHR + NaOH \longrightarrow ArSO_2NHR + NaCl + H_2O$$

The availability of benzenesulfonyl chloride and of *p*-toluenesulfonyl chloride has made their reaction with amines important as a means of distinguishing between primary, secondary, and tertiary amines, and of separating mixtures of different types of amines. This procedure is known as the *Hinsberg reaction*. It is based on the fact that sulfonamides prepared from primary amines are sufficiently acidic to form sodium salts in aqueous solution and hence are soluble in dilute aqueous alkali (p. 269). Those from secondary amines cannot form a salt and do not dissolve in dilute alkali. Tertiary amines do not react with sulfonyl chlorides in the presence of aqueous alkali.

$$C_6H_5SO_2Cl + H_2NR + 2\,NaOH \longrightarrow [C_6H_5SO_2\bar{N}R]Na^+ + NaCl + 2\,H_2O$$

Soluble in water; reacts with dilute acid to give water-insoluble sulfonamide

$$C_6H_5SO_2Cl + HNR_2 + NaOH \longrightarrow C_6H_5SO_2NR_2 + NaCl + H_2O$$

Insoluble in dilute alkali or dilute acid

$$C_6H_5SO_2Cl + NR_3 \longrightarrow$$ No reaction; tertiary amine soluble in dilute acids

Sulfonamides are hydrolyzed with difficulty. The best reagent is 30 per cent hydrogen bromide in acetic acid containing phenol.

$$ArSO_2NHR + 2\,HBr \longrightarrow ArSO_2Br + RNH_2{}^{+-}Br$$

The phenol prevents side reactions by reacting with the bromine formed by the oxidizing action of the sulfonyl bromide on hydrogen bromide.

Like amines and other amides (p. 225), sulfonamides having hydrogen on the nitrogen atom react with alkaline hypochlorite solutions to give *N*-halo derivatives.

$$ArSO_2NHR + NaOCl \rightleftharpoons ArSO_2N(Cl)R + NaOH$$

p-Toluenesulfonamide gives the sodium salt, which is soluble in water. It commonly is called **Chloramine-T.**

$$p\text{-}CH_3C_6H_4SO_2NH_2 + NaOCl \rightleftharpoons [p\text{-}CH_3C_6H_4SO_2\bar{N}Cl]Na^+ + H_2O$$

Sodium salt of *N*-chloro-*p*-toluenesulfonamide (*Chloramine-T*)

Since the product is stable when dry and the reaction is reversible in aqueous solution, the *N*-chlorosulfonamides have the antiseptic properties of hypochlorite solutions. The analogous compound from benzenesulfonamide is known as **Chloramine-B.** If the chlorination is carried further, a dichloro derivative is formed, which is soluble in oils and salves.

$$p\text{-}CH_3C_6H_4SO_2NH_2 + 2\,NaOCl \longrightarrow p\text{-}CH_3C_6H_4SO_2NCl_2 + 2\,NaOH$$

N,N-Dichloro-*p*-toluenesulfonamide (*Dichloramine-T*)

Sulfonyl chlorides can be reduced to sulfinic acids and to mercaptans (p. 266). These reactions are useful in the aromatic series as a means of obtaining **sulfinic acids** and **thiophenols.**

$$2\,C_6H_5SO_2Cl + 2\,Zn\ \text{(in ether)} \longrightarrow \underset{\text{Zinc benzene-sulfinate}}{(C_6H_5SO_2)_2Zn} + ZnCl_2$$

$$ArSO_2Cl + Na_2SO_3 + 2\,NaOH \longrightarrow \underset{\text{Sodium arene-sulfinate}}{ArSO_2Na} + Na_2SO_4 + NaCl + H_2O$$

$$2\,C_6H_5SO_2Cl + 6\,Zn + 5\,H_2SO_4\ \text{(in water)} \longrightarrow \underset{\text{Thiophenol}}{2\,C_6H_5SH} + ZnCl_2 + 5\,ZnSO_4 + 4\,H_2O$$

Sodium sulfinates react with ethylene oxide (p. 605) to give **β-hydroxyethyl sulfones,** which can be converted to the hydrogen sulfates.

$$ArSO_2Na + \underset{\diagdown\ O\ \diagup}{CH_2{-}CH_2} + H_2O \longrightarrow ArSO_2CH_2CH_2OH + NaOH$$

$$ArSO_2CH_2CH_2OH + H_2SO_4 \longrightarrow ArSO_2CH_2CH_2OSO_3H + H_2O$$

In alkaline solution β elimination readily takes place to give the **vinyl sulfone.**

$$ArSO_2\underset{\substack{|\\H}}{CH}{-}CH_2{-}O{-}SO_3Na + NaOH \longrightarrow ArSO_2CH{=}CH_2 + Na_2SO_4 + H_2O$$

Like that in α,β-unsaturated carbonyl compounds and α,β-unsaturated nitriles (p. 618), the double bond under alkaline conditions readily adds compounds with reactive hydrogen such as amines, alcohols, or thiols, a process that has been termed *sulfonoethylation.*

$$ArSO_2CH{=}CH_2 \xrightarrow{RNH_2} ArSO_2CH_2CH_2NHR$$

$$ArSO_2CH{=}CH_2 \xrightarrow{ROH} ArSO_2CH_2CH_2OR$$

These reactions are important technically in the application of vinyl sulfone reactive dyes (p. 565).

When arenesulfinic acids are heated with mercuric chloride, the **arylmercuric chloride** is formed.

$$C_6H_5SOOH + HgCl_2 \longrightarrow C_6H_5HgCl + SO_2 + HCl$$

Sulfonyl chlorides undergo the Friedel-Crafts reaction with aromatic compounds in the presence of anhydrous aluminum chloride to yield **sulfones.**

$$CH_3SO_2Cl + C_6H_6 \xrightarrow{AlCl_3} \underset{\text{Methyl phenyl sulfone}}{C_6H_5SO_2CH_3} + HCl$$

Because of the strong electron-attracting effect of the sulfone group, hydrogen on carbon adjacent to it is sufficiently acidic to react with Grignard reagents.

$$ArSO_2CH_3 + CH_3MgI \longrightarrow ArSO_2CH_2MgI + CH_4$$

The resulting product undergoes all of the reactions of Grignard reagents and can be used to prepare a wide variety of compounds that contain the sulfone group.

PROBLEMS

22–*1.* Give equations for the preparation of the following compounds: (*a*) *p*-tolyl sulfone, (*b*) *o*-methylthiophenol, (*c*) *N*-methylbenzenesulfonamide, (*d*) *n*-butyl *p*-toluenesulfonate, (*e*) *p*-bromobenzenesulfonyl chloride, (*f*) 2,4-dimethylbenzene-sulfinic acid, (*g*) β-hydroxyethyl *p*-tolyl sulfone.

22–2. Give equations for the following reactions: (*a*) ethyl *p*-toluenesulfonate and benzylmagnesium chloride, (*b*) *N*-chlorobenzenesulfonamide and water, (*c*) cumene and *p*-toluenesulfonyl chloride in the presence of aluminum chloride, (*d*) 3,5-dibromobenzenesulfonic acid boiled with aqueous hydrochloric acid, (*e*) 2,4-dinitrophenyl disulfide, chlorine, and water, (*f*) fusion of sodium *p*-toluenesulfonate with sodium hydroxide, (*g*) *n*-propyl benzenesulfonate and ammonia.

22–3. Give a series of reactions for converting (*a*) *m*-xylene to 4-cyano-*m*-xylene, (*b*) toluene to pure *p*-toluenesulfonic acid, (*c*) benzyl chloride to *n*-propylbenzene, (*d*) 2,4-dinitrochlorobenzene to 2,4-dinitrobenzenesulfonic acid, (*e*) methanesulfonic acid to methyl *p*-tolyl sulfone, (*f*) methyl phenyl sulfone to carboxymethyl phenyl sulfone, (*g*) benzenesulfonyl chloride to phenylmercuric chloride, and (*h*) sodium *β*-benzenesulfonylethyl sulfate to *β*-(methylamino)ethyl phenyl sulfone.

22–4. Describe a test for distinguishing between the members of the following pairs of compounds: (*a*) methyl *p*-chlorobenzenesulfonate and benzenesulfonyl chloride, (*b*) sodium benzenesulfonate and sodium sulfate, (*c*) benzenesulfonyl chloride and *p*-chlorophenyl sulfone, (*d*) *N*-methylbenzenesulfonamide and *N,N*-dimethylbenzenesulfonamide, (*e*) *N*-chlorobenzenesulfonamide and *p*-chlorobenzenesulfonamide, (*f*) benzenesulfonic acid and phenyl sulfone, (*g*) benzyl chloride and benzenesulfonyl chloride, (*h*) ethyl *p*-toluenesulfonate and *N,N*-dimethyl-*p*-toluenesulfonamide, (*i*) ethyl *p*-nitrobenzenesulfonate and *N,N*-dimethylbenzenesulfonamide.

CHAPTER TWENTY-THREE

AROMATIC NITRO COMPOUNDS

Although aliphatic nitro compounds have been manufactured commercially only since 1940 (p. 243), and still are produced in relatively small volume, aromatic nitro compounds long have been technically important. They have been used as intermediates for the manufacture of dyes since the discovery of mauve by Perkin in 1856. Other technical developments have led to their use as explosives and as intermediates for the manufacture of pharmaceuticals and many other aromatic compounds of commercial importance.

Preparation of Nitro Compounds

The factor contributing most to the widespread use of aromatic nitro compounds is the ease of preparation by direct nitration of aromatic nuclei. If a nitro group is present in a benzene ring, the rate of substitution of hydrogen is decreased even more than when halogen is present (Table 20–*1*, p. 380). Consequently more concentrated acids and higher temperatures are needed to obtain appreciable quantities of a dinitrated product. When substitution of a second hydrogen takes place, the nitro group enters chiefly at the *meta* position.

$$C_6H_5NO_2 + HONO_2 \xrightarrow{(H_2SO_4)} m\text{-}C_6H_4(NO_2)_2 + H_2O$$

m-Dinitrobenzene

It is very difficult to introduce a third nitro group into benzene by direct nitration (cf. p. 417).

Because of the activating effect of the methyl group (p. 383), toluene nitrates more readily than benzene. The principal products are first a mixture of *o*- and *p*-nitrotoluene, then 2,4-dinitrotoluene, and finally 2,4,6-trinitrotoluene (*TNT*). *o*-Nitrotoluene gives in addition some 2,6-dinitrotoluene, which also nitrates further to *TNT*.

$$C_6H_5CH_3 \xrightarrow[(H_2SO_4)]{HONO_2} o\text{-}CH_3C_6H_4NO_2 \text{ and } p\text{-}CH_3C_6H_4NO_2$$

o-Nitrotoluene and *p*-Nitrotoluene

$\downarrow$ $HONO_2$ (H_2SO_4)

2,6-Dinitrotoluene and 2,4-Dinitrotoluene $\xrightarrow[(H_2SO_4)]{HONO_2}$ 2,4,6-Trinitrotoluene

Physical Properties of Nitro Compounds

Aromatic nitro compounds usually are solids, either colorless or yellow. Only a few mononitro hydrocarbons are liquids at room temperature. Because of the presence of the semipolar bond, the nitro group has a high dipole moment, and the nitro compounds have high boiling points. The lowest boiling aromatic nitro compound, nitrobenzene, has a dipole moment of 4.3 D and boils at 209°. The ability of nitro groups to produce crystallinity and high melting point in aromatic compounds has led to the extensive use of nitro compounds, such as 2,4-dinitrophenylhydrazine (p. 400) and 3,5-dinitrobenzoyl chloride (p. 471), for the preparation of derivatives for identification purposes.

Aromatic nitro hydrocarbons are practically insoluble in water. The liquids are good solvents for most organic compounds and fair solvents for many inorganic salts, especially those such as zinc chloride or aluminum chloride that can accept an unshared pair of electrons and form a complex.

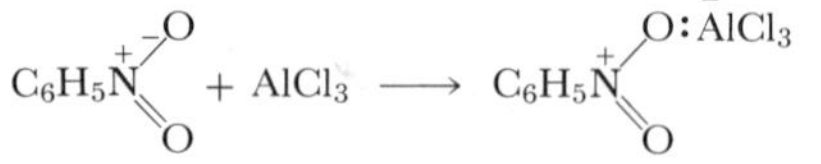

$$C_6H_5\overset{+}{N}(\overset{-}{O})O + AlCl_3 \longrightarrow C_6H_5\overset{+}{N}(O{:}\overset{-}{A}lCl_3)O$$

Nitrobenzene absorbs ultraviolet light strongly just beyond the visible (Fig. 31–*11*, p. 556). Electronic disturbances, produced by other groups in the molecule or by the proximity of other molecules, may shift the absorption band to longer wavelengths, causing the compounds or their solutions to be yellow, orange, or red (p. 556).

Although nitro groups increase the complexity of the infrared absorption spectrum of aromatic compounds (p. 393), the C—N absorption bands lie in the same ranges as for aliphatic compounds, namely 1565–1500 cm.$^{-1}$ and 1380–1320 cm.$^{-1}$. Additional strongly electronegative groups in the *para* position or large groups in the *ortho* position that prevent the nitro group from becoming planar cause absorption in the higher portions of the ranges, whereas strongly electron-donating groups in the *ortho* or *para* positions lead to absorption at the lower frequencies.

Physiological Action

Nitro compounds having a sufficiently high vapor pressure have strong characteristic odors. For the most part they are highly toxic substances. Even those compounds with low vapor pressures are dangerous because they are absorbed readily through the skin, particularly from solutions. Symptoms of poisoning are dizziness, headache, irregular pulse, and cyanosis (blue lips and finger tips caused by a change in the hemoglobin of the blood). Prolonged contact or exposure leads to death.

Reactions of Aromatic Nitro Compounds

Reactions at the Nucleus. The deactivating and *meta* directive effects of the nitro group on electrophilic substitution of nuclear hydrogen have been noted previously (p. 383). The same electron-withdrawing power of the nitro group, however, leads to activation of the nucleus for **nucleophilic substitution** of hydrogen. Moreover the intermediate σ complexes produced by electron-donating reagents have an unshared pair of electrons and hence a negative charge instead of a positive charge in the resonance structures represented on page 383. Thus electron-attracting groups lead to *ortho-para* substitution. The nitro group is particularly activating for these positions because both the inductive effect and the resonance effect operate in the same direction.

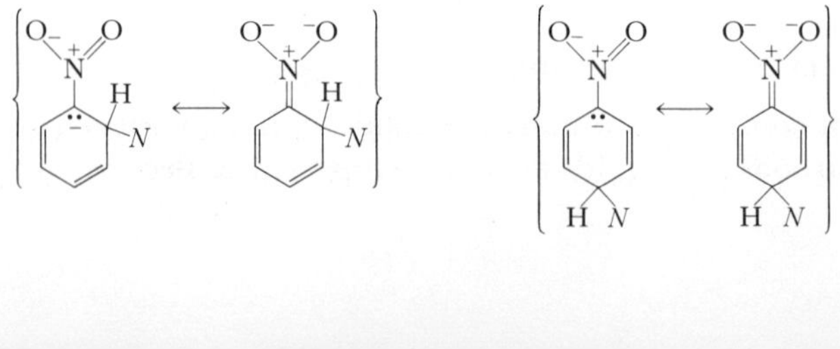

[Resonance structures of σ complexes: ortho attack (H, N on ring carbon) and para attack (H N on ring carbon), each shown as two contributing structures ⟷]

An example of nucleophilic substitution is the formation of *o*-nitrophenol (*o-hydroxynitrobenzene*) when nitrobenzene is heated with potassium hydroxide in the presence of air.

$$C_6H_5NO_2 + {}^-\!:\ddot{O}H \longrightarrow o\text{-}C_6H_4(NO_2)OH + {}^-\!:H$$

$$2\,H\!:^- + O_2 \longrightarrow 2\,{}^-\!:\ddot{O}H$$

More common is the **nucleophilic displacement of nitro groups,** especially if more than one nitro group is present.

$$o\text{-}C_6H_4(NO_2)_2 \xrightarrow{\text{Aq. KOH}} o\text{-}C_6H_4(NO_2)OH + KNO_2$$

$$o\text{-}C_6H_4(NO_2)_2 \xrightarrow[+\,C_2H_5OH]{KOH} o\text{-}C_6H_4(NO_2)OC_2H_5 + KNO_2 + H_2O$$

$$o\text{-}C_6H_4(NO_2)_2 \xrightarrow{NH_3} o\text{-}C_6H_4(NO_2)NH_2 + NH_4NO_2 \longrightarrow N_2 + 2\,H_2O$$

p-Dinitrobenzene behaves similarly but not *m*-dinitrobenzene. On the other hand 1,3,5-trinitrobenzene reacts with sodium methoxide in methanol solution to give 1,3-dinitro-5-methoxybenzene. The effect of nitro groups on the nucleophilic displacement of halogen has been discussed (p. 400).

Reduction of the Nitro Group. Like the aliphatic nitro compounds (p. 245), the aromatic nitro compounds are reduced readily to primary amines. The reduction may be brought about by hydrogen and a hydrogenation catalyst at room temperature and atmospheric pressure. Because of the large amount of heat evolved and the speed of the reaction, proper precautions should be taken to keep the reaction always under control when large quantities of nitro compounds are hydrogenated in a closed system.

$$C_6H_5NO_2 + 3\,H_2 \xrightarrow[\text{Raney Ni}]{\text{Pt or}} \underset{\text{Aniline}}{C_6H_5NH_2} + 2\,H_2O$$

Hydrazine may be used conveniently as a source of hydrogen.

$$2\,C_6H_5NO_2 + 3\,H_2NNH_2 \xrightarrow{\text{Raney Ni}} 2\,C_6H_5NH_2 + 4\,H_2O + 3\,N_2$$

The ease of reduction of nitro groups is indicated by their selective hydrogenation in the presence of less active catalysts such as rhodium sulfide, without hydrogenolysis of carbon-sulfur or carbon-halogen bonds that may be present.

Ordinarily reduction to the amine is brought about by active metals in acid solution. The reagents commonly used are iron or tin in the presence of hydrochloric acid. When an excess of tin is used, the product is the arylammonium chlorostannite from which the amine is liberated by the addition of alkali.

$$2\,C_6H_5NO_2 + 6\,Sn + 24\,HCl \longrightarrow \underset{\text{Phenylammonium chlorostannite}}{[C_6H_5\overset{+}{N}H_3]_2[SnCl_4]^{2-}} + \underset{\text{Chlorostannous acid}}{5\,H_2SnCl_4} + 4\,H_2O$$

$$\downarrow 6\,NaOH$$

$$\underset{\text{Aniline}}{2\,C_6H_5NH_2} + \underset{\text{Sodium stannite}}{Na_2SnO_2} + 4\,NaCl + 4\,H_2O$$

Stannous chloride in the presence of hydrochloric acid reduces nitro compounds to amines, the actual product being the chlorostannate.

$$2\,C_6H_5NO_2 + 6\,SnCl_2 + 24\,HCl \longrightarrow \underset{\text{Phenylammonium chlorostannate}}{[C_6H_5\overset{+}{N}H_3]_2[SnCl_6]^{2-}} + \underset{\text{Chlorostannic acid}}{5\,H_2SnCl_6} + 4\,H_2O$$

Nitro groups may be estimated quantitatively by means of standardized solutions of stannous chloride. Zinc and hydrochloric acid is not a satisfactory reducing agent because considerable amounts of chloroanilines are formed (p. 415). Sodium hydrosulfite is a good reducing agent in alkaline solution.

$$C_6H_5NO_2 + 3\,Na_2S_2O_4 + 6\,NaOH \longrightarrow C_6H_5NH_2 + 6\,Na_2SO_3 + 2\,H_2O$$

Commercially nitrobenzene is reduced to aniline by scrap iron and water in the presence of about one-fortieth of the calculated amount of hydrochloric acid. Hence the reduction is brought about essentially by iron and water, the iron being converted to black oxide of iron, Fe_3O_4 (p. 425).

$$4\,C_6H_5NO_2 + 9\,Fe + 4\,H_2O \xrightarrow{(HCl)} 4\,C_6H_5NH_2 + 3\,Fe_3O_4$$

If two nitro groups are present in the ring, it is possible to reduce one group without reducing the other. Ammonium or sodium sulfide is the preferred reagent.

$$C_6H_4(NO_2)_2 + 3\,(NH_4)_2S \longrightarrow C_6H_4(NO_2)(NH_2) + 6\,NH_3 + 3\,S + 2\,H_2O$$

This behavior is understandable, since the electron-attracting properties of the nitro groups mutually increase their strength as oxidizing agents and enable one of them to be reduced more easily than if a single nitro group were present. After one nitro group is reduced, the electron-donating power of the amino group decreases the ease of reduction of the remaining nitro group.

Nitro compounds are not reduced readily by lithium aluminum hydride, sodium borohydride, or diborane. Hence these reagents may be used to reduce other groups selectively in the presence of nitro groups.

Although primary amines are the final reduction products of both aliphatic and aromatic nitro compounds, it is possible with aromatic nitro compounds to isolate a number of intermediate reduction products, which may be obtained in good yield by the use of milder and controlled reducing conditions. Thus nitrobenzene, when boiled with an aqueous solution of ammonium chloride and zinc dust, gives **N-phenylhydroxylamine** (also called *β-phenylhydroxylamine*).

$$C_6H_5NO_2 + 2\,Zn + 4\,NH_4Cl \longrightarrow \underset{N\text{-Phenylhydroxylamine}}{C_6H_5NHOH} + 2\,ZnCl_2 + H_2O + 4\,NH_3$$

N-Phenylhydroxylamine is a colorless solid that melts at 81°. It is not stable, however, the molecules undergoing mutual intermolecular oxidation and reduction and condensation to give a mixture of the various reduction products of nitrobenzene (p. 416). When a solution of β-phenylhydroxylamine in ether is treated with dry ammonia and an alkyl nitrite, the ammonium salt of *N*-phenyl-*N*-nitrosohydroxylamine precipitates.

$$C_6H_5NHOH + NH_3 + C_4H_9ONO \longrightarrow C_6H_5N(NO)O^{-}\,{}^{+}NH_4 + C_4H_9OH$$

It is known as **cupferron** and is used in analytical chemistry for the separation of certain metal ions, such as those of copper, iron, and titanium, that react with it to form insoluble chelate coordination compounds (p. 604).

$$2\ C_6H_5N(N{=}O){-}O^-{}^+NH_4 + Cu^{2+} \longrightarrow (C_6H_5N(N{=}O){-}O)_2Cu + 2\ NH_4^+$$

If zinc and hydrochloric acid are used to reduce nitrobenzene, the main product is aniline, but a considerable amount of *o*- and *p*-chloroaniline is formed, presumably by rearrangement of *N*-chloroaniline produced from *N*-phenylhydroxylamine.

$$C_6H_5NHOH + HCl \longrightarrow H_2O + C_6H_5NHCl \longrightarrow o\text{-}ClC_6H_4NH_2 \text{ and } p\text{-}ClC_6H_4NH_2$$

Although **nitrosobenzene** has been shown to be an intermediate reduction product of nitrobenzene, it has not been isolated because it is reduced very rapidly to *N*-phenylhydroxylamine (p. 414). It can, however, be prepared by oxidation of the latter compound.

$$3\ C_6H_5NHOH + Na_2Cr_2O_7 + 4\ H_2SO_4 \xrightarrow{0^\circ} \underset{\text{Nitrosobenzene}}{3\ C_6H_5NO} + Na_2SO_4 + Cr_2(SO_4)_3 + 7\ H_2O$$

Several intermediate bimolecular reduction products can be obtained using mild reducing agents in alkaline medium. Sodium methoxide in alcohol, or aqueous sodium arsenite, convert nitrobenzene to **azoxybenzene.**

$$4\ C_6H_5NO_2 + \underset{(\text{or } 6\ Na_3AsO_3)}{3\ NaOCH_3} \longrightarrow \underset{\text{Azoxybenzene}}{2\ C_6H_5\overset{+}{N}(O^-){=}NC_6H_5} + \underset{\substack{\text{Sodium formate}\\(\text{or } 6\ Na_3AsO_4)}}{3\ NaOCHO} + 3\ H_2O$$

Oxidation of azobenzene with hydrogen peroxide in acetic acid also yields azoxybenzene.

$$C_6H_5N{=}NC_6H_5 + H_2O_2 \longrightarrow C_6H_5\overset{+}{N}(O^-){=}NC_6H_5 + H_2O$$

The oxygen atom in azoxybenzene is believed to be linked to nitrogen by a semipolar bond as in the amine oxides (p. 229). In accordance with this view oxidation of unsymmetrical azo compounds yields two structurally isomeric products.

$$C_6H_5N{=}NC_6H_4NO_2 \xrightarrow{H_2O_2} C_6H_5\overset{+}{N}(O^-){=}NC_6H_4NO_2 \text{ and } C_6H_5N{=}\overset{+}{N}(O^-)C_6H_4NO_2$$

Azobenzene results when azoxybenzene is heated with iron filings,

$$3\ C_6H_5\overset{+}{N}(O^-){=}NC_6H_5 + 2\ Fe \longrightarrow 3\ C_6H_5N{=}NC_6H_5 + Fe_2O_3$$

but the usual laboratory method of preparation is by the oxidation of hydrazobenzene with hypobromite solution.

$$\underset{\text{Hydrazobenzene}}{C_6H_5NHNHC_6H_5} + NaOBr \longrightarrow C_6H_5N{=}NC_6H_5 + NaBr + H_2O$$

The oxidation may be brought about also by passing air through an alkaline alcoholic solution of hydrazobenzene.

The nitrogen-nitrogen double bond in azo compounds can give rise to *cis-trans* isomerism analogous to that caused by a carbon-carbon double bond (p. 76). The ordinary stable form of azobenzene has the *trans* configuration. When it is irradiated by light, it is converted partially to the *cis* isomer, which can be isolated by crystallization or by adsorption on alumina.

C_6H_5—N=N—C_6H_5 *trans*-Azobenzene

C_6H_5—N=N—C_6H_5 *cis*-Azobenzene

Many azo dyes (p. 563) undergo a reversible color change when exposed to light (*phototropism*) because the *cis* form has a different color than the *trans* form. Numerous other types of compounds exhibit phototropism for which explanations other than *cis-trans* isomerism must be invoked.

When nitrobenzene is heated with zinc dust and aqueous sodium hydroxide, the product is **hydrazobenzene.**

$$2\,C_6H_5NO_2 + 5\,Zn + 10\,NaOH \longrightarrow C_6H_5NHNHC_6H_5 + 5\,Na_2ZnO_2 + 4\,H_2O$$

Any of the intermediate reduction products can be reduced to aniline by means of a strong reducing agent. The reactions of nitrobenzene involving reduction are summarized in Fig. 23–*1*.

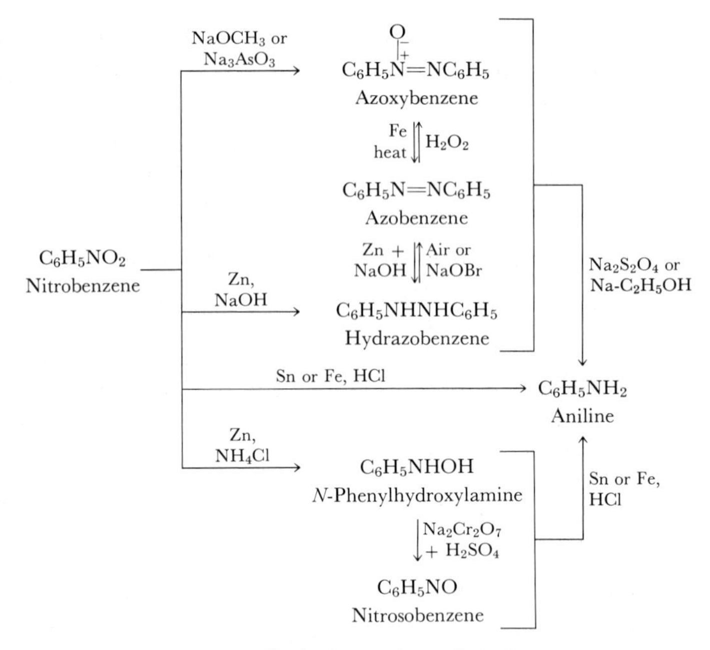

Figure 23–*1*. Reduction products of nitrobenzene.

Important Nitro Hydrocarbons

Nitrobenzene was prepared first in 1834 by Mitscherlich, who called it *nitrobenzid.* It is the most important of the nitro aromatic hydrocarbons. Production in the United States was 239 million pounds in 1964, and the selling price was about 9 cents per pound. Very little nitrobenzene is used as such, practically all being converted to aniline. Because of its odor, nitrobenzene was called *oil of mirbane* or *artificial oil of bitter almonds.* It once was used as a flavoring principle and as a perfume for soaps, and to adulterate oil of bitter almonds. Its high toxicity soon led to a discontinuance of this practice. Because it penetrates leather, it long was used as a solvent for shoe dyes, but here again poisoning occurred by adsorption of the vapors through the skin, and such formulations now are prohibited in most countries.

Some of the di- and trinitroalkylbenzenes have a musk-like odor and are used in perfumery. Methoxyl or acetyl groups also may be present. These compounds are known as **"synthetic musks"** or **"nitro musks,"** although they are not related chemically to the true musks, which are macrocyclic ketones (p. 656).

Musk xylene	Musk tibetine	Musk ambrette	Musk ketone
1-CH_3, 2,6-$(NO_2)_2$, 3-CH_3, 5-$C(CH_3)_3$	1-CH_3, 2,6-$(NO_2)_2$, 3,4-$(CH_3)_2$, 5-$C(CH_3)_3$	1-CH_3, 2,6-$(NO_2)_2$, 3-OCH_3, 4-$C(CH_3)_3$	1-CH_3, 2-$COCH_3$, 3-CH_3, 4-NO_2, 5-$C(CH_3)_3$, 6-NO_2

1,3,5-Trinitrobenzene is a more powerful explosive than TNT, but it cannot be made in satisfactory yield by direct nitration of benzene because of the difficulty of introducing a third nitro group in the absence of an activating group. It is prepared on a small scale for reagent purposes by decarboxylating 2,4,6-trinitrobenzoic acid, which is obtained by the oxidation of TNT.

$$CH_3C_6H_2(NO_2)_3 \xrightarrow[H_2SO_4]{Na_2Cr_2O_7 +} HOOCC_6H_2(NO_2)_3 \xrightarrow{Heat} C_6H_3(NO_2)_3 + CO_2$$

2,4,6-Trinitrobenzoic acid; 1,3,5-Trinitrobenzene

2,4,6-Trinitrotoluene, commonly known as TNT, is an important military explosive. It is used for filling bombs, shells, and hand grenades either alone or with other explosives (pp. 214, 609). Since it melts at 81° and does not explode until 280°, it can be poured into shells in a liquid state and allowed to solidify. It is relatively insensitive to shock and must be exploded by a detonator. Production in the United States during World War II probably reached a rate of one million tons per year. The first H-bomb exploded is believed to have had a force equivalent to the explosion of around 14 million tons of TNT. Some TNT is used for nonmilitary demolition purposes.

PROBLEMS

23–*1*. Give equations for the following reactions: (*a*) *p*-nitrotoluene with iron and hydrochloric acid, (*b*) *o*-nitrocumene with sodium methoxide, (*c*) *p,p'*-dimethylazobenzene with hydrogen peroxide, (*d*) *p*-nitroethylbenzene with zinc dust and sodium hydroxide, (*e*) *m*-chloronitrobenzene with zinc dust and ammonium chloride, (*f*) *p*-hydrazocumene with sodium hypobromite, (*g*) *N*-*o*-tolylhydroxylamine with dichromate and sulfuric acid.

23–*2*. Give the correct pairing of the formulas (*a*) $C_6H_5N{=}NC_6H_5$, (*b*) C_6H_5NHOH, (*c*) $C_6H_5NHNHC_6H_5$, (*d*) $C_6H_5NO{=}NC_6H_5$, and (*e*) C_6H_5NO with the names (*1*) azoxybenzene, (*2*) azobenzene, (*3*) hydrazobenzene, (*4*) *N*-phenylhydroxylamine, and (*5*) nitrosobenzene.

23–*3*. Give a series of reactions for the conversion of (*a*) *m*-xylene to 4-nitroso-*m*-xylene, (*b*) cumene to *sym*-di-*p*-cumylhydrazine, (*c*) toluene to 4-amino-2-nitrotoluene, and (*d*) 2,4-dinitrochlorobenzene to 2,4-dinitrobenzenesulfonic acid.

23–*4*. In the reduction of azobenzene to aniline with an alkaline solution of sodium hydrosulfite, what is the theoretical ratio of moles of azobenzene to moles of hydrosulfite?

23–*5*. How can one distinguish by a chemical reaction between the members of the following pairs of compounds: (*a*) nitrobenzene and 1-nitrohexane, (*b*) azoxybenzene and hydrazobenzene, (*c*) *m*-nitrochlorobenzene and 2,4-dinitrochlorobenzene, (*d*) nitrobenzene and *N*-phenylhydroxylamine, (*e*) *m*-chloronitrobenzene and 2-chloro-4-nitrotoluene.

23–*6*. Arrange the following compounds in the order of increasing ease with which halogen is displaced by a nucleophile: (*a*) bromobenzene, (*b*) *o*-nitrochlorobenzene, (*c*) *m*-nitrochlorobenzene, (*d*) chlorobenzene, (*e*) 3,5-dinitrochlorobenzene, (*f*) 2,4-dinitrochlorobenzene, (*g*) benzyl chloride, (*h*) 2,4,6-trinitrochlorobenzene.

23–*7*. Give a plausible explanation of the fact that methoxide ion displaces one nitro group from *ortho* or *para* dinitrobenzene, or 1,3,5-trinitrobenzene, but not from *m*-dinitrobenzene.

CHAPTER TWENTY-FOUR

AROMATIC AMINES AND PHOSPHINES. DIAZONIUM SALTS

Compounds classed as aromatic amines have an amino group or an alkyl- or aryl-substituted amino group attached directly to an aromatic nucleus. Usually they are made by a procedure different from those for aliphatic amines and undergo additional reactions. Aromatic phosphines resemble the amines only in structure. Their methods of preparation and reactions are entirely different.

AMINES

Nomenclature

Aromatic amines may be primary, secondary, or tertiary, and in the secondary or tertiary amines, the second or third hydrocarbon group may be alkyl or aryl. Usually the primary amines are named as amino derivatives of the aromatic hydrocarbon or as aryl derivatives of ammonia, but some are known best by common names such as aniline or toluidine.

NH_2 — Aniline

CH_3, NH_2 — *o*-Toluidine

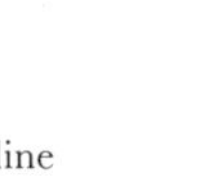

NH_2, NH_2 — *m*-Phenylenediamine

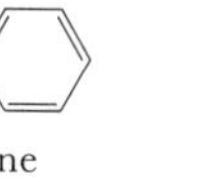

Secondary and tertiary amines are named as derivatives of the primary amine, or as derivatives of ammonia.

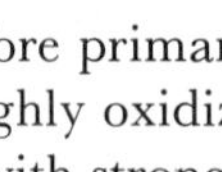

$N(CH_3)_2$ — *N,N*-Dimethylaniline

—NH— — Diphenylamine

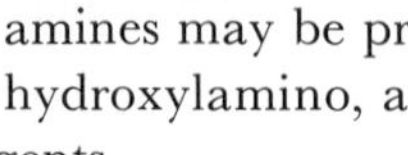

Preparation

1. ***By Reduction of More Highly Oxidized Nitrogen Compounds.*** Aromatic nitro compounds yield a series of reduction products, the final product being the primary amine (Fig. 23–*1*, p. 416). Therefore primary aromatic amines may be prepared from nitro compounds or from the less highly oxidized nitroso, hydroxylamino, azoxy, azo, and hydrazo compounds, by reduction with strong reducing agents.

2. ***By Ammonolysis of Halogen Compounds.*** Halogen attached to an aromatic nucleus usually is very stable to hydrolysis or ammonolysis, and rather drastic conditions are required to bring about reaction, which may occur with rearrangement (p. 399). If, however, electron-attracting groups are present in the *ortho* and *para* positions, the halogen

is more easily displaced. Thus 2,4,6-trinitrochlorobenzene (*picryl chloride*) reacts readily with ammonia to yield 2,4,6-trinitroaniline (*picramide*) by an $S_{N_{Ar}}2$ mechanism (p. 400).

$$\text{Cl-C}_6\text{H}_2(\text{NO}_2)_3 + 2\,\text{NH}_3 \longrightarrow \text{NH}_2\text{-C}_6\text{H}_2(\text{NO}_2)_3 + \text{NH}_4\text{Cl}$$

2,4,6-Trinitrochlorobenzene (*picryl chloride*) — 2,4,6-Trinitroaniline (*picramide*)

Physical Properties

The physical properties of the aromatic amines are about what would be expected. Just as benzene (b.p. 80°) boils at a higher temperature than *n*-hexane (b.p. 69°), so aniline (b.p. 184°) has a higher boiling point than *n*-hexylamine (b.p. 130°). The greater difference in the boiling points of the second pair may be ascribed to the fact that aniline has a higher dipole moment ($\mu = 1.6$) than *n*-hexylamine ($\mu = 1.3$). *N*-Methylaniline (b.p. 195°) boils at a higher temperature than aniline, but *N,N*-dimethylaniline (b.p. 193°) boils at a lower temperature than methylaniline despite the increase in the number of electrons because proton bonding is not possible for dimethylaniline.

Aniline is considerably more soluble in water (3.6 g. per 100 g. of water) than *n*-hexylamine (0.4 g. per 100 g. of water). Water dissolves in aniline to the extent of about 5 per cent. Aniline is miscible with benzene but not with *n*-hexane.

As is true for all of the disubstituted benzenes, the *para*-substituted anilines, being the most symmetric, have the highest melting point. Thus *p*-toluidine is a solid at room temperature whereas both the *ortho* and *meta* isomers are liquids.

Physiological Properties

The aromatic amines, like the aromatic hydrocarbons and their halogen and nitro derivatives, are highly toxic. The liquids are absorbed readily through the skin, and low concentrations of the vapors produce symptoms of toxicity when inhaled for prolonged periods. Aniline vapors may produce symptoms of poisoning after several hours of exposure to concentrations as low as 7 parts per million. Aniline affects both the blood and the nervous system. Hemoglobin of the blood is converted into methemoglobin with reduction of the oxygen-carrying capacity of the blood and resultant cyanosis. A direct depressant action is exerted on heart muscle. Continued exposure leads to mental disturbances. Aromatic amines appear to be responsible also for bladder irritation and the formation of tumors in workers engaged in the manufacture of dye intermediates.

The chloro and nitro nuclear-substituted amines, the *N*-alkylated and acylated amines, and the diamines all are highly toxic. The *N*-phenylamines are considerably less toxic than the *N*-alkyl derivatives. The phenolic hydroxyl group also decreases the toxicity somewhat. Toxicity is greatly reduced by the presence of free carboxylic or sulfonic acid groups in the ring.

Reactions of the Nucleus

1. ***Hydrogen Exchange.*** Electrophilic substitution of deuterium for hydrogen of benzene takes place only with strong acids under anhydrous conditions (p. 375). The amino group is so strongly activating, however, that exchange with hydrogen in the *ortho* and *para* positions takes place readily in aqueous solutions, although less readily than with the amino hydrogen. As would be expected, these exchange reactions are catalyzed by acids.

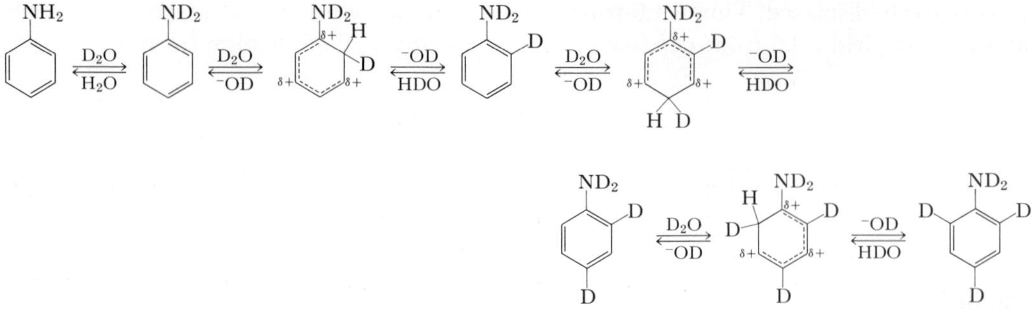

2. ***Oxidation.*** Aliphatic amines are fairly stable to oxidation, but many aromatic amines oxidize readily. Unless carefully purified, they soon darken on standing in air. Stronger oxidizing agents produce highly colored products. Even the simplest aromatic amine, aniline, can give rise to numerous and frequently complex oxidation products. It is not surprising that, depending on the oxidizing agent used, azobenzene, azoxybenzene, phenylhydroxylamine, nitrosobenzene, and nitrobenzene have been isolated (p. 416), since aniline is a reduction product of these compounds. In addition to the amino group, however, the hydrogen atoms of the benzene ring that are *ortho* and *para* to the amino group can be oxidized to hydroxyl groups because the amino group increases the electron density at the *ortho* and *para* positions. Thus when sodium hypochlorite solution is added to aniline, *p*-aminophenol is formed along with azobenzene and other products.

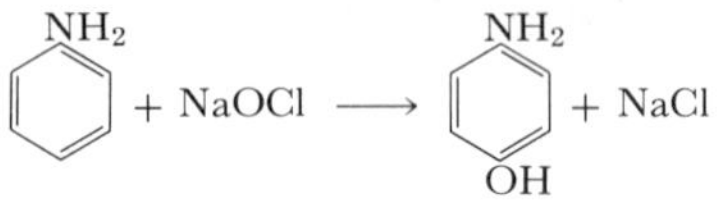

$$C_6H_5NH_2 + NaOCl \longrightarrow p\text{-}HOC_6H_4NH_2 + NaCl$$

These hydroxy amines are oxidized very readily to quinones (p. 450), which undergo further oxidation and condensation reactions. For example, the violet color produced when aniline is mixed with a solution of bleaching powder is due to a series of reactions that form a blue compound known as *indoaniline.*

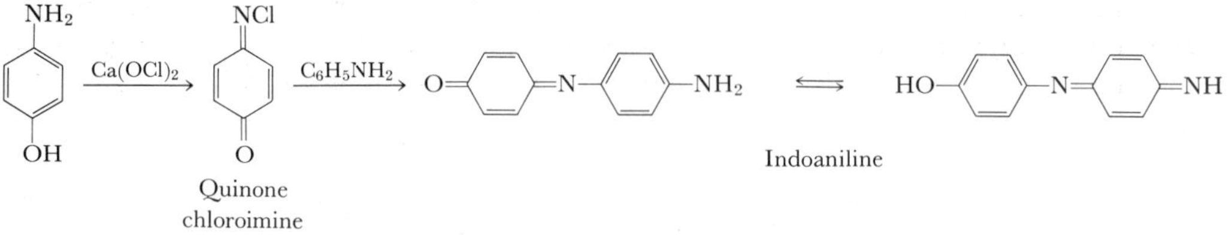

Quinone chloroimine

Indoaniline

More complicated reactions are involved in the formation of the Aniline Blacks (p. 576).

Amine salts are much less readily oxidized than the free amines because the positive charge makes the group electron-attracting rather than electron-donating. Similarly, electronegative substituents such as the nitro group decrease the electron density of the ring and greatly reduce the ease of oxidation.

3. ***Halogenation.*** Because of the strong activating effect of the amino group, no catalyst is required in the halogenation of the nucleus. Furthermore, halogenation takes place in aqueous solution and is so rapid that the only product readily isolated is 2,4,6-trichloro- or 2,4,6-tribromoaniline. The three halogen atoms in the *ortho* and *para* positions reduce the basicity of the amino group, and the salt does not form in aqueous solution.

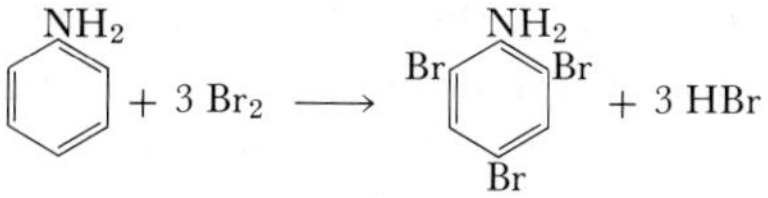

$$C_6H_5NH_2 + 3\,Br_2 \longrightarrow 2,4,6\text{-}Br_3C_6H_2NH_2 + 3\,HBr$$

2,4,6-Tribromo-aniline

Actually trichloroaniline or tribromoaniline is formed even when chlorine or bromine is added to an aqueous solution of an aniline salt. This behavior seems anomalous at first, since salt formation should lead to deactivation and *meta* orientation. The experimental results can be explained by the presence of free amine in equilibrium with the salt in aqueous solution. This view is confirmed by the fact that aniline dissolved in concentrated sulfuric acid is not chlorinated or brominated at room temperature. At higher temperatures the *meta* substitution product is formed.

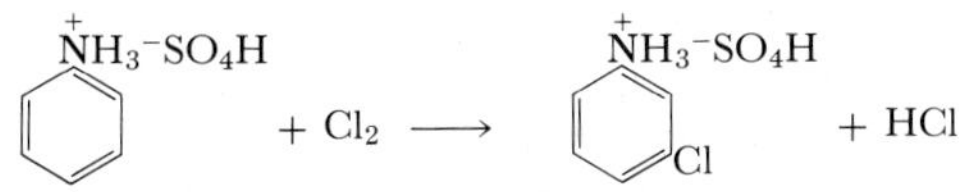

$$C_6H_5\overset{+}{N}H_3\,^{-}SO_4H + Cl_2 \longrightarrow m\text{-}ClC_6H_4\overset{+}{N}H_3\,^{-}SO_4H + HCl$$

Aniline acid sulfate — *m*-Chloroaniline acid sulfate

4. ***Nitration.*** Because of the ease of oxidation of free aniline (p. 420), only the salt can be nitrated efficiently, and nitration is carried out in concentrated sulfuric acid solution. Hence the chief product is *m*-nitroaniline.

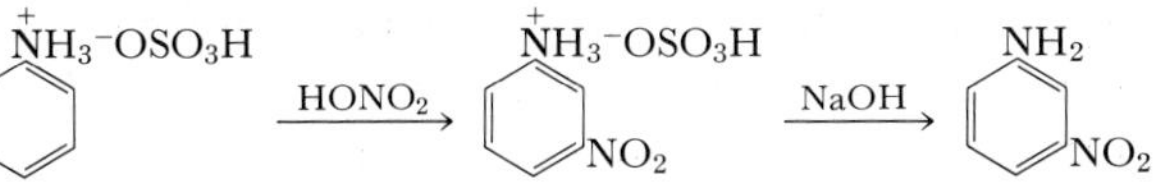

$$C_6H_5\overset{+}{N}H_3\,^{-}OSO_3H \xrightarrow{HONO_2} m\text{-}O_2NC_6H_4\overset{+}{N}H_3\,^{-}OSO_3H \xrightarrow{NaOH} m\text{-}O_2NC_6H_4NH_2$$

m-Nitroaniline

Some *o*- and *p*-nitroaniline also are formed, probably by nitration of the small amount of free amine in equilibrium with its salt, since the amount of *meta* increases with the concentration of the sulfuric acid (cf. p. 422). The three nitroanilines differ in basicity and can be separated by fractional precipitation from their salts with alkali. The order of precipitation is *ortho,* then *para,* then *meta.* *m*-Nitroaniline usually is made by the partial reduction of *m*-dinitrobenzene (p. 414).

If salt formation is prevented by the conversion of the basic amino group to the neutral acetamido group, nitration in acetic acid takes place almost exclusively in the *para* position. If the nitration is carried out in acetic anhydride, the *ortho* isomer is the chief product.

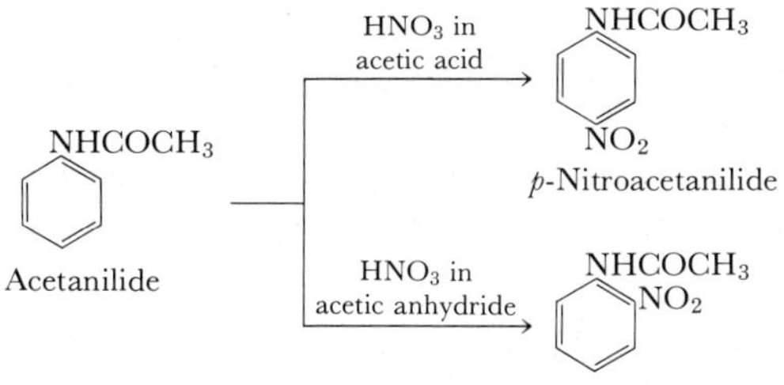

$$C_6H_5NHCOCH_3 \xrightarrow{HNO_3\ \text{in acetic acid}} p\text{-}O_2NC_6H_4NHCOCH_3$$

$$C_6H_5NHCOCH_3 \xrightarrow{HNO_3\ \text{in acetic anhydride}} o\text{-}O_2NC_6H_4NHCOCH_3$$

Acetanilide — *p*-Nitroacetanilide — *o*-Nitroacetanilide

Saponification of the nitroacetanilides with sodium hydroxide solution gives the nitroanilines.

5. ***Sulfonation.*** Sulfonation of aniline at room temperature with fuming sulfuric acid gives a mixture of *o*-, *m*-, and *p*-aminobenzenesulfonic acids. Since the effect of the $^{+}NH_3$ group should be comparable to that of the $^{+}NR_3$ group, it should lead to pure *meta* substi-

tution. Hence, as in nitration (p. 421), the *ortho* and *para* isomers probably arise from the sulfonation of the small amount of free amine in equilibrium with the salt. When aniline is heated with concentrated sulfuric acid for several hours at 180° (*baking process*), the sole product is the *para* isomer, sulfanilic acid. There is some evidence that here the initial product is the sulfamic acid, which should be *ortho,para*-directing.

$$C_6H_5\overset{+}{N}H_3\,^{-}SO_4H \xrightarrow{\text{Heat}} C_6H_5NHSO_3H \xrightarrow{H_2SO_4} p\text{-}HO_3SC_6H_4NHSO_3H \xrightarrow{H_2O} p\text{-}H_2NC_6H_4SO_3H \text{ or better } p\text{-}H_3\overset{+}{N}C_6H_4SO_3^-$$

Phenylsulfamic acid; Sulfanilic acid

N,N-Dimethylaniline, however, behaves in the same way as aniline, and since it cannot form a sulfamic acid, *para* substitution is assumed to result from sulfonation of the free amine at both low and high temperatures.

Although the formulas for the sulfonated amines frequently are written as aminosulfonic acids, they actually are inner salts or dipolar ions (cf. p. 355). Thus sulfanilic acid decomposes at 280–300° without melting, whereas aniline is a liquid, benzenesulfonic acid is a low melting solid, and both can be distilled. Whereas the amino carboxylic acids are more soluble in either strong base or strong acid than in water, sulfanilic acid is more soluble only in strong bases because the sulfonic acid group is as strong as any of the mineral acids in aqueous solution.

The common names for *o*-, *m*-, and *p*-aminobenzenesulfonic acids are *orthanilic, metanilic,* and *sulfanilic* acids respectively. Metanilic acid is prepared by the reduction of *m*-nitrobenzenesulfonic acid. Orthanilic acid is not readily available but can be obtained by removal of the bromine atom in 2-amino-5-bromobenzenesulfonic acid by reduction, or by reduction of *o*-nitrobenzenesulfonic acid made from *o*-nitrophenyl disulfide (p. 400) by oxidation.

Reactions of the Amino Group

1. ***Basicity.*** When an amino group is attached to an aromatic nucleus, the unshared pair of electrons on nitrogen interacts with the π orbital system of the nucleus (p. 381) and makes the unshared pair less available for bonding with other groups. Moreover the electronegativity of the phenyl group is greater than that of an alkyl group because the greater *s* character of the aryl sp^2 orbital pulls the electrons closer to the nucleus. Hence the basicity of aromatic amines is less than that of aliphatic amines, although it is still greater than that of amides. Thus the acidity constants (pK_a's) of methylamine, aniline, and acetamide are 10.6, 4.6, and −1.5 respectively. The introduction of a second aromatic nucleus on the nitrogen atom decreases the basicity still further, the pK_a for diphenylamine being 0.9. On the other hand the introduction of alkyl groups increases the basicity, the pK_a's for *N*-methylaniline and *N,N*-dimethylaniline being 4.8 and 5.1 respectively.

2. ***Alkylation and Arylation.*** Like the aliphatic amines, the primary aromatic amines react with alkyl halides to give secondary and tertiary amines and quaternary ammonium salts.

$$C_6H_5NH_2 + RX \longrightarrow [C_6H_5\overset{+}{N}H_2R]X^- \xrightarrow{NaOH} C_6H_5NHR + NaX + H_2O$$

N-Alkylaniline

$$C_6H_5NHR + RX \longrightarrow [C_6H_5\overset{+}{N}HR_2]X^- \xrightarrow{NaOH} C_6H_5NR_2 + NaX + H_2O$$

N,N-Dialkylaniline

$$C_6H_5NR_2 + RX \longrightarrow [C_6H_5\overset{+}{N}R_3]X^-$$

Phenyltrialkylammonium halide

Simple aryl halides react with difficulty. Although diphenylamine is a minor coproduct of the commercial production of aniline from chlorobenzene (p. 426), it is made best by heating aniline with aniline hydrochloride.

$$C_6H_5NH_2 + [C_6H_5\overset{+}{N}H_3]Cl^- \xrightarrow{220°} (C_6H_5)_2NH + NH_4Cl$$

Diphenylamine

Reaction of the lithium salt of a diarylamine with an aryl iodide in the presence of catalytic amounts of cuprous iodide yields the *triarylamine.* The lithium salt is prepared from the amine and phenyllithium (p. 398).

$$Ar_2NH + C_6H_5Li \longrightarrow Ar_2N^{-+}Li + C_6H_6$$

$$Ar_2N^{-+}Li + IAr' \xrightarrow{CuI} Ar_2NAr' + LiI$$

3. ***Acylation.*** Acid anhydrides and acyl halides convert primary and secondary amines into the amides.

$$C_6H_5NH_2 + (CH_3CO)_2O \longrightarrow C_6H_5NHCOCH_3 + CH_3COOH$$

Acetanilide

$$2\ C_6H_5NHCH_3 + CH_3COCl \longrightarrow C_6H_5N(CH_3)COCH_3 + [C_6H_5\overset{+}{N}H_2CH_3]Cl^-$$

N-Methylaniline — *N*-Methylacetanilide — *N*-Methylaniline hydrochloride

Acylation can be brought about also by heating the amine salts of carboxylic acids (p. 166).

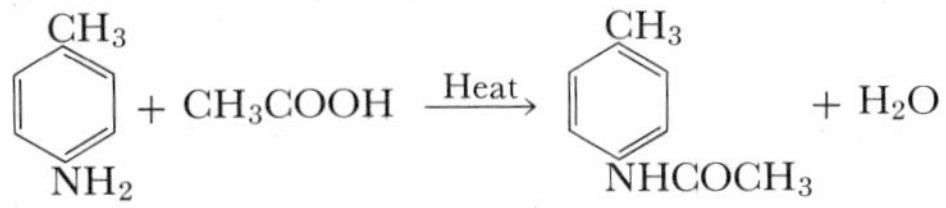

$$p\text{-}CH_3C_6H_4NH_2 + CH_3COOH \xrightarrow{Heat} p\text{-}CH_3C_6H_4NHCOCH_3 + H_2O$$

p-Toluidine — *p*-Acetotoluidide

Reaction of aniline with phosgene gives phenylcarbamyl chloride. When it is heated, hydrogen chloride is lost and **phenyl isocyanate** is produced.

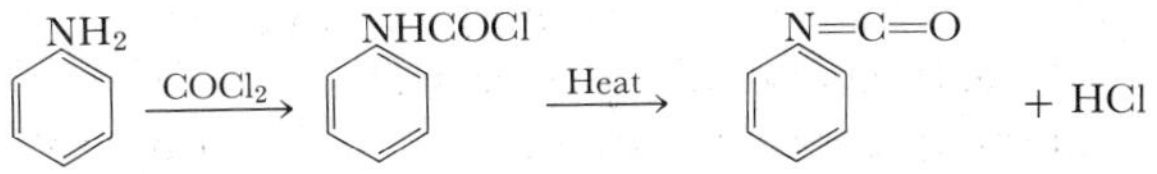

$$C_6H_5NH_2 \xrightarrow{COCl_2} C_6H_5NHCOCl \xrightarrow{Heat} C_6H_5N{=}C{=}O + HCl$$

Phenyl isocyanate

Phenyl isocyanate is useful for the identification of alkyl halides. The latter can be converted to Grignard reagents, which add to phenyl isocyanate. Hydrolysis of the addition product gives a solid anilide.

$$C_6H_5N{=}C{=}O + RMgX \longrightarrow C_6H_5N{=}\overset{OMgX}{\overset{|}{C}}R \xrightarrow{H_2O} \left[C_6H_5N{=}\overset{OH}{\overset{|}{C}}R\right] \longrightarrow C_6H_5NH\overset{O}{\overset{\|}{C}}R$$

When phenyl isocyanate is used for the preparation of derivatives of alcohols and amines (p. 287), all moisture must be excluded. Otherwise the phenylcarbamic acid that is formed loses carbon dioxide, and the resulting aniline reacts with more phenyl isocyanate to give insoluble **diphenylurea.**

$$C_6H_5N{=}C{=}O \xrightarrow{H_2O} [C_6H_5NHCOOH] \longrightarrow CO_2 + C_6H_5NH_2 \xrightarrow{C_6H_5N{=}C{=}O} C_6H_5NHCONHC_6H_5$$

Diphenylurea is one of the substances in coconut milk that stimulates the growth of plant cells. Various 1-aryl-1,3-dialkylureas and alkyl *N*-arylcarbamates, such as *i*-propyl *N*-phenylcarbamate (*IPC*), are made commercially from aryl isocyanates for use as selective herbicides. An 80:20 mixture of **2,4- and 2,6-tolylene diisocyanate** (so-called *toluene diisocyanate*) made from the mixed diaminotoluenes is of commercial importance for the manufacture of urethan plastics (p. 605). Production of isocyanates in 1964 was 138 million pounds, 82 per cent of which was toluene diisocyanate.

4. ***Reaction with Nitrous Acid.*** The behavior of aromatic amines toward nitrous acid, like that of the aliphatic amines, depends on whether the amine is primary, secondary, or tertiary. The reactions of primary and tertiary aromatic amines, however, differ from those of primary and tertiary aliphatic amines (p. 228).

(*a*) PRIMARY AMINES. At temperatures below 0° in strongly acid solution, nitrous acid reacts with the primary aromatic amine salts to give water-soluble compounds known as *diazonium salts.*

$$[C_6H_5\overset{+}{N}H_3]Cl^- + HONO\,(NaNO_2 + HCl) \longrightarrow [C_6H_5\overset{+}{N}_2]Cl^- + 2\,H_2O$$

Aniline hydrochloride → Benzenediazonium chloride

The properties and uses of these important compounds are described on pages 429–433.

(*b*) SECONDARY AMINES. Secondary aromatic amines behave like secondary aliphatic amines and yield *N*-nitroso derivatives.

$$C_6H_5NHCH_3 + HONO \longrightarrow C_6H_5N(NO)CH_3 + H_2O$$

N-methylaniline → *N*-Nitroso-*N*-methylaniline

(*c*) TERTIARY AMINES. Tertiary aromatic amines having an unsubstituted *para* position yield *p*-nitroso derivatives.

$$C_6H_5N(CH_3)_2 + HONO \longrightarrow p\text{-}ON{-}C_6H_4{-}N(CH_3)_2 + H_2O$$

N,N-Dimethylaniline → *p*-Nitroso-*N,N*-dimethylaniline

This reaction takes place because of the strong activating effect of the dimethylamino group. Although most of the dimethylaniline is present as the salt in the acid solution, and the dimethylammonium group is deactivating and *meta*-directing, sufficient free dimethylaniline is in equilibrium with the salt to react with nitrous acid, and the equilibrium shifts until nitrosation is complete. Nitrous acid does not bring about the nitrosation of benzene or even of toluene or mesitylene.

Activation by the dimethylamino group depends on the resonance effect (p. 381), which requires that the dimethylamino group must be able to take up a position coplanar with the benzene ring.

[Resonance structures of N,N-dimethylaniline: neutral $(H_3C)_2\ddot{N}$–benzene ↔ $(H_3C)_2\overset{+}{N}$= ring with negative charge at ortho ↔ negative charge at para ↔ negative charge at other ortho]

If groups larger than hydrogen occupy the *ortho* positions, coplanarity cannot be attained, and activation of the ring is not possible. Thus 2,6,*N,N*-tetramethylaniline does not undergo reactions that require strong activation of the nucleus such as nitrosation and coupling with diazonium salts (p. 431).

5. ***Other Reactions.*** Aromatic amines undergo most of the reactions described for aliphatic amines. Thus they give condensation products with aldehydes and ketones. Intermediate condensation products frequently are more stable than those of the aliphatic

amines. For example, the products of reaction of an aldehyde with one or two moles of aniline can be isolated.

$$C_6H_5NH_2 + OCHR \longrightarrow C_6H_5N{=}CHR + H_2O$$

$$2\ C_6H_5NH_2 + OCHR \longrightarrow (C_6H_5NH)_2CHR + H_2O$$

The products from one mole each of amine and aldehyde are known as *Schiff bases* or *anils.* These intermediates undergo further polymerization and condensation. The condensation products have been used as rubber accelerators and antioxidants (p. 585).

Unlike the aliphatic amines, aniline does not react with carbon disulfide at room temperature to give the dithiocarbamate (p. 292). When a solution of aniline and carbon disulfide in alcohol is refluxed, hydrogen sulfide is evolved with the formation of **thiocarbanilide.**

$$2\ C_6H_5NH_2 + CS_2 \longrightarrow \underset{\text{Thiocarbanilide } (\textit{diphenylthiourea})}{C_6H_5NHCSNHC_6H_5} + H_2S$$

Thiocarbanilide at one time was an important rubber accelerator. It now is used chiefly for the preparation of 2-mercaptobenzothiazole, which has supplanted it (p. 523).

When thiocarbanilide is boiled with strong hydrochloric acid, **phenyl isothiocyanate** (*phenyl mustard oil*), a very pungent compound, is produced.

$$C_6H_5NHCSNHC_6H_5 + HCl \longrightarrow C_6H_5N{=}C{=}S + C_6H_5NH_3^{+-}Cl$$

Phenyl isothiocyanate reacts readily with primary and secondary amines to give thioureas, which are useful for the identification of amines.

$$C_6H_5N{=}C{=}S + H_2NR \longrightarrow C_6H_5NHCSNHR$$

Reaction with ammonia gives **phenylthiourea,** $C_6H_5NHCSNH_2$, which is of interest in that it is extremely bitter to some persons and tasteless to others. The ability to taste the compound has been shown to be hereditary. ***p*-Ethoxyphenylurea** (*Dulcin*), *p*-$C_2H_5OC_6H_4NHCONH_2$, on the other hand, is about 100 times sweeter than sucrose. Its toxicity is too great for use in foods.

In the presence of ammonia, aniline reacts with carbon disulfide to give **ammonium phenyldithiocarbamate.**

$$C_6H_5NH_2 + CS_2 + NH_3 \longrightarrow C_6H_5NHCSS^{-+}NH_4$$

Removal of hydrogen sulfide from the salt by reaction with lead nitrate (p. 293) gives phenyl isothiocyanate.

$$C_6H_5NHCSS^{-+}NH_4 + Pb(NO_3)_2 \longrightarrow C_6H_5N{=}C{=}S + PbS + NH_4NO_3 + HNO_3$$

Technically Important Aromatic Amines and Their Derivatives

Aniline is by far the most important amine from the technical viewpoint. Almost 170 million pounds were produced in the United States in 1964, the selling price being about 14 cents per pound. Aniline was discovered in 1826 in the products of the destructive distillation of indigo (p. 570) and given the name *krystallin* because it readily formed crystalline salts. It was detected in coal tar in 1834 and called *kyanol,* because it gave a blue color with bleaching powder. It was rediscovered in the distillation products of indigo in 1841 and called *aniline* from *añil,* the Spanish word for indigo. In the same year it was produced by the reduction of nitrobenzene with ammonium sulfide and called *benzidam.* In 1843 Hofmann (p. 222) proved that all four substances are identical.

Both the reduction of nitrobenzene and the ammonolysis of chlorobenzene are used in the commercial production of aniline. In the reduction process scrap cast-iron turnings and water are placed in a cast-iron vessel fitted with a stirrer and a reflux condenser. A small amount of hydrochloric acid or ferric chloride is added, and the mixture is heated to remove oxides from the surface of the iron, the hydrochloric acid or ferric chloride being converted to ferrous chloride. Nitrobenzene then is added with vigorous stirring. The iron is converted to black iron oxide, Fe_3O_4, which is recovered and used as a pigment (p. 414). The aniline is distilled with steam, and the mixed vapors are condensed. The aniline layer of the distillate is separated from the water layer and purified by distillation at reduced pressure. Since aniline is soluble in water to the extent of about 3 per cent, it must be recovered from the aqueous layer of the distillate. In order to avoid extraction with a solvent and recovery of the solvent, the aniline-saturated aqueous layer is returned to the steam generator for processing a subsequent batch. In another procedure the aniline is extracted from the water with nitrobenzene, and the extract put through the reduction process. Operation of a continuous vapor-phase catalytic hydrogenation process using a fluidized catalyst bed began in 1956.

Since 1926 aniline has been prepared on a large scale by the reaction of chlorobenzene with ammonia. The chlorobenzene is heated in a pressure system with 28 per cent aqueous ammonia (mole ratio 1:6) in the presence of cuprous chloride (introduced as cuprous oxide) at 190–210°.

$$C_6H_5Cl + 2\,NH_3 \xrightarrow[190-210^\circ]{CuCl} C_6H_5NH_2 + NH_4Cl$$

A pressure of around 900 p.s.i. develops. The process is continuous, the reactants entering at one end of the system and the products leaving the other end. About 5 per cent of phenol and 1 to 2 per cent of diphenylamine are formed as coproducts.

$$C_6H_5Cl + H_2O + NH_3 \longrightarrow \underset{\text{Phenol}}{C_6H_5OH} + NH_4Cl$$

$$C_6H_5Cl + H_2NC_6H_5 + NH_3 \longrightarrow \underset{\text{Diphenylamine}}{C_6H_5NHC_6H_5} + NH_4Cl$$

These side reactions would take place to a greater extent were it not for the presence of the large excess of ammonia. At the end of the reaction the liquid is blown into a column. The free ammonia and aniline vaporize and are condensed. Caustic soda is added to the residue to liberate ammonia and aniline from their hydrochlorides, convert the phenol into its sodium salt, and precipitate the copper salts.

The first technical use for aniline was in 1856 for the production of mauve, the first commercial synthetic dye (p. 559). Aniline still is used almost exclusively as an intermediate in the production of other compounds. About 65 per cent of the total production is used in the manufacture of rubber accelerators and antioxidants (p. 584), 15 per cent for dyes and dye intermediates, 6 per cent for drug manufacture and 2 per cent for photographic developers (p. 450).

The **toluidines, xylidines, phenylenediamines,** and most other primary aromatic amines are prepared by similar procedures involving reduction of the nitro compounds. ***m*-Nitroaniline** is prepared commercially by the partial reduction of *m*-dinitrobenzene using sodium sulfide as the reducing agent (p. 414). ***o*- or *p*-Nitroaniline** may be prepared by the ammonolysis of *o*- or *p*-nitrochlorobenzene. This reaction takes place more readily than the ammonolysis of chlorobenzene because of the activating effect of nitro groups in the *ortho* or *para* position (p. 400).

Acetanilide was produced to the extent of about 13 million pounds in 1943, but output in 1964 was only about one third of this amount because sulfa drugs (p. 427) have been largely superseded by penicillin (p. 524) and other antibiotics. A small amount is used as a dye intermediate. Acetanilide was introduced as an antipyretic in 1886 under the name *antifebrine,* and at one time it was used widely for this purpose and as an analgesic. It is highly toxic, however, being similar to aniline in its action, and it has been displaced largely by the relatively safer salicylates (p. 473), especially aspirin, which was introduced in 1899. Because acetanilide is cheap, it still is used in some proprietary headache and pain-killing remedies.

Lidocaine (*Xylocaine*), a local anesthetic (footnote 4, p. 142) that now is used widely instead of Novocaine (p. 472), is the hydrochloride of 2,6-dimethyl-α-diethylaminoacetanilide and is prepared by the following series of reactions.

$$2,6\text{-}(CH_3)_2C_6H_3NH_2 \xrightarrow{ClCOCH_2Cl} 2,6\text{-}(CH_3)_2C_6H_3NHCOCH_2Cl \xrightarrow{NH(C_2H_5)_2} \left[2,6\text{-}(CH_3)_2C_6H_3NHCOCH_2\overset{+}{N}H(C_2H_5)_2\right] Cl^-$$

About 11 million pounds of ***N,N*-dimethylaniline** was produced in 1964. It is made from aniline and methyl alcohol in the presence of hydrochloric or sulfuric acid in a pressure reactor at 220°.

$$C_6H_5NH_2 + 2\,CH_3OH \xrightarrow[220^\circ]{H_2SO_4} C_6H_5N(CH_3)_2 + 2\,H_2O$$

It can be made also from aniline and methyl ether over activated alumina at 260°.

$$C_6H_5NH_2 + (CH_3)_2O \xrightarrow[260^\circ]{Al_2O_3} C_6H_5N(CH_3)_2 + H_2O$$

Dimethylaniline is used as a dye intermediate (pp. 569, 576) and in the manufacture of tetryl (p. 427). ***p*-Nitrosodimethylaniline** is made commercially by the nitrosation of dimethylaniline (p. 424). Reduction of *p*-nitrosodimethylaniline gives ***p*-dimethylaminoaniline,** which is used in the manufacture of Methylene Blue (p. 576). *p*-Dimethylaminoaniline is useful also for the separation of aldehydes and ketones, since only aldehydes react to give the anils (p. 425). The zinc chloride double salts of diazotized **p-dimethylaminoaniline** and of diazotized **p-diethylaminoaniline** are used in diazotype photographic processes. *p*-Diethylaminoaniline is used also as a developer in color photography. Nitrosation of *N*-methylaniline gives **N,4-dinitroso-N-methylaniline** (*N,4-DNMA*), which is used in the compounding of rubber (p. 585). **N,N′-Di-s-butyl-p-phenylenediamine** (p. 447), is one of the more widely used antioxidants for preventing the polymerization of the unsaturated components of cracked gasoline. Long-chain alkyl derivatives, such as **N,N′-di-2-octyl-p-phenylenediamine,** are used as antioxidants for synthetic rubber.

Diphenylamine is the principal stabilizer for smokeless powder (p. 343), being added in amounts of 1 to 8 per cent of the finished product. Its function is to combine with any oxides of nitrogen that are liberated, which otherwise would catalyze further decomposition. Large quantities of diphenylamine are used also in the manufacture of phenothiazine (p. 527), an intestinal disinfectant for animals. When a solution of diphenylamine in concentrated sulfuric acid reacts with nitrous or nitric acid or with their salts or esters, a deep blue color is formed. The reaction can be used as a test for diphenylamine or for nitrous or nitric acid or their salts or esters.

Sulfanilic acid is used principally as a dye intermediate. **Sulfanilamide** was shown in 1935 to be effective against streptococcal and other cocci infections such as pneumonia and gonorrhea, and other bacterial infections. These results led to the synthesis and testing of over five thousand sulfonamides by 1947. Sulfanilamide is synthesized from acetanilide by a series of reactions.

$C_6H_5NHCOCH_3 \xrightarrow{ClSO_3H} p\text{-}CH_3CONHC_6H_4SO_2Cl \xrightarrow{NH_3} p\text{-}CH_3CONHC_6H_4SO_2NH_2 \xrightarrow{H_2O,\ H^+} p\text{-}H_2NC_6H_4SO_2NH_2$

Sulfanilamide

Most of the derivatives of sulfanilamide that have proved to be superior to it differ from it in structure only in that one of the hydrogens of the sulfonamide group is replaced by a more complex organic group. These derivatives are made by substituting another amine for ammonia in the step in which the sulfonamide is formed. One exception is *Marfanil* in which the nuclear amino group is separated from the benzene ring by a methylene group. It is more effective than the sulfanilamides against anaerobic bacteria, such as the anthrax bacillus, and is used for topical treatment of open wounds and superficial infections.

Sulfadiazine — Sulfamerazine — Sulfamethazine — Marfanil

The disadvantage of the sulfanilamides has been their low solubility in water, which has led to deposition in the kidneys and renal damage. This difficulty has been alleviated by administering a mixture of three different compounds. The combined dose is as effective as an equal amount of any one drug, but the concentration of each is only one third. The production of **sulfa drugs,** which amounted to about 6 million pounds in 1945, dropped to 2.5 million pounds in 1948 because of the increased use of penicillin and other antibiotics. Since then production has increased again to over 5 million pounds in 1964.

p-Aminophenyl sulfone [*diaminodiphenyl sulfone*, *p*-$(H_2NC_6H_4)_2SO_2$] and its derivatives are effective in the treatment of Hansen's disease (*leprosy*). It is useful as a hardening agent for epoxy resins (p. 608).

Tetryl, 2,4,6-trinitrophenylmethylnitramide, is the standard booster charge for high explosive shells. The explosion of a shell is initiated by the primary explosive such as mercury fulminate which is sensitive to heat or shock. Detonation of the primary charge causes the explosion of the less sensitive booster charge, which in turn detonates the still less sensitive main charge such as TNT Tetryl usually is manufactured by the nitration of dimethylaniline in concentrated sulfuric acid. During the course of the reaction one of the methyl groups is removed by oxidation.

$C_6H_5N(CH_3)_2 \xrightarrow{HNO_3,\ H_2SO_4} 2,4\text{-}(NO_2)_2C_6H_3N(CH_3)_2 \longrightarrow 2,4\text{-}(NO_2)_2C_6H_3N(CH_3)COOH \longrightarrow 2,4\text{-}(NO_2)_2C_6H_3N(CH_3)H \longrightarrow 2,4,6\text{-}(NO_2)_3C_6H_2N(CH_3)NO_2$

Tetryl

PHOSPHINES

The most important arylphosphine is **triphenylphosphine,** which can be made by the reaction of phosphorus trichloride with phenylmagnesium chloride (cf. p. 110) and now is available commercially. Technically it usually is called *triphenylphosphorus.* It is a white solid that melts at 79° and is insoluble in water but soluble in aromatic hydrocarbons. Oxidation by air takes place much less readily than oxidation of the trialkylphosphines (p. 273). Although triphenylphosphine is not basic, it is a good nucleophile and displaces halogen readily from reactive halides to give quaternary salts. Those quaternary salts that have α hydrogen react with base to give **ylides** (p. 231). The general reactions may be represented as

$$(C_6H_5)_3P + XCH(Y)(Z) \longrightarrow \left[(C_6H_5)_3\overset{+}{P}{-}CH(Y)(Z)\right] X^- \xrightarrow{^-:B} (C_6H_5)_3P^{\pm}{=}C(Y)(Z) + HB + X^-$$

Y and *Z* may be any groups that are not affected by the base. The base used depends on the ease of removal of the proton, which in turn depends on the electronegativity of *Y* and *Z*. Among the bases commonly employed are butyllithium, phenyllithium, sodium or potassium alkoxides, sodium carbonate, or ammonia. Benzene, ethers, alcohols, or *N,N*-dimethylformamide are the usual solvents. Dihalides give **diylides.**

$$2\,(C_6H_5)_3P + XCH_2{-}A{-}CH_2X \longrightarrow [(C_6H_5)_3\overset{+}{P}{-}CH_2{-}A{-}CH_2{-}\overset{+}{P}(C_6H_5)_3]\,2\,X^- \xrightarrow{2\,^-:B}$$

$$(C_6H_5)_3P^{\pm}{=}CH{-}A{-}CH{=}^{\pm}P(C_6H_5)_3 + 2\,HB + 2\,X^-$$

A may be aliphatic or aromatic.

The importance of the triphenylphosphorus ylides[1] lies in the *Wittig reaction* in which all types of carbonyl groups undergo functional exchange of the oxygen for the ylidene group with the formation of a carbon-carbon double bond.

$$(C_6H_5)_3P^{\pm}{=}CYZ + O{=}CY'Z' \longrightarrow (C_6H_5)_3P^{\pm}{=}O + YZC{=}CY'Z'$$

Some specific applications of the Wittig reaction illustrate its general usefulness. When *Y* and *Z* are hydrogen or hydrocarbon groups, olefins result. If the ylide has the general formula $(C_6H_5)_3P^{\pm}{=}CH(CH_2)_nY$, where *Y* may be COR, C≡N, COOEt, or $CONH_2$, unsaturated ketones, nitriles, esters, or amides are formed in which the position of the double bond depends on the number of carbon atoms separating *Y* from the ylide carbon atom. If $n = 0$ and $Y = OCH_3$, an α,β-unsaturated ether is formed. Hydrolysis gives a saturated aldehyde.

$$(C_6H_5)_3P^{\pm}{=}CHOCH_3 + R_2CO \longrightarrow (C_6H_5)_3\overset{+}{P}{-}CHOCH_3 \;(\text{with } {}^-O{-}CR_2) \longrightarrow (C_6H_5)_3P + R_2C{=}CHOCH_3 \xrightarrow{H_2O,\,H^+}$$

$$CH_3OH + [R_2C{=}CHOH] \longrightarrow R_2CHCHO$$

Nucleophilic displacement of halogen from acyl halides in the presence of a tertiary amine gives a 2-oxoalkylphosphonium halide, which is converted to an acylated ylide. Thermal decomposition of the ylide gives a substituted acetylene.

---

[1] Phosphorus ylides frequently are represented as having a phosphorus-carbon double bond, for example $(C_6H_5)_3P{=}CH_2$ (cf. p. 271), and are named as derivatives of the hypothetical *phosphorane,* PH_5. Thus phosphorus ylides are called *alkylidenephosphoranes.*

$$(C_6H_5)_3P^{\pm}{=}CHY + RCOCl \longrightarrow \left[(C_6H_5)_3P^{\pm}{-}\underset{\displaystyle O{=}C{-}R}{CHY}\right]Cl^- \xrightarrow{(C_2H_5)_3N}$$

$$\left\{(C_6H_5)_3P^{\pm}{=}\underset{\displaystyle O{=}C{-}R}{CY} \longleftrightarrow (C_6H_5)_3P^{\pm}{-}\underset{\displaystyle {}^-O{-}C{-}R}{CY}\right\} \xrightarrow{\text{Heat}} (C_6H_5)_3PO + RC{\equiv}C{-}Y$$

Hydrolysis of an acylated ylide in basic aqueous alcohol yields a ketone.

$$(C_6H_5)_3P^{\pm}{=}\underset{\displaystyle O{=}C{-}R}{C}{-}Y \xrightarrow{H_2O,\ {}^-OH} (C_6H_5)_3PO + H_2\underset{\displaystyle O{=}C{-}R}{C}{-}Y$$

If R is aryl, better yields of ketone are obtained by reduction with zinc and acetic acid.

$$(C_6H_5)_3P^{\pm}{=}\underset{\displaystyle O{=}C{-}Ar}{C}{-}Y \xrightarrow{Zn,\ HOAc} (C_6H_5)_3P + H_2\underset{\displaystyle O{=}C{-}Ar}{C}{-}Y$$

Diylides react with carbonyl compounds to give diunsaturated compounds.

$$(C_6H_5)_3P^{\pm}{=}CH{-}A_n{-}CH{=}^{\pm}P(C_6H_5)_3 + 2\ R_2CO \longrightarrow 2\ (C_6H_5)_3PO + R_2C{=}CH{-}A_n{-}CH{=}CR_2$$

Because of its generality and the mild conditions under which it takes place, the Wittig reaction can be put to a great variety of uses.

DIAZONIUM SALTS

Diazonium salts and Grignard reagents constitute the two most versatile types of reagents known to organic chemists. The value of Grignard reagents is limited by the fact that for the most part they can be prepared only from the halogen derivatives of aliphatic or aromatic hydrocarbons; that is, few other functional groups may be present. This type of limitation does not exist for the formation of diazonium salts. On the other hand the latter can be prepared only if an amino group is attached to an aromatic nucleus.

Diazonium salts were prepared first by Peter Griess[2] in 1858 by the action of nitrous acid on the salt of an aromatic amine (p. 424). The importance of these compounds soon was recognized, and within the course of the next five years, their reactions had been widely investigated, and azo dyes derived from them were being manufactured commercially. Moreover the investigations concerning the structure of aromatic diazo compounds have played an important part in the development of the theoretical aspects of organic chemistry.

Structure, Physical Properties, and Nomenclature

The products of the reaction of aromatic primary amines with nitrous acid in strongly acid solutions show all of the properties of salts. They are solids, soluble in water, and insoluble in organic solvents. Electrical conductivity measurements show that they are completely ionized in dilute solution. The only reasonable structure is that in which one of the nitrogen atoms is quaternary as in ammonium salts and in which the nitrogen atoms are joined by a triple bond.

$$\left[Ar{:}\overset{+}{N}{\vdots}N{:}\right]X^- \quad \text{or} \quad [Ar{-}\overset{+}{N}{\equiv}N]X^-$$

---

[2] Johan Peter Griess (1829–1888) discovered the diazonium salts while working on a problem suggested by Kolbe. He continued his investigations first as an assistant to Hofmann and then throughout his life in whatever time he could spare while working as a chemist for an English brewery.

In naming these compounds, the name of the hydrocarbon from which they are derived is affixed to *diazonium,* and the name of the anion is added.

$$[C_6H_5\overset{+}{N}{\equiv}N]\,Cl^-$$

Benzenediazonium chloride

$$[O_2N{-}C_6H_4{-}\overset{+}{N}{\equiv}N]\,\overset{-}{O}SO_3H$$

p-Nitrobenzenediazonium acid sulfate

Preparation

In general aromatic diazonium compounds are not isolated but are prepared and used in aqueous solution. The reaction of the amine salt with nitrous acid is known as *diazotization* and must be carried out in strongly acid solution to prevent the diazonium salt from coupling with unreacted amine (p. 431). The nitrous acid is generated by the addition of a solution of sodium nitrite to the suspension of amine salt in excess mineral acid. Most diazonium salts are unstable at room temperature. Hence the reaction is carried out at 0°, and the solution is used immediately.

A detailed study of the kinetics of diazotization has shown that it involves nucleophilic attack by the amine on dinitrogen trioxide, which is in equilibrium with nitrous acid (p. 228).

$$HONO \rightleftharpoons H^+ + ONO^-$$

$$HONO + H^+ \rightleftharpoons H_2\overset{+}{O}{-}NO$$

$$H_2\overset{+}{O}{-}NO + {}^-ONO \longrightarrow H_2O + ON{-}O{-}NO$$

$$ArNH_2 + ON{-}O{-}NO \longrightarrow Ar\overset{+}{N}H_2(N{=}O) + {}^-ONO$$

The conjugate acid of the nitrosoamine, by a series of proton transfers and loss of water, gives the diazonium ion (cf. p. 228).

$$Ar\overset{+}{N}H_2(N{=}O) \longrightarrow Ar\ddot{N}{=}\ddot{N}{-}\overset{+}{O}H_2 \longrightarrow H_2O + Ar\overset{+}{N}{\equiv}N{:}$$

Solid diazonium salts may be obtained by dissolving the amine salt in an acid solution of alcohol and adding an alkyl nitrite.

$$[Ar\overset{+}{N}H_3]Cl^- + C_2H_5ONO \longrightarrow [Ar\overset{+}{N}_2]Cl^- + C_2H_5OH + H_2O$$

In this procedure no inorganic salts are introduced or formed, and the diazonium salt can be precipitated by the addition of ether. The solid salts are crystalline and colorless but darken in air. They explode when heated or when subjected to mechanical shock. The stability of diazonium salts varies greatly with the structure of the molecule. *p*-Nitrobenzenediazonium chloride is considerably more stable than benzenediazonium chloride. Double salts with zinc chloride and the salts of some acids such as 1,5-naphthalenedisulfonic acid (p. 490) are more stable than salts of the mineral acids. Their wet pastes may be kept for some time at room temperature without decomposition and are used in certain types of dyeing and in diazotype printing.

Reactions

The diversity of the reactions of aromatic diazo compounds arises in part from the fact that they can take place by either ionic or free radical mechanisms. Those reactions in which nitrogen is not evolved result from electrophilic substitution by the positive diazonium ion.

$$\{Ar\overset{+}{N}{\equiv}N \longleftrightarrow ArN{=}N^+\} + HAr \longrightarrow ArN{=}NAr + H^+$$

Those taking place in acid solution with the evolution of nitrogen appear to involve an S_N1 mechanism with formation of an aryl carbonium ion intermediate (cf. p. 432).

$$Ar{-}\overset{+}{N}{\equiv}N \longrightarrow Ar^+ + N_2$$

Reactions catalyzed by cuprous ions or those that take place in neutral or basic solutions appear to involve free radicals.

$$Ar{-}N{=}N{-}OH \longrightarrow Ar\cdot + N_2 + \cdot OH$$

1. ***The Coupling Reaction.*** The diazonium ion is a weak electrophile and substitution in an aromatic ring occurs only if the nucleus is strongly activated by a hydroxyl or amino group. The products are highly colored azo compounds, and the coupling reaction gives rise to the commercially important azo dyes (p. 563).

$$[C_6H_5\overset{+}{N}{\equiv}N]Cl^- + H{-}C_6H_4{-}OH + NaOH \longrightarrow C_6H_5N{=}N{-}C_6H_4{-}OH + NaCl + H_2O$$

p-Hydroxyazobenzene

$$[C_6H_5\overset{+}{N}{\equiv}N]Cl^- + H{-}C_6H_4{-}N(CH_3)_2 + NaOH \longrightarrow C_6H_5N{=}N{-}C_6H_4{-}N(CH_3)_2 + NaCl + H_2O$$

p-Dimethylaminoazobenzene

Coupling takes place only in weakly acid, neutral, or alkaline solutions. In strongly acid solutions the hydroxyl group is undissociated, and the amino group is present as its salt. Only the free amino group and the oxide ion are sufficiently activating to permit coupling (p. 379).

In the coupling reaction the diazonium salt is known as the *primary component,* and the compound with which it couples is called the *secondary component.* Coupling takes place in the unhindered *para* position of the secondary component. If the *para* position is occupied, coupling takes place in the *ortho* position. If the *para* position and both *ortho* positions are blocked, coupling usually does not take place, although occasionally the group in the *para* position is displaced.

Coupling takes place with primary and secondary *aliphatic* amines to give *diazoamino compounds.*

$$[ArN_2^+]Cl^- + 2\,HNHR \longrightarrow ArN{=}N{-}NHR + RNH_3Cl$$

Some of the primary and secondary aromatic amines also undergo this reaction. For example, aniline couples with benzenediazonium chloride buffered with sodium acetate to give diazoaminobenzene.

$$C_6H_5N_2Cl + H_2NC_6H_5 + NaOCOCH_3 \longrightarrow C_6H_5N{=}N{-}NHC_6H_5 + NaCl + CH_3COOH$$

Diazoaminobenzene

It is to prevent this reaction that diazotization is carried out in strongly acid solution, since the amine salt does not couple.

If the diazoamino compound is heated in the presence of an amine salt to catalyze the reaction, rearrangement to the aminoazo compound takes place.

$$C_6H_5N{=}N{-}NHC_6H_5 \xrightarrow{\text{Heat } (+\ C_6H_5NH_3Cl)} C_6H_5N{=}N{-}C_6H_4{-}NH_2$$

p-Aminoazobenzene

From many primary aromatic amines, for example the naphthylamines (p. 492), the diazoamino compounds cannot be isolated, the nuclear-substituted azo compounds being formed directly.

2. ***Reduction to Hydrazines.*** When a diazonium salt is reduced with zinc dust and acetic acid, with sulfur dioxide, with sodium hydrosulfite, or with stannous chloride, an arylhydrazine is formed.

$$[C_6H_5\overset{+}{N}{\equiv}N]^{-}OSO_3H + 2\,H_2SO_3 + 2\,H_2O \longrightarrow [C_6H_5NH\overset{+}{N}H_3]^{-}OSO_3H + 2\,H_2SO_4$$

Since the reaction is general, many of these valuable reagents are easily available. It was the discovery of phenylhydrazine by Emil Fischer in 1875 that led to his work on the structure of the sugars (p. 332).

3. ***Replacement by Hydroxyl or Alkoxyl.*** If an aqueous solution of a diazonium sulfate is heated, nitrogen is evolved and a phenol is formed. The reaction is first order, and an electron-withdrawing group such as the nitro group in the *para* position decreases the rate of the reaction. Hence an S_N1 mechanism is indicated (cf. p. 431).

$$ArN_2^+ \longrightarrow N_2 + Ar^+ \xrightarrow{H_2O} Ar\overset{+}{O}H_2 \longrightarrow ArOH + H^+$$

The great stability of the nitrogen molecule contributes to the driving force for its elimination and accounts for the instability of diazo compounds in general.

Reaction of benzenediazonium chloride with methanol in the presence of air, which inhibits a free radical reduction, gives anisole (*methyl phenyl ether*).

$$C_6H_5N_2^+Cl^- + CH_3OH \longrightarrow C_6H_5OCH_3 + N_2 + HCl$$

Other alcohols give ethers in varying yields, which depend on the structure of the alcohol and of the diazonium salt.

4. ***Replacement by Hydrogen.*** Although alcohols have been used to remove a diazonium group reductively, a more general and reliable reducing agent is hypophosphorus acid, catalyzed by traces of cuprous, ferrous, or manganous ions.

$$ArN_2^{+-}OSO_3H + HO\overset{O^-}{\overset{|+}{P}}H_2 \longrightarrow H_2SO_4 + Ar{-}N{=}N{-}O\overset{O^-}{\overset{|+}{P}}H_2 \longrightarrow Ar\cdot + N_2 + \cdot\overset{O^-}{\overset{|+}{P}}H(OH) \quad \textit{Initiation}$$

$$Ar\cdot + HPO_2H_2 \longrightarrow ArH + \cdot\overset{O^-}{\overset{|+}{P}}H(OH)$$

$$(HO)H\overset{O^-}{\overset{|+}{P}}\cdot + ArN_2^+ \longrightarrow Ar\cdot + N_2 + (HO)H\overset{O}{\overset{\|}{P}}{}^+ \xrightarrow{H_2O} (HO)H\overset{O^-}{\overset{|+}{P}}OH + H^+$$

Propagation

The over-all reaction is

$$ArN_2^{+-}OSO_3H + H_3PO_2 + H_2O \longrightarrow ArH + N_2 + H_2SO_4 + H_3PO_3$$

Formaldehyde in alkaline solution frequently is an equally satisfactory reducing agent.

$$ArN_2^{+-}OSO_3H + HCHO + 3\,NaOH \longrightarrow ArH + N_2 + Na_2SO_4 + NaOCHO + 2\,H_2O$$

5. ***Replacement by Halogen.*** The diazonium group can be replaced readily by any of the halogens, although different conditions may be required. For replacement by **chlorine** or **bromine** the aqueous solution of the corresponding salt is heated either with copper bronze (*Gattermann reaction*) or with cuprous chloride or cuprous bromide (*Sandmeyer reaction*). In general the cuprous halides give better yields.

$$[ArN_2^+]Cl^- \xrightarrow[\text{heat}]{\text{Cu or CuCl}} ArCl + N_2 \qquad [ArN_2^+]Br^- \xrightarrow[\text{heat}]{\text{Cu or CuBr}} ArBr + N_2$$

Cuprous chloride is believed to bring about a free radical decomposition involving one-electron transfers, with the oxidation state of copper going from Cu(I) to Cu(II) and back to Cu(I).

$$ArN_2{}^{+-}Cl + CuCl \longrightarrow ArN_2{}^{+-}[CuCl_2] \longrightarrow Ar\cdot + N_2 + CuCl_2$$
$$Ar\cdot + CuCl_2 \longrightarrow ArCl + CuCl$$

Although free radicals are indicated as intermediates, there is evidence that in reactions catalyzed by cuprous ion, the radicals never are actually free in the sense that ions are free in aqueous solution. The reactants, complexed with cuprous ion, merely undergo homolysis within the complex. In the Sandmeyer reaction, the rate of reaction decreases with the square of the chloride ion concentration indicating that the formation of the $[CuCl_3]^{2-}$ complex ion prevents the formation of the complex with the diazonium ion.

When a diazonium salt is treated with copper bronze, some biaryl is formed.

$$2\,ArN_2Cl + 2\,Cu \longrightarrow Ar_2 + N_2 + 2\,CuCl$$

The yield of biphenyl from benzenediazonium chloride is about 20 per cent, but the amount of coupling is greater for some diazonium salts.

To replace the diazonium group with **iodine,** it is necessary only to add potassium iodide to the aqueous solution of the sulfate and heat.

$$[ArN_2{}^{+}]^{-}OSO_3H + KI \longrightarrow ArI + N_2 + KHSO_4$$

Fluorine compounds are prepared by adding a solution of the diazotized aromatic amine to a solution of fluoroboric acid (*Balz-Schiemann reaction*). The fluoroborate precipitates and is washed and dried. When heated, it decomposes to yield the fluoro derivative and boron fluoride.

$$ArN_2{}^{+-}Cl + HBF_4 \longrightarrow HCl + ArN_2{}^{+-}BF_4 \xrightarrow{\text{Heat}} ArF + N_2 + BF_3$$

From these reactions it is evident that for replacement by groups other than halogen, it is preferable to use the diazonium sulfates rather than the diazonium halides, since the latter always yield some of the halogen substitution product. Although chlorine and bromine compounds can be prepared by direct substitution, the preparation from the diazonium salt frequently has an advantage because the halogen usually enters only at the position formerly occupied by the diazonium group. Hence single products are obtained and isomers can be made that may not be available by direct halogenation because of unfavorable directive influences.

6. ***Replacement by the Cyano (and Carboxyl) Group.*** If a neutral solution of a diazonium salt is added to a solution of cuprous cyanide-sodium cyanide complex, a precipitate is formed that decomposes to the nitrile when heated.

$$[ArN_2{}^{+}]^{-}OSO_3Na + NaCu(CN)_2 \longrightarrow ArCN + N_2 + CuCN + Na_2SO_4$$

Since the nitrile can be hydrolyzed to the acid, the reaction affords a method for replacing the diazonium group by a carboxyl group as well.

7. ***Other Replacement Reactions.*** Numerous other types of replacement reactions are known that yield, for example, thiocyanates, sulfides, disulfides, sulfinic acids, thiophenols, nitro compounds, mercury compounds, and arsonic and stibonic acids.

PROBLEMS

24–*1*. Give equations for the preparation of the following compounds: (*a*) *p*-toluidine from toluene, (*b*) *N*-nitrosodiphenylamine from aniline, (*c*) *p*-chlorophenyl isocyanate from *p*-chloroaniline, (*d*) *p*-nitroacetanilide from aniline.

24–*2*. Give equations for the following reactions and name the product formed: (*a*) ethylmagnesium bromide with phenyl isocyanate, (*b*) nitrous acid with 4-amino-*m*-xylene hydrochloride, (*c*) *p*-toluidine heated with *p*-toluidine hydrochloride, (*d*) phenyl isothiocyanate with dimethylamine, (*e*) 2,6-dinitrotoluene with ammonium sulfide.

24–*3*. Pair each of the compounds, (*a*) 1,2-diphenylhydrazine, (*b*) 1,3-diaminobenzene, (*c*) *p*-aminotoluene, (*d*) *sym*-diphenylthiourea, (*e*) phenyl isothiocyanate, (*f*) *p*-aminobenzenesulfonic acid, and (*g*) *m*-aminobenzenesulfonic acid, with the proper common name, (*1*) phenyl mustard oil, (*2*) *p*-toluidine, (*3*) sulfanilic acid, (*4*) thiocarbanilide, (*5*) hydrazobenzene, (*6*) metanilic acid, and (*7*) *m*-phenylenediamine.

24–*4*. Give a series of reactions for preparing the following compounds starting with triphenylphosphine: (*a*) 4-octene, (*b*) ethyl 4-methyl-3-pentenoate, (*c*) decanal, (*d*) phenyl-*n*-propylacetylene, (*e*) 3,6-nonadiene.

24–*5*. Describe a procedure for distinguishing between the members of the following pairs of compounds: (*a*) thiocarbanilide and diphenylurea, (*b*) aniline and *n*-hexylamine, (*c*) acetanilide and acetamide, (*d*) *o*-toluidine and *o*-nitrotoluene, (*e*) phenyl isocyanate and phenyl isocyanide, (*f*) *N*-methylaniline and *N,N*-dimethylaniline, (*g*) *o*-chloroaniline and aniline hydrochloride, (*h*) diphenylamine and acetanilide, (*i*) thiocarbanilide and *p*-toluenesulfonamide, (*j*) *m*-toluidine and *N*-methylaniline.

24–*6*. Which member of the following pairs of compounds is the stronger base: (*a*) diphenylamine or methylphenylamine, (*b*) *p*-nitroaniline or 2,4-dinitroaniline, (*c*) *p*-toluidine or *N,N*-dimethylaniline, (*d*) aniline or *p*-toluidine, (*e*) aniline or *n*-amylamine, (*f*) *m*-chloroaniline or *p*-chloroaniline, (*g*) aniline or phenylphosphine?

24–*7*. Five compounds, *A*, *B*, *C*, *D*, and *E*, have the molecular formula C_7H_9N. All are soluble in dilute hydrochloric acid. All react with benzenesulfonyl chloride in the presence of aqueous sodium hydroxide to give soluble products except *A*, which gives a product that is not soluble in dilute alkali. When subjected to vigorous oxidation, *B* yields an acid having a neutralization equivalent of 122, but the others give only products of low molecular weight. All of the compounds are liquids at room temperature except *E*, which is a solid. When *C* and *D* are diazotized, the diazonium groups replaced by chlorine, and the products oxidized with sodium dichromate and sulfuric acid, they yield acids having neutralization equivalents of 157. The acid from *C* melts at 158° and that from *D* at 142°. Give possible structural formulas for the five compounds and equations for the reactions that they undergo.

24–*8*. Give reactions for the synthesis of the following compounds from the proper amine by way of a diazonium salt: (*a*) *p*-hydroxytoluene, (*b*) *o*-chlorobromobenzene, (*c*) phenyl cyanide, (*d*) *p*-hydroxy-*p*′-nitroazobenzene, (*e*) *m*-bromophenylhydrazine.

24–*9*. Give a series of reactions for the preparation of the following compounds starting with an aromatic hydrocarbon: (*a*) 2-cyano-4-chlorotoluene, (*b*) *m*-dibromobenzene, (*c*) *p*-bromofluorobenzene, (*d*) 3,3′-dinitrobiphenyl, (*e*) *m*-iodonitrobenzene, (*f*) *m*-chlorophenol, (*g*) 2,5-dimethylphenylhydrazine, (*h*) 1,3,5,-trichlorobenzene, (*i*) *m*-toluidine.

24–*10*. Assuming that the properties of the three phthalic acids are known, devise a procedure for determining the orientation of (*a*) the three nitroanilines, and of (*b*) the three nitrotoluenes.

24–*11*. Assuming that the properties of the three nitroanilines are known, and that *Y* and *Z* are two groups that can be introduced by way of the diazonium salts, list the disubstituted benzenes in terms of *Y* and *Z* whose orientation readily can be determined.

24–*12*. Aniline was dissolved in one equivalent of aqueous hydrochloric acid and the solution diazotized at 0° by the addition of a solution of sodium nitrite. Instead of a colorless solution of the diazonium salt, a yellow precipitate was formed. Explain.

24–*13*. Devise a probable mechanism for the reduction of a diazonium group by formaldehyde in alkaline solution.

CHAPTER TWENTY-FIVE

PHENOLS, AMINOPHENOLS, AND QUINONES

PHENOLS

The phenols are compounds that have a hydroxyl group attached directly to an aromatic nucleus. In general their methods of preparation and reactions differ from those of the alcohols.

Occurrence and Nomenclature

The name *phenol* for hydroxybenzene is derived from *phene,* an old name for benzene (p. 373). The hydroxy derivatives of toluene have the common name *cresols,* and those of xylene are called *xylenols.* Phenols in general are named as derivatives of phenol.

OH | CH_3 OH | OH NH_2

Phenol | *o*-Cresol | *p*-Aminophenol[1]

Phenol, cresols, and xylenols, along with other phenolic compounds, occur in coal tar, wood tar, and petroleum distillates and are known as *tar acids.* The mixture of phenols from the cresol fraction is known technically as *cresylic acid.* Derivatives of phenols frequently occur as plant products (p. 446).

Preparation

The general reactions leading to the production of phenols have been discussed previously, namely the fusion of sodium arenesulfonates with sodium hydroxide (p. 405), the displacement of the diazonium group by hydroxyl (p. 432), and the hydrolysis of aryl halides (p. 400). The ease with which the last reaction takes place depends on the reactivity of the halogen, that is, the electrophilicity of the carbon atom to which the halogen is attached. Chlorobenzene requires heating with aqueous sodium hydroxide at 350°, and the uncatalyzed reaction takes place largely by way of dehydrobenzene (p. 399). The

[1] If a compound has more than one kind of function, only one kind should be represented by the suffix. This function is called the *principal function.* CHEMICAL ABSTRACTS determines the principal function from the following order of precedence based on general usage: "onium" compounds, peroxide, hydroperoxide, acid (*carboxylic, arsonic, arsonous, sulfonic, sulfinic, sulfenic, stibonic, stibonous, phosphonic, phosphonous, phosphinic, phosphinous, boronic, borinic, others*), acid halide, amide, imide, amidine, aldehyde, nitrile, isocyanide, ketone, thione, alcohol, phenol, thiol, amine, imine, organometallic compounds, ether, sulfide, sulfoxide, sulfone. Thus *p*-aminophenol is preferred to *p*-hydroxyaniline. The halo, nitro, nitroso, azo, and azoxy functions are not expressed as name endings and do not appear in the list. The introduction to the Subject Index of Volume 56 of CHEMICAL ABSTRACTS deals with nomenclature and carries lists of names for organic groups and suffixes.

copper-catalyzed reaction (p. 400) may involve the homolytic decomposition of a complex. Activated halogen undergoes an $S_{N_Ar}2$ displacement (p. 400). Thus, 2,4-dinitrochlorobenzene is hydrolyzed easily by dilute sodium carbonate solution. The halogen of aryl halides that can form Grignard reagents can be replaced easily by hydroxyl by oxidation of the Grignard reagent with bromomagnesium *t*-butyl hydroperoxide (cf. p. 127).

Physical Properties

The pure phenols are colorless solids or liquids, although as usually encountered they are colored red by oxidation products. Like the aromatic amines (p. 419), they boil higher than the normal aliphatic analogs of the same molecular weight. For example, phenol boils at 181° whereas 1-hexanol boils at 157°. Phenol is soluble in water to the extent of 9 g. per 100 g. of water at 25° and becomes miscible at 65°. Water is soluble in phenol to the extent of 29 g. per 100 g. Since the melting point of phenol is only 42°, a small amount of water lowers the melting point below room temperature. This liquid form containing about 5 per cent water is called **carbolic acid.** Of the monosubstituted phenols, the *para* isomer has the highest melting point.

Physiological Action

All phenols having a sufficiently high vapor pressure have a characteristic odor. Phenol is highly toxic, killing all types of cells. It precipitates proteins, and when applied to the skin, produces a white spot that soon turns red; later the dead skin sloughs. If allowed to remain in contact with the skin, it penetrates to the deeper tissues and severe burns result. It also is absorbed into the blood stream and acts as a systemic poison. Phenol is eliminated in the urine as sodium phenyl sulfate, $C_6H_5OSO_3Na$. Frequently other toxic aromatic compounds, for example benzene, bromobenzene, and naphthalene, are detoxified in the body by oxidation to the phenol and elimination as the ester of sulfuric acid.

The phenols in general are toxic to microorganisms. Although many substances had been discovered empirically to have a preservative and healing action long before the nature of bacterial infection was known, phenol itself was the first compound to be used widely for the avowed purpose of antisepsis. It was introduced by Lister in 1867. Much more effective and less toxic compounds have been developed since then, but antiseptic activity still is reported in terms of the *phenol coefficient,* a number that compares the effectiveness of a preparation with that of a 5 per cent solution of phenol against *Staphylococcus aureus.*

Reactions of the Hydroxyl Group

1. ***Acidity.*** Unlike the alkoxide ion, the phenoxide ion is stabilized by resonance (p. 384). Hence phenols are considerably more acidic than alcohols, although less so than carboxylic acids or even than carbonic acid. The negative charge on a carboxylate ion is more strongly stabilized than that on a phenoxide ion because the resonance structures of the carboxylate ion are identical and because the second oxygen atom increases the inductive effect. The acidity constant (pK_a) for acetic acid is 4.8; for carbonic acid (plus CO_2), 6.5; for hydrocyanic acid, 9.1; for phenol, 9.9; and for water, 15.7. Hence phenols react with sodium hydroxide solutions to form water-soluble salts but not with aqueous sodium bicarbonate. Moreover water-insoluble phenols are precipitated from their salts by carbonic acid.

$$ArOH + NaOH \longrightarrow ArO^{-+}Na + H_2O$$
$$ArO^{-+}Na + CO_2 + H_2O \longrightarrow ArOH + NaHCO_3$$

These reactions are used to distinguish phenols and to separate them from alcohols or carboxylic acids.

These statements apply to phenols that do not contain strongly electron-attracting groups in the nucleus. Thus the acidity constants of some of the nitrophenols are: *o*-nitrophenol, 7.2; *m*-nitrophenol, 8.3; *p*-nitrophenol, 7.2; 2,4-dinitrophenol, 4.1; and 2,4,6-trinitrophenol (picric acid), 0.4. In the mononitrophenols the effect of the nitro group is greater in the *ortho* or *para* position than in the *meta* position because the resonance effect adds to the inductive effect. Two nitro groups in the *ortho* and *para* positions give an acid approximately as strong as a carboxylic acid. The strength of picric acid with three nitro groups in the *ortho* and *para* positions approaches that of the mineral acids.

2. ***Colored Complexes with Ferric Chloride.*** Enols in general give colored water-soluble complexes with ferric chloride. The exact nature of these colored compounds is uncertain, although it seems likely that a coordination compound is formed in which the iron is hexacovalent. The reaction with phenols probably is complicated by the formation of colored oxidation products (p. 438). Whereas with simple enols the color produced is a burgundy red, the phenols give colors that are less pure and usually purplish or greenish. This reaction is a convenient test for certain phenols and other enols, but phenols with electronegative substituents may give no visible color because the complex absorbs in the ultraviolet region of the spectrum.

3. ***Ester Formation.*** Phenols do not esterify directly with carboxylic acids because the position of equilibrium is unfavorable. For example, the equilibrium constant for the reaction of phenol with acetic acid is 0.009 compared to 4.0 for ethanol. Esters can be prepared by reaction with anhydrides or acid chlorides.

$$\underset{\text{Phenol}}{C_6H_5OH} + (CH_3CO)_2O \longrightarrow \underset{\text{Phenyl acetate}}{C_6H_5OCOCH_3} + HOCOCH_3$$

$$C_6H_5OH + CH_3COCl \longrightarrow C_6H_5OCOCH_3 + HCl$$

4. ***Ether Formation.*** Phenols form ethers very easily by the Williamson synthesis (p. 139). Since sodium aryl oxides are hydrolyzed only partially in aqueous solution, the reaction can be carried out in this medium.

$$ArONa + XR \longrightarrow ArOR + NaX$$

The methyl ethers are most important because they can be prepared readily by agitating an alkaline solution of the phenol with methyl sulfate.

$$\underset{\substack{\text{Sodium}\\\text{phenoxide}}}{C_6H_5O^{-+}Na} + (CH_3)_2SO_4 \longrightarrow \underset{\substack{\text{Anisole}\\(\textit{methyl phenyl ether})}}{C_6H_5OCH_3} + CH_3SO_4{}^{-+}Na$$

Ethyl phenyl ether is known as *phenetole. Phenyl ether* is a coproduct of one of the commercial methods for the synthesis of phenol (p. 442).

5. ***Replacement by Halogen.*** The hydroxyl group of phenols, unlike that of alcohols, is difficult to replace by halogen for the same reasons that it is difficult to replace aromatic halogen by other groups (p. 399). Halogen acids are without action, and phosphorus trihalides yield only phosphorous esters. If phosphorus pentachloride or pentabromide is used, some replacement occurs, but the reaction is not used for preparative purposes.

$$ArOH + PX_5 \longrightarrow ArX + POX_3 + HX$$

6. ***Replacement by Hydrogen.*** When phenols are heated with zinc dust, the oxygen is removed from the molecule.

$$ArOH + Zn \xrightarrow{\text{Heat}} ArH + ZnO$$

The reaction, although of no value for synthesis, has been useful in arriving at the fundamental structure of oxygenated aromatic compounds. The method was used first by Baeyer in his elucidation of the structure of indigo.

7. ***Replacement by Amino Groups.*** Ordinarily the replacement of a hydroxyl group by an amino group is brought about by heating the phenol with the amine and zinc chloride or calcium chloride.

$$4\ ArOH + Zn(NH_3)_4Cl_2 \longrightarrow 4\ ArNH_2 + Zn(H_2O)_4Cl_2$$

In the naphthalene series (p. 492), ammonium sulfite is the preferred reagent for ammonolysis.

Reactions of the Nucleus

1. ***Hydrogen Exchange.*** As with the aromatic amines (p. 419), the hydrogen in the *ortho* and *para* positions of phenols is exchanged readily for deuterium or tritium. Whereas exchange with amines is catalyzed by acids, exchange with phenols is catalyzed by bases. Conversion to the phenoxide ion so greatly increases the electron density on the nucleus that the water molecule can act as an electrophilic reagent.

2. ***Oxidation.*** Like the amino group in aromatic amines, the hydroxyl group can supply electrons to the nucleus, especially in alkaline solution, and permits ready oxidation. Complex mixtures of oxidation products are formed by either air or other oxidizing agents. One of the oxidation products of phenol by air is *quinone* (p. 450), which forms a brilliant red addition product with phenol known as *phenoquinone*.

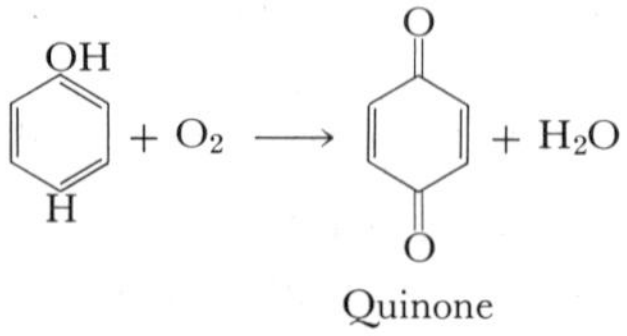

$$C_6H_5OH\ (p\text{-}H) + O_2 \longrightarrow \text{Quinone} + H_2O$$

Quinone

$$C_6H_4O_2 + 2\ C_6H_5OH \longrightarrow C_6H_4O_2 \cdot 2\ C_6H_5OH$$

Phenoquinone

The X-ray diffraction pattern of crystalline phenoquinone indicates that the quinone molecule is sandwiched between the two phenol molecules with the planes of the benzene rings parallel. The quinone acts as the electron-acceptor (π *acid,* p. 451) and the phenol as the electron-donor to form a 1:2 charge-transfer complex (cf. p. 387).

3. ***Sulfonation.*** Because of the strong activating effect of the hydroxyl group, sulfonation takes place very readily. At room temperature concentrated sulfuric acid yields chiefly the *ortho* isomer, whereas at 100° the more stable *para* isomer predominates.

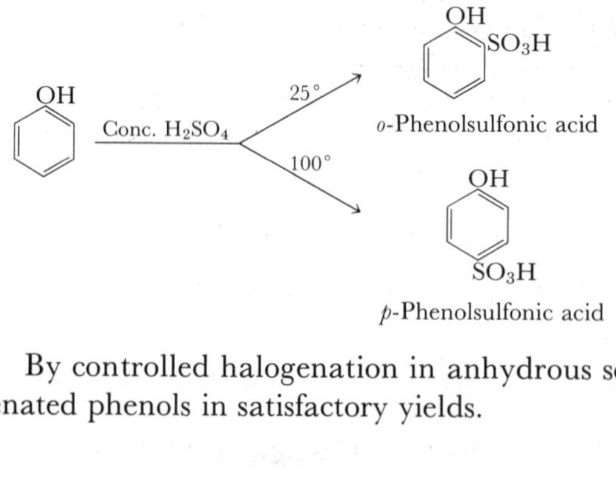

$$C_6H_5OH \xrightarrow{\text{Conc. } H_2SO_4} \begin{cases} 25°: & o\text{-}HOC_6H_4SO_3H \\ 100°: & p\text{-}HOC_6H_4SO_3H \end{cases}$$

o-Phenolsulfonic acid

p-Phenolsulfonic acid

4. ***Halogenation.*** By controlled halogenation in anhydrous solvents, it is possible to obtain the monohalogenated phenols in satisfactory yields.

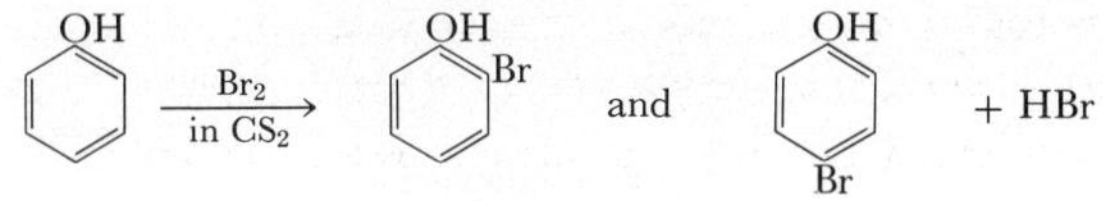

OH $\xrightarrow[\text{in } CS_2]{Br_2}$ OH, Br and OH, Br + HBr

o-Bromophenol *p*-Bromophenol

When bromine water is added to an aqueous solution of phenol, a quinonoid tetrabromo derivative precipitates. Reduction with sodium bisulfite produces 2,4,6-tribromophenol.

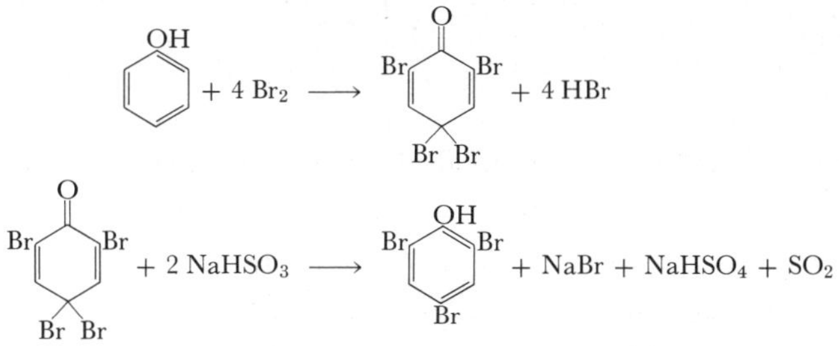

OH + 4 Br_2 ⟶ O, Br, Br, Br Br + 4 HBr

O, Br, Br, Br Br + 2 $NaHSO_3$ ⟶ OH, Br, Br, Br + NaBr + $NaHSO_4$ + SO_2

5. ***Nitrosation.*** The activating effect of the hydroxyl group is so great that phenol reacts with nitrous acid to give **p-nitrosophenol.** This compound is tautomeric with the monoxime of quinone (p. 450).

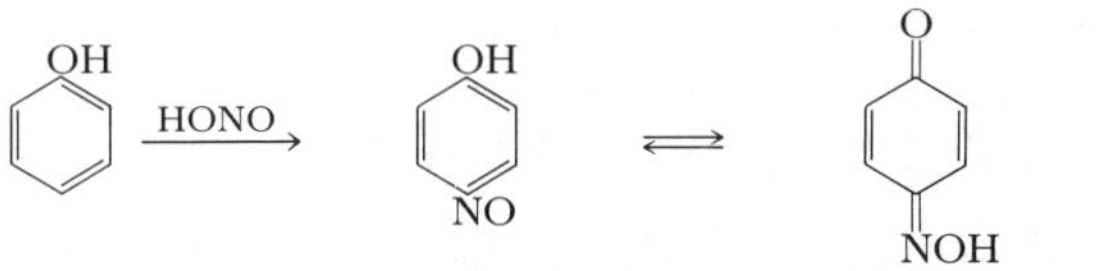

OH $\xrightarrow{HONO}$ OH, NO ⇌ O, NOH

p-Nitrosophenol Quinone monoxime

If the reaction is carried out in the presence of concentrated sulfuric acid, further condensation with phenol takes place to give a dark blue solution of indophenol acid sulfate. When the acid solution is diluted with water, the red **indophenol** (also called *phenolindophenol*) is liberated. When excess sodium hydroxide is added, the deep blue sodium salt is formed.

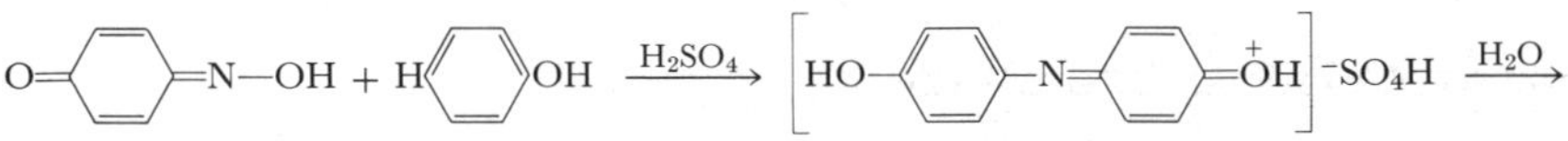

O═⟨ ⟩═N—OH + H⟨ ⟩OH $\xrightarrow{H_2SO_4}$ [HO—⟨ ⟩—N═⟨ ⟩═$\overset{+}{O}H$] $^{-}SO_4H$ $\xrightarrow{H_2O}$

Acid sulfate of indophenol
(deep blue)

O═⟨ ⟩═N—⟨ ⟩—OH $\xrightarrow{NaOH}$ [O═⟨ ⟩═N—⟨ ⟩—$\bar{O}$] Na^+

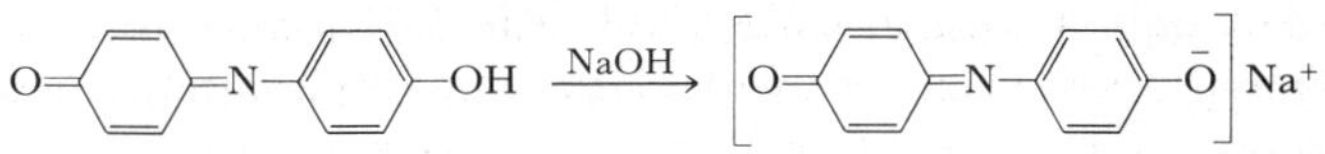

Indophenol
(red)

Sodium salt of indophenol
(deep blue)

The reaction is used as a test for phenols and for nitrites and is known as the **Liebermann nitroso reaction.**[2] Even some nitrates and aliphatic nitro compounds give the test, since reduction to nitrite takes place in the concentrated sulfuric acid solution. The deep blue color in either strongly basic or strongly acidic solutions is due to the resonance of equivalent structures and their high dipole moment (p. 558).

---

[2] Carl Theodor Liebermann (1842–1914), professor at the Berlin Technische Hochschule. His work dealt chiefly with natural dyes and with the chemistry of anthracene (p. 496).

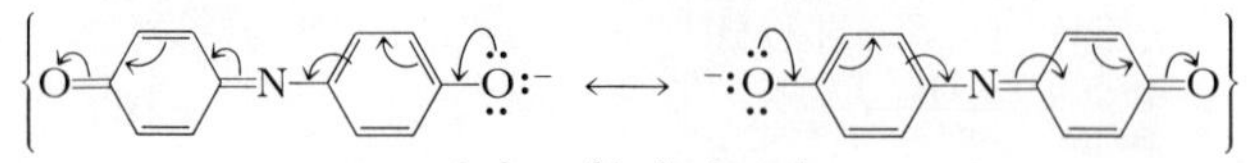

Anion of indophenol

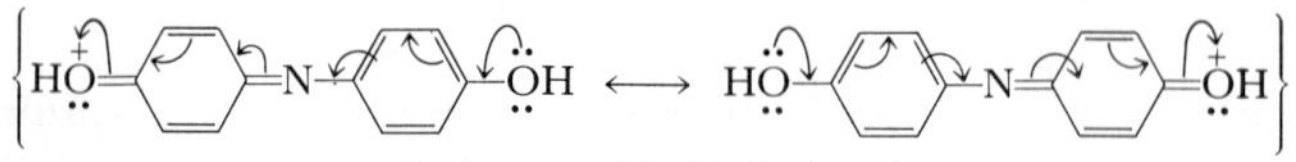

Conjugate acid of indophenol

6. ***Nitration.*** Phenol nitrates so rapidly that if mononitration is desired, dilute nitric acid at room temperature is used. Under these conditions the concentration of nitronium ion, $^{+}NO_2$, is low, and most of the reaction results from nitrosation by nitrosonium ion, ^{+}NO, followed by oxidation to the nitro compound. Some nitrous acid always is present because of oxidation of the organic compound, and this nitrous acid catalyzes the reaction.

$$HONO + 2\,HNO_3 \longrightarrow {}^{+}NO + H_3O^{+} + 2\,NO_3^{-}$$

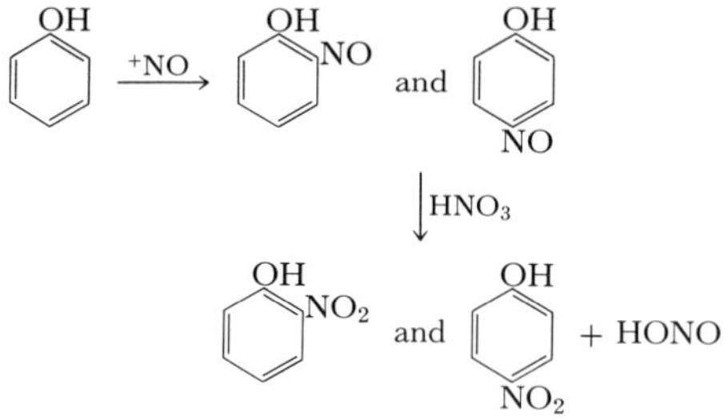

If urea is added to the nitric acid to remove the nitrous acid (p. 284), no reaction takes place.

The isomers can be separated by steam distillation. The *ortho* isomer is much more volatile because it can undergo intramolecular hydrogen bonding, whereas the *para* isomer is intermolecularly hydrogen bonded.

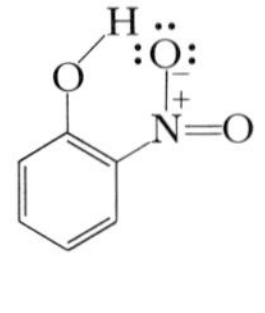

Nonassociated *o*-nitrophenol

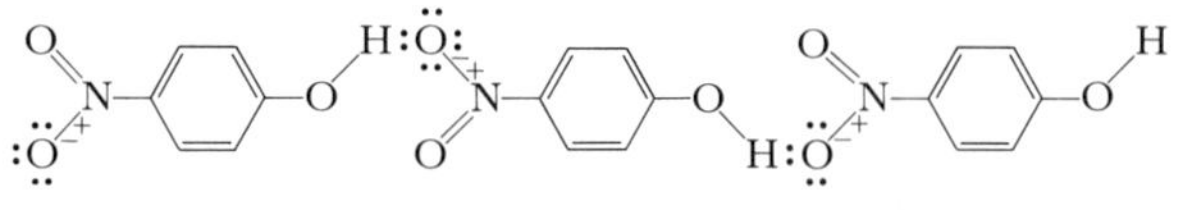

Associated *p*-nitrophenol

Most *ortho, meta,* and *para* isomers boil within 10° of each other, but *o*-nitrophenol boils at 214° and *p*-nitrophenol at 245°. The *ortho* isomer also is less soluble in water than the *meta* or *para* compound. Thus the *ortho* isomer not only has less tendency to associate with itself but also with other hydroxylic compounds.

Nitration of phenol with concentrated nitric acid gives 2,4,6-trinitrophenol (*picric acid*), but the amount of oxidation is excessive, and indirect methods of preparation are used. The nitrophenols are colorless or pale yellow. Their salts are deep yellow (p. 556).

7. ***Ring Alkylation and Acylation.*** Because of the activating effect of the hydroxyl group, phenols alkylate and acylate in the aromatic ring much more readily than hydrocarbons. Usually warming the phenol with an olefin or an alcohol in the presence of sulfuric acid is sufficient for alkylation.

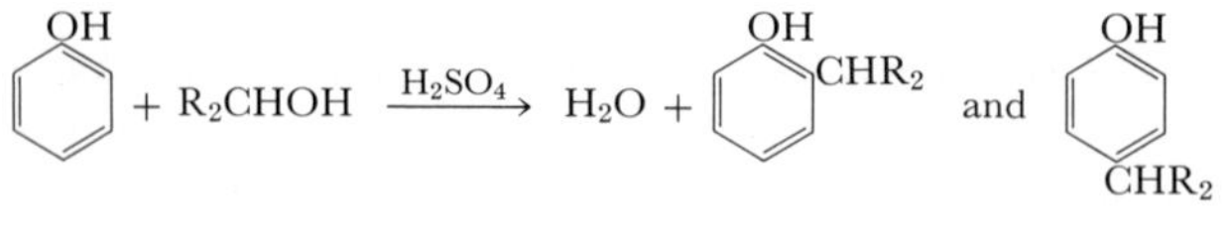

$$C_6H_5OH + R_2CHOH \xrightarrow{H_2SO_4} H_2O + o\text{-}HOC_6H_4CHR_2 \text{ and } p\text{-}HOC_6H_4CHR_2$$

Ring acylation takes place when phenols are heated with a carboxylic acid and zinc chloride (p. 464).

8. ***Mercuration.*** Readily substituted aromatic compounds undergo direct mercuration when heated with mercuric acetate in alcohol.

$$C_6H_5OH + Hg(OAc)_2 \longrightarrow o\text{- and } p\text{-}HOC_6H_5HgOAc + HOAc$$

9. ***Condensation with Aldehydes and Ketones*** (*Lederer-Manasse Reaction*). Electrophilic substitution of phenols by aldehydes is catalyzed by either alkali or acid. Under alkaline conditions the phenoxide ion is sufficiently activated to react with the weakly electrophilic free aldehyde.

$$^{-}O\text{—}C_6H_5 + RCH\text{=}O \rightleftharpoons O\text{=}C_6H_5(H)(CHO^{-}R) \rightleftharpoons {}^{-}O\text{—}C_6H_4\text{—}CHOH\text{—}R$$

In acid solution, protonation of the aldehyde makes it a stronger electrophile that can bring about substitution of the undissociated phenol.

$$RCH\text{=}O \overset{H^+}{\rightleftharpoons} \left\{RCH\text{=}\overset{+}{O}H \longleftrightarrow R\underset{+}{C}H\text{—}OH\right\} \xrightarrow[H^+]{C_6H_5OH} RCH(OH)\text{—}C_6H_4\text{—}OH$$

These reactions are closely related to aldol additions (p. 201).

Formaldehyde reacts with phenol in the presence of dilute alkali to give *o*- and *p*-hydroxybenzyl alcohols.

$$C_6H_5OH + HCHO \xrightarrow{\text{Dil. NaOH}} p\text{-}HOC_6H_4CH_2OH \text{ and } o\text{-}HOC_6H_4CH_2OH$$

p-Hydroxybenzyl alcohol — *o*-Hydroxybenzyl alcohol (*saligenin*)

Unless the reaction is controlled carefully, these products undergo further condensation to yield phenol-formaldehyde resins (p. 442).

Phenol and Some Important Derivatives

Phenol in 1964 held third place among the synthetic aromatic chemicals,[3] the volume of production being exceeded only by that of styrene (p. 483) and ethylbenzene (p. 391). The quantity produced in the United States by synthesis was 8 million pounds in 1928, 274 million pounds in 1948, and 1063 million pounds in 1964. In the same years the amount of natural phenol obtained from coal tar was 2, 23, and 35 million pounds. About 15 million pounds was obtained in 1964 from the refining of petroleum products.

Production of Phenol. Numerous methods have been developed for the synthesis of phenol. At least five different processes are operated in the United States.

(*a*) SULFONATION AND CAUSTIC FUSION. This process is the oldest. It was discovered independently in 1867 by Kekulé, Wurtz, and Dusart. The commercial application originally involved sulfonation of benzene with fuming sulfuric acid, conversion to the sodium salt by liming out (p. 404), and fusion with sodium hydroxide. In 1918 a continuous process was developed in which a large excess of benzene runs counter-current to the sulfonating acid, which starts out at a concentration of 98 per cent and ends up as 77 per cent. The sulfonic acid remains dissolved in the benzene at a concentration of 2 per cent. It is washed out with water and neutralized with sodium sulfite recovered from the caustic fusion. The solution of sodium salt is concentrated and run into fused sodium hydroxide at 320–350°. The melt is treated with a minimum of water to dissolve the sodium phenoxide, leaving the sodium sulfite undissolved. The phenol is liberated or *sprung* by passing carbon dioxide into the solution. The sodium bicarbonate is reconverted to caustic soda by the addition of lime, and the calcium carbonate that precipitates is converted to carbon dioxide and lime in the lime kilns. Alternatively the phenol may be liberated with sulfur dioxide formed when the benzenesulfonic acid is converted to the sodium salt, or the benzenesulfonic acid itself may be used.

[3] Excluding benzene, toluene, and the xylenes derived from petroleum.

With the exception of losses, only benzene, sodium hydroxide, and sulfuric acid are consumed in the process, the products being phenol and sodium bisulfite (or sulfur dioxide and sodium sulfite). The over-all process therefore is the oxidation of benzene by sulfuric acid. Since sulfuric acid is made by the oxidation of sulfur dioxide with air, the latter is the ultimate oxidizing agent.

(*b*) HYDROLYSIS OF CHLOROBENZENE (*Dow Process*). Since 1928 phenol has been manufactured in large quantities by the hydrolysis of chlorobenzene, which is emulsified with 10 per cent sodium hydroxide solution and then preheated in a heat exchanger and passed through the reactor where it is kept at 370° for 20 minutes. The product is bled out of the system at the rate at which the original emulsion enters, making the process continuous. Almost 100 per cent reaction occurs. Besides phenol, about 20 per cent of phenyl ether and smaller amounts of *o*- and *p*-phenylphenol are formed.

$$C_6H_5Cl + 2\,NaOH \longrightarrow C_6H_5ONa + NaCl + H_2O$$

$$C_6H_5Cl + NaOC_6H_5 \longrightarrow \underset{\text{Phenyl ether}}{C_6H_5OC_6H_5} + NaCl$$

$C_6H_5Cl + C_6H_5ONa + NaOH \longrightarrow$ [o-phenylphenoxide: ONa on ring bearing phenyl] and [p-phenylphenoxide]–ONa $+ NaCl + H_2O$

When cool, the mixture separates into two layers, and the upper layer containing the phenyl ether and unreacted chlorobenzene is separated and recycled. The aqueous layer is acidified with hydrochloric acid to liberate the phenol, which is separated from the phenylphenols and purified by distillation.

Some phenyl ether is used in the perfume industry, and a large quantity is used to form with biphenyl (p. 485) the eutectic mixture, which serves as a heat transfer medium in industrial operations. This mixture of 74 parts of phenyl ether and 26 parts of biphenyl is stable up to 400° at 135 p.s.i. If the amount of phenyl ether produced during the manufacture of phenol is greater than the demand, it can be recycled, since at 370° in the presence of sodium hydroxide it is in equilibrium with sodium phenoxide.

$$C_6H_5OC_6H_5 + 2\,NaOH \rightleftharpoons 2\,C_6H_5ONa + H_2O$$

(*c*) RASCHIG PROCESS. This process was introduced into the United States from Germany about 1940. As in the Dow process two stages are involved, the chlorination of benzene and hydrolysis. The differences are that hydrogen chloride and air replace electrolytic chlorine, water replaces caustic soda, and the process is carried out in the vapor phase. In the first stage, benzene, hydrogen chloride, and air are passed over a catalyst.

$$C_6H_6 + HCl + \tfrac{1}{2}\,O_2\,(\text{air}) \xrightarrow[CuCl_2-FeCl_3]{230^\circ} C_6H_5Cl + H_2O$$

Although carried out in one step, the process undoubtedly involves the intermediate formation of chlorine by the old Deacon process from hydrogen chloride and air in the presence of cupric chloride. The conversion is about 10 per cent per pass, but the chlorobenzene is separated readily and the unchanged reagents are recycled.

In the second stage the chlorobenzene is hydrolyzed with water. Here again the conversion per pass is about 10 per cent.

$$C_6H_5Cl + H_2O \xrightarrow{425^\circ\ +\ \text{cat.}} C_6H_5OH + HCl$$

The recovery of hydrogen chloride for use in the first stage is about 97 per cent. About 10 per cent of coproducts are formed including 6 per cent of dichlorobenzene in the first stage. This process, like the first, is an indirect air oxidation of benzene.

(*d*) FROM CUMENE. Since 1954 phenol has been synthesized from cumene, prepared from benzene and propylene (p. 391). Oxidation of cumene by air gives cumene hydroperoxide (cf. p. 141). Strong acids catalyze the decomposition of the hydroperoxide into phenol and acetone.

$$\underset{\text{Cumene}}{C_6H_5CH(CH_3)_2} + O_2\,(\text{air}) \longrightarrow \underset{\text{Cumene hydroperoxide}}{C_6H_5C(CH_3)_2(O{-}OH)} \xrightarrow{H_2SO_4} C_6H_5OH + (CH_3)_2CO$$

Small amounts of methanol, acetophenone (p. 464), and α-methylstyrene also are formed.

The acid-catalyzed decomposition takes place by an ionic mechanism. The strong acid removes hydroxide ion from the hydroperoxide to give an intermediate ion with a positive charge on oxygen. Migration of the phenyl group and acquisition of a molecule of water gives the oxonium salt of a hemiacetal, which decomposes into phenol, acetone, and a proton.

$$C_6H_5C(CH_3)_2(O{-}OH) \underset{}{\overset{H^+}{\rightleftharpoons}} \left[C_6H_5{-}C(CH_3)_2(O{-}\overset{+}{O}H_2)\right] \underset{H_2O}{\rightleftharpoons} \left[{}^{+}C(CH_3)_2(OC_6H_5)\right] \overset{H_2O}{\rightleftharpoons}$$

$$H_2\overset{+}{O}C(CH_3)_2(OC_6H_5) \rightleftharpoons H{-}O{-}C(CH_3)_2(H\overset{+}{O}C_6H_5) \longrightarrow H^+ + HOC_6H_5 + O{=}C(CH_3)_2$$

(*e*) BY OTHER PROCESSES. In 1962 Dow Chemical Company began operating a process that starts with toluene, a less expensive raw material than benzene. Oxidation of toluene in the liquid phase by air, using a soluble cobalt salt as catalyst, gives benzoic acid. The purified benzoic acid is vaporized and is passed with air over a bed of cupric oxide at 300°. *o*-Hydroxybenzoic acid (*salicyclic acid,* p. 472) is formed, which decarboxylates to phenol. Most of the latter esterifies with the excess benzoic acid to phenyl benzoate, which then is hydrolyzed to phenol and benzoic acid. The benzoic acid is recycled.

$$C_6H_5CH_3 + 1\tfrac{1}{2}\,O_2 \longrightarrow C_6H_5COOH + H_2O$$
$$C_6H_5COOH + \tfrac{1}{2}\,O_2 \xrightarrow{CuO} C_6H_4(OH)COOH$$
$$C_6H_4(OH)COOH \longrightarrow C_6H_4OH + CO_2$$
$$C_6H_4OH + C_6H_5COOH \longrightarrow C_6H_4OCOC_6H_5 + H_2O$$
$$C_6H_5OCOC_6H_5 + H_2O \longrightarrow C_6H_5OH + C_6H_5COOH$$

Another plant to be built will use a process that is reported to involve catalytic hydrogenation of benzene to cyclohexane, oxidation to cyclohexanol (p. 647), and dehydrogenation to phenol.

Of the total synthetic production about 37 per cent is made from cumene, 27 per cent from chlorobenzene by the Dow process, and 18 per cent each through sulfonation and by the Raschig process. Approximately 55 per cent is used for plastics and adhesives (p. 444), 11 per cent to make caprolactam for nylon 6 (p. 649), 9 per cent for alkylphenols (p. 444), 6 per cent for chlorinated phenols (p. 444), 5 per cent each for bisphenol-A (p. 444) and adipic acid (p. 630), 2 per cent for medicinals, especially salicylic acid (p. 472), and 7 per cent for miscellaneous uses including the refining of lubricating oils (p. 95).

Uses of Phenol and Its Derivatives. The chief use for phenol is for the manufacture of phenol-formaldehyde resins and plastics. Production of these *phenolics* in the United States amounted to 833 million pounds in 1964. Although the reaction of phenol and formaldehyde to give resins was reported by Baeyer in 1872, it was not until 1909 that a commercially useful product was developed by Baekeland, a Belgian-born American chemist, who was looking for a substitute for shellac. For a long time the term *Bakelite* was synonymous with phenol-formaldehyde plastics. The trade name now is owned by the Union Carbide Chemicals Company and is used to designate all of their plastics except the vinyl polymers (p. 599). Phenol-formaldehyde plastics and resins now usually are referred to as phenolics.

Two general processes of manufacture are in use. The *one-step process* produces cast phenolics, which usually are converted into useful objects by machining. Phenol and somewhat more than one molecular equivalent of aqueous formaldehyde are heated for a short time with a basic catalyst such as ammonia or sodium hydroxide. A *resole* or *A-stage resin* is formed, which is fusible and soluble in organic solvents. It is a linear polymer, condensation having taken place in the *ortho* and *para* positions. This molten resin is drawn off and cast into molds, which then are heated at 75–85° until the reaction is complete. During this stage the linear polymeric chains are cross linked by the excess formaldehyde, and the infusible, insoluble resin is formed. The following formulas merely indicate possible modes of linkage. In the actual resin these various types of linkage undoubtedly are distributed randomly.

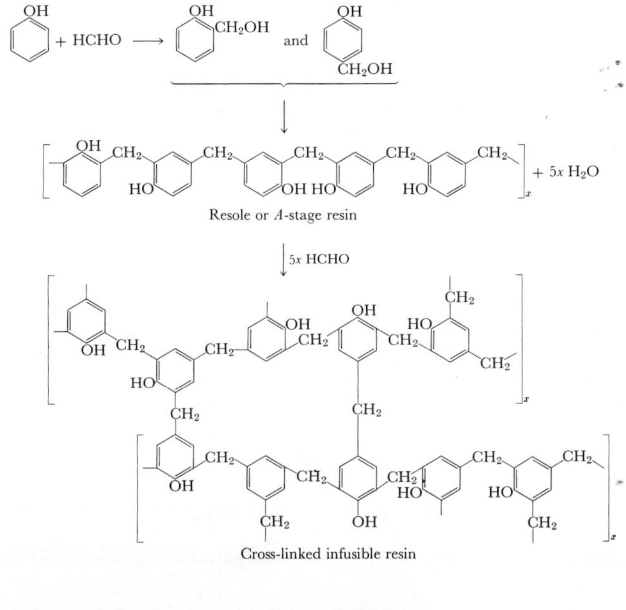

Phenol + HCHO ⟶ *o*-hydroxymethylphenol (CH_2OH) and *p*-hydroxymethylphenol (CH_2OH) ⟶ + $5x\,H_2O$

Resole or *A*-stage resin

↓ $5x$ HCHO

Cross-linked infusible resin

The *two-step process* is used for the compression molding of objects. Phenol and about 0.8 molecular equivalent of formaldehyde is heated with about 0.1 per cent of hydrogen chloride as an acid catalyst. After two hours the water is removed by distillation at reduced pressure, and the resole is run into pans or onto the floor, where it cools to a glassy solid. It is broken into pieces and ground in ball mills with lime to neutralize the acid. It then is mixed with hexamethylenetetramine (p. 213), a mold lubricant such as zinc stearate, a filler such as sawdust, and a brown or black coloring matter. After compacting and granulating, it is known as *molding powder.* When the proper amount of the powder is placed in a mold and subjected to heat and pressure, the material flows to fill the interstices of the mold, and the hexamethylenetetramine supplies the additional formaldehyde and ammonia necessary to produce cross linking and setting of the resin.

The phenol-formaldehyde plastics usually are dark colored. They are brittle, the strength of the finished product being dependent largely on the filler. They are cheap, however, and have a good finish, high heat resistance, and high dielectric strength. If *para*-substituted phenols are used, for example *p*-cresol or *p-t*-butylphenol, the cross-chain reaction is blocked and the product is thermoplastic. About one third of the phenolic resin production is used for molding purposes. Most of the remainder is used as a bonding agent in the manufacture of plywood and other laminated materials, and as water-proof adhesives for other purposes.

Bisphenol-A is the condensation product from acetone and two moles of phenol.

$$2\,C_6H_5OH + (CH_3)_2CO \xrightarrow{HCl} HO\text{-}C_6H_4\text{-}C(CH_3)_2\text{-}C_6H_4\text{-}OH + H_2O$$

Production amounted to over 93 million pounds in 1964. Reaction with phosgene gives the very hard and strong **polycarbonate plastics,** which can be molded and extruded.

$$x\,HOC_6H_4C(CH_3)_2C_6H_4OH + x\,ClCOCl \longrightarrow 2x\,HCl + [\text{—OCO—O—}C_6H_4C(CH_3)_2C_6H_4\text{—}]_x$$

The polymer can be made also by reaction of bisphenol-A with phenyl carbonate (cf. alcoholysis, p. 173), phenol being removed by distillation.

$$n\,HOC_6H_4C(CH_3)_2C_6H_4OH + n\,(C_6H_5)_2CO_3 \longrightarrow 2n\,C_6H_5OH + [\text{—OCO—O—}C_6H_4C(CH_3)_2C_6H_4\text{—}]_n$$

At present most of the bisphenol-A (94 per cent) is used for the production of **epoxy resins** (p. 608). **Tetrachlorobisphenol-A,** made by chlorination of bisphenol-A, gives fire-retardant resins. If phenol is condensed with levulinic acid (p. 330), the compound known as **diphenolic acid** [*4,4-bis(p-hydroxyphenyl)valeric acid*] is formed.

$$2\,C_6H_5OH + CH_3CO(CH_2)_2COOH \longrightarrow H_2O + HO\text{-}C_6H_4\text{-}C(CH_3)(CH_2CH_2COOH)\text{-}C_6H_4\text{-}OH$$

It is used as an intermediate in the production of various types of plastics and resins.

So-called **nonylphenol** is made by the condensation of phenol with propylene trimer (p. 87).

$$CH_3CH(CH_3)CH_2CH(CH_3)CH_2CH{=}CH_2 + C_6H_4OH \xrightarrow{H_2SO_4} CH_3CH(CH_3)CH_2CH(CH_3)CH_2CH(CH_3)C_6H_4OH$$

The 61 million pounds produced in 1964 was used chiefly for the production of nonionic detergents (p. 606). Because of the necessity of producing only biodegradable detergents by 1966, propylene trimer is being replaced by straight-chain alkenes (cf. p. 406). **Pentachlorophenol** dissolved in oil is used extensively for treating wood to prevent destruction by fungi and termites. The sodium salt is used to treat industrial water to prevent the growth of slime and algae. The salts and esters of the halogenated phenoxy aliphatic acids are the most important of the selective weed killers. **2,4-Dichlorophenoxyacetic acid** (*2,4-D*) is made from sodium 2,4-dichlorophenoxide and sodium chloroacetate. Total production of the acid and its salts and esters in 1964 was 108 million pounds.

$$2,4\text{-}Cl_2C_6H_3ONa + ClCH_2COONa \longrightarrow 2,4\text{-}Cl_2C_6H_3OCH_2COONa + NaCl$$

2,4,5-Trichlorophenoxyacetic acid (*2,4,5-T*) is made in the same way from 2,4,5-trichlorophenol. The latter compound results from the hydrolysis of 1,2,4,5-tetrachlorobenzene obtained by direct chlorination of benzene.

$$C_6H_6 + 4\,Cl_2 \xrightarrow{FeCl_3} 1,2,4,5\text{-}C_6H_2Cl_4 \xrightarrow{NaOH} 2,4,5\text{-}Cl_3C_6H_2OH$$

Treatment of rubber trees with 2,4-D increases the flow and yield of latex (p. 582) from two to three times.

p-Nitrophenol is an intermediate for the synthesis of **parathion,** an important insecticide.

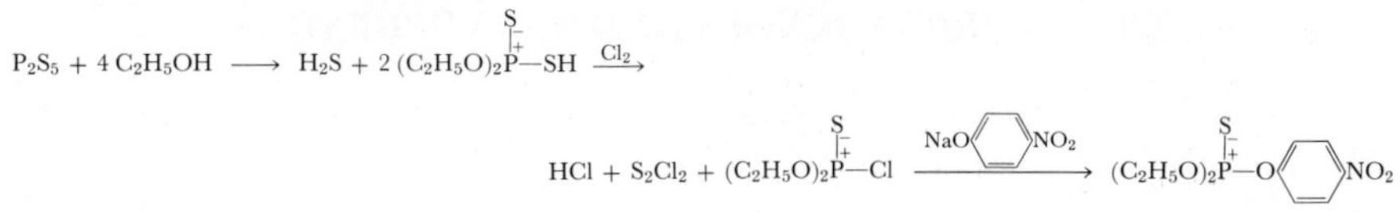

$$P_2S_5 + 4\,C_2H_5OH \longrightarrow H_2S + 2\,(C_2H_5O)_2\overset{S}{P}{-}SH \xrightarrow{Cl_2}$$

$$HCl + S_2Cl_2 + (C_2H_5O)_2\overset{S}{P}{-}Cl \xrightarrow{NaO{-}C_6H_4{-}NO_2} (C_2H_5O)_2\overset{S}{P}{-}O{-}C_6H_4{-}NO_2$$

The corresponding methyl ester is called **methyl parathion.** Combined production in 1964 was over 28 million pounds. They are extremely toxic to mammals and must be used with great care. **2,4,6-Trinitrophenol** (p. 440), because of its high acidity and its bitter taste, was given the name **picric acid** (Gr. *pikros* bitter). At times it has been used as a yellow dye for silk and as a military explosive. It forms salts with amines and complexes with polynuclear aromatic compounds that have been used for the purpose of isolation or characterization. The aromatic ring of trinitrophenol is oxidized by alkaline hypochlorite to give **chloropicrin** (*trichloronitromethane,* p. 246).

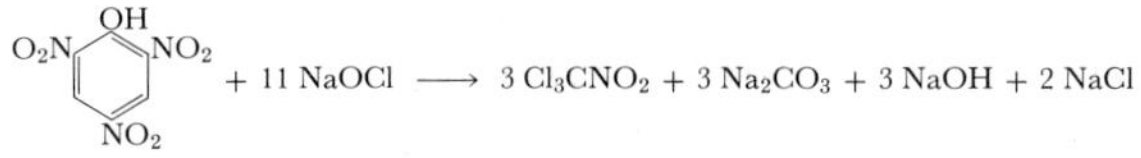

$$2,4,6\text{-}(O_2N)_3C_6H_2OH + 11\,NaOCl \longrightarrow 3\,Cl_3CNO_2 + 3\,Na_2CO_3 + 3\,NaOH + 2\,NaCl$$

The *aryl phosphates* are made by the reaction of the phenol with phosphorus oxychloride.

$$3\,ArOH + POCl_3 \longrightarrow (ArO)_3PO + 3\,HCl$$

They are used extensively as plasticizers and flame retardants. **Phenyl phosphate** (*triphenyl phosphate*) is used in photographic film base to increase flexibility, to produce flat sheets, and to decrease flammability. **Cresyl diphenyl phosphate** is added to gasoline to prevent preignition and the fouling of spark plugs. Technical **tricresyl phosphate,** prepared from the mixed cresols, is an important plasticizer, particularly for vinyl polymers. It is used also as a nonflammable hydraulic fluid and as an additive for lubricating oils. Production of the three esters in the United States in 1964 was over 9, 16, and 32 million pounds respectively.

Creosote oil, a coal, wood, or petroleum tar fraction boiling at 225–270°, contains considerable amounts of the **cresols** and is used on a large scale for wood preservation. ***p*-Cresol** is synthesized from the hydroperoxide formed by the air-oxidation of *p*-cymene (cf. phenol from cumene, p. 442), by the nuclear methylation of phenol (p. 377), and by the caustic fusion of sodium *p*-toluenesulfonate. Alkylation of *p*-cresol with isobutylene gives **2,6-di-*t*-butyl-4-methylphenol,** which is used widely as an antioxidant for gasoline, lubricating oils, rubber, and edible fats and oils (p. 186). It is known by various names such as **Ionol** or **BHT** (*butylated hydroxy toluene*). Almost 20 million pounds was produced in 1964. Phenols inhibit autoxidation by terminating the chain reaction. Although they are converted to free radicals, resonance stabilization prevents them from being sufficiently reactive to start a new chain.

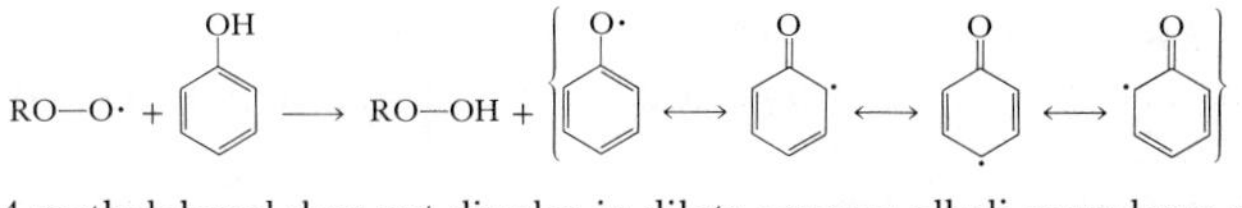

$$RO{-}O\cdot + C_6H_5OH \longrightarrow RO{-}OH + \{C_6H_5O\cdot \longleftrightarrow \longleftrightarrow \longleftrightarrow\}$$

2,6-Di-*t*-butyl-4-methylphenol does not dissolve in dilute aqueous alkali or undergo ring oxidation, and does not give a color with ferric chloride. Phenols that have large groups in the *ortho* positions and that do not give the usual reactions of phenols have been called *cryptophenols.*

When air or oxygen is passed through 2,6-dimethylphenol in the presence of a cuprous chloride–amine complex, oxidative coupling takes place.

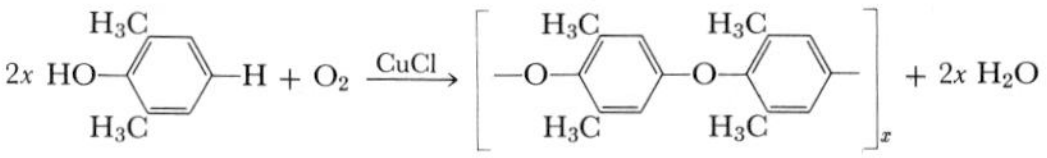

$$2x\,HO{-}C_6H_3(CH_3)_2{-}H + O_2 \xrightarrow{CuCl} \left[{-}O{-}C_6H_2(CH_3)_2{-}O{-}C_6H_2(CH_3)_2{-}\right]_x + 2x\,H_2O$$

The resulting **poly(phenylene oxide)** of high molecular weight has excellent mechanical strength and resistance to chemical action from −170° to +190° and is expected to replace glass and stainless steel for many purposes.

Thymol, (*3-hydroxy-4-isopropyltoluene*) occurs in thyme oil. In the absence of a sufficient supply of the natural product, it is synthesized from *m*-cresol and propylene (p. 440). Thymol is antiseptic in high dilutions. It has a more pleasant aromatic odor than phenol or the cresols and frequently is used in proprietary antiseptic preparations. Thymol is the starting point for the synthesis of menthol (p. 660). **Carvacrol** (*2-hydroxy-4-isopropyltoluene*) also occurs in some essential oils. Many other phenols or phenol ethers are responsible for the aromatic properties of essential oils. **Anethole,** the chief component of anise oil, is 4-propenylanisole.

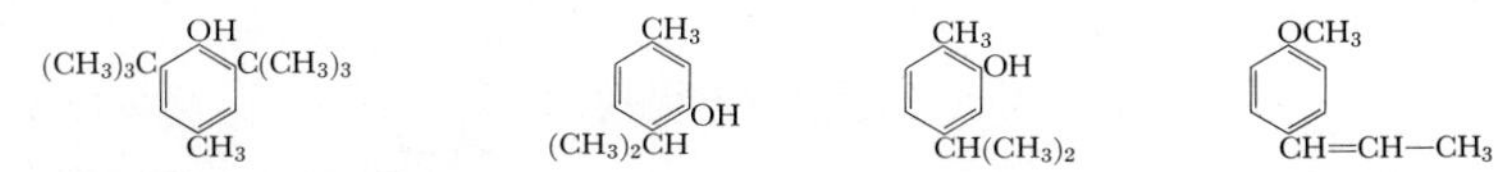

2,6-Di-*t*-butyl-4-methylphenol — Thymol — Carvacrol — Anethole

POLYHYDRIC PHENOLS AND AMINOPHENOLS

The *o*-, *m*-, and *p*-dihydroxybenzenes are known as *catechol, resorcinol,* and *hydroquinone* respectively. **Catechol** (also called *pyrocatechol*) is so named because it is one of the distillation products of *gum catechu,* obtained from certain Asiatic tropical plants. It can be prepared by the general method of synthesis, namely the fusion of sodium *o*-phenolsulfonate with caustic soda, or by scission of its monomethyl ether (*guaiacol*) with hydrogen bromide.

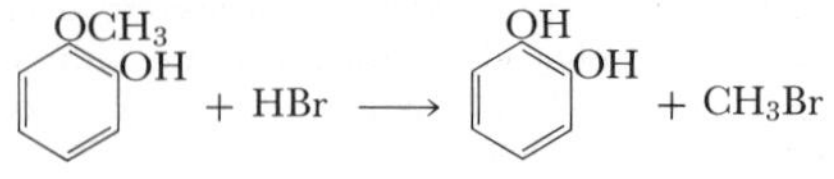

$$\text{Guaiacol} + HBr \longrightarrow \text{Catechol} + CH_3Br$$

Guaiacol Catechol

Guaiacol occurs in the distillation products of guaiac, the resin from American tropical trees of the genus *Guaiacum* but is produced commercially from wood tar. The technical product, because of its antioxidant properties, is used as an anti-skinning agent for paints. **Guaiacol carbonate** is used as an expectorant in cough remedies. **Eugenol** from oil of cloves is 2-methoxy-4-allylphenol, and **safrole** from oil of sassafras and from camphor oil is the formaldehyde acetal of 4-allylcatechol. Being an acetal, safrole is hydrolyzed easily by dilute acids.

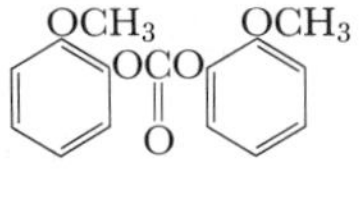

Guaiacol carbonate 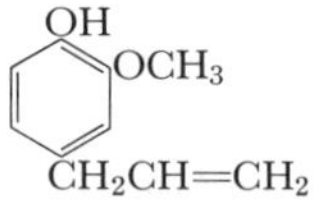 Eugenol ($CH_2CH{=}CH_2$) Safrole ($CH_2CH{=}CH_2$)

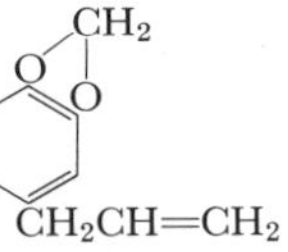

The toxic irritants of poison ivy (*Rhus toxicodendron*), poison oak (*Rhus diversiloba*), and certain related plants of the *Anacardiaceae* are mixtures of catechols having unbranched 15-carbon side chains in the 3 position. Four compounds have been isolated from poison ivy in which the side chain is saturated or contains one, two, or three double bonds in the 8, the 8,11, and the 8,11,14 positions. Three of the components of **urushiol,** the toxic principle of one of the lac trees (*Rhus vernicifera*), are identical with three of the compounds obtained from poison ivy, but the triply unsaturated compound has two of the double bonds conjugated in the 12,14 positions.

Nordihydroguaiaretic acid (*NDGA*) is a catechol derivative isolated from the creosote bush (*Larrea divaricata*), a native of the deserts of southwestern United States and of Mexico.

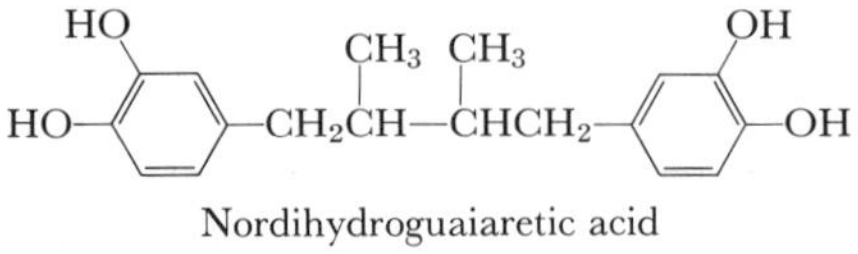

$$(HO)_2C_6H_3-CH_2CH(CH_3)-CH(CH_3)CH_2-C_6H_3(OH)_2$$

Nordihydroguaiaretic acid

It is a powerful antioxidant and is permitted for use in foods to the extent of 0.01 per cent. It prevents the development of rancidity in lard or in frozen cream for up to one year.

Resorcinol is a product of the distillation of certain natural resins but is manufactured by the fusion of sodium *m*-benzenedisulfonate with caustic soda. It undergoes substitution reactions readily in the 4 position. The condensation products with formaldehyde are used as cold-setting adhesives. Resorcinol also is an intermediate for the preparation of dyes (p. 570), and of ***n*-hexylresorcinol.** The last compound is a popular antiseptic and is synthesized by Clemmensen reduction of hexanoylresorcinol (cf. pp. 464, 208).

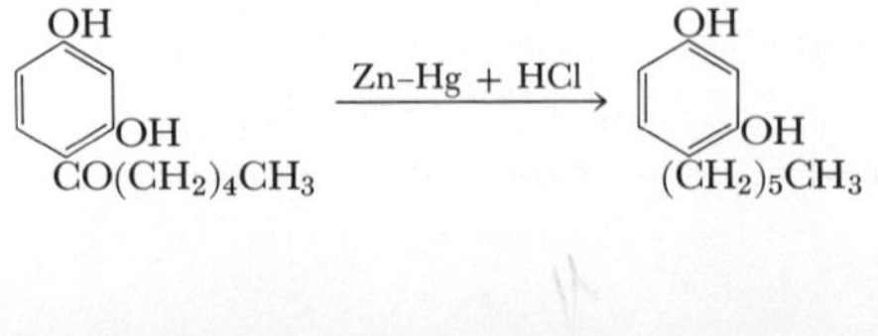

$$(HO)_2C_6H_3CO(CH_2)_4CH_3 \xrightarrow{Zn\text{-}Hg\ +\ HCl} (HO)_2C_6H_3(CH_2)_5CH_3$$

The antiseptic power of phenol is increased greatly by the substitution of alkyl groups into the nucleus. Thus the cresols are nearly as toxic as phenol, but their phenol coefficients (p. 436) are about 3. As the length of the alkyl group is increased, the effectiveness increases up to six carbon atoms, and then decreases. *n*-Hexylphenol is 500 times more effective than phenol. Evidently a hydrocarbon chain of six carbon atoms corresponds to the optimum solubility in water and in fats, both of which are present in cells. The effect of the compound in lowering the surface tension of water also is important.

Reduction of resorcinol by sodium amalgam in aqueous solution or by high-pressure catalytic hydrogenation yields **dihydroresorcinol,** which is tautomeric with 1,3-cyclohexanedione. As a 1,3-dicarbonyl compound, it is a useful intermediate for the synthesis of other compounds (p. 647).

$$\text{Resorcinol} \xrightarrow[H_2/CuO\text{-}Cr_2O_3]{Na\text{-}Hg\ +\ H_2O,\ \text{or}} \text{dihydroresorcinol} \rightleftharpoons \text{1,3-cyclohexanedione}$$

2,4,6-**Trinitroresorcinol** is known as *styphnic acid* and is used like picric acid to prepare derivatives of organic compounds (p. 445). **Orcinol,** which can be obtained from certain lichens and aloes, is *5-methylresorcinol.* It is used as a reagent to distinguish among pentoses, methylpentoses, and hexoses (p. 330).

Hydroquinone is manufactured by the reduction of quinone (p. 450) with iron and water.

$$4\ \text{Quinone} + 3\ Fe + 4\ H_2O \longrightarrow 4\ \text{Hydroquinone} + Fe_3O_4$$

As the result of hydrogen bonding, crystalline hydroquinone consists of two interpenetrating three-dimensional networks that enclose roughly spherical cavities in the ratio of two cavities for each three molecules of hydroquinone. These cavities can hold molecules of argon, krypton, xenon, nitrogen, oxygen, hydrogen chloride, sulfur dioxide, methane, acetylene, or methanol. The clathrate of argon can be kept indefinitely, since the gas cannot escape without breaking hydrogen bonds.

Hydroquinone and its monomethyl ether in amounts of 0.1 per cent or less are used extensively to prevent unwanted autoxidation and polymerization of organic compounds (p. 445). The mixture of 2- and 3-*t*-butyl-4-methoxyphenol (**BHA** = *butylated hydroxyanisole*) is used to prevent the development of rancidity in foods. Hydroquinone and some of its derivatives, for example the monobenzyl ether, have the undesirable property of causing permanent depigmentation of the skin. Aminolysis of hydroquinone with primary amines (p. 438) yields substituted *p*-phenylenediamines that are used as antioxidants for gasoline and rubber products (p. 427).

$$RNH_2 + HO\text{—}C_6H_4\text{—}OH + H_2NR_2 \xrightarrow[\text{heat}]{ZnCl_2,} RNH\text{—}C_6H_4\text{—}NHR + 2\ H_2O$$

The chief use for hydroquinone is as a photographic developer (p. 449).

Pyrogallol (*pyrogallic acid* or *1,2,3-trihydroxybenzene*) is prepared by decarboxylating gallic acid, obtained by hydrolysis of gallotannin (p. 473).

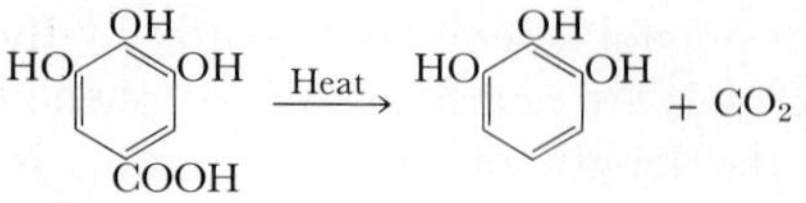

Gallic acid $\xrightarrow{\text{Heat}}$ Pyrogallol + CO_2

Benzoic acids that have hydroxyl or amino groups in the *ortho* and *para* positions decarboxylate more readily than the unsubstituted acid. The resonance effect increases the electron density at the carbon atom bearing the carboxyl group and facilitates displacement of the carboxyl group by a proton.

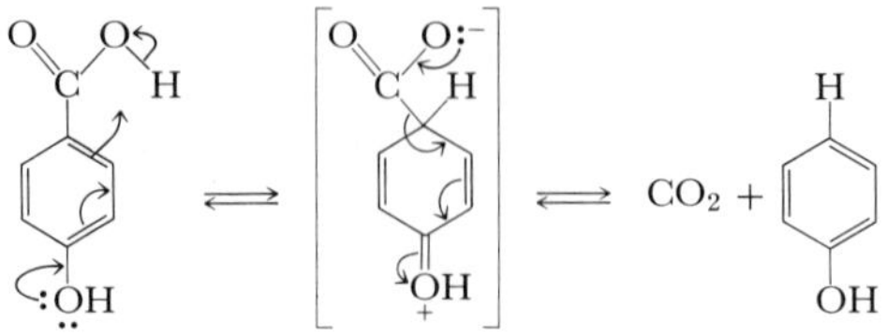

$\rightleftharpoons$ CO_2 +

Pyrogallol is an important photographic developer (p. 449). Alkaline pyrogallol solutions absorb oxygen very readily and are used to remove oxygen from mixtures with other gases.

Phloroglucinol (*sym-trihydroxybenzene* or *1,3,5-trihydroxybenzene*) is a useful reagent for the estimation of furfural and hence of pentoses (p. 512). It is made by a series of reactions from trinitrotoluene,

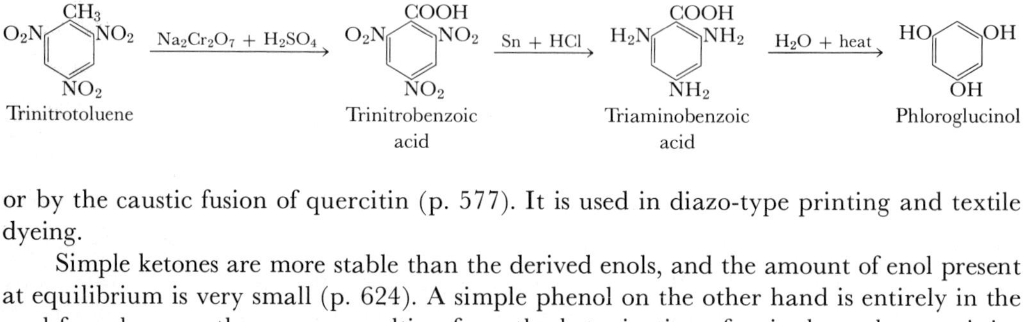

Trinitrotoluene $\xrightarrow{Na_2Cr_2O_7 + H_2SO_4}$ Trinitrobenzoic acid $\xrightarrow{Sn + HCl}$ Triaminobenzoic acid $\xrightarrow{H_2O + \text{heat}}$ Phloroglucinol

or by the caustic fusion of quercitin (p. 577). It is used in diazo-type printing and textile dyeing.

Simple ketones are more stable than the derived enols, and the amount of enol present at equilibrium is very small (p. 624). A simple phenol on the other hand is entirely in the enol form because the energy resulting from the ketonization of a single enol group is insufficient to overcome the resonance energy of the nucleus. The energy gained by ketonization of three hydroxyl groups, however, appears to approach the aromatic resonance energy. Although interatomic distances as determined by electron diffraction show that phloroglucinol is 1,3,5-trihydroxybenzene, its reactions indicate that the various tautomeric forms are in mobile equilibrium with each other.

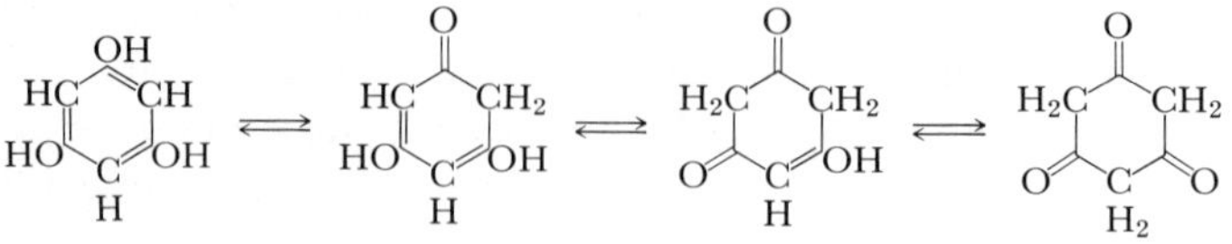

$\rightleftharpoons$ $\rightleftharpoons$ $\rightleftharpoons$

Thus it gives a blue-violet color with ferric chloride and forms a trimethyl ether with diazomethane (p. 164).

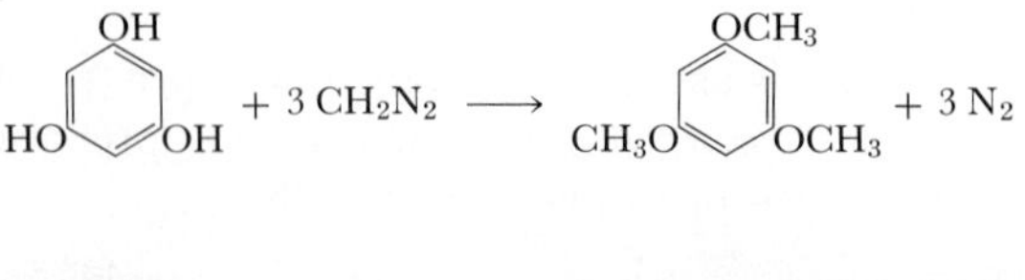

$$C_6H_3(OH)_3 + 3\,CH_2N_2 \longrightarrow C_6H_3(OCH_3)_3 + 3\,N_2$$

On the other hand it reacts with ammonia and with hydroxylamine, reactions characteristic of ketones.

$$\text{cyclo-}[C(=O)CH_2C(=O)CH_2C(=O)CH_2] + 3\,H_2NOH \longrightarrow \text{cyclo-}[C(=NOH)CH_2C(=NOH)CH_2C(=NOH)CH_2] + 3\,H_2O$$

Of the aminophenols, **p-aminophenol** (*P.A.P.*) is the most important because of its use as a photographic developer. It is made by the nitrosation of phenol followed by reduction.

$$C_6H_5OH \xrightarrow{HONO} p\text{-}ON{-}C_6H_4{-}OH \xrightarrow{Fe,\ H_2O,\ HCl} p\text{-}H_2N{-}C_6H_4{-}OH$$

It can be made also by the electrolytic reduction of nitrobenzene in acid solution. *N*-Phenylhydroxylamine is formed, which rearranges in acid solution to the salt of *p*-aminophenol (cf. p. 415).

$$C_6H_5NO_2 \xrightarrow{\text{Elect. red.}} C_6H_5NHOH \xrightarrow{H_2SO_4} [p\text{-}H_3\overset{+}{N}{-}C_6H_4{-}OH]_2\ SO_4^{2-}$$

p-Hydroxyphenylglycine (*photographer's Glycine*) is made by the condensation of *p*-aminophenol with sodium chloroacetate.

$$p\text{-}H_2N{-}C_6H_4{-}OH + ClCH_2COONa \longrightarrow p\text{-}HOOCCH_2HN{-}C_6H_4{-}OH + NaCl$$

When *p*-hydroxyphenylglycine is heated in a mixture of cresols, decarboxylation takes place to give **p-(methylamino)phenol.** This compound can be made also by heating an aqueous solution of methylamine and hydroquinone in an autoclave at 100° (cf. p. 438). The sulfate, known as *Metol* or *Elon,* is another commercial photographic developer. The widely used MQ developer is a mixture of hydroquinone and *p*-(methylamino)phenol.

The common black and white photographic plate, film, or printing paper consists of a support on which is coated an emulsion of mixed silver bromide and iodide in gelatin solution. Silver halides darken on exposure to light. The light energy dissociates the silver halide into silver and halogen atoms. The number of silver atoms formed depends on the intensity of the light falling on the silver halide and the time of exposure. If a photographic plate were exposed long enough to a light image, enough silver would be formed to produce a silver image on the plate. A more satisfactory procedure is to give a short exposure and produce an invisible latent image on the plate. Each silver particle thus formed then can act as a nucleus for the deposition of more silver when the plate is subjected to mild chemical reduction. During this reduction a visible image develops because the density of the silver deposited is proportional to the number of silver nuclei, which in turn is proportional to the intensity of the light that fell on the plate. After the desired amount of development, the unreduced silver halide is removed by dissolving it with sodium thiosulfate solution (*photographer's hypo*), leaving the silver image on the plate, film, or print.

The most important developers are hydroquinone, pyrogallol, and *p*-aminophenol and its derivatives, because they bring about the chemical reduction of the silver halide at the desired rate. The introduction of amino and hydroxyl groups into the benzene nucleus increases the ease of oxidation by increasing the availability of the unsaturation electrons (pp. 420, 438). The ease of oxidation is increased greatly in alkaline solution and decreased in acid solution because the electron density on an oxide ion or on a free amino group is much greater than that on a hydroxyl group or an ammonium ion. If two hydroxyl groups or amino groups are *ortho* or *para* to each other, the ease of oxidation is increased greatly because such compounds can be oxidized to quinones (p. 450). Different developers produce different types of deposition of silver and hence influence the characteristics of the developed image. Usually a combination of developers is used to produce the desired effect.

The methyl ethers of the aminophenols are known as *anisidines* and the ethyl ethers as *phenetidines*. **p-Phenetidine,** its acetyl derivative known as **acetophenetidine** or **phenacetin,** and **p-hydroxyacetanilide** (*APAP* for *acetyl-p-aminophenol*) have antipyretic and analgesic action. Although less toxic than acetanilide, they also reduce the oxygen-combining power of the blood.

QUINONES

Of the three dihydroxybenzenes, or the diamines, or the aminophenols, the *ortho* and *para* isomers oxidize much more easily than the *meta* isomers. The reason is that the *ortho* and *para* isomers can lose two hydrogen atoms from oxygen or nitrogen to give stable compounds known as *quinones*.

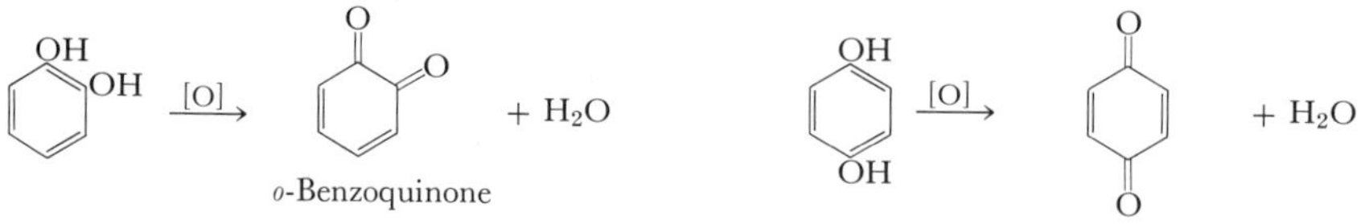

OH, OH —[O]→ O, O + H_2O
o-Benzoquinone

OH, OH —[O]→ O, O + H_2O
p-Benzoquinone
(*quinone*)

This type of oxidation is not possible for the *meta* isomers because no stable structure can exist for a *meta* quinone. Quinones are formed also by the oxidation of aminophenols and diamines because the intermediate quinonimines and quinonediimines are hydrolyzed rapidly in aqueous solution.

OH, NH_2 —[O]→ O, NH —H_2O→ O, O + NH_3

NH_2, NH_2 —[O]→ NH, NH —2 H_2O→ O, O + 2 NH_3

Quinone is a generic term for the above class of compounds, but it frequently is used as a specific name for **p-benzoquinone,** a bright yellow solid with a sharp odor. The name *quinoyl* was assigned to this compound when it first was obtained by the oxidation of quinic acid extracted from cinchona bark (p. 536). Berzelius later changed the name to *quinone*. It is prepared commercially by the oxidation of aniline with manganese dioxide and sulfuric acid.

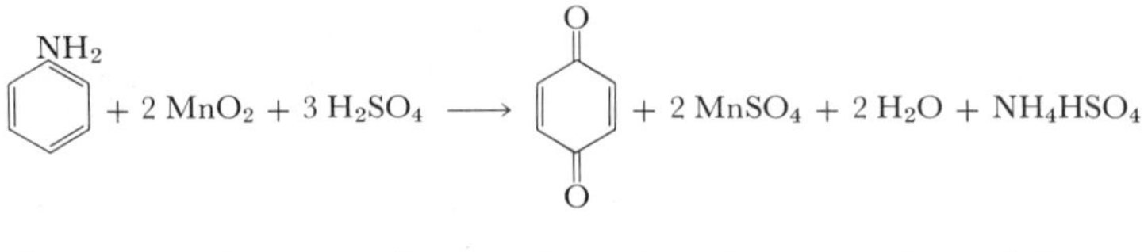

NH_2 + 2 MnO_2 + 3 H_2SO_4 ⟶ O, O + 2 $MnSO_4$ + 2 H_2O + NH_4HSO_4

In general, groups that are electron-donating decrease the oxidizing power of the quinone and increase the reducing power of the quinol. Electron-attracting groups increase the oxidizing power of the quinone and decrease the reducing power of the quinol. **Tetrachloro-1,4-benzoquinone** (*chloranil*) is obtained in practically quantitative yield by the reaction of either quinone or hydroquinone with chlorine in hydrochloric acid. It has been used as a dehydrogenating agent for organic compounds that have hydrogen α and β to a double bond. Thus the cyclohexadiene ring is converted to a benzene ring. The initial step is removal of a hydride ion by the quinone, which is followed by removal of a proton.

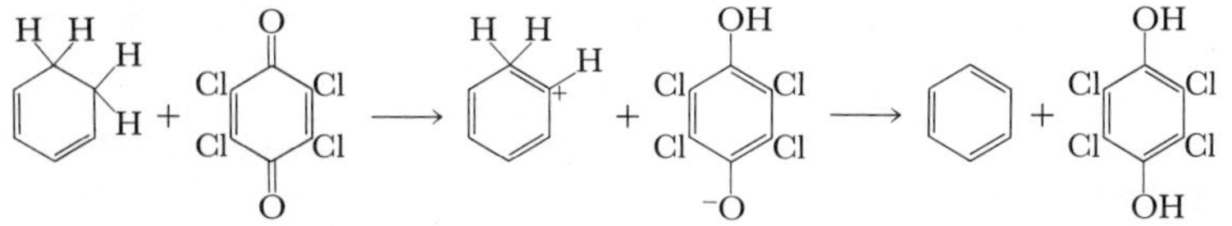

Tetrachloro-1,2-benzoquinone and **2,3-dichloro-5,6-dicyano-1,4-benzoquinone** have higher oxidation potentials and are four and five thousand times more reactive than chloranil.

Compounds that are strongly electron-attracting and have unfilled π orbitals form highly colored charge-transfer complexes with electron-donor molecules (cf. p. 387) and have been called *π acids.* **1,4-Benzoquinonetetracarboxylic anhydride** is reported to be a stronger π acid than tetracyanoethylene (p. 636). It gives a deep red 1 : 1 complex with benzene. The solution in benzene is orange, in toluene, red, in *m*-xylene, magenta, and in mesitylene, violet.

Quinones are widely distributed in plants and animals, the majority occurring as pigments in fungi and flowering plants (pp. 495, 499). Over 160 different quinones have been isolated from members of the plant kingdom and 30 from the animal kingdom, the latter mostly from insects and sea urchins. Practically all are completely substituted *para* quinones that vary greatly in structure. **Polyporic acid,** isolated from *Polyporus* species, is representative of benzoquinone derivatives present as pigments in fungi.

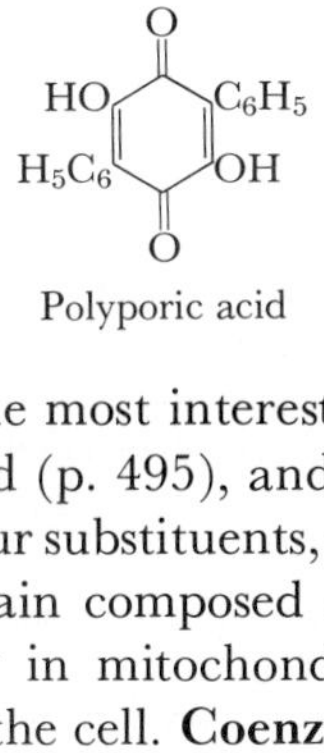

Polyporic acid

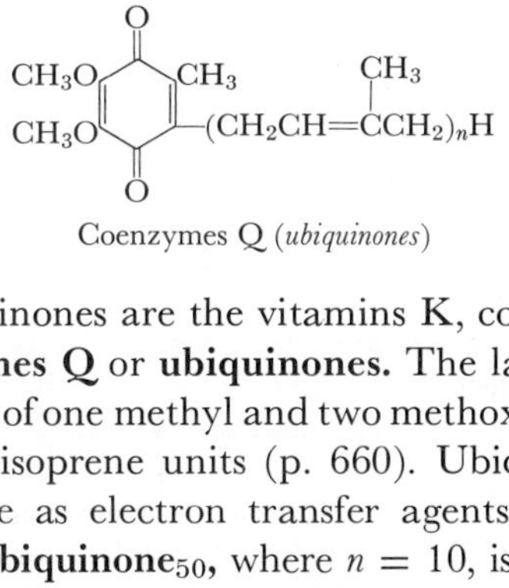

Coenzymes Q (*ubiquinones*)

$n = 6, 7, 8, 9$, or 10

The most interesting of the quinones are the vitamins K, concerned with the clotting of blood (p. 495), and the **coenzymes Q** or **ubiquinones.** The latter are *p*-benzoquinones with four substituents, which consist of one methyl and two methoxyl groups, and a branched side chain composed of six to ten isoprene units (p. 660). Ubiquinones are present universally in mitochondria and serve as electron transfer agents in oxidative metabolism within the cell. **Coenzyme Q_{10}** or **ubiquinone$_{50}$,** where $n = 10$, is most commonly encountered in the higher plants and animals and is isolated best from tissues of high respiratory function such as heart tissue.

PROBLEMS

25–*1*. Place the following compounds in the order of increasing acidity: (*a*) acetic acid, (*b*) 1-hexanol, (*c*) benzenesulfonic acid, (*d*) nitrobenzene, (*e*) nitroethane, (*f*) carbonic acid, (*g*) benzene, (*h*) cyclohexane, (*i*) phenol.

25–*2*. Which compound is the most convenient starting point for the synthesis of *m*-chlorophenol: (*a*) phenol, (*b*) chlorobenzene, (*c*) benzenesulfonic acid, (*d*) *m*-chloroaniline, or (*e*) *m*-dichlorobenzene?

25–*3*. For each of the following compounds give a synthesis that starts with an aromatic hydrocarbon: (*a*) *p*-methoxytoluene, (*b*) *o*-nitrophenyl acetate, (*c*) *m*-nitrophenol, (*d*) 2,4-dinitroanisole (*e*) 2,4-dimethylphenyl benzyl ether, (*f*) *m*-chlorophenol, (*g*) *m*-phenolsulfonic acid, (*h*) 4,6-dimethylresorcinol, (*i*) *p*-bromophenyl carbonate, (*j*) *p*-methoxybenzenesulfonamide.

25–*4*. Which reactions can be carried out more readily with phenol than with ethanol: (*a*) direct esterification, (*b*) formation of alkyl ether, (*c*) replacement of hydroxyl by hydrogen, (*d*) oxidation to a carbonyl compound, (*e*) conversion to the anion, (*f*) dehydration, (*g*) bromination?

25–*5*. Which of the reagents given can be used to distinguish between (*a*) *n*-hexanol and phenol, (*b*) *N*-methylaniline and *p*-toluidine, (*c*) acetamide and acetanilide, (*d*) *n*-hexylamine and aniline, (*e*) aromatic azo and diazo compounds, (*f*) sulfuric acid and benzenesulfonic acid, (*g*) *n*-hexane and benzene, (*h*) chlorobenzene and *n*-hexyl chloride: (*1*) fuming H_2SO_4, (*2*) hot aq. NaOH, (*3*) $FeCl_3$, (*4*) $Ba(OH)_2$, (*5*) NaI-acetone, (*6*) cold HNO_2 + HCl, (*7*) $C_6H_5SO_2Cl$ + NaOH, (*8*) phenol?

25–*6*. Pair the following compounds with the proper name: (*a*) 1,2-dihydroxybenzene, (*b*) 1,3-dihydroxybenzene, (*c*) 1,4-dihydroxybenzene, (*d*) 1,2,3-trihydroxybenzene, (*e*) 1,3,5-trihydroxybenzene, (*f*) 2-methoxyphenol, (*g*) 4-methoxypropenylbenzene, (*h*) 3-methoxy-4-hydroxyallylbenzene, (*i*) 3-hydroxy-4-isopropyltoluene, (*j*) 4-hydroxytoluene; (*1*) anethole, (*2*) pyrogallol, (*3*) eugenol, (*4*) guaiacol, (*5*) thymol, (*6*) hydroquinone, (7) resorcinol, (8) catechol, (*9*) phloroglucinol, (*10*) *p*-cresol.

25–*7*. Give equations for the synthesis of the following compounds from readily available chemicals: (*a*) *o*-cresyl phosphate, (*b*) thymol, (*c*) *p*-methylaminophenol, (*d*) *p*-diethylaminophenyl di-*n*-propylcarbamate, (*e*) acetophenetidine, (*f*) styphnic acid, (*g*) 2-*n*-butyl-5-methylphenol, (*h*) *p*-chloromercuriphenol.

25–*8*. For each of the following mixtures, describe a procedure for its separation that will yield each component in a relatively pure state: (*a*) *p*-cymene, *p*-cresol, benzenesulfonic acid, and aniline; (*b*) *N*-methylaniline, *N,N*-dimethylaniline, *o*-chlorophenol, and benzoic acid; (*c*) acetanilide, *n*-butyric acid, phenol, and sulfanilic acid; (*d*) *o*-dichlorobenzene, benzenesulfonamide, phenol, and *p*-phenolsulfonic acid.

25–*9*. Write the structural formulas for the components of (*a*) the toxic principles of poison ivy and poison oak, and (*b*) urushiol.

25–*10*. Compound *A* has the molecular formula $C_9H_{12}O_2$. It gives a positive iodoform test and evolves two moles of methane when treated with methylmagnesium iodide. When heated with sodium bisulfate, it gives compound *B*, $C_9H_{10}O$, which is reduced catalytically to *C*, $C_9H_{12}O$. When *C* is distilled with zinc dust, compound *D*, C_9H_{12}, is obtained, which is converted by vigorous oxidation to terephthalic acid. Compound *A* reacts with thionyl chloride to give a product having the molecular formula $C_9H_{10}O_3S$. Give the structural formula for *A* and equations for the reactions that take place.

25–*11*. Compound *A* has the molecular formula $C_{15}H_{14}O_5$. It is insoluble in cold dilute alkali but dissolves completely when boiled with aqueous sodium hydroxide. When the alkaline solution is distilled only water comes over. When the alkaline solution is acidified, effervescence takes place and an oil separates. The gas gives a white precipitate when passed into a solution of barium hydroxide. The oil gives a color with ferric chloride. When the oil is boiled with 48 per cent hydrobromic acid, methyl bromide is evolved. Distillation of the excess hydrogen bromide under reduced pressure leaves an oil that solidifies at room temperature. It proves to be identical with the compound obtained by fusing sodium *m*-benzenedisulfonate with sodium hydroxide and acidifying the melt. Give a structural formula for *A* and equations for the reactions that take place.

25–*12*. Compound *A* has the molecular formula $C_{10}H_{11}NO_3$. It is insoluble in dilute alkali or dilute acid. When boiled with dilute sulfuric acid and distilled, the distillate is acid to litmus. When a portion of the residue in the flask is made alkaline no precipitate forms. When the remainder of the acid solution is cooled to 0° and sodium nitrite is added, there is no evolution of gas, but when this solution is warmed, nitrogen is evolved. When the solution is cooled, a solid crystallizes. An aqueous solution of the solid reduces Tollens reagent and gives a transient green color with ferric chloride. As more ferric chloride is added, a yellow precipitate forms. Give the structural formula for *A* and equations for the reactions that take place.

25–*13*. A water-insoluble compound *A*, having the molecular formula C_7H_9NO, is soluble in dilute sodium hydroxide and in dilute hydrochloric acid but insoluble in sodium bicarbonate solution. It reacts with an excess of acetic anhydride to give compound *B*, which is insoluble in dilute sodium hydroxide and in dilute hydrochloric acid. Compound *A* decolorizes bromine water. When it is dissolved in an excess of dilute hydrochloric acid and the cold solution is treated with sodium nitrite, a new product *C* separates without the evolution of nitrogen. Give a possible formula for *A* and the reactions it undergoes.

CHAPTER TWENTY-SIX

AROMATIC ALCOHOLS, ARALKYLAMINES, ALDEHYDES, AND KETONES. STEREOCHEMISTRY OF THE OXIMES

ALCOHOLS AND ARALKYLAMINES

The methods for preparing compounds containing hydroxyl or amino groups in alkyl side chains of aromatic nuclei, and the reactions of these groups, are the same as the methods of preparation and reactions of aliphatic alcohols and amines. The chief difference is a considerably greater reactivity of the groups that are attached to the carbon atom adjacent to the ring. The reason for this enhanced reactivity is that a benzyl group, whether a free radical, cation, or anion, is stabilized by resonance with the nucleus.

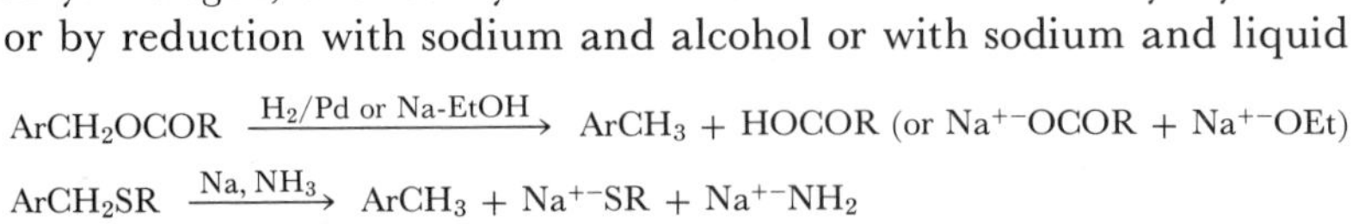

The stabilization of the benzyl free radical is evident from the bond dissociation energies of corresponding alkyl, aryl, and benzyl analogs as given in Table 26–*1*. Thus the benzyl-oxygen, benzyl-nitrogen, and benzyl-sulfur bonds are cleaved easily by either catalytic reduction or by reduction with sodium and alcohol or with sodium and liquid ammonia.

$$ArCH_2OCOR \xrightarrow{H_2/Pd \text{ or } Na\text{-}EtOH} ArCH_3 + HOCOR \text{ (or } Na^{+-}OCOR + Na^{+-}OEt)$$

$$ArCH_2SR \xrightarrow{Na,\ NH_3} ArCH_3 + Na^{+-}SR + Na^{+-}NH_2$$

$$ArCH_2NHR \xrightarrow{H_2/Ni} ArCH_3 + H_2NR$$

Hence a benzyl group can be used to protect a hydroxyl, mercapto, or amino group while another reaction is carried out and then can be removed by hydrogenolysis.

TABLE 26–*1*. BOND DISSOCIATION ENERGIES IN KCAL./MOLE FOR SOME ALKYL, ARYL, AND BENZYL COMPOUNDS

BOND BROKEN \ R	C_2H_5	C_6H_5	$C_6H_5CH_2$
R—H	96	102	77
R—CH_3	85	87	63
R—NH_2	80 (R = CH_3)	—	59
R—SH	69	—	53
R—Br	65	71	51

The high reactivity of benzyl halides has been discussed (p. 402). Benzyl esters of strong acids undergo displacement reactions with alkyl-oxygen scission.

$$ArCH_2\text{—}O\text{—}SO_2Ar + {}^-I \longrightarrow ArCH_2I + ArSO_3^-$$

All of these effects are greatly enhanced in the benzylhydryl and triphenylmethyl systems, where the additional aryl groups provide increased resonance stabilization of the transition states and intermediates.

Benzyl alcohol is made by the hydrolysis of benzyl chloride (p. 402).

$$C_6H_5CH_2Cl + H_2O + Na_2CO_3 \longrightarrow C_6H_5CH_2OH + NaHCO_3 + NaCl$$

Over 3 million pounds was produced in 1964. It is used chiefly to make benzyl esters. Benzyl alcohol is added to the extent of 1 to 3 per cent to solutions or suspensions intended for intramuscular or subcutaneous injections in order to minimize pain at the site of the injection. **Benzyl acetate** is used for the preparation of perfumes of the jasmine or gardenia type, and **benzyl benzoate** (p. 462) is used as a miticide. The antipyretic (fever-reducing) power of willow bark (*Salix alba*) was known to the ancients and is due to the bitter glucoside **salicin,** first isolated in 1827. Hydrolysis yields glucose and **saligenin** (*o-hydroxybenzyl alcohol,* p. 441).

$$o\text{-}(CH_2OH)C_6H_4OC_6H_{11}O_5 \xrightarrow{H_2O} o\text{-}(CH_2OH)C_6H_4OH + C_6H_{12}O_6$$

Salicin Saligenin Glucose

Diphenylcarbinol (*benzhydrol*) is made by the reduction of phenyl ketone (*benzophenone,* p. 464).

$$C_6H_5COC_6H_5 + Zn + 2\,NaOH \longrightarrow C_6H_5CHOHC_6H_5 + Na_2ZnO_2$$

It dehydrates with extreme ease to form **benzhydryl ether,** dilute acids being sufficient to cause reaction.

$$2(C_6H_5)_2CHOH \longrightarrow (C_6H_5)_2CHOCH(C_6H_5)_2 + H_2O$$

Benadryl, $(C_6H_5)_2CHOCH_2CH_2N(CH_3)_2$, the β-dimethylaminoethyl ether of benzhydrol, was one of the first synthetic chemicals to be used widely in the treatment of histamine allergies (p. 523). It also is one of the more effective preventives and cures for motion sickness (p. 529). It is synthesized by a modified Williamson synthesis from diphenylmethyl bromide and β-dimethylaminoethanol. Because of the high reactivity of the halide, sodium carbonate in an inert solvent suffices to cause reaction.

$$(C_6H_5)_2CHBr + HOCH_2CH_2N(CH_3)_2 \xrightarrow[140^\circ]{Na_2CO_3} (C_6H_5)_2CHOCH_2CH_2N(CH_3)_2 + NaBr + NaHCO_3$$

Triphenylcarbinol usually is prepared by the action of phenylmagnesium bromide on ethyl benzoate.

$$C_6H_5COOC_2H_5 + 2\,C_6H_5MgBr \longrightarrow MgBr(OC_2H_5) + (C_6H_5)_3COMgBr \xrightarrow{H_2O} (C_6H_5)_3COH + MgBr(OH)$$

When colorless triphenylcarbinol is dissolved in concentrated sulfuric acid, a yellow solution is obtained. When the solution is poured into water, colorless triphenylcarbinol precipitates. Moreover the depression of the freezing point of 100 per cent sulfuric acid by triphenylcarbinol shows that four particles are produced in solution. These results indicate that solution of triphenylcarbinol in sulfuric acid yields triphenylmethyl cations.

$$(C_6H_5)_3COH + 2\,H_2SO_4 \longrightarrow (C_6H_5)_3C^+ + H_3O^+ + 2\,HSO_4^-$$

The triphenylmethyl cation forms readily because it is stabilized by resonance.

$$\left\{ \begin{array}{ccc} (C_6H_5)_2\overset{+}{C}\text{—}C_6H_5 & \longleftrightarrow & (C_6H_5)_2C\text{=}C_6H_5^{+}\ (ortho) \\ \updownarrow & & \updownarrow \\ (C_6H_5)_2C\text{=}C_6H_5^{+}\ (ortho) & \longleftrightarrow & (C_6H_5)_2C\text{=}C_6H_5^{+}\ (para) \end{array} \right\}$$

Highly colored compounds formed by the reaction of organic compounds with anhydrous acids are known as **halochromic salts.** Other examples are the deep blue solutions obtained in the Liebermann nitroso reaction (p. 439) and the red to purple solutions of dibenzylideneacetone and its derivatives in concentrated sulfuric acid (p. 463). The same color is obtained whether the reagent is sulfuric acid or solutions of anhydrous aluminum chloride, zinc chloride, or boron fluoride. The ease with which these highly colored solutions are formed depends on the stability of the resulting carbonium ion and on the "acidity" of the reagent, that is, its attraction for a pair of electrons. Because of the indicator action of the organic compounds, their relative basicity or the relative acidity of the Lewis acids can be easily determined. Thus it can be shown that the ability of the reagents to acquire a pair of electrons is in the order anhydrous aluminum chloride in nitrobenzene > concentrated sulfuric acid > anhydrous zinc chloride in nitrobenzene, which order is the same as the catalytic activity of these reagents in condensation reactions.

Reaction of triphenylcarbinol with concentrated aqueous hydrochloric acid gives **triphenylmethyl chloride** (*trityl chloride*).

$$(C_6H_5)_3COH + HCl \longrightarrow (C_6H_5)_3CCl + H_2O$$

Triphenylmethyl chloride in the presence of pyridine reacts with primary alcohols to give triphenylmethyl ethers, a process frequently called *tritylation.*

$$RCH_2OH + ClC(C_6H_5)_3 + C_5H_5N \longrightarrow RCH_2OC(C_6H_5)_3 + C_5H_5NHCl$$

Primary amino groups also are alkylated readily. Since the trityl group is removed easily by acid hydrolysis, tritylation provides a means for the protection of hydroxyl or amino groups while reactions are carried out on other functional groups.

Phenethyl alcohol (*β-phenylethyl alcohol*) is a component of the essential oil of the flowers of roses grown by the perfume industry. It is synthesized commercially either by the sodium-alcohol reduction of ethyl phenylacetate, or by a Friedel-Crafts reaction between benzene and ethylene oxide (p. 605).

$$C_6H_5CH_2COOC_2H_5 + 4\,Na + 3\,C_2H_5OH \longrightarrow C_6H_5CH_2CH_2OH + 4\,NaOC_2H_5$$

$$C_6H_6 + \underset{O}{CH_2\text{—}CH_2} \xrightarrow{AlCl_3} C_6H_5CH_2CH_2OAlCl_2 \xrightarrow{H_2O} C_6H_5CH_2CH_2OH$$

In the acetolysis of β,β-dimethylphenethyl benzenesulfonate, rearrangement takes place to give α,α-dimethylphenethyl acetate. It is postulated that the transition state is a carbonium ion in which the phenyl group is bound to two adjacent carbon atoms.

$$(CH_3)_2\overset{C_6H_5}{C}\text{—}CH_2\text{—}OBs \longrightarrow {}^-OBs + (CH_3)_2C\overset{C_6H_5^{+}}{\text{——}}CH_2 \xrightarrow{HOAc} (CH_3)_2\underset{OAc}{C}\text{—}\overset{C_6H_5}{CH_2} + H^+$$

Such postulated intermediates are referred to as *nonclassical carbonium ions.* The reaction is an example of *neighboring group participation.* That such transition states or intermediates can exist has been proved for a number of reactions. For example, acetolysis of optically active $\alpha S,\beta S$-α,β-dimethylphenethyl *p*-toluenesulfonate[1] gives the optically active acetate with 94 per cent retention of configuration.

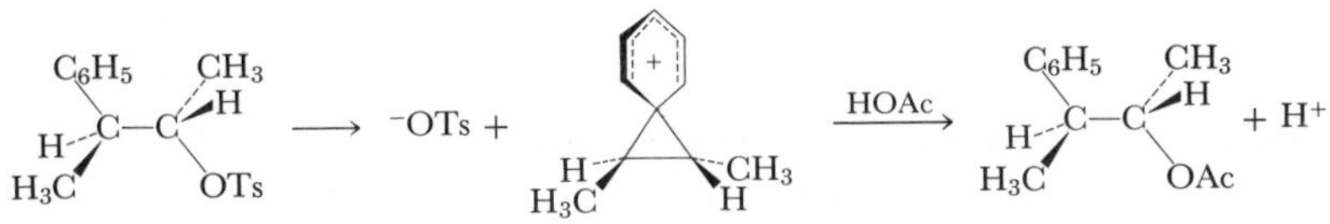

$$C_6H_5(H)(H_3C)C{-}C(CH_3)(H)(OTs) \longrightarrow {}^{-}OTs + \text{[bridged phenonium ion: } H, H_3C \text{ and } CH_3, H\text{]} \xrightarrow{HOAc} C_6H_5(H)(H_3C)C{-}C(CH_3)(H)(OAc) + H^+$$

If an ordinary S_N1 reaction had taken place with initial formation of a carbonium ion at C-1, this carbon would have become planar and a mixture of the αS and αR isomers would have been formed. An S_N2 reaction would have given only the αR isomer. Direct evidence has been obtained for the existence of a bridged phenonium ion. The proton magnetic resonance spectrum (p. 549) of a solution of 2,3-dimethyl-3-phenyl-2-butanol in liquid sulfur dioxide containing antimony pentafluoride and fluorosulfonic acid (cf. footnote 4, p. 104) at $-60°$ shows that all four methyl groups are equivalent as expected for a symmetrical intermediate.

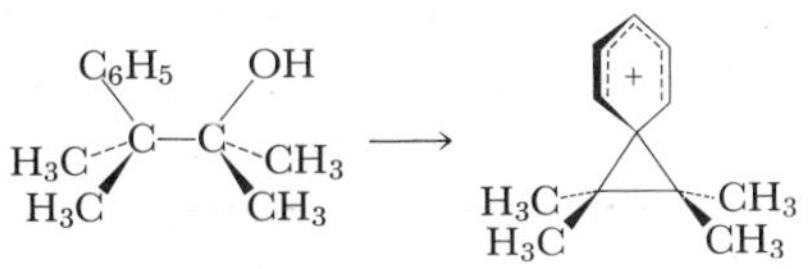

$$C_6H_5(H_3C)_2C{-}C(OH)(CH_3)_2 \longrightarrow \text{[bridged phenonium ion: } (H_3C)_2 \text{ and } (CH_3)_2\text{]}$$

Cinnamyl alcohol is made from cinnamaldehyde (p. 462) by the Meerwein-Ponndorf reduction (p. 206), which reduces the aldehyde group without affecting the double bond.

$$C_6H_5CH{=}CHCHO + (CH_3)_2CHOH \xrightarrow{Al(OPr\text{-}i)_3} C_6H_5CH{=}CHCH_2OH + (CH_3)_2CO$$

It has a hyacinth-like odor. The alcohol and its esters are used in perfumery.

Coniferin, which has been isolated from the cambial sap of conifers, is the glucoside of **coniferyl alcohol.**

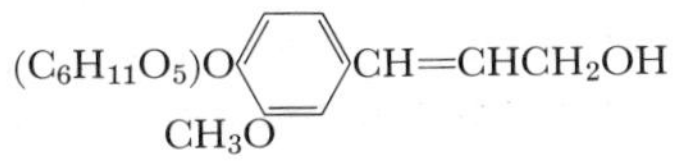

$(C_6H_{11}O_5)O$–$C_6H_3(OCH_3)$–$CH{=}CHCH_2OH$

Coniferin

HO–$C_6H_3(OCH_3)$–$CH{=}CHCH_2OH$

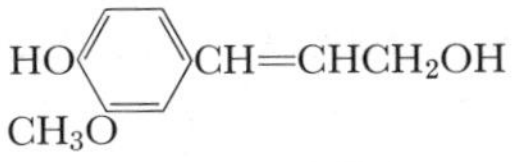

Coniferyl alcohol

The lignin of conifers (p. 342) appears to be an oxidative polymerization product of coniferyl alcohol. Alkaline oxidation of the lignin yields up to 25 per cent of its weight as vanillin (p. 463). The lignin of deciduous trees yields a mixture of vanillin and syringic aldehyde, the amount of syringic aldehyde frequently equaling the amount of vanillin. Accordingly the lignin of deciduous trees appears to be derived not only from coniferyl alcohol but also from **syringenin,** which first was obtained from the glycoside **syringin** present in bark of the lilac, *Syringa vulgaris.*

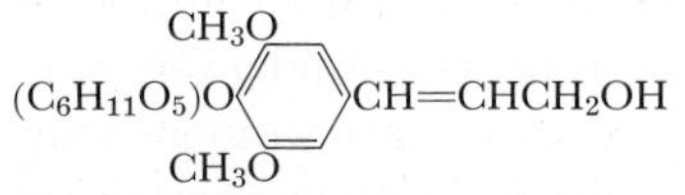

$(C_6H_{11}O_5)O$–$C_6H_2(OCH_3)_2$–$CH{=}CHCH_2OH$

Syringin

HO–$C_6H_2(OCH_3)_2$–$CH{=}CHCH_2OH$

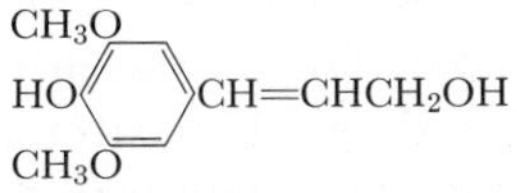

Syringenin

[1] For configurational nomenclature see page 312.

Benzylamine, dibenzylamine, and **tribenzylamine** are formed when ammonia reacts with benzyl chloride. They are only slightly less basic than the aliphatic amines. Primary and secondary aralkylamines can be made also from aromatic aldehydes or ketones by catalytic reduction of Schiff bases (p. 462). **α-Phenylethylamine** (*α-methylbenzylamine*) usually is made in the laboratory from acetophenone by reductive amination with ammonium formate (*Leuckart reaction*).

$$C_6H_5COCH_3 + HCOONH_4 \longrightarrow C_6H_5CH(CH_3)NH_2 + CO_2 + H_2O$$

It can be resolved readily into the two active forms, which are useful for the resolution of racemic acids. An important advantage of the Leuckart reaction over other reductive methods is that halogen and nitro groups that may be present in the aromatic nucleus are not affected.

Phenethylamine (*β-phenylethylamine*) may be regarded as the parent substance of a large group of medicinally important compounds known as *sympathomimetic amines.* The name was coined to indicate that they produce effects in the animal organism similar to those that result from stimulation of the postganglionic sympathetic nerve fibers. For example, they dilate the pupil of the eye (*mydriatic action*), strengthen the heart beat, and increase blood pressure (*pressor activity*). It now is believed that the sympathetic system performs these functions by the elaboration of epinephrine (*Adrenalin*) and norepinephrine (*Arterenol*), which are substituted phenethylamines. Formulas for only a few of the important derivatives of phenethylamine are given.

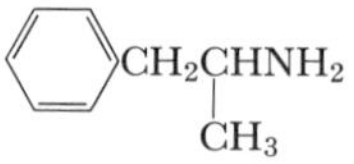

$C_6H_5CH_2CH(CH_3)NH_2$

Amphetamine
(*Benzedrine*)

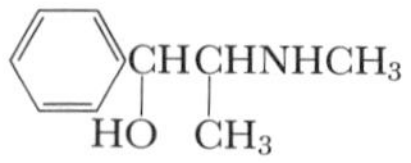

$C_6H_5CH(OH)CH(CH_3)NHCH_3$

Ephedrine

$C_6H_5CH(OH)CH(CH_3)NH_2$

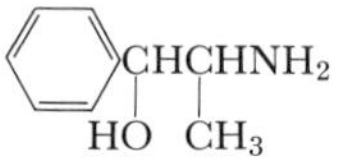

Phenylpropanolamine
(*Propadrine*)

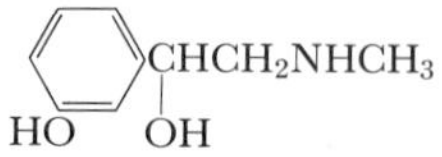

$HOC_6H_4CH(OH)CH_2NHCH_3$

Phenylephrine
(*Neo-Synephrine*)

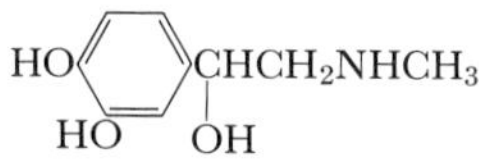

$(HO)_2C_6H_3CH(OH)CH_2NHCH_3$

Epinephrine
(*Adrenalin*)

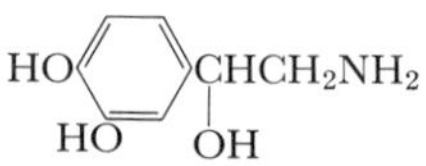

$(HO)_2C_6H_3CH(OH)CH_2NH_2$

Norepinephrine
(*Arterenol*)

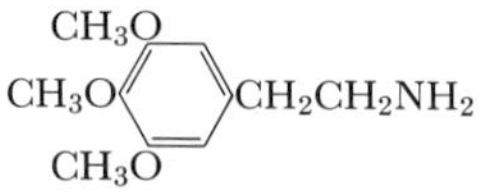

$(CH_3O)_3C_6H_2CH_2CH_2NH_2$

Mescaline

$O_2NC_6H_4CH(OH)CH(CH_2OH)NHCOCHCl_2$

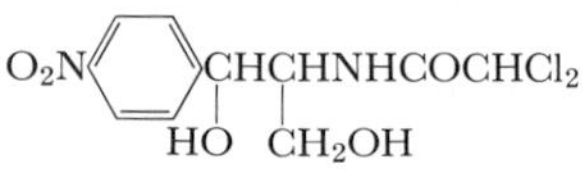

Chloramphenicol
(*Chloromycetin*)

Benzedrine has a powerful action on the central nervous system leading to temporary increase in alertness, lessened fatigue, and increased irritability and sleeplessness. This action is followed by fatigue and mental depression; hence considerable danger lies in promiscuous use of the drug. The dextrorotatory form is two to four times more active than racemic Benzedrine and is sold under the trade name **Dexedrine.** It reduces gastric contractions and is prescribed as an anorexic to prevent overeating. Catalytic reduction of the benzene ring of the *N*-methyl derivative of Benzedrine gives **Benzedrex,** $C_6H_{11}CH_2CH(CH_3)NHCH_3$. It constricts capillaries with minimal stimulation of the central nervous system and is sufficiently volatile to be used in nasal inhalers.

The closely related compound **ephedrine** [(−)-*αR,βR-N,α-dimethyl-β-hydroxyphenethylamine*][2] is present in the herb *Ma Huang* (*Ephedra vulgaris*) used medicinally by Chinese physicians for thousands of years. The active principle was isolated by Japanese workers in 1885, but it did not become well known in the western world until about 1925 after the investigations of Chen and Schmidt in the United States. It is administered for the treatment of bronchial asthma, and for a few years was used in nose drops for contracting the capillaries and relieving nasal congestion caused by colds. For the latter purpose it has been replaced largely by other synthetic phenethylamines such as **Neo-Synephrine,** which is the (−)-*R* isomer. **Propadrine** is used orally for this purpose. It is of interest that Propadrine differs from Benzedrine only in having an additional hydroxyl group; yet it is much less likely to cause central stimulation.

Epinephrine (*Adrenalin*) and **norepinephrine**[3] (*Arterenol*), which also are the (−)-*R* isomers, are the active principles produced by the adrenal medulla. Norepinephrine is liberated also by the postganglionic nerve endings and is regarded as one of the chemical transmitters of nerve impulses (p. 232). Epinephrine was the first hormone to be isolated in crystalline form (Abel, 1897) and the first hormone to be synthesized (Stolz, 1904, and Dakin, 1905). Hormones (Gr. *hormaein* to excite) are chemical substances produced by the cells of one part of an organism and transported by the fluids of the organism to another site where they exert their specific action. Ordinary epinephrine from adrenal glands (*U.S.P. epinephrine*) contains from 12 to 18 per cent of norepinephrine. Epinephrine in very small amounts increases the blood pressure by increasing the force and the rate of the heart beat and by constricting the arteries. This pressor effect is diminished by a dilating action on the peripheral capillary blood vessels. Norepinephrine is a powerful vasoconstrictor of all types of blood vessels and has about 1.5 times the pressor activity of epinephrine. U.S.P. epinephrine is administered with local anesthetics (p. 472) to prolong their action by constricting the blood vessels locally and preventing the anesthetic from being carried away from the site of injection. Epinephrine also is a powerful bronchodilator and is used in the treatment of bronchial asthma. It is about 100 times more effective than norepinephrine for this purpose. Both epinephrine and norepinephrine contain an asymmetric carbon atom, and the naturally occurring (−) forms are around twenty times more active than the (+) forms. It is of interest that norepinephrine is present in banana to the extent of 2 μg. per gram in the pulp and over 1 mg. per gram in the peel. Banana pulp also contains 8 μg. per gram and the peel 7 mg. per gram of **dopamine** (*3,4-dihydroxyphenethylamine*), a biological precursor of norepinephrine. All of the naturally occurring phenethylamines and numerous other compounds have their origin in the amino acid tyrosine (p. 353).

Mescaline has been isolated from mescal buttons, the tops of tubercles on the dumpling cactus (*Lophophora williamsii*). Mescal is employed by the Indians of southwestern United States and northern Mexico in religious ceremonies for the psychic effects and hallucinations that it produces. **Chloromycetin** [*chloramphenicol,* (−)-*αR, βR-N-dichloroacetyl-α-hydroxymethyl-β-hydroxy-4-nitrophenethylamine*], isolated from a species of *Streptomyces,* a soil organism, is the first antibiotic to be synthesized by a practical procedure. It is a relatively simple mole-

---

[2] For configurational nomenclature see page 312.

[3] The prefix *nor* is used to designate a homolog with one less methylene group. For example, norepinephrine is one of the next lower homologs of epinephrine. The prefix originated when a chemist of the Bayer Company in Germany labeled a compound *norbase* to indicate that the original base had a substituent on nitrogen. *Nor* was his abbreviation for "N ohne Radical."

cule and contains an aromatic nitro group and a dichloroacetyl group. Neither structural feature had been found previously in a natural product. Chloromycetin is especially effective against typhus and Rocky Mountain fever, but it is highly toxic and must be used with care.

Reaction of phenethylamine hydrochloride with dicyandiamide (p. 288) gives **1-phenethylbiguanide hydrochloride** (*DBI*).

$$C_6H_5CH_2CH_2NH_3{}^{+-}Cl + N{\equiv}C{-}NH\underset{\underset{NH}{\|}}{C}NH_2 \longrightarrow C_6H_5CH_2CH_2NH\underset{\underset{NH}{\|}}{C}{-}NH\underset{\underset{{}^{+}NH_2{}^{-}Cl}{\|}}{C}NH_2$$

It is effective orally in lowering the blood sugar of diabetics. **Phenethylhydrazine hydrogen sulfate** (*phenelzine*) is a monoamine oxidase inhibitor. It is reported to be an antidepressant that acts by blocking an enzyme that breaks down brain-stimulating substances.

ALDEHYDES

Preparation

The aromatic aldehydes can be prepared by any of the general methods used to prepare aliphatic aldehydes (p. 193). Additional methods are available that are applicable only to aromatic compounds. Usually some one procedure is preferable for a particular aromatic aldehyde. The more commonly used procedures are summarized below.

1. ***Hydrolysis of Dihalides.*** A methyl side chain may be oxidized indirectly by halogenation to the dihalide (p. 401) and hydrolysis.

$$ArCH_3 + 2\,Cl_2 \xrightarrow{h\nu} ArCHCl_2 + 2\,HCl$$

$$ArCHCl_2 \xrightarrow{H_2O(Na_2CO_3)} HCl + [ArCH(OH)Cl] \longrightarrow ArCHO + HCl$$

2. ***Gattermann[4] Carbon Monoxide Synthesis (Gattermann-Koch Reaction).*** Formyl chloride is stable only at liquid air temperature, but a mixture of carbon monoxide and hydrogen chloride in the presence of anhydrous aluminum chloride and cuprous chloride behaves like formyl chloride in its reactions with aromatic compounds (p. 377).

$$C_6H_5CH_3 + \underbrace{CO + HCl}_{[ClCHO]} \xrightarrow{AlCl_3\text{-}CuCl} p\text{-}CH_3C_6H_4CHO + HCl$$

An equimolecular mixture of carbon monoxide and hydrogen chloride can be prepared conveniently by the addition of chlorosulfonic acid to formic acid.

$$HCOOH + ClSO_3H \longrightarrow CO + HCl + H_2SO_4$$

3. ***Gattermann Hydrogen Cyanide Synthesis.*** This procedure is used chiefly on phenols and phenol ethers, which substitute with greater ease than hydrocarbons. It again is a Friedel-Crafts type of reaction (p. 377), an addition product of hydrogen chloride and hydrogen cyanide probably being the active reagent.

$$m\text{-}C_6H_4(OH)_2 + \underbrace{HCN + HCl}_{[ClCH{=}NH]} \xrightarrow{ZnCl_2} C_6H_3(OH)_2CH{=}NH_2{}^{+-}Cl \xrightarrow{H_2O} C_6H_3(OH)_2CHO + NH_4Cl$$

Zinc chloride is a sufficiently active catalyst for resorcinol. Zinc cyanide and hydrogen

[4] Ludwig Gattermann (1860–1920), professor at Heidelberg. He is best known for his work with diazonium salts (p. 432), and for his syntheses of aromatic aldehydes.

chloride may be used in place of anhydrous hydrogen cyanide, or the hydrogen cyanide may be replaced by bromocyanogen. Formerly the reaction was believed to be useful only with phenols and phenol ethers, but it has been shown to give 85 to 100 per cent yields with toluene or xylene if carried out at 100°.

4. ***Reimer-Tiemann Reaction.*** When strongly alkaline solutions of phenols react with chloroform, a formyl group is introduced *ortho* to the hydroxyl group. If the *ortho* position is occupied, substitution takes place in the *para* position. As with other reactions of chloroform in alkaline solution (p. 178), dichlorocarbene is formed, which acts as an electrophile. Substitution gives the benzylidene chloride, which is hydrolyzed rapidly to the aldehyde. Acidification yields the hydroxy aldehyde.

$$CHCl_3 + {}^-OH \longrightarrow H_2O + {}^-{:}CCl_3 \longrightarrow Cl^- + {:}CCl_2$$

$$C_6H_5O^- + {:}CCl_2 \longrightarrow [\text{cyclohexadienone with } H \text{ and } {}^-CCl_2] \longrightarrow o\text{-}(CHCl_2)C_6H_4O^- \xrightarrow{2\ {}^-OH} 2\,Cl^- + H_2O + o\text{-}(CHO)C_6H_4O^- \xrightarrow{H^+} o\text{-}(CHO)C_6H_4OH$$

Salicylaldehyde

5. ***Through Aldehyde-Ammonia Intermediates.*** Several methods for preparing aldehydes involve the intermediate formation of an aldehyde-ammonia that cannot undergo dehydration, which decomposes to the aldehyde and amine. Thus an activated aromatic nucleus reacts with *N,N*-dimethylformamide or *N*-methylformanilide in the presence of phosphorus oxychloride to give the aromatic aldehyde (*Vilsmeier reaction*). The reaction frequently is called *formylation.*

$$(CH_3)_2NC_6H_4H + OCHN(CH_3)_2 \xrightarrow{POCl_3} \left[(CH_3)_2NC_6H_4CH(OH)N(CH_3)_2\right] \longrightarrow (CH_3)_2NC_6H_4CHO + HN(CH_3)_2$$

p-Dimethylaminobenzaldehyde

Reactions

Aromatic aldehydes undergo most of the general addition reactions of aliphatic aldehydes such as reduction and oxime formation, but they do not polymerize. Some of the reactions deserve special consideration.

1. ***Oxidation.*** Aromatic aldehydes are oxidized by air to the acid, by way of the peroxy acid (p. 205). They are not oxidized as readily as aliphatic aldehydes, however, by solutions of oxidizing agents. For example, benzaldehyde does not reduce Fehling solution.

2. ***Cannizzaro Reaction.*** Like all aldehydes that do not have hydrogen on an α carbon atom, aromatic aldehydes undergo the Cannizzaro reaction (p. 209).

$$2\,C_6H_5CHO + NaOH \longrightarrow \underset{\text{Benzyl alcohol}}{C_6H_5CH_2OH} + \underset{\text{Sodium benzoate}}{C_6H_5COONa}$$

Because formaldehyde is oxidized more easily than the aromatic aldehydes, the latter can be converted completely to the alcohol by heating a mixture of the aromatic aldehyde with an excess of formaldehyde in the presence of concentrated aqueous sodium hydroxide (*crossed Cannizzaro reaction*).

$$C_6H_5CHO + HCHO + NaOH \longrightarrow C_6H_5CH_2OH + HCOONa$$

If a hydroxyl group is present in the *ortho* or *para* position, the Cannizzaro reaction takes place only when catalyzed by finely divided metals, especially silver.

Benzaldehyde is converted into benzyl benzoate by the catalytic action of aluminum ethoxide (Tishchenko reaction, p. 209).

$$2\ C_6H_5CHO \xrightarrow{Al(OC_2H_5)_3} C_6H_5COOCH_2C_6H_5$$

3. ***Condensation with Primary Amines.*** The initial products of addition of primary aliphatic or aromatic amines to aromatic aldehydes lose water spontaneously to give imino derivatives known as *Schiff bases*. The products from aromatic amines are known also as *anils*.

$$C_6H_5CHO + H_2NCH_3 \longrightarrow C_6H_5\underset{\displaystyle OH}{\underset{|}{C}}HNHCH_3 \longrightarrow \underset{\text{N-Methylbenzylidenimine}}{C_6H_5CH{=}NCH_3} + H_2O$$

$$C_6H_5CHO + H_2NC_6H_5 \longrightarrow C_6H_5\underset{\displaystyle OH}{\underset{|}{C}}HNHC_6H_5 \longrightarrow \underset{\text{Benzylideneaniline}}{C_6H_5CH{=}NC_6H_5} + H_2O$$

Schiff bases in general are more stable than *N*-alkyl alkylidenimines, RCH=NR, because the conjugation of the double bond with the aromatic nucleus greatly reduces its reactivity. Benzylideneanilines are still more stable because the double bond is conjugated with two aryl groups.

Schiff bases can be reduced easily by hydrogen and Raney nickel to the secondary amines. It is not necessary to isolate the Schiff bases. An alcoholic solution of aromatic aldehyde and primary amine is shaken with hydrogen in the presence of the catalyst (p. 458).

4. ***Aldol-Type Condensations with Aliphatic Aldehydes and Ketones.*** Aromatic aldehydes condense with other aldehydes or ketones having two α hydrogen atoms. The intermediate aldols usually cannot be isolated. They lose water even more readily than the aliphatic aldols (p. 202) because the double bond formed by loss of water is conjugated not only with the carbonyl group but also with the aromatic ring.

$$C_6H_5CHO + CH_3CHO \xrightarrow{\text{Dil. NaOH}} \left[C_6H_5\underset{\displaystyle OH}{\underset{|}{C}}HCH_2CHO\right] \longrightarrow \underset{\text{Cinnamaldehyde}}{C_6H_5CH{=}CHCHO} + H_2O$$

$$C_6H_5CHO + CH_3COCH_3 \xrightarrow[\text{10 per cent}]{\text{NaOH}} \left[C_6H_5\underset{\displaystyle OH}{\underset{|}{C}}HCH_2\underset{\displaystyle O}{\underset{\|}{C}}CH_3\right] \longrightarrow \underset{\text{Benzylideneacetone}}{C_6H_5CH{=}CH\underset{\displaystyle O}{\underset{\|}{C}}CH_3} + H_2O$$

$$C_6H_5CH{=}CH\underset{\displaystyle O}{\underset{\|}{C}}CH_3 + OCHC_6H_5 \xrightarrow[\text{10 per cent}]{\text{NaOH}} \underset{\text{Dibenzylideneacetone}}{C_6H_5CH{=}CH\underset{\displaystyle O}{\underset{\|}{C}}CH{=}CHC_6H_5} + H_2O$$

The yield of cinnamaldehyde is not so good as that of either benzylidene- or dibenzylideneacetone because of the polymerizing action of the alkali on the aliphatic aldehyde. Aldehyde condensations with ketones brought about by the use of 10 per cent sodium hydroxide are known as *Claisen* or *Claisen-Schmidt reactions*. The benzylideneacetones give highly colored halochromic salts with strong acids in anhydrous media (p. 456).

$$C_6H_5\text{—}CH{=}CH\text{—}\underset{\underset{:\ddot{O}:}{\|}}{C}CH_3 \xrightarrow{H^+} \left\{ C_6H_5\text{—}CH{=}CH\text{—}\underset{\underset{\ddot{O}^{+}:H}{\|}}{C}CH_3 \longleftrightarrow \overset{+}{C_6H_5}{=}CH\text{—}CH{=}\underset{\underset{:\ddot{O}:H}{|}}{C}CH_3 \right\}$$

5. ***Perkin Synthesis.*** An aldol-type addition of anhydrides to aromatic aldehydes was discovered by Perkin (p. 559). The basic catalyst for the reaction usually is the carboxylate ion corresponding to the anhydride used. The final product is an α,β-unsaturated acid formed by hydrolysis of the intermediate anhydride.

$$ArCHO + (RCH_2CO)_2O \xrightarrow[100^\circ]{Na^{+-}OCOCH_2R} \left[ArCH{=}\overset{R}{\overset{|}{C}}\text{—}\overset{O}{\overset{\|}{C}}\text{—}O\text{—}\overset{O}{\overset{\|}{C}}\text{—}CH_2R + H_2O \right] \longrightarrow ArCH{=}\overset{R}{\overset{|}{C}}COOH + RCH_2COOH$$

6. ***Benzoin Condensation.*** When benzaldehyde is shaken with an aqueous alkali cyanide solution, two molecules react to give a keto alcohol known as **benzoin.**

$$2\,C_6H_5CHO \xrightarrow{KCN} \underset{\text{Benzoin}}{C_6H_5\underset{\underset{O}{\|}}{C}CHOHC_6H_5}$$

Formally this reaction appears to take place by the addition of the hydrogen of one aldehyde group to the carbonyl group of the other and union of the carbon atoms. That the reaction is not simply an aldol-type condensation, however, is indicated by the fact that it is not catalyzed by ordinary bases but specifically by alkali cyanides. Lapworth (p. 197) proposed that the first step is the addition of cyanide ion followed by proton transfer to give the cyanohydrin. The cyanohydrin contains hydrogen α to a nitrile group and hence can undergo a base-catalyzed addition to a second molecule of aldehyde. The resulting cyanohydrin of benzoin then loses hydrogen cyanide.

$$ArCHO \xrightleftharpoons{^-:CN} Ar\underset{\underset{CN}{|}}{C}HO:^- \xrightleftharpoons[^-:OH]{H_2O} Ar\underset{\underset{CN}{|}}{C}HOH \xrightleftharpoons[H_2O]{^-:OH} Ar\underset{\underset{CN}{|}}{\ddot{C}}OH \xrightleftharpoons{ArCHO} Ar\text{—}\underset{CN\ \ OH}{C\text{—}\overset{O:^-}{\overset{|}{C}}HAr} \xrightleftharpoons[^-:OH]{H_2O}$$

$$Ar\text{—}\underset{CN\ \ OH}{C}\text{—}CHOHAr \xrightleftharpoons[H_2O]{^-:OH} \left[Ar\text{—}\underset{CN\ \ O:^-}{C}\text{—}CHOHAr \right] \xrightleftharpoons[^-:CN]{} Ar\underset{\underset{O}{\|}}{C}CHOHAr$$

Important Aromatic Aldehydes

Benzaldehyde is one of the products of hydrolysis of the cyanogenetic glycoside, *amygdalin* (Gr. *amygdalon* almond), which is present in the seeds of members of the prune family. It once was called *oil of bitter almonds.*

$$\underset{\text{Amygdalin}}{C_6H_5\underset{\underset{OC_{12}H_{21}O_{10}}{|}}{C}HCN} \xrightarrow{H_2O} 2\,\underset{\text{Glucose}}{C_6H_{12}O_6} + \underset{\text{Benzaldehyde cyanohydrin}}{C_6H_5CHOHCN} \rightleftharpoons \underset{\text{Benzaldehyde}}{C_6H_5CHO} + HCN$$

Benzaldehyde played an important part in the work that laid the foundations of structural organic chemistry because Liebig and Woehler showed that the benzoyl group, C_6H_5CO, could be transported intact through a large number of chemical transformations. Benzaldehyde is prepared commercially by the hydrolysis of benzylidene chloride, one of the products of the side-chain chlorination of toluene (p. 401). It is used to some extent as a flavoring agent and in the formulation of perfumes, but chiefly for the synthesis of other organic compounds.

Cinnamaldehyde, $C_6H_5CH{=}CHCHO$, is the chief component of cassia oil and oil of cinnamon, the volatile oils of the bark of *Cinnamomum cassia* and of *Cinnamomum ceylonicum.* It is synthesized by the aldol condensation of benzaldehyde with acetaldehyde (p. 462).

Vanillin is the principal odorous component of vanilla beans, the long podlike capsules of a tropical climbing orchid, *Vanilla planifolia.* It probably is the most widely used flavoring material with the exception of salt, pepper, and vinegar. Annual production of the synthetic product in the United States is over half a million pounds. Besides being used to produce a desired odor or flavor of vanilla, it has a pronounced effect in masking undesirable odors. For example, one part in 2000 will mask the undesirable odor of fresh paint. The masking and neutralizing of the odors of articles manufactured from rubber, textiles, and plastics is an important phase of the perfumer's art.

Numerous processes have been developed for the synthesis of vanillin. Currently it is obtained by the alkaline air oxidation of lignin sulfonates from waste sulfite liquors (pp. 342, 457). **Ethavan** is the trade name of a synthetic product containing an ethoxy group in place of the methoxy group. It has 3.5 times the flavoring power of vanillin.

Methylation of vanillin with methyl sulfate and alkali yields **veratral** (*3,4-dimethoxybenzaldehyde*), a material useful in organic synthesis. **Piperonal** (*heliotropin*) is 3,4-methylenedioxybenzaldehyde and is made from safrole (p. 446) by isomerization of the side chain and oxidation with permanganate. It has a pleasant odor and is used in perfumery. It also is valuable in organic synthesis. In contrast to benzaldehyde and cinnamic aldehyde, which rapidly oxidize in air to the acids, anisaldehyde, vanillin, and piperonal are highly stable to autoxidation.

KETONES

Ketones that do not have the carbonyl group adjacent to an aromatic ring are prepared by the general procedures described for aliphatic ketones (p. 193). Those ketones having a carbonyl group adjacent to the ring usually are prepared by the Friedel-Crafts reaction (p. 377). Only one acyl group is introduced because the carbonyl group is sufficiently deactivating to prevent a second substitution. For the same reason halogen and *meta*-directing groups prevent the Friedel-Crafts reaction from taking place. Nitrobenzene frequently is used as a solvent for these reactions. On the other hand, activating groups permit reactions of the Friedel-Crafts type to take place that do not occur with aromatic hydrocarbons. Resorcinol, for example, gives ketones by condensation with carboxylic acids in the presence of zinc chloride.

$$C_6H_4(OH)_2 + HOCOR \xrightarrow{ZnCl_2} C_6H_3(OH)_2COR + H_2O$$

Acetophenone is methyl phenyl ketone and **benzophenone** is phenyl ketone. Both have some use in perfumery and are valuable intermediates for organic synthesis. They have been made by the Friedel-Crafts reaction from benzene and acetyl chloride or acetic anhydride, and from benzene and benzoyl chloride respectively. The Claisen reaction of benzaldehyde with acetophenone (p. 462) gives **benzylideneacetophenone,** $C_6H_5COCH{=}CHC_6H_5$. It was given the name *chalcone* because its hydroxy derivatives have a reddish yellow color (Gr. *chalkos* copper). **Michler ketone** (*p-dimethylaminophenyl ketone*) is made from dimethylaniline and phosgene. It is used as a dye intermediate (p. 569).

$$4\,(CH_3)_2NC_6H_5 + COCl_2 \xrightarrow{ZnCl_2} (CH_3)_2NC_6H_4{-}\overset{\overset{O}{\|}}{C}{-}C_6H_4N(CH_3)_2 + 2\,[C_6H_5\overset{+}{N}H(CH_3)_2]Cl^-$$

Michler ketone

Chlorination or bromination of acetophenone gives **ω-chloro-** or **ω-bromoacetophenone** (*phenacyl chloride* or *phenacyl bromide*). Both are relatively harmless but potent lachrymators, and chloroacetophenone (*CN*) is used extensively as a tear gas for dispelling mobs.

Benzil, a 1,2 diketone, is obtained easily by the oxidation of benzoin (p. 463). Even mild oxidizing agents such as Fehling solution or copper sulfate in pyridine bring about the reaction because the new carbonyl group is conjugated with the original carbonyl group.

$$\underset{\text{Benzoin}}{C_6H_5COCHOHC_6H_5} + 2\,CuSO_4 + \underset{\text{Pyridine}}{2\,C_5H_5N} \longrightarrow \underset{\text{Benzil}}{C_6H_5COCOC_6H_5} + Cu_2SO_4 + (C_5H_5NH)_2SO_4$$

The cupric sulfate can be regenerated by passing air into the cuprous solution. Benzil when heated with aqueous or alcoholic alkali undergoes the **benzilic acid rearrangement** to give sodium benzilate.

$$C_6H_5COCOC_6H_5 + NaOH \longrightarrow (C_6H_5)_2COHCOONa$$

The rearrangement involves addition of hydroxide ion to a carbonyl group followed by a 1,2 shift of a phenide ion, with removal of a proton from the resulting carboxyl group and addition to the oxide ion. Acidification yields benzilic acid.

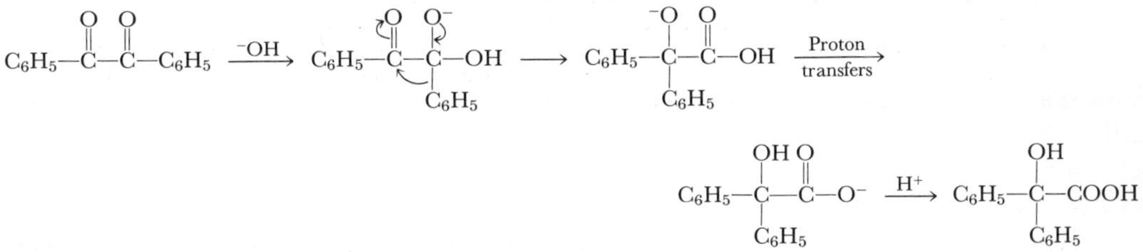

$$C_6H_5{-}\overset{O}{\overset{\|}{C}}{-}\overset{O}{\overset{\|}{C}}{-}C_6H_5 \xrightarrow{^{-}OH} C_6H_5{-}\overset{O}{\overset{\|}{C}}{-}\underset{C_6H_5}{\underset{|}{\overset{O^-}{\overset{|}{C}}}}{-}OH \longrightarrow C_6H_5{-}\underset{C_6H_5}{\underset{|}{\overset{^{-}O}{\overset{|}{C}}}}{-}\overset{O}{\overset{\|}{C}}{-}OH \xrightarrow[\text{transfers}]{\text{Proton}}$$

$$C_6H_5{-}\underset{C_6H_5}{\underset{|}{\overset{OH}{\overset{|}{C}}}}{-}\overset{O}{\overset{\|}{C}}{-}O^- \xrightarrow{H^+} C_6H_5{-}\underset{C_6H_5}{\underset{|}{\overset{OH}{\overset{|}{C}}}}{-}COOH$$

STEREOCHEMISTRY OF THE OXIMES

When benzaldehyde reacts with hydroxylamine, an oxime called α-benzaldoxime, melting at 34°, is formed. If this oxime is dissolved in ether and dry hydrogen chloride is added, a hydrochloride precipitates. Decomposition of the hydrochloride with sodium carbonate solution yields an oxime called β-benzaldoxime, which melts at 128–130°. Both oximes on hydrolysis yield benzaldehyde and hydroxylamine. Benzophenone gives only a single oxime, but mixed ketones give two oximes. For example, phenyl *p*-tolyl ketone gives an α-ketoxime, m.p. 153–154°, and a β-ketoxime, m.p. 115–116°.

Hantzsch[5] and Werner[6] in 1890 proposed a stereochemical explanation for the isomerism. They assumed that a carbon-nitrogen double bond can give geometrical isomers in the same way as a carbon-carbon double bond (p. 76). Hantzsch proposed the terms *syn* and *anti* for the two forms (Gr. *syn* with, *anti* opposite). For the aldoximes the prefixes refer to the relative positions of the hydrogen and hydroxyl group, and for the ketoximes they refer to the relative positions of the hydroxyl group and the group adjacent to the prefix.

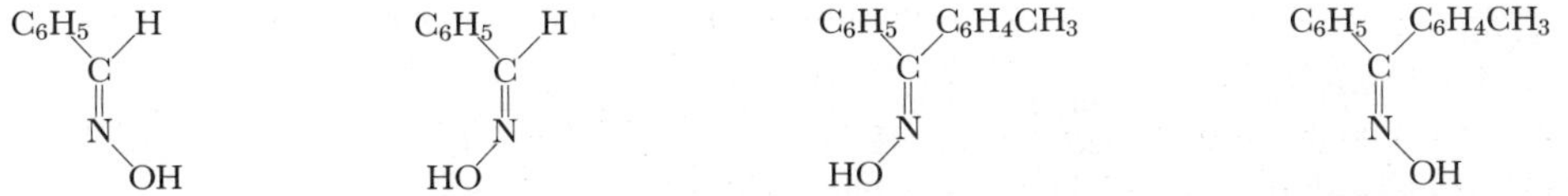

$$\begin{array}{cccc}
C_6H_5\ \ \ H & C_6H_5\ \ \ H & C_6H_5\ \ \ C_6H_4CH_3 & C_6H_5\ \ \ C_6H_4CH_3 \\
\diagdown\ \diagup & \diagdown\ \diagup & \diagdown\ \diagup & \diagdown\ \diagup \\
C & C & C & C \\
\| & \| & \| & \| \\
N & N & N & N \\
\ \ \diagdown OH & HO\diagup\ \ & HO\diagup\ \ & \ \ \diagdown OH
\end{array}$$

α- or *syn*-Benzaldoxime | β- or *anti*-Benzaldoxime | *syn*-Phenyl tolyl ketoxime or *anti*-tolyl phenyl ketoxime | *syn*-Tolyl phenyl ketoxime or *anti*-phenyl tolyl ketoxime

[5] Arthur Rudolf Hantzsch (1857–1935) made his Habilitationsschrift in 1882 on the synthesis of pyridines. He succeeded Victor Meyer at the Polytechnicum in Zurich as a full professor at the age of 28, and then succeeded Emil Fischer at Wuerzburg in 1893, and Wislicenus at Leipzig in 1903. Up to 1890 his research was concerned largely with the synthesis of heterocyclic compounds. After Meyer had proved that the benzil monoximes were not structural isomers, Hantzsch and Werner put forward their stereochemical explanation. This view led to the investigation of diazo compounds. Later Hantzsch investigated the tautomeric forms of aliphatic nitro compounds. Under the influence of the strong school of physical chemistry at Leipzig, he was led to the application of physical chemical methods, such as conductivity and freezing point depression, to the study of organic chemical problems. Following 1906 he concerned himself largely with the absorption spectra of organic compounds and the relation between color and chemical constitution.

[6] Alfred Werner (1866–1919), Swiss chemist, professor at the University of Zuerich. He was noted for his studies of coordination compounds and for his contributions to stereochemistry. He was awarded the Nobel Prize in Chemistry in 1913.

The configurations of the aldoximes can be determined by the behavior of their acetates when warmed with sodium carbonate solution. The aldoxime is regenerated from *syn* acetates, whereas *anti* acetates yield the nitrile.

When ketoximes are treated with a variety of acidic reagents such as concentrated sulfuric acid, acetyl chloride, or phosphorus pentachloride in ether solution, rearrangement to an amide takes place.

$$\underset{\text{Benzophenone oxime}}{\begin{array}{c} C_6H_5CC_6H_5 \\ \| \\ N \\ \diagdown OH \end{array}} \xrightarrow[H_2O]{PCl_5,\ \text{then}} \left[\begin{array}{c} HO \quad C_6H_5 \\ \diagdown \ \diagup \\ C \\ \| \\ C_6H_5{-}N \end{array} \right] \longrightarrow \underset{\text{Benzanilide}}{\begin{array}{c} C_6H_5 \\ \diagup \\ O{=}C \\ | \\ C_6H_5{-}NH \end{array}}$$

This reaction is known as the **Beckmann[7] rearrangement.** The isomeric ketoximes yield different products. The groups that exchange places in the rearrangement are those that are *anti* to each other. Since the structure of the products can be determined readily by hydrolysis, the Beckmann rearrangement can be used to determine the configurations of ketoximes. For example, the α isomer of phenyl tolyl ketoxime yields an anilide that can be hydrolyzed to *p*-toluic acid and aniline, and is assigned the *anti*-phenyl tolyl ketoxime structure, whereas the β isomer yields an anilide that can be hydrolyzed to benzoic acid and toluidine, and is assigned the *syn*-phenyl tolyl ketoxime structure.

$$\begin{array}{c} C_6H_5CC_6H_4CH_3 \\ \| \\ NOH \end{array} \xrightarrow[H_2O]{PCl_5,\ \text{then}} \left[\begin{array}{c} HOCC_6H_4CH_3 \\ \| \\ NC_6H_5 \end{array} \right] \longrightarrow \begin{array}{c} O{=}CC_6H_4CH_3 \\ | \\ NHC_6H_5 \end{array} \xrightarrow{H_2O} CH_3C_6H_4COOH + H_2NC_6H_5$$

$$\begin{array}{c} C_6H_5CC_6H_4CH_3 \\ \| \\ HON \end{array} \xrightarrow[H_2O]{PCl_5,\ \text{then}} \left[\begin{array}{c} C_6H_5COH \\ \| \\ CH_3C_6H_4N \end{array} \right] \longrightarrow \begin{array}{c} C_6H_5C{=}O \\ | \\ CH_3C_6H_4NH \end{array} \xrightarrow{H_2O} C_6H_5COOH + H_2NC_6H_4CH_3$$

PROBLEMS

26–*1*. Pair the following names with the appropriate reaction: (*a*) Gattermann, (*b*) Beckmann, (*c*) Perkin, (*d*) Claisen, (*e*) Reimer-Tiemann, (*f*) Cannizzaro, (*g*) Lapworth, (*h*) Hantzsch; (*1*) stereochemical explanation of isomeric oximes, (*2*) synthesis of α,β-unsaturated acids, (*3*) conversion of ketoximes to amides, (*4*) synthesis of α,β-unsaturated ketones, (*5*) synthesis of polyhydroxybenzaldehydes, (*6*) reaction of phenol with chloroform, (*7*) mechanism of benzoin condensation, (*8*) intermolecular oxidation and reduction of aldehydes.

26–*2*. Pair the following compounds with the proper common name: (*a*) 3-phenyl-2-propenal, (*b*) 1-phenyl-2-aminopropane, (*c*) α-hydroxy-β-methylamino-3,4-dihydroxyethylbenzene, (*d*) methyl phenyl ketone, (*e*) 2-hydroxybenzaldehyde, (*f*) 3-methoxy-4-hydroxybenzaldehyde, (*g*) dioxodiphenylethane, (*h*) diphenylhydroxyacetic acid; (*1*) salicylaldehyde, (*2*) cinnamaldehyde, (*3*) benzil, (*4*) Benzedrine, (*5*) vanillin, (*6*) epinephrine, (*7*) benzilic acid, (*8*) acetophenone.

26–*3*. Give equations for the following preparations: (*a*) *p*-chlorobenzaldehyde from toluene, (*b*) ethylphenylcarbinol using a Grignard reagent, (*c*) *p*-methylacetophenone from an acyl halide, (*d*) *o*-methylbenzylamine from *o*-xylene, (*e*) *p,p'*-dimethoxybenzil from anisaldehyde, (*f*) *o*-nitrobenzaldehyde from *o*-nitrobenzyl chloride, (*g*) piperonylideneacetone from safrole, (*h*) 2,4-dimethylbenzaldehyde from a hydrocarbon, (*i*) diethylbenzylamine from benzaldehyde.

---

[7] Ernst Beckmann (1853–1923), first director of the Kaiser Wilhelm Institut fuer Chemie. He was trained as a pharmacist and then studied under Kolbe and Wislicenus. He first observed the rearrangement of benzophenone oxime in 1886. Beckmann had a wide range of interests and occupied chairs in organic, physical, and pharmaceutical chemistry, and in food technology and nutrition. He developed many pieces of laboratory apparatus such as the sodium press, the differential thermometer, electromagnetic stirrers, and electrical heating apparatus. Beckmann redesigned the laboratories at Leipzig, and the Kaiser Wilhelm Institut was built largely according to his plans.

26–*4*. Give equations for the conversion of benzaldehyde into the following compounds: (*a*) $C_6H_5CH_2OH$, (*b*) C_6H_5COOH, (*c*) $C_6H_5CHOHCH_3$, (*d*) $C_6H_5CH{=}CHCHO$, (*e*) $C_6H_5CHOHCOC_6H_5$, (*f*) *m*-BrC_6H_4CHO, (*g*) C_6H_5CN, (*h*) $C_6H_5CHOHCN$, (*i*) $C_6H_5CH{=}N{-}NHCONH_2$, (*j*) $C_6H_5CH{=}NNHC_6H_5$, (*k*) $C_6H_5N{=}CHC_6H_5$, (*l*) $CH_3COCH{=}CHC_6H_5$, (*m*) $C_6H_5CHCl_2$, (*n*) $C_6H_5CH_2OCOC_6H_5$, (*o*) $C_6H_5CH{=}CHCOOH$, (*p*) $C_6H_5CH{=}CHNO_2$.

26–*5*. Give reactions for the following syntheses: (*a*) triphenylmethyl chloride from ethyl benzoate, (*b*) phenethyl chloride from benzyl chloride, (*c*) anisaldehyde from phenol, (*d*) coniferyl alcohol from vanillin.

26–*6*. Indicate a possible series of reactions for the synthesis of the following compounds that starts with hydrocarbons only as the source of the aromatic nuclei: (*a*) α-methylbenzyl ether, (*b*) phenethylamine, (*c*) *p*-chlorobenzaldehyde, (*d*) Michler ketone, (*e*) *m*-nitrobenzaldehyde, (*f*) ω-chloroacetophenone, (*g*) *p*-dimethylaminobenzaldehyde, (*h*) 4,6-dibenzoylresorcinol.

26–*7*. How can one distinguish by a chemical reaction between the members of the following pairs of compounds: (*a*) phenylethyl alcohol and phenetole, (*b*) anethole and anisaldehyde, (*c*) vanillin and piperonal, (*d*) *p*-tolualdehyde and anisaldehyde, (*e*) benzylaniline and diphenylamine, (*f*) acetophenone and isobutyrophenone, (*g*) nitrobenzene and benzaldehyde, (*h*) biphenyl and benzophenone, (*i*) benzyl phenyl ether and phenyl ether, (*j*) benzylamine and dibenzylamine.

26–*8*. Write the structural formulas and configurations for the oximes that give the following products when the amides formed by Beckmann rearrangement are hydrolyzed: (*a*) benzoic acid and *p*-toluidine, (*b*) *p*-bromobenzoic acid and *p*-aminocumene, (*c*) *m*-toluic acid and *p*-chloroaniline, (*d*) *o*-nitrobenzoic acid and aniline.

26–*9*. It is desired to prepare the following compounds with radioactive carbon at the starred position. Give the series of reactions by which this may be accomplished, starting with radioactive barium carbonate, BaC^*O_3, and any other readily available materials: (*a*) $C_6H_5C^*H_2CH_2OH$, (*b*) $C_6H_5C^*H{=}CHCOOH$, (*c*) $C_6H_5COOC^*H_2CH_3$, (*d*) $C_6H_5C^*HOHCOOH$, (*e*) $(C_6H_5)_2C^*(OH)C_2H_5$, (*f*) *m*-$C_6H_4(C^*OOH)COOH$.

26–*10*. Indicate which of the following compounds are oxidized by Fehling solution: (*a*) methyl α-glucoside, (*b*) benzil, (*c*) benzoin, (*d*) sucrose, (*e*) fructose, (*f*) lactose, (*g*) benzaldehyde, (*h*) glucose, (*i*) hydroxyacetic acid, (*j*) acetylmethylcarbinol.

26–*11*. Write perspective formulas that show the configuration of the natural forms of (*a*) epinephrine, (*b*) ephedrine, and (*c*) chloramphenicol.

26–*12*. An optically active substance *A*, $C_{10}H_{14}O$, is insoluble in dilute sodium hydroxide and in dilute hydrochloric acid. On treatment with excess methylmagnesium iodide, one mole of methane is produced per mole of *A*. When *A* is heated at 300° over alumina, it is converted into an optically inactive hydrocarbon *B*, $C_{10}H_{12}$, which decolorizes bromine water as well as permanganate solution. Ozonolysis of *B* gives formaldehyde and *D*, $C_9H_{10}O$. Substance *D* gives a silver mirror with Tollens reagent, and on being boiled with zinc-amalgam and hydrochloric acid, it is reduced to *E*, C_9H_{12}. Vigorous oxidation of *E* with boiling alkaline potassium permanganate yields an acid identical with that obtained by the oxidation of *m*-xylene. Deduce the structural formulas of all these compounds.

26–*13*. (*a*) Compound *A* is a colorless, neutral substance that gives a positive test with Tollens reagent. Treatment of *A* with refluxing 3 per cent potassium cyanide in aqueous alcohol solution gives *B*, $C_{16}H_{16}O_2$. Mild oxidation of *B* gives a pale yellow solid, $C_{16}H_{14}O_2$. Reaction of *B* with amalgamated zinc and hot concentrated hydrochloric acid gives a compound *C* that adds one mole of bromine to give *D*. Vigorous oxidation of *C* with alkaline permanganate gives an acid, *E*, that is identical with the acid obtained by the oxidation of *p*-xylene. Give the structure of compound *A* and the reactions that it undergoes. (*b*) Give the probable configurations of compounds *C* and *D*.

CHAPTER TWENTY-SEVEN

AROMATIC CARBOXYLIC ACIDS AND THEIR DERIVATIVES

The effect of an aromatic nucleus on the properties of a carboxyl group attached to it is less pronounced than the effect on other groups such as halogen, amino, and hydroxyl groups. Nevertheless special attention should be given to some of the methods of preparation and reactions of aromatic carboxylic acids. Many of their derivatives are important compounds.

Preparation

Strictly aromatic acids, in which the carboxyl group is attached to the ring, can be prepared by the general methods available for aliphatic carboxylic acids, such as the carbonation of Grignard reagents (p. 153), and the hydrolysis of nitriles (p. 238). They can be prepared also by the oxidation of carbon side chains (p. 387). Since the trichloromethyl group frequently can be formed by direct chlorination (p. 401), hydrolysis of this group becomes an important procedure for preparing aromatic acids.

$$\underset{\text{Benzotrichloride}}{C_6H_5CCl_3} + 2\,Na_2CO_3\text{ (aqueous)} \longrightarrow \underset{\text{Sodium benzoate}}{C_6H_5COONa} + 3\,NaCl + 2\,CO_2$$

Reactions

For the most part the reactions of aromatic acids are identical with those of aliphatic acids, the more important differences being in degree rather than in kind.

1. ***Acidity.*** The unsubstituted benzene ring gives rise to a slightly stronger acid than an unsubstituted alkyl group (benzoic acid, $pK_a = 4.2$, acetic acid, 4.8). The effect of substituents in the nucleus on the ionization of benzoic acid has been studied extensively. Acidity constants for a few compounds are listed in Table 27–*1*.

2. ***Esterification.*** If no substituents are present in the *ortho* positions, direct esterification of the carboxyl group proceeds as with straight-chain aliphatic acids. If, however, one

TABLE 27–*1*. ACIDITY CONSTANTS (pK_a) OF BENZOIC ACID AND DERIVATIVES

SUBSTITUENT	POSITION		
	meta	*ortho*	*para*
H	4.2	4.2	4.2
CH_3	4.3	3.9	4.4
OH	4.1	3.0	4.5
OCH_3	4.1	4.1	4.5
Br	3.8	2.9	4.0
Cl	3.8	2.9	4.0
NO_2	3.5	2.2	3.4

of the *ortho* positions is substituted, the rate of esterification is greatly decreased, and if both *ortho* positions are occupied, esterification does not take place. This behavior was noted first by Victor Meyer (p. 242) and sometimes is called the *Victor Meyer esterification law.* The esters of *ortho*-substituted benzoic acids can be prepared by the reaction of the silver salts with alkyl halides. Once they are formed, however, they cannot be hydrolyzed easily.

These effects are observed regardless of the nature of the substituents. Apparently a **steric effect** (*steric hindrance,* p. 103) is involved analogous to that noted for secondary and tertiary aliphatic carboxylic acids (p. 164). Groups larger than hydrogen so effectively occupy the space surrounding the carbon atom of the carboxyl group that there is insufficient room to form the intermediate or transition state necessary for the formation or the saponification of the ester. Reaction between the acid and alcohol can be effected by a mechanism that does not involve addition with sp^3 hybridization of the carbon atom of the carbonyl group. Thus mesitoic acid, which cannot be esterified readily by the usual acid-catalyzed procedure, can be converted to the carbonium ion by solution in concentrated sulfuric acid. When this solution is poured into alcohol, the ester results.

$$2,4,6\text{-}(CH_3)_3C_6H_2\text{—}COOH + 2\,H_2SO_4 \longrightarrow 2,4,6\text{-}(CH_3)_3C_6H_2\text{—}\overset{+}{C}O + H_3O^+ + 2\,HSO_4^-$$

$$2,4,6\text{-}(CH_3)_3C_6H_2\text{—}\overset{+}{C}O + ROH + HSO_4^- \longrightarrow 2,4,6\text{-}(CH_3)_3C_6H_2\text{—}COOR + H_2SO_4$$

In the same way when a solution of a hindered ester in concentrated sulfuric acid is poured into water, the hydrolysis products are formed.

3. ***Acid Halide Formation.*** In general the hydroxyl group of aromatic acids is replaced with more difficulty than that of aliphatic acids, and phosphorus pentachloride is the preferred reagent instead of phosphorus trichloride.

$$ArCOOH + PCl_5 \longrightarrow ArCOCl + POCl_3 + HCl$$

Thionyl chloride also can be used. Sometimes the acid chloride can be prepared conveniently by the chlorination of the aldehyde.

The aromatic acid chlorides, frequently called *aroyl chlorides,* are insoluble in water and hence react with it very slowly. As a result, alcohols and amines can be acylated in aqueous solution. When dilute alkali is used to catalyze the reaction and to combine with the hydrogen chloride formed, the reaction is called the *Schotten-Baumann reaction.*

$$ArCOCl + ROH + NaOH \longrightarrow ArCOOR + NaCl + H_2O$$

$$ArCOCl + H_2NR + NaOH \longrightarrow ArCONHR + NaCl + H_2O$$

Often a tertiary amine such as pyridine (p. 514) is used as both solvent and base.

$$ArCOCl + HOR + C_5H_5N \longrightarrow ArCOOR + C_5H_5NHCl$$

4. ***Decarboxylation.*** When the salt of an aromatic carboxylic acid is fused with alkali, the carboxyl group is replaced by hydrogen. The usual reagent is *soda lime,* a mixture of sodium hydroxide and calcium hydroxide.

$$ArCOO^- + {}^-OH \xrightarrow{\text{Heat}} Ar\text{—}C(O^-)(O^-)\text{—}OH \longrightarrow ArH + CO_3^{2-}$$

This reaction takes place with aliphatic carboxylic acids also, but except for sodium acetate, which yields methane, the side reactions are so complex that the reaction has no practical value. Certain substituted aliphatic (pp. 243, 622, 628), aromatic (pp. 417, 448), and heterocyclic (p. 517) carboxylic acids decarboxylate cleanly and even more easily than benzoic acid.

5. ***Reduction of the Carboxyl Group.*** Aromatic acids are reduced to benzyl alcohols in 80 to 99 per cent yields by lithium aluminum hydride. As with aliphatic compounds, reduction of the esters or acyl halides is preferred (p. 170) because they are more soluble in ether, and one third less reagent is required. Reduction of aromatic esters with sodium and alcohol (p. 175) is not satisfactory for the preparation of benzyl alcohols because the benzyl alcohol is reduced further to a methyl group (p. 454).

$$C_6H_5COOC_2H_5 + 6\,Na + 4\,C_2H_5OH \longrightarrow C_6H_5CH_3 + 5\,C_2H_5ONa + NaOH$$

Some Important Aromatic Acids and Their Derivatives

Benzoic acid was described in 1560 as a product of the distillation of Siam gum benzoin, an aromatic resin. The term *benzoin* is a corruption of the Arabic *luban jawi,* which means the *frankincense of Java*. The composition of benzoic acid was established by Liebig and Woehler in 1832. It has been manufactured technically by the oxidation of toluene, by the hydrolysis of benzotrichloride, and by the partial decarboxylation of phthalic acid (p. 475). The benzoyl derivative of glycine, $C_6H_5CONHCH_2COOH$, occurs in the urine of horses and other herbivora and is known as **hippuric acid** (Gr. *hippos* horse). When benzoic acid is ingested by animals, including humans, it is detoxified by combination with glycine and eliminated in the urine as hippuric acid. **Sodium benzoate** frequently is added to foods as a preservative and accounts for about half of the 16 million pounds of benzoic acid produced in 1964. Since only free benzoic acid is effective in inhibiting the growth of microorganisms, the *p*H of the food must be less than 4.5.

Benzoyl chloride is made by the partial hydrolysis of benzotrichloride. It is a liquid having a characteristic odor and strong lachrymatory action, and is used as a benzoylating agent (p. 469). When benzoyl chloride is shaken with sodium peroxide in water or with an alkaline solution of hydrogen peroxide, **benzoyl peroxide** is formed.

$$2\,C_6H_5COCl + Na_2O_2 \longrightarrow \underset{\text{Benzoyl peroxide}}{C_6H_5\overset{O}{\overset{\|}{C}}{-}O{-}O{-}\overset{O}{\overset{\|}{C}}C_6H_5} + 2\,NaCl$$

Benzoyl peroxide is used as a bleaching agent for edible oils and fats and for flour and as an initiator for polymerization reactions (pp. 484, 593). When it reacts with sodium methoxide, methyl benzoate and sodium peroxybenzoate are formed. Removal of the methyl benzoate, acidification, and extraction with chloroform gives a chloroform solution of **peroxybenzoic acid.**

$$(C_6H_5COO)_2 + {}^{-}OCH_3 \longrightarrow CH_3OOCC_6H_5 + C_6H_5\overset{O}{\overset{\|}{C}}{-}O{-}O^{-} \xrightarrow{H^+} \underset{\text{Peroxybenzoic acid}}{C_6H_5\overset{O}{\overset{\|}{C}}{-}O{-}OH}$$

Peroxybenzoic acid reacts quantitatively with nonconjugated double bonds to form the epoxide, a useful preparative procedure.

$$RCH{=}CHR + C_6H_5CO_3H \longrightarrow \underset{\diagdown O \diagup}{RCH{-}CHR} + C_6H_5COOH$$

Since the peroxybenzoic acid liberates iodine from potassium iodide, the chloroform solution can be standardized and the reaction used for the *quantitative estimation of double bonds.*

The methylbenzoic acids are called **toluic acids.** They are obtained by the partial oxidation of *o*-, *m*-, and *p*-xylene. ***N,N*-Diethyl-*m*-toluamide** (*deet*) is an effective insect repellent.

The sodium salts of **3-acetamido-2,4,6-triiodobenzoic acid** and **3,5-diacetamido-2,4,6-triiodobenzoic acid** are opaque to X-rays because of their high iodine content. They are excreted by the kidneys and are used for X-ray examination of the urinary tract.

Nitration of benzoic acid yields chiefly ***m*-nitrobenzoic acid.** The *ortho* or *para* derivative is made by the oxidation of *o*- or *p*-nitrotoluene. **3,5-Dinitrobenzoic acid** is made by the nitration of benzoic acid. Reaction with phosphorus pentachloride gives **3,5-dinitrobenzoyl chloride,** a valuable reagent for the identification of alcohols. The latter react with it even in aqueous solution by the Schotten-Baumann reaction (p. 469) to give solid 3,5-dinitrobenzoates. The introduction of three nitro groups is difficult (p. 411), and **2,4,6-trinitrobenzoic acid** is made by the oxidation of 2,4,6-trinitrotoluene (p. 417).

Anthranilic acid, *o*-aminobenzoic acid, is made by the action of alkaline sodium hypochlorite on phthalimide. The reaction involves hydrolysis to the sodium salt of phthalamic acid and a Hofmann rearrangement (p. 223).

$$C_6H_4(CO)_2NH \xrightarrow{NaOH} C_6H_4(COONa)(CONH_2) \xrightarrow{NaOH,\ NaOCl} C_6H_4(COONa)(NH_2)$$

Sodium anthranilate

Diazotized anthranilic acid is a convenient source of dehydrobenzene (*benzyne,* p. 399) that avoids the use of strong bases. Decomposition of the aqueous solution yields salicylic acid, but when the dry salt is decomposed in the presence of anhydrous *t*-butyl alcohol, *t*-butyl phenyl ether is formed.

$$C_6H_4(CO_2^-)(N_2^+) \longrightarrow CO_2 + N_2 + C_6H_4\!: \xrightarrow{HOC(CH_3)_3} C_6H_5OC(CH_3)_3$$

At one time a large quantity of anthranilic acid was used for the synthesis of indigo (p. 570). It still is used as an intermediate for the synthesis of thioindigo and other dyes, but production is only around 500,000 pounds. **Methyl anthranilate** is present in several odorous oils and contributes to the odor and flavor of grapes. The synthetic product is used in grape flavors. Production was 412,000 pounds in 1964.

***p*-Aminobenzoic acid** (*PABA*) is made by the reduction of *p*-nitrobenzoic acid. It is present in the mixture of substances known as the vitamin B complex and makes up a portion of the folic acid molecule (p. 529), but its necessity in the human diet has not been established. Certain bacteria, however, require it for the synthesis of folic acid, which is necessary for growth. Apparently this accounts for the effectiveness of sulfa drugs in combating infections by such organisms. The bacteria synthesize folic acid by means of an enzyme. If sites on the enzyme normally occupied by *p*-aminobenzoic acid are occupied by the structurally related sulfanilamide, the production of folic acid is prevented. Compounds that are related in structure to normal metabolic products but that inhibit essential metabolic processes are called *antimetabolites.* They are playing an increasingly important role in chemotherapy (p. 530).

Certain *p*-aminobenzoic esters have a local anesthetic action. The ethyl ester is known as **Anesthesin** (*benzocaine*) and the *n*-butyl ester as **Butesin.** They are used for relieving the pain of burns and open wounds. The most important derivatives are the aminoalkyl *p*-aminobenzoates, which on injection at the proper site anesthetize nerve fibers or endings, or block the transmission of pain by the nerve trunks. Their development resulted from attempts to find agents less toxic than cocaine, the first local anesthetic (p. 535). **Novocain** (*procaine,* β-diethylaminoethyl *p*-aminobenzoate hydrochloride) was synthesized by Einhorn in 1905. Literally thousands of related compounds have been synthesized since then and many placed on the market. Its method of synthesis is that generally used for all of the compounds of this type.

$$O_2N{-}C_6H_4{-}COCl + HOCH_2CH_2N(C_2H_5)_2 \longrightarrow$$

$$\left[O_2N{-}C_6H_4{-}COOCH_2CH_2\overset{+}{N}H(C_2H_5)_2\right]Cl^- \xrightarrow{\text{Fe, HCl}} \left[H_2N{-}C_6H_4{-}COOCH_2CH_2\overset{+}{N}H(C_2H_5)_2\right]Cl^-$$

Novocain (*procaine*)

Butyn (*butacaine*) is more active but also is more toxic than Novocain. **Pontocaine** (*tetracaine*) is used for surface anesthesia of the eye and for spinal anesthesia.

$$p\text{-}H_2N{-}C_6H_4{-}COOCH_2CH_2CH_2\overset{+}{N}H(n\text{-}C_4H_9)_2\ Cl^-$$

Butyn (*butacaine*)

$$p\text{-}n\text{-}C_4H_9NH{-}C_6H_4{-}COOCH_2CH_2\overset{+}{N}H(CH_3)_2\ Cl^-$$

Pontocaine (*tetracaine*)

Until recently procaine was the most widely used local anesthetic, but it has been largely displaced by lidocaine (*Xylocaine,* p. 426). Over 550,000 pounds of procaine was used in 1964, however, in the production of the salt of penicillin G (p. 524).

Of the hydroxy acids, the *ortho* isomer, **salicylic acid,** is by far the most important. It is prepared by the action of carbon dioxide on sodium phenoxide at 150° (*Kolbe synthesis*). The reaction is an electrophilic attack by carbon dioxide at the electron-rich *ortho* position.

$$\left\{C_6H_5O^- \longleftrightarrow O{=}C_6H_5{:}^-\right\} \xrightarrow{CO_2} \left[O{=}C_6H_5(H)COO^-\right] \longrightarrow o\text{-}HOC_6H_4COO^- \xrightarrow{H^+} o\text{-}HOC_6H_4COOH$$

Salicylic acid was prepared first by Piria in 1838 from salicylaldehyde, which derived its name from the fact that it was obtained by the oxidation of saligenin from the glucoside salicin (p. 455). Salicylic acid also was known as *spirsaeure* in the older German literature because salicylaldehyde is present in the volatile oil from the blossoms and leaves of various species of *Spiraea.*

The importance of salicylic acid and its derivatives lies in their antipyretic and analgesic action. **Sodium salicylate** was used first for this purpose in 1875. In the following year it was used for the treatment of rheumatic fever. The irritating action of sodium salicylate on the lining of the stomach led to the investigation of the action of various derivatives. **Salol** (*phenyl salicylate*) was introduced in 1886. It passes unchanged through the stomach and is hydrolyzed to phenol and salicylic acid in the alkaline juices of the intestines. Since, however, the weight of phenol liberated is almost equal to that of salicylic acid, there is considerable danger of phenol poisoning. Salol now is used as an enteric coating for medicinals that otherwise would be destroyed by the secretions of the stomach. When the pill reaches the alkaline intestines the salol is hydrolyzed and dissolved, and the medicinal is liberated. Salol absorbs ultraviolet light, and its main use is as a sun-screening agent and stabilizer of plastics. Salicylic acid now is administered chiefly as the acetyl derivative, which is known as **aspirin** (from the German *acetylspirsaeure*). Like salol it passes through the stomach unchanged and is hydrolyzed to salicylic acid in the intestines.

OH, $COOC_6H_5$ — Salol; $OCOCH_3$, COOH — Aspirin; OH, $COOCH_3$ — Methyl salicylate; $SHgC_2H_5$, COONa — Merthiolate

Salicylates lower body temperature rapidly and effectively in subjects having fever (*antipyretic action*) but have little effect if the temperature is normal. They are mild analgesics relieving certain types of pain such as headaches, neuralgia, and rheumatism, and compare favorably with the corticosteroids (p. 666) for the treatment of rheumatic fever and arthritis. The extensive use of aspirin is indicated by the production in the United States in 1964 of over 28 million pounds or 39 tons per day. The yearly production was enough for two hundred and four 5-grain pills for every member of the entire population. Although the toxic dose is large, promiscuous use of salicylates is not without danger, especially to children. Single doses of 5 to 10 grams have caused death, and 12 grams taken over a period of twenty-four hours causes symptoms of poisoning. Moreover in some persons salicylates cause stomach ulcers or violent allergic reactions.

The chief component of oil of wintergreen (*Gaultheria procumbens*) was identified as **methyl salicylate** in 1843. It is used to a considerable extent as a flavoring agent and in rubbing liniments. It has a mild irritating action on the skin and acts as a counterirritant for sore muscles. Production in 1964 was 3.8 million pounds. **Merthiolate,** a popular antiseptic, is the ethylmercuri derivative of sodium 2-mercaptobenzoate.

2-Hydroxy-4-aminobenzoic acid, commonly called *p-aminosalicylic acid* or *PAS,* can be made by the Kolbe synthesis (p. 472) from *m*-aminophenol. It is being used along with streptomycin (p. 350) and isoniazid (p. 518) in the treatment of tuberculosis.

Gallic acid (*3,4,5-trihydroxybenzoic acid*) is found free in sumach, tea, and many other plants. It is prepared by the hydrolysis of the **tannin** (*tannic acid*) present in galls (*gall nuts* or *nutgall*), which are the excrescences on young twigs of oaks and other plants caused by parasitic insects. This particular tannin is a mixture of gallic acid esters of glucose. A basic core of β-penta-*O*-galloylglucose has the phenolic hydroxyl groups esterified with three to five additional galloyl groups. The phenolic ester linkages are called *depside linkages* (Gr. *depsein* to tan). Tannins in general are substances that have the property of rendering the gelatin of hides insoluble, thereby converting the hide into leather. They are divided into the hydrolyzable tannins, which are esters of phenolic acids and are subdivided into gallotannins and ellagic tannins, and the nonhydrolyzable or condensed tannins, which are converted by acids into insoluble amorphous products and appear to be related to the flavans (p. 577). Most natural tannins used by the leather industry belong to the latter type.

Gallic acid and tannic acid are used in the manufacture of permanent writing inks. They form colorless water-soluble ferrous salts, which oxidize in air to black insoluble ferric salts. The latter are more permanent to light than the dye used to make the ink initially visible. Gallic acid is decarboxylated to produce pyrogallol (p. 447). **Propyl gallate** has been approved as an antioxidant for foods (cf. p. 445).

The *o*-, *m*-, and *p*-benzenedicarboxylic acids are known as *phthalic, isophthalic,* and *terephthalic acids.* **Phthalic acid** when heated above 180° rapidly loses water to form the volatile cyclic anhydride.

$$C_6H_4(COOH)_2 \xrightarrow{\text{Heat } 180°} C_6H_4(CO)_2O + H_2O$$

Phthalic acid — Phthalic anhydride

Hence in the usual methods for synthesizing phthalic acid, namely the high temperature oxidation of *o*-xylene or naphthalene (p. 487), it is the anhydride that is obtained.

$$C_6H_4(CH_3)_2 + 3\,O_2\,(\text{air}) \xrightarrow{V_2O_5,\ 360°} C_6H_4(CO)_2O + 3\,H_2O + 220\text{ kcal.}$$

$$\underset{\text{Naphthalene}}{C_{10}H_8} + 4\tfrac{1}{2}\,O_2\,(\text{air}) \xrightarrow{V_2O_5,\ 360°} C_6H_4(CO)_2O + 2\,CO_2 + 2\,H_2O + 511\text{ kcal.}$$

The oxidation of *o*-xylene is more readily controlled than the oxidation of naphthalene because less than half as much heat is evolved per mole of anhydride formed. Prior to the development of this process by Gibbs in 1917, sulfuric acid in the presence of mercuric sulfate had been used as the oxidizing agent.

Over 557 million pounds of phthalic anhydride was produced in 1964 at a price of about $0.10 per pound. About one third of the phthalic anhydride is converted to the **methyl, ethyl, *n*-butyl, 2-ethylhexyl,** and **higher alkyl esters** of phthalic acid, which are used as plasticizers of synthetic polymers, especially polyvinyl chloride. Production of the esters in millions of pounds in 1964 was as follows: methyl, 4; ethyl, 15; butyl, 18; 2-ethylhexyl, 189; "diisooctyl," 135; "diisodecyl," 78; all others, 162. **Methyl phthalate** (*dimethyl phthalate, DMP*) is an effective insect repellent.

Over 50 per cent of the phthalic anhydride is used for the manufacture of synthetic resins of which the **glyptal** (*glycerol* and *phthalic anhydride*) type is the simplest. Since both phthalic acid and glycerol are polyfunctional, heating a mixture gives polymeric esters. If three moles of phthalic anhydride to two of glycerol are used, a fusible resin first is obtained, which on further heating gives an infusible solid, insoluble in organic solvents. The cross links occur not only between two polymeric chains as indicated, but between large numbers of chains to give a three-dimensional network.

$$C_6H_4(CO)_2O + HOCH_2CHOHCH_2OH \longrightarrow HOOC\text{-}C_6H_4\text{-}COOCH_2CHOHCH_2OH \xrightarrow{x\text{ moles}}$$

$$x\,H_2O + \left[\text{—CO-}C_6H_4\text{-}COOCH_2CHOHCH_2O\text{—}\right]_x \xrightarrow{x\,C_6H_4(CO)_2O}$$

Fusible resin

$$\left[\text{—CO-}C_6H_4\text{-}COOCH_2\underset{|}{C}HCH_2O\text{—}\right]_x \text{ cross-linked through } \text{—O—CO-}C_6H_4\text{-CO—O—} \text{ to } \left[\text{—CO-}C_6H_4\text{-}COOCH_2\overset{|}{C}HCH_2O\text{—}\right]_x$$

Infusible resin

If one mole of phthalic anhydride, one mole of glycerol, and one mole of a monocarboxylic acid (a *modifying agent*) is heated, one of the hydroxyl groups is esterified with the monocarboxylic acid and cross linking is prevented, thus forming a fusible solid, soluble in organic solvents. If the monocarboxylic acids are polyunsaturated, such as those from the drying oils, the resin, when coated on a surface and heated, undergoes further oxidative polymerization like a drying oil to give very tough, elastic, weather-resistant films. It is these synthetic baking and drying enamels that have been used so extensively for the finishing of automobiles and household appliances. Water emulsions are used as the vehicle for interior wall paints. These resins belong to a more general class of polymeric substances derived from polyhydric alcohols and polybasic acids known as **alkyd resins.** Other types of polyester resins are used for laminating fiber glass (p. 635) and for making rubber-like products (p. 605).

About 5 per cent of the phthalic anhydride produced is used for the manufacture of anthraquinone (p. 497) and its derivatives, which are intermediates for the synthesis of anthraquinone dyes (p. 572). Smaller amounts are used for miscellaneous purposes such as the manufacture of phthalein and xanthene dyes (p. 569), benzoic acid, and phthalimide.

$$C_6H_4(CO)_2O + H_2O \xrightarrow[220^\circ]{\text{Cr–Na salt}} C_6H_5COOH + CO_2$$

Benzoic acid

$$C_6H_4(CO)_2O + NH_3 \xrightarrow{\text{Heat}} C_6H_4(CO)_2NH + H_2O$$

Phthalimide

Phaltan, an agricultural fungicide closely related to Captan (p. 649), is made by the reaction of potassium phthalimide with trichloromethanesulfenyl chloride (p. 269).

$$C_6H_4(CO)_2N^{-+}K + ClSCCl_3 \longrightarrow KCl + C_6H_4(CO)_2NSCCl_3$$

Reaction of phthalic anhydride with glutamic acid (p. 353) yields phthalylglutamic acid, which can be cyclized to the imide by heating with ammonia at 150°.

$$C_6H_4(CO)_2O + H_2NCH(COOH)CH_2CH_2COOH \longrightarrow H_2O + C_6H_4(CO)_2N{-}CH(COOH){-}CH_2{-}CH_2COOH \xrightarrow[150^\circ]{NH_3}$$

$$C_6H_4(CO)_2N{-}CH\langle CO{-}NH{-}CO{-}CH_2{-}CH_2\rangle + 2\,H_2O$$

The product, known as **thalidomide,** is an effective sedative and sleep producer but when taken during the early months of pregnancy leads to malformation of the fetus (*teratogenic activity*).

Isophthalic and **terephthalic acids** have become of technical importance only since 1950. They cannot be made by the same process used for phthalic anhydride because bond lengths and bond angles do not permit the formation of volatile monomeric anhydrides. Usually the oxidation of *m*- or *p*-xylene is brought about in several steps. The xylene, in the liquid phase and in the presence of a soluble cobalt or manganese salt, is oxidized by air to the toluic acid. The toluic acid is converted to the methyl ester, which is oxidized to the methyl hydrogen phthalate. The latter is isolated as the dimethyl ester.

$$\underset{m\text{- or }p\text{-Xylene}}{C_6H_4(CH_3)_2} \xrightarrow[140^\circ]{O_2,\text{ Co or Mn salt}} \underset{m\text{- or }p\text{-Toluic acid}}{C_6H_4(CH_3)COOH} \xrightarrow{CH_3OH,\ H^+}$$

$$\underset{\text{Methyl }m\text{- or }p\text{-toluate}}{C_6H_4(CH_3)COOCH_3} \xrightarrow[\text{salt}]{O_2,\text{ Co or Mn}} \underset{\text{Methyl hydrogen iso- or terephthalate}}{C_6H_4(COOH)COOCH_3} \xrightarrow{CH_3OH,\ H^+} \underset{\text{Methyl iso- or terephthalate}}{C_6H_4(COOCH_3)_2}$$

Alternatively, the toluic acids can be oxidized further in alkaline solution, or the xylenes can be oxidized in acetic acid solution. For most purposes methyl isophthalate and methyl terephthalate are preferable to the free acids, although the latter can be obtained by hydrolysis.

British investigators, extending some earlier work of Carothers (p. 588), found that fibers spun from the polymeric ester of terephthalic acid with the dihydric alcohol, ethylene glycol (p. 604), had superior properties. Textile fibers made from poly(ethylene terephthalate) are called **Terylene, Fortrel,** or **Dacron.** Transparent sheets are sold as **Mylar** and photographic film base as **Cronar.** The product is made by alcoholysis of methyl terephthalate with ethylene glycol to give hydroxyethyl terephthalate. Further alcoholysis with removal of ethylene glycol by distillation gives the polymer for which *n* has a value of 80 to 130.

$$CH_3OCO{-}C_6H_4{-}COOCH_3 + 2\ HOCH_2CH_2OH \xrightarrow{^-OCH_3}$$

$$2\ CH_3OH + HOCH_2CH_2OCO{-}C_6H_4{-}COOCH_2CH_2OH \xrightarrow[\text{heat}]{n\ \text{Moles,}}$$

$$n\ HOCH_2CH_2OH + ({-}CO{-}C_6H_4{-}COOCH_2CH_2O{-})_n$$

Terylene or Dacron

Catalytic hydrogenation of methyl terephthalate gives (1,4-dihydroxymethyl)cyclohexane.

$$CH_3OOC{-}C_6H_4{-}COOCH_3 + 7\ H_2 \xrightarrow{CuO\text{-}Cr_2O_3} HOCH_2(H)C_6H_{10}(H)CH_2OH + 2\ CH_3OH$$

Reaction of this dihydric alcohol with methyl terephthalate gives a polymeric ester having the structure

$$\left({-}CO{-}C_6H_4{-}COOCH_2{-}(H)C_6H_{10}(H){-}CH_2O{-}\right)_n$$

Fibers spun from it are sold as **Kodel.** The corresponding film base is called **Kodar.**

***N,N′*-Dimethyl-*N,N′*-dinitrosoterephthalamide** (*NTA*) is used as a blowing agent for foamed rubbers and plastics. When heated, it decomposes to nitrogen and methyl terephthalate. It is useful particularly for thick sections because of the small amount of heat evolved on decomposition, but its use is limited to those plastics that soften well below its decomposition point of 105°.

$$CH_3{-}N(NO){-}CO{-}C_6H_4{-}CO{-}N(NO){-}CH_3 \xrightarrow{\text{Heat}} N_2 + CH_3OCO{-}C_6H_4{-}COOCH_3$$

Because NTA is commercially available and stable at room temperature, it is the most convenient material for the preparation of diazomethane (p. 249).

$$CH_3{-}N(NO){-}CO{-}C_6H_4{-}CO{-}N(NO){-}CH_3 + 2\ NaOH\ (aq.) \longrightarrow 2\ CH_2N_2 + NaOOC{-}C_6H_4{-}COONa + 2\ H_2O$$

Oxidation of *o*-toluenesulfonic acid gives *o*-sulfobenzoic acid, which is cyclized by heat to **sulfobenzoic anhydride.** The anhydride is used in the synthesis of sulfonphthalein indicators (p. 569). **Saccharin,** first made in Remsen's laboratory,[1] is the cyclic imide of *o*-sulfobenzoic acid. It usually is stated to have a sweetness from 550 to 750 times that of cane sugar. Relative sweetness, however, depends on the method of determination and on the individual (p. 331). For most tastes a $\frac{1}{2}$ grain (0.03 gram) tablet replaces a heaping teaspoon (10 grams) of sucrose, indicating a sweetening power about 300 times that of sucrose. Saccharin has no food value and is used only when it is desirable to reduce the consumption of sugar. Saccharin has been made from toluene by a series of reactions.

$$C_6H_5CH_3 \xrightarrow{HOSO_2Cl} p\text{-}CH_3C_6H_4SO_2Cl \text{ and } o\text{-}CH_3C_6H_4SO_2Cl \xrightarrow{NH_3} o\text{-}CH_3C_6H_4SO_2NH_2 \xrightarrow{KMnO_4}$$

$$o\text{-}C_6H_4(COOH)(SO_2NH_2) \xrightarrow{Heat} C_6H_4\langle{}^{CO}_{SO_2}\rangle NH \xrightarrow{NaOH} C_6H_4\langle{}^{CO}_{SO_2}\rangle N^{-+}Na$$

Saccharin — Saccharin soluble

The imide is converted to the sodium salt to increase the solubility in water.

Mandelic acid (*α-hydroxyphenylacetic acid*) is prepared by the hydrolysis of benzaldehyde cyanohydrin (*mandelonitrile*).

$$C_6H_5CHO \xrightarrow{HCN} \underset{\text{Mandelonitrile}}{C_6H_5CHOHCN} \xrightarrow{H_2O,\ H^+} \underset{\text{Mandelic acid}}{C_6H_5CHOHCOOH}$$

Its name arises from the fact that it first was obtained by hydrolysis of an extract of bitter almonds (Ger. *mandel* almond) with hydrochloric acid (p. 463). The levorotatory form has the *R* configuration. Mandelic acid administered orally is excreted unchanged in the urine. Since it is bactericidal in acidic medium, it is used in the treatment of urinary infections. It may be administered as its salt with hexamethylenetetramine (p. 213), which also is eliminated in the urine and has antiseptic properties. ***p*-Bromomandelic acid** is used to determine the ratio of hafnium to zirconium in mixtures of their salts.

Cinnamic acid is made by the Perkin synthesis (p. 463).

$$C_6H_5CHO + (CH_3CO)_2O \xrightarrow{NaOCOCH_3} [C_6H_5CH{=}CHCOOCOCH_3 + H_2O] \longrightarrow$$

$$\underset{\text{Cinnamic acid}}{C_6H_5CH{=}CHCOOH} + HOOCCH_3$$

When salicylaldehyde is heated with acetic anhydride and sodium acetate, the resulting free acid is unstable and cyclizes spontaneously to the lactone known as **coumarin.**

$$o\text{-}C_6H_4(CHO)(OH) \xrightarrow[+\ NaOCOCH_3]{(CH_3CO)_2O} o\text{-}C_6H_4(OH)CH{=}CHCOOH \longrightarrow \underset{\text{Coumarin}}{C_6H_4\langle{}^{CH=CH}_{O-C=O}\rangle} + H_2O$$

Coumarin is present in the tonka bean, the seed of a tropical South American tree (*Dipteryx odorata*) known to the natives as *cumaru,* and has been isolated also from numerous other plants. It has the odor of newly mown hay and is used in perfumery. Coumarin was the

[1] Ira Remsen (1846–1927), professor of chemistry at Johns Hopkins University, later president of the university. He established the first adequate graduate school of chemistry in the United States. He was an outstanding teacher, and his textbooks in both inorganic and organic chemistry were used throughout the world. He established in 1879 the American Chemical Journal, which later was combined with the Journal of the American Chemical Society. Remsen was interested in the effect of substituents on the reactivity of other groups, and it was during an investigation of the effect of substituents on the oxidation of side chains that saccharin was discovered.

first natural perfume to be synthesized from a coal tar chemical (p. 559). Its use as a flavoring agent has been discontinued because of its toxicity. An interesting property is the ability to inhibit the germination of seeds.

Dicumarol (*dicoumarin*) prevents the coagulation of blood and is the agent responsible for the hemorrhagic disease that results when cattle eat spoiled sweet clover. **Coumadin,** another coumarin derivative, is used medicinally to reduce the possibility of blood clot formation, for example, after surgery or in the treatment of phlebitis and coronary thrombosis. Under the name *Warfarin,* it is sold as an effective rat poison.

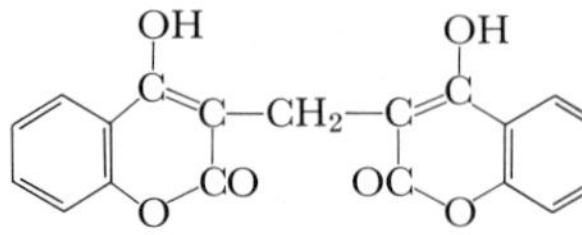

Dicumarol (*dicoumarin*)

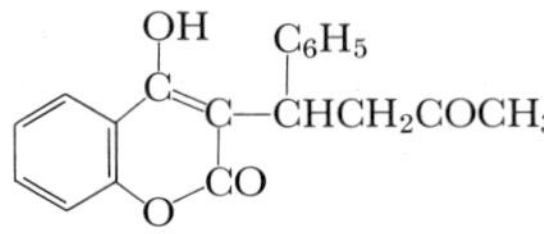

Coumadin (*Warfarin*)

PROBLEMS

27–*1*. Arrange the following compounds in the order of increasing boiling point: (*a*) benzoic acid, (*b*) benzamide, (*c*) phenol, (*d*) ethylbenzene, (*e*) thiophenol, (*f*) anisole.

27–*2*. Give reactions for the following preparations: (*a*) *o*-toluic acid from *o*-nitrotoluene, (*b*) benzanilide from benzotrichloride, (*c*) methyl *o*-toluate from *o*-xylene, (*d*) *m*-chlorobenzyl alcohol from *m*-chlorobenzoic acid, (*e*) benzoic acid from bromobenzene.

27–*3*. Give equations illustrating the steps necessary for the replacement of the indicated group in each of the following compounds by hydrogen: (*a*) COOH in *p*-toluic acid, (*b*) NH_2 in 3,4-dichloroaniline, (*c*) SO_3H in 3,5-dinitrobenzenesulfonic acid, (*d*) OH in 2,4,6-trimethylphenol, (*e*) Br in *p*-bromo-*s*-butylbenzene, (*f*) NO_2 in 2,4-dimethylnitrobenzene, (*g*) CH_3 in 2,4,6-trinitrotoluene, (*h*) SO_2NH_2 in sulfanilamide, (*i*) C_2H_5O in ethyl *p*-tolyl ether, (*j*) $C_6H_5CH_2$ in *N*-benzyl-*p*-chloroaniline.

27–*4*. Give a series of reactions for preparing the following compounds starting with an aromatic hydrocarbon as the source of the aromatic nucleus: (*a*) *o*-iodohippuric acid, (*b*) peroxybenzoic acid, (*c*) 3,5-dinitrobenzoyl chloride, (*d*) aspirin, (*e*) Anesthesin, (*f*) 2-hydroxy-4-aminobenzoic acid, (*g*) phenylacetic acid, (*h*) phenyl salicylate, (*i*) *p*-*t*-butylbenzoic acid, (*j*) 3-acetylamino-2,4,6,triiodobenzoic acid, (*k*) *N,N*-diethyltoluamide.

27–*5*. Give a series of reactions for the following preparations: (*a*) methyl anthranilate from phthalic anhydride, (*b*) Butyn from toluene and trimethylene chlorohydrin, (*c*) 3,4,5-trimethoxybenzaldehyde from gallic acid.

27–*6*. Give reactions for four procedures for the synthesis of *p*-toluic acid starting with toluene or *p*-xylene.

27–*7*. For the members of each of the following groups of compounds, predict the order of increasing acid strength: (*a*)—(*1*) benzenesulfonic acid, (*2*) hexanoamide, (*3*) benzoic acid, (*4*) *p*-nitrobenzoic acid, (*5*) *p*-toluic acid, (*6*) benzamide, and (*7*) phenol; (*b*)—(*1*) benzoic acid, (*2*) 2,4,6-trinitrobenzoic acid, (*3*) diphenylamine, (*4*) phthalimide, (*5*) mesitoic acid, (*6*) saccharin, and (*7*) 2,4,6-trichlorobenzoic acid; (*c*)—(*1*) phenylacetic acid, (*2*) 2,4-dichlorobenzoic acid, (*3*) 2,3-dimethylbenzoic acid, (*4*) *p*-chlorobenzoic acid, (*5*) *p*-toluamide, (*6*) benzenesulfonamide, and (*7*) 2,4,6-trichlorobenzoic acid.

27–*8*. Which of the following compounds will not give immediate evidence of reaction with cold concentrated hydrochloric acid: (*a*) $C_6H_5NH_2$, (*b*) $(C_2H_5)_4N^{+-}Cl$, (*c*) C_6H_5Cl, (*d*) aqueous C_6H_5COONa, (*e*) $(CH_3)_3COH$, (*f*) C_6H_5OH, (*g*) $(C_2H_5)_3N$, (*h*) $C_6H_5NO_2$ (*i*) C_6H_5COOH, (*j*) $C_6H_5N(CH_3)_3{}^{+-}OH$.

27–*9*. The following compounds all have the molecular formula C_9H_{12}. From the indicated behavior deduce the structural formula of each compound: (*a*) nitration gives only two mononitro derivatives and oxidation gives an acid having a neutralization equivalent of 83, (*b*) nitration gives only two mononitro derivatives and oxidation gives an acid having a neutralization equivalent of 70, (*c*) nitration gives only three mononitro derivatives and oxidation gives an acid with a neutralization equivalent of 122, (*d*) nitration gives only three mononitro derivatives and oxidation gives an acid having a neutralization equivalent of 70, (*e*) undergoes the same reactions as (*c*) but boils lower.

27–*10*. Compound *A* has the molecular formula $C_{18}H_{18}O_3$. It is insoluble in dilute sodium hydroxide, but when boiled with the alkali, it goes into solution. The saponification equivalent is 141. When the alkaline solution is distilled, only water comes over. Acidification of the alkaline solution gives a precipitate, *B*, which has a neutralization equivalent of 150. Strong oxidation of *B* gives *C*, which is an acid having a neutralization equivalent of 83. When *C* is heated above its melting point, it evolves a gas and forms a compound that dissolves only slowly in aqueous alkali. Write a structural formula for *A* and give equations for the reactions that take place.

27–*11*. Compound *A* has the molecular formula $C_{14}H_{10}N_2O_5$. It is soluble in aqueous bicarbonate and insoluble in dilute hydrochloric acid. When *A* is shaken with hydrogen in the presence of platinum catalyst, a product *B* is obtained that is soluble in either dilute acid or dilute alkali. When *B* is boiled with strong hydrochloric acid for some time and the solution is neutralized to *p*H 7, a single compound *C* is obtained that is soluble in both acid and alkali and has the molecular formula $C_7H_7NO_2$. Write a possible structural formula for *A* and give equations for the reactions that take place.

27–*12*. Compound *A* has the molecular formula $C_9H_{12}ClNO_2$. It is soluble in water and gives a precipitate with silver nitrate. When the aqueous solution is made alkaline, a precipitate *B* is formed that is free of halogen. When *B* is refluxed with alkali, it dissolves. When the alkaline solution is distilled, the distillate gives a positive iodoform test. When the alkaline residue is neutralized to *p*H 7, a precipitate *C* is formed that is soluble in either acid or alkali. When a cold acid solution of *C* is treated with sodium nitrite, no nitrogen is evolved. Addition of a cuprous cyanide–sodium cyanide solution yields a product that evolves ammonia when it is boiled with alkali. After ammonia no longer is formed, the alkaline solution is acidified. A precipitate is obtained that is identical with terephthalic acid. Write a structural formula for *A* and give equations for the reactions that take place.

27–*13*. Compound *A*, containing carbon, hydrogen, and oxygen, has a neutralization equivalent of 181 ± 1. When *A* is boiled with aqueous sodium hydroxide and the solution cooled and acidified, a solid, *B*, separates. *B* is soluble in sodium bicarbonate solution, decolorizes bromine, and gives a positive test with ferric chloride. Distillation of the filtrate from *B* gives an acid *C* that is isolated as the silver salt, which contains 64.7 per cent silver. What structures are possible for compound *A*?

27–*14*. A yellow solid, *A*, that gives a positive test for nitrogen, is soluble in water. Addition of cold alkali liberates a water-soluble, ether-soluble compound, *B*, that has an ammoniacal odor and reacts with benzenesulfonyl chloride to give a derivative that is insoluble in alkali. Addition of hydrochloric acid to a dilute aqueous solution of *A* gives a solid acid, *B*, that has a neutralization equivalent of 167 ± 2. After it is boiled with tin and hydrochloric acid, it becomes soluble in acid as well as alkali. What structures are possible for *A*?

27–*15*. Compound *A*, $C_8H_5ClO_2$, when heated with absolute alcohol, gives *B*, $C_{14}H_{20}O_4$. Vigorous oxidation of *A* with chromic acid gives an acid *C*, $C_8H_6O_4$. Reaction of *A* with aniline gives *D*, $C_{20}H_{16}N_2O$. When *C* is allowed to react with ammonia and heated to decomposition, compound *E*, $C_8H_5NO_2$, is obtained, which forms a potassium salt of the composition $C_8H_4NO_2K$. Give structural formulas for compounds *A* through *E*.

CHAPTER TWENTY-EIGHT

ARYLALKANES. ARYLALKENES. BIPHENYL AND ITS DERIVATIVES

ARYLALKANES

The simplest of the arylalkanes is **toluene** (p. 372). **Diphenylmethane** can be synthesized by the Friedel-Crafts reaction from benzene and either benzyl chloride or methylene chloride.

$$C_6H_6 + C_6H_5CH_2Cl \xrightarrow{AlCl_3} C_6H_5CH_2C_6H_5 + HCl$$

$$2\,C_6H_6 + CH_2Cl_2 \xrightarrow{AlCl_3} \underset{\text{Diphenylmethane}}{C_6H_5CH_2C_6H_5} + 2\,HCl$$

Oxidation of diphenylmethane yields benzophenone, which is relatively stable to further oxidation because of the lack of α hydrogen.

$$C_6H_5CH_2C_6H_5 \xrightarrow{Na_2Cr_2O_7,\ H_2SO_4} C_6H_5\overset{O}{\overset{\|}{C}}C_6H_5 + H_2O$$

Triphenylmethane may be prepared from benzene and chloroform.

$$3\,C_6H_6 + CHCl_3 \xrightarrow{AlCl_3} (C_6H_5)_3CH + 3\,HCl$$

As the hydrogens of methane are replaced successively by phenyl groups, the acidity of the remaining hydrogens increases. Toluene, diphenylmethane, and triphenylmethane have been assigned pK_a's of 37, 34, and 32.5, respectively, compared to 58 for methane. Removal of a proton leaves a negatively charged carbanion. The arylalkanes are more acidic than the alkanes because of the electronegative inductive effect of the aromatic nucleus and because the anions are stabilized by resonance. The more widely the charge can be distributed, the more stable the carbanion and the more acidic the hydrocarbon.

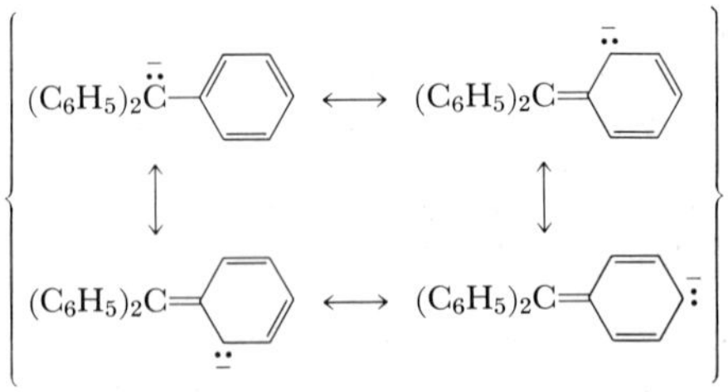

$(C_6H_5)_2\ddot{C}^{-}$–C₆H₅ ⟷ $(C_6H_5)_2C$=C₆H₅ (ortho carbanion)

↕ ↕

$(C_6H_5)_2C$=C₆H₅ (ortho carbanion) ⟷ $(C_6H_5)_2C$=C₆H₅ (para carbanion)

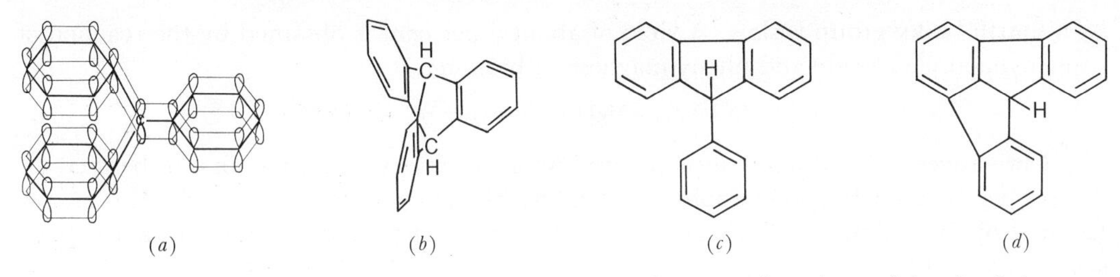

Figure 28–*1*. (*a*) Molecular orbital representation of resonance in a planar triphenylmethyl carbonium ion, carbanion, or free radical; (*b*) triptycene; (*c*) 9-phenylfluorene; (*d*) fluoradene.

Triphenylmethane is a stronger acid than ammonia ($pK_a = 35$) and reacts with potassium amide in liquid ammonia to give a red solution of potassium triphenylmethide.

$$(C_6H_5)_3CH + K^{+-}NH_2 \longrightarrow (C_6H_5)_3C^{-+}K + NH_3$$

The anion can be formed more conveniently by reaction of triphenylmethane with lithium aluminum hydride in tetrahydrofuran.

Resonance in the triphenylmethide ion results from overlap of the *p* orbital of the sp^2-hybridized central carbon with the π orbitals of the benzene nuclei and would be at a maximum when the rings and the central carbon atom are coplanar (Fig. 28–*1a*). Actually the ion cannot become entirely coplanar because steric interference of the *ortho* hydrogen atoms causes the rings to be twisted out of the plane and resonance accordingly is somewhat reduced. In **triptycene** (*9,10-benzeno,-9,10-dihydroanthracene,* p. 497) the methane carbon atoms are part of a cage structure (Fig. 28–*1b*), and the aromatic rings and methane carbons cannot approach planarity. Hence resonance stabilization of the carbanion is not possible and triptycene does not form a potassium salt in liquid ammonia. **9-Phenylfluorene** (Fig. 28–*1c*), in which two of the rings are forced to be planar, is a much stronger acid ($pK_a = 18.5$) than triphenylmethane, and **fluoradene** (Fig. 28–*1d*), in which three rings are planar, is a still more acidic hydrocarbon ($pK_a = 13.5$). Less than half of the acidity of triphenylmethane appears to be due to resonance stabilization of the anion. The remainder results from the electronegative inductive effect of the three benzene rings. Some of the increased acidity of the fluorenes is associated with the cyclopentadiene portion (p. 644).[1] Electron-attracting groups in the *ortho* and *para* positions of the ring increase the acidity. For example, **tri-*p*-nitrophenylmethane** gives a blue salt with potassium hydroxide in alcoholic solution. Here the nitro groups not only have an inductive effect but extend the resonance.

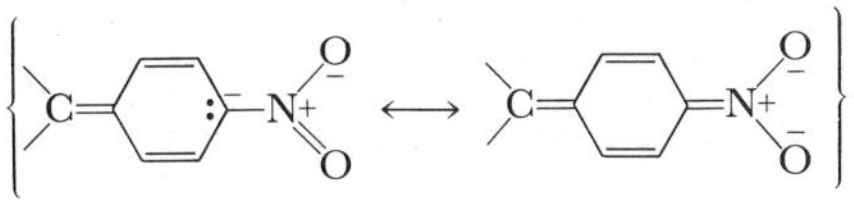

Direct oxidation of triphenylmethane yields triphenylcarbinol. The reaction stops at this point because the tertiary alcohol cannot dehydrate and be oxidized further without destroying a benzene ring.

$$(C_6H_5)_3CH \xrightarrow{Na_2Cr_2O_7,\ H_2SO_4} \underset{\text{Triphenylcarbinol}}{(C_6H_5)_3COH}$$

Tetraphenylmethane is difficult to prepare because the replacement of a halogen atom

[1] These considerations do not hold for carbonium ions, where other factors come into play. Thus the 9-phenylfluorenyl cation is less stable than the triphenylmethyl cation.

by a fourth bulky group is slow. A yield of about 5 per cent is obtained by the reaction of triphenylmethyl chloride and phenylmagnesium bromide.

$$(C_6H_5)_3CCl + C_6H_5MgBr \longrightarrow (C_6H_5)_4C + MgBrCl$$

Derivatives of **1,1-diphenylethane** are important insecticides. The first to be used for this purpose was 1,1-bis(*p*-chlorophenyl)-2,2,2-trichloroethane, known as **DDT.** It is manufactured by the action of sulfuric acid on a mixture of chlorobenzene and chloral (*trichloroacetaldehyde,* p. 121).

$$\underset{\text{Chloral}}{2\,ClC_6H_5 + OCHCCl_3} \xrightarrow{H_2SO_4} \underset{\text{DDT}}{(p\text{-}ClC_6H_4)_2CHCCl_3} + H_2O$$

Although first prepared in 1874, its insecticidal properties did not become known until 1942. During World War II it was used in delousing powders to prevent the spread of typhus, and as a mosquito larvicide to render swampy areas habitable. Since then it has been used widely as an agricultural insecticide and in household sprays. By 1963 the body fat of the general population of both Great Britain and the United States contained from 0.85 to 18.4 p.p.m. of DDT and its metabolic product, DDE [*1,1-dichloro-2,2-bis(p-chlorophenyl)ethylene*]. Production in the United States was around 124 million pounds in 1964. Many related compounds have been prepared and their properties investigated. **DDD** or **TDE** is 1,1-bis(*p*-chlorophenyl)-2,2-dichloroethane, and **methoxychlor** is 1,1-bis(*p*-methoxyphenyl)-2,2,2-trichloroethane.

Hexaphenylethane was prepared by Gomberg[2] in 1900 by the action of finely divided metals on a benzene solution of triphenylmethyl chloride.

$$2\,(C_6H_5)_3CCl + 2\,Ag\ (\text{or Zn}) \longrightarrow \underset{\text{Hexaphenylethane}}{(C_6H_5)_3CC(C_6H_5)_3} + 2\,AgCl\ (\text{or } ZnCl_2)$$

Hexaphenylethane is a colorless crystalline solid, but its solutions are deep yellow. The solutions rapidly absorb oxygen from the air to give triphenylmethyl peroxide.

$$(C_6H_5)_3CC(C_6H_5)_3 + O_2 \longrightarrow (C_6H_5)_3C{-}O{-}O{-}C(C_6H_5)_3$$

They decolorize iodine with the formation of triphenylmethyl iodide and react with alkali metals to give brick red metallic salts.

$$(C_6H_5)_3CC(C_6H_5)_3 + I_2 \longrightarrow 2\,(C_6H_5)_3CI$$
$$+ 2\,K \longrightarrow 2\,(C_6H_5)_3C^{-+}K$$

None of these reactions would be expected from the structure of the hydrocarbon. The explanation lies in the dissociation of hexaphenylethane in solution into two free **triphenylmethyl** groups, each with an unpaired electron.

$$(C_6H_5)_3C{:}C(C_6H_5)_3 \rightleftharpoons 2\,(C_6H_5)_3C\cdot$$

These unpaired electrons can pair with the unpaired electrons of the oxygen molecule or with that of an iodine atom to form covalent bonds, or can accept an electron from a metallic atom to form an ionic bond.

As with the triphenylmethyl cation (p. 456) and the triphenylmethyl anion (p. 480), the greater stability of the triphenylmethyl free radical as compared to that of free alkyl radicals (p. 248) is the result, at least in part, of resonance.

---

[2] Moses Gomberg (1866–1947), Russian-born American chemist who obtained both undergraduate and graduate training at the University of Michigan and then worked with Baeyer and Victor Meyer. After synthesizing tetraphenylmethane, he returned to Michigan and attempted to synthesize hexaphenylethane with the result that he discovered stable free radicals.

$(C_6H_5)_2\dot{C}—C_6H_5 \longleftrightarrow (C_6H_5)_2C=C_6H_5\cdot$ (ortho)

$\updownarrow \qquad\qquad \updownarrow$

$(C_6H_5)_2C=C_6H_5\cdot$ (ortho) $\longleftrightarrow (C_6H_5)_2C=C_6H_5\cdot$ (para)

The decrease in bond dissociation energy from 83 kcal. for ethane to 11 kcal. for hexaphenylethane is attributable possibly half to resonance and half to steric repulsion of the aryl groups. Dissociation energies for ethane and for hexaphenylethane are not strictly comparable, however, since one is for the gas phase and the other for solution.

Since electronic interaction takes place with all three benzene nuclei, there are ten possible positions for the unpaired electron, and a total of 44 valence bond structures can be written. If more aromatic nuclei are conjugated with the system, the possibility for resonance increases and the free radicals become still more stable. Thus replacement of the phenyl groups of hexaphenylethane successively by *p*-biphenyl groups (*xenyl groups*) increases the degree of dissociation. In 3 per cent solutions in benzene, triphenylmethyl exists in equilibrium with its dimer to the extent of about 2 per cent, xenyldiphenylmethyl 15 per cent, dixenylphenylmethyl 79 per cent, and trixenylmethyl 100 per cent.

The development of apparatus for the detection of electron paramagnetic resonance absorption (p. 550) has provided a means for detecting unpaired electrons, and hence free radicals, in concentrations as low as 10^{-8} molar. These studies have confirmed the existence of free radicals and have shown the presence of free radicals not previously detected. Moreover the distribution of the unpaired electron over the whole group in a free radical such as triphenylmethyl has been confirmed, since the interaction of the unpaired electron with all of the hydrogen atoms, whose nuclei also have a magnetic moment (p. 547), has been observed.

ARYLALKENES AND ARYLALKYNES

From a practical viewpoint the simplest of the arylalkenes, **phenylethylene,** is a very important substance. Three billion pounds was produced in the United States in 1964, which exceeded the production of any other synthetic aromatic chemical except ethylbenzene, from which it is made. The common name **styrene** was given to it because it was obtained first in 1831 by the distillation of *liquid storax,* a fragrant balsam from *Liquidambar styraciflua* and *Liquidambar orientalis.* It probably is formed from the cinnamic acid present, which decomposes on slow distillation.

$$C_6H_5CH{=}CHCOOH \xrightarrow{\text{Heat}} \underset{\text{Styrene}}{C_6H_5CH{=}CH_2} + CO_2$$

Styrene is present also in coal tar, which was the first commercial source. At present it is made by dehydrogenating ethylbenzene (p. 391) catalytically. The catalyst is either iron oxide or zinc oxide with small amounts of promoters such as chromic oxide. Superheated steam is used to supply the heat for the endothermic reaction and to shift the equilibrium by reducing the partial pressure of the products.

$$C_6H_5CH_2CH_3 \xrightarrow[650^\circ]{Fe_2O_3 \text{ or } ZnO} C_6H_5CH{=}CH_2 + H_2$$

For storage and shipment of styrene, an antioxidant such as *t*-butylcatechol is added, and the temperature is kept low by refrigeration.

The importance of styrene lies in its easy polymerization, which was observed as early as 1839. As late as 1930, however, polystyrene had little commercial use because the product was very brittle. Subsequently the brittle polymers were avoided by proper purification of the monomer to remove the small amounts of impurities that cause cross linking. Very pure styrene, polymerized at room temperature, gives a fibrous, tough solid having a molecular weight of about 500,000, which is a **thermoplastic linear polymer,** soluble in benzene. To carry out the polymerization, the antioxidant is removed by washing with alkali, benzoyl peroxide is added as a catalyst, and the solution is heated to 85–100°.

$$x\, C_6H_5CH{=}CH_2 \xrightarrow{(C_6H_5CO)_2O_2} \left[-CH_2\underset{\displaystyle C_6H_5}{\underset{|}{C}}H- \right]_x$$

If as little as 0.01 per cent of divinylbenzene, $CH_2{=}CHC_6H_4CH{=}CH_2$, is present, the product no longer is thermoplastic and only swells in benzene because cross linking of the linear chains has taken place.

$$\left[\begin{array}{c} -CH_2CHCH_2CHCH_2CH- \\ \quad | \qquad\quad | \qquad\quad | \\ C_6H_5 \quad C_6H_4 \quad C_6H_5 \\ \qquad\quad | \\ -CH_2CHCH_2CHCH_2CH- \\ \quad | \qquad\qquad\quad | \\ C_6H_5 \qquad\quad C_6H_5 \end{array} \right]_x$$

Polystyrene has high transparency and strength and, being a hydrocarbon, it is a good electrical insulator, is chemically resistant, and is light in weight.

Beads of cross-linked copolymers of styrene with divinylbenzene are the starting point for the production of most **ion-exchange resins.** Sulfonation with sulfuric acid or chlorosulfonic acid introduces into each benzene ring a sulfonic acid group, which may be converted into the sodium salt. The chief use is for softening or for deionizing water. Hard water contains alkaline earth and heavy metal salts, which leave deposits in water heaters and boilers and which form precipitates with soaps (p. 190). When such water is passed over the sodium salt of a **cation-exchange material,** these undesirable ions are replaced by sodium ions, the salts of which are more soluble.

$$\left[-C_6H_4-SO_3^{-+}Na \right]_x + x\, M^{+-}A \rightleftharpoons \left[-C_6H_4-SO_3^{-+}M \right]_x + x\, Na^{+-}A$$

In the above equilibrium, M^+ is any cation and A^- any anion, which need not be monovalent. When exchange has taken place to the extent that the undesirable cations no longer are removed sufficiently from solution, the ion exchanger is regenerated by allowing it to stand with a concentrated salt solution, which shifts the equilibrium to the left.

If the ion-exchange resin is used as the free acid, it exchanges hydrogen ions for other cations and can be regenerated with concentrated hydrochloric acid.

$$\left[-C_6H_4-SO_3^{-+}H \right]_x + x\, M^{+-}A \rightleftharpoons \left[-C_6H_4-SO_3^{-+}M \right]_x + x\, H^{+-}A$$

Synthetic organic resins give rise to the only useful **anion-exchange materials.** The more common ones are polystyrenes that have quaternary ammonium groups. Chloromethylation (p. 401) of the resin followed by reaction with a tertiary amine such as trimethylamine gives the quaternary salt.

$$\left[-C_6H_5 \right]_x + x\, HCHO + x\, HCl \xrightarrow{ZnCl_2} \left[-C_6H_4-CH_2Cl \right]_x \xrightarrow{x\, (CH_3)_3N} \left[-C_6H_4-CH_2\overset{+}{N}(CH_3)_3Cl^- \right]_x$$

The chloride ion can be exchanged for hydroxide ion by allowing the resin to stand with a solution of sodium hydroxide. Water that has exchanged its metallic ions for hydrogen ions can be passed over the basic anion-exchange resin to give ion-free water.

Numerous types of ion-exchange resins are now available for special purposes. Other addition and condensation polymers may be used as well as other types of ion-exchange groups such as carboxyl, phosphonic, and phosphinic groups, amino, phosphonium, and sulfonium groups, and metal chelating groups. Of the total production of styrene about 50 per cent is used for plastics (p. 484), 30 per cent for synthetic rubber (p. 587), 7 per cent for latex paints (p. 589), 4 per cent for polyester resins (p. 635), and 2 per cent for other copolymers such as those with acrylonitrile (*SAN*, p. 589) and acrylonitrile-butadiene (*ABS*, p. 589).

Three diphenylethylenes are possible. **1,1-Diphenylethylene** can be made by the reaction of acetophenone and phenylmagnesium bromide. The intermediate carbinol can be dehydrated readily to the olefin.

$$C_6H_5COCH_3 \xrightarrow{C_6H_5MgBr} (C_6H_5)_2C(OMgBr)CH_3 \xrightarrow{H_2O} (C_6H_5)_2C(OH)CH_3 \xrightarrow{H_2SO_4} (C_6H_5)_2C{=}CH_2 + H_2O$$

The two 1,2-diphenylethylenes are known as *cis-* and *trans-stilbene* (p. 76). ***trans*-Stilbene** is the stable isomer and is obtained by the dehydration of benzylphenylcarbinol.

$$C_6H_5CHOHCH_2C_6H_5 \xrightarrow{H_2SO_4} \underset{\textit{trans}\text{-Stilbene}}{(C_6H_5)HC{=}CH(C_6H_5)} + H_2O$$

Irradiation with ultraviolet light converts it to ***cis*-stilbene** (cf. p. 416).

$$\underset{\text{(trans)}}{(C_6H_5)HC{=}CH(C_6H_5)} \overset{h\nu}{\rightleftarrows} \underset{\textit{cis}\text{-Stilbene}}{(C_6H_5)HC{=}CH(C_6H_5)}$$

BIPHENYL AND ITS DERIVATIVES

Biphenyl and its derivatives form the largest group of compounds having benzene rings directly united. **Biphenyl** is obtained when benzene is heated to a high temperature. In the laboratory preparation the vapor from refluxing benzene is allowed to come in contact with a spiral of electrically heated Nichrome wire, but in the commercial method the vapor is passed through molten lead or hot tubes.

$$2\ C_6H_6 \xrightarrow{700\text{–}800^\circ} C_6H_5C_6H_5 + H_2$$

Since the initial stage in the formation of biphenyl undoubtedly is the decomposition of benzene into phenyl radicals and hydrogen atoms, it is not surprising that small amounts of the terphenyls, ***m***- and ***p*-diphenylbenzene,** and the quaterphenyls, **1,3,5-triphenylbenzene** and **4,4′-bis(biphenyl),** also are produced.

Besides the use of biphenyl in the eutectic with phenyl ether as a heat transfer medium (p. 442), some of its substitution products are important. The halogenated biphenyls (*Arochlors*) are used as transformer oils and as nonflammable heat exchange mediums, and some of the aminobiphenyls are used as dye intermediates. Direct substitution reactions lead to *ortho* and *para* monosubstitution products, but since the groups introduced usually are deactivating, a second substitution takes place in the second ring.

The most important derivative of biphenyl, *p,p′*-diaminobiphenyl or **benzidine,** is not made from biphenyl but by the *benzidine rearrangement* of hydrazobenzene (p. 416). It is an important intermediate for the synthesis of direct dyes for cotton (p. 566).

$$C_6H_5NHNHC_6H_5 + 2\ HCl \xrightarrow{\text{Heat}} [H_3\overset{+}{N}C_6H_4{-}C_6H_4\overset{+}{N}H_3]\ 2\ Cl^-$$

Benzidine hydrochloride

o-Tolidine (*p,p′-diamino-m,m′-dimethylbiphenyl*) is made similarly from *o*-hydrazotoluene, and **dianisidine** (*p,p′-diamino-m,m′-dimethoxybiphenyl*) from *o*-hydrazoanisole.

Benzidine is oxidized under certain conditions to give the intensely colored **Benzidine Blue.** This reaction has been used as the basis of tests for hydrogen peroxide and peroxidases; of sensitive spot tests for readily reduced ions of metals such as cerium, manganese, cobalt, and copper; and for bromide, iodide, cyanide, and thiocyanate ions. The test for copper is even more sensitive if tolidine is used instead of benzidine.

Biphenyl derivatives result when iodobenzenes are heated with finely divided copper (*Ullmann reaction*). The reaction is valuable because it permits the synthesis of derivatives of known constitution and of compounds that cannot be prepared readily in other ways.

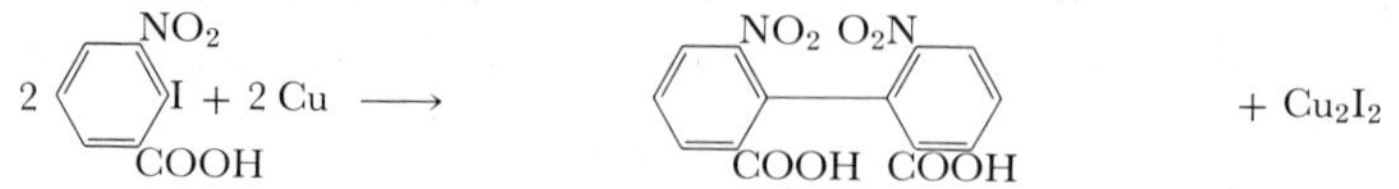

$$2\ (NO_2)(COOH)C_6H_3I + 2\ Cu \longrightarrow (NO_2)(COOH)C_6H_3{-}C_6H_3(NO_2)(COOH) + Cu_2I_2$$

2,2′-Dinitro-6,6′-biphenyldicarboxylic acid

The derivatives of biphenyl have been of interest from the theoretical viewpoint. 2,2′-Dinitro-6,6′-biphenyldicarboxylic acid, for example, can be resolved into two optically active isomers, although it does not contain an asymmetric carbon atom. Many derivatives of biphenyl have been studied, and only those having substituents in the *ortho* positions are capable of resolution. Figure 28–2*a* shows that if two *ortho* groups have an apparent diameter greater than that of the aromatic carbon atom (1.39 A), these groups cannot pass each other, and free rotation about the bond joining the two phenyl groups is prevented. If the two phenyl groups are restricted to a nonplanar conformation and the resulting molecule lacks a plane, center, or alternating axis of symmetry (p. 302), then two configurations are possible that are nonsuperposable mirror images (Fig. 28–2*b*).

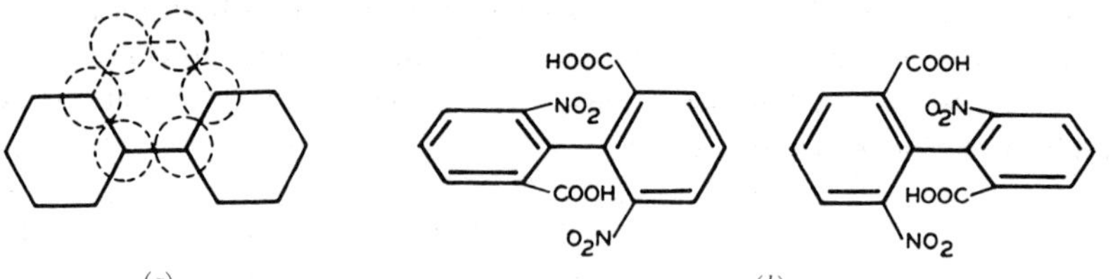

Figure 28–2. (*a*) Shortest internuclear distance between *ortho* groups that will permit rotation in biphenyl derivatives; (*b*) active forms of 2,2′-dinitrobiphenyl-6,6′-dicarboxylic acid.

PROBLEMS

28–*1*. Give a series of reactions for the following syntheses: (*a*) methoxychlor from phenol, (*b*) *o*-tolidine from toluene, (*c*) triphenylmethyl from benzene, (*d*) diphenylacetylene from toluene, (*e*) *p,p′*-biphenyldicarboxylic acid from toluene, (*f*) *p,p′*-diiodobiphenyl from nitrobenzene, (*g*) styrene from acetophenone.

28–*2*. How can one distinguish between the members of the following pairs of compounds by chemical reactions: (*a*) *p*-methylbiphenyl and diphenylmethane, (*b*) biphenyl and triphenylmethane, (*c*) *sym*-diphenylethylene and *p,p′*dimethylbiphenyl, (*d*) styrene and phenylacetylene, (*e*) *sym*-diphenylethylene and *unsym*-diphenylethylene?

28–*3*. List the following compounds in the order of increasing acidity: (*a*) water, (*b*) methane, (*c*) triphenylmethane, (*d*) toluene, (*e*) ammonia, (*f*) tri-(*p*-nitrophenyl)methane, and (*g*) benzene.

CHAPTER TWENTY-NINE

AROMATIC COMPOUNDS WITH CONDENSED RINGS

Compounds with condensed rings are compounds in which two or more carbon atoms are shared in common by two or more rings. The most familiar aromatic compounds of this type are naphthalene and anthracene and their derivatives. Numerous systems that are more complex are known (p. 501).

NAPHTHALENE AND DERIVATIVES

Occurrence and Structure

Naphthalene, isolated some time before 1820, was the first pure compound to be obtained from the distillation products of coal. The reason for its early discovery is that it is a beautiful crystalline solid that sublimes readily. It was noticed first as a deposit in the condensers during the distillation of the naphtha fraction and hence was called naphthalene. Naphthalene is obtained from coal tar by crystallization from the carbolic and creosote fractions (p. 390). Production amounted to 426 million pounds in 1964. It was produced also from the C_{10} fraction of petroleum reformates (p. 94) to the extent of 315 million pounds.

The empirical formula, C_5H_4, was established by Faraday in 1826 by the analysis of barium naphthalenesulfonate, but it was not until after Kekulé propounded his theory of aromatic structure that Erlenmeyer[1] proposed in 1866 a satisfactory structure for the molecular formula, $C_{10}H_8$. Erlenmeyer's formula contains two aromatic nuclei having two carbon atoms in common.

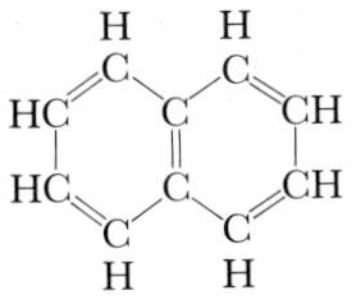

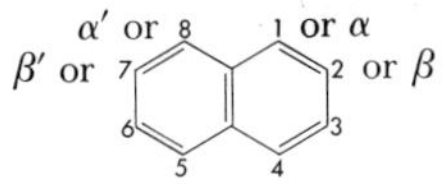

This type of compound is said to have a *fused* or a *condensed* ring system.

Graebe[2] first showed in 1869 that two different benzene rings are present in naphthalene, but a later proof is more direct. When 1-nitronaphthalene is oxidized, 3-nitrophthalic acid is formed, but if the nitro group first is reduced to the amino group, the product of subsequent oxidation is phthalic acid.

[1] Richard August Karl Emil Erlenmeyer (1825–1909), professor at the University of Munich. He is well known for the still popular conical flask that he devised.

[2] Carl Graebe (1841–1927), student of Bunsen and Baeyer and later professor at the University of Geneva. He is known chiefly for determining the structure of alizarin and synthesizing it, but he did important work in the field of polynuclear compounds and dyes in general.

Phthalic acid $\xleftarrow{KMnO_4}$ 1-Naphthylamine $\xleftarrow{Fe,\ HCl}$ 1-Nitronaphthalene $\xrightarrow{KMnO_4}$ 3-Nitrophthalic acid

The reason for the difference in behavior on oxidation is that the nitro group makes an aromatic ring harder to oxidize than an unsubstituted benzene ring, whereas the amino group increases the ease of oxidation of the ring to which it is attached (p. 420).

Several syntheses of naphthalene substantiate the Erlenmeyer formula. One of the more unequivocal syntheses is that of Fittig, which involves the cyclization of β-benzylidenepropionic acid to 1-naphthol and reduction to naphthalene (p. 437).

$C_6H_5CH{=}CHCH_2COOH \xrightarrow{H_2SO_4}$ [keto form] $\longrightarrow$ 1-Naphthol $\xrightarrow{\text{Zn dist.}}$ naphthalene

The formula accounts for the existence of two monosubstitution products, frequently designated as α and β, and ten disubstitution products when both groups are alike. Up to ten hydrogen atoms can be added to naphthalene indicating five potential double bonds.

As with benzene, the unsaturation electrons are in molecular orbitals characteristic of closed conjugated systems (p. 371). Naphthalene, like benzene, usually is represented as a resonance hybrid of the conventional bond structures.

The resonance energy of naphthalene is 61 kcal. Since the resonance energy of benzene is 36 kcal., the additional resonance energy contributed by the second ring is only 25 kcal. This decreased resonance energy is reflected in a lower ionization potential of naphthalene (8.3 e.v.) compared to benzene (9.6 e.v.), and in greater chemical reactivity.

Reactions

Addition Reactions. One ring of naphthalene undergoes addition reactions more readily than benzene. Reaction with sodium and ethyl alcohol produces a 1,4-dihydro derivative. At the higher temperature of boiling *i*-amyl alcohol, the 1,2,3,4-tetrahydro derivative (*tetralin*) is formed.

Naphthalene $\xrightarrow{Na,\ C_2H_5OH}$ 1,4-Dihydronaphthalene

Naphthalene $\xrightarrow{Na,\ C_5H_{11}OH}$ 1,2,3,4-Tetrahydronaphthalene (*tetralin*)

Catalytically the reduction may be carried to the tetrahydro or the decahydro stage.

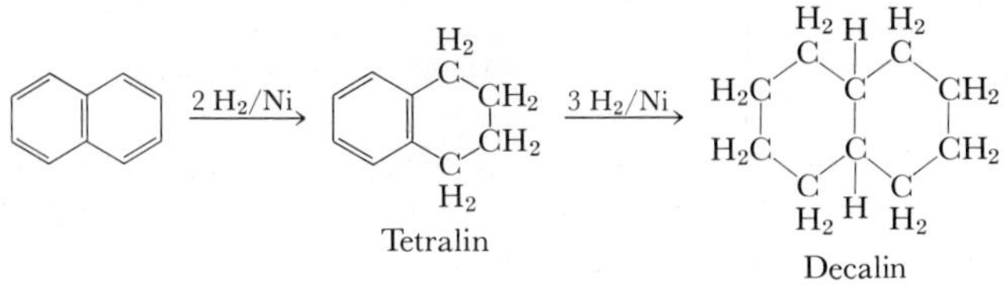

$\xrightarrow{2\,H_2/Ni}$ Tetralin $\xrightarrow{3\,H_2/Ni}$ Decalin

Tetralin

Decalin

Tetralin and **decalin** find some use as solvents. Decalin is of interest in that it exists in two stereoisomeric configurations, *cis*-decalin in which the methylidyne (CH) hydrogen atoms are on the same side of the ring union, and *trans*-decalin in which they are on opposite sides (p. 651).

Substitution Reactions. Naphthalene undergoes all of the substitution reactions of benzene by use of the same reagents and catalysts. **Direct halogenation** in the presence of iron gives 95 per cent of α-chloronaphthalene and about 5 per cent of the β isomer.

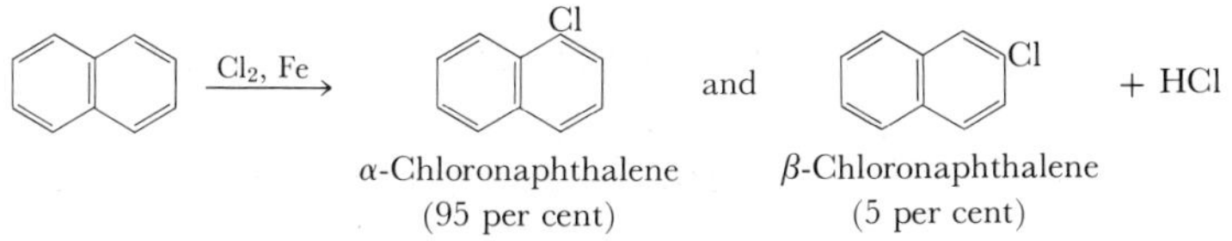

$\xrightarrow{Cl_2,\ Fe}$ and $+ HCl$

α-Chloronaphthalene (95 per cent)

β-Chloronaphthalene (5 per cent)

Similarly nitration with a mixture of nitric and sulfuric acids yields 95 per cent of α-nitronaphthalene together with about 5 per cent of β-nitronaphthalene. Sulfonation of naphthalene with 98 per cent sulfuric acid at 40° gives a product containing 96 per cent of the α isomer, whereas at 160° the mixture of sulfonic acids produced contains 85 per cent of the β isomer.

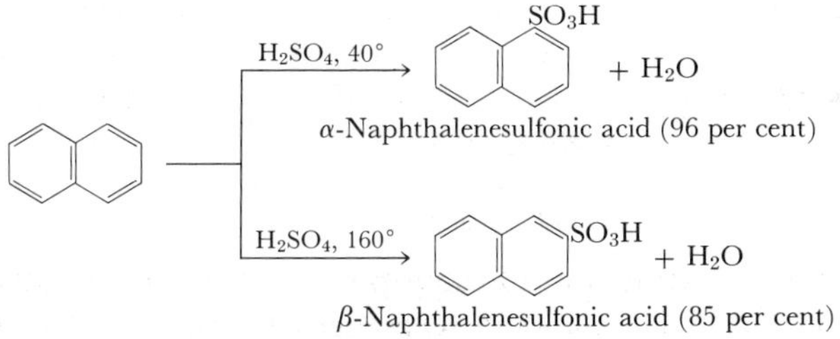

$\xrightarrow{H_2SO_4,\ 40°}$ SO_3H $+ H_2O$

α-Naphthalenesulfonic acid (96 per cent)

$\xrightarrow{H_2SO_4,\ 160°}$ SO_3H $+ H_2O$

β-Naphthalenesulfonic acid (85 per cent)

This behavior is important because it provides the principal method for introducing a functional group into the β position. Most β-substituted naphthalenes are prepared by way of the β sulfonic acid. The reason for this behavior appears to be that the rate of sulfonation in the α position is very much faster than in the β position, but the position of equilibrium for the β sulfonic acid is more favorable. Since the reaction is reversible (p. 405), the β isomer eventually predominates. Mixtures of the α and β acids are separated by crystallization of the calcium salts, the calcium salt of the α isomer being the more soluble. Alternatively the α sulfonic acid, which hydrolyzes more rapidly than the β, may be removed by steam blown through the mixture (p. 405).

The Friedel-Crafts reaction also gives mixtures of α and β substitution products, the relative amounts of which can be controlled by varying the conditions of the reaction. Reaction with acetyl chloride in carbon disulfide gives about three fourths α and one fourth β substitution, whereas in nitrobenzene solution the β isomer is formed almost exclusively.

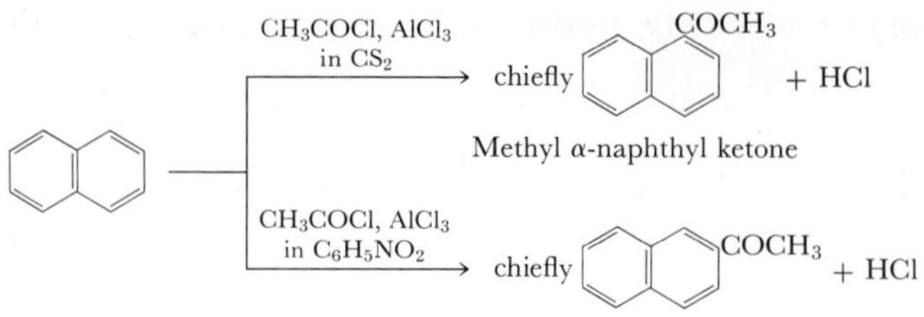

CH_3COCl, $AlCl_3$ in CS_2 → chiefly [naphthalene-1-$COCH_3$] + HCl

Methyl α-naphthyl ketone

CH_3COCl, $AlCl_3$ in $C_6H_5NO_2$ → chiefly [naphthalene-2-$COCH_3$] + HCl

Methyl β-naphthyl ketone

Disubstitution follows essentially the same rules as for benzene, but the reversibility of some reactions frequently permits rearrangement of the initial product to a more stable isomer. If the group in a monosubstituted naphthalene is deactivating, a second substituent usually enters the second ring in one of the α′ positions, although sulfonation can take place in a β′ position.

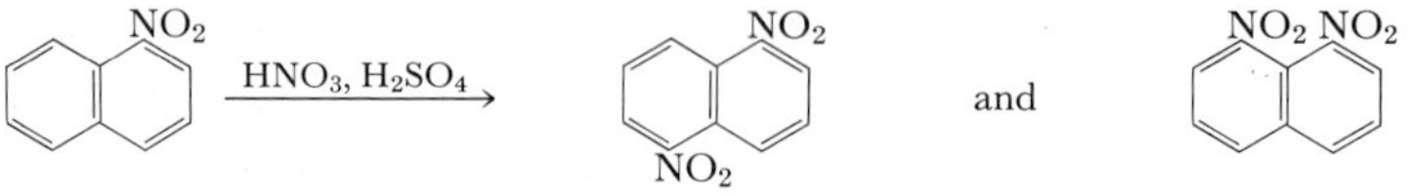

NO_2 —(HNO_3, H_2SO_4)→ 1,5-Dinitronaphthalene and 1,8-Dinitronaphthalene

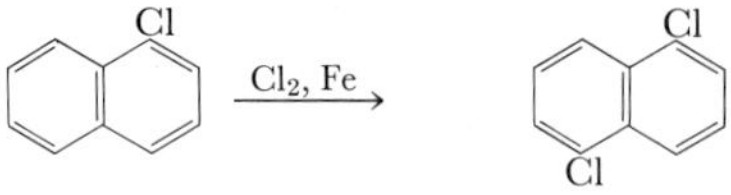

Cl —(Cl_2, Fe)→ 1,5-Dichloronaphthalene

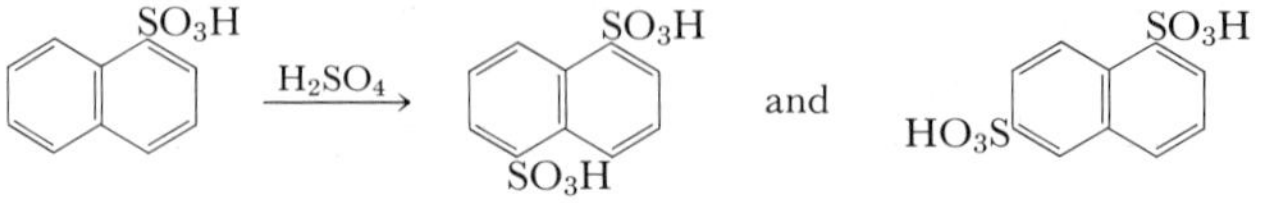

SO_3H —(H_2SO_4)→ 1,5-Naphthalenedisulfonic acid and 1,6-Naphthalenedisulfonic acid

SO_3H (β) —(H_2SO_4)→ 2,6-Naphthalenedisulfonic acid and 2,7-Naphthalenedisulfonic acid

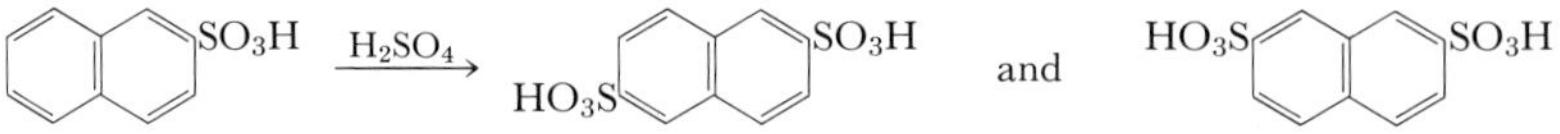

If the group present is activating and in the α position, a second substituent enters the 4 (*para*) position, except that in sulfonation a rearranged product may result. If an activating group is in the β position, a second substituent enters the α position. Again sulfonation may give anomalous products (p. 493).

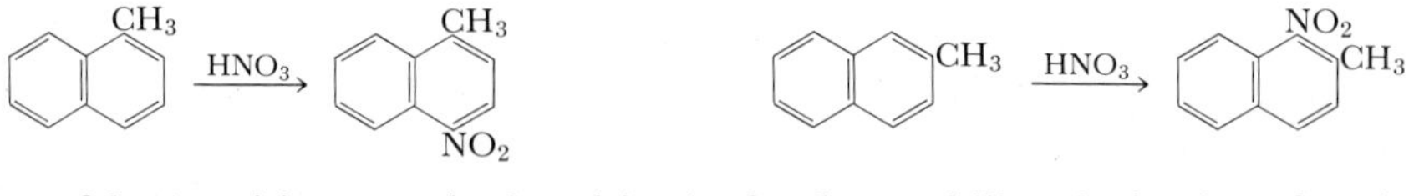

CH_3 —(HNO_3)→ CH_3, NO_2 (1,4) CH_3 (β) —(HNO_3)→ NO_2, CH_3 (1,2)

Preference of the 1 position over the 3 position in the electrophilic substitution of naphthalenes having an activating group in the 2 position can be explained by noting that two of the resonance structures in the σ intermediate for substitution at the 1 position have a benzene nucleus, whereas in that for substitution at the 3 position only one structure has an aromatic ring. Hence the first is more stable and more readily formed.

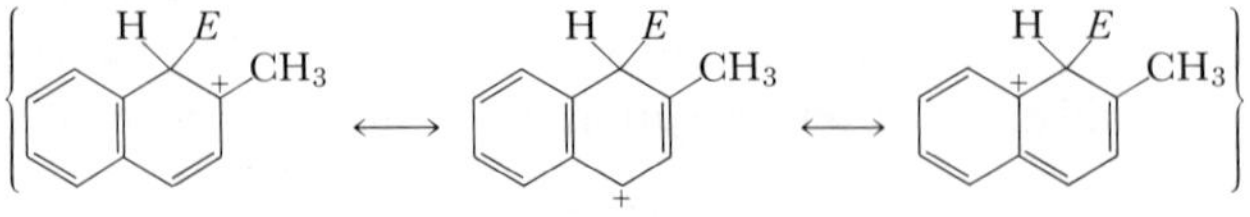

{ H E, +, CH_3 ⟷ H E, CH_3, + ⟷ H E, +, CH_3 }

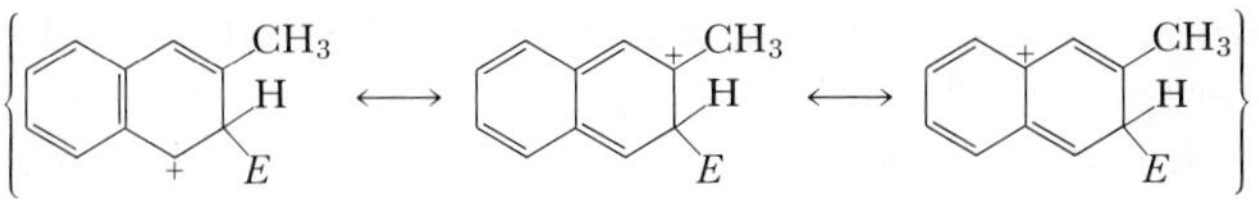

CH_3 H E ⟷ CH_3 H E ⟷ CH_3 H E

An exception to the rule is the reaction of sodium β-naphthoxide with carbon dioxide, which yields the 3-hydroxy-2-naphthoate (p. 495) at elevated temperature.

Naphthalene Derivatives

Alkyl Derivatives. α- and β-Methylnaphthalene are isolated commercially from coal tar. **α-Methylnaphthalene** has been selected as the standard fuel with zero rating in testing Diesel fuels because of its poor burning characteristics in the Diesel engine. During shortages of naphthalene, the methylnaphthalenes are used for the production of phthalic anhydride (p. 474).

Naphthols. α- and β-Naphthol are prepared from the corresponding sodium sulfonates by fusion with sodium hydroxide. They not only are important dye intermediates themselves but are used for the synthesis of other dye intermediates. β-Naphthol is the more important, production being forty to fifty times that of the α isomer. Pure α-naphthol is made by the hydrolysis of α-naphthylamine (cf. p. 493). Reaction of α-naphthol with phosgene followed by reaction with methylamine, or direct reaction of α-naphthol with methyl isocyanate (p. 287), gives **1-naphthyl *N*-methylcarbamate,** α-$C_{10}H_7OCONHCH_3$ (trade name *Sevin*). It is highly toxic to a wide variety of insects but unusually nontoxic to mammals and may be used on agricultural products as late as one to three days before harvest. Unlike many insecticides, it does not contain either halogen or phosphorus.

Halogenated Naphthalenes. α-Chloro- and α-bromonaphthalene are made by direct halogenation. The β isomers are only of research interest. They can be made by the action of phosphorus pentahalides on the naphthol or by the Sandmeyer reaction from diazotized β-naphthylamine.

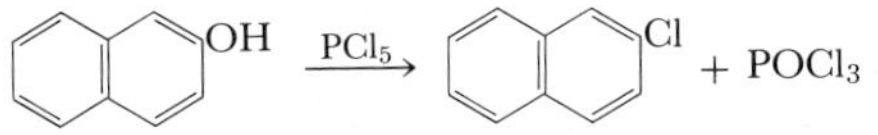

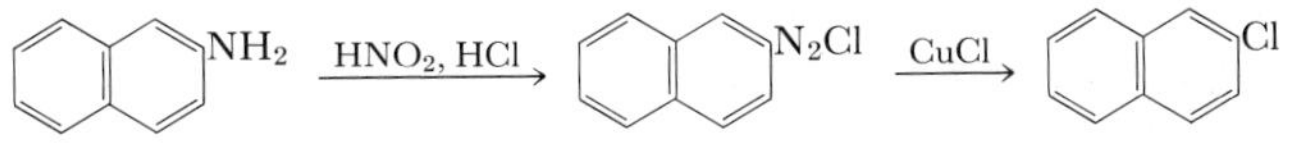

OH $\xrightarrow{PCl_5}$ Cl + $POCl_3$

NH_2 $\xrightarrow{HNO_2,\ HCl}$ N_2Cl $\xrightarrow{CuCl}$ Cl

The polychloronaphthalenes made by direct chlorination are solids that find use as nonflammable impregnating agents for electrical insulating materials under the trade name *Halowax.*

The reaction of 1- or 2-chloro-, bromo-, or iodonaphthalene with lithium amides gives about the same mixture of 30 per cent α and 70 per cent β isomers. Hence these reactions are believed to take place by way of a 1,2-dehydronaphthalene (*naphthalyne*) intermediate (cf. p. 399).

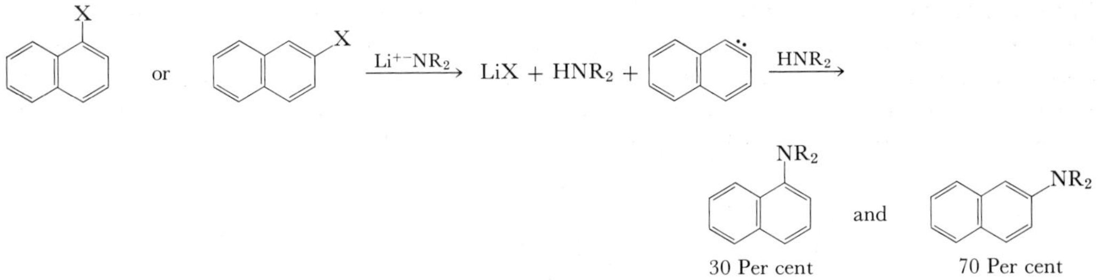

X or X $\xrightarrow{Li^{+}\,{}^{-}NR_2}$ LiX + HNR_2 + $\xrightarrow{HNR_2}$

NR_2 and NR_2

30 Per cent 70 Per cent

Naphthylamines. α-Naphthylamine is made by the reduction of α-nitronaphthalene with iron and water (p. 414). β-Naphthylamine is made by the ammonolysis of β-naphthol. This reaction, which is slow in the benzene series (p. 438), can be brought about easily in the naphthalene series by aqueous ammonia and ammonium sulfite, and is known as the **Bucherer reaction.**

$$C_{10}H_7OH \underset{}{\overset{NH_3,\ (NH_4)_2SO_3}{\rightleftarrows}} C_{10}H_7NH_2 + H_2O$$

β-Naphthylamine has been withdrawn from the open market because prolonged exposure to it leads to bladder tumors.

Amines other than ammonia can be used to give substituted aminonaphthalenes. The reaction is reversible and is used to prepare naphthols from naphthylamines if the naphthylamine is more readily available. For example, pure α-naphthol is obtained best from pure α-naphthylamine.

It usually is assumed that the Bucherer reaction proceeds through the keto form and that the function of the sulfite is to stabilize this form by converting it to the bisulfite addition compound, although evidence has been presented for the addition of bisulfite to the ring double bond.

$$C_{10}H_7OH \rightleftarrows \left[\text{keto form } (H_2C,\ C{=}O)\right] \overset{HOSO_2NH_4}{\rightleftarrows} \left[H_2C{-}C(OH)(SO_3NH_4)\right] \underset{H_2O}{\overset{NH_3}{\rightleftarrows}}$$

$$\left[H_2C{-}C(NH_2)(SO_3NH_4)\right] \underset{(NH_4)_2SO_3}{\overset{NH_3}{\rightleftarrows}} \left[H_2C{-}C{=}NH\right] \rightleftarrows C_{10}H_7NH_2$$

The keto form of the naphthols and the imino form of the naphthylamines are more stable than those of phenol and aniline because the resonance stabilization lost by ketonization is only the difference between that of naphthalene and that of benzene (p. 488).

Sulfonated Naphthols and Naphthylamines. Naphthols and naphthylamines couple with diazonium salts to give azo compounds that absorb light of longer wavelength than the simpler azo compounds derived from benzene derivatives, because the difference between the energy levels of the highest occupied orbital and the lowest unoccupied orbital decreases with increasing conjugation with the azo group (p. 554). In order that these colored compounds be water-soluble, however, strongly water-solubilizing groups must be present. Hence azo dyes for direct dyeing always contain sulfonic acid groups, and the sulfonated naphthols and naphthylamines are used widely as dye intermediates. Usually these intermediates have common names that were assigned by dye manufacturers because the constitution of the compound was unknown when it first came into use, or because the originator did not care to be overly helpful to his competitors.

Naphthionic acid is made from α-naphthylamine by the baking process described for the preparation of sulfanilic acid (p. 422).

$$C_{10}H_7\overset{+}{N}H_3\overset{-}{S}O_4H \xrightarrow{\text{Heat}} NH_2C_{10}H_6SO_3H + H_2O$$

Naphthionic acid

5-Dimethylamino-1-naphthalenesulfonyl chloride (*dansyl chloride*) is a useful reagent for identifying the amino-end residue of peptides (p. 362).

Naphthionic acid is hydrolyzed to 1-hydroxy-4-naphthalenesulfonic acid **(Nevile-Winther acid)** when boiled with aqueous sodium bisulfite.

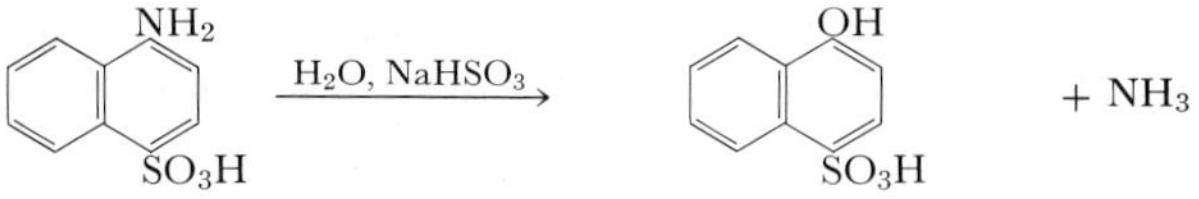

NH_2 / SO_3H $\xrightarrow{H_2O,\ NaHSO_3}$ OH / SO_3H + NH_3

1-Naphthol-4-sulfonic acid
(*Nevile-Winther acid*)

This reaction was known but not openly published eight years before Bucherer discovered it in 1904, but Bucherer demonstrated the general usefulness and the reversibility of the reaction.

Sulfonation of β-naphthol at a low temperature gives the 1 sulfonic acid, which rapidly rearranges to the 8 sulfonic acid, and at higher temperatures to the 6 sulfonic acid **(Schaeffer acid).** With an excess of sulfuric acid, the 3,6 and 6,8 disulfonic acids **(*R*-acid** and **G-acid)** are formed.

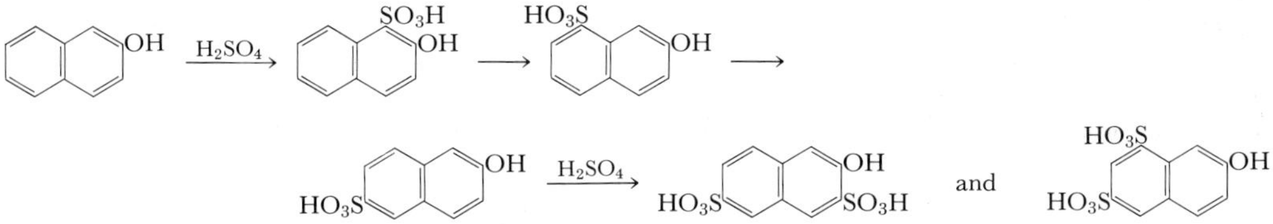

OH $\xrightarrow{H_2SO_4}$ SO_3H / OH $\longrightarrow$ HO_3S / OH $\longrightarrow$

HO_3S / OH $\xrightarrow{H_2SO_4}$ HO_3S / OH / SO_3H and HO_3S / HO_3S / OH

Schaeffer acid   *R*-Acid   *G*-Acid

The designations *R* and *G* come from the German words *rot* and *gelb,* which refer to the red and yellow shades of the azo dyes produced by coupling. The acids are isolated as the sodium salts, which are known as Schaeffer salt, *R*-salt, and *G*-salt.

All three acids are important but *G*-acid is the most valuable. The relative amounts of the isomers can be varied by changing the conditions of sulfonation. Thus 1 part of naphthol to 2 parts of 100 per cent sulfuric acid at 80–110° gives 1 part of Schaeffer salt and 0.5 part of *R*-salt, whereas 1 part of naphthol to 3 parts of sulfuric acid at 30–35° for two to three days gives 1 part of *G*-salt, and 1 part of *R*-salt.

If β-naphthol is sulfonated at 5° in nitrobenzene with one mole of chlorosulfonic acid, the 2-hydroxy-1-naphthalenesulfonic acid is obtained. It is converted to the technically important **Tobias acid** (4.5 million pounds, 1964) by the Bucherer reaction (p. 492).

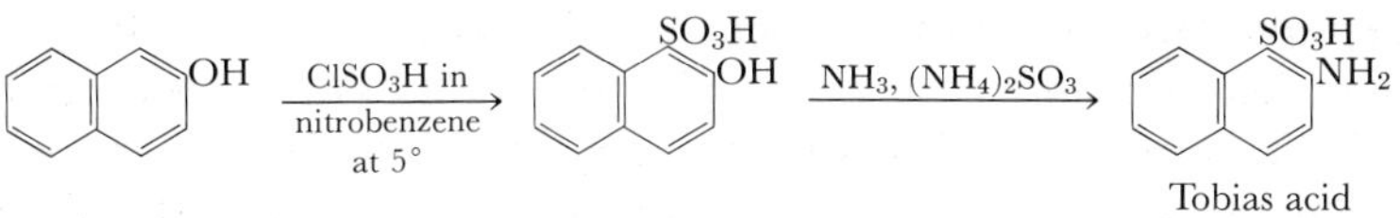

OH $\xrightarrow[\text{nitrobenzene at } 5^\circ]{ClSO_3H\ \text{in}}$ SO_3H / OH $\xrightarrow{NH_3,\ (NH_4)_2SO_3}$ SO_3H / NH_2

Tobias acid

Aminonaphthols. 4-Amino-1-naphthol and 1-amino-2-naphthol are prepared best by coupling α- or β-naphthol with diazotized sulfanilic acid and reducing the azo dye with sodium hydrosulfite.

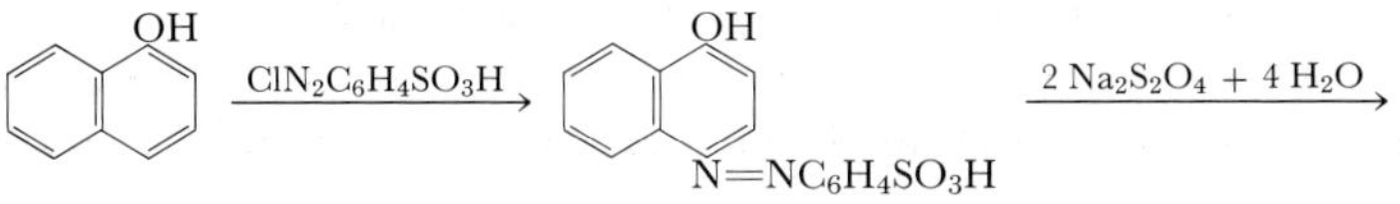

OH $\xrightarrow{ClN_2C_6H_4SO_3H}$ OH / $N{=}NC_6H_4SO_3H$ $\xrightarrow{2\ Na_2S_2O_4\ +\ 4\ H_2O}$

Acid Orange 20 (*Orange I*)

OH / NH_2 + $H_2NC_6H_4SO_3H$ + 4 $NaHSO_3$

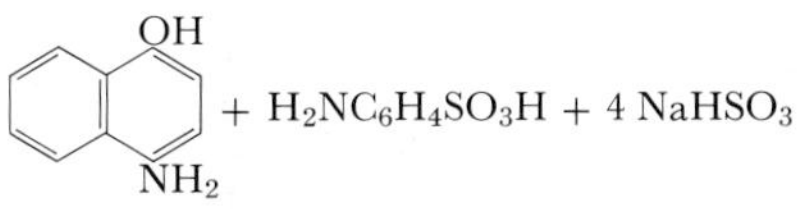

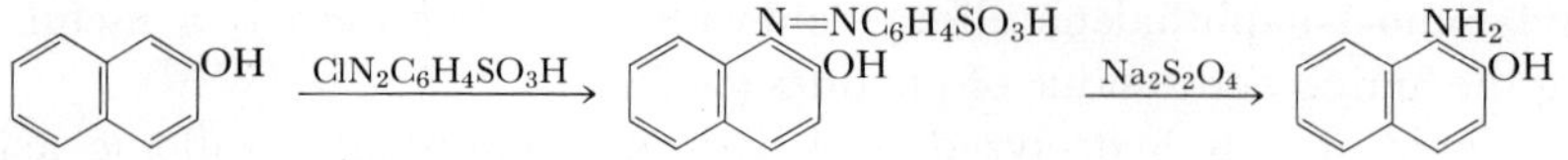

$$\text{2-naphthol} \xrightarrow{ClN_2C_6H_4SO_3H} \text{1-}(N{=}NC_6H_4SO_3H)\text{-2-naphthol} \xrightarrow{Na_2S_2O_4} \text{1-amino-2-naphthol}$$

Acid Orange 7 (*Orange II*)

Other aminonaphthols can be made by the fusion of salts of amino sulfonic acids with sodium hydroxide, or of hydroxy sulfonic acids with sodium amide.

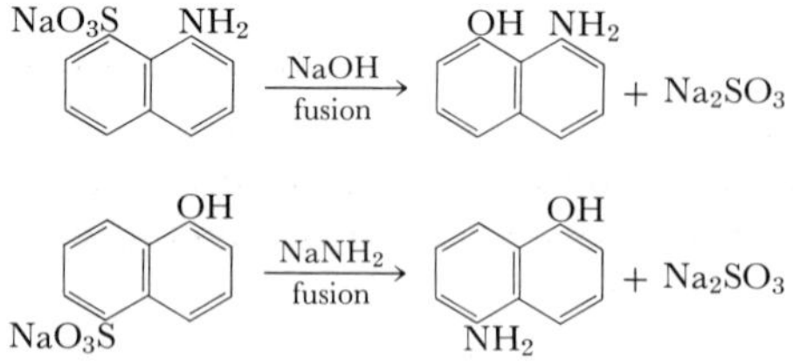

$$\text{8-}NaO_3S\text{-1-}NH_2\text{-naphthalene} \xrightarrow[\text{fusion}]{NaOH} \text{8-OH-1-}NH_2\text{-naphthalene} + Na_2SO_3$$

$$\text{5-}NaO_3S\text{-1-naphthol} \xrightarrow[\text{fusion}]{NaNH_2} \text{5-}NH_2\text{-1-naphthol} + Na_2SO_3$$

Aminonaphtholsulfonic Acids. The most important derivative of an aminonaphthol is ***H*-acid,** which is made by the controlled sodium hydroxide fusion of 1-amino-3,6,8-naphthalenetrisulfonic acid **(Koch acid),** the sulfonic acid group in the 8 position being easily replaced.

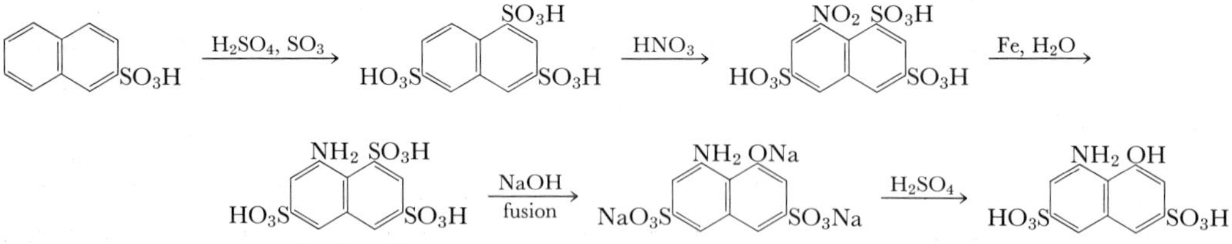

$$SO_3H\text{-naphthalene} \xrightarrow{H_2SO_4,\ SO_3} HO_3S/SO_3H/SO_3H\text{-naphthalene} \xrightarrow{HNO_3} NO_2,\ SO_3H,\ HO_3S,\ SO_3H \xrightarrow{Fe,\ H_2O}$$

$$\underset{\text{Koch acid}}{NH_2,\ SO_3H,\ HO_3S,\ SO_3H} \xrightarrow[\text{fusion}]{NaOH} NH_2,\ ONa,\ NaO_3S,\ SO_3Na \xrightarrow{H_2SO_4} \underset{H\text{-Acid}}{NH_2,\ OH,\ HO_3S,\ SO_3H}$$

Production of the monosodium salt amounted to 3.5 million pounds in 1964.

Sulfonation of β-naphthylamine gives a mixture of 2-amino-1,5,7-naphthalenetrisulfonic acid and 2-amino-6,8-naphthalenedisulfonic acid. Controlled caustic fusion of the disulfonic acid gives **γ-acid,** 2-amino-8-hydroxy-6-naphthalenesulfonic acid.

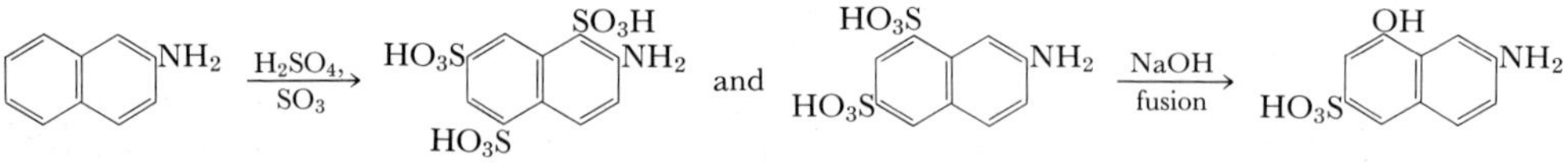

$$\text{2-naphthylamine} \xrightarrow[SO_3]{H_2SO_4,} \text{2-}NH_2\text{-1,5,7-}(SO_3H)_3 \text{ and } \text{2-}NH_2\text{-6,8-}(HO_3S)_2 \xrightarrow[\text{fusion}]{NaOH} \underset{\gamma\text{-Acid}}{\text{2-}NH_2\text{-8-OH-6-}HO_3S}$$

Naphthoquinones. Whereas the benzene ring permits the existence of only two benzoquinones, the *ortho* and *para,* the naphthalene ring system permits more. α- and β-Naphthoquinones are prepared by the oxidation of 4-amino-1-naphthol and 1-amino-2-naphthol respectively. The latter are obtained more readily than the dihydroxy or diamino derivatives. A third naphthoquinone, called *amphi,* is obtained by the oxidation of 2,6-naphthalenediol, which is made by the alkali fusion of the 2,6 disulfonic acid. *amphi*-Naphthoquinone is a brick-red solid that is stable for several hours in benzene solution but rapidly decomposes in other solvents to dark insoluble products. Structural formulas can be written also for 1,5-, 1,7-, and 2,3-naphthoquinones. Although evidence for the transient formation of 1,5-naphthoquinone has been obtained, none of these compounds has been isolated.

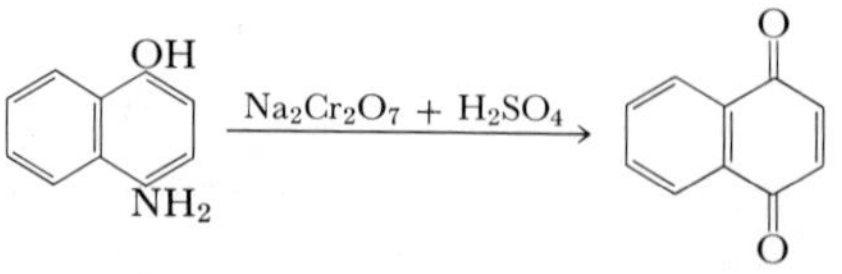

$$\text{4-amino-1-naphthol} \xrightarrow{Na_2Cr_2O_7\ +\ H_2SO_4} \text{1,4-naphthoquinone}$$

α-Naphthoquinone (*1,4-naphthoquinone*)

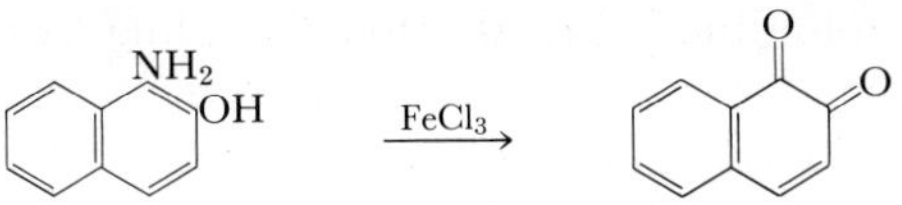

$\xrightarrow{FeCl_3}$ β-Naphthoquinone (*1,2-naphthoquinone*)

$\longrightarrow$ *amphi*-Naphthoquinone (*2,6-naphthoquinone*)

Numerous derivatives of α-naphthoquinone have been isolated from natural sources, where they early attracted attention because of their color. The fleshy hulls of walnuts (*Juglans regia*) contain 1,4,5-naphthalenetriol, which oxidizes in air to give the dark colored quinhydrone from which **juglone** was isolated in 1856.

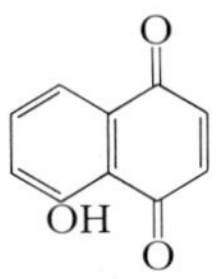

Juglone 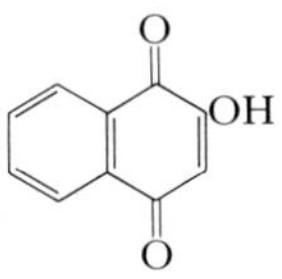 Lawsone 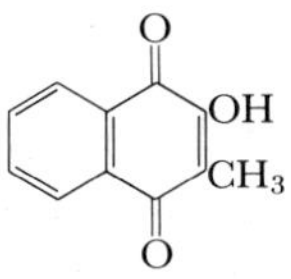 Phthiocol

Lawsone is a yellow pigment isolated from Indian henna (*Lawsonia alba*). A paste of henna leaves and an extract of *Acacia catechu* was used for dyeing the hair red and gave rise to the term *henna* for this particular hue. **Phthiocol** is obtained by the saponification of the ether-soluble fraction (*lipid fraction*) of tubercle bacilli.

The most interesting of the naphthoquinones are two forms of **vitamin K.** Vitamin K_1 is found in green plants such as alfalfa, and K_2 in putrefied fish meal. One of the K vitamins must be supplied by the diet for the adequate clotting of blood. Hence they are termed *antihemorrhagic factors.*

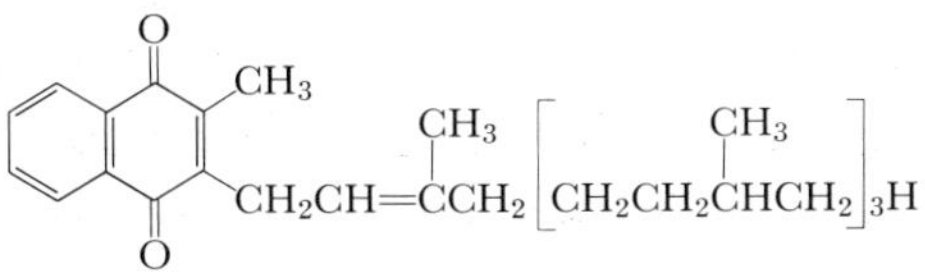

$CH_2CH{=}C(CH_3)CH_2[CH_2CH_2CH(CH_3)CH_2]_3H$

Vitamin K_1

$[CH_2CH{=}C(CH_3)CH_2]_7H$

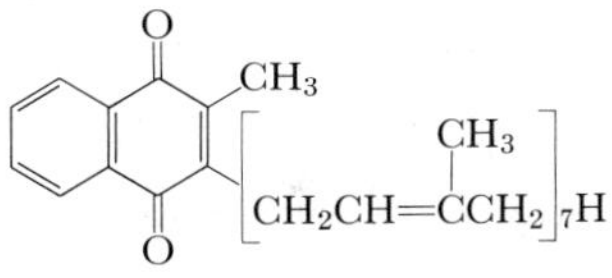

Vitamin K_2

Carboxylic Acids. α- or β-Naphthoic acid can be made by hydrolysis of the corresponding nitrile. The latter is prepared from the sulfonic acid by fusion with sodium cyanide, or from the amine through the diazo reaction (pp. 405, 433). α- or β-Naphthoic acid results also from the oxidation of α- or β-methylnaphthalene with dilute nitric acid. α-Naphthoic acid can be obtained from α-bromonaphthalene through the Grignard reagent.

3-Hydroxy-2-naphthoic acid (*β-oxynaphthoic* or *BON acid*) is made on a large scale by a Kolbe synthesis (p. 472) from β-naphthol.

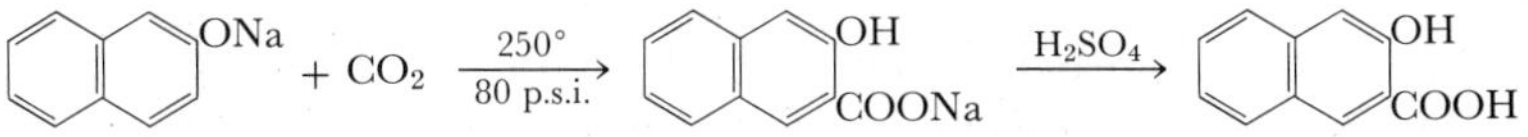

$$C_{10}H_7ONa + CO_2 \xrightarrow[80\ p.s.i.]{250^\circ} C_{10}H_6(OH)COONa \xrightarrow{H_2SO_4} C_{10}H_6(OH)COOH$$

It is used to make the anilide known as Azoic Coupling Component 7 (*Naphtol AS*), an important intermediate (over 1 million pounds, 1964) for the preparation of azoic dyes (p. 563).

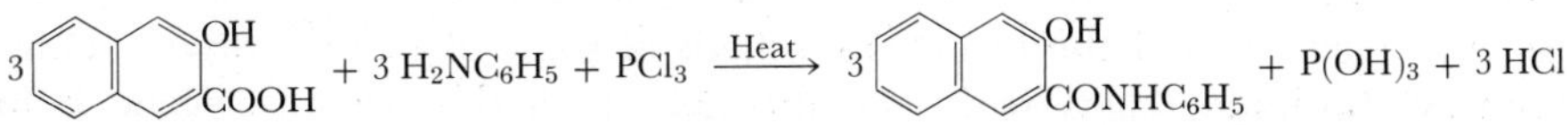

$$3\ C_{10}H_6(OH)COOH + 3\ H_2NC_6H_5 + PCl_3 \xrightarrow{Heat} 3\ C_{10}H_6(OH)CONHC_6H_5 + P(OH)_3 + 3\ HCl$$

Naphtol AS

The corresponding product from *o*-toluidine is called Azoic Coupling Component 18 (744 thousand pounds, 1964).

The most important dibasic acid is **naphthalic acid** prepared by the oxidation of acenaphthene, a component of coal tar (p. 391).

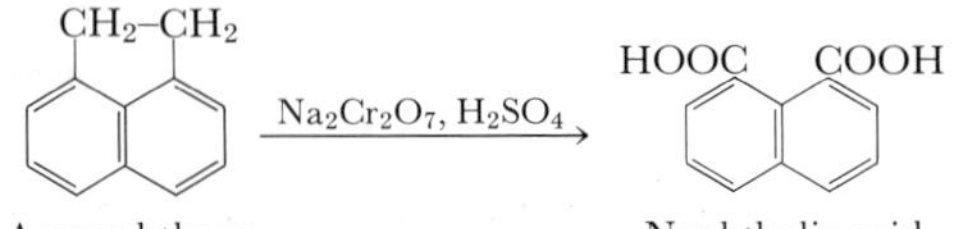

CH_2-CH_2 $\xrightarrow{Na_2Cr_2O_7,\ H_2SO_4}$ HOOC COOH

Acenaphthene Naphthalic acid

The spatial relation of the two carboxyl groups is such that naphthalic acid closely resembles phthalic acid in its properties. When heated, it yields an anhydride, which can be converted to the imide.

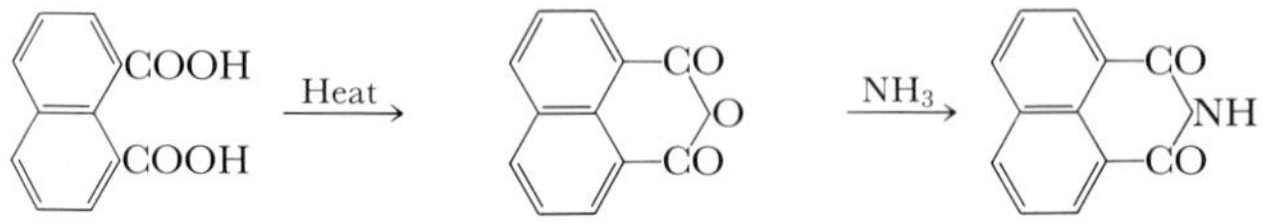

COOH COOH $\xrightarrow{Heat}$ CO O CO $\xrightarrow{NH_3}$ CO NH CO

Naphthalic anhydride Naphthalimide

Certain chemicals present in plants accelerate the growth of cells and are known as **auxins.** Many synthetic compounds have similar properties and are used by nurserymen to increase the ease of rooting of cuttings and by orchardists to prevent premature bud formation and premature dropping of fruit. One of the more widely used substances for this purpose is **α-naphthylacetic acid,** which can be made by the chloromethylation of naphthalene, conversion to the nitrile, and hydrolysis.

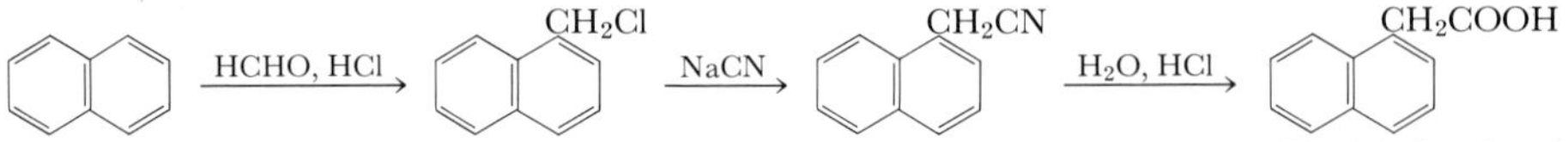

$\xrightarrow{HCHO,\ HCl}$ CH_2Cl $\xrightarrow{NaCN}$ CH_2CN $\xrightarrow{H_2O,\ HCl}$ CH_2COOH

α-Naphthylacetic acid

All of the various derivatives of naphthalene and its miscellaneous uses account for only 20 per cent of production. The remainder is used to make phthalic anhydride (p. 474).

ANTHRACENE AND DERIVATIVES

The coal tar fraction boiling at 300–350° is known as **anthracene oil,** or as **green oil** because of its dark green fluorescence. It is run into tanks and allowed to crystallize over a period of one to two weeks and filtered. The nearly dry cakes are hot-pressed at 50,000 to 70,000 pounds pressure to remove liquid and lower-melting components and give a mixture of anthracene, phenanthrene, and carbazole. The press cake is ground and washed with coal tar naphtha to remove most of the phenanthrene and then with pyridine to remove the carbazole. Anthracene was the second compound to be isolated from coal tar (p. 390) although only about 1 per cent is present. At first it was believed to be an isomer of naphthalene and was called paranaphthalene, but later the name was changed to anthracene (Gr. *anthrax* coal). At one time anthracene was the sole source of the anthraquinone dyes, but intermediates for the latter now are synthesized also from benzene.

Anthracene has the molecular formula $C_{14}H_{10}$. It has both unsaturated and aromatic properties. Thus it readily adds one mole of hydrogen, chlorine, or bromine, and two atoms of sodium. It also sulfonates directly and can be oxidized to anthraquinone, $C_{14}H_8O_2$. Its structure is arrived at from the rational synthesis of some of its derivatives. Thus when *o*-bromobenzyl bromide reacts with metallic sodium, a dihydroanthracene is formed, identical with that obtained by the reduction of anthracene. Mild oxidation converts it to anthracene.

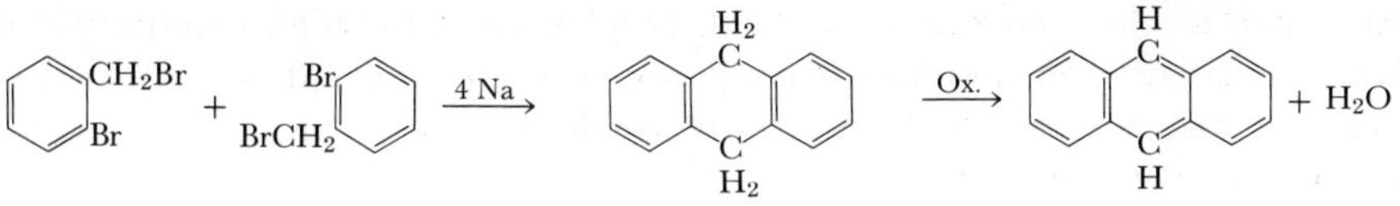

CH_2Br, Br + Br, $BrCH_2$ —4 Na→ H_2C, CH_2 —Ox.→ HC, CH + H_2O

9,10-Dihydroanthracene

Anthracene

9,10-Anthraquinone is obtained in good yield by condensing benzene with phthalic anhydride and cyclizing with sulfuric acid (cf. p. 488).

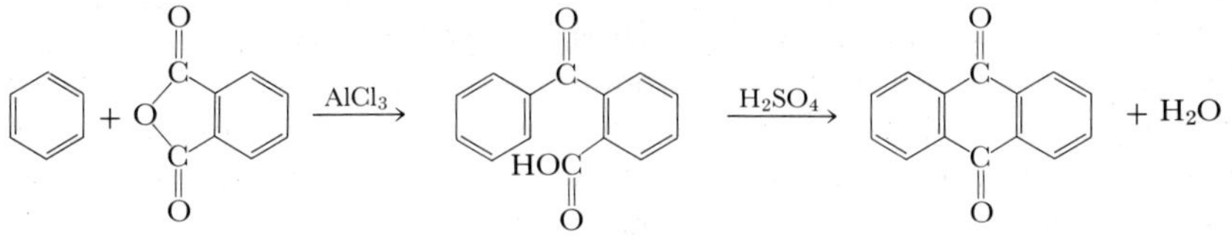

Benzene + phthalic anhydride (O=C, O, C=O) —$AlCl_3$→ (C=O, HOC=O) —H_2SO_4→ (C=O, C=O) + H_2O

o-Benzoylbenzoic acid

9,10-Anthraquinone

From these syntheses it is clear that anthracene contains three fused rings and that the addition and oxidation reactions can be formulated as taking place at the 9,10 positions.

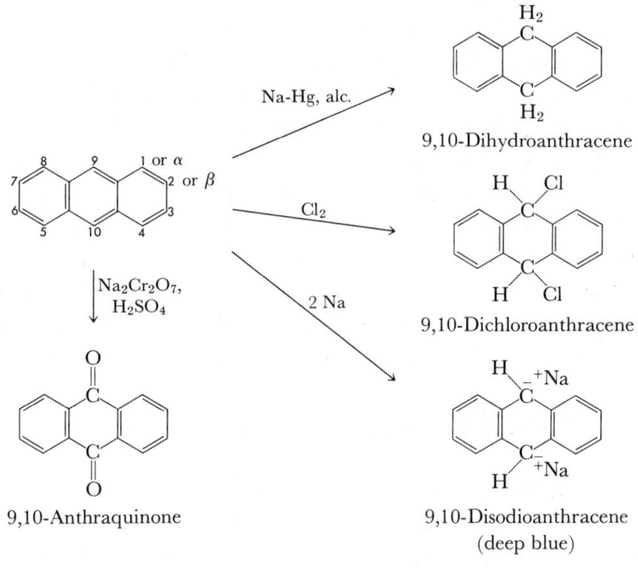

Anthracene (positions 1 or α, 2 or β, 3, 4, 5, 6, 7, 8, 9, 10)

—Na-Hg, alc.→ H_2C, CH_2

9,10-Dihydroanthracene

—Cl_2→ H, Cl, C; C, H, Cl

9,10-Dichloroanthracene

—2 Na→ H, C$^-$ $^+$Na; C$^-$ $^+$Na, H

9,10-Disodioanthracene (deep blue)

—$Na_2Cr_2O_7$, H_2SO_4→ O=C, C=O

9,10-Anthraquinone

Dehydrobenzene (*benzyne*), formed by the decomposition of diazotized anthranilic acid (p. 471), adds to anthracene to give 9,10-benzeno-9,10-dihydroanthracene (*triptycene,* p. 481).

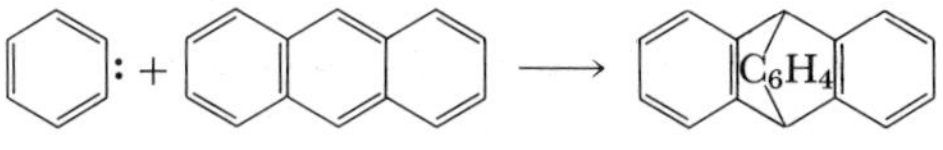

Benzyne : + anthracene ⟶ (C_6H_4)

With various other reagents, anthracene reacts at the 9,10 positions in a manner analogous to conjugated polyenes (p. 634).

Anthracene, like other aromatic hydrocarbons, is considered to be a resonance hybrid.

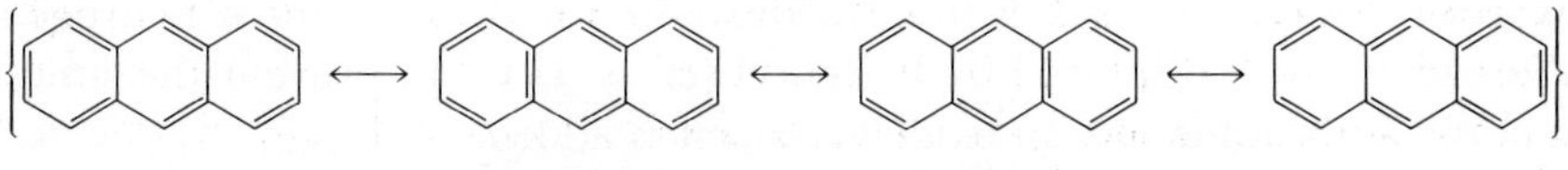

{ ⟷ ⟷ ⟷ }

The resonance energy of anthracene is 84 kcal. Since the resonance energy of two benzene rings is 72 kcal., the resonance energy for the additional ring is only 12 kcal., which is in accordance with the low ionization potential (7.6 e.v.) and the high reactivity of anthracene. Reaction at the 9,10 positions leaving two benzene rings takes place more readily than at an end ring because the latter course would leave a naphthalene nucleus with a resonance energy of only 61 kcal.

When anthracene is pure, it is colorless and has a strong, pale-blue fluorescence when exposed to ultraviolet light. This blue fluorescence is quenched by the presence of as little as one part of naphthacene (p. 501) in 3 million parts of anthracene. Ordinary anthracene has a pale yellow color and exhibits a strong greenish-yellow fluorescence.

Of the several anthraquinones, the 9,10 isomer is the only one of importance, and usually it is designated simply as **anthraquinone.** Its derivatives are of great importance in the dye industry (p. 572). Formerly anthraquinone was made by oxidation of anthracene with chromic acid, but now it is made by vapor phase air oxidation or by the cyclization of *o*-benzoylbenzoic acid (p. 497). About 2.8 million pounds was produced in 1964. Since halogen or nitro groups may be introduced into phthalic anhydride by direct substitution, and since it or its derivatives may be condensed with any aromatic compound capable of undergoing the Friedel-Crafts reaction, this general method of synthesis can lead to a large number of derivatives of anthraquinone.

Because of the presence of the deactivating carbonyl groups, anthraquinone does not undergo substitution readily by electrophilic reagents. Thus halogenation is difficult and Friedel-Crafts reactions do not take place. On the other hand, the carbonyl groups increase the ease of displacement of hydrogen and of other groups by nucleophilic reagents (p. 412).

Sulfonation catalyzed by mercuric sulfate at 120° gives chiefly **α-anthraquinonesulfonic acid,** whereas the uncatalyzed sulfonation at 140° gives chiefly β-anthraquinonesulfonic acid. Rapid mercuration followed by reaction with sulfuric acid appears to account for the catalysis of α sulfonation. At the higher temperature, the more stable β isomer results (cf. sulfonation of naphthalene, p. 489). The sodium salt of the β isomer, obtained by the liming out process (p. 404), is known as **silver salt** because of its silvery gray appearance.

H_2SO_4, SO_3, $HgSO_4$, 120° → α-Anthraquinonesulfonic acid (SO_3H)

H_2SO_4, SO_3, 140° → β-Anthraquinonesulfonic acid (SO_3H) → Liming out → Silver salt (SO_3Na)

Nitration of anthraquinone yields only **α-nitroanthraquinone.**

α- or β-Anthraquinonesulfonates yield the corresponding **α-** or **β-hydroxyanthraquinones** when heated with milk of lime at 180°. Under the more drastic conditions of fusion with sodium hydroxide used to make phenols, the α sulfonate breaks down to benzene derivatives. When silver salt is fused with sodium hydroxide, the α hydrogen as well as the sulfonate group is displaced by hydroxyl (cf. p. 413). To prevent the simultaneous reduction of the anthraquinone, an oxidizing agent is added.

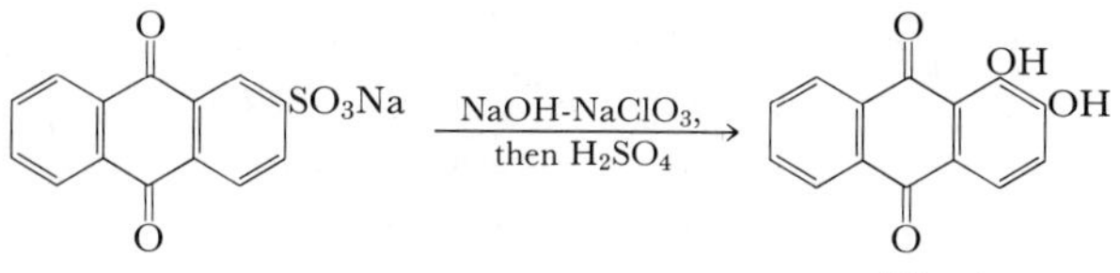

$\xrightarrow[\text{then } H_2SO_4]{NaOH\text{-}NaClO_3,}$

Alizarin

The product is **alizarin,** formerly an important mordant dye (p. 572).

The 1,4 isomer, **quinizarin,** is an important dye intermediate (1.4 million pounds, 1964). It is made by the reaction of phthalic anhydride with *p*-chlorophenol. Condensation, cyclization, and hydrolysis take place in a single operation.

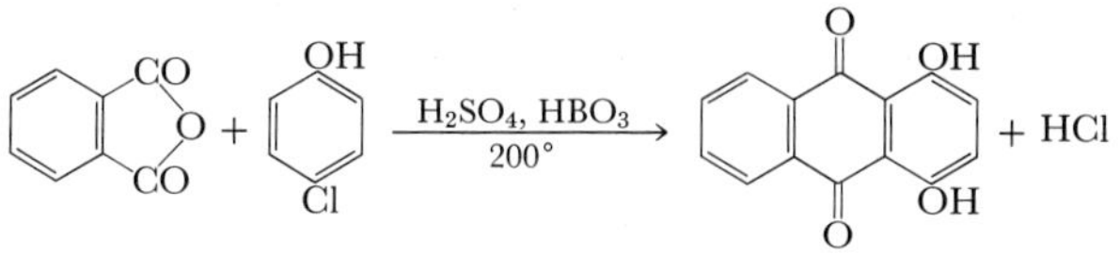

$\xrightarrow[200°]{H_2SO_4,\ HBO_3}$ + HCl

Quinizarin

Over fifty anthraquinone derivatives have been identified as pigments in plants, fungi, lichens, and insects. Many are hydroxymethylanthraquinones. **Emodin** has strong cathartic properties and is an active principle of cascara, senna, aloes, and rheum (*rhubarb*). **Physcion** occurs widely in molds and lichens.

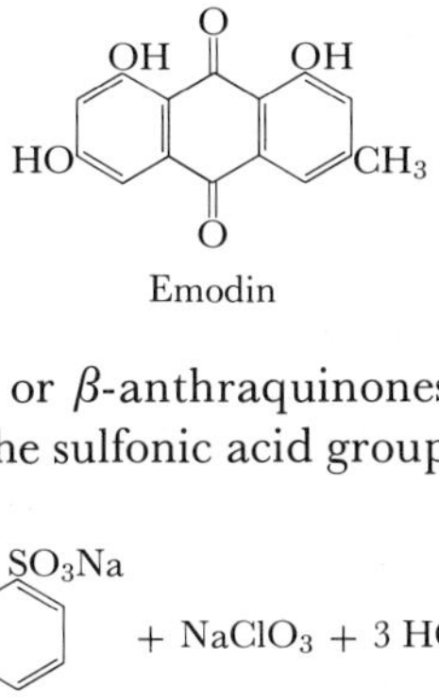

Emodin

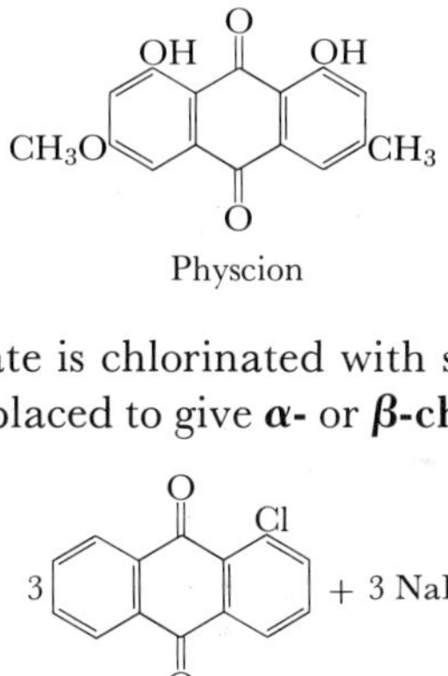

Physcion

When either α- or β-anthraquinonesulfonate is chlorinated with sodium chlorate and hydrogen chloride, the sulfonic acid group is displaced to give **α- or β-chloroanthraquinone.**

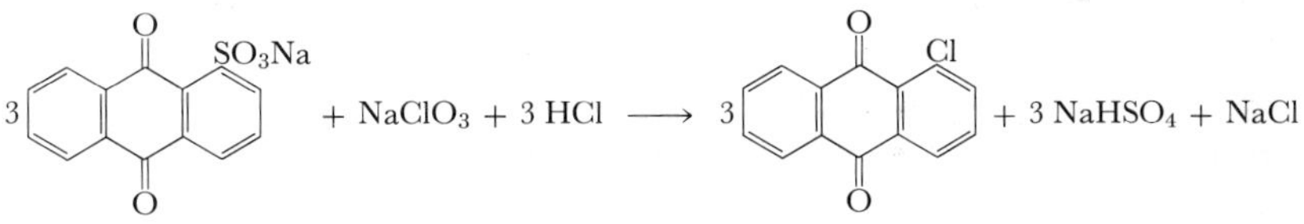

$+ NaClO_3 + 3\ HCl \longrightarrow$ $+ 3\ NaHSO_4 + NaCl$

β-Chloroanthraquinone is prepared commercially, however, from chlorobenzene and phthalic anhydride.

α-Aminoanthraquinone can be made by reduction of α-nitroanthraquinone or by nucleophilic displacement of the α sulfonate group with ammonia in the presence of barium chloride. Production was 1.2 million pounds in 1964.

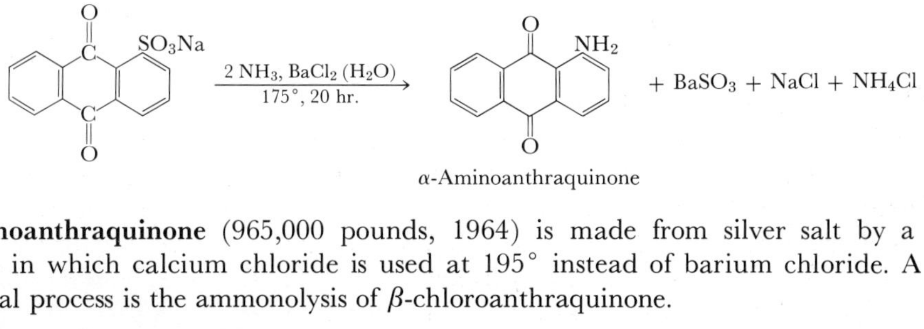

$\xrightarrow[175°,\ 20\ hr.]{2\ NH_3,\ BaCl_2\ (H_2O)}$ $+ BaSO_3 + NaCl + NH_4Cl$

α-Aminoanthraquinone

β-Aminoanthraquinone (965,000 pounds, 1964) is made from silver salt by a similar process in which calcium chloride is used at 195° instead of barium chloride. A second technical process is the ammonolysis of β-chloroanthraquinone.

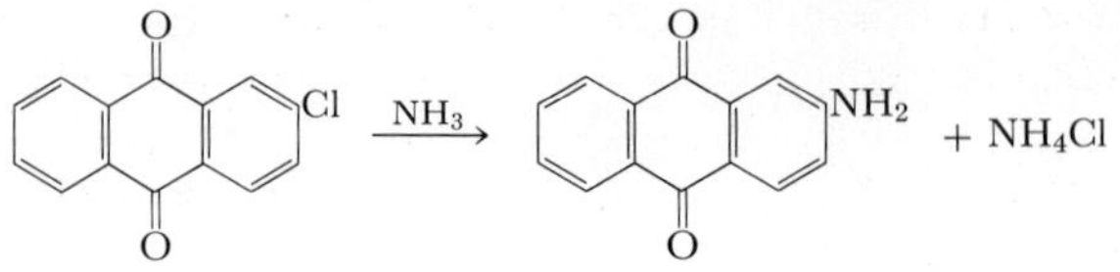

$$\text{2-Chloroanthraquinone} \xrightarrow{NH_3} \text{2-Aminoanthraquinone} + NH_4Cl$$

β-Aminoanthraquinone

The most important reaction of anthraquinones is their easy reduction to dihydro derivatives. The latter are phenols and are soluble in alkali. The reduced form on exposure to air is reoxidized to the water-insoluble anthraquinone.

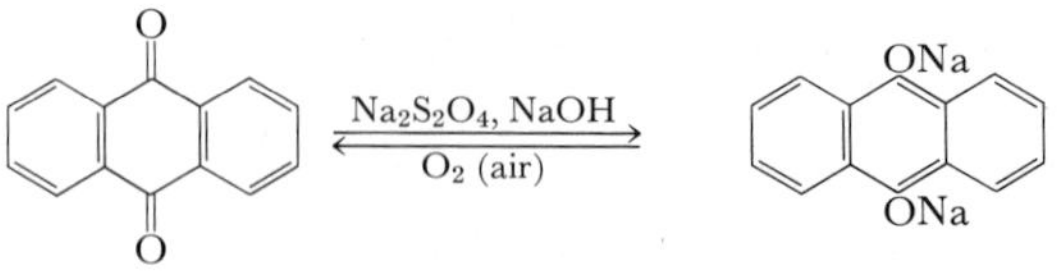

$$\text{Anthraquinone} \underset{O_2\ (air)}{\overset{Na_2S_2O_4,\ NaOH}{\rightleftharpoons}} \text{9,10-di(ONa)anthracene}$$

Water-insoluble anthraquinone — Water-soluble sodium salt of dihydroanthraquinone

These reactions are the basis for the application of the anthraquinone vat dyes to cotton cloth (p. 573).

OTHER CONDENSED RING SYSTEMS

Phenanthrene is isomeric with anthracene. Its structure can be deduced from its oxidation to 9,10-phenanthraquinone and diphenic acid (*2,2′-biphenyldicarboxylic acid*).

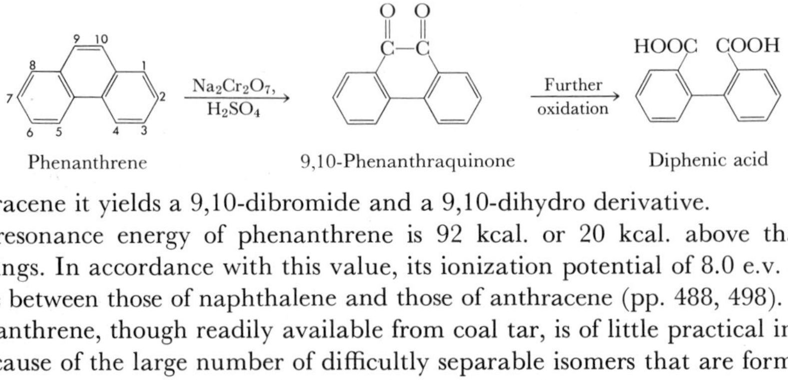

$$\text{Phenanthrene} \xrightarrow[H_2SO_4]{Na_2Cr_2O_7,} \text{9,10-Phenanthraquinone} \xrightarrow[\text{oxidation}]{\text{Further}} \text{Diphenic acid}$$

Phenanthrene (positions 1–10 numbered) — 9,10-Phenanthraquinone — Diphenic acid (HOOC COOH)

Like anthracene it yields a 9,10-dibromide and a 9,10-dihydro derivative.

The resonance energy of phenanthrene is 92 kcal. or 20 kcal. above that for two benzene rings. In accordance with this value, its ionization potential of 8.0 e.v. and its reactivity lie between those of naphthalene and those of anthracene (pp. 488, 498).

Phenanthrene, though readily available from coal tar, is of little practical importance, chiefly because of the large number of difficultly separable isomers that are formed in substitution reactions. The carbon skeleton of phenanthrene is present as a nonaromatic ring system in many natural products. When these products are heated with sulfur or selenium, dehydrogenation takes place and phenanthrene derivatives are formed. Since the aromatic hydrocarbon usually can be purified easily and its constitution determined more readily than that of the original compound, these dehydrogenation products have played an important part in elucidating the constitution of complex natural products. For example, the ring system in abietic acid, the chief component of pine rosin (p. 662), was identified by the isolation of **retene,** 1-methyl-7-isopropylphenanthrene, from the dehydrogenation products.

$$\text{Abietic acid} \xrightarrow{5\ S,\ 300^\circ} \text{Retene} + CO_2 + CH_3SH + 4\ H_2S$$

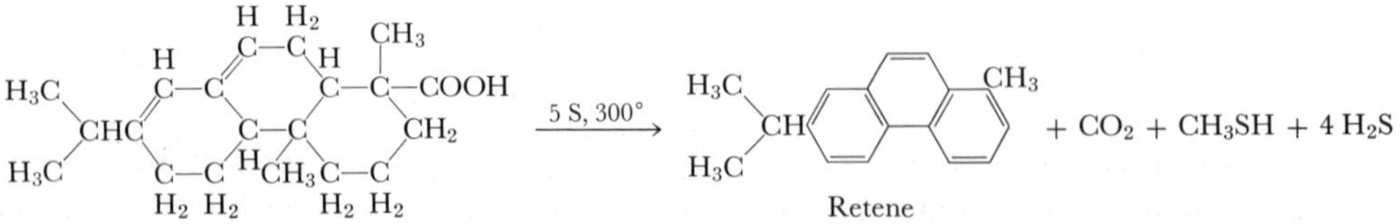

Abietic acid — Retene

Anthracene has an absorption band in the near ultraviolet and is colorless, but naphthacene is orange-yellow and pentacene is purple. All polynuclear aromatic hydrocarbons form π complexes with π electron-acceptor molecules, the stability of the complex being proportional to the ionization potential of the hydrocarbon and the electron affinity of the acceptor molecule (cf. pp. 387, 451, 637).

Two important antibiotics are complex derivatives of a partially hydrogenated naphthacene nucleus. **Terramycin** is obtained from *Streptomyces rimosus* and **Aureomycin** from *Streptomyces aureofaciens.*

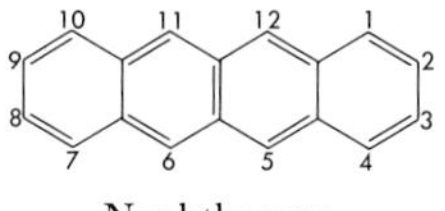

Naphthacene

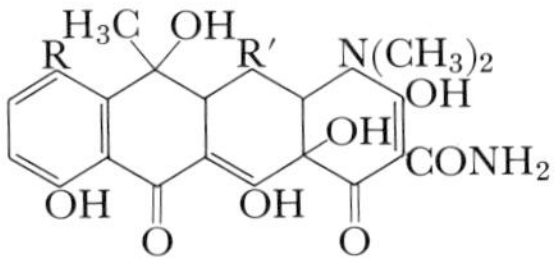

Tetracycline R = H, R′ = H

Terramycin R = H, R′ = OH

Aureomycin R = Cl, R′ = H

They are called *broad-spectrum antibiotics* because they are effective against a large variety of organisms. **Tetracycline,** obtained by the replacement of either a hydroxyl group or a chlorine atom by hydrogen, also is an effective antibiotic. Aureomycin is used to delay the spoilage of fish and poultry (*Acronizing process*).

1,2 : 5,6-Dibenzanthracene, 1,2-benzopyrene, and **3-methylcholanthrene** are powerful carcinogenic hydrocarbons; that is, they produce skin cancers in mice when applied over a considerable period of time, and sarcomas when injected subcutaneously. Numerous other polynuclear aromatic hydrocarbons are less active carcinogens. 1,2-Benzopyrene has been isolated from coal tar and is believed to be the chief agent responsible for the production of skin cancers in chimney sweeps and workers in the coal tar industry. It is present to the extent of up to one per cent in carbon blacks obtained as a by-product when petroleum fractions are cracked to produce isoprene (p. 588). Fifteen different polynuclear aromatic hydrocarbons have been isolated from the outer layer of charcoal-broiled steaks, including 8 μg. of 1,2-benzopyrene per kilo of steak. They are believed to arise from decomposition of fat that drips on the glowing charcoal and vaporization of the hydrocarbons to the steak.

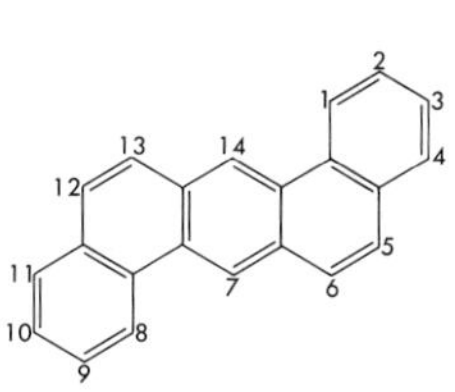

1,2:5,6-Dibenzanthracene

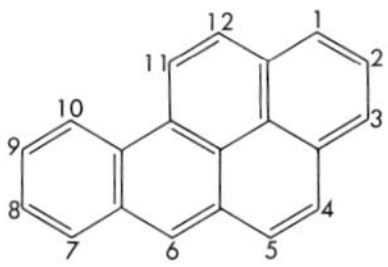

1,2-Benzopyrene

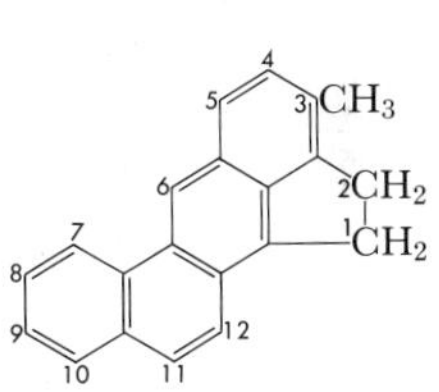

3-Methylcholanthrene

PROBLEMS

29–*1*. Give structural formulas for the theoretically possible isomers, including stereoisomers, for each of the following groups of compounds: (*a*) dinitronaphthalenes, (*b*) trichloronaphthalenes, (*c*) dinaphthylethylenes, (*d*) bromo-β-nitronaphthalenes.

29–*2*. Predict the chief product or products formed when one more group is introduced by the following reactions: (*a*) nitration of 1-bromonaphthalene, (*b*) bromination of β-naphthalenesulfonic acid, (*c*) nitration of β-naphthol, (*d*) chlorination of α-methylnaphthalene, (*e*) bromination of β-chloronaphthalene, (*f*) chlorination of β-naphthylamine, (*g*) sulfonation of 8-amino-1-naphthol.

29–*3*. Give reactions for the synthesis of the following compounds starting with naphthalene: (*a*) α-naphthylamine, (*b*) β-naphthylamine, (*c*) α-naphthol, (*d*) β-naphthol, (*e*) α-naphthoic acid, (*f*) β-naphthoic acid, (*g*) β-naphthoquinone, (*h*) β-isopropylnaphthalene.

29–*4*. What would be the chief product expected on vigorous oxidation of each of the following compounds: (*a*) α-naphthoic acid, (*b*) β-naphthol, (*c*) β-chloroanthracene, (*d*) β-fluoronaphthalene, (*e*) phenanthrene, (*f*) 2,4-diaminonaphthalene?

29–*5*. Give a practical synthesis for each of the following compounds starting with an aromatic hydrocarbon: (*a*) 3-chlorophthalic acid, (*b*) 4-nitrophthalic anhydride, (*c*) β-chloroanthraquinone, (*d*) 1,4-dianilinoanthraquinone, (*e*) 2-methyl-1-nitroanthraquinone, (*f*) 2-methyl-5-nitroanthraquinone.

29–*6*. Suggest the probable intermediate steps that take place in the synthesis of (*a*) alizarin from silver salt, and (*b*) quinizarin from phthalic anhydride and *p*-chlorophenol.

CHAPTER THIRTY

HETEROCYCLIC COMPOUNDS. NUCLEIC ACIDS. ALKALOIDS

Compounds in which three or more atoms are joined to form a closed ring are known as cyclic compounds. If all of the ring atoms are carbon, the compound is said to be *carbocyclic,* but if different kinds of atoms compose the ring, the compound is said to be *heterocyclic.* Theoretically any atom capable of forming at least two covalent bonds can be a member of a ring, but the heterocyclic compounds encountered most frequently contain nitrogen, oxygen, and sulfur as the hetero atoms. Recently rings containing phosphorus, boron, or silicon as the hetero atom have received considerable attention.

Nomenclature

The second edition of the *Ring Index* and the first and second supplements list 11,524 different carbocyclic and heterocyclic ring systems reported in the literature up to January 1, 1962. Clearly it is not useful to attempt to outline here the rules for systematic nomenclature. Ring systems that are encountered frequently have common names. For others the *Ring Index* is consulted. A few simple general rules should be noted. For monocyclic compounds the hetero atom of the ring is numbered one, and the other ring atoms are numbered consecutively to give substituents the smaller possible numbers. If two or more hetero atoms are present, they are given the lowest numbers. If they are different, the lowest number is given to oxygen, then sulfur, and then nitrogen.

3-Methylfuran 4-Methylpyrimidine

Occasionally several tautomeric structures may be possible, and it becomes necessary to specify the position of the wandering hydrogen.

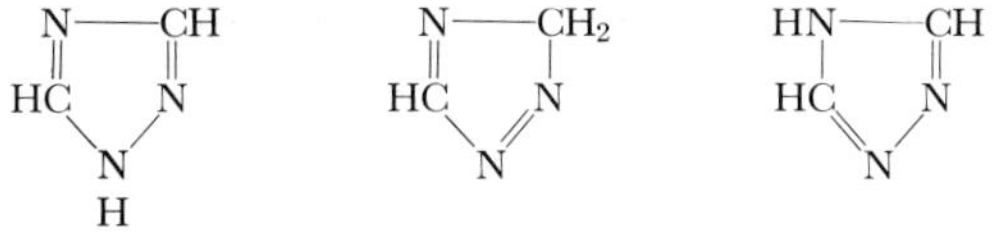

1,2,4,1H-Triazole 1,2,4,3H-Triazole 1,2,4,4H-Triazole

The difficulty of remembering the names of heterocycles has led to increasing use of the *oxa-aza* system whereby the hetero rings are named from the homocyclic hydrocarbon that has the same number of ring atoms. *Oxa, thia, aza, phospha,* or *bora,* for example, denote replacement of CH or CH_2 by O, S, N or NH, P or PH, or B or BH.

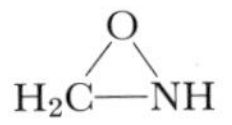

H_2C—NH (with O bridging)
Oxaazacyclopropane

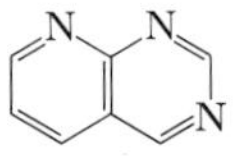

1,3,8-Triazanaphthalene

The classification of heterocyclic compounds used in this chapter is based on the number and kind of hetero atoms and size of the ring. As each new ring system is introduced, the common name or systematic name or both will be given together with the numbering of the ring. The naming of derivatives, however, usually will be limited to common names and occasionally the oxa-aza system.

RING COMPOUNDS CONTAINING ONE HETERO ATOM

THREE- AND FOUR-MEMBERED RINGS

The members of this group that are of most importance are the ethylene oxides and ethylenimines. Their methods of preparation and reactions are discussed in Chapter 35.

FIVE-MEMBERED RINGS

Thiophenes

In 1879 Baeyer reported that when benzene is mixed with isatin (p. 572) and concentrated sulfuric acid, a blue color is produced. This behavior was called the *indophenin reaction* (p. 572) and was believed to be characteristic of benzene until Victor Meyer in 1882 attempted to demonstrate the reaction on a sample of benzene that had been prepared by the decarboxylation of benzoic acid. The lecture demonstration failed, and the resulting investigation as to the cause of the failure led to the discovery that the indophenin reaction is not characteristic of benzene. It is due to a sulfur compound with physical and chemical properties resembling those of benzene so closely that the few tenths of a per cent present in coal tar benzene had not been detected previously. This compound was named *thiophene.* Subsequent development of thiophene chemistry was so rapid that five years after its isolation in 1883, Meyer published a 300-page book on the subject.

Thiophene has the molecular formula C_4H_4S and forms two monosubstitution products. It has been assigned a cyclic structure with two double bonds, which is supported by the methods of synthesis. Until 1946 thiophene was synthesized by heating sodium succinate (p. 629) with tetraphosphorus heptasulfide, sometimes called phosphorus trisulfide.

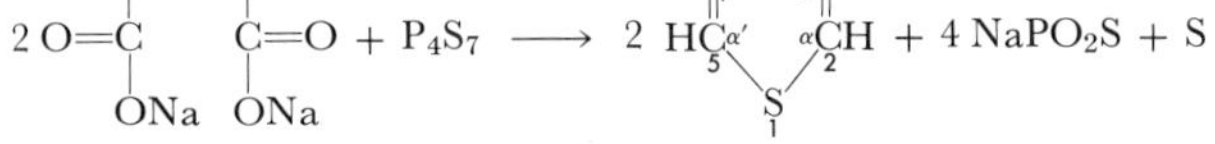

$$2\ \mathrm{O{=}C(ONa)CH_2{-}CH_2C(ONa){=}O} + P_4S_7 \longrightarrow 2\ \mathrm{C_4H_4S} + 4\ NaPO_2S + S$$

Sodium succinate → Thiophene[1] (ring: S 1, α CH 2, β CH 3, β' CH 4, α' CH 5)

Homologs of thiophene can be made by the reaction of 1,4 diketones with phosphorus pentasulfide (*Paal-Knorr synthesis*).

[1] *Chemical Abstracts* and the *Ring Index* orient the formulas of heterocyclic compounds with the hetero atom at the top. By custom, however, it usually is placed at the bottom for the simpler molecules, or in whatever position it assumes if other features of the molecule have a customary orientation. As an aid to memory it seems better to follow custom even though the Ring Index system may have some points of logic in its favor.

$$R{-}CO{-}CH_2{-}CH_2{-}CO{-}R \xrightarrow{P_2S_5} \text{(2,5-R}_2\text{-thiophene: } HC{=}C(R){-}S{-}C(R){=}CH) + P_2O_2S_3 + H_2S$$

A 1,4 diketone

The reaction probably involves replacement of one oxygen atom or both by sulfur, enolization, cyclization by addition, and elimination of water or hydrogen sulfide.

$$O{=}CR{-}CH_2{-}CH_2{-}CR{=}O \;(+\; {>}P^{+}{-}S{:}^{-}) \longrightarrow {>}P^{+}{-}O^{-} + S{=}CR{-}CH_2{-}CH_2{-}CR{=}O \rightleftharpoons RC(SH){=}CH{-}CH_2{-}CR{=}O \rightleftharpoons \text{(cyclic: } RC{=}CH{-}CH_2{-}C(OH)(R){-}S) \longrightarrow H_2O + \text{(thiophene: } RC{=}CH{-}CH{=}CR{-}S)$$

Renewed interest in the chemistry of thiophene and its derivatives has developed with the commercial production of thiophene from butane and sulfur.

$$CH_3CH_2CH_2CH_3 + 4\,S \xrightarrow{650^\circ} \text{(thiophene: } HC{=}CH{-}CH{=}CH{-}S) + 3\,H_2S$$

The butane and sulfur vapor are preheated separately to 600° and mixed in the reaction tube at 650° for a contact time of 0.07 second, after which the exit gases are cooled rapidly. The unreacted materials are recycled. A longer reaction time leads to a more complex mixture of products.

Thiophene boils at 84°, and the boiling points of its homologs and derivatives are close to those of the benzene analogs. The chemical reactions of thiophene also are remarkably similar to those of benzene. The replacement of a benzene ring in physiologically active compounds by a thiophene ring has little effect on their activity. Thus the thiophene analogs of cocaine or atropine (p. 535) have similar local anesthetic or mydriatic action. 2-Thiophenecarboxylic acid when ingested is eliminated in the urine as the amide of glycine, just as benzoic acid is eliminated as hippuric acid (p. 470).

The resemblance of thiophene to benzene can be ascribed to similar molecular weights, similar shapes of the molecules, and most of all to the similar electronic interactions. The p orbitals of the carbon atoms of the benzene ring overlap to form molecular π orbitals above and below the plane of the ring (Fig. 20–*1*, p. 371); likewise a p orbital of the sulfur atom overlaps with the p orbitals of the carbon atoms to form molecular orbitals above and below the plane of the ring (Fig. 30–*1*). In both benzene and thiophene, six electrons occupy three π orbitals, an arrangement that frequently is referred to as an *aromatic sextet*. The electronic interactions may be represented also as resonance among valence-bond structures.

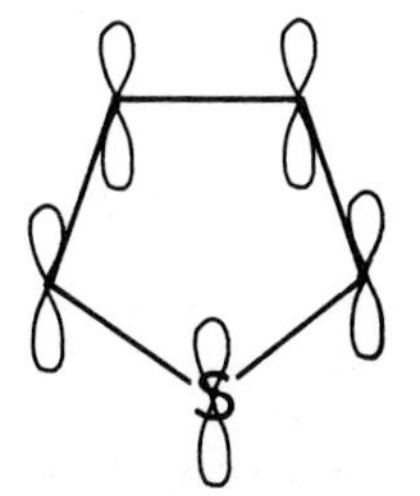

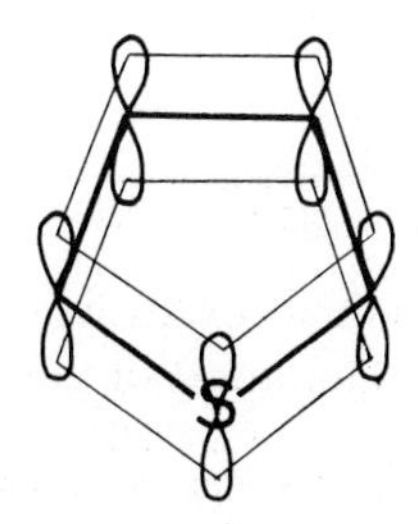

Figure 30–*1*. Resonance in the thiophene molecule.

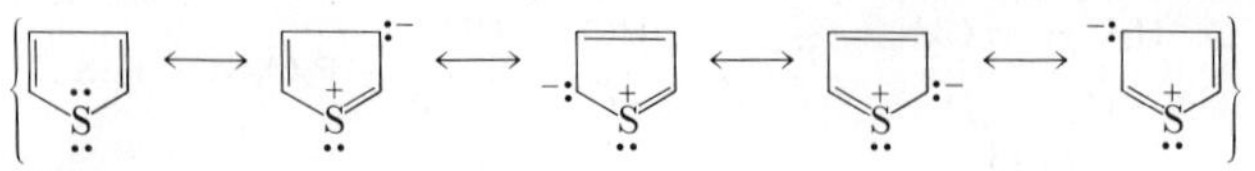

Thiophene undergoes the typical electrophilic substitution reactions of aromatic compounds. The electron density at the carbon atoms, however, is considerably higher than that at benzene carbon atoms because the sulfur atom has contributed a pair of electrons to the π system rather than a single electron. Such systems have been referred to as *π-excessive* aromatic compounds, that is, having an excess of π electrons. The result of the greater electron density at the ring carbon atoms is that thiophene is considerably more reactive to electrophilic substitution than benzene. The situation is analogous to the effect of the hydroxyl or amino group in phenol or aniline where interaction of an unshared pair of electrons on oxygen or nitrogen with the π electrons of the nucleus increases the electron density at the ring carbon atoms.

When possible, electrophilic substitution in thiophene usually takes place almost exclusively in the 2 or 5 position. On the approach of electrophilic reagents, σ complexes can be formed more readily at the 2 position than at the 3 position because the positive charge can be dispersed more widely over the ring.

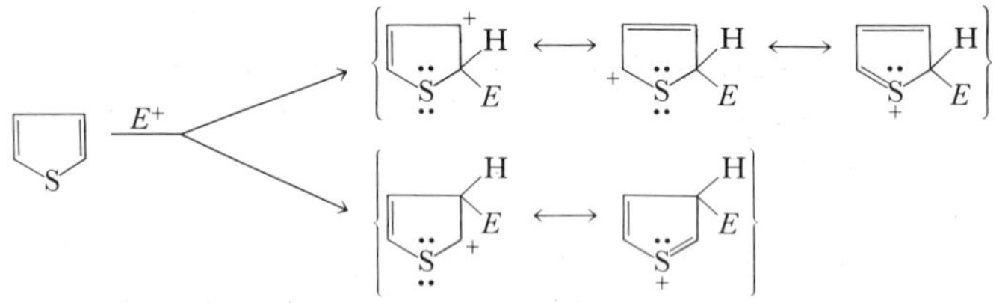

E^+

If either a *meta*- or an *ortho,para*-directing group is present in the 2 position, a second group enters the 5 position. When substituents are in the 3 (or 4) position, mixtures of isomers usually are formed.

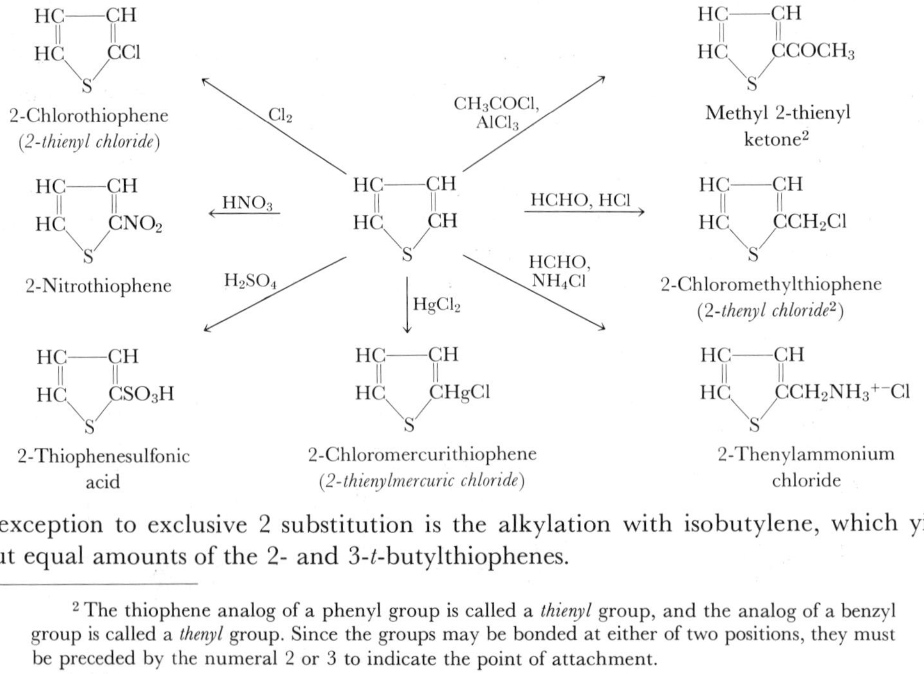

2-Chlorothiophene (*2-thienyl chloride*) ← Cl_2

Methyl 2-thienyl ketone[2] ← CH_3COCl, $AlCl_3$

2-Nitrothiophene ← HNO_3

2-Chloromethylthiophene (*2-thenyl chloride*[2]) ← HCHO, HCl

2-Thiophenesulfonic acid ← H_2SO_4

2-Chloromercurithiophene (*2-thienylmercuric chloride*) ← $HgCl_2$

2-Thenylammonium chloride ← HCHO, NH_4Cl

An exception to exclusive 2 substitution is the alkylation with isobutylene, which yields about equal amounts of the 2- and 3-*t*-butylthiophenes.

[2] The thiophene analog of a phenyl group is called a *thienyl* group, and the analog of a benzyl group is called a *thenyl* group. Since the groups may be bonded at either of two positions, they must be preceded by the numeral 2 or 3 to indicate the point of attachment.

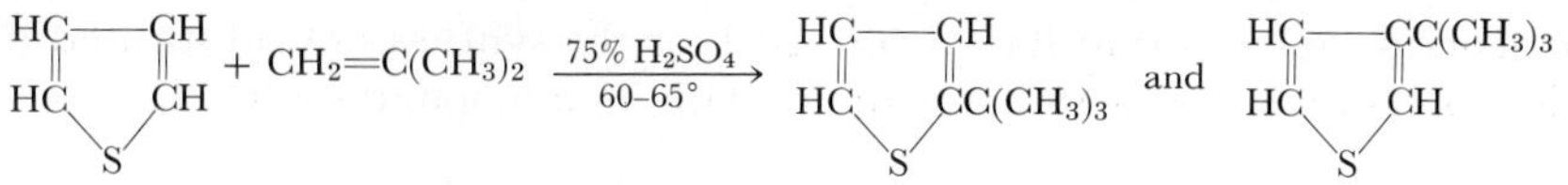

HC—CH, HC CH, S (thiophene) + $CH_2{=}C(CH_3)_2$ $\xrightarrow[60\text{–}65°]{75\%\ H_2SO_4}$ HC—CH, HC $CC(CH_3)_3$, S and HC—$CC(CH_3)_3$, HC CH, S

Thiophene reacts with butyllithium to give **2-thienyllithium,** which undergoes the usual reactions of reactive organometallic compounds.

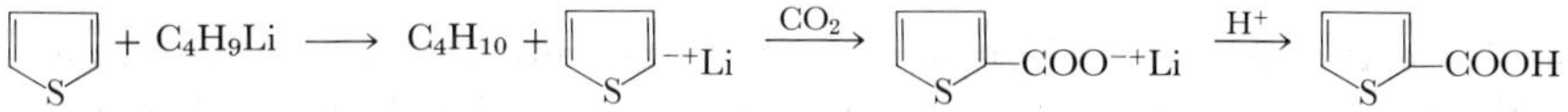

$$\text{Thiophene} + C_4H_9Li \longrightarrow C_4H_{10} + \text{2-thienyl}{-}^{-+}Li \xrightarrow{CO_2} \text{2-thienyl}{-}COO^{-+}Li \xrightarrow{H^+} \text{2-thienyl}{-}COOH$$

The greater reactivity of thiophene over benzene permits the reactions to be carried out under milder conditions that are more nearly like those used for phenols. Thus halogenation takes place without a catalyst. Unless special conditions are observed, direct nitration gives 2,5-dinitrothiophene. Concentrated sulfuric acid reacts at room temperature, preferably in an inert solvent. The Friedel-Crafts reaction is carried out in petroleum ether solution, or a milder catalyst such as stannic chloride is used rather than aluminum chloride. The greater ease of sulfonation is the basis for the removal of thiophene from benzene by extraction with concentrated sulfuric acid.

Thiophene and its homologs do not behave like sulfides. Alkyl halides, for example, do not give sulfonium salts, and oxidation does not produce the sulfoxide or sulfone. The sulfone of 2,5-dihydrothiophene (*sulfolene*) results from the addition of sulfur dioxide to 1,3-butadiene (p. 582) at ordinary temperature. When heated at 140°, it decomposes to sulfur dioxide and butadiene.

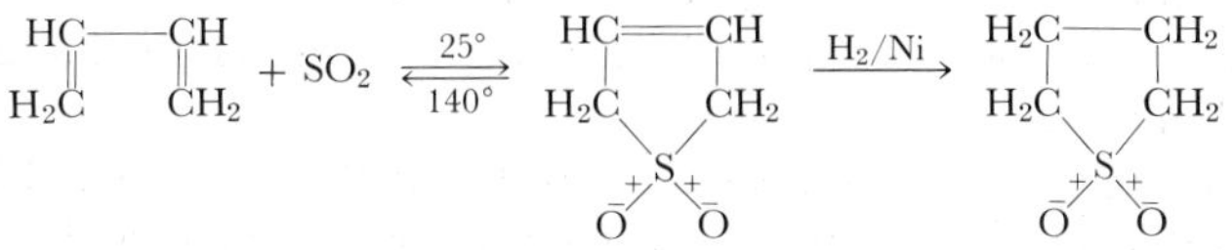

HC—CH, H_2C CH_2 + SO_2 $\underset{140°}{\overset{25°}{\rightleftharpoons}}$ HC=CH, H_2C CH_2, $\overset{+}{S}$, $\bar{O}$ $\bar{O}$ $\xrightarrow{H_2/Ni}$ H_2C—CH_2, H_2C CH_2, $\overset{+}{S}$, $\bar{O}$ $\bar{O}$

Sulfolene Sulfolane

Sulfolene has been used to purify butadiene and has been recommended as a controlled source of sulfur dioxide. Hydrogenation of sulfolene gives the sulfone of tetrahydrothiophene (*sulfolane*). It is used for the extraction of aromatics from catalytic reformates (p. 391) and as a high-dielectric aprotic solvent (p. 102).

Pyrroles

Pyrrole, C_4H_5N, is the nitrogen analog of thiophene. It is present in coal tar and in the tars obtained by the distillation of waste animal matter such as bones (*bone oil* or *Dippel oil*), horn, and scrap leather. Its presence in bone oil was detected in 1834 by the red color that is produced when its vapors come in contact with a pine splint dipped in concentrated hydrochloric acid (Gr. *pyrros* red). It was not isolated in a pure state from these sources until 1858.

The **structure** of pyrrole is most obvious from its formation by distillation of succinimide (p. 629) with zinc dust.

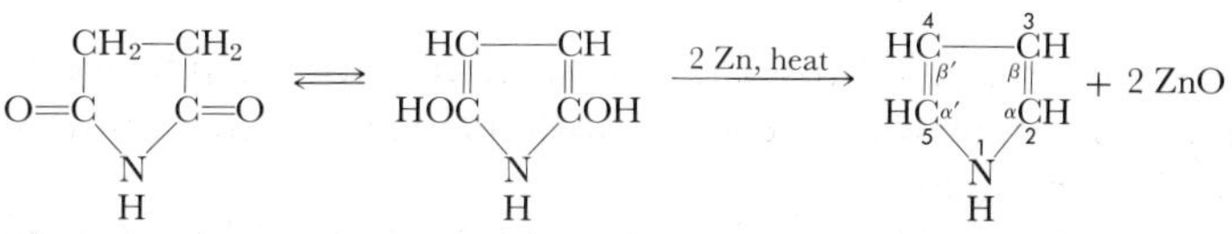

CH_2—CH_2, O=C C=O, N, H $\rightleftharpoons$ HC—CH, HOC COH, N, H $\xrightarrow{2\ Zn,\ heat}$ $H\overset{4}{C}$—$\overset{3}{C}H$ (β', β), $HC\alpha'$ (5) αCH (2), N (1), H + 2 ZnO

Succinimide Pyrrole

Pyrrole is produced commercially by the reaction of furan with ammonia (p. 513). One of the more general methods that have been used for the synthesis of substituted pyrroles is the reaction of 1,4 diketones with ammonia or with an ammonium salt.

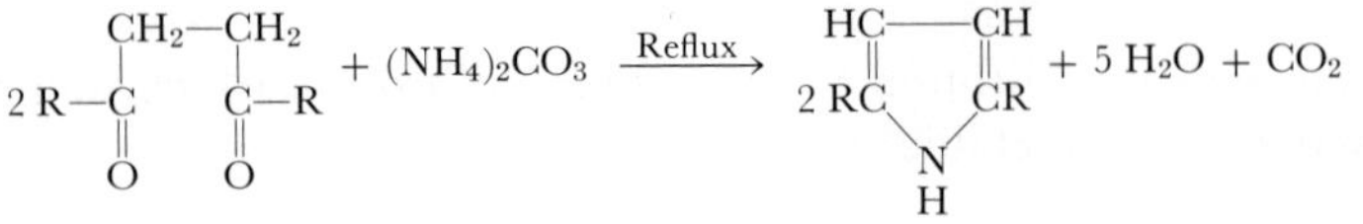

$$2\,R{-}\underset{\underset{O}{\|}}{C}{-}CH_2{-}CH_2{-}\underset{\underset{O}{\|}}{C}{-}R + (NH_4)_2CO_3 \xrightarrow{\text{Reflux}} 2\,\underset{\text{(2,5-disubstituted pyrrole: } HC{=}CR{-}NH{-}CR{=}CH\text{)}}{RC{=}CH{-}CH{=}CR{-}NH} + 5\,H_2O + CO_2$$

The **physical** and **chemical properties** of pyrrole differ markedly from those of thiophene. The nitrogen of pyrrole, like the carbon atoms, is sp^2-hybridized, leaving a pair of electrons in a p orbital that can form an aromatic sextet with the four electrons of the p orbitals of the carbon atoms. Because a $2p$ orbital of nitrogen is used instead of a $3p$ orbital of sulfur, overlap with the p orbitals of carbon is better than in thiophene, and the ring carbon atoms have an even higher excess of π electrons. These factors account for both the physical and the chemical properties of pyrrole. The considerable transfer of negative charge to the carbon atoms gives pyrrole a high dipole moment of 1.8 D compared to 0.5 D for thiophene. The high dipole moment causes the high boiling point of 131°, compared to 84° for thiophene and 55° for diethylamine. Since the unshared pair of electrons on nitrogen is part of the aromatic sextet, it is not available for hydrogen bonding. Thus the boiling point is little different from that of *N*-methylpyrrole, and the solubility is only 1 part in 17 of water whereas diethylamine is miscible with water.

Withdrawal of charge from nitrogen by the aromatic sextet makes pyrrole a much weaker base ($pK_a = 0.4$) than ammonia ($pK_a = 9.2$) and a much stronger acid ($pK_a = 15$ compared to 33 for ammonia). It is not, however, so strong an acid as phenol ($pK_a = 9.9$). Thus it reacts with anhydrous potassium hydroxide, but the potassium salt is almost completely hydrolyzed in water.

Because of the high nuclear π electron density, pyrrole is oxidized by air even more rapidly than aniline or phenol to give dark-colored resins. Electrophilic substitution takes place with extreme ease. The formation of red resinous polymers in the presence of mineral acids may be explained if it is assumed that the σ complex with a proton is sufficiently stable to initiate a chain reaction.

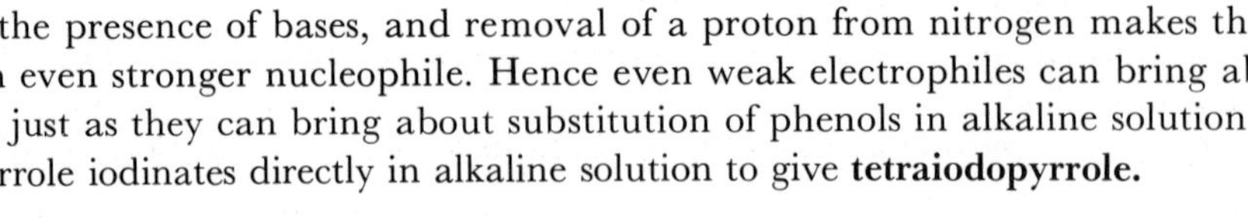

$$\text{pyrrole (N–H)} \underset{B^-}{\overset{HB}{\rightleftharpoons}} \left[\text{2H-pyrrolium (H, H at C-2; N–H; +)}\right] \overset{\text{pyrrole}}{\rightleftharpoons} \left[\text{dimer cation (H, H; N–H; H H; N; +; H)}\right] \xrightarrow{x\ \text{pyrrole}} \text{Polymers}$$

Because of the sensitivity to acids, reactions such as halogenation, nitration, sulfonation, and Friedel-Crafts under the usual conditions are not applicable. Pyrrole is stable, however, in the presence of bases, and removal of a proton from nitrogen makes the pyrrole anion an even stronger nucleophile. Hence even weak electrophiles can bring about substitution just as they can bring about substitution of phenols in alkaline solution (p. 439). Thus pyrrole iodinates directly in alkaline solution to give **tetraiodopyrrole.**

$$C_4H_4NH + 4\,I_2 + 4\,NaOH \longrightarrow C_4I_4NH + 4\,NaI + 4\,H_2O$$

Various other electrophilic substitutions common with the phenoxide ion take place with equal or greater ease than with phenol. As with thiophene, monosubstitution usually takes place initially at the 2 position.

Pyrrole reacts with methylmagnesium bromide with evolution of methane to give the bromomagnesium amide, which reacts in much the same way as other reactive organometallic compounds. At 0° alkyl halides give the *N*-alkyl derivative and acyl chlorides give the *N*-acyl derivative. At higher temperatures the 2-substituted alkyl- or acylpyrroles are obtained. Carbon dioxide gives the 2-carboxylic acid, ethyl chloroformate gives ethyl 2-pyrrolecarboxylate, and ethyl formate gives 2-pyrrolecarboxaldehyde.

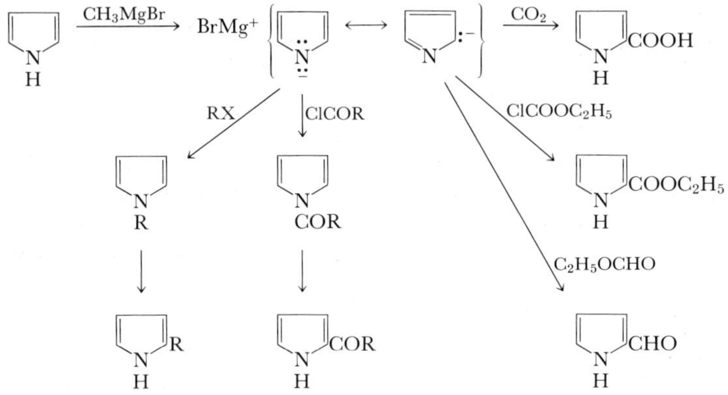

CH_3MgBr BrMg+ CO_2 COOH
RX ClCOR $ClCOOC_2H_5$
R COR $COOC_2H_5$
C_2H_5OCHO
R COR CHO

The aldehyde can be made also by the *Gattermann hydrogen cyanide synthesis* (p. 460).

Reduction of pyrrole with zinc and acetic acid yields **3,4-pyrroline** (*2,5-dihydropyrrole*), and catalytic reduction yields the tetrahydro derivative, **pyrrolidine.**

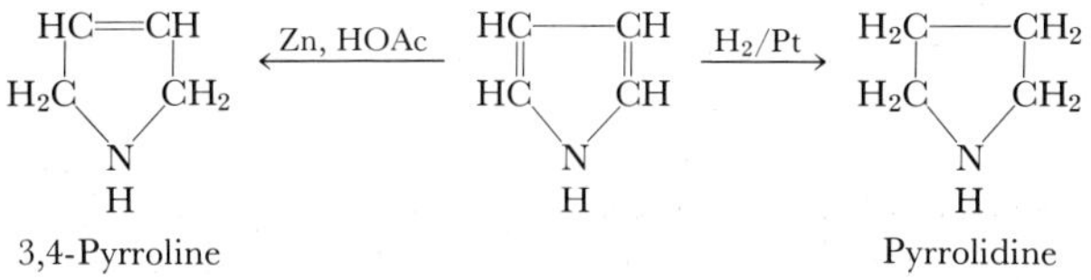

Zn, HOAc H_2/Pt

3,4-Pyrroline Pyrrolidine

The formation of pyrrolidine hydrochloride when 1-chloro-4-aminobutane is heated is a confirmation of the structure of pyrrolidine.

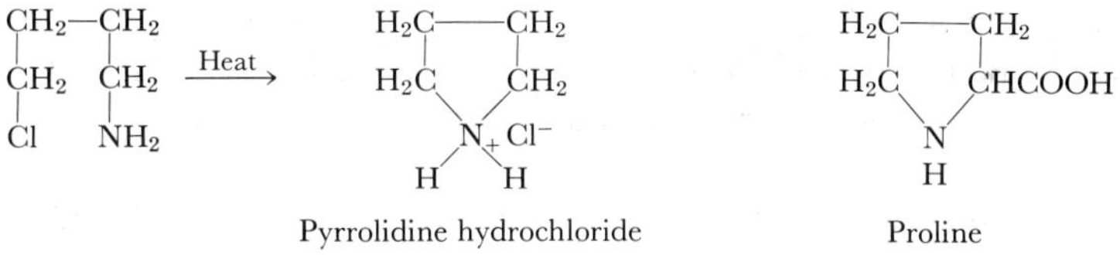

Heat

Pyrrolidine hydrochloride Proline

Pyrrolidine is a typical secondary aliphatic amine. It is miscible with water, has about the same basicity ($pK_a = 11.3$) as diethylamine ($pK_a = 11.0$), and boils at 88°. **Proline** (*2-pyrrolidinecarboxylic acid*) and **hydroxyproline** (*4-hydroxy-2-pyrrolidinecarboxylic acid*) are natural amino acids (pp. 353, 354). The **pyrrolidones** are lactams of γ-amino acids (p. 618).

Porphin Derivatives. Alkylated pyrrole nuclei form the building units for many biologically important pigments, for example those of bile and blood and the green coloring matter of plants. Hence pyrrole and the alkylated pyrroles are present in bone oil (p. 507), which arises from the decomposition of bone marrow, the source of the blood pigments. These pigments have a **porphin nucleus,** which is a flat 16-membered ring. The **porphyrins** derived from natural pigments (Gr. *porphyra* purple) have substituents in the eight β positions of the pyrrole nuclei. The natural pigments themselves are metal chelate complexes (p. 604) of the porphyrins. Thus reaction of **protoporphyrin** with ferric chloride in alkaline solution gives **hemin.** The synthesis of hemin has been reported, but the method of synthesis is not sufficiently rational to be an unequivocal proof of structure. The reduced compound lacking the chloride ion is the **heme** of hemoglobin (p. 363).

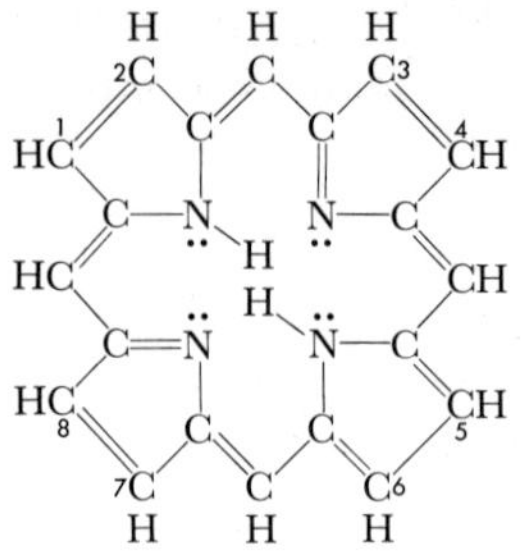

Porphin nucleus

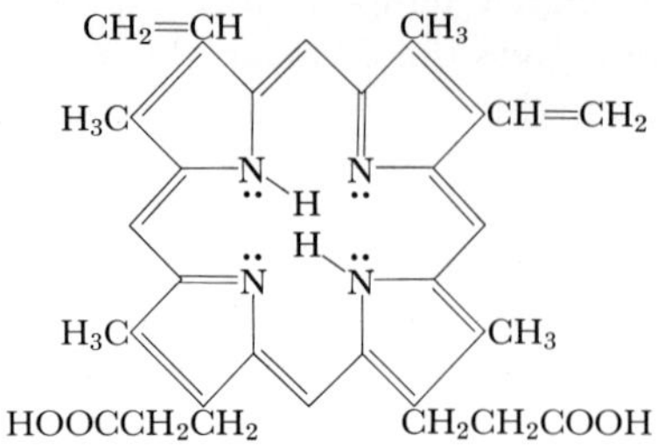

Protoporphyrin

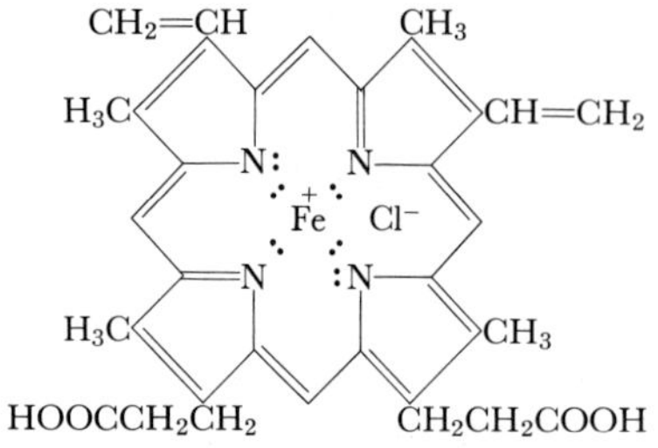

Hemin

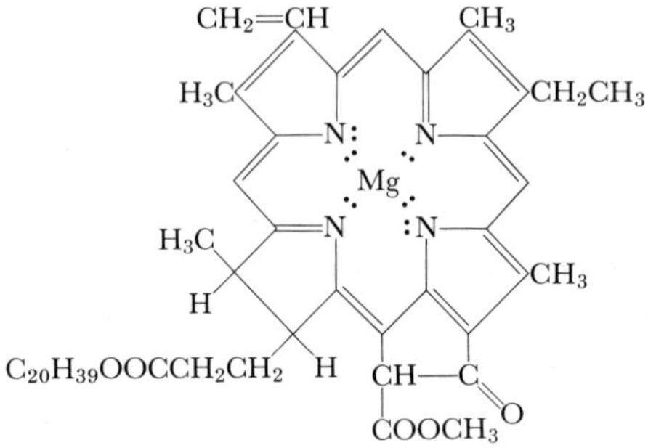

Chlorophyll *a*

The **chlorophylls** are magnesium complexes of porphyrins esterified with the long-chain alcohol phytol, $C_{20}H_{39}OH$ (p. 661). Chlorophyll *b* differs from chlorophyll *a* in that it has an aldehyde group replacing the methyl group in the 3 position. By the use of isotopically labeled molecules it has been shown that the porphyrin nucleus of both heme and chlorophyll is synthesized biologically from glycine (p. 353) and acetic acid and that the steps of the synthesis by the red blood cell and by the chloroplast of the plant are identical. A rational laboratory synthesis of chlorophyll *a* was reported in 1960. Nickel and vanadium complexes of porphyrins occur in petroleum and lead to difficulty in the processing operations.

Vitamin B_{12}, isolated from liver, is a cobalt coordination complex that contains a substituted *corrin* nucleus, which differs from the porphin nucleus in that two of the pyrrole nuclei are linked directly. Its complete structure has been established by chemical and X-ray analysis. With a molecular formula of $C_{63}H_{88}N_{14}O_{14}PCo$, it is the most complicated nonprotein organic compound of known structure. Vitamin B_{12} is effective in the treatment of pernicious anemia. The normal daily requirement is estimated to be about 1 μg. Commercial production in 1963 amounted to 2000 pounds valued at $8,519 per pound.

Benzopyrroles. Benzo[*b*]pyrrole is known as **indole** because it was obtained first by Baeyer[3] in 1866 by distilling oxindole, a degradation product of indigo (p. 571), with zinc dust. It first was synthesized in 1869 and was not found in coal tar until 1910. Numerous syntheses of indole and its derivatives have been developed. One that clearly indicates the structure of indole is the intramolecular condensation of *N*-*o*-tolylformamide (*Madelung synthesis*).

[3] Johann Friedrich Wilhelm Adolf von Baeyer (1835–1917), student of Bunsen and Kekulé and successor to Liebig at the University of Munich. He was awarded the Nobel Prize in Chemistry in 1905. Of the many who studied under Baeyer may be mentioned Emil and Otto Fischer, Perkin Jr., Friedlaender, Bamberger, Curtius, Rupe, and Willstaetter.

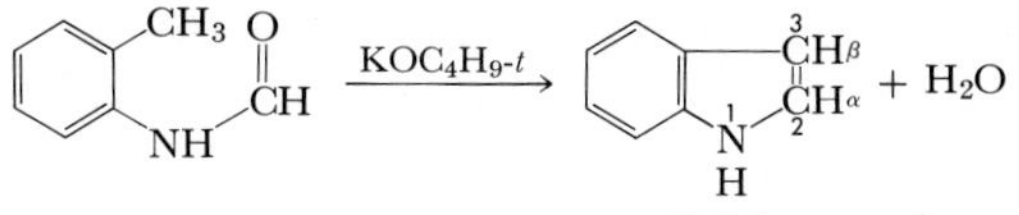

$\xrightarrow{KOC_4H_9\text{-}t}$ $+ H_2O$

Indole

Indole and 3-methylindole (*skatole*) are formed during the putrefaction of proteins. They contribute to the characteristic odor of feces. In contrast, pure indole in high dilution has a flowery odor and is used in the preparation of jasmine, orange blossom, and lilac blends. In fact it is present in natural jasmine, orange blossom, and jonquil extracts. **Skatole** can be synthesized by the *Fischer indole synthesis,* which consists of heating the phenylhydrazones of aldehydes, ketones, or keto acids with zinc chloride or alcoholic sulfuric acid.

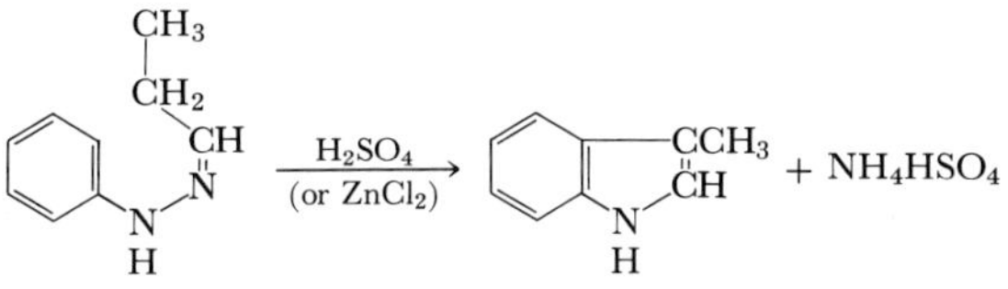

$\xrightarrow[\text{(or } ZnCl_2)]{H_2SO_4}$ $+ NH_4HSO_4$

Phenylhydrazone of propionaldehyde — Skatole

The most important derivative of indole is **tryptophan,** an essential amino acid (p. 365). The name arises from the fact that it is destroyed by the acid hydrolysis of proteins, but can be obtained by enzymic hydrolysis with trypsin. Practical methods of synthesis have been developed. **Serotonin** (*5-hydroxytryptamine*), a vasoconstrictor, is present in the serum of mammals, in brain tissue, and in other body tissues. Its close relationship to hallucinogenic compounds such as bufotenine, isolated from toads and toadstools, has led to the postulation that schizophrenia may result from its abnormal metabolism.

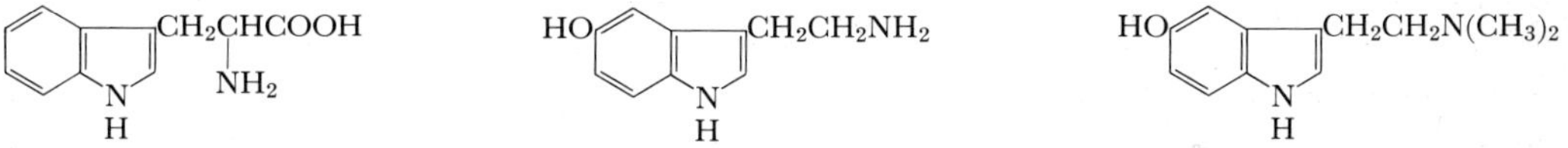

Tryptophan — Serotonin — Bufotenine

Carbazole, 2,4-dibenzopyrrole, is present to the extent of over 1 per cent in coal tar. It can be isolated from the anthracene fraction as the potassium salt. The demand, however, exceeded the supply from coal tar with the result that a synthetic process from *o*-aminobiphenyl was developed that made the recovery from coal tar uneconomical.

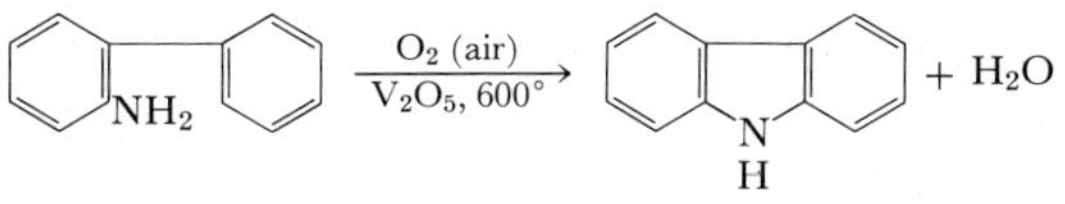

$\xrightarrow[V_2O_5,\ 600^\circ]{O_2\text{ (air)}}$ $+ H_2O$

The chief use for carbazole is in the preparation of blue sulfur dyes (p. 575).

Furans

Furan, C_4H_4O, is the oxygen analog of thiophene. The most available derivative of furan is the α aldehyde, **furfural,** from which most other furans are prepared. Furfural first was obtained in 1840 by the distillation of bran (L. *furfur*) with dilute sulfuric acid. It results from the dehydration of pentoses formed by the hydrolysis of the pentosans in the bran (p. 348).

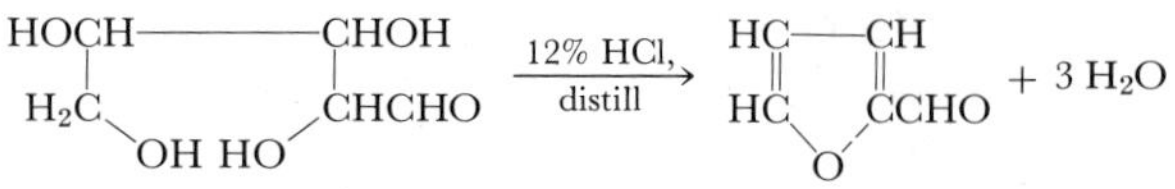

$$\underset{\text{A pentose}}{\begin{matrix} HOCH\text{———}CHOH \\ | \qquad\qquad | \\ H_2C \qquad CHCHO \\ \diagdown OH \; HO \diagup \end{matrix}} \xrightarrow[\text{distill}]{12\%\ HCl,} \underset{\text{Furfural}}{\begin{matrix} HC\text{—}CH \\ \| \quad \| \\ HC \quad CCHO \\ \diagdown O \diagup \end{matrix}} + 3\,H_2O$$

The commercial production since 1922 by the hydrolysis of oat hulls stemmed from an attempt to produce an improved cattle feed from them. The availability of cheap furfural led to several large-scale applications, and now it is made also from other agricultural wastes that contain pentosans, such as corncobs and straw. One general method for the synthesis of furans is the dehydration of 1,4 diketones.

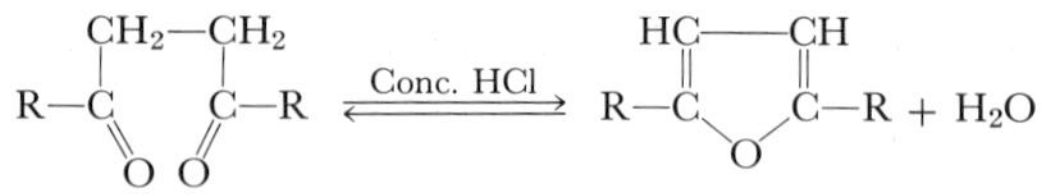

$$\begin{matrix} CH_2\text{—}CH_2 \\ | \qquad | \\ R\text{—}C \qquad C\text{—}R \\ \| \qquad \| \\ O \qquad O \end{matrix} \xrightleftharpoons{\text{Conc. HCl}} \begin{matrix} HC\text{—}CH \\ \| \quad \| \\ R\text{—}C \quad C\text{—}R \\ \diagdown O \diagup \end{matrix} + H_2O$$

Furfural is a colorless liquid with a pleasant characteristic odor. It is oxidized by air, the color changing through shades of yellow and brown to almost black. These changes can be prevented by storing in the absence of oxygen or by adding an antioxidant (p. 447). Furfural can be detected by the brilliant red color that it gives with aniline in the presence of acetic acid. Apparently two moles of aniline condense with one of furfural with opening of the ring.

$$C_6H_5NH_2 + \begin{matrix} HC\text{—}CH \\ \| \quad \| \\ HC \quad CCHO \\ \diagdown O \diagup \end{matrix} + H_2NC_6H_5 \xrightarrow{HOCOCH_3} C_6H_5NHCH{=}CHCH{=}\underset{\underset{OH}{|}}{C}CH{=}NC_6H_5$$

Reaction of furfural with an acid solution of phloroglucinol (p. 448) gives a dark green precipitate of unknown and probably variable composition. The weight of precipitate formed has been related empirically to the weight of furfural from which it was produced. Hence the reaction can be used for the quantitative estimation of furfural, and indirectly of pentoses and pentosans (p. 330).

Furfural is used on a large scale as a solvent for refining lubricating oils (p. 95) and for removing butadiene from its mixtures with butenes and butane (p. 587). It can replace some of the formaldehyde in phenolic resins (p. 443). The common reactions are summarized in Fig. 30–*2*. It will be noted that furfural gives all of the reactions characteristic of benzaldehyde. Tetrahydrofuran, tetramethylene chloride, pyrrole, and pyridine are made commercially by the series of reactions indicated.

Interaction of an unshared pair of electrons of the hetero atom with those of the carbon atoms to give an aromatic sextet is more important with furan than with thiophene because of the better overlap of the $2p$ orbital. Interaction is less, however, with furan than with pyrrole because the greater electronegativity of oxygen holds the pair more strongly. The dipole moment of furan is only 0.7 D, and it boils at 31°. It is about one half as soluble in water as ethyl ether, indicating that it forms less strong hydrogen bonds and probably is somewhat less basic because of the greater electronegativity of the sp^2-hybridized carbon atoms. Furan, like pyrrole, reacts with acids and forms a dark brown polymer. The ring exhibits its excess of π electrons by undergoing electrophilic substitution, but reagents that are strong acids or that produce strong acids cannot be used unless electronegative groups are present.

Tetrahydrofuran (*THF*), which has been synthesized on an industrial scale from acetylene as well as from furfural, has a dipole moment of 1.6 D and boils at 67°. It is more basic ($pK_a = -2.1$) than ethyl ether ($pK_a = -3.6$) and forms more stable complexes with Grignard reagents. Moreover, its higher boiling point makes it safer to use.

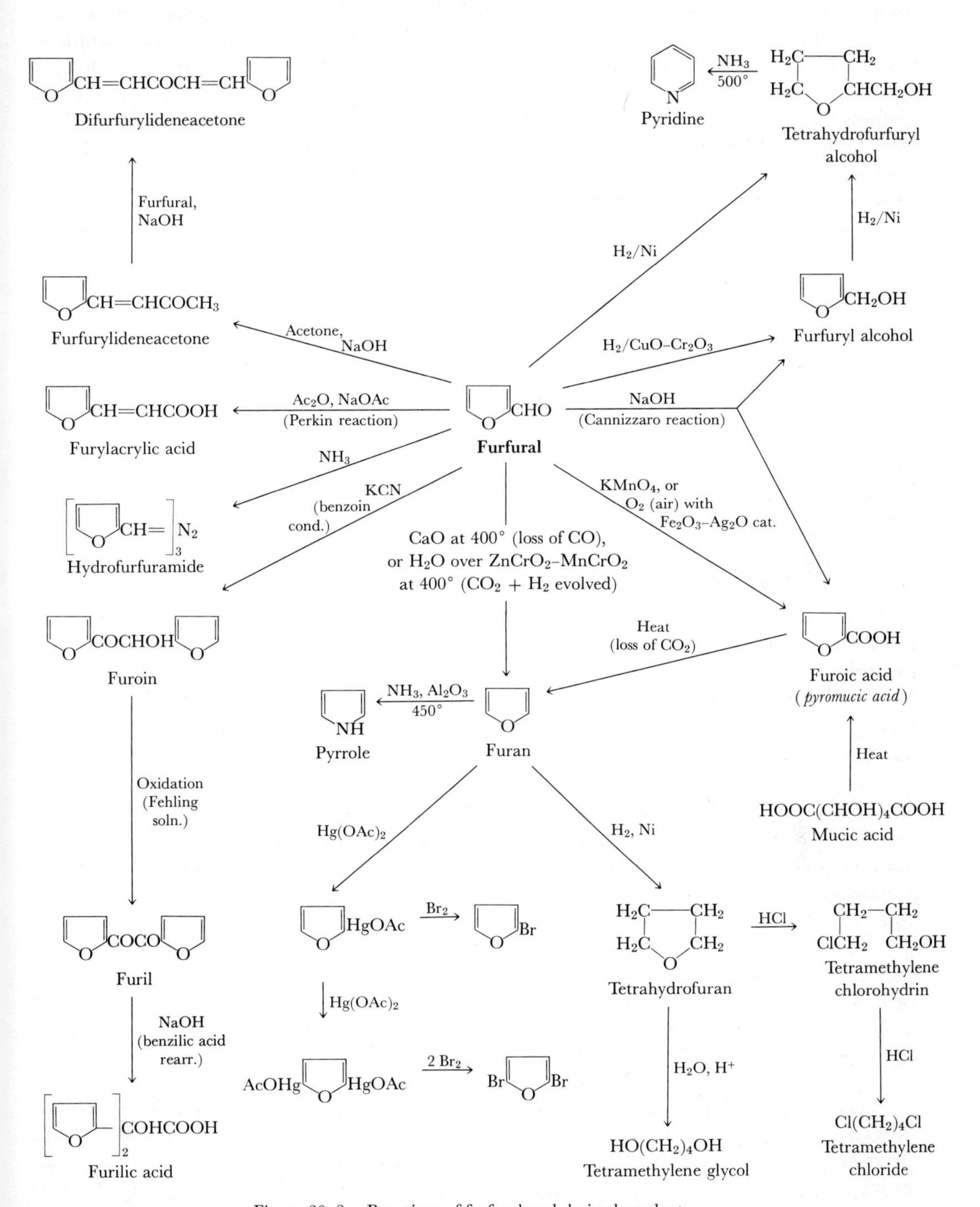

Figure 30–2. Reactions of furfural and derived products.

Because of its greater basicity, tetrahydrofuran forms stronger hydrogen bonds with water complexes, which makes it miscible with water. Besides being used as a solvent, tetrahydrofuran has been converted through tetramethylene chloride and tetramethylene cyanide to hexamethylenediamine, an intermediate for the production of nylon 6-6 (p. 631). Hydrolysis gives tetramethylene glycol used in the production of polyurethans (p. 605). **Tetrahydrofurfuryl alcohol** is used not only for the commercial synthesis of pyridine but also as an intermediate for the production of a variety of organic chemicals (p. 521). The **γ-butyrolactones** are **2-oxotetrahydrofurans** (p. 616). **Muscarine,** the toxic principle of one class of poisonous mushrooms (*Amanita muscaria*) is the trimethyl quaternary salt of 2-(aminomethyl)-4-hydroxy-5-methyltetrahydrofuran.

H_2C—CH_2 / H_2C, $C{=}O$ / O

γ-Butyrolactone

HOCH—CH_2 / CH_3CH $CHCH_2\overset{+}{N}(CH_3)_3$ Cl^- / O

Muscarine (chloride)

Benzo[*b*]furan

SIX-MEMBERED RINGS

Pyridines

Pyridine, C_5H_5N, was isolated first from bone oil in 1851 and then from coal tar in 1854. Until around 1950, coal tar was the sole commercial source, although less than 0.1 per cent of pyridine is present (Table 20–*3*, p. 291). Its chief use had been as a denaturant for ethyl alcohol. The increasing demand for high purity pyridine for the synthesis of medicinals led to its commercial synthesis in which acetylene, formaldehyde hemimethylal, and ammonia are passed into a fluidized bed of alumina-silica at 500°.

$$2\,HC{\equiv}CH + 2\,CH_2(OH)OCH_3 + NH_3 \xrightarrow{Al_2O_3\text{-}SiO_2} C_5H_5N + H_2O + 3\,CH_3OH$$

The acetylene may be replaced by acetaldehyde. Almost as much 3-methylpyridine (p. 516), a less useful chemical, is obtained along with pyridine. A more recent commercial synthesis of pyridine starts with tetrahydrofurfuryl alcohol (Fig. 30–*2*, p. 513).

Reduction of pyridine with sodium and alcohol gives a hexahydro derivative known as *piperidine* (p. 519). Three monosubstitution products of pyridine are known. Hence it can be represented best as an analog of benzene in which a methylidyne (CH) group is replaced by nitrogen.

$$\text{Pyridine} \xrightarrow[\text{Na, alcohol}]{6\,[H]} \text{Piperidine}$$

Pyridine Piperidine

The nitrogen atom in pyridine is sp^2-hybridized. Two orbitals are used to form σ bonds with carbon atoms, and an unshared electron pair is in the third sp^2 orbital. The remaining electron in a *p* orbital forms with the five electrons in the *p* orbitals of the carbon atom an aromatic sextet. In contrast to pyrrole, however, pyridine is π deficient compared with benzene because nitrogen, like each carbon atom, contributes only one electron to the aromatic sextet but is more electronegative than carbon. The dipole moment of pyridine of 2.3 D is high and accounts for the high boiling point of 115° relative to benzene, but the moment has the direction opposite to that of pyrrole. Since the nitrogen has an unshared pair of electrons not involved with the aromatic sextet, pyridine forms hydrogen bonds with water molecules and is miscible in water. It coordinates also with metal ions and dissolves many inorganic salts as well as other organic compounds. Like other volatile amines, it has a very disagreeable odor.

Pyridine, although basic ($pK_a = 5.2$), is much less so than aliphatic tertiary amines ($pK_a = 10\text{–}11$). The sp^2 orbital has more s character than an sp^3 orbital and the pair of electrons is held closer to the nucleus (p. 148). Like aniline ($pK_a = 4.6$), it is sufficiently basic to react with alkyl iodides to give quaternary salts.

$$C_5H_5N\colon + RI \longrightarrow C_5H_5\overset{+}{N}\colon R\ {}^{-}I$$

Pyridine finds important use in organic synthesis as a basic solvent, whereby it not only can exert a catalytic action but also can combine with acids produced in reactions. For example, acetylations and benzoylations take place smoothly in pyridine solution.

$$ROH + ClCOC_6H_5 + C_5H_5N \longrightarrow ROCOC_6H_5 + C_5H_5NH^{+}\ {}^{-}Cl$$

For this purpose the pyridine must be anhydrous since water in the presence of pyridine hydrolyzes the reagent. The presence of interfering amounts of water in pyridine can be detected easily by adding pure benzoyl chloride free of benzoic acid. If water is present an immediate precipitate of the slightly soluble benzoic anhydride is formed.

$$C_6H_5COCl + H_2O + C_5H_5N \longrightarrow C_6H_5COOH + C_5H_5NH^{+}\ {}^{-}Cl$$
$$C_6H_5COOH + ClCOC_6H_5 + C_5H_5N \longrightarrow (C_6H_5CO)_2O + C_5H_5NH^{+}\ {}^{-}Cl$$

Because of the deficiency of π electrons, pyridine is stable to oxidation. Solutions of potassium permanganate in pyridine can be boiled without affecting the ring. Pyridine inherently is less reactive to electrophilic reagents than benzene. It is particularly unreactive under acidic conditions because salt formation places a positive charge on the nitrogen, thus making it very strongly electronegative. When substitution is forced, the group enters the 3 position in preference to the 2 or 4 position because the σ intermediate for attack at the 2 or 4 position would require greater increase in the positive charge on the nitrogen atom.

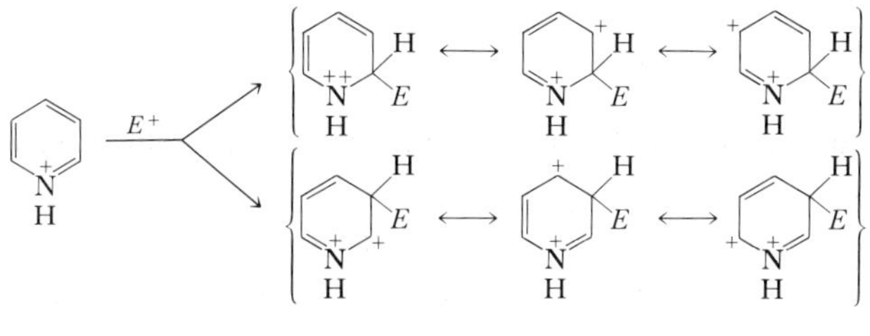

Pyridine does not react with hot nitric, sulfuric, or chromic acids. Pyridinium bromide reacts with bromine to give the solid **pyridinium tribromide** (*pyridinium bromide perbromide*), $C_5H_5NH^{+}\ {}^{-}Br_3$, which is a convenient brominating agent for other compounds.

Although the decrease in π electron density makes electrophilic substitution difficult, it makes pyridine more susceptible than benzene to nucleophilic substitution (cf. p. 412). Thus the amino group can be introduced directly into pyridine by heating with sodium amide (*Chichibabin*[4] *reaction*) to give the sodium salt of **2-aminopyridine.** Addition of water gives the free amine.

[4] Aleksei Eugenievitsch Chichibabin (1871–1945), professor of chemistry at the Imperial College of Technology of Moscow until 1929. From 1931 until his death he worked at the Collège de France in Paris. He is noted chiefly for his studies of the chemistry of pyridine compounds.

$$C_5H_5N \xrightarrow{^{-}:NH_2} \sigma\text{-complex (H, NH}_2) \longrightarrow \text{2-}C_5H_4N\text{–}NH_2 + {}^{-}:H \longrightarrow H_2 + \text{2-}C_5H_4N\text{–}NH^{-} \xrightarrow[^{-}OH]{H_2O} \text{2-}C_5H_4N\text{–}NH_2$$

Substitution does not take place at the 3 position because in the σ intermediate the negative charge would be concentrated only on carbon, which is less able to accommodate it than the more electronegative nitrogen. 2-Aminopyridine is an important intermediate for the synthesis of medicinals such as sulfapyridine and antihistaminics (p. 519).

Pyridine reacts with peroxy acids or with 30 per cent hydrogen peroxide in acetic acid solution to give good yields of **pyridine *N*-oxide** (p. 229).

$$C_5H_5N + H_2O_2 \longrightarrow C_5H_5\overset{+}{N}\text{–}\overset{-}{O} + H_2O$$

Electrophilic substitution in the 2 and 4 positions of the oxide takes place much more easily than with pyridine because the unshared electrons on oxygen can contribute to the π electron density and stabilize the σ intermediates at the 2 or 4 positions but not at the 3 position. Thus potassium nitrate and fuming sulfuric acid at 100° gives **4-nitropyridine *N*-oxide** in 90 per cent yield.

$$C_5H_5\overset{+}{N}\text{–}\overset{-}{O} + {}^{+}NO_2 \longrightarrow \text{(H, NO}_2\text{ σ intermediate, }\overset{+}{N}\text{=O)} \longrightarrow H^{+} + \text{4-}O_2N\text{–}C_5H_4\overset{+}{N}\text{–}\overset{-}{O}$$

Reaction of the nitro oxide with a five-fold excess of phosphorus trichloride in chloroform gives a 65 per cent yield of **4-nitropyridine.**

$$\text{4-}O_2N\text{–}C_5H_4\overset{+}{N}\text{–}\overset{-}{O} + PCl_3 \longrightarrow \text{4-}O_2N\text{–}C_5H_4N + POCl_3$$

4-Aminopyridine can be obtained from the nitro *N*-oxide by reduction with iron and acetic acid.

$$\text{4-}O_2N\text{–}C_5H_4\overset{+}{N}\text{–}\overset{-}{O} + 4\,Fe + 8\,HOCOCH_3 \longrightarrow \text{4-}H_2N\text{–}C_5H_4N + 3\,H_2O + 4\,Fe(OCOCH_3)_2$$

The three methylpyridines are known as **α-, β-,** and **γ-picolines** (L. *pix, picis* pitch). They occur in bone oil, but the commercial source is coal tar. The α isomer is present in largest amount and was isolated first in 1846. The β isomer was isolated in 1879, and the γ isomer, which is present in smallest amount, in 1887. Special extractive distillation procedures are necessary to separate β- and γ-picoline because of their close boiling points (143.5° and 143.1°). The picolines are obtained also by synthesis, since modifications in procedure can alter the relative amounts of α-, β-, and γ-picolines that are formed during the synthesis of pyridine (p. 514).

Oxidation of the three isomers with permanganate yields the three **carboxylic acids.**

α-Picoline $\xrightarrow{KMnO_4}$ α-Picolinic acid (*picolinic acid*)

β-Picoline $\xrightarrow{KMnO_4}$ β-Picolinic acid (*nicotinic acid*)

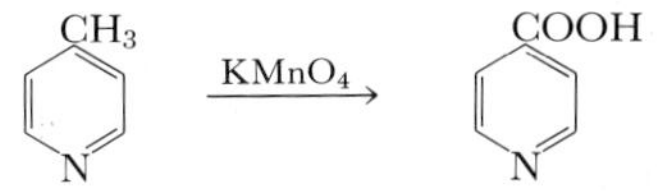

γ-Picoline $\xrightarrow{KMnO_4}$ γ-Picolinic acid (*isonicotinic acid*)

The α and γ acids when heated lose carbon dioxide with the formation of pyridine. Loss of carbon dioxide from a carboxyl group in a 2 or 4 position takes place by way of the dipolar ion, the intermediate being stabilized by the ability of the nitrogen to accept a pair of electrons. This mechanism is not available to carboxyl groups in the 3 position.

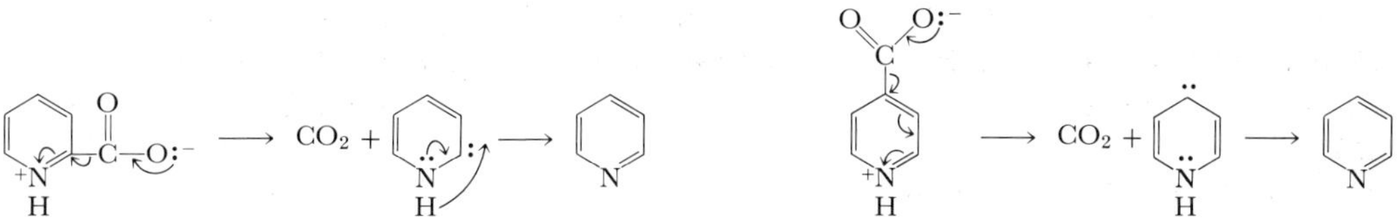

$\longrightarrow CO_2 + \longrightarrow$ $\qquad$ $\longrightarrow CO_2 + \longrightarrow$

The name **nicotinic acid** for the β isomer was given to it because it first was obtained in 1867 by the oxidation of nicotine. This reaction elucidated the structure of half of the nicotine molecule.

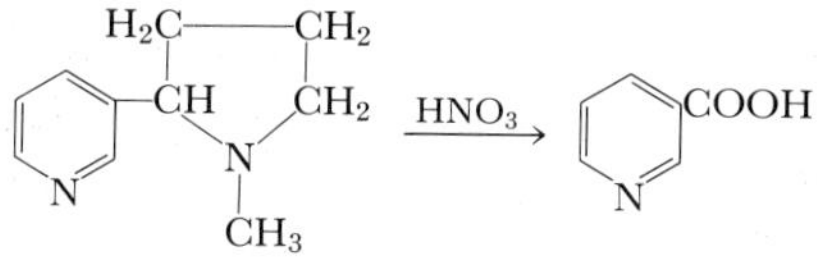

Nicotine $\xrightarrow{HNO_3}$ (nicotinic acid, COOH)

It was not until 1937 that it was recognized that the absence of nicotinic acid or its amide from the diet resulted in the deficiency disease known as *pellagra* in humans and *black tongue* in dogs. Nicotinic acid is required for the production of nicotinamide adenine dinucleotide (*NAD*),[5] which acts as a hydrogen acceptor in the presence of dehydrogenating enzymes.

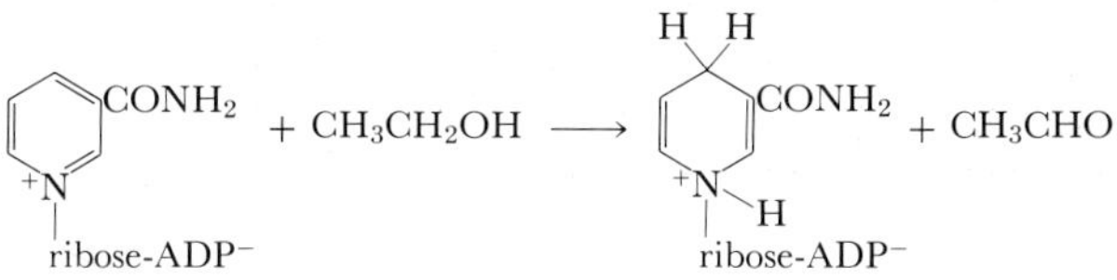

$$\text{NAD (CONH}_2\text{, ribose-ADP}^-) + CH_3CH_2OH \longrightarrow \text{NADH (CONH}_2\text{, ribose-ADP}^-) + CH_3CHO$$

Over 2.3 million pounds of the acid and amide valued at $3.4 million was manufactured in the United States in 1964 for use in fortified wheat flour and vitamin preparations. The names *niacin* and *niacinamide* have been coined for the acid and its amide because they are considered to be more acceptable to the public. They are made not only from β-picoline, but also from 2-methyl-5-ethylpyridine and from quinoline (p. 520).

6-Aminonicotinamide is a teratogen that, when administered to mice during the early

[5] Formerly called coenzyme I or diphosphopyridine nucleotide (*DPN*). *ADP* = adenosine diphosphate (p. 533).

stages of gestation, causes cleft palate in 95 per cent of the fetuses. Moreover chromosomal changes are found even in cells remote from the defect.

Reaction of ethyl isonicotinate with hydrazine gives **isonicotinhydrazide** (*isoniazid*). It is one of the more effective tuberculostatic drugs, although it allows the emergence of resistant strains. The mood-elevating effect of isoniazid and its derivatives, such as **iproniazid,** has led to the term *psychic energizer.* These compounds are used to treat depressive states and are believed to act by inhibiting an enzyme that brings about the destruction of brain-stimulating substances (cf. pp. 460, 511).

$CONHNH_2$ (4-pyridyl) — Isonicotinhydrazide (*isoniazid*)

$CONHNHCH(CH_3)_2$ (4-pyridyl) — Iproniazid

$CH{=}NOH$ (2-pyridyl, N^+–CH_3 I^-) — 2-Pyridinecarboxaldoxime methiodide (*PAM*)

Oxidation of α- or γ-picoline by air in the vapor phase gives **2-** or **4-pyridinecarboxaldehyde. 2-Pyridinecarboxaldoxime methiodide** (*PAM*) appears to be an effective antidote for poisons such as the nerve gases and insecticides (p. 272) that inhibit choline esterase (p. 232).

α- and γ-Picolines add to aldehydes.

$$C_5H_4N\text{-}CH_3 + RCH{=}O \longrightarrow C_5H_4N\text{-}CH_2CH(OH)R$$

These reactions are nothing more than aldol additions, the electron-attracting effect of the nitrogen atom in the picoline taking the place of that of a doubly bound oxygen atom in an aldehyde or ketone. The effect can be transmitted by conjugation to the γ position but not to the β position. Addition of α- or γ-picoline to formaldehyde gives the **hydroxyethylpyridines,** which can be dehydrated to the **vinylpyridines.**

$$C_5H_4N\text{-}CH_3 \xrightarrow{HCHO} C_5H_4N\text{-}CH_2CH_2OH \longrightarrow C_5H_4N\text{-}CH{=}CH_2$$

2-Vinylpyridine

2- and 4-Vinylpyridine are analogs of and undergo typical addition reactions of α,β-unsaturated carbonyl compounds (p. 618). **2-Vinylpyridine** when copolymerized with butadiene gives an elastomer that adheres to rayon and nylon cord and makes possible cord tires. When copolymerized with acrylonitrile, it introduces a basic group, which permits polyacrylonitrile fibers (p. 619) to be dyed more readily.

α- and γ-Picolines condense with aromatic aldehydes to give benzylidene derivatives, but β-picoline does not.

$$C_5H_4N\text{-}CH_3 + OCHC_6H_5 \xrightarrow{Ac_2O} C_5H_4N\text{-}CH{=}CHC_6H_5 + 2\ HOAc$$

Benzylidene-α-picoline

$$4\text{-}CH_3C_5H_4N + OCHC_6H_5 \xrightarrow{Ac_2O} 4\text{-}C_6H_5CH{=}CHC_5H_4N + 2\ HOAc$$

Benzylidene-γ-picoline

Since the benzylidene derivatives boil higher than the picolines, the reaction can be used to separate β-picoline from its isomers.

The greater ease with which the methyl groups in the α and γ positions lose a proton is illustrated by the ease of formation of organometallic derivatives. Thus phenyllithium (p. 398) reacts to give lithium derivatives, which react with carbon dioxide to give lithium salts of the **pyridylacetic acids.**

$$\text{C}_5\text{H}_4\text{N-CH}_3 \xrightarrow{C_6H_5Li} \text{C}_5\text{H}_4\text{N-CH}_2\text{Li} \xrightarrow{CO_2} \text{C}_5\text{H}_4\text{N-CH}_2\text{COO}^{-+}\text{Li}$$

Pyridoxal (*vitamin* B_6) in conjunction with certain chelating metal ions is required for the enzymatic decarboxylation and transamination of α-amino and α-keto acids. **Pyribenzamine** and **Chlor-Trimeton** are among the more widely used antihistaminics (p. 523). All of the antihistamine activity of the latter resides in the dextrorotatory isomer, which is marketed as **Polaramine.**

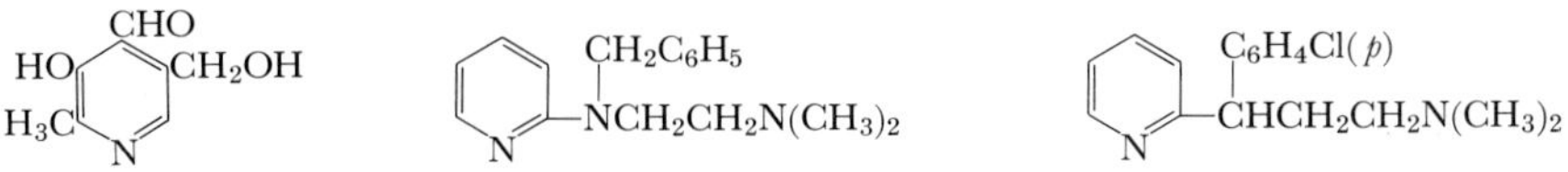

Pyridoxal (ring bearing HO, CHO, CH_2OH, H_3C; N)

Pyribenzamine (pyridine–$N(CH_2C_6H_5)CH_2CH_2N(CH_3)_2$)

Chlor-Trimeton (±), Polaramine (+) (pyridine–$CH(C_6H_4Cl(p))CH_2CH_2N(CH_3)_2$)

Histadyl is a thiophene analog of Pyribenzamine that has a 2-thienyl group instead of a phenyl group. Like many antihistaminics, it has mild sedative properties and is used in proprietary sleeping pills (*Dormin* and others).

The dimethylpyridines were called **lutidines** because they are isomeric with the toluidines. All six possible isomers are known. The trimethylpyridines are called **collidines** (Gr. *kolla* glue) and occur with the other pyridine and pyrrole bases in coal tar and in the decomposition products of animal matter.

Piperidine is a typical secondary aliphatic amine ($pK_a = 11.2$). It is made commercially by the ruthenium-catalyzed hydrogenation of pyridine. The assigned structure is supported by its synthesis from pentamethylenediamine hydrochloride (p. 612).

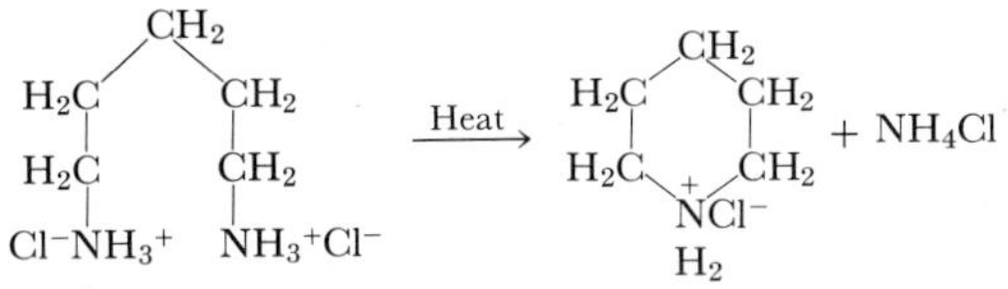

$$Cl^-NH_3^+CH_2CH_2CH_2CH_2CH_2NH_3^+Cl^- \xrightarrow{\text{Heat}} (CH_2)_5\overset{+}{N}H_2Cl^- + NH_4Cl$$

Pentamethylenediamine hydrochloride — Piperidine hydrochloride

Quinolines

Quinoline is 2,3-benzopyridine. It first was isolated from coal tar in 1834. It was obtained also by the distillation of quinine alkaloids (p. 537) with alkali in 1842. Both products were accompanied by impurities that gave different color reactions, and they were not proven to be identical until 1882.

The structure of quinoline is indicated clearly by the *Friedlaender synthesis* from *o*-aminobenzaldehyde and acetaldehyde in the presence of dilute alkali.

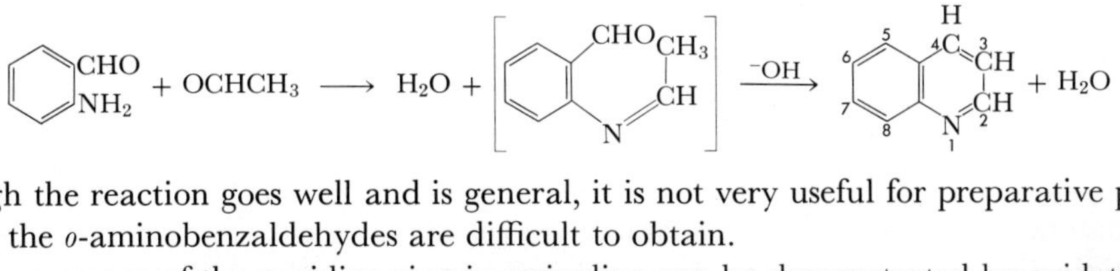

$$o\text{-}C_6H_4(CHO)(NH_2) + OCHCH_3 \longrightarrow H_2O + \left[C_6H_4(CHO)(N{=}CHCH_3)\right] \xrightarrow{^-OH} \text{quinoline} + H_2O$$

(Quinoline ring numbered: N 1, C 2, C 3, C 4, 5, 6, 7, 8)

Although the reaction goes well and is general, it is not very useful for preparative purposes because the *o*-aminobenzaldehydes are difficult to obtain.

The presence of the pyridine ring in quinoline can be demonstrated by oxidation with permanganate. Because of the lower electron density of the pyridine ring, the benzene ring is oxidized preferentially. The chief product is **quinolinic acid** (*α,β-pyridinedicarboxylic acid*), which can be decarboxylated to nicotinic acid or pyridine.

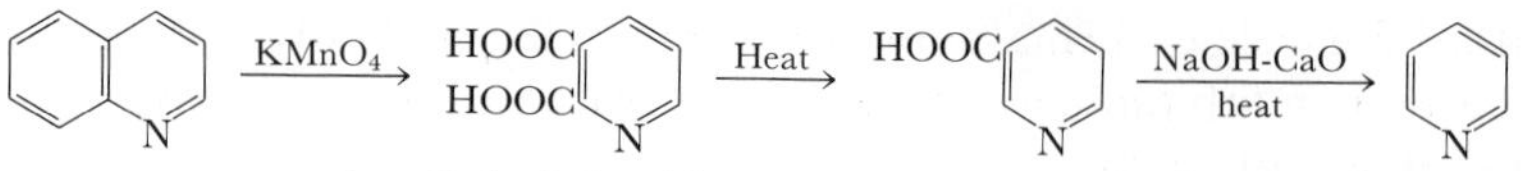

$\xrightarrow{KMnO_4}$ HOOC, HOOC $\xrightarrow{\text{Heat}}$ HOOC $\xrightarrow[\text{heat}]{\text{NaOH-CaO}}$

Quinolinic acid

Pure quinoline is obtained best by the *Skraup synthesis,* in which a mixture of glycerol and aniline is heated with concentrated sulfuric acid and a mild oxidizing agent such as arsenic acid or nitrobenzene. Ferrous sulfate and boric acid usually are added to moderate the reaction. Presumably the first stage is the dehydration of the glycerol to the unsaturated aldehyde, acrolein. Aniline then adds 1,4 to the conjugated unsaturated aldehyde. Subsequent ring closure and oxidation yields the quinoline. The method can be used also for the synthesis of quinoline derivatives.

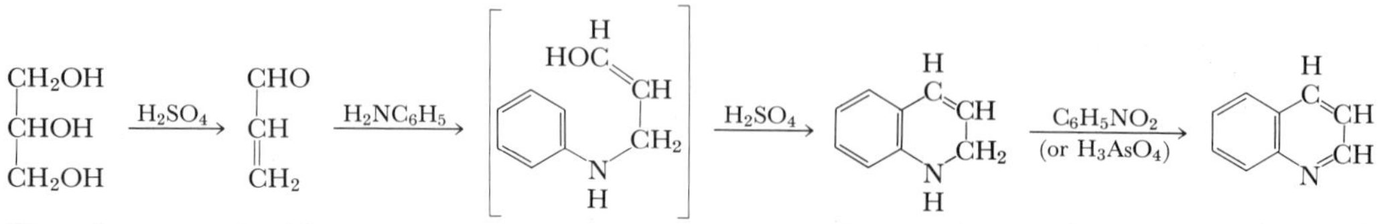

CH_2OH–CHOH–CH_2OH $\xrightarrow{H_2SO_4}$ CHO–CH=CH_2 $\xrightarrow{H_2NC_6H_5}$ [HOC(H)=CH–CH_2–NH–C_6H_5] $\xrightarrow{H_2SO_4}$ 1,2-dihydroquinoline $\xrightarrow[\text{(or } H_3AsO_4)]{C_6H_5NO_2}$ quinoline

Glycerol — Acrolein — 1,2-Dihydroquinoline — Quinoline

2-Methylquinoline, known as **quinaldine,** occurs in coal tar. **Lepidine** (*4-methylquinoline*) accompanies quinoline in the decomposition products of quinine alkaloids. Both quinaldine and lepidine can be made by variations of the Skraup synthesis.

8-Hydroxyquinoline (*oxine*), made from *o*-aminophenol by the Skraup reaction, is a valuable reagent in analytical chemistry. The space relationship of the hydroxyl group and the unshared pair of electrons on the nitrogen atom is such that insoluble chelate (p. 604) coordination complexes are formed with metallic ions.

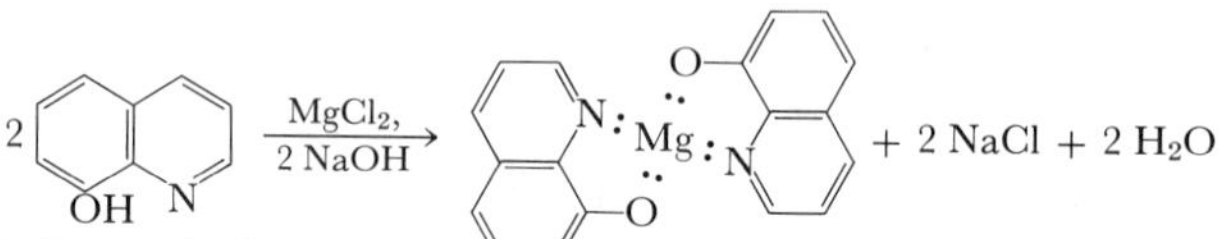

2 (8-Hydroxyquinoline) $\xrightarrow[2\,NaOH]{MgCl_2,}$ Mg chelate + 2 NaCl + 2 H_2O

8-Hydroxyquinoline

It is also bactericidal and fungicidal and is used in pharmaceutical preparations.

3,4-Benzopyridine is known as **isoquinoline.** It accompanies quinoline in coal tar and is available commercially from this source. Oxidation with permanganate yields both phthalic acid and **cinchomeronic acid** (*3,4-pyridinedicarboxylic acid*).

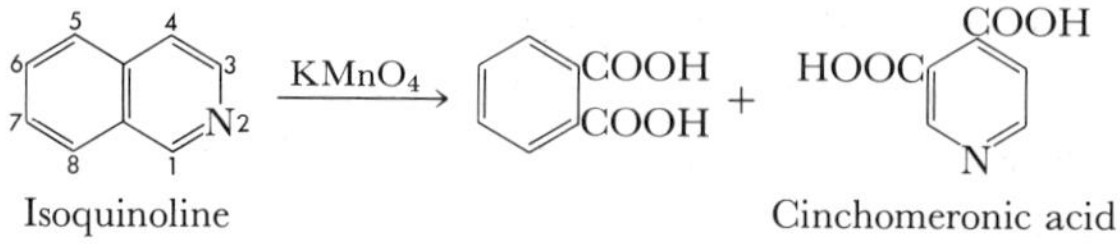

Isoquinoline $\xrightarrow{KMnO_4}$ $C_6H_4(COOH)_2$ + Cinchomeronic acid

Isoquinoline — Cinchomeronic acid

The most general synthesis of isoquinolines is the *Bischler-Napieralski reaction,* which starts with β-phenylethylamines.

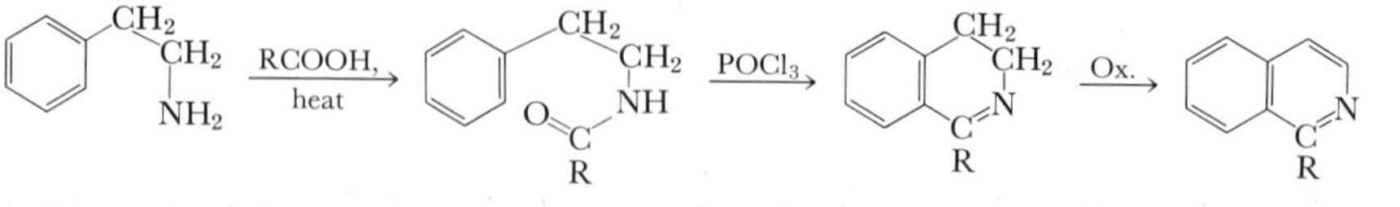

$C_6H_5CH_2CH_2NH_2$ $\xrightarrow[\text{heat}]{\text{RCOOH,}}$ $C_6H_5CH_2CH_2NHCOR$ $\xrightarrow{POCl_3}$ dihydroisoquinoline $\xrightarrow{\text{Ox.}}$ 1-R-isoquinoline

Several natural alkaloids are benzylisoquinolines (p. 535).

Acridine is benzo[*b*]quinoline. It is present in coal tar and received its name from its irritating action on the skin and mucous membranes. Benzo[*c*]quinoline is known as **phenanthridine** because of its relation to phenanthrene (p. 500).

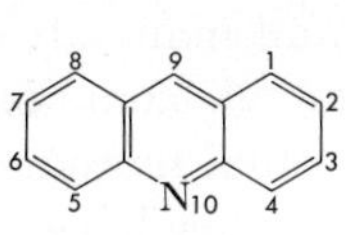

Acridine

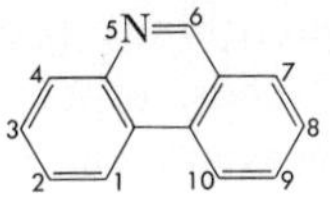

Phenanthridine

o-Phenanthroline contains two nitrogen atoms replacing two methylidyne (CH) groups of the phenanthrene nucleus and can be made by a Skraup synthesis from *o*-phenylenediamine. The position of the nitrogen atoms is such that *o*-phenanthroline can form stable complex cations with metal ions. The *iron chelate complex* (p. 604), in which iron has a coordination number of six, is a useful oxidation-reduction indicator. It is intensely red in the reduced form and faintly blue in the oxidized form.

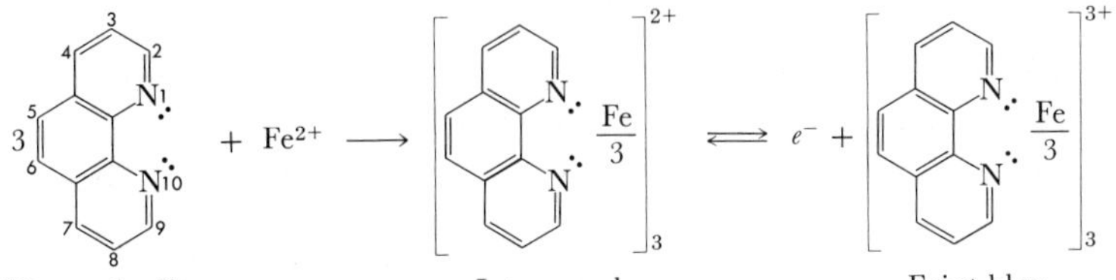

$$o\text{-Phenanthroline} + Fe^{2+} \longrightarrow [\ldots]_3^{2+} \rightleftharpoons e^- + [\ldots]_3^{3+}$$

o-Phenanthroline   Intense red   Faint blue

Pyrans

The six-membered ring containing a single oxygen atom and two double bonds is known as the *pyran ring*. The oxonium salts containing an aromatic system are known as *pyrylium salts*.

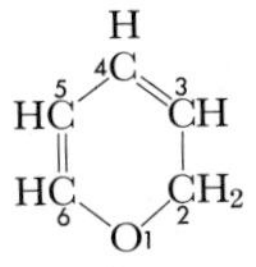

1,2- or α-Pyran 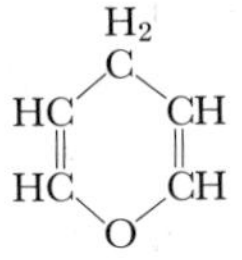   1,4- or γ-Pyran 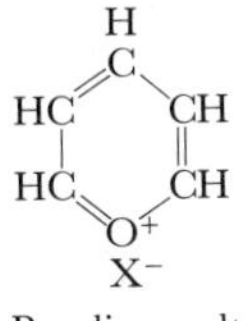   Pyrylium salt

No simple pyran or pyrylium salt is known. **Dihydropyran** is made by the catalytic dehydration with ring enlargement of tetrahydrofurfuryl alcohol (p. 513). Catalytic hydrogenation of dihydropyran gives **tetrahydropyran.**

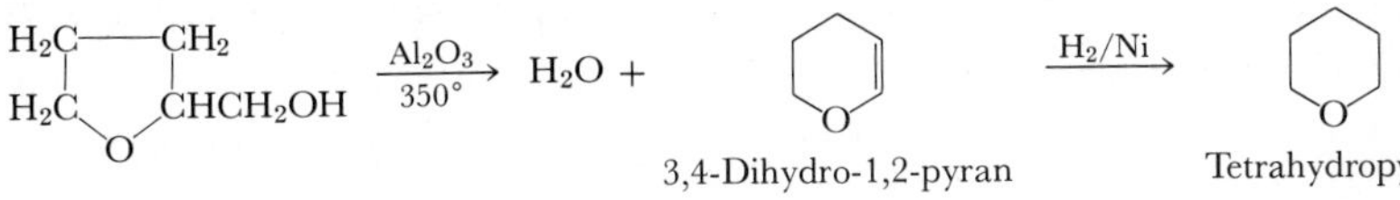

$$\text{Tetrahydrofurfuryl alcohol} \xrightarrow[350^\circ]{Al_2O_3} H_2O + \text{3,4-Dihydro-1,2-pyran} \xrightarrow{H_2/Ni} \text{Tetrahydropyran}$$

3,4-Dihydro-1,2-pyran   Tetrahydropyran

Numerous natural products contain the pyran ring. **Kojic acid,** formed in the fermentation of starch by *Aspergillus oryzae,* **maltol,** first isolated from larch bark, and **chelidonic acid,** from the greater celandine (*Chelidonium majus*), are γ-pyrones.

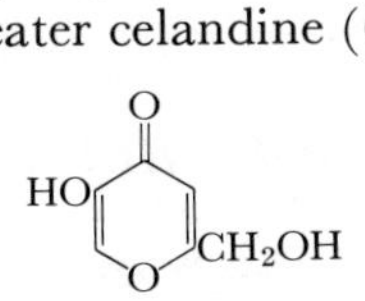

Kojic acid 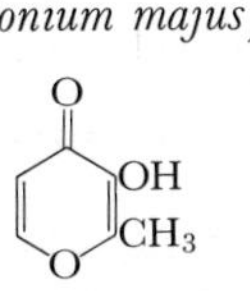   Maltol 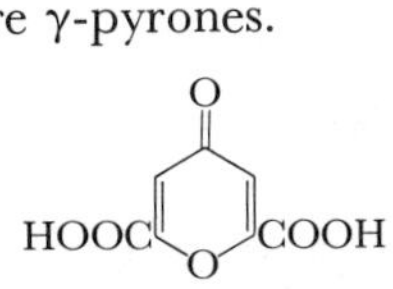   Chelidonic acid

The weakly acidic properties of kojic acid are due to the enolic hydroxyl. Maltol (*Veltol, Palatone*) is a product also of the dry distillation of carbohydrates and hardwood and of a process involving fermentation. It has the property of enhancing nonnitrogenous flavors much as monosodium glutamate enhances nitrogenous flavors (p. 358). The δ lactones are 2-oxotetrahydropyrans (p. 616).

Benzopyrans frequently are present in plants, and many are of considerable interest. Thus the germ oil of seeds, especially wheat germ oil, contains substances designated as *vitamin E* that are necessary for the growth and normal reproduction of the rat. At least four compounds are present that have this activity, and they have been called the *tocopherols* (Gr. *tokos* childbirth, *pherein* to bear). The most active is **α-tocopherol,** the racemic form of which has been synthesized from trimethylhydroquinone and phytyl bromide (p. 661).

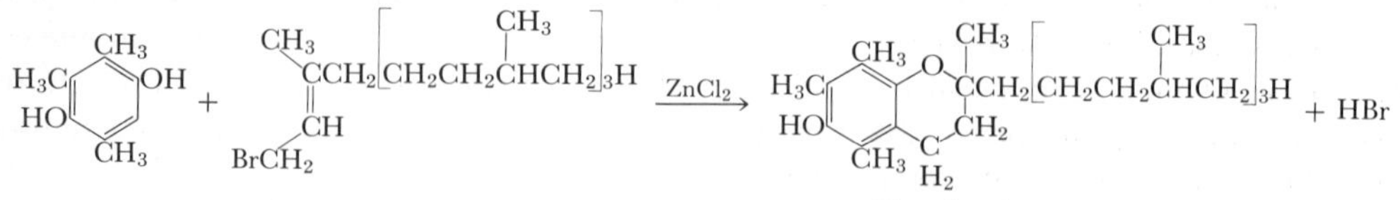

$\xrightarrow{ZnCl_2}$ + HBr

α-Tocopherol

The tocopherols are distributed widely in food. Their use as a supplement to the diet of man and animals for therapeutic purposes has been controversial. They have a marked antioxidant action and decrease the rate at which rancidity develops in foods containing fats (p. 186).

The resin from the flowering tops of hemp (*Cannabis sativa*) has been used since antiquity for its physiological effects. The resin is known as *hashish* or *bhang* and the dried tops as *marijuana*. The active components are believed to be isomeric **tetrahydrocannabinols,** the structure and configuration of one of which has been established.

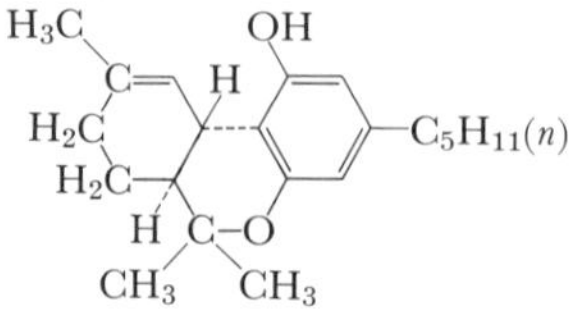

trans-Tetrahydrocannabinol

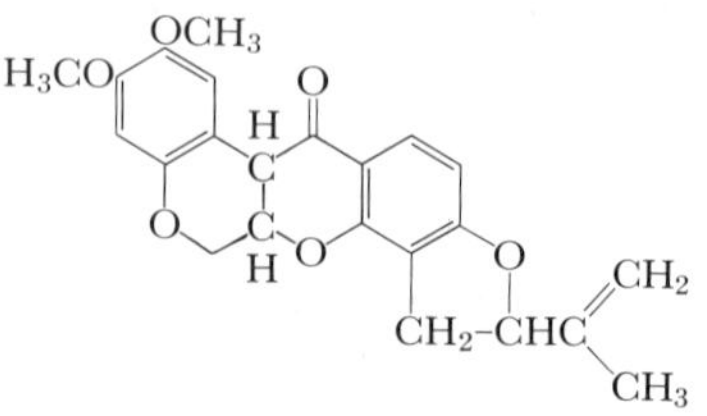

Rotenone

Rotenone contains two dihydropyran rings and a dihydrofuran ring. It is one of the active principles of several plants that are used as insecticides. They are highly toxic to fish also but are relatively harmless to human beings. Hence they are safe to use on plants bearing food. Rotenone and related compounds have been identified in sixty-seven species of plants, but the chief commercial sources are the roots of tuba (*Derris elliptica*) cultivated in Malaya and the East Indies, and of timbo or cube (*Lonchocarpus nicou*) grown in South America.

Many of the brilliant coloring matters of flowers are derivatives of benzopyran. They occur as glucosides and are discussed in Chapter 32 (pp. 577–579).

RING COMPOUNDS CONTAINING TWO OR MORE HETERO ATOMS

The number of possible compounds falling into this group is very large. Only a few of the more important ones are considered.

FIVE-MEMBERED RINGS

Pyrazole and **imidazole** each have two heterocyclic nitrogens. **Histidine,** which contains the imidazole nucleus, is an essential amino acid (p. 365).

```
 4     3          4     3
HC————CH         HC————N         HC====CCH2CHCOOH         HC====CCH2CH2NH2
||    ||         ||    ||        |     |   |              |     |
HC    N2         HC    CH        N     NH  NH2            N     NH
5 \  /           5 \  / 2         \\  /                    \\  /
   N1               N1              C                        C
   H                H               H                        H
Pyrazole         Imidazole        Histidine                Histamine
```

Histamine is derived from histidine by decarboxylation and is present in all tissues of the body. It is extremely toxic when administered parenterally, that is, by means other than absorption through the intestines, and hence it must be combined with protein in the tissues. No other chemical has such a wide variety of actions. Nearly every tissue responds to it in some way. Excessive amounts of free histamine are thought to result from allergic reactions. Since 1941 many synthetic organic compounds have been found to relieve allergic symptoms such as those caused by hay fever, poison ivy and poison oak, and the common cold. They are known as *antihistaminics* (pp. 455, 519, 526, 527).

The *hydantoins* also contain the imidazole ring system. **Hydantoin** may be prepared by the cyclization of hydantoic acid obtained by boiling glycine with urea.

```
                                           CH2————NH                      CH2—NH
HOOCCH2NH2 + H2NCONH2  ——→  NH3 +          |        \CO   —Heat——→        |     \CO
                                           COOH H2N/      dil. acid       CO—NH /
                                           Hydantoic acid                 Hydantoin
```

The name hydantoin was assigned by Baeyer because he obtained it by reduction of allantoin with hydrogen iodide. Hydantoic acid was prepared first by the hydrolysis of hydantoin. The sodium salt of diphenylhydantoin, known as **Dilantin Sodium,** is used in the treatment of epilepsy. It has an anticonvulsant action equal to that of phenobarbital (p. 525) but is less depressant.

```
(C6H5)2C——NH             (CH3)2C——NBr             H2C———CH(CH2)4COOH
|          \CO-+Na       |          \CO           |     |
CO—N//                   CO—NBr/                  H2C   S
                                                     \ /
                                                      S
Dilantin Sodium          1,3-Dibromo-5,5-di-      Lipoic acid
                         methylhydantoin          (thioctic acid)
```

1,3-Dibromo-5,5-dimethylhydantoin is more stable than *N*-bromosuccinimide (p. 629) and can be used for the same purposes. The chlorine analog reacts with water to give hypochlorous acid and can be used as a stable, dry bleaching agent in laundry preparations.

Lipoic acid (*thioctic acid*) contains the *1,2-dithiole* ring. It plays a part in the enzyme systems involved in photosynthesis and carbohydrate metabolism (p. 526).

The five-membered ring containing one oxygen and one nitrogen atom in adjacent positions is known as the **isoxazole** ring, and that in which they occupy the 1,3 positions is called the **oxazole** ring. The sulfur-nitrogen analogs are the **isothiazoles** and the **thiazoles.**

```
 4     3          4     3          4     3          4     3
HC————CH         HC————N          HC————CH         HC————N
||    ||         ||    ||         ||    ||         ||    ||
HC    N          HC    CH         HC    N          HC    CH
5 \  / 2         5 \  / 2         5 \  / 2         5 \  / 2
   O                O                S                S
   1                1                1                1
Isoxazole        Oxazole          Isothiazole      Thiazole
```

2-Mercaptobenzothiazole (*Captax*) is an important rubber accelerator (p. 585). It is formed by heating thiocarbanilide (p. 425) with carbon disulfide and sulfur.

```
C6H5NHCNHC6H5 + CS2 + 2 S  —Heat→  2 C6H4<(N=)(S)>CSH + H2S
       ||
       S
Thiocarbanilide                    2-Mercaptobenzothiazole (Captax)
```

Luciferin, the compound responsible for the chemiluminescence of the firefly, is a bisthiazole derivative. Oxygen, in the presence of the enzyme luciferase, adenosine monophosphate (p. 533), and magnesium ion, converts it to dehydroluciferin with the production of light.

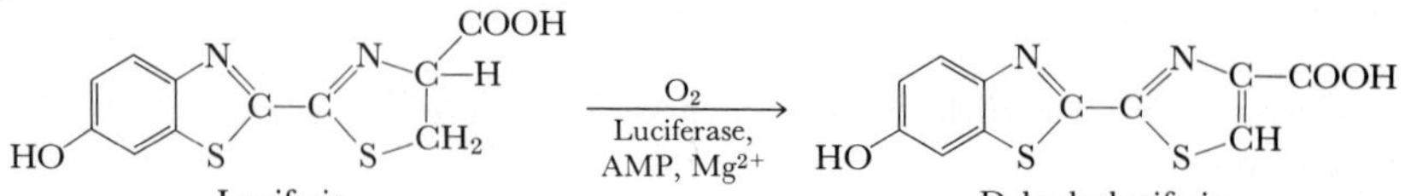

$$\text{Luciferin (HO-benzothiazolyl-C}-\text{C}\!\!<\!\text{N}-\text{CH(COOH)}-\text{CH}_2-\text{S)} \xrightarrow[\text{Luciferase, AMP, } Mg^{2+}]{O_2} \text{Dehydroluciferin (HO-benzothiazolyl-C}-\text{C}\!\!<\!\text{N}-\text{C(COOH)}=\text{CH}-\text{S)}$$

Luciferin Dehydroluciferin

The natural **penicillins,** the first antibiotics to be used in medicine, contain the thiazole ring system. Over 1.5 million pounds of penicillin salts valued at 106 million dollars was produced in the United States in 1964, chiefly from the mold *Penicillium chrysogenum.* Over half was used as the procaine salt. It has been estimated that during World War II a thousand chemists in thirty-nine laboratories in Great Britain and the United States worked on the problem of the structure and synthesis of penicillin at a cost of $20,000,000. A successful synthesis was not developed, however, until 1957. Commercial production still is from the mold, although modifications in the structure of the natural product have yielded compounds with improved properties.

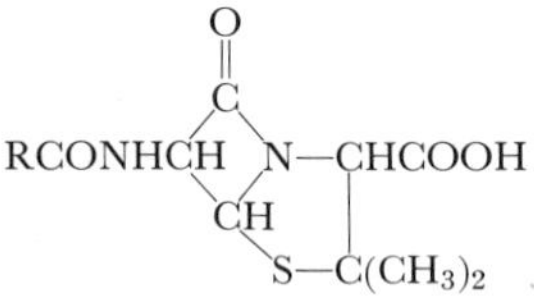

RCONHCH—CH(—C(=O)—N—CHCOOH)—S—C(CH₃)₂

Natural Penicillins

PENICILLIN	R
G or II	$C_6H_5CH_2$—
X or III	*p*-$HOC_6H_4CH_2$—
F or I	$CH_3CH_2CH{=}CHCH_2$—
Dihydro F	$CH_3(CH_2)_4$—
Flavicidin	$CH_3CH{=}CH(CH_2)_2$—
K or IV	$CH_3(CH_2)_6$—
V	$C_6H_5OCH_2$—

Arylhydrazones couple with diazonium salts (p. 431) to give highly colored water-insoluble *formazans.* The latter can be oxidized to colorless water-soluble **tetrazolium salts.**

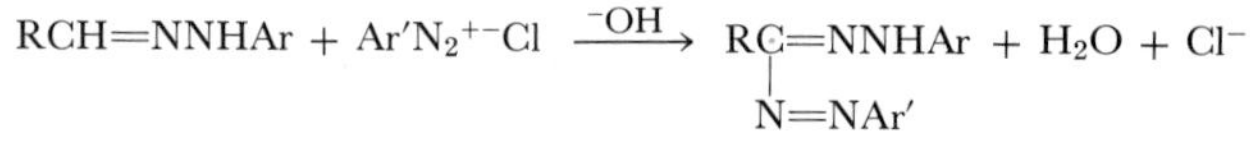

$$RCH{=}NNHAr + Ar'N_2^{+-}Cl \xrightarrow{^-OH} RC({=}NNHAr)(N{=}NAr') + H_2O + Cl^-$$

Colored formazan

$$RC({=}NNHAr)(N{=}NAr') + H_2O_2 + HCl \xrightarrow{V_2O_5} \left[\begin{array}{c} RC{=}N \\ | \quad\quad NAr \\ N{=}\overset{+}{N} \\ Ar' \end{array}\right] Cl^- + 2\,H_2O$$

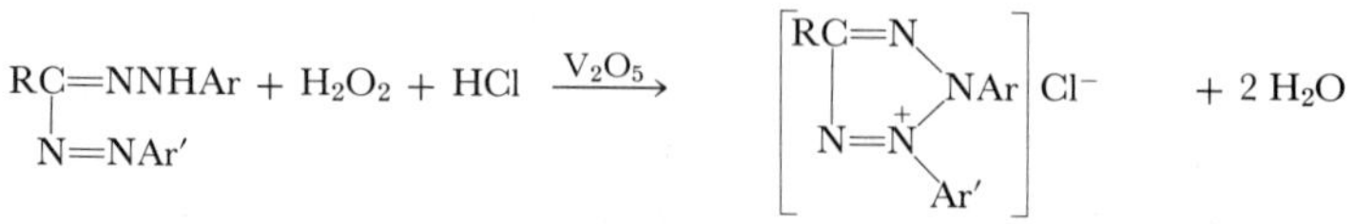

Colorless tetrazolium chloride

The reduction of tetrazolium salts to the formazan can be brought about readily by various reagents including biological systems. When tissues are treated with a tetrazolium salt, the colored insoluble formazan is deposited at the site where reduction is taking place. **Triphenyltetrazolium chloride,** which is most used, is made from benzaldehyde phenylhydrazone and benzenediazonium chloride and gives a red formazan. The diformazan from benzaldehyde phenylhydrazone and diazotized *o*-dianisidine is called **Tetrazolium Blue.**

SIX-MEMBERED RINGS

Three six-membered ring systems containing two nitrogen atoms are known, namely the 1,2-diazines or pyridazines, the 1,3-diazines or pyrimidines, and the 1,4-diazines or pyrazines.

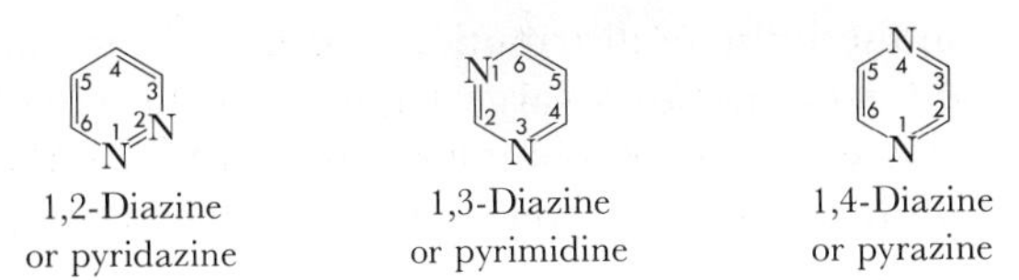

1,2-Diazine or pyridazine	1,3-Diazine or pyrimidine	1,4-Diazine or pyrazine

The 1,2-, 1,3-, and 1,4-diazines are very much weaker bases (pK_a = 2.3, 1.3, and 0.7) than pyridine (pK_a = 5.2) because of the additional electronegativity of the second nitrogen. The diazines are less stable to acid or alkaline hydrolysis than pyridine. Electron-donating substituents both increase the basicity and increase the stability to hydrolysis. As would be expected, electrophilic substitution of the parent diazines is even more difficult than for pyridine. Chlorination can be brought about, however, if one strongly electron-donating group (NH_2 or OH) is present, and nitration if two are present. Two electron-releasing groups also permit nitrosation and coupling with diazonium salts.

Compounds that contain the **pyrimidine** ring are present in all living cells. Hydrolysis of nucleic acids obtained from nucleoproteins (p. 358) yields the pyrimidines **cytosine, 5-methylcytosine, uracil,** and **thymine,** along with purines (p. 528), D-ribose or 2-deoxy-D-ribose (p. 319), and phosphoric acid.

Cytosine *(2-hydroxy-4-amino-pyrimidine)*	5-Methylcytosine *(2-hydroxy-4-amino-5-methylpyrimidine)*	Uracil *(2,4-dihydroxy-pyrimidine)*	Thymine *(2,4-dihydroxy-5-methylpyrimidine)*

Although systematic nomenclature leads to a preference for the structural representations and names as given, biochemists usually place the 4-amino group or the 4-hydroxyl group at the top, which they designate as the 6 position, presumably because of the structural relation of the pyrimidines to the purines (p. 528).

The **barbiturates** also form an important group of pyrimidine derivatives. They are prepared by condensing urea or thiourea with disubstituted malonic esters (p. 633). As cyclic diimides they form water-soluble sodium salts.

$$R_1R_2C(COOC_2H_5)_2 + (H_2N)_2CO \xrightarrow{NaOC_2H_5} R_1R_2C(CO{-}N^{-}\,{}^{+}Na)(CO{-}NH)CO + 3\ C_2H_5OH$$

Although these compounds as a class are called **barbiturates** by the medical profession, the free compounds as well as their sodium salts are used medicinally. The ending commonly used for the free compound is *al,* and *sodium* is added as a separate word to indicate the sodium salt, for example, *barbital* and *barbital sodium.*

The barbiturates have a depressant action on the central nervous system and are valuable sedatives and soporifics. **Barbital** (R_1 and R_2 = C_2H_5) was synthesized first in 1882 by the reaction of ethyl iodide with the silver salt of barbituric acid. In 1903 von Mering discovered its hypnotic properties and called it *Veronal* because he considered Verona to be the most restful city in the world. Barbital is its nonproprietary name. The synthesis from diethylmalonate was developed by Emil Fischer. **Phenobarbital** (R_1 = ethyl, R_2 = phenyl) was introduced several years after Veronal under the trade name *Luminal.* It has a specific action in preventing epileptic seizures. **Amobarbital** (*Amytal,* R_1 = ethyl, R_2 = isoamyl), and **pentobarbital** (*Nembutal,* R_1 = ethyl, R_2 = 1-methylbutyl) act more quickly but have a shorter duration of action than either barbital or phenobarbital. **Secobarbital** (*Seconal,* R_1 = allyl, $CH_2{=}CHCH_2{-}$, and R_2 = 1-methylbutyl) acts still more quickly and for a relatively short period.

Unfortunately the indiscriminate distribution and use of barbiturates and the danger of overdosage make them a hazard to a large segment of the population. Total production of barbiturates in 1964 was 779,000 pounds, valued at $4.7 million. This amount was equivalent to ten 0.2-gram doses per capita. About 250,000 pounds of phenobarbital alone was produced.

Thiopental sodium (*Pentothal Sodium*) is a thiobarbiturate that is used for general anesthesia by intravenous injection. The use of barbiturates for general anesthesia requires

C_2H_5, $CO{-}N^{-+}Na$, C, CS, $CH_3CH_2CH_2CH$, $CO{-}NH$, CH_3

Pentothal Sodium (*thiopental sodium*)

$CO{-}NH$, OC, CO, $CO{-}NH$

Alloxan

careful technique because the anesthetic dose is 50 to 70 per cent of the lethal dose. Their ease of administration and rapid action, and the rapid recovery of the patient, led to their use for major surgical operations in the front battle lines during World War II, and the experience gained was carried over to civilian practice.

Alloxan was isolated from the oxidation products of uric acid (p. 528) in 1817. The discovery in 1943 that either oral or parenteral administration to animals brings about destruction of the islets of Langerhans in the pancreas, resulting in diabetes, led to renewed interest in the compound.

Thiamine (*vitamin B_1, aneurin*) contains a pyrimidine nucleus joined through a methylene group to a thiazole nucleus. Its absence from the diet causes the deficiency diseases known as beriberi in man and polyneuritis in birds. The pyrophosphate is **cocarboxylase,** which is necessary for the enzymic decarboxylation of pyruvic acid, $CH_3COCOOH$, and other α-keto acids. There is evidence that the active portion of the enzyme is the lipoic acid amide of cocarboxylase (p. 523).

NH_2, N, $CH{-}S$, $CH_2\overset{+}{N}$, CH_3, N, ^{-}Cl $C{=}CCH_2CH_2OH$, CH_3

Thiamine (*vitamin B_1, aneurin*)

N, N, CH_2, OCH_3, N, $CH_2CH_2N(CH_3)_2$

Thonzylamine (*Neohetramine*)

Thonzylamine (*Neohetramine, Anahist*) was among the first of the antihistaminics (p. 523) that the U.S. Federal Drug Administration permitted to be sold without a physician's prescription.

Hexahydropyrazines are called **piperazines.** Their hydrochlorides are formed when the hydrochlorides of 1,2 diamines are heated.

$$2\ \begin{array}{l}CH_2NH_3^+Cl^-\\ |\\ CH_2NH_3^+Cl^-\end{array} \xrightarrow{\text{Heat}} \begin{array}{c} H_2\\ N^+Cl^-\\ H_2C \quad CH_2\\ H_2C \quad CH_2\\ N^+Cl^-\\ H_2\end{array} + 2\,NH_4Cl$$

Piperazine dihydrochloride

Piperazine is used extensively (6.9 million pounds, 1964) as a deworming agent for poultry and swine.

A few six-membered heterocycles containing oxygen or sulfur are important, **1,4-Dioxane** is made by the dehydration of ethylene glycol or of diethylene glycol (p. 605).

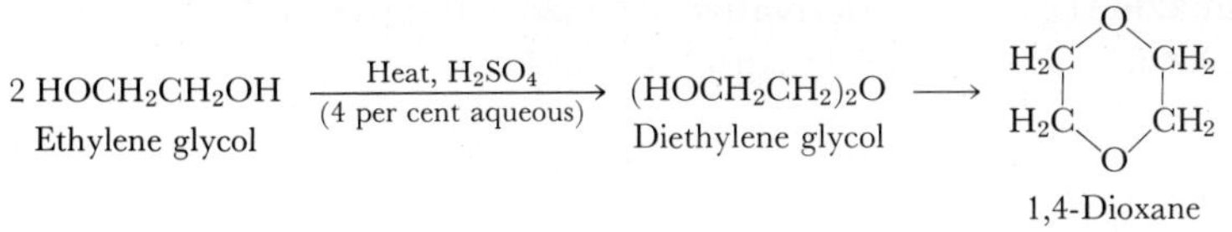

$$2\ HOCH_2CH_2OH \xrightarrow[\text{(4 per cent aqueous)}]{\text{Heat, } H_2SO_4} (HOCH_2CH_2)_2O \longrightarrow$$

Ethylene glycol — Diethylene glycol — 1,4-Dioxane

It is used chiefly as a solvent and paint remover, but its toxicity requires adequate ventilation during its use. **Morpholine** is made by the dehydration of diethanolamine (p. 610).

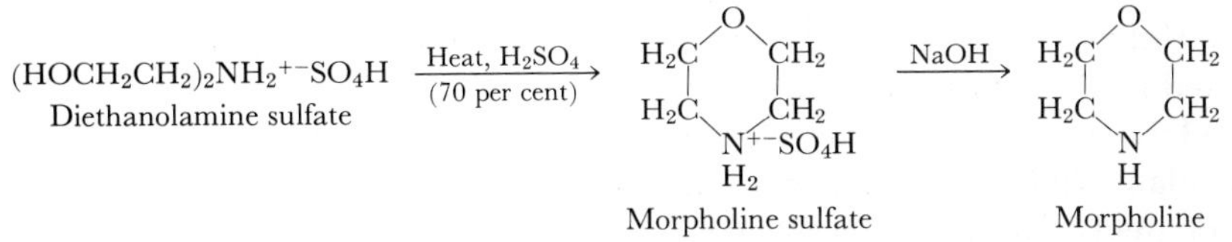

$$(HOCH_2CH_2)_2NH_2{}^{+\,-}SO_4H \xrightarrow[\text{(70 per cent)}]{\text{Heat, } H_2SO_4} \text{Morpholine sulfate} \xrightarrow{NaOH} \text{Morpholine}$$

Diethanolamine sulfate — Morpholine sulfate — Morpholine

The cyclic esters of δ-hydroxy sulfonic acids and the cyclic amides of δ-amino sulfonic acids contain the 1,2-oxathiane and the 1,2-thiazane rings. They are known as **δ sultones** and **δ sultams** (cf. pp. 616, 617).

A few dibenzo-1,4-thiazines are important. **Phenothiazine,** sometimes called *thiodiphenylamine,* is prepared by heating diphenylamine with sulfur. It is used extensively as an intestinal antiseptic for cattle and poultry.

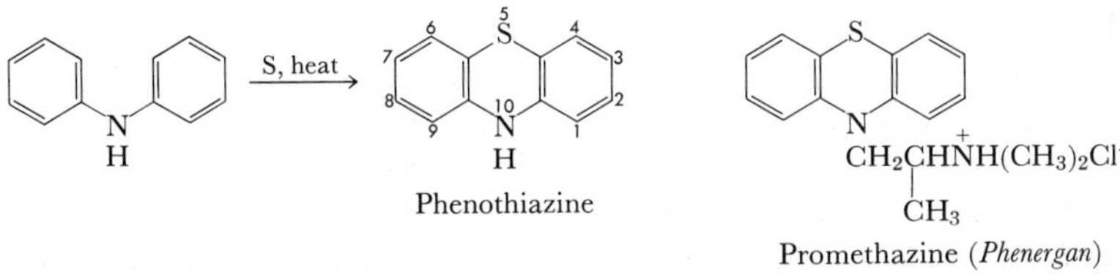

S, heat

Phenothiazine

$CH_2\underset{CH_3}{CH}\overset{+}{N}H(CH_3)_2Cl^-$

Promethazine (*Phenergan*)

N-(2-Methyl-2-dimethylaminoethyl)phenothiazine (*promethazine, Phenergan*) is a potent antihistaminic. **Promazine,** the *N*-(3-dimethylaminopropyl) derivative, however, has very little antihistaminic power but is a tranquilizer. **Chlorpromazine,** the 2-chloro derivative, is a still more effective tranquilizer used in psychiatric treatments. If the sulfur of promazine is replaced by a dimethylene group, the resulting product, named **imipramine,** is a psychic energizer instead of a tranquilizer. It is used to treat depressive states, but is suspected of having teratogenic properties.

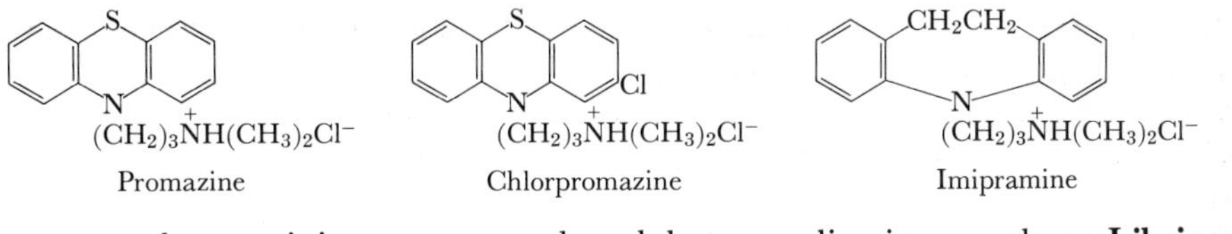

$(CH_2)_3\overset{+}{N}H(CH_3)_2Cl^-$ — Promazine

$(CH_2)_3\overset{+}{N}H(CH_3)_2Cl^-$ — Chlorpromazine

$(CH_2)_3\overset{+}{N}H(CH_3)_2Cl^-$ — Imipramine

Other compounds containing seven-membered heterocyclic rings, such as **Librium** and **Valium,** are important tranquilizers.

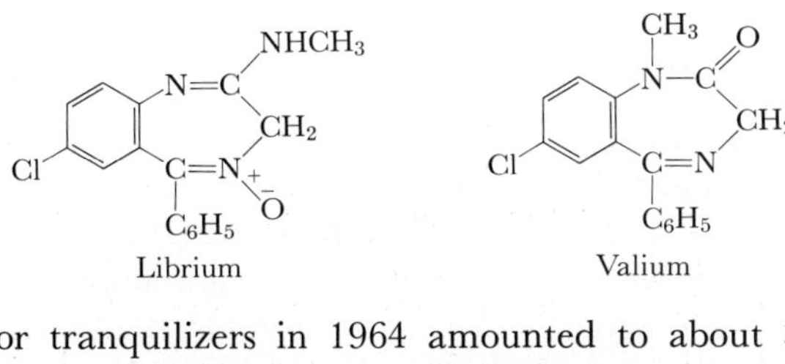

Librium

Valium

The total market for tranquilizers in 1964 amounted to about $220,000,000 of which phenothiazine derivatives accounted for around one third.

Chlorothiazide (*Diuril*) is a derivative of benzo-4-thiapyrimidine made by the cyclization of 2-chloro-4-formamido-1,5-benzenedisulfonamide.

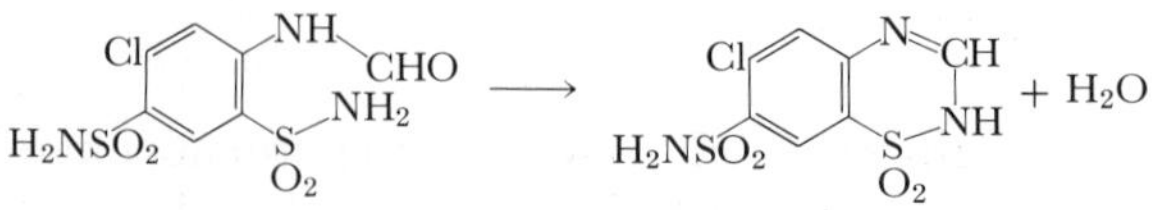

Diuril

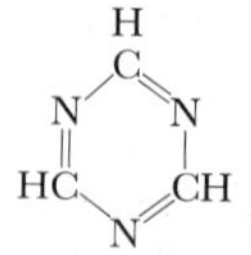

1,3,5-Triazine

It is a powerful diuretic useful in the treatment of high blood pressure and edema. Although first reported in 1957, sales in 1964 amounted to 54,000 pounds valued at $6,777,000.

1,3,5-Triazine (*s-triazine*) derivatives can be made easily because cyanuryl chloride is readily available (p. 288). Its halogens are displaced easily by nucleophiles such as amines, alcohols, or thiols. Numerous herbicides (p. 288), insecticides (p. 611), fiber-reactive dyes (p. 565), and fluorescent brightening agents (p. 567) are prepared in this way.

COMPOUNDS WITH CONDENSED HETERO RINGS

The **purines** are the most important class of compounds containing two condensed heterocyclic rings. They contain both a pyrimidine ring and an imidazole ring.

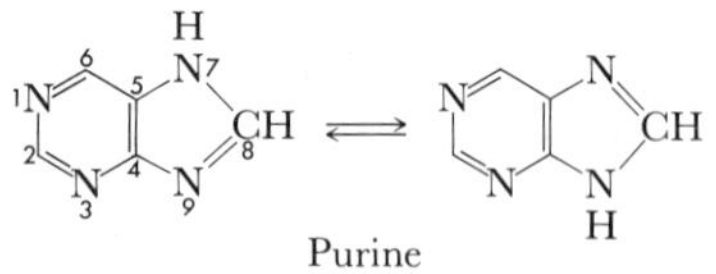

Purine

Amino and hydroxy derivatives of purine accompany pyrimidines in the hydrolysis products of nucleic acids (p. 530). They occur also free in the fluids and tissues of animals and plants. **Adenine** is 6-aminopurine, **hypoxanthine** is 6-hydroxypurine, **guanine** is 2-amino-6-hydroxypurine, and **xanthine** is 2,6-dihydroxypurine. Because of its important role in life processes (p. 533), it is of interest that adenine has been detected along with amino acids in the products resulting from the bombardment of a mixture of methane, ammonia, hydrogen, and water with electrons from a 4.5 m.e.v. accelerator. Adenine, guanine, and hypoxanthine have been reported to be formed also when a concentrated solution of ammonium cyanide is heated, and uracil when malic acid is heated with urea. Hence it appears possible for these rather complex organic compounds to have been produced under conditions existing on earth before life began.

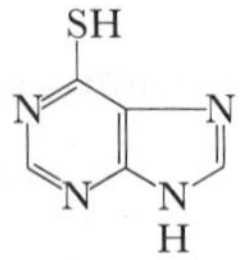

6-Mercaptopurine

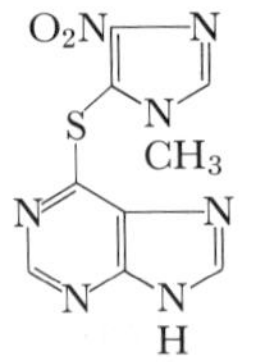

Azathioprine (*Imuran*)

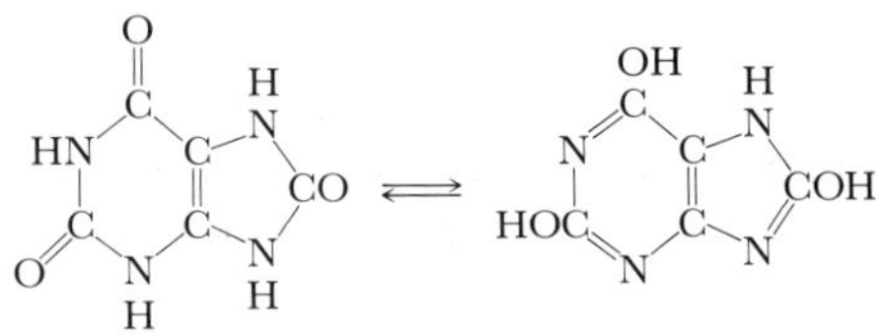

Uric acid

6-Mercaptopurine, the thio analog of hypoxanthine, is an antimetabolite (p. 471) used in the treatment of acute leukemia. It is administered also as an immunosuppressive agent to prevent the rejection of transplanted organs. **Azathioprine** (*Imuran*) and its 2-amino derivative, **aminothioprine,** were developed specifically as immunosuppressive agents.

Uric acid is 2,6,8-trihydroxypurine. It is present in blood and urine and can cause the formation of urinary calculi, from which it first was isolated by Scheele in 1776. Crystals of the monosodium salt deposited in the joints cause the painful condition known as gout. Although uric acid is eliminated only in small amounts by mammals, it is the chief product of nitrogen metabolism by caterpillars, birds, and reptiles. Guano, which contains about 25 per cent uric acid, is one of the best sources.

The compounds responsible for the stimulating action of coffee, tea, and cocoa are methyl derivatives of xanthine. **Theophylline** is present in tea (*Thea sinensis*), **theobromine** in cocoa (*Theobroma cacao*), and **caffeine** in tea, coffee (*Coffea arabica*), cola nuts (*Cola acuminata*), maté (*Ilex paraguayensis*), and many other plants.

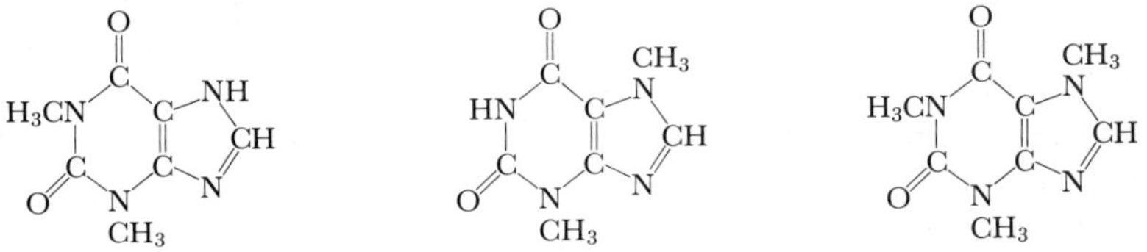

Theophylline

Theobromine

Caffeine

Numerous legends are told concerning the origin of the use of coffee and tea as beverages. It is of interest that wherever plants having a high caffeine content are indigenous to an area, the natives use extracts of the plant as a beverage. All of these compounds stimulate the central nervous system, caffeine being most active and theobromine least. The effective dose of caffeine is 150 to 250 mg., corresponding to 1 to 2 cups of coffee or tea. About 80 per cent is broken down in the body to urea, and the remainder is excreted unchanged or partially demethylated. The fatal dose of caffeine has been estimated as 10 grams, but no deaths from it have been reported. In recent years the use of caffeine-containing beverages has increased enormously, largely because of their stimulating effect. Production of natural and synthetic caffeine in the United States in 1964 was 2.8 million pounds valued at $4,890,000. This amount is equivalent to over 8 billion 150-mg. doses or about 47 per capita.

Theophylline is a stronger acid ($pK_a = 8.8$) than phenol ($pK_a = 9.9$) and is soluble in aqueous ammonia. The salt of 8-chlorotheophylline and Benadryl (p. 455) is called **Dramamine** and is used for the prevention of motion sickness. Its effectiveness appears to be due entirely to the Benadryl content.

Several biologically important compounds have the *pteridine nucleus,* in which a pyrimidine ring is fused to a pyrazine ring. The **pterins** (Gr. *pteron* wing) are colorless or yellow compounds found in the wings of various species of butterflies and wasps, in the skin and eyes of fish, amphibia, and reptiles, and in the liver and urine of mammals. They are characterized by marked fluorescence in neutral solution.

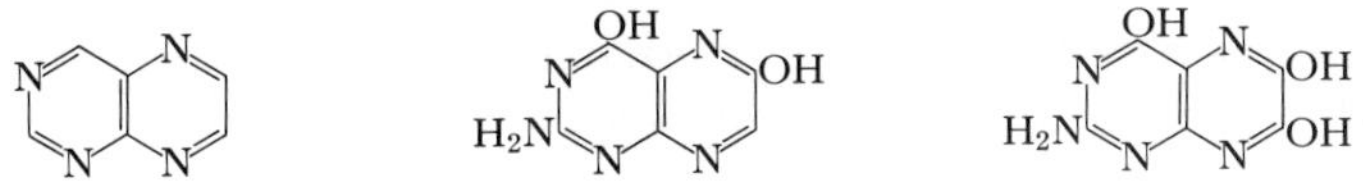

Pteridine

Xanthopterin

Leucopterin

The **folic acids** are necessary for the growth of certain micro-organisms. Because the portion remaining after removal of the glutamic acid units by hydrolysis is called **pteroic**

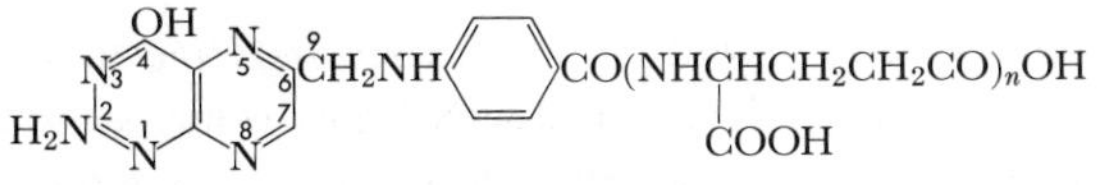

Folic acids

acid, the folic acids have been termed *pteroylglutamic acids.* **Pteroylglutamic acid** ($n = 1$) is synthesized commercially for use in the treatment of certain types of anemia. It appears to be a precursor of the so-called *citrovorum factor,* which is a 5,6,7,8-tetrahydro *N*-formyl derivative and is required for cell division in mammals. **Aminopterin** has an amino group instead of a hydroxyl group in the 4 position. **Amethopterin** (*methotrexate*) has the side-chain aromatic amino group of aminopterin methylated. Both compounds are antimetabolites (p. 471) that inhibit cell division and are used in the treatment of acute and subacute leukemia. Amethopterin is administered also as an immunosuppressive agent to aid the retention of organ transplants.

Riboflavin (*vitamin* B_2) is the prosthetic group of the various *yellow enzymes* or *flavoproteins.* They play a part in the metabolism of α-amino acids and in aerobic carbohydrate metabolism. Riboflavin is yellow and its solutions are highly fluorescent in ultraviolet light. The polyhydroxy side chain has the D-ribose configuration (p. 319). Production of riboflavin in the United States in 1964 amounted to 662,000 pounds valued at 6 million dollars. **Biotin,** another member of the vitamin B complex that may be involved in the metabolism of pyruvic acid, contains an imidazole ring fused to a thiophene ring.

OH N N CH_3 CH_3 O C N N
$CH_2(CHOH)_3CH_2OH$
Riboflavin (*vitamin* B_2)

CH_2—CH—NH S CO CH—CH—NH
$(CH_2)_4COOH$
Biotin

Hypothetically any multivalent element can be the hetero atom in a ring. Many cyclic compounds are known that have hetero atoms other than nitrogen, oxygen, or sulfur, for example, phosphorus, boron, or silicon.

NUCLEIC ACIDS

Nucleoproteins are essentially salt-like combinations of proteins (p. 358) with nucleic acids. Their names result from the early belief that they were present only in cell nuclei, but it now is known that they occur in all parts of the cell. The nucleic acids originally were isolated by extraction of nucleoproteins with base, but this procedure degrades the nucleic acids and they now are separated from the protein portion by extraction with salt solutions, anionic detergents, phenol, and other agents. Tobacco mosaic virus (*TMV*) is almost all, if not entirely, nucleoprotein. It has been separated by fractional precipitation with salt solutions into the protein component and the nucleic acid component. The protein portion has none of the infective property of the virus and the nucleic acid portion has only a few per cent of the activity of the virus. The two fractions can be recombined, however, to give a product as infective as the original virus. Thus the separation procedure does not appear to have altered the two components.

Structure

Complete hydrolysis of **nucleic acids** gives pyrimidines (p. 525), purines (p. 528), a sugar, and phosphoric acid. It early was recognized that the nucleic acids fall into two classes, those that give ribose (p. 318) as the sugar constituent and those that give deoxyribose (p. 319). Accordingly they were classed as **ribonucleic acids** (*RNA*'s) and **deoxyribonucleic acids** (*DNA*'s). Often RNA and DNA are referred to as if they were single compounds. It is important to remember that there are many ribonucleic acids and many deoxyribonucleic acids. With better purification and more accurate analyses, it became recognized that RNA's and DNA's differ not only in the sugar constituent but also in the purine constituents. Whereas RNA's yield adenine, guanine, cytosine, and uracil, DNA's yield adenine, guanine, cytosine, and thymine. Moreover in the latter some of the cytosine may be replaced by methylcytosine, and the DNA's of certain bacteriophages yield instead of cytosine only hydroxymethylcytosine. The latter is bound through the hydroxyl group to glucose by a glycosidic link.

DNA's appear to be present only in the cell nucleus and are localized in the chromosomes. RNA's are present in all parts of the cell with the highest concentration in the microsomes of the cytoplasm. The RNA's present in the microsomes are of higher molecular weight than those in cell fluid (*soluble RNA's*), but both appear to be involved in protein synthesis.

The nucleic acids are polymeric compounds with molecular weight up to several million. Partial hydrolysis gives the **nucleotides** composed of one molecule of either pyrimidine or purine, one molecule of ribose or deoxyribose, and one of phosphoric acid. The pyrimidine is linked through N-3 and the purine through N-9 to the ribose or deoxyribose as a β-glycoside, and the $5'$-hydroxyl of the sugar is esterified with phosphoric acid. Both the ribose and the deoxyribose have the furanose structure.

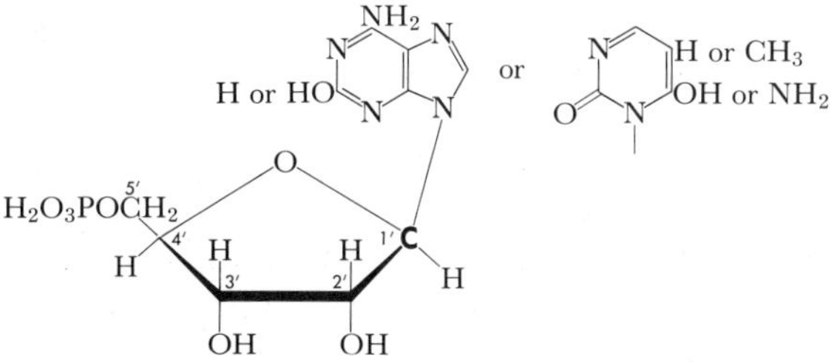

NH_2 / H or HO / or / H or CH_3 / OH or NH_2 / $H_2O_3POCH_2$ / H / OH / OH

Ribonucleotides

NH_2 / H or HO / or / H or CH_3 / OH or NH_2 / $H_2O_3POCH_2$ / H / OH / H

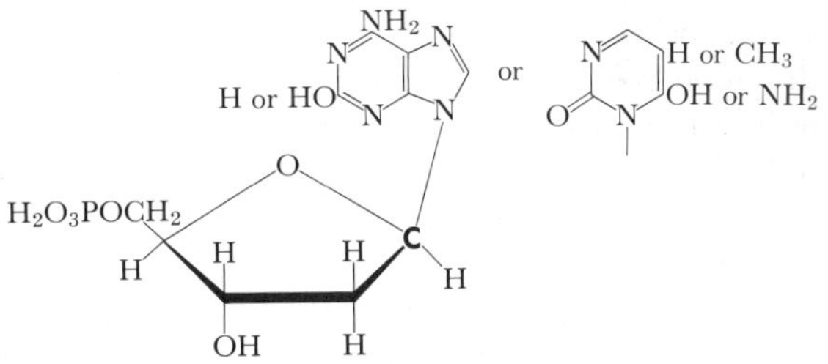

Deoxyribonucleotides

The ribonucleotides that yield adenine, guanine, cytosine, thymine, and uracil are called respectively *adenylic, guanylic, cytidylic, thymidylic,* and *uridylic* acids. Further hydrolysis removes the phosphoric acid to give the **ribonucleosides,** known as *adenosine, guanosine, cytidine, thymidine,* and *uridine,* or the corresponding deoxy compounds. In the polymer the nucleotides are bonded to each other through phosphoric acid ester links between the $3'$-hydroxyl of one sugar and the $5'$-hydroxyl of the next. Hence the structure may be represented schematically as

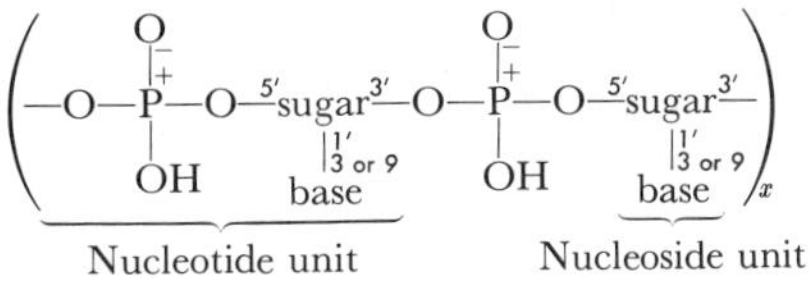

$$\left(-O-\overset{O}{\underset{OH}{\overset{|}{P}}}-O-{}^{5'}\underset{\text{base}}{\text{sugar}}{}^{3'}-O-\overset{O}{\underset{OH}{\overset{|}{P}}}-O-{}^{5'}\underset{\text{base}}{\text{sugar}}{}^{3'}-\right)_x$$

(sugar–base link: $1'$ to 3 or 9)

Nucleotide unit   Nucleoside unit

At the *p*H of cellular fluids, protons have been removed from the phosphate hydroxyl groups, and the nucleic acid is present as the polynegative ion. In the nucleoproteins the anionic portions of the nucleic acid are held to the cationic portions of the protein by electrostatic attraction.

Most is known about the structure of DNA's. Analyses indicate that within a given species the purine and pyrimidine composition is the same in all tissues and at all ages of a given organism, but that the composition varies with different organisms. For all DNA's

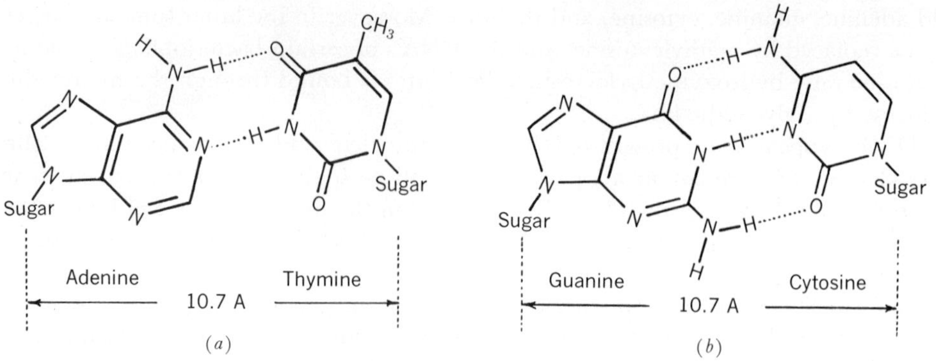

Figure 30–3. Hydrogen bonding (*a*) between adenine and thymine and (*b*) between guanine and cytosine.

the number of moles of adenine equals the number of moles of thymine, and the number of moles of guanine equals the number of moles of cytosine, or of cytosine + methylcytosine. The ratio of adenine + thymine to guanine + cytosine, however, is different for different organisms.

The correspondence noted indicates a pairing of adenine and thymine and of guanine and cytosine. One member of each pair has an oxo group adjacent to an NH group, and the other has an amino group on a carbon adjacent to a nitrogen with an unshared pair of electrons. Hence there is the possibility of strong hydrogen bonding analogous to that present in carboxylic acids (p. 157). Scale models show that the over-all dimensions of an adenine-thymine pair is identical with that of a guanine-cytosine pair as illustrated in Fig. 30–3.

Photographs taken with the electron microscope show that the molecules of DNA are unbranched threads, and X-ray analysis and other evidence indicate that the planar purine and pyrimidine portions are stacked parallel to each other and roughly perpendicular to the length of the thread. Furthermore X-ray analysis shows that the molecule is a regular helix, but that the purines and pyrimidines are arranged in an irregular sequence. These facts have led to the postulation that the molecules of DNA consist of two intertwined spirals held together by hydrogen bonds. Since every adenine unit is bonded to a thymine unit and every guanine to a cytosine, the two spirals are complementary to each other. This relationship is illustrated schematically in Fig 30–4. It will be noted that although hydrogen bonding accounts in part for the helical structure of both proteins and deoxyribonucleic acids, the former are intramolecularly bonded along the length of the fiber (p. 363), whereas the latter are intermolecularly bonded across the fiber.

The model for DNA not only accounts for its composition and physical properties but also for biological replication. If in preparation for cell division the two spirals begin to un-

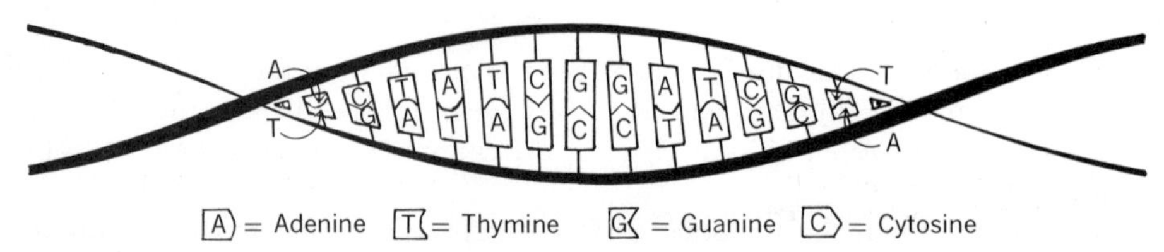

Figure 30–4. Representation of the double-stranded spiral structure of a hypothetical deoxyribonucleic acid.

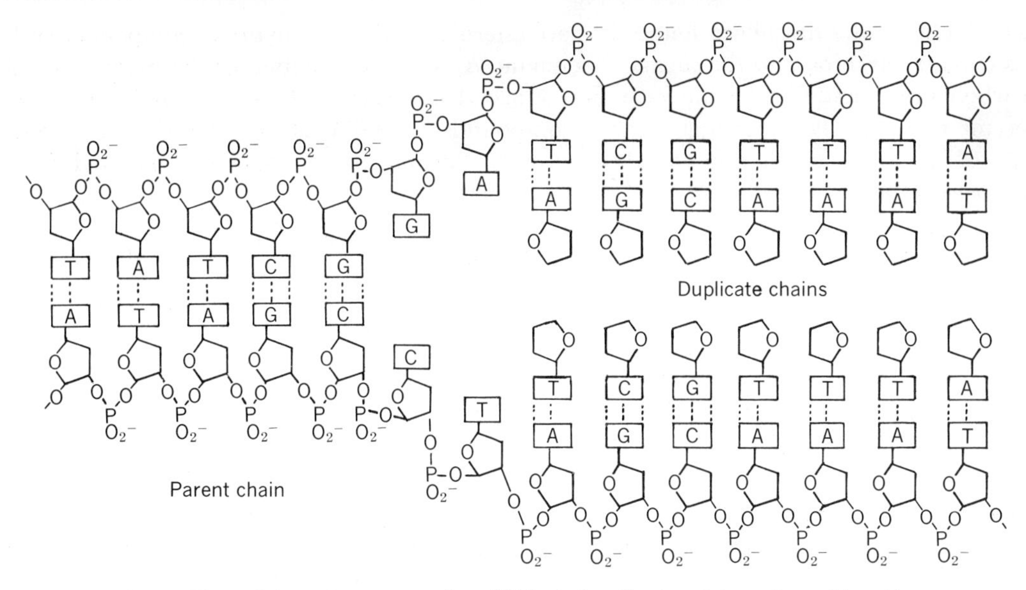

Figure 30–5. Schematic representation of biological replication of deoxyribonucleic acids.

twine, the separated strands can act as templates for the production of two identical molecules, as indicated schematically in Fig. 30–5. Experimental evidence for this postulation is the isolation of an enzyme that catalyzes the synthesis of DNA from its components but only if some DNA is added as a primer. Not only can DNA's duplicate themselves; it is thought that they carry the genetic code for the synthesis of RNA, which in turn can direct the synthesis of the specific enzymes and other proteins required by the animal organism. Major efforts now are being made to establish the purine and pyrimidine sequences in DNA.

Less is known about the structure of RNA's. As with DNA's the sugar units are joined by phosphate ester linkages at the 3′,5′ positions. The purine-pyrimidine content does not show the regularities present in DNA's but this may be due to degradation of the more reactive RNA's during isolation. Within the same tissue, RNA's of the nuclei appear to differ from RNA's of the cytoplasm. X-ray data show a resemblance to the DNA pattern, and doubly stranded helical portions have been postulated. The RNA's of high molecular weight are believed to be concerned with replication and those of low molecular weight with transfer of amino acids during protein synthesis.

Numerous other biologically active compounds contain the nucleotide unit. Among the earliest recognized were adenosine diphosphate (*ADP*) and adenosine triphosphate (*ATP*). Since then pyrophosphates of all the nucleosides have been isolated (B = purine or pyrimidine base).

$$\mathrm{HO\overset{O^-}{\underset{OH}{\overset{|}{\underset{|}{P^+}}}}{-}O{-}\overset{O^-}{\underset{OH}{\overset{|}{\underset{|}{P^+}}}}{-}OCH_2}\text{–(ribose ring: O; H, H, H; OH, OH; C(H)–}B\text{)}$$

Ribonucleoside-5′ diphosphate
(*ADP if B = adenine residue*)

$$\mathrm{HO\overset{O^-}{\underset{OH}{\overset{|}{\underset{|}{P^+}}}}{-}O{-}\overset{O^-}{\underset{OH}{\overset{|}{\underset{|}{P^+}}}}{-}O{-}\overset{O^-}{\underset{OH}{\overset{|}{\underset{|}{P^+}}}}{-}OCH_2}\text{–(deoxyribose ring: O; H, H, H; OH, H; C(H)–}B\text{)}$$

Deoxyribonucleoside-5′ triphosphate

Nicotinamide adenine dinucleotide (*NAD*) has been mentioned earlier (p. 517), and riboflavin is combined with ADP in **flavin adenine dinucleotide** (*FAD*). Moreover nucleoside diphosphates have been found that are esterified with other hydroxy compounds such as choline, glycerol, and the sugars. **Coenzyme A,** which is essential for the metabolism of carbohydrates and fats, for the formation of acetylcholine (p. 232), and for the biosynthesis of many naturally occurring organic compounds (p. 667), is an ester of pantetheine (p. 618) and a triphosphorylated adenosine. **Puromycin,** an antibiotic isolated from *Streptomyces alboniger,* is related to adenosine (p. 531).

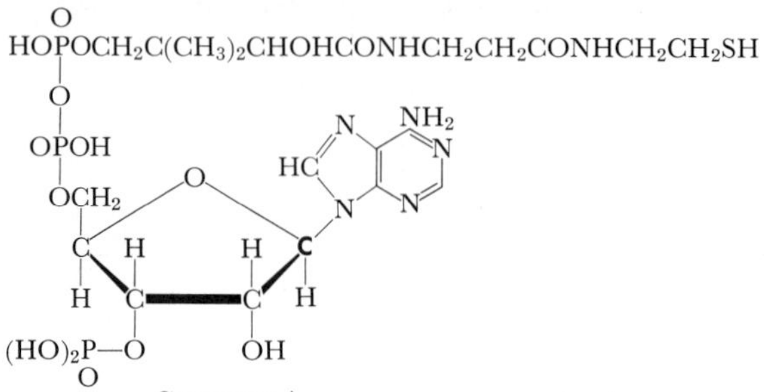

$HOPOCH_2C(CH_3)_2CHOHCONHCH_2CH_2CONHCH_2CH_2SH$

Coenzyme A

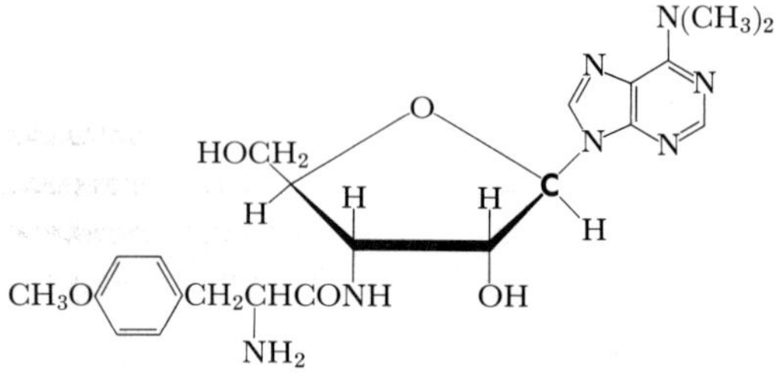

Puromycin

ALKALOIDS

The term *alkaloid* means *like an alkali,* and alkaloids usually are defined as basic nitrogenous plant products having a marked physiological action when administered to animals. Some compounds, however, are included under the term that do not conform to this definition. For example piperine, the alkaloid of pepper, is not basic and has practically no physiological activity. On the other hand some compounds such as caffeine definitely are alkaloids, but are so innocuous that they frequently are not considered as such. Furthermore, some compounds either are so closely related in structure to the alkaloids or have such similar physiological action that it is natural to think of them along with the alkaloids even though they do not come within the usual definition. Thus epinephrine and ephedrine are closely related but only ephedrine is a plant product; opium and hashish (*marijuana*) are both habit-forming drugs having similar action, yet the active principle of the latter is not basic and does not contain nitrogen (p. 522).

The simplest alkaloid **coniine,** is α-*n*-propylpiperidine. It is present in all parts of the poison hemlock (*Conium maculatum;* Gr. *konas* to whirl around).

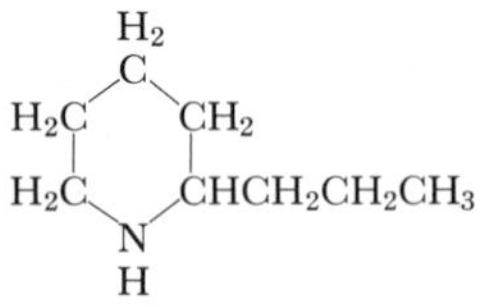

Coniine

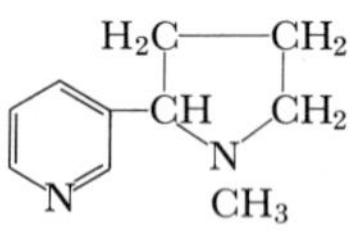

Nicotine

At least ten alkaloids are present in tobacco (*Nicotiana tabacum*). About three fourths of the total alkaloids is **nicotine.** It is highly toxic to animals but in small amounts causes an initial and transient stimulation followed by depression. It is used extensively as a contact insecticide.

Piperine is the alkaloid of black pepper (*Piper nigrum*). The piperine content varies from 5 to 9 per cent. Hydrolysis yields piperidine and piperic acid, which indicates an amide linkage. Piperic acid, $C_{12}H_{10}O_4$, contains two double bonds, and oxidation gives piperonal.

```
      H2 H2                                   O—CH2
      C—C                                     |   |
  H2C      NCOCH=CHCH=CH—(C6H3)———————————————O
      C—C
      H2 H2
                  Piperine

                     | Hydrolysis
                     ↓

      H2 H2                          O—CH2                         O—CH2
      C—C                            |   |                          |   |
  H2C      NH + HOOCCH=CHCH=CH—(C6H3)—O    ——KMnO4——>  OCH—(C6H3)—O
      C—C
      H2 H2
  Piperidine            Piperic acid                          Piperonal
```

Numerous alkaloids have a **tropane ring system,** in which two methylene groups bridge the 2,6 positions of a piperidine ring.

```
        H2                     OCOCHC6H5                    OCOC6H5
        C                      CH  CH2OH                    CH
  H2C        CH2         H2C              CH2        H2C          CHCOOCH3
     CH2CH2                  CH2CH2                     CH2CH2
  HC         CH          HC               CH         HC           CH
        N                       N                           N
        H                       CH3                         CH3
     Tropane             (−)-Hyoscyamine                  Cocaine
                          and atropine
```

(−)-Hyoscyamine is the chief alkaloid of many plants of the family *Solanaceae,* especially henbane (*Hyoscyamus niger*), belladonna (*Atropa belladona*), and the deadly nightshade (*Datura stramonium*). It is racemized readily to **atropine,** which probably does not occur naturally in more than traces. **Cocaine** is the chief alkaloid in the leaves of several varieties of the coca bush (*Erythroxylon*) native to the Peruvian and Bolivian Andes.

All of the alkaloids of the tropane group are characterized by their mydriatic action. Atropine causes dilation of the pupil of the eye at a dilution of 1 in 130,000 parts of water. Cocaine is noted particularly for its stimulating action on the central nervous system, permitting great physical endurance, and for its local anesthetic action. Before the dangers of addiction were understood, it was self-administered widely in Europe and was a component of many patent medicines. The development of the local anesthetics of the procaine type (p. 472) resulted from the observation that the toxic effects of cocaine are associated with the methoxycarbonyl group, whereas the anesthetic action is due to the portion that is the benzoic ester of an amino alcohol.

At least twenty-four different alkaloids are present in **opium** (Gr. *opion* poppy juice), the dried latex of the species of poppy (*Papaver somniferum*) that is indigenous to Asia Minor. These alkaloids fall chiefly into two groups, the benzylisoquinoline group and the phenanthrene group. In the former are **papaverine, narcotine,** and **laudanine,** and in the latter **morphine, codeine,** and **thebaine.**

Papaverine   Morphine

One of the hydroxyl groups of morphine is phenolic and can be methylated readily. The resulting methyl ether is identical with natural **codeine. Thebaine** is the dimethyl derivative. Acetylation of morphine gives the diacetate, **heroin,** which does not occur in opium.

The effects of opium were known before recorded history, and it has been and remains one of the most valuable drugs at the disposal of the physician. Morphine, which constitutes about 10 per cent of opium, was isolated by Sertuerner in 1805, and the isolation of other pure components soon followed. Since then the individual alkaloids have been used medicinally. The benzylisoquinoline alkaloids have very little action on the central nervous system but relax smooth muscle. Papaverine is the most important member of the group and is a valuable antispasmodic. Morphine exerts a simultaneous depressing and stimulating action on the central nervous system, producing drowsiness and sleep, yet causing excitation of smooth muscle with resulting nausea and vomiting. The most important use for morphine is for the relief of pain. Its synthesis was completed in 1952.

In 1939 it was discovered accidentally that a relatively simple synthetic compound, which has been named **meperidine** (*pethidine, Demerol*), when injected into rats caused them to hold their tails in a position such as that assumed when morphine is injected. This observation led to the discovery that meperidine has a marked analgesic action although less than that of morphine. It has the advantage that it does not cause nausea.

Meperidine   Methadone   Propoxyphene hydrochloride (*Darvon*)

This discovery turned attention to the fact that morphine contains a phenylpiperidine structure. Accordingly many derivatives of phenylpiperidine were synthesized and tested, and some of them were found to have marked analgesic activity. About 1941 a compound, which has been named **methadone** (*Amidone*), was found to be even more effective than morphine as an analgesic. Although certain structural features are present in methadone that may be said to be present in morphine, the relationship is rather remote. Unfortunately continued use of either meperidine or methadone also leads to habituation. Methadone, however, does not cause physical and mental deterioration, and heroin addicts can be rehabilitated by daily doses of methadone. Search for a chemical that will relieve deep-seated pain and yet not lead to habituation has continued. **Propoxyphene hydrochloride** (*Darvon*) is as effective as codeine but has shown little or no liability to cause addiction, although its structure is similar to that of methadone. It is synthesized by a series of standard reactions.

Another important group of alkaloids is that derived from **cinchona bark.** The name commonly is thought to be derived from that of Countess Anna del Chinchon, wife of the Spanish Viceroy to Peru, who was cured of malaria by treatment with it in 1638. It has been suggested also that the name was derived from the Inca word *kinia* meaning *bark*. The bark probably reached Europe about 1632, and its use was widespread by 1640. The genus *Cinchona* was established by Linnaeus in 1742, and the tree known as *Cinchona officinalis,* which is native to the high eastern slopes of the Andes, was described by him in 1753. The alkaloids **quinine** and **cinchonine** were not isolated until 1820 by Pelletier and Caventou. By about 1860 the near extinction of the native trees caused such a rise in price of the drug that attempts were made to cultivate cinchona elsewhere. It was grown successfully in India, Ceylon, and Java. Today more than 90 per cent of the quinine produced comes from plantations in Java. The total alkaloid content of the bark of the cultivated trees is 6 per cent, of which 70 per cent is quinine. The unusual feature of the structure of the cinchona alkaloids is the presence of a quinuclidine ring. The last phase of a total synthesis of quinine was completed successfully in 1944.

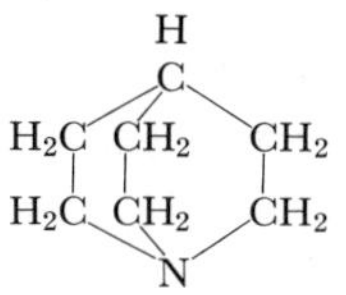

Quinuclidine

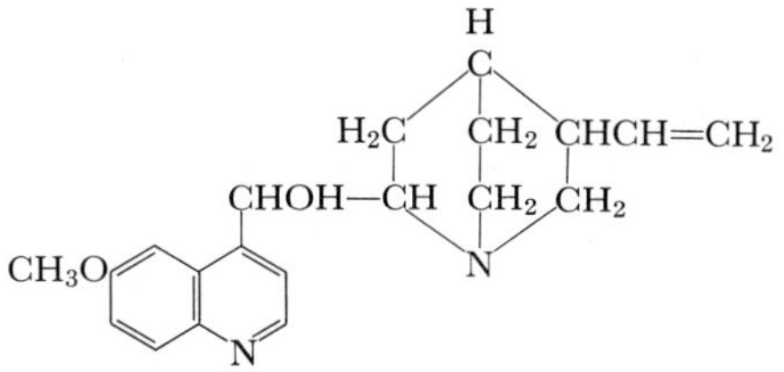

Quinine

Over twenty other alkaloids have been isolated from various species of *Cinchona* and *Cuprea.* **Epiquinine, quinidine,** and **epiquinidine** are stereoisomers of quinine. **Cinchonine, cinchonidine,** and **cinchonicine** are stereoisomeric with each other, and differ from quinine and its isomers in that they lack the methoxyl group.

The value of quinine lies in its specific action in the treatment of malaria. It is estimated that throughout the world there are several hundred million cases of malaria per year, resulting in 3 million deaths and partial or total debilitation for the remainder. Although annual production of quinine has reached 600 tons it is only a fraction of what would be needed to treat all cases even if the cost were not prohibitive for most of the victims. Starting from the observation of Ehrlich[6] in 1891 that the dye Methylene Blue (p. 576) has antimalarial action, chemists at the I. G. Farbenindustrie of Germany began an extensive program of synthesis and testing, which resulted in the introduction of **pamaquine** (*Plasmochin*) in 1926. It had the desirable property not possessed by quinine of being able to kill the sexual form of the parasite, thus preventing the spread of malaria by mosquitoes. Its toxic effects precluded its general use, however, and **quinacrine** (*Atabrine*) was introduced in 1930.

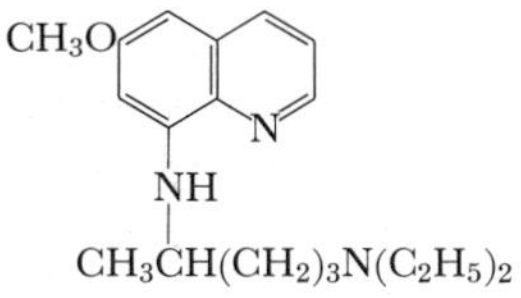

Pamaquine (*Plasmochin*)

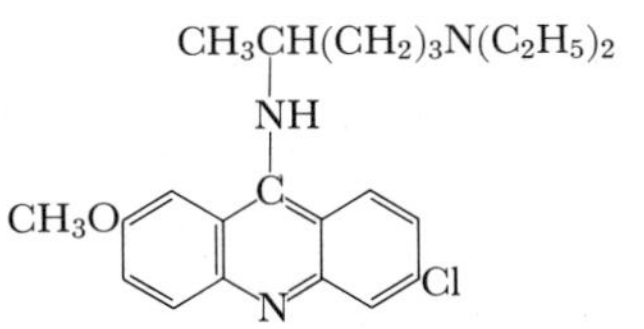

Quinacrine (*Atabrine*)

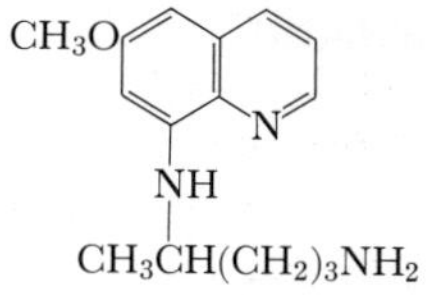

Primaquine

Quinacrine, like quinine, acts on the schizont stage of the parasite and reduces the formation of the gametocytes whose asexual reproduction causes the destruction of red blood corpuscles and leads to the symptomatic chills and fever. Quinacrine was not used exten-

[6] Paul Ehrlich (1854–1915), physiologist and professor of experimental therapy at the University of Frankfurt. He was the first to attempt to synthesize chemical compounds that are more toxic for a pathogenic organism than for the host, a field of investigation to which he gave the name *chemotherapy.* He was awarded the Nobel Prize in Medicine in 1908.

sively, however, until World War II cut off supplies of quinine from Java. Having become familiar with quinacrine, physicians preferred it to quinine. Its use was abandoned, however, because it causes the skin to become yellow.

Between 1920 and 1930, during the development of pamaquine and quinacrine, over 12,000 compounds were prepared and tested. With the advent of tropical warfare in World War II, an intensive search for better antimalarials was undertaken in the United States, Canada, and England. In this program over 14,000 compounds were screened. Several compounds that were developed, especially **primaquine,** are more satisfactory than quinacrine.

Despite the effectiveness of these synthetic antimalarials, concern is being expressed because strains of the malaria parasites are developing that are resistant to the drugs most commonly used. Since there is no evidence that malaria plasmodia become resistant to quinine, its availability has acquired renewed interest.

The **ergot alkaloids** are derived from ergot, a fungus that grows on rye and other cereals. They are responsible for a disease known as *ergotism,* which was prevalent in the Middle Ages. All are amides of **lysergic acid. Ergonovine,** which is the amide of 2-amino-1-propanol, is used medicinally in the treatment of migraine headache and to induce uterine contractions at childbirth. The oral administration of as little as 50 μg. (50 millionths of a gram) of the synthetic **diethylamide** of lysergic acid (*LSD-25*) produces a psychosis resembling schizophrenia.

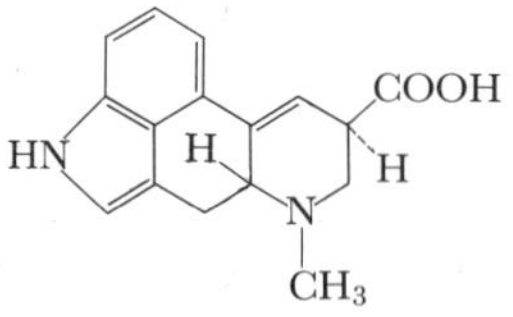

Lysergic acid

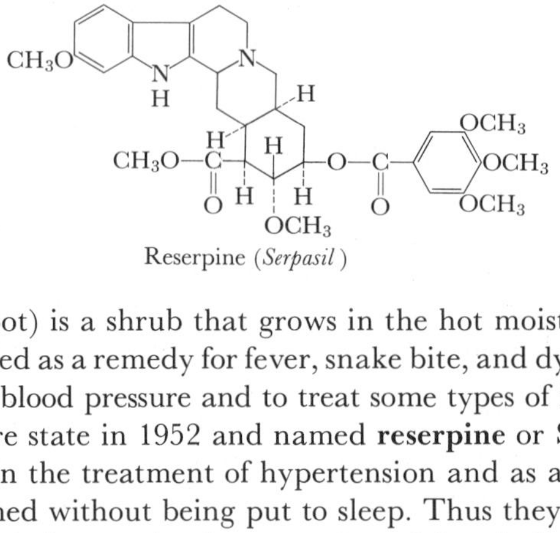

Reserpine (*Serpasil*)

Rauwolfia serpentina (Indian snake root) is a shrub that grows in the hot moist regions of India. Extracts of the root have been used as a remedy for fever, snake bite, and dysentery. In recent times it has been used to lower blood pressure and to treat some types of insanity. The active alkaloid was isolated in a pure state in 1952 and named **reserpine** or **Serpasil.** Since then it has been used extensively in the treatment of hypertension and as a general sedative. Violent schizophrenics are calmed without being put to sleep. Thus they become amenable to psychiatric treatment, and frequently the necessity of hospitalization is avoided.

PROBLEMS

30–*1*. Pair each of the following nuclei with the related compound or class of compounds: (*a*) pyrrole, (*b*) purine, (*c*) pyridine, (*d*) isoquinoline, (*e*) tropane, (*f*) quinuclidine, (*g*) pteridine, (*h*) imidazole, (*i*) indole, (*j*) pyrimidine; (*1*) nicotinic acid, (*2*) quinine, (*3*) barbiturates, (*4*) histidine, (*5*) porphins, (*6*) papaverine, (*7*) caffeine, (*8*) tryptophan, (*9*) cocaine, (*10*) folic acids.

30–*2*. Give a series of reactions for the following preparations: (*a*) 2-thiophenecarboxylic acid from thiophene, (*b*) thienylacetic acid from thiophene, (*c*) methyl α-pyrryl ketone from pyrrole, (*d*) 5-oxo-2-pyrrolecarboxylic acid from glutamic acid, (*e*) β-methylindole from aniline, (*f*) α-methylindole from *o*-toluidine, (*g*) 2-hydroxypyridine from pyridine, (*h*) 1-phenyl-2-α-pyridylethane from α-picoline, (*i*) 2-ethylpyridine from α-picoline.

30–*3*. Tell how one may distinguish readily by chemical reactions between the members of the following pairs of compounds: (*a*) benzaldehyde and furfural, (*b*) pyridine and piperidine, (*c*) α-picoline and β-picoline, (*d*) thiophene and benzene, (*e*) furoic acid and benzoic acid, (*f*) α-picolinic acid and β-picolinic acid, (*g*) quinoline and quinaldine, (*h*) quinoline and tetralin, (*i*) phthalic acid and quinolinic acid, (*j*) phenanthridine and *o*-phenanthroline, (*k*) 6-hydroxyquinoline and 8-hydroxyquinoline, (*l*) pyrrole and pyrrolidine, (*m*) coniine and nicotine, (*n*) meperidine and methadone.

30–*4*. Give reactions for the following syntheses: (*a*) proline from tetrahydrofuran, (*b*) isonicotinic acid hydrazide from γ-picoline, (*c*) 2-pyridinecarboxaldoxime methiodide from α-picoline, (*d*) Pyribenzamine from pyridine and other readily available chemicals, (*e*) 2-chloroquinoline from *o*-nitrobenzaldehyde, (*f*) 1,3-dichloro-5,5-dimethylhydantoin from acetone, (*g*) Dilantin from benzaldehyde, (*h*) triphenyltetrazolium chloride from benzaldehyde and aniline, (*i*) Tetrazolium Blue from benzaldehyde and *o*-nitroanisole, (*j*) 8-hydroxyquinoline from *o*-nitroaniline, (*k*) tetrahydropyran from furfural.

30–*5*. Devise a synthesis for chlorothiazide starting with *m*-chloroaniline.

30–*6*. Explain why the α halogen of 2,3-dichlorotetrahydropyran undergoes nucleophilic displacement more readily than the β halogen.

30–*7*. Indicate the reactions expected to occur when morphine is treated with (*a*) dilute hydrochloric acid, (*b*) dilute sodium hydroxide, (*c*) excess bromine in carbon tetrachloride, (*d*) methyl iodide, then heated with silver oxide, (*e*) excess acetic anhydride, (*f*) aluminum *t*-butoxide and acetone, (*g*) benzenediazonium chloride and alkali, and (*h*) excess methylmagnesium bromide.

30–*8*. Anthranilic acid is treated in hydrochloric acid solution at 0° with sodium nitrite, cuprous chloride is added, and the solution warmed. The resulting compound *A* is heated with a mixture of nitric and sulfuric acids to give *B*, which when heated with aniline gives *C*, $C_{13}H_{10}N_2O_4$. *C* is treated with tin and hydrochloric acid to give *D*, which, after isolation, is dissolved in hydrochloric acid and cooled. Sodium nitrite first is added and then hypophosphorous acid and the solution warmed. The product *E* has the formula $C_{13}H_{11}NO_2$. When *E* is heated with concentrated sulfuric acid, compound *F*, $C_{13}H_9NO$, is obtained which, when heated with zinc dust, yields *G*, $C_{13}H_9N$. Give the reactions involved, with structural formulas for the organic products.

30–*9*. Compound *A* is optically active and has the molecular formula $C_8H_{14}O_2S_2$. It dissolves readily in aqueous bicarbonate and is precipitated unchanged by acidifying the alkaline solution. Reduction with zinc and hydrochloric acid gives *B*, $C_8H_{16}O_2S_2$, which is reconverted to *A* with sodium hypoiodite solution. When *A* is heated with Raney nickel containing adsorbed hydrogen, it is converted to octanoic acid. Give the likely structures for *A* and equations for the reactions that it undergoes.

30–*10*. Compound *A*, $C_{10}H_{13}N$, is soluble in dilute hydrochloric acid and reacts with benzenesulfonyl chloride to give a derivative insoluble in dilute aqueous sodium hydroxide. When subjected to the Hofmann exhaustive methylation and decomposition of the quaternary hydroxide, *A* is converted into compound *B*, $C_{12}H_{17}N$. *B* is soluble in dilute acid but does not react with acetyl chloride. When the exhaustive methylation and decomposition procedure is repeated on *B*, compound *C*, $C_{10}H_{10}$, is formed. *C* decolorizes bromine in carbon tetrachloride rapidly without the evolution of hydrogen bromide. Vigorous oxidation converts either *A*, *B*, or *C* into compound *D*. *D* is soluble in dilute alkali and has a neutralization equivalent of 83. When heated, *D* is converted into compound *E*, $C_8H_4O_3$. Give a structural formula for *A* and equations for the reactions involved.

30–*11*. Give equations for the following reactions with structural formulas for the organic compounds: Vanillin treated with methyl sulfate and sodium hydroxide gives compound *A*. When an alcoholic solution of *A* is shaken with hydrogen and platinum catalyst, compound *B* is formed. *B* reacts with dry hydrogen chloride to give *C*, which reacts with sodium cyanide in the presence of cuprous cyanide to give *D*. When *D* is boiled with alcohol containing sulfuric acid, *E* is obtained. When *D* is shaken with hydrogen and platinum catalyst in the presence of ammonia, compound *F* is obtained. Heating *E* and *F* together in tetralin at the boiling point gives *G*. Refluxing *G* in toluene solution with phosphorus oxychloride gives *H*. When *H* is heated with platinum black, compound *I* is obtained, which has the molecular formula $C_{20}H_{21}NO_4$.

30–*12*. Give equations for the following transformations: Vigorous sulfonation of benzene, liming out, concentration, and fusing with sodium hydroxide gives a melt from which, after acidification, compound *A*, $C_6H_6O_2$, can be isolated. When hydrogen chloride is passed into a mixture of *A* and zinc cyanide in benzene and the reaction product is decomposed with water, compound *B*, $C_7H_6O_3$, is obtained. *B* heated with acetic anhydride and sodium acetate gives *C*, $C_{13}H_{12}O_6$. Saponification of *C* and acidification gives compound *D*, $C_9H_6O_3$.

CHAPTER THIRTY-ONE

MASS SPECTROMETRY. NUCLEAR MAGNETIC AND ELECTRON PARAMAGNETIC RESONANCE. ELECTRONIC ABSORPTION SPECTRA

The application of physical methods always has been important for the characterization and the determination of structure of organic compounds. The routine use of these methods by organic chemists has depended on the complexity of the apparatus, the time and patience required for the measurements, the training and knowledge required for the operation of the equipment and for the interpretation of the results, and the usefulness of the information that is obtained. Among the simplest procedures may be mentioned the determination of melting point, boiling point, refractive index, and optical rotation. The usefulness of the last procedure has been extended by making measurements with ultraviolet as well as with visible light (p. 311). Apparatus now is available also for measuring circular dichroism. Among the more complicated procedures should be mentioned electron diffraction by organic molecules in the vapor state, X-ray diffraction by crystals, mass spectrometry, and the absorption of electromagnetic radiation from the ultraviolet through radio frequencies.

Electron and X-ray diffraction techniques permit the calculation of bond lengths and bond angles. Electron diffraction is of limited application, and a difficulty with the X-ray analysis of complex organic compounds has been the extremely time-consuming calculations necessary to interpret X-ray diffraction patterns. With the advent of modern computers, however, this burden has been reduced considerably. As a result, the X-ray analysis of organic crystals has undergone a rejuvenation, and in the hands of X-ray crystallographers has resulted in striking triumphs such as the elucidation of the structure of hemoglobin (p. 363) and of vitamin B_{12} (p. 510). The procedure still is too complicated, however, for routine use. An exception to this statement is the comparison of the X-ray diffraction pattern of a known and an unknown compound for the purpose of establishing identity or nonidentity of the samples.

The mass spectrometer was developed around 1900, but for many years it was used only to establish the existence and to determine the relative abundance of isotopes. The first extensive application to organic compounds was by the petroleum industry, which began using it to analyze hydrocarbon gas streams about 1940. Its use as a tool for the determination of the structure of organic molecules dates from the development of heated inlet systems around 1955.

It long has been recognized that valuable information concerning the structure of organic molecules can be obtained by determining the intensity of the absorption of electromagnetic radiation at different wavelengths. The absorption of radiation by a molecule may increase its electronic energy, its vibrational energy, its rotational energy, or the energy of its nuclear and electronic spin states, all changes being quantized.

The energy required to promote an electron from the ground state to an orbital of higher energy can be supplied by radiation in the ultraviolet region of the spectrum. Although absorption in the ultraviolet was investigated for many years by photochemists using photographic processes, it was not until around 1940 that the commercial production of simply operated photoelectric spectrometers made it a convenient tool for organic chemists.

Changes in molecular vibrational energy are smaller than changes in electronic energy and can be brought about by radiation in the near infrared. With the development after 1945 of commercial infrared spectrometers that are easy to operate, the use of infrared spectra has become one of the most valuable physical aids for the determination of the structure of organic compounds (cf. pp. 133, 392).

Changes in the rotational energies of molecules are small and can be brought about by absorption of radiation in the microwave region. Microwave absorption spectra permit the calculation of the principal moments of inertia of the molecule from which internuclear distances and bond angles can be calculated for simple molecules. Although the absorption spectrum in this region cannot be used to identify functional groups in a molecule, it is characteristic of the molecule as a whole and could be used to identify the molecule. Investigations in this area, however, are made at the present time only by specialists in microwave spectroscopy.

Atomic nuclei that possess magnetic moments behave like minute spinning bar magnets and tend to line up parallel to an applied magnetic field. Under appropriate conditions these nuclei can absorb energy from a magnetic field oscillating in the radio frequency region and give rise to what has come to be known as nuclear magnetic resonance spectroscopy. The phenomenon first was detected in bulk matter independently by E. M.

TABLE 31-*1*. REGIONS OF THE ELECTROMAGNETIC SPECTRUM

REGION	WAVELENGTH*,†	FREQUENCY EXPRESSED AS WAVENUMBER‡ ($cm.^{-1}$)	ENERGY IN KCAL. PER CHEM. MOLE§
Cosmic rays	0.00005 mμ		
Gamma rays	0.001–0.14 mμ		
X-rays	0.01–15 mμ		
Far ultraviolet	15–200 mμ	666,667–50,000	1907–143
Near ultraviolet	200–400 mμ	50,000–25,000	143–71.5
Visible	400–800 mμ	25,000–12,500	71.5–35.7
Near infrared	0.8–2.5 μ	12,500–4,000	35.7–11.4
Vibrational infrared	2.5–25 μ	4,000–400	11.4–1.14
Far infrared	0.025–0.5 mm.	400–200	1.14–0.57
Microwave radar	0.5–300 mm.	200–0.033	0.57–0.000094
Short radio	0.3–200 m.		
Broadcast radio	200–550 m.		
Long radio	550–10^5 m.		
Electric power	10^5–10^9 m.		

* 1 Millimicron (mμ) = 0.001 micron (μ); 1 micron = 0.001 millimeter (mm.); 1 millimeter = 0.001 meter (m.). 1 Angstrom (A) = 0.1 mμ.

† The various regions are not sharply defined. Each overlaps the adjacent portions at both ends.

‡ Wavenumber ($\bar{\nu}$) = number of cycles per centimeter of length along the path of the wave = $1/\lambda$ in cm.; cyclic frequency = c/λ in cm., where c = velocity of light = 3×10^{10} cm./sec.

§ 1000 $cm.^{-1}$ = 2.86 kcal./mole = 0.124 e.v.; 100 kcal./mole = 34,965 $cm.^{-1}$ = 4.38 e.v.; 1 e.v. = 23.06 kcal./mole = 8063 $cm.^{-1}$; $2.86/\lambda$ in μ = kcal./mole.

Purcell and F. Bloch and their co-workers in 1945 and within a remarkably short time commercial equipment for its measurement became available. As a result, this new research tool, which is used mainly to give information about the electronic and hence the atomic environment of hydrogen nuclei, rapidly has come into general use for arriving at the structure of organic compounds. Electron paramagnetic resonance is a similar phenomenon connected with spinning unpaired electrons and has been of considerable theoretical interest.

Table 31–*1* defines the regions of the electromagnetic spectrum. The unit of radiant energy is the *photon,* the energy of a photon being given by the relation $E = h\nu = hc/\lambda$ where h is the Planck constant, ν the frequency of the radiation, c the velocity of light, and λ the wavelength of the radiation. Thus the magnitude of a radiation quantum is directly proportional to the frequency and inversely proportional to the wavelength of the radiation. The various unit and energy interconversions are listed under the table.

MASS SPECTROMETRY

When an electron that possesses high kinetic energy collides with a molecule, the electron usually knocks an outer electron from the molecule to give a positive ion.

$$A + e \longrightarrow A^{+} + 2\,e$$

The moving positive ion can be deflected from its path by a magnetic field. The amount of the deflection for a given field strength is dependent on the momentum (= mass × velocity) of the ion. If a monoenergetic collimated beam of these ions is produced, ions having different masses and therefore differing in momentum will be deflected by a magnetic field to an extent dependent only on the masses of the ions per unit charge. Since practically all of the ions have unit charge, the original beam will be separated and spread out into as many beams as there are ions having different mass numbers. A **mass spectrometer** is an instrument that converts molecules into ions, which it separates into a spectrum according to their mass to charge ratio, and that determines the relative number of each species of ion present.

Figure 31–*1* is a simplified diagram of a mass spectrometer, which is evacuated to a pressure of around 10^{-6} mm. of mercury. Electrons given off by a heated filament are

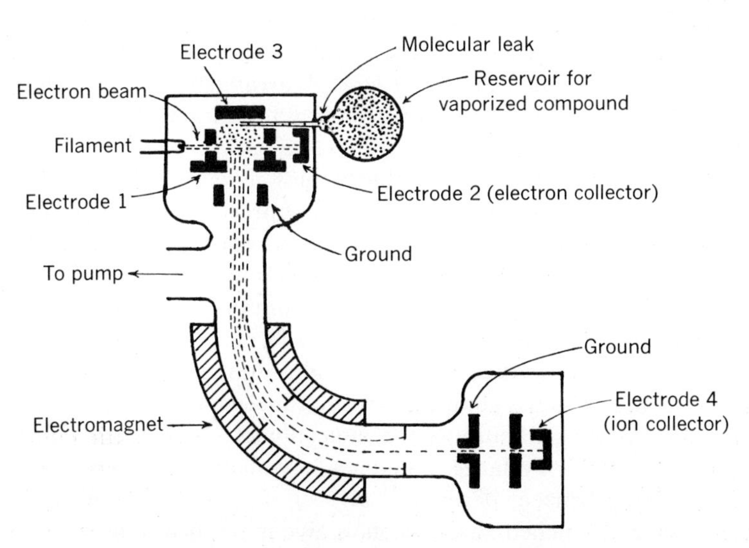

Figure 31–*1*. Schematic diagram of one form of mass spectrometer.

accelerated into the ionization chamber by a potential difference between the filament and electrode 1. The electron beam is collected at electrode 2. The sample, which has been vaporized into a reservoir at about 10^{-2} mm., diffuses through a minute hole, called a *molecular leak,* into the ionization chamber, and the ions formed are pushed out by a small difference in voltage between electrodes 3 and 1. The ions then are accelerated by a potential difference of 300–4000 v. between electrode 1 and the ground. The ion beam is deflected and separated by the magnetic field, and a single beam of ions, all having the same mass to charge ratio, is collected at electrode 4, where the ion current is measured by a vacuum-tube electrometer having a sensitivity of 10^{-14} to 10^{-15} ampere. By changing the magnetic field (*magnetic scanning*) or the accelerating voltage (*voltage scanning*), different beams may be made to impinge on the detector and the intensity of each beam measured. The **mass spectrum** is a graph of the intensity of each beam against the mass number.

For investigation with a mass spectrometer, an organic compound must be stable at whatever temperature is necessary to give it a vapor pressure of about 10^{-2} mm. of mercury. Although hydrocarbons may be heated to temperatures as high as 350° without thermal decomposition, the optimum maximum temperatures for less stable compounds is in the 150–250° range. A three-liter reservoir requires only one micromole of compound. Present commercial instruments give unit resolution up to around 375, but ions of still higher mass number give useful information.

If the ionized molecule were stable, all that could be determined with a mass spectrometer would be the exact molecular weight divided by the charge on the ion. The usual ionizing electron beam, however, has an energy of 70 electron volts, and since only 9 to 15 e.v. is required to remove an electron, the molecule ion has a large excess of energy. As a result the ion undergoes immediate fragmentation into smaller ions and neutral free radicals.

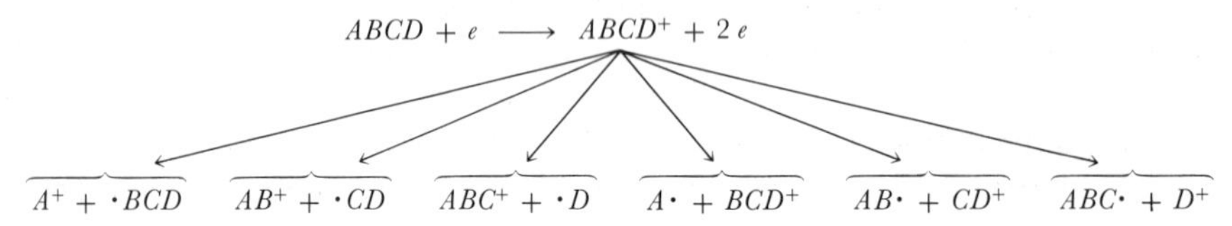

The smaller polyatomic ions can undergo further fragmentation. Thus even simple molecules give a whole spectrum of masses. The relative heights of the peaks are proportional to the relative number of the different kinds of ions. Usually the strongest peak, called the *base peak,* is given the value of 100, and the intensities of the other peaks are expressed as the per cent of this highest intensity. The peak corresponding to the exact numerical weight of the compound is called the *parent peak.* Rearrangements and combination of ions with neutral fragments also can occur but usually do not interfere with the interpretation of the main fragmentation pattern.

Because of the presence of isotopes, the highest mass number is not necessarily the molecular weight of the compound since even if an appreciable quantity of the original ions escapes fragmentation, small peaks with mass numbers one, two, or three units higher may be obtained depending on the elements present. Mass numbers corresponding to the molecular weights must be even for carbon compounds containing hydrogen, oxygen, halogen, or an even number of nitrogen atoms. All fragments formed by the cleavage of single bonds have odd mass numbers, unless they contain an odd number of nitrogen atoms or unless a neutral molecule such as water is eliminated from a molecule ion.

The order of decreasing stability of the molecule ions is aromatic compounds $>$ conjugated polyenes $>$ olefins $>$ cyclanes $>$ ketones $>$ *n*-alkanes $>$ ethers $>$ esters $>$ acids $>$ alcohols $>$ amines. Thus benzene derivatives give strong peaks at the molecular weight and few fragments, whereas branched-chain alcohols give many peaks, none of which is as high as the molecular weight.

The point at which the ions cleave and the particular fragment that carries the positive charge correspond to current ideas regarding the stability of the resulting fragments; that is, cleavage takes place most readily at the point that yields the most stable positive ion of a stable neutral molecule. Thus for carbonium ions that are not stabilized by resonance, the order of stability is tertiary > secondary > primary (p. 83), and cleavage takes place to give the more highly branched ion. Isobutane gives peaks at mass 43 and 57 but none at 15 and 29. Methyl and ethyl groups are produced also but they are neutral radicals because the secondary ions are more stable.

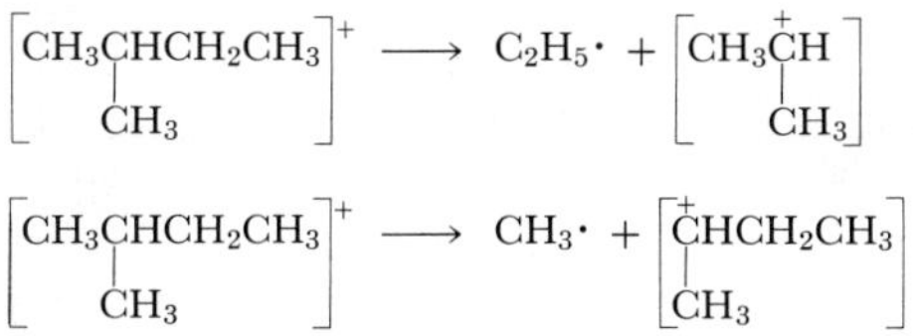

$$\left[\begin{array}{l}CH_3\underset{|}{C}HCH_2CH_3\\ \quad CH_3\end{array}\right]^+ \longrightarrow C_2H_5\cdot + \left[\begin{array}{l}CH_3\overset{+}{C}H\\ \quad CH_3\end{array}\right]$$

$$\left[\begin{array}{l}CH_3\underset{|}{C}HCH_2CH_3\\ \quad CH_3\end{array}\right]^+ \longrightarrow CH_3\cdot + \left[\begin{array}{l}\overset{+}{C}HCH_2CH_3\\ CH_3\end{array}\right]$$

Ions stabilized by resonance are formed in preference to those not so stabilized. Thus olefins yield allyl cations (p. 594).

$$[CH_2{=}CHCH_2CH_3]^+ \longrightarrow CH_3\cdot + \{CH_2{=}CH\overset{+}{C}H_2 \longleftrightarrow \overset{+}{C}H_2CH{=}CH_2\}$$

All primary alcohols give strong peaks at mass 31 because the $^+CH_2OH$ ion is resonance stabilized.

$$[R{-}CH_2OH]^+ \longrightarrow R\cdot + \{\overset{+}{C}H_2OH \longleftrightarrow CH_2{=}\overset{+}{O}H\}$$

2-Alkanols give a peak at mass 45 and 3-alkanols at mass 59.

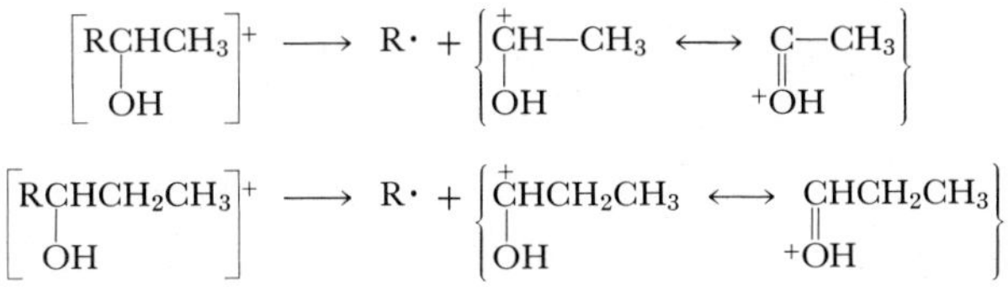

$$\left[\begin{array}{l}R\underset{|}{C}HCH_3\\ \;OH\end{array}\right]^+ \longrightarrow R\cdot + \left\{\begin{array}{l}\overset{+}{C}H{-}CH_3\\ OH\end{array} \longleftrightarrow \begin{array}{l}C{-}CH_3\\ \|\\ ^+OH\end{array}\right\}$$

$$\left[\begin{array}{l}R\underset{|}{C}HCH_2CH_3\\ \;OH\end{array}\right]^+ \longrightarrow R\cdot + \left\{\begin{array}{l}\overset{+}{C}HCH_2CH_3\\ OH\end{array} \longleftrightarrow \begin{array}{l}CHCH_2CH_3\\ \|\\ ^+OH\end{array}\right\}$$

The latter, however, give a strong peak at mass 31 also because of the easy elimination of a molecule of ethylene.

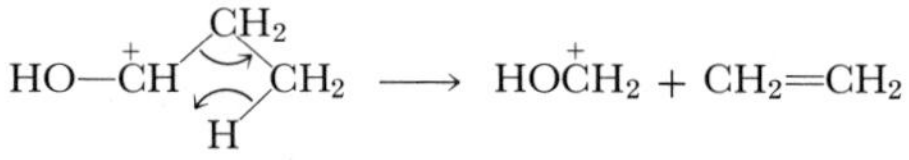

$$HO{-}\overset{+}{C}H\overset{CH_2}{\frown}\underset{H}{\smile}CH_2 \longrightarrow HO\overset{+}{C}H_2 + CH_2{=}CH_2$$

2-Methyl-2-alkanols also give a peak at mass 59.

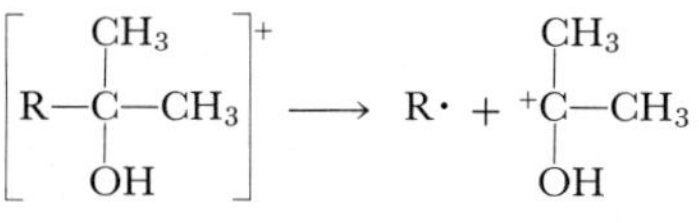

$$\left[\begin{array}{c}CH_3\\ |\\ R{-}C{-}CH_3\\ |\\ OH\end{array}\right]^+ \longrightarrow R\cdot + \begin{array}{c}CH_3\\ |\\ ^+C{-}CH_3\\ |\\ OH\end{array}$$

This peak, however, is much stronger than that from 2-alkanols because the tertiary carbonium ion is more stable than the secondary carbonium ion. Ethyl ether gives a strong peak at mass 59 but the strongest peak is at mass 31 because of the loss of ethylene.

$$[CH_3CH_2OCH_2CH_3]^+ \longrightarrow CH_3\cdot + \{CH_3CH_2O\overset{+}{C}H_2 \longleftrightarrow CH_3CH_2\overset{+}{O}{=}CH_2\}$$

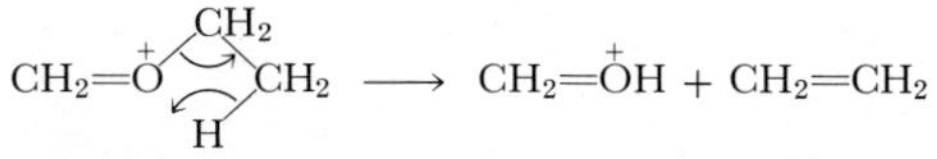

$$CH_2{=}\overset{+}{O}\overset{CH_2}{\frown}\underset{H}{\smile}CH_2 \longrightarrow CH_2{=}\overset{+}{O}H + CH_2{=}CH_2$$

In addition to ethylene, other small, stable, neutral molecules such as carbon monoxide, carbon dioxide, water, ammonia, hydrogen sulfide, or hydrogen cyanide may be eliminated from appropriate ions.

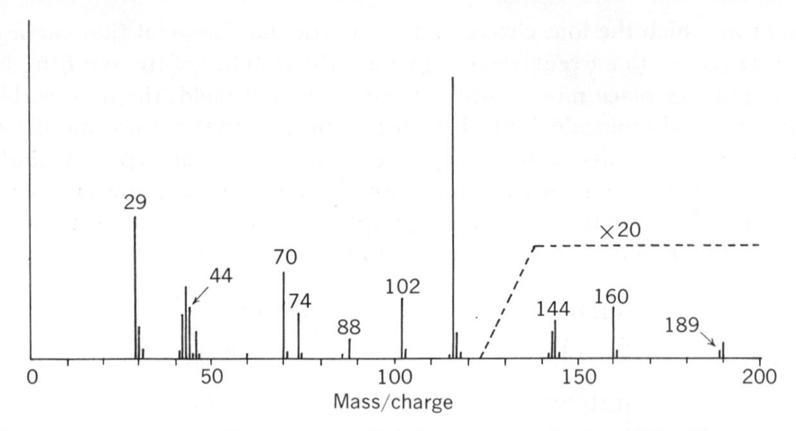

Figure 31–*2*. Mass spectrum of ethyl aspartate, molecular weight 189.

The behavior of amines, thiols, and sulfides is analogous to that of the corresponding oxygen compounds. If more than one hetero atom is present, the ion carrying the more electron-releasing element is the more stable. Thus 1,2-amino alcohols give a more intense peak for the nitrogen fragment than for the oxygen fragment because the nitrogen is less electronegative than oxygen and can stabilize the positive charge more readily.

$$\left[\begin{array}{c}\mathrm{RCH{-}CHR} \\ \mathrm{OH\quad NH_2}\end{array}\right]^+ \longrightarrow \begin{cases} \mathrm{RCH(OH)\cdot} + \mathrm{\overset{+}{C}HR(NH_2)} \\ \mathrm{R\overset{+}{C}H(OH)} + \mathrm{\cdot CHR(NH_2)} \end{cases}$$

Figure 31–*2* is a chart of the mass spectrum of ethyl aspartate, the fragmentation pattern of which has been interpreted as follows:

Primary fragmentation

$$\mathrm{C_2H_5{-}O{-}CO{-}CH_2{-}CH(NH_2){-}CO{-}O{-}C_2H_5}$$

(cleavages marked: 102, 144, 160, 116)

Secondary fragmentations

$$\underset{116}{\mathrm{COOC_2H_5{-}CH_2{-}\overset{+}{C}HNH_2}} \longrightarrow \mathrm{CH_2{=}CH_2} + \underset{88}{\mathrm{COOH{-}CH_2{-}\overset{+}{C}HNH_2}} \longrightarrow \mathrm{H_2O} + \underset{70}{\begin{array}{c}\mathrm{CH_2{-}CO} \\ \mathrm{{}^{+}CH{-}NH}\end{array}}$$

$$\mathrm{(88)}\ \longrightarrow \mathrm{CO_2} + \underset{44}{\mathrm{CH_2{=}CH{-}\overset{+}{N}H_3}}$$

$$\underset{102}{\mathrm{COOC_2H_5{-}\overset{+}{C}HNH_2}} \longrightarrow \mathrm{CH_2{=}CH_2} + \underset{74}{\mathrm{COOH{-}\overset{+}{C}HNH_2}}$$

The chief primary fragmentation products result from breaking the bond on either side of the $CHNH_2$ group because the amino group permits stabilization of the positive ion.

$$\{R\overset{+}{C}H{-}NH_2 \longleftrightarrow RCH{=}\overset{+}{N}H_2\}$$

The fragment of mass 102 is less readily formed than that of mass 116 because the electron-attracting power of the carbonyl group attached directly to the methylidyne (CH) carbon decreases the ease with which this carbon can accept a positive charge. Loss of the neutral ethylene molecule from the chief primary fragments yields ions of mass 88 and 74. Loss of water or carbon dioxide from the mass 88 ion yields the ions of masses 70 and 44.

The number of ions of mass above 116 is very small because of the extensive fragmentation, and this portion of the chart has been enlarged twenty times. The principal ions are accompanied by small amounts of ions one mass number higher chiefly because of the presence of carbon-13. The greater number of ions of mass 190 than mass 189 results from the acquisition of a hydrogen atom by the molecular ion by collision with another molecule.

Many more generalizations have been made concerning the behavior of organic molecules when bombarded by electrons. By the application of these principles, the mass spectrometer has become a valuable tool for the elucidation of the structures of even complex organic compounds.

NUCLEAR MAGNETIC RESONANCE

The nuclei of certain isotopes such as H^1, F^{19}, or P^{31} act like charged spinning bodies. The moving charge creates a magnetic field, and the nuclei have a magnetic moment. In the absence of an external magnetic field the nuclear magnetic moments are randomly oriented, but in the presence of a uniform magnetic field they tend to become aligned with the field. This tendency is opposed by thermal agitation, but because of quantum restrictions, magnetic moments that are not aligned with the field must be aligned against the field. Those nuclei whose moments are aligned with the field have a lower energy than those aligned against the field, but the difference in energy is so small that at room temperature and at field strengths conveniently obtainable, the equilibrium concentration of the two states is nearly equal. There is, however, a slight excess of nuclei at the lower energy level, the number in excess being dependent on the strength of the applied magnetic field.

The difference in energy of the two states, ΔE, equals $2\,\mu H$ where μ is the magnetic moment of the nucleus and H is the field strength at the nucleus. To bring about a transition from the lower to the higher energy level, that is, to cause the nuclei aligned with the field to be aligned against the field, radiation energy of frequency ν is required, where $\nu h = 2\,\mu H$. For hydrogen nuclei in a field of 9400 gauss, this frequency is approximately 40 megacycles, which is in the radio frequency range. The particular frequency at which energy is absorbed is called the *resonance frequency*. The phenomenon commonly is referred to as **nuclear magnetic resonance** (*n.m.r.*), although **nuclear spin resonance** (*n.s.r.*) is more descriptive, and one of the discoverers of the phenomenon, F. Bloch, prefers to call it **nuclear inductance.**

Figure 31–*3* illustrates schematically the essential features of one form of an instrument, known as a *nuclear magnetic resonance spectrometer,* for detecting and measuring the absorption of energy that results from the transition from the lower to the higher energy state. The sample, as a liquid or in solution in a solvent such as deuteriochloroform that does not contain protons, is put in a narrow thin-walled glass tube, and the tube is placed in a "probe" between the poles of an electromagnet. The tube is spun by an air turbine to

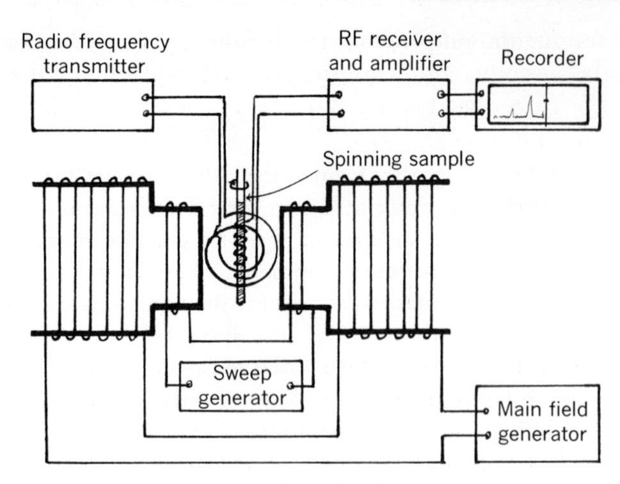

Figure 31–3. Schematic diagram of one form of nuclear magnetic resonance spectrometer.

average out slight inhomogeneities of the applied field. Energy of radio frequency is transmitted to the sample by means of a coil so placed in the probe that it produces a radio-frequency magnetic field perpendicular to the main magnetic field, H_0. Any absorption of energy by the sample is detected by a receiver coil that surrounds the sample and is amplified to operate a recorder. The axis of the receiver coil is perpendicular both to that of the transmitter coil and to the main magnetic field. Hence only radiation that has been modified by the absorption is detected.

Because of the way most commercial spectrometers operate and because the frequency is directly proportional to field strength, the basic applied field of an instrument is specified in terms of the equivalent frequency, one cycle per second (1 c.p.s.) being equivalent to 0.235 milligauss. Thus an instrument for the study of protons that uses an applied field strength of 7050 gauss operates at a fundamental frequency of 30 megacycles and is called a 30 megacycle instrument; 40 Mc is equivalent to 9400 gauss, 60 Mc to 14,100 gauss, and 100 Mc to 23,500 gauss.

To determine the amount of absorption and at what frequency it occurs, the n.m.r. spectrometer scans a sample over the range of frequencies expected for the nucleus under examination. In practice it is easier to keep the applied radio frequency constant and to vary the field strength by means of an auxiliary winding on the magnet known as a *sweep coil.* As the applied field is varied, the current from the receiver coil is amplified and recorded on a calibrated chart in the form of a graph that can be interpreted as a plot of the absorption of energy against frequency at a constant applied magnetic field. The area under an absorption peak of the graph is proportional to the number of nuclei absorbing at that particular frequency.

Of the abundant isotopes that have a strong nuclear magnetic moment, hydrogen is the most important to the organic chemist. Carbon-12, oxygen-16, and sulfur-32 have zero nuclear magnetic moment and cannot be detected, and the moment of nitrogen-14 is weak. Most elements have at least one isotope with a reasonably strong moment, for example C^{13} and O^{17}, that can be introduced into organic compounds if necessary. Except for special applications, however, routine determination of nuclear magnetic resonance spectra is confined ordinarily to hydrogen, and unless otherwise specified *n.m.r.* usually refers to proton magnetic resonance (*p.m.r.*).

If at a fixed frequency all protons absorbed energy at the same applied field strength, the observation of nuclear magnetic resonance would be of little value. The applied field, however, induces circulation of the electrons in the molecule, which produces a secondary magnetic field about the nuclei that is opposed to the applied field. Hence the local field about a proton usually is smaller than the applied field, by an amount that depends on the electronic environment and hence the chemical environment of the proton. The greater the shielding or screening of the protons by the induced magnetic field, the higher the field strength necessary to bring about resonance. Therefore the position of the signal depends on the environment of each individual proton in the molecule. Since it is not practical to use an absolute measure of field strength when the range covered is only around 120 milligauss out of a total of 14×10^6 milligauss, measurements are made relative to some reference compound, which usually is added to the sample. The displacement of a particular signal relative to that of the reference compound is called the *chemical shift.*

Chemical shifts for protons usually are measured relative to some standard that has only one kind of hydrogen and gives a single sharp signal. Tetramethylsilane, which is isotropic and absorbs energy at a higher field strength than most other organic compounds, commonly is used. A small amount of tetramethylsilane is added to the liquid used to dissolve the sample, and signals are measured in cycles per second below that for the signal for tetramethylsilane. The chemical shift, however, is directly proportional to the field strength or to the fundamental frequency. Thus the distance between signals at 40 Mc (9400 gauss) is one and one third times that at 30 Mc (7050 gauss), at 60 Mc (14,100 gauss) it is twice that at 30 Mc and at 100 Mc (23,500 gauss) it is three and one third times that at 30 Mc. Hence when recording chemical shifts in milligauss or cycles per second it is necessary to specify the fundamental frequency of the field as well. The trend is to use instruments that operate at higher and higher field strengths because of the better separation of signals. Accordingly it is desirable to express the chemical shifts in a unit that is independent of the applied magnetic field strength. One such unit, *delta* (δ), is defined as the chemical shift from the reference compound in cycles per second per million cycles of fixed frequency. Hence for a chemical shift of 150 c.p.s. downfield from the signal for internal tetramethylsilane at a fixed frequency of 60 megacycles,

$$\delta^{\text{int.}}_{\text{Si(CH}_3)_4} = \frac{150}{60} = 2.5 \text{ p.p.m. (parts per million).}$$

By this definition increasing values for δ correspond to decreasing field strength and less magnetic shielding of protons. Moreover for signals at field strengths higher than that for tetramethylsilane, δ values are negative. In another commonly used system, a unit designated as *tau* (τ) is defined as 10 minus δ. Thus proton signals ordinarily have positive τ values, including signals at field strengths higher than that for the TMS signal. Moreover τ values increase with increasing shielding.

The following specific example illustrates the application of n.m.r. spectroscopy. Chemical evidence indicated that a new compound had either structure *I* or structure *II.*

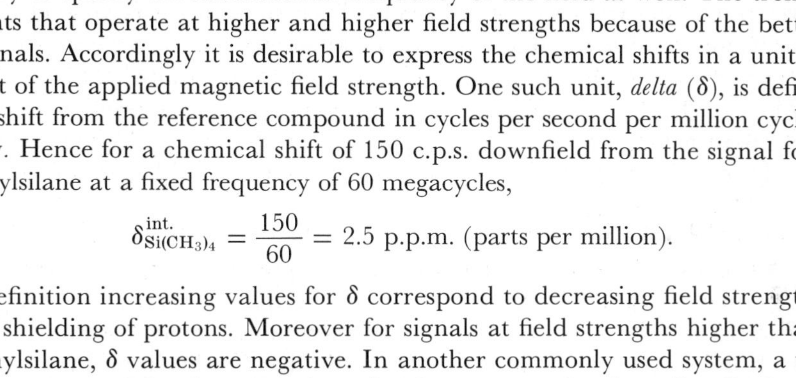

I *II*

The proton n.m.r. spectrum is given in Fig. 31–*4*, together with the integration curve for the areas under the peaks. It shows clearly that formula *I* is the correct structure and confirms completely the chemical work. From left to right, each doublet and the next two peaks represent two protons each, the next peak corresponds to three protons, and the last peak corresponds to six protons, a total of 17 hydrogens. If the compound had structure *II,*

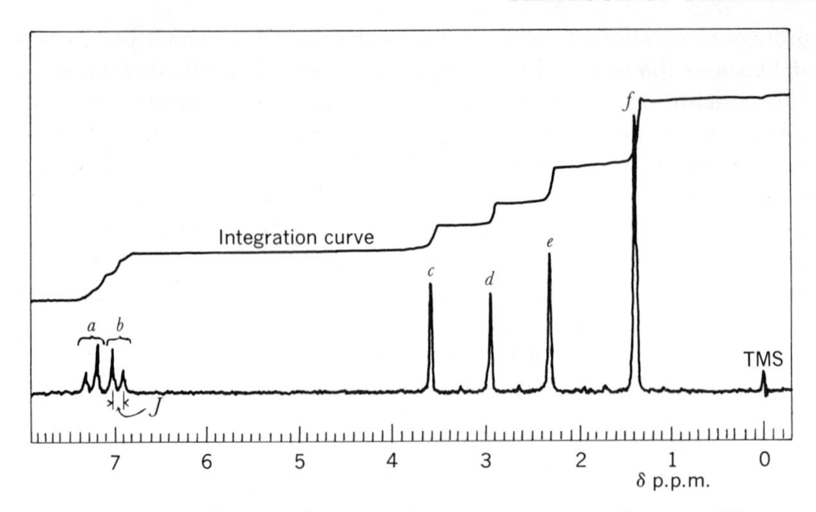

Figure 31–4. Proton nuclear magnetic resonance spectrum of compound I.

two methyl signals for three protons each would have been obtained instead of one for three protons at (*e*) and one for six protons at (*f*). Moreover a methyl group attached to a doubly bonded carbon atom would give a peak further downfield near $\delta = 1.7$ rather than $\delta = 1.4$ because an sp^2 orbital is more electronegative than an sp^3 orbital. The three protons at *e* are those of the aromatic methyl group. Their signal appears downfield from that for the *gem*-dimethyl groups because of the deshielding effect of the benzene ring currents. The signals at (*d*) and (*c*) are those for the two methylene groups on either side of the nitrogen. They are further downfield than that for (*e*) because the protons are deshielded by the electron-attracting power of the nitrogen; that at (*c*) is the methylene group between nitrogen and the carbonyl group, since the electronegativity of the carbonyl group causes further deshielding.

Although only two kinds of hydrogen are attached to the benzene ring, four signals are observed because of the phenomenon known as **spin-spin coupling.** When two non-equivalent protons are on adjacent carbon atoms, the spin of one proton, (*a*), produces a magnetic field that can be transferred by way of intervening bonding electrons to the other proton, (*b*). It either adds to or subtracts from the field of the second proton, (*b*), and causes a splitting of the signal. The distance between peaks of the split signal, known as the *coupling constant, J*, is independent of the applied field, whereby it can be distinguished from a chemical shift. It is measured in cycles per second and is characteristic of the structure of the compound. The spin of the proton (*b*) likewise couples with that of (*a*) and splits the signal for (*a*). The doublet for the two (*a*) protons is at lower field strength than that for the (*b*) protons because the sulfone group is electron-attracting whereas the methyl group is electron-releasing.

ELECTRON PARAMAGNETIC RESONANCE

The spin of an unpaired electron, such as that present in free radicals, also produces a magnetic moment and can give rise to a resonance phenomenon analogous to that caused by nuclear spin. It commonly is referred to as *electron paramagnetic resonance* (*e.p.r.*) or *electron spin resonance* (*e.s.r.*). The e.p.r. spectrum is more complicated than that of n.m.r. The number of absorption lines observable is equal to one plus twice the magnetic quantum number of the nucleus with which the unpaired electron is associated. Thus an unpaired electron in the neighborhood of a C^{12} nucleus, for which the nuclear spin is zero, gives a single line. For H^1, with nuclear spin 1/2, splitting of the absorption line into two components occurs, and N^{14}, with nuclear spin 1, gives three lines. Moreover if the electron interacts with several "equivalent" nuclei, the total magnetic quantum number applies. Thus if interaction is with two nitrogen nuclei, five lines [2(1 + 1) + 1] are observed, if with three hydrogen nuclei, four lines [2(1/2 + 1/2 + 1/2) + 1]. The observation that for stable free radicals the absorption line is split into numerous components is experimental evidence for the delocalization of the electron. At high resolution triphenylmethyl gives a number of groups of multiple lines, indicating that the unpaired electron is interacting with all the nuclei of the molecule (cf. p. 483).

The semiquinone ion gives a group of five signals in its e.p.r. spectrum indicating interaction of the unpaired electron with the four hydrogen nuclei.

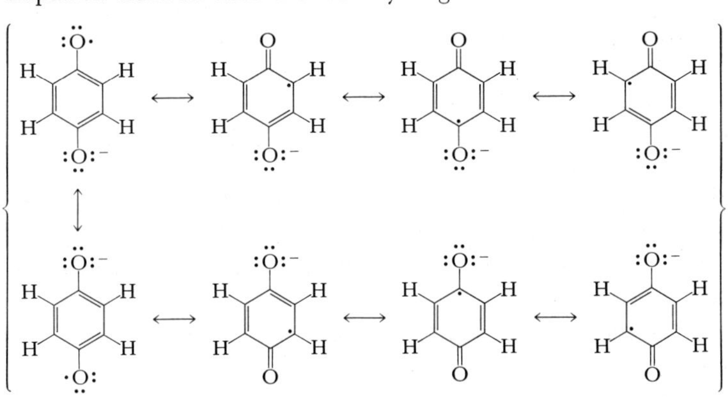

The observation of a group of 13 signals for the tetramethylsemiquinone ion, indicating interaction of the unpaired electron with all 12 hydrogen nuclei, has been hailed as direct evidence for so-called *hyperconjugation* or *no-bond resonance* represented by structures such as *III*.

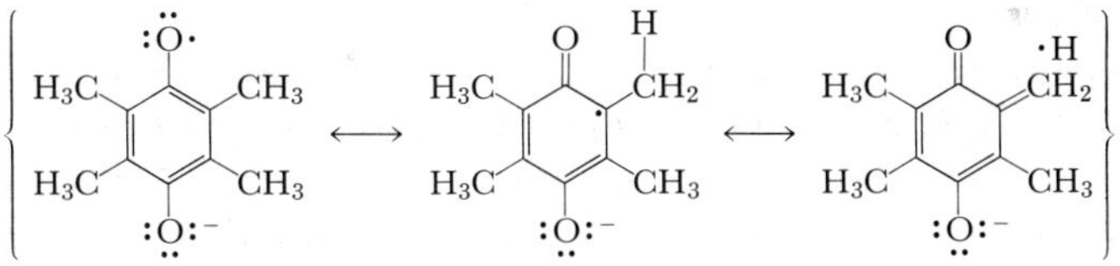

I *II* *III*

It is not necessary, however, to postulate this type of resonance in methyl groups. The unpaired electron in structure *II* can interact with the protons by way of the electrons in the carbon-carbon bond and those in the carbon-hydrogen bonds, just as spin-spin coupling of two protons on adjacent carbon atoms takes place by way of the electrons in the intervening bonds and gives rise to the splitting of the nuclear magnetic resonance signal (p. 550).

Unpaired electrons can be detected at extremely low concentrations by e.p.r. Thus 5×10^{-11} moles of diphenylpicrylhydrazyl has been detected in a volume of approximately 0.1 cc. or a molar concentration of 5×10^{-7}. Hence it is possible to establish the presence of free radicals in concentrations too low to detect by static susceptibility methods. Free radicals also have been detected in polymers formed by free radical polymerizations and in substances that have been irradiated by high-energy particles. Moreover e.p.r. absorption has been observed in substances not previously considered to contain molecules with unpaired electrons. Thus carbonium ions (p. 456) and carbanions (p. 480) appear to contain some ions in the biradical state.

ELECTRONIC (ULTRAVIOLET AND VISIBLE) SPECTRA

In molecules at ordinary temperatures the electrons are in orbitals having the lowest possible energy, a condition known as the *ground state.* Higher orbitals are empty, but it requires energy to cause the transfer (*transition*) of an electron from a lower orbital to a higher orbital, and the kinetic energy of the molecules is not sufficiently high to do this except at high temperatures. It happens, however, that electromagnetic vibrations having wavelengths between 100 and 1300 mμ, that is, in the ultraviolet, visible, and near infrared, have just the right energy range to cover the energy range of the lower electronic transitions of molecules. Hence the energy of a light quantum can be used to transfer an electron from its ground state to the next higher orbital, during which process light of that particular wavelength is absorbed.

The graphical representation ordinarily used for ultraviolet and visible spectra differs from that used for infrared spectra. The unit of wavelength is the millimicron (1 mμ = 0.001 μ). The absorption is expressed as the **absorbance,** A (formerly called *optical density,* D, or *extinction,* E), which is the logarithm of the ratio of the intensity, I_0, of the incident light to the intensity, I, of the transmitted light or the logarithm of $1/T$, where T, the **transmittance,** = I/I_0. The intensity of ultraviolet and visible light can be measured accurately more readily than the intensity of infrared radiation, and absorption usually is expressed on a quantitative rather than on a qualitative basis. The **absorbtivity,** a, is the absorbance divided by the product of the concentration of the solution in grams per liter, c, and the sample path length in centimeters, b.

$$a = \frac{A}{cb}$$

If the molecular weight of the compound is known, the absorption is expressed as the **molar absorbtivity,** ϵ (formerly called *molar extinction coefficient*), which is the product of the absorbtivity and the molecular weight.

$$\epsilon = \frac{A \times \text{Mol. wt.}}{cb} = \log \frac{I_0}{I} \times \frac{\text{Mol. wt.}}{cb}$$

The maximum value of ϵ to be expected is about 10^5. Large values of a or ϵ sometimes are expressed as log a or log ϵ. A plot of the absorbtivity or of the molar absorbtivity against wavelength is referred to by organic chemists as the *ultraviolet* or *visible absorption spectrum* of the compound. Since absorption is plotted rather than transmittance, the absorption peaks are maxima in the curves for the visible and ultraviolet regions rather than minima as in the usual curves for infrared absorption (p. 135).

Electrons in the ground state of organic molecules are in three main levels of increasing energy; those in bonding σ orbitals, those in bonding π orbitals, and those that are nonbonding. Examples of the last class, which are designated as n electrons, are the unshared electrons on nitrogen, oxygen, or halogen. The orbitals of next higher energy, which usually are empty, are the antibonding π orbitals designated as π^* (*pi starred*), and then the antibonding σ orbitals designated as σ^*. Hence transitions may be designated as $\sigma \longrightarrow \sigma^*$, $\pi \longrightarrow \pi^*$, $n \longrightarrow \sigma^*$, and $n \longrightarrow \pi^*$ (Fig. 31–5).[1]

Although Fig. 31–5 indicates a single level for each electronic state, molecules at room temperature have rotational and vibrational levels that increase in number and decrease in energy difference with increasing complexity of the molecule. Hence absorption is not confined to a single wavelength but extends over a broad spectral region. With ordinary

[1] In another notation, $\sigma \longrightarrow \sigma^*$ and $\pi \longrightarrow \pi^*$ transitions are designated as $N \longrightarrow V$ transitions, while $n \longrightarrow \pi^*$ are called $N \longrightarrow Q$.

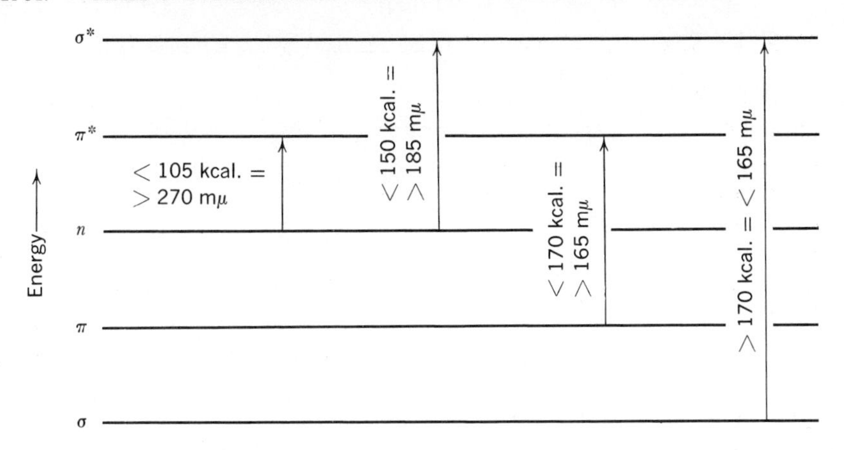

Figure 31–5. Energy or wavelength required for an electronic transition between a bonding or nonbonding orbital and an antibonding orbital.

resolution, the plot of intensity of absorption against wavelength usually appears as a smooth curve. The position and intensity of the maximum is recorded as being characteristic of the transition, for example $\lambda_{max.} = 250$ mμ, $\epsilon = 12{,}500$. The various attractive and repulsive forces between molecules also affect the energy levels of the molecule under observation and tend to increase further the number of energy levels. Hence the spectrum varies with the physical state of the compound and the solvent used. Fine structure is best observable in the vapor phase, the coalescence of the various peaks being greatly increased in the liquid state and in solution. The spectra of solutions are most highly resolved in hydrocarbon solvents and least in high-dielectric protic solvents such as water and alcohol. In practice the choice of solvent is governed by its transparency in the region under investigation and the solubility characteristics of the compound.

The intensity of the absorption is proportional to the probability that the transition will occur, which in turn depends on the magnitude of the change in molecular dipole moment brought about by the transition and is related to certain considerations of symmetry. In general $n \longrightarrow \pi^*$ bands are weak ($\epsilon_{max.} = < 100$), the $n \longrightarrow \sigma^*$ bands have moderate intensity, and the $\pi \longrightarrow \pi^*$ bands are very strong. The weak bands are the result of transitions "forbidden" according to certain selection rules that are not considered here.

The lowest energy transitions for molecules that contain only single bonds and no atoms with unshared valence electrons are $\sigma \longrightarrow \sigma^*$ transitions, which are at wavelengths shorter than 165 mμ. The longest wavelength absorption of ethane is at 140 mμ. Since air absorbs below 200 mμ, this region is not readily accessible. If a carbon-carbon double bond is present, the difference in energy between the bonding π orbital and the antibonding π^* orbital is less than that for the σ,σ^* orbitals and absorption takes place at longer wavelengths. Thus ethylene absorbs at 165 mμ (Fig. 31–*6a*).

If two isolated double bonds are present, the absorption simply is the sum of the separate absorptions. If, however, the double bonds are conjugated as in 1,3-butadiene, $CH_2{=}CH{-}CH{=}CH_2$, interaction of the two pairs of π orbitals leads to four orbitals, two of which, π_1 and π_2, are bonding and have energies lower and higher than the π orbital of ethylene, and the other two, π_3^* and π_4^*, are antibonding and have energies lower and higher than the π^* orbital of ethylene (Fig. 31–*6b*). In the ground state π_1 and π_2 are occupied. The lowest energy transition now is $\pi_2 \longrightarrow \pi_3^*$ which requires only 136 kcal. compared to 176 kcal. for the $\pi \longrightarrow \pi^*$ transition. As a result, the absorption in the gas

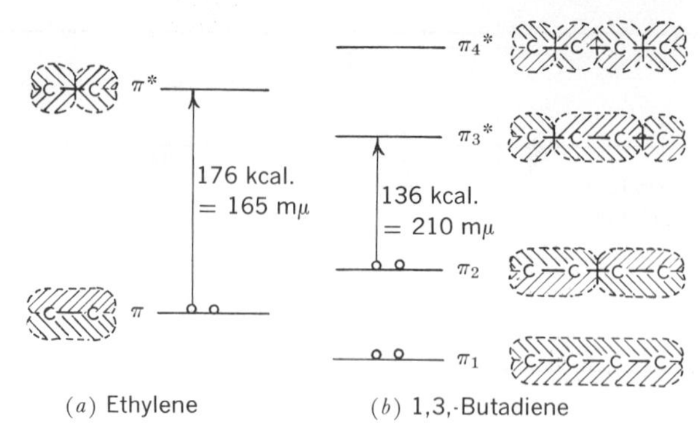

Figure 31–6. Energy levels for π orbitals of (a) ethylene and (b) 1,3-butadiene.

phase shifts to 210 mμ, which now is in the accessible portion of the ultraviolet. Moreover the intensity of the absorption is very strong ($\epsilon \sim 10^4$). Alkyl substitution at the double bonds shifts the absorption to longer wavelengths. If 217 mμ is taken as the maximum for 1,3-butadiene in alcohol solution, the position of the maximum for alkyl-substituted conjugated dienes in solution can be calculated by adding 5 mμ for each substituent. In addition to the $\pi_2 \longrightarrow \pi_3{}^*$ transition, an electron can be promoted from π_2 to $\pi_4{}^*$ and from π_1 to $\pi_3{}^*$ or $\pi_4{}^*$, giving rise to a total of four band systems. Increasing energy is required for these transitions, however, and the last three absorb only at wavelengths less than 200 mμ.

As the number of double bonds in conjugation increases, the number of band systems increases, and the energy difference between the highest occupied orbital in the ground state and the lowest antibonding orbital decreases. Hence the first absorption band moves to longer and longer wavelengths (Fig. 31–7). By the time eight or nine double bonds are in conjugation, absorption is taking place in the blue region of the spectrum and the compound appears yellow. As conjugation continues to increase, the absorption progresses through the visible, causing color changes through orange, red, and purple.

Carbon-oxygen and carbon-nitrogen double bonds resemble carbon-carbon double bonds in that the two pairs of electrons forming the double bond are in a σ orbital and a π orbital. Hence similar electronic transitions occur. Formaldehyde absorbs strongly at 185 mμ, corresponding to the absorption at 165 mμ by ethylene because of a $\pi \longrightarrow \pi^*$ transition. Formaldehyde, other aldehydes, and ketones absorb weakly also at about 280 mμ

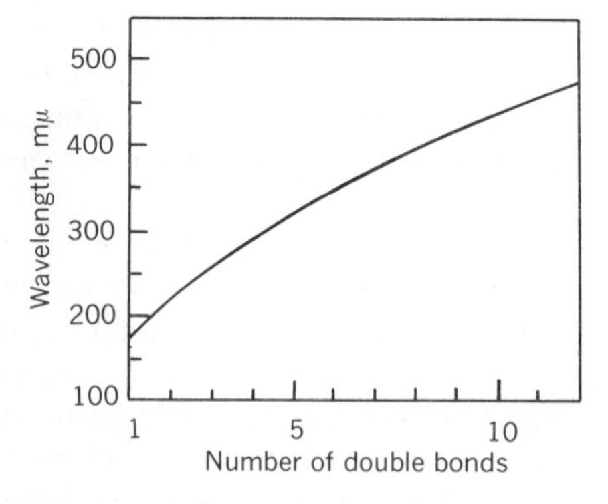

Figure 31–7. Longest wavelength absorption by conjugated polyenes.

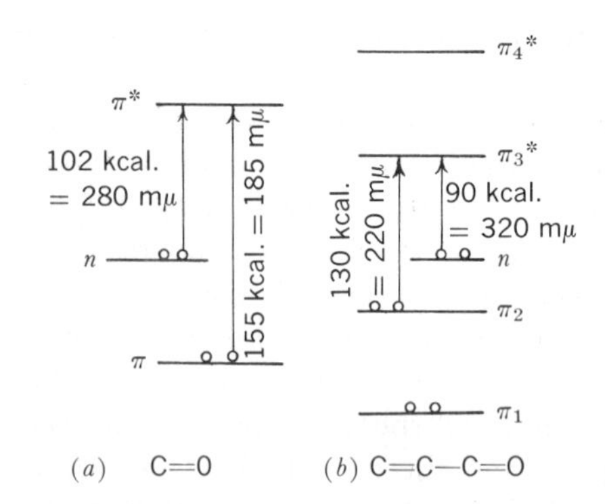

Figure 31–8. Energy levels for (a) an isolated carbonyl group and (b) a conjugated carbonyl group.

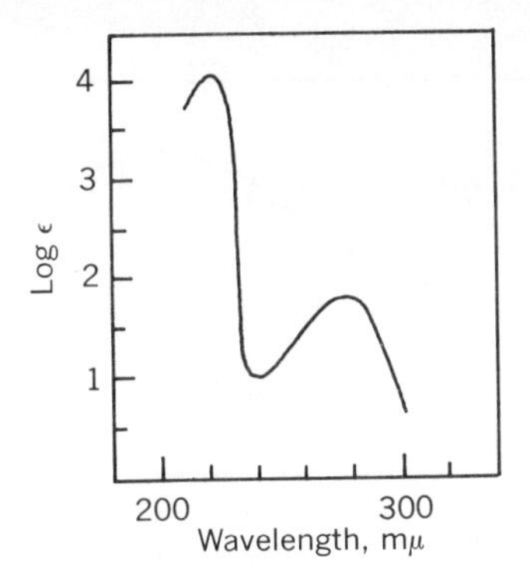

Figure 31–*9*. Absorption spectrum of mesityl oxide, $(CH_3)_2C{=}CHCOCH_3$, a conjugated unsaturated ketone.

Figure 31–*10*. Absorption spectrum of diacetyl, $CH_3COCOCH_3$, a conjugated diketone.

($\epsilon = 10^2$), whereas ethylene and its homologs do not. This band appears to be the result of a "forbidden" $n \longrightarrow \pi^*$ transition (Fig. 31–*8a*). Similarly azomethane, $CH_3N{=}NCH_3$, gives a high-intensity absorption at $\lambda_{max.} = 245$ mμ and a low-intensity band at $\lambda_{max.} = 345$ mμ. The latter band is sufficiently broad to extend well into the visible, and azomethane is yellow. Neither the carboxyl group nor its derivatives show appreciable absorption in the near ultraviolet.

A carbon-carbon double bond conjugated with a carbonyl group gives rise to a high-intensity absorption ($\epsilon \sim 10^4$) at about the same place ($\lambda_{max.} = 220$ mμ) as two conjugated carbon-carbon bonds because of a $\pi_2 \longrightarrow \pi_3^*$ transition (Figs. 31–*8b* and 31–*9*). Again the position of the maximum is shifted to the red with increasing substitution at the carbon-carbon double bond. The lower intensity absorption resulting from the $n \longrightarrow \pi^*$ transition of the carbonyl group moves to about 320 mμ.

If two carbonyl groups are in conjugation as in diacetyl, $CH_3COCOCH_3$, two absorption bands occur below 200 mμ, one at 280 mμ and the other in the visible extending from about 400 to 460 mμ (Fig. 31–*10*), and the compound is yellow. Similarly, conjugation of N=N with phenyl groups in azobenzene (p. 415) shifts the long wavelength band of yellow azomethane further towards the red ($\lambda_{max.} = 448$ mμ), and the compound is orange. The short wavelength band at 313 mμ corresponds to that for colorless *trans*-stilbene ($\lambda_{max.} = 300$ mμ).

Benzene with three conjugated double bonds might be expected to absorb strongly in the near ultraviolet. Instead it exhibits strong absorption at 186 mμ, and only weak absorption occurs at 200 and 260 mμ (Fig. 31–*11*). This band system requires a more detailed treatment than that used to discuss the simpler systems, and its interpretation is not attempted here.

Substituents on the benzene ring perturb the electronic system by both a resonance and an inductive effect. The latter shifts all levels to the same extent in the same direction and has little effect on the position of the absorption. Thus the anilinium ion absorbs at almost the same wavelength and only slightly more strongly than benzene. Substituents having an unshared pair of electrons have a marked effect because of resonance with the benzene ring. The mixing of the *p* orbital with the π orbitals of benzene decreases the energy difference between the highest energy level of the ground state and the lowest excited state. Moreover the intensity is greatly increased because of the increase in the transition dipole moment (Fig. 31–*11*). Thus for aniline the 260 mμ band of benzene is shifted to 300 mμ and the intensity is ten times that for benzene. In *N,N*,2,6-tetramethylaniline, resonance is inhibited because the steric effect of the two *ortho* methyl groups prevents the *p* orbital of

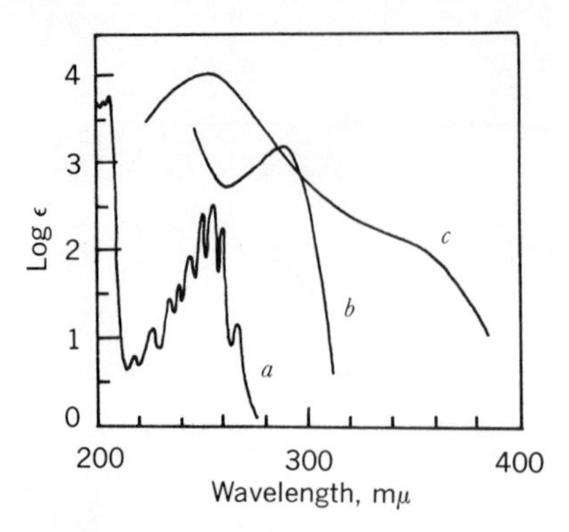

Figure 31–*11*. Absorption spectra of (*a*) benzene, (*b*) aniline, and (*c*) nitrobenzene.

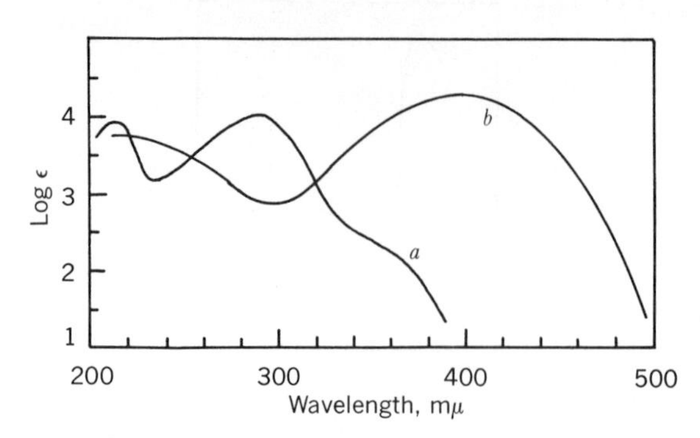

Figure 31–*12*. Absorption spectra of *p*-nitrophenol in (*a*) hexane, (*b*) 0.1*N* sodium hydroxide solution.

nitrogen from being perpendicular to the plane of the ring (p. 424). Accordingly both the wavelength and the intensity of the absorption are decreased.

If an *ortho,para*-directing group and a *meta*-directing group are *para* to each other, the shift is greater than that of the sum of the shifts for the monosubstituted benzenes because of the resonance over the whole system. Thus the high-intensity band at 250 mμ for nitrobenzene (Fig. 31–*11*) is shifted to 290 mμ for *p*-nitrophenol (Fig. 31–*12*). In alkaline solution the absorption band of the latter is shifted still further to the visible to give a yellow salt. The removal of the proton leaves a negative charge on oxygen and makes it easier for the unshared electrons to interact with the nucleus and the nitro group, thus permitting longer wavelengths to bring about the transition. The intensity of the absorption remains unchanged. Portions of a molecule that are fundamentally responsible for the electronic transition are known as **chromophores.** Groups such as the amino group or the hydroxyl group that modify the chromophore, usually by shifting the absorption to longer wavelength and intensifying it, are called **auxochromes.**

High-intensity absorption in the visible is characteristic of organic dyes. They all have (*1*) a large number of conjugated double bonds, giving rise to a large number of π orbitals differing little in energy, and (*2*) two or more polar groups a considerable distance apart, which causes a large transition dipole.

Complexity of Spectra

The preceding abbreviated discussion may give the impression that it is an easy matter to identify the absorption bands of organic compounds with specific structural features of the molecule. Such is the case only for relatively simple chromophoric systems under certain conditions. Almost any external influence modifies the absorption spectrum of a molecule. Solvent molecules may produce electric fields about the molecule, which change the energy differences between the ground state and excited states, they may form solvates, or they may act as an acid or a base and alter the proportions in which a compound exists as a base and its conjugated acid. The solute itself may dissociate, polymerize, or isomerize in solution. All of these factors alter and increase the complexity of molecular spectra. Moreover many different electronic systems can give rise to absorption in the same region of the spectrum.

CHAPTER THIRTY-TWO

COLOR. DYES AND DYEING

Color always has played an important role in the life of man, even though its significance is almost wholly esthetic. Throughout man's history dyes and pigments, both natural and synthetic, have been an important article of commerce.

COLOR

Color Sensation

The human eye is sensitive only to electromagnetic vibrations having a wavelength between 400 and 800 mμ. This region of the spectrum is known as the *visible* (cf. Table 32–*1*). The mixture of all wavelengths in the visible having the relative intensities produced by a body at a white heat is known as *white light.* If the light striking the retina of the eye does not contain all of the wavelengths of the visible spectrum, or if their intensities differ considerably, the sensation of color results.

Light may be colored because only a limited region of the spectrum is emitted by a light source, as for example the yellow light of the sodium flame. Or light may be colored because of the separation or removal of certain wavelengths of the visible light. Because light of different wavelengths is refracted (velocity reduced) to different degrees on passing through a transparent medium, it is possible to separate the different wavelengths of white light by the use of a prism and produce a colored spectrum. Another way in which portions of the spectrum may be removed is by interference. When light is reflected from the two surfaces of a thin film, the thickness of the film may be such that a light wave reflected from the far surface travels a sufficiently longer path to be thrown out of phase with a light wave reflected from the near surface. Cancellation of this wavelength results, and if white light is incident, the reflected light is colored. Colored bird feathers and colors of soap bub-

TABLE 32–*1*. RELATION BETWEEN ABSORPTION AND VISUAL COLOR

WAVELENGTHS ABSORBED (mμ)	COLOR ABSORBED	VISUAL COLOR
400–435	violet	yellow-green
435–480	blue	yellow
480–490	green-blue	orange
490–500	blue-green	red
500–560	green	purple
560–580	yellow-green	violet
580–595	yellow	blue
595–605	orange	green-blue
605–750	red	blue-green

bles are examples of this phenomenon. Finally certain wavelengths of white light may be removed by absorption, which is by far the most common cause of color. The color may be observed as the light transmitted through a solution of the substance in a transparent medium, or as the light reflected from the surface of a substance.

The visual color is complementary to the color absorbed; that is, it is the color sensation produced by all of the wavelengths minus the wavelengths absorbed. Table 32–*1* gives the observed colors when relatively narrow bands of the visible spectrum are absorbed. If the absorption bands are broader or if more than one absorption band is present, the visible color is altered.

Color and Chemical Structure

Until recently chemists contented themselves with attempts to correlate visual color with structural features of the molecule. As early as 1868 Graebe and Liebermann discussed the importance of unsaturation in producing color and noted that reduction of a colored compound always led to a colorless product.

In 1876 Witt pointed out that two types of groups usually are present in highly colored compounds, unsaturated groups, which he called **chromophores** (Gr. *chroma* color, *phoros* from *pherein* to bear), and groups that intensified the color, which he called **auxochromes** (Gr. *auxein* to increase). To the chromophores he assigned the groups NO_2, C=O, and N=N. Later C=C, C=N, C=S, and N=O were added to the list. In 1888 Armstrong[1] added the very important chromophoric quinonoid structure, $=\!\langle\,\rangle\!=$, pointing out that the structures of most highly colored substances containing an aromatic ring could be written in such a way that they contained a quinonoid structure. The important auxochromes were the hydroxyl group, the amino group, and the alkylated amino group. Witt associated the effect of the auxochromes with their salt-forming properties because acetylation of the amino group or methylation of the hydroxyl group destroys the effect, but methylation of the amino group does not. Moreover the salts of the phenolic compounds are more strongly colored than the free phenols. Numerous other empirical observations have been made. For example, auxochromes do not affect the color when in the *meta* position to the chromophore.

At the time that Witt discussed the effect of salt-forming groups on color, he mentioned also the necessity for their presence if the colored compound was to act as a dye, that is, to have the ability to fix itself to a fiber. This dual property of auxochromes has led to some confusion, since both properties are not always exhibited simultaneously.

The inconsistencies and inadequacy of the chromophore-auxochrome theory long have been recognized, and in recent years it has been reinterpreted and extended in terms of current electronic theory as discussed under electronic spectra in Chapter 31 (pp. 552–556). There it is shown that the entire conjugated system is responsible for color, and that either nitro groups or amino groups shift the absorption to longer wavelengths (p. 556). The more conjugation is extended, the greater is the number of molecular orbitals and the closer their energy levels are spaced. Hence less energy is required for electronic transitions, and absorption is shifted to longer wavelengths (p. 554). The interaction of auxochromes with the conjugated system not only extends the conjugation but leads to large dipole moments and large transition dipole moments with resulting high-intensity absorption (p. 553). Acylation of an amino or a hydroxyl group merely decreases the availability of an unshared

[1] Henry Edward Armstrong (1848–1937), professor of chemistry at Central Technical College, South Kensington, England. He is remembered not only for his contributions to the theory of color and the structure of benzene, but also for his sharp criticisms of contemporary thinking in the field of chemistry.

pair of electrons for interaction with the conjugated system. The nature of fixation of a dye to the fiber, and especially to mordants, also now is better understood, and the function of hydroxyl and amino groups in this respect has been separated from their effect on color.

DYES AND DYEING

History

Natural coloring matters have been used by man since the beginnings of civilization. The first synthetic dye was picric acid, made by Woulfe in 1771 by the action of nitric acid on natural indigo. Not until 1855 was a technical method introduced to prepare it from coal tar. The first dye prepared from coal tar was *aurin* or *rosolic acid* reported by Runge[2] in 1834. He noted that with the usual mordants it produced red colors and lakes that rivaled those produced from the natural dyes, cochineal and madder. Since little was known about the components of coal tar at that time and since Kekulé's theory of the structure of benzene was not proposed until 1865, Runge's observations were not extended.

As early as 1843, Hofmann had observed that aniline, as prepared at that time, gave red colors under certain conditions. In 1856 Perkin[3] oxidized aniline sulfate with potassium dichromate and obtained a purple dye called *mauve* (p. 576), which became the first synthetic coal tar dye to be manufactured commercially. In 1859 a process was patented in France for the oxidation of aniline with stannic chloride to give a dye having a color resembling that of fuchsia flowers and named *fuchsin*. After Hofmann showed that fuchsin is a derivative of triphenylmethane, this class of dyes was investigated extensively and came into widespread use.

Meanwhile the azo dyes were discovered by Griess in 1862, the theory of aromatic structure became established, alizarin was synthesized by Graebe and Liebermann in 1868, and indigo was synthesized by Baeyer in 1879. Direct dyes for cotton were introduced in 1880, sulfur colors from coal tar derivatives were produced in 1893, anthraquinone vat dyes in 1901, azoic dyes in 1911, acetoacetanilides in 1923, phthalocyanins in 1934, fluorescent brightening agents in 1940, and fiber-reactive dyes in 1956.

The synthetic coal tar dye industry developed first in England under Perkin, Nicholson, and others, but gradually passed into German hands. By 1913, just before World War I, Germany made three fourths of the world production of dyes and 90 per cent of the dyes used in England and the United States. France and Switzerland also had flourishing dye industries, but they were connected with the German cartel. Among the reasons for German predominance were (*1*) favor-

---

[2] Friedlieb Ferdinand Runge (1794–1867), professor of chemistry at the University of Breslau and later technical director of the Chemical Products factory at Oranienburg. He was the first to isolate aniline, quinoline, pyrrole, and phenol from coal tar in 1834. He discovered Aniline Black (p. 576) and pioneered in the use of absorbent paper for spot tests.

[3] William Henry Perkin (1838–1907) entered the Royal College of Chemistry in London in 1853 to study chemistry under Hofmann. He discovered anthraquinone while trying to nitrate anthracene, but did no further work with this compound until 1869, when he devised a commercial synthesis of alizarin from it (p. 572).

In his report on the work of the Royal College of Chemistry in 1849, Hofmann had remarked that the synthesis of quinine would be very desirable. Seven years later Perkin, then 18 years old and research assistant to Hofmann, was impressed by this statement. At that time the chief lead to the structure of a compound was the difference between its molecular formula and that of a known compound. Perkin thought that quinine might result by the oxidation of allyltoluidine.

$$\underset{\text{Allyltoluidine}}{2\,C_{10}H_{13}N} + 3\,[O] \longrightarrow \underset{\text{Quinine}}{C_{20}H_{24}N_2O_2} + H_2O$$

During the Easter vacation of 1856, he made allyltoluidine in his home laboratory and oxidized the sulfate with potassium dichromate. He obtained only a dirty reddish-brown precipitate, but this behavior interested him, and he decided to try the reaction on a simpler base. On treating an impure aniline sulfate with dichromate, he obtained a black precipitate from which he extracted a purple compound that had the property of a dye and was fast to light. After submitting samples to dyers and receiving favorable comments, he resigned his post at the Royal College and with the aid of his father and brother began manufacture in 1857. The dye was known as *aniline purple* or *mauve*. Material dyed with it became so popular that it gave its name to the *mauve decade*. It is of interest that the total synthesis of quinine was not accomplished until 1944.

Perkin became a successful industrialist but never gave up scientific research. In 1867 he published his first paper on what now is known as the Perkin reaction (p. 463), and in 1868 announced the synthesis of coumarin (p. 478), the first natural perfume to be synthesized from a coal tar component. In 1874 at the age of 36, Perkin retired from business to devote full time to scientific research, especially to the study of the effect of the magnetic field on optical rotation.

able American patent laws that permit a foreign country to patent processes and products without requiring manufacture in the United States, (*2*) vigorous research programs by German manufacturers and close collaboration with university laboratories, (*3*) low tariff restrictions, and (*4*) cartel control of prices with price-cutting and dumping to prevent competition. Along with control of the dye industry went control of the manufacture of synthetic medicinals and all other organic chemicals.

During World War I Germany lost control of the organic chemical market because the United States and England had to make their own dyes and pharmaceuticals, and these countries realized that the dye plants with their facilities for nitration and for the manufacture of chlorine and phosgene were potential munitions factories. The Chemical Foundation took over German patents in this country and licensed them to firms in the United States, and after the war adequate tariff laws were passed to ensure the continuance and growth of the organic chemical industry. Since then the general industrial development and aggressive research programs have permitted the organic chemical industry to flourish. Total sales of chemical products in 1963 amounted to 35 billion dollars. No single firm or cartel controls the industry. The largest company accounted for only 7.5 per cent of the total and the five largest companies accounted for less than 25 per cent.

Types of Material Dyed

A **dye** may be defined as an organic compound used to impart color to other substances. The property of a dye to retain its color on prolonged exposure to light or to resist removal under conditions to which the substance normally is exposed, such as washing, is referred to as the *fastness* of the dye. **Dyeing** is the process by which the dye is applied to the substance being dyed. For textile fibers this process usually involves immersion of the yarn or cloth in an aqueous solution or dispersion of the dye. It is believed that, in general, the dye initially is adsorbed on the surface of the fiber and that adsorption is followed by solution and diffusion of the dye into the swollen fiber. The mechanisms of the initial adsorption and of the diffusion process differ with the nature of the material being dyed, that is, whether it is protein, cellulose, or some synthetic substance. Dyes used for coloring liquids or plastics ordinarily are soluble in the medium.

Leather and **animal fibers** such as wool and silk contain potentially free amino and carboxyl groups and are dyed readily by dyes that have potential acidic or basic groups, which aid the initial adsorption. The **nylons** contain amide groups and terminal amino groups (p. 631) and can be dyed by many acid dyes at a temperature near the boiling point of the dye bath. The **acrylic fibers** (*Orlon, Acrilan*) have hydrogen α to nitrile groups (p. 619), and hence they are very weakly acid and are dyed by most basic dyes. Monomers lacking acidic or basic groups may be copolymerized with a small amount of a monomer having such groups to give products dyeable with basic or acid dyes.

Cotton and viscose rayon and other **cellulosic materials** such as linen and other bast fibers contain only hydroxyl groups that are relatively unreactive and can form only hydrogen bonds with most dyes. Although hydrogen bonding assists in the adsorption of the dye by the fiber, it is believed that London forces (p. 22) also are important and that long flat molecules that can approach the cellulose molecules closely are required. **Cellulose acetate** lacks even free hydroxyl groups, but it will take up from an aqueous dispersion water-insoluble dyes that are soluble in cellulose acetate. The polyester fibers (*Dacron, Terylene, Kodel,* p. 476) and the nylons and acrylics can be dyed in the same way by dyes that are soluble in the polymer. The dye also may be dissolved in a solution or melt of a substance such as cellulose acetate (*Chromspun*) or a vinylidene chloride polymer (p. 593) before it is spun into fibers. **Paper** may be colored by almost any water-soluble dye because the dye need not be adsorbed strongly to the fiber. Since esters and amides are hydrolyzed by aqueous strong acids and alkalies, the azoic, vat, and sulfur dyes, which require such conditions, usually are applied only to cellulose fibers. **Glass fibers** and **aluminum** can be dyed after the surface has been treated to provide an adherent oxide layer that can adsorb the dye (cf. p. 561).

Because of their commercial importance, a very large number of dyes have been synthesized, and many have been placed on the market. The latest *Colour Index* (1956), sponsored jointly by the Society of Dyers and Colourists (Great Britain) and the American Association of Textile Chemists and Colorists, lists about 3800 different dyes and pigments. They have been assigned names according to **method of application,** and given a *Colour Index* number according to their **structures.** Each manufacturer, however, usually labels his products with registered trademarks, and hence the same dye may appear on the market under several names. Frequently the original or classical name is in common use. Since all names are cross-indexed in the *Colour Index,* it is possible to find the structure of any named dye. Dyes considered in this text bear the *Colour Index* name and frequently in addition the classical name.

Methods of Application

Direct or Substantive Dyes. The direct or substantive dyes are those that are applied to the fiber or cloth from a hot solution of the dye in water. **Acid dyes** are sodium salts of sulfonic acids and are applied from a bath acidified with sulfuric or acetic acid. **Basic dyes** are the hydrochlorides or zinc chloride complexes of dyes that have basic groups, but they frequently contain also sodium sulfonate groups to make them more soluble in water. They are applied from a neutral bath, usually on a fiber that has been treated with tannic acid. The so-called **direct dyes** are substantive to cotton and other cellulose fibers. These dyes have a molecular weight sufficiently high to give colloidal solutions that are adsorbed strongly. They are called *salt colors* also because adsorption on the fiber usually is assisted by the addition of a salt such as sodium sulfate.

Mordant or Adjective Dyes, and Chrome Dyes. A mordant is any substance that can be fixed to the fiber (L. *mordere* to bite) and that later can be dyed. Thus albumin was used as a mordant for printing cotton cloth to produce calico. The protein was coagulated on the cotton fiber by heat, and then dyed with an acid dye. Tannic acid was used as a mordant for basic dyes. The terms mordant dye and chrome dye, however, are reserved for those dyes that can form coordination chelate complexes with metal ions. Coordination increases the complexity of the dye and its adsorbability by the fiber.

Mordanting with metal ions probably had its origin in a prehistoric observation that certain clays had the property of adsorbing strongly certain natural coloring matters. Although aluminum hydroxide used in the dyeing of madder was one of the earliest mordants (p. 572), it now is used only to produce certain organic pigments. With few exceptions the currently important metal ions are those of copper, cobalt, and especially chromium. Chroming with unmetallized dyes is carried out in one of three ways. In *prechroming,* wool is treated with an excess of sodium dichromate and then the dye is added. In *metachroming* the dichromate is added after the fiber has been added to the dye bath but before it is fully dyed. In *afterchroming* the dichromate is added only after the dye bath has become fully exhausted. These operations are carried out by the dyer. In 1919, 1:1 dye:metal complexes became available that could be used as direct dyes, and in 1940, 2:1 dye:metal chromium and cobalt complexes were introduced that are useful for dyeing wool and nylon (p. 631). These premetallized dyes are complexes with *o,o'*-dihydroxy azo dyes (p. 564). They are easier to apply than the older mordant dyes and their importance is increasing.

Azoic Dyes. Azoic dyes are water-insoluble azo dyes that are formed on the fiber. They usually are used on cotton. In the application of **ice colors,** the cloth is impregnated with a compound capable of coupling with a diazonium salt and then the cloth is immersed in an ice-cold solution of a diazotized amine. Thus a water-insoluble azo dye is formed within the fibers. In **developed dyeing** the cloth is dyed with a direct dye containing a free amino group. The dye then is diazotized on the fiber and developed by coupling with an amine or phenol. The new dye on the fiber is faster to washing because of its higher molecular weight. It usually has a deeper color.

Sulfur Dyes. This class includes those sulfur-containing dyes that are applied to cotton from their solution in aqueous sodium sulfide. The soluble reduced form of the dye is substantive to cotton. After dyeing, the cloth is exposed to air or chemical oxidizing agents, which regenerate the insoluble dye on the fibers.

Vat Dyes. Vat dyes are water-insoluble but can be rendered water-soluble by reduction in alkaline solution. Cotton adsorbs the reduced form of the dye from this solution. Reconversion of the adsorbed dye to the oxidized form is accomplished as in sulfur dyeing. The reduction formerly was carried out by fermentation in large vats and gave rise to the name *vat dye.*

Disperse Dyes. Dyeing of cellulose acetate, polyester, or other synthetic fibers can be brought about by colloidal aqueous dispersions of azo or anthraquinone dyes that are of low molecular weight, lack sulfonic acid groups, and are soluble in the organic polymer. The colors must be very finely divided to increase the rate of solution in water because the dyeing mechanism appears to involve the removal by the fiber of the small amount of dye that is dissolved in the water. Disperse dyes are not very fast to washing. They are faster the more drastic the conditions required for their application, since these conditions are not duplicated in the washing process. A peculiar weakness is that they are subject to fading even in the dark. The fading is due to oxides of nitrogen and of ozone in the atmosphere and is called *gas fading.*

Fiber-Reactive Dyes. Simple acid and basic dyes are substantive to wool and silk and yield brilliantly colored fibers. The large complex molecules necessary for the direct dyeing of cotton or those formed in the chroming of dyes for wool tend to be dull in color. Hence much research has been devoted to devising a simple dye that can be attached to the reactive groups of the fiber by a chemical bond. Two such types were placed on the market in 1956 and 1957. One class consists essentially of colored acid chlorides that react with the hydroxyl groups of cellulose or the amino groups of protein to form ester or amide linkages (p. 565). Members of the other class are vinyl sulfones to which hydroxyl or amino groups add to give ether or secondary amine linkages (p. 565). Such dyes have been termed *reactive* or *fiber-reactive* dyes.

Oil- and Spirit-Soluble Dyes. Many colored compounds lacking sulfonic acid groups are soluble in organic solvents and are used to dye gasoline, plastics, fats, oils and waxes, spirit printing inks, and stains. They may be added also to solutions or melts of synthetics before the spinning operation.

Fluorescent Brightening or Optical Whitening Agents. Colorless compounds that have an affinity for fibers and have a blue fluorescence when exposed to near ultraviolet light are used extensively as whitening agents. They are known as *direct whites, optical bleaches,* or *Blankophors.*

Food, Drug, and Cosmetic Colors. Certain dyes are toxic and some even promote the growth of tumors. Hence most nations permit the coloring of foods, drugs, and cosmetics only with those dyes considered to be harmless. Dyes listed in this chapter as FDC dyes are those permitted for such use in the United States by the U.S. Food and Drug Administration.

Production and Value. The total production of synthetic dyes in 1964 was 184,387,000 pounds valued at $273,000,000. On a quantity basis, the per cent distribution was as follows: vat, 28; direct for cotton, 17; acid, 10; sulfur, 10; fluorescent brightening, 9; disperse, 7; azoic, 5; basic, 5; solvent, 4.5; mordant, 2; FDC, 5; miscellaneous, 1.

CHEMICAL CLASSES OF DYES

Azo Dyes

This class constitutes the largest single group of dyes, making up over half of the total number of synthetic colors of known structure. It accounted for about 32 per cent of the total production by weight and 37 per cent by value in 1964. Almost 58 million pounds were produced at an average unit price of $1.68 per pound. Most azo dyes are sodium salts of sulfonic acids. Those of high molecular weight containing two, three, and four azo groups (*disazo, trisazo,* and *tetrakisazo dyes*) are substantive for cotton.

Azo dyes are prepared by coupling a diazotized aromatic amine, the *primary component,* with a phenol or an aromatic amine, the *secondary component* (p. 431). In the benzene series coupling takes place *para* to the hydroxyl or amino group, or *ortho* if the *para* position is occupied. If all *ortho* and *para* positions are occupied, no coupling takes place. Occasionally a group such as the carboxyl group in the *para* position may be displaced. Phenols when coupled in strongly alkaline solution undergo some *ortho* substitution. With diamines and dihydroxy compounds only the *meta* isomers couple.

In the naphthalene series, α-naphthol and α-naphthylamine couple in the 4 position. If the 4 position is occupied or if a sulfonate group is in the 3 or 5 position, coupling takes place in the 2 position. β-Naphthol and β-naphthylamine couple only in the 1 position. If both an amino group and a hydroxyl group are present, the amino group directs in weakly acid solution, and the hydroxyl group directs in alkaline solution (cf. p. 384).

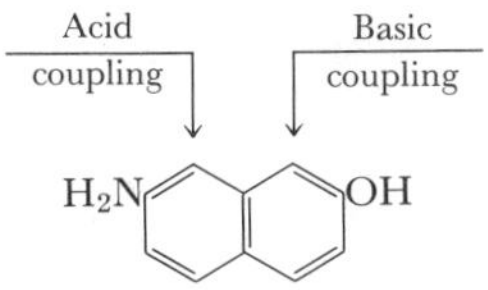

Acid coupling   Basic coupling

H_2N … OH

Monoazo Dyes. **(*a*)** BASIC. Only a few monoazo basic dyes still are in use. **Basic Orange 2** (*Chrysoidine Y,* 581[4]), the first azo dye made commercially in 1875, still is used for dyeing leather and paper. It is made by coupling diazotized aniline with *m*-phenylenediamine.

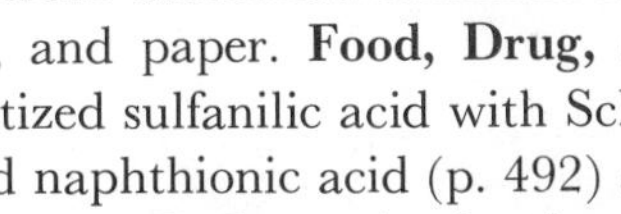

Basic Orange 2 (*Chrysoidine Y*)

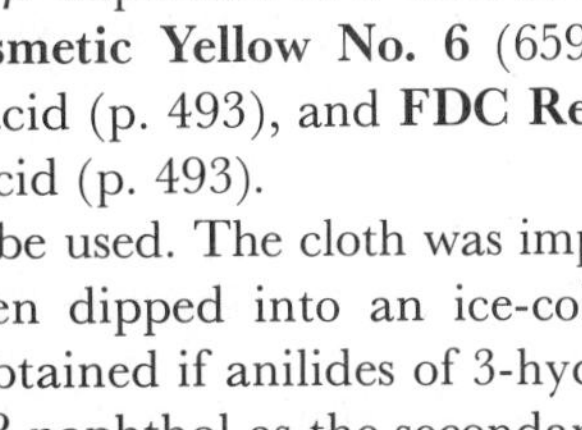

Acid Orange 7 (*Orange II*)

(*b*) ACID. The acid monoazo dye produced in largest amount is **Acid Orange 7** (*Orange II,* 763) made from diazotized sulfanilic acid and β-naphthol. It is used to dye wool, silk, nylon, leather, and paper. **Food, Drug, and Cosmetic Yellow No. 6** (659) is made by coupling diazotized sulfanilic acid with Schaeffer acid (p. 493), and **FDC Red No. 2** (786) from diazotized naphthionic acid (p. 492) and R acid (p. 493).

(*c*) AZOIC. **Para Red** was the first ice color to be used. The cloth was impregnated first with an alkaline solution of β-naphthol and then dipped into an ice-cold solution of diazotized *p*-nitroaniline. Much faster colors are obtained if anilides of 3-hydroxy-2-naphthoic acid (BON acid, p. 495) are used instead of β-naphthol as the secondary component. These compounds are known as the **Naphtol AS** series (from the German *Naphtol Anilid Saeure*). The free amines are sold as *Fast Bases* and the stabilized diazotized amines as *Fast*

[4] U.S. production in 1964 in thousands of pounds, when known, is given in parentheses following the first occurrence of the name of the dye.

Salts. About 30 Naphtols and 50 bases or salts are available commercially, making 1500 combinations possible. Not all combinations are used, however, because many of the colors overlap and not all of the colors produced have the best possible properties. The primary component produced in largest amount is **Azoic Diazo Component 13** (*2-methoxy-5-nitroaniline,* 419). The two most widely used secondary components are **Azoic Coupling Component 18** (*BON acid o-toluidide,* 744) and **Coupling Component 7** (*BON acid β-naphthylamide,* 1041). **Coupling Component 2** (*BON acid anilide,* 227) is the original Naphtol AS. Azoic dyes are used not only for solid dyeing of cloth but also for print-dyeing. Most of the colors do not have specific names since they are sold in the form of their components rather than as the coupled product. **Fast Scarlet R** is the color produced from Diazo Component 13 and Coupling Component 2.

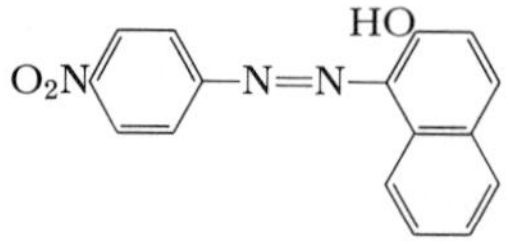

Para Red

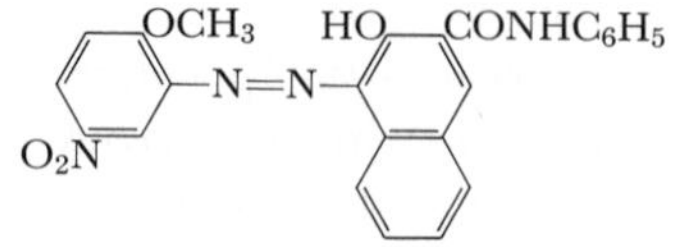

Fast Scarlet R

Since the azoic colors are insoluble in water, many are used as pigments for paints and printing inks.

(*d*) DISPERSE and OIL-SOLUBLE DYES. **Disperse Yellow 3** (933) is an example of a widely-used monoazo disperse dye for cellulose acetate. **Solvent Yellow 14** (*Sudan I,* 747) is representative of oil-soluble colors.

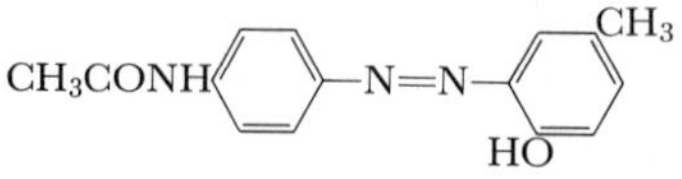

Disperse Yellow 3

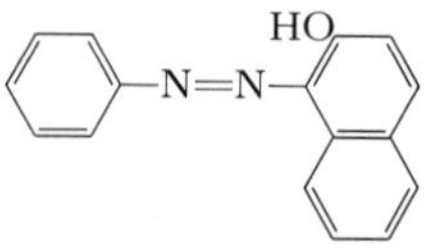

Solvent Yellow 14 (*Sudan I*)

(*e*) MORDANT. **Mordant Black 11** (*Eriochrome Black T,* 2054) is one of five dyes, all blacks, each of which is produced in larger amount than any dye other than one of the five. Its formation is unusual in that it represents one of the two known cases in which α-naphthol couples in the 2 position. Like other mordant azo dyes, it has two hydroxyl groups in the positions *o,o′* to the azo group. Only one nitrogen of the azo group is involved in chelation. The fiber may be treated with chroming solution before, after, or during the dyeing (p. 561). The chroming deepens the color and produces on the fiber a molecule of much higher molecular weight and greater fastness.

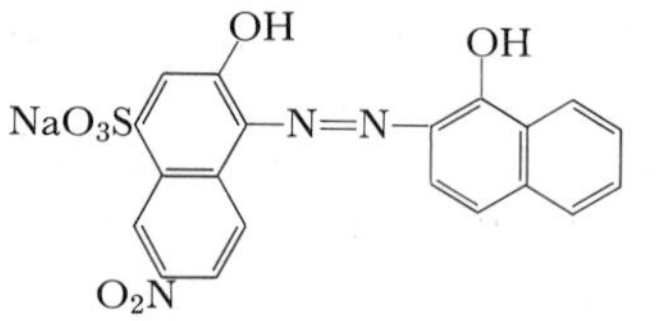

Mordant Black 11 (*Eriochrome Black T*)

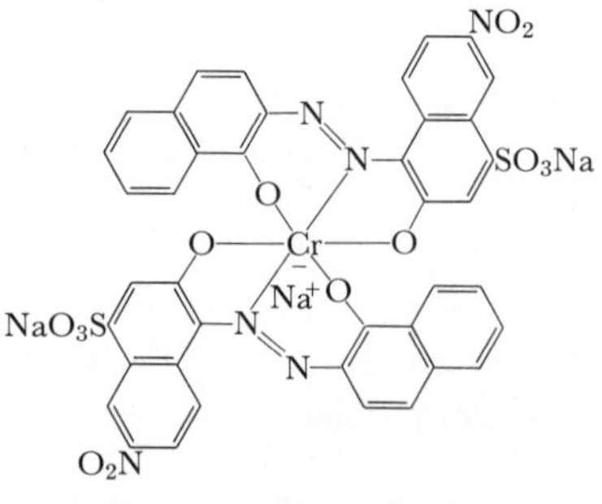

Chromed Mordant Black 11

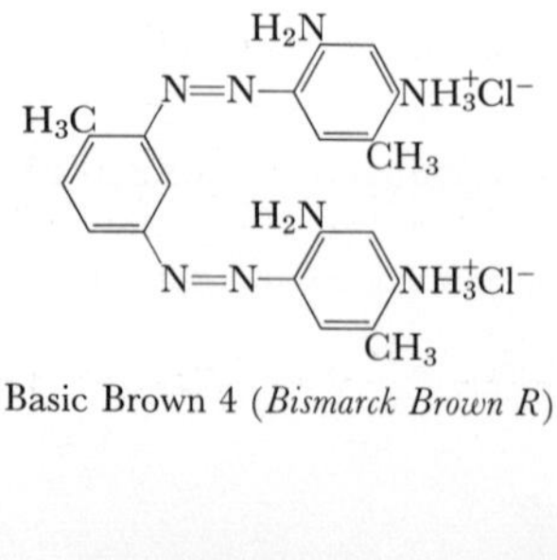

Basic Brown 4 (*Bismarck Brown R*)

Mordant Black 11, under the name *Eriochrome Black T,* is used as an indicator in chelatometric titrations. The chelate complexes that it forms with metal ions in dilute aqueous solutions are red whereas solutions of the free dye are blue. If a solution of polyvalent metal ions and indicator is titrated with a standardized solution of EDTA (p. 612), which coordinates more strongly with the metal ions than does the indicator, a change in color takes place when sufficient EDTA has been added to complex with all of the metal ion.

(*f*) FIBER-REACTIVE. Thus far two types of fiber-reactive dyes are being marketed. The **s-triazine derivatives** are made by allowing a dye containing free amino groups to react with cyanuryl chloride (p. 288) to give the dichlorotriazine.

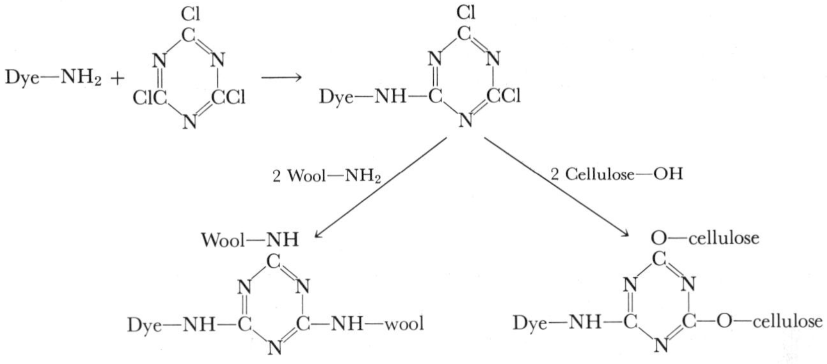

Dye—NH2 + (2,4,6-trichloro-s-triazine: ClC, CCl, C–Cl) ⟶ Dye—NH—C (triazine with Cl, CCl)

2 Wool—NH2 ⟶ Dye—NH—C (triazine with C—NH—wool, C—NH—wool)

2 Cellulose—OH ⟶ Dye—NH—C (triazine with C—O—cellulose, C—O—cellulose)

When wool or cotton is dyed with such derivatives and the dye bath then made slightly alkaline with sodium carbonate and warmed, the remaining halogens are displaced by the amino groups of wool or the hydroxyl groups of cellulose to give essentially an amide bond with the wool or an ester bond with the cellulose. Alternatively the cloth can be padded with a cold solution of the dye and sodium bicarbonate and then dried at a higher temperature to fix the dye. An example of an azo fiber-reactive dye is **Procion Red.**

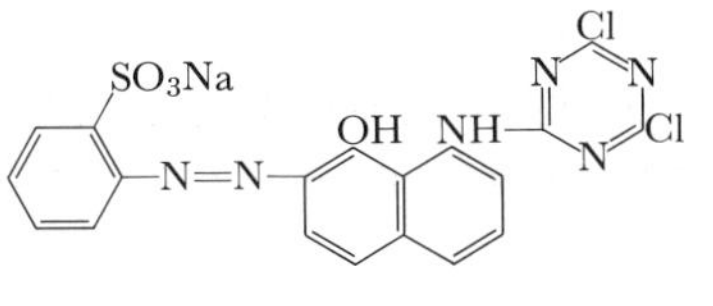

(Structure: SO_3Na-substituted phenyl —N=N— naphthalene with OH and NH—(chloro-triazine with Cl, Cl))

Procion Red

(Structure: naphthalene with OH and SO_3Na —N=N— phenyl with OCH_3 and $SO_2CH{=}CH_2$)

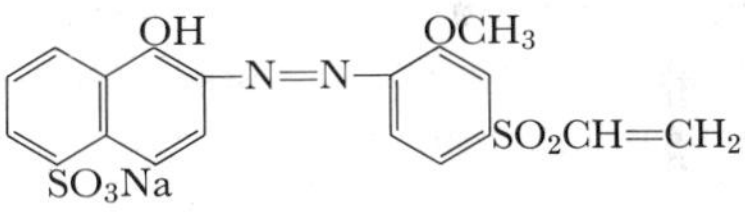

A vinyl sulfone (*VS*) dye

These simple dyes are more brilliant than the complex mordanted and polyazo dyes and have a fastness approaching that of the anthraquinone vats (p. 573). Other colored compounds that contain amino groups, such as the substituted anthraquinones, also can be converted to fiber-reactive dyes. Moreover, since the halogen atoms of cyanuryl chloride are displaced with decreasing ease, it is possible to prepare by stepwise reaction disubstituted triazines having two different dye groupings with one halogen remaining (*Cibacron series*). These dyes still are fiber-reactive but require more concentrated alkali or a higher temperature to fix the dye to the fiber.

A second type of fiber-reactive dyes, called **VS dyes,** contains the vinyl sulfone group, which can be introduced by way of the sulfonyl chloride (p. 409) into either the primary or the secondary component. An example is the red dye obtained by coupling diazotized 2-methoxy-5-vinylsulfonylaniline with 5-hydroxy-1-naphthalenesulfonic acid. The VS dyes react with wool or cotton by sulfonoethylation (p. 409) of the amino or hydroxyl groups.

$$ArSO_2CH_2CH_2{-}NH{-}wool \xleftarrow{Wool-NH_2} ArSO_2CH{=}CH_2 \xrightarrow{Cellulose-OH} Ar{-}SO_2CH_2CH_2{-}O{-}cellulose$$

In practice the sulfonylethyl sulfates (p. 409) are preferred because they confer water solubility on the dye without the necessity of introducing other water-solubilizing groups and are converted to the vinyl sulfones under the mildly alkaline conditions used to fix the dye.

$$ArSO_2CH_2CH_2OSO_3Na + NaOH \longrightarrow ArSO_2CH{=}CH_2 + Na_2SO_4 + H_2O$$

Like the triazine dyes, the fastness of the VS dyes lies between that of after-chromed or direct dyes and the anthraquinone vat dyes. It should be mentioned that VS disperse, mordant, and anthraquinone dyes also can be prepared.

Disazo Dyes. ***(a)*** BASIC. **Basic Brown 4** (*Bismarck Brown R,* 605, p. 564) is made by the action of nitrous acid on 2,4-diaminotoluene (*m-toluylenediamine*). The original Bismarck Brown from *m*-phenylenediamine was discovered by Martius in 1863.

(b) ACID. **Acid Black 1** (1381) is made by coupling *H*-acid (p. 494) first with diazotized *p*-nitroaniline in weakly acid solution and then with diazotized aniline in alkaline solution.

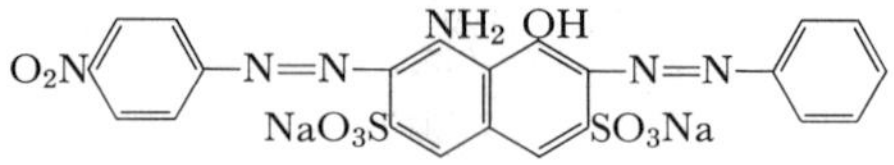

Acid Black 1

(c) DIRECT FOR COTTON. The discovery that the disazo acid dyes derived from benzidine are direct dyes for cotton gave enormous impetus to the synthetic dye industry. **Congo Red,** derived from benzidine and naphthionic acid, was the first dye of this class, but it is sensitive to acids. The direct cotton dye that established financially the Badische Anilin und Soda Fabrik in Germany was **Direct Red 2** (*Benzopurpurin 4B,* 381) made from *o*-tolidine (p. 486) and naphthionic acid (p. 492).

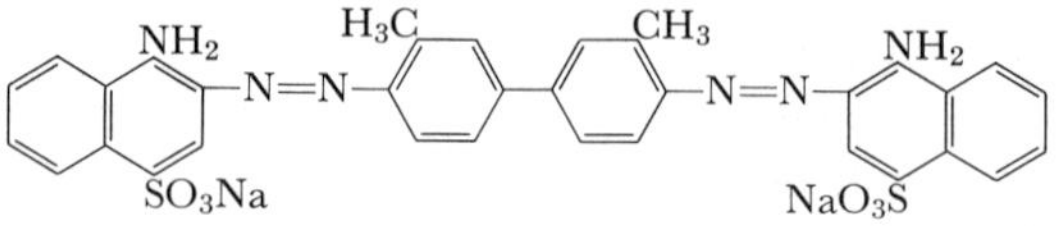

Direct Red 2 (*Benzopurpurin 4B*)

Many other important dyes are derived from benzidine. Thus **Direct Blue 2** (1828) is made by coupling diazotized benzidine first with one mole of γ-acid (p. 494) and then with *H*-acid, both in alkaline solution.

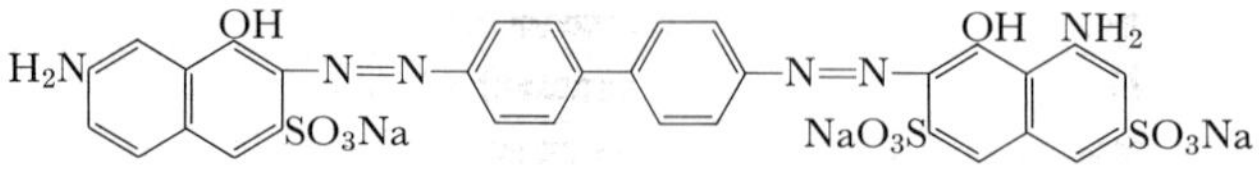

Direct Blue 2

(d) DEVELOPED. Direct Blue 2 is known also as **Developed Black BH.** Direct Blue 2 dyes cotton a bright blue. If, after dyeing, the remaining amino groups are diazotized on the cloth and then coupled with β-naphthol, a navy blue is produced; if coupled with *m*-phenylene diamine, the color is black.

Trisazo Dye. The dye produced in largest amount is **Direct Black 38** (*Direct Black EW,* 6338), discovered in the United States in 1901 by Oscar Mueller. It is made by coupling one mole of *H*-acid first with one mole of diazotized benzidine in acid solution, and then with one mole of diazotized aniline in alkaline solution. Finally the second diazonium group of the benzidine portion is coupled with *m*-phenylenediamine.

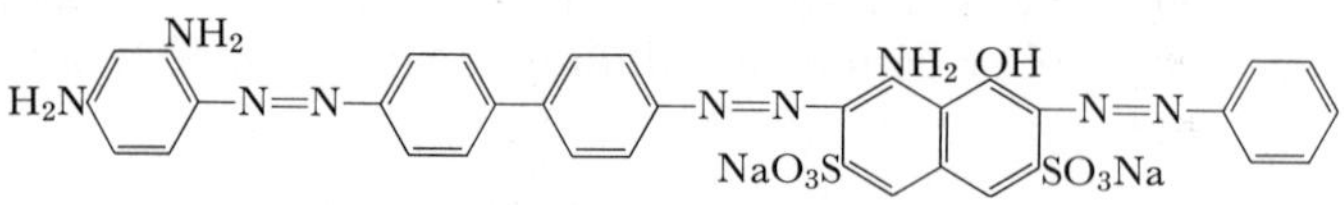

Direct Black 38 (*Direct Black EW*)

Stilbene Dyes. These yellow to orange direct dyes for cotton are azo or azoxy compounds that are not made by the usual coupling reaction but are derived from 4-nitro-2-toluenesulfonic acid (*para acid*) by boiling it with dilute aqueous sodium hydroxide. The initial product from para acid is the sodium salt of 4,4′-dinitroso-2,2′-stilbenedisulfonic acid.

$$2\ O_2N\text{-}C_6H_3(SO_3H)\text{-}CH_3 \xrightarrow{2\ NaOH} ON\text{-}C_6H_3(SO_3Na)\text{-}CH{=}CH\text{-}C_6H_3(SO_3Na)\text{-}NO + 4\ H_2O$$

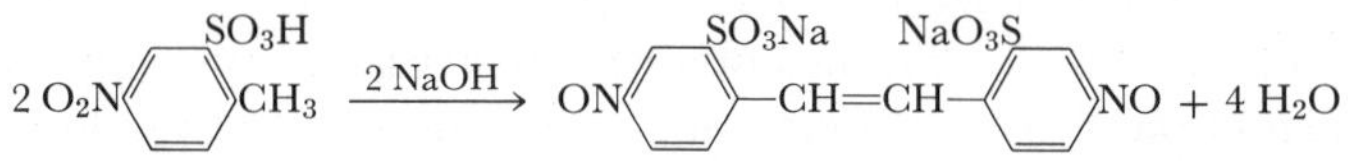

Para acid — Sodium 4,4′-Dinitroso-2,2′-stilbenedisulfonate

The constitutions of the dyes resulting from intermolecular oxidation and reduction are uncertain.

The most important stilbene dyes now are not azo dyes but are acyl derivatives of 4,4′-diamino-2,2′-stilbenedisulfonic acid, which is made by reduction of the dinitroso compound. They account for over 80 per cent of the fluorescent brightening agents (p. 562) and hold fifth place in both volume and value among synthetic dyes. Those most used at present, especially in detergents, are *s*-triazine derivatives prepared from cyanuryl chloride (p. 288).

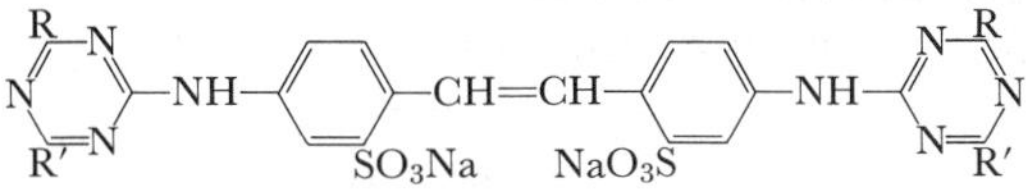

$$(R)(R')C_3N_3\text{-}NH\text{-}C_6H_3(SO_3Na)\text{-}CH{=}CH\text{-}C_6H_3(SO_3Na)\text{-}NH\text{-}C_3N_3(R)(R')$$

Products with R = C_6H_5NH or CH_3NH and R′ = OH or $NHCH_2CH_2OH$ are used for cotton. Products with R = Cl have affinity for wool and nylon.

Triphenylmethane Dyes

Triphenylmethane dyes are basic dyes for wool or silk, or for cotton mordanted with tannic acid. From the time of their discovery in 1859 to the development of the anthraquinone vat dyes, the triphenylmethane dyes were regarded highly because of their brilliant colors; that is, they not only absorb strongly some parts of the spectrum, but they reflect strongly other parts of the spectrum. They are not fast to light or washing, however, except when applied to acrylic fibers.

Malachite Green Series. Dyes of this group are derivatives of bis(*p*-aminophenyl)-phenylmethane. **Basic Green 4** (*Malachite Green,* 489), used for dyeing acrylic fibers and leather and for coloring bast fibers, paper, and lacquers, is made by condensing benzaldehyde with dimethylaniline to give bis(*p*-dimethylaminophenyl)phenylmethane, which is known as the *leuco base* (Gr. *leukos* white). Oxidation converts it to the carbinol, which also is colorless and is known as the *color base* or the *carbinol base.* Strong acids convert the color base into the colored dye.

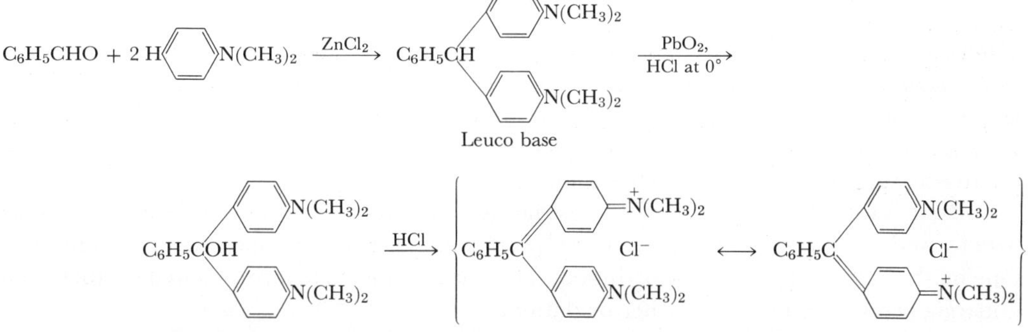

$$C_6H_5CHO + 2\ H\text{-}C_6H_4\text{-}N(CH_3)_2 \xrightarrow{ZnCl_2} C_6H_5CH[C_6H_4N(CH_3)_2]_2 \xrightarrow[HCl\ at\ 0^\circ]{PbO_2,}$$

Leuco base

$$C_6H_5C(OH)[C_6H_4N(CH_3)_2]_2 \xrightarrow{HCl} \left\{ C_6H_5C[{=}C_6H_4{=}\overset{+}{N}(CH_3)_2][C_6H_4N(CH_3)_2]\ Cl^- \longleftrightarrow C_6H_5C[C_6H_4N(CH_3)_2][{=}C_6H_4{=}\overset{+}{N}(CH_3)_2]\ Cl^- \right\}$$

Color (or carbinol) base — Basic Green 4 (*Malachite Green*)

Rosaniline Series. The rosanilines are derivatives of tris(*p*-aminophenyl)methane. **Basic Red 9** (*pararosaniline*), patented by Verguin in France in 1859, was the first triphenylmethane dye. Like mauve, it was prepared by the oxidation of aniline, but with stannic chloride, nitrobenzene, or arsenic oxide, instead of dichromate. The dye was manufactured by the Society for Chemical Industry in Basle (Ciba) until the plant was forced to shut down because the aniline imported from France no longer gave satisfactory yields. Hofmann found that the formation of the dye depended on the presence of toluidine in the aniline, and the results of his investigations established the structure of rosaniline. The methyl group of the *p*-toluidine supplies the methylidyne (CH) carbon atom of the leuco base.

$$\begin{matrix} H-C_6H_4-NH_2 \\ CH_3-C_6H_4-NH_2 \\ H-C_6H_4-NH_2 \end{matrix} \xrightarrow[\text{or } As_2O_5]{C_6H_5NO_2} \underset{\text{Leuco base}}{HC[-C_6H_4-NH_2]_3} \xrightarrow[\text{ox.}]{\text{Further}}$$

$$\underset{\text{Color base}}{HOC[-C_6H_4-NH_2]_3} \xrightarrow{HCl} \underset{\text{Basic Red 9 (pararosaniline)}}{[H_2N-C_6H_4-]_2C{=}C_6H_4{=}\overset{+}{N}H_2\overset{-}{Cl}}$$

The commercial **fuchsins, rosanilines,** and **magentas** usually are mixtures of pararosaniline with its methyl homologs. The names *fuchsin* and *rosaniline* were given because of the fuchsia or rose colors of the dyes, but *magenta* was named in honor of the victory of Napoleon III in the battle of Magenta, Italy, in 1859.

The decolorization of fuchsin by sulfur dioxide appears to involve the formation of the leuco sulfonic acid and also the reaction of sulfur dioxide with the amino groups. Subsequent reaction with an aldehyde gives an addition product that loses sulfurous acid to form a new colored compound.

$$\underset{X^-}{H_2\overset{+}{N}{=}}C_6H_4{=}C[-C_6H_4-NH_2]_2 \xrightarrow{H_2SO_3} \underset{\text{Colorless Schiff reagent}}{\underset{X^-}{HO_2S\overset{+}{N}H_2}-C_6H_4-\underset{SO_3H}{C}[-C_6H_4-NHSO_2H]_2} \xrightarrow{RCHO}$$

$$\underset{\text{Colored aldehyde addition product}}{\underset{X^-}{RCHOHSO_2\overset{+}{N}H_2{=}}C_6H_4{=}C[-C_6H_4-NHSO_2CHOHR]_2}$$

The reaction is used for the colorimetric determination of long-chain aldehydes. Here the molar ratio of aldehyde to fuchsin in the colored product appears to be 3:1. The reaction is more complicated than the simple equation indicates. Paper chromatography has shown that formaldehyde gives from three to seven different products depending on the relative amounts of dye, formaldehyde, and sulfur dioxide.

Basic Violet 1 (*Methyl Violet,* 914) is the dye commonly used in purple inks, indelible pencils, and typewriter ribbons. It is made by oxidizing dimethylaniline with cupric chloride. One methyl group of a molecule of dimethylaniline is oxidized to formaldehyde, which then undergoes condensation and further oxidation.

$$C_6H_5N(CH_3)_2 + [O] \xrightarrow[NaCl]{CuSO_4,} C_6H_5NHCH_3 + HCHO$$

$$HCHO + C_6H_5NHCH_3 + 2\,C_6H_5N(CH_3)_2 \xrightarrow{[O]} HC(C_6H_4NHCH_3)(C_6H_4N(CH_3)_2)_2 \xrightarrow{[O]}$$

$$HOC(C_6H_4NHCH_3)(C_6H_4N(CH_3)_2)_2 \xrightarrow{HCl} (C_6H_4NHCH_3)(C_6H_4N(CH_3)_2)C{=}C_6H_4{=}\overset{+}{N}(CH_3)_2\overset{-}{Cl}$$

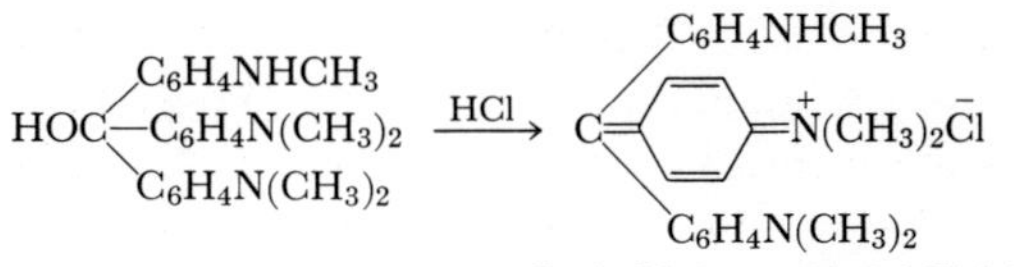

Basic Violet 1 (*Methyl Violet*)

Basic Violet 3 (*Crystal Violet,* 1056) is the completely methylated compound obtained by condensing Michler ketone (p. 464) with dimethylaniline.

$$[(CH_3)_2NC_6H_4]_2CO + C_6H_5N(CH_3)_2 \xrightarrow{POCl_3} HOC[C_6H_4N(CH_3)_2]_3 \xrightarrow{HCl}$$

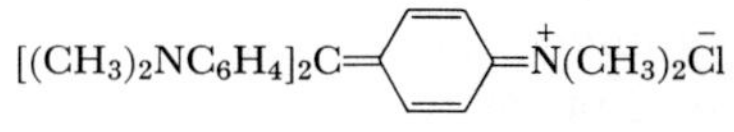

$$[(CH_3)_2NC_6H_4]_2C{=}C_6H_4{=}\overset{+}{N}(CH_3)_2\overset{-}{Cl}$$

Basic Violet 3 (*Crystal Violet*)

Gentian Violet, which is used as an antiseptic, is a mixture of Methyl Violet and Crystal Violet.

Phthaleins. Although not used as a dye, **phenolphthalein** is the most important member of this group, of which it is typical. It is prepared by condensing phthalic anhydride and phenol in the presence of an anhydrous acid catalyst.

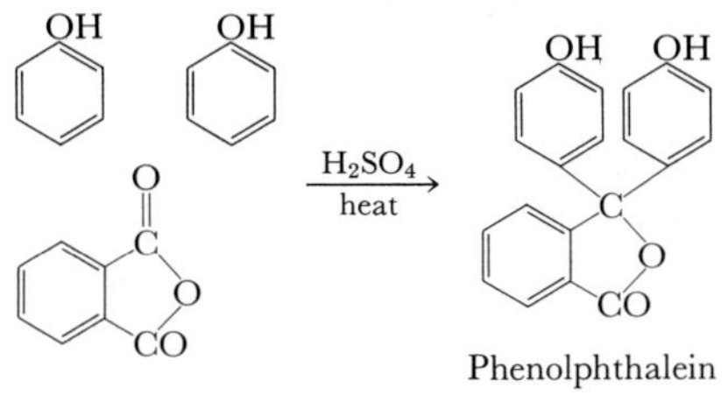

$$2\,C_6H_5OH + C_6H_4(CO)_2O \xrightarrow[heat]{H_2SO_4} (HOC_6H_4)_2C\text{—}O\text{—}CO(C_6H_4)$$

Phenolphthalein

Although widely used as an acid-base indicator (p. 579), the chief commercial importance of phenolphthalein is as a medicinal. It is the usual active ingredient of the candy-type laxatives. **Tetraiodophenolphthalein** is made by the direct iodination of phenolphthalein in alkaline solution. It is used in the X-ray examination of the gallbladder because it accumulates there and the heavy iodine atoms are opaque to X-rays.

The **sulfonphthaleins** result from the condensation of phenols with sulfobenzoic anhydride (p. 476). Like phenolphthalein and its derivatives, the sulfonphthaleins are used as acid-base indicators.

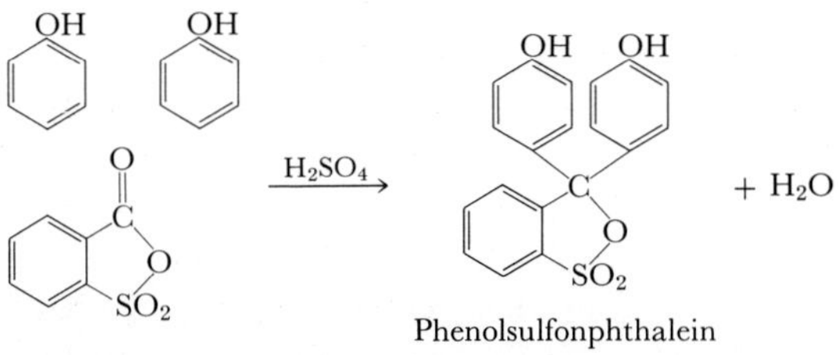

$$2\,C_6H_5OH + C_6H_4(CO)(SO_2)O \xrightarrow{H_2SO_4} (HOC_6H_4)_2C\text{—}O\text{—}SO_2(C_6H_4) + H_2O$$

Phenolsulfonphthalein

Xanthenes. The xanthenes are related to the phthaleins and are made in the same way. They are derived, however, from *m*-dihydroxy compounds or *m*-hydroxy amines, which permit the formation of the xanthene (*2,5-dibenzopyran*) ring system. A typical example is **Acid Yellow 73** (*fluorescein,* 219), the sodium salt of which is called **Uranine.**

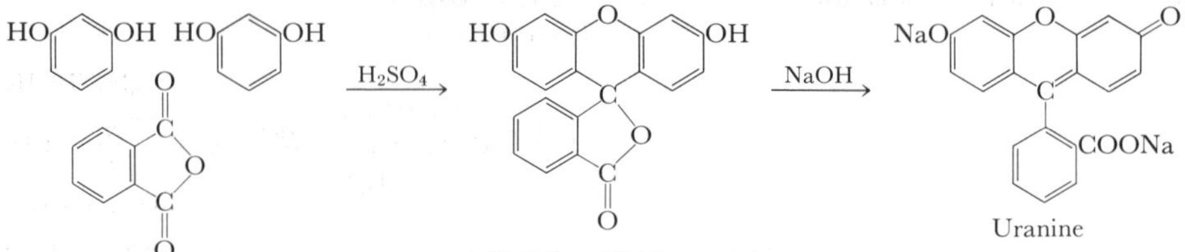

$\xrightarrow{H_2SO_4}$ $\xrightarrow{NaOH}$

Acid Yellow 73 (*fluorescein*)

Uranine

Aqueous solutions of the sodium salt of fluorescein, even at very low concentration, have an intense yellow-green fluorescence when exposed to sunlight. There is distinct visual evidence of its presence at one part in 40 million parts of water. It has been used to trace the course of underground waters and to detect the source of contamination of water supplies. Production in normal times is small, but over a million pounds was manufactured in the United States in 1943. It was supplied in packets to airmen during World War II for use as a sea marker, and many rescued men owe their lives to this dye. Other dyes that fluoresce brilliantly also found wartime use. For example, planes were landed at night on aircraft carriers by signal men whose dyed flags and clothing were made to glow by illumination with ultraviolet light.

Tetrabromofluorescein, prepared by direct bromination of fluorescein, is known as **Acid Red 87** (*eosin,* 465). The sodium salt is the usual dye in red ink. The insoluble lead salt is sold as **Pigment Red 90** (1243). **Mercurochrome,** an antiseptic dye, is the sodium salt of hydroxymercuridibromofluorescein. **Acid Red 51,** also known as **FDC Red No. 3** (*erythrosin,* 49), used as a food color and photographic sensitizer, is tetraiodofluorescein.

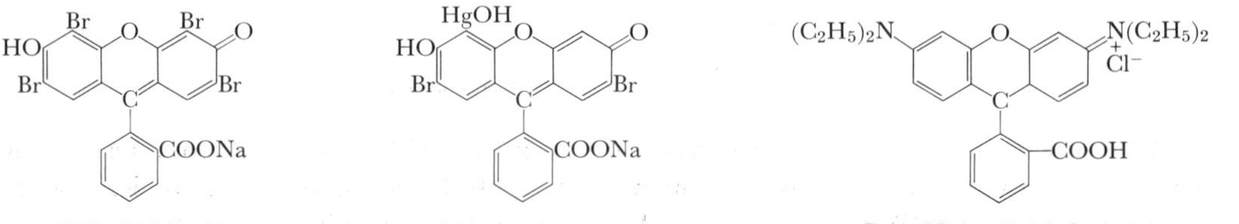

Acid Red 87 (*eosin*)

Mercurochrome

Basic Violet 10 (*rhodamine*)

Basic Violet 10 (*rhodamine,* 260) results from the condensation of phthalic anhydride with *m*-diethylaminophenol. It yields a luminescent pigment when incorporated in a resin.

Indigoid Dyes

Indigos. The oldest known recorded use of an organic dye is that of the vat dye, **indigo** (*Vat Blue 1*). Egyptian mummy cloths estimated to be over four thousand years old were dyed with it. It is present in many plants as a glucoside and has been obtained in the western world from woad (*Isatis tinctoria*) and from plants of the *Indigofera* species. After the twelfth century it became an important article of commerce. Production reached a maximum of about five million pounds in 1890, valued at about $3 per pound. In India alone, 250,000 acres were planted to indigo. The development of commercial syntheses for the natural dyes, alizarin (p. 572) and indigo, were severe blows to the agricultural economies of those times.

In the usual synthesis, aniline reacts with chloroacetic acid to give *N*-phenylglycine, which cyclizes to indoxyl when heated with sodium amide. Air oxidation of indoxyl gives indigo, which has the *trans* configuration.

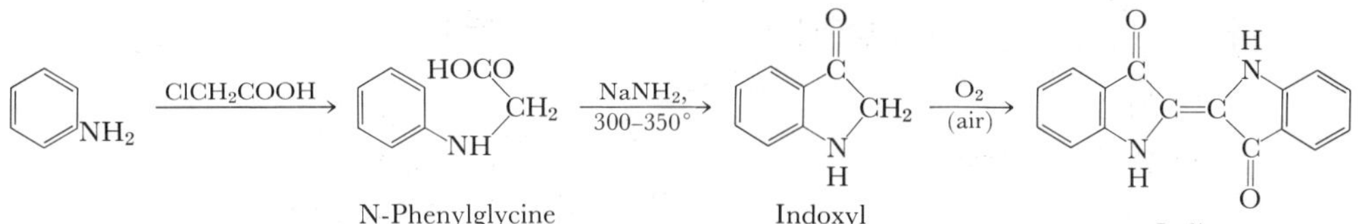

$C_6H_5NH_2 \xrightarrow{ClCH_2COOH}$ N-Phenylglycine $\xrightarrow[300\text{–}350°]{NaNH_2,}$ Indoxyl $\xrightarrow[(air)]{O_2}$ Indigo

N-Phenylglycine Indoxyl Indigo

Production in the United States has been decreasing and has not been published since 1960. Sales in 1961 amounted to 1.3 million pounds on a 100 per cent basis,[5] the price being $1.30 per pound.

Indigo is a deep-blue water-insoluble substance having a bronze reflex. Its application to textiles depends on its easy reduction to a bright-yellow dihydroxy compound known as *indigo white*. The latter is soluble in alkali because of the acidic nature of the hydroxyl groups. The cloth is dyed in vats of the hot indigo white solution and then exposed to air, which rapidly oxidizes the indigo white and deposits the insoluble indigo within the fiber.

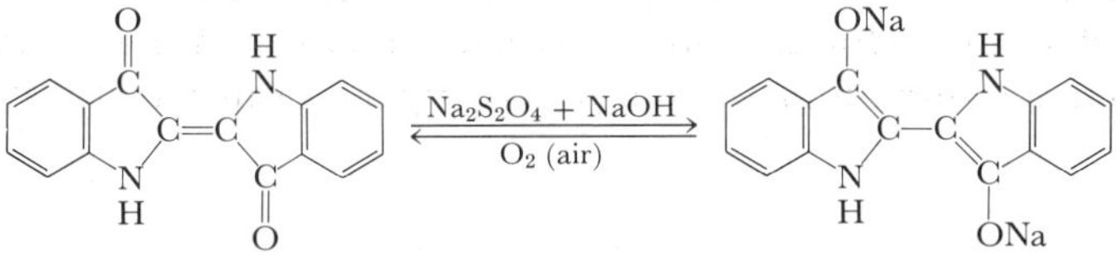

Vat Blue 1 (*indigo*) (water-insoluble) $\underset{O_2\ (air)}{\overset{Na_2S_2O_4 + NaOH}{\rightleftarrows}}$ Indigo white (water-soluble)

The *cis* form of indigo was unknown until 1939. It is of interest that the indigo white is adsorbed on the fiber in the *cis* conformation, and the product first formed on air oxidation is *cis* indigo, which goes over to the *trans* form in the solid state on the fiber.

Another natural dye used by the ancients is **Tyrian Purple.** It is believed to have been used in Crete as early as 1600 B.C. It was derived from several species of mollusk of the genus *Murex*. The term *royal purple* and the phrase *born to the purple* testify to its limited use, 9000 mollusks being required to yield one gram of dye. Modern investigations have shown that the chief component of the dye is 6,6′-dibromoindigo. 5,5′,7,7′-Tetrabromoindigo (**Vat Blue 5,** *Bromindigo Blue 2BD*, 39) sold for $6.20 per pound on a 100 per cent basis in 1963.

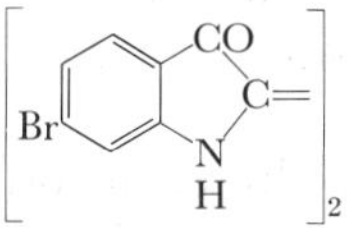

Tyrian Purple 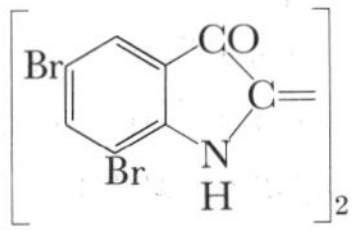 Vat Blue 5 (*Bromindigo Blue 2BD*)

The blue color of the **indophenin reaction** (p. 504) is due to the formation of compounds related to indigo. Two products have been isolated, α- and β-indophenin.

[5] Although production, sales, and price per pound of vat dyes are reported for the pastes, which contain from 6 to 20 per cent dye, these figures have been converted for this chapter to a 100 per cent basis.

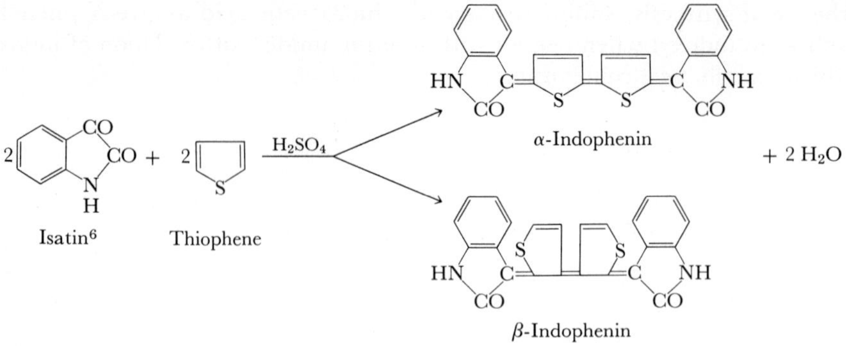

2 Isatin[6] + 2 Thiophene $\xrightarrow{H_2SO_4}$ α-Indophenin (or β-Indophenin) + 2 H_2O

Anthraquinone Dyes

Mordant Dyes. The best known member of the mordant anthraquinones is **alizarin,** another natural dye used by the ancient Egyptians and Persians. It occurs in the root of the madder (*Rubia tinctorum:* Fr. *garance,* Ger. *krapp,* Ar. *alizari*). The red caps and trousers of French soldiers and the red fox-hunting and army coats of the British were required by royal decree to be dyed with madder in support of agricultural interests. Cultivation in Europe resulted in a production of 70,000 tons of madder in 1868. In 1869 patents were issued in England to Caro, Graebe, and Liebermann, and almost simultaneously to Perkin, for the production of alizarin from sodium β-anthraquinonesulfonate (p. 498).

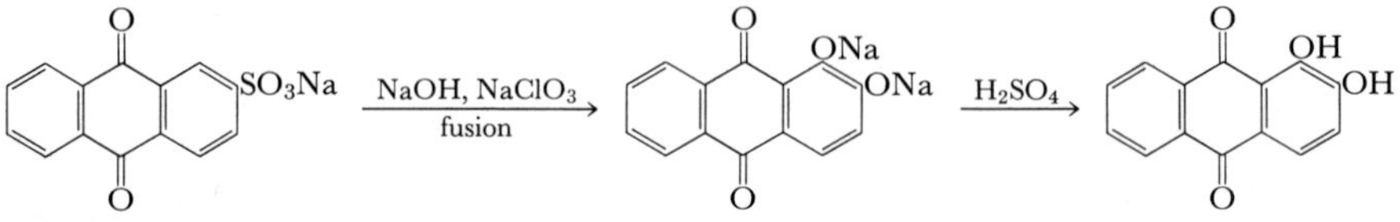

$$\text{(sodium }\beta\text{-anthraquinonesulfonate, }SO_3Na) \xrightarrow[\text{fusion}]{NaOH,\ NaClO_3} \text{(1,2-di-ONa anthraquinone)} \xrightarrow{H_2SO_4} \text{Alizarin (1,2-di-OH anthraquinone)}$$

The process immediately was successful technically and soon drove the natural product from the market. Needless to say, a considerable disturbance to the agricultural economy of Western Europe resulted. The cost of alizarin (100 per cent) dropped from $15 per pound in 1870 to $0.55 per pound in 1914.

Alizarin is *polygenetic,* yielding different colors with different mordants. Thus a magnesium mordant gives a violet color, calcium a purple-red, barium a blue, aluminum a rose-red, chromium a brown-violet, and ferrous iron a black-violet. Alizarin was used chiefly to produce the color known as *Turkey Red* on cotton mordanted with aluminum hydroxide (p. 561) in the presence of sulfated castor or olive oil. Although the dye now is obsolete, sulfated oils still are known as Turkey Red oil (p. 191).

Disperse Dyes. Disperse anthraquinone dyes for cellulose acetate, nylon, and polyester fibers generally are simple aminoanthraquinones or derivatives having one or more hydrogen atoms of the amino groups replaced by other groups. **Disperse Red 15** and **Disperse Blue 3** (1741) are examples.

---

[6] In 1963 it was reported that the oral administration of the β-thiosemicarbazone of *N*-methylisatin prevents more effectively than revaccination or γ-globulin the development of smallpox in persons who have been in contact with active cases. If supported by further evidence, it is the first instance of the successful prophylaxis of a virus disease by a chemical agent. The report has led to intensive investigation of the therapeutic properties of thiosemicarbazones.

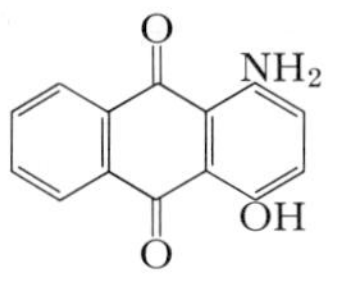

O, NH_2, OH, O

Disperse Red 15

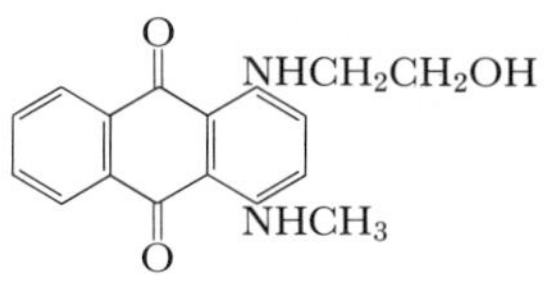

O, $NHCH_2CH_2OH$, $NHCH_3$, O

Disperse Blue 3

Fiber-reactive Dyes. **Procion Blue** is a fiber-reactive dye of the *s*-triazine type (p. 565). Vinyl sulfone (p. 565) derivatives of anthraquinones also are marketed.

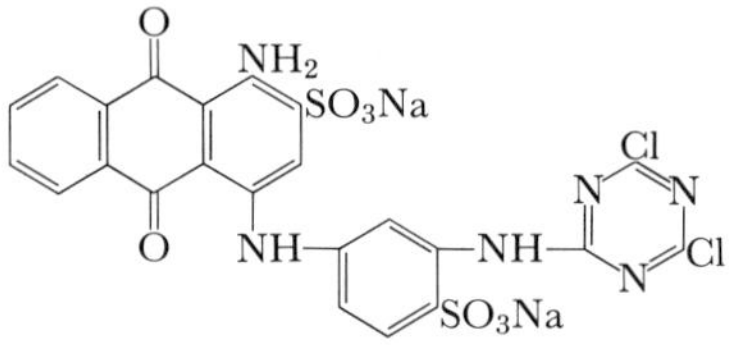

O, NH_2, SO_3Na, Cl, N, N, O, NH—, —NH—, N, Cl, SO_3Na

Procion Blue

O, NH_2, SO_3H, O, NH—, CH_3O, $SO_2CH_2CH_2OSO_3H$

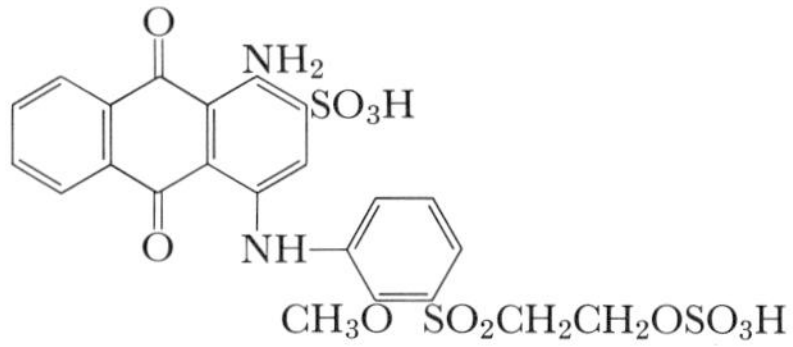

A VS anthraquinone dye

Vat Dyes. The anthraquinones, like indigo, give on reduction dihydro derivatives that are soluble in alkali and are oxidized back to the insoluble anthraquinone on exposure to air or by chemical oxidizing agents (p. 500). The simple quinones are not fixed to animal or vegetable fibers, but the more complex compounds are.

(a) (ACYLAMINO)ANTHRAQUINONES. Although the vats of the aminoanthraquinones have no affinity for cotton, they become substantive if an amino group in the α position is acylated with an aromatic acyl group. The resulting products are the simplest of the vat dyes. Examples are **Vat Red 42** (*Indanthrene Red 5GK*) and **Vat Violet 17** (*Indanthrene Brilliant Violet RK*).

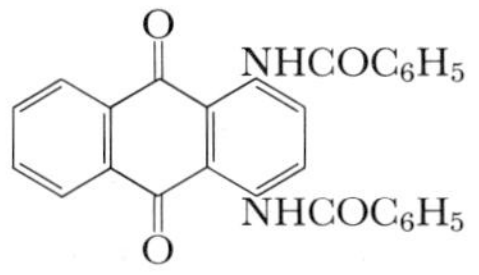

O, $NHCOC_6H_5$, $NHCOC_6H_5$, O

Vat Red 42 (*Indanthrene Red 5GK*)

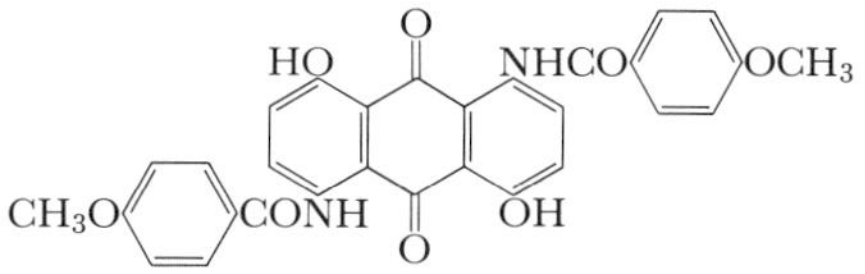

HO, O, NHCO, OCH_3, CH_3O, CONH, O, OH

Vat Violet 17 (*Indanthrene Brilliant Violet RK*)

(b) HYDROAZINES. This group is the oldest of the anthraquinone vat dyes. **Vat Blue 4** (*Indanthrene Blue R,* 7), the first anthraquinone vat dye, was discovered accidentally by Bohn in 1901. He was trying to make diphthaloylindigo by the alkali fusion of the glycine derived from β-aminoanthraquinone. A blue vat dye was obtained that proved to be a dehydrogenation product of β-aminoanthraquinone. He found that the dye could be made by the action of alkali on β-aminoanthraquinone in the presence of an oxidizing agent.

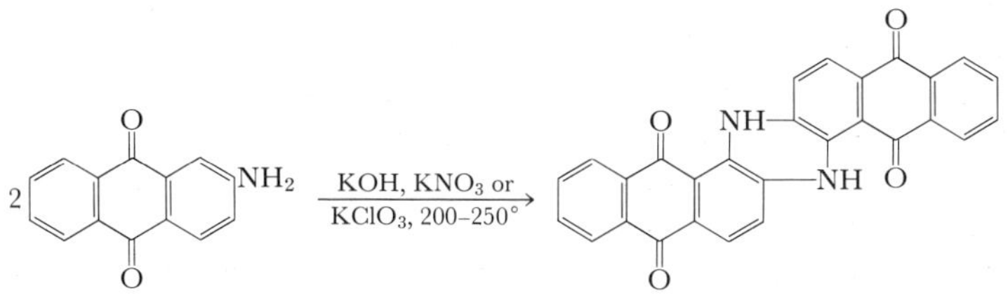

$$2\ \beta\text{-aminoanthraquinone} \xrightarrow[KClO_3,\ 200\text{–}250^\circ]{KOH,\ KNO_3\ \text{or}} \text{Vat Blue 4}$$

Vat Blue 4 (*Indanthrene Blue R*)

Indanthrene Blue is one of the most stable organic compounds known. It can be heated in air at 470°, with strong hydrochloric acid at 400°, and with potassium hydroxide at 300° without decomposition. When applied to cloth, it is extremely fast to washing and light, a property shared by most anthraquinone vat colors. It has been superseded by **Vat**

Blue 6 (*Anthraquinone GCD,* 296) prepared by introducing two chlorine atoms into Vat Blue 4 by direct chlorination.

(*c*) CARBAZOLE DERIVATIVES. These dyes are made by condensing an aminoanthraquinone with a chloroanthraquinone and cyclizing to form a carbazole nucleus, the latter step involving an oxidation or dehydrogenation.

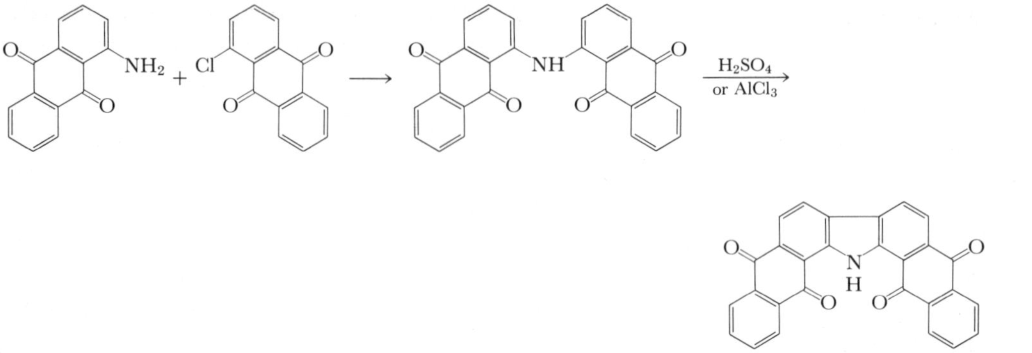

NH_2 + Cl → NH $\xrightarrow[\text{or } AlCl_3]{H_2SO_4}$

Vat Yellow 28 (*Indanthrene Yellow FFRK*)

Vat Brown 1 (*Indanthrene Brown BR,* 112) is made from one mole of 1,4-diaminoanthraquinone and two moles of 1-chloroanthraquinone. **Vat Green 8** (*Indanthrene Khaki GG,* 96) was one of the major products of the dyestuff industry during World War II. It is made from one mole of 1,4,5,8-tetrachloroanthraquinone and four moles of 1-aminoanthraquinone.

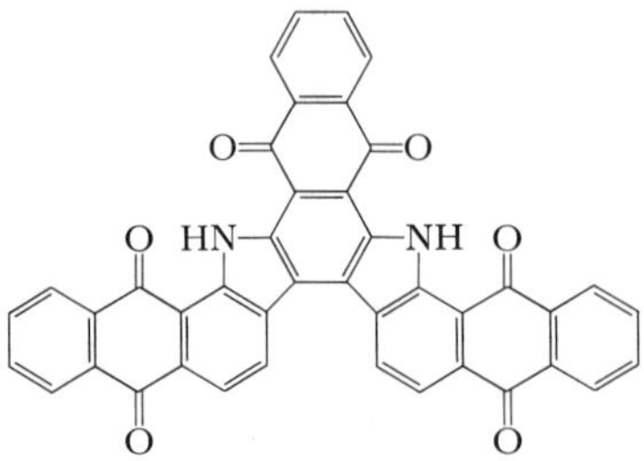

Vat Brown 1 (*Indanthrene Brown BR*)

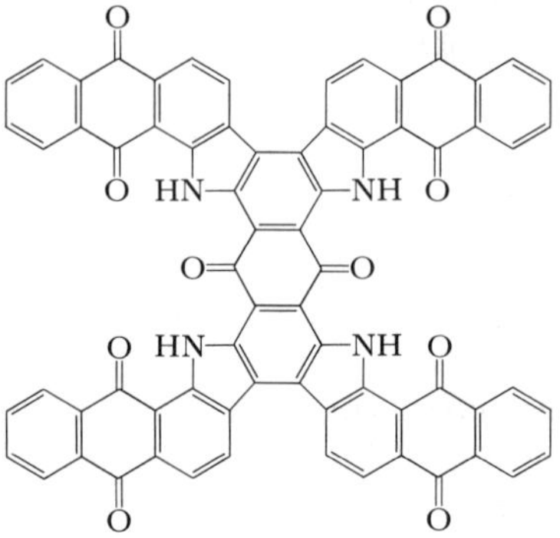

Vat Green 8 (*Indanthrene Khaki GG*)

(*d*) COMPLEX CARBOCYCLIC COMPOUNDS. Anthraquinone is not colored and not fixed to fibers, but complex anthraquinones are. It is not necessary that the two carbonyl groups be present in a single ring provided that they are connected through aromatic rings by a conjugated system. Such complex ring systems can be made either by dehydration or by dehydrogenation reactions.

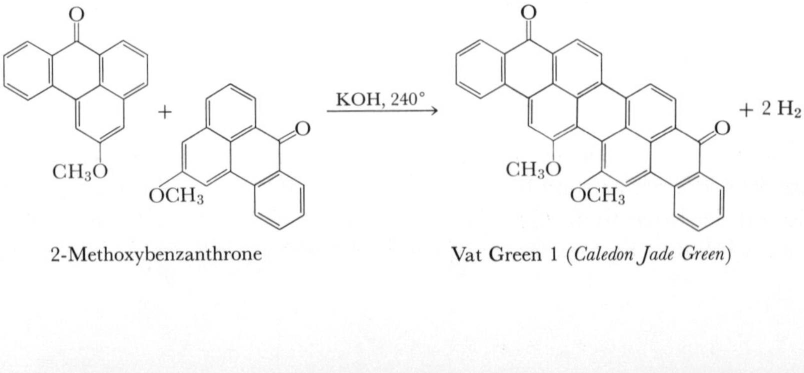

2-Methoxybenzanthrone $\xrightarrow{\text{KOH, } 240°}$ + 2 H_2

Vat Green 1 (*Caledon Jade Green*)

Vat Green 1 (*Caledon Jade Green,* 354), a British discovery first produced in 1920, is considered by dye chemists to be the finest cotton dye.

Anthraquinone vat dyes are applied from a hydrosulfite bath (p. 500), which must be considerably more strongly alkaline than that for dyeing indigo. Hence in the past they have been used only on cotton and viscose rayon. More recently processes have been developed for applying them to wool at a lower temperature than is used for cotton. Strong alkali can be avoided by the use of *solubilized vat dyes,* which are sodium or potassium salts of the reduced dye. Application to the fiber, acidification, and oxidation regenerates the dye.

The greater difficulty of manufacture has made the anthraquinone vats more expensive than most other dyes. Thus the average price per pound for various classes of dyes (100 per cent basis) in 1964 was sulfur, $0.57; azo, $1.68; indigoid, $2.70; anthraquinone vat, $13.60. Nevertheless, the superior fastness of anthraquinone vats to light and washing, the brilliance of their colors, and their high tinctorial power have resulted in a rapid expansion in their use in recent years. The value of their annual sales in 1964 was $66,889,000 compared to $96,579,000 for azo dyes, $29,166,000 for stilbene dyes, $12,682,000 for triphenylmethane dyes, $9,798,000 for sulfur dyes, and $3,302,000 for indigoid dyes.

Sulfur or Sulfide Dyes

This important class of dyes ordinarily includes those dyes made by heating organic materials with sulfur and sodium sulfide, a process known as *thionation,* and does not include other sulfur-containing dyes such as the thiazines, thioindigos, and the thiazoles. The first sulfur dyes were yellows and browns produced by heating sawdust, bran, or manure with sulfur. In 1893 Vidal introduced the use of derivatives of benzene and naphthalene to produce black dyes, and later blues, greens, yellows, and oranges were developed. Sulfur dyes are insoluble in water but are reduced by sodium sulfide to water-soluble products that are substantive to cotton and from which the insoluble dye is regenerated by air or chemical oxidation. They are used only for cotton because they are applied from the sodium sulfide solution, which attacks protein and ester fibers.

The exact chemical structures of the sulfur dyes is unknown, but they are complex compounds and may contain thiazole (p. 523), phenothiazine (p. 527), and 9,10-dithia-anthracene ring systems. Thionation also introduces mercapto or disulfide groups, which accounts for the solubilization by alkaline reduction and precipitation by oxidation.

$$\underset{\text{Water-insoluble}}{\text{DH}} \xrightarrow{\text{S}} \text{DSH} \xrightarrow{\text{NaOH}} \underset{\text{Water-soluble}}{\text{DSNa}} \xrightarrow{O_2,\ H_2O} \underset{\text{Water-insoluble}}{\text{DS—SD}}$$

Since little is known about the constitution of the dyes and since the products of various manufacturers differ, they usually are grouped together according to color. The most important members are **Sulfur Black 1** (1076) and **Sulfur Black 2** (2776), both of which are made from 2,4-dinitrophenol. At $0.34 and $0.40 per pound, they are the cheapest dyes. **Blues** are made from indophenols (p. 439), from diphenylamine derivatives, and from carbazole and *p*-nitrosophenol. **Greens** are made by adding copper salts to melts for blue dyes. **Yellows, oranges,** and **browns** are made from compounds that have reactive groups in the *meta* position such as *m*-toluylenediamine.

Azine Dyes

These dyes are oxidized amino derivatives of phenoxazine, phenothiazine, and dihydrophenazine, and are known as oxazines, thiazines, and phenazines. They can be reduced to colorless compounds and the color can be regenerated by oxidation.

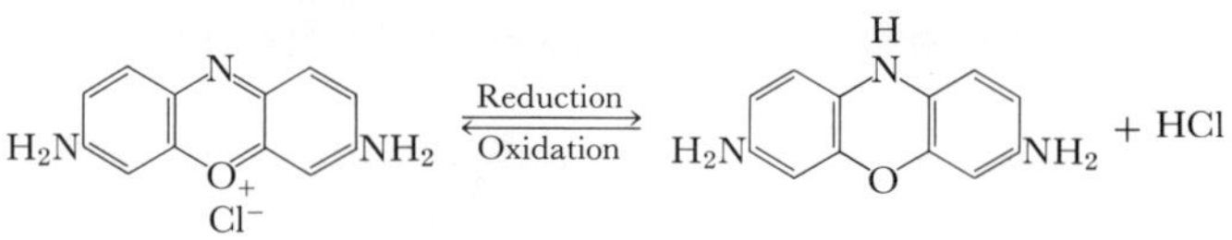

Reduction / Oxidation + HCl

Perkin's **mauve** is a phenazine derivative. The methyl groups are present because the aniline used by Perkin (cf. p. 568) contained *o*- and *p*-toluidines as impurities. **Basic Red 2** (*Safranine T,* 176) also is a phenazine. **Basic Blue 9** (*Methylene Blue,* 403), which has been used extensively in biological oxidation-reduction experiments, is a thiazine.

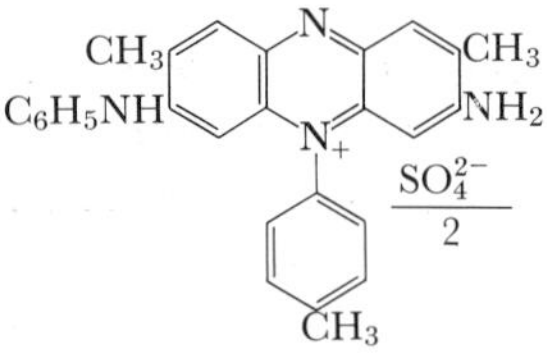

Mauve

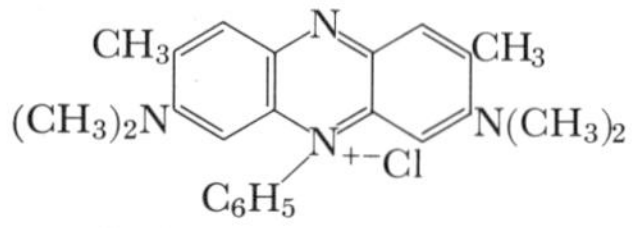

Basic Red 2 (*Safranine-T*)

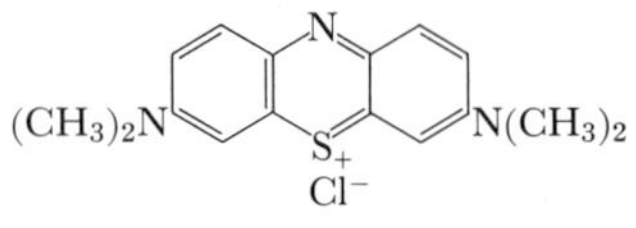

Basic Blue 9 (*Methylene Blue*)

One of the most important black dyes is a phenazine derivative called **Aniline Black.** It does not appear in statistics on dyes because it never is isolated as such. It is produced directly on cotton fiber by the oxidation of aniline salts with oxidizing agents such as potassium chlorate, sodium dichromate, or ferric chloride in the presence of vanadium, copper, or iron salts as catalysts. The first Aniline Black was produced by Lightfoot in 1863. The early Aniline Blacks turned green with age. They are believed to have the structure assigned to **Nigraniline.** Later an "ungreenable" Aniline Black was developed, which is called **Pernigraniline** and is believed to be a polyphenazine derivative.

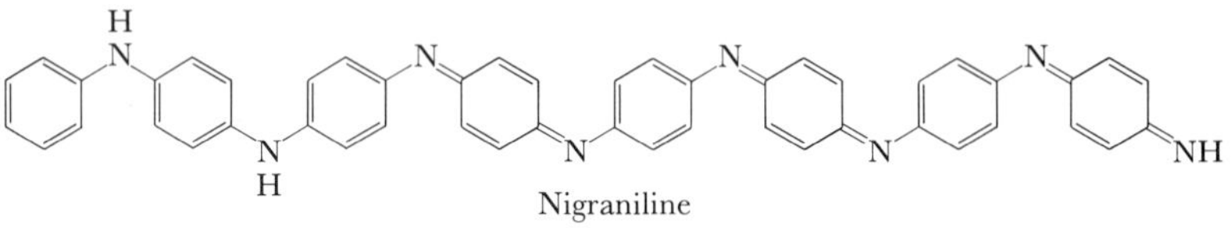

Nigraniline

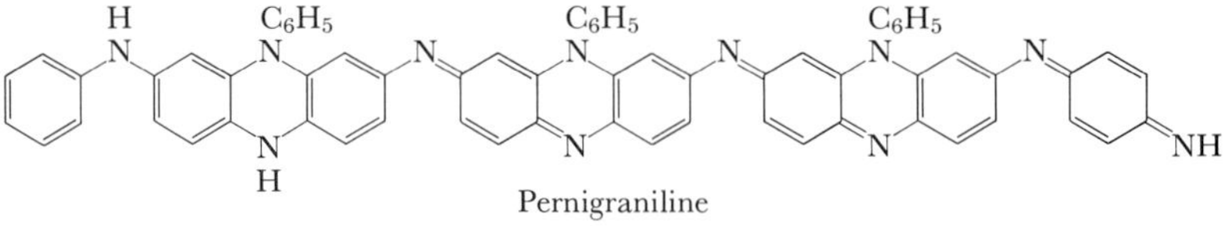

Pernigraniline

Hydrolysis and further oxidation of the Aniline Blacks yields quinone (p. 450).

Nigrosine is a cheap black dye made by oxidizing aniline and aniline hydrochloride with nitrobenzene or nitrophenol in the presence of ferric chloride at 180°. The free base, **Solvent Black 7,** is soluble in oils and waxes, the hydrochloride in alcohol, and the sulfonated product, **Acid Black 2,** in water. Around 3 million pounds per year is made for dyeing shoe polish, printing inks, leather, and paper.

Phthalocyanines

The **phthalocyanines** form an important group of pigments reported in 1934. They are metal complexes of a compound having a 16-membered ring analogous to the porphin nucleus (p. 510) but containing four nitrogen atoms in place of the four methylidyne groups. They can be prepared by a strongly exothermic reaction between phthalonitrile and a metal or metallic salt.

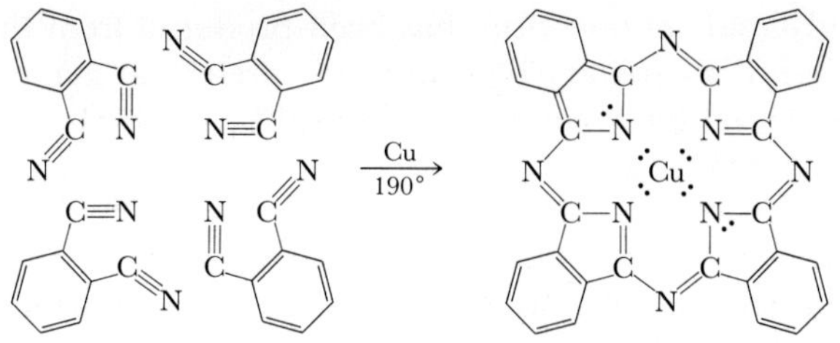

Cu
190°

Copper phthalocyanine
Pigment Blue 15 (*Monastral Blue*)

Copper phthalocyanine, known as **Pigment Blue 15** (*Monastral Blue,* 5871), is made more cheaply by heating phthalic anhydride with urea in the presence of cuprous chloride and small amounts of a catalyst such as vanadium oxide, aluminum oxide, or molybdenum oxide. The only other phthalocyanine pigments that have found important use are the chlorinated copper phthalocyanines. As the number of chlorine atoms substituted in the benzene nuclei increases, the color becomes greener. The hexadecachloro compound, **Pigment Green 7** (4004), now is made commercially from tetrachlorophthalic anhydride, urea, cuprous chloride, and catalyst, together with zirconium or titanium tetrachloride as an additional condensing agent.

Flavans and Flavylium Salts

Flavans, flavanols, flavanones, flavones, and flavonols are derivatives of 2-phenylchroman.

Flavan

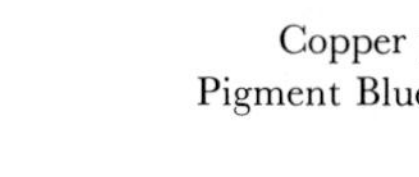

Flavanol

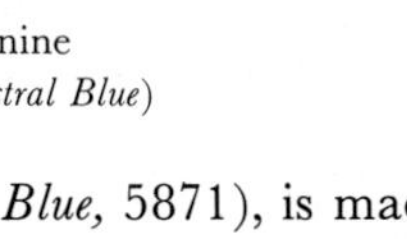

Flavanone

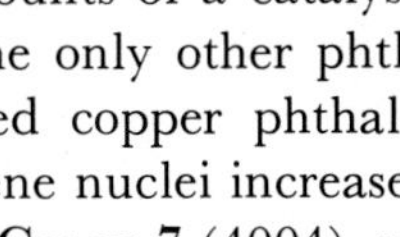

Flavone

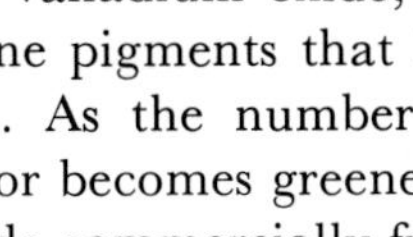

Flavonol

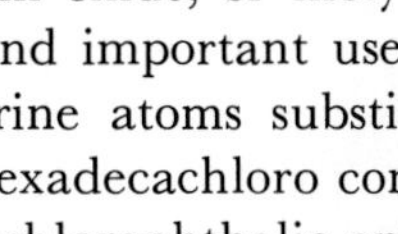

Isoflavan

The corresponding *iso* compounds are derivatives of 3-phenylchroman. Numerous plant products, especially pigments and colors (L. *flavus* yellow), are polyhydroxy derivatives of these basic structures. Usually they are present in the plant as glycosides and may occur in any part of the plant.

Flavonols. Quercitrin is distributed widely in plants but was isolated first from the bark of the black oak, *Quercus tinctoria.* It is a 3-glycoside. Hydrolysis gives L-rhamnose and the aglycone **quercitin,** which has been used since earliest times as a yellow mordant dye. It can be produced in any desired amount by the air oxidation of dihydroquercitin, extracted by hot water from Douglas fir bark.

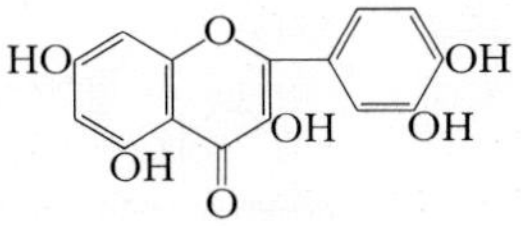

Quercitin

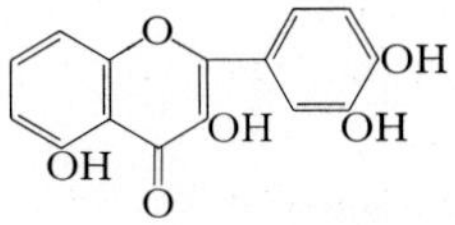

Morin

Rutin, another glycoside of quercitin, has been produced from the buckwheat plant and used in the treatment of capillary bleeding. It is present in many other plants and is the yellow coloring matter on the stems and leaves of the tomato plant. **Morin** is a sensitive reagent for aluminum.

Flavylium Salts. Reaction of an *o*-hydroxy aromatic aldehyde with an aldehyde or ketone in the presence of acid yields a benzopyran derivative known as a *benzopyrylium salt.* A probable intermediate in the reaction is the cyclic hemiacetal, which is called a pseudo base because it is in equilibrium with the oxonium hydroxide.

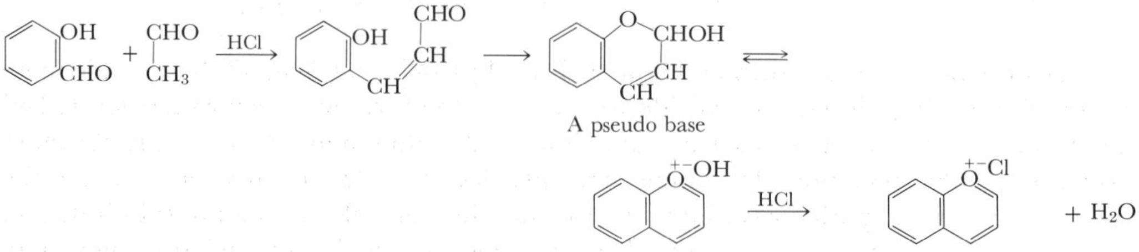

A pseudo base

$+ H_2O$

Benzopyrylium chloride

Most of the red and blue pigments of flowers are derivatives of 2-phenylbenzopyrylium salts, which are known as *flavylium salts.* They occur in the plants as glucosides known as **anthocyanins,** the aglycone being called an **anthocyanidin.** The anthocyanidins derived from natural anthocyanins belong to three groups. All have hydroxyl groups in the 3, 5, and 7 positions. **Pelargonidin chloride** (scarlet pelargonium, orange dahlia) contains an additional hydroxyl group in the 4′ position; **cyanidin chloride** (red rose, blue cornflower, red dahlia, black cherry, plum) has two hydroxyl groups in the 3′ and 4′ positions; **delphinidin chloride** (delphinium, violet pansy, purple grape) contains three hydroxyl groups in the 3′, 4′, and 5′ positions. The anthocyanidins have been synthesized from the properly substituted aldehydes and ketones by the general method for the synthesis of benzopyrylium salts.

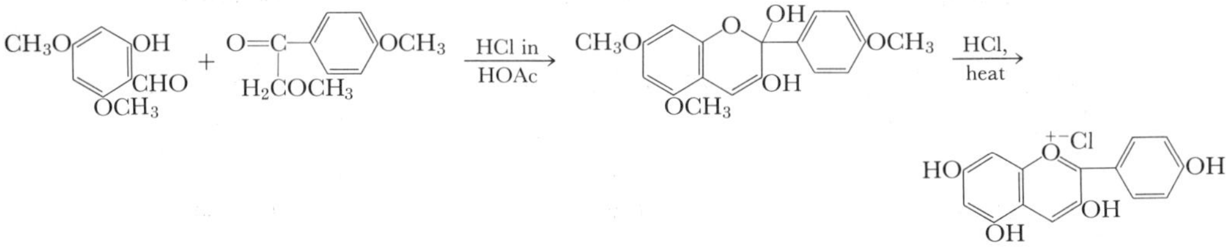

Pelargonidin chloride

The color of anthocyanins, which usually are 3,5-diglucosides, depends not only on their constitution but also on whether the anthocyanin is complexed with a metal ion. Thus the rose is red because the anthocyanin is not coordinated with a metal. The cornflower is blue because the pigment is present as a complex with ferric or aluminum ion. The anion and water molecule and the four oxygen atoms attached to the rings are distributed octahedrally about the metal (p. 14).

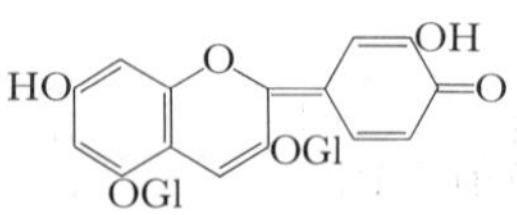

Red rose pigment

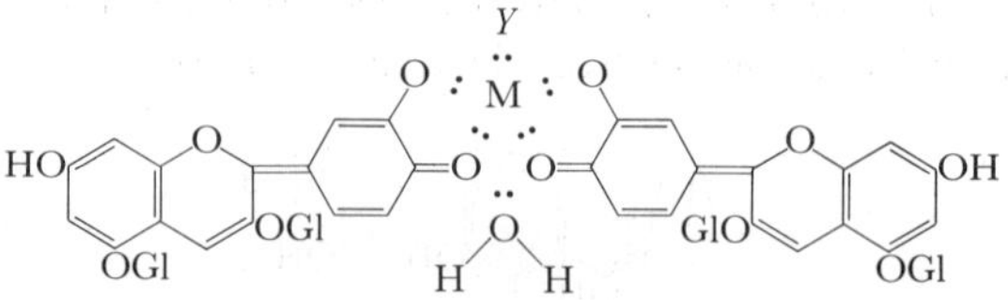

Blue cornflower pigment
$M = Fe^{3+}$ or Al^{3+}, Y = some anion

Oxygen in both the 3′ and 4′ positions is necessary for complex formation. Flowers that contain only pelargonidin as the anthocyanidin cannot have blue varieties.

INDICATOR ACTION

Many compounds have different colors at different hydrogen ion concentrations. These color changes occur because the compounds themselves are acids or bases that enter into proton transfer reactions, and the undissociated acid or the free base has a different color than the ionized salt. The hydrogen ion concentration at which the color change takes place depends on the strength of the compound as an acid or a base.

Methyl Orange in solutions that are more basic than *p*H 4.4 exists almost entirely as the yellow negative ion. In solutions more acidic than *p*H 3.1 it combines almost completely with a proton and forms the red dipolar ion.

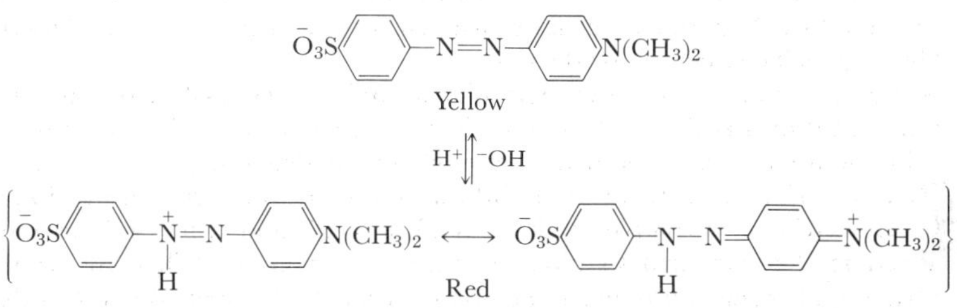

$\bar{O}_3S$–C₆H₄–N=N–C₆H₄–$N(CH_3)_2$

Yellow

$H^+ \uparrow\downarrow {}^-OH$

$\bar{O}_3S$–C₆H₄–$\overset{+}{N}$(H)=N–C₆H₄–$N(CH_3)_2$ ⟷ $\bar{O}_3S$–C₆H₄–N(H)–N=C₆H₄=$\overset{+}{N}(CH_3)_2$

Red

Phenolphthalein is colorless in solutions having a *p*H less than 8.3, where it exists almost entirely as the phenolic lactone. At *p*H greater than 10 it is in the form of a salt that is red. In very strongly alkaline solutions it is converted slowly to the carbinol, which again is colorless.

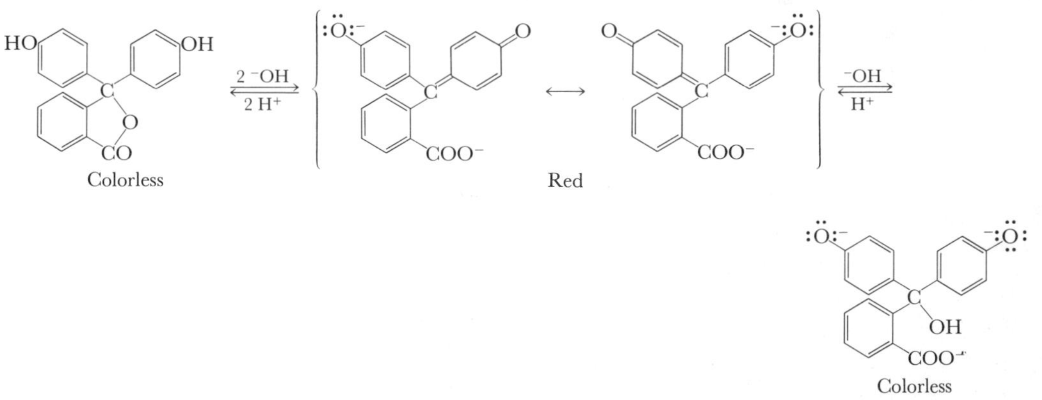

HO, OH, C, O, CO — Colorless $\underset{2\,H^+}{\overset{2\,{}^-OH}{\rightleftharpoons}}$ [:Ö:⁻, C, O, COO⁻ ⟷ O, C, :Ö:⁻, COO⁻] Red $\underset{H^+}{\overset{{}^-OH}{\rightleftharpoons}}$:Ö:⁻, :Ö:⁻, C, OH, COO⁻ — Colorless

The reason that Methyl Orange changes color on the acid side whereas phenolphthalein changes on the alkaline side is that the salt from Methyl Orange and an acid is a much stronger acid than phenolphthalein, just as an amine hydrochloride is a stronger acid than phenol. The conversion of the red ion of phenolphthalein to the colorless carbinol is characteristic of all triphenylmethane dyes, and merely is the reverse of the formation of the dye from the carbinol base (p. 567).

PROBLEMS

32–*1*. Which of the following classes of dyes, (*a*) anthraquinone vat, (*b*) indigoid, (*c*) benzidine, (*d*) mordant, (*e*) fiber-reactive, (*f*) sulfur, (*g*) triphenylmethane, and (*h*) disperse, are (*1*) direct dyes for cotton, (*2*) in general fastest to light and to washing, (*3*) vat dyes other than anthraquinone derivatives, (*4*) the cheapest dyes, (*5*) used extensively for synthetic fibers, (*6*) the newest commercial class, (*7*) esteemed for brilliance but are not very fast to light, and (*8*) applied with the aid of heavy-metal salts?

32–*2*. Starting with compounds obtainable from coal tar, give a series of reactions for the following syntheses: (*a*) Acid Orange 7, (*b*) Fast Scarlet R, (*c*) Disperse Yellow 3, (*d*) Direct Red 2, (*e*) Basic Green 4.

32–*3*. Starting with anthracene, give the reactions that Bohn was attempting to complete that led to the discovery of Indanthrene Blue.

32–*4*. A dye was decolorized when boiled with tin and hydrochloric acid. From the colorless solution two products were isolated, both of which contained nitrogen, but only one of which contained sulfur. Fusion of the sulfur-containing compound with sodium hydroxide gave a product identical with the compound that was free of sulfur. Oxidation of the latter with chromic acid gave 1,4-naphthoquinone. Give a possible structure for the dye.

32–*5*. A red compound, *A*, has the molecular formula $C_{28}H_{18}N_2O_4$. It is insoluble in dilute acid or alkali, but dissolves when warmed with alkaline sodium hydrosulfite solution. *A* is saponified when boiled with alcoholic potassium hydroxide. Removal of most of the alcohol and dilution with water gives a precipitate, *B*, $C_{14}H_{10}N_2O_2$. *B* is soluble in dilute sulfuric acid, and when sodium nitrite is added to the cold acid solution no nitrogen is evolved. When the solution is warmed, nitrogen is produced and compound *C* is obtained. When *C* is distilled with zinc dust, anthracene sublimes. When the alkaline solution from which compound *B* was removed is acidified, compound *D* is obtained, which has a neutralization equivalent of 122. When *D* is heated with sodalime, benzene distills. Give a likely structural formula for *A* and equations for the reactions that take place.

CHAPTER THIRTY-THREE

POLYENES. RUBBER AND SYNTHETIC RUBBERS

POLYENES

The chemical properties of polyenes differ according to the relative positions of the double bonds. If the double bonds are isolated, that is, separated by two or more single bonds, each double bond reacts independently, and the reactions are no different from those when only a single double bond is present. If three or more consecutive carbon atoms are joined by double bonds, the latter are known as *cumulative* double bonds. Hydrocarbons that have two cumulative double bonds, sometimes called *twin* double bonds, are known as *allenes* after the name of the first member of the series, $CH_2{=}C{=}CH_2$. Compounds having more than two double bonds joining adjacent carbon atoms are called *cumulenes.* If double bonds alternate with single bonds, they are said to be *conjugated.*

Cumulenes

Allene (*propadiene*) is the simplest hydrocarbon that contains cumulative double bonds. It can be made by a series of reactions from glycerol.

$$\begin{array}{c}CH_2OH\\ |\\ CHOH\\ |\\ CH_2OH\end{array} \xrightarrow{HBr} \begin{array}{c}CH_2Br\\ |\\ CHBr\\ |\\ CH_2Br\end{array} \xrightarrow{\text{Alc. KOH}} \begin{array}{c}CH_2\\ \|\\ CBr\\ |\\ CH_2Br\end{array} \xrightarrow[\text{alcohol}]{\text{Zn in}} \begin{array}{c}CH_2\\ \|\\ C\\ \|\\ CH_2\\ \text{Allene}\end{array}$$

A general synthesis of allenes involves addition of dibromocarbene, generated from bromoform and potassium *t*-butoxide (p. 478), to an olefin (p. 641) followed by reaction of the *gem*-dibromocyclopropene with magnesium in ether or various organometallic reagents.

$$RCH{=}CHR + :CBr_2 \longrightarrow \underset{\substack{\diagdown\ \diagup\\ C\\ \diagup\ \diagdown\\ Br\quad Br}}{RCH{-}CHR} \xrightarrow[\text{ether}]{\text{Mg in}} RCH{=}C{=}CHR + MgBr_2$$

The allenes readily undergo rearrangement. Thus reaction of allene with sodium gives sodium methylacetylide.

$$CH_2{=}C{=}CH_2 \xrightarrow[\text{ether}]{\text{Na in}} CH_3C{\equiv}CNa$$

Dihalides having the halogen atoms on adjacent carbon atoms yield chiefly acetylenes on reaction with alcoholic alkali (p. 148). For example, the gas obtained by the reaction of propylene bromide with hot alcoholic potassium hydroxide solution is about 95 per cent methylacetylene and 5 per cent allene.

Conjugated Polyenes

1,3 Dienes can be prepared by the acid-catalyzed dehydration of 1,2, 1,3, or 1,4 diols, which are available by pinacol reduction of aldehydes or ketones (p. 207), reduction of aldols (p. 202), or reduction of 1,4-dihydroxy-2-butynes (p. 200).

Conjugation in polyenes can be detected by their strong absorption of light (p. 554). A characteristic chemical behavior is 1,4 addition. Thus Thiele (p. 75) found that if one mole of bromine is added to 1,3-butadiene, the chief product is 1,4-dibromo-2-butene.

$$CH_2{=}CH{-}CH{=}CH_2 + Br_2 \longrightarrow \underset{80\text{ per cent}}{\underset{Br}{CH_2}{-}CH{=}CH{-}\underset{Br}{CH_2}} \rightleftharpoons \underset{20\text{ per cent}}{CH_2{=}CH{-}\underset{Br}{CH}{-}\underset{Br}{CH_2}}$$

This behavior is in accord with a stepwise mechanism in which the intermediate carbonium ion is a resonance hybrid.

$$CH_2{=}CH{-}CH{=}CH_2 \underset{Br^-}{\overset{Br_2}{\rightleftharpoons}} \left\{CH_2{=}CH{-}\underset{+}{CH}{-}CH_2Br \longleftrightarrow \underset{+}{CH_2}{-}CH{=}\dot{C}H{-}CH_2Br\right\} \xrightarrow{Br_2}$$

$$CH_2{=}CHCHBrCH_2Br \quad \text{and} \quad BrCH_2{-}CH{=}CH{-}CH_2Br + Br^+$$

Actually 1,2 and 1,4 addition take place at about equal rates but the *trans* 1,4 addition product is the more stable, and the equilibrium composition rapidly is established. The most important reaction of conjugated dienes is their polymerization to rubber-like products by 1,4 addition under the influence of initiators of the free radical type (pp. 587–589).

$$x\,CH_2{=}CH{-}CH{=}CH_2 \xrightarrow[\text{initiator}]{\text{Free radical}} ({-}CH_2{-}CH{=}CH{-}CH_2{-})_x$$

1,3-Butadiene can be converted to a cyclic dimer (p. 654) and cyclic trimers (p. 655) by Ziegler-type catalysts.

Numerous naturally occurring compounds are conjugated polyenes. They are considered under di- and tetraterpenes (pp. 661, 663).

RUBBER

Sources

Rubber was introduced to Europe shortly after the discovery of America. Early Spanish explorers found that South and Central American natives used the substance to waterproof household utensils and to make balls for their games. The name *rubber* was given to it by Joseph Priestley, who used it to rub out pencil marks.

Rubber is distributed extensively in the plant kingdom. It usually occurs as a colloidal solution in a white fluid known as *latex*. If the milky fluid from goldenrod or dandelion is rubbed between the fingers, a small ball of rubber soon is formed. Many such sources have been investigated, but the principal commercial production is from the rubber tree, *Hevea brasiliensis,* native to the Amazon valley. In 1940, 1.4 million tons was produced of which 97 per cent was grown on plantations, 1 per cent was from wild *Hevea,* and 2 per cent was from other sources. After recovery from the disruption of World War II, production reached 2 million tons in 1955 and was 2.4 million tons in 1963.

Latex is not the sap of the rubber tree. It occurs in microscopic tubules distributed throughout the plant and is obtained from those in the phloem between the bark and the cambium layer. A sloping V-shaped incision is made one third of the way around the trunk starting three feet above the ground, and the latex is drained into a cup containing a small amount of preservative and attached to the trunk at the end of the incision. The latex is carried to the collecting station, where it is strained to remove bark and dirt, and then is transported to the central factory. There it is diluted to 15 per cent rubber and coagulated by the addition of salt and acetic acid. The precipitate is rolled into sheets, washed, and smoked to preserve it against mold. This product is dark brown. In the preparation of the light-colored crepe rubber, bisulfite is added before precipitation to prevent oxidation, and the product is washed more thoroughly to remove the serum and prevent spoilage.

Increasing quantities of rubber are shipped as latex, which is stabilized by the addition of ammonia and concentrated to 60 to 75 per cent solids in one of three ways: (*1*) by centrifuging, (*2*) by *creaming,* in which the addition of a small amount of a hydrated colloid such as Irish moss or gum tragacanth causes a more concentrated layer to separate, or (*3*) by evaporation.

Constitution

Crude plantation rubber contains 2 to 4 per cent protein and 1 to 4 per cent acetone-soluble material consisting of resins, fatty acids, and sterols. The remainder is the *rubber hydrocarbon,* which has the empirical formula C_5H_8 as established by Faraday in 1826. Many attempts have been made to determine its molecular weight. One investigation indicates that the molecular weights of individual molecules range from 50,000 to 3,000,000, with 60 per cent of the molecules having molecular weights greater than 1,300,000.

Destructive distillation of rubber yields among other products a hydrocarbon called **isoprene,** which has the molecular formula C_5H_8 and is 2-methyl-1,3-butadiene, $CH_2{=}C(CH_3)CH{=}CH_2$ (p. 660). The fact that isoprene reverts gradually to a rubber-like product led to the view that rubber is a polymerization product of isoprene. Rubber is unsaturated. It adds one mole of hydrogen catalytically, one mole of bromine, or one mole of hydrogen chloride for each five carbon atoms. Hence one double bond is present for each isoprene unit. Harries[1] in 1904 prepared the ozonide of rubber and isolated from the hydrolysis products levulinic aldehyde and levulinic acid. A painstaking quantitative investigation of the products of ozonolysis reported in 1936 accounted for 95 per cent of the carbon content of the rubber molecule, and 90 per cent of the products isolated can be considered as derived from levulinic aldehyde. These results leave little doubt that the rubber hydrocarbon is a linear polymer of isoprene having the structure first postulated by Pickles in 1910. The nature of the end groups has not been determined.

$$\left[-CH_2\underset{\displaystyle CH_3}{\underset{|}{C}}{=}CHCH_2(CH_2\underset{\displaystyle CH_3}{\underset{|}{C}}{=}CHCH_2)_xCH_2\underset{\displaystyle CH_3}{\underset{|}{C}}{=}CHCH_2-\right] \xrightarrow{\text{Ozonolysis}} (x+2)\ O{=}CHCH_2CH_2\underset{\displaystyle CH_3}{\underset{|}{C}}{=}O$$

Rubber hydrocarbon — Levulinic aldehyde

Vulcanization

Because there is little if any cross linking of the chains of the molecules, rubber is thermoplastic and becomes soft and sticky when heated. When cooled to low temperatures, it becomes hard and brittle. These properties were undesirable even in the early use of rubber, which was chiefly for the waterproofing of textiles. In 1834 Charles Goodyear[2] began experiments attempting to overcome this disadvantage. Mixtures with sulfur had been tried previously, and in 1839, while attempting to improve these mixtures, Goodyear

[1] Carl Dietrich Harries (1866–1923), professor at the University of Kiel. He is known for his extensive work on the ozonolysis of unsaturated organic compounds.

[2] Charles Goodyear (1800–1860) was a New England inventor. His first patent was granted in 1844 and was followed in subsequent years by over sixty more. Many honors were bestowed on Goodyear, but he was kept in poverty because of litigations arising from infringements of his patents.

accidentally dropped one of his preparations on a hot stove, thus discovering the process that he called **vulcanization.** Development of the process led to the production of a material with much greater toughness and elasticity than natural rubber, and one which withstood relatively high temperatures without softening and which retained its elasticity and flexibility at low temperatures.

Vulcanization is a chemical reaction of the rubber hydrocarbon with sulfur, which produces cross links between the chains of rubber molecules. If only a few cross links are present, the molecules can be aligned and elongated to a considerable extent by stretching but cannot slip past one another. When the tension is removed, thermal agitation returns the molecules to their original random orientation (Fig. 33–*1*). In the stretching process, the entropy of the system is decreased; on contraction it is regained (p. 28). Cross linking probably involves disulfide or polysulfide linkages between carbon atoms α to the double bonds rather than by addition to the double bonds. As little as 0.3 per cent of sulfur effects a cure. Commercial rubber is either low in sulfur (1 to 3 per cent) for soft rubber, or high in sulfur (23 to 35 per cent) for hard rubber or ebonite. Rubbers containing intermediate amounts are intractable and of no value. The vulcanization of natural and synthetic rubbers consumes around 100,000 tons of sulfur per year.

Accelerators and Other Additives

Many types of compounds, both inorganic and organic, increase the rate of vulcanization and permit vulcanization to be carried out rapidly at a lower temperature with less sulfur. These compounds are known as **accelerators.** Some of the more important accelerators in current use are sodium and zinc dimethyl-, diethyl-, and dibutyldithiocarbamates (p. 293), 2-mercaptobenzothiazole (Captax, p. 523), and the corresponding disulfide, and tetramethylthiuramdisulfide (*Tuads,* p. 293). Compounds prepared by the oxidative condensation of 2-mercaptobenzothiazole and aliphatic amines are especially useful for milled rubber goods and synthetic rubbers.

$$C_6H_4(S)(N)CSH + H_2NC_6H_{11} \xrightarrow{NaOCl} C_6H_4(S)(N)CSNHC_6H_{11} + NaCl + H_2O$$

2-Mercaptobenzothiazole — Cyclohexylamine

Most organic accelerators work best in the presence of **accelerator activators,** the most commonly used being zinc oxide, along with stearic acid to increase its solubility in rubber.

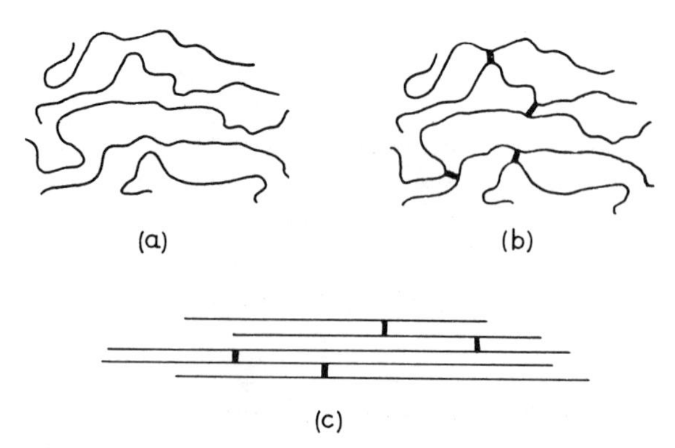

Figure 33–*1*. Rubber molecules: (*a*) unvulcanized; (*b*) vulcanized but unstretched; (*c*) vulcanized and stretched.

One of the outstanding developments in the rubber industry has been the use of **antioxidants** (p. 447) to prolong the life of rubber articles. The ageing of rubber is due to autoxidation of the unsaturated centers with subsequent scission of bonds and reduction in the molecular weight. This reaction is autocatalytic and can be prevented by the addition of secondary aromatic amines such as *N*-phenyl-β-naphthylamine. The aldehyde-aromatic amine condensation products, such as the mixture of condensation products from acetaldehyde or *n*-butyraldehyde and aniline (p. 425), not only are antioxidants but also have an accelerating action.

Reinforcing agents increase the stiffness, tensile strength, and resistance to abrasion. *Carbon black* (p. 93) is used most for this purpose (950,000 tons, 1964). *N*,4-Dinitroso-*N*-methylaniline (p. 427) enhances this property of carbon black. **Fillers** such as barium sulfate, calcium carbonate, and diatomaceous earth decrease the strength but are used to reduce the cost of articles where strength is not important. **Softeners** such as fatty acids or pine oil also may be added.

Manufacturing Operations

Rubber is compounded by incorporating the various ingredients on mixing rolls, an operation known as *milling*. The rubber is squeezed through two large metal rolls rotating slowly in opposite directions at different speeds. The rubber mass is so handled that it encircles one of the rolls and is passed continuously between them. The rubber thus is subjected to a shearing action, which causes it to become warm. Air oxidation takes place, the chain length of the rubber molecules is reduced, and the mass becomes sticky and plastic. The rubber is said to be *broken down* or *masticated* by this operation. In this condition the various solids and liquids that are to be added to the rubber can be worked in on the rolls. Currently the trend is to use Banbury mixers, which by means of powerful revolving arms can compound a batch of rubber in as little as eight minutes. The thoroughly mixed compounded rubber is rolled into sheets. The sheets are used to line a mold, which then is subjected to heat and pressure to form the finished product. The mix also can be spread onto fabric, or extruded into tubing or around wire for insulation.

Latex can be compounded with the various finely powdered ingredients of the rubber mix by simple mixing. Articles are shaped by gelation on a mold and then vulcanized. The mix also can be deposited by electrodeposition. Latex can be spun into a coagulating bath to give thread. The use of latex is advantageous because expensive heavy machinery and high power consumption are not required in the manufacturing operations. Moreover the products are stronger because the rubber is not broken down in a milling operation.

Reclaimed Rubber

Vulcanized rubber is reclaimed on a large scale, especially from discarded automobile tires. The tire is shredded and heated with 4 to 8 per cent sodium hydroxide solution and pine, petroleum, or coal tar oils at 180–200° to plasticize the rubber and to disintegrate the fabric, which then can be washed out. After straining, the mixture is sheeted. No sulfur is removed during the process, but since the initial sulfur content was low, there still are sufficient double bonds present for revulcanization. The price usually is somewhat less than half that of SBR, but it contains only about 50 per cent of rubber hydrocarbon. A certain amount of reclaim is used in most rubber articles. Production in the United States is around 300,000 tons.

Foamed Rubber

Most foamed rubber now is produced by the use of dinitrosopentamethylenetetramine (p. 214) as a foaming agent. When incorporated with the rubber mix and heated, nitrogen is evolved, and vulcanization takes place to give the foamed product.

$$C_5H_{12}N_6O_2 \longrightarrow C_5H_{12}N_2O_2 + 2\ N_2$$

SYNTHETIC RUBBERS

History

Isoprene was obtained first by the distillation of rubber by Gregory in 1835. In 1879 Bourchardt reported the polymerization of isoprene to an elastic product that again gave isoprene on distillation. Continuously since then attempts have been made to develop commercial processes for the synthesis of rubber or rubber substitutes. For years some of the most able English and German chemists, both industrial and academic, vied with each other in their efforts to solve the problem. In 1910 Harries discovered the catalytic effect of sodium on the polymerization of isoprene and found that homologs of isoprene such as butadiene and 2,3-dimethylbutadiene also polymerized to rubber-like products. Although the synthetic rubber industry usually is thought to be a recent development, and in some ways rightly so, it is of interest that Matthews and Perkin, Jr., had worked out a process based on sodium polymerization in 1910, and the Badische Company exhibited a pair of automobile tires made of synthetic rubber at the Eighth International Congress of Pure and Applied Chemistry held in New York in 1912.

With the loss of the East Indies and the Malay Peninsula as a source of rubber during World War II, American chemists and chemical engineers were faced with the problem of creating a synthetic rubber industry within two years to replace the total output of 9 million acres of plantations employing over 2 million people and resulting from 60 years of development. Fortunately far-sighted industries had built experimental pilot plants before the war, and by the pooling of knowledge and resources the remarkable feat was accomplished. By 1945 total synthetic rubber production was at the rate of 700,000 tons per year and reached 1.7 million tons in 1964, or four times the consumption of natural rubber. Total world production of natural and synthetic rubber in 1963 was around 2.4 million tons and 2.8 million tons respectively.

In 1955 a synthetic polyisoprene with properties almost indistinguishable from those of natural rubber was announced and has been available commercially since 1959. The price is around $0.25 per pound. Although producers of natural rubber undoubtedly will continue to hold their market by keeping their price near the cost of synthesis, the latter value should control the price in the future.

Manufacture

The term *elastomer* has been introduced for all substances that have rubber-like properties, but the synthetic products usually are called *synthetic rubbers*. Actually there is no synthetic product identical with natural rubber, which has a regular head to tail arrangement of isoprene units with entirely *cis* configuration at the double bonds. Most so-called synthetic rubbers are not polymers of isoprene, do not have a regular arrangement of monomer units, and the double bonds have mainly *trans* configurations. Polyisoprenes, made with stereoselective catalysts (p. 88), resemble natural rubber very closely but are not identical with it (p. 588).

For a material to have rubber-like properties, it must (*1*) be a linear polymer of high molecular weight, (*2*) have either some short side chains or a steric configuration that precludes strong London forces by preventing the main chains from approaching too closely, (*3*) be relatively nonpolar to avoid strong dipole interaction between chains, and (*4*) have a few functional groups that permit cross linking of the chains after the material has been applied or shaped to the desired product. Clearly many structurally different polymers can satisfy these basic requirements. Although a synthetic material absolutely identical with natural rubber is unknown, many different "synthetic rubbers" are made, each having its own desirable characteristics, which frequently are superior to those of natural rubber. It should be emphasized also that each of the main classes considered in this chapter may have numerous variants that are designed to have optimum properties for a specific purpose. For example, although all neoprenes are based on chloroprene, eleven types differing considerably in composition and properties were available commercially in 1960. United States production of the main classes in 1964 in thousands of tons was styrene-butadiene,

1153; *cis*-butadiene, 182; neoprene, 158; Butyl, 111; nitrile, 59; and silicone, 4.1. The price per pound was \$0.23 for styrene-butadiene rubber, \$0.48 for nitrile rubber, and \$3.77 for silicone rubber.

Styrene-Butadiene Rubber. This workhorse of the rubber industry (*SBR, cold rubber, GRS, Buna S*) is made by a free radical copolymerization of one part of styrene (p. 483) and three parts of 1,3-butadiene by weight, or a mole ratio of approximately 1:6. The butadiene is made most cheaply by the catalytic dehydrogenation of *n*-butane or of 1- and 2-butenes.

$$CH_3CH_2CH_2CH_3 \xrightarrow[600°,\ 1\ at.]{Al_2O_3\text{-}Cr_2O_3} CH_3CH{=}CHCH_3 \xrightarrow[650°,\ 0.1\ at.]{Ca_8Ni(PO_4)_6\text{-}Cr_2O_3} CH_2{=}CHCH{=}CH_2$$

A reduced partial pressure is achieved by admixture with steam. The conversion of butenes to butadiene in 85 per cent yield reaches 30 per cent per pass. The chief problem is the purification of the product. Extraction with ammoniacal cuprous acetate solution usually is used. United States production for all purposes was 1,162,000 tons in 1963. During emergencies when the supply of the straight-chain butanes and butenes is insufficient, butadiene can be made from ethanol, acetaldehyde, or acetylene.

Butadiene adds by both 1,4 and 1,2 addition. Hence the units present in SBR are $—CH_2CH{=}CHCH_2—$, $—\underset{\displaystyle CH{=}CH_2}{\underset{|}{C}}HCH_2—$, and $—CH_2\underset{\displaystyle C_6H_5}{\underset{|}{C}}H—$. About 20 per cent of the butadiene units are incorporated by 1,2 addition, and the various units appear to be randomly distributed. About one fifth of the double bonds resulting from 1,4 addition have the *cis* configuration and about four fifths are *trans*. Some branching and cross linking of molecules also takes place.

In order to keep cross linking at a minimum, the polymerization is carried out at as low a temperature as possible that is consistent with a practical rate of reaction. By the use of oxidation-reduction systems, free radicals can be generated at a sufficient rate to permit the polymerization to be run at 5°. Numerous complex formulations have been used, but in general the process consists of emulsification of the monomers in water by means of a soap or synthetic dispersing agent in the presence of an oxidation-reduction system, a sequestering agent, a buffer, and a molecular weight modifier. The oxidation-reduction system consists of an organic hydroperoxide that is soluble in the monomers, such as α,α'-dihydroperoxy-*p*-diisopropylbenzene, a water-soluble reducing agent, such as sodium formaldehyde sulfoxalate (*SFS*, p. 214), and ferrous sulfate. After addition of an antioxidant such as *N*-phenyl-β-naphthylamine, the emulsion is stripped of unreacted monomer, coagulated by addition of an acidified salt solution, washed, and dried.

Tire treads made from polymers of very high molecular weight have better wear resistance than those made from products of lower molecular weight, but their plasticity is too low for easy processing. Since 1951 this difficulty has been overcome by adding 25 to 50 parts of high-aromatic petroleum oils to each 100 parts of SBR. The oil is added as an emulsion to the polymerized rubber emulsion before coagulation. The extender not only increases workability but, at one sixth the price of SBR, considerably decreases cost. Most synthetic rubbers generate more heat on flexing than natural rubber. Hence the latter is preferred for the casings of heavy duty truck tires, but SBR is used in the tread.

***Neoprene* (CR, GRM).** During his investigations of the chemistry of acetylene, Nieuwland[3] discovered the formation of a dimer, **vinylacetylene,** by the action of cuprous

[3] Julius Arthur Nieuwland (1879–1936), Catholic priest and professor of chemistry at the University of Notre Dame. Besides being well known for his investigations of the chemistry of acetylene, he was a distinguished botanist.

salts. Carothers[4] found that the dimer adds hydrogen chloride to yield 2-chloro-1,3-butadiene, known as **chloroprene.** The initial 1,4 addition product rapidly undergoes an allylic rearrangement (p. 594) in the presence of acid.

$$2\,HC{\equiv}CH \xrightarrow[NH_4Cl]{Cu_2Cl_2} \underset{\text{Vinylacetylene}}{H_2C{=}CH{-}C{\equiv}CH}$$

$$H_2C{=}CH{-}C{\equiv}CH \xrightarrow{HCl} ClCH_2{-}CH{=}C{=}CH_2 \longrightarrow \underset{\text{Chloroprene}}{H_2C{=}CH{-}\underset{Cl}{C}{=}CH_2}$$

Emulsion polymerization of chloroprene with potassium persulfate as the initiator gives **Type W neoprenes.** The molecular weight of **Type G neoprenes** is controlled by the addition of sulfur as a modifier during the emulsion polymerization. The neoprenes are good general purpose rubbers but the high cost of manufacture has limited their use to those applications that require their special properties, such as resistance to oils, chemicals, air, light, heat, and flame.

***cis*-Polybutadiene (BR).** Although polymerization of butadiene by the usual catalysts leads to useless mixed polymers, the stereoselective catalysts of the Ziegler type (p. 113), especially the isobutylaluminum-titanium tetrachloride complex, give a product that is almost entirely the *cis*-1,4 addition polymer, which has rubber-like properties. It rapidly has come into use for blending with natural rubber and with SBR rubber. Although *cis*-polybutadiene first was introduced in 1961, production climbed to third place in 1962 and was second only to SBR in 1963. Other coordination catalysts give a product that is essentially 100 per cent *trans*. It is being marketed as a replacement for gutta-percha and balata in golf ball covers.

***cis*-Polyisoprene (IR).** Early attempts to prepare satisfactory synthetic rubbers from isoprene were unsuccessful. Moreover no reasonably cheap and abundant source of isoprene was available. In 1955 it was announced that stereoselective polymerization of high purity isoprene using Ziegler-type catalysts, finely dispersed lithium,[5] or butyllithium, with rigorous exclusion of air and moisture, gives products that are chiefly *cis*-1,4-polyisoprene. They differ from natural rubber only in that a small amount of 1,2 addition has taken place. Their gross physical properties are almost indistinguishable from natural rubber. Although isoprene is being made by the cracking of naphtha fractions (p. 643), a synthetic process starting with propylene appears to offer a distinct advantage. The propylene is dimerized to 2-methyl-1-pentene with *n*-propylaluminum as catalyst, isomerized to 2-methyl-2-pentene, and cracked in the presence of hydrogen bromide to isoprene.

$$2\,CH_3CH{=}CH_2 \longrightarrow CH_3CH_2CH_2\underset{CH_3}{C}{=}CH_2 \rightleftharpoons CH_3CH_2CH{=}\underset{CH_3}{C}CH_3 \longrightarrow CH_4 + CH_2{=}CH\underset{CH_3}{C}{=}CH_2$$

Thus the long-standing problem of the technical synthesis of natural rubber has been solved, but the extent to which it will be used is questionable. Its chief importance lies in placing a ceiling on the price of natural rubber and in providing insurance against any future possibility that supplies of natural rubber may be cut off.

---

[4] Wallace Hume Carothers (1896–1937), American-trained chemist and director of a laboratory for fundamental research in organic chemistry at the du Pont Experimental Station. In addition to his work on neoprene, he conducted an investigation of the reactions of polyfunctional molecules that led to the discovery of nylon (p. 631) and Dacron (p. 476) and to the production of many-membered ring compounds (p. 656).

[5] Harries (p. 583) had reported the polymerization of isoprene by lithium in 1913, but its specific properties apparently were not recognized until forty years later. The less expensive sodium was used in the early commercial processes, which then gave way to emulsion polymerization.

***Butyl Rubber* (IIR, GRI).** The polymerization of isobutylene catalyzed by a Lewis acid (p. 34) gives a viscous to rubber-like product that is completely saturated and hence cannot be vulcanized (p. 87). If it is copolymerized with a small amount of a diene such as isoprene, one double bond remains for each molecule of the diene used. When about 2 per cent of the diene is used, sufficient double bonds are present to give on vulcanization a rubber-like product that is practically saturated and hence has good resistance to chemicals and oxidation. Polymerization is carried out at $-100°$ in methyl chloride with aluminum chloride as a catalyst and liquid ethylene as a refrigerant. Because Butyl rubber is very impermeable to gases, its chief use has been for the manufacture of inner tubes for automobile tires and as a liner for tubeless tires. New developments in the processing of Butyl rubber have improved its other physical properties, and it shows promise of becoming a general purpose rubber.

Miscellaneous Elastomers. **Nitrile rubber** (*NBR, GRN, Buna N*) is a copolymer of two parts of 1,3-butadiene with one part of acrylonitrile, $CH_2{=}CHCN$ (p. 619). It is hard to mill but has high resistance to aromatic oils, fuels, and solvents. One of the latest commercial elastomers is **ethylene-propylene terpolymer** (*EPT*), a sulfur-vulcanizable material made by the copolymerization of ethylene, propylene, and a small amount of a nonconjugated diene such as dicyclopentadiene (p. 643) or 1,4-hexadiene. **Polysulfide rubbers** such as Thiokols *A, ST,* and *FA* are made by condensing a dichloro compound with sodium polysulfide. **Silicone rubbers** are linear siloxanes (p. 278) of high molecular weight with carbon-carbon cross links produced by the action of peroxides on methyl side chains. **Hypalon** is a chlorinated polyethylene containing a small amount of sulfonyl chloride groups. Hydrolysis of the latter to sulfonic acid groups and salt formation with magnesium oxide brings about cross linking of the chains. **Polyacrylate rubbers** result from copolymerization of ethyl acrylate, $CH_2{=}CHCOOC_2H_5$ (p. 619), with vinyl chloroethyl ether, $CH_2{=}CHOCH_2CH_2Cl$, and cross linking with polyamines such as diethylenetriamine (p. 611). **Urethan rubbers** are discussed on page 605 and **fluorocarbon rubbers** on page 597. Many other rubber-like products result from plasticizing nonrubber-like substances. For example, polyvinyl chloride (p. 593) can be plasticized with cresyl phosphate (p. 445) or alkyl phthalates (p. 474) (*Koroseal, Geon, Tygon*).

Chemically Related Products. **Solid rocket propellants** usually are mixtures of an organic binder with a finely divided inorganic oxidizing agent such as ammonium perchlorate or ammonium nitrate. The binder is a fluid organic material, for example a partially polymerized elastomer or a polysulfide polymer made fluid by reduction of disulfide bonds, that can be mixed with the oxidizing agent and poured into the rocket motor case where it is converted to solid polymer by controlled heating.

The so-called **rubber-base latex paints** are water emulsions of a pigment and a resin vehicle made by copolymerizing one part of butadiene with three parts of styrene. Copolymers of around 75 per cent styrene and 25 per cent acrylonitrile are known as **SAN plastics.** They are used for dinnerware, household articles, and battery cases, and as monofilament for brushes. If styrene and acrylonitrile are added to the latex of an emulsion-polymerized butadiene and polymerization is continued, some of the styrene and acrylonitrile copolymerize with the polybutadiene as side chains to give a *graft polymer* along with SAN. The graft polymer blends well with the SAN and confers greater impact strength to the molded product. Such materials are called **ABS plastics.** They are used for telephone hand sets, pipe, automotive and household appliances, and women's shoe heels. Combined production of SAN and ABS plastics reached 264 million pounds in 1964 and is expanding rapidly.

PROBLEMS

33–*1*. Pair the compounds (*a*) $CH_3CH{=}C{=}CH_2$, (*b*) $CH_3CH{=}CH{-}CH_2CH{=}CH_2$, (*c*) $CH_3C{\equiv}CCH_3$, and (*d*) $CH_3CH_2CH{=}CH{-}CH{=}CH_2$ with the appropriate type: (*1*) conjugated diene, (*2*) allene, (*3*) nonconjugated diene, (*4*) acetylene.

33–*2*. Give reactions for the synthesis of (*a*) 2,3-pentadiene from 2-butene, (*b*) 1,3-butadiene from acetaldehyde.

33–*3*. Pair the following types of cross linkage, (*a*) C—S—S—C, (*b*) C—C, (*c*) $C{-}SO_3{}^{-+}Mg^{+-}O_3S{-}C$, and (*d*) $C{-}NH(CH_2)_2NH(CH_2)_2NH{-}C$, with the elastomer in which they usually are present in the vulcanized product: (*1*) Hypalon, (*2*) natural rubber, (*3*) polyacrylate rubber, (*4*) silicone rubber.

CHAPTER THIRTY-FOUR

CHLORINATED AND FLUORINATED ALIPHATIC HYDROCARBONS

CHLORINATED HYDROCARBONS

Over 6 million tons of chlorine was produced in the United States in 1964, and 65 per cent was used for the preparation of organic compounds. The total production of noncyclic halogenated hydrocarbons alone amounted to over 4 million tons. Both substitution and addition reactions are used in their preparation.

The manufacture and uses of methyl, ethyl, and amyl chlorides are discussed on page 105. **Methylene chloride,** b.p. 40°, is made only by the chlorination of methane or methyl chloride. It is the least toxic of the chlorinated methanes. One third of production is used in paint and varnish removers, 30 per cent as solvent and diluent in aerosol paints and sprays and to reduce the flammability of other solvents, and 10 per cent as a solvent for cellulose triacetate in the manufacture of film and fiber (p. 345). **Chloroform,** b.p. 61°, is made almost exclusively by the chlorination of methane or methyl chloride. Its principal use is as an intermediate in the synthesis of chlorodifluoromethane (p. 597). **Carbon tetrachloride,** b.p. 77°, is the most important of the chlorinated methanes. About one fourth of the total production is by the chlorination of methane. One half results from the two-stage reaction of chlorine with carbon disulfide (p. 291).

$$CS_2 + 3\,Cl_2 \longrightarrow CCl_4 + S_2Cl_2$$
$$2\,S_2Cl_2 + CS_2 \longrightarrow CCl_4 + 6\,S$$

The sulfur is reconverted to carbon disulfide. High temperature chlorinolysis of propane accounts for the remaining fourth. At 500–700° with a large excess of chlorine and rapid quench, the chief products are carbon tetrachloride and tetrachloroethylene.

$$C_3H_8 + 8\,Cl_2 \xrightarrow{600^\circ} CCl_4 + Cl_2C{=}CCl_2 + 8\,HCl$$

Small amounts of hexachloroethane and hexachlorobenzene also are produced, the former being recycled. Both the first and the third methods utilize only one half of the chlorine, and hydrogen chloride must be recovered for use in some other process (cf. p. 592). Most of the carbon tetrachloride is converted to tetrachloroethylene (p. 592), trichlorofluoromethane (p. 597), and dichlorodifluoromethane (p. 596).

Ethylene chloride (*1,2-dichloroethane*), b.p. 84°, is produced in larger amount than any other organic chlorine compound. It is made by the addition of chlorine to ethylene at 40° in liquid ethylene chloride containing ferric chloride as catalyst. It can be prepared also by "oxychlorination" of ethylene. To utilize hydrogen chloride formed in the production of vinyl chloride (p. 592), mixtures of hydrogen chloride with ethylene and oxygen are passed over a supported copper chloride catalyst at 300° (cf. p. 442).

$$2\,C_2H_4 + 4\,HCl + O_2 \longrightarrow 2\,C_2H_4Cl_2 + 2\,H_2O$$

The chief use for ethylene chloride is the production of vinyl chloride (p. 592). Smaller amounts are used for the manufacture of Thiokols (p. 589) and of the ethylene amines (p. 611), as a solvent, and in Ethyl Fluid (p. 94).

Various polyhalogenated compounds are produced from acetylene. **1,1,2,2-Tetrachloroethane** (*acetylene tetrachloride*) is prepared by mixing solutions of acetylene and chlorine in tetrachloroethane at 70–95° with ferric chloride as a catalyst. To prevent explosions the tetrachloroethane is circulated downwards, acetylene mixed with it, and the chlorine bubbled in at a point below, where it reacts with a dilute solution of the acetylene in tetrachloroethane. Tetrachloroethane, b.p. 146°, is an excellent solvent, but it is highly toxic and corrodes metals in the presence of moisture. Most of it is used for the production of **trichloroethylene** by pyrolysis over supported barium chloride, followed by treatment with an aqueous slurry of lime to complete the reaction.

$$CHCl_2CHCl_2 \xrightarrow[250^\circ]{BaCl_2 \text{ on act. C}} CHCl{=}CCl_2 + HCl$$

$$2\ CHCl_2CHCl_2 + Ca(OH)_2 \longrightarrow 2\ CHCl{=}CCl_2 + CaCl_2 + 2\ H_2O$$

Trichloroethylene boils at 87° and is one of the more important of the chlorinated solvents. It is stable and noncorrosive and is about as toxic as carbon tetrachloride. It is used chiefly for degreasing metal parts. The cold metal part is passed through the hot vapors, which condense on the metal and wash away the oil and dirt.

Tetrachloroethylene (*perchloroethylene*[1]), b.p. 118°, is made chiefly by the chlorinolysis of propane (p. 591) or by the pyrolysis of carbon tetrachloride.

$$2\ CCl_4 \xrightarrow{800\text{–}900^\circ} Cl_2C{=}CCl_2 + 2\ Cl_2$$

This reaction was discovered by Kolbe (p. 166) in 1843 but has been used commercially only recently. Around 80 per cent of production is used as a nonflammable solvent in dry-cleaning plants and coin-operated machines. It is preferred to trichloroethylene because it does not extract disperse dyes from synthetic fibers (p. 562).

The relative importance of some of these compounds is indicated by their 1964 United States production in millions of pounds: ethylene chloride, 2199; ethyl chloride, 666; carbon tetrachloride, 535; trichloroethylene, 370; tetrachloroethylene, 366; methylene chloride, 180; methyl chloride, 134; and chloroform, 119. The prices ranged from $0.05 per pound for ethylene chloride to $0.09 per pound for tetrachloroethylene.

Vinyl chloride (*chloroethylene*) is made chiefly by the thermal dehydrohalogenation of ethylene chloride and by the addition of hydrogen chloride to acetylene.

$$ClCH_2CH_2Cl \xrightarrow{400^\circ} CH_2{=}CHCl + HCl$$

$$HC{\equiv}CH + HCl \xrightarrow[200^\circ]{Hg_2Cl_2/\text{act. C}} CH_2{=}CHCl$$

Frequently the two processes are operated by the same company to utilize all of the chlorine.

The halogen atom in vinyl chloride is about as unreactive as that in chlorobenzene (p. 399). This reduced reactivity is characteristic of halogen on a doubly bonded carbon atom, and is ascribed to the interaction of the unshared electrons in a *p* orbital of the halogen atom with the electrons in the π orbital of the double bond, which leads to greater bond strength and decreased bond length and reactivity.

[1] The prefix *per* (L. *per* through) is used to indicate not only a high state of oxidation but also the maximum amount of substitution or addition.

$$\left\{CH_2{=}CH\ddot{\underset{..}{Cl}}: \longleftrightarrow \bar{C}H_2CH{=}\overset{+}{\underset{..}{Cl}}:\right\}$$

The decrease in the C—Cl bond distance is observable, the interatomic distance for ethyl chloride being 1.77 A and for vinyl chloride 1.69 A.

All of the huge production of vinyl chloride (1.6 billion pounds in 1964) is converted to resins and plastics. Vinyl chloride polymerizes readily in the presence of peroxides to a hard brittle resin in which the units are linked regularly in a head to tail fashion.

$$3x\ CH_2{=}CHCl \longrightarrow [-CH_2CHCl|CH_2CHCl|CH_2CHCl-]_x$$

Addition of plasticizers such as cresyl phosphate (p. 445) or alkyl phthalates (p. 474) gives tough, long-wearing, leather- or rubber-like materials that are used in hundreds of consumer products such as floor tile, shoe soles, raincoats, purses, wire insulation, pipe, tubing, protective coatings, and phonograph records. Many other commercial plastics, such as *Vinylite* (p. 599) and *Dynel* (p. 619), are copolymers with vinyl chloride.

Vinylidene chloride (*1,1-dichloroethylene*) can be prepared from acetylene and chlorine at 135° in the presence of ferric chloride, a reaction that may involve substitution and addition of hydrogen halide.

$$HC{\equiv}CH + Cl_2 \xrightarrow[135°]{FeCl_3} CH_2{=}CCl_2$$

Vinylidene chloride polymerizes to a material that is characterized by chemical inertness, high tensile strength, and resistance to abrasion. The commercial products, known as *sarans,* usually are copolymers with vinyl chloride, acrylonitrile, or other substituted ethylenes to improve their working properties. Emulsions of vinylidene chloride polymers have the unusual property of forming tough coherent sheets when spread on a smooth surface and allowed to dry.

Although olefins that yield secondary alkyl derivatives on the addition of unsymmetric addenda (*secondary-base olefins*) add halogen readily in the liquid phase by a polar mechanism (p. 80), they do not do so in the vapor phase. At a sufficiently high temperature (500–600°) rapid substitution takes place by a free radical mechanism. Thus propylene gives 85–90 per cent of **allyl chloride.**

$$Cl_2 \longrightarrow 2\,Cl\cdot$$

$$Cl\cdot + HCH_2CH{=}CH_2 \longrightarrow ClH + \{\cdot CH_2CH{=}CH_2 \longleftrightarrow CH_2{=}CHCH_2\cdot\} \xrightarrow{Cl_2} CH_2{=}CHCH_2Cl + Cl\cdot$$

With olefins that can give tertiary alkyl derivatives (*tertiary-base olefins*), the ratio of substitution to addition is not affected by temperature, and substitution by a polar mechanism is extremely rapid in the liquid phase or at higher temperatures in the vapor phase in contact with porous materials. At 300° and a contact time of 1 sec., up to 87 per cent **methallyl chloride** (*2-methylallyl chloride*) is obtained from isobutylene.

$$Cl_2 + CH_2{=}\underset{\displaystyle CH_3}{\underset{|}{C}}{-}CH_3 \longrightarrow Cl^- + ClCH_2{-}\underset{\displaystyle CH_3}{\underset{|}{\overset{+}{C}}}{-}CH_3 \longrightarrow \underset{\text{Methallyl chloride}}{ClCH_2\underset{\displaystyle CH_3}{\underset{|}{C}}{=}CH_2} + H^+$$

The combination C=C—C is known as the **allylic system.** Like the benzyl system (p. 454), it confers high reactivity on halogen attached to the carbon atom that is not doubly bound because free radical, carbonium ion, or carbanion intermediates are stabilized by resonance (p. 494).

Frequently allylic systems react with rearrangement. Thus solvolysis of 3-chloro-1-butene takes place readily with ethanol, but 82 per cent of the product is ethyl crotyl ether.

Here an S_N1 mechanism is involved. Nucleophilic attack by the alcohol can take place at either end of the allylic system but does so chiefly at the less hindered carbon.

$$CH_2{=}CHCH(CH_3)Cl \longrightarrow Cl^- \left\{ CH_2{=}CHCH^+(CH_3) \longleftrightarrow {}^+CH_2{-}CH{=}CH(CH_3) \right\} \xrightarrow{C_2H_5OH}$$

$$H^+ + \underset{18\text{ per cent}}{CH_3CHCH(CH_3)OC_2H_5} \quad \text{and} \quad \underset{82\text{ per cent}}{C_2H_5OCH_2CH{=}CHCH_3}$$

This behavior of allyl derivatives is called *allylic rearrangement.*

The principal use for allyl chloride is as a starting point for the preparation of allyl alcohol (p. 600), glycerol (p. 607), and epichlorohydrin (p. 608). Addition of hydrogen bromide in the absence of antioxidants gives **1-chloro-3-bromopropane** (*trimethylene chlorobromide*) used for the manufacture of cyclopropane (p. 641).

$$CH_2{=}CHCH_2Cl + HBr \xrightarrow{\text{Peroxides}} BrCH_2CH_2CH_2Cl$$

FLUORINATED HYDROCARBONS

The extensive contributions to the chemistry of fluorine compounds in recent years tend to obscure the fact that their fundamental chemistry has been known for a long time. The pioneering work of Moissan[2] was completed about 1900 and that of Swarts[3] by about 1925. The discovery of commercial uses for organic fluorine compounds, however, has led to the entrance of a large number of workers into the field, particularly chemists connected with industrial research laboratories. The commercial production of liquid hydrogen fluoride and fluorine, the development of methods for handling them safely, and the ready availability of commercially manufactured chlorinated organic compounds have led to a rapid extension of the work of the early investigators.

Chemical and Physical Properties

Fluorine compounds are characterized by extremes and opposites. Some are the most reactive of organic compounds and others are the most inert. Some are extremely toxic and others are as nontoxic as nitrogen or water. The introduction of fluorine may raise or lower the boiling point, and the progressive introduction of fluorine not only reduces the solubility in water but also in other organic solvents. Thus completely fluorinated hydrocarbons, known as *fluorocarbons,* are soluble in ether and in chlorofluorocarbons but are insoluble in most other solvents.

Primary alkyl fluorides are stable to aqueous or alcoholic alkali, do not form Grignard reagents, and are not affected by reducing agents. Nucleophiles in general do not react with them. Secondary and tertiary alkyl fluorides are more reactive. Many of them decompose on distillation and lose hydrogen fluoride when treated with alkali. They also undergo the Wurtz type of coupling reaction with alkylmagnesium halides.

Compounds that have one fluorine on each of two adjacent carbon atoms are very

[2] Ferdinand Frederic Henri Moissan (1852–1907), professor of chemistry at the University of Paris. He was the first to isolate fluorine in 1886, and in 1892 he prepared calcium carbide by heating lime and carbon in the electric arc furnace. He investigated many other reactions at high temperatures and was awarded the Nobel Prize in Chemistry in 1906.

[3] Frederic-Jean Edmond Swarts (1866–1940), professor at the University of Ghent. He developed methods for the synthesis of many organic fluorine compounds, particularly the reaction of antimony fluoride and mercurous fluoride with chloro derivatives, and studied extensively the thermochemistry and refractometry of fluorine compounds. He also prepared many organic chlorine compounds for purposes of comparison.

reactive. Thus 1,2-difluoroethane spontaneously loses hydrogen fluoride to give vinyl fluoride and is converted to ethylene glycol (p. 604) when shaken with water. On the other hand, compounds with two fluorines on the same carbon atom are even less reactive than the primary alkyl fluorides, being extremely inert to all chemical reactions. Fluorocarbons are surpassed in stability only by the inert gases. They are decomposed only at a red heat, the products being carbon and carbon tetrafluoride. They react with sodium or potassium at 300–400° and with sodium in liquid ammonia. Above 400° they react with silica to form silicon tetrafluoride. It is the reactions with sodium and silica that must be used for analytical purposes.

Two fluorine atoms also reduce the reactivity of other halogen atoms attached to the same carbon atom. Thus the reactivity of the chlorine atoms in CCl_2F_2 is less than that in CH_2Cl_2 or CCl_4. Even the usual course of a reaction may be changed. For example, methyl iodide is hydrolyzed easily to methyl alcohol, but trifluoromethyl iodide gives fluoroform and hypoiodite.

$$CF_3I + KOH \longrightarrow CF_3H + KOI$$

Whereas monofluorides are toxic compounds, *gem*-difluorides[4] usually are nontoxic. Even CCl_2F_2 lacks the toxicity and anesthetic action of CH_2Cl_2 or CCl_4. Exceptions to this rule are some extremely toxic polyfluorocyclopropanes and some fluorinated alkenes such as perfluoroisobutylene.

Organic fluorine compounds show unusual physical properties as well as unexpected chemical properties. The progressive replacement of hydrogen by chlorine causes a continual increase in the boiling point, but progressive replacement by fluorine causes an initial rise, which then is followed by a decrease. Thus the boiling points of methane, methyl fluoride, methylene fluoride, fluoroform, and carbon tetrafluoride are respectively −161°, −78°, −52°, −83°, and −128°. Similarly the boiling points of chloro-, chlorofluoro-, chlorodifluoro-, and chlorotrifluoromethane are −24°, −9°, −41°, and −81°. When the nuclear hydrogen atoms of aromatic compounds are replaced by fluorine, little change in boiling point takes place. Thus benzene, fluorobenzene, *o*-, *m*-, and *p*-difluorobenzene, trifluorobenzene, and hexafluorobenzene all boil within the range 80–91°.

A striking difference between mono- and polyfluoro compounds is observed in the interatomic distances. Although there is no detectable difference in the C—Cl bond distances in the chlorinated methanes, the C—F distance in methyl fluoride is 1.42 A, whereas that in *gem*-difluoro compounds is 1.36 A. Moreover the bond distances to other atoms attached to the same carbon atom are detectably decreased. Thus the C—Cl distance of 1.76 A in carbon tetrachloride is decreased to 1.70 A in dichlorodifluoromethane, and the C—C distance of 1.53 A in ethane is decreased to 1.48 A in 1,1,1-trifluoroethane.

Preparation

The direct fluorination of alkanes and the difficulties encountered are discussed on page 67. Hydrogen fluoride adds to double bonds to give alkyl fluorides, and to acetylenes to give vinyl fluorides or difluoroalkanes. One of the most useful procedures is the *Swarts reaction* in which chlorine is exchanged for fluorine. Various metallic fluorides may be used, but the most important reagent is hydrogen fluoride in the presence of antimony pentafluoride. The active reagent undoubtedly is antimony pentafluoride, which continuously is regenerated by the hydrogen fluoride.

$$RCl + SbCl_5 \longrightarrow RF + SbCl_4F$$
$$SbCl_4F + HF \longrightarrow SbCl_5 + HCl$$

[4] The prefix *gem-* (L. *geminus* twin) refers to two like atoms or groups attached to the same carbon atom.

The most recent method for preparing fluorinated hydrocarbons is by the reaction of sulfur tetrafluoride with oxygenated compounds. Sulfur tetrafluoride is a toxic gas, b.p. $-38°$, that can be prepared by refluxing sulfur chloride with a suspension of sodium fluoride in acetonitrile.

$$3\,SCl_2 + 4\,NaF \longrightarrow SF_4 + S_2Cl_2 + 4\,NaCl$$

In pressure vessels at temperatures from 120 to 180°, it brings about the replacement of hydroxyl and oxo groups by fluorine. Thus the primary alcohols are converted to fluoromethyl derivatives, ketones and quinones to difluoromethylene derivatives, and carboxylic acids, acid chlorides, acid anhydrides, esters, and amides to trifluoromethyl derivatives.

$$RCH_2OH + SF_4 \longrightarrow RCH_2F + SOF_2 + HF$$

$$R_2C{=}O + SF_4 \longrightarrow R_2CF_2 + SOF_2$$

$$R\underset{\underset{O}{\|}}{C}{-}OH \longrightarrow R{-}\underset{\underset{O}{\|}}{C}{-}F \longrightarrow R{-}CF_3$$

The reactions are catalyzed by Lewis acids. Double and triple bonds are not affected, nor do ethers react.

An electrochemical process for the synthesis of fluorine compounds is important commercially. Solutions of organic oxygen or nitrogen compounds in liquid hydrogen fluoride conduct the electric current. If a current is passed through the solution using a potential of 5 to 6 volts, the organic compound is converted without the evolution of fluorine to a perfluorohydrocarbon or a perfluoro derivative, depending on the conditions. Acetic acid, for example, can give chiefly carbon tetrafluoride or trifluoroacetyl fluoride. Sometimes rearranged products are obtained. The actual product or products formed must be determined by experiment. The perfluorinated products are insoluble in liquid hydrogen fluoride and are removed from the top of the cell as gases or from the bottom of the cell as liquids.

Effect of the Trifluoromethyl Group on Other Functions

The strong electron-withdrawing effect of the trifluoromethyl group markedly alters the behavior of other functional groups, particularly by increasing their acidity and decreasing their basicity. Thus 2,2,2-trifluoroethanol forms a salt with aqueous sodium hydroxide, and the acidity of 1,1,1,3,3,3-hexafluoro-2-propanol ($pK_a = 6.7$) is about like that of carbonic acid. Perfluoromethyl ether does not dissolve boron fluoride. Tris(trifluoromethyl)amine does not form salts even with strong acids. Hexafluoroacetone forms a stable hydrate (p. 199), and trifluoromethyl phenyl ketone is soluble in 10 per cent sodium hydroxide because of the acidity of its hydrate. Immediate acidification regenerates the ketone. On standing, however, the usual haloform decomposition takes place.

$$C_6H_5COCF_3 \xrightarrow{NaOH} C_6H_5\overset{\overset{O^{-+}Na}{|}}{\underset{\underset{OH}{|}}{C}}CF_3 \longrightarrow C_6H_5COO^{-+}Na + CHF_3$$

Commercially Important Products

Interest in fluorine compounds in recent years arises from their technical applications. The most important compound is **dichlorodifluoromethane** (*Freon-12*), which is used as a refrigerant for household and commercial refrigerators and air-conditioning equipment, and as a solvent-propellant for aerosol-type spray preparations. It is entirely noncorrosive, nontoxic, and nonflammable. Its cheap production depends on the reaction of carbon tetrachloride with anhydrous hydrogen fluoride in the presence of antimony pentafluoride, on the relatively high boiling point of hydrogen fluoride, on the decrease in boiling point of 40–50° for each replacement of chlorine by fluorine, and by the fact that the reaction does not proceed readily beyond the desired stage.

$$CCl_4 \xrightarrow{HF,\ SbF_5} CCl_3F \xrightarrow{HF,\ SbF_5} CCl_2F_2$$

The boiling points of the reactants and products in decreasing order are carbon tetrachloride, 76°; **trichlorofluoromethane** (*Freon-11*), 24°; hydrogen fluoride, 20°; dichlorodifluoromethane, −30°; and hydrogen chloride, −85°. Hence hydrogen fluoride and carbon tetrachloride can be added continuously to a reactor, and hydrogen chloride and dichlorodifluoromethane removed through a column and condenser, the hydrogen chloride being absorbed in water. The conditions of reaction can be modified to yield mainly F-11, which finds use in industrial cooling systems with centrifugal compressors, as a combination with F-12 to increase the solvent properties of the latter and to reduce the pressure in aerosol containers, and as a blowing agent for polyurethan foams (p. 605).

Chlorodifluoromethane (*Freon-22*), b.p. −41°, is made from chloroform by the Swarts reaction. It is used in refrigeration equipment having reciprocating compressors, and as an intermediate for the production of tetrafluoroethylene. **Bromotrifluoromethane,** b.p. −58°, is made by brominating fluoroform at 600–700°. It is very efficient in extinguishing fires and, because of its low toxicity, is used advantageously in closed areas. It has replaced carbon dioxide systems in commercial aircraft because the reduction in weight permits increase in payload by one passenger.

1,1,2,2-Tetrachloro-1,2-difluoroethane (*Freon-112*), b.p. 93°, **1,1,2-trichloro-1,2,2-trifluoroethane** (*Freon-113*), b.p. 48°, and **1,2-dichloro-1,1,2,2-tetrafluoroethane** (*Freon-114*), b.p. 3.8°, are made from hexachloroethane by the Swarts reaction. Because fluorine decreases the reactivity of chlorine on the same carbon atom, the reaction always takes place to produce a symmetric product and stops when four chlorines have been replaced.

$$\underset{}{CCl_3CCl_3} \longrightarrow \underset{\text{F-111}}{CCl_3CCl_2F} \longrightarrow \underset{\text{F-112}}{CCl_2FCCl_2F} \longrightarrow \underset{\text{F-113}}{CCl_2FCClF_2} \longrightarrow \underset{\text{F-114}}{CClF_2CClF_2}$$

Freon-114 is the common refrigerant for household refrigerators with rotary-type compressors.

1-Bromo-1-chloro-2,2,2-trifluoroethane (*halothane, Fluothane*) has been used widely as an inhalation anesthetic since 1956 because, in contrast to ether and cyclopropane, it is nonflammable. Its safety as regards toxicity, however, was questioned in 1963.

Pyrolysis of chlorodifluoromethane gives **tetrafluoroethylene. Hexafluoropropylene** is a coproduct.

$$2\ CHClF_2 \xrightarrow[\text{1 sec.}]{700°\text{ for}} F_2C{=}CF_2 + 2\ HCl$$

$$3\ CHClF_2 \longrightarrow F_3CCF{=}CF_2 + 3\ HCl$$

Tetrafluoroethylene is a surprisingly reactive compound, easily adding halogen acids and halogen, including iodine. At 200° in the absence of catalysts it gives the cyclic dimer, **perfluorocyclobutane.** The mechanism appears to be a simple 1,2 addition of the four-center type.

$$\begin{matrix} F_2C \\ \| \\ F_2C \end{matrix} + \begin{matrix} CF_2 \\ \| \\ CF_2 \end{matrix} \longrightarrow \begin{matrix} F_2C\cdots CF_2 \\ \vdots \quad \vdots \\ F_2C\cdots CF_2 \end{matrix} \longrightarrow \begin{matrix} F_2C-CF_2 \\ | \quad\ | \\ F_2C-CF_2 \end{matrix}$$

It has been approved as a propellant for canned food products such as whipped cream and cake toppings. Polymerization of tetrafluoroethylene at 700 p.s.i. in the presence of peroxide catalysts gives a product of molecular weight 500,000 to 2,000,000 known as **Teflon.**

$$x\ CF_2{=}CF_2 \longrightarrow (-CF_2CF_2-)_x$$

This polymer is characterized by extreme chemical inertness. It withstands the attack of all reagents except molten alkali metals. Aqueous alkalies, concentrated acids, oxidizing agents, and organic solvents have no effect on it. It can be used in the temperature range −70 to 250°. It softens above 250°, changes to a rubbery state at 325°, and depolymerizes at 600–800° without charring. Although it thus is thermoplastic, it is much more difficult to work than most plastics. A property useful in the laboratory is the ability of small pieces to prevent bumping in boiling operations. Introduced in 1944 at $18 per pound, it sold for $3.25 per pound in 1963. Production was 12.5 million pounds.

Copolymerization of tetrafluoroethylene with hexafluoropropylene gives a product called **Teflon-100** that has the same chemical resistance, electrical insulating, and antifrictional properties of Teflon but softens at a lower temperature and can be molded or extruded more readily. The maximum temperature at which it is serviceable, however, is about 50° lower than for Teflon.

Of the approximately 410 million pounds of fluorinated hydrocarbons reported in 1963, F-12 accounted for 53 per cent; F-11, 34 per cent; F-22, 10 per cent; F-114, 3 per cent. Aerosol sprays consumed 49 per cent; refrigerants, 29 per cent; and all other uses such as plastics, films, elastomers, coatings, lubricants, textile treatment, solvents, and foaming agents, 22 per cent.

PROBLEMS

34–*1*. Construct a chart indicating the steps for the preparation of the following compounds starting with acetylene, including necessary reagents, catalysts, and conditions: acetylene tetrachloride, vinyl chloride, vinylidene chloride, ethylidene chloride, vinyl fluoride, 1,1-difluoroethane, 1-chloro-1,1-difluoroethane, trichloroethylene, pentachloroethane, hexachloroethane, 1,2-dichloroethylene, and perchloroethylene.

34–*2*. What halogenated hydrocarbons containing one and two carbon atoms can be made starting with the chlorination of methane?

34–*3*. Give reactions for the conversion of propylene into (*a*) 1,2-dichloropropane, (*b*) allyl chloride, (*c*) 1-bromo-3-chloropropane, and (*d*) 1,2-dibromo-3-chloropropane.

34–*4*. Give reactions for the following syntheses: (*a*) chlorotrifluoroethylene from methane, (*b*) tetrafluoroethylene from acetaldehyde, (*c*) 1,2-dichloro-2-fluoropropane from allyl chloride, (*d*) trifluoroiodomethane from acetic acid, (*e*) tribromofluoromethane from acetone, (*f*) halothane from trichloroethylene.

34–*5*. Explain the fact that carbonation of cinnamylmagnesium chloride gives chiefly 2-phenyl-3-butenoic acid.

34–*6*. Compound *A* has the molecular formula $C_7H_{13}Cl$. It readily decolorizes a solution of bromine in carbon tetrachloride. Ozonolysis gives two products, each of which gives a color with Schiff reagent, but only one gives a positive iodoform test. After *A* is boiled with water, the aqueous layer gives a precipitate with silver nitrate. When *A* is heated, two products are obtained that can be separated by careful fractional distillation. One is identical with *A*, but the other, *B*, is isomeric with *A*. *B* likewise decolorizes bromine but hydrolyzes somewhat more slowly than *A*. Ozonolysis of *B* also gives two products, one of which gives a color with Schiff reagent but a negative iodoform test, whereas the other gives a negative test with Schiff reagent and a positive iodoform test. Give a formula for *A* and equations for the reactions that take place.

CHAPTER THIRTY-FIVE

UNSATURATED ALCOHOLS, EPOXIDES, POLYHYDRIC ALCOHOLS, AMINO ALCOHOLS, AND POLYAMINES

UNSATURATED ALCOHOLS

The first member of this series, **vinyl alcohol,** is unknown in the monomeric state because acetaldehyde, the carbonyl form, is more stable.

$$\underset{\text{Vinyl alcohol}}{[CH_2{=}CHOH]} \longrightarrow \underset{\text{Acetaldehyde}}{CH_3CHO}$$

Vinyl acetate can be made by the catalyzed addition of acetic acid to acetylene in either the liquid or the vapor phase, by reaction of ethylene chloride and sodium acetate, or by the reaction of acetaldehyde with acetic anhydride.

$$HC{\equiv}CH + HOCOCH_3 \xrightarrow[\text{or Zn (OAc)}_2\text{ at }210\text{–}250°]{HgSO_4\text{ at }75\text{–}80°} \underset{\text{Vinyl acetate}}{H_2C{=}CHOCOCH_3}$$

$$ClCH_2CH_2Cl + 2\,NaOCOCH_3 \longrightarrow H_2C{=}CHOCOCH_3 + 2\,NaCl + CH_3COOH$$

$$CH_3CHO \rightleftharpoons [CH_2{=}CHOH] \xrightarrow{Ac_2O} CH_2{=}CHOCOCH_3 + CH_3COOH$$

If the palladium chloride–copper chloride process for making acetaldehyde from ethylene (p. 214) is carried out in acetic acid and sodium acetate, the product is vinyl acetate. The first method is used most commercially. Vinyl acetate boils at 72° and can be polymerized readily with peroxide catalysts to give a tough, thermoplastic resin,[1] soluble in aromatic hydrocarbons.

$$2x\,CH_2{=}CHOCOCH_3 \longrightarrow \underset{\text{Poly(vinyl acetate)}}{\left[\begin{array}{c}-CH_2\underset{|}{C}H-CH_2-\underset{|}{C}H- \\ \quad OCOCH_3 \quad OCOCH_3\end{array}\right]_x}$$

Poly(vinyl acetate) emulsions are used as adhesives and in latex paints. Copolymerization of vinyl acetate and vinyl chloride in various proportions and to varying degrees of polymerization gives products known as *Vinylite* resins, which have a wide range of properties.

[1] Although the term *vinyl plastics* originally was applied only to those made from monomers containing the simple vinyl group, such as vinyl acetate and vinyl chloride, it now refers to all polymers of negatively substituted ethylenes, for example polystyrene (p. 484) and poly(methyl methacrylate) (p. 620).

They are very inert to chemical agents and weathering, and can be used to produce rigid sheets, flexible sheeting and films, textile fibers (*Vinyon*), molded and extruded articles, and surface coatings that are long-wearing and scuff- and stain-resistant. Floor coverings, upholstering materials, and shoe soles are made from it.

Saponification of poly(vinyl acetate) gives **poly(vinyl alcohol),** commonly called **PVA.**

$$\left[-CH_2CH(OCOCH_3)-CH_2-CH(OCOCH_3)-\right]_x \xrightarrow{NaOH} \underset{\text{Poly(vinyl alcohol)}}{\left[-CH_2CH(OH)CH_2CH(OH)-\right]_x}$$

The physical properties of the product depend on the extent of hydrolysis. Products in which about 90 per cent of the acetyl groups have been removed dissolve readily in water. As with partially hydrolyzed cellulose acetate and partially methylated cellulose (p. 344), the presence of a few larger groups keeps the chains apart and permits hydration of the hydroxyl groups. The fully hydrolyzed material is water resistant. It can be made completely insoluble in water by treatment with an aldehyde that will form acetal cross links. With this wide range of properties, various types of poly(vinyl alcohol) have found uses from water-soluble films and adhesives to strong fibers used in fish nets.

Because most of the hydroxyl groups of poly(vinyl alcohol) occupy 1,3 positions, it reacts readily with aldehydes to give cyclic acetals having six-membered rings (cf. p. 603).

$$\left[-CH_2CH(OH)CH_2CH(OH)-\right]_x + x\,RCHO \xrightarrow[\text{acid}]{\text{Dil.}} \left[-CH_2CH\overset{CH_2}{\frown}CH-\ (\text{ring: } -O-CH(R)-O-)\right]_x + x\,H_2O$$

The reaction product with formaldehyde is known as **poly(vinyl formal)** or *Formvar* and is used as an insulating enamel for electric wire. **Poly(vinyl butyral)** or *Butacite,* from PVA and *n*-butyraldehyde, is sandwiched between two sheets of glass to make a very strong safety glass.

Alcohols add to acetylene in the presence of potassium hydroxide at elevated temperatures to give *vinyl alkyl ethers.*

$$HC{\equiv}CH + HOR \xrightarrow[150\text{–}180^\circ]{KOH} H_2C{=}CHOR$$

Polymerization of **methyl vinyl ether** gives **poly(vinyl methyl ether),** which may be obtained as a sticky thick liquid or as a soft solid. It is soluble in organic solvents and in cold water but is precipitated from aqueous solutions when heated to 35°. It is used as an adhesive coagulant in aqueous media.

Allyl alcohol is made on a large scale commercially by the hydrolysis of allyl chloride (p. 593) at *p*H 8 to 11. It is used as an intermediate for the synthesis of glycerol (p. 607).

$$\underset{\text{Allyl chloride}}{CH_2{=}CHCH_2Cl} \xrightarrow[Na_2CO_3,\ NaOH]{H_2O} \underset{\text{Allyl alcohol}}{CH_2{=}CHCH_2OH}$$

The *cis* form of **1-hydroxy-3-hexene** (*leaf alcohol*), $CH_3CH_2CH{=}CHCH_2CH_2OH$, is responsible for the characteristic odor of green grass and leaves. **Bombykol,** the sex attractant secreted by the abdominal cells of the silkworm moth (*Bombyx mori*), is 1-hydroxy-*trans*-10-*cis*-12-hexadecadiene; 12 mg. was obtained from 500,000 virgin females.

EPOXIDES

The cyclic ethers that have three and four atoms in the ring commonly are called *1,2* and *1,3 epoxides.* They are considered here because their reactions lead to the glycols and structurally related compounds. The simplest representative has the common name *ethylene oxide* and may be called *epoxyethane.* The official name for the heterocycle is *oxirane.* A general method of preparation is by the reaction of alkali with 1,2 chlorohydrins.

$$\underset{\text{OH}\ \ \text{Cl}}{\text{RCH—CH—R}} \xrightarrow{^-\text{OH}} H_2O + \underset{\text{O}^-\ \ \text{Cl}}{\text{RCH—CHR}} \longrightarrow Cl^- + \underset{\text{O}}{\text{RCH—CHR}}$$

The chlorohydrins are obtained by the reaction of alkenes with chlorine and water.

$$\text{RCH=CHR} \xrightarrow{Cl_2} Cl^- + \overset{+}{\text{RCH}}\text{—}\underset{\text{Cl}}{\text{CHR}} \xrightarrow{H_2O} \underset{H_2O^+\ \ \text{Cl}}{\text{RCH—CHR}} \longrightarrow H^+ + \underset{\text{OH}\ \ \text{Cl}}{\text{RCH—CHR}}$$

This reaction frequently is spoken of as the addition of hypohalous acid. That hypohalous acid is not an intermediate is indicated by the fact that if the reaction is carried out in methanol, the methoxy halide is formed. 1,2 Epoxides can be prepared directly from unsaturated compounds and peroxy acids. In the laboratory peroxybenzoic acid (p. 470) usually is used, but for large scale operations peroxyacetic acid is preferred (p. 165).

Although unprotonated ethers do not react with nucleophiles (cf. p. 123), the 1,2 epoxides are very susceptible to nucleophilic attack because of the release of strain in the three-membered ring when it is opened. Alcohols, mercaptans, carboxylic acids, or amines react readily, especially in the presence of base. The point of attack is determined by the relative electron density and relative steric hindrance at the ring carbon atoms. Thus monoalkyl-substituted ethylene oxides give secondary β-hydroxy compounds.

$$\underset{\text{O}}{\text{RCH—CH}_2} \xrightarrow{^-:SR'} \underset{\text{O}^-}{\text{RCHCH}_2\text{SR}'} \xrightarrow{HSR'} {}^-SR' + \underset{\text{OH}}{\text{RCHCH}_2\text{SR}'}$$

$$\underset{\text{O}}{\text{RCH—CH}_2} + \text{HOR}' \longrightarrow \text{RCHOHCH}_2\text{R}'$$

$$+ \text{HNHR}' \longrightarrow \text{RCHOHCH}_2\text{NHR}$$

$$+ \text{HOCOR}' \longrightarrow \text{RCHOHCH}_2\text{OCOR}'$$

The reaction of Grignard reagents with ethylene oxide is very useful in organic synthesis because it permits the building of a carbon chain two atoms at a time.

$$\text{RMgX} + (CH_2)_2O \longrightarrow RCH_2CH_2OMgX \xrightarrow{HX} RCH_2CH_2OH + MgX_2$$

In reactions with nucleophiles at a pH less than about 4, acid catalysis takes over, and the initial attack is on oxygen. Here the ease with which the oxygen can leave with the bonding pair of electrons determines which bond breaks. The chief product from a monoalkyl oxide is the primary alcohol.

$$\underset{\text{O}}{\text{RCH—CH}_2} \xrightarrow{H^+} \underset{+\text{OH}}{\text{RCH—CH}_2} \xrightleftharpoons{R'OH} \overset{\overset{+}{\text{HOR}'}}{\underset{\text{OH}}{\text{RCHCH}_2}} \underset{H^+}{\rightleftharpoons} \overset{\text{OR}'}{\underset{\text{OH}}{\text{RCHCH}_2}}$$

If R is a strongly electron-attracting group, the mode of addition may be reversed. Mixtures or predominantly one or the other isomer may be obtained depending on the relative importance of the various factors. In either the acid- or base-catalyzed reactions, the nucleophilic attack phase causes Walden inversion at one of the carbon atoms. Thus if both groups are alike in a 1,2-disubstituted alkoxide, hydrolysis of *cis* compounds gives racemic glycols, and *trans* compounds yield *meso* glycols (p. 305).

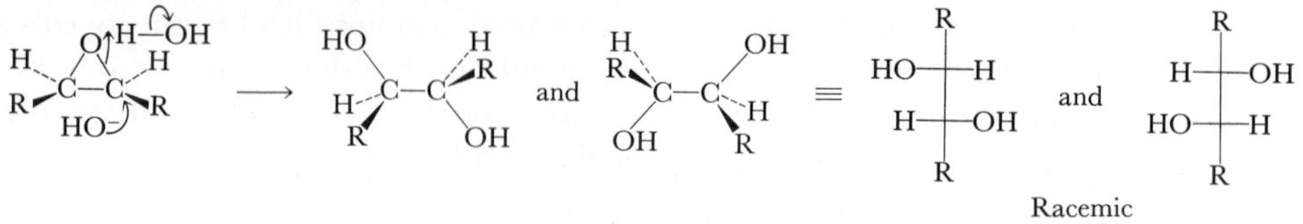

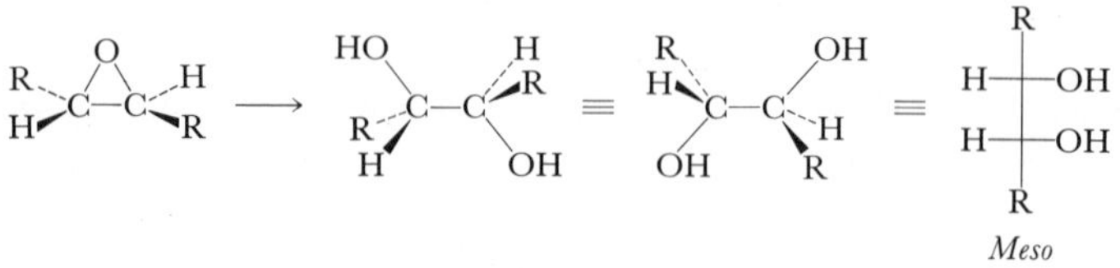

Racemic

Meso

For the *cis* isomer, the second configuration results from attack on the other carbon. For the *trans* isomer, the same configuration results from attack on either carbon atom.

The **1,3 epoxides** commonly are called *trimethylene oxides* or *oxetanes.* They usually are made by way of the 1,3 chlorohydrins. Although the four-membered ring, once it is formed, is less strained than the three-membered ring, the probability of forming the four-membered ring by intramolecular cyclization is less, and the yields are low (20–30 per cent). **Trimethylene oxide** is obtained in good yield if the glycol is dissolved in sulfuric acid, and the solution neutralized and distilled with an excess of sodium hydroxide.

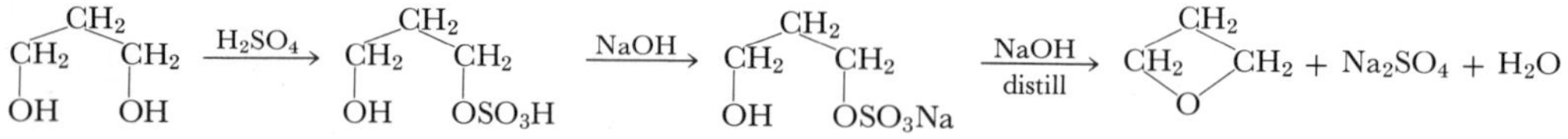

$$HOCH_2CH_2CH_2OH \xrightarrow{H_2SO_4} HOCH_2CH_2CH_2OSO_3H \xrightarrow{NaOH} HOCH_2CH_2CH_2OSO_3Na \xrightarrow[\text{distill}]{NaOH} (CH_2)_3O + Na_2SO_4 + H_2O$$

Oxetanes undergo the same types of reactions as oxiranes but more slowly. The five- and six-membered oxygen heterocycles, such as tetrahydrofuran (p. 512) and tetrahydropyran (p. 521), and the still higher-membered rings are unstrained and show the typical inertness of acyclic ethers.

POLYHYDRIC ALCOHOLS

Aldehyde Hydrates

Compounds that have two hydroxyl groups on the same carbon atom usually are unstable and lose water to form the carbonyl derivative. Methanediol, the hydrate of formaldehyde, appears to exist only in aqueous solution (p. 199). The presence of electron-attracting groups, however, increases the stability of the hydrate. Thus dichloroacetaldehyde and trichloroacetaldehyde form monohydrates melting at 55° and 52° respectively (p. 215). Hexafluoroacetone and triketohydrindene also form stable hydrates (pp. 596, 646). Derivatives of 1,1-alkanediols such as the hemiacetals (p. 199) are unstable, but the acetals (p. 200) and acylals are stable compounds.

1,2 Glycols

Preparation. Simple diols usually are called *glycols*. Several general methods for the preparation of 1,2 glycols commonly are used.

(*a*) OXIDATION OF UNSATURATED COMPOUNDS. Suitable oxidizing agents for the hydroxylation of the double bond are dilute aqueous permanganate (p. 90) and hydrogen peroxide in the presence of osmium tetroxide.

$$RCH{=}CHR + OsO_4 \longrightarrow \underset{\substack{| \quad\ \ | \\ O \quad O \\ \diagdown \ \diagup \\ Os \\ \diagup \ \diagdown \\ O \quad O}}{RCH{-}CHR} \xrightarrow{H_2O} \underset{\substack{| \quad\ \ | \\ OH \ \ OH}}{RCH{-}CHR} + OsO_3$$

$$OsO_3 + H_2O_2 \longrightarrow OsO_4 + H_2O$$

These hydroxylations take place by way of *cis* addition to give *cis* cyclic esters (p. 90). Since hydrolysis does not involve the carbon-oxygen bonds, they yield the so-called *cis* glycols; that is, if the R groups are identical and not hydrogen, the *cis* olefin gives the *meso* form and the *trans* olefin gives the racemic form (cf. p. 602).

(b) HYDROLYSIS OF 1,2 EPOXIDES. The hydrolysis of 1,2 epoxides (*ethylene oxides,* p. 601) to 1,2 glycols is catalyzed by either acids or bases. Here the products are *trans* glycols; that is, *cis* epoxides give racemic glycols, and *trans* epoxides give *meso* glycols (p. 602).

(c) HYDROLYSIS OF 1,2 DIHALIDES OR HALOHYDRINS. Dihalides are converted to the glycols best by way of the diacetates.

$$\underset{\substack{| \ \ | \\ X \ X}}{RCHCHR} \xrightarrow{NaOCOCH_3} \underset{\substack{| \qquad\qquad | \\ OCOCH_3 \ OCOCH_3}}{RCH{-}{-}{-}CHR} \xrightarrow{NaOH} \underset{\substack{| \quad\ \ | \\ OH \ \ OH}}{RCH{-}CHR}$$

The halohydrins hydrolyze very easily to the glycols.

$$\underset{\substack{| \quad\ \ | \\ OH \ \ X}}{RCH{-}CHR} + H_2O + Na_2CO_3 \longrightarrow \underset{\substack{| \ \ | \\ HO \ OH}}{RCHCHR} + NaX + NaHCO_3$$

1,2 Glycols result also from the pinacol reduction of ketones (p. 207) and from the reduction of acyloins (p. 176), and have been isolated from natural products (p. 182).

Reactions. 1,2 Glycols exhibit several characteristic reactions not given by simple alcohols.

(a) FORMATION OF CYCLIC COMPOUNDS. Because of the 1,2 position of the hydroxyl groups, reactions leading to the formation of 5-membered rings are common. Thus reaction with aldehydes or ketones yields **cyclic acetals.**

$$\begin{matrix} RCHOH \\ | \\ RCHOH \end{matrix} + O{=}CHC_6H_5 \xrightarrow{HCl} \begin{matrix} RCH{-}O \\ | \qquad\ \ \ \\ RCH{-}O \end{matrix}\! \qquad\qquad\quad \rangle CHC_6H_5$$

Benzylidene derivative

$$\begin{matrix} RCHOH \\ | \\ RCHOH \end{matrix} + O{=}C(CH_3)_2 \xrightarrow{HCl} \begin{matrix} RCH{-}O \\ | \qquad\ \ \ \\ RCH{-}O \end{matrix}\! \qquad\qquad\quad \rangle C(CH_3)_2$$

Isopropylidene derivative

If the carbonyl compound is relatively nonvolatile, the high boiling point of the glycol permits removal of the water by distillation, thus driving the reaction to completion. Benzene or toluene may be used as an entrainer. The products frequently are called *ethylene ketals* or *dioxolanes.* The reaction with ethylene glycol (p. 604) is used to protect carbonyl groups while other reactions are carried out under neutral or basic conditions, since the protecting group can be removed readily by acid hydrolysis.

1,2 Glycols increase the conductivity of boric acid solutions because, after formation of the borate, the unshared pair of electrons of the fourth hydroxyl group fills the empty orbital of the boron atom, permitting ionization of a proton.

$$\begin{matrix}\text{RCHOH}\\ | \\ \text{RCHOH}\end{matrix} + B(OH)_3 + \begin{matrix}\text{HOCHR}\\ | \\ \text{HOCHR}\end{matrix} \longrightarrow 3\,H_2O + \left[\begin{matrix}\text{RCH—O} & & \text{O—CHR}\\ | & B & | \\ \text{RCH—O} & & :\ddot{O}\text{—CHR}\\ & & | \\ & & H\end{matrix}\right] \longrightarrow \left[\begin{matrix}\text{RCH—O} & & \text{O—CHR}\\ | & B & | \\ \text{RCH—O} & & \text{O—CHR}\end{matrix}\right]^- H^+$$

Cyclic compounds in which the ring is closed by coordination with an unshared pair of electrons are known as **chelate compounds** (Gr. *chele* claw), the ring closure being thought of as a pincer-like action. The process is called **chelation.**

(b) PINACOL-PINACOLONE REARRANGEMENT. In the acid-catalyzed rearrangement of 1,2 glycols (p. 207), hydrogen migrates in preference to a hydrocarbon group.

$$RCHOHCH_2OH \xrightarrow{H^+} RCH_2CHO + H_2O$$
$$RCHOHCHOHR \longrightarrow RCH_2COR$$
$$R_2COHCOHR_2 \longrightarrow R_3CCOR$$

The easy preparation of isobutyraldehyde by hydrolysis of methallyl chloride and of methyl isopropyl ketone by hydrolysis of 2,3-dichloro-2-methylbutane is the result of a pinacol-pinacolone rearrangement.

$$CH_2{=}C(CH_3)CH_2Cl \xrightarrow[\text{heat}]{H_2O,\ H_2SO_4} (CH_3)_2COHCH_2Cl \xrightarrow{H_2O} (CH_3)_2COHCH_2OH \xrightarrow{H^+} (CH_3)_2CHCHO$$

$$(CH_3)_2CClCHClCH_3 \xrightarrow[\text{heat}]{H_2O,\ H_2SO_4} (CH_3)_2COHCHOHCH_3 \xrightarrow{H^+} (CH_3)_2CHCOCH_3$$

(c) OXIDATIVE SCISSION. The bond between the two hydroxylated carbon atoms can be split readily by oxidation. If reagents such as permanganate or acid dichromate are used, two moles of acid are formed. With certain other reagents, the oxidation stops at the aldehyde stage. Lead tetraacetate is the preferred reagent for this purpose in anhydrous solvents (*Criegee reaction*), and periodic acid is the preferred reagent in aqueous solutions (*Malaprade reaction*).

$$RCHOHCHOHR + Pb(OCOCH_3)_4 \longrightarrow 2\,RCHO + Pb(OCOCH_3)_2 + 2\,HOCOCH_3$$
$$RCHOHCHOHR + HIO_4 \longrightarrow 2\,RCHO + HIO_3 + H_2O$$

Since both lead tetraacetate and periodic acid can be estimated readily by iodimetric methods, the reactions can be used for the quantitative determination of 1,2 glycols.

Important 1,2 Glycols. After deciding that the structural formula for glycerol is 1,2,3-propanetriol, Wurtz reasoned that an analogous 1,2-ethanediol should be possible. In 1859 he reported its synthesis by the saponification of the acetate, which he prepared by the action of silver acetate on ethylene iodide. Because the product is sweet and resembles glycerol in its properties, it was called *glycol.*

Since 1925 **ethylene glycol,** $HOCH_2CH_2OH$, has been technically important, and it was one of the first synthetic organic chemicals to be produced commercially from petroleum. Its development was not planned, although it resulted from a research program. One of the first industrial fellowships to be established at Mellon Institute was that by the Prest-O-Lite Company for the development of a cheaper process for the manufacture of acetylene. A process was developed for the thermal cracking of natural gas or petroleum to acetylene, but a large amount of ethylene also was obtained. Attempts to utilize the ethylene led to the production of ethylene glycol. When glycol first became available in quantity in 1922 there were no important uses for it. Production in 1964 was over 1.8 billion pounds.

Most of the ethylene glycol is made by the hydrolysis of ethylene oxide with dilute sulfuric acid at 60° or with water at 200°.

$$(CH_2)_2O + H_2O \xrightarrow{200^\circ} HOCH_2CH_2OH$$

It boils at 197° and is separated from water and the coproducts, diethylene and triethylene glycols

(p. 605) by distillation. The **ethylene oxide** (2.2 billion pounds, 1964), a toxic flammable gas, is obtained chiefly by direct oxidation of ethylene with air or oxygen in the presence of a silver catalyst.

$$CH_2{=}CH_2 + \tfrac{1}{2}\,O_2 \xrightarrow[250^\circ]{\text{Ag cat.,}} \underset{\diagdown O \diagup}{CH_2{-}CH_2}$$

About three fourths of the ethylene glycol is used as a nonvolatile antifreeze for automobile radiators, and as a coolant for airplane motors. The next largest uses are for the synthesis of polyester fiber (3 per cent) and glycol nitrate (2 per cent). Numerous miscellaneous uses account for the balance. Like glycerol it is hygroscopic and can replace glycerol for many technical uses. Ethylene glycol should not be used in foods or cosmetics, however, because it is relatively toxic. Large doses depress the central nervous system and lead to cyanosis and respiratory failure. Administration of sublethal doses to rats over a long period leads to the deposition of calcium oxalate in the renal tubules and uremic poisoning. The **dinitrate** is an explosive and is used to lower the freezing point of nitroglycerin (p. 608).

When ethylene oxide reacts with an alcohol or a phenol, a monoalkyl or monoaryl ether of ethylene glycol is formed.

$$(CH_2)_2O + ROH \longrightarrow HOCH_2CH_2OR$$

$$(CH_2)_2O + ArOH \longrightarrow HOCH_2CH_2OAr$$

The **monoethyl ether** was called *Cellosolve* because it is a solvent for cellulose nitrate. It is used in the formulation of lacquers. Other glycol ethers and their acetates also are useful for this purpose. The **monomethyl ether** is added to jet fuel to prevent the formation of ice crystals. The **dimethyl ether,** commonly called *glyme,* is used as a solvent medium for organic reactions.

The alkali-catalyzed hydration of ethylene oxide yields **di(ethylene glycol), tri(ethylene glycol),** higher condensation products, and finally high molecular weight **poly(ethylene glycols),** the molecular weight depending on the amount of water initially added.

$$(CH_2)_2O \xrightarrow{H_2O,\,^-OH} HOCH_2CH_2OH \xrightarrow{(CH_2)_2O} \underset{\text{Di(ethylene glycol)}}{HOCH_2CH_2OCH_2CH_2OH} \xrightarrow{(CH_2)_2O}$$

$$\underset{\text{Tri(ethylene glycol)}}{HOCH_2CH_2OCH_2CH_2OCH_2CH_2OH} \xrightarrow{x\,(CH_2)_2O} \underset{\text{Poly(ethylene glycols)}}{HO(CH_2CH_2O)_{x+2}CH_2CH_2OH}$$

Di(ethylene glycol) and tri(ethylene glycol) are used to absorb moisture in the processing of natural gas. The poly(ethylene glycols) vary in properties from sticky viscous liquids to wax-like or tough solids (*Carbowaxes*) depending on their molecular weight (20,000 to several million), and all are soluble in water.

If the polymerization of ethylene oxide is initiated by the trihydric alcohol, glycerol (p. 607), a branched polyether triol is obtained.

$$\begin{array}{l} CH_2OH \\ | \\ CHOH \\ | \\ CH_2OH \end{array} + \begin{array}{l} x\,(CH_2)_2O \\ \\ y\,(CH_2)_2O \\ \\ z\,(CH_2)_2O \end{array} \longrightarrow \begin{array}{l} CH_2O(CH_2CH_2O)_xH \\ | \\ CHO(CH_2CH_2O)_yH \\ | \\ CH_2O(CH_2CH_2O)_zH \end{array}$$

Similarly sorbitol (p. 349) yields a polyether hexol. These polymeric polyols now are the chief resins used in the manufacture of *polyurethan foams.* The basic chemistry of the **polyurethans** is the reaction of a polyfunctional resin that has terminal hydroxyl groups with a diisocyanate to give a product of high molecular weight.

$$x\,OCNArNCO + x\,HO(CH_2CH_2O)_yH \longrightarrow [-CONHArNHCOO(CH_2CH_2O)_y-]_x$$

The amount of branching and cross linking depends on the number of hydroxyl groups in the original resin and can be varied to give elastic fibers (*spandex*), elastomers, or elastomeric, semirigid, or rigid foams. The diisocyanate most used is the technical "toluene diisocyanate" (p. 423). Catalysts are amines, such as diethylenetriamine (p. 611). Foamed products are obtained by adding a low-boiling liquid, such as methylene chloride or trichlorofluoromethane, which is volatilized by heating during the curing period. Alternatively water can be added, which reacts with some of the diisocyanate with the formation of carbon dioxide (p. 287). This process leads to foaming at room temperature but is more expensive because of the high cost of the diisocyanate. The cell size and structure of the foams are controlled by the addition of liquid silicones (p. 278). Other types of polyurethans can be made from other resins such as poly(tetramethylene glycol) (p. 514) or the adipic acid-ethylene glycol polyesters (p. 635). Poromeric materials such as *Corfam* are polyurethans reinforced with polyester. They are permeable to water vapor and other gases and are used to replace leather in shoe uppers.

Reaction of monoalkyl ethers of ethylene glycol with ethylene oxide gives the **monoalkyl ethers of di(ethylene glycol),** which are known as *Carbitols* and are used in lacquer formulation. The **dimethyl ether of di(ethylene glycol),** commonly referred to as *diglyme,* is a frequently used aprotic solvent medium for organic reactions. When alkylated phenols react with an excess of ethylene oxide, the **alkylaryl ethers** of **poly(ethylene glycol)** are formed.

$$RC_6H_4OH + (x + 1)(CH_2)_2O \longrightarrow RC_6H_4O(CH_2CH_2O)_xCH_2CH_2OH$$

Products of this type in which the alkyl group has 8 to 10 carbon atoms and x is 8 to 12 are good **nonionic detergents.** The nonylphenol obtained by the reaction of propylene trimer with phenol (p. 444) commonly has been used, but straight-chain alkyl phenols now are used to increase biodegradability. The reaction of ethylene oxide with fat acids such as those from tall oil (p. 342) gives **esters of poly(ethylene glycol),** which also are used as nonionic detergents in the formulation of low-sudsing products for automatic washers.

$$RCOOH + (x + 1)(CH_2)_2O \longrightarrow RCOO(CH_2CH_2O)_xCH_2CH_2OH$$

Propylene oxide (539 million pounds, 1964) is made almost entirely by way of the chlorohydrin.

$$CH_3CH{=}CH_2 \xrightarrow[H_2O]{Cl_2,} \left\{\begin{matrix} CH_3CHOHCH_2Cl \\ \text{90 per cent} \\ CH_3CHClCH_2OH \\ \text{10 per cent} \end{matrix}\right\} \xrightarrow{Ca(OH)_2} \underset{\text{Propylene oxide}}{CH_3CH{-}CH_2 \ (\text{-O-})}$$

A direct oxidation process was reported in 1965 but no details were given. Hydrolysis yields **propylene glycol.**

$$CH_3CH{-}CH_2\ (\text{-O-}) + H_2O \xrightarrow{H^+} CH_3\underset{OH}{C}HCH_2OH$$

It has properties similar to those of ethylene glycol. Unlike ethylene glycol, however, it is nontoxic and can replace glycerol in food products and cosmetics. It is used as a coolant in refrigerator systems in dairies, breweries, and food packaging plants where toxicity may be a factor. About one fourth of the 1964 production of 236 million pounds is the U.S.P. grade. Polyesters (p. 635) consume over one third and 16 per cent is used to prevent cellophane from drying out and becoming brittle. Aerosols of propylene glycol have been used in hospitals and schools to reduce the incidence of air-borne infections. Apparently the droplets take up moisture from the air and condense on bacteria, which are carried to the floor.

The **poly(propylene glycols)** are more oil-soluble and less hygroscopic than the poly(ethylene glycols). Their chief use has been in the manufacture of polyurethans (p. 605).

1,3 Glycols

Many 1,3 glycols can be prepared by reduction of β-hydroxy ketones or β-keto esters. The latter compounds are the product of aldol additions (p. 202) or Claisen ester condensations (p. 620). Thus aldol addition of n-butyraldehyde, followed by reduction, gives **2-ethyl-1,3-hexanediol.**

$$n\text{-}C_3H_7CHO + \underset{C_2H_5}{C}H_2CHO \xrightarrow{\text{Dil. NaOH}} C_3H_7CHOH\underset{C_2H_5}{C}HCHO \xrightarrow{H_2/Ni} \underset{\text{2-Ethyl-1,3-hexanediol}}{C_3H_7CHOH\underset{C_2H_5}{C}HCH_2OH}$$

It is an effective insect repellent, marketed as "6–12." **2-Methyl-2-propyl-1,3-propanediol,** the intermediate for the synthesis of meprobamate (p. 286), is obtained by the reaction of α-methylvaleraldehyde with formaldehyde.

α,ω Glycols

Glycols in general may be made by the hydrolysis of the corresponding dihalide. Frequently the reaction of the dihalide with sodium acetate in acetic acid solution to produce the diacetate proceeds better, and the ester can be hydrolyzed to the glycol.

$$Br(CH_2)_xBr + 2\,NaOCOCH_3 \longrightarrow CH_3COO(CH_2)_xOCOCH_3 \xrightarrow{NaOH} HO(CH_2)_xOH$$

If the proper dicarboxylic acid is available, the desired glycol can be prepared by the reduction of the diester.

$$CH_3OOC(CH_2)_xCOOCH_3 \xrightarrow[\text{Na + ROH}]{H_2/Ni,\text{ or}} HOCH_2(CH_2)_xCH_2OH$$

Tetramethylene glycol (*1,4-butanediol*) is made by the hydrolysis of tetrahydrofuran (p. 512) or by reduction of 1,4-dihydroxy-2-butyne (p. 200). Tetramethylene halohydrins are useful intermediates in laboratory syntheses and tetramethylene chloride is used technically for the synthesis of adiponitrile (p. 631). These halogen derivatives can be obtained directly from tetrahydrofuran by reaction with halogen acids (p. 513).

Tri- to Hexahydric Alcohols

The trihydric alcohol of most importance is **glycerol.** It was isolated first by Scheele in 1779 as a saponification product of fats (p. 184). He called it *oelsuess* because of its sweet taste. It was named *glycerin* (Gr. *glykeros* sweet) by Chevreul (p. 182). Berthelot (p. 121) showed it to be a trihydric alcohol in 1854, and its structural formula was assigned to it by Wurtz in 1855. Until 1948 it was obtained almost exclusively as a coproduct of the manufacture of soap, and its price varied greatly. In 1939 it sold for $0.12 and in 1946 for as high as $0.75 per pound.

Many synthetic processes for glycerol manufacture have been developed, but it was not until 1948 that a plant was built utilizing a process announced in 1938. In this process allyl alcohol made from allyl chloride (p. 600) is converted with aqueous chlorine to glycerol α-chlorohydrin, a product that can be hydrolyzed easily to glycerol.

$$CH_2{=}CHCH_2OH \xrightarrow{Cl_2,\ H_2O} ClCH_2CHOHCH_2OH \xrightarrow[NaOH]{Na_2CO_3,} HOCH_2CHOHCH_2OH$$

A contract was made for the entire output of the first plant for the synthesis of glycerol at the reputed price of $0.19 per pound when natural glycerol was selling for $0.33 per pound. In 1955 a new process was announced in which allyl alcohol made from acrolein (p. 216) is hydroxylated by aqueous hydrogen peroxide in the presence of tungstic acid.

$$CH_2{=}CHCH_2OH + H_2O_2 \xrightarrow[65^\circ]{WO_3} HOCH_2CHOHCH_2OH$$

In 1962 a plant was completed to produce glycerol, along with ethylene glycol, by the catalytic reduction and hydrogenolysis of hexoses from molasses or other sources of carbohydrate.

Production of glycerol amounted to 287 million pounds in 1964 of which about half was synthetic. About 24 per cent was used for the manufacture of alkyd resins (p. 474), 20 per cent each as a softening agent for cellophane (p. 345) and in drugs and cosmetics, 16 per cent as a humectant for tobacco, and 6 per cent for the manufacture of nitroglycerin for explosives. The remaining 14 per cent was used in printing inks, textile processing, food products, and in the manufacture of emulsifying agents (p. 191) and other chemicals.

When glycerol is heated with sodium bisulfate, dehydration takes place to yield acrolein.

$$HOCH_2CHOHCH_2OH \xrightarrow[\text{heat}]{KHSO_4,} [HOCH_2CH{=}CHOH] \longrightarrow HOCH_2CH_2CHO \longrightarrow \underset{\text{Acrolein}}{CH_2{=}CHCHO}$$

Acrolein results also from the pyrolysis of fats and is responsible for the lachrymatory properties of the fumes from overheated fats.

Glyceryl nitrate, commonly called *nitroglycerin,* is an important explosive made by the reaction of glycerol with nitric acid in the presence of sulfuric acid.

$$\begin{array}{l} CH_2OH \\ | \\ CHOH \\ | \\ CH_2OH \end{array} + 3\,HNO_3 \xrightarrow{(H_2SO_4)} \begin{array}{l} CH_2ONO_2 \\ | \\ CHONO_2 \\ | \\ CH_2ONO_2 \end{array} + 3\,H_2O$$

Glyceryl nitrate
(*nitroglycerin*)

It is an oil that freezes at 13°. Formerly it was the chief explosive ingredient of **dynamite,** being mixed in amounts up to 40 per cent with a combustible mixture such as powdered wood pulp and sodium nitrate in the ratio of about 1 to 3. At present, dynamites contain up to 55 per cent ammonium nitrate mixed with about 15 per cent each of sodium nitrate, wood pulp, and *explosive oil.* The last is a mixture of glycol nitrate and glyceryl nitrate, which is added merely as a sensitizer for the ammonium nitrate. **Gelatin dynamite** is a mixture of wood pulp, sodium nitrate, and nitroglycerin gelatinized with 2 to 6 per cent of cellulose nitrate. It is plastic and can be loaded solidly into bore holes. It has a high water resistance, a requirement for work in wet places. Dynamite has many useful applications. Around 1.2 billion pounds was consumed in the United States in 1962, of which 74 per cent was used in mining operations and 21 per cent for construction purposes. Double-base military smokeless powders such as *Ballistite* and *Cordite* consist of about 60 per cent cellulose nitrate gelatinized with 40 per cent nitroglycerin. **Glycerol α-monochlorohydrin** and **glycerol α,γ-dichlorohydrin** are made by the reaction of aqueous chlorine with allyl alcohol or allyl chloride respectively.

$$CH_2{=}CHCH_2OH \xrightarrow{Cl_2,\ H_2O} ClCH_2CHOHCH_2OH$$

$$CH_2{=}CHCH_2Cl \xrightarrow{Cl_2,\ H_2O} ClCH_2CHOHCH_2Cl$$

The chlorohydrins when heated with alkali give the epoxy derivatives, which have the common names **glycidol** and **epichlorohydrin** respectively.

$$ClCH_2CHOHCH_2OH + NaOH \xrightarrow{Heat} \underset{\diagdown O \diagup}{CH_2{-}CH}CH_2OH + NaCl + H_2O$$

Glycidol

$$ClCH_2CHOHCH_2Cl + NaOH \xrightarrow{Heat} \underset{\diagdown O \diagup}{CH_2{-}CH}CH_2Cl + NaCl + H_2O$$

Epichlorohydrin

Epichlorohydrin is one of the two important raw materials for the manufacture of **epoxy resins,** the other being bisphenol-A (p. 444). The condensation of the two reactants in the presence of base is believed to take place by attack on the oxide ring by phenoxide ion rather than by direct displacement of chloride ion.

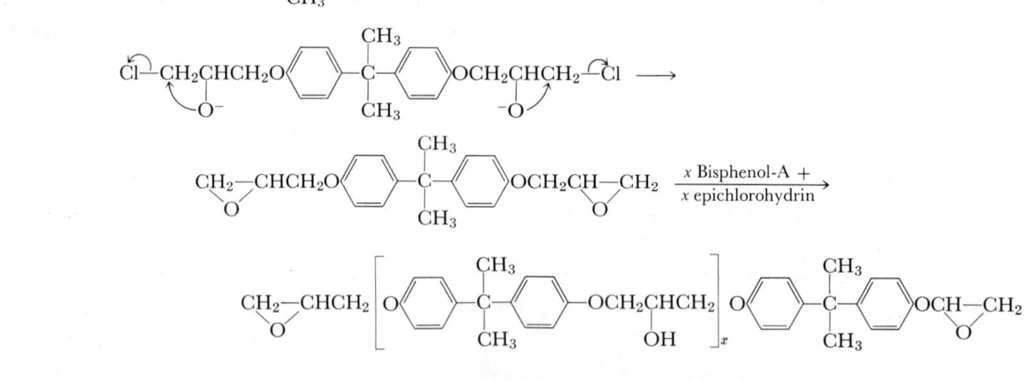

$$ClCH_2\underset{\diagdown O \diagup}{CH{-}CH_2} + {}^{-}O{-}C_6H_4{-}\underset{CH_3}{\overset{CH_3}{C}}{-}C_6H_4{-}O^{-} + \underset{\diagdown O \diagup}{CH_2{-}CH}CH_2Cl \longrightarrow$$

$$Cl{-}CH_2\underset{O^{-}}{CH}CH_2O{-}C_6H_4{-}\underset{CH_3}{\overset{CH_3}{C}}{-}C_6H_4{-}OCH_2\underset{{}^{-}O}{CH}CH_2{-}Cl \longrightarrow$$

$$\underset{\diagdown O \diagup}{CH_2{-}CH}CH_2O{-}C_6H_4{-}\underset{CH_3}{\overset{CH_3}{C}}{-}C_6H_4{-}OCH_2\underset{\diagdown O \diagup}{CH{-}CH_2} \xrightarrow[x\ \text{epichlorohydrin}]{x\ \text{Bisphenol-A} +}$$

$$\underset{\diagdown O \diagup}{CH_2{-}CH}CH_2\left[O{-}C_6H_4{-}\underset{CH_3}{\overset{CH_3}{C}}{-}C_6H_4{-}OCH_2\underset{OH}{CH}CH_2\right]_x O{-}C_6H_4{-}\underset{CH_3}{\overset{CH_3}{C}}{-}C_6H_4{-}O\underset{\diagdown O \diagup}{CH{-}CH_2}$$

The use of an excess of epichlorohydrin leads to terminal epoxy groups, and the amount of the excess governs the molecular weight. Commercial resins vary from thin liquids of low average molecular weight used as casting resins, through viscous adhesives, to solids used as surface coatings where the average value of x is as high as 25. The linear polymer can be modified by the replacement of all or part of the bisphenol-A with other polyhydroxy compounds, and by the use of other epoxides such as those obtained by the epoxidation of unsaturated fatty acids (p. 186).

The linear polymers are converted to the final products of high molecular weight by addition of a curing agent, which may be any compound having nucleophilic properties.

$$-\underset{\diagdown O \diagup}{CH-CH_2} + :NR_3 \longrightarrow -\underset{O^-}{\underset{|}{C}}H-CH_2\overset{+}{N}R_3 \xrightarrow{\underset{\diagdown O \diagup}{CH_2-CH-}}$$

$$-\underset{OCH_2CHO^-}{\underset{|}{C}}HCH_2\overset{+}{N}R_3 \xrightarrow{x\,\underset{\diagdown O \diagup}{CH_2-CH-}} -\underset{(OCH_2CHO)_xCH_2CHO^-}{\underset{|}{C}}HCH_2\overset{+}{N}R_3$$

$$(x + 2)\ -\underset{\diagdown O \diagup}{CH-CH_2} + HOOCR \longrightarrow -\underset{(OCH_2CHO)_xCH_2CHOH}{\underset{|}{C}}HCH_2OCOR$$

The epoxy resins have outstanding adhesive properties and are used for bonding to metal, glass, and ceramics. They also are used as casting resins in electrical assemblies. Their chief use has been in protective coatings because of their good adhesion, inertness, hardness, and unusual flexibility.

Pentaerythritol is the most important tetrahydric alcohol. It is prepared by the reaction of an aqueous solution of acetaldehyde with an excess of paraformaldehyde or formalin in the presence of calcium hydroxide. The reaction consists of aldol additions followed by a crossed Cannizzaro reaction (p. 462).

$$3\,HCHO + CH_3CHO \xrightarrow{Ca(OH)_2} (HOCH_2)_3CCHO \xrightarrow{HCHO,\ Ca(OH)_2} \underset{\text{Pentaerythritol}}{(HOCH_2)_4C} + \underset{\text{Calcium formate}}{\tfrac{1}{2}\,Ca(OCHO)_2}$$

The highly symmetric structure accounts for the high melting point of 262°. Over 69 million pounds was produced in 1964, most of which was used for the preparation of alkyd and other polyester resins and for the up-grading of drying oils. By esterification with unsaturated fat acids an ester is produced that has a higher molecular weight than the glycerides of natural fats. Hence a lower degree of polymerization is needed to produce films having the necessary toughness and hardness. Soybean oil, for example, is not a very good drying oil (p. 188). If it is hydrolyzed and the fat acids are reesterified with pentaerythritol, the product "dries" more rapidly. Another use for pentaerythritol is for the preparation of the nitrate known as **PETN.** Mixed with 30 per cent TNT it is used as a high explosive charge for bombs, torpedoes, and mines, and for demolition purposes. Straight-chain **penta-** and **hexahydric alcohols** result from the reduction of sugars (p. 348).

AMINO ALCOHOLS AND POLYAMINES

Like the aldehyde hydrates, the **1-hydroxy-1-amino compounds** (*aldehyde ammonias*) and **1,1-diamino compounds** usually are unstable, undergoing dehydration or deamination, and polymerization (p. 203).

1-Hydroxy-2-amino compounds can be prepared by the general methods, but those compounds most readily available are prepared by the reaction of ethylene oxides with ammonia or primary or secondary amines (p. 601) or by the reduction of nitro alcohols prepared by aldol addition of nitroalkanes to aldehydes or ketones (p. 246). Reaction of ethylene oxide with ammonia gives 2-aminoethanol, commonly called **ethanolamine** (sometimes *colamine*). A second molecule of ethylene oxide gives bis-(2-hydroxyethyl)amine or **diethanolamine,** and a third yields tris(2-hydroxyethyl)amine or **triethanolamine.**

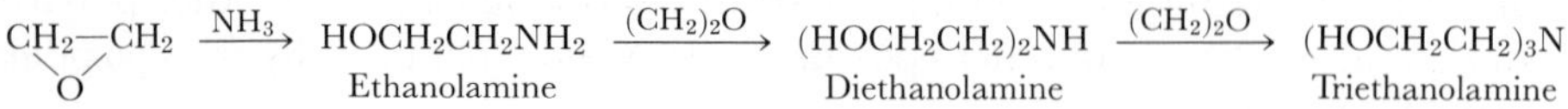

$$\underset{\text{O}}{CH_2{-}CH_2} \xrightarrow{NH_3} \underset{\text{Ethanolamine}}{HOCH_2CH_2NH_2} \xrightarrow{(CH_2)_2O} \underset{\text{Diethanolamine}}{(HOCH_2CH_2)_2NH} \xrightarrow{(CH_2)_2O} \underset{\text{Triethanolamine}}{(HOCH_2CH_2)_3N}$$

Salts of the ethanolamines with fat acids are soluble in both water and hydrocarbons and are good emulsifying agents. Thus kerosene or paraffin oil containing a small amount of triethanolamine oleate can be mixed with water to give stable emulsions useful as agricultural sprays, or as lubricating coolants during high speed metal-cutting operations. Ethanolamine and diethanolamine are used for the removal and recovery of acidic gases such as carbon dioxide and hydrogen sulfide from natural gas and other industrial gases. The gases are absorbed by the cold amine and are liberated by heating the solution.

Ethanolamine and its trimethylammonium salts, the cholines (p. 232), constitute a portion of an important class of biological substances known as the *phospholipids* or *phosphatides.* Thus the **lecithins** are mixed esters of glycerol and choline with fat acids and phosphoric acid. The **cephalins** (or *kephalins*) are esters of ethanolamine or serine (p. 353) instead of choline. **Sphingomyelin** is a derivative of the unsaturated dihydroxy amine *sphingosine* instead of glycerol. **Sphingosine** is (+)-2*S*,3*R*-1,3-dihydroxy-2-amino-*trans*-4-octadecene (also designated as D-(+)-*erythro*-). In the formulas, COR represents a fat acid acyl group.

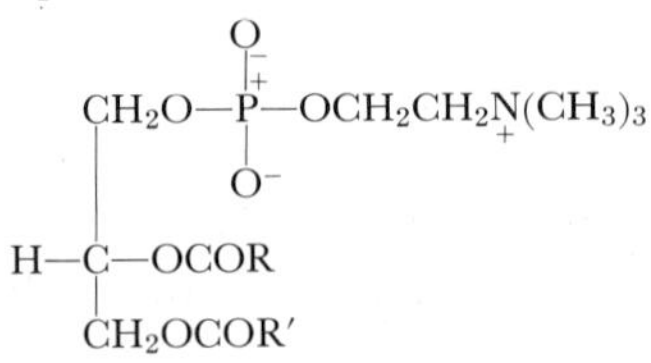

$$\begin{array}{l} CH_2O{-}\overset{O^-}{\underset{O^-}{\overset{|+}{P}}}{-}OCH_2CH_2\overset{+}{N}(CH_3)_3 \\ | \\ H{-}C{-}OCOR \\ | \\ CH_2OCOR' \end{array}$$

Lecithins

$$\begin{array}{l} CH_2O{-}\overset{O^-}{\underset{O^-}{\overset{|+}{P}}}{-}OCH_2CH_2\overset{+}{N}(CH_3)_3 \\ | \\ H{-}C{-}NHCOR \\ | \\ H{-}C{-}OH \\ \quad \backslash \qquad H \\ \quad C{=}C \\ H \qquad (CH_2)_{12}CH_3 \end{array}$$

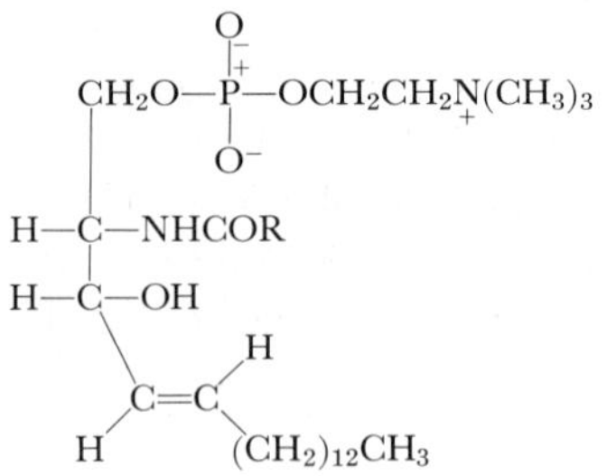

Sphingomyelins

The phospholipids are components of all animal and vegetable cells and are abundant in the brain, spinal cord, eggs, and soybeans. They possibly function as emulsifying agents for fats and are important in the metabolism of fats. *Soybean lecithin* is used in large quantities for the stabilization of emulsified food fats such as oleomargarine and mayonnaise. The importance of choline and acetylcholine salts has been discussed (p. 232). In the **brain gangliosides,** the phosphate portion of the sphingomyelins is replaced by a complex carbohydrate unit made up of glucose, galactose, galactosamine, and neuraminic acid (p. 364). In the sulfolipids, sulfate or sulfonic acid groups are present instead of phosphate groups.

Reaction of methyldiethanolamine hydrochloride with thionyl chloride or phosphorus trichloride gives **methylbis(2-chloroethyl)amine hydrochloride.**

$$(HOCH_2CH_2)_2NHCH_3{}^{+-}Cl + 2\,SOCl_2 \longrightarrow (ClCH_2CH_2)_2NHCH_3{}^{+-}Cl + 2\,HCl + 2\,SO_2$$

The free base is a nitrogen analog of mustard gas (p. 269) and belongs to the class of compounds known as **nitrogen mustards.** Many nitrogen mustards were prepared and investigated as toxic agents during World War II. They have a local vesicant action similar to that of mustard gas and in addition penetrate the skin and exert a generalized systemic action on living cells similar to the action of X-rays. Exposure to very low concentrations may cause opacity of the cornea.

Reaction of ethanolamine with sulfuric acid yields **2-aminoethyl hydrogen sulfate,** which is an inner salt.

$$H_2NCH_2CH_2OH + H_2SO_4 \xrightarrow{\text{Heat}} H_3\overset{+}{N}CH_2CH_2OSO_3^- \xrightarrow{\text{NaOH}} H_2C\overset{N}{\underset{H}{—}}CH_2$$

2-Aminoethyl hydrogen sulfate (inner salt structure) — Ethylenimine

When distilled with aqueous sodium hydroxide it gives **ethylenimine** (*aziridine*), a toxic liquid, b.p. 55°. This reaction is an intramolecular alkylation analogous to the ethylation of an amine by ethyl sulfate.

The aziridines are somewhat weaker bases ($pK_a = 8$ to 9) than other similarly substituted aliphatic amines but are stronger than pyridine. They undergo acid- and base-catalyzed nucleophilic displacement reactions with opening of the ring strictly analogous to those of the oxiranes (p. 601). Ethylenimine reacts with aqueous sulfur dioxide to give **taurine** (*2-aminoethanesulfonic acid*).

$$(CH_2)_2NH + H_2SO_3 \longrightarrow H_3\overset{+}{N}CH_2CH_2SO_3^-$$

Taurine

The acid-catalyzed polymerization of ethylenimine yields **polyethylenimine.**

$$H_2C\overset{N}{\underset{H}{—}}CH_2 + HN\begin{matrix}CH_2\\|\\CH_2\end{matrix} \xrightarrow{H^+} H_2NCH_2CH_2N\begin{matrix}CH_2\\|\\CH_2\end{matrix} \xrightarrow[\text{condensation}]{\text{Further}} H_2NCH_2CH_2(NHCH_2CH_2)_xN\begin{matrix}CH_2\\|\\CH_2\end{matrix}$$

Diethylenimine — Polyethylenimine

Ethylenimine and polyethylenimine react with the hydroxyl groups of cellulose and are used to impart wet strength and abrasion resistance to paper and to decrease the tendency of rayon and cotton fibers to swell in water. Reaction of aziridine with phosphorus oxychloride gives **triaziridylphosphine oxide** (*APO*), $(C_2H_4N)_3PO$, which also reacts with the hydroxyl groups of cellulose and makes textiles fire resistant. This compound under the name *tepa* is used as a sterilant for male insects. The 2,4,6-triaziridyl derivative of 1,3,5-triazine (p. 528) made from aziridine and cyanuryl chloride (p. 288) is called *tretamine* and also is an effective agent for this purpose.

In the reaction of propylene oxide with ammonia the point of attack is the 1 position and gives rise to **1-amino-2-propanol.**

$$CH_3CH\overset{O}{—}CH_2 + NH_3 \longrightarrow CH_3CHOHCH_2NH_2$$

2-Amino-1-hydroxy compounds are available by reduction of the corresponding nitro alcohols obtained by the addition of aliphatic nitro compounds to aldehydes (p. 245).

$$(CH_3)_2\underset{NO_2}{\underset{|}{C}}CH_2OH \xrightarrow{H_2/Ni} (CH_3)_2\underset{NH_2}{\underset{|}{C}}CH_2OH$$

Diaminoalkanes can be obtained by the usual methods for preparing amines. Thus **ethylenediamine** results from the reaction of ethylene chloride with ammonia.

$$ClCH_2CH_2Cl + 4\,NH_3 \longrightarrow H_2NCH_2CH_2NH_2 + 2\,NH_4Cl$$

Further reaction with ethylene chloride and ammonia gives the coproducts **diethylenetriamine,** $H_2NCH_2CH_2NHCH_2CH_2NH_2$, **triethylenetetramine,** $H_2N(CH_2CH_2NH)_2CH_2$-CH_2NH_2, and **tetraethylenepentamine,** $H_2N(CH_2CH_2NH)_3CH_2CH_2NH_2$. All of these compounds are manufactured commercially and are used as basic catalysts for epoxy resins

(p. 608) and polyurethans (p. 605), and in the synthesis of pharmaceuticals, textile finishing agents, emulsifying agents, and fungicides. **Ethylenediammonium tartrate** crystals are piezo-electric, and thin plates cut from them are used instead of quartz plates for the control of high-frequency electric currents in telephony, radio, radar, and television.

The reaction of ethylenediamine with formaldehyde and sodium cyanide in alkaline solution yields the sodium salt of **ethylenediaminetetraacetic acid** (*EDTA, Sequestrene, Versene*).

$$H_2NCH_2CH_2NH_2 + 4\,HCHO + 4\,NaCN + 4\,H_2O \xrightarrow[80^\circ]{NaOH} (NaOOCCH_2)_2NCH_2CH_2N(CH_2COONa)_2 + 4\,NH_3$$

It can be made also from ethylenediamine and sodium chloroacetate. EDTA is a strong chelating (or complexing) agent for alkaline earth and heavy metal ions, the relative position of the unshared pairs of electrons on the four carboxylate ions and the two nitrogen atoms being such that it can form stable five-membered chelate rings with the metal ions. With six unshared pairs of electrons available, EDTA would be expected to complex completely with a metal ion having a coordination number of six. Usually, however, only four rings form, presumably for steric reasons, and the sixth pair of electrons is supplied by a water molecule.

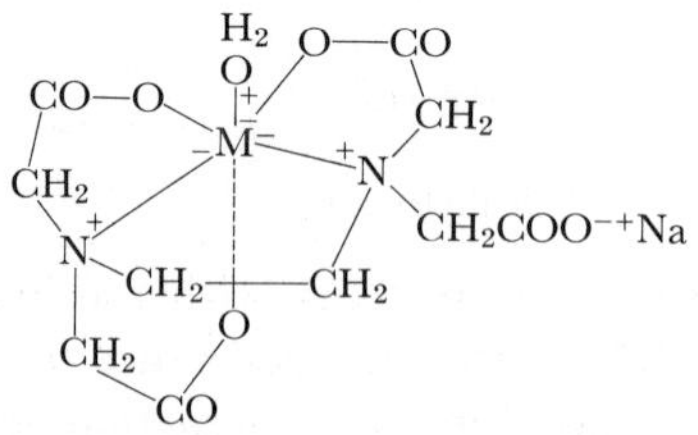

```
                H2  O—CO
                O  /    \
      CO—O     |+/      CH2
     /     \  -M=      /
    /        /   \ + N
  CH2       /     |    \
     \ +   /      |     CH2COO-+Na
      N           |   /
      | \CH2——————+——CH2
      |           O
     CH2         /
        \      /
          CO
```

Ethylenediaminetetraacetic acid and its salts have found important uses in analytical chemistry. Their chief use, however, is to bind traces of alkaline earth and heavy metal ions in a nonionic form, thus preventing undesirable catalytic effects or precipitation by other components of aqueous solutions. A food grade is used in canned foods and to inhibit autoxidation of salad dressings and oleomargarine. The iron complex is used to combat plant chlorosis caused by insufficient availability of iron in the soil. Although first introduced in the United States after World War II, production of the acid and its salts in 1955 was over 5 million pounds and over 25 million pounds in 1964.

Tetramethylenediamine, known as *putrescine,* and **pentamethylenediamine,** known as *cadaverine,* occur among the bacterial decomposition products of proteins. They arise from the decarboxylation of ornithine and lysine, respectively.

$$\underset{\text{Ornithine}}{H_2N(CH_2)_3CHNH_2COOH} \xrightarrow{\text{Bacteria}} \underset{\text{Putrescine}}{H_2N(CH_2)_4NH_2} + CO_2$$

$$\underset{\text{Lysine}}{H_2N(CH_2)_4CHNH_2COOH} \xrightarrow{\text{Bacteria}} \underset{\text{Cadaverine}}{H_2N(CH_2)_5NH_2} + CO_2$$

Spermine, $H_2N(CH_2)_3NH(CH_2)_4NH(CH_2)_3NH_2$, is widely distributed in the organs of mammals and has been isolated also from yeast. **Hexamethylenediamine** is an intermediate for the synthesis of *nylon* 6-6 (p. 631) and is made commercially by the catalytic reduction of tetramethylene cyanide (*adiponitrile,* p. 631) in the presence of ammonia to prevent the formation of imino derivatives (p. 223).

$$\underset{\text{Adiponitrile}}{NC(CH_2)_4CN} + 4\,H_2 \xrightarrow{Ni(NH_3)} \underset{\text{Hexamethylenediamine}}{H_2N(CH_2)_6NH_2}$$

PROBLEMS

35–*1.* Give reactions for the following syntheses: (*a*) crotyl alcohol from acetaldehyde, (*b*) 1-hydroxy-3-hexene from acetylene, (*c*) 2-hydroxy-2-methyl-3-hexyne from acetylene, (*d*) 1-hydroxy-2-hexene from acetylene.

35–*2.* Give the reagents required to bring about the reactions indicated in the following chart.

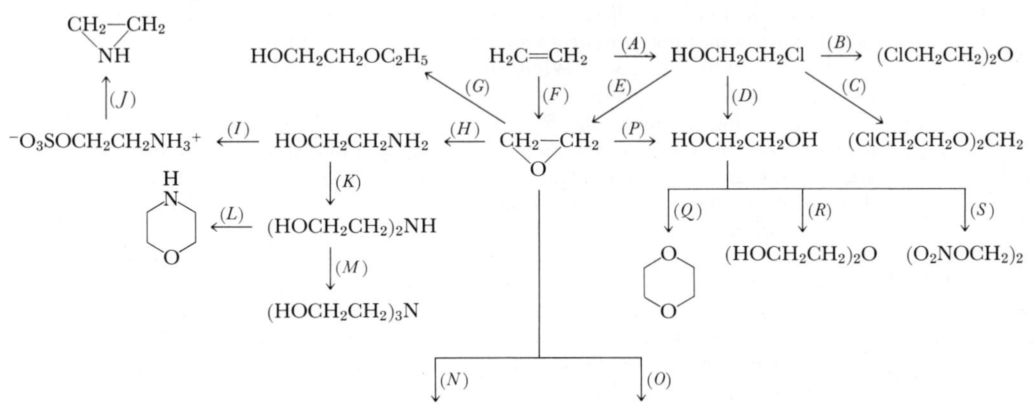

35–*3.* Give reactions for the following conversions: (*a*) propylene to propylene glycol diacetate, (*b*) styrene to phenylacetaldehyde, (*c*) *n*-butyl alcohol to *n*-hexyl alcohol, (*d*) *n*-butyl alcohol to 2,2-bis-(hydroxymethyl)-1-butanol, (*e*) tetrahydrofurfuryl alcohol to 1,5-pentanediol, (*f*) acetaldehyde to 1,3-butanediol, (*g*) allyl alcohol to glycidol, (*h*) ethylene oxide to mustard gas.

35–*4.* Construct a chart illustrating the reactions, reagents, and conditions for converting acetylene into the following compounds: (*A*) propargyl alcohol, $HC\equiv CCH_2OH$, (*B*) tetrahydrofuran, (*C*) 1,4-dihydroxy-2-butyne, (*D*) ethyl alcohol, (*E*) tetramethylene glycol, (*F*) acetaldehyde, (*G*) *N*-vinylcarbazole, (*H*) 3-hydroxy-3-methyl-1-butyne, (*I*) vinyl methyl ether, and (*J*) 1,4-dichloro-2-butyne.

35–*5.* Give reactions for the preparation of: (*a*) pyrrolidine from tetramethylene glycol, (*b*) spermine from trimethylene and tetramethylene glycols, (*c*) hexamethylenediamine from tetramethylene chloride, (*d*) trimethylenediamine from ethylene cyanohydrin, (*e*) 2-amino-2-methyl-1,3-propanediol from 1-nitroethane, (*f*) 1-amino-2-propanol from propylene, (*g*) 1,4-dimethylpiperazine from ethylene chloride.

35–*6.* Devise a practical synthesis for the glycol intermediate necessary for the preparation of meprobamate (p. 286).

35–*7.* How can one distinguish readily by chemical reactions between the members of the following pairs of compounds: (*a*) glycerol and ethylene glycol, (*b*) propylene glycol and 1-amino-2-propanol, (*c*) 1,4-butanediol and 2-3-butanediol, (*d*) pentaerythritol nitrate and 2,4,6-trinitrotoluene, (*e*) Cellosolve and ethylene glycol.

35–*8.* During the preparation of a batch of 2-methyl-2,4-pentanediol by the catalytic reduction of diacetone alcohol, the Raney nickel was not completely removed before purification by distillation. During the distillation through a column, only acetone distilled. Give an explanation for this behavior.

35–*9.* Indicate the reactions that take place in the preparation of triaziridylphosphine oxide and in its cross-linking reaction with cellulose.

35–*10.* Xylose boiled with 12 per cent hydrochloric acid gives *A*, which can be hydrogenated in the presence of Raney nickel to *B*. When *B* is passed over hot alumina, *C* is obtained. Catalytic hydrogenation of *C* gives *D*, which reacts with hydrogen chloride to give *E*. *E* reacts with ammonia to give *F*, which, when pyrolyzed, gives *G*. Give equations for the reactions that take place.

35–*11*. Three hydrocarbons have the empirical formula CH. Two are liquids and one is a solid. One of the liquids decolorizes bromine, but the other does not. A solution of the solid in benzene does not decolorize bromine. When the reactive liquid is allowed to stand with a chloroform solution of peroxybenzoic acid, a new product is obtained. When this product is heated with hydrochloric acid, another compound is isolated that gives a color with Schiff reagent. When either this product, the reactive liquid hydrocarbon, or the solid hydrocarbon is vigorously oxidized, the same acid is obtained. When this acid is heated with soda-lime, a hydrocarbon is formed that is identical with the unreactive liquid hydrocarbon. What are the three hydrocarbons and what are the reactions that take place?

35–*12*. A compound has the molecular formula $C_7H_{12}O_5$. It is neutral to moist litmus, is hydrolyzed by alkali, and gives a saponification equivalent of 88. When the alkaline solution after saponification is distilled, no organic compound can be detected in the distillate. Acidification with sulfuric acid and distillation gives a volatile acid that has a neutralization equivalent of 60. Neutralization of the sulfuric acid solution remaining in the flask after the distillation and evaporation gives a sticky mass of salts. When this residue is mixed with sodium bisulfate and heated strongly, a volatile substance with a sharp odor is formed. Collection of the volatile compound in a small amount of water and addition of Schiff reagent gives a red color. Give a possible structural formula for the original compound and equations for the reactions that take place.

CHAPTER THIRTY-SIX

SUBSTITUTED MONOCARBOXYLIC ACIDS

If more than one functional group is present in a molecule and if they are widely separated, they react independently of each other. If they are adjacent or separated by one, two, or three carbon atoms, they may behave differently. Hence it is convenient to discuss substituted carboxylic acids from this viewpoint.

Hydroxy Acids

α-Hydroxy acids may be prepared by the hydrolysis of α-halogen acids, or by the hydrolysis of the cyanohydrins of aldehydes or ketones.

$$RCHO \xrightarrow{HCN} RCH\begin{matrix} OH \\ CN \end{matrix} \xrightarrow{H_2O,\ HCl} RCHOHCOOH + NH_4Cl$$

They undergo bimolecular esterification with the formation of a six-membered ring. Such cyclic esters are known as *lactides.*

$$\begin{matrix} RCHOHCOOH \\ + \\ HOOCCHOHR \end{matrix} \longrightarrow \begin{matrix} RCH{-}CO \\ O \qquad\quad O \\ CO{-}CHR \end{matrix} + 2\,H_2O$$

A lactide

This reaction takes place so readily that it is not possible to keep α-hydroxy acids in their monomolecular state except in the form of their salts.

When α-hydroxy acids are boiled with dilute sulfuric acid, carbon monoxide and water are lost with the formation of an aldehyde, a behavior called *decarbonylation.* The carbonium ion formed on loss of carbon monoxide is merely the conjugate acid of the aldehyde.

$$RCHOHCOOH \xrightarrow{H^+} RCHOHCO\overset{+}{O}H_2 \xrightarrow[H_2O]{} RCHOH\overset{+}{C}O \xrightarrow[CO]{}$$

$$\{R\overset{+}{C}HOH \longleftrightarrow RCH{=}\overset{+}{O}H\} \xrightarrow[H^+]{} RCHO$$

This reaction is valuable for the synthesis of higher aldehydes from acids through the α-bromo acid. Since one carbon atom is lost in the process, the series of reactions may be used for the stepwise degradation of a carbon chain.

β-Hydroxy acids may be made by the catalytic reduction of β-keto esters (p. 620) followed by hydrolysis. The β-hydroxy esters are obtained also by the **Reformatsky reaction.** This reaction is brought about by the addition of zinc to a mixture of an α-halogen ester, usually the α-bromo ester, with an aldehyde or ketone in ether or aromatic hydrocarbon solution.

$$RCHO + BrCH_2COOC_2H_5 + Zn \longrightarrow \underset{\displaystyle OZnBr}{RCHCH_2COOC_2H_5} \xrightarrow{HCl} RCHOHCH_2COOC_2H_5$$

β-Hydroxy acids lose water, especially when heated with sulfuric acid, to give α,β-unsaturated acids, frequently mixed with β,γ-unsaturated acids.

γ- and δ-Hydroxy acids are stable only in the form of their salts. The free acids spontaneously cyclize to lactones, which are monomeric cyclic esters.

$$RCHOHCH_2CH_2COOH \longrightarrow \underset{\text{(ring: }CH_2\text{—}CH_2\text{, RCH—O—C=O)}}{\gamma\text{-lactone}} + H_2O$$

$$RCHOHCH_2CH_2CH_2COOH \longrightarrow \underset{\text{(ring: }CH_2\text{—}CH_2\text{—}CH_2\text{, RCH—O—C=O)}}{\delta\text{-lactone}} + H_2O$$

Glycolic acid, $HOCH_2COOH$, can be made by the hydrolysis of chloroacetic acid or by the oxidation of ethylene glycol with dilute nitric acid. The commercial product, used by tanners and dyers, is made by the acid-catalyzed reaction of formaldehyde, carbon monoxide, and water at high pressure.

$$HCHO + CO + H_2O \xrightarrow{H^+} HOCH_2COOH$$

Thioglycolic acid is made from sodium chloroacetate and sodium hydrosulfide.

$$HSNa + ClCH_2COONa \longrightarrow HSCH_2COONa$$

The ammonium salt is the active agent in preparations used for the cold permanent-waving of hair (p. 364).

Lactic acid, $CH_3CHOHCOOH$, the acid formed when milk turns sour because of the action of *Lactobacillus* organisms on the lactose, was isolated from sour milk by Scheele in 1780. It is manufactured by the fermentation of lactose from whey, of molasses, or of starch hydrolysates, in the presence of an excess of calcium carbonate, or synthesized by the hydrolysis of acetaldehyde cyanohydrin. About 8 million pounds was consumed in the United States in 1964. In the past, the main food use has been as an acidulant, but lately large amounts are used in the synthesis of the calcium salt of **(stearyllactyl)lactic acid,** $\left[CH_3(CH_2)_{16}COO\underset{CH_3}{CH}COO\underset{CH_3}{CH}COO^-\right]_2 Ca^{2+}$. Under the trade name *Verv-Ca,* it is added to bread dough to the extent of 0.5 per cent of the weight of flour to improve the baking properties and thus give a more uniform product. Esters of lactic acid are valuable high-boiling solvents for the formulation of lacquers, and can be used for the manufacture of acrylic esters (p. 619). Lactic acid contains an asymmetric carbon atom, and the lactic acid formed on muscular contraction is dextrorotatory. It is known as **sarcolactic acid** (Gr. *sarx* flesh).

γ-Butyrolactone is made commercially by the catalytic dehydrogenation of tetramethylene glycol (p. 607).

$$\underset{HOCH_2\quad CH_2OH}{CH_2\text{—}CH_2} \xrightarrow[200^\circ]{Cu\text{-}SiO_2} \underset{CH_2\text{—}O\text{—}CO}{CH_2\text{—}CH_2} + 2\,H_2$$

It is an intermediate for the preparation of pyrrolidone (cf. p. 617). Both γ-hydroxybutyrate ion and γ-butyrolactone, when administered orally or intravenously, produce sleep. The anion is the active compound.

Of the polyhydroxy acids, the most interesting is **mevalonic acid.** The racemic form is synthesized best from 4-hydroxy-2-butanone, the aldol addition product of acetone and formaldehyde, which is commercially available. The proton-catalyzed reaction with ketene yields the acetate, which reacts with a second molecule of ketene in the presence of boron fluoride to give the β lactone of mevalonic acid. Hydrolysis of the latter gives the hydroxy acid.

$$CH_3\underset{\|}{\overset{}{C}}CH_2CH_2OH \xrightarrow{CH_2=C=O,\ H^+} CH_3\underset{\|}{C}CH_2CH_2OAc \xrightarrow{CH_2=C=O,\ BF_3}$$
$$\qquad\ \ O \qquad\qquad\qquad\qquad\qquad\qquad O$$

$$CH_3\underset{O\text{—}CO\text{—}CH_2}{C}CH_2CH_2OAc \xrightarrow{2\,H_2O} CH_3\underset{CH_2COOH}{C}HOHCH_2CH_2OH + HOAc$$

The optically active form with the R configuration (p. 312) is a growth factor for lactobacilli. It is also a precursor in the biosynthesis, through the 5-pyrophosphate, of terpenes (p. 667), of polyterpenes such as rubber and squalene, and of cholesterol (p. 668). Numerous other hydroxy and polyhydroxy acids have been isolated from natural sources. They occur widely distributed in the fats and waxes of animals, plants, and microorganisms.

Amino Acids

The synthesis of **α-amino acids** is discussed on pages 246 and 357. **β-Amino acids** can be made by the addition of ammonia or amines to α,β-unsaturated acids (p. 618).

$$RCH{=}CHCOOH + NH_3 \longrightarrow RCH(NH_2)CH_2COOH$$

Amino acids in general exist as dipolar ions and are less likely to undergo the types of reactions noted for the hydroxy acids. Nevertheless comparable reactions take place under more drastic conditions. Thus α-amino acids, when heated in glycerol solution to 170°, lose water and form cyclic amides known as *2,5-dioxopiperazines.*

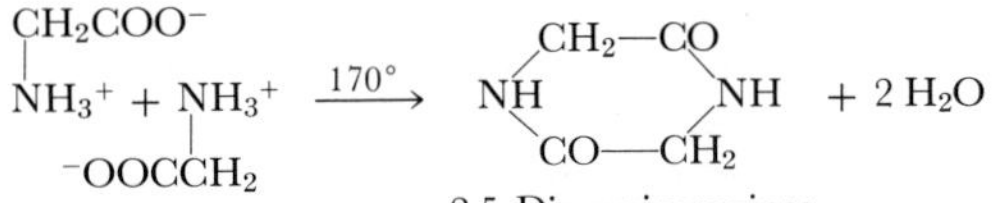

$$CH_2(NH_3^+)COO^- + {}^-OOCCH_2NH_3^+ \xrightarrow{170^\circ} \text{cyclo}(NH{-}CH_2{-}CO{-}NH{-}CH_2{-}CO) + 2\,H_2O$$

2,5-Dioxopiperazines

When the salts of β-amino acids are heated to decomposition, α,β-unsaturated acids are formed.

$$RCH(NH_3^{+\,-}PO_4H_2)CH_2COOH \xrightarrow{\text{Heat}} RCH{=}CHCOOH + (NH_4)H_2PO_4$$

γ- and δ-Amino acids when heated yield cyclic amides, which are known as *lactams.*

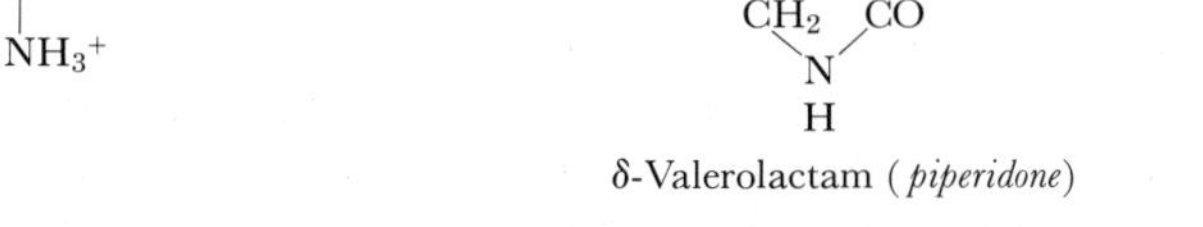

$$CH_2(NH_3^+)CH_2CH_2COO^- \xrightarrow{\text{Heat}} \text{cyclo}(CH_2{-}CH_2{-}CO{-}NH{-}CH_2) + H_2O$$

γ-Butyrolactam (*pyrrolidone*)

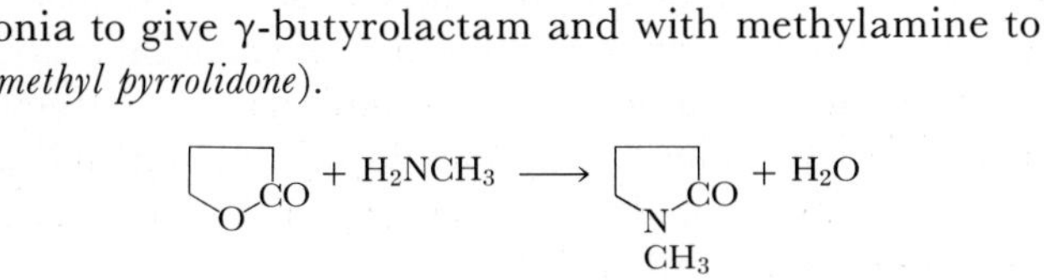

$$CH_2(NH_3^+)CH_2CH_2CH_2COO^- \xrightarrow{\text{Heat}} \text{cyclo}(CH_2{-}CH_2{-}CH_2{-}CO{-}NH{-}CH_2) + H_2O$$

δ-Valerolactam (*piperidone*)

γ Lactams result also from the reaction of γ lactones with amines. Thus γ-butyrolactone reacts with ammonia to give γ-butyrolactam and with methylamine to give *N*-methyl-γ-butyrolactam (*N-methyl pyrrolidone*).

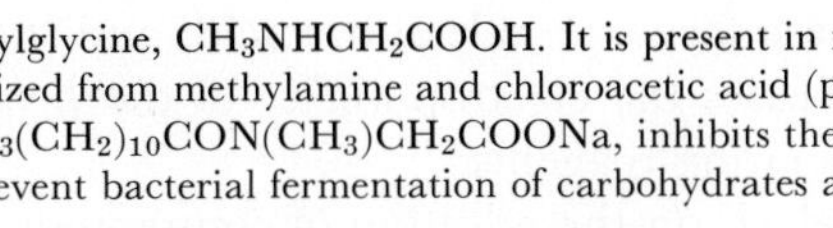

$$\text{cyclo}(CH_2CH_2CH_2{-}CO{-}O) + H_2NCH_3 \longrightarrow \text{cyclo}(CH_2CH_2CH_2{-}CO{-}N(CH_3)) + H_2O$$

The latter compound is a useful high-dielectric aprotic solvent (p. 102).

Sarcosine is *N*-methylglycine, CH_3NHCH_2COOH. It is present in muscle and hence in meat extract. It can be synthesized from methylamine and chloroacetic acid (p. 347). The sodium salt of **N-lauroyl sarcosine,** $CH_3(CH_2)_{10}CON(CH_3)CH_2COONa$, inhibits the action of enzymes and is used in toothpastes to prevent bacterial fermentation of carbohydrates and the formation of acids in the mouth.

The internal quaternary ammonium salts of amino acids are known as *betaines* after betaine itself, which is the simplest representative. **Betaine,** $(CH_3)_3\overset{+}{N}CH_2COO^-$, is present in the juice of beets (*Beta vulgaris*), and the residue from the manufacture of beet sugar is an abundant source. A quaternary ammonium derivative of aminoacetic acid that has proved to be very valuable is the **Girard T-reagent,** the hydrazide of carboxymethyltrimethylammonium chloride. It is made by the reaction of ethyl chloroacetate with trimethylamine, followed by reaction with hydrazine.

$$(CH_3)_3N + ClCH_2COOC_2H_5 \longrightarrow \left[(CH_3)_3\overset{+}{N}CH_2COOC_2H_5\right]\overset{-}{Cl} \xrightarrow{H_2NNH_2} \underset{\text{Girard T-reagent}}{\left[(CH_3)_3\overset{+}{N}CH_2CONHNH_2\right]\overset{-}{Cl}} + C_2H_5OH$$

Water-insoluble ketones give water-soluble hydrazones with this reagent because of the presence of the quaternary ammonium group, thus enabling the separation from other nonketonic water-insoluble compounds. The ketone is regenerated easily by hydrolysis. The reagent has been invaluable for the separation of ketonic from nonketonic steroidal hormones (p. 665).

β-Alanine can be made from succinimide (p. 629). It is of special interest because it comprises a portion of **pantetheine,** which in turn makes up a portion of coenzyme A (p. 534). **Pantothenic acid,** first isolated from liver extracts and considered to be a member of the vitamin B complex (B_5), can be derived from pantetheine.

$$\overbrace{HOCH_2C(CH_3)_2CHOHCO}^{\textit{Pantoic acid portion}}\overbrace{NHCH_2CH_2CO}^{\textit{β-Alanine portion}}\overbrace{NHCH_2CH_2SH}^{\textit{Thioethanolamine portion}}$$

Pantothenic acid portion: $HOCH_2C(CH_3)_2CHOHCONHCH_2CH_2COOH$ part

Pantetheine

Production of synthetic pantothenic acid and its derivatives in 1964 was over 1.7 million pounds valued at $4.3 million.

The long-chain *N*-alkyl-β-amino acids are commercial *amphoteric surfactants.* The sodium salts result from the addition of alkylamines, made from fat acid amides, to methyl acrylate (p. 619) followed by saponification.

$$RNH_2 + CH_2{=}CHCOOCH_3 \longrightarrow RNHCH_2CH_2COOCH_3 \xrightarrow{NaOH} RNHCH_2CH_2COO^{-+}Na$$

The active agent is anionic in alkaline solution and cationic in acid solution.

Condensation of pyrrolidone (p. 617) with acetylene gives **N-vinylpyrrolidone** which polymerizes to **polyvinylpyrrolidone.** The latter gives colloidal solutions in water and has been used as a blood plasma substitute under the name *Periston.* It now is the most widely used film-forming ingredient of hair sprays.

Pyrrolidone (N–H, C=O) $\xrightarrow{HC\equiv CH}$ Vinylpyrrolidone (N–$CH{=}CH_2$, C=O) $\longrightarrow$ Polyvinylpyrrolidone $[-CH-CH_2-]_x$

Unsaturated Acids

Only α,β-unsaturated acids are readily available. The methods for the preparation of these compounds are numerous. They usually are made by the oxidation of α,β-unsaturated aldehydes, which are available through the aldol addition of aldehydes (p. 202). The β-aryl-substituted α,β-unsaturated acids may be obtained by the Perkin reaction (p. 463).

The most important behavior of α,β-unsaturated acids and their derivatives is the increased reactivity of the double bond to the addition of chemical reagents. Thus halogen acid, hydrogen cyanide, hydrogen sulfide, and ammonia give β-halogen, β-cyano, β-mercapto, and β-amino acids. Mercaptans and amines give the corresponding sulfides and substituted amino derivatives, and alcohols and phenols give β-alkoxy and β-phenoxy acids. Sodium bisulfite gives the sodium β-sulfonate.

α,β-Unsaturated esters and nitriles react with these reagents with even greater ease than α,β-unsaturated acids, especially when the reaction is catalyzed by bases. The free acids in the presence of a base form the salt of the acid, and the resulting negative charge on the carboxylate ion decreases the electrophilicity of the β carbon atom and its ability to combine with electron-donating reagents.

The effect of basic catalysis on the addition of compounds containing reactive hydrogen may be very striking. For example, a mixture of α,β-unsaturated ester and mercaptan may show no sign of reaction at room temperature, but the addition of a small amount of pyridine induces a vigorous reaction. Undoubtedly the reaction occurs by the attack of a negative ion at the β position.

$$RSH \underset{}{\overset{C_5H_5N}{\rightleftharpoons}} C_5H_5\overset{+}{N}H + RS\overset{-}{:} \overset{CH_2{=}CH\overset{O}{\overset{\|}{C}}OR}{\rightleftharpoons} RS{-}CH_2{-}\overset{-}{C}H\overset{O}{\overset{\|}{C}}OR \overset{C_5H_5\overset{+}{N}H}{\rightleftharpoons} RSCH_2CH_2\overset{O}{\overset{\|}{C}}OR$$

α,β-Unsaturated esters also undergo the Michael addition of reactive methylene compounds (p. 633).

Acrylonitrile, methyl acrylate, and methyl methacrylate are the most important α,β-unsaturated compounds from a technical viewpoint. **Acrylonitrile** is made chiefly from either acetylene or propylene.

$$HC{\equiv}CH + HCN \xrightarrow[90^\circ]{CuCl,\ NH_4Cl,\ HCl} H_2C{=}CHCN$$

In the newer processes, a mixture of propylene and ammonia is oxidized by air at 450° in a fluidized bed of molybdenum oxide promoted with bismuth oxide. Acrolein probably is an intermediate in the reaction (p. 216).

$$CH_2{=}CHCH_3 + O_2 \xrightarrow[450^\circ]{MoO_3\text{-}Bi_2O_3} CH_2{=}CHCHO \xrightarrow{NH_3} CH_2{=}CHCH{=}NH \xrightarrow{O_2,\ MoO_3\text{-}Bi_2O_3} CH_2{=}CHC{\equiv}N$$

Similar processes start with acrolein made from propylene (p. 216).

Acrylonitrile long has been used as a comonomer in the production of synthetic rubbers and plastics (p. 589). When polymerized alone in an aqueous medium containing ammonium persulfate along with sodium bisulfite as an activator, it gives a solid resin that is infusible and insoluble in the common organic solvents.

$$x\ CH_2{=}CHCN \xrightarrow{Persulfate} \left[\begin{array}{c}{-}CH_2CH{-}\\ \ \ \ |\\ \ \ \ CN\end{array}\right]_x$$

Polyacrylonitrile

The discovery of solvents in which it could be dissolved and from which it could be spun into threads led to the production of the so-called **acrylic fibers.** The solvents most commonly used for spinning are *N,N*-dimethylformamide and *N,N*-dimethylacetamide. The acrylic fibers such as *Orlon, Acrilan, Creslan,* and *Zefran* are copolymers or graft polymers with 10–15 per cent of one or more other polymerizable monomers, such as methyl acrylate, vinyl acetate, vinylpyridine (p. 518), or vinylpyrrolidone, that increase the affinity of the fiber for either acid or basic dyes. Fibers formed by cospinning polyacrylonitrile with cellulose acetate also are dyed readily. *Dynel* is a copolymer of 50 to 60 per cent acrylonitrile with 50 to 40 per cent vinyl chloride and can be spun from a solution in acetone. *Verel* is a similar copolymer with a higher proportion of vinyl chloride.

Acrylonitrile undergoes base-catalyzed addition reactions with all compounds containing reactive hydrogen, such as alcohols, mercaptans, hydrogen cyanide, acids, amines, phosphines, and even with ketones.

$$HZ \underset{HB}{\overset{B^-}{\rightleftharpoons}} Z^- \xrightarrow{CH_2{=}CHCN} \left\{ZCH_2\overset{-}{C}H{-}C{\equiv}N \longleftrightarrow ZCH_2CH{=}C{=}N{:}^-\right\} \underset{B^-}{\overset{HB}{\rightleftharpoons}} ZCH_2CH_2C{\equiv}N$$

A particularly useful catalyst for these reactions is a quaternary ammonium base such as trimethylbenzylammonium hydroxide (*Triton-B*). Because a cyanoethyl group appears in the product, the process has been called *cyanoethylation.*

Total United States production of acrylonitrile in 1964 was 594 million pounds. Acrylic fibers consumed 49 per cent and nitrile rubber (*NBR*) and *ABS-SAN* plastics (p. 589) about 8 per cent each. The remainder was about equally divided between miscellaneous uses and exports.

Acrylic acid usually is made commercially by the hydracarbonylation of acetylene in the presence of nickel carbonyl.

$$HC{\equiv}CH + CO + H_2O \xrightarrow{Ni(CO)_4,\ HCl} H_2C{=}CHCOOH$$

It is used primarily to make methyl, ethyl, *n*-butyl, and 2-ethylhexyl esters, which are copolymerized with other monomers or converted by peroxide initiators to homopolymers known as *acrylic* or *acryloid resins.* **Methyl acrylate** can be made directly by the carbonylation of acetylene in aqueous methanol.

$$HC{\equiv}CH + CO + CH_3OH \xrightarrow[\text{aq. HCl}]{Ni(CO)_4,} H_2C{=}CHCOOCH_3$$

Other esters are prepared by using other alcohols or by alcoholysis of the methyl ester.

Methyl α-methylacrylate, commonly called *methyl methacrylate,* is made from acetone.

$$(CH_3)_2CO \xrightarrow{HCN} (CH_3)_2COHCN \xrightarrow{98\%\ H_2SO_4} CH_2{=}\underset{\displaystyle CH_3}{\underset{|}{C}}CONH_2 \cdot H_2SO_4 \xrightarrow[H_2SO_4]{CH_3OH,} CH_2{=}\underset{\displaystyle CH_3}{\underset{|}{C}}COOCH_3$$

Methyl methacrylate

Polymerization of methyl methacrylate using peroxide initiators gives **poly(methyl methacrylate),** a strong thermoplastic solid that is highly transparent and has a high refractive index. It is sold under the names *Lucite, Crystallite, Plexiglas,* or *Perspex* and is used in place of glass and for molding transparent objects. Alkyd baking enamels for automobile finishes and household appliances (p. 474) were being replaced in 1961 by lacquers based on poly(methyl methacrylate), but by 1963 the latter were being replaced by thermosetting enamels consisting of poly(methyl methacrylate) and melamine (p. 289). Baking produces cross linking by reaction of the ester side chains of the polymer with the amino groups of melamine to give amide linkages. The result is a finish that does not require polishing.

Numerous long-chain unsaturated acids such as oleic, vaccenic, petroselenic, linoleic, linolenic, eleostearic, and erucic acid can be obtained by the saponification of natural fats (p. 183). *Royal jelly,* the only food of the queen bee, is a complex mixture of proteins, fats, and carbohydrates that contains about 3 per cent of **10-hydroxy-*trans*-2-decenoic acid,** along with smaller amounts of several other C_{10} acids. It possibly is the precursor of the queen-bee sex attractant, 9-oxo-*trans*-2-decenoic acid.

Keto Acids

Of the keto acids, those with the oxo group in the β position are of most interest, chiefly because of the properties of their esters. In 1863 Geuther attempted to prepare a sodium salt of ethyl acetate by reaction with metallic sodium. He observed the evolution of hydrogen, but the solid products proved to be sodium ethoxide and a compound with the molecular formula $C_6H_9O_3Na$. Acidification of the latter compound gave a liquid that, though neutral to litmus, could be reconverted to the original sodium salt. The liquid proved to be ethyl β-oxobutyrate, commonly called **ethyl acetoacetate** or *acetoacetic ester,* formed by the elimination of ethyl alcohol from two molecules of ethyl acetate. The presence of some alcohol is necessary for the reaction to start easily. The explanation of the reaction is that ethoxide ion catalyzes an aldol-type condensation.

$$CH_3COOC_2H_5 \underset{C_2H_5OH}{\overset{^-OC_2H_5}{\rightleftarrows}} {^-}{:}CH_2COOC_2H_5 \overset{CH_3COOC_2H_5}{\rightleftarrows} CH_3\overset{\displaystyle O^-}{\overset{|}{\underset{\displaystyle OC_2H_5}{\underset{|}{C}}}}{-}CH_2COOC_2H_5 \underset{^-OC_2H_5}{\rightleftarrows}$$

$$CH_3\overset{\displaystyle O}{\overset{\|}{C}}CH_2COOC_2H_5 \underset{C_2H_5OH}{\overset{^-OC_2H_5}{\rightleftarrows}} CH_3\overset{\displaystyle O}{\overset{\|}{C}}{-}\underset{\cdot\cdot}{C}HCOOC_2H_5{}^-$$

With most simple aliphatic esters and ethoxide ion, the position of equilibrium of the first step is far on the side of the nonionized ester because ethanol is the stronger acid. The reaction goes to completion only because the final product, ethyl acetoacetate, is a much stronger acid ($pK_a = 10.7$) than ethanol ($pK_a = \sim 16$), and the equilibrium position for the last step is far on the side of the salt.

Claisen[1] and others showed that esters can be condensed by means of sodium ethoxide with a wide variety of compounds having hydrogen α to a carbonyl group or a nitrile group. The reaction usually is called the *Claisen ester condensation* and should not be confused with the *Claisen reaction* (p. 462). Other alkoxides such as magnesium alkoxides can be used.

[1] Ludwig Claisen (1851–1930), professor at the University of Kiel. He is noted chiefly for his work on the condensation of aromatic aldehydes with ketones, on ester condensations, on the rearrangement of allyl phenyl ethers, and on tautomeric compounds.

To avoid the formation of mixed esters, the alkyl group of the alkoxide usually corresponds to that of the ester. At the end of the reaction, the keto ester is liberated from its salt by the addition of dilute acetic acid.

The best yields are obtained in the condensation of like molecules of esters having two α hydrogen atoms or in mixed condensations in which one of the esters lacks α hydrogen atoms.

$$\underset{\text{Ethyl formate}}{HC(=O)OC_2H_5} + CH_3COOC_2H_5 \xrightarrow{^-OC_2H_5} \underset{\text{Ethyl formylacetate (\textit{formylacetic ester})}}{HC(=O)CH_2COOC_2H_5} + C_2H_5OH$$

$$\underset{\text{Ethyl oxalate}}{C_2H_5OCOCOOC_2H_5} + CH_3COOC_2H_5 \xrightarrow{^-OC_2H_5} \underset{\text{Ethyl oxaloacetate (\textit{oxaloacetic ester})}}{C_2H_5OCOCOCH_2COOC_2H_5} + C_2H_5OH$$

$$\underset{\text{Ethyl benzoate}}{C_6H_5COOC_2H_5} + CH_3COOC_2H_5 \xrightarrow{^-OC_2H_5} \underset{\text{Ethyl benzoylacetate (\textit{benzoylacetic ester})}}{C_6H_5COCH_2COOC_2H_5} + C_2H_5OH$$

If in a dibasic ester the hydrogen α to one ester group is δ or ϵ to the other, intramolecular condensation may occur with the formation of a five- or six-membered ring. This type of reaction is called a *Dieckmann condensation.*

$$\underset{\text{Ethyl adipate}}{\begin{array}{l} \quad C(=O)OC_2H_5 \\ CH_2 \quad CH_2COOC_2H_5 \\ | \qquad | \\ CH_2\text{—}CH_2 \end{array}} \xrightarrow{^-OC_2H_5} \underset{\text{Ethyl 2-oxocyclopentanecarboxylate}}{\begin{array}{l} \qquad C=O \\ CH_2 \quad CHCOOC_2H_5 \\ | \qquad | \\ CH_2\text{—}CH_2 \end{array}} + C_2H_5OH$$

Esters condense also with ketones to give **1,3 diketones.**

$$RCOCH_3 \xrightarrow{^-OC_2H_5} RCOCH_2^- \xrightarrow{R'COOC_2H_5} RCOCH_2COR' + {}^-OC_2H_5$$

If ethyl carbonate is the ester, the product is a β-keto ester. The preferred condensing agent here is sodium hydride.

$$RCOCH_3 \xrightarrow[H_2]{^-:H} RCOCH_2:^- \xrightarrow[^-OC_2H_5]{(C_2H_5O)_2CO} RCOCH_2COOC_2H_5$$

Hydrogen on carbon between two carbonyl groups is more acidic than hydrogen on carbon attached to a single carbonyl group because of the greater electron-withdrawing effect of two carbonyl groups and because of the increased resonance stabilization of the anion resulting from ionization.

$$RCOCH_2COR' \rightleftharpoons H^+ + \left\{ RC(O^-)\text{=}CH\text{—}C(=O)R' \longleftrightarrow RC(=O)\text{—}\ddot{C}H^-\text{—}C(=O)R' \longleftrightarrow RC(=O)\text{—}CH\text{=}C(O^-)R' \right\}$$

Thus ethyl acetate ($pK_a \sim 26$) is a weaker acid than water ($pK_a = 15.7$), but ethyl acetoacetate ($R' = OC_2H_5$) is much stronger ($pK_a = 10.7$). Acidic methylene groups commonly are referred to as *active* or *reactive* methylene groups. They react readily with alkoxide ion in alcohol solution. The sodium salt of ethyl acetoacetate frequently is called *sodioacetoacetic ester.*

The carbanion is a good nucleophile and can bring about the usual S_N2 displacement reactions (p. 100). Nucleophilic displacement of halogen leads to *carbon alkylation,* which provides a method for the production of α-substituted β-keto esters.

$$RCOCH_2COOC_2H_5 \xrightarrow{^-OC_2H_5} RCO\bar{C}HCOOC_2H_5 \xrightarrow{R'X} RCOCHR'COOC_2H_5 \xrightarrow{^-OC_2H_5} RCO\bar{C}R'COOC_2H_5 \xrightarrow{R''X} RCOCR'(R'')COOC_2H_5$$

The importance of these reactions lies in their use for the preparation of other compounds. Hydrolysis and decarboxylation of the carbon-alkylated esters gives good yields of the **substituted ketones.** This reaction frequently is referred to as *ketonic cleavage.*

$$RCOCHR'COOC_2H_5 \xrightarrow[\text{then HCl}]{\text{Dil. NaOH,}} RCOCHR'COOH \xrightarrow{\text{Heat}} RCOCH_2R' + CO_2$$

$$RCOCR'(R'')COOC_2H_5 \xrightarrow[\text{then HCl}]{\text{Dil. NaOH,}} RCOCR'(R'')COOH \xrightarrow{\text{Heat}} RCOCHR'(R'') + CO_2$$

The free β-keto acids decarboxylate easily because a cyclic planar transition state is possible that permits electronic shifts to take place readily with the formation of more stable products.

$$RC(=O)CR_2C(=O)O{-}H \text{ (cyclic)} \longrightarrow CO_2 + \left[RC(OH){=}CR_2\right] \longrightarrow RC(=O)CHR_2$$

Since the Claisen ester condensation is reversible, bases catalyze cleavage between the α carbon and the β carbonyl group. With concentrated aqueous alkali, the reaction is irreversible because of the formation of salts of the carboxylic acids.

$$RCOCR_2COOR' + 2\ OH^- \rightleftharpoons RCOOH + {}^-CR_2COOH + {}^-OR'$$

$$RCOOH \xrightarrow{OH^-} RCOO^- \qquad {}^-CR_2COOH \longrightarrow R_2CHCOO^-$$

This behavior frequently is referred to as the *acid cleavage* of β-keto esters because the products are salts of carboxylic acids. Although these reactions may give satisfactory yields, substituted acetic acids usually are prepared from ethyl malonate (p. 633).

Aldol-type additions take place readily when catalyzed by secondary amines. If the β-keto ester is unsubstituted at the α position, the initial hydroxy compound loses water to give an α,β-unsaturated β-keto ester. The over-all process is known as the Knoevenagel[2] reaction.

$$R'CHO + H_2C(COCH_3)(COOC_2H_5) \xrightarrow{R_2NH} \left[R'CHOHCH(COCH_3)(COOC_2H_5)\right] \longrightarrow H_2O + R'CH{=}C(COCH_3)(COOC_2H_5)$$

The resulting α,β-unsaturated carbonyl compounds may undergo the Michael[3] reaction, which is the addition of a compound with a reactive methylene group to an α,β-unsaturated carbonyl compound (cf. p. 619).

$$R'CH{=}C(COCH_3)(COOC_2H_5) \xrightarrow{CH_3COCH_2COOC_2H_5} (CH_3CO)(C_2H_5OOC)CHCHR'CH(COCH_3)(COOC_2H_5)$$

[2] Heinrich Emil Albert Knoevenagel (1865–1921), professor at Heidelberg. He is known chiefly for his work on the condensation reactions of aldehydes and ketones.

[3] Arthur Michael (1853–1942), professor of chemistry at Harvard University. His work, which was published for the most part in the German chemical literature, was concerned chiefly with the theoretical aspects of addition to the double bond and the behavior of active methylene compounds.

Hydrolysis of these products gives rise to β-keto acids that decarboxylate to **α,β-unsaturated ketones** or **1,5 diketones.**

$$R'CH{=}C\begin{matrix}COCH_3\\COOC_2H_5\end{matrix} \xrightarrow[\text{then HCl}]{\text{Dil. NaOH,}} R'CH{=}C\begin{matrix}COCH_3\\COOH\end{matrix} \xrightarrow{\text{Heat}} R'CH{=}CHCOCH_3 + CO_2$$

$$\begin{matrix}CH_3CO\\C_2H_5OOC\end{matrix}CHCHR'CH\begin{matrix}COCH_3\\COOC_2H_5\end{matrix} \xrightarrow[\text{then HCl}]{\text{Dil. NaOH,}} \begin{matrix}CH_3CO\\HOOC\end{matrix}CHCHR'CH\begin{matrix}COCH_3\\COOH\end{matrix} \xrightarrow{\text{Heat}}$$

$$CH_3COCH_2CHR'CH_2COCH_3 + CO_2$$

1,3 Diketones also have reactive methylene groups (p. 621). They undergo reactions entirely analogous to those of β-keto esters.

Tautomerism. Ethyl acetoacetate is the classical example of tautomerism, which usually is defined as the ability of a substance to possess or to react as if it possesses more than one structure. This definition is satisfactory except for the implication that the substance concerned is a single compound, the view held by Laar who coined the term tautomerism (Gr. *tauto* the same) in 1885. Actually tautomerism is the dynamic equilibrium existing between two spontaneously interconvertible isomers. In 1887 Wislicenus reported the isolation of two isomeric ethyl formylphenylacetates, $C_6H_5\underset{\displaystyle CHO}{\underset{|}{C}}HCOOC_2H_5$, a liquid that gave a color with ferric chloride and a solid that did not. The solid slowly changed to the liquid on standing. In 1893 Claisen reported two forms of acetyldibenzoylmethane. One melted at 85–90°, was soluble in dilute carbonate solution, gave in alcoholic solution a red color with ferric chloride, and reacted at once with cupric acetate to give an insoluble blue copper salt. Crystallization from alcohol of the product melting at 85–90° gave a compound melting at 109–112°, which at first was completely insoluble in dilute alkali but which slowly dissolved. Alcoholic solutions gave no color immediately with ferric chloride and no blue precipitate with cupric acetate, but both reactions took place slowly. In an article published in 1896 Claisen postulated that the lower-melting form had an enol structure and that the higher-melting form was completely ketonic.

$$CH_3\underset{\displaystyle OH}{\underset{|}{C}}{=}C(COC_6H_5)_2 \rightleftharpoons CH_3COCH(COC_6H_5)_2$$

Enolic form of acetyldibenzoylmethane — Ketonic form of acetyldibenzoylmethane

In the same year Hantzsch isolated two forms of phenylnitromethane. The solid form had acidic properties and changed spontaneously to the nonacidic liquid form (cf. p. 244).

$$C_6H_5CH_2\overset{+}{N}\begin{matrix}O^-\\O\end{matrix} \rightleftharpoons C_6H_5CH{=}\overset{+}{N}\begin{matrix}O^-\\OH\end{matrix}$$

It was not until 1911 that Knorr[4] succeeded in isolating two forms of ethyl acetoacetate. The ketonic form separated when solutions in alcohol, ether, or petroleum ether were cooled to −78°. It did not give a color immediately with ferric chloride and did not decolorize bromine. When dry hydrogen chloride was passed into a solution of the sodium salt of ethyl acetoacetate at −78°, a glassy solid was obtained that reacted instantaneously

[4] Ludwig Knorr (1859–1921), professor at the University of Jena. He is known chiefly for his syntheses of heterocyclic compounds and for his separation of the tautomers of ethyl acetoacetate.

with ferric chloride and with bromine. When either isomer was permitted to reach room temperature, the equilibrium mixture was obtained.

$$CH_3COCH_2COOC_2H_5 \rightleftharpoons CH_3COH{=}CHCOOC_2H_5$$

This interconversion is catalyzed by traces of acids or bases. By using specially treated quartz apparatus, K. H. Meyer in 1920 succeeded in separating the two forms by distillation. Since alcohols boil higher than ketones, it is surprising that the enol form of ethyl acetoacetate boils lower than the keto form. An explanation is that the enol form contains an internal hydrogen bond that reduces intermolecular dipolar association.

```
            H
            C
   CH3C          C—OC2H5
     |           ||
     O—H   :O:
```

Knorr determined the refractive index of the keto form to be $n_D^{10} = 1.4225$ and of the enol form $n_D^{10} = 1.4480$. From the refractive index of the equilibrium mixture, $n_D^{10} = 1.4232$, the enol content was estimated to be 3 per cent. Meyer found that the keto form does not isomerize too rapidly to prevent the estimation of the enol content by reaction with bromine if the procedure is carried out quickly. He mixed a solution of the ester with an excess of a solution of bromine at 0° and removed the excess bromine by the addition of a solution of β-naphthol within a period of only fifteen seconds.

$$CH_3C(OH){=}CHCOOC_2H_5 + Br_2 \longrightarrow CH_3COCHBrCOOC_2H_5 + HBr$$

$$\beta\text{-}C_{10}H_7OH + Br_2 \longrightarrow 1\text{-}Br\text{-}2\text{-}C_{10}H_6OH + HBr$$

Since α-bromo ketones are reduced by hydrogen iodide, they can be estimated by acidifying, adding sodium iodide, and titrating the iodine with standard thiosulfate solution.

$$CH_3COCHBrCOOC_2H_5 + 2\,HI \longrightarrow CH_3COCH_2COOC_2H_5 + I_2 + HBr$$

By this procedure the amount of enol form in the pure ester was estimated to be 8 per cent.

The difference between simple aldehydes and ketones, β diketones and β-keto esters, and phenols is purely one of degree. For the aldehydes and ketones, the position of equilibrium is far on the keto side. Determinations by a modification of Meyer's procedure indicate that only 0.00025 per cent of pure acetone is in the enol form. For β diketones (p. 621) and β-keto esters appreciable quantities of both the keto and enol forms are present at equilibrium. With phenols the position of equilibrium is far on the enol side.

The existence of an appreciable amount of the enol form in β diketones and of β-keto esters results from stabilization by conjugation of the double bond with the second carbonyl group, and by internal hydrogen bonding of the enolic hydroxyl group with the carbonyl group (p. 624). β-Keto esters are less highly enolized than β diketones because the electronegativity of the ethoxyl group reduces the availability of electrons on the carbonyl group for hydrogen bonding. Thus pure acetylacetone is 80 per cent enolized compared to 8 per cent for ethyl acetoacetate. On the other hand ethyl α-phenylacetoacetate is 30 per cent enolized because the electronegative phenyl group increases the acidity of the remaining α hydrogen. The importance of internal hydrogen bonding is indicated by the decrease in enolization of acetylacetone to 15 per cent in water where the carbonyl groups can be hydrogen-bonded to water molecules. Phenols are entirely in the enol form because of resonance stabilization of the aromatic ring.

The term *tautomers,* as applied thus far, has referred to structures that differ in the position of hydrogen, and tautomerization has involved transfer of a proton from one part of the molecule to another. This type of tautomerism has been called more specifically **prototropy** (Gr. *trope* a turning). A less frequent type of tautomerism is **anionotropy** as exemplified by the allylic rearrangement (p. 594) of certain allyl halides. Thus either 1-chloro-2-butene (*crotyl chloride*) or 3-chloro-1-butene can be converted into an equilibrium mixture of the two, presumably by an S_N1 mechanism involving a resonance-stabilized carbonium ion.

$$CH_3CH{=}CHCH_2Cl \rightleftharpoons \left\{CH_3CH{=}CH{-}\overset{+}{C}H_2 \longleftrightarrow CH_3\overset{+}{C}H{-}CH{=}CH_2\right\} Cl^- \rightleftharpoons CH_3\underset{\displaystyle Cl}{\underset{|}{C}H}{-}CH{=}CH_2$$

When one of the tautomers is a ring compound and the other an open-chain compound, the phenomenon is known as **ring-chain tautomerism.** Examples of prototropic ring-chain tautomerism are the equilibria between cyclic hemiacetals and δ-hydroxy aldehydes or ketones as in the sugars (p. 325).

An entirely different type of tautomerism that involves only a redistribution of bonds and changes in internuclear distances and bond angles is referred to as **valence tautomerism.** Thus 1,3,5-cyclooctatriene and bicyclo[4.2.0]-2,4-octadiene[5] exist in equilibrium with each other at 100°. Catalytic reduction yields cyclooctane, whereas permanganate oxidation yields *cis*-cyclobutane-1,2-dicarboxylic acid.

$$\text{cyclooctane} \xleftarrow{H_2/Pt} \text{1,3,5-cyclooctatriene} \rightleftharpoons \text{bicyclo[4.2.0]-2,4-octadiene} \xrightarrow{KMnO_4} \begin{array}{l} HOOCCH{-}CH_2 \\ \quad\;\; | \quad\;\; | \\ HOOCCH{-}CH_2 \end{array}$$

This example illustrates a necessary criterion for resonance. If these electronic shifts could occur without a movement of nuclei, a resonance hybrid would result.

It should be noted that there is no sharp distinction among tautomerization, molecular rearrangement, and isomerization. Usually the chemist thinks of *tautomerization* as an isomerization that is rapid and reversible and either autocatalyzed or not requiring catalysis. Molecular rearrangements and isomerizations may or may not be reversible and usually require external catalysts or rather drastic thermal treatment. The term *molecular rearrangement* frequently is applied to reactions in which a change in the carbon skeleton or in the functional group takes place during the conversion of one compound into another (pp. 207, 223). *Isomerizations,* on the other hand, involve a change in structure with no change in the nature of the functional group (pp. 69, 89, 387).

PROBLEMS

36–*1.* What can be said concerning the structure of each of the following compounds: (*a*) when an aqueous solution of the sodium salt of a hydroxy acid is acidified, a neutral compound is obtained (*b*) an amino acid hydrochloride when heated gives an unsaturated acid; (*c*) an unsaturated acid reacts with hydrogen sulfide to give a mercapto acid.

36–*2.* Give reactions for the following preparations: (*a*) tridecanol from myristic acid, (*b*) phenylalanine from benzaldehyde, (*c*) 3,4-methylenedioxycinnamic acid from piperonal, (*d*) 5-chloropentanoic acid from tetrahydrofuran, (*e*) β-hydroxypro-

---

[5] In the Baeyer system for naming bridged rings, the numbers in the bracket indicate the number of carbons in the bridges in descending order. Numbering starts at a bridgehead and follows the same order, proceeding around the larger ring in the way that gives functions the smaller numbers.

pionic acid from ethylene, (*f*) β-mercaptobutyric acid from acetaldehyde, (*g*) leucine from *i*-valeraldehyde, (*h*) γ-bromovaleric acid from levulinic acid, (*i*) *N*-lauroylsarcosine from lauric acid and acetic acid, (*j*) 13-aminotridecanoic acid from erucic acid.

36–*3*. Devise a reasonable synthesis of pantoic acid from readily available materials.

36–*4*. Construct a chart indicating the reactions, including reagents and conditions, for the synthesis of the following compounds starting with acrylonitrile: (*A*) methyl acrylate, (*B*) ethylene cyanide, (*C*) tetramethylenediamine, (*D*) γ-nitrobutyronitrile, (*E*) β-mercaptoethyl cyanide, (*F*) β-cyanopropionaldehyde, (*G*) γ-aminobutyric acid, (*H*) pyrrolidine, (*I*) β-aminopropionitrile, (*J*) β-mercaptopropionic acid, (*K*) β-alanine, (*L*) 2-phenoxyethyl cyanide, (*M*) γ-butyrolactam, (*N*) trimethylenediamine, (*O*) γ-aminobutyronitrile, (*P*) 3-phenoxypropylamine.

36–*5*. Which of the given reagents commonly is used to bring about the cyclization of (*a*) β-phenylpropionyl chloride, (*b*) phthalic acid, (*c*) *o*-benzoylbenzoic acid, (*d*) *N*-(β-phenylethyl)acetamide, (*e*) *o*-hydroxycinnamic acid, (*f*) adipic acid, (*g*) 1,5-diaminopentane: (*1*) none, (*2*) concentrated H_2SO_4, (*3*) $Ba(OH)_2$, (*4*) $AlCl_3$, (*5*) $POCl_3$, (*6*) heat, (*7*) hydrochloric acid and heat.

36–*6*. Starting with ethyl acetoacetate, give reactions for the preparation of the following compounds: (*a*) methyl *i*-hexyl ketone, (*b*) α,β-dimethylbutyric acid, (*c*) 4-*i*-propyl-2,6-heptanedione, (*d*) δ-oxocaproic acid.

36–*7*. Give reactions for the following preparations: (*a*) 1-oxo-1,3-diphenylpropane from ethyl benzoate, (*b*) 2-methyl-1-oxocyclopentane using a Dieckmann condensation.

CHAPTER THIRTY-SEVEN

POLYCARBOXYLIC ACIDS

Of the polycarboxylic acids, the dicarboxylic acids are encountered most frequently, and some are of considerable importance. Thus the ethyl ester of malonic acid is a valuable intermediate for organic syntheses. Adipic acid and the unsaturated maleic acid are technically important intermediates. Some of the hydroxy polycarboxylic acids, such as malic, tartaric, and citric acids, occur in fruit juices.

Nomenclature

The unsubstituted polycarboxylic acids have common names that are in general use and serve as family names for the substituted acids. The names of the normal dibasic acids having 2 to 10 carbon atoms are oxalic, malonic, succinic, glutaric, adipic, pimelic, suberic, azelaic, and sebacic acids.

Dicarboxylic Acids

Preparation and Reactions. Several general methods for the preparation of dibasic acids are available.

(*a*) OXIDATION OF α,ω GLYCOLS.

$$(CH_2)_x(CH_2OH)_2 \xrightarrow{HNO_3,\ KMnO_4,\ or\ HCrO_4} (CH_2)_x(COOH)_2$$

(*b*) HYDROLYSIS OF DINITRILES.

$$(CH_2)_x(CN)_2 \xrightarrow{H_2O,\ HCl} (CH_2)_x(COOH)_2 + 2\ NH_4Cl$$

(*c*) ELECTROLYSIS OF SALTS OF ACID ESTERS OF LOWER DICARBOXYLIC ACIDS (*Kolbe synthesis*, p. 166).

$$2\ CH_3OOC(CH_2)_xCOONa \xrightarrow{Electrolysis} CH_3OOC(CH_2)_x(CH_2)_xCOOCH_3 + 2\ CO_2 (+\ NaOH\ and\ H_2\ at\ cathode)$$

Special methods frequently are used for the preparation of the more common dicarboxylic acids, such as the oxidation of unsaturated acids (p. 632) or the synthesis from malonic esters (p. 633).

Like all polyfunctional compounds, the dicarboxylic acids have certain characteristic behaviors that depend on the relative positions of the functional groups. Since a carboxyl group is electron-attracting, the presence of one carboxyl close to another increases the ease of ionization of the first hydrogen ion. This inductive effect rapidly decreases as the carboxyl groups become more separated. Thus oxalic acid is somewhat stronger than phosphoric acid, the pK_a's for the ionization of the first proton being 1.3 and 2.1 respectively. In malonic acid, the carboxyl groups are farther apart, and it is weaker ($pK_a = 2.8$), although still stronger than the unsubstituted monocarboxylic acids. When two or more methylene groups intervene, the two carboxyl groups have little effect on each other. Succinic acid, $pK_a = 4.2$, is only slightly stronger than acetic acid, $pK_a = 4.8$. More striking differences in chemical behavior that are dependent on the length of the carbon chain that separates the carboxyl groups are discussed under representative members of each series.

Ethanedioic Acid. This acid with the common name **oxalic acid** was known at an early date. The presence of its potassium acid salt in the sorrels (various species of *Rumex* and *Oxalis,* Gr. *oxys* sharp or acid) was observed at the beginning of the seventeenth century. It is present in many other plants such as spinach, rhubarb, sweet potatoes, cabbage, grapes, and tomatoes. When these fruits and vegetables are eaten, microscopic star-shaped crystals of the insoluble calcium oxalate may appear in the urine.

Nitric acid oxidation of any α,β-oxygenated or aminated substance such as the carbohydrates or the amino acids produces oxalic acid. Commercially it is manufactured by heating sodium formate (p. 167) and liberating the free acid from the sodium salt with sulfuric acid.

$$2\,HCOONa \xrightarrow{400^\circ} H_2 + NaOOCCOONa \xrightarrow{H_2SO_4} HOOCCOOH$$

When oxalic acid is heated slowly to 150°, it sublimes unchanged, but rapid heating to a higher temperature decomposes it into carbon dioxide and formic acid, and the latter decomposes further into carbon monoxide and water.

$$HOOCCOOH \xrightarrow{Heat} CO_2 + HCOOH \longrightarrow CO + H_2O$$

These reactions take place more readily when oxalic acid is warmed with concentrated sulfuric acid.

Propanedioic Acids. **Malonic acid** was so named because it first was obtained in 1858 as an oxidation product of malic acid (p. 635). It ordinarily is prepared by a series of reactions from sodium chloroacetate, first carried out by Kolbe (p. 166) in 1864. The intermediate cyanoacetate may be hydrolyzed to the acid or alcoholyzed to the ester.

$$NaOOCCH_2Cl \xrightarrow{NaCN} \underset{\text{Sodium cyanoacetate}}{NaOOCCH_2CN} \begin{cases} \xrightarrow{H_2O,\ H^+} \underset{\text{Malonic acid}}{HOOCCH_2COOH} \\ \xrightarrow{C_2H_5OH,\ H^+} \underset{\text{Ethyl malonate}}{C_2H_5OOCCH_2COOC_2H_5} \end{cases}$$

Esters of malonic acid undergo a wide variety of reactions, which are considered separately (p. 632). Substituted malonic acids can be obtained by the hydrolysis of substituted esters made by alkylation of ethyl malonate (p. 633), or by a Claisen ester condensation (p. 633).

Malonic acid and substituted malonic acids, when heated above the melting point, lose carbon dioxide to give the monocarboxylic acid. As was true for β-keto acids (p. 622), a cyclic transition state permits easy decarboxylation.

$$\text{HOC(=O)–CR}_2\text{–C(=O)–O–H (cyclic)} \xrightarrow{Heat} CO_2 + \left[HO\overset{OH}{\underset{}{C}}{=}CR_2 \right] \longrightarrow R_2CHCOOH$$

When a pyridine solution of malonic acid and an aldehyde is heated, the salt of the initial aldol addition product loses carbon dioxide and water to give an α,β-unsaturated acid.

$$RCHO + H_2C(COOH)_2 + C_5H_5N \longrightarrow \text{RCH(OH)–CH(COOH)–C(=O)O}^- \cdots H\overset{+}{N}C_5H_5 \longrightarrow RCH{=}CHCOOH + CO_2 + C_5H_5N + H_2O$$

Malononitrile can be made by the dehydration of cyanoacetamide.

$$2\,NCCH_2CONH_2 + POCl_3 \longrightarrow NCCH_2CN + HPO_3 + 3\,HCl$$

It is reported to trimerize explosively in the presence of traces of alkali, although metallic salts have been prepared.

Butanedioic Acids. **Succinic acid** was known in the sixteenth century as a distillation product of amber, a fossil resin (*L. succinum* amber). It can be obtained by the hydrolysis of ethylene cyanide, but is manufactured now by the catalytic reduction of maleic acid or the electrolytic reduction of fumaric acid (p. 635).

$$\underset{\text{Maleic or fumaric acid}}{\text{HOOCCH=CHCOOH}} + 2\,[\text{H}] \xrightarrow[\text{electrolytic reduction}]{\text{Cat. or}} \underset{\text{Succinic acid}}{\text{HOOCCH}_2\text{CH}_2\text{COOH}}$$

Derivatives of succinic acid can be made by the hydrolysis of β-cyano esters formed by the addition of hydrogen cyanide to α,β-unsaturated esters (p. 618).

$$\text{RCH=CHCOOC}_2\text{H}_5 \xrightarrow{\text{HCN}} \text{RCH(CN)CH}_2\text{COOC}_2\text{H}_5 \xrightarrow{\text{H}_2\text{O, HCl}} \text{RCH(COOH)CH}_2\text{COOH}$$

Succinonitrile, $NCCH_2CH_2CN$, can be made from ethylene bromide and sodium cyanide, but it is made commercially by the addition of hydrogen cyanide to acrylonitrile (p. 619). Substituted succinonitriles result from the decomposition of α,α'-dicyanoazoalkanes (p. 248).

Succinic acid and its substitution products lose water when pyrolyzed to give the stable five-membered cyclic anhydrides.

$$\begin{matrix}\text{CH}_2\text{COOH}\\ |\\ \text{CH}_2\text{COOH}\end{matrix} \xrightarrow{\text{Heat}} \underset{\text{Succinic anhydride}}{\begin{matrix}\text{CH}_2\text{—CO}\\ | \quad\quad \text{O}\\ \text{CH}_2\text{—CO}\end{matrix}} + \text{H}_2\text{O}$$

This reaction takes place at much lower temperatures in the presence of dehydrating agents. Thus if a solution of succinic acid in acetyl chloride is heated and allowed to cool, a good yield of succinic anhydride crystallizes. Succinic acids do not decarboxylate readily because no pathway with a low activation energy, such as the cyclic mechanism for malonic acids (p. 628), is available.

When succinic anhydride is heated with ammonia **succinimide** is formed.

$$\begin{matrix}\text{CH}_2\text{—CO}\\ | \quad\quad \text{O}\\ \text{CH}_2\text{—CO}\end{matrix} + \text{NH}_3 \xrightarrow{\text{Heat}} \underset{\text{Succinimide}}{\begin{matrix}\text{CH}_2\text{—CO}\\ | \quad\quad \text{NH}\\ \text{CH}_2\text{—CO}\end{matrix}} + \text{H}_2\text{O}$$

When succinimide is treated with bromine and alkali under the usual conditions of the Hofmann reaction, the product is β-alanine.

$$\begin{matrix}\text{CH}_2\text{—CO}\\ | \quad\quad \text{NH}\\ \text{CH}_2\text{—CO}\end{matrix} \xrightarrow{\text{H}_2\text{O}} \begin{matrix}\text{CH}_2\text{COOH}\\ |\\ \text{CH}_2\text{CONH}_2\end{matrix} \xrightarrow{\text{Br}_2\text{, NaOH}} \underset{\beta\text{-Alanine}}{\begin{matrix}\text{CH}_2\text{COOH}\\ |\\ \text{CH}_2\text{NH}_2\end{matrix}}$$

If, however, bromine is added to an ice-cold alkaline solution of succinimide, ***N*-bromosuccinimide** precipitates in almost quantitative yield.

$$(\text{CH}_2\text{CO})_2\text{NH} + \text{Br}_2 + \text{NaOH} \xrightarrow{0^\circ} \underset{N\text{-Bromosuccinimide}}{(\text{CH}_2\text{CO})_2\text{NBr}} + \text{NaBr} + \text{H}_2\text{O}$$

This valuable reagent has the property of being able, in boiling carbon tetrachloride solutions, to brominate unsaturated or aromatic compounds α to the double bond or the ring to give allyl or benzyl bromides.

$$RCH_2CH{=}CHCH_3 + (CH_2CO)_2NBr \longrightarrow RCHBrCH{=}CHCH_3 + (CH_2CO)_2NH$$
$$ArCH_2R + (CH_2CO)_2NBr \longrightarrow ArCHBrR + (CH_2CO)_2NH$$

Catalysis of the bromination by light and peroxides indicates that the reaction takes place by a free radical mechanism. The hydrogen-abstracting agent is the bromine atom. The *N*-bromosuccinimide appears to act chiefly as a source of a low concentration of bromine for reaction with the hydrogen bromide by a polar mechanism.

$$ZNBr \rightleftharpoons ZN\cdot + Br\cdot$$
$$RH + Br\cdot \rightleftharpoons R\cdot + HBr$$
$$ZNBr + HBr \rightleftharpoons ZNH + Br_2$$
$$R\cdot + Br_2 \rightleftharpoons RBr + Br\cdot$$

Pentanedioic Acids. **Glutaric acid** first was prepared from the readily obtainable glutamic acid (p. 358), which was converted to α-hydroxyglutaric acid and reduced with hydrogen iodide. It is not a commercial product but is made in the laboratory by the hydrolysis of trimethylene cyanide, from malonic ester (p. 633), or by the nitric acid oxidation of cyclopentanone (p. 631).

$$\underset{\text{Cyclopentanone}}{(CH_2)_4C{=}O} \xrightarrow{HNO_3} \underset{\text{Glutaric acid}}{HOOC(CH_2)_3COOH}$$

Like the succinic acids, the glutaric acids lose water when pyrolyzed to give the cyclic anhydrides.

$$CH_2(CH_2COOH)_2 \xrightarrow{\text{Heat}} \underset{\text{Glutaric anhydride}}{CH_2(CH_2CO)_2O} + H_2O$$

Hexanedioic Acids. **Adipic acid** (L. *adeps* fat) is one of the compounds formed when many unsaturated fats or fat acids are oxidized. It is the chief product from the oxidation of cyclohexanol or cyclohexanone (pp. 647, 649) with nitric acid.

$$\underset{\text{Cyclohexanol}}{(CH_2)_5CHOH} \xrightarrow{HNO_3} \underset{\text{Cyclohexanone}}{(CH_2)_5C{=}O} \xrightarrow{HNO_3} \underset{\text{Adipic acid}}{HOOC(CH_2)_4COOH}$$

It formerly was made from cyclohexanol but now is produced by the catalytic air oxidation of cyclohexane (p. 647) in the liquid phase. Sufficient acetic acid is added to render the cobalt acetate catalyst soluble.

$$\underset{\text{Cyclohexane}}{(CH_2)_6} \xrightarrow[95^\circ,\ 150\ \text{p.s.i.}]{O_2,\ CoAc_2} HOOC(CH_2)_4COOH + H_2O$$

Adipic acids and other acids that have the carboxyl groups more widely separated do not give cyclic anhydrides except under special conditions that favor intramolecular reaction. When heated with dehydrating agents, they give linear polymeric anhydrides.

$$(x+1)\ HOOC(CH_2)_4COOH + x\,(CH_3CO)_2O \xrightarrow{\text{Heat}} \underset{\text{Poly(adipic anhydride)}}{HOOC(CH_2)_4[COOCO(CH_2)_4]_xCOOH} + 2x\ CH_3COOH$$

If the monomeric cyclic anhydride is sufficiently volatile, it can be formed by distillation of the polymer under reduced pressure. When adipic acids are heated, especially in the presence of a small amount of barium hydroxide, five-membered cyclic ketones are formed.

$$\begin{matrix} CH_2CH_2COOH \\ | \\ CH_2CH_2COOH \end{matrix} \xrightarrow{Ba(OH)_2,\ heat} \begin{matrix} CH_2{-}CH_2 \\ | \quad\quad \diagdown \\ | \quad\quad\quad CO \\ | \quad\quad \diagup \\ CH_2{-}CH_2 \end{matrix} + H_2O + CO_2$$

Cyclopentanone

Adipic acid is the most important of the aliphatic dicarboxylic acids commercially. It is an intermediate for the synthesis of **nylon 6-6,** which is a polyamide formed when the hexamethylenediamine (p. 612) salt of adipic acid is heated. The word *nylon* is a generic term coined to designate any synthetic polyamide with recurring units in the main chain. The number 6-6 indicates that this particular nylon has two six-carbon components.

$$(x+1)\ {}^-OOC(CH_2)_4COO^{-}\,{}^+NH_3(CH_2)_6NH_3^+ \xrightarrow[200\text{–}300°]{Heat}$$

Nylon salt

$$^-OOC(CH_2)_4CO[NH(CH_2)_6NHCO(CH_2)_4CO]_xNH(CH_2)_6NH_3^+ + (2x+1)\ H_2O$$

Nylon 6-6

The molecular weight of nylon 6-6 is about 10,000 and the melting point around 260°. It is spun from a melt, and the filaments are cold-drawn to four times their original length to orient the molecules along the axis of the fiber. The resulting fibers are elastic and lustrous and either dry or wet have a higher tensile strength than silk. Of all the nylons, nylon 6-6 because of its high melting point has the best properties for textile fibers and accounts for 80 per cent of production. Most of the remainder is nylon 6 (p. 649). A limited amount of nylon 6-10 is made from hexamethylenediamine and sebacic acid for the manufacture of synthetic bristles.

Hexamethylenediamine is made by the reduction of **adiponitrile** (p. 612), which can be made from adipic acid by passing the vapors with an excess of ammonia over a catalyst such as boron phosphate at 350°.

$$\underset{\text{Ammonium adipate}}{(CH_2)_4(COONH_4)_2} \longrightarrow 2\ H_2O + \underset{\text{Adipamide}}{(CH_2)_4(CONH_2)_2} \longrightarrow 2\ H_2O + \underset{\text{Adiponitrile}}{(CH_2)_4(CN)_2}$$

With a growing scarcity of benzene, a process was developed for preparing adiponitrile from furfural by way of tetrahydrofuran and tetramethylene chloride (Fig. 30–2, p. 513).

$$(CH_2)_4Cl_2 + 2\ NaCN \longrightarrow (CH_2)_4(CN)_2 + 2\ NaCl$$

Adiponitrile can be made also from acetylene by way of tetrahydrofuran (p. 512) or from 1,3-butadiene (p. 582).

$$CH_2{=}CHCH{=}CH_2 \xrightarrow{Cl_2} ClCH_2CH{=}CHCH_2Cl \xrightarrow{NaCN} NCCH_2CH{=}CHCH_2CN \xrightarrow{H_2/Ni} NC(CH_2)_4CN$$

The latest process reported uses the electrolytic bimolecular reduction of acrylonitrile at lead or mercury cathodes.

$$2\ CH_2{=}CHCN + 2\ H^+ + 2\ e \longrightarrow (-CH_2CH_2CN)_2$$

These reactions illustrate the diverse raw materials that frequently are available to the manufacturer of organic chemicals.

Production of adipic acid was estimated to be about 720 million pounds in 1964 with 90 per cent being used to make nylon 6-6. The remainder is used in the production of plasticizers, lubricants, polyurethans (p. 605), and polyester resins (p. 635).

Higher Alkanedioic Acids. **Pimelic acid** (Gr. *pimele* fat) also is an oxidation product of unsaturated fats. It is not available commercially, but it can be synthesized by the hydrolysis of pentamethylene cyanide or by several other procedures.

$$\underset{\text{Pentamethylene cyanide}}{CN(CH_2)_5CN} \xrightarrow{H_2O,\ HCl} \underset{\text{Pimelic acid}}{HOOC(CH_2)_5COOH}$$

Pimelic acids pyrolyze to six-membered cyclic ketones.

$$\begin{matrix} \diagup CH_2CH_2COOH \\ CH_2 \\ \diagdown CH_2CH_2COOH \end{matrix} \xrightarrow{Ba(OH)_2,\ heat} \begin{matrix} \diagup CH_2{-}CH_2 \diagdown \\ CH_2 \quad\quad\quad CO \\ \diagdown CH_2{-}CH_2 \diagup \end{matrix} + H_2O + CO_2$$

The generalization that succinic and glutaric acids give cyclic anhydrides, whereas adipic and pimelic acids yield cyclic ketones, is known as the **Blanc rule.** It has been of considerable value in determining whether oxygenated or unsaturated rings in compounds of unknown constitution are five- or six-membered. Thus oxidation of a five-membered ring gives a glutaric acid that cyclizes to an anhydride, whereas a six-membered ring gives an adipic acid that cyclizes to a ketone. The rule is not infallible, however, since at least one example is known in which a nonterminal six-membered ring in a polycyclic compound gives on oxidation an adipic acid that forms a seven-membered cyclic anhydride.

Suberic acid (L. *suber* cork oak), $HOOC(CH_2)_6COOH$, is obtained in small amounts by the oxidation of cork with nitric acid. It can be prepared by the nitric acid oxidation of castor oil, although the yield still is low. It has been postulated that suberic acid arises by oxidation of trihydroxystearic acid, but it is conceivable that a methylene group α to the double bond is the initial point of attack (p. 630). It is made commercially by the oxidation of cyclooctane (p. 654).

Azelaic acid is the chief product of the oxidation of unsaturated fat acids with nitric acid (F. *azote* nitrogen, and Gr. *elaion* olive oil) or with potassium permanganate.

$$\underset{\text{Oleic acid}}{CH_3(CH_2)_7CH{=}CH(CH_2)_7COOH} \xrightarrow{\text{Ox.}} \underset{\text{Pelargonic acid}}{CH_3(CH_2)_7COOH} + \underset{\text{Azelaic acid}}{HOOC(CH_2)_7COOH}$$

Oxidation of the ozonide of oleic acid with chromic acid gives better yields. Esters of azelaic acid are used as plasticizers and as synthetic lubricants.

Sebacic acid (L. *sebum* tallow) is a waxy solid. The sodium salt is one of the products of the destructive distillation of the sodium soap from castor oil with excess alkali. The other chief product is 2-octanol (*capryl alcohol*).

$$\underset{\text{Sodium ricinoleate}}{CH_3(CH_2)_5CHOHCH_2CH{=}CH(CH_2)_7COONa} + NaOH + H_2O \xrightarrow{\text{Heat}}$$

$$\underset{\text{2-Octanol}}{CH_3(CH_2)_5CHOHCH_3} + \underset{\text{Sodium sebacate}}{NaOOC(CH_2)_8COONa} + H_2$$

Sebacic acid is an intermediate for the synthesis of nylon 6-10 (p. 631), which is used to manufacture filament for synthetic bristles. Its esters are used as synthetic lubricants.

Reactions of Malonic Esters

The reactions of the esters of malonic acid are sufficiently important to require special attention. Usually the ethyl ester is used, and the term *malonic ester* ordinarily means ethyl malonate. For certain purposes **ethyl *t*-butyl malonate** and ***t*-butyl malonate** are useful. They are made by the reaction of ethyl hydrogen malonate or malonic acid with isobutylene in the presence of a small amount of sulfuric acid (p. 164). **Ethyl cyanoacetate,** made by the esterification of cyanoacetic acid (p. 628), undergoes many of the reactions of malonic ester. It often can lead to the same final products and may advantageously replace ethyl malonate.

The methylene group joined to two alkoxycarbonyl groups or to one alkoxycarbonyl and one cyano group is unusually reactive (cf. β-keto esters, p. 621). Bromination takes place with extreme ease, although ethyl malonate does not give a color with ferric chloride, and other tests indicate an enol content of only 0.008 per cent. Because of resonance stabilization of the carbanion, esters of malonic acid are considerably more acidic ($pK_a = \sim 13$) than alcohols ($pK_a = \sim 16$), and they readily react with alkoxide ion in alcohol solution. The sodium salt of ethyl malonate commonly is called *sodiomalonic ester*. Like that from β-keto esters, the carbanion is a good nucleophile and can bring about S_N2 displacement reactions and undergo base-catalyzed addition to aldehydes and to α,β-unsaturated carbonyl compounds. Nucleophilic displacement of halogen, usually referred to as *carbon alkylation*, leads to a variety of products and is the most important reaction of malonic esters. Simple alkyl halides yield the alkyl-substituted malonic ester. A second alkyl group may be introduced if desired. Since the esters can be saponified and the free acids easily decarboxylated (p. 628), the reaction leads to the preparation of substituted acetic acids.

$$CH_2(COOC_2H_5)_2 \xrightarrow[HOC_2H_5]{^-OC_2H_5} :\bar{C}H(COOC_2H_5)_2 \xrightarrow[X^-]{RX}$$

$$\underset{\text{Ethyl alkylmalonate}}{RCH(COOC_2H_5)_2} \xrightarrow{\text{NaOH, then HCl}} \underset{\text{Alkylmalonic acid}}{RCH(COOH)_2} \xrightarrow{\text{Heat}} \underset{\text{Alkylacetic acid}}{RCH_2COOH} + CO_2$$

$$RCH(COOC_2H_5)_2 \xrightarrow[HOC_2H_5]{^-OC_2H_5} R\bar{\underset{\cdot\cdot}{C}}(COOC_2H_5)_2 \xrightarrow[X^-]{R'X}$$

$$\underset{\text{Ethyl dialkylmalonate}}{RR'C(COOC_2H_5)_2} \xrightarrow{\text{NaOH, then HCl}} \underset{\text{Dialkylmalonic acid}}{RR'C(COOH)_2} \xrightarrow{\text{Heat}} \underset{\text{Dialkylacetic acid}}{RR'CHCOOH} + CO_2$$

Methylene halides react with two moles of ethyl sodiomalonate to give **tetracarboxylic esters.** If it is possible to form a three- to six-membered ring, a cyclic derivative also may be formed.

$$ClCH_2CH_2Cl + 2\,\bar{C}H(COOC_2H_5)_2 \longrightarrow \underset{\text{Ethyl 1,1,4,4-butanetetracarboxylate}}{(C_2H_5OOC)_2CHCH_2CH_2CH(COOC_2H_5)_2} + 2\,Cl^-$$

$$ClCH_2CH_2Cl + \bar{C}H(COOC_2H_5)_2 \longrightarrow Cl^- + ClCH_2CH_2CH(COOC_2H_5)_2 \xrightarrow{^-OC_2H_5}$$

$$ClCH_2CH_2\underline{C}(COOC_2H_5)_2 \longrightarrow \underset{\text{Ethyl 1,1-cyclopropanedicarboxylate}}{(CH_2)_2C(COOC_2H_5)_2} + Cl^-$$

Similarly trimethylene, tetramethylene, and pentamethylene bromides give **cyclobutane-, cyclopentane-,** and **cyclohexanedicarboxylic esters.** If the halogen atoms are farther apart, only the tetracarboxylic esters are obtained.

Arylmalonic esters cannot be made from nonactivated aryl halides and sodiomalonic ester in alcohol, because the halogen is not readily displaced. Usually they are made by a Claisen ester condensation (p. 621). For example, ethyl phenylmalonate results from the reaction of ethyl phenylacetate with ethyl carbonate.

$$C_6H_5CH_2COOC_2H_5 + C_2H_5OCOOC_2H_5 \xrightarrow{NaOC_2H_5} C_6H_5CH(COOC_2H_5)COOC_2H_5 + C_2H_5OH$$

Like other reactive methylene compounds, ethyl malonate undergoes the **Knoevenagel reaction** (p. 622) with aldehydes. Diethylamine or piperidine is the usual catalyst.

$$RCH{=}O + H_2C(COOC_2H_5)_2 \xrightarrow[\text{or } C_5H_{10}NH]{(C_2H_5)_2NH} [RCHOHCH(COOC_2H_5)_2] \longrightarrow RCH{=}C(COOC_2H_5)_2 + H_2O$$

If R is aliphatic and an excess of ethyl malonate is used, the initial reaction usually is followed by a Michael addition (p. 622) to give the tetracarboxylic ester.

$$RCH{=}C(COOC_2H_5)_2 + H_2C(COOC_2H_5)_2 \xrightarrow{R_2NH} RCH[CH(COOC_2H_5)_2]_2$$

The synthesis of **pyrimidines** by the reaction of malonic esters with urea and related compounds has been discussed (p. 525).

Unsaturated Dicarboxylic Acids

A few unsaturated dicarboxylic acids are of special interest. **Maleic** and **fumaric acids** (*cis-* and *trans-ethylene*-1,2-*dicarboxylic acids*) are the classic examples of geometric isomerism (p. 76). Maleic acid readily yields a monomeric cyclic anhydride when heated, which indicates that the carboxyl groups are on the same side of the double bond. Fumaric acid on the other hand cannot be converted to a monomeric fumaric anhydride. When it is heated to a sufficiently high temperature (250–300°), isomerization takes place, and maleic anhydride is formed.

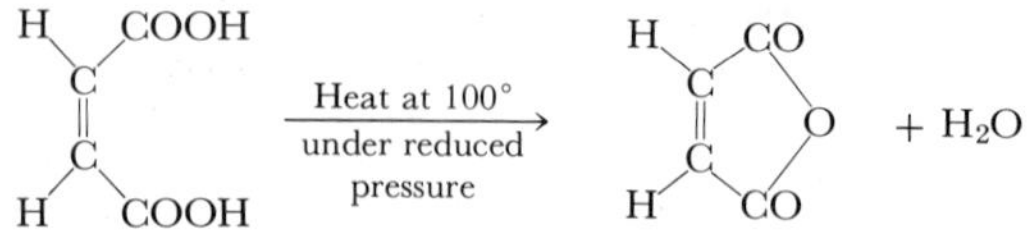

$$\underset{\text{Maleic acid}}{\text{H(COOH)C=C(COOH)H}} \xrightarrow[\text{under reduced pressure}]{\text{Heat at } 100^\circ} \underset{\text{Maleic anhydride}}{\text{H(CO)C=C(CO)H (–O–)}} + H_2O$$

$$\underset{\text{Fumaric acid}}{\text{HOOC(H)C=C(H)COOH}} \xrightarrow{\text{Heat at } 200^\circ} \text{Sublimes unchanged}$$

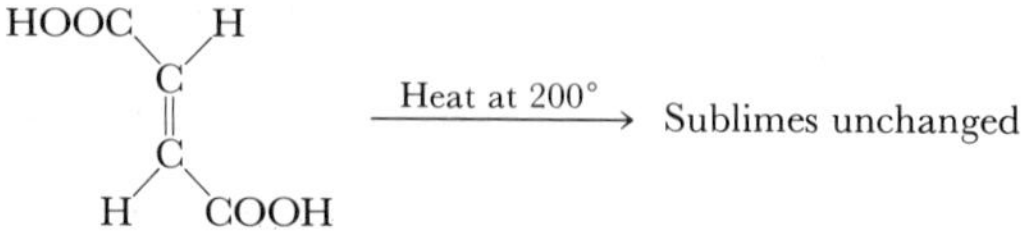

The most characteristic reaction of maleic anhydride is its 1,4 addition to conjugated dienes (p. 582).

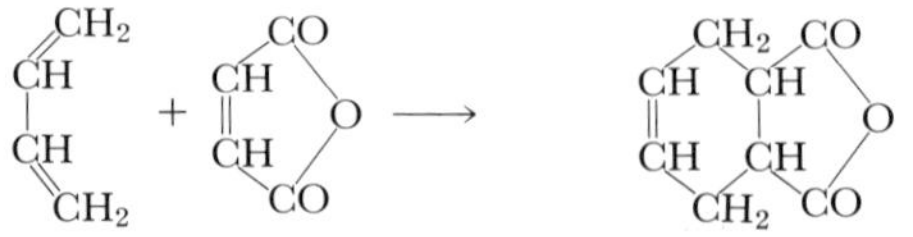

$$CH_2{=}CH{-}CH{=}CH_2 + \text{CH(CO)=CH(CO)>O} \longrightarrow \underset{\text{Tetrahydrophthalic anhydride}}{\text{cyclo-}[CH_2{-}CH{=}CH{-}CH_2{-}CH(CO){-}CH(CO)]\text{>O}}$$

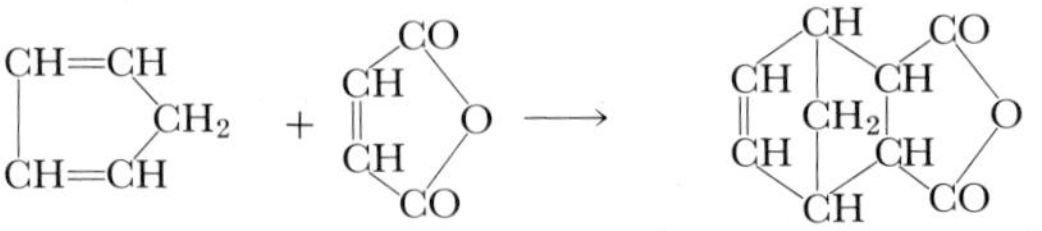

$$\underset{\text{Cyclopentadiene}}{\text{CH=CH–CH}_2\text{–CH=CH (ring)}} + \text{CH(CO)=CH(CO)>O} \longrightarrow \underset{\text{3,6-Methylenecyclohexene-4,5-dicarboxylic anhydride}}{\text{adduct}}$$

Other compounds containing a double or triple bond conjugated with a carbonyl group or a nitrile group, such as acrolein, crotonaldehyde, acrylic acid, crotononitrile, acetylenedicarboxylic esters, ethyl azoformate, $C_2H_5OOC{-}N{=}N{-}COOC_2H_5$, quinones, and vinyl ethers, also add to a system of conjugated carbon-carbon multiple bonds. The first group of compounds are called *dienophiles,* the second are the *dienes,* and the product of the reaction of dienophile with a diene is called the *adduct.* These reactions commonly are known as **Diels-Alder diene syntheses.**[1]

Maleic anhydride is made commercially by the catalytic oxidation of benzene with air.

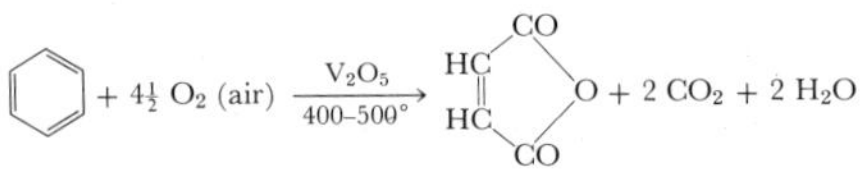

$$C_6H_6 + 4\tfrac{1}{2}\ O_2\ (\text{air}) \xrightarrow[400\text{–}500^\circ]{V_2O_5} \text{HC(CO)=HC(CO)>O} + 2\ CO_2 + 2\ H_2O$$

[1] Otto Diels (1876–1954), professor of chemistry at the University of Kiel. He discovered carbon suboxide, C_3O_2, and was the first to use selenium for the dehydrogenation of natural products, but he is best known for the reaction that bears his name. He was awarded the Nobel Prize in Chemistry in 1950.

It is formed also to the extent of 5 to 8 per cent in the production of phthalic anhydride (p. 474) and is separated as a coproduct. Production began in the United States in 1933 and in 1964 amounted to 118 million pounds. About 60 per cent was used in the formulation of **polyester resins.** Alkyd resins (p. 475), polyethylene terephthalate (p. 476), and polyurethans (p. 605) also have the ester linkage as the most important type of bonding in the polymer, but technically the term *polyester* or *polyester resin* refers to a solution of an unsaturated linear polymer in a liquid monomer that is capable of copolymerizing with the linear polymer. Most polyester resins consist of a solution of an alkyd resin, prepared for example from propylene glycol, maleic anhydride, and adipic acid, or from diethylene glycol, tetrahydrophthalic anhydride, and fumaric acid, in 30 per cent of its weight of styrene. An inhibitor, such as a quaternary ammonium salt, is added to prevent polymerization before use. Just prior to use, a peroxide initiator such as benzoyl peroxide or *t*-butylhydroperoxide together with a cobalt or manganese salt as a promoter is added. The liquid may be cast or applied to a reinforcing agent such as glass fiber. Copolymerization of the maleic anhydride units of the alkyd resin with the monomer takes place at room temperature or higher depending on the formulation. Wide variations in the composition of the alkyd are possible and other liquid monomers, such as allyl phthalate or mixtures of styrene with vinyl acetate, methyl methacrylate, or vinyltoluene, may be used as solvent. Other alkyd resins consume about 20 per cent and agricultural chemicals 10 per cent of the production of maleic anhydride.

Reagents that add to the double bond of other α,β-unsaturated acids (p. 618) also add to maleic and fumaric acids and their derivatives. Addition of sodium bisulfite to 2-ethylhexyl maleate gives a wetting agent that is called *dioctyl sodium sulfosuccinate* when used medicinally or *Aerosol-OT* for technical use.

$$\begin{matrix} CHCOOC_8H_{17} \\ \| \\ CHCOOC_8H_{17} \end{matrix} + NaHSO_3 \longrightarrow Na^{+-}O_3S-\begin{matrix} CH_2COOC_8H_{17} \\ | \\ CHCOOC_8H_{17} \end{matrix}$$

It makes water more effective as a softening agent for fecal matter and is used with potentially constipating agents such as barium sulfate, which is used in X-ray examination of the alimentary tract.

Fumaric acid is made by the isomerization of maleic acid and also by a fermentation process from starch or other carbohydrate using molds of the genus *Rhizopus*. It is used technically in alkyd and polyester resins. In contrast to maleic acid which is toxic, fumaric acid is permitted as an acidulant in food products without restriction. The cost of providing the same flavor equivalent is less than half that of other acidulants such as citric acid, and its use is limited chiefly by the low solubility in water of 0.7 per cent at 20°.

Hydroxy Dicarboxylic Acids

Malic acid is hydroxysuccinic acid, $HOOCCH_2CHOHCOOH$. It is present in many fruit juices and was isolated by Scheele in 1785 from unripe apples (L. *malum* apple). Calcium acid (−)-malate separates during the concentration of maple sap and is known as *sugar sand.* Malic acid has the property of attracting plant spermatozoids; that is, they migrate toward the point of highest concentration. Surprisingly D- and L-malic acid are equally effective. Racemic malic acid is manufactured by the hydration of maleic or fumaric acid and is used as a food acidulant.

Thiomalic acid is made by the addition of hydrogen sulfide to maleic acid.

$$HOOCCH{=}CHCOOH + H_2S \longrightarrow HOOCCH_2CH(SH)COOH$$

Sodium thiomalate is reported to be effective as an antidote for heavy metal poisoning. Addition of *O,O*-diethyldithiophosphate to ethyl maleate gives the mixed dithiophosphate known as **malathion.** It is an important insecticide because of its high toxicity to a variety of insects and its low toxicity to mammals.

$$4\,C_2H_5OH + P_2S_5 \longrightarrow 2\,(C_2H_5O)_2P(S)SH + H_2S$$

$$\begin{matrix} (C_2H_5O)_2P{-}SH \\ \overset{+}{\underset{-}{|}} \\ S \end{matrix} + \begin{matrix} CHCOOC_2H_5 \\ \| \\ CHCOOC_2H_5 \end{matrix} \longrightarrow \underset{\text{Malathion}}{\begin{matrix} (C_2H_5O)_2P{-}S{-}CHCOOC_2H_5 \\ \overset{+}{\underset{-}{|}} \qquad\quad | \\ S \qquad CH_2COOC_2H_5 \end{matrix}}$$

The **tartaric acids** are dihydroxysuccinic acids, $HOOCCHOHCHOHCOOH$. The (+) form is one of the most widely distributed plant acids. Its potassium acid salt is present in grape juice and is the chief component of the lees of wine (p. 300). The crude product is called *argol,* whereas the purified material is called *cream of tartar.* It is used as the acid component in some baking powders. Neutralization of cream of tartar with sodium hydroxide yields sodium potassium tartrate, which is known as *Rochelle salts* (after Rochelle, France) and is used as a purgative. Tartar was known to the ancients, but tartaric acid was isolated first by Scheele in 1769.

Fehling solution (p. 321) is prepared from copper sulfate, sodium hydroxide, and Rochelle salts. The tartrate ion forms a chelate complex (p. 604), which decreases the cupric ion concentration below that necessary for the precipitation of cupric hydroxide.

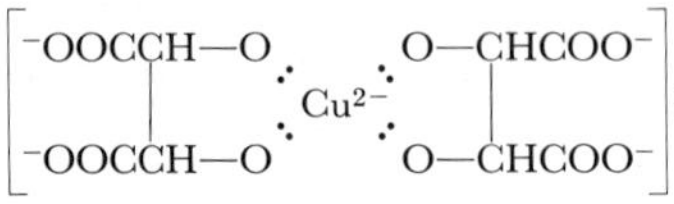

$$\left[\begin{array}{ccc} {}^-OOCCH-O & & O-CHCOO^- \\ | & Cu^{2-} & | \\ {}^-OOCCH-O & & O-CHCOO^- \end{array}\right] 6\,\overset{+}{Na}$$

Sodium cupritartrate

$$\left[\begin{array}{ccccc} {}^-OOCCH_2 & O & & O-CO & CH_2COO^- \\ & C & Cu^{2-} & C & \\ {}^-OOCCH_2 & OC-O & & O & CH_2COO^- \end{array}\right] 6\,\overset{+}{Na}$$

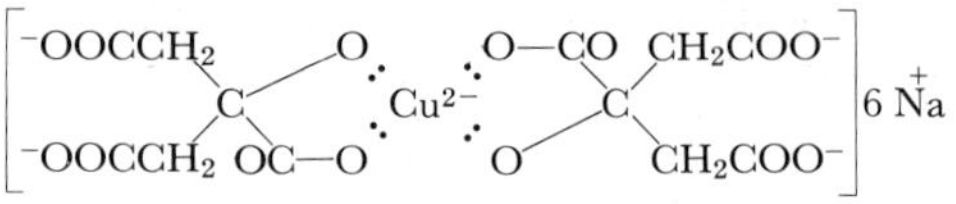

Sodium cupricitrate

Polycarboxylic Acids

Citric acid, $HOOCCH_2C(OH)(COOH)CH_2COOH$, is 2-hydroxy-1,2,3-propanetricarboxylic acid. It is the chief acid of citrus fruits, amounting to 6 to 7 per cent of lemon juice. It is present also in currants, gooseberries, and many other fruits, as well as in the roots and leaves of many plants. It was obtained in a crystalline form from unripe lemons by Scheele in 1784. Commercial manufacture is from cull lemons, or by the fermentation of molasses or starch with *Aspergillus niger* at *p*H 3.5. It is used as an acidulant in foods and soft drinks and accounts for about three-fourths of the 60 million pounds of organic acids, other than acetic acid, used for this purpose. Its ability to complex with metal ions accounts for its use to clean metals.

Benedict solution (p. 321) is prepared from copper sulfate, sodium carbonate, and sodium citrate. The structure of the complex is similar to that of the tartrate complex except that a carboxyl group has entered into complex formation instead of a hydroxyl group (p. 636).

As an α-hydroxy acid, citric acid loses carbon monoxide and water when it reacts with fuming sulfuric acid at 0° (p. 615) to give **acetonedicarboxylic acid.** Esterification of the latter gives ethyl acetonedicarboxylate, which is a β-keto ester that has two reactive methylene groups (p. 621) and is a useful intermediate in organic syntheses.

$$HOOCCH_2\underset{\displaystyle COOH}{\underset{|}{C}}OHCH_2COOH \xrightarrow[0^\circ]{H_2SO_4,\ SO_3} HOOCCH_2COCH_2COOH + CO + H_2O \xrightarrow{C_2H_5OH,\ HCl} C_2H_5OOCCH_2COCH_2COOC_2H_5$$

Ethyl acetonedicarboxylate

Ethyl ethylenetetracarboxylate is obtained when ethyl bromomalonate is heated with anhydrous sodium carbonate.

$$2\,(C_2H_5OOC)_2CHBr + Na_2CO_3 \longrightarrow (C_2H_5OOC)_2C{=}C(COOC_2H_5)_2 + 2\,NaBr + H_2O + CO_2$$

Because of the four strongly electronegative cyano groups, **tetracyanoethylene,** the nitrile corresponding to ethylenetetracarboxylic acid, is a very unusual compound. First made in 1957, it is prepared best by the reaction of dibromomalononitrile with copper powder.

$$2\,Br_2C(CN)_2 + 4\,Cu \longrightarrow (NC)_2C{=}C(CN)_2 + 4\,CuBr$$

Hydrocarbons that have had all of the hydrogen replaced by cyano groups are called *cyanocarbons.* In contrast to ethylene, which is weakly nucleophilic, tetracyanoethylene is strongly electrophilic. Thus it forms highly colored charge-transfer complexes (p. 387) with aromatic compounds and has been used as an indicator for their detection in paper chromatography. Its electrophilic property puts it in the class of strong π acids (p. 451).

Tetracyanoethylene is attacked readily by nucleophiles with displacement of cyanide ion.

$$(NC)_2C{=}C(CN)_2 + H_2O \longrightarrow (NC)_2C{=}C\begin{matrix}CN\\OH\end{matrix} + HCN$$

In contrast to ordinary vinyl alcohols, **tricyanovinyl alcohol** is stable in the enol form. It is a moderately strong acid ($pK_a = 2.0$) comparable to phosphoric acid ($pK_a = 2.1$).

PROBLEMS

37–*1*. Give reactions for the synthesis of the following compounds from malonic acid or ethyl malonate: (*a*) cyclobutanecarboxylic acid, (*b*) β-phenylglutaric acid, (*c*) 2,5-dimethylcaproic acid, (*d*) 4-methyl-2-pentenoic acid.

37–*2*. Give the series of reactions necessary for the following conversions: (*a*) chloroacetic acid to malononitrile, (*b*) phenol to glutaric acid, (*c*) chloroacetic acid to glutaric acid, (*d*) benzaldehyde to phenylsuccinic acid, (*e*) acetylene to adipic acid, (*f*) benzene to malic acid, (*g*) citric acid to β-bromoglutaric acid.

37–*3*. Group the following products according to (*a*) those that are essentially linear polymers, (*b*) those that are moderately cross-linked, and (*c*) those in which the final product is highly cross-linked: (*1*) nylon, (*2*) proteins, (*3*) polystyrene, (*4*) phenol-formaldehyde resins, (*5*) cellulose, (*6*) poly(vinyl chloride), (*7*) polyurethan rubbers, (*8*) vulcanized rubber, (*9*) polyacrylonitrile, (*10*) urea-formaldehyde resins, (*11*) polyisobutylene, (*12*) alkyd resins, (*13*) polyethylene, (*14*) polyformaldehyde, (*15*) epoxy resins, (*16*) Dacron, (*17*) polyurethan rigid foams, (*18*) polyester resins.

37–*4*. Give the formulas for the products of the following Diels-Alder additions: (*a*) furan and maleic anhydride, (*b*) 1,3-butadiene and acrylic acid, (*c*) acrolein and methyl vinyl ether, (*d*) cyclopentadiene and crotonaldehyde, (*e*) anthracene and maleic anhydride, (*f*)1,3-butadiene and quinone, (*g*) dimerization of acrolein.

37–*5*. Give reactions for the synthesis of quinquephenyl, $C_6H_5{-}(C_6H_4)_3{-}C_6H_5$, making use of the following steps: (*1*) synthesis of *p*-bis(4-phenyl-1,3-butadienyl)benzene from *p*-bis(chloromethyl)benzene and cinnamaldehyde making use of the Wittig reaction, (*2*) Diels-Alder reaction with ethyl acetylenedicarboxylate, (*3*) saponification, decarboxylation, and oxidation with ferricyanide to quinquephenyl.

37–*6*. Compound *A* containing only carbon, hydrogen, and oxygen, is insoluble in water but soluble in sodium bicarbonate solution. Titration with standard alkali gives a neutralization equivalent of 256 ± 3. It decolorizes bromine in carbon tetrachloride solution without the evolution of hydrogen bromide. Vigorous oxidation gives a neutral compound, *B*, and an acidic compound, *C*. Compound *C* is a nonvolatile solid with a neutralization equivalent of 72 ± 1. When *C* is distilled with barium hydroxide, a ketone is formed. Compound *B* reacts with phenylhydrazine and gives iodoform when treated with iodine and sodium hydroxide. Write a possible structural formula for compound *A* and give equations for the reactions that it undergoes.

CHAPTER THIRTY-EIGHT

NONBENZENOID CARBOCYCLIC COMPOUNDS

Carbocyclic compounds that have the same properties as aliphatic (*acyclic*) compounds are called *alicyclic compounds*. A few compounds closely related in structure to alicyclic compounds show resonance stabilization similar to that of benzene, although they do not contain a benzene nucleus. They have been termed *nonbenzenoid aromatic* compounds. Alicyclic compounds and nonbenzenoid aromatic compounds can be grouped together conveniently as *nonbenzenoid carbocyclic compounds*. The saturated alicyclic hydrocarbons frequently are called *cycloparaffins* or *cyclanes*. Petroleum technologists usually call them *naphthenes* because cyclopentane (*pentamethylene*) and cyclohexane (*hexamethylene*) and their homologs have been isolated from the naphtha fraction of petroleum. Those compounds obtainable by the hydrogenation of aromatic rings frequently are called *hydroaromatic* compounds.

General Theory Regarding Cyclic Compounds

Previous to 1881 only five- and six-membered ring compounds were known. They could be accounted for without too much difficulty, since the internal angles of a regular pentagon are 108 degrees and of a regular hexagon 120 degrees and are not greatly different from the tetrahedral angle of 109°28′. A four-membered ring compound, however, was synthesized by Markovnikov in 1881, three-membered ring compounds by Freund in 1882, and both types by Perkin Jr.[1] in 1883. The difficulty in forming and the ease of opening these small rings led Baeyer in 1885 to propose a **strain theory.** Baeyer postulated that the ease of formation of a ring depends on the amount that the bond must deviate from its normal tetrahedral angle of 109°28′ in order to form the bond. The amount of deviation was designated as the *strain* in the ring. The greater the amount of strain, the easier it should be to open the ring, that is, the more reactive the compound should be. Thus in the formation of the highly reactive double bond, which in older theory consisted of two identical single bonds, each bond must be bent through one-half of the tetrahedral angle or 54°44′; for a cyclopropane ring, $\frac{1}{2}(109°28' - 60°) = 24°44'$; for cyclobutane, $\frac{1}{2}(109°28' - 90°) = 9°44'$; for cyclopentane, $\frac{1}{2}(109°28' - 108°) = 0°44'$; and for cyclohexane, $\frac{1}{2}(109°28' - 120°) = -5°44'$. Since rings having more than seven atoms were unknown, Baeyer assumed that all of the atoms must be in a plane, which would require increasing negative strain for larger rings. This assumption was questioned at once by Werner, and other discrepancies were obvious. For example, although olefins are highly reactive, they are obtained easily in excellent yield. Moreover rings containing from seven to over thirty carbon atoms have been synthesized, all of which are approximately as stable as cyclopentane or cyclohexane. Currently it is assumed that the ease of formation of cyclic compounds, that is, the tendency for intramolecular reaction, depends on the proximity of the atoms being joined

[1] William Henry Perkin, Jr. (1860–1929), student of Wislicenus and Baeyer, and professor of chemistry at the University of Oxford. He is noted for his synthetic and degradative work in the field of natural products, particularly the terpenes and alkaloids.

in the reaction. This tendency is high for the formation of a double bond, where the two atoms are adjacent to each other. Moreover it no longer is assumed that the double bond is a two-membered ring. Instead it is believed that one pair of electrons occupies a π orbital, which involves no strain in the original sense, and the high reactivity is accounted for by the higher energy level of the π orbital (p. 74).

If the atoms were forced to remain in an extended chain, the chance for intramolecular reaction would decrease as more and more carbon atoms separated the reacting groups. Rotation about the single bonds, however, permits the assumption of a spiral structure as indicated in Fig. 38–*1*. If in this figure the distance between C-1 and C-2 = 1.53 A, then C-1 to C-3 = 2.51 A, C-1 to C-4 = 2.52 A, and C-1 to C-5 = 1.67 A. Hence the double bond and the five-membered ring can be formed readily, but it is more difficult to form three- and four-membered rings. Because of the flexibility of the molecule, C-1 and C-6 can approach each other to any desired distance and still retain tetrahedral bond angles. Therefore once the bond is formed the ring should be entirely free of angle strain. The same situation exists for longer chains, but above six carbon atoms other atoms in the chain begin to get in the way of the reacting groups, and a greater amount of maneuvering of the chain is necessary to bring the reacting groups into the proper space relationship for reaction to take place. As a result, intermolecular reaction to give polymeric products becomes predominant, and the yields of cyclic compounds are very low. In confirmation of this theory, large ring compounds can be obtained in excellent yields if the reaction is carried out at extremely high dilution where the chance of intramolecular reaction is again greater than the chance of intermolecular reaction (p. 656), or if the reactive groups can be brought close to each other by the reagent (p. 655).

If only angle strain were involved, cyclopentane should be practically strainless and rings above C_5 entirely so. Since all of the cyclanes are composed only of methylene groups joined to each other and since the products of combustion for all are carbon dioxide and water in the molar ratio of 1:1, the heat of combustion per methylene group is a measure of the stability of the cyclane. The heat of combustion of the higher normal alkanes approaches 157.5 kcal. per CH_2 group per mole. For cyclanes the values are for C_3, 166.5; C_4, 163.8; C_5, 158.7; C_6, 157.5; C_7, 158.2; C_8, 158.5; and C_9, 158.7. Thus the stability is at a maximum at C_6, and both cyclopentane and the higher cyclanes are somewhat strained. The explanation for this behavior lies in *bond-opposition* or *eclipsing strain.* If ethane were forced to exist in the eclipsed conformation, it would have an energy content about 3 kcal. higher than in the staggered conformation (p. 56), or about 1 kcal. per eclipsed pair of hydrogens. If all of the carbon atoms of all cyclanes were planar, all hydrogen atoms would be eclipsed. This situation applies to cyclopropane because a ring of three atoms must be planar. The heat of combustion per CH_2 group is 9 kcal. higher than that for alkanes, indicating a total ring strain of 27 kcal. With six hydrogen bonds in opposition, approximately 6 kcal. of this amount can be attributed to eclipsing strain and 21 kcal. to angle strain. Actually the eclipsing strain probably is somewhat less because the smaller

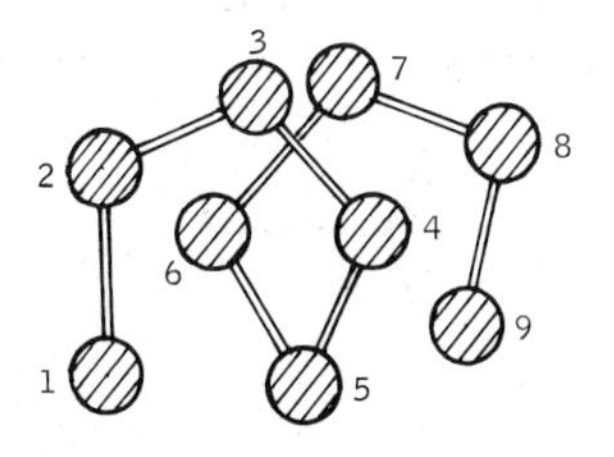

Figure 38–*1*. Spiral arrangement of a carbon chain.

C—C—C angle of the ring carbons keeps the hydrogens farther apart than would be true in eclipsed ethane.

The conformational behavior of cyclohexane has been discussed (p. 326). Here the ring atoms need not be planar, and numerous conformations can be assumed in which all C—C—C angles are a strainless 109.5 degrees. Moreover in the chair form (Fig. 38–*2*) all of the hydrogens are completely staggered, and there is no eclipsing strain. Accordingly the stability is the same as that of the alkanes.

The introduction of a double bond into a ring system increases the strain, the effect being greatest for cyclopropene and decreasing with increasing ring size. Only the *cis* configuration is possible for the lower members. The first known *trans* cyclene has an eight-carbon ring (p. 654).

Under certain circumstances, however, unsaturation can lead to resonance stabilization. The outstanding examples are benzene and its congeners, which now usually are referred to as *benzenoid compounds*. In 1931 E. Hueckel noted that most of the aromatic compounds had $4n + 2\pi$ electrons, where n is an integer. Thus benzene has six π electrons, naphthalene has ten, and anthracene has fourteen. Consequently he postulated that the bonding orbitals can be grouped into shells analogous to those of atomic orbitals and that the filling of the shell confers stability on the molecule, just as the stable and unreactive atoms of the rare gases are obtained when an atomic shell is full. It should be recalled that for appreciable interaction of π electrons, the p orbitals of adjacent sp^2-hybridized carbon atoms must have good overlap (pp. 424, 481). Hence the axes of the p orbitals must be parallel, which requires that the sp^2-hybridized ring atoms be in the same plane or very nearly so.

These basic ideas have been extended to other types of compounds, and physical chemists have referred to the extent of resonance stabilization of a cyclic compound as its *aromaticity*. Compounds that do not contain a benzene nucleus but show considerable resonance stabilization are called *nonbenzenoid aromatic compounds*. Thus the term *aromatic* has at least three different meanings. Originally it referred to a classification based on the physiological property of odor. The recognition that many of these compounds had distinct chemical properties led to the appropriation of the term for all compounds that have these properties whether odorous or not. For 70 years the term *aromatic compounds* meant to the chemist only those compounds that underwent the electrophilic substitutions characteristic of benzene and related compounds. With the advent of the electronic theory these chemical properties were ascribed to the "aromatic sextet" of electrons. Now there is a physical viewpoint, which, since it refers to the energy of the ground state of the molecule and not to the energy of a transition state, does not necessarily have any appreciable bearing on the chemical reactivity of the compound. Although many of the so-called nonbenzenoid aromatic compounds have certain chemical properties common to benzenoid compounds, as do also certain aliphatic compounds such as the enols, only the metallocenes (p. 644) approach the truly aromatic benzenoids in the chemical sense.

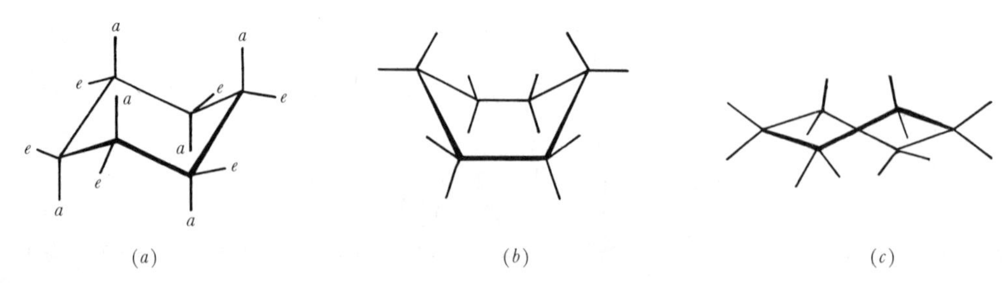

Figure 38–*2*. Conformations of cyclohexane: (*a*) chair form with axial and equatorial bonds indicated; (*b*) boat form; (*c*) right-hand twist or skew boat form.

Three-carbon Rings

Preparation. Cyclopropane rings can be formed by the reaction of 1,3-dihalogenated compounds with zinc dust in alcohol or other solvents.

$$\begin{matrix} R{-}CHX \\ CH_2 \\ R{-}CHX \end{matrix} + Zn \longrightarrow \begin{matrix} RCH{-}CHR \\ CH_2 \end{matrix} + ZnX_2$$

The reaction is general but is limited by the accessibility of the halogen compounds. **Cyclopropane** is prepared commercially from 1-bromo-3-chloropropane (*trimethylene chlorobromide,* p. 594). It is a gas, b.p. −34°, that is used as a general anesthetic, although the explosion hazard is greater than with other inhalation anesthetics.

1,1-Cyclopropanedicarboxylic esters may be made by the reaction of one mole of a 1,2 dihalide with one mole of sodiomalonic ester, followed by the addition of a second mole of sodium ethoxide (p. 633). Saponification and decarboxylation gives the **cyclopropanecarboxylic acid.**

$$\triangleright(COOC_2H_5)_2 \longrightarrow \triangleright(COOH)_2 \longrightarrow \triangleright{-}COOH$$

Better yields of the acid are obtained, however, by hydrolysis of the nitrile. γ-Halo ketones and γ-halo nitriles cyclize on reaction with base to give **cyclopropyl ketones** and **cyclopropyl cyanides.**

$$\begin{matrix} CH_2CH_2COCH_3 \\ | \\ CH_2Cl \end{matrix} \xrightarrow{^-OH} H_2O + \triangleright{-}COCH_3 + Cl^-$$

$$\begin{matrix} CH_2CH_2CN \\ | \\ CH_2Cl \end{matrix} \xrightarrow{^-OH} H_2O + \triangleright{-}CN + Cl^- + H_2O$$

One of the good general methods for the synthesis of cyclopropanes is the reaction of unsaturated compounds with carbenes (p. 250). The reaction is stereoselective, *cis* olefins yielding chiefly *cis* cyclopropanes, and *trans* olefins yielding chiefly *trans* cyclopropanes. If the carbene carbon is bound to two different groups, two isomers are formed.

$$\begin{matrix} R & & R \\ & C{=}C & \\ H & & H \end{matrix} + :C\begin{matrix} A \\ B \end{matrix} \longrightarrow \text{(cyclopropane: } A \text{ up}, B \text{ down; } R,R \text{ cis; } H,H) + \text{(cyclopropane: } B \text{ up}, A \text{ down; } R,R \text{ cis; } H,H)$$

Many derivatives of cyclopropane have been prepared by interconversion of groups present in derivatives made by ring formation. Thus **cyclopropylcarbinol** is obtained by the reduction of the carboxylic ester, **secondary alcohols** by reduction of cyclopropyl alkyl ketones, and **tertiary alcohols** by reaction of ketones with Grignard reagents. **Cyclopropylamine** results from the amide by the Hofmann rearrangement (p. 223) or from the carboxylic acid by the Schmidt reaction (p. 224).

Cyclopropene has been prepared by exhaustive methylation of cyclopropylamine and pyrolysis of the quaternary hydroxide (p. 231).

$$\triangleright{-}\overset{+}{N}(CH_3)_3\,^-OH \xrightarrow{\text{Heat}} \triangleright + N(CH_3)_3 + H_2O$$

Derivatives of cyclopropene are obtained readily by the reaction of acetylenes with carbenes.

$$R{-}C{\equiv}C{-}R' + :CR''R''' \longrightarrow \begin{matrix} R{-}C{=}C{-}R' \\ CR'R'' \end{matrix}$$

Several natural products that contain a 3-carbon ring have been isolated. **Lactobacillic acid** is present to the extent of 30 per cent and 19 per cent respectively in the fat acids from *Lactobacillus arabinosus* and from *L. casei,* and **sterculic acid** from the kernel oil of *Sterculia foetida.*

$$CH_3(CH_2)_5CH{-}CH(CH_2)_9COOH \text{ (ring } CH_2\text{)}$$
Lactobacillic acid

$$CH_3(CH_2)_7C{=}C(CH_2)_7COOH \text{ (ring } CH_2\text{)}$$
Sterculic acid

$$CH_2{=}C{-}CHCH_2CH(NH_2)COOH \text{ (ring } CH_2\text{)}$$
Hypoglycin *A*

The cyclopropane ring also makes up a portion of the structure of the active principles of pyrethrum (p. 646). **Hypoglycin A** is from the seed of the akee tree (*Blighia sapida*).

Physical and Chemical Properties. The three-membered ring is highly strained (p. 639) and resembles the double bond in certain aspects. Thus the ionization potential of 10.2 e.v. for cyclopropane is between that of ethylene (10.6 e.v.) and that of propylene (9.8 e.v.) and considerably less than that of ethane (11.7 e.v.). The C—H stretching absorption for cyclopropanes having a CH_2 group lies in the range 3058–3040 cm.$^{-1}$ compared to 3040–3010 cm.$^{-1}$ for ethylenes and 2936–2916 cm.$^{-1}$ for the CH_2 group in alkanes. Thus the CH bonds in cyclopropanes resemble those in ethylenes rather than those in alkanes. Cyclopropanes show characteristic absorption also in the range 1020–1000 cm.$^{-1}$ owing to the three-membered ring.

The strain energy of cyclopropane is around 27 kcal. (p. 639). Since this strain is relieved when the ring is opened, the bond dissociation energy of the carbon-carbon bond is expected to be around 27 kcal. less than the 83 kcal. for the carbon-carbon bond in normal alkanes, or 56 kcal. This value is approximately the same as that for the π bond in ethylene (p. 74). Accordingly the most characteristic property of cyclopropanes is the ease with which the ring is opened. Cyclopropane and many of its derivatives give open-chain compounds with most of the reagents that react with olefins.

$$(CH_2)_3 + H_2SO_4 \longrightarrow CH_3CH_2CH_2OSO_3H$$
$$(CH_2)_3 + HBr \longrightarrow CH_3CH_2CH_2Br$$
$$(CH_2)_3 + Br_2 \longrightarrow BrCH_2CH_2CH_2Br$$
$$(CH_2)_3 + H_2 \xrightarrow[80^\circ]{Pt} CH_3CH_2CH_3$$

Four-carbon Rings

Cyclobutane and its simple derivatives usually are made by way of the carboxylic acids obtained from ethyl malonate (p. 633), as illustrated by the following series of reactions.

$$C_4H_7COOC_2H_5 \longleftarrow \square\text{-}COOH \longrightarrow C_4H_7COOAg \xrightarrow{Br_2} C_4H_7Br$$
$$C_4H_7COOC_2H_5 \xrightarrow{LiAlH_4} C_4H_7CH_2OH$$
$$\square\text{-}COOH \xrightarrow{HN_3,\ H_2SO_4} C_4H_7NH_2$$
$$\square\text{-}COOH \longrightarrow C_4H_7CONH_2 \longrightarrow C_4H_7CN \longrightarrow C_4H_7CH_2NH_2$$
$$C_4H_7CONH_2 \xrightarrow{Br_2,\ NaOH} C_4H_7NH_2$$
$$C_4H_7NH_2 \xrightarrow{CH_3I,\ AgOH} C_4H_7\overset{+}{N}(CH_3)_3\,{}^{-}OH \longrightarrow C_4H_6$$
$$C_4H_7Br \xrightarrow{Mg} C_4H_7MgBr \xrightarrow{C_4H_9OH} C_4H_8$$

The cyclobutane ring is much less reactive than the cyclopropane ring. It is not opened with sulfuric acid, hydrobromic acid, or bromine. It can be reduced catalytically at 120° to *n*-butane and hence is more reactive than the higher cyclanes, which are not reduced at temperatures up to 200°.

The cyclobutadiene ring system is of interest in that molecular orbital theory predicts that it should not be aromatic despite the fact that the valence-bond representation might indicate considerable resonance stabilization.

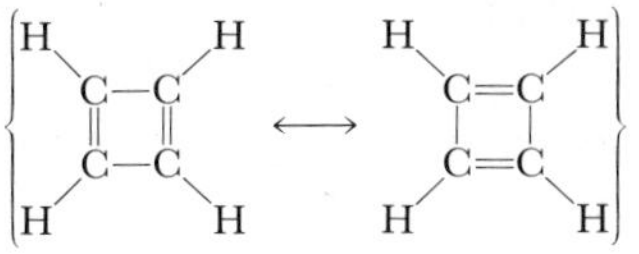

After many ingenious but unsuccessful attempts were made to synthesize unsubstituted cyclobutadiene, it finally was shown in 1965 to be capable of transient existence as a highly reactive gas that gives no evidence of resonance stabilization.

Five-carbon Rings

Ring closure to produce cyclopentane derivatives can be brought about in several ways, such as the reaction of sodiomalonic ester with tetramethylene halides (p. 633), the pyrolysis of hexanedioic acids (p. 631), or the Dieckmann condensation of their esters (p. 621).

Since the cyclopentane ring is only slightly strained (p. 639), there is no more tendency for the ring to open than there is for scission of an alkane chain. One point should be noted concerning the **infrared spectra** of **cyclic ketones.** Although cyclohexanones and higher ring ketones show a carbonyl stretching absorption in the same range as aliphatic ketones (1725–1705 cm.$^{-1}$), the carbonyl group in a five-membered ring absorbs at 1750–1740 cm.$^{-1}$ This difference has been especially useful in establishing the presence of five-membered rings in polycyclic compounds.

The five-membered carbocyclic compound most readily available is **cyclopentadiene,** C_5H_6, b.p. 41°. It is a component of coal gas and is a coproduct in the manufacture of isoprene and butadiene by the cracking of petroleum naphtha or gas oil fractions (p. 588). Cyclopentadiene is a typical conjugated diene. The resonance energy is 3 kcal. per mole, the same as that for 1,3-butadiene. It readily undergoes 1,4 addition to the double bonds. Thus maleic anhydride gives the Diels-Alder adduct (p. 634). Cyclopentadiene dimerizes slowly on standing to **dicyclopentadiene,** m.p. 33°, which dissociates at its boiling point, 170°, to the monomer.

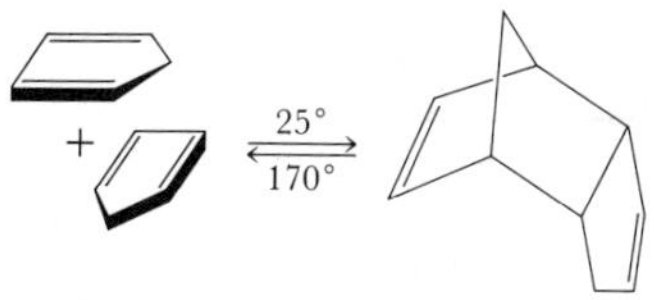

Dicyclopentadiene

The properties of the methylene group are unusual, however, because it is considerably more acidic than methylene groups adjacent to an open-chain conjugated system. Thus cyclopentadiene reacts with potassium in benzene with the evolution of hydrogen and evolves methane on treatment with methylmagnesium bromide.

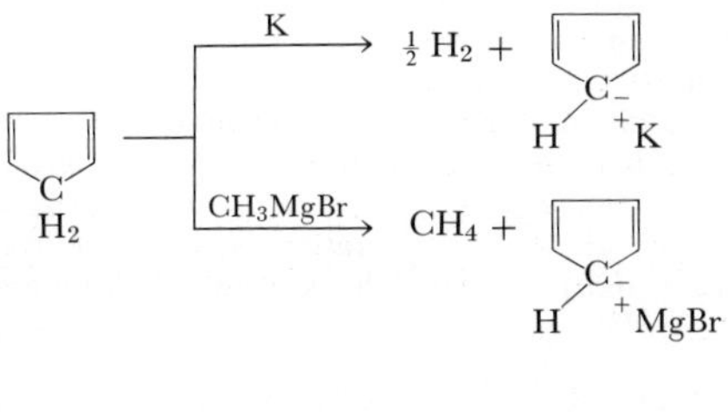

The high acidity ($pK_a = \sim 16$) results from the resonance stabilization of the cyclopentadienyl anion, which obeys the Hueckel $4n + 2$ rule and has an aromatic sextet of electrons (p. 640). Although the same number of valence-bond resonance structures can be written both for the anion and for the cation, the latter would have only $4n$ π electrons and has not been detected in solution.

The most interesting behavior of the cyclopentadienyl anion is its reaction with metallic ions. Ferrous ion gives an orange solid, m.p. 174°, having the molecular formula $Fe(C_{10}H_{10})$.

$$2\,C_5H_5^{-+}Na + FeCl_2 \longrightarrow Fe(C_2H_5)_2 + 2\,NaCl$$

The compound even is formed and sublimes when cyclopentadiene vapor is passed over activated iron at 300°. It boils at 249°, is insoluble in water and soluble in organic solvents, and is not affected by boiling acid or alkali. It does not behave like an unsaturated compound but shows typical aromatic chemical properties. Thus it undergoes electrophilic substitutions such as the Friedel-Crafts reaction, sulfonation, and formylation, and the hydrogen exchanges readily with deuterium. Metalation takes place readily, indicating that the hydrogens are more acidic than those in benzene.

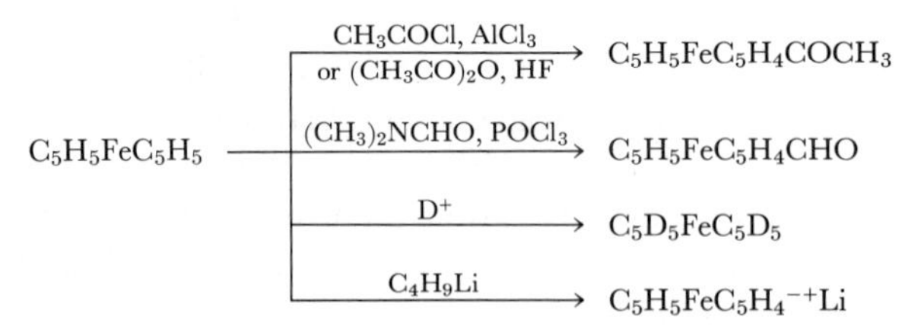

Because of these reactions, the compound was called **ferrocene.**

As to the structure of ferrocene and its analogs, all of the chemical and physical evidence indicates two parallel cyclopentadienyl rings bonded to the metal atom by two delocalized single covalent bonds, with the inner lobes of the carbon π orbitals overlapping the $3d_{xz}$ and $3d_{yz}$ orbitals of the iron. Since the metal atom is located symmetrically between the rings, they have been called *sandwich molecules* (Fig. 38–3).

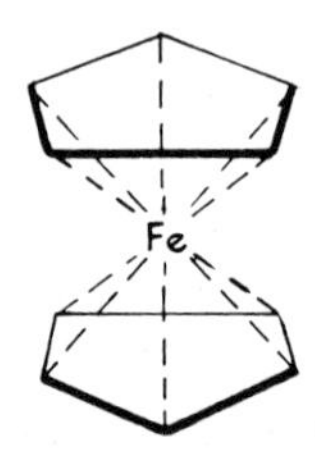

Figure 38–*3*. Ferrocene.

A variety of polychlorinated cyclic compounds have been prepared from cyclopentadiene, several of which are important insecticides. Vapor phase chlorination of cyclopentadiene gives **hexachlorocyclopentadiene.** It can be prepared also by the pyrolysis of mixed polychlorinated pentanes. The Diels-Alder addition of hexachlorocyclopentadiene to cyclopentadiene yields **chlordene,** which adds chlorine in carbon tetrachloride solution to give the important insecticide **chlordan,** or chlorinates at 5° in the presence of diatomaceous earth to give **heptachlor,** another insecticide. These products have the *endo* configuration, in which the five-membered ring is inclined toward the inside (Gr. *endo* within) of the folded six-membered ring and away from the dichloromethylene group.

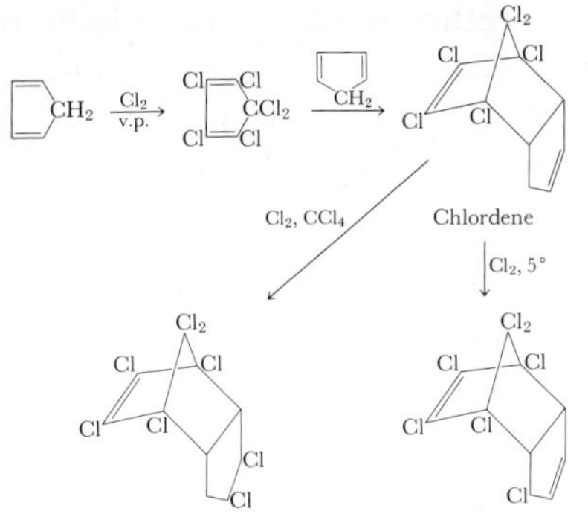

Cl_2, v.p. → ; CH_2 (cyclopentadiene) → Chlordene

Cl_2, CCl_4 → Chlordan

Cl_2, 5° → Heptachlor

Chlordan

Heptachlor

Diels-Alder addition of acetylene to cyclopentadiene gives **bicyclo[2.2.1]-2,5-heptadiene** (*norbornadiene*[2]), which adds to hexachlorocyclopentadiene to give the insecticide **aldrin.** When aldrin is converted to the epoxide with peroxyacetic acid or with hydrogen peroxide and tungstic oxide, still another insecticide known as **dieldrin** is formed. Here the chlorinated five-membered ring is *endo* with respect to the two central carbon atoms whereas the nonchlorinated ring is *exo.* The oxide ring in dieldrin is *exo.*

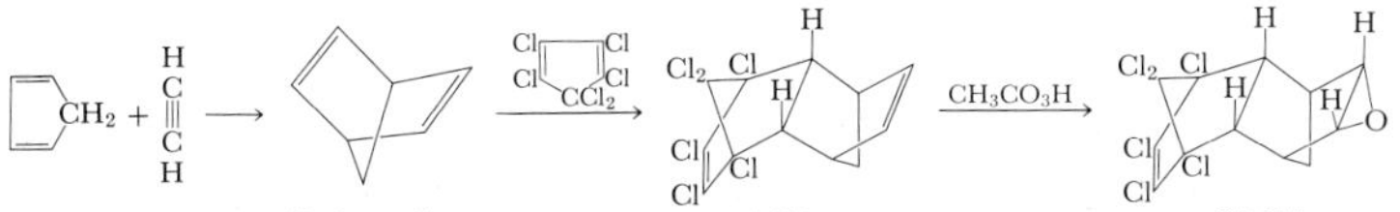

CH_3CO_3H

Norbornadiene Aldrin Dieldrin

Diels-Alder addition of acetylene to hexachlorocyclopentadiene gives 1,2,3,4,7,7-hexachlorobicyclo[2.2.1]-2,5-heptadiene, which adds to cyclopentadiene to give **isodrin,** the *endo-endo* isomer of aldrin. Epoxidation gives **endrin,** the *endo-endo* isomer of dieldrin.

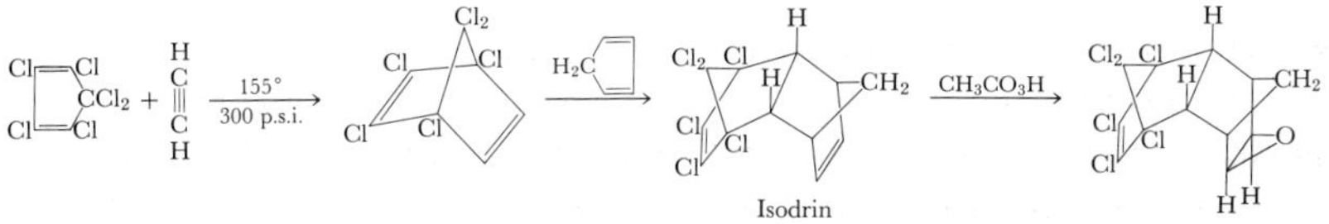

155°, 300 p.s.i. ; CH_3CO_3H

Isodrin

Endrin

The use of highly chlorinated insecticides, especially heptachlor and endrin, has been severely criticized because of their persistence and high toxicity to animal life, particularly to fish.

The term **naphthenic acids** is applied to the mixture of carboxylic acids obtained from the alkali washes of petroleum fractions. Judging from the pure compounds that have been isolated, they are complex mixtures of normal and branched aliphatic acids, alkyl derivatives of cyclopentane- and cyclohexanecarboxylic acids, and cyclopentyl and cyclohexyl derivatives of aliphatic acids. The crude naphthenic acids are available in large amounts and are used chiefly in the form of metallic salts, which are soluble in oils and organic solvents. The copper salts are used in wood preservatives and the lead, manganese, zinc, and iron salts as driers (*oxidation catalysts*) for paints and varnishes (p. 188). **Chaulmoogric** and **hydnocarpic acids,** the characteristic fat acids of chaulmoogra oil, long used in the treatment of leprosy, contain a terminal cyclopentene ring.

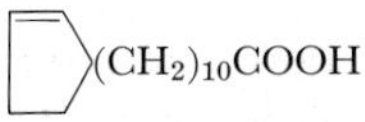

$(CH_2)_{10}COOH$ — Hydnocarpic acid

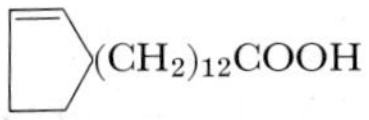

$(CH_2)_{12}COOH$ — Chaulmoogric acid

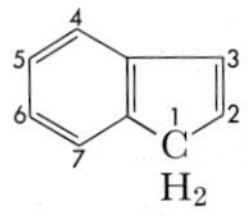

Indene

[2] This common name is derived from *norbornane,* which is a contraction of a more correct name, *trisnorbornane* (bicyclo[2.2.1]heptane). For systematic nomenclature, see footnote 5, page 625.

The active principles of **pyrethrum,** the commercially important insecticide derived from the flower heads of *Chrysanthemum cinerariaefolium,* and a few other varieties, are four esters of cyclopropanecarboxylic acids with cyclopentyl alcohols, called *pyrethrin I, pyrethrin II, cinerin I,* and *cinerin II.* The relative toxicity to houseflies is 100:23:71:18.

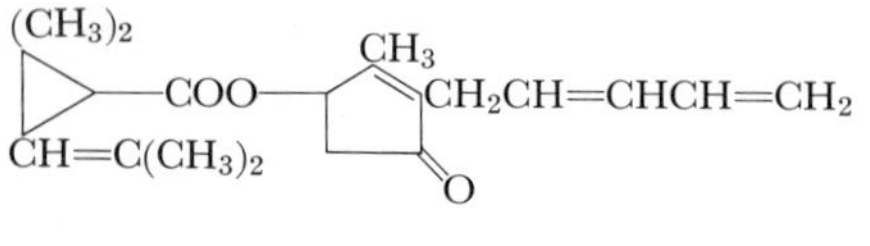

$(CH_3)_2$ —COO— CH_3 $CH_2CH{=}CHCH{=}CH_2$ $CH{=}C(CH_3)_2$ O

Pyrethrin *I*

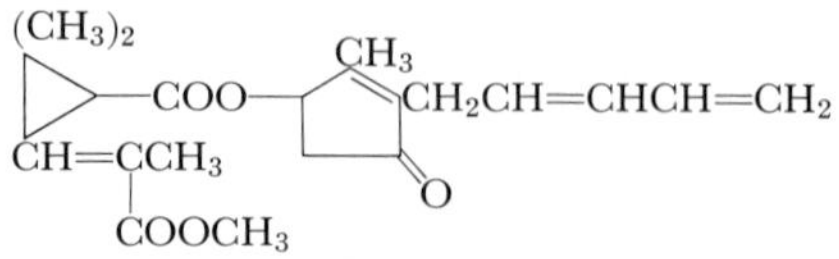

$(CH_3)_2$ —COO— CH_3 $CH_2CH{=}CHCH{=}CH_2$ $CH{=}CCH_3$ $COOCH_3$ O

Pyrethrin *II*

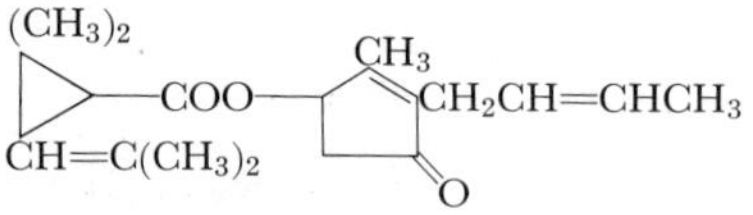

$(CH_3)_2$ —COO— CH_3 $CH_2CH{=}CHCH_3$ $CH{=}C(CH_3)_2$ O

Cinerin *I*

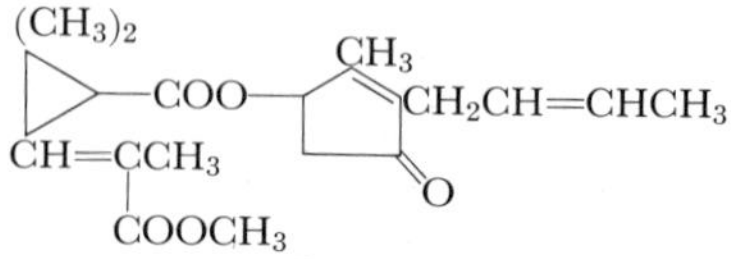

$(CH_3)_2$ —COO— CH_3 $CH_2CH{=}CHCH_3$ $CH{=}CCH_3$ $COOCH_3$ O

Cinerin *II*

The acid formed by hydrolysis of pyrethrin *I* and cinerin *I* is called *chrysanthemum monocarboxylic* or *chrysanthemic acid,* and that from pyrethrin *II* and cinerin *II* is called *chrysanthemum dicarboxylic acid* or *pyrethric acid.* The alcohol from pyrethrin *I* and pyrethrin *II* is called *pyrethrolone* and that from cinerin *I* and cinerin *II* is called *cinerolone.* A commercial synthetic analog of pyrethrin *I,* called **allethrin,** has an allyl side chain instead of the pentadienyl side chain.

Indene (*benzocyclopentadiene*) occurs in coal tar. 2,3-Dihydroindene is called **indane** or **hydrindene.** Its most important derivative is the stable hydrate of 1,2,3-indanetrione, which is known as triketohydrindene hydrate or **ninhydrin.** A recent synthesis of ninhydrin involves a Claisen ester condensation of ethyl phthalate with methyl sulfoxide. The product undergoes an internal oxidation and reduction in the presence of hydrochloric acid to give an α-chloro methyl sulfide. Hydrolysis by boiling with water gives ninhydrin in approximately 80 per cent over-all yield.

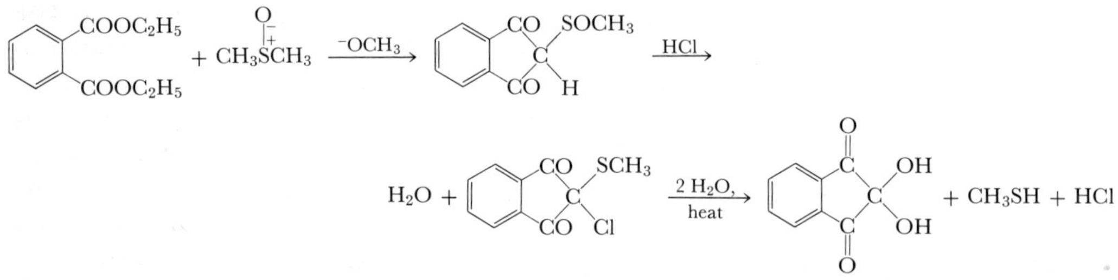

$C_6H_4(COOC_2H_5)_2 + CH_3\overset{+}{S}(\overset{-}{O})CH_3 \xrightarrow{^-OCH_3} C_6H_4(CO)_2C(H)SOCH_3 \xrightarrow{HCl}$

$H_2O + C_6H_4(CO)_2C(Cl)SCH_3 \xrightarrow[heat]{2\,H_2O,} C_6H_4(C{=}O)_2C(OH)_2 + CH_3SH + HCl$

Ninhydrin

Ninhydrin is an important reagent for the colorimetric detection and estimation of α-amino acids (p. 356). In this reaction the Schiff base initially formed decomposes with loss of carbon dioxide and hydrolyzes to 2-amino-1,3-diketohydrindene and an aldehyde (*Strecker degradation*). Reaction of the amino diketone with ninhydrin gives a new Schiff base whose anion has a deep blue color.

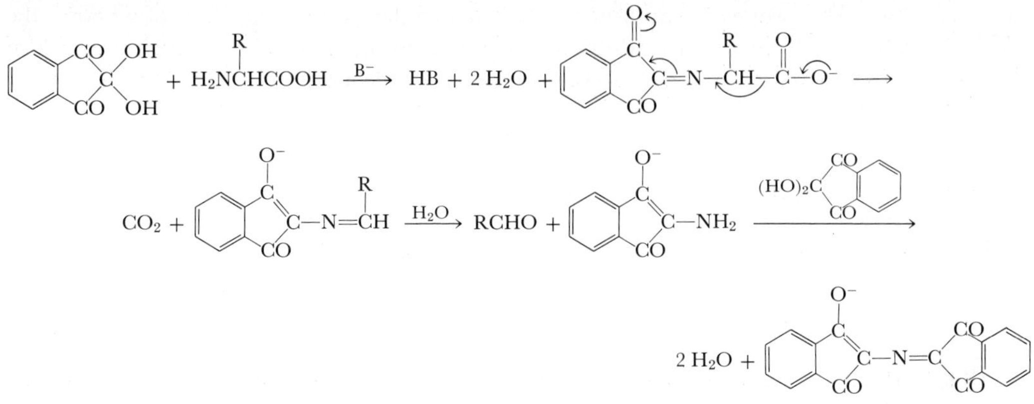

(Blue)

For quantitative estimation of amino acids, the absorption is measured at 570 mμ. The reaction takes a different course with proline and hydroxyproline. The products are yellow in acid solution and red-purple in neutral solution. To estimate these amino acids, absorption is measured at 440 mμ at *p*H 3–4.

Fluorene (*dibenzocyclopentadiene*) also occurs in coal tar. Colorless fluorene can be oxidized directly to yellow **fluorenone.** Nitration of fluorenone gives **2,4,5,7-tetranitrofluorenone,** which forms addition complexes with polynuclear aromatic hydrocarbons and their derivatives that are useful for the purpose of purification and identification.

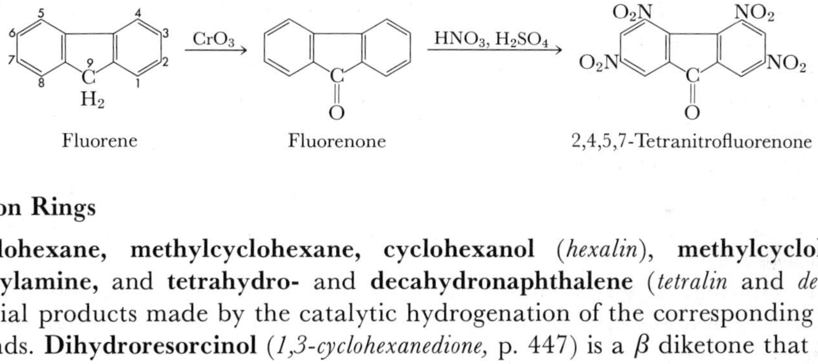

Fluorene → (CrO_3) Fluorenone → (HNO_3, H_2SO_4) 2,4,5,7-Tetranitrofluorenone

Six-carbon Rings

Cyclohexane, methylcyclohexane, cyclohexanol (*hexalin*), **methylcyclohexanols, cyclohexylamine,** and **tetrahydro-** and **decahydronaphthalene** (*tetralin* and *decalin*) are commercial products made by the catalytic hydrogenation of the corresponding aromatic compounds. **Dihydroresorcinol** (*1,3-cyclohexanedione,* p. 447) is a β diketone that is a valuable intermediate for synthesizing other compounds. It can be used, for example, to lengthen a chain by six carbon atoms.

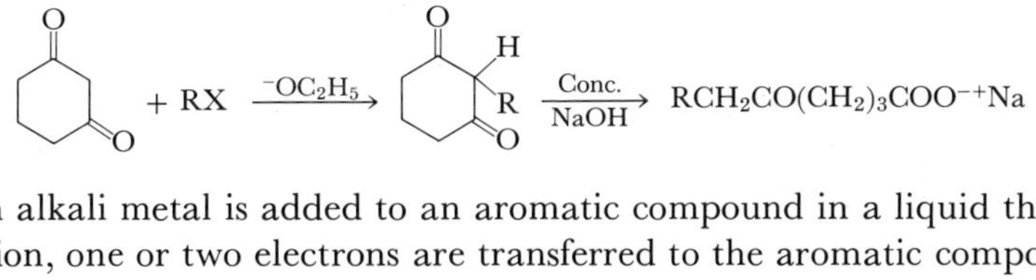

$$\text{+ RX} \xrightarrow{^{-}OC_2H_5} \xrightarrow[\text{NaOH}]{\text{Conc.}} RCH_2CO(CH_2)_3COO^{-}\,{}^{+}Na$$

When an alkali metal is added to an aromatic compound in a liquid that can solvate the metal cation, one or two electrons are transferred to the aromatic compound to give a mono- or divalent salt. Solvents that have been used are methyl ether, dimethoxyethane (*glycol methyl ether* or *glyme*), tetrahydrofuran, liquid ammonia, and certain amines. Ethyl ether is not suitable, presumably because the alkyl groups are too large, and it is not as basic as tetrahydrofuran (p. 512). In dilute solutions of the other ethers, the monovalent anion predominates, whereas in more concentrated solutions in liquid ammonia, the main species probably is the dianion. Hydrocarbons such as benzene that have a low electron affinity are expected also to form only the monovalent anion. Reaction of the salt with a proton donor of the proper strength gives the dihydro derivative regardless of whether the mono- or dianion is formed, since addition of a proton to the monoanion, which is a radical ion, permits the formation of a new anion that adds the second proton.

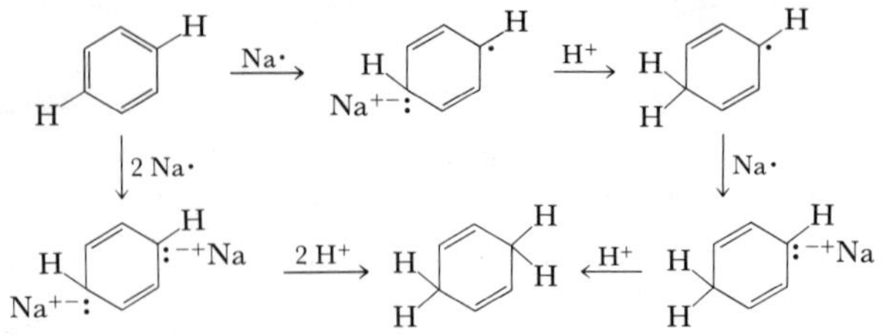

The reducing system most commonly used is sodium in liquid ammonia with absolute ethanol as the proton donor, a process commonly referred to as a *Birch reduction.* When alkoxyl groups are present, the negative charge stays away from the concentration of electrons on oxygen, and the 2,5-dihydro derivative is formed. Acid hydrolysis yields the β,γ-unsaturated ketone, which rearranges to the conjugated α,β-unsaturated ketone.

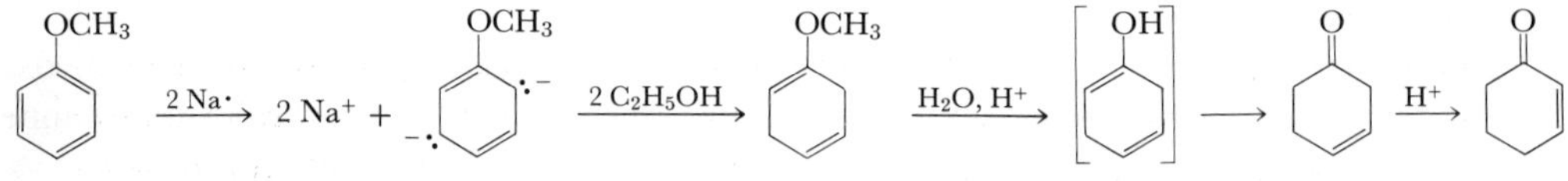

Electron-donating alkyl groups have an effect similar to that of alkoxyl groups. If a carboxyl group is attached to the ring, reduction takes place at this position because resonance stabilizes the intermediate anion.

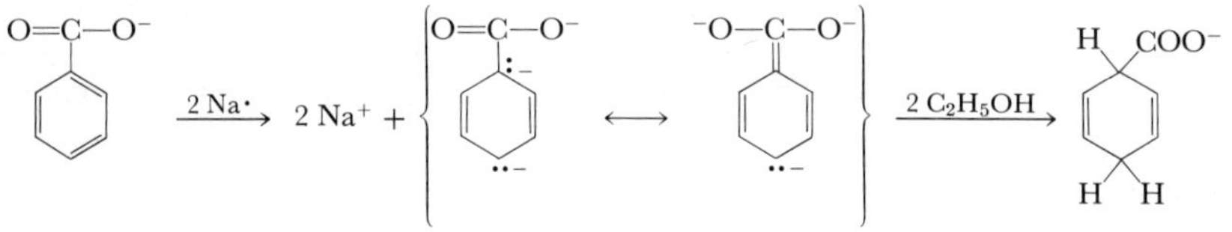

In the naphthalene series, an alkoxyl or hydroxyl group in the α position makes the oxygenated ring less susceptible to reduction than the unsubstituted ring, whereas the opposite holds for the β isomers.

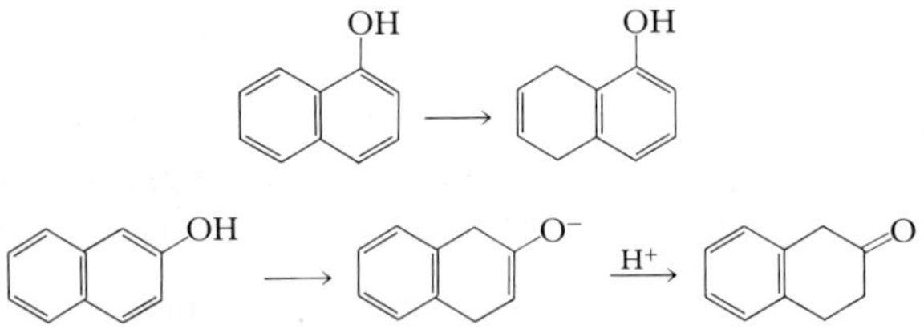

This behavior is not unexpected because 2,5 addition to the α-substituted compound would require complete destruction of the aromatic resonance.

Any of the standard reactions can be used on derivatives of cyclohexanes to produce other derivatives. Cyclohexanols, for example, can be converted to the cyclohexyl halides, dehydrated to cyclohexenes, or oxidized to the cyclohexanones. In contrast to the acyclic alkenes (p. 84), cyclohexene adds hydrogen bromide *trans* to the extent of at least 94 per cent.

Cyclohexanone is made commercially on a large scale by the dehydrogenation of cyclohexanol or by the catalyzed oxidation of cyclohexane with air. It is used chiefly for the production of **nylon 6.** Conversion to the oxime and Beckmann rearrangement (p. 466) gives **caprolactam** (*ε-aminocaproic acid lactam*), which contains a seven-membered ring and is converted to a linear polymer by heating with a trace of water.

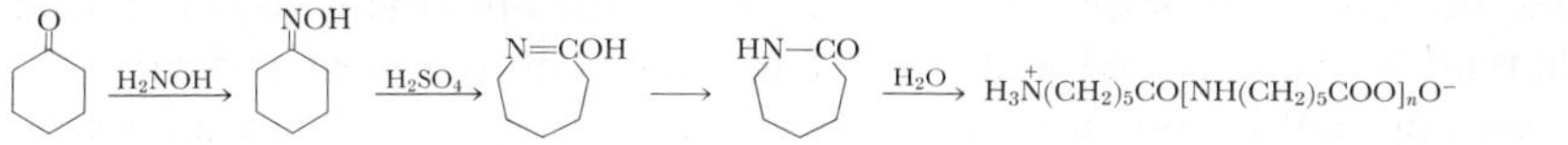

$$\xrightarrow{H_2NOH} \quad \xrightarrow{H_2SO_4} \quad N{=}COH \quad \longrightarrow \quad HN{-}CO \quad \xrightarrow{H_2O} H_3\overset{+}{N}(CH_2)_5CO[NH(CH_2)_5COO]_nO^-$$

Various other technical processes for the production of caprolactam have been developed. Nylon 6 melts around 215° and is less suitable for textile fibers than nylon 6-6 (m.p. 264°, p. 631). Nylon 6 is cheaper, however, and is being manufactured on a large scale for conversion into tire cord, filament for carpets and brushes, and molded articles.

Cyclohexylsulfamic acid and its salts, called *cyclamates,* are about 30 times sweeter than cane sugar and are used as noncarbohydrate sweetening agents. **Dicyclohexylammonium nitrite** under the name VPI (*vapor-phase inhibitor*) is used to impregnate bags and wrapping materials for machine parts to retard rusting (p. 228). **Captan,** a powerful fungicide, contains a cyclohexene ring (cf. p. 475).

$[C_6H_{11}NHSO_3^-]_2\ Ca^{2+}$

Calcium cyclohexylsulfamate
(*calcium cyclamate*)

$[C_6H_{11}]_2\overset{+}{N}H_2\overset{-}{N}O_2$

Dicyclohexylammonium nitrite
(*VPI*)

CO, N—$SCCl_3$, CO

Captan

The number of isomers of substituted cyclohexanes that can be formed readily often is limited. Just as steric interaction of groups in the axial positions plays a part in the behavior of tetrahydropyrans as exemplified by the sugars (p. 328), so does it influence the behavior of substituted cyclohexanes. All-*cis* 1,2,3,4,5,6-hexasubstituted cyclohexanes are difficult to prepare because three groups must be crowded into axial positions on the same side of the ring in either chair conformation (cf. p. 327). Thus far, all-*cis*-inositol (Fig. 38–*4a*) is the only such compound that has been prepared. No naturally occurring inositol has even two axial hydroxyl groups on the same side of the ring (p. 349). Group opposition is greatest in the boat form of cyclohexane (Fig. 38–*2b*, p. 640), but since this form is flexible (p. 326), strain can be relieved to a certain extent if a twisted form is assumed, which can be either right-handed or left-handed (Fig. 38–*2c,* p. 640). In bicyclic and polycyclic compounds the cyclohexane ring often is held or forced into a boat (p. 650) or twist conformation, since these forms may have an energy only a few kilocalories higher than that of the chair form.

Groups in the equatorial position of the chair form are less subject to steric hindrance than those in the axial positions. Hence it is to be expected that reactions involving addition intermediates will be faster when the group is in the equatorial position. Thus hydroxyl and carboxyl groups are more easily esterified and ester groups are more easily hydrolyzed when in the equatorial than when in the axial position. Secondary alcohols are more easily oxidized to ketones when the hydroxyl group is in the axial position, presumably because the rate-determining step is the breaking of the carbon-hydrogen bond, which then is equatorial. On the other hand, the simple displacement reaction cannot operate on groups in the equatorial position because the approach from the back side is blocked by the groups

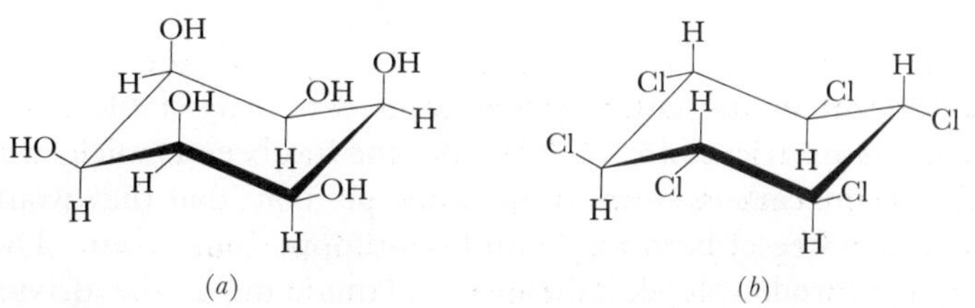

(*a*) (*b*)

Figure 38–*4.* Conformations of (*a*) all-*cis*-inositol and (*b*) *β*-benzene hexachloride.

in the axial positions. Before reaction can take place, the group must be forced into the energetically less favorable axial position. In this position, however, it is *trans* to hydrogen on an adjacent carbon atom, a situation that is favorable to elimination of hydrogen along with the functional group with the formation of a double bond. Of the nine benzene hexachlorides, the all-*trans* β isomer (Fig. 38–4*b*) is the only one that does not eliminate hydrogen chloride readily when treated with base. Since no axial hydrogen and axial chlorine can be present on adjacent carbon atoms, *trans* elimination by the *E*2 mechanism is not possible. The rate of reaction is only one ten-thousandth of that of the other isomers, whose rates are all of the same order of magnitude. Considerations such as these, called *conformational analysis,* have been very useful in arriving at constitution and configuration, particularly in condensed ring systems such as those present in the terpenes and steroids.

Fused and Bridged Rings. **Decalin** (*decahydronaphthalene*) exists in two isomeric forms that differ in the space relationship of the 9 and 10 hydrogen atoms.

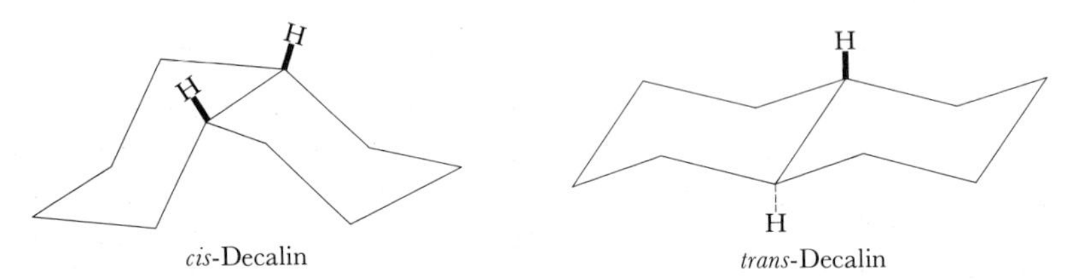

cis-Decalin *trans*-Decalin

These two fused-ring forms were postulated by Mohr in 1918 as being possible but were not isolated until 1927. Since isomers could not exist if the carbon atoms of the rings were planar, because of the amount of strain that would be involved, the isolation of the two forms by W. Hueckel was a proof that carbocyclic rings need not be planar (p. 638). Although *cis*-decalin is dissymmetric, it cannot be resolved because, as with cyclohexane, the easy inversion of the chair conformations converts one enantiomorph into the other. The interconvertibility of chair conformations also permits substituents in *cis*-decalin to occupy either an axial or an equatorial position regardless of their isomeric configuration, the form with the large groups equatorial being the more stable. *trans*-Decalin is a rigid molecule because the central bond must be diequatorial. For isomeric compounds the substituents are either axial or equatorial but not interconvertible. The same principles apply when more than two cyclohexane rings are fused as in the terpenes and steroids, (pp. 662, 665) and have an important application in the determination of configuration.

Adamantane, $C_{10}H_{16}$, m.p. 268°, was isolated first from a petroleum of Eastern Silesia (*Hodonin, Czechoslovakia*). It since has become fairly easily available by the acid-catalyzed rearrangement of tetrahydrodicyclopentadiene.

$\xrightarrow{H_2/Ni}$ $\xrightarrow{AlCl_3}$

1 2 3 4 5 6 7 8 9 10

Adamantane crystallizes in the cubic system. It is extremely stable both thermally and chemically. The structure tricyclo[3.3.1^{1,5}.1^{3,7}]decane has been confirmed by X-ray analysis. The molecule has the carbon atoms in the same positions that they would occupy in the diamond lattice and is free of both angle and bond-opposition strain. The relief of strain in going from tetrahydrodicyclopentadiene to adamantane is the driving force for the rearrangement.

Adamantane provides a recent example of the value of fundamental pure research with no regard for possible practical applications. Nothing could appear to be of more esoteric interest than an investigation of adamantane and its chemical behavior. Yet in 1964 it was reported that 1-adamantanamine (*amantadine*), used as the water-soluble hydrochloride, inhibits infections by several strains of influenza virus. It thus is one of the few chemicals with demonstrated antiviral activity (cf. footnote, p. 572).

Widespread interest in unusual polycyclic systems, which have been given whimsically descriptive common names, has led to the synthesis of **barrelene** in 1960, **cubane** in 1964, and **diphenyltetrahedrane** in 1965. Tetrahedrane, the parent hydrocarbon, C_4H_4, has four CH groups at the corners of a tetrahedron.

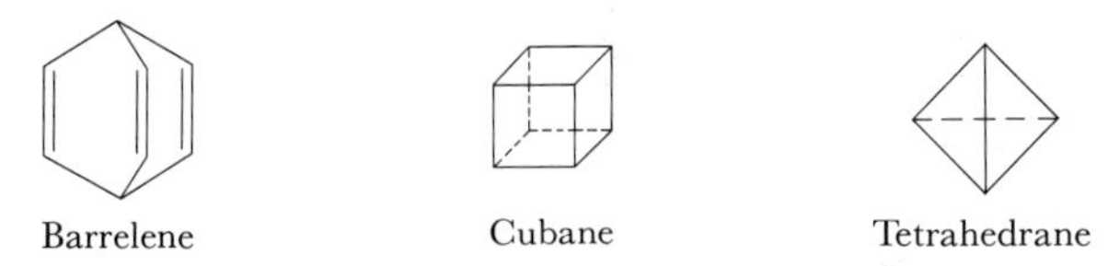

Barrelene Cubane Tetrahedrane

Congressane was so named when its synthesis was set as a goal for organic chemists in 1963 by the organizers of the XIXth Congress of the International Union of Pure and Applied Chemistry. Like adamantane, its carbon atoms have the diamond arrangement, but it has two kinds of bridgehead hydrogens instead of one. The goal was achieved in 1965 by a procedure analogous to that used to synthesize adamantane, and the structure of the product was confirmed by X-ray diffraction.

2 — $h\nu$ → — $AlCl_3$ →

Norbornane Congressane

Renewed interest has been shown in some older suggested polycyclic structures. The so-called "Dewar structure" for benzene has two double bonds and a bond joining the *para* carbon atoms (*I*). Actually it was not advocated by Dewar as a structure for benzene but merely was given in a paper published in 1867 as an example of a type of structure that could be built by a set of mechanical models that he had designed. Nevertheless the structure has been perpetuated in the minds of chemists, and in 1962 a tri-*t*-butyl derivative (*II*) was obtained by the irradiation of 1,2,4-tri-*t*-butylbenzene (t = t-C_4H_9).

I *II*

Several syntheses of other derivatives of the Dewar structure have since been reported.

In 1869, Ladenburg[3] put forward his historically famous prism formula for benzene (*III*, now named *prismane*). In 1964 it was reported that spontaneous trimerization of *t*-butylfluoroacetylene gives a derivative (*IV*) of this structure, together with derivatives of the Dewar structure (*V*) and of a fourth valence-bond isomer of benzene (*VI*, named *benzvalene*).

---

[3] Albert Ladenburg (1842–1911), student of Bunsen, Friedel, and Kekulé, and later professor at the University of Breslau. He is known for his proof of the equivalence of the hydrogen atoms of benzene, for early work on organosilicon compounds (p. 278), and for his investigations of heterocyclic compounds and alkaloids.

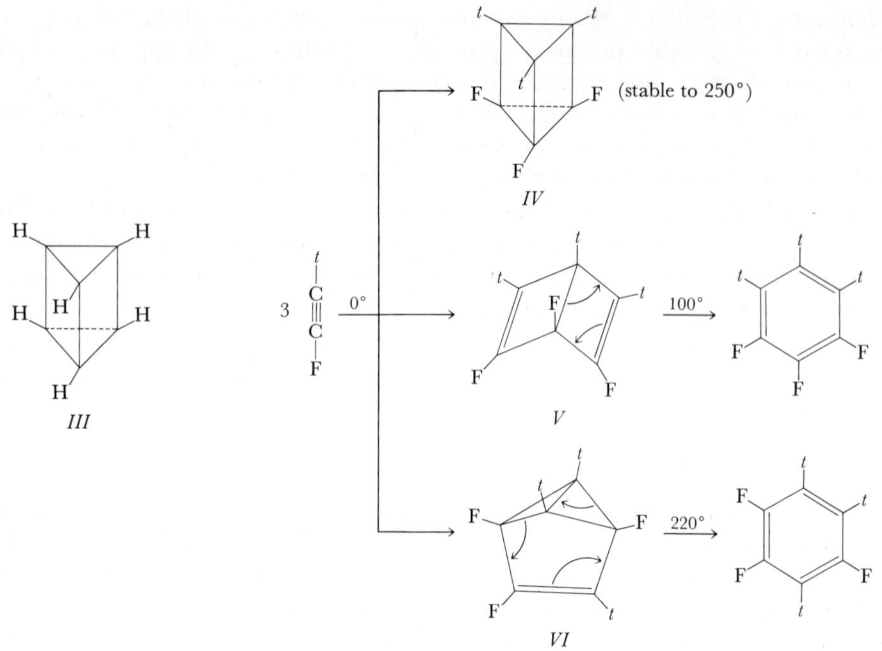

(stable to 250°)

III *IV* *V* *VI*

3 t—C≡C—F 0° 100° 220°

Seven-carbon Rings

Cycloheptanone is the most readily available seven-membered ring compound and is the usual starting point for the simpler derivatives of cycloheptane. It is made by ring enlargement of cyclohexanone by means of the diazomethane procedure for inserting a methylene group adjacent to a carbonyl group (p. 249).

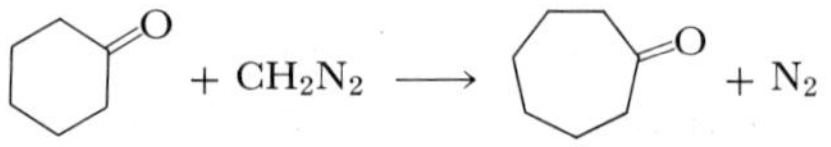

$$+ CH_2N_2 \longrightarrow + N_2$$

Approximately 50 per cent yields of cycloheptanone can be obtained by the pyrolysis of salts of suberic acid, which recently has become available (p. 632).

The chief interest in seven-membered rings has been in the polyunsaturated derivatives. **Cycloheptatriene** is obtained in 87 per cent yield by the cuprous chloride–catalyzed reaction of benzene with diazomethane.

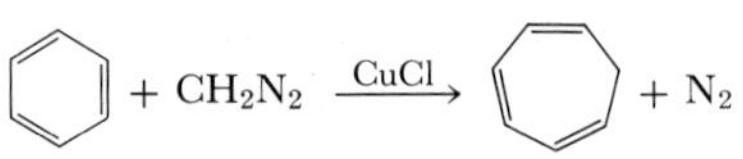

$$+ CH_2N_2 \xrightarrow{CuCl} + N_2$$

Oxidation of cycloheptatriene with selenium dioxide in aqueous dioxane gives the oxo derivative known as **tropone** and permanganate oxidation gives a small yield of 2-hydroxytropone, which is called **tropolone.** Addition of one mole of bromine and pyrolytic elimination of hydrogen bromide gives **cycloheptatrienyl bromide** (*tropylium bromide*).

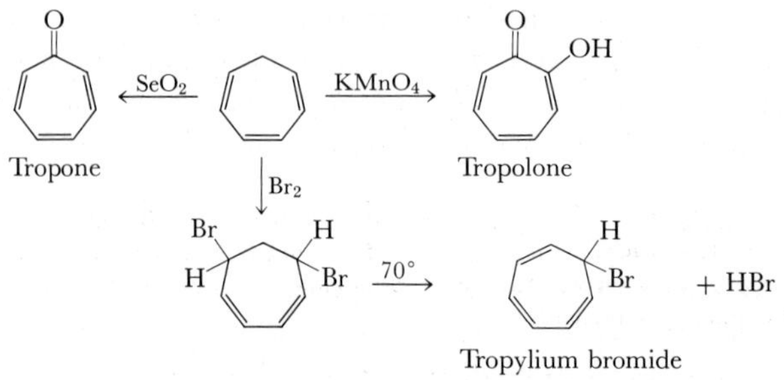

SeO_2 $KMnO_4$ Br_2 $\xrightarrow{70°}$ + HBr

Tropone Tropolone

Tropylium bromide

All of these compounds have unusual properties. Estimations of the acidity of cycloheptatriene vary from $pK_a = 25$ to 36, indicating that resonance stabilization of the anion may be near that for the anion of triphenylmethane ($pK_a = 33$). Tropone has a high dipole moment (4.3 D), high boiling point (113°/15 mm.), and abnormal stretching absorption in the infrared (1638 cm.$^{-1}$) for a ketone. It does not form a phenylhydrazone, although the oxime and the semicarbazone have been prepared under forcing conditions. Tropolone is devoid of carbonyl reactivity. Both tropone and tropolone form crystalline hydrochlorides Tropolone is a stronger acid ($pK_a = 7$) than enols or phenols, its acidity approaching that of carbonic acid. Tropylium bromide is a high-melting deliquescent salt, soluble in water, and insoluble in aprotic low-dielectric solvents.

These properties indicate that the classical formulas do not represent these compounds. They are explained adequately by resonance stabilization of a cycloheptatrienyl cation, a cycle that obeys the Hueckel rule (p. 640).

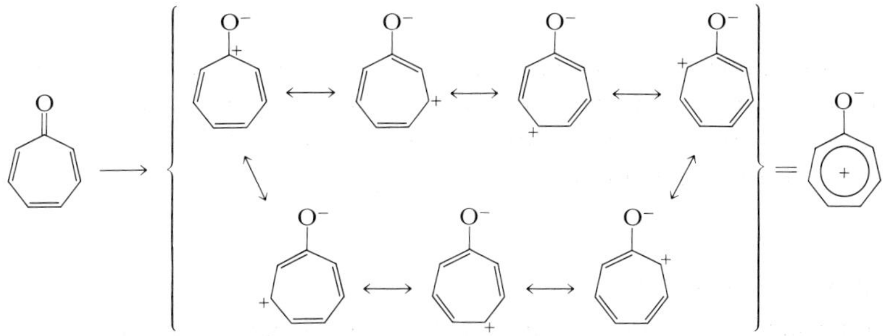

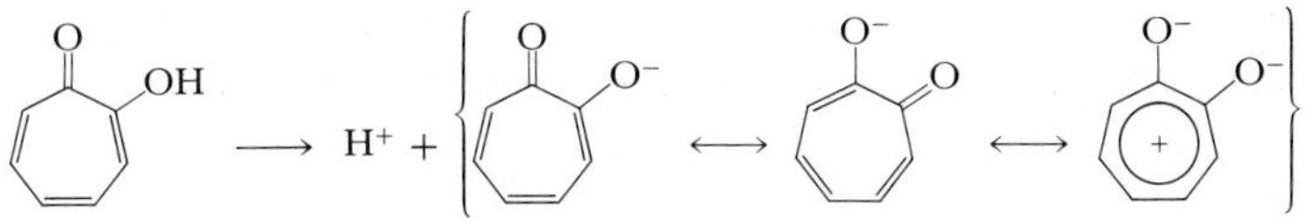

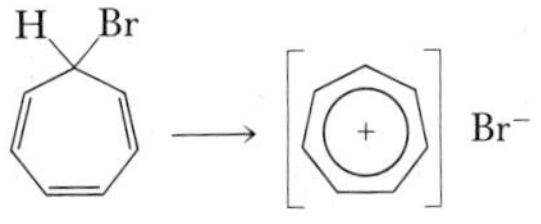

The tropylium ion is the only simple carbonium ion that is stable in aqueous solution. Tropylium salts do not show any of the electrophilic substitution reactions characteristic of benzenoid hydrocarbons. The positive charge makes them extremely resistant to nitration, sulfonation, and Friedel-Crafts reactions. Tropone has both unsaturated and aromatic properties. It adds a mole of chlorine or bromine, but it also couples in the α position with benzenediazonium chloride. Tropolone is resistant to electrophilic substitution in strongly acid solution because of salt formation.

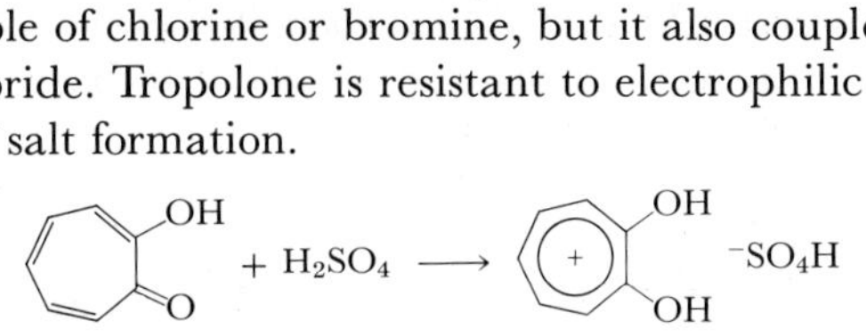

In weakly acid or neutral solution, however, it behaves like phenol, undergoing coupling with diazonium salts, nitrosation, tribromination in the 3,5,7 positions, and the Reimer-Tiemann reaction (p. 461). 5-Nitrosotropolone can be reduced to the amine, which can be diazotized. The diazonium group can be replaced by other groups as in the Sandmeyer reactions (p. 432). The properties of the hydroxyl group are intermediate between those of phenols and carboxylic acids. Thus it can be alkylated but not acylated readily, and the alkoxyl group is more easily hydrolyzed than phenolic alkoxyl. Presumably the adjacent oxygen can interact with the hydroxyl electronically through the π system in the same way that the carbonyl interacts with the hydroxyl in a carboxyl group.

Numerous natural products contain seven-membered carbon rings and several contain the tropolone system. **Stipitatic acid,** the first compound for which a tropolone structure was proposed, was isolated from the mold *Penicillium stipitatum*. The **α-, β-,** and **γ-thujaplicins** are the three isopropyl tropolones. They are present in the heartwood of western red cedar (*Thuja plicata*), and their high fungicidal activity is responsible for the resistance of red cedar to decay. **Purpurogallin** is a red crystalline compound formed by the oxidation of pyrogallol. It occurs naturally as a diglucoside in oak galls.

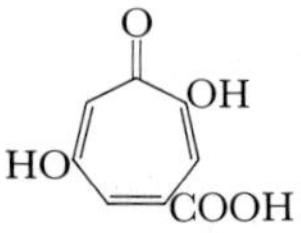

Stipitatic acid 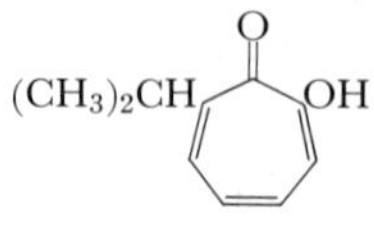   α-Thujaplicin 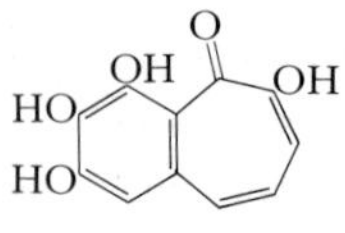   Purpurogallin 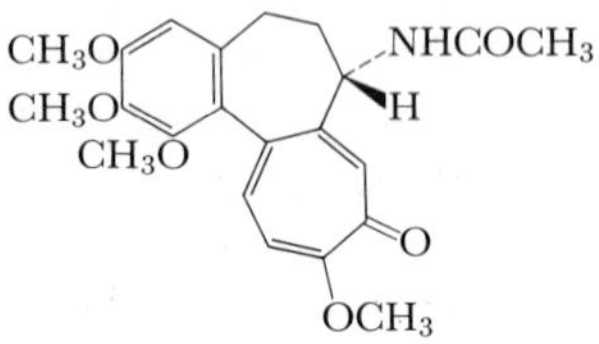   (−)-Colchicine

Colchicine is an important alkaloid obtained from the corms or seeds of the so-called autumn crocus (*Colchicum autumnale*). It is used to relieve acute attacks of gout. It also has the ability to increase the number of chromosomes during cell division in plants and is used commercially to produce new plant races.

Medium Rings (*8 to 12 Carbons*)

Cyclooctanone is prepared by ring enlargement of cycloheptanone with diazomethane. The yield by pyrolysis of salts of azelaic acid is less than 20 per cent. It is made commercially by a series of reactions starting with the dimerization of butadiene by a Ziegler catalyst.

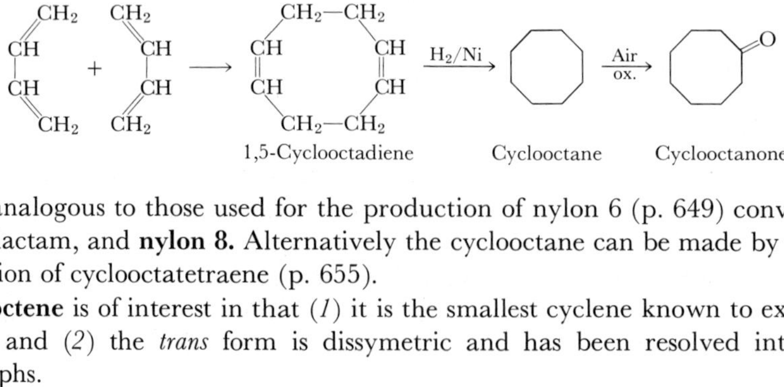

$$2\ CH_2{=}CH{-}CH{=}CH_2 \longrightarrow \text{1,5-Cyclooctadiene} \xrightarrow{H_2/Ni} \text{Cyclooctane} \xrightarrow[\text{ox.}]{\text{Air}} \text{Cyclooctanone}$$

1,5-Cyclooctadiene   Cyclooctane   Cyclooctanone

Reactions analogous to those used for the production of nylon 6 (p. 649) converts it to the oxime, the lactam, and **nylon 8.** Alternatively the cyclooctane can be made by the catalytic hydrogenation of cyclooctatetraene (p. 655).

Cyclooctene is of interest in that (*1*) it is the smallest cyclene known to exist in *cis* and *trans* forms and (*2*) the *trans* form is dissymetric and has been resolved into the active enantiomorphs.

R(−)-*trans*-Cyclooctene   *S*(+)-*trans*-Cyclooctene

Rotation of the ethylenic group through the inside of the ring would result in interconversion and racemization, but this is not possible because it would require excessive steric and angle strain. The resolving agent used is the coordination complex, ethylene-(+)-α-phenylethylaminedichloroplatinum, which exchanges the ethylene ligand for a cyclooctene ligand and gives diastereoisomers separable by fractional crystallization.

Cyclooctatetraene has been of considerable interest for many years. It was prepared first by Willstaetter[4] by a rational synthesis from the alkaloid pseudopelletierine. More recently it has been prepared in quantity by the tetramerization of acetylene.

$$4\ HC{\equiv}CH \xrightarrow[\text{tetrahydrofuran, } 65^\circ \text{ and 250 p.s.i.}]{Ni(CN)_2 \text{ in}} \text{cyclo-}(CH{=}CH)_4$$

Cyclooctatetraene

Originally cyclooctatetraene was prepared to determine whether its chemical properties would resemble those of benzene (*cyclohexatriene*). They do not, since the compound readily adds four moles of halogen or four moles of halogen acid and is oxidized by cold permanganate. Its resonance energy is estimated from heats of hydrogenation to be only 2 kcal. and from its heat of combustion between 4 and 6 kcal. That it is much less stable than benzene is shown by its ready and complete rearrangement to styrene. The nonaromatic character no longer is surprising since cyclooctatetraene is not planar but tub-shaped and does not obey the $4n + 2$ rule (p. 640).

Although rings that contain a double bond have been made in all sizes, the linear distribution of acetylenic bonds apparently prevents the formation of rings that contain a triple bond and have less than eight carbons. **Cyclooctyne** is the smallest cyclic acetylene whose synthesis has been reported (cf. p. 400).

Until 1941, rings with 9 to 11 carbons were very difficult to synthesize. Yields by the usual ring-closure methods are only a few per cent or less. Intramolecular acyloin condensation (p. 176) of α,ω-dicarboxylic esters, however, gives yields of 30 to 60 per cent. Presumably adsorption of both ends of the molecule on the surface of the metal brings them close together and facilitates reaction.

$$(CH_2)_n(COOC_2H_5)_2 + 4\ Na \longrightarrow 2\ C_2H_5ONa + (CH_2)_n\left<\begin{matrix} CO^{-+}Na \\ \| \\ CO^{-+}Na \end{matrix}\right. \xrightarrow{H_2O} \left[(CH_2)_n\left<\begin{matrix} COH \\ \| \\ COH \end{matrix}\right.\right] \longrightarrow (CH_2)_n\left<\begin{matrix} CO \\ | \\ CHOH \end{matrix}\right.$$

Twelve-carbon rings now are made commercially by the cyclotrimerization of butadiene with Ziegler catalysts. Chromic chloride-trialkylaluminum complex gives the *trans-trans-trans* isomer, whereas titanium trichloride-dialkylaluminum chloride gives the *cis-trans-trans* isomer.

$$3\ CH_2{=}CH{-}CH{=}CH_2 \xrightarrow{CrCl_3\text{-}AlR_3} \text{cyclododecatriene } (trans\text{-}trans\text{-}trans)$$

$$3\ CH_2{=}CH{-}CH{=}CH_2 \xrightarrow{TiCl_3\text{-}AlR_2Cl} \text{cyclododecatriene } (cis\text{-}trans\text{-}trans)$$

Catalytic reduction gives cyclododecane, which can be converted to **cyclododecanone,** the oxime, and **nylon 12** (cf. p. 649).

Large Rings (*Macrocycles*)

The reason for Baeyer's postulation that large rings are planar was that large-ring compounds were unknown. In 1918 a ten-membered ring containing nitrogen and bridging

[4] Richard Willstaetter (1872–1942), successor to Baeyer at the University of Munich. He was one of the outstanding investigators of the constitution of natural products such as the alkaloids, anthocyanins, carotenes, and chlorophylls. He was awarded the Nobel Prize in Chemistry in 1915.

the *meta* positions of a benzene ring was prepared by v. Braun[5] (cf. p. 657), and in 1926 it was shown that **muscone** from the secretion of the musk deer and **civetone** from the secretion of the civet cat are fifteen- and seventeen-membered ring ketones respectively. In the following year the plant musks, **pentadecanolide** from angelica root and **ambrettolide** from ambrette seed, were found to be lactones containing sixteen- and seventeen-membered rings.

$CH_3CH\langle CH_2, (CH_2)_{12}\rangle CO$ — Muscone

$CH{-}(CH_2)_7 / CH{-}(CH_2)_7$ (CH=CH) $\rangle CO$ — Civetone

$(CH_2)_{14}$ CO—O — Pentadecanolide

$(CH_2)_5$, CH=CH, CO—O, $(CH_2)_8$ — Ambrettolide

Since 1956 several antibiotics isolated from various species of *Streptomyces* have been shown to contain highly oxygenated large lactone rings. Thus **methymycin, erythromycin,** and **carbomycin** contain 12-, 14-, and 17-membered rings respectively. This class of natural products has been designated as **macrolides.**

Over a period of years several types of compounds containing up to thirty-four atoms in a ring were synthesized. It is of interest that all cyclic ketones, lactones, carbonates, imines, and formals having fourteen to seventeen atoms in a ring have a musk odor. **Cyclopentadecanone,** known as *Exaltone,* is manufactured commercially for use in perfumery in place of the natural musks. It is used in the laboratory as solvent for molecular weight determinations, since it has a high cryoscopic constant and melts lower than camphor (Table 4–2, p. 50). Cyclopentadecanone was prepared first by the decomposition of the thorium salt of 1,14-tetradecanedicarboxylic acid (cf. p. 631). The yield is less than 1 per cent. Better yields of large-ring ketones have been obtained by carrying out an intramolecular Thorpe[6] condensation at high dilution. Hydrolysis gives a β-keto acid, which loses carbon dioxide when heated.

$$(CH_2)_{13}\langle C{\equiv}N, CH_2CN\rangle \xrightarrow{Li^+{}^-NR_2} (CH_2)_{13}\langle C{=}NH, CHCN\rangle \xrightarrow{Hydrolysis} (CH_2)_{13}\langle C{=}O, CHCOOH\rangle \xrightarrow{Heat} (CH_2)_{14}CO + CO_2$$

Much synthetic work in the field of polyunsaturated macrocyclic compounds has resulted from attempts to synthesize hydrocarbons that obey the $4n + 2$ rule for values of n greater than 1. The first such compound to be prepared is **cyclooctadecanonaene,** which contains a conjugated π system of 18 electrons ($n = 4$). It was made by the oxidative coupling of three moles of 1,5-hexadiyne to give a cyclooctadecahexayne, hydrogen migration to the conjugated hexaenetriyne, and selective reduction.

$$3\ HC{\equiv}C(CH_2)_2C{\equiv}CH \xrightarrow[\text{pyridine}]{Cu(OAc)_2\text{ in}} \text{cyclooctadecahexayne } (-C{\equiv}C-C{\equiv}C-) \xrightarrow{{}^-OC_4H_9\text{-}t} \text{Tridehydro[18]annulene} \xrightarrow{H_2/Pd} \text{[18]Annulene}$$

Tridehydro[18]annulene

[18]Annulene

[5] Julius von Braun (1875–1939), professor at Frankfurt University. His work dealt chiefly with organic nitrogen compounds. Several methods of opening nitrogen heterocycles bear his name.

[6] Jocelyn Field Thorpe (1872–1940), professor of organic chemistry at the Imperial College of Science and Technology in London. He was noted particularly for his investigations of the tautomerism of the glutaconic acids and for the effect of substitution on ring formation.

Compounds of this type have been given the common name *annulenes.* A number in brackets indicates the number of carbon atoms in the ring. [18]Annulene is fairly stable, but crowding of the hydrogens inside the ring prevents the molecule from being completely planar. It has no aromatic chemical properties. On the other hand, the proton n.m.r. spectrum shows that there is sufficient cyclic delocalization of the π electron system in the ground state of the molecule to sustain a magnetically induced ring current (p. 550).

Bridged Benzene Rings

Since the carbon-hydrogen bonds of benzene lie in the plane of the ring, small external rings can be formed only across the *ortho* positions. Thus only *o*-phthalic acid forms a cyclic anhydride. With large enough alicyclic rings, however, it is possible to include the *meta* and *para* positions in a cycle as well as the *ortho* positions.

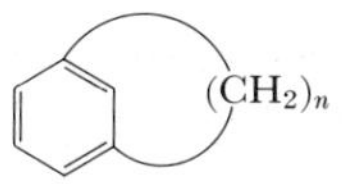

$(CH_2)_n$ $(CH_2)_n$

Such structures have been called *ansa* compounds (L. *ansa* handle). For entirely methylene groups in the bridge, it appears that n must be at least 6 in *meta* rings to retain the characteristic aromatic properties of the benzene ring.

Structures incorporating two benzene rings likewise are of interest. In the *para* series, known as *paracyclophanes,* X-ray diffraction measurements show that when $n = 2$, the strain is sufficient to bend the benzene rings into boat shapes.

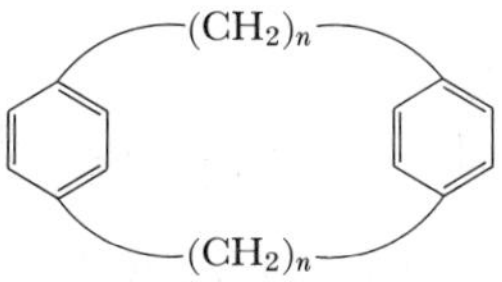

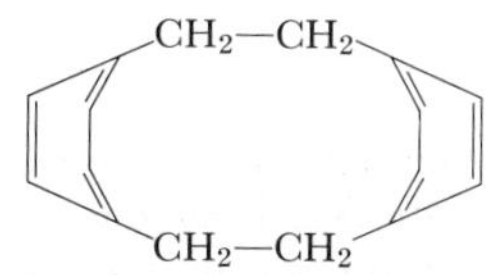

$(CH_2)_n$ $CH_2{-}CH_2$

$(CH_2)_n$ $CH_2{-}CH_2$

Since they no longer are planar, they do not show the characteristic ultraviolet absorption of the benzene rings. When $n = 4$, the compound has the normal ultraviolet absorption spectrum.

Catena Compounds

Compounds that have interlocking rings in the manner of a linked chain have been termed **catena compounds** (L. *a chain*). A rational synthesis of the following compound, which contains a 28-membered ring linked with a 26-membered ring, was reported in 1964.

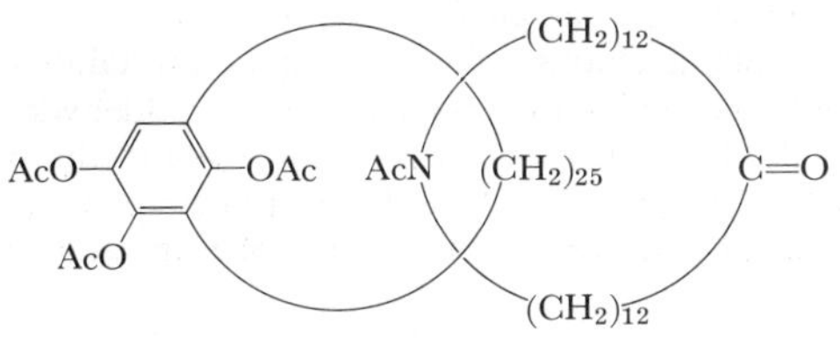

$(CH_2)_{12}$

AcO— —OAc AcN $(CH_2)_{25}$ C=O

AcO

$(CH_2)_{12}$

PROBLEMS

38–*1*. Starting with ethylene bromide, give a consecutive series of reactions that includes the preparation of 1,1-cyclopropanedicarboxylic acid, cyclopropyldimethylamine, ethyl cyclopropanecarboxylate, cyclopropanecarboxylic acid, cyclopropylamine, cyclopropene, cyclopropanecarboxamide, cyclopropyltrimethylammonium hydroxide, and ethyl 1,1-cyclopropanedicarboxylate.

38–*2*. Give equations illustrating the preparation of cyclopropanecarboxylic acid and of cyclopropyl methyl ketone starting with ethyl acetoacetate.

38–*3*. Give the structural formulas for the compounds designated by the italicized capital letters in the following chart.

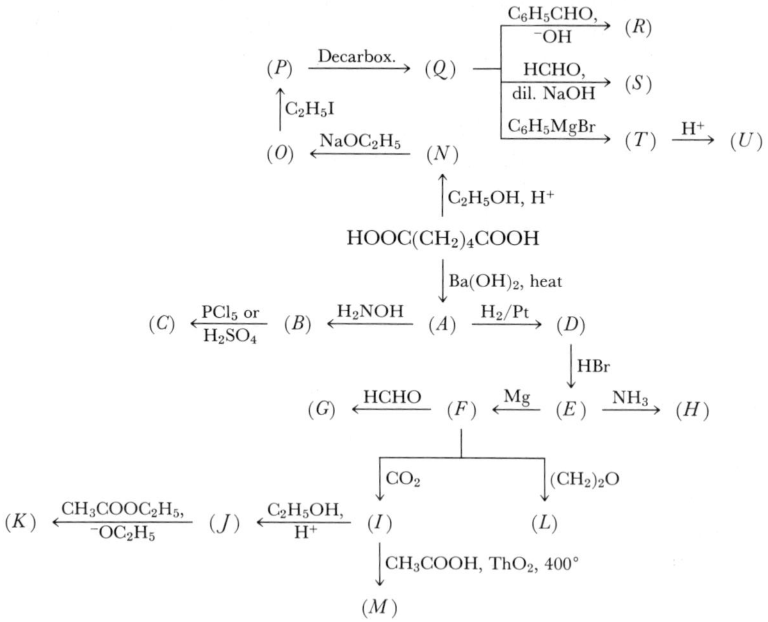

$$(Q) \xrightarrow[\text{-OH}]{C_6H_5CHO,} (R)$$

$$(P) \xrightarrow{\text{Decarbox.}} (Q) \xrightarrow[\text{dil. NaOH}]{\text{HCHO,}} (S)$$

$$(Q) \xrightarrow{C_6H_5MgBr} (T) \xrightarrow{H^+} (U)$$

$$(O) \xrightarrow{C_2H_5I} (P)$$

$$(O) \xleftarrow{NaOC_2H_5} (N)$$

$$HOOC(CH_2)_4COOH \xrightarrow{C_2H_5OH,\ H^+} (N)$$

$$HOOC(CH_2)_4COOH \xrightarrow{Ba(OH)_2,\ \text{heat}} (A)$$

$$(C) \xleftarrow[H_2SO_4]{PCl_5 \text{ or}} (B) \xleftarrow{H_2NOH} (A) \xrightarrow{H_2/Pt} (D)$$

$$(D) \xrightarrow{HBr} (E)$$

$$(G) \xleftarrow{HCHO} (F) \xleftarrow{Mg} (E) \xrightarrow{NH_3} (H)$$

$$(F) \xrightarrow{CO_2} (I) \qquad (F) \xrightarrow{(CH_2)_2O} (L)$$

$$(K) \xleftarrow[{}^{-}OC_2H_5]{CH_3COOC_2H_5,} (J) \xleftarrow[H^+]{C_2H_5OH,} (I)$$

$$(I) \xrightarrow{CH_3COOH,\ ThO_2,\ 400^\circ} (M)$$

38–*4*. 1-Ethynylcyclohexyl carbamate is a short-acting soporific. Devise a synthesis starting with cyclohexanone.

38–*5*. Devise a synthesis for the fungicide Captan starting with 1,3-butadiene (cf. pp. 634, 475).

38–*6*. Predict the configuration of the chief isomer formed on the reduction of *m*-hydroxybenzoic acid with sodium and alcohol to hexahydro-*m*-hydroxybenzoic acid.

38–*7*. When either the *cis* or *trans* isomer of 1,3-cyclohexanediol is treated with hydrogen bromide only a single 1,3-dibromocyclohexane is obtained. Similarly only a single 1,4-dibromocyclohexane is obtained from either *cis* or *trans* 1,4-cyclohexanediol. Predict the configuration of the 1,3- and the 1,4-dibromocyclohexanes.

38–*8*. Starting with erucic acid, give equations for the synthesis of 2-hydroxycyclotetracosanone and for its conversion into the diol, the diketone, the ketone, the halide, the hydrocarbon, the alcohol, and the amine.

38–*9*. Give the probable mechanism for the magnesium halide-catalyzed rearrangement of cyclohexene oxide to cyclopentanecarboxaldehyde.

38–*10*. A substance *A*, $C_{15}H_{22}O_2$, reacts with hydrogen in the presence of platinum catalyst to give *B*, $C_{15}H_{28}O_2$, which does not take up any more hydrogen. What can be concluded concerning the number of rings and reducible bonds in *A* and *B*?

38–*11*. A pure organic compound, *A*, $C_7H_{12}O$, is optically active and gives a positive permanganate test. Hydrogenation of *A* gives *B*, $C_7H_{14}O$, which is optically inactive and gives a negative permanganate test. Oxidation of *A* with chromic acid gives *C*, $C_7H_{10}O$, which also is optically inactive. Ozonation of *C* and oxidation of the ozonide with chromic acid gives *D*, $C_7H_{10}O_5$, which is soluble in sodium bicarbonate solution. When *D* is heated, a gas is evolved and compound *E*, $C_6H_{10}O_3$, results. *E* also is soluble in sodium bicarbonate solution and gives a positive iodoform test. What structures for *A*, *B*, *C*, *D*, and *E* satisfy all of the experimental data?

38–*12*. Give a likely interpretation for each of the following rearrangements.

(*a*) 2-chlorocyclohexanone $\xrightarrow{^{-}OC_2H_5}$ cyclopentyl–$COOC_2H_5$

(*b*) camphor oxime (=N–OH) $\xrightarrow{H^+}$ dimethylcyclopentene–CH_2CN

(*c*) 1-ethynylcyclohexanol (HO, C≡CH) $\xrightarrow{H^+}$ 1-cyclohexenyl–$COCH_3$

CHAPTER THIRTY-NINE

TERPENES AND STEROIDS. BIOGENESIS

TERPENES

The odorous components of plants are volatile with steam and usually are separated from the plant material by steam distillation. They are known as the **volatile** or **essential oils.** They consist of hydrocarbons, alcohols, ethers, aldehydes, ketones, and lactones. Some of these substances such as anethole, cinnamaldehyde, and methyl salicylate belong to the aromatic series (pp. 445, 463, 473). In the exudations of conifers and in the oils from citrus fruits and from eucalyptus trees, alicyclic hydrocarbons of the composition $C_{10}H_{16}$ are especially abundant, and it is to these compounds that the term *terpene* (Gr. *terebinthos* turpentine tree) was applied in the restricted sense. Closely related open-chain hydrocarbons having ten carbon atoms also were included under this term. The oxygenated terpenes were known as *camphors.* It soon became evident, however, that compounds containing 15, 20, 30, and 40 carbon atoms also are closely related to the terpenes. The one common characteristic of all of these compounds is that their carbon skeletons are evenly divisible into iso-C_5 units, frequently referred to as isoprene or isopentane units. These divisions often are indicated by dotted lines through the formulas. The term *terpene* in its broadest sense now includes all such compounds, whether hydrocarbons or not. Terpene in the limited sense still refers to compounds containing two iso-C_5 units. Hence the broad class of terpenes is divided into hemiterpenes, C_5; terpenes, C_{10}; sesquiterpenes, C_{15}; diterpenes, C_{20}; triterpenes, C_{30}; tetraterpenes, C_{40}; and polyterpenes, C_{5x}.

Citral is a mixture of the *cis* and *trans* isomers of an unsaturated aldehyde that constitutes 80 per cent of East Indian lemongrass oil, the essential oil of *Cymbopogon flexuosus.* It is important as the starting point for the synthesis of vitamin A (p. 661) and of β-carotene (p. 663). Because of the uncertain availability, varying quality, and fluctuating price of natural citral, processes for its commercial synthesis have been developed.

$$(CH_3)_2C{=}CHCH_2 \vdots CH_2\underset{\displaystyle CH_3}{\underset{|}{C}}{=}CHCHO$$

Citral

$$(CH_3)_2C{=}CHCH_2CH_2\underset{\displaystyle CH_3}{\underset{|}{C}}HCH_2CHO$$

Citronellal

***R*-(+)-Citronellal** is the chief component of the oil from citronella, a Ceylon lemongrass (*Cymbopogon nardus*). It differs from citral only in that it lacks the double bond that is conjugated with the carbonyl group, and hence it contains an asymmetric carbon atom. **S-(−)-Citronellol,** the enantiomorphic alcohol, occurs in rose oil (p. 456).

Limonene is the main terpene component of lemon, orange, and many other oils. **(−)-Menthol** is produced chiefly from Japanese peppermint oil. The racemic compound is synthesized by hydrogenating thymol (p. 445).

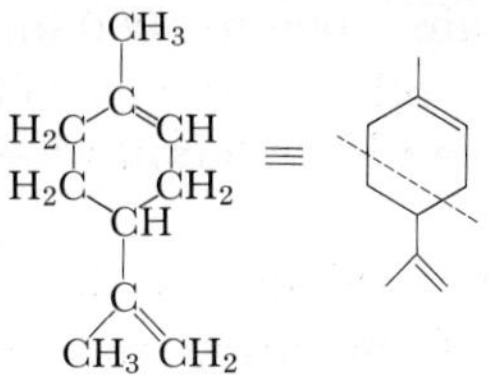

Limonene

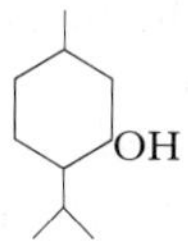

Menthols

α-Pinene, the principal component of oil of turpentine from the exudate of pine trees, is a bicyclic hydrocarbon. Turpentine is an important paint thinner (p. 188) and pinene is used for the synthesis of other chemicals.

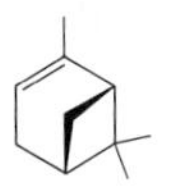

(+)-α-Pinene

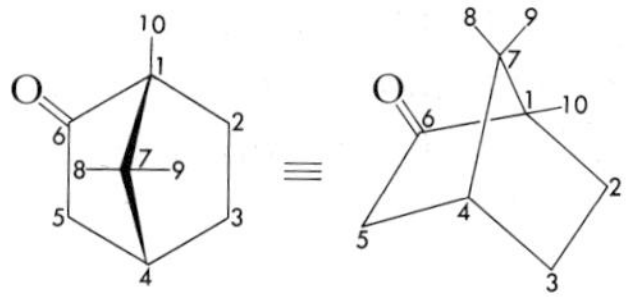

(+)-Camphor

Camphor is a bicyclic ketone, the dextro form of which occurs in the wood of the camphor tree, *Cinnamomum camphora.* Although camphor contains two asymmetric carbon atoms, only one pair of enantiomorphs is known. An examination of models shows that the second pair of isomers cannot exist because of the extreme distortion of bond angles that would be required.

Camphor has been known and valued for medicinal purposes since earliest times, although modern medicine has found it to have no therapeutic value. Its chief industrial importance has been as a plasticizer for the manufacture of celluloid and photographic film base (p. 343). Since celluloid largely has been replaced by other plastics and since nitrate film base no longer is manufactured in the United States (p. 343), the commercial importance of camphor has greatly decreased.

Phytol, an acyclic diterpene, $C_{20}H_{39}OH$, constitutes about one third of the chlorophyll molecule (p. 510), from which it is obtained by saponification. The same alcohol has been isolated from the chlorophyll of over 200 species of plants.

$$(CH_3)_2CHCH_2CH_2(CH_2\underset{CH_3}{\underset{|}{C}}HCH_2CH_2)_2CH\underset{CH_3}{\underset{|}{C}}=CHCH_2OH$$

Phytol

Vitamin A is a fat-soluble vitamin that is necessary for the growth of rats, that plays a part in the resistance of the animal organism to infection, and that is required for the production of visual purple, a pigment necessary for sight. Its structure has been determined by degradation reactions.

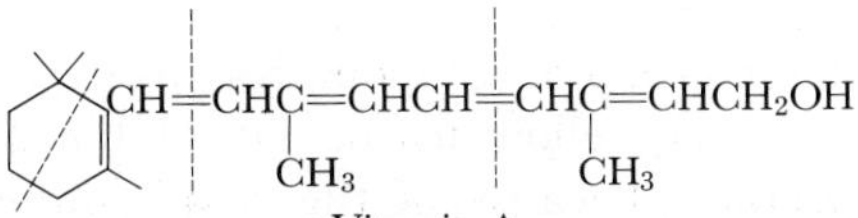

$$\text{—CH=CH}\underset{CH_3}{\underset{|}{C}}\text{=CHCH=CH}\underset{CH_3}{\underset{|}{C}}\text{=CHCH}_2\text{OH}$$

Vitamin A

The pure alcohol melts at 64° and has a biological potency of 4.3×10^6 U.S.P. units per gram or 3.3×10^6 International units per gram. Before commercial production of the acetate by synthesis, which began in 1950, the chief source was the fish liver oils, which vary greatly in potency. Thus cod liver oil contains 3000 to 5000 units per gram, halibut liver oil 10,000 to 15,000, and soupfin shark liver oil 15,000 to 500,000 (average 350,000) units per gram. Production of synthetic vitamin A and its esters in 1964 was 694,000 pounds, valued at $25,300,000.

The most important diterpene commercially is **abietic acid,** $C_{20}H_{30}O_2$, the chief component of rosin or colophony, the resin obtained from various species of pine (L. *abies* fir). Abietic acid is an original component of the tree secretions, and is formed also by the isomerization of other acids during the distillation of the turpentine.

Abietic acid was one of the first resin acids to be investigated, and the chief features of its carbon skeleton have been known since 1910 when retene, the product obtained by Vesterberg in 1903 by dehydrogenation with sulfur, was shown to be 1-methyl-7-isopropylphenanthrene (p. 500). Determination of the location of the remaining carbon atoms and the position of the two double bonds was considerably more difficult, and the verification of the currently accepted structure was not made until 1941.

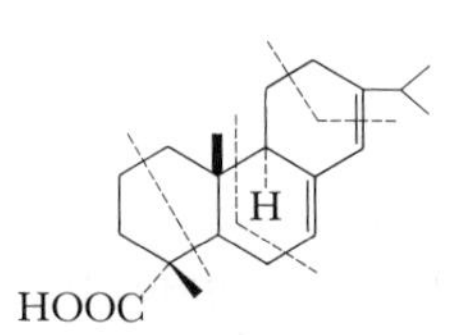

HOOC

Abietic acid

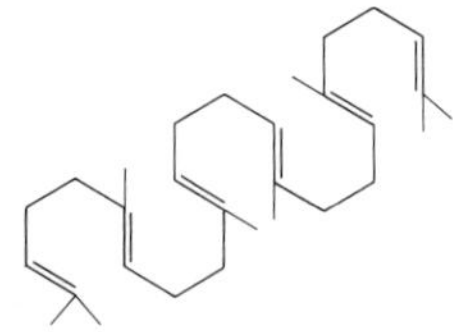

Squalene

Although hydrocarbons usually have not been considered to play an important part in animal metabolism, **squalene,** an acyclic triterpene, $C_{30}H_{50}$, makes up as high as 90 per cent of the liver oil of certain species of sharks of the family *Squalidae.* It now is known that squalene is an intermediate in the biological synthesis of cholesterol (p. 668) and probably is present in all animals. Numerous tetracyclic triterpenes are known, the most important of which is **lanosterol,** $C_{30}H_{50}O$. It occurs along with other triterpenes and the steroid, cholesterol, in wool grease (*lanolin*) and is an intermediate in the biological conversion of squalene to cholesterol (p. 668). The pentacyclic triterpenes are present in many natural resins. Some of the **saponins,** which are responsible for the foam-producing properties of the so-called soap plants, are glycosides of pentacyclic triterpenes. **Oleanolic acid,** $C_{30}H_{48}O_3$, is a representative member of this class. It occurs as the aglycone of a saponin in guaiac bark, sugar beet, and calendula flowers, and free in olive leaves, clove buds, mistletoe, and grape skins.

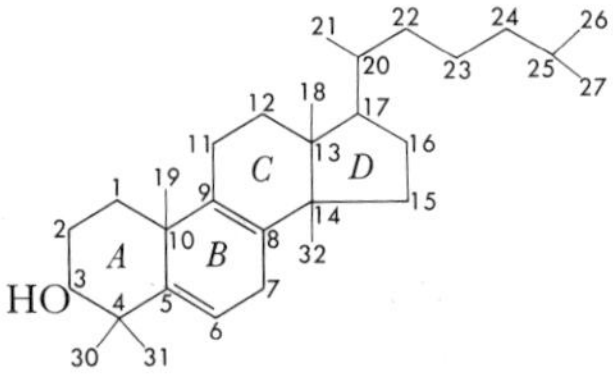

HO

Lanosterol

HO COOH

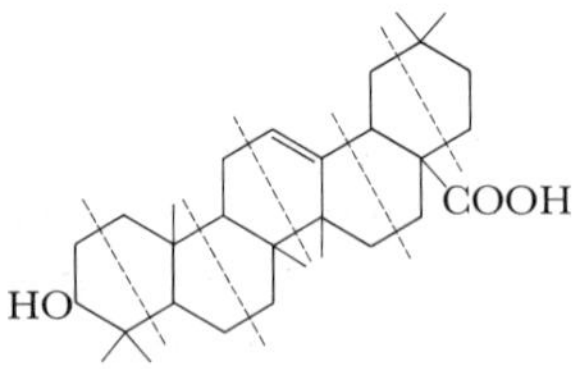

Oleanolic acid

Most members of the large group of compounds known as *carotenoids* may be classed as tetraterpenes. They constitute the yellow to red fat-soluble pigments of plants. Usually several pigments occur together. Because of the small amounts present and the close similarities in structure, isolation and purification by the usual crystallization procedures have been difficult. Rapid progress in the chemistry of the carotenoids began with the use of chromatographic adsorption (p. 44) on alumina, magnesia, or calcium carbonate. It was in the separation of carotenoids that this technique first was highly developed. Because the mixtures separated as colored bands on the column of adsorbent, the procedure was called *chromatography*. This term now is applied to the differential distribution of any mixture of compounds between any two phases, one of which is stationary.

Lycopene, $C_{40}H_{56}$, is the red pigment in the ripe fruit of the tomato (*Lycopersicum esculentum*) and of the watermelon (*Cucumis citrullus*).

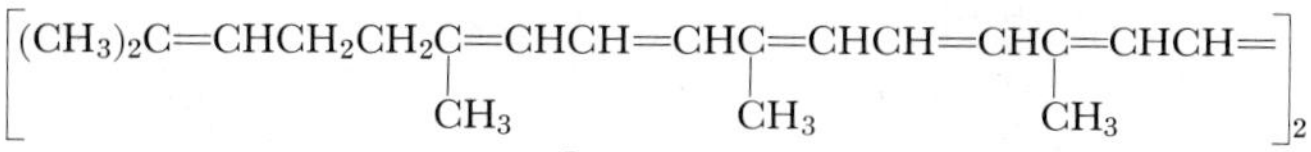

$$\left[(CH_3)_2C{=}CHCH_2CH_2\underset{CH_3}{C}{=}CHCH{=}CH\underset{CH_3}{C}{=}CHCH{=}CH\underset{CH_3}{C}{=}CHCH{=}\right]_2$$

Lycopene

β-Carotene is the chief pigment of the carrot (*Daucus carota*). It can be converted by the animal organism into vitamin A and hence has vitamin A activity. The commercial synthesis of β-carotene makes available a yellow coloring matter for foods that can replace the carcinogenic azo dyes.

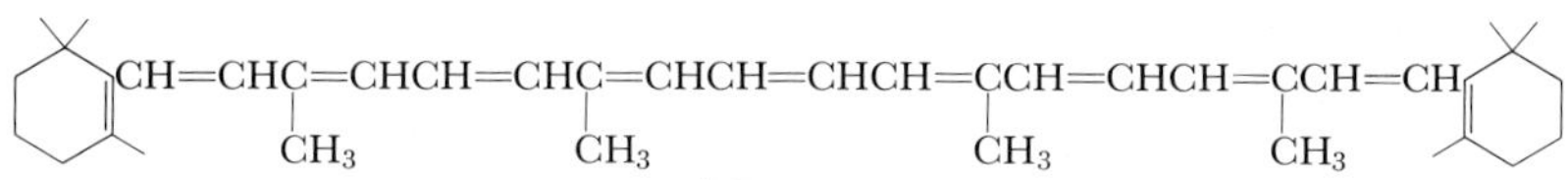

$$CH{=}CH\underset{CH_3}{C}{=}CHCH{=}CH\underset{CH_3}{C}{=}CHCH{=}CHCH{=}\underset{CH_3}{C}CH{=}CHCH{=}\underset{CH_3}{C}CH{=}CH$$

β-Carotene

Lutein (*leaf xanthophyll*), $C_{40}H_{56}O_2$, is a yellow pigment present in leaves, yellow flowers, and egg yolk.

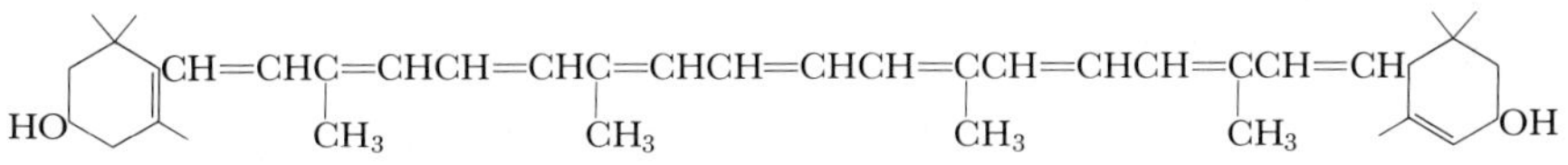

$$HO\ \ CH{=}CH\underset{CH_3}{C}{=}CHCH{=}CH\underset{CH_3}{C}{=}CHCH{=}CHCH{=}\underset{CH_3}{C}CH{=}CHCH{=}\underset{CH_3}{C}CH{=}CH\ \ OH$$

Lutein

Rubber hydrocarbon, $(C_5H_8)_x$, is a polyterpene (p. 583).

STEROIDS

Steroids may be defined as those compounds that contain a ring system like that present in cholesterol. They are characterized by the fact that they yield methylcyclopentenophenanthrene (*Diels hydrocarbon*) on dehydrogenation with selenium.

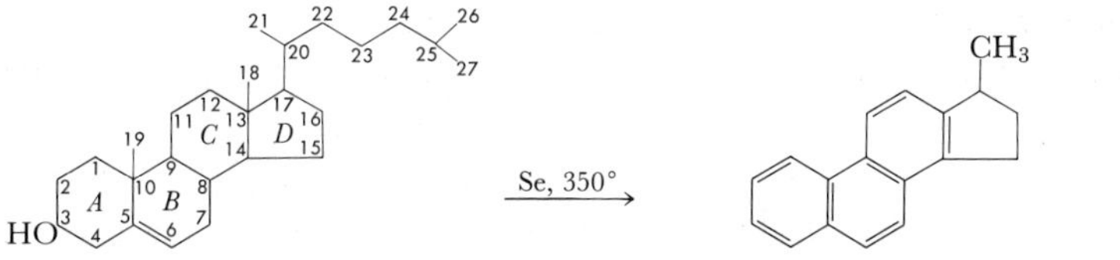

Cholesterol $\xrightarrow{\text{Se, 350°}}$ Methylcyclopentenophenanthrene (*Diels hydrocarbon*)

To this group belong the sterols, the bile acids, the cardiac aglycones, the sex hormones, the adrenal steroids, the toad poisons, and the steroid sapogenins. In view of the complexity of the chemistry and the large number of compounds in the group, the formulas of only a few representatives of the various subgroups are given.

Cholesterol, $C_{27}H_{46}O$, is present in the blood of animals and hence in all parts of the body. It is concentrated in the spinal cord, the brain, skin secretions, and gallstones. It was isolated first from gallstones by Conradi in 1775 and was named cholesterine (Gr. *chole* bile, *stereos* solid) in 1816 by Chevreul (p. 182), who showed that unlike the fats it is not saponifiable. Berthelot (p. 121) recognized in 1859 that it is an alcohol, but the correct molecular formula, $C_{27}H_{46}O$, was not proposed until 1888 by Reinitzer. The currently accepted structure given above was not arrived at until 1932 after over eighty years of active chemical investigation, the last thirty years of which was conducted mainly by Windaus[1] and his collaborators. With eight asymmetric carbon atoms, 256 active isomers are possible. Not only has the configuration of each asymmetric carbon atom been established, but the last steps in the total synthesis of cholesterol having the exact configuration of the natural product were completed in 1951.

Ergosterol, $C_{28}H_{44}O$, was isolated first from ergot but is obtained more readily from yeast. Irradiation by ultraviolet light transforms it into **previtamin D_2** (*precalciferol*), which is converted into **vitamin D_2** (*calciferol*) by heat. The further action of light on previtamin D_2 converts it into a variety of other products.

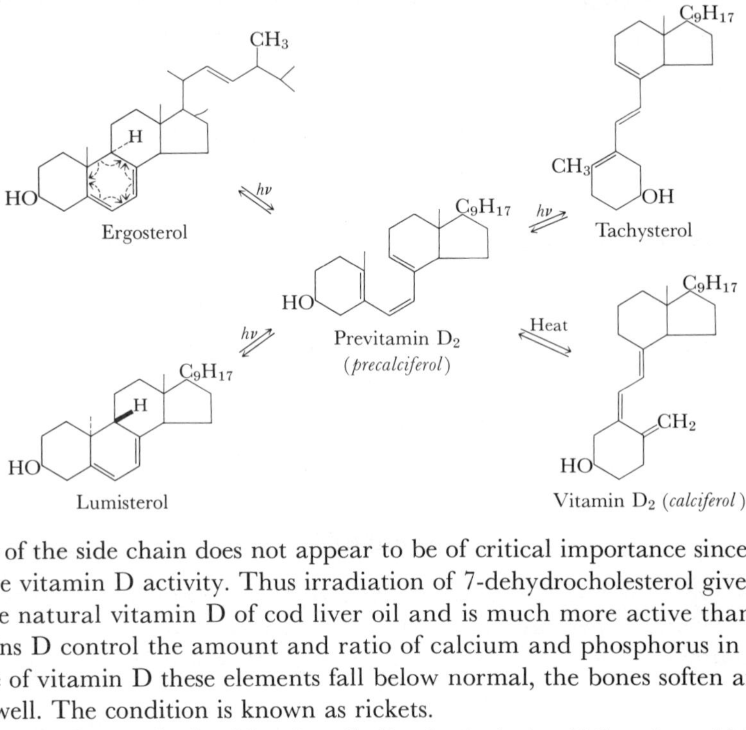

CH_3 H HO Ergosterol $h\nu$ C_9H_{17} HO Previtamin D_2 (*precalciferol*) $h\nu$ C_9H_{17} CH_3 OH Tachysterol $h\nu$ C_9H_{17} H HO Lumisterol Heat C_9H_{17} CH_2 HO Vitamin D_2 (*calciferol*)

The nature of the side chain does not appear to be of critical importance since several compounds have vitamin D activity. Thus irradiation of 7-dehydrocholesterol gives vitamin D_3, which is the natural vitamin D of cod liver oil and is much more active than vitamin D_2. The vitamins D control the amount and ratio of calcium and phosphorus in the blood. In the absence of vitamin D these elements fall below normal, the bones soften and bend, and the joints swell. The condition is known as rickets.

The **bile acids** are obtained by the alkaline hydrolysis of bile salts, which are present in the bile of various animals. In the bile salts, the bile acids are combined by an amide linkage between their carboxyl group and the amino group of glycine, H_2NCH_2COOH, or of taurine, $H_2NCH_2CH_2SO_3H$. Thus **glycocholic acid** on hydrolysis gives cholic acid and glycine, whereas **taurocholic acid** gives cholic acid and taurine. The function of the bile salts is to act as emulsifying agents for fats and hence to promote the hydrolysis and absorption of fats from the intestinal tract. The four bile acids occurring in human and ox bile are **cholic acid, deoxycholic acid, chenodeoxycholic acid,** and **lithocholic acid.** Wieland's investigations[2] of the bile acids, begun in 1912, were carried on concurrently with those of Windaus on cholesterol. The two fields supplemented each other and laid the basis for structural determination in the whole area of steroids.

---

[1] Adolf Windaus (1876–1959), professor of chemistry at the University of Goettigen. His investigations dealt entirely with the sterols, especially the structure of cholesterol and of ergosterol and its irradiation products. He was awarded the Nobel Prize in Chemistry in 1928.

[2] Heinrich Wieland (1877–1957), professor of chemistry at the Bayerische Akademie der Wissenschaften. He made important contributions to many fields of organic chemistry such as oxidation mechanisms, free radicals, alkaloids, butterfly pigments, toad poisons, and especially the bile acids. The 1927 Nobel Prize in Chemistry was awarded to him in 1928.

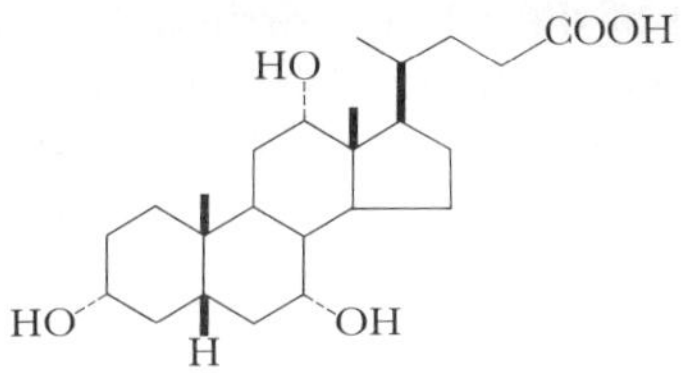

Cholic acid

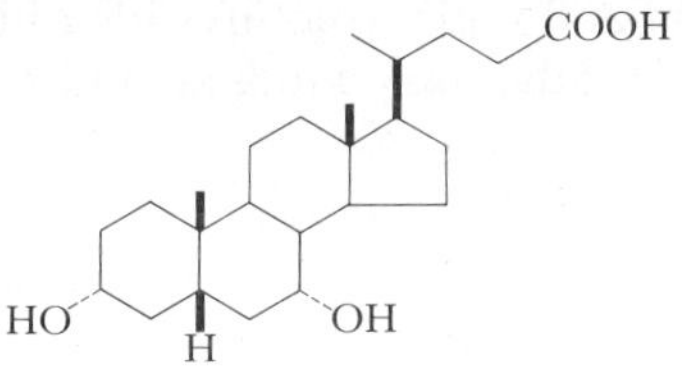

Chenodeoxycholic acid

The **sex hormones** are substances responsible for the sex characteristics and the sexual processes of the animal organism. They are formed in the testes and ovaries, which are stimulated by the gonadotropic hormones secreted by the anterior lobe of the pituitary gland. **Testosterone** is secreted by the testes and controls the development of the genital tract, accessory male organs, and secondary male characteristics such as the comb and wattles of a rooster. **Estradiol** is produced in the ovaries, probably in the ripening follicles.

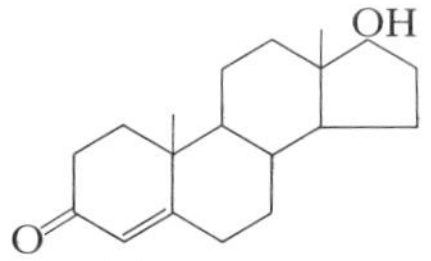

Testosterone

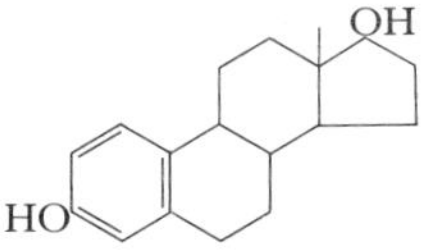

Estradiol

It controls the development of female characteristics and initiates the first phase in the menstrual cycle, namely the proliferation of cells in the uterus. The estrogenic effect is not very specific and a synthetic compound, **stilbestrol** (*diethylstilbestrol*), is used more commonly than natural estradiol to alleviate trouble arising from a deficiency of the hormone and to arrest prostatic cancer. Its use as a growth stimulant for cattle has been discontinued in the United States.

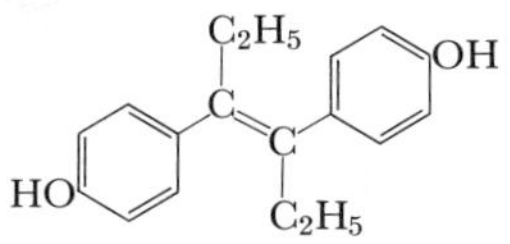

Stilbestrol

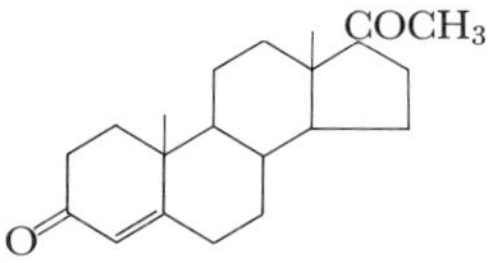

Progesterone

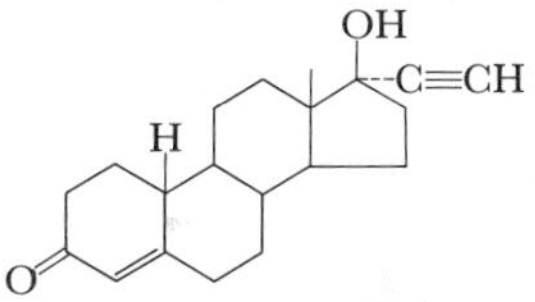

Norlutin

Progesterone is secreted by the corpus luteum (*yellow body*) formed after the expulsion of the ovum. This hormone prepares the bed of the uterus for the implantation of the fertilized ovum and suppresses further ovulation. It is used clinically to prevent abortion. Certain synthetic steroids such as 17α-ethynyl-19-nortestosterone (*Norlutin*) and the isomeric compound with the double bond in the 5,10 position (*Enovid*) are more effective in suppressing ovulation and are used as oral contraceptives. Progestins are used also to control the breeding cycle of sheep and cattle.

The **adrenal steroids** constitute another important group of hormones. The adrenals are two small glands, one above each kidney, that have two important functions, namely the secretion of epinephrine and norepinephrine (p. 459) and the secretion of cortin. Both secretions are essential to life, but the secretion of cortin is the more important because it is secreted only by the adrenals, whereas epinephrine and norepinephrine are secreted by other organs as well. A deficit of cortin leads to a bronzing of the skin, muscular weakness, increased excretion of sodium and chloride ions, and an increase in blood urea (*Addison's disease*). An excess in children produces precocious sex development. It is involved also in carbohydrate and protein metabolism.

Cortin activity resides in the steroidal fraction present in the adrenal cortex. Over forty different compounds, a few of which probably are artifacts, have been isolated from this fraction and their structures identified. Eight of them have cortin activity. The structures of three of the compounds are indicated, together with those of three synthetic products.

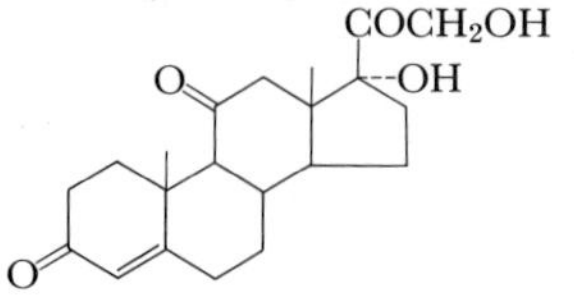

Cortisone 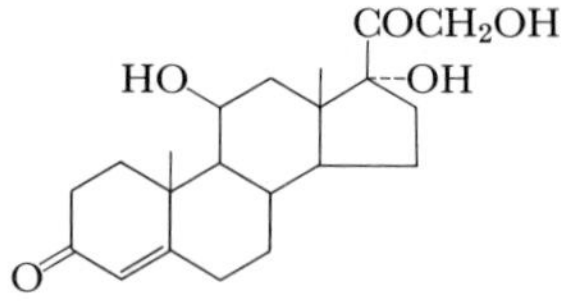   Hydrocortisone (*cortisol*) 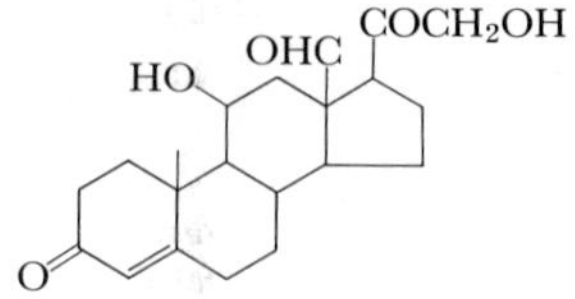   Aldosterone

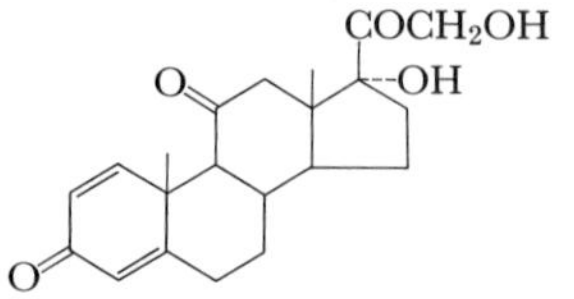

Prednisone (*unnatural*) 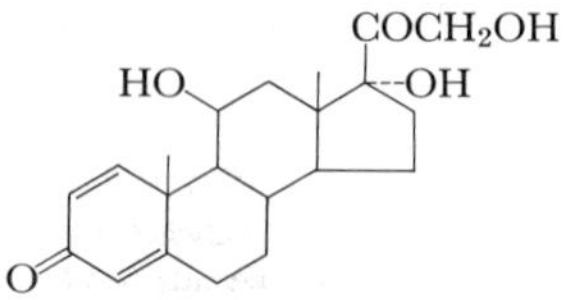   Prednisolone (*unnatural*) 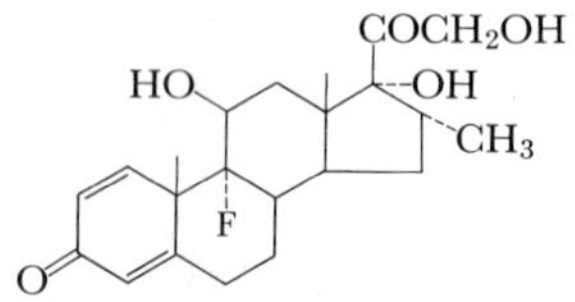   Dexamethasone (*unnatural*)

After 1948 considerable interest centered on a component called **cortisone** because of its beneficial effects in the treatment of various maladies, especially rheumatoid arthritis. **Hydrocortisone** (*cortisol*), however, is one and one half times more active and is the most important of the adrenal steroids. It is used extensively in the treatment of acute inflammation of various tissues and to relieve allergies and the effects of numerous other diseases. Many modifications in the structure of cortisone have been made, such as the introduction of unsaturation, halogen, methyl groups, and hydroxyl groups, with the object of enhancing its desirable properties or reducing undesirable action. **Prednisone** and **prednisolone,** for example, surpass cortisone in antirheumatic and antiallergic activity and cause less of the undesirable side effects of cortisone and hydrocortisone. **Dexamethasone** is seven times more effective than hydrocortisone and twenty times more effective than prednisolone as an anti-inflammatory agent and causes little increase in salt retention. **Aldosterone** has an aldehyde group at C-13, which gives rise to hemiacetal formation with the hydroxyl group at C-11. Aldosterone is from twenty to several hundred times more active than the next most active substance, deoxycorticosterone, in causing the retention of sodium, and it probably is the principal hormonal factor in maintaining electrolyte balance in the animal organism.

BIOSYNTHESIS OF ORGANIC COMPOUNDS

As the interrelationships in the structures of the terpenes and the steroids became clear, it appeared almost certain that some unifying principle accounted for their biogenesis. For many years chemists postulated various possible precursors and speculated about the steps involved in their conversion to other products. Speculation gave way to facts only after a number of developments had taken place. Enzyme systems became better understood; more selective methods for the detection and isolation of biologically produced intermediates were developed; mutant strains of microorganisms were produced that were unable to carry out a particular step in a normal synthesis and allowed an intermediate to accumulate; and isotopes of hydrogen, nitrogen, and carbon, especially radioactive carbon-14, became readily available for labeling molecules in various positions. The fundamental observation made was that the administration of labeled acetate ion to microorganisms, plants, and animals led to the appearance of isotopes in metabolic end products such as the sterols and rubber. The acetyl group in acetylcoenzyme A (*acetyl-CoA* or *$CH_3COSCoA$*, p. 534) now is accepted as the ultimate source of carbon for all of the terpenes and sterols, as well as for many other biological products. The acetyl group may be derived from acetate ion or by a series of reactions from fats or from carbohydrates. The several origins usually are referred to as the *acetate metabolic pool* since they all are interconnected by reversible processes.

The key compound in the biosynthesis of terpenes and sterols is **mevalonic acid** (p. 616). It is produced from acetyl-CoA by a series of reactions analogous to Claisen ester condensations (p. 620) that are terminated by reduction with the reduced form of nicotinamide adenine dinucleotide pyrophosphate (*NADPH,* p. 517).

$$CH_3COSCoA + CH_3COSCoA \longrightarrow HSCoA + \underset{\text{Acetoacetylcoenzyme A}}{CH_3COCH_2COSCoA} \xrightarrow{CH_3COSCoA}$$

$$HSCoA + \underset{\text{3-Hydroxy-3-methylglutarylcoenzyme A}}{CH_3{-}\overset{\overset{CH_2COOH}{|}}{\underset{\underset{CH_2COSCoA}{|}}{C}}{-}OH} \xrightarrow{2\,NADPH} HSCoA + 2\,NADP + \underset{\text{Mevalonic acid}}{CH_3{-}\overset{\overset{CH_2COOH}{|}}{\underset{\underset{CH_2CH_2OH}{|}}{C}}{-}OH}$$

Coenzyme A is not involved in the further reactions of mevalonic acid. The next step is the enzymatic conversion of mevalonic acid by means of adenosine triphosphate (*ATP*, p. 533) and metal ion cofactors into 1-hydroxy-3-methyl-3-butene pyrophosphate, commonly called **isopentenyl pyrophosphate** (*IPP*). The series of reactions involves the formation of the phosphate, the pyrophosphate, probably the pyrophosphate phosphate, and decarboxylation, thereby arriving at the five-carbon fragment responsible for the isoprene rule.

$$HOOCCH_2\overset{\overset{CH_3}{|}}{\underset{\underset{OH}{|}}{C}}CH_2CH_2OH \xrightarrow[\text{metal ions}]{\text{ATP,}} ADP + HOOC\overset{\overset{CH_3}{|}}{\underset{\underset{OH}{|}}{C}}CH_2CH_2O\overset{\overset{O^-}{|}}{\underset{\underset{OH}{|}}{P^+}}OH \xrightarrow[\text{metal ions}]{\text{ATP,}}$$

$$ADP + HOOCCH_2\overset{\overset{CH_3}{|}}{\underset{\underset{OH}{|}}{C}}CH_2CH_2O\overset{\overset{O^-}{|}}{\underset{\underset{OH}{|}}{P^+}}{-}O{-}\overset{\overset{O^-}{|}}{\underset{\underset{OH}{|}}{P^+}}OH \xrightarrow[\text{metal ions}]{\text{ATP,}} ADP + H{-}O{-}\overset{\overset{O}{\|}}{C}{-}CH_2{-}\overset{\overset{CH_3}{|}}{\underset{\underset{OPO_3H_2}{|}}{C}}CH_2CH_2OPP \longrightarrow$$

$$H_3PO_4 + CO_2 + \underset{\substack{\text{Isopentenyl}\\ \text{pyrophosphate}}}{CH_2{=}\overset{\overset{CH_3}{|}}{C}CH_2CH_2OPP} \underset{}{\overset{\text{Isomerase}}{\rightleftharpoons}} \underset{\substack{\text{3,3-Dimethylallyl}\\ \text{pyrophosphate}}}{CH_3{-}\overset{\overset{CH_3}{|}}{C}{=}CHCH_2OPP}$$

Isopentenyl pyrophosphate has a terminal methylene group whereas the terpenes usually have a terminal isopropylidene group. An enzyme is available, however, that brings about the isomerization of IPP to **3,3-dimethylallyl pyrophosphate** (*DMAPP*). Pyrophosphate ion, because of its allylic position, is lost easily from this intermediate, and nucleophilic attack by IPP gives **geranyl pyrophosphate.** The geranyl pyrophosphate so formed is an allyl pyrophosphate also and subject to a second attack by IPP to give **farnesyl pyrophosphate.**

$$(CH_3)_2C{=}CHCH_2(OPP) + CH_2{=}C(CH_3){-}CH(H)CH_2OPP \longrightarrow$$

$$HOPP + (CH_3)_2C{=}CHCH_2CH_2C(CH_3){=}CHCH_2OPP \xrightarrow{CH_2{=}C(CH_3)CH_2CH_2OPP}$$

Geranyl pyrophosphate

$$HOPP + (CH_3)_2C{=}CHCH_2CH_2C(CH_3){=}CHCH_2CH_2C(CH_3){=}CHCH_2OPP$$

Farnesyl pyrophosphate

Repetition of these condensations with IPP leads to polyisoprenes such as rubber.

The exact mechanism for the end-to-end union of two farnesyl chains to give squalene is not yet certain. Even before the early steps in the synthesis of farnesyl pyrophosphate were worked out in detail, however, the pattern of synthesis for squalene and for cholesterol had been determined by pinpointing the origin of each carbon atom. Acetate ion labeled with C^{14} in either the methyl group (*m*) or the carboxyl group (*c*) was fed to rats or incorporated in the medium used to grow animal tissue, especially rat liver. After the desired time had elapsed, the product of interest was isolated, usually after dilution with the nonradioactive compound to act as a carrier for the small amount of the metabolic product. Stepwise chemical degradation then was carried out, and the location of the C^{14} determined.

When acetate ion labeled at either C-1 or C-2 was used, the same distribution pattern was found, namely that of *I* for cholesterol and of *II* for squalene. It then was postulated that, under the influence of enzymes, the concerted electron transfers and methyl migrations indicated in *II*, together with hydration and reduction, can convert squalene to lanosterol. Later the distribution of acetate methyl and carboxyl in lanosterol was found to be that predicted by this postulation (*III*).

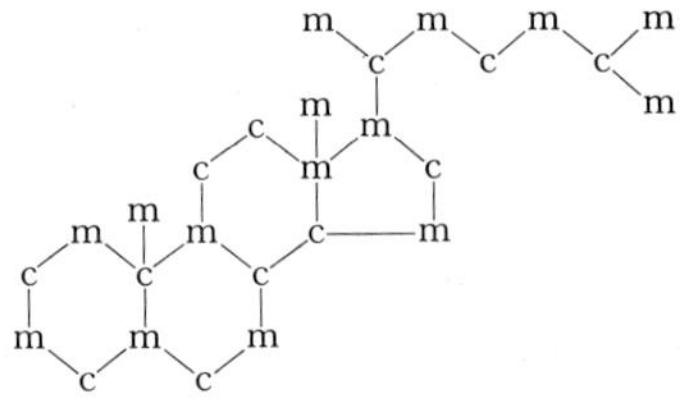

I (*Cholesterol*)

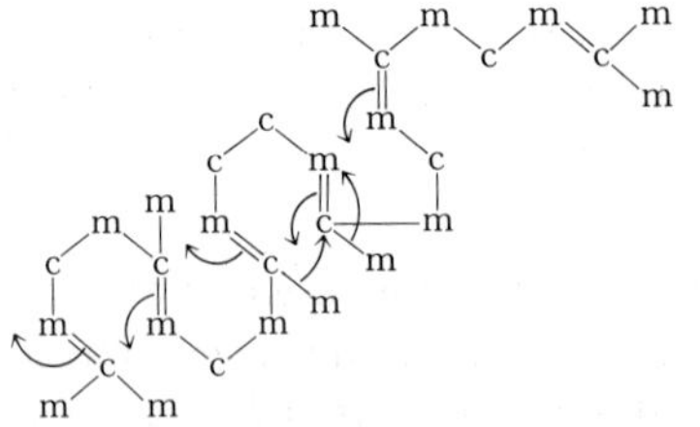

II (*Squalene*) 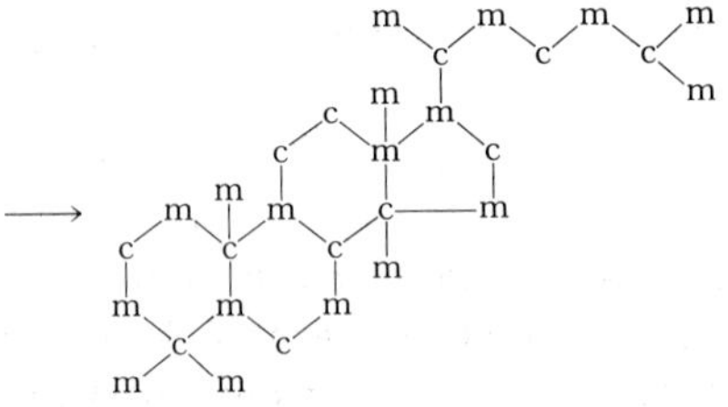 $\longrightarrow$ *III* (*Lanosterol*)

That the sequence acetate $\longrightarrow$ squalene $\longrightarrow$ lanosterol $\longrightarrow$ cholesterol is correct was indicated by feeding labeled acetate to rats and determining the relative amounts of radioactivity in the squalene, lanosterol, and cholesterol after different lengths of time.

After ten minutes, the per cent of activity in each compound was 60.5, 20, and 6.3, whereas after 60 minutes it was 1, 1.8, and 83; that is, the results are compatible with the view that the relative rates of appearance are squalene > lanosterol > cholesterol, and that cholesterol is formed at the expense of lanosterol and squalene.

Attempts to identify intermediates of greater complexity than acetate ion showed that mevalonic acid is a more efficient precursor of cholesterol. Furthermore when mevalonic acid is incubated with rat liver homogenate in the absence of molecular oxygen, squalene is formed but no cholesterol. When squalene or cholesterol from mevalonic acid labeled with C^{14} at C-2 is degraded, C^{14} is found only at the six positions indicated by the black dots in squalene (*IV*) and at five in cholesterol (*V*); that is, the carboxyl groups that were attached to these radioactive centers are lost entirely as carbon dioxide, in agreement with the mechanism for the formation of isopentenyl pyrophosphate.

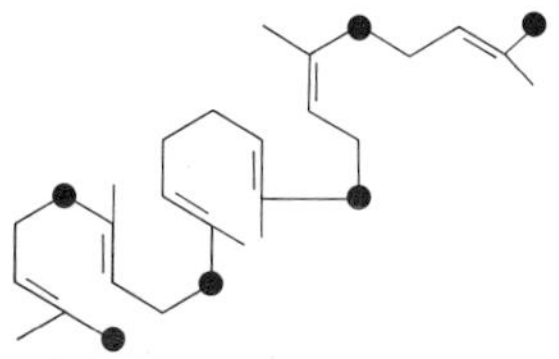

IV 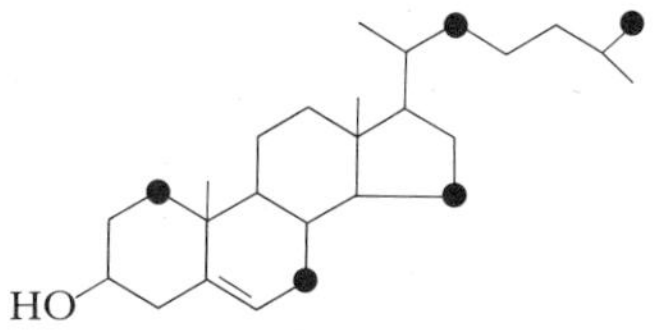 *V* 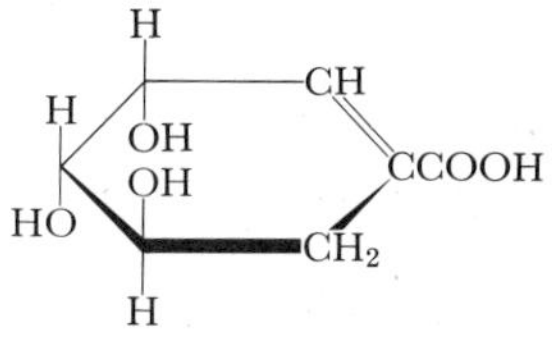 Shikimic acid

Mevalonic acid is an efficient precursor also for carotenoids and rubber.

The steps involving cyclization, methyl migrations, and methyl eliminations that lead to the triterpenes and the sterols are less well established, but reasonable concerted mechanisms that give proper stereochemistry have been proposed and are being actively investigated. All of the sex hormones, corticosteroids, and bile acids are derived from cholesterol.

Analogous procedures are being used to follow the metabolic pathways to other natural products. Fatty acids long have been known to be synthesized from acetate, and most of the intermediate steps have been established. Many aromatic compounds, including gallic acid, lignin, and the amino acids that have an aromatic ring, arise from carbohydrate by way of shikimic acid. This pathway in general holds for *ortho* dihydroxy compounds and carboxylic acids. On the other hand, *meta* dihydroxy compounds, anthraquinones, and the tropolones appear to be synthesized from acetate. Half of the molecule of flavones and of anthocyanidines (p. 577) results from the acetate route and half from the carbohydrate route. Many alkaloids are derived from amino acids. The pyrrolidine ring of nicotine is formed from glutamic acid, probably by way of putrescine (p. 612), whereas phenylalanine and tyrosine are precursors of papaverine and the morphine alkaloids (p. 535). Mevalonic acid, however, is a precursor of lysergic acid (p. 538), and shikimic acid of the strychnos alkaloids. The porphyrins (p. 510) arise from acetate and glycine by way of succinoylcoenzyme A and δ-aminolevulinic acid, $HOOCCH_2CH_2COCH_2NH_2$. Thus no single pathway accounts for any particular group of compounds. Just as a likely mechanism proposed for a particular nonenzymatic reaction must be considered only as tentative until it has been established by physical and chemical methods, so no likely pathway for the biological synthesis of a natural product is more than an assumption until all of the intermediates have been isolated and identified. The interpretation of the mode of action of the enzyme systems that bring about the biological transformations still is largely in the purely speculative stage. Recent work indicates, however, that rapid advances in this field can be expected.

Appendix

ANSWERS TO PROBLEMS

These answers are not always detailed. Enough is given for the student to be able to determine if his answer is correct, or to correct it if he is in error. Syntheses usually give only the reagents for carrying out consecutively the correct steps. Use of the text may be necessary for the student to follow the answers. Sometimes the answer given may not be the only correct answer. For example, several alternative syntheses may be possible. It is to be understood that a useful reaction need not necessarily be one that yields a single product; that where salts of weak acids or weak bases are products, the free acid or base can be obtained by addition of a strong acid or a strong base; that special conditions such as temperature or concentration may be necessary to bring about a reaction although these conditions are not always indicated. The answers to "road map" problems usually give only the structure of the initial product. The student should be able to follow the steps if he has not arrived at the correct answer.

CHAPTER 2

2–*1.* (*a*) −10.79 and −4.02 e.v. (*b*) −7.78 and −16.30 e.v.

2–*2.* H:S:H   :C:O: H:O:C:O:H (with O above C)   H:O:N:O

{:N:N:O: ⟷ :N:N:O:}   ·N:O

{O:N:O ⟷ O:N:O}   {O:N:O:N:O ⟷ O:N:O:N:O}

O:N:N:O (each N bearing an additional O) and three other resonance forms

O:N:O:N:O (each N bearing an additional O) and three other resonance forms

H:O:P:O (P bearing O–H above and H below)   :Cl:P:Cl: (P bearing Cl above)   {:O:S:O: ⟷ :O:S:O:}

{:O:S:O: ⟷ :O:S:O: ⟷ :O:S:O:} (S bearing O above)   O:S:Cl (S bearing Cl above)   O:S:O (S bearing Cl above and below)

2–*3.* (*a*) H_2SO_4, H_3PO_4, $POCl_3$, CO, $(HO)_2PH(O)$, $SOCl_2$, SO_2Cl_2 (*b*) CO_2, H_2CO_3, HNO_2 (*c*) NO, NO_2

2–*4.* (*a*) 1.0 (*b*) 13 (*c*) 22 (*d*) 64.5 (*e*) 32 (*f*) 58.5 (*g*) 10 (*h*) −24

2–*5.* (*a*) 3.25 (*b*) 1.90 (*c*) 2.20

2–*6.* (*a*) $\mathrm{N^{-}{-}^{+}H}$ (*b*) $\mathrm{N^{+}{-}^{-}F}$ (*c*) $\mathrm{B^{+}{-}^{-}H}$ (*d*) $\mathrm{B^{+}{-}^{-}Cl}$ (*e*) $\mathrm{C^{-}{-}^{+}Si}$ (*f*) $\mathrm{B^{+}{-}^{-}C}$ (*g*) $\mathrm{C^{+}{-}^{-}N}$

2–*7.* (*a*) $sp^3 < sp^2 < sp$ (*b*) sp

CHAPTER 3

3–*1.* $\Delta G = 3198$ cal., $K = 4.6 \times 10^{-3}$

3–*2.* (*a*) 1.23 (*b*) 13.82 (*c*) 4.58 (*d*) 21.1 (*e*) −0.5

3–*3.* (*a*) The first (*b*) The first

3–*4.* (*a*) The second (*b*) The second (*c*) The second

CHAPTER 4

4–*1.* (*a*) 81.7, 18.3 (*b*) 52.2, 13.1, 34.7 (*c*) 30.6, 3.9, 45.1, 20.4 (*d*) 40.0, 13.4, 46.6 (*e*) 32.0, 6.7, 18.7, 42.6 (*f*) 25.7, 6.5, 45.7, 22.1 (*g*) 30.2, 5.1, 44.5, 20.2 (*h*) 10.7, 5.4, 71.4, 12.5 (*i*) 10.4, 2.6, 87.0 (*j*) 33.2, 8.3, 32.7, 25.8

4–*2.* (*a*) 0.04, 0.04 (*b*) 0.31, 0.01 (*c*) 0.02, 0.10 4–*3.* 10.9 cc. 4–*4.* C_2H_6S

4–*5.* C_2H_4O, 88 4–*6.* 67

4–*7.* (*a*) 59.9 (*b*) 754 (*c*) 121 (*d*) 1500 (*e*) 16,900

4–*8.* C_3H_7BrO 4–*9.* $C_4H_8N_2O_2$

CHAPTER 5

5–*1.* (*a*)

```
C—C—C—C—C—C—C
```

Heptane

```
C—C—C—C—C—C
            |
            C
```

2-Methylhexane
Dimethyl-*n*-butylmethane

```
C—C—C—C—C—C
      |
      C
```

3-Methylhexane
Methylethyl-*n*-propylmethane

```
C—C—C—C—C
    |  |
    C  C
```

2,3-Dimethylpentane
Methylethyl-*i*-propylmethane

```
C—C—C—C—C
  |   |
  C   C
```

2,4-Dimethylpentane
Dimethyl-*i*-butylmethane

```
         C
         |
C—C—C—C—C
         |
         C
```

2,2-Dimethylpentane
Trimethyl-*n*-propylmethane

```
    C
    |
C—C—C—C—C
    |
    C
```

3,3-Dimethylpentane
Dimethyldiethylmethane

```
C—C—C—C—C
    |
    C—C
```

3-Ethylpentane
Triethylmethane

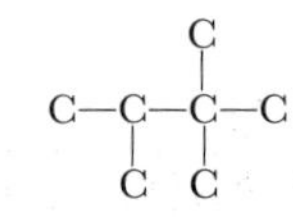

2,2,3-Trimethylbutane
Trimethyl-*i*-propylmethane

(*b*)

```
C—C—C—C—C
  |  |  |
  C  C  C
```
2,3,4-Trimethylpentane
Methyldi-*i*-propylmethane

```
         C
         |
C—C—C—C—C
      |  |
      C  C
```
2,2,3-Trimethylpentane
Trimethyl-*s*-butylmethane

```
          C
          |
C—C—C—C—C
  |       |
  C       C
```
2,2,4-Trimethylpentane
Trimethyl-*i*-butylmethane

```
C—C—C—C—C
     |  |
   C—C  C
```
2-Methyl-3-ethylpentane
Diethyl-*i*-propylmethane

```
      C
      |
C—C—C—C—C
      |  |
      C  C
```
2,3,3-Trimethylpentane
Dimethylethyl-*i*-propylmethane

```
      C
      |
C—C—C—C—C
      |
      C—C
```
3-Methyl-3-ethylpentane
Methyltriethylmethane

5–**2.** (*a*) $(CH_3)_2CH(CH_2)_6CH_3$ (*b*) $CH_3CH_2C(CH_3)(C_2H_5)CH_2CH(CH_3)_2$
(*c*) $CH_3(CH_2)_4CH(CH_2)_5CH_3$ with $CH(CH_3)CH(CH_3)_2$ on the CH (*d*) $CH_3(CH_2)_{14}CH_3$
(*e*) $CH_3CH(CH_3)CH(CH_3)CH(CH_3)CH_3$ (*f*) $CH_3(CH_2)_3CH(CH_2)_3CH_3$ with $CH_2C(CH_3)_2CH_3$ on the CH
(*g*) $CH_3(CH_2)_2CH{-}CH(CH_2)_2CH_3$ with $(CH_3)_2CH$ and $C(CH_3)_3$ on the two CH groups
(*h*) $CH_3(CH_2)_3CHCH_2CH(CH_2)_4CH_3$ with $(CH_3)_3CCH_2$ and $CH(CH_3)CH(C_2H_5)CH_2CH_3$ on the two CH groups

5–**3.** (*a*) 2-Methyldecane, isoundecane (*b*) 2,4-Dimethylpentane, di-*i*-propylmethane
(*c*) 2,2,3-Trimethylpentane, methylethyl-*t*-butylmethane
(*d*) 2-Methyl-4-ethylheptane, ethyl-*n*-propyl-*i*-butylmethane
(*e*) 3,4-Dimethyl-4-ethyloctane, methylethyl-*n*-butyl-*s*-butylmethane

5–**4.** (*a*) 1,2,3-Trimethylpentyl (*b*) 1,1,2,2-Tetramethylpropyl (*c*) 2,2,3,3-Tetramethylbutyl

5–**5.** (*a*) 5-(1,1-Dimethylpropyl)decane (*b*) 2,2,3,3-Tetramethyl-5-(2,3-dimethylbutyl)decane

5–**6.** (*a*) Longest chain, 4-ethylheptane (*b*) Smaller number, 2-methylpentane
(*c*) Numeral position, 3-methylpentane (*d*) Most branched carbon, dimethylethylmethane
(*e*) Most branched chain, 2-methyl-3-ethylhexane
(*f*) Longest side chain, 4-(1-methylpropyl)heptane

5–**7.** (*a*) 2-Methylpentyl, *I*
4-Methylpentyl, *I*
1,1-Dimethylbutyl, *III*
1,3-Dimethylbutyl, *II*
1-Ethyl-2-methylpropyl, *II*

(*b*) 2,3-Dimethyl-3-ethylpentyl, *I*
3,4-Dimethyl-3-ethylpentyl, *I*
1,1,2-Trimethyl-2-ethylbutyl, *III*
1,2,3-Trimethyl-2-ethylbutyl, *II*
2,2-Diethyl-3-methylbutyl, *I*

5–**8.** (*a*) Five (*b*) One (*c*) Four (*d*) Three (*e*) Seven

5–**9.** (*a*) 5, 1, 1, 1 (*b*) 6, 0, 0, 2 (*c*) 3, 3, 1, 0 (*d*) 6, 2, 2, 1

5–**10.** $c < e < d < b < a$ 5–**11.** (*a*) Six and one half (*b*) Nine

5–**12.** (*a*) Longer (*b*) $E^{\ddagger} = 46$ kcal. $\Delta H = +15$, -22, and -8 kcal.
(*c*) No change in the number of molecules
(*d*) $K = 4.0 \times 10^{17}$, 7.4×10^{5}, and 2.9×10^{-10} (*e*) 2.5 and 5.4 kcal.

5–**13.** 19.2% *i*-butane 5–**14.** 92.6% *i*-pentane

CHAPTER 6

6–***1.*** (***a***) C—C—C—C=C
1-Pentene

C—C—C=C—C
2-Pentene

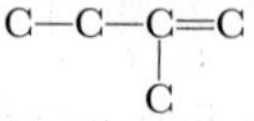
2-Methyl-1-butene

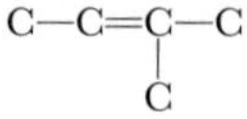
2-Methyl-2-butene

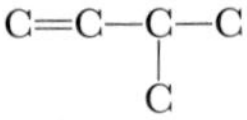
3-Methyl-1-butene

(***b***) C—C—C—C=C—C—C—C
4-Octene

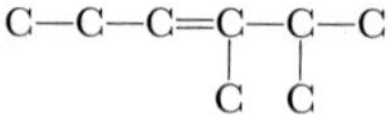
2,3-Dimethyl-3-hexene

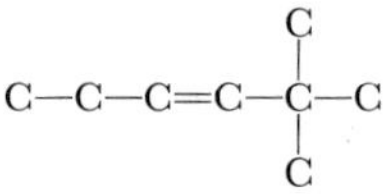
2,2-Dimethyl-3-hexene

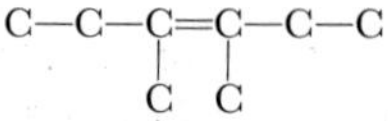
3,4-Dimethyl-3-hexene

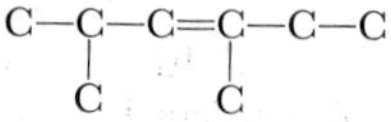
2,4-Dimethyl-3-hexene

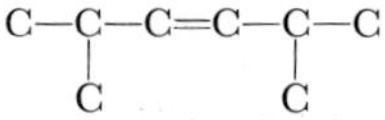
2,5-Dimethyl-3-hexene

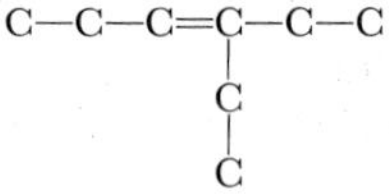
3-Ethyl-3-hexene

6–***2.*** (***a***) $CH_3CH_2CH{=}CHC(CH_3)_2CH_3$ (***b***) $CH_3CH_2CH(CH_3)CH{=}CH_2$
(***c***) $C_2H_5CH{=}CHC(CH_3)_3$ (***d***) 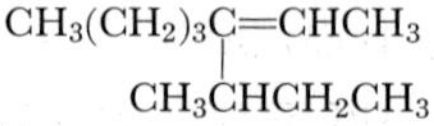 (***e***) 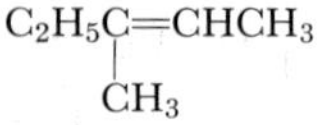

6–***3.*** (***a***) 4,4-Dimethyl-2-pentene, *sym*-methyl-*t*-butylethylene
(***b***) 5-Methyl-3-heptene, *sym*-ethyl-*s*-butylethylene
(***c***) 2,4-Dimethyl-3-ethyl-2-pentene, (*unsym*-dimethyl)ethyl-*i*-propylethylene
(***d***) 3,4-Dimethyl-3-nonene, (*sym*-dimethyl)ethyl-*n*-amylethylene

6–***4.*** (***a***) Longest chain, 5-methyl-3-heptene
(***b***) Smaller number for function, 5-methyl-3-heptene (***c***) Two isomers, use *sym* or *unsym*
(***d***) Number precedes parent name, 2,3-dimethyl-2-pentene
(***e***) Mixed systems, *i*-propylethylene or 3-methyl-1-butene

6–***5.*** (***a***) $CH_3CH_2CHClCHClCH_3$, 2,3-dichloropentane
(***b***) $CH_3CH_2CH_2CH_2CH(OSO_3H)CH_3$, 1-methylpentyl hydrogen sulfate
(***c***) $CH_3CH_2CH_2CCl(CH_3)_2$, 2-chloro-2-methylpentane
(***d***) $(C_2H_5)_3CH$, triethylmethane (***e***) $(CH_3)_2CHCH_2CH_2CH_2Br$, isohexyl bromide
(***f***) $[CH_3CH_2CH(CH_3)CH_2CH_2]_3B$, tris(3-methylpentyl)borane

6–***6.*** (***a***) Any 2,3- or 3,4-dimethylpentene + H_2/Pt (***b***) 2-Methyl-1(or 2)-butene + HCl
(***c***) 2,3-Dimethyl-1(or 2)-butene + H_2SO_4 (***d***) 2-Methyl-2-hexene + Br_2
(***e***) 2-Methyl-1-pentene + HBr + peroxides

6–***7.*** (***a***) $CH_2{=}C(CH_3)_2$ (***b***) $CH_3CH{=}C(CH_3)C_2H_5$ (***c***) $C_2H_5C(CH_3){=}C(CH_3)C_2H_5$
(***d***) $CH_3CH{=}CHCH(CH_3)_2$ (***e***) $(CH_3)_2C{=}C(CH_3)C_2H_5$

6–***8.*** (***a***) 2,4-Dimethyl-3-hexene (***b***) 3,3-Dimethyl-1-butene (***c***) 3-Hexene
(***d***) 2,3,4-Trimethyl-2-pentene

6–***9.*** $(CH_3)_2C{=}C(C_2H_5)_2$, $(CH_3)_2C{=}C(CH_3)CH(CH_3)_2$, and $(CH_3)_2C{=}C(CH_3)CH_2CH_2CH_3$

6–***10.*** *a4*, *b3*, *c7*, *d6*, *e5*, *f1*, *g2* 6–***11.*** $c > b > a > e > d$ 6–***12.*** $d < a < b < e < c$

6–***13.*** $b < d < a < c$ 6–***14.*** 80 6–***15.*** 57 6–***16.*** C_7H_{14}

6–*17.* Three, C_7H_{10} 6–*18.* 60% 6–*19.* 49% 6–*20.* 135.6 liters

6–*21.* (*a*) $CH_3CH{=}CH_2 \xrightarrow{H^+} CH_3\overset{+}{C}HCH_3 \xrightarrow{(CH_3)_2CHOH}$

$$(CH_3)_2CH{-}\underset{H}{\overset{+}{O}}{-}CH(CH_3)_2 \longrightarrow (CH_3)_2CH{-}O{-}CH(CH_3)_2 + H^+$$

(*b*) $(CH_3)_2C{=}CH_2 \xrightarrow{H^+} (CH_3)_3C^+ \xrightarrow{HOOH} (CH_3)_3C{-}\underset{H}{\overset{+}{O}}{-}OH \longrightarrow$

$$(CH_3)_3C{-}OOH + H^+$$

6–*22.* 4,6,8-Trimethyl-1-nonene and 4,6,8-trimethyl-2-nonene

6–*23.* $(CH_3)_3\overset{+}{C} + CH_2{=}C(CH_3)_2 \longrightarrow (CH_3)_3CCH_2\overset{+}{C}(CH_3)_2 \xrightarrow{(CH_3)_3CH} (CH_3)_3CCH_2CH(CH_3)_2 + (CH_3)_3C^+$

$(CH_3)_3CCH_2\overset{+}{C}(CH_3)_2 \xrightarrow{H^-\ \text{shift}} (CH_3)_3C\overset{+}{C}HCH(CH_3)_2$

$(CH_3)_3C\overset{+}{C}HCH(CH_3)_2 \xrightarrow{CH_3^-\ \text{shift}} (CH_3)_2\overset{+}{C}CH(CH_3)CH(CH_3)_2 \xrightarrow{(CH_3)_3CH} (CH_3)_2CHCH(CH_3)CH(CH_3)_2 + (CH_3)_3C^+$

$(CH_3)_3C\overset{+}{C}HCH(CH_3)_2 \xrightarrow{CH_3^-\ \text{shift}} (CH_3)_3CCH(CH_3)\overset{+}{C}HCH_3 \xrightarrow{(CH_3)_3CH} (CH_3)_3CCH(CH_3)CH_2CH_3 + (CH_3)_3C^+$

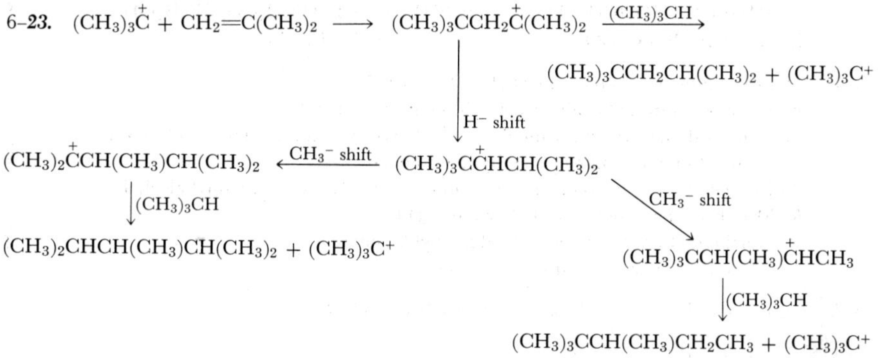

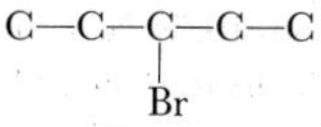

CHAPTER 7

7–*1.*

C—C—C—C—CBr — *n*-Pentyl bromide, 1-Bromopentane

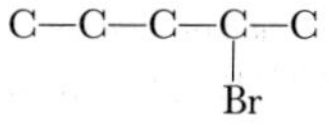

C—C—C—C(Br)—C — 2-Bromopentane

C—C—C(Br)—C—C — 3-Bromopentane

C—C(C)—C—CBr — *i*-Pentyl bromide, 1-Bromo-3-methylbutane

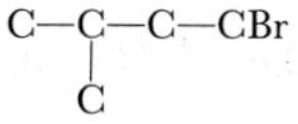

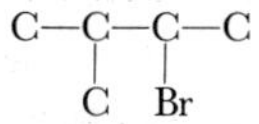

C—C(C)—C(Br)—C — 2-Bromo-3-methylbutane

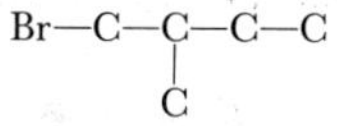

Br—C—C(C)—C—C — 1-Bromo-2-methylbutane

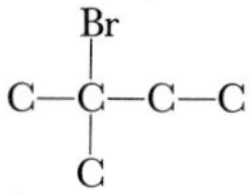

C—C(Br)(C)—C—C — *t*-Pentyl bromide, 2-Bromo-2-methylbutane

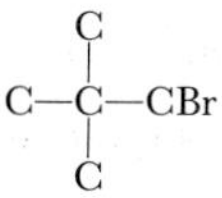

C—C(C)(C)—CBr — Neopentyl bromide, 1-Bromo-2,2-dimethylpropane

7–*2.* 2 *I*, 3 *II*, 1 *III* 7–*4.* $e < b < a < d < c$ 7–*5.* $d < a < e < b < c$

7–*6.* Tertiary halides hydrolyze easily. 7–*7.* $I^{\pm}{=}Cl$, $CH_3CH_2CHClCH_2I$

7–*8.* (*a*) H_2O or ROH (*b*) O_2 (*c*) I_2 (*d*) CO_2 (*e*) $HgCl_2$ (*f*) PCl_3

7–*9.* (*a*) Free radical (*b*) S_N2 (*c*) S_N1 (*c*) *E*1 (*e*) *E*2 7–*10.* Ten

7–*11.* 49.8 cc. 7–*12.* Two

CHAPTER 8

8–*1.*

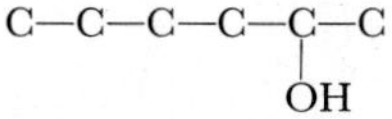

C—C—C—C—C(OH)—C — 2-Hexanol

C—C—C—C(OH)—C—C — 3-Hexanol

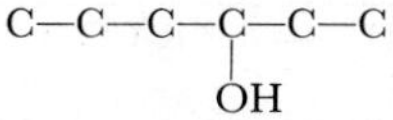

C—C—C(HO)—C(C)—C — 2-Methyl-3-pentanol

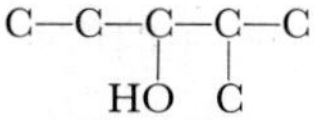

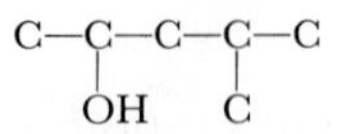

```
                                                        C
                                                        |
C—C—C—C—C             C—C—C—C—C            C—C—C—C
  |     |                   |  |                |  |
  OH    C                   C  OH               HO C
4-Methyl-2-pentanol   3-Methyl-2-pentanol   3,3-Dimethyl-2-butanol
```

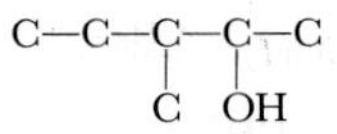

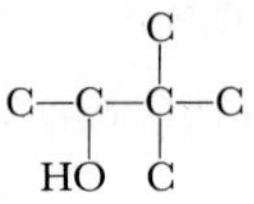

8–**2.** (***a***) Five (***b***) Seven (***c***) Three (***d***) Three (***e***) Three

8–**3.** (***a***) 3, 1, 1 (***b***) 2, 4, 1 (***c***) 2, 1, 0 (***d***) 1, 2, 0 (***e***) 1, 1, 1

8–**4.** (***a***) $CH_3CH_2CH_2CH_2CH_2OH$ (***b***) $C_2H_5C(CH_3)_2OH$ (***c***) $CH_3CH_2CHOHC(CH_3)_3$
(***d***) $(CH_3)_2CH(CH_2)_2CH_2OH$ (***e***) $CH_3CH_2CH(CH_3)C(C_2H_5)_2OH$
(***f***) $CH_3(CH_2)_2CH(C_2H_5)CHOHCH_3$ (***g***) $(CH_3)_3CCH_2OH$
(***h***) $(CH_3)_2CHCHOHCH(CH_3)_2$ (***i***) $(CH_3)_2CHCH_2C[CH(CH_3)_2]CH_2OH$
(***j***) $C_2H_5CH(CH_3)CH_2OH$ 8–**5.** *b* and *f*

8–**6.** (***a***) Mixed systems, isopropyl alcohol or 2-propanol
(***b***) Use most branched chain, 4-methyl-3-ethyl-1-pentanol
(***c***) One word, methylethylcarbinol (***d***) Mixed systems, 2,4-dimethyl-2-hexanol
(***e***) Numeral precedes name, 2-methyl-1-hexanol
(***f***) Not definitive, 2-pentanol or 3-pentanol (***g***) Two words, *i*-amyl alcohol
(***h***) Mixed systems, *s*-butyl alcohol or 2-butanol
(***i***) Numbered from wrong end, 2,2-dimethyl-3-pentanol
(***j***) Use longest chain, 3-methyl-2-pentanol

8–**7.** $c < d < e < b < a$ 8–**8.** $c > b > a > e > d > f$

8–**9.** $b < g < a < d < e < f < c$

8–**10.** (***a***) 1- or 2-Pentene, H_2SO_4, H_2O (***b***) 2-Methyl-1(or 2)-pentene
(***c***) 2-Methyl-1(or 2)-butene (***d***) 3-Ethyl-2-pentene

8–**11.** (***a***) Na, then RCH_2X. *n*-Nonyl alkyl ether or 1-alkoxynonane
(***b***) H_2SO_4, heat. $(CH_3)_2CHCH{=}CHCH_3$, 4-methyl-2-pentene
(***c***) HNO_2. 2,3-Dimethylpentyl nitrite

8–**12.** (***a***) H_2SO_4, heat $\longrightarrow$ 2-butene $\xrightarrow{\text{Cold } H_2SO_4}$
s-butyl hydrogen sulfate $\xrightarrow{H_2O}$ *s*-butyl alcohol
(***b***) H^+, heat (***c***) H_2SO_4, heat, then HCl
(***d***) H_2SO_4, heat $\longrightarrow$
propene, then HBr + peroxides followed by AgOH or BH_3 followed by alkaline H_2O_2

8–**13.** (***a***) $3 < 4 < 2 < 5 < 1$ (***b***) $1 < 5 < 2 < 4 < 3$

8–**14.** (***a***) *n*-Butyl phosphate (***b***) Ethyl chloride (***c***) Methanol

8–**15.** (***a***) *i*-Propyl alcohol slowly reacts with Lucas reagent.
(***b***) Ethyl nitrite gives oxides of nitrogen with mineral acid.
(***c***) Aluminum alkoxides react with water and give insoluble aluminum hydroxide.
(***d***) Alcohols evolve H_2 with Na. (***e***) Alkyl sulfates hydrolyze to give sulfate ions.
(***f***) Test for halogen. (***g***) Alkenes decolorize bromine in CCl_4.

8–**16.** 74 8–**17.** *a3*, *b4*, *c1*, *d5*, *e2*

8–**18.** 54.2 g. of alcohol, 112 g. of sodium bromide

8–**19.**

$$
\begin{array}{rcl}
Cr_2O_7^{2-} + H_2O + 2\,H^+ & \longrightarrow & 2\,H_2CrO_4 \\
2\,R_2CHOH + 2\,H_2CrO_4 & \longrightarrow & 2\,R_2CO + 2\,H_2CrO_3 + 2\,H_2O \\
2\,H_2CrO_3 & \longrightarrow & HCrO_3 + HCrO_2 + H_2O \\
R_2CHOH + HCrO_3 & \longrightarrow & R_2CO + HCrO_2 + H_2O \\
2\,HCrO_2 + 6\,H^+ & \longrightarrow & 2\,Cr^{3-} + 4\,H_2O \\
\hline
Cr_2O_7^{2-} + 8\,H^+ + 3\,R_2CHOH & \longrightarrow & 3\,R_2CO + 7\,H_2O + 2\,Cr^{3-}
\end{array}
$$

8–**20.** $(RO)_2BOR + H_2O \longrightarrow (RO)_2\overset{-}{B}{-}OR$ (with $H{-}\overset{+}{O}{-}H$ attached to B) $\longrightarrow (RO)_2BOH + HOR$

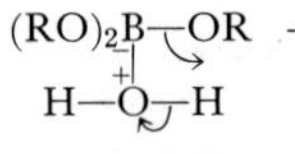

CHAPTER 9

9–*1.* Three ethers and four alcohols

9–*2.* Sodium ethoxide and *n*-propyl halides, or sodium *n*-propoxide and ethyl halides

9–*3.* (*a*) C_2H_5Cl and *n*-C_3H_7OH (*b*) C_2H_5Br and *n*-C_3H_7OH

9–*4.* $AlCl_3$, HBr, HI, warm H_2SO_4, and Al_2O_3 at 300° 9–*5.* *c*

9–*6.* (*a*) Sodium and Lucas reagent (*b*) Bromine and conc. aq. HCl (ether dissolves)
(*c*) Sodium iodide in acetone and aqueous silver nitrate
(*d*) Cold conc. H_2SO_4 and sodium fusion test for halogen

9–*7.* (*a*) 35.2% (*b*) 88 9–*8.* Methyl *t*-butyl ether and *t*-amyl alcohol

9–*9.* Ethyl ether and *n*-butyl alcohol

CHAPTER 10

10–*1.* (*a*) 2 *i*-PrBr + 2 Na (*b*) 1,2-Dibromo-3-methylpentane + Zn
(*c*) *n*-Heptyl bromide + Mg, then H_2O (*d*) *t*-BuOH + H^+ and heat

10–*2.*

C—C—C—C—C≡C
1-Hexyne
n-Butylacetylene

C—C—C—C≡C—C
2-Hexyne
Methyl-*n*-propylacetylene

C—C—C≡C—C—C
3-Hexyne
Diethylacetylene

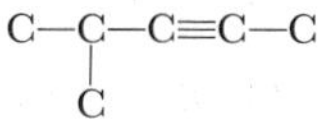

C—C(C)—C≡C—C
4-Methyl-2-pentyne
Methyl-*i*-propylacetylene

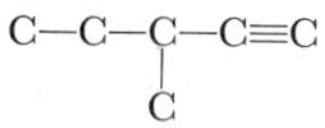

C—C—C(C)—C≡C
3-Methyl-1-pentyne
s-Butylacetylene

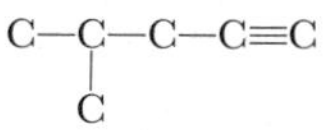

C—C(C)—C—C≡C
4-Methyl-1-pentyne
i-Butylacetylene

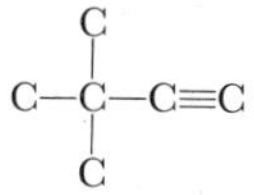

C—C(C)(C)—C≡C
Dimethylbutyne
t-Butylacetylene

10–*3.*

H—C≡C, 180°	C≡C—C, 180°	C—C—H, 109.5°
C—C—C, 109.5°	C—C=C, 120°	H—C=C, 120°
C=C—C, 120°	C—C—O, 109.5°	C—O—H, 105°

10–*4.* $c < d < a < b$

10–*5.* (*a*) HC≡CH + $NaNH_2$, then *n*-PrBr (*b*) EtC≡CH + 2 HBr (*c*) H_2/Pd
(*d*) 2 $NaNH_2$, then MeI (*e*) Br_2, then alc. KOH

10–*6.* (*a*) Trace H_2SO_4, heat (*b*) H_2SO_4, heat, then H_2/Pt
(*c*) Hot soda-lime, then Br_2 followed by alc. KOH (*d*) HBr, then Na
(*e*) Alc. KOH, then 2 HBr (*f*) Alc. KOH, H_2/Pd, Br_2

10–*7.* *c* 10–*8.* Four

10–*9.* (*a*) Fusion test for halogen and $Ag(NH_3)_2NO_3$ or Br_2 (*b*) Br_2 and $Ag(NH_3)_2NO_3$
(*c*) Br_2 and $Ag(NH_3)_2NO_3$

10–*10.* (*a*) Absorb acetylene in $Ag(NH_3)_2NO_3$ and liberate with HNO_3, absorb ethylene in Br_2 and liberate with Zn.
(*b*) Absorb 1-butyne in $Ag(NH_3)_2NO_3$ and 2-butyne in Br_2 and liberate as in (*a*).
(*c*) Same as (*b*).
(*d*) Absorb ethylacetylene in $Ag(NH_3)_2NO_3$, allow remainder to react with excess bromine and separate dibromide from tetrabromide by distillation, liberate as in (*a*).

10–*11.* $\Delta H = -11$ kcal.

CHAPTER 11

11–*1.*

C—C—C—C—C—COOH
Hexanoic acid
Caproic acid

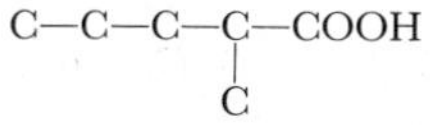

C—C—C—C(C)—COOH
2-Methylpentanoic acid
Methyl-*n*-propylacetic acid

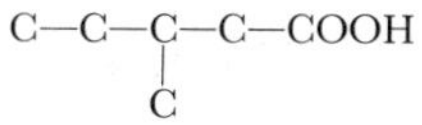

C—C—C(C)—C—COOH
3-Methylpentanoic acid
s-Butylacetic acid

```
C—C—C—C—COOH          C—C—C—COOH          C—C—C(C)(C)—COOH
  |                     |  |
  C                     C  C
```

4-Methylpentanoic acid, Isocaproic acid; 2,3-Dimethylbutanoic acid, Methyl-*i*-propylacetic acid; 2,2-Dimethylbutanoic acid, Dimethylethylacetic acid

```
C—C(C)(C)—C—COOH          C—C—C—COOH
                                |
                                C—C
```

3,3-Dimethylbutanoic acid, *t*-Butylacetic acid; 2-Ethylbutanoic acid, Diethylacetic acid

11–2. $CH_3C(=O)—OH$ (755) $O=CHCH_2OH$ (755) $H_3C—O—CH(=O)$ (744)

$CH_2—CHOH$ (epoxide, ring O) (743) $CH_2=C(OH)_2$ (735) $HOCH=CHOH$ (735)

$CH_2—O—CH_2—O$ (four-membered ring) (732) $CH_2—CH_2—O—O$ (four-membered ring) (680) $O—CHCH_3$ (ring with O) (680)

11–3. *a* < *c* < *b* < *d*

11–4. (***a***) $3\ CH_3CH_2CH{=}O_3{=}CHCH_3 + Na_2Cr_2O_7 + 4\ H_2SO_4 \longrightarrow 3\ CH_3CH_2COOH + 3\ HOOCCH_3 + Na_2SO_4 + Cr_2(SO_4)_3 + 4\ H_2O$

(***b***) $3\ (CH_3)_2CHCH_2OH + 4\ KMnO_4 \longrightarrow 3\ (CH_3)_2CHCOOH + 4\ KOH + 4\ MnO_2 + H_2O$

(***c***) $CH_3(CH_2)_6CH_2OH + 4\ HNO_3 \longrightarrow CH_3(CH_2)_6COOH + 4\ NO_2 + 3\ H_2O$

11–5. (***a***) *4* (***b***) *5* and *6* by HBr, *1* by HBr + peroxides

11–6. *f* < *d* < *e* < *c* < *g* < *a* < *i* < *h* < *b*

11–7. (***a***) $CH_3(CH_2)_5COOH + PBr_3$

(***b***) $CH_3(CH_2)_2COOH + (CH_3CO)_2O$; $CH_3(CH_2)_2COCl + NaOCO(CH_2)_2CH_3$

(***c***) $(CH_3)_2CH(CH_2)_2MgBr + CO_2$; $(CH_3)_2CH(CH_2)_2CN + H_2O + H^+$

(***d***) $CH_3(CH_2)_3COONH_4$ + heat; $CH_3(CH_2)_3COCl$, $[CH_3(CH_2)_3CO]_2O$, or $CH_3(CH_2)_3COOCH_3 + NH_3$

(***e***) $CH_3CH_2CHOHCH_3 + CH_3CH_2COOH$ and H^+; $(CH_3CH_2CO)_2O$ or CH_3CH_2COCl

(***f***) $CH_3COOH + Ca(OH)_2$ or CaO

11–8. Five **11–9.** *d* **11–*10*.** *b* **11–*12*.** *e* > *c* > *b* > *a* > *d*

11–*13*. (***a***) H_2O, NaOH (***b***) H_2O, fusion test for halogen

(***c***) Acidity, Na (***d***) H_2O, aq. NaOH (***e***) Acid indicator or aq. NaOH, Na

(***f***) Conc. H_2SO_4, aq. NaOH, Br_2

11–*14*. *a4*, *b5*, *c3*, *d2*, *e1*

11–*15*. (***a***) NaCN, then H_2O, H^+, or Mg, then CO_2

(***b***) $Na_2Cr_2O_7 + H_2SO_4$, then C_2H_5OH, H^+

(***c***) Na, C_2H_5OH, or $LiAlH_4$, or H_2/Ni, then HBr

(***d***) $Na_2Cr_2O_7$, H_2SO_4, convert to Na salt and electrolyze

(***e***) C_2H_5OH, H^+, then Na, C_2H_5OH, or $LiAlH_4$, or H_2/Ni

(***f***) Reduction, then dehydration over neutral Al_2O_3

(***g***) Alcohol ⟶ acid ⟶ silver salt, then Br_2 (***h***) $LiAlH_4$, PI_3

(***i***) Esterify, then excess C_2H_5MgBr (***j***) HBr, Mg, CO_2

(***k***) $Na_2Cr_2O_7 + H_2SO_4$, then ThO_2 at 300° (***l***) $KMnO_4$, H^+, then $Br_2 + PBr_3$

(***m***) PCl_3, then H_2O_2 + NaOH (***n***) $H_2O_2 + H_2SO_4$, then CH_3COOH, H^+

11–*16*. (***a***) −0.82 kcal. (***b***) 97% **11–*17*.** 102 g. **11–*18*.** 60

11–*19*. *n*-Propyl and *i*-propyl propionate **11–*20*.** 9.2

11–**21.** Litmus changes color on either side of neutrality. Should acidify to *p*H 1 to ensure complete liberation of the carboxylic acid.

11–**22.** An anhydride

11–**23.** It is the ethyl ester of one of the four C_5 saturated carboxylic acids.

11–**24.** $CH_3(CH_2)_7CH{=}CH(CH_2)_7COOH$

11–**25.** (*a*) C_3H_5O (*b*) $C_6H_{10}O_2$, 114
(*c*) $(CH_3)_2CHCOOH$ and $(CH_3)_2CHCH{=}CHCOOH$

11–**26.** (*a*) $R{-}C({=}O){-}O{-}C(CH_3)_3 + HBr \longrightarrow R{-}C({=}\overset{+}{O}H){-}O{-}C(CH_3)_3 + {:}Br^- \longrightarrow RC(OH){=}O + BrC(CH_3)_3$

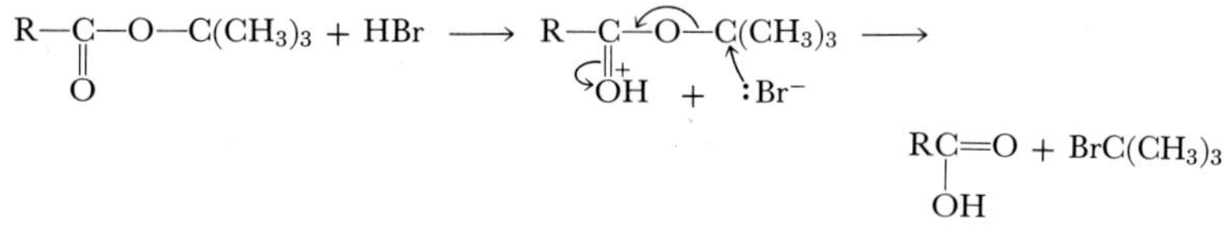

(*b*) $RC({=}O){-}OR' \underset{NH_3}{\overset{NH_4^+}{\rightleftharpoons}} RC({=}\overset{+}{O}H){-}OR' \overset{:NH_3}{\rightleftharpoons} RC(\overset{+}{N}H_3)(OH){-}OR' \rightleftharpoons RC(H_2N)(O{-}H){-}\overset{+}{O}(H){-}R' \rightleftharpoons$

$RC({=}O){-}NH_2 + HOR' + H^+ \xrightarrow{NH_3} NH_4^+$

(*c*) $R{-}C(OR')_2({:}\ddot{O}{-}R') \xrightarrow{MgX_2} R{-}C(OR')_2(X_2Mg{:}\ddot{O}{-}R') \xrightarrow{R''{-}MgX}$

$R{-}C(R'')(OR')_2 + \underbrace{{}^+MgX + X_2\bar{M}gOR'}_{MgX_2 + XMgOR'}$

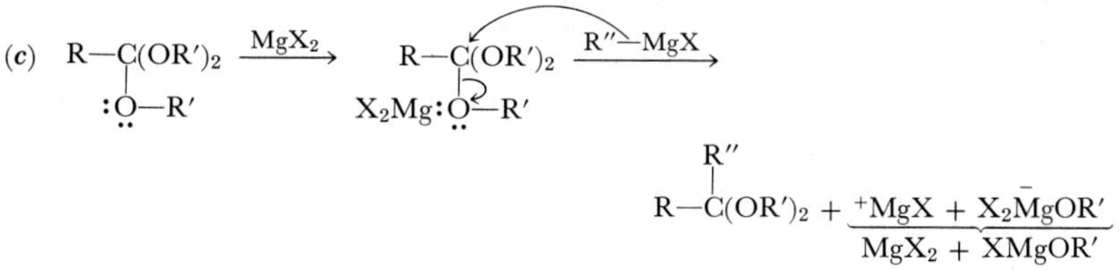

CHAPTER 12

12–**1.** *c* 12–**2.** (*a*) *6* (*b*) *4* (*c*) *1* and *5* (*d*) *6* (*e*) *5* (*f*) *4*

12–**3.** (*a*) *2* (*b*) *4* 12–**4.** 39.7 liters

12–**5.** (*a*) Palmitoleic acid (*b*) Compare IR spectrum, thin-layer chromatogram, and melting point of a solid derivative with those of an authentic sample.

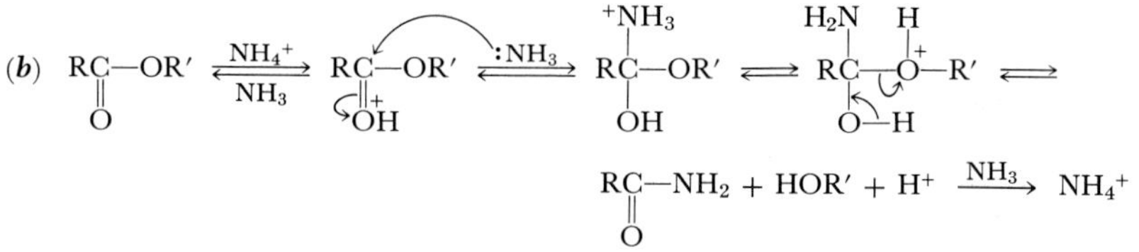

12–**6.** 1.39 12–**7.** (*a*) 1.07 (*b*) 10.3 12–**8.** (*a*) 41.1 (*b*) 128

CHAPTER 13

13–**1.** (*a*)

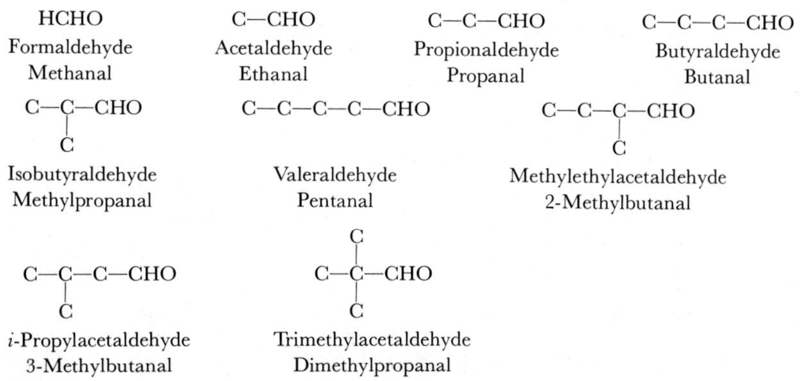

HCHO — Formaldehyde, Methanal

C—CHO — Acetaldehyde, Ethanal

C—C—CHO — Propionaldehyde, Propanal

C—C—C—CHO — Butyraldehyde, Butanal

C—C(C)—CHO — Isobutyraldehyde, Methylpropanal

C—C—C—C—CHO — Valeraldehyde, Pentanal

C—C—C(C)—CHO — Methylethylacetaldehyde, 2-Methylbutanal

C—C(C)—C—CHO — *i*-Propylacetaldehyde, 3-Methylbutanal

C—C(C)(C)—CHO — Trimethylacetaldehyde, Dimethylpropanal

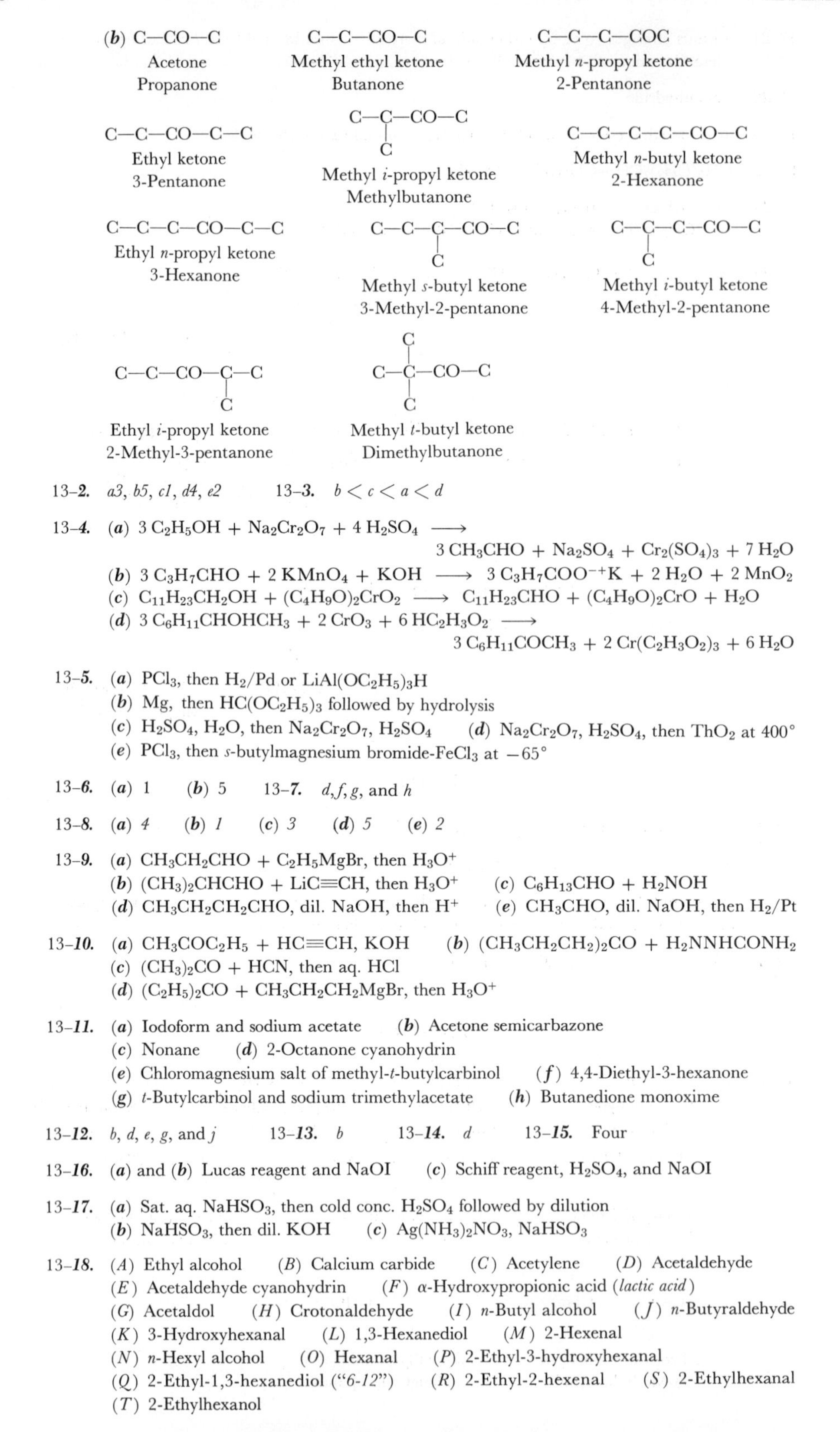

(*b*) C—CO—C
Acetone
Propanone

C—C—CO—C
Methyl ethyl ketone
Butanone

C—C—C—COC
Methyl *n*-propyl ketone
2-Pentanone

C—C—CO—C—C
Ethyl ketone
3-Pentanone

C—C(C)—CO—C
Methyl *i*-propyl ketone
Methylbutanone

C—C—C—C—CO—C
Methyl *n*-butyl ketone
2-Hexanone

C—C—C—CO—C—C
Ethyl *n*-propyl ketone
3-Hexanone

C—C—C(C)—CO—C
Methyl *s*-butyl ketone
3-Methyl-2-pentanone

C—C(C)—C—CO—C
Methyl *i*-butyl ketone
4-Methyl-2-pentanone

C—C—CO—C(C)—C
Ethyl *i*-propyl ketone
2-Methyl-3-pentanone

C—C(C)(C)—CO—C
Methyl *t*-butyl ketone
Dimethylbutanone

13–**2.** *a3, b5, c1, d4, e2*   13–**3.** $b < c < a < d$

13–**4.** (*a*) $3\ C_2H_5OH + Na_2Cr_2O_7 + 4\ H_2SO_4 \longrightarrow 3\ CH_3CHO + Na_2SO_4 + Cr_2(SO_4)_3 + 7\ H_2O$
(*b*) $3\ C_3H_7CHO + 2\ KMnO_4 + KOH \longrightarrow 3\ C_3H_7COO^{-+}K + 2\ H_2O + 2\ MnO_2$
(*c*) $C_{11}H_{23}CH_2OH + (C_4H_9O)_2CrO_2 \longrightarrow C_{11}H_{23}CHO + (C_4H_9O)_2CrO + H_2O$
(*d*) $3\ C_6H_{11}CHOHCH_3 + 2\ CrO_3 + 6\ HC_2H_3O_2 \longrightarrow 3\ C_6H_{11}COCH_3 + 2\ Cr(C_2H_3O_2)_3 + 6\ H_2O$

13–**5.** (*a*) PCl_3, then H_2/Pd or $LiAl(OC_2H_5)_3H$
(*b*) Mg, then $HC(OC_2H_5)_3$ followed by hydrolysis
(*c*) H_2SO_4, H_2O, then $Na_2Cr_2O_7$, H_2SO_4 (*d*) $Na_2Cr_2O_7$, H_2SO_4, then ThO_2 at 400°
(*e*) PCl_3, then *s*-butylmagnesium bromide-$FeCl_3$ at $-65°$

13–**6.** (*a*) 1   (*b*) 5   13–**7.** *d, f, g*, and *h*

13–**8.** (*a*) *4*   (*b*) *1*   (*c*) *3*   (*d*) *5*   (*e*) *2*

13–**9.** (*a*) $CH_3CH_2CHO + C_2H_5MgBr$, then H_3O^+
(*b*) $(CH_3)_2CHCHO + LiC{\equiv}CH$, then H_3O^+ (*c*) $C_6H_{13}CHO + H_2NOH$
(*d*) $CH_3CH_2CH_2CHO$, dil. NaOH, then H^+ (*e*) CH_3CHO, dil. NaOH, then H_2/Pt

13–**10.** (*a*) $CH_3COC_2H_5 + HC{\equiv}CH$, KOH   (*b*) $(CH_3CH_2CH_2)_2CO + H_2NNHCONH_2$
(*c*) $(CH_3)_2CO + HCN$, then aq. HCl
(*d*) $(C_2H_5)_2CO + CH_3CH_2CH_2MgBr$, then H_3O^+

13–**11.** (*a*) Iodoform and sodium acetate   (*b*) Acetone semicarbazone
(*c*) Nonane   (*d*) 2-Octanone cyanohydrin
(*e*) Chloromagnesium salt of methyl-*t*-butylcarbinol   (*f*) 4,4-Diethyl-3-hexanone
(*g*) *t*-Butylcarbinol and sodium trimethylacetate   (*h*) Butanedione monoxime

13–**12.** *b, d, e, g*, and *j*   13–**13.** *b*   13–**14.** *d*   13–**15.** Four

13–**16.** (*a*) and (*b*) Lucas reagent and NaOI   (*c*) Schiff reagent, H_2SO_4, and NaOI

13–**17.** (*a*) Sat. aq. $NaHSO_3$, then cold conc. H_2SO_4 followed by dilution
(*b*) $NaHSO_3$, then dil. KOH   (*c*) $Ag(NH_3)_2NO_3$, $NaHSO_3$

13–**18.** (*A*) Ethyl alcohol   (*B*) Calcium carbide   (*C*) Acetylene   (*D*) Acetaldehyde
(*E*) Acetaldehyde cyanohydrin   (*F*) α-Hydroxypropionic acid (*lactic acid*)
(*G*) Acetaldol   (*H*) Crotonaldehyde   (*I*) *n*-Butyl alcohol   (*J*) *n*-Butyraldehyde
(*K*) 3-Hydroxyhexanal   (*L*) 1,3-Hexanediol   (*M*) 2-Hexenal
(*N*) *n*-Hexyl alcohol   (*O*) Hexanal   (*P*) 2-Ethyl-3-hydroxyhexanal
(*Q*) 2-Ethyl-1,3-hexanediol ("*6-12*")   (*R*) 2-Ethyl-2-hexenal   (*S*) 2-Ethylhexanal
(*T*) 2-Ethylhexanol

13–**19.** Petroleum or Natural gas $\xrightarrow{\text{Cracking}}$ (*A*) $\xrightarrow{H_2O,\ H^+}$ (*B*)

(*B*) $\xrightarrow{\text{Cu-Zn},\ 325°}$ (*C*) (+ H_2) or (*B*) $\xrightarrow{\text{Air},\ 450°}$ (*C*) (+ H_2O_2)

Carbohydrates $\xrightarrow{\textit{Clostridium}}$ (*C*) (+ H_2)

(*C*) $\xrightarrow{NaNH_2}$ (*K*); (*C*) $\xrightarrow{ZnCl_2}$ (*L*); (*C*) $\xrightarrow{Ca(OH)_2}$ (*F*); (*C*) $\xrightarrow{HCN}$ (*D*)

(*J*) $\xleftarrow{H_2/Ni}$ (*F*)

(*F*) $\xrightarrow{H^+}$ (*G*); (*D*) $\xrightarrow{H^+,\ CH_3OH}$ (*E*)

(*I*) $\xleftarrow{H_2/Ni}$ (*H*) $\xleftarrow{H_2/Ni}$ (*G*)

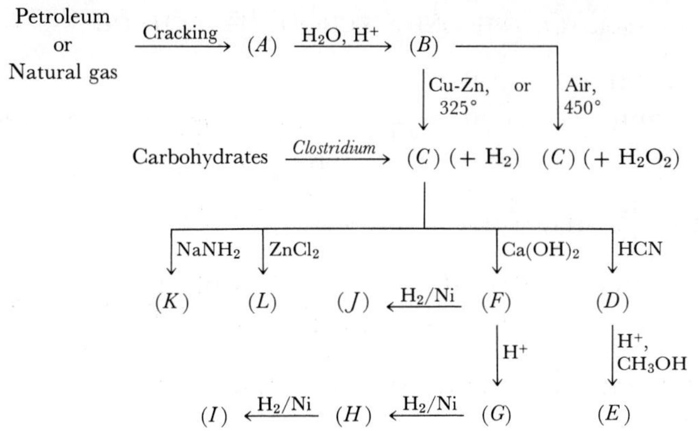

13–**20.** (***a***) C_2H_5MgBr, H^+, H_2/Pt (***b***) Mg, $HC(OC_2H_5)_3$, H_3O^+
(***c***) PCl_3, C_2H_5MgBr, H_2/Pt (***d***) HI, Mg, HCHO (***e***) H_2/Pt, HI
(***f***) Na + C_2H_5OH or $LiAlH_4$, *t*-butyl chromate (***g***) ThO_2 at 450°, CH_3MgBr
(***h***) HBr, Mg, HCHO (***i***) *i*-C_3H_7MgBr, H^+ (***j***) H_2/Pt, H_2SO_4
(***k***) CrO_3, C_2H_5MgBr (***l***) H_2CrO_4, $C_2H_5OH + H^+$ (***m***) NaOH, ThO_2 at 450°
(***n***) H_2O, Hg_2SO_4, H_2SO_4, then H_2/Pt (***o***) H_2CrO_4, Mg-Hg
(***p***) H_2O, Hg_2SO_4, H_2SO_4, then H_2/Pt (***q***) O_3, H_2/Pt (***r***) Mg, HCHO, HBr

13–**21.** (***a***) $CH_3OCH_2CHOHCHO$, $CH_3OCH_2COCH_2OH$
(***b***) $CH_3OCOCH_2CH_2OH$, $CH_3OCOCHOHCH_3$, $C_2H_5OCOCH_2OH$

13–**22.** $C_2H_5C(CH_3){=}CHCH_2CH_2CH_3$ or $C_2H_5C(CH_3){=}CHCH(CH_3)_2$, *cis* or *trans*

13–**23.** $C_2H_5C(CH_3){=}CHC_2H_5$ 13–**24.** $CH_3CHOHCH_2CH_2CHO$

13–**25.** $(CH_3)_2CHC(CH_3)(OH)CH(CH_3)_2$ or $(CH_3)_2C(OH)CH(CH_3)CH(CH_3)_2$

13–**26.** $C_2H_5{-}O{-}CH(CH_3)_2$ 13–**27.** $C_2H_5COCH(CH_3)_2$ 13–**28.** $(CH_3)_2CHCH_2I$

13–**29.** $CH_3CH_2CH_2{-}O{-}CH(CH_3)CH_2CH_3$ 13–**30.** $CH_3(CH_2)_4COOC_2H_5$

13–**31.** (***a***) $RCH(OEt)_2 \underset{B^-}{\overset{HB}{\rightleftharpoons}} RCH(OEt){-}\overset{+}{O}H{-}Et \underset{HOEt}{\overset{H_2O}{\rightleftharpoons}} RCH(OEt)(\overset{+}{O}H_2) \rightleftharpoons RCH(OH)(\overset{+}{O}H{-}Et) \underset{EtOH}{\rightleftharpoons}$

$RCH{=}\overset{+}{O}H \underset{HB}{\overset{B^-}{\rightleftharpoons}} RCHO$

(***b***) $R_2C{=}O \underset{B^-}{\overset{HB}{\rightleftharpoons}} R_2C{=}\overset{+}{O}H \underset{H\overset{+}{C}(OEt)_2}{\overset{HC(OEt)_3}{\rightleftharpoons}} R_2C(OEt){-}OH \underset{HOCH(OEt)_2}{\overset{H\overset{+}{C}(OEt)_2}{\rightleftharpoons}}$

$R_2\overset{+}{C}(OEt) \overset{HOCH(OEt)_2}{\rightleftharpoons} R_2C(OEt){-}\overset{+}{O}(Et){-}CH(OH)(OEt) \underset{HB}{\overset{B^-}{\rightleftharpoons}} R_2C(OEt)_2 + O{=}CHOEt$

(***c***) $R_2C{=}NOH \underset{B^-}{\overset{HB}{\rightleftharpoons}} R_2C{=}\overset{+}{N}H{-}OH \overset{H_2O}{\rightleftharpoons} R_2C(H_2O^+){-}\ddot{N}H{-}OH \rightleftharpoons$

$R_2C(OH){-}\overset{+}{N}H_2{-}OH \underset{HB}{\overset{B^-}{\rightleftharpoons}} R_2C{=}O + H_2NOH \underset{B^-}{\overset{HB}{\rightleftharpoons}} H_3\overset{+}{N}OH$

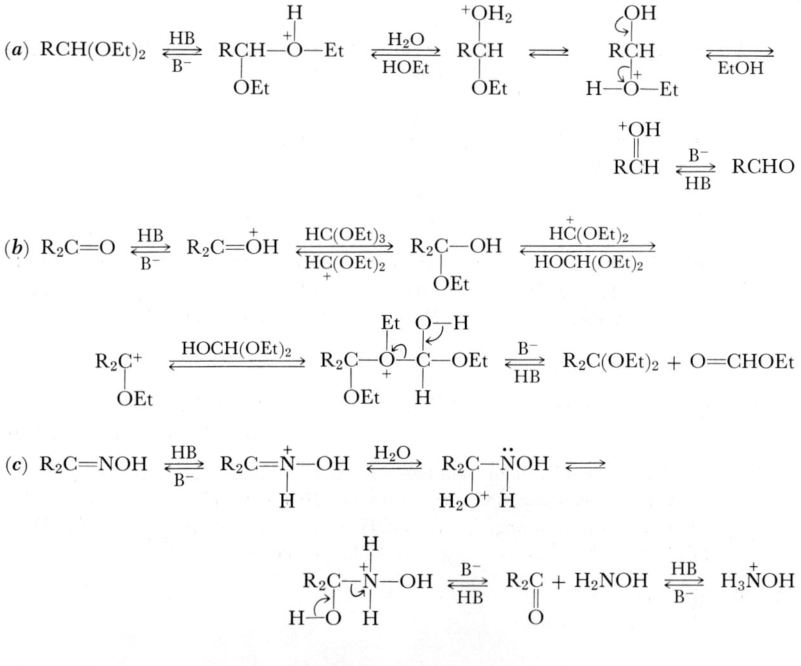

(*d*) ($H_2N—U$ equals $H_2N—NHCONH_2$) $RCH{=}O \underset{B^-}{\overset{HB}{\rightleftharpoons}} RCH{=}\overset{+}{O}H \xrightarrow{H_2N—U}$

$$\begin{matrix} OH \\ | \\ RCH \\ |+ \\ H_2N—U \end{matrix} \rightleftharpoons \begin{matrix} {}^{+}OH_2 \\ | \\ RCH \\ | \\ :NH—U \end{matrix} \xrightarrow[H_2O]{} \begin{matrix} RCH \\ \| \\ H—\underset{+}{N}—U \end{matrix} \underset{HB}{\overset{B^-}{\rightleftharpoons}} RCH{=}N—NHCONH_2$$

(*e*) $HCHO \underset{B^-}{\overset{HB}{\rightleftharpoons}} \{H_2C{=}\overset{+}{O}H \longleftrightarrow H_2\overset{+}{C}—OH\} \xrightarrow{R_2C{=}NOH}$

$$\left\{\begin{matrix} R_2C{=}\overset{+}{N}—OH \\ | \\ H—O—CH_2 \end{matrix} \longleftrightarrow \begin{matrix} R_2\overset{+}{C}—\ddot{N}—OH \\ | \\ H—O—CH_2 \end{matrix}\right\} \underset{HB}{\overset{B^-}{\rightleftharpoons}} \begin{matrix} R_2C \\ \| \\ O \end{matrix} + CH_2{=}NOH$$

(*f*) $CH_3CH{=}NH \underset{B^-}{\overset{HB}{\rightleftharpoons}} \{CH_3CH{=}\overset{+}{N}H_2 \longleftrightarrow CH_3\overset{+}{C}H—\ddot{N}H_2\}$

$$\begin{matrix} & \overset{+}{C}HCH_3 & \\ HN & & NH_2 \\ \| & & \downarrow \\ CH_3CH & & CHCH_3 \\ & N & \\ & H & \end{matrix} \rightleftharpoons \begin{matrix} & CHCH_3 & \\ HN & & \overset{+}{N}H_2 \\ | & & | \\ CH_3CH & & CHCH_3 \\ & N & \\ & H & \end{matrix} \underset{HB}{\overset{B^-}{\rightleftharpoons}} \begin{matrix} & CHCH_3 & \\ HN & & NH \\ | & & | \\ CH_3CH & & CHCH_3 \\ & N & \\ & H & \end{matrix}$$

(*g*) $RCH{=}N—\ddot{N}HC_6H_5 \underset{B^-}{\overset{HB}{\rightleftharpoons}} RCH_2—N{=}\overset{+}{N}HC_6H_5 \underset{HB}{\overset{B^-}{\rightleftharpoons}} RCH_2—N{=}N—C_6H_5$

CHAPTER 14

Amines

14–*1*. Four alcohols, eight amines

14–*2*. (*a*) Propionamide + NaOBr (*b*) *n*-Butyraldoxime + Na + C_2H_5OH
(*c*) *i*-Propylmagnesium bromide + chloro-di-*n*-butylamine
(*d*) Palmitic acid + NaN_3 + H_2SO_4 (*e*) *i*-Propyl cyanide + Na + C_2H_5OH
(*f*) C_2H_5Br, then aq. HCl or NaOH

14–*3*. (*a*) $C_2H_5{:}\ddot{N}{:}H$ with H below N (*b*) $\left[H_3C{:}\underset{CH_3}{\overset{CH_3}{N}}{:}CH_3\right]^+ :\ddot{Cl}:^-$ (*c*) $(CH_3)_3N{:}\ddot{O}{:}$ (*d*) $(C_2H_5)_3N{:}\underset{F}{\overset{F}{B}}{:}F$

(*e*) $H_3C{:}\underset{CH_3}{\ddot{N}}{:}\ddot{N}{:}\ddot{O}$ (*f*) $\left[H_3C{:}\underset{H}{\overset{H}{N}}{:}H\right]^+_2 \left[:\ddot{O}{:}\underset{:O:}{\overset{:O:}{S}}{:}\ddot{O}:\right]^{2-}$

14–*4*. *f* > *d* > *c* > *a* > *e* > *b* 14–*5*. *c* > *f* > *b* > *g* > *a* > *e* > *d*

14–*6*. (*a*) NH_3, then H_2O_2 (*b*) NH_3, then NaOBr (*c*) $LiAlH_4$, then $(CH_3CO)_2O$
(*d*) H_2NOH, then Na + C_2H_5OH (*e*) HBr, then NH_3 (*f*) HBr, then AgCN
(*g*) NaOH, then H_2SO_4 + NaN_3 (*h*) Mg, HCHO, HBr, NH_3

14–*7*. *a5*, *b1*, *c4*, *d3*, *e6*, *f2*

14–*8*. $CH_3C{\equiv}CH + HCHO + HN(CH_3)_2 \longrightarrow CH_3C{\equiv}CCHOHN(CH_3)_2 + H_2O$
1-Hydroxy-1-dimethylamino-2-butyne

14–*9*. (*a*) Water-solubility, odor, acid-base indicator, or dilute acid, then HNO_2
(*b*) Odor or acid-base indicator (*c*) Dil. HCl, HNO_2
(*d*) Fusion test for nitrogen, cold NaOH (*e*) Water-solubility, cold NaOH
(*f*) NaOH, then HNO_2 (*g*) Dil. HCl, hot NaOH

14–*10*. *a4*, *b1*, *c3*, *d2*

14–11. (*a*) To separate, use dil. HCl, sat. $NaHSO_3$, cold conc. H_2SO_4; to recover, add NaOH to solution of amine salt, HCl to bisulfite addition compound, and pour H_2SO_4 into cold water.
(*b*) Add HCl and distill; add NaOH to distillate and distill; remove methanol from hexane with cold conc. H_2SO_4; regenerate from salts.
(*c*) Distill to remove volatiles, add HCl to distillate and distill to remove ethyl alcohol, liberate amines from residue with NaOH and distill; treat distillate with acetic anhydride, neutralize to phenolphthalein and distill to remove tertiary amine; boil residue with NaOH to liberate secondary amine.

14–12. The ammonia competes with the amine in adding to the aldimine.

14–13. $C_4H_9NH_2 + HONO \xrightarrow{H^+} CH_3CH_2CH_2\overset{+}{C}H_2 + N_2 + 2\ H_2O$ (cf. p. 228).

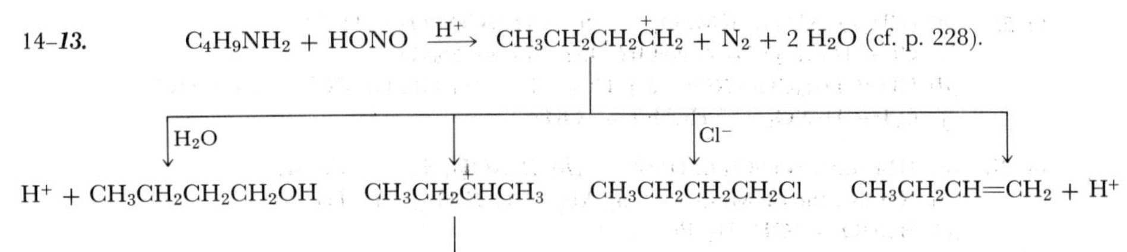

H_2O | Cl^-

$H^+ + CH_3CH_2CHOHCH_3$ $\quad CH_3CH_2CHClCH_3$ $\quad CH_3CH{=}CHCH_3 + H^+$

14–14. $RNH_2 + O{=}N{-}Br \longrightarrow Br^- + R\overset{+}{N}H_2NO \longrightarrow Br{-}R + NH_2NO \longrightarrow N_2 + H_2O$

Amides and Imidic Acids

14–15. (*a*) $CH_3CH_2COCl + C_2H_5NH_2$ (*b*) $CH_3(CH_2)_3COOC_2H_5 + NH_3$
(*c*) $(CH_3CH_2CH_2CO)_2O + (CH_3)_2NH$ (*d*) $CH_3(CH_2)_{10}COOH + NH_3$ + heat

14–16. *b*, *e*, and *h*

14–17. (*a*) NaOBr, Ac_2O (*b*) $C_2H_5OH + H^+$ (*c*) NaOH, HNO_2
(*d*) P_2O_5, then Na, EtOH (*e*) C_4H_9COOH, heat
(*f*) CrO_3, then $CH_3CH_2CH(NH_2)CH_3$ + heat (*g*) Mg, then $(CH_3)_2NCHO$
(*h*) C_2H_5OH, HCl in ether, then C_2H_5OH + heat (*i*) PCl_5, then NH_3

14–18. (*a*) $CH_3CH_2CONH_2$ and $(CH_3)_2NCH{=}O$ or $C_2H_5NHCH{=}O$
(*b*) $CH_3\underset{\underset{NOH}{\|}}{C}C_4H_9$ (four isomers), $C_5H_{11}CONH_2$ (eight isomers), and $C_4H_9CONHCH_3$ (four isomers; other structurally possible compounds give water-soluble acids)
(*c*) $HOCH_2CH_2NHCH_3$ [$HOCH_2N(CH_3)_2$ and $CH_3OCH_2NHCH_3$, unstable and not included] and $(CH_3)_3NO$

14–19. $R{-}C({=}O)NH_2 + :\overset{+}{S}(O^-)(Cl){-}Cl \longrightarrow$ [cyclic intermediate: $R{-}C(O{-}S(O^-)(Cl)Cl){=}N{-}H$] $\longrightarrow HCl +$ [$R{-}C(O{-}\overset{+}{S}(O^-)Cl){=}N{-}H$] $\longrightarrow$

$RC{\equiv}N + O{=}\ddot{S}{=}O + HCl$

Cyanides and Isocyanides

14–20. (*a*) C_2H_5I + NaCN (*b*) $CH_3CH_2CH(CH_3)CN + H_2O, H^+$
(*c*) $(CH_3)_2CHNH_2 + CHCl_3$ + KOH (*d*) $(CH_3)_2CHCH_2CH_2CN$ + Na, C_2H_5OH
(*e*) $(CH_3)_2CHCH{=}CH_2$ + HCN over Al_2O_3, 350° (*f*) $CH_3C{\equiv}N + H_2O_2, OH^-$
(*g*) $(CH_3)_2CHCH_2CONH_2 + POCl_3$, pyridine (*h*) $(CH_3)_3CC{\equiv}N + CH_3OH, H^+$
(*i*) $CH_3(CH_2)_5CH{=}NOH + P_2O_5$ (*j*) $CH_3(CH_2)_3N{\equiv}C + H_2/Pt$
(*k*) $(CH_3)_2CHCH_2NHCHO + POCl_3$ + pyridine

14–21. (*a*) NaOBr, then $CHCl_3$ + KOH (*b*) H_2NOH, then Na + C_2H_5OH
(*c*) HCN over Al_2O_3 at 350° (*d*) HCN + H_2SO_4, then H_2O

(*e*) $LiAlH_4$, HBr, NaCN (*f*) HBr, NaCN, Na + C_2H_5OH
(*g*) NH_3, heat, then $SOCl_2$ followed by $NaNH_2$ and CH_3I

14–*22*. (*a*) *N*-Ethylformamide (*b*) Methyl *N*-ethylformimidate (*c*) Triethylformamidine
(*d*) *N*-Ethylacetaldimine (probably as cyclic trimer)

14–*23*. (*a*) *1, 7* (*b*) *4, 6*

14–*24*. (*a*) Dil. HCl and fusion test for nitrogen (*b*) HNO_2 and hot NaOH (NH_3 odor)
(*c*) Dil. NaOH and hot NaOH (odor) (*d*) Cold NaOH (odor) and HNO_2

14–*25*. (*a*) *2, 6, 9* (*b*) *1, 3, 4, 5* (*c*) All except *1* (*d*) *2, 10*

14–*26*. (*a*) $(CH_3)_3CNH_2 + KMnO_4$ (*b*) $(CH_3)_2CHNO_2 + H_2/Pt$
(*c*) $CH_3CH_2CH_2NO_2$ + NaOH, then 25% aq. H_2SO_4
(*d*) $CH_3(CH_2)_5CH{=}NOH + F_3CCO_3H$ (*e*) $CH_3CH_2NO_2$ + conc. HCl
(*f*) $C_2H_5CH_2NO_2 + 2\,HCHO + {}^{-}OH$

14–*27*. (*a*) HBr, then $NaNO_2$ in DMF (*b*) H_2NOH, then F_3CCO_3H
(*c*) NaOBr, then $KMnO_4$ (*d*) $Mg(OCOOCH_3)_2$, H^+, H_2/Pt
(*e*) HCHO + ^{-}OH, H_2/Pt

14–*28*. (*a*) Dil. NaOH and dil. HCl (*b*) Fusion test for nitrogen, then HNO_2
(*c*) Test for nitrogen, then conc. H_2SO_4 (*d*) Dil. NaOH, then Br_2
(*e*) HCl, dil. NaOH

14–*29*. $H_3C{:}O{:}N{:}O$ $H_3C{:}N(O){:}O$ $H_2C{:}N(O){:}O{:}H$ $H_3C{:}O{:}N(O){:}O$ $H_2C{:}N(O){:}O$ / $:N{:}O$

14–*30*. $HOCH_2(C_3H_6)CONH_2$ (four isomers) and $C_3H_7CHNO_2CH_3$ (two isomers)

Other Nitrogen Compounds

14–*31*. $H_3\overset{+}{N}CH_2COOC_2H_5 + {}^{-}NO_2 \longrightarrow HONO + H_2NCH_2COOC_2H_5 \longrightarrow$

$H_2O + O{=}N{-}NHCH_2COOC_2H_5 \longrightarrow$:N(–O–H)=N–CHCOOC$_2H_5$ (with H) $\longrightarrow$

$H_2O + {}^{-}{:}\ddot{N}{=}\overset{+}{N}{=}CHCOOC_2H_5$

14–*32*. (*a*) CH_2N_2 (*b*) NaN_3, then H_2/Pt or $LiAlH_4$ (*c*) HNO_2, then H_2O, heat

14–*33*. $CH_3CONHNHCOCH_3$ is more likely than $(CH_3CO)_2NNH_2$ because acylation decreases the basicity of nitrogen, and diacylation of hydrazine would give the symmetrical compound.

CHAPTER 15

Sulfur Compounds

15–*1*. (*a*) $C_4H_9SH + Na_2Cr_2O_7 + 4\,H_2SO_4 \longrightarrow$
$C_4H_9SO_3H + Na_2SO_4 + Cr_2(SO_4)_3 + 4\,H_2O$
(*b*) $C_3H_7S{-}SC_3H_7 + 10\,HNO_3 \longrightarrow 2\,C_3H_7SO_3H + 10\,NO_2 + 4\,H_2O$
(*c*) $3\,C_2H_5SC_2H_5 + 4\,KMnO_4 + 2\,H_2O \longrightarrow 3\,C_2H_5SO_2C_2H_5 + 4\,KOH + 4\,MnO_2$

15–*2*. (*a*) $2\,(CH_3)_2CHI + Na_2S$ (*b*) $C_2H_5I + Na^{+-}SC_4H_9$ (*c*) $(CH_3)_2S + KMnO_4$
(*d*) $CH_3CH{=}O + 2\,HSC_2H_5$ (*e*) Zn, H_2SO_4 (*f*) $2\,C_5H_{11}Cl + Na_2S_2$
(*g*) NaSH (*h*) I_2, KOH (*i*) H_2/Ni

15–*3*. (*a*) $C_2H_5COOH + P_2S_5$ (*b*) $C_3H_7CONH_2 + P_2S_5$ (*c*) $CH_3COCl + HSCH_3$
(*d*) $CH_3SO_2Cl + HOC_2H_5$ (*e*) $(C_3H_7)_2S + C_3H_7I$ (*f*) $C_3H_7SO_2Cl + NH_3$
(*g*) HNO_3 (*h*) Mg, SO_2 (*i*) Cl_2, H_2O (*j*) Mg, CS_2 (*k*) HNO_3
(*l*) PCl_3, then Zn + H_2SO_4 (*m*) $NaHSO_3$ (*n*) $(CH_3)_3CMgCl$

15–*4.* (*a*) $C_2H_5{:}\overset{:\ddot{O}:}{\ddot{S}}{:}C_2H_5$ (*b*) $\left[C_2H_5{:}\overset{C_2H_5}{\underset{C_2H_5}{S}}\right]^+ \; :\ddot{I}:^-$ (*c*) $K^{+-}{:}\ddot{O}{:}\overset{:\ddot{O}:}{\underset{:\ddot{O}:}{S}}{:}\ddot{O}{:}CH_3$

(*d*) $C_2H_5{:}\overset{:\ddot{O}:}{\underset{:\ddot{O}:}{S}}{:}C_3H_7$ (*e*) $C_4H_9{:}\ddot{O}{:}\overset{:\ddot{O}:}{S}{:}\ddot{O}{:}C_4H_9$ (*f*) $C_4H_9{:}\overset{:\ddot{O}:}{\underset{:\ddot{O}:}{S}}{:}\ddot{O}{:}C_2H_5$

(*g*) $CH_3{:}\overset{:\ddot{S}}{C}{:}\ddot{S}{:}H$ (*h*) $Na^{+-}{:}\ddot{O}{:}\overset{:\ddot{O}:}{S}{:}C_2H_5$ (*i*) $CH_3{:}\overset{:\ddot{S}}{C}{:}\underset{H}{\ddot{N}}{:}H$

15–*5.* (*a*) $(CH_3)_2S + H_2O + CH_2{=}CHCH_2CH_3$ (*b*) $C_2H_5CS{-}S_2{-}CSC_2H_5 + H_2O$
(*c*) $C_3H_7S{-}SC_3H_7$ (*d*) $C_4H_9SO_2Cl + C_4H_9Cl + HCl$

15–*6.* (*a*) Boil with HCl. Sulfites give SO_2. Sulfates give H_2SO_4; test for sulfate ion with $Ba(OH)_2$.
(*b*) Mercaptans give precipitate with lead acetate; sulfides are oxidized to neutral sulfones, disulfides to sulfonic acids.
(*c*) Sulfide with CH_3I gives sulfonium salt; sulfoxide is oxidized to sulfone.
(*d*) Sulfonium chloride is soluble in water; test for S.
(*e*) Hydrolyze and test for sulfate ion, then test for S.
(*f*) Cold NaOH liberates NH_3 from salt; boiling with NaOH liberates NH_3 from amide.

15–*7.* (*a*) Add NaOH and distill, add HCl to residue and distill, distill residue at reduced pressure.
(*b*) Extract ammonium salt with water, extract sulfonamide with dilute NaOH and regenerate.
(*c*) Add NaOH and distill, remove water from distillate, and fractionally distill. Liberate mercaptan from salt.
(*d*) Extract mercaptan with aq. NaOH, dry insoluble layer and extract with cold conc. H_2SO_4. (*e*) Extract sulfonic acid with water, extract sulfonamide with NaOH.

15–*8.* (*a*) HNO_3, then P_2O_5 (*b*) Na_2S, then H_2O_2 (*c*) Na, then CH_3I
(*d*) $Na + C_2H_5OH$, then HBr, then NaSH (*e*) $C_2H_5C{\equiv}N + H_2O_2$, then P_2S_5
(*f*) $SOCl_2$, then $Zn + H_2SO_4$ (*g*) H_2/Pt, HBr, Na_2S_2

15–*9.* (*a*) $R\overset{O}{\overset{\|}{C}}{-}O{-}\overset{O}{\overset{\|}{C}}R + HSR' \longrightarrow R\overset{O}{\overset{\|}{C}}{-}O^- + H\overset{R'}{\underset{+}{S}}{-}\overset{O}{\overset{\|}{C}}R \longrightarrow RCOOH + R'S\overset{O}{\overset{\|}{C}}R$

(*b*) $RSO_2OR' + HOR'' \longrightarrow RSO_2O^- + R'{-}\overset{+}{\underset{H}{O}}{-}R'' \underset{}{\overset{ROH}{\rightleftharpoons}} R'OR'' + R\overset{+}{O}H_2$

Phosphorus, Silicon, and Tin Compounds

15–*10.* (*a*) $C_4H_9OH + POCl_3 + R_3N$
(*b*) $3\,C_2H_5OH + PCl_3 \longrightarrow (C_2H_5O)_2PH(O) + C_2H_5Cl + 2\,HCl$
(*c*) $6\,(C_2H_5O)_3PO + 2\,POCl_3 \longrightarrow$
$3\,(C_2H_5O)_2P(O){-}O{-}P(O)(OC_2H_5)_2 + 6\,C_2H_5Cl + P_2O_5$
(*d*) $CH_3OH + P_2S_5$ (*e*) $C_2H_5MgBr + PCl_3$

15–*11.* (*a*) $PCl_3 + (C_3H_7)_2Hg$, then $NaOC_2H_5$ (*b*) Na, then C_2H_5Br
(*c*) $PCl_3 + R_3N$, then n-C_3H_7Br
(*d*) Final step, $(HOCH_2)_3P + [H_2C{=}\overset{+}{O}H]Cl^- \longrightarrow (HOCH_2)_4P^{+-}Cl$

15–*12.* 73 kcal.

15–*13.* (*a*) Na, C_2H_5Br, then $HSiCl_3/Pt$ (*b*) H_2O, NaOH

CHAPTER 16

16–*1.* (*A*) $COCl_2$ (*B*) H_2NCONH_2 (*C*) $CH_3CONHCONHCOCH_3$
(*D*) H_2NCOOR (*E*) HNCO (*F*) $H_2NCONHCONH_2$
(*I*) $H_2NCONHNO_2$ (*J*) $H_2NCONHNH_2$
(*K*) $H_2NCONHCH_2OH$ (*L*) $(HOCH_2NH)_2CO$ (*M*) ClCOOR
(*N*) $OC(OR)_2$ (*O*) RNHCOCl (*P*) RNHCOOR (*Q*) RN=C=O
(*R*) RNHCONHR (*S*) $RN{=}C(OCH_3)NHR$ (*T*) RNHCHO (*U*) RN≡C

16–**2.** *c*, *f*, *h*, and *j* 16–**3.** *b*, *h*, *j*, and *l*

16–*4.* (***a***) $COCl_2 + CH_3OH$ (***b***) $COCl_2 + 2\,C_4H_9OH$
(***c***) $CO(NH_2)_2 + (C_2H_5O)_2SO_2 + NaOH$ (***d***) $CS(NH_2)_2 + C_4H_9Br + NaOH$
(***e***) $(CH_3)_2CHBr + NaSCN$ (***f***) $C_2H_5NHCSNHC_2H_5 + C_2H_5I + NaOH$
(***g***) $C_5H_{11}OH + CS_2 + KOH$ (***h***) $CH_3NH_2 + CS_2 + NaOH$
(***i***) $C_3H_7NHCONHC_3H_7 + C_7H_7SO_2Cl + (C_2H_5)_3N$ (***j***) $(CH_3)_2NH + COCl_2$

16–*5.* (***a***) $(CH_3)_2SO_4$, then $C_2H_5NH_2$ (***b***) $C_3H_7NH_2$, then $Pb(NO_3)_2$
(***c***) CS_2, then NaOCl (***d***) CS_2, then heat (***e***) $COCl_2$, heat, then C_2H_5OH

16–*6.* (***a***) Heat with aq. H_2SO_4 and distill, test distillate for CO_2 or CH_3COOH.
(***b***) By odor or add H_2O. (***c***) Heat with HgO.
(***d***) Distill with H_2O and test distillate for CO_2, C_2H_5OH, or CH_3COOH. (***e***) HNO_2
(***f***) Distill with NaOH and test for C_2H_5OH. (***g***) RNH_2 (***h***) H_2O (***i***) H_2O

16–*7.* (***a***) Two; $-CH_2NH{-}C({=}NH)NH_2 \longrightarrow NH_3 + {-}CH_2NHCONH_2 \longrightarrow$
$NH_3 + {-}CH_2NHCOOH \longrightarrow {-}CH_2NH_2 + CO_2$
(***b***) Four; $HOOCCH(NH_2)(CH_2)_3NHC({=}NH)NH_2 \longrightarrow$
$2\,N_2 + HOOCCH(OH)(CH_2)_3NHC({=}NH)OH \rightleftharpoons {-}NHCONH_2 \longrightarrow$
$N_2 + {-}NHCOOH \longrightarrow CO_2 + {-}NH_2 \longrightarrow$
$N_2 + HOOCCHOH(CH_2)_3OH$ (or dehydration products)

16–*8.* $$RN{=}C{=}O \underset{B^-}{\overset{HB}{\rightleftharpoons}} R\overset{+}{N}(H){=}C{=}O \xrightarrow{H_2O} R\ddot{N}(H){-}C(^{+}OH_2){=}O \underset{HB}{\overset{B^-}{\rightleftharpoons}} RN(H){-}C(O{-}H){=}O \underset{HB}{\overset{B^-}{\longrightarrow}}$$
$$CO_2 + R\ddot{\underset{..}{N}}\bar{H} \underset{B^-}{\overset{HB}{\rightleftharpoons}} RNH_2 \xrightarrow{HB} RNH_3{}^{+-}B$$

16–**9.** Compound *A* is $(CH_3)_2CH{-}O{-}COCl$. 16–*10.* Urea nitrate

CHAPTER 17

17–*1.* *a3, b1, c2* 17–**2.** *a2, b1, c3*

17–**3.** None, *e* and *q*; two active, *a*, *b*, *f*, *i*, *k*, *l*, and *n*; two active and one *meso*, *c*, *h*, and *m*; four active, *g*, *j*, *o*, *p*, and *t*; sixteen active, *d*. Geometric, 8*s*, 16*r*.

17–*4.* (***a***) Eight active; four active; four active; two active, one *meso*.
(***b***) Two active; four active; eight active; eight active.
(***c***) Four active; eight active; eight active; four active; eight active.

17–*5.* (***a***)

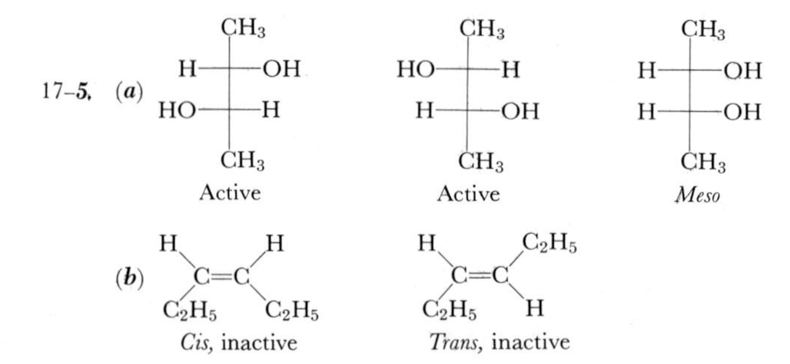

(c) 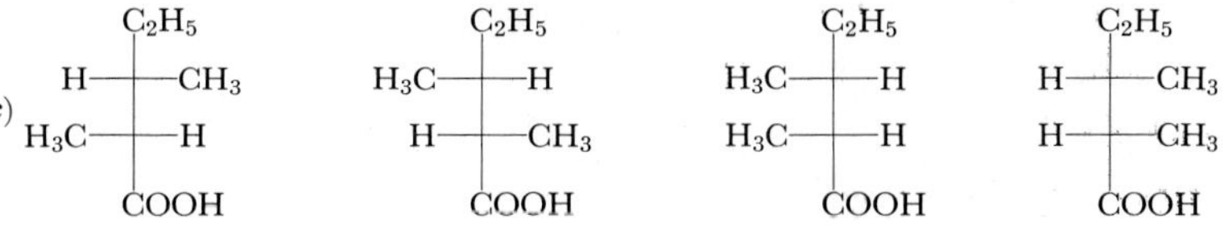 Fischer projections (top C_2H_5, bottom COOH):

	Upper carbon	Lower carbon	
1	H—C—CH_3	H_3C—C—H	Active
2	H_3C—C—H	H—C—CH_3	Active
3	H_3C—C—H	H_3C—C—H	Active
4	H—C—CH_3	H—C—CH_3	Active

(d) 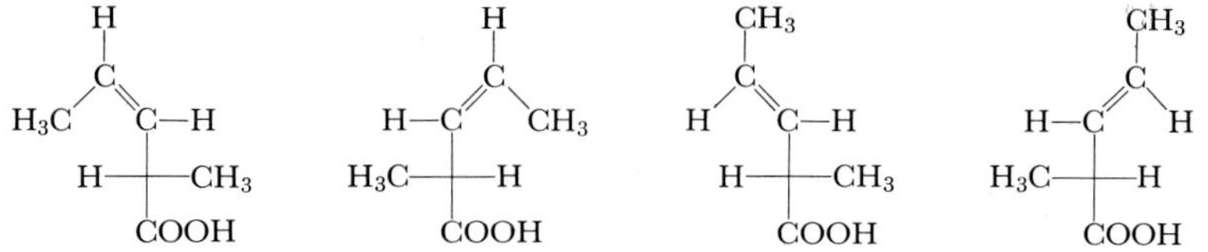 Structures with CH(H or CH_3)=CH— groups on the carbon above COOH:

	Double bond	Lower carbon	
1	H_3C(H)C=C—H	H—C—CH_3, COOH	Active
2	H—C=C(H)CH_3	H_3C—C—H, COOH	Active
3	H(CH_3)C=C—H	H—C—CH_3, COOH	Active
4	H—C=C(CH_3)H	H_3C—C—H, COOH	Active

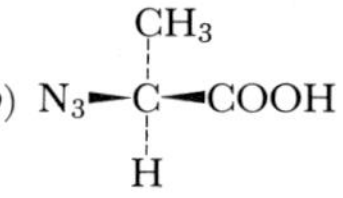

(e) 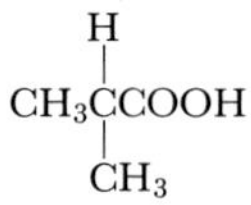 $CH_3CH(CH_3)COOH$
Inactive

17–**6.** *a4, b1, c6, d2, e3, f9, g5, h7, i8* 17–**7.** *a2, b1, c6, d5, e3, f4, g7*

17–**8.** (**a**) 66.5° (**b**) 8.08 g./100 cc. 17–**9.** (**a**) 5.9° (**b**) 18.9 per cent

17–**10.** (**a**) *S* (**b**) *S* (**c**) *1S, 2S, 3R, 4R* (**d**) *1R, 2S*

17–**11.** (**a**) H---C(OH)((CH$_2$)$_5$CH$_3$)---CH$_2$—CH=CH—(CH$_2$)$_7$COOH (cis, H and H on same side) (**b**) N_3—C(CH_3)(H)—COOH

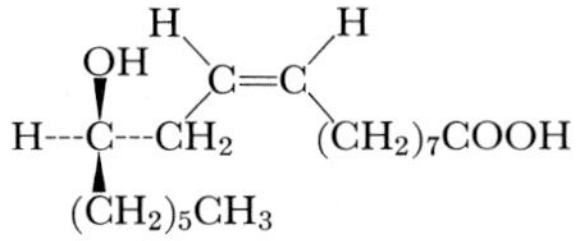

(**c**) HC(=O)—CH(OH)—CH(OH)—CH(OH)—CH(OH)—CH_2OH (zigzag: OH up, H down; H up, OH down; OH up, H down; H up, OH down)

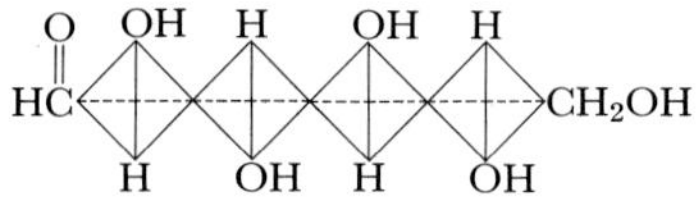

CHAPTER 18

18–**1.** (**a**) 32 (**b**) 128

18–**2.** (**a**) *4* (**b**) *2, 10* (**c**) *8* (**d**) *3* (**e**) *7* (**f**) *2, 3, 6, 10, 11*
(**g**) *1, 3, 6, 10, 11* (**h**) *4, 5, 7* (**i**) *10* (**j**) *11* (**k**) *1*
(**l**) *1, 2, 3, 5, 6, 8, 9, 10, 11*

18–**3.** *a4, b3, c2, d1*

18–**4.** (**a**) Arabinose gives furfural when distilled with 12% HCl.
(**b**) Ribose is oxidized by nitric acid to an inactive dibasic acid.
(**c**) Mannose gives an insoluble colorless phenylhydrazone.
(**d**) Rhamnose gives only a mono carboxylic acid when oxidized with nitric acid.
(**e**) Xylose gives furfural when distilled with 12% HCl.
(**f**) The phenylosazone of maltose is soluble in hot water, that of glucose is not; the osazones differ also in crystal structure.
(**g**) Lactose gives inactive mucic acid when oxidized with nitric acid.
(**h**) Sucrose is nonreducing.
(**i**) Take rotation after hydrolysis.
(**j**) Nitric acid oxidation of pectic acid gives inactive mucic acid.
(**k**) Test for nitrogen or flammability. (**l**) Acetate dissolves in acetone.
(**m**) Ethylcellulose gives ethyl iodide with HI (Zeisel determination); saponification of acetate, acidification, and distillation yields acetic acid.

18–**5.** (***a***)

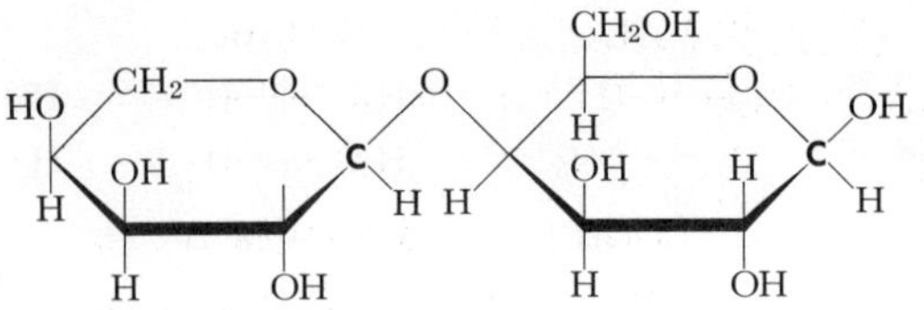

(***b***)

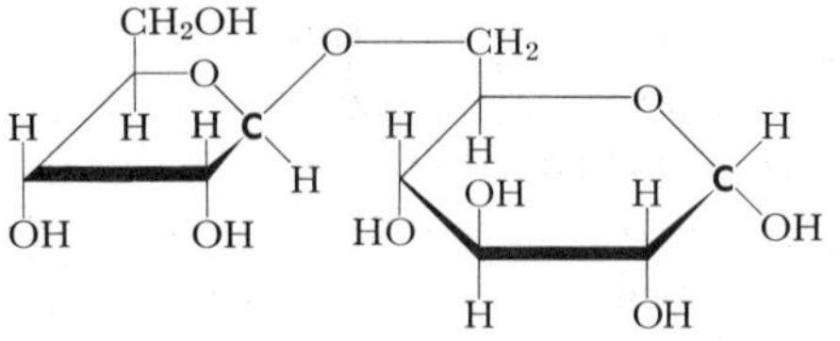

(***c***)

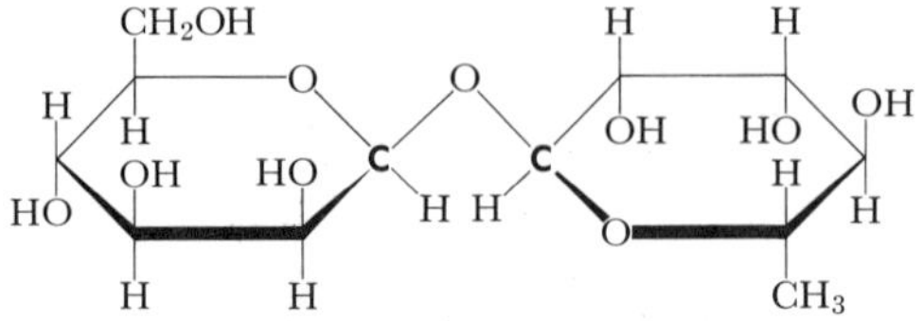

18–**6.**

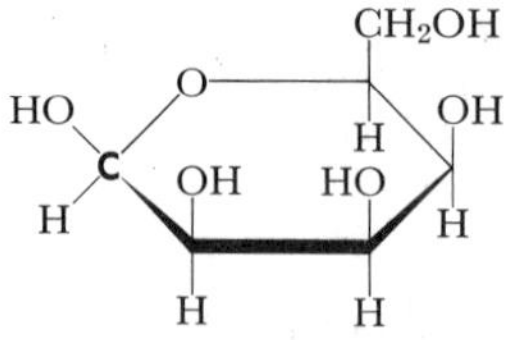

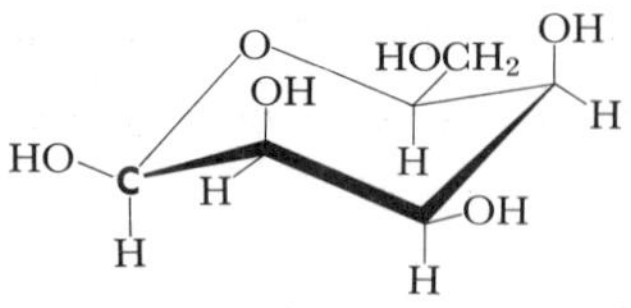

18–**7.** D-Ribose

18–**8.** CHO---C----C----C----C----C---CH_2OH

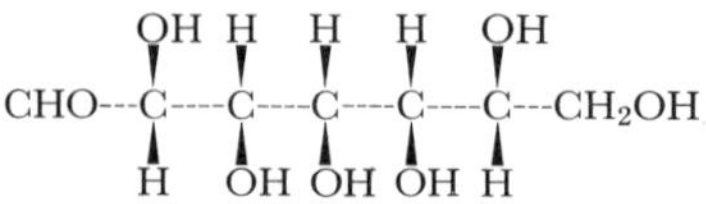

CHAPTER 19

19–**1.** (***a***) *1* (***b***) *5* (***c***) *4* 19–**2.** *a2, b4, c7, d3, e9, f1, g6, h8, i5*

19–**3.** (***a***) Br_2 + P, then excess NH_3, or NaN_3 followed by catalytic reduction
(***b***) NaCN + NH_4Cl, then acid hydrolysis
(***c***) $Mg(OCOOCH_3)_2$, then catalytic reduction
(***d***) Mg, HCHO, HBr, Mg, CO_2, Cl_2 + PCl_3, excess NH_3

19–**4.** (***A***) H_2N-Gly-Ser-Gly-Ser-Gly-COOH (***B***) H_2N-Cys-Leu-Cys-Cys-Leu-COOH
(***C***) H_2N-Ileu-Ileu-Asp-Asp-Ileu-COOH (***D***) H_2N-Met-Ser-Met-Met-Gly-COOH

CHAPTER 20

20–**1.** *a2, b4, c3, d5, e1* 20–**2.** *b, e, g,* and *h*

20–**3.** $b < c < f < d < e < a$ 20–**4.** $e < b < d < c < a$

20–**5.** (***a***)

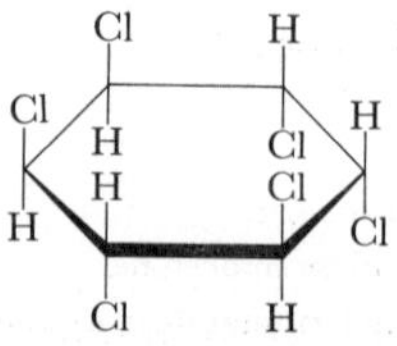

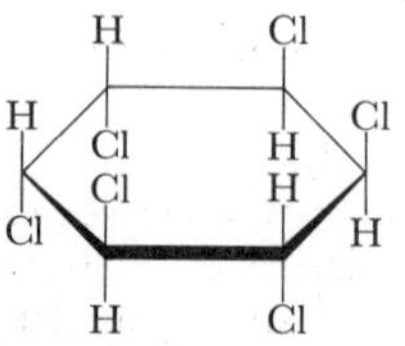

(***b***) *eeeeaa, eeeeee, eeeaaa, eeeeea, eeaeea, eeaaea, eeeaea, eaeaea*

20–*6*. (***a***) 1,2,3-, 1,2,4-, and 1,3,5-Trimethylbenzenes
(***b***) 2-Chloro-3-, 4-, 5-, and 6-nitrotoluenes, 3-chloro-2-, 4-, 5-, and 6-nitrotoluenes, and 4-chloro-2- and 3-nitrotoluenes
(***c***) 2,3-, 2,4-, 2,5-, 2,6-, 3,4-, and 3,5-Dinitrochlorobenzenes
(***d***) 2,3,4-, 2,3,5-, 2,3,6-, 3,4,5-, 2,4,5-, and 2,4,6-Trimethylbromobenzenes

20–*7*. Both the *ortho* and *meta* isomers can yield four compounds when one more group is introduced.

20–*8*. (***a***) Four butylbenzenes (***b***) *m*-Diethylbenzene
(***c***) *o*-Diethylbenzene, *p*-methyl-*n*-propylbenzene, or *p*-methyl-*i*-propylbenzene
(***d***) 2,6-Dimethylethylbenzene or 3,5-dimethylethylbenzene

20–*9*. 1,2,3-, m.p. 87.4°; 1,2,4-, m.p. 44°; 1,3,5-, m.p. 119°

20–*10*. *Ortho*, b.p. 225°; *meta*, b.p. 219°; *para*, m.p. 87°

20–*11*. *A* is *para*, *B* is *ortho*, and *C* is *meta*

20–*12*. (***a***) 2- and 4-Bromo-1-nitrobenzene (***b***) *m*-Chloronitrobenzene
(***c***) *o*- and *p*-Toluenesulfonic acid (***d***) *m*-Bromobenzoic acid
(***e***) *m*-Nitrobenzenesulfonic acid (***f***) 4-Chloro-*m*-xylene
(***g***) 1,3,5-Trinitrobenzene (***h***) 3-Nitro-4-*i*-propylbenzenesulfonic acid
(***i***) *o*- and *p*-Bromochlorobenzene

20–*13*. (***a***) $C_6H_6 + HNO_3 + H_2SO_4$, then $Br_2 + FeBr_3$
(***b***) *m*-Xylene + Cl_2 + Fe, then $HNO_3 + H_2SO_4$
(***c***) Ethylbenzene + Br_2 + Fe, then $HNO_3 + H_2SO_4$
(***d***) $C_6H_6 + Cl_2$ + Fe, then $H_2SO_4 + SO_3$
(***e***) Toluene + $HNO_3 + H_2SO_4$, then $Cl_2 + FeCl_3$
(***f***) Toluene + $KMnO_4$, or $Na_2Cr_2O_7 + H_2SO_4$, then $HNO_3 + H_2SO_4$
(***g***) Toluene + Br_2 + Fe, then $Na_2Cr_2O_7 + H_2SO_4$
(***h***) Benzene + Cl_2 + Fe, then $HNO_3 + H_2SO_4$

CHAPTER 21

21–*1*. (***a***) $C_6H_6 + Br_2 + h\nu$ (***b***) Cumene + Cl_2 + Fe (***c***) *o*-Xylene + $Br_2 + h\nu$
(***d***) Chlorobenzene + Br_2 + Fe (***e***) Iodobenzene + *p*-iodotoluene + Cu
(***f***) *p*-Xylene + $Cl_2 + h\nu$ (***g***) Ethylbenzene + HCHO + HCl + $ZnCl_2$
(***h***) 1-Chloro-2,4-dinitrobenzene + H_2NNH_2 (***i***) *p*-Bromocumene + CuCN in MSO
(***j***) Chlorobenzene + ethyl mercaptan + Cu_2O

21–*2*. $c < e < b < d < a$

21–*3*. (***a***) $Cl_2 + h\nu$, then Mg, followed by CO_2; or C_6H_5Li, then CO_2
(***b***) Propylene + H_3PO_4, then HCHO + HCl
(***c***) Br_2 + Fe, then $KMnO_4$, or $Na_2Cr_2O_7 + H_2SO_4$
(***d***) Br_2 + Fe, then *n*-C_3H_7Br + Na
(***e***) Conc. H_2SO_4, then I_2 + HgO or HNO_3

21–*4*. (***a***) Fuming H_2SO_4 (***b***) Alc. $AgNO_3$ (***c***) Test for chlorine
(***d***) Hydrolyze and test for benzaldehyde. (***e***) Bases convert *t*-amyl chloride to alkene.
(***f***) Na fusion, then HNO_3 and aq. $AgNO_3$ (***g***) Oxidize with NaOCl.

21–*5*. $Br-C_6H_4-CHClCH_3$ or $Br-C_6H_4-CH_2CH_2Cl$

CHAPTER 22

22–*1*. (***a***) *p*-$CH_3C_6H_4SO_2Cl + CH_3C_6H_5 + AlCl_3$ (***b***) *o*-$CH_3C_6H_4SO_2Cl$ + Zn + H_2SO_4
(***c***) $C_6H_5SO_2Cl + H_2NCH_3$ + NaOH (***d***) *p*-$CH_3C_6H_4SO_2Cl$ + *n*-C_4H_9OH + NaOH
(***e***) $BrC_6H_5 + 2\ ClSO_3H$ (***f***) 2,4-$(CH_3)_2C_6H_3SO_2Cl$ + Zn in ether, then H^+

22–**2.** The products: (***a***) *n*-Propylbenzene + $MgCl_2$ + $Mg(OSO_2C_6H_4CH_3)_2$
(***b***) Benzenesulfonamide + HOCl (***c***) *p*-Tolyl *p*-cumyl sulfone
(***d***) *m*-Dibromobenzene (***e***) 2,4-Dinitrobenzenesulfonic acid
(***f***) Sodium *p*-methylphenoxide (***g***) *n*-Propylammonium benzenesulfonate

22–**3.** (***a***) H_2SO_4 + SO_3, liming out, then NaCN fusion
(***b***) $ClSO_3H$, crystallize, boil with water, and evaporate to dryness
(***c***) Mg, then ethyl *p*-toluenesulfonate (***d***) Na_2S_2, then oxidation with HNO_3
(***e***) PCl_3, then toluene + $AlCl_3$ (***f***) CH_3MgBr, then CO_2
(***g***) Zn in ether, then HCl, followed by $HgCl_2$ (***h***) NaOH, then CH_3NH_2

22–**4.** (***a***) Boil with water and test for halogen (***b***) Test with barium chloride
(***c***) As in (*a*) (***d***) *N*-Methylbenzenesulfonamide is soluble in dilute NaOH
(***e***) Aq. $AgNO_3$ (***f***) The sulfonic acid is soluble in water and strongly acid
(***g***) Allow to react with ammonia; the sulfonamide is soluble in dilute NaOH whereas benzylamine is soluble in dilute acid (***h***) Fusion test for nitrogen
(***i***) Distill with sodium hydroxide and test distillate for ethyl alcohol or boil with 25% HCl, make alkaline, distill and test distillate for dimethylamine

CHAPTER 23

23–**1.** (***a***) $CH_3C_6H_4NO_2 + 3\ Fe + 7\ HCl \longrightarrow CH_3C_6H_4NH_3^{+-}Cl + 3\ FeCl_2 + 2\ H_2O$
(***b***) $4\ C_3H_7C_6H_4NO_2 + 3\ NaOCH_3 \longrightarrow$
$2\ C_3H_7C_6H_4NO{=}NC_6H_4C_3H_7 + 3\ NaOCHO + 3\ H_2O$
(***c***) $CH_3C_6H_4N{=}NC_6H_4CH_3 + H_2O_2 \longrightarrow CH_3C_6H_4NO{=}NC_6H_4CH_3 + H_2O$
(***d***) $2\ C_2H_5C_6H_4NO_2 + 5\ Zn + 10\ NaOH \longrightarrow$
$C_2H_5C_6H_4NHNHC_6H_4C_2H_5 + 5\ Na_2ZnO_2 + 4\ H_2O$
(***e***) $ClC_6H_4NO_2 + 2\ Zn + 4\ NH_4Cl \longrightarrow ClC_6H_4NHOH + 2\ Zn(NH_3)_2Cl_2 + H_2O$
(***f***) $C_3H_7C_6H_4NHNHC_6H_4C_3H_7 + 2\ NaOBr \longrightarrow$
$C_3H_7C_6H_4N{=}NC_6H_4C_3H_7 + 2\ NaBr + H_2O$
(***g***) $3\ CH_3C_6H_4NHOH + Na_2Cr_2O_7 + 4\ H_2SO_4 \longrightarrow$
$3\ CH_3C_6H_4NO + Na_2SO_4 + Cr_2(SO_4)_3 + 7\ H_2O$

23–**2.** *a2*, *b4*, *c3*, *d1*, *e5*

23–**3.** (***a***) HNO_3 + H_2SO_4, then Zn + NH_4Cl, followed by $Na_2Cr_2O_7$ + H_2SO_4
(***b***) HNO_3 + H_2SO_4, then Zn + NaOH (***c***) HNO_3 + H_2SO_4, then $(NH_4)_2S$
(***d***) Na_2S_2, then HNO_3 oxidation

23–**4.** One to two

23–**5.** (***a***) Primary nitroalkanes are soluble in dil. NaOH.
(***b***) NaOBr oxidizes hydrazobenzene to orange azobenzene.
(***c***) 2,4-Dinitrochlorobenzene gives a precipitate with alcoholic $AgNO_3$.
(***d***) Conc. HCl rearranges *N*-phenylhydroxylamine to *p*-chloroaniline, which is soluble in dil. HCl.
(***e***) 2-Chloro-4-nitrotoluene can be oxidized to the carboxylic acid, which is soluble in dil. NaOH.

23–**6.** $d < a,c < b,e < f,g < h$

23–**7.** A nitro group in the *ortho* or *para* position can provide additional resonance stabilization for the negative charge on the intermediate, but a group in the *meta* position cannot. In 1,3,5-trinitrobenzene the inductive effect of three electronegative groups provides sufficient withdrawal of electrons from the benzene ring to permit formation of the σ intermediate.

CHAPTER 24

24–**1.** (***a***) HNO_3 + H_2SO_4, then Fe + HCl (***b***) Heat with aniline hydrochloride, then HNO_2
(***c***) HNO_2, then NaOH (***d***) $COCl_2$, then heat (***e***) CH_3COOH + heat, then HNO_3

24–2. (*a*) Propionanilide (*b*) 2,4-Dimethylbenzenediazonium chloride
(*c*) *p*-Ditolylamine (*d*) *N*-*N*-Dimethyl-*N′*-phenylthiourea (*e*) 6-Nitro-*o*-toluidine

24–3. *a5, b7, c2, d4, e1, f3, g6*

24–4. (*a*) *n*-Butyl bromide, base, *n*-butyraldehyde (*b*) Ethyl β-chloropropionate, base, acetone
(*c*) Chlormethyl methyl ether, base, nonanal, hydrolysis
(*d*) Benzyl chloride, base, butyryl chloride + $(C_2H_5)_3N$, heat
(*e*) Trimethylene bromide, base, propionaldehyde

24–5. (*a*) Heat with HgO or other test for S (*b*) HCl + HNO_2 at 0° (*c*) NaOH, heat
(*d*) Dil. HCl
(*e*) Isocyanide + H_2O gives formanilide, which reduces Tollens reagent; isocyanate gives carbanilide. (*f*) HNO_2 + HCl (*g*) Aq. $AgNO_3$
(*h*) Distill with aq. H_2SO_4 and test distillate for acetic acid. (*i*) Dil. NaOH
(*j*) HNO_2 + HCl

24–6. (*a*) *N*-Methylaniline (*b*) *p*-Nitroaniline (*c*) *N,N*-Dimethylaniline
(*d*) *p*-Toluidine (*e*) *n*-Amylamine (*f*) *p*-Chloroaniline (*g*) Aniline

24–7. *A*, *N*-Methylaniline *B*, Benzylamine *C*, *m*-Toluidine *D*, *o*-Toluidine
E, *p*-Toluidine

24–8. (*a*) *p*-Toluidine, H_2SO_4 + $NaNO_2$, heat (*b*) *o*-Chloroaniline, HBr + $NaNO_2$, CuBr
(*c*) Aniline, H_2SO_4 + $NaNO_2$, $NaCu(CN)_2$
(*d*) *p*-Nitroaniline, HCl + $NaNO_2$, C_6H_5OH + $NaOCOCH_3$
(*e*) *m*-Bromoaniline, H_2SO_4 + $NaNO_2$, $NaHSO_3$, then NaOH

24–9. (*a*) Toluene, Cl_2 + $FeCl_3$, HNO_3 + H_2SO_4, Fe + HCl, H_2SO_4 + $NaNO_2$, $NaCu(CN)_2$
(*b*) Benzene, HNO_3 + H_2SO_4, Br_2 + $FeBr_3$, Fe + HCl, HBr + $NaNO_2$, CuBr
(*c*) Benzene, Br_2 + Fe, HNO_3 + H_2SO_4, Fe + HCl, HCl + $NaNO_2$, HBF_4, heat
(*d*) Benzene ⟶ *m*-dinitrobenzene ⟶ *m*-nitroaniline, then HCl + $NaNO_2$, Cu
(*e*) Benzene, HNO_3 + H_2SO_4 to *m*-dinitro, $(NH_4)_2S_2$, H_2SO_4 + $NaNO_2$, KI
(*f*) Benzene ⟶ nitrobenzene ⟶ *m*-chloronitrobenzene, then H_2SO_4 + $NaNO_2$, H_2O and heat
(*g*) *p*-Xylene, HNO_3 + H_2SO_4, HCl + Fe, H_2SO_4 + $NaNO_2$, $NaHSO_3$
(*h*) Benzene, HNO_3 + H_2SO_4, Fe + HCl, Cl_2, H_2SO_4 + $NaNO_2$, H_3PO_2
(*i*) Toluene, HNO_3 + H_2SO_4, Fe + HCl, CH_3COOH + heat, HNO_3 + H_2SO_4, NaOH, H_2SO_4 + $NaNO_2$, H_3PO_2 (or HCHO + NaOH), Fe + HCl

24–10. (*a*) Reduce to the diamines, diazotize both amino groups, convert to the dicyanobenzenes with $NaCu(CN)_2$, and hydrolyze to the dicarboxylic acids.
(*b*) Treat as in (*a*) and carry out as an additional step oxidation with permanganate.

24–11. The *ortho*, *meta*, and *para* isomers of $C_6H_4NO_2Y$, $C_6H_4NO_2Z$, $C_6H_4NH_2Y$, $C_6H_4NH_2Z$, $C_6H_4Y_2$, C_6H_4YZ, and $C_6H_4Z_2$. (Also all of the corresponding derivatives of NH_2, *Y*, or *Z*.)

24–12. For each mole of diazonium chloride formed, a mole of aniline is liberated, which couples at once with the diazonium salt to give insoluble diazoaminobenzene.

24–13.

$$C_6H_5N_2{}^{+-}Cl + {}^{-}OH \longrightarrow Cl^- + C_6H_5{:}N{\equiv}N{:}OH \longrightarrow C_6H_5\cdot + N_2 + \cdot OH$$

$$C_6H_5\cdot + HCHO \longrightarrow C_6H_5H + \cdot CHO$$

$$HO\cdot + \cdot CHO \longrightarrow HOCHO \xrightarrow{{}^{-}OH} H_2O + {}^{-}OCHO$$

CHAPTER 25

25–1. $h < g < d < b < i < e < f < a < c$ **25–2.** *d*

25–3. (*a*) Toluene, H_2SO_4, NaOH fusion, $(CH_3)_2SO_4$
(*b*) Benzene, H_2SO_4, NaOH fusion, $(CH_3CO)_2O$, HNO_3
(*c*) Benzene, HNO_3 + H_2SO_4, H_2SO_4 + SO_3, NaOH fusion

(*d*) Benzene ⟶ sodium phenoxide, $(CH_3)_2SO_4$, HNO_3
(*e*) *m*-Xylene ⟶ sodium 2,4-dimethylphenoxide, then benzyl chloride
(*f*) Benzene, $H_2SO_4 + SO_3$, Cl_2 + Fe, NaOH fusion
(*g*) Benzene, $HNO_3 + H_2SO_4$, $H_2SO_4 + SO_3$, Fe + HCl, $H_2SO_4 + NaNO_2$, H_2O + heat
(*h*) *m*-Xylene, H_2SO_4, NaOH fusion
(*i*) Benzene ⟶ phenol, $(CH_3CO)_2O$, Br_2, NaOH, $COCl_2$
(*j*) Benzene ⟶ phenol, then $ClSO_3H$, NH_3

25–**4.** *b*, *e*, and *g* 25–**5.** *a3*, *b7*, *c2*, *d6*, *e8*, *f4*, *g1*, *h5*

25–**6.** *a8*, *b7*, *c6*, *d2*, *e9*, *f4*, *g1*, *h3*, *i5*, *j10*

25–**7.** (*a*) *p*-Cresol + $POCl_3$ + NaOH (*b*) *m*-Cresol + $CH_3CH{=}CH_2 + H_3PO_4$
(*c*) Phenol, HNO_2, Fe + HCl, $ClCH_2COOH$, heat
(*d*) *p*-Aminophenol + 2 EtI, NaOH + $COCl_2$, $(C_3H_7)_2NH$
(*e*) $C_6H_5ONa + (C_2H_5)_2SO_4$, then nitration, reduction, and acetylation
(*f*) Resorcinol + HNO_3 (*g*) *m*-Cresol + $C_4H_9Cl + AlCl_3$
(*h*) Phenol + $Hg(OAc)_2$, then HCl

25–**8.** (*a*) Extract with aq. NaOH, then dil. HCl to give *p*-cymene; add NaOH to HCl extract to free aniline; pass CO_2 into first alkaline extract to free phenol; separate phenol and evaporate aqueous layer to dryness, extract with anhydrous alcohol to remove sodium benzenesulfonate, pass dry HCl into alcohol solution to precipitate NaCl, filter and evaporate filtrate to dryness.
(*b*) Extract with aq. NaOH, pass CO_2 into alkaline extract to free chlorophenol and acidify with HCl to free benzoic acid; acetylate amine mixture with $(CH_3CO)_2O$, dilute, filter *N*-methylacetanilide and saponify to recover *N*-methylaniline; liberate *N,N*-dimethylaniline from filtrate with NaOH.
(*c*) Make alkaline with dil. NaOH and filter acetanilide, pass in CO_2 to liberate phenol, acidify with HCl and extract butyric acid with ether, evaporate aqueous solution to dryness and heat to drive off HCl.
(*d*) Steam-distill to remove dichlorobenzene and phenol, make distillate alkaline and remove dichlorobenzene, acidify filtrate and extract with ether to obtain phenol; filter still residue to remove benzenesulfonamide, evaporate filtrate to dryness to obtain phenolsulfonic acid.

25–**9.** (*a*) Catechol with the following side chains in the 3 position: $-(CH_2)_7CH{=}CH(CH_2)_5CH_3$, $-(CH_2)_7CH{=}CHCH_2CH{=}CHCH_2CH_2CH_3$, and $-(CH_2)_7CH{=}CHCH_2CH{=}CHCH_2CH{=}CH_2$
(*b*) Like (*a*) except that the triply unsaturated side chain is $-(CH_2)_7CH{=}CHCH_2CH_2{-}CH{=}CH{-}CH{=}CH_2$

25–**10.** CH_3—[benzene ring]—$CHOHCH_3$, OH

25–**11.** $(CH_3O$—[benzene ring]—$O)_2CO$

25–**12.** CH_3COO—[benzene ring]—$NHCOCH_3$

25–**13.** CH_3NH—[benzene ring]—OH

CHAPTER 26

26–**1.** *a5*, *b3*, *c2*, *d4*, *e6*, *f8*, *g7*, *h1* 26–**2.** *a2*, *b4*, *c6*, *d8*, *e1*, *f5*, *g3*, *h7*

26–**3.** (*a*) Cl_2 + Fe, then 2 moles $Cl_2 + h\nu$, followed by hydrolysis
(*b*) Benzaldehyde + C_2H_5MgBr (*c*) Toluene + $CH_3COCl + AlCl_3$
(*d*) $Br_2 + h\nu$, then NH_3 (*e*) Dil. KCN, then $CuSO_4$ + pyridine
(*f*) Hydrolysis, then *t*-C_4H_9OCl, or $Cl_2 + h\nu$, then hydrolysis
(*g*) KOH, $KMnO_4$, acetone + NaOH (*h*) *m*-Xylene + CO + HCl + $AlCl_3$ + CuCl
(*i*) $C_2H_5NH_2 + H_2/Ni$, then C_2H_5I

26–4. (*a*) H_2/Pt (*b*) $KMnO_4$ or H_2CrO_4 (*c*) CH_3MgBr (*d*) CH_3CHO + dil. NaOH
(*e*) Aq. KCN (*f*) $Br_2 + FeBr_3$ (*g*) $KMnO_4$, NH_3 + heat, P_2O_5 (*h*) HCN
(*i*) $H_2NNHCONH_2$ (*j*) $C_6H_5NHNH_2$ (*k*) $C_6H_5NH_2$
(*l*) CH_3COCH_3 + 10% NaOH (*m*) PCl_5 (*n*) $Al(OC_2H_5)_3$
(*o*) $(CH_3CO)_2O + NaOCOCH_3$ (*p*) CH_3NO_2 + dil. NaOH

26–5. (*a*) C_6H_5MgBr, then conc. HCl (*b*) Mg, HCHO, HCl + $ZnCl_2$
(*c*) $(CH_3)_2SO_4$ + NaOH, $OCHN(CH_3)_2 + POCl_3$
(*d*) CH_3CHO + dil. NaOH, $i\text{-}C_3H_7OH + Al(OC_3H_7\text{-}i)_3$

26–6. (*a*) $C_6H_6 + CH_3COCl + AlCl_3$, H_2/Pt, H_2SO_4
(*b*) $C_6H_5CH_3 + Cl_2 + h\nu$, NaCN, $H_2/Pt + NH_3$
(*c*) $C_6H_5CH_3 + Cl_2$ + Fe, 2 moles $Cl_2 + h\nu$, hydrolysis
(*d*) $C_6H_6 + HNO_3 + H_2SO_4$, Fe + HCl, CH_3OH + HCl, $COCl_2 + ZnCl_2$
(*e*) $C_6H_5CH_3 + Cl_2 + h\nu$, hydrolysis, $HNO_3 + H_2SO_4$
(*f*) $C_6H_6 + ClCOCH_3 + AlCl_3$, then Cl_2
(*g*) $C_6H_6 + HNO_3 + H_2SO_4$, Fe + HCl, CH_3OH + HCl, $OCHN(CH_3)_2 + POCl_3$
(*h*) C_6H_6, $H_2SO_4 + SO_3$, NaOH fusion, then $C_6H_5COCl + AlCl_3$

26–7. (*a*) Na (*b*) $C_6H_5NHNH_2$ (*c*) Dil. NaOH (*d*) HBr (*e*) Dil. HCl
(*f*) NaOI (*g*) Conc. H_2SO_4 (*h*) Conc. H_2SO_4 (*i*) H_2/Pt or HBr (*j*) HNO_2

26–8. (*a*) *syn*-Phenyl *p*-tolyl ketoxime (*b*) *syn-p*-Bromophenyl *p*-cumyl ketoxime
(*c*) *syn-m*-Tolyl *p*-chlorophenyl ketoxime (*d*) *syn-o*-Nitrophenyl phenyl ketoxime

26–9. (*a*) $C_6H_5MgBr \xrightarrow{C^*O_2} C_6H_5C^*OOH \xrightarrow{CH_2N_2} C_6H_5C^*OOCH_3 \xrightarrow{LiAlH_4}$
$C_6H_5C^*H_2OH \xrightarrow{HBr} C_6H_5C^*H_2Br \xrightarrow{Mg} C_6H_5C^*H_2MgBr \xrightarrow{HCHO}$
$C_6H_5C^*H_2CH_2OH$

(*b*) C^*O_2 + C at 800° $\longrightarrow C^*O$; $C_6H_6 + C^*O + HCl + AlCl_3\text{-}CuCl \longrightarrow$
$C_6H_5C^*HO \xrightarrow{(CH_3CO)_2,\ NaOCOCH_3} C_6H_5C^*H{=}CHCOOH$

(*c*) $CH_3MgBr \xrightarrow{C^*O_2} CH_3C^*OOH \xrightarrow{CH_2N_2} CH_3C^*OOCH_3 \xrightarrow{LiAlH_4}$
$CH_3C^*H_2OH \xrightarrow{C_6H_5COCl,\ NaOH} CH_3C^*H_2OCOC_6H_5$

(*d*) $C_6H_5C^*HO \xrightarrow{HCN} C_6H_5C^*HOHCN \xrightarrow{HCl,\ H_2O} C_6H_5C^*HOHCOOH$

(*e*) $C_6H_5C^*OOH \xrightarrow{PCl_5} C_6H_5C^*OCl \xrightarrow{C_6H_6,\ AlCl_3} C_6H_5C^*OC_6H_5 \xrightarrow{C_2H_5MgBr}$
$C_6H_5C^*(OH)C_6H_5$
$\quad |$
C_2H_5

(*f*) $C_6H_5C^*OOH \xrightarrow{HNO_3,\ H_2SO_4} m\text{-}O_2NC_6H_4C^*OOH \xrightarrow{Fe,\ HCl}$ amine $\xrightarrow{HNO_2,\ HCl}$
diazonium chloride $\xrightarrow{NaCu(CN)_2}$ nitrile $\xrightarrow{H_2O,\ HCl} m\text{-}HOOCC_6H_4C^*OOH$

26–10. *c*, *e*, *f*, *h*, and *j*

26–11. (*a*)
$C_6H_3(OH)_2$
HO—C—CH_2NHCH_3
H

(*b*)
H
C_6H_5—C—OH
CH_3NH—C—CH_3
H

(*c*)
H
$O_2NC_6H_4$—C—OH
$Cl_2CHCONH$—C—CH_2OH
H

26–12. *A*, $m\text{-}C_2H_5C_6H_4CHOHCH_3$

26–13. (*a*) *A*, $p\text{-}CH_3C_6H_4CHO$ (*b*) *C*, *trans* and *D*, *meso*

CHAPTER 27

27–1. $d < f < e < c < a < b$

27–2. (*a*) Fe + HCl, $H_2SO_4 + NaNO_2$, $NaCu(CN)_2$, hydrolysis
(*b*) Hydrolysis, heat with $C_6H_5NH_2$ (*c*) Partial oxidation with $KMnO_4$, $CH_3OH + H^+$
(*d*) $LiAlH_4$ (*e*) Mg, CO_2

27–*3.* (*a*) NaOH-CaO + heat (*b*) H_2SO_4 + $NaNO_2$, then H_3PO_2, or HCHO + NaOH
(*c*) Boil with 25% HCl (*d*) Distill with Zn dust (*e*) Mg, then H_2O
(*f*) Fe + HCl, then as in (*b*) (*g*) $Na_2Cr_2O_7$ + H_2SO_4, then heat
(*h*) Boil with 25% HCl (*i*) HBr, then distill with Zn dust (*j*) H_2/Pt

27–*4.* (*a*) Toluene + HNO_3 + H_2SO_4, $KMnO_4$, Fe + HCl, H_2SO_4 + $NaNO_2$, KI, $SOCl_2$, H_2NCH_2COOH (*b*) Toluene, $KMnO_4$, PCl_5, H_2O_2 + NaOH, $NaOCH_3$, H^+
(*c*) Benzoic acid + HNO_3 + H_2SO_4, PCl_5
(*d*) C_6H_6 + H_2SO_4, NaOH fusion, CO_2 at 150°, $(CH_3CO)_2O$
(*e*) $C_6H_5CH_3$ + HNO_3 + H_2SO_4, $Na_2Cr_2O_7$ + H_2SO_4, C_2H_5OH + H^+, Fe + HCl
(*f*) C_6H_6 + HNO_3 + H_2SO_4, H_2SO_4 + SO_3, NaOH fusion, Fe + HCl, NaOH + CO_2 at 150° (*g*) $C_6H_5CH_3$ + Cl_2 + *hv*, Mg, CO_2 (*h*) Salicylic acid + $SOCl_2$, C_6H_5OH
(*i*) $C_6H_5CH_3$ + $(CH_3)_2C{=}CH_2$ + HF, $KMnO_4$
(*j*) $C_6H_5CH_3$ + $Na_2Cr_2O_7$ + H_2SO_4, HNO_3 + H_2SO_4, Fe + HCl, I_2 + NaOH, $(CH_3CO)_2O$ (*k*) *m*-$C_6H_4(CH_3)_2$ + $KMnO_4$ (partial oxidation), $(C_2H_5)_2NH$ + heat

27–*5.* (*a*) NH_3, NaOCl, CH_3OH + H^+
(*b*) HNO_3 + H_2SO_4, $Na_2Cr_2O_7$ + H_2SO_4, PCl_5, $HO(CH_2)_3Cl$, $(C_4H_9)_2NH$, Fe + HCl
(*c*) $(CH_3)_2SO_4$ + NaOH, PCl_5, H_2/Pd

27–*6.* (*1*) Toluene, H_2SO_4, NaOH, NaCN fusion, H_2O, H^+.
(*2*) Toluene, HNO_3 + H_2SO_4, Fe + HCl, H_2SO_4 + $NaNO_2$, $NaCu(CN)_2$, H_2O, H^+.
(*3*) Toluene, Br_2 + Fe, Mg, CO_2.
(*4*) Partial oxidation of *p*-xylene with $KMnO_4$, or $H_2Cr_2O_7$, or air + Co-Mn salts.

27–*7.* (*a*) $2 < 6 < 7 < 5 < 3 < 4 < 1$ (*b*) $3 < 4 < 6 < 1 < 7 < 5 < 2$
(*c*) $5 < 6 < 1 < 3 < 4 < 2 < 7$

27–*8.* *b*, *c*, *f*, *h*, and *i*

27–*9.* (*a*) *p*-Ethyltoluene (*b*) 1,2,3-Trimethylbenzene (*c*) *n*-Propylbenzene
(*d*) 1,2,4-Trimethylbenzene (*e*) *i*-Propylbenzene

27–*10.* (*o*-$CH_3CH_2C_6H_4CO)_2O$

27–*11.* One of nine possible isomers represented by $O_2NC_6H_4CONHC_6H_4COOH$

27–*12.* *p*-$C_2H_5OCOC_6H_4NH_3^{+-}Cl$ 27–*13.* *o*-, *m*-, or *p*-$CH_3COOC_6H_4COOH$

27–*14.* *o*-, *m*-, or *p*-$O_2NC_6H_4COO^{-+}NH_2RR'$ where R = R′ = C_2H_5, or R = CH_3 and R′ = *n*- or *i*-C_3H_7 27–*15.* *A*, *o*-$OCHC_6H_4COCl$

CHAPTER 28

28–*1.* (*a*) $(CH_3)_2SO_4$ + NaOH, Cl_3CCHO + H_2SO_4
(*b*) HNO_3 + H_2SO_4 ⟶ *o*-nitrotoluene, then Zn + NaOH, conc. HCl
(*c*) Br_2 + Fe, Mg, CO_2, C_2H_5OH + H^+, C_6H_5MgBr, conc. HCl, Zn dust
(*d*) Cl_2 + *hv*, Mg, C_6H_5CHO, H^+ + heat, Br_2, alc. KOH
(*e*) I_2 + HgO, Cu, $KMnO_4$ (*f*) Zn + NaOH, conc. HCl, $NaNO_2$ + H_2SO_4, KI
(*g*) H_2/Ni, H^+ + heat

28–*2.* (*a*) $(C_6H_5)_2CH_2$ + H_2CrO_4 ⟶ $(C_6H_5)_2CO$; $C_6H_5{-}C_6H_4{-}CH_3$ + H_2CrO_4 ⟶ $C_6H_5C_6H_4{-}COOH$
(*b*) K in liquid NH_3, or H_2CrO_4 ⟶ $(C_6H_5)_3COH$ (*c*) Br_2 (*d*) $Ag(NH_3)_2NO_3$
(*e*) *unsym* + H_2CrO_4 ⟶ $(C_6H_5)_2CO$, *sym* ⟶ C_6H_5COOH

28–*3.* $b < g < d < c < e < a < f$

CHAPTER 29

29–*1.* (*a*) Ten isomers, 1,2- through 1,8-, 2,3-, 2,6-, and 2,7-
(*b*) Fourteen isomers, 1,2,3- through 1,2,8-, 1,3,5- through 1,3,8-, 1,4,5- and 1,4,6-, 1,6,7-, and 2,3,7-

(*c*) Nine isomers, $\alpha_2C{=}CH_2$, $\beta_2C{=}CH_2$, $\alpha\beta C{=}CH_2$, *cis* and *trans* $\alpha CH{=}CH\alpha$, $\beta CH{=}CH\beta$, $\alpha CH{=}CH\beta$ (*d*) Seven isomers

29–2. (*a*) 8-Nitro (*b*) 5-Bromo (*c*) 1-Nitro (*d*) 4-Chloro (*e*) 8-Bromo (*f*) 1-Chloro (*g*) 5-Amino-4-hydroxy-1-naphthalenesulfonic acid

29–3. (*a*) $HNO_3 + H_2SO_4$, Fe + HCl (*b*) H_2SO_4 at 160°, NaOH fusion, $NH_3 + (NH_4)_2SO_3$
(*c*) H_2SO_4 at 80°, NaOH fusion (*d*) As in (*b*) (*e*) Br_2 + Fe, Mg, CO_2
(*f*) H_2SO_4 at 160°, NaCN fusion, $H_2O + H^+$
(*g*) β-Naphthol + $^-O_3SC_6H_4N_2^+$, $Na_2S_2O_4$, $Na_2Cr_2O_7 + H_2SO_4$
(*h*) $CH_3CH{=}CH_2 + AlCl_3$ in $C_6H_5NO_2$

29–4. (*a*) 1,2,3-Benzenetricarboxylic acid (*b*) Phthalic acid (*c*) β-Chloroanthraquinone
(*d*) 4-Fluorophthalic acid (*e*) 9,10-Phenanthraquinone, then diphenic acid
(*f*) α-Naphthoquinone

29–5. (*a*) $C_{10}H_8 + Cl_2$ + Fe, $Na_2Cr_2O_7 + H_2SO_4$
(*b*) $C_{10}H_8 + Na_2Cr_2O_7 + H_2SO_4$, $HNO_3 + H_2SO_4$, heat
(*c*) $C_6H_6 + Cl_2$ + Fe, $C_8H_4O_3 + AlCl_3$, conc. H_2SO_4
(*d*) $C_6H_5Cl + H_2SO_4$, NaOH fusion, $C_8H_4O_3 + H_2SO_4 + HBO_3$, $C_6H_5NH_2$ + heat
(*e*) $C_6H_5CH_3 + C_8H_4O_3 + AlCl_3$, conc. H_2SO_4, $HNO_3 + H_2SO_4$
(*f*) As in (*e*), using 3-nitrophthalic anhydride instead of phthalic anhydride

29–6. (*a*) The electronegative carbonyl group and sulfonic acid group permit nucleophilic substitution at the α position by hydroxide ion, which is followed by displacement of the sulfonic acid group.
(*b*) The water formed in the cyclization of the benzoylbenzoic acid causes hydrolysis of the α chlorine, which is readily displaced because of the electronegative carbonyl group on the adjacent carbon atom.

CHAPTER 30

30–1. *a5, b7, c1, d6, e9, f2, g10, h4, i8, j3*

30–2. (*a*) $CH_3COCl + AlCl_3$, NaOCl (*b*) HCHO + HCl, NaCN, $H_2O + H^+$
(*c*) CH_3MgBr, CH_3COCl
(*d*) Heat (*e*) $H_2SO_4 + NaNO_2$, SO_2, CH_3CH_2CHO, H_2SO_4
(*f*) CH_3COOH + heat, $K^{+-}OC_4H_9$-*t* (*g*) $NaNH_2$, conc. $H_2SO_4 + NaNO_2$, H_2O
(*h*) $C_6H_5CHO + (CH_3CO)_2O$, H_2/Ni (*i*) HCHO + dil. NaOH, H^+ + heat, H_2/Ni

30–3. (*a*) Aniline acetate or phloroglucinol (*b*) C_6H_5COCl + NaOH
(*c*) $C_6H_5CHO + (CH_3CO)_2O$ (*d*) Conc. H_2SO_4 at 25°
(*e*) Neutralization equivalent (*f*) Decarboxylation (*g*) $C_6H_5CHO + (CH_3CO)_2O$
(*h*) Dil. HCl (*i*) CaO-NaOH + heat (*j*) $FeSO_4$
(*k*) Alkaline-earth or heavy-metal salt solution (*l*) Dil. HCl or pine splint test
(*m*) HNO_3 oxidation (*n*) Saponify

30–4. (*a*) HCl, NaCN, Br_2, NH_3, $H_2O + H^+$ (*b*) $KMnO_4$, H_2NNH_2 + heat
(*c*) Vapor-phase air oxidation, H_2NOH, CH_3I
(*d*) $NaNH_2$, $C_6H_5CH_2Cl$, $ClCH_2CH_2N(CH_3)_2$
(*e*) Ac_2O + NaOAc, Fe + HCl, heat, $POCl_3$
(*f*) NH_4CN, $CH_3OH + H^+$, H_2CONH_2 + heat, Cl_2
(*g*) Aq. KCN, $CuSO_4$ + pyridine, NaOH, HBr, H_2NCONH_2 + heat
(*h*) $C_6H_5NHNH_2$, $C_6H_5N_2^{+-}Cl$, $H_2O_2 + HCl + V_2O_5$
(*i*) Zn + NaOH, HCl + heat, HNO_2, $C_6H_5NHNH_2$ $H_2O_2 + HCl + V_2O_5$
(*j*) $H_2SO_4 + NaNO_2$, H_2O + heat, Fe + HCl, then Skraup synthesis
(*k*) H_2/Ni, Al_2O_3 at 350°, H_2/Ni

30–5. HCOOH + heat, 4 moles $ClSO_3H$, NH_3, $C_2H_5O^{-+}Na$

30–6. As with other α-halogenated ethers, the electron-withdrawing effect of the oxygen decreases the electron density on the carbon atom to which it is bonded.

30–7. (***a***) Hydrochloride formation (***b***) Monosodium salt formation
(***c***) Monobromination in the aromatic ring and addition to the double bond in the alicyclic ring
(***d***) Quaternary salt formation, conversion to the quaternary hydroxide, opening of nitrogen ring (***e***) Formation of diacetate (*heroin*)
(***f***) Conversion of secondary alcoholic hydroxyl to carbonyl group
(***g***) Coupling *ortho* to the phenolic hydroxyl (***h***) Evolution of two moles of methane

30–8. *C* is 5-nitro-2-(phenylamino)benzoic acid and *G* is acridine.

30–9. *A* is lipoic acid. **30–10.** *A* is 1-methyl-1,2,3,4-tetrahydroisoquinoline.

30–11. *E* is ethyl 3,4-dimethoxyphenylacetic acid, *F* is 3,4-dimethoxyphenethylamine, and *I* is papaverine.

30–12. *B* is resorcinol-4-carboxaldehyde. *D* is 7-hydroxycoumarin.

CHAPTER 32

32–1. *a2, b3, c1, d8, e6, f4, g7, h5*

32–2. (***a***) C_6H_6, $HNO_3 + H_2SO_4$, Fe + HCl, H_2SO_4 at 180°, HCl + $NaNO_2$ ⟶ *A*. $C_{10}H_8 + H_2SO_4$ at 160°, NaOH fusion ⟶ *B*. *A* + *B* + NaOH ⟶ Acid Orange 7.
(***b***) $C_6H_6 + Cl_2$ + Fe, $HNO_3 + H_2SO_4$, NH_3 + heat, H_2SO_4, NaOH fusion, $(CH_3)_2SO_4$ + NaOH, HCl + $NaNO_2$ ⟶ *C*. *B* from (***a***) + CO_2 + NaOH at 150°, $C_6H_5NH_2$ + heat ⟶ *D*. *C* + *D* + NaOH ⟶ Fast Scarlet R.
(***c***) $C_6H_6 + HNO_3 + H_2SO_4$, Fe + HCl, CH_3COOH + heat, $HNO_3 + H_2SO_4$, Fe + HCl, HCl + $NaNO_2$ ⟶ *F*. *F* + *p*-cresol + NaOH ⟶ Disperse Yellow 3.
(***d***) $C_6H_5CH_3 + HNO_3 + H_2SO_4$, Zn + NaOH, conc. HCl, HCl + $NaNO_2$ ⟶ *G*. $C_{10}H_8 + HNO_3 + H_2SO_4$, Fe + HCl, H_2SO_4 at 180° ⟶ *H*. *G* + *H* + NaOH ⟶ Direct Red 2.
(***e***) $C_6H_5CH_3 + Cl_2 + h\nu$, $H_2O + Na_2CO_3$ ⟶ *J*. $C_6H_6 + HNO_3 + H_2SO_4$, Fe + HCl, CH_3OH + HCl at 220° ⟶ *K*. *J* + *K* + $ZnCl_2$, PbO_2 + HCl ⟶ Basic Green 4.

32–3. $C_{14}H_{10} + Na_2Cr_2O_7 + H_2SO_4$, $H_2SO_4 + SO_3$ at 140°, NH_3, $CaCl_2$ at 195°, $ClCH_2COOH$ ⟶ [anthraquinone-NH–CH$_2$–COOH] $\xrightarrow{\text{KOH fusion}}$ [anthraquinone-fused ring with CO, CH_2, N–H] $\xrightarrow{O_2}$ [dimer with CO, C=C, CO, N–H, N–H]

32–4. HO–(naphthyl)–N=N–(naphthyl)–SO_3Na ⟶ HO–(naphthyl)–NH_2 + H_2N–(naphthyl)–SO_3Na

32–*5.* [structure: anthraquinone bearing two $NHCOC_6H_5$ groups at positions 1 and 4]

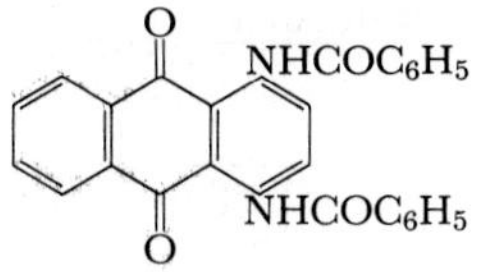

CHAPTER 33

33–*1.* *a2, b3, c4, d1*

33–*2.* (*a*) $CHBr_3 + K^{+-}OC_4H_9\text{-}t$, Mg (*b*) Dil. NaOH, H_2/Ni, H^+ + heat

33–*3.* *a2, b4, c1, d3*

CHAPTER 34

34–*1.* See text for methods of preparation.

34–*2.* Methyl chloride, methylene chloride, chloroform, carbon tetrachloride, tetrachloroethylene, tetrachloroethane, and hexachloroethane, and the various fluorine compounds that can be derived from them.

34–*3.* $CH_3CH{=}CH_2 \xrightarrow[600^\circ]{Cl_2} ClCH_2CH{=}CH_2 \xrightarrow{Br_2} ClCH_2CHBrCH_2Br$

$CH_3CH{=}CH_2 \xrightarrow{Cl_2,\ \text{liq. phase}} CH_3CHClCH_2Cl$

$ClCH_2CH{=}CH_2 \xrightarrow{HBr,\ \text{peroxides}} Cl(CH_2)_3Br$

34–*4.* (*a*) $Cl_2 \longrightarrow CCl_4 \xrightarrow{900^\circ} Cl_2C{=}CCl_2 \xrightarrow{Cl_2} Cl_3C{-}CCl_3 \xrightarrow{HF,\ SbF_5}$ $Cl_2FC{-}CClF_2 \xrightarrow{Zn} ClFC{=}CF_2$ (*b*) NaOCl, HF + SbF_5, pyrolysis
(*c*) HCl, alc. KOH, HF, Cl_2 (*d*) Electrolysis in HF, H_2O, AgOH, I_2
(*e*) NaOBr, CoF_3 (*f*) Br_2, HF + SdF_5

34–*5.* $\{C_6H_5CH{=}CH{-}\ddot{C}H_2^- \longleftrightarrow C_6H_5\ddot{C}H^-{-}CH{=}CH_2\}^+MgCl$

34–*6.* $CH_3C(Cl)(C_2H_5){-}CH{=}CHCH_3 \rightleftharpoons CH_3C(C_2H_5){=}CH{-}CHClCH_3$

A *B*

CHAPTER 35

35–*1.* (*a*) Dil. NaOH, H^+ + heat, $(CH_3)_2CHOH + Al(OC_3H_7\text{-}i)$
(*b*) Na, C_2H_5Br, CH_3MgBr, $(CH_2)_2O$, H_2/Pd
(*c*) Na, C_2H_5Br, $(CH_3)_2CO$ + NaOH (*d*) Na, $n\text{-}C_3H_7Br$, HCHO, H_2/Pd

35–*2.* (*A*) $Cl_2 + H_2O$ (*B, I, L, Q*) H_2SO_4 (*C*) HCHO + H^+ (*D*) H_2O, Na_2CO_3
(*E*) $Ca(OH)_2$ (*F*) O_2/Ag at 250° (*G*) $C_2H_5OH + H^+$ (*H*) NH_3
(*J*) NaOH (*K, M, R*) $(CH_2)_2O$ (*N*) HCN (*O*) $(C_2H_5)_2NH$
(*P*) $H_2O + H^+$ (*S*) $HNO_3 + H_2SO_4$

35–*3.* (*a*) Cl_2, $Na^{+-}OCOCH_3$ (*b*) $C_6H_5CO_3H$, $H_2O + H^+$ + heat
(*c*) HBr, Mg, $(CH_2)_2O$ (*d*) Cu-Zn at 325°, HCHO + $Ca(OH)_2$
(*e*) Al_2O_3 at 350°, H_2/Ni, $H_2O + H^+$ (*f*) Dil. NaOH, H_2/Ni
(*g*) $Cl_2 + H_2O$, NaOH (*h*) H_2S, HCl

35–*4.*

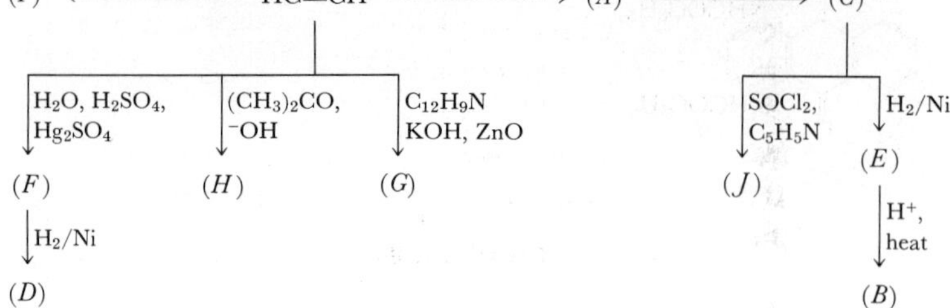

35–*5.* (***a***) HCl, NH_3, pyrolysis
(***b***) $HO(CH_2)_3OH$, HBr, NH_3 ⟶ *A*. $HO(CH_2)_4OH$ + HBr ⟶ *B*. 2 *A* + *B* ⟶ spermine.
(***c***) NaCN, H_2/Ni + NH_3 (***d***) H_2/Ni + NH_3, $SOCl_2$, NH_3
(***e***) HCHO + dil. NaOH, Fe + HCl (***f***) Cl_2 + H_2O, NH_3
(***g***) 2 CH_3NH_2 + 2 $ClCH_2CH_2Cl$

35–*6.* $CH_3CH_2CH_2CHBrCH_3$ + Mg, $(C_2H_5O)_3CH$, H_2O + H^+, excess HCHO + $Ca(OH)_2$

35–*7.* (***a***) $NaHSO_4$ + heat (***b***) HNO_2 (***c***) $Pb(OCOCH_3)_4$ or HIO_4
(***d***) Saponification (***e***) HI or $Pb(OCOCH_3)_4$ or HIO_4

35–*8.* The metal caused dehydrogenation to diacetone alcohol, which reverted to acetone.

35–*9.* CH_2—CH_2 (N:, H) + $Cl_2P^{+}(Cl)O^{-}$ ⟶ HCl + (CH_2, CH_2)N—$P^{\pm}$=O (Cl)(Cl) $\xrightarrow{2\,(CH_2)_2NH}$

$[(CH_2)_2N]_3PO$ + 2 HCl

Cell—CH_2—O(H): + $[(CH_2)_2N]_3P^{+}$—O^{-} + :O(H)—CH_2—Cell + :O(H)—CH_2—Cell ⟶

Cell—$CH_2OCH_2CH_2NH$—$P^{+}(O^{-})$—$NHCH_2CH_2OCH_2$—Cell
NH
$CH_2CH_2OCH_2$—Cell

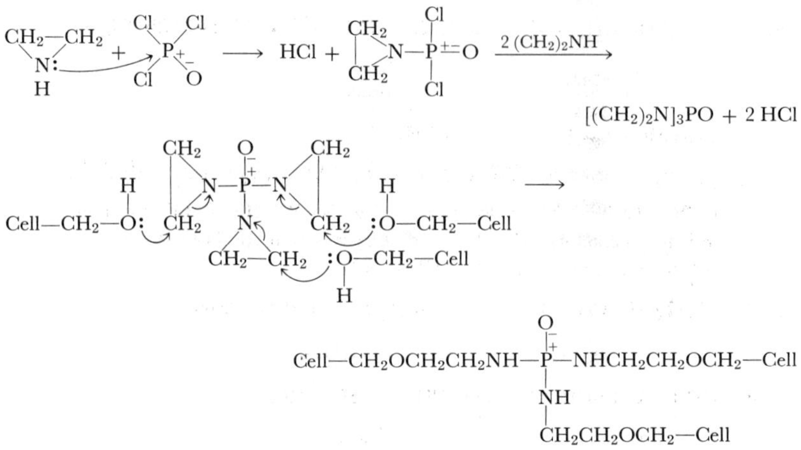

35–*10.* Xylose ⟶ furfural ⟶ tetrahydrofurfuryl alcohol ⟶ dihydropyran ⟶ tetrahydropyran ⟶ pentamethylene chloride ⟶ pentamethylenediamine hydrochloride ⟶ piperidine

35–*11.* Benzene, styrene, and 1,2-diphenylethane 35–*12.* Glycerol diacetate

CHAPTER 36

36–*1.* (***a***) The sodium salt is that of a γ-hydroxy lactone, which spontaneously cyclizes to the γ lactone. (***b***) A β-amino acid gives an α,β-unsaturated acid.
(***c***) An α,β-unsaturated acid.

36–*2.* (***a***) Br_2 + PBr_3, KOH, dil. H_2SO_4 + heat, H_2/Pt
(***b***) Ac_2O + NaOAc, Na-Hg, Br_2(+PBr_3), NH_3 (***c***) $(CH_3CO)_2O$ + CH_3COONa
(***d***) HCl, NaCN, H_2O + H^+ (***e***) H_2O + Cl_2, KOH, HCN, H_2O + H^+
(***f***) Dil. NaOH, CrO_3 in CH_3COOH, H_2S + pyridine

(g) Strecker synthesis (h) H_2/Pt, HBr
(i) $CH_3COOH + Cl_2 + PCl_3 \longrightarrow ClCH_2COOH$, $CH_3NH_2 \longrightarrow A$;
$CH_3(CH_2)_{10}COOH + PCl_3 \longrightarrow B$; $A + B \longrightarrow$ product
(j) Ozonolysis, H_2/Pt, HBr, NH_3

36–*3.* $(CH_3)_2CHCHO + HCHO +$ dil. $NaOH \longrightarrow (CH_3)_2C(CH_2OH)CHO \xrightarrow{HCN}$

$$(CH_3)_2C(CH_2OH)CHOH(CN) \xrightarrow{H_2O,\ H^+} (CH_3)_2C(CH_2OH)CHOHCOOH$$

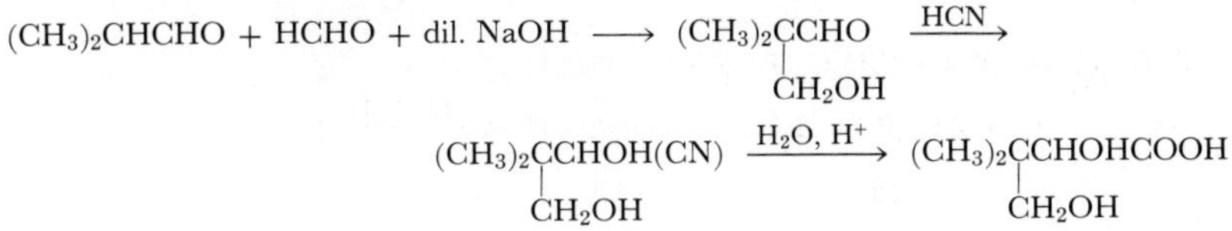

36–*4.*

$(A) \xleftarrow{CH_3OH,\ H^+}$ $CH_2{=}CHC{\equiv}N$ $\xrightarrow{H_2S} (E) \xrightarrow{H_2O,\ H^+} (J)$

$CH_2{=}CHC{\equiv}N \xrightarrow{C_6H_5OH} (L) \xrightarrow[C_2H_5OH]{Na,} (P)$

$CH_2{=}CHC{\equiv}N \xrightarrow{HCN} (B) \xrightarrow{Na,\ C_2H_5OH} (C) \xrightarrow{HCl,\ heat} (H)$

$CH_2{=}CHC{\equiv}N \xrightarrow{NH_3} (I) \xrightarrow[C_2H_5OH]{Na,} (N)$; $(I) \xrightarrow{H_2O,\ H^+} (K)$

$CH_2{=}CHC{\equiv}N \xrightarrow{CH_3NO_2} (D) \xrightarrow[H_2SO_4]{NaOH,} (F)$; $(D) \xrightarrow{Fe,\ HCl} (O) \xrightarrow{H_2O,\ H^+} (G) \xrightarrow{Heat} (M)$

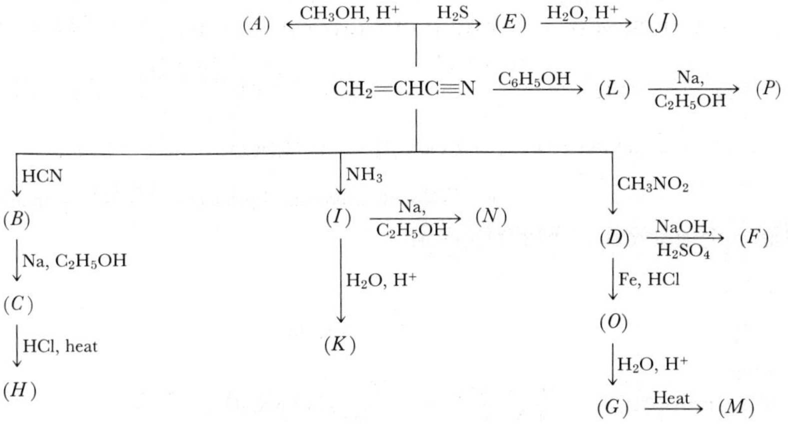

36–*5.* *a4, b6, c2, d5, e1, f3, g7*

36–*6.* (a) $NaOC_2H_5 + i\text{-}C_5H_{11}Br$, decarboxylation
(b) $NaOC_2H_5 + CH_3Br$, $NaOC_2H_5 + i\text{-}C_3H_7Br$, conc. NaOH
(c) $(CH_3)_2CHCHO + (C_2H_5)_2NH$, decarboxylation
(d) $NaOC_2H_5 + ClCH_2CH_2COOC_2H_5$, decarboxylation

36–*7.* (a) $CH_3COOC_2H_5 + NaOC_2H_5$, $C_6H_5CH_2Cl$, decarboxylation
(b) $(CH_2)_4(COOC_2H_5)_2 + NaOC_2H_5$, CH_3Br, decarboxylation

CHAPTER 37

37–*1.* (a) Ester + $Br(CH_2)_3Br$ + 2 $NaOC_2H_5$, then decarboxylate.
(b) Ester + ethyl cinnamate + $NaOC_2H_5$, then decarboxylate.
(c) Ester + $NaOC_2H_5 + CH_3Br$, then $NaOC_2H_5 + i\text{-}C_5H_{11}Br$, followed by decarboxylation. (d) Malonic acid + $i\text{-}C_3H_7CHO$ + pyridine, then heat.

37–*2.* (a) NaCN, NH_3 + heat, P_2O_5 (b) H_2/Ni, HNO_3, $Ba(OH)_2$ + heat, HNO_3
(c) NaCN, $C_2H_5OH + H_2SO_4$, $HCHO + (C_2H_5)_2NH$, decarboxylation
(d) $(CH_3CO)_2O + CH_3COONa$, $C_2H_5OH + H^+$, HCN, $H_2O + H^+$
(e) $HCHO/Cu_2C_2$ on SiO_2, H_2/Pt, $HCl + ZnCl_2$, NaCN, $H_2O + H^+$
(f) $O_2 + V_2O_5$ at 360°, $H_2O + H^+$ (g) Conc. H_2CO_4, H_2/Pt, HBr

37–*3.* (a) *1, 2, 3, 5, 6, 9, 11, 13, 14, 16* (b) *7, 8* (c) *4, 10, 12, 15, 17, 18*

37–*4.* (a) [bicyclic ring with O bridge; CO, O, CO anhydride] (b) [cyclohexene ring] COOH (c) [dihydropyran ring, O] OCH_3 (d) [bicyclic ring with CH_2 bridge] CH_3, CHO (*e*) [9,10-dihydroanthracene-9,10-endo-succinic anhydride: bridged CHCO–O–CHCO]   (*f*) [tetrahydronaphthoquinone: O, O]   (*g*) [tetrahydropyran-2-carboxaldehyde: O, CHO]

37–*5*. $(C_6H_5)_3P + ClCH_2C_6H_4CH_2Cl + P(C_6H_5)_3 \longrightarrow$

$$(C_6H_5)_3\overset{+}{P}-CH_2C_6H_4CH_2-\overset{+}{P}(C_6H_5)_3 \; \overset{-}{Cl} \; \overset{-}{Cl} \xrightarrow{LiOC_2H_5}$$

$$(C_6H_5)_3P^{\pm}{=}CHC_6H_4CH{=}^{\pm}P(C_6H_5)_3 \xrightarrow{C_6H_5CH{=}CHCHO}$$

$$C_6H_5CH{=}CH{-}CH{=}CHC_6H_4CH{=}CH{-}CH{=}CHC_6H_5 \xrightarrow{2\,C_2H_5OCOC{\equiv}CCOOC_2H_5}$$

$$C_6H_5{-}\underset{\text{(ring: CH=CH, C=C)}}{CH}\;CH{-}C_6H_4{-}CH\;CH{-}C_6H_5 \quad (C_2H_5OCO,\ COOC_2H_5,\ C_2H_5OCO,\ COOC_2H_5) \xrightarrow{KOH}$$

$$\text{salt} \xrightarrow{\text{Heat}} \text{decarboxylated product} \xrightarrow{K_3Fe(CN)_6} \text{quinquephenyl}$$

37–*6*. $C_8H_{17}C(CH_3){=}CH(CH_2)_4COOH$

CHAPTER 38

38–*1*. $BrCH_2CH_2Br \xrightarrow[+\ NaOC_2H_5]{CH_2(COOC_2H_5)_2} (H_2C)_2C(COOC_2H_5)_2 \xrightarrow{\text{Sap.}}$

$$(H_2C)_2C(COOH)_2 \xrightarrow{\text{Heat}} (H_2C)_2CHCOOH \xrightarrow{C_2H_5OH,\ H^+}$$

$$(H_2C)_2CHCOOC_2H_5 \xrightarrow{NH_3} (H_2C)_2CHCONH_2 \xrightarrow[NaOH]{NaOBr,} (H_2C)_2CHNH_2 \xrightarrow[NaOH]{2\,CH_3I,}$$

$$(H_2C)_2CHN(CH_3)_2 \xrightarrow[AgOH]{CH_3I,} (H_2C)_2CHN(CH_3)_3{}^-OH \xrightarrow{\text{Heat}} HC{=}CH{-}H_2C \text{ (cyclopropene)}$$

38–*2*. $CH_3COCH_2COOC_2H_5 + BrCH_2CH_2Br + 2\,NaOC_2H_5 \longrightarrow$

$$CH_3CO\underset{H_2C-CH_2}{C}COOC_2H_5 \xrightarrow{\text{Conc. NaOH}} CH_3COONa + (H_2C)_2CHCOONa + C_2H_5OH$$

↓ Dil. KOH, then HCl + heat

$$CH_3COCH(CH_2)_2 + C_2H_5OH + CO_2$$

38–*3*. (*A*) Cyclopentanone (*B*) Cyclopentanone oxime (*C*) δ-Valerolactam
(*D*) Cyclopentanol (*E*) Cyclopentyl bromide (*F*) Cyclopentylmagnesium bromide
(*G*) Cyclopentylcarbinol (*H*) Cyclopentylamine (*I*) Cyclopentanecarboxylic acid
(*J*) Ethyl cyclopentanecarboxylate (*K*) Ethyl β-oxo-β-cyclopentylpropionic acid
(*L*) 2-Cyclopentylethanol (*M*) Cyclopentyl methyl ketone (*N*) Ethyl adipate
(*O*) Sodio-2-ethoxycarbonylcyclopentanone
(*P*) 2-Ethoxycarbonyl-2-ethylcyclopentanone (*Q*) 2-Ethylcyclopentanone
(*R*) 2-Benzylidene-5-ethylcyclopentanone (*S*) 2-Ethyl-6-hydroxymethylcyclopentanone
(*T*) 2-Ethyl-1-phenylcyclopentanol (*U*) 1-Ethyl-2-phenylcyclopentene

38–*4*. $NaC{\equiv}CH$, HNCO, or $COCl_2$, followed by NH_3

38–**5.** Maleic anhydride, NH_3 + heat, trichloromethanesulfenyl chloride

38–**6.** *cis* 38–**7.** 1,3 is *cis*, 1,4 is *trans*.

38–**8.** $HNO_3 \longrightarrow$ brassylic acid $\xrightarrow{C_2H_5OH\ +\ H^+}$ acid ester $\xrightarrow[\text{potassium salt}]{\text{Electrolysis of}}$

ester of C_{24} dibasic acid $\xrightarrow[\text{ether}]{\text{Na,}}$ cyclic acyloin $\xrightarrow{H_2/Pt}$ diol $\xrightarrow[\text{rearr.)}]{H^+ \text{ (pinacol}}$

cyclotetracosanone $\xrightarrow{H_2/Pt}$ the alcohol $\xrightarrow{HBr}$ bromide $\xrightarrow[NaN_3,\ H_2/Pt]{NH_3 \text{ or}}$ amine.

Acyloin $\xrightarrow[CH_3COOH]{CrO_3 \text{ in}}$ dione. Cyclotetracosanone $\xrightarrow[HCl]{\text{Zn-Hg,}}$ cyclotetracosane.

38–**9.** [structures: cyclohexane epoxide–MgX_2 complex (H, $\overset{+}{O}$:$\overset{-}{Mg}X_2$, H) → cyclopentyl–$\overset{+}{C}$(H)–$\ddot{O}$:$\overset{-}{Mg}X_2$ (H) → cyclopentyl(H)–CH=O + MgX_2]

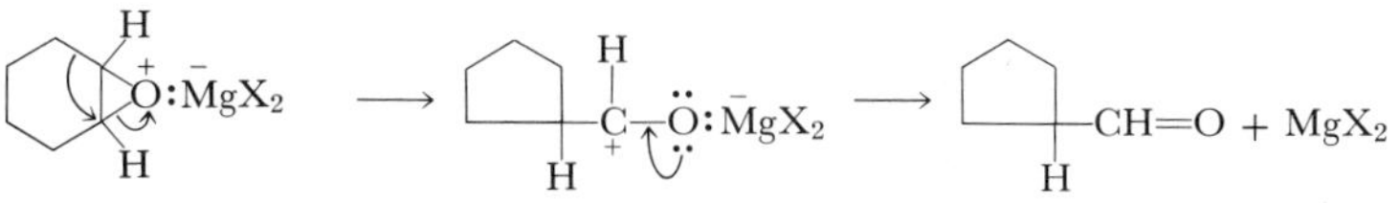

38–**10.** *A* has two rings and three reducible double bonds, either C=C or C=O or combinations; or two rings, one triple bond, and one double bond; or one ring and three reducible bonds and a carboxyl or ester group; or one ring, one triple bond, one double bond, and a carboxyl or ester group. *B* has two rings or one ring and a carboxyl or ester group.

38–**11.** *A* is 4-hydroxycycloheptene.

38–**12.** (*a*) [structures: $C_2H_5\overset{-}{O}$: H (2-chlorocyclohexanone, O, Cl) → C_2H_5OH + [bicyclic cyclopropanone, O] + :Cl^-]

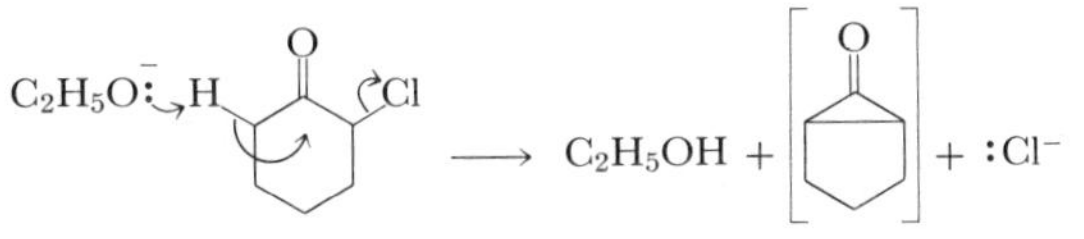

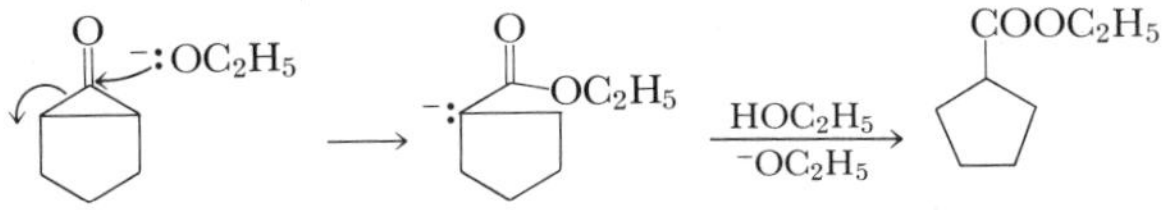

[structures: bicyclic cyclopropanone (O) + $^-$:OC_2H_5 → $^-$: ring carbanion (O, OC_2H_5) $\xrightarrow[^-OC_2H_5]{HOC_2H_5}$ cyclopentane–$COOC_2H_5$]

(*b*) [structures: B:$^-$ H … bicyclic oxime (N, $\overset{+}{O}$H, H) → substituted cyclopentene with CH_2CN + H_2O]

(*c*) [structures: B:$^-$ H … 1-ethynylcyclohexanol (HO, C≡CH, H^+) → B:$^-$ H–O … 1-cyclohexenyl–C=CH_2, H^+ → 1-cyclohexenyl–$COCH_3$]

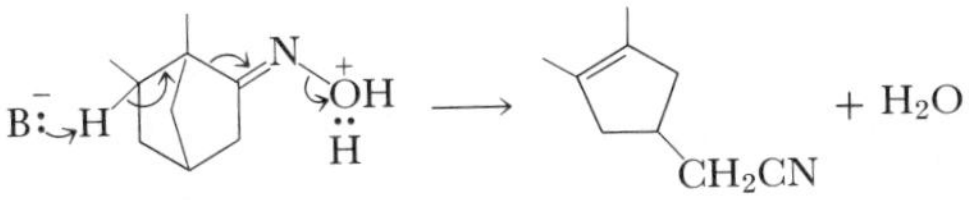

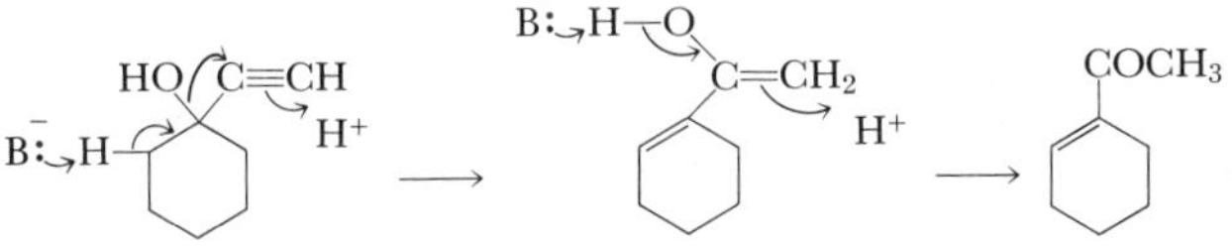

Index